The Chromosomes in Human Cancer and Leukemia

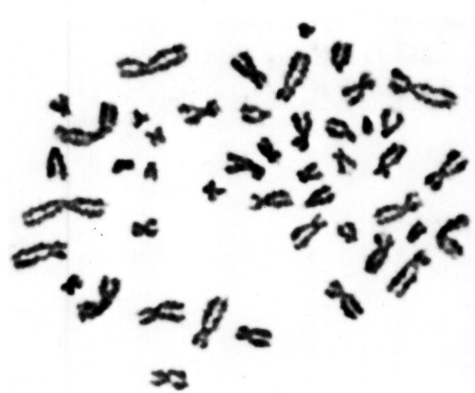

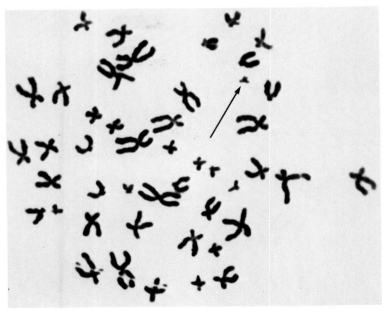

1 In this frontispiece are shown two metaphases representing milestones in human cytogenetics. The top picture shows the first metaphase published (Tjio and Levan, 1956) in which the correct number of chromosomes in the human was established. The lower metaphase shows (arrow) a Ph¹-chromosome, to date the most characteristic and consistent karyotypic change in human cancer and leukemia.

The Chromosomes in Human Cancer and Leukemia

Avery A. Sandberg, M.D.
Roswell Park Memorial Institute
Buffalo, New York, U.S.A.

Elsevier
New York • Amsterdam

© 1980 by Elsevier North Holland, Inc.

Published by:
Elsevier North Holland, Inc.
52 Vanderbilt Avenue, New York, New York 10017

Sole distributors outside of the United States and Canada:
Elsevier/North-Holland Biomedical Press
335 Jan van Galenstraat, P.O. Box 211
Amsterdam, The Netherlands

ISBN: 0-444-00289-8

Managing Editor Barbara A. Conover
Copy Editor Bradley Hundley
Design Edmée Froment
Art Editor Virginia Kudlak
Production Director Ethel G. Langlois
Production Manager Joanne Jay
Compositor The Clarinda Co.
Printer Halliday Lithograph

Manufactured in the United States of America

*To my wife, Maryn,
without whom nothing would have been possible*

Contents

Foreword

Preface

Introduction

1 **Historical Background** 1

2 **Terminology, Classification, Symbols, and Nomenclature of Human Chromosomes** 22

The Normal Human Karyotype 27

Conspectus of Human Mitotic Chromosomes 30
Group A (Chromosomes #1 – #3) 30
Group B (Chromosomes #4, #5) 31
Group C (Chromosomes #6 – #12 and the X-Chromosome) 31
Group D (Chromosomes #13 – #15) 31
Group E (Chromosomes #16 – #18) 31
Group F (Chromosomes #19, #20) 31
Group G (Chromosomes #21, #22) 31
Y-Chromosome 31

Chromosome Measurements 31

Autoradiography 33

Chromosome Banding Techniques 33
Methods and Terminology 34

Characterization of Chromosomes by Fluorescent Banding Techniques 35
Q-Banded Pattern of Individual Chromosomes 35

Characterization of Chromosomes by Other Banding Techniques 39
C-Bands 39
G- and R-Bands 41
T-Banding 41

Localization of Nucleolar Organizers (NOR) 42

Chromosome Band Nomenclature 43
Identification of Chromosome Landmarks and Bands 43
　Definitions 43
　Designation of arms, regions, and bands 44

Diagrammatic representation of landmarks and bands 45
   Subdivision of an existing landmark or band 45

System for Designating Break Points Within Bands 45
   Examples 45

Designating Structural Chromosome Abnormalities by Breakage Points and Band Composition 45

Specification of Chromosome Rearrangements 46
   Deletions 46
   Translocations 46
   Three-break arrangements 46

Specification of Break Points 46
Short System 47
   Two-break arrangements 47
   Three-break arrangements 47
   Rearrangements affecting two or more chromosomes 47

Detailed System 47
   Additional symbols 47
   Designating the band composition of a chromosome 47
   Examples 48

Four-Break Rearrangements 50
   Examples 50
   Terminal rearrangements 50
   Whole-arm translocations 50
   Duplication of a chromosome segment 51

Description of Heteromorphic Chromosomes 51

Code to Describe Banding Techniques 51
   Short terminology 51
   Complete description 51
   Examples 52
   Numerical aberrations 52
   Structural alterations 53

Marker Chromosomes 55
Mosaics and Chimeras 55
   Examples 55

Usage of + and − Signs 55
   Examples 56

Length Changes of Secondary Constrictions 56
   Examples 56

Structurally Abnormal Chromosomes 56
   Examples 56

Abbreviating Lengthy Descriptions 56
Special Terminology 57

Terminology of Acquired Aberrations 57

Chromatid Aberrations 57
Chromosome Aberrations 58
Complex Chromosome Rearrangements 59
Scoring of Aberrations 59
Cell Populations with Acquired Abnormalities 60
Tumor Cell Populations 60

3 Chromosome Structure 63

The Ultrastructure of a Chromosome 64

Morphology of a Chromosome Observed with Electron Microscopy 64
The Fibrils 65
DNA Packing Ratio 66
Models of Chromosome Structure 67
Recent Considerations of Chromosome Structure 67
Heterochromatin 71

4 Some Facets of the Cell Cycle and Unique Aspects of Chromosome Structure 74

Mitosis 75

The Morphology of Human Chromosomes 76

The Chromosome Complement 76
Ploidy 76
Karyotype 76
Characteristic Features Used for Chromosome Identification 76
Secondary Constrictions 77
Satellites 77
Visibility of the Satellites 77
Size of the Satellites 77
Satellite Association 80
Identification of Chromosomes with Autoradiographic Studies and DNA Replications 80
   Technique 81
Labeling Pattern of Specific Chromosomes 81
   Group A (Chromosomes #1 – #3) 81

Group B (Chromosomes #4, #5)   81
Group C (Chromosomes #6–#12)   82
Group D (Chromosomes #13–#15)   82
Group E (Chromosomes #16–#18)   82
Group F (Chromosomes #19, #20)   82
Group G (Chromosomes #21, #22)   82
Y-Chromosome   83
X-Chromosome   83

Identification of Chromosomes by Transverse Bands   83
Quinacrine Fluorescence or Q-Bands   83
Mechanism of Q-Banding   87
Centromeric Bands or C-Bands   88
  Mechanism   88
Giemsa or G-Bands   90
Comparison of Q- and G-Banding Techniques   91
Reverse Bands or R-Bands (Including T-Bands)   91
Differential Staining of the Nucleolus Organizers in Human Chromosomes   91

Some Comments on Polymorphism   94
Lateral Asymmetry   94

Sister Chromatid Exchange   95

5 Methods   98

Techniques Used in My Laboratory: Details of Specific Methods in Cytogenetics   99

Chromosome Preparations   100
Leukocyte Culture—Microtechnique   101
Leukocyte Culture—Macrotechnique   101
Chromosome Preparation from Bone Marrow   102
Harvesting of Bone Marrow or Blood Culture Cells   102
Chromosome Preparation from Effusions   103
Chromosome Preparation from Lymph Nodes and Spleen   103
Chromosome Preparation from Solid Tumors (Primary and Metastatic)   103

Banding Methods   103
Q-Banding   103
  Staining procedure   103
C-Banding   104
G-Banding   104
  Giemsa staining solution   104
R-Banding   104
T-Banding   104
  Method 1   104
  Method 2   105
Staining Technique of Nucleolus Organizer Regions   105
Silver Staining Method for NOR   105

Method for Sister Chromatid Exchange   106
BRdU-Dye Methods for Detecting DNA Synthesis   106
  Sister chromatid exchanges   106
  Modification for detection of late-replicating regions   108
  Modification for detection of regions containing DNA with thymine asymmetry   108
Concluding Remarks   109
Autoradiography   109

High-Resolution Banding of Chromosomes   109

High Resolution of G-Banded Chromosomes   109
Amethopterin-Synchronized Lymphocyte Culture Technique   114
G-Banding of Chromosome Preparations with Wright Stain   115
  Wright's stock solution   115
  Phosphate buffer   115
  Procedure   115

Actinomycin D–Treated Chromosomes   115
  Mitogens   116

6 Congenital Chromosome Anomalies and Neoplasia   118

Autosomal Anomalies   119
Down's Syndrome (Mongolism)   119
Other Autosomal Trisomies   120
Other Autosomal Anomalies   121
Retinoblastoma   123

Sex-Chromosome Anomalies   124
XO Karyotype and Its Variants (Gonadal Dysgenesis, Including Turner's Syndrome)   124

XXY and Its Variants (Klinefelter's
   Syndrome) 129
XXX Syndrome 129
Gonadal Dysgenesis Without Chromosomal
   Changes 129
XYY Male 129
Relation of Congenital and Familial
   Chromosome Abnormalities to Cancer and
   Leukemia 132

7 Effects of Noxious Agents and Viruses on the Human Karyotype 136

Effects of X-Ray and Other Forms of Irradiation 137

Effects of Chemical Agents 140
Benzene Poisoning and Leukemia 141

Effects of Viruses 143

DNA Viruses 145
Adenoviruses 145
Herpes Viruses 145
   HSV type 1 145
   HSV type 2 146
   Cytomegalovirus 146
   Herpes zoster virus 146
   EB virus 146
   Papovaviruses 146
   Polyoma virus 146
   SV40 146
   Poxviruses 147

RNA Viruses 147
Paramyxovirus 147
   Measles virus 147
   Mump virus 147
   Sendai virus 147
   Myxoviruses, arboviruses,
      picornaviruses, and unclassified
      viruses 147
Oncornaviruses 147
   Avian leukemia-sarcoma complex 147
   Murine leukemia-sarcoma complex 148
Chromosomal Changes upon Exposure to
   X-Ray or to Transformation by Certain
   Viruses 148

Survey of Cancer Patients for Chromosomal Anomalies 149

8 Chromosome Breakage Syndromes 152

Fanconi's Anemia (Congenital Pancytopenia) 155
Early Chromosome Studies of Fanconi's
   Anemia 155
Recent Studies of Fanconi's Anemia 156
Acute Leukemia in Fanconi's Anemia 157

Bloom's Syndrome 158
Studies of Bloom's Syndrome Cells in My
   Laboratory 160
DNA Replication in Bloom's Syndrome 161

Louis-Barr Syndrome (Ataxia Telangiectasia Syndrome) 161

Xeroderma Pigmentosum 165
Studies of DNA Defects in Xeroderma
   Pigmentosum 166

New Chromosome Instability Syndromes 167
Incontinentia Pigmenti (Bloch-
   Sulzberger) Syndrome 167
Scleroderma (Progressive Systemic
   Sclerosis) 167
Porokeratosis of Mibelli 168
Glutathione Reductase Deficiency
   Anemia 168
Kostmann's Agranulocytosis 168
Basal Cell Nevus Syndrome 168

9 Preleukemia 171

Idiopathic Aplastic Anemia 174

Pure Red Cell Aplasia 175

Refractory Anemias 176

Sideroblastic Anemia 177

10 The Leukemias: Chronic Granulocytic Leukemia 183

Introduction 184

Chronic Myelocytic Leukemia (CML) 185

Definition of the Ph¹   187
Nature of the Ph¹   190
Possible Mechanism for the Genesis of the Ph¹-Chromosomes   194
The Ph¹ and the Genesis of CML   195
Some Novel Views of the Ph¹   196
Pathogenetic Aspects of the Ph¹ in CML and Related Disorders   198
Polyploid Cells with Ph¹-Chromosomes   198
Nature of Precursor Cells of CML and the Ph¹   198
Extramedullary Origin of Ph¹-Positive Cells, Including Those in the Blastic Phase   200
Therapy and the Ph¹   200
Ph¹-Negative CML   201
Value of Detailed Chromosome Studies on Large Numbers of Cells in CML   204
Clinical Implications of Chromosomal Findings in CML   205
Unusual and Complex Ph¹-Translocations   207
Ph¹ Without Evidence of Translocations   214
Ph¹-Positive CML and the Missing Y-Chromosome   216
Ph¹-Negative CML with a Missing Y-Chromosome   220
Further Aspects of CML and the Missing Y-Chromosome   221
Additional Chromosome Changes in the Chronic Phase of CML   222
CML in Children   225
Eosinophilic Leukemia   227
Chromosome Findings in Eosinophilic Leukemia   230
Basophilic Leukemia   230
Leukocyte Alkaline Phosphates and the Ph¹   231
Peroxidase Activity, Periodic Acid–Schiff, and Sudan Black B Reactions of Blasts in CML   232
Cell Surface and Other Markers in the Blastic Phase of Ph¹-Positive CML   232
Ph¹ in Hematopoietic Disorders Other than CML   234
Chromosomal Findings in the Blastic Phase of CML   235
Multiple Ph¹ Chromosomes in the Blastic Phase   247
Double-Ph¹, Blastic Phase, and Cell Source in CML   248
Ph¹-Negative, Diploid Cells in CML   249
Karyotypic "Staging" of CML and Its Prognostic Implications   250
Clonal Origin and Evolution of CML   252
Extramedullary Aspects of Ph¹-Positive CML   254
Miscellaneous Aspects of Ph¹-Positive CML   256

Chronic Myelomonocytic Leukemia   257

11 The Leukemias: Acute Leukemia   262

Acute Leukemia (AL)   263

Significance of the Ph¹ in Various Acute Leukemias: Personal Experience   267
Ph¹-Positive Acute Lymphoblastic Leukemia   268
Ph¹-Positive Acute Myeloblastic Leukemia   270
Ph¹-Positive Erythroleukemia   270
General Remarks Regarding the Ph¹ in AL   274
Marrow Transplantation in AL and Chromosome Markers   275
Nuclear Blebs and Aneuploidy in AL   276
Cellular Markers in Differentiating Various Leukemias   276

Acute Lymphoblastic Leukemia (ALL)   276

Studies Before Banding   276
Frequency of Abnormalities and Number of Chromosomes   284
Common Abnormalities in ALL, Including the 6q−   284
Karyotypic Evolution   290
Cell Surface Markers and Chromosomal Abnormalities   290
General Comments   290
Cytogenetic Experience in ALL   291
Congenital Chromosome Abnormalities and ALL   292

Acute Myeloblastic Leukemia (AML)   294

Common Chromosomal Changes in AML   294

Chromosomal Changes and Prognosis of AML 303
The Ph¹ in AML 308
Chromosomes and the Genesis of AML 309
Chromosome Patterns in AML Revealed by Banding 310
Classifications of AML by the Karyotypic Patterns and Their Relation to Prognosis 314
AML Complicating Other Conditions 319
The Missing Y-Chromosome in AML 320
Monosomy-7 and Trisomy-8 in AML 322

Congenital Leukemia 322

Erythroleukemia (EL) 323
Chromosomal Changes in EL 323
MIKA and MAKA Acute Leukemias 327
The MAKA Group—A Type of Erythroleukemia? 328
Polyploidy—A Function of Abnormal Erythroid Cells and Karyotypic Instability 329
A Survey—Three Chromosomal Types of EL? 329
Further Aspects of MAKA and EL 331
Prophasing and EL 331

Down's Syndrome and Leukemia 332

Acute Promyelocytic Leukemia 337

Acute Myelomonocytic Leukemia 337

Acute Megakaryocytic Leukemia 340

Acute Monocytic Leukemia 340

"Hairy Cell" Leukemia 340

Near-Haploidy in Acute Leukemia 341

12 Chronic Lymphocytic Leukemia 349

13 Myeloproliferative Disorders 354

Myeloid Metaplasia with Myelofibrosis and/or Osteomyelosclerosis 358

Acute Myelofibrosis 362

Myeloproliferative Syndrome in Children with #7 Monosomy 362

Essential (Primary or Idiopathic) Thrombocythemia 363

Megaloblastic Anemia 363

Polycythemia Vera 365
Experience Before Banding 365
Chromosome Changes in PV 367
Chromosome Changes and Progression of Polycythemia Vera 369
The Ph¹ in Polycythemia Vera 371
Banding Studies in Polycythemia Vera 372

Comments on Chromosomal Changes in Myeloproliferative Disease 375

14 The Lymphomas 377

Hodgkin's Disease 378
Some Cytogenetic Findings in Hodgkin's Disease 380
Acute Leukemia in Hodgkin's Disease 381

Reticulum-Cell Sarcoma, Lymphosarcoma, and Follicular Lymphoma 383
Chromosome Studies Before Banding 384
Banding Studies in Lymphoma 387
Chromosome Findings in Follicular Lymphoma 392
Acute Leukemia in Patients with Lymphoma 393
The 14q+ Anomaly in Lymphoma 304

Burkitt Lymphoma 395
Nonendemic Burkitt Lymphoma 398

Comparison of Karyotypic Findings in Various Lymphomas 399
Studies in Established Cell Lines 399
The 14q+ Anomaly in Burkitt Lymphoma 400

Chromosome Changes in Nonendemic Burkitt Lymphoma 404
Some In Vitro Observations 405

Sézary Syndrome and Mycosis Fungoides 405

Angioimmunoblastic
Lymphadenopathy   409

Histiocytic Medullary Reticulosis   409
General Comments   410

15 Plasma Cell Dyscrasis   413

Waldenstrom's
Macroglobulinemia   414

Multiple Myeloma   416
*Chromosome Studies in Multiple
   Myeloma Before Banding   417
Banding Studies in Multiple
   Myeloma   419
Leukemia Complicating Multiple
   Myeloma   420
Comments on Chromosome Findings in
   Multiple Myeloma   423*

Plasma Cell Dyscrasias of Unknown
Significance   423

16 Chromosomes and Cancers   426

Chromosomes in Cancers: Some
"Practical" Aspects   427

Significance of Chromosome
Abnormalities in Cancer   432
*Chromosome Changes in Malignant
   Transformation   437
DNA Measurements in Tumors   439
Changes in Blood Leukocytes   439*

Chromosomes and Causation of
Human Cancer and Leukemia   439
*The Chromosomes in Animal Tumors and
   Leukemia   440*
   Spontaneous tumors and leukemia   440
   Tumors due to nutritional or endocrine
      imbalance   440
   Chemical carcinogens   441
   Oncogenic viruses   442
   Contagious tumors   444
   Minimal deviation hepatomas   444
   Further comments on some experimental
      tumors   445

Chromosomes and Progression of
Human Neoplasia   447

Chromosomes in the Diagnosis,
Therapy, and Prognosis of Cancer and
Leukemia   453

Chromosomal Findings in
Preneoplastic Lesions   455

17 Solid Tumors and Metastatic Cancer   458

Benign Tumors   459
*Benign Tumors Without Chromosome
   Anomalies   461
Benign Tumorous Conditions with
   Chromosome Anomalies   461
Cancer: Primary and Metastatic   462*

Tumors of the Alimentary Tract   468
*Tumors of the Oral Cavity and the
   Esophagus   468
Tumors of the Stomach   469
Some Recent Studies of Cancer of the
   Stomach   471
Tumors of the Colon, Cecum, Appendix,
   Rectum, and Anus   471
Cancer of the Colon, Cecum, Appendix,
   Rectum, and Anus   473
Banding Studies of Primary Tumors of the
   Large Bowel   481
Tumors of Liver, Pancreas, and
   Peritoneum   484*

Tumors of the Female   485
*Tumors of the Breast   485
Tumors of the Ovary   490*
   Early studies of cancer of the ovary   491
   Chromosome studies of cancer of the ovary based
      on banding analyses   492
   Ovarian teratomas   496
   Further comments on the origin of benign or
      variant teratomas   496
*Cancer of the Corpus Uteri and
   Endometrium   497*
   Studies of cancer of the uterus before
      banding   499
   Banding studies in cancer of the uterus   499
*Chorionic Villi, Hydatidiform Moles, and
   Chorionepithelioma   500
Miscellaneous Gynecologic Tumors   501*
   Hydatidiform moles   501
   Trophoblastic tumors   501

Tumors of the Urinary Tract   503

Chromosomes in Cancer of the Bladder   503
    Spectrophotometric DNA studies   504
    Cytogenetic studies: chromosome numbers and morphology   505
    Studies with banding   509
Cancer of the Kidney   511

Tumors of the Male   511

Tumors of the Testis   511
    Further comments on tumors of the testis   514
Cancer of the Prostate   515

Tumors of the Lung   516

Cancer of the Lung (Bronchus) and Larynx   516

Tumors of the Thyroid and Adrenal Glands   518

Further Comments on Tumors of the Thyroid   519

Malignant Melanoma   520

Some Chromosome Findings in Malignant Melanoma   520
Chromosome #1, Malignant Melanoma, and Other Cancers   521

Miscellaneous Tumors   523

Cancer of the Uterine Cervix   526

Pathology of Cervical Lesions   528
Problems Related to Cytogenetic Studies in Cancer of the Cervix   529
Chromosomal Findings in Cervical Lesions   529
    Invasive carcinoma   529
    Microcarcinoma (microinvasive carcinoma)   530
    Carcinoma in situ and dysplasia   531
    Mild dysplasia   532
    Evidence for clonal evolution in cervical lesions   532
Certain Aspects of Chromosomal Changes in the Uterine Cervix   533
    Polyploidy   533
    Aneuploidy   533
    Chromosome breakage   534

Summary of Chromosome Findings in Lesions of the Uterine Cervix   534
    Some facets of chromosomes in cervical cancer   534

Brain Tumors   535

Meningiomas   535
    Studies prior to banding   536
    Banding studies   541
    Summary of cytogenetic findings in meningioma   542
Malignant (Astrocytic) Gliomas   543
Oligodendrogliomas   546
Ependymomas   546
Medulloblastomas and Neuroblastomas   547
Retinoblastomas and Optical Gliomas   548
Neurinomas (Schwannomas)   548
Pituitary Adenomas   549
Metastatic Tumors to the Nervous System   549
Double-Minute Chromosomes (DMS)   550
Comments on DMS Observed in Human Tumors   552
Homogeneously Staining Regions and Double-Minute Chromosomes   558

18 Synoptic View of Specific Chromosome Changes in Human Cancer, Leukemia, and Gene Loci   66

Involvements of Certain Chromosomes in Human Neoplasia, Particularly Lympho- and Myeloproliferative Disorders   567

Gene Mapping and Cancer Causation   590

References   597

Indexes   731

Foreword

Cancer is a malignant form of uncontrollable cellular growth occurring in an organism through a continuous series of cell generations. Both the origin of malignant development and its management require a full understanding of cytological principles based on the use of a variety of methods, as abnormal growth and its inhibition are intimately connected with the mechanisms of cell division. About 100 years ago, microscopic studies of cancer in domestic animals were undertaken; however, because of technical limitations at that time, gross histological aspects were primarily presented. Not long thereafter emphasis was given to cytogenetic studies of malignant material, since cancer originates from pre-existing somatic cells, and cellular phenomena involving various mitotic events must have an essential bearing upon the prime elements of many cancer problems. The genetic constitution of individual cancer cells reflects the genetic pattern of the tumor; it is these individual cells that maintain a distinct genetic pattern for each tumor and determine its genetic nature. Thus, successful investigations of the genetic nature of cellular changes in tumors, as represented by the chromosome constitution, are important in solving the mechanism of malignant transformation.

Much interest has been generated by the many types of chromosome changes observed in almost all cases of cancer studied to date. Theodor Boveri advanced the theory that mutation in the genetic constitution of cells, particularly in the chromosomes, may explain the change from normal to malignant status.

Until relatively recently, the data on the chromosome constitution of cancer were utterly confused. Much of the older literature merely described high mitotic rates, a remarkable frequency of mitotic abnormalities, and striking aberrations of chromosome numbers as universal features of neoplastic cells. Extensive studies of transplantable tumors in rodents subsequently revealed that specific cells have a characteristic chromosome constitution contributing to the growth and development of the tumor in a new host, generating a stemline lineage. Following these studies of animal tumors, and based on them, much critical information was gathered regarding the significance of chromosomal mechanisms in malignant transformation. Furthermore, current advances in the technical methods in the field of mammalian cytogenetics have afforded a precise and reliable analysis of the chromosomes in the cells of mammals, including man, *in vivo* and *in vitro*. These methodologies have provided karyological data essential for the understanding of some of the etiological aspects of various types of diseases and established the importance of chromosomal data in cancer as helpful criteria for clinical and pathological considerations, and in the understanding of the mechanisms of malignant growth. The cytogenetic findings in human tumors have been shown to be comparable in many ways to those obtained in experimental and spontaneous tumors in animals; however, valuable information, supplementing the knowledge gained through animal studies, has been provided through the acquisition of karyotypic data in human neoplastic cells. Furthermore, the introduction of modern cytogenetic techniques, and the subsequent discovery of the association of many syndromes with specific types of abnormal karyotypes have led to this field becoming one of considerable clinical importance.

Dr. Avery A. Sandberg, the author of this book, has long been involved in cytogenetics, as a leading investigator in the field of human neoplasia as well as of congenital disorders. "The Chromosomes in Human Cancer and Leukemia", based on his experience, consists of some 850 pages and contains over 200 figures, more than 100 tables, and nearly 4,000 references. The book presents the chromosome findings in various cancerous conditions of man and a detailed, inclusive overview of many major areas of chromosome findings, both clinical and investigative, of human neoplasia, with emphasis on the contribution of these findings to a better understanding of the pathology and clinical aspects of human cancer and leukemia. To a large extent, the volume is a chronicle of his own detailed work, made through joint efforts with his colleagues and well-trained students, all providing an enormous amount of information in the relevant fields mentioned above. It reflects the ebullient energy of Dr. Sandberg.

This book contains a well-organized presentation in the following areas: chromosome breakage syndromes, lymphomas, plasma cell disorders, primary and metastatic cancer, cancers of specific sites such as those of the alimentary tract, female organs, urinary tracts, male organs, lung, thyroid and adrenal, melanoma, brain and nervous system, and so on. Details of many of the methodologies used in cancer cytogenetics, as well as those in congenital disorders, and the effects of noxious agents and viruses are presented in a well-arranged fashion. These elements give the book a special value as a reference volume. In addition, the newly developed chromosome banding procedures, each useful in the characterization and identification of normal and abnormal chromosomes, as well as a large body of data of new and significant findings derived from their application, are presented. Synoptic views of specific chromosome changes in human cancer and leukemia, together with the historical background, and the citing of the most essential publications, constitute an impressive and outstanding presentation and a ready source of cancer cytogenetic material for the reader. All these facets will be advantageous to a great extent not only to scientists in clinical and medical fields, but also to those in a variety of disciplines such as biochemistry, molecular biology, evolutionary genetics, and practitioners, as well as for lay persons.

In my opinion, the publication of this book reflects the vast knowledge in the fields of both human oncology and cytogenetics and bears witness to the vast production and devotion of workers involved in this intellectual discipline and scientific endeavor.

Over the years, I have come to be associated with Dr. Sandberg, both through our chromosome research work, as well as through the collaboration of many of my students, who have worked with him in Buffalo. It is a pleasure and an honor to have this opportunity to express my appreciation for his outstanding contribution to the science of cancer biology and cytogenetics.

Sajiro Makino, D.Sc.
Professor Emeritus
Hokkaido University
Sapporo, Japan
Member of Japan Academy

Preface

The evolution of my interest in cytogenetics may be worthy of record, because not only is my work in steroid metabolism often assumed to be that of another individual, but, in fact, this work was responsible for leading me into the field of cytogenetics. In the late 1950s we were engaged in studying testosterone synthesis by testes of subjects with Klinefelter's syndrome and other forms of gonadal dysgenesis or disease. To ensure the nature of the abnormalities affecting the testes being utilized in our studies, I committed myself to the establishment of a cytogenetic laboratory in my department, necessitating a period of intense study and training in the field. In addition, the then existing needs of the medical community and the various hospitals led to an expansion of my cytogenetic laboratory, and the subsequent study of chromosomes in human cancer and leukemia. Roswell Park Memorial Institute was an ideal place for the study of the latter disease, because not only is it a recognized center for the treatment of cancer and leukemia, but also its staff is deeply committed to understand as much as possible about these diseases. Thus, it is not surprising that over the years I have enjoyed the utmost cooperation from the clinicians and surgeons of the various services at Roswell Park Memorial Institute. In addition, I have been fortunate to have had associates, many of them from Japan, who have contributed to the field of cytogenetics, and in so doing have enlarged my knowledge of and appreciation for the chromosomal changes in human disease. To all of these individuals I offer my sincere thanks and gratefulness.

Even though a large number of references are given in this book, the list is not all inclusive. I have included only references to papers I had the opportunity to read, excluded some which duplicated papers of authors published in more than one journal but did not contain any additional information, and obviously could not include those papers of which I was not cognizant or had been unable to obtain for reading.

I wish to thank my secretarial staff, in particular Mrs. Cathy Russin, for taking care of the various clerical aspects of the book; a number of authors who have kindly supplied

Preface

me with manuscripts and data prior to publication and/or pictorial material for use in this book; the Departments of Medical Illustrations and Medical Photography for their help in so many different ways; and Dr. Surabhi Kakati of my department for her help in writing several of the chapters.

Avery A. Sandberg, M.D.
Buffalo, New York

Introduction

The description of an impressive number of human developmental disorders and diseases associated with *gross* chromosomal changes during the 20 years since the establishment of the correct number of chromosomes in man represents an epochal period in medicine and genetics. The field of cytogenetics, including human cytogenetics, has become an established discipline in its own right, as evidenced not only by the appearance of an array of journals and books dealing with this area, but also by the fact that the material published has been beyond the scientific appetite or comprehension of any one individual. Hence, it is the major aim of this book to focus almost exclusively on the correlation of karyotypic changes with specific human cancers and leukemias. Even this limited approach to human cytogenetics may be too much for one individual to accomplish, but it is the only way in which my views, opinions, and, possibly, prejudices in this field can be presented unabashedly, clearly, and, I hope, responsibly, and within the context of a critical and comprehensive correlation and summation of chromosomal changes with every malignant disease described to date.

Developments in the area of cytogenetics have overlapped with those on the molecular basis of genetics, i.e., the structure and function of DNA, and the architectural scheme of chromosomes in relation to the newly acquired knowledge of DNA.

This book will concern itself with *visibly recognizable* chromosomal changes in human cancer and leukemia. Thus, until recently, these changes consisted exclusively of readily ascertained morphologic and/or numerical changes of the chromosomes involving, in genetic terms, very large amounts of DNA. Newer techniques of fluorescent staining and banding patterns of chromosomes have already revealed finer karyotypic features in cancer and leukemia, too delicate to have been realized with the older methods. No available method, however, is capable of visibly showing changes at the gene level. Inasmuch as such a change is probably an essential part of carcinogenesis and leukemogenesis, our inability to examine chromosomes at that level will continue to be an egregious shortcoming of oncologic cytogenetics. Thus, even though

Introduction

emphasis has been put in the following chapters on *gross* (visibily recognizable) changes as seen with microscopy, including electron microscopy, it must be remembered that until we have means of reliably recognizing functional or molecular changes at the gene level, which may be the most common, if not sole, site of genetic changes (mutations) resulting in cancer or leukemia, our understanding of the causation and role of chromosomal changes in these conditions will continue to be incomplete.

The chromosomal alterations in human cancer and leukemia are almost always confined to the cells of the neoplastic tissues. For example, the chromosomal changes in acute leukemia are present only in the leukemic cells of the marrow or blood; those in various cancers are present only in the involved tissue and are not reflected in the karyotypes of the blood lymphocytes of cultured skin cells, which almost always reveal a diploid pattern. Thus, the karyotypic picture of the cancerous and leukemic cells is of value to clinicians, biologists, cytogeneticists, pathologists, and researchers interested in cancer and leukemia.

In this book an attempt has been made to present a comprehensive evaluation of the cytogenetic findings in human cancer and leukemia, with emphasis on those areas pertinent to medical oncology and pathology; and, additionally, on those areas not covered in previous reviews and books. In a number of cases the author presents his personal views, views, as often happens, that may appear putatively erroneous as evidence is obtained with new methodologies or approaches. In these, he has drawn generally on his own experience, though he has relied heavily on those publications that contain comprehensive and sufficient information for reliable interpretation. However, the guiding principle in and the special emphasis of this book will be a correlation of specific diseases and their subdivisions with their chromosomal picture, so that, busy clinicians, medical students, and clinical and basic science researchers will not need to search the widely dispersed and voluminous literature for their data.

The Chromosomes in Human Cancer and Leukemia

1

Historical Background

We have now to speak in greater detail of certain points which have hitherto been only cursorily touched upon, and to add others which have not yet been mentioned. In the first place I must beg leave to propose a separate technical name "chromosome" for those things which have been called by Boveri "chromatic elements," in which there occurs one of the most important acts in karyokinesis, viz. the longitudinal splitting. The name "primary loops" does not do, since these things have not always the form of a loop. "Chromatic elements" is too long. On the other hand, they are so important that a special and shorter name appears useful. . . . If the one I propose is practically applicable it will become familiar, otherwise it will soon sink into oblivion.

Waldeyer, W. 1890. Karyokinesis and its relation to the process of fertilization. Q. J. Micr. Sci., 30: 159–281. (Translated by W. B. Benham, from the original paper by W. Waldeyer in 1888. Arch. Mikr. Anat. 32:1).

More than a century has elapsed since the first descriptions appeared on the detailed events involved in nuclear division in animal cells and the search for terms to describe these events concisely and accurately. Thus, for example, the term "karyokinesis" was introduced by Schleicher in 1878–1879, "mitosis," "chromatin," and "equatorial plate" by Flemming in the 1880s and, as indicated above, the term "chromosome" by Waldeyer in 1888. The keenness and accuracy of these observers of cellular biology are witnessed by the persistent use of the terms coined by them, many of which we now use as if they had always been part of the scientific language. Certainly, Waldeyer need not have worried about the term "chromosome" disappearing into oblivion, though it is interesting that in an authoritative book on the history of medicine (Garrison, 1929), he is credited with a number of contributions to descriptive anatomy but no mention is made of his work on karyokinesis and the introduction of the term "chromosome."

Advances in science in general, and in cytogenetics in particular, have been and continue to be at the mercy of methodology. No better example can be given than the establishment of the correct number of chromosomes in the human. Even though the introduction of vital staining (affinity of the chromosome for basophilic dyes) and improvements in microscopy afforded the visualization and presentation of animal chromosomes, including those of human tumors (Arnold, 1879a,b; Galeotti, 1893; Farmer, More et al., 1906; Walker and Whittingham, 1911) and normal tissues of man (Flemming, 1882, 1889; Wieman, 1912), as well as the behavior of the meiotic chromosomes in *Ascaris* (van Beneden, 1883), the methods and materials utilized precluded a determination of the exact chromosome number in man. Until the 1920s the method commonly used for chromosome analysis was based on fixation of the tissue in Carnoy-Flemming's solution or a variant thereof, preparation of microtome sections by the paraffin method, and staining with iron hematoxylin. Most of the chromosome counts in man were based on testicular tissue obtained from patients with tuberculosis of the epididymis or from executed criminals. Thus it seems likely that the processing of these tissues, i.e., fixation, dehydration, cleaving, paraffin embedding under high-temperature conditions, sectioning, staining, and often postmortem changes, may have played a role in the variability of chromosome counts obtained by different workers. At best, these methods yielded mitotic figures with considerable overlapping of the chromosomes in a crowded metaphase plate, with the morphology of the chromosomes often being indistinct. Thus, it is not surprising to find that Flemming, the first to describe and depict human chromosomes in 1882, thought the chromosome number in human somatic cells (corneal epithelium) ranged from 22 to 24, though most workers felt that the haploid number (as observed in spermatocytes) was either 16, 24, 32, 34 (on sections of fetal tissue used by Wieman in 1912), 38 or 40, the most commonly found number being 24 chromosomes.* Flemming's figure of 22–24 chromosomes was accepted and "reaffirmed" repeatedly for 30 years.

A few years before World War I, the Belgian Winiwarter, working in Liège, published a paper (1912a,b) indicating that the number of chromosomes in man was 48 for females ($n = 23 + XX$) and 47 for males ($n = 23 + XO$). After the war, he reiterated his concept in a number of papers (Winiwarter and Oguma, 1925, 1926, 1930). Apparently, a similar conclusion was reached by the Japanese Oguma, working at Hokkaido University, in 1922 and, subsequently, with his co-worker Kihara (Makino, 1973). Oguma joined Winiwarter at Liège for a few years in the early 1920s, and they jointly published several papers reaffirming the chromosome number for the human as basically 47 ($n = 23 + XO$).

In 1921 Painter, an American, published a paper in *Science* indicating that the human had 48 chromosomes, an XX constitution characterizing the female and an XY characterizing the male. In subsequent papers (1923–1925), Painter restressed the existence of the XY pair in spermatogonia; in this he was joined by Evans and Swezy, working in California (1929), who also felt that the chromosome number for man was 48 with an XY and XX constitution being present in male and female cells, respectively. Thus, American workers appear to have been essentially responsible for determining

---

*For an extensive listing and references regarding this area, see Table 1 in Makino's (1975) book on *Human Chromosomes*.

the chromosomal sex of the human, even though these concepts were not universally accepted. Oguma in 1930 and again in 1937 denied the existence of the Y, believing that what the American workers were looking at was part of the tripartite components of the X-chromosomes. However, the Japanese Minouchi and Ohta (1932, 1934) and Iriki (1936) agreed with the concepts of Painter (1930). A similar acceptance of these concepts was expressed by Koller in 1937. Interestingly, Painter (1921) was somewhat ambiguous about the exact chromosome number in the human; though agreeing in his publication with Winiwarter's finding of 48, he nevertheless stated that "in the clearest equatorial plates so far studied only 46 chromosomes have been found." By 1921 the techniques for chromosome analysis had improved greatly, and it was a pity that Painter (and for that matter, a number of others over the next 35 years) trusted the spurious conclusions of the past rather than the probity of their observatory powers.

For nearly 45 years after Winiwarter's publication (1912a,b), the chromosome number of human somatic cells was believed to be 48. However, certain methodologic developments during that period ultimately led to and were responsible for the establishment of the correct number of chromosomes in the human. These included the development of tissue culture techniques, the introduction of colchicine for mitotic arrest, and hypotonic treatment of the cells before chromosome analysis.

The controversy between those who believed that there was remarkable "inconstancy" in the chromosome number,[1] not only between various normal tissues, but also within the same tissue specimen, and those who held that the chromosome number was constant ultimately was settled in favor of the latter group when more reliable techniques were utilized. Nevertheless, this did not prevent the statement, applied to the constancy of the chromosome number (in the normal human uterus), that the "chromosome counts were made on 50 cells and these have shown no variations from the normal ($2n = 48$) diploid chromosome number" (Sachs, 1953, 1954).

Attempts at growing human cells in culture date back to 1929, when Kemp "grew" marrow and skin cells in vitro and determined the chromosome number in these cells. Apparently, successful culture of human leukocytes was obtained by Chrustschov and associates in the 1930s (Chrustschov, Andres, et al., 1931; Chrustschov and Berlin, 1935). In 1938 Levan introduced colchicine as a mitotic inhibitor of mammalian cells, thus increasing the number of metaphase cells for chromosome analysis. Hughes (1952) and Makino and Nishimura (1952) used hypotonic pretreatment for swelling of the cells, thus making possible the more facile spreading of the chromosomes within the swollen cytoplasm of the cell. Combining these various developments, T. C. Hsu (1952) and T. C. Hsu and Pomerat (1953) determined the chromosome number (erroneously, as 48) in cultured human cells exposed to colchicine and hypotonicity before processing the cells for cytogenetic analysis. It was obvious that it was merely a matter of time before the exact number of chromosomes in the human would be established with these new and refined techniques. The breakthrough came in 1956, when Tjio and Levan published their paper on the correct number of chromosomes in the human ($2n = 46$) on the basis of cultured embryo lungs from four fetuses. This was quickly confirmed by C. E. Ford and Hamerton (1956) in the cells of three subjects, including the haploid number ($n = 23$). Interestingly, a reexamination of past materials by Levan and Hsu (1959) revealed the chromosome counts to be exclusively 46 and not 48 as previously reported (T. C. Hsu and Pomerat, 1953). Makino and Sasaki (1959, 1960) and Kodani (1958) confirmed the diploid number of 46 in the Japanese literature only a few years after the correct chromosome number in man was established by Tjio and Levan.[2]

Progress in cytogenetics of animal tumors antedated that of the human and, in many ways, paved the way methodologically and conceptually for similar studies in human neoplasia. The study of chromosomes in animals was facilitated by the development of the ascites sarcoma in rats by Yoshida (1949) and the ascites phase of induced or spontaneous tumors in mice by Klein (1951). However, studies based on the ascites form of cancer in the mouse date back to 1905, when it was introduced by Ehrlich and Apolaut. These served as favorite materials for chromosome analysis by Makino and his co-workers (1951; Yoshida, 1952) and by Hauschka (1952, 1953). The malignant cells in the ascites offered suitable cells for cytogenetic analysis:[3] they contained a significant number of mitotic cells, their number could be readily increased by the instillation of colchicine into the peritoneal cavity of the animal, and the ascitic fluid could be removed at appropriate and optimal periods for chromosome analysis.

The rapid evolution of tissue culture afforded the application of such experience to special fields of interest. Osgood and Brooke (1955) reported on the gradient culture system for normal and leukemic leukocytes. Importantly, they used phytohemagglutinin (PHA), an extract from the red kidney bean *(Phaseolus vulgaris)*, to agglutinate the erythrocytes before harvesting of the leukocytes in the blood. In 1960 Nowell reported that PHA actually stimulated lymphocyte growth in vitro, yielding many more dividing cells than afforded by any other method. Thus, the determination of the chromosome constitution of somatic cells was greatly enhanced by this lymphocyte culture technique, requiring no more than 48–72 hr of in vitro cultivation. The spreading of the chromosomes for easy analysis was further facilitated by the introduction of the air- or flame-drying technique (Rothfels and Siminovitch, 1958; Tjio and Puck, 1958), the latter replacing the cumbersome and not always successful or reproducible squash technique. Thus, the culture of various somatic cells, the PHA-blood lymphocyte method, and the utilization of bone marrow cells, the latter tissue containing a relatively large number of mitotic cells and usually not requiring culture, afforded the cytogeneticist a reliable armamentarium in studying the chromosome number in normal and pathologic conditions in the human. Hence, it was not unexpected that in one year (1959) at least four new and basic chromosomal abnormalities associated with human disease were described: i.e., Down's syndrome (mongolism), associated with trisomy of chromosome #21 by Lejeune, Gautier, et al. (1959); and Lejeune, Turpin, et al. (1959), (predicted as a chromosomal abnormality in 1932 by Waardenburg); Turner's syndrome (45 chromosomes with an XO sex constitution) by C. E. Ford, Jones, et al. (1959); Klinefelter's syndrome with an XXY chromosome constitution (predicted by Painter in 1923) by Jacobs and Strong (1959); and the XXX female (triple-X) by Jacobs, Baikie, et al (1959*b*). Within a few years the other "classical" sex chromosome syndrome (XYY males) was described by Sandberg, Koepf, et al. (1961); Sandberg, Ishihara, et al. (1963*a*); and Hauschka, Hasson, et al. (1962). The chromosome picture in testicular feminization was also clarified (Jacobs, Baikie, et al., 1959*a*).

Refinements of and expansion in the analysis of human chromosomes were introduced over the years and include autoradiography (German and Bearn, 1961), based on methodologies developed in lower animals and plants (J. H. Taylor, 1958, 1963; J. H. Taylor, Woods, et al., 1957; Lima-di-Faria, 1959), and, more recently, the various banding techniques. Autoradiography, primarily performed with tritiated thymidine, has not only revealed the way in which chromosomes replicate, but has also demonstrated an asynchronous replication of the DNA of the various chromosomes and within individual chromosomes. Thus, the late labeling of one X-chromosome in female cells (or any cells containing more than two X-chromosomes), resulting in a heavily labeled X, has been demonstrated with autoradiography and used as a means for identifying and characterizing the X-chromosome.

The various banding techniques, starting with the fluorescent method introduced by Caspersson, Zech, et al. (1970*a*–*c*), utilizing quinacrine mustard, have greatly improved our confidence in the identification of individual chromosomes. This Q-banding method is now used in addition to a number of others (e.g., G-banding [Giemsa], R-banding, T-banding, C-banding, N-banding, etc.). All of these methods have afforded the cytogeneticist reliable and unusual tools in recognizing, classifying, and interpreting the normal and abnormal karyotypic pictures in the human. No doubt other even more refined and informative methods will become available in the future, giving the biologist a more intimate knowledge of the organization of the individual chromosomes, with the ultimate aim being the recognition of genes and the changes they may undergo. More recently, the incorporation of bromodeoxyuridine (BUdR) into chromosomes followed by appropriate staining has revealed sister chromatid exchanges, as a result of differential staining between the chromatid containing BUdR and that devoid of it. This approach will undoubtedly allow for a much better understanding of certain hereditary conditions and the effects of physical and biologic agents on chromosomal changes leading to karyotypic abnormalities.

It was stated previously that basic advances in cytogenetics are at the mercy of methodologic developments. The vast literature since 1956 in human cytogenetics is indicative of the strides made in the application of new methodologies to the understanding and diagnosis of many conditions and diseases associated with chromosomal abnormalities. Inasmuch as this is but the beginning of our appreciation of the relation of chromosomal changes in human pathology, it seems reasonable to hope for future unraveling of the role of and changes in the human genome in conditions hitherto not well established etiologically and the subtle chromosomal alterations leading to abnormalities of various types.

* * *

The crucial role of chromosomes in the heredity and reproduction of the cell was well appreciated by a number of biologists about a century ago. The most comprehensive view of the role played by cellular chromosomes was probably held by Boveri (1902, 1912). It was his work with sea urchin eggs and *Ascaris* that led him early in his career (1887) to correlate the Mendelian theory of segregation of specific characteristics with the reduction phenomenon of the chromosomes at meiosis. This was further supported by Sutton (1902*a,b*), based on his results with grasshopper chromosomes, at the time when the genetic theories of Mendel were being reemphasized. The élan of the cytologists was due, in large measure, to the fact that they could visually follow the behavior of the genetic material, instead of having to rely on the statistical approaches of the so-called classical geneticist. Hence, it is not surprising to find that eventually the chromosomes would be implicated in the causation of various cancers in the human, with Boveri playing the most outstanding and provocative role in the arena of cytogenetics.

Before the concepts of Boveri on the role of chromosomes in tumor development are discussed, it may be worthwhile to summarize the theories and observations preceding the publication of his book on *Zur Frage der Entstehung maligner Tumoren (On the Problem of the Origin of Malignant Tumors)* in 1914.[4]

The chromosomes in human malignant tumors were described by Arnold in 1879. Almost all workers before Boveri's publication pointed to the role of abnormal, i.e., asymmetrical, mitosis in the genesis of cancer. Although Arnold's (1879) was the first paper on this subject, it was Hansemann (1890) who postulated that all carcinomas are characterized by asymmetrical karyokinesis (mitosis), which leads to an unequal chromatin distribution resulting in cancer. This relationship was soon questioned by Stroebe (1892). He did not believe it was possible to estimate reliably and reproducibly the chromatin content of daughter cells, and he could not agree that asymmetrical karyokinesis was restricted to carcinomas, because he found it also in sarcomas, benign tumors, and regenerative tissues. Stroebe's findings were confirmed by others. Subsequently, students of cancer paid attention to the measurement of the nuclear size of cancer cells. It was Heiberg and Kemp (1929; T. Kemp, 1929, 1930) who pioneered this work and who found that cancer cells usually had larger nuclei than normal cells, but that there were some carcinomas with normal-sized nuclei and some with nuclei even smaller than normal. Being confronted with such conflicting evidence, shortly before his death in 1920 Hansemann stated that he was wrong, and that asymmetrical mitosis does not cause the development of cancer. As we shall see later, there was no reason for Boveri to withdraw his hypothesis, because he had insisted not on quantitative but on qualitative chromosomal differences between normal and malignant cells.

Because Boveri had never experimented with tumors or their chromosomes, his decision to write his treatise on the role of chromosomes in cancer causation must be regarded as a courageous and somewhat scientifically reckless undertaking. Boveri had devoted most of his scientific life to the study of normal and abnormal mitosis of sea urchin eggs and, to a lesser extent, of *Ascaris*. He obviously understood, and certainly sensed, that the chromosomes control the life of the cell and its functions and ability to divide, and that each chromosome had resident in it material that dictated specific and varied functions of the cell. Thus, it is possible, though he did not know it then, that he had formulated the concept of the gene. Boveri knew that the "hereditary units" are located in the chromosomes and that an equilibrium among them is the prerequisite for the normal function of cells and the organism. For some reason, much more emphasis has been given to his theory regarding the role of numerical chromosomal changes in cancer causation than to his clairvoyant concept of changed genic function (mutation) without concomitant chromosomal changes as a cause of neoplasia.

* * *

When Theodor Boveri published his *Zur Frage der Entstehung maligner Tumoren* in 1914, he most probably did not imagine that more than 60 years later the problems and questions related to the role of chromosomal changes in the genesis of neoplasia would still occasion deep controversy. Oncogenesis has become an immensely sophisticated multifaceted research area in which karyotypic alterations represent a small but important, if not crucial, aspect. Amazingly enough, this aspect of oncogenesis has attracted more attention than any other, yet the validity of Boveri's hypothesis remains to be conclusively established. What, in essence, is Boveri's hypothesis? It states, first, that the cell of a malignant tumor has an abnormal chromosome constitution, and, second, that any event *leading* to an abnormal chromosome constitution will result in a malignant tumor. Having proved the individuality and genetic diversity of chromosomes in his experiments with dispermic sea urchin eggs, he concluded that there might be cell division-enhancing or -suppressing chromosomes; and malignant growth would occur if loss of suppressing chromosomes occurred or the enhancing chromosomes became predominant. Boveri emphasized that he was a zoologist, and that he did not have personal experience with karyological aspects of cancer cells. His hypothesis was largely based on his experiments with sea urchin eggs, and he could find only scanty support for his hypothesis in the literature. Though he stressed the development of an unbalanced chromosome constitution as a possible cause of cancer, particularly emphasizing mitotic disturbances and asymmetry, he did advance a number of theories and visualized the possible occurrence of chromosomal nondisjunction and pseudodiploidy as causes of cancer. He also anticipated the concept of the monoclonal origin of cancer when he stated that "typically each tumor takes its origin from one and a single cell."[5] This original cell must have acquired "a certain abnormal chromatin constitution" of whatever cause and must already be endowed with all the properties of the tumor. It must be emphasized that although Boveri stressed chromosomal imbalance resulting from mitotic disturbances as a cause of cancer, he did consider other possibilities in which the genetic equilibrium must be disturbed in some special way to make a cell malignant.

* * *

To evaluate Boveri's hypothesis and test it in human neoplasia, it was necessary to know the exact chromosome number of normal human somatic cells. Because it took more than 40 years after Boveri's publication to establish the normal chromosome number in man, let alone the morphologic characteristics of these chromosomes, the cytogenetic data on human cancer before 1956 have to be interpreted with caution. Thus, Belling (1927) found the chromosome number in human tumors unchanged; Heiberg and Kemp (1929) and Levine (1930, 1931) observed chromosome numbers in the diploid, triploid, and haploid ranges. Levine advanced the notion that "the initial aberrant nuclear and chromosomal phenomena are responses of the cell to some chemical or physical influences."[6] He questioned the significance of polyploid metaphases in the etiology of cancer, a point which had been stressed by others using experimental tumors.[7] Important contributions were published by Andres in 1932 and 1934. He, for the first time, compared chromosomal changes of primary cancer and its metastases. In both cases the chromosome number ranged between hypodiploid and near triploid, but in the metastases the degree of anomalies with regard to the number and morphology of chromosomes was much more pronounced than in the primary tumor. The morphological changes he described were fuzziness and stickiness of chromosomes, as well as fragmentation. Similar morphological changes were observed by him in cases of chronic myelocytic leukemia (CML) and in the blastic phase of the disease (Andres and Shiwago, 1933). He stated that these *morphological* changes were restricted to the myeloid cells, whereas the chromosome number was normal, i.e., 48, in both the erythroid and myeloid cell series. From his observations and some evidence available in the literature to the effect that in cancer and leukemia no consistent chromosomal anomaly existed and that there were malignant cancer cells with a normal chromosome number, Andres concluded that chromosomal changes were not the cause of cancer but only "Folgeerscheinungen" ("consequences"). He suggested that chromosomal changes in neoplasia are the consequence of irregularities of spindle formation caused by an abnormal physicochemical state of the cytoplasm. However, he conceded that a perfectly normal chromosome constitution would not exclude a gene mutation as the cause of neoplasia, as proposed by Bauer in his most interesting thesis on *"Mutations-Theorie der Geschwulstentstehung (Mutation Theory of Tumor Origin)*, published in 1928. In 1947 Koller, who at that time had studied the chromosomes of more tumors than any other worker, divided the mitotic anomalies into three classes: (1) structural alterations of chromosomes; (2) numerical changes, and (3) complete or partial suppression of the spindle. Stickiness, causing nonsegregation, nondisjunction, and chromosome lagging, was the most common alteration observed at mitosis. The latter alteration and suppression of the spindle, which he attributed to lack of food supply and toxic breakdown products, would then lead to numerical chromosomal changes. Having studied the cytological and chromosomal behavior of an adenocarcinoma in great detail, he made two statements: (1) cells with a deficient chromosomal complement do not have a shortened life-span; they can continue to divide, and (2) from this it follows that in a tumor the cell behavior is under cytoplasmic and not nuclear control. He concluded that Boveri's hypothesis was wrong (see also, Koller, 1960). His view was echoed by Timonen and Therman (1950) three years later. Another "historical" paper worthy of mention is that of Fritz-Niggli (1955). She made the important point that two histologically identical tumors may have different abnormal chromosome sets and suggested a three-phase evolution in cancer development with different "noxen" causing different chromosome anomalies. Among cells with different chromosome anomalies, selection occurs, and the most vital cells then cause cancer.

In 1956 Tjio and Levan, and shortly thereafter C. E. Ford and Hamerton (1956), described the definite chromosome number of man. By combining the acetoorcein squash technique (Hauschka and Levan, 1951), hypotonic treatment (T. C. Hsu, 1961), and colchicine pretreatment, chromosome analysis became relatively easy, reliable, and accurate, and the modern era of cytogenetics began, and with it the analysis of chromosomes in human cancer and leukemia (E. H. Y. Chu, 1960).

* * *

Even though the significance of sex-chromatin bodies in tumors remains obscure and no attempt will be made to present in detail the findings on sex-chromatin in human cancers, some discussion of this area is indicated.

Sex-chromatin was first described by Barr and Bertram (1949) in the nuclei of the interphase cells of the nervous system of female cats. This sex-body was absent in the cells of the male animals. These findings were soon confirmed in a number of other species and tissues (Mittwoch, 1964). The presence of a sex-chromatin body (X-body, Barr body) in female cells is most readily ascertained in buccal mucosa scrapings or in cultured skin cells. The X-body is usually 1–2 μm in size, rests compactly against the nuclear membrane, and has certain rather specific staining characteristics (Figure 2). In degenerating cells the X-body can usually not be well visualized. In determining the sex of an individual, each laboratory has to establish the percent of cells with X-bodies leading to the unequivocal determination of the female sex.

Several important developments related to the X-body have increased our understanding of the function and significance of the X-chromosomes in cells and have been important for the light this, in turn, has shed on the heredity of a number of disorders. Because the number of X-bodies in a cell is the number of X-chromosomes minus one, it was rightly assumed that the X-bodies were functionally inactivated X-chromosomes. This was corroborated by autoradiography with tritiated thymidine in which all X-chromosomes but one incorporate the labeled compound at a rapid rate late in the S-period, thus making it "dark" upon photographic development.

The "inactive" X-chromosome, represented

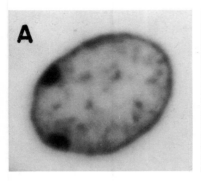 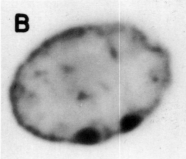 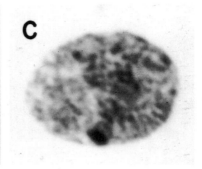

2 Sex-chromatin (Barr) bodies in the nuclei of buccal mucosa cells of a female with an XXX chromosome constitution (A and B) and of a normal female (C). Two "inactive" X-chromosomes in the former case result in two bodies being seen, whereas the single body in the latter case is the result of only one "inactive" X-chromosome.

by the X-body, is thought not to have a transcriptional function in the cell (Lyon, 1971). However, this may not apply during the early stages of embryogenesis when both X-chromosomes in female cells, one of paternal and another of maternal origin, are probably functional. At some stage of differentiation only one of the X-chromosomes remains functional and is so for the life of the particular cell. Whether it is the paternal or maternal X-chromosome that remains active is apparently decided by chance. This has been corroborated by the distribution among the cells of certain isoenzyme or enzyme deficiencies (e.g., glucose-6-phosphate dehydrogenase, G6PD) associated with genes on the X-chromosome.

The significance and reliability of clublike projections called "drumsticks" (Figure 3) in neutrophilic and other granulocytic cells of blood and marrow of females as representative of an inactive X-chromosome, similar to the X-body in other cells, is open to question. The percent of such cells having a drumstick varies from <3% to about 6% of female blood leukocytes. Barr and Carr (1960, 1962a,b) expressed the opinion that the most plausible interpretation of the leukocytic drumstick is an inherent tendency for some female granulocytic leukocytes toward nuclear lobulation in which the X-body induces the formation of a small lobule containing the body. Our experience with the drumsticks as indicators of an individual's phenotypic sex has been rather disappointing, including the observation of such drumsticks in the marrow cells of elderly males. In determining the sex of a cell or an individual, the utilization of the presence of drumsticks in granulocytic leukocytes should be discouraged. For that matter, the exact sex-chromosomal picture of any individual is best determined on the basis of a chromosomal karyotype because the absence or presence of an X-body fails to yield the important information required in complicated cases. For example, the absence of the X-body in the cells of patients with Turner's syndrome could be wrongly interpreted as

3 Two polymorphonuclear neutrophils of a female with one of them (arrow) containing a body thought to be related to the X-chromosome.

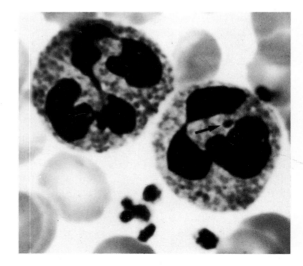

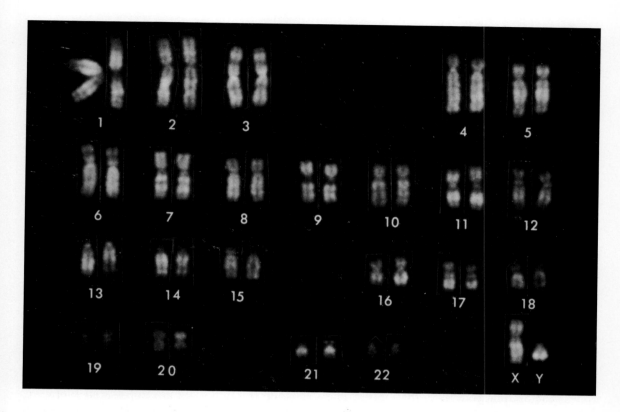

4 Q-banded karyotype of a normal male marrow cell. Note the prominent fluorescence of the long arms of the Y-chromosome.

indicating a male (XY) chromosome constitution, when, in fact, these subjects have 45 chromosomes with an XO sex-chromosome constitution; the presence of an X-body in the cells of males does not necessarily indicate the presence of an XX chromosome constitution, e.g., the XXY or Klinefelter's syndrome is accompanied by an X-body, but these individuals do have a Y-chromosome in the cells.

The introduction of the fluorescent technique for the visualization of chromosomal structure revealed the Y-chromosome to be the most brightly fluorescing chromosome in the human set, particularly intense in the long arm of the Y. Furthermore, in interphase nuclei this fluorescence persists as a heteropyknotic body (Y-body) of intense fluorescence. In contrast to the unique localization of the X-body in interphase nuclei and the prerequisites for the presence of more than one X-chromosome in a cell for the X-body to appear, the Y-body does not appear to have any particular localization in the nucleus, and all morphologically normal Y-chromosomes within a cell show this unique and intense fluorescence (Figures 4 and 5).

Studies of sex-bodies in human tumor cells have yielded limited information because the information supplied is related only to the number of sex-chromosomes present in the cell. Furthermore, chromosomal rearrangements in malignant cells may lead to the formation of abnormal chromosomes ("markers") which may simulate the X-body or the fluorescence of the Y and, thus, yield spurious results on the basis of such examinations. Of course, detailed karyotypic analysis will reveal the exact number of sex-chromosomes and the possible nature of the abnormal chromosomes.

Sex-chromatin analysis in tumor cells may reveal chromosomal changes when it indicates an anomalous pattern, e.g., triple sex-chromatin (Atkin, 1967a, 1974a). The presence of an X-body in a malignant melanoma of a male patient whose metastases regressed (Atkin, 1971c) and in a Burkitt lymphoma in a boy whose tumor cells had a female karyotype (Manolov, Levan, et al., 1970) raises a question regarding

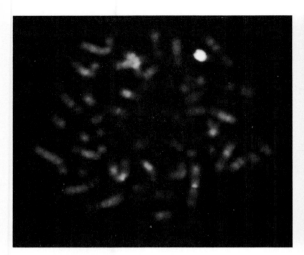 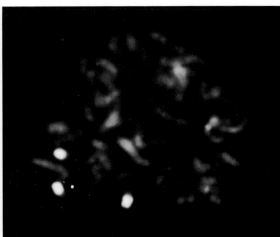

5 Two metaphases from a brain tumor, showing a highly fluorescent Y-body (Y-chromosome) (left) and three such bodies (right).

the origin of these X-bodies and their possible role in the unusual courses in these cases. The significance and frequency of the Y-body in human tumors will be discussed in a subsequent section of this book.

Some comment is indicated about the measurement of the DNA content of tumor cells, as a reflection of their ploidy. It must be realized that though DNA measurements are usually made of cells in interphase, synthesis of DNA goes on throughout the S-period without visibly affecting the microscopic appearance of the cell. Thus, spectrophotometric measurements of the nuclear DNA at the beginning of S will reveal a lower value than in late S. Ideally, spectrophotometric measurements of nuclear DNA should be applied to telophase nuclei, when the total amount of DNA is equal to $2n$. Even when DNA measurement reveals a near-diploid mode, it does not necessarily mean that the cell is diploid, for a pseudodiploid chromosome constitution or slight hypo- or hyperdiploidy, conditions not uncommon in human cancer and leukemia, would be recorded as being near-diploid by the DNA measurement method. Furthermore, it is doubtful whether slight losses or gains of chromosomal material, e.g., loss or gain of a chromosome of group #21 or #22, would be determined with the usual DNA methodologies. Although spectrophotometric DNA measurement in severely hypodiploid or hyperdiploid cancer cells may be of some value, the utilization and application of this method are very limited in the study of human tumors.

The scope of the present book does not include the important and substantial literature on the cytogenetics of human cells grown in long-term cultures. Only when such findings bear directly on the interpretation and meaning of chromosomal results obtained with direct or short-term techniques on bone marrow or blood cells have the cytogenetic data on long-term human cell lines been referred to. Furthermore, the complexity of the karyotypic data on these cell lines precludes a concise and meaningful application to the theme of the present book.

References

1. Andres, 1934; Andres and Jiv, 1936; Hungerford, 1978; Timonen, 1950; Walker and Boothroyd, 1954.
2. Ishihara and Makino, 1960; Makino and Ishihara, 1960; Makino, Kikuchi, et al., 1962.
3. Hauschka, 1953; Makino, 1951, 1957b; Makino and Kano, 1951.
4. Boveri, 1929 (English translation).
5. Boveri 1914.
6. Levine 1931.
7. Goldschmidt and Fischer, 1930; R. M. Lewis and Lockwood, 1929; Winge, 1927, 1930.

GLOSSARY OF CYTOGENETIC TERMS[1]

Acentric chromosome	A chromosome lacking a centromere and, hence, incapable of attaching to the spindle at mitosis and thus left out in the cytoplasm at telophase where it tends to degenerate.
Acentric fragment	Chromosome material lacking a centromere.
Acrocentric	Describes a chromosome with the centromere near one end.
Anaphase	The next stage of cell division after metaphase. Begins when the centromere of each chromosome divides longitudinally, so that the two halves of each chromosome, the chromatids, are completely separated from each other. The two sets of chromatids then begin to move away from each other fairly rapidly, under the influence of the spindle fibers, to opposite poles of the cell. When the two sets of chromatids have reached the opposite poles of the cell, anaphase is complete and the next stage, telophase, begins.
Aneuploidy	Deviation of the chromosome number from that characteristic for the species. In man, a chromosome complement that is not an exact multiple of the haploid (n) number (23 chromosomes).
Asymmetrical division	A mitosis that leads to the formation of two visibly different daughter cells often with abnormal chromosome complements.
Autosome	Any chromosome other than the sex-chromosomes.
Banding	The bringing out of various intrachromosomal bands and/or regions of varying intensity by procedures of differential staining. The most commonly used procedures are Q-, G-, C-, and T-banding.
Binucleate cells	Cells with two nuclei, resulting from incomplete mitosis when no cell membrane is formed between daughter cells. At the following mitosis, the two nuclei usually divide synchronously, a common equatorial plate is formed where the two sets of chromosomes mix together and the new daughter cells each contain twice the diploid chromosome number. If the process is repeated, "giant" cells can be formed which may have several hundred chromosomes.
Bivalents	In meiosis, paired homologous chromosomes.
Breaks (Figure 6). See Gaps	Chromatid breaks occur only in one of the two chromatids. They are usually the result of breakage that has taken place after DNA replication ("post-split-breaks"). When a break is present at the same site on both chromatids, it is termed a chromosome, isochromatid, or isolocus break.
Centric fusion	A process of translocation usually involving the long arms of acrocentric or subtelocentric chromosomes which join their proximal ends and lead to the appearance of a large metacentric chromosome.

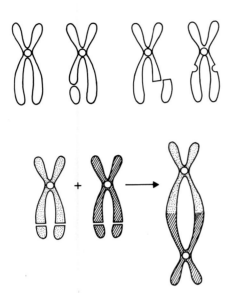

6 Schematic presentation of a gap, break, and secondary constriction (upper row) and the formation of a dicentric (lower row) chromosome.

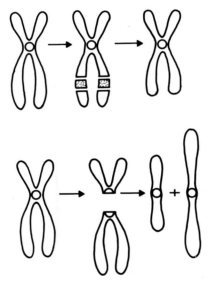

7 Schematic presentation of an interstitial deletion (stippled area) leading to a shortened chromosome (upper row) and the formation of an isochromosome (lower row). In the latter process, instead of the longitudinal separation of the chromatids at the centromere, the separation occurs in a horizontal fashion leading to the formation of the isochromosomes.

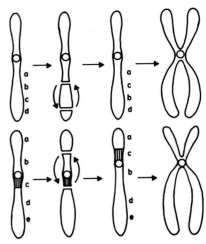

8 Schematic presentation of inversions. In the top row is shown an interstitial inversion resulting in no change of the chromosomal structure even though the band sequences have been changed from abcd to acbd. In the bottom row is shown a pericentric inversion; in this type there is a change in chromosomal morphology due to shifting of the centromere location. Both types of inversions may lead to abnormal phenotypes as a result of changes of *intrachromosomal* structure.

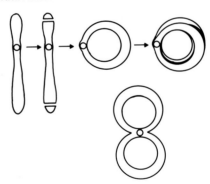

9 Schematic presentation of the formation of a ring chromosome with a centromere, followed by its duplication and an occasional figure 8 appearance of such a chromosome in metaphase spreads.

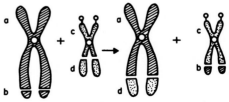

10 Schematic presentation of a reciprocal translocation between two chromosomes.

Centromere (Kinetochore)	An area of chromosomal constriction, especially visible in metaphase, where the two chromatids are held together, and by which attachment to the spindle is accomplished. The location of the centromere is very useful in the identification of individual chromosomes. An early term for the centromere was "primary constriction."
Chromatid	Two sister chromatids, held together at the centrometric region, make up a chromosome and, hence, each sister chromatid represents the replicated form of the genetic material of the other. One chromatid of each chromosome migrates to each new cell when cellular division is completed.
Chromatid breaks (Figure 6)	In which only one chromatid is broken, the other remaining intact.
Chromatin	The substance of which chromosomes are composed.
Chromocenters	Discrete, highly condensed regions of the chromatin in the interphase nucleus. The sex-chromatin body in female cells is an example of a chromocenter.
Chromomeres	Dark staining areas of enlargements along the chromosomes.
Chromosome	Arrangement of the nuclear genetic material (chromatin, DNA) into small and discrete bodies, typically containing a centromere and two chromosome arms, in preparation for cell division. They contain the genes, or hereditary factors, and are constant in number in each species. The normal number in man is 46, with 22 pairs of autosomes and 2 sex-chromosomes (XX or XY). Many chromosomal aberrations may occur, some of them being typically associated with the various abnormalities.
Chromosome bridges	See Stickiness.
Chromosome gaps and breaks (Figure 6)	The same terminology applies as chromatid lesions, except that the event is usually the result of breakage that had taken place before DNA replication (presplit breaks). At replication, the lesion is reproduced in both chromatids. Similar appearing lesions called isochromatid breaks are due to the chance occurrence of two independent chromatid breaks at the same site. Even though chromosome and isochromatid breaks are morphologically indistinguishable, they can be distinguished on the basis of the amount of time that has elapsed between exposure of the breaks to the breaking agent and the mitosis at which they are observed.
Clonal evolution	A continuous step-wise evolution of new chromosomal patterns by successive acquisitions and duplications of extra chromosomes.
C-mitosis	An abnormal mitosis in which anaphase is entirely omitted so that both groups of daughter chromosomes are included in the same nucleus leading to a $4n$ number of chromosomes. Colchicine, thought to act as a spindle poison, may lead to C-mitosis.

Cytogenetics	The study of cell genetics.
Deletion	Loss of a chromosome segment.
Dicentric (Figure 6)	A chromosome having two centromeres, formed by breakage and reunion involving two chromosomes.
Diploidy ($2n$)	The number of chromosomes present in the somatic and primary germ cells. Each species has a characteristic number ($2n$) with the haploid number (n) being present in the gamete (sperm, ovum). The $2n$ number for man is 46.
Duplications or repeats	These may be *tandem* or *reversed*. A tandem duplication may be represented by the sequence ABCDCDEF, a reversed one by ABCDDCEF. The former may result from mutual translocations between homologous chromosomes, or by a similar type of rearrangement involving the two chromatids of a single chromosome. The latter probably arise mainly through a fusion between an acentric fragment and a chromatid mechanically broken in an anaphase bridge after sister strand union.
	More complex types of rearrangements may result from three or more chromosome or chromatid breaks in the same nucleus or from various combinations of chromosome and chromatid breaks. Three breaks in the same chromosome may lead to tandem inversions or to the region between breaks 1 and 2 being *inserted* into break 3 (either inverted or noninverted). Four breaks in the same chromosome may lead to *independent, included,* or *overlapping* inversions. A special type of reversed duplication is where we find a metacentric chromosome with the two arms homologous throughout (e.g., xyx-Cxyx, where C represents the centromere). Such isochromosomes, as they have been called, may arise in various ways, following on breaks through or adjacent to the centromere. Attached X-chromosomes in *Drosophila melanogaster* are isochromosomes that have been extensively studied in genetic experiments.
Endoreduplication (endopolyploidy)	Endoreduplication of chromosomes occurs during interphase resulting in the formation of metaphase chromosomes with four sister chromatids (diplochromosomes) and exact karyotype doubling. Arises through a process of endomitosis whereby all the chromosomes in a nucleus reproduce, the daughter chromosomes separating from one another within the intact nuclear membrane without the formation of any spindle or mitotic apparatus in the ordinary sense. When endoreduplication occurs in a $2n$ cell, it leads to the appearance of a cell with 92 chromosomes ($4n$). Selective endoreduplication of a single chromosome, e.g., the Ph[1], or parts of chromosomes has been postulated.
Euchromatin	See Heterochromatin.

Euploid	The diploid number of chromosomes or an exact multiple thereof.
Exchanges	A simple break results in a chromosome deletion in a chromosome fragment. If two breaks occur in reasonable proximity to each other, exchange or reunion takes place. If the fusion occurs between both proximal ends or both distal ends, a U-type exchange occurs. If all four ends involved in a potential exchange are reunited, the exchange is termed complete. If none or only two are reunited, it is incomplete. In the U-type exchange, these may be divided into incomplete proximal and incomplete distal.
Fragmentation	Fragmentation of chromosomes is an example of one of the most common anomalies shown by tumor cells, the primary event taking place during chromosome replication of interphase, but the result can only be observed at the following mitosis. The chromosome fragments are scattered in the cells and, because they lack a centromere, fail to move toward the poles at mitosis. The fragments may form micronuclei or they may disintegrate in the cytoplasm and consequently be lost in the next division. The presence of such micronuclei in the cytoplasm is evidence that the nucleus of such cells has lost chromosomal material and the deficient genome may cause the death of the cell.
Fragment	A broken part of a chromosome: an acentric fragment is one in which no centromere is present, whereas a centric fragment is one containing a centromere.
G_1-period	The period between the end of the previous cell division and the beginning of chromosomal replication.
G_2-period	The interval between the end of the S-period and the beginning of the next cell division.
Gaps	Severe attenuation of the chromatids, producing an achromatic lesion or gap. When there is displacement of the distal fragment, the lesion can be positively identified as a true break.
Genome	The total genetic constitution of a cell or organism.
Gonosomes	The sex-chromosomes, i.e., X and Y.
Haploidy	Cells, usually germ cells, with the n number of chromosomes. In the human $n = 22 + X$ in the mature ovum and $n = 22 + Y$ chromosomes in the mature sperm.
Heterochromatin	A chromosome or part of it may appear in a condensed or decondensed state during the course of development. The condensed regions stain dark and are called heterochromatin; the lightly stained regions are called euchromatin. The heterochromatic regions can be divided into two categories: facultative and constitutive heterochromatins. Facultative heterochromatins contain structural genes, which are inactivated, depending upon the phys-

iological and developmental process. An example is one of the X-chromosomes in female cells. During the early stages of development both the X-chromosomes are active, but after a certain stage most (i.e., the long arm) of one of the X-chromosomes of female cells is inactivated.

Constitutive heterochromatin is permanent and is present in all human chromosomes in variable quantities. In man, it is localized in the centromeric region, in the secondary constrictions of chromosomes #1, #9, and #16 on their long arm and the distal part of the long arm of the Y, which are characteristically stained by certain techniques and are generally enriched in highly repeated DNA sequences.

Heteropyknosis	More or less condensed appearance of chromosomes or their segments when compared to the rest of the karyotype; there may be positive heteropyknosis ("overcondensation") or negative heteropyknosis ("undercondensation"). Chromosomal material that shows heteropyknosis at some stage is referred to as *heterochromatin,* the "standard" regions which do not show heteropyknosis being *euchromatin.* The same chromosome or chromosome region may exhibit positive heteropyknosis at one stage of its cycle and no heteropyknosis or even negative heteropyknosis at another stage.
Hexaploid ($6n$)	$6n$ number of chromosomes, in the human 148. See Ploidy.
Hyperdiploid	More than the diploid number of chromosomes, e.g., more than 46 in man.
Hypodiploid	Less than the diploid number of chromosomes, e.g., less than 46 in man.
Idiogram	Diagrammatic representation of a karyotype, which may be based on measurements of the chromosomes in several or many cells.
Insertion	A transfer of a chromosome segment into another chromosome. This involves two breakages in each of the two chromosomes, where a segment of one chromosome is inserted into a breakage point in another.
Intercalary	Chromosome segments not located terminally.
Interchange	An interchromosomal exchange involving reunion between two different chromosomes. Every chromosome is involved in chromosomal interchanges in man. An intrachromosomal exchange involving reunion between different types of the same chromosome is an interchange.
Interphase	In the cell cycle, when the cell is not in mitosis. Interphase includes the G_1-, S-, and G_2-phases. During interphase, the chromosomes are active in RNA synthesis. At a certain, rather fixed time before the cell next divides, the chromosomes replicate themselves and thus double the amount of chromosomal material in the nucleus. The

	time at which they are doing this is known as the S- (synthetic) period.
Interstitial deletion (Figure 7)	Loss of chromosome material, not involving the terminal ends of the chromosome.

Inversions (Figure 8)
  Paracentric
  Pericentric
	These result from two breaks in the same chromosome with rotation of the segment between them through 180° and fusion of the broken ends. A chromosome with the sequence of regions ABCDEF, if broken between B and C and between D and E, becomes ABDC-EF after the inversion. Inversions may originate from either chromatid or chromosome breaks.
	
	In paracentric inversion the breaks occur on the same arm on one side of the centromere, in contrast to pericentric inversion in which the breaks occur on both sides of the centromere. In paracentric inversion the intra-arm exchange will lead to no apparent altered morphology.
	
	A pericentric inversion is one that occurs as a result of breakage followed by relocation of a segment within a chromosome about the centromere. If the breaks are equidistant from the centromere, no apparent change in morphology will occur; when they are not of equal distance, an abnormal chromosome (morphologically) will result.

Isochromatid break or deletion	A discontinuity in both chromatids of a chromosome at identical sites.
Isochromosome (Figure 7)	A symmetrical chromosome composed of duplicated long or short arms formed after misdivision of the centromere in a transverse plane.
Karyotype	A systematic arrangement of the chromosomes into various groups according to size, centromere location, and other morphologic features. In man, the karyotype consists of 22 pairs of autosomes and the 2 sex-chromosomes (XX in females, XY in males).
Lagging	Lagging of chromosomes is another anomaly which results in the loss of chromosomal material, such chromosomes usually lie outside the mitotic spindle and at the end of mitosis they may form micronuclei in the cytoplasm.
Marker chromosome	An abnormal chromosome easily identified by its peculiar morphology. Marker chromosomes are always abnormal.
Meiosis	A process occurring in the germinal cells of the testes (spermatogenesis) or ovary (ovogenesis) in diploid organism which results in the $2n$ chromosome number being reduced to the haploid number (n), e.g., in man to 23.
Metacentric (median)	Describes a chromosome with the centromere located midway between the two extremities.
Metaphase	One of the phases of cellular division characterized by the disappearance of the nuclear membrane and the nucleo-

	li, the chromosomes becoming arranged in one plane at the equator of the cell, and this arrangement is called a metaphase plate. In this phase, the chromosomes are maximally condensed, most easily seen, and least genetically active, and become attached to the spindle at the region along their length called the centromere (kinetochore).
Mitosis	The process of cellular division resulting in the appearance of two daughter cells with a diploid number of chromosomes under normal circumstances.
Modal chromosome number	The predominent number in cells with a wide range of chromosome numbers.
Monosomic	The presence of only one chromosome of a homologous pair, instead of the normal number of two.
Monosomy	The absence of one member of a homologous pair of chromosomes.
Multipolar mitotic spindle	A common occurrence in cells of malignant tissue and responsible for the irregular distribution of chromosomes into several nuclei. Multipolar mitosis is often incomplete and chromosome bridges hold the various daughter nuclei together. The sizes of the nuclei vary, indicating that their chromosome content is different.
n	Haploid number of chromosomes, in the human 23.
$2n$	Diploid number of chromosomes, in the human 46.
Nondisjunction	Failure of two daughter chromosomes to separate at anaphase, thus migrating together to the same pole along the spindle and as a result one daughter nucleus contains more and the other less chromosomes than normal. Nondisjunction can also be caused by stickiness of chromosomes.
Nucleolar organizers	Chromosomes allegedly involved in the organization of the nucleolus during and after mitosis.
Octoploidy ($8n$)	Cells containing an $8n$ number of chromosomes, i.e., in the human 184. See Ploidy.
Pairing	A process by which two homologous chromosomes pair starting at the chromosome ends and then proceeding zipperlike along the length of each chromosome pair until it is complete; this does not apply to the X and Y.
Paracentric inversion	See Inversions.
Pericentric inversion	See Inversions.
Ploidy	Indicates a specific multiple of a set of chromosomes or deviations from it, i.e., diploid ($2n$), triploid ($3n$), tetraploid ($4n$), etc.
Polymorphism	The presence in a population of coexistent alternative phenotypes. Applied in cytogenetics to differences in chromosome morphology, particularly banding patterns.
Polyploidization	Conversion to a polyploid stage, e.g., by endoreduplication. In tumors, also used to signify one or more dou-

	blings of the chromosome complement of aneuploid cells.
Polyploidy	Cells containing a multiple of the n number of chromosomes but more than $2n$; in the case of a tumor, multiple of S.
Prophase	The first stage of cell division initiated by the chromosomes beginning to become visible in the nucleus as fine threads which progressively shorten, become thicker and so more obvious as prophase proceeds.
Prophasing (premature chromosome [or chromatin] condensation, PCC)	A process occurring in a multinucleate cell, i.e., interphase-metaphase binucleate cell, in which the presence of a metaphase cell leads to dissolution of the nuclear membrane of the interphase nucleus resulting in a state similar to prophase in a mononucleate cell.[2]
Pseudodiploidy	In man, the presence of 46 chromosomes but with an abnormal karyotype.
Restitution	*Reunion* of broken chromosomes in such a way as to restore the original sequence.
Ring chromosome (Figure 9). See Fragments	Two breaks occur in the same chromosome, on opposite sides of the centromere; the ends of the centric fragment fuse, to form a *centric ring chromosome*, leaving the two acentric terminal fragments to their fate (it is immaterial whether they fuse together or undergo sister strand reunion); being acentric they will degenerate in the cytoplasm anyhow.
S	Number of chromosomes characterizing the modal cells in a cancer or leukemia, i.e., the stemline cells.
Satellite association	A tendency for the 10 acrocentric chromosomes to group together during metaphase with their short arms and satellites in association; there is some evidence that the association in mitosis is due to the fact that all the secondary constrictions concerned are closely related to the nucleolus up to the end of prophase and that it is probably the remnants of the nucleolus that hold these chromosomes together in this way. Satellite association is thus a secondary phenomena which relates to the state of the cell before fixation and spreading, and therefore may be dependent on such factors as variability in numbers and size of nucleoli present, as well as on the technical methods employed in making the preparation.
Satellite DNA	Highly repetitive DNA segments of the genome that are not transcribed into RNA for protein synthesis, i.e., inactive DNA.
Satellites	Small rounded bodies attached to the short arm of the acrocentric chromosomes by a thin stalk (groups D and G); they are thought to be related to nucleolar physiology and integrity. Rarely, they can be seen on the short arms of #17.
Secondary constriction (Figure 6)	Any constricted region other than the centromere.

Sex-chromosomes	In man, the X- and Y-chromosomes.
Single chromosome break	Both chromatids are broken at the same level resulting in the appearance of a centric fragment consisting of the proximal part of the chromosome and an acentric fragment consisting of the distal part. Such single breaks may and frequently do undergo restitution. Alternatively, the "sticky end" of the two acentric chromatids may fuse to give a U-shape fragment ("distal sister strand union"), and the corresponding end of the centric chromatids may do likewise ("proximal sister strand union"). Sister strand union occurs rather regularly in the case of breaks produced by mechanical stretching of the chromosomes on the spindle at anaphase-telophase, but less frequently after breaks induced at certain mitotic stages by radiation. Acentric fragments will be lost in the cytoplasm regardless of whether or not they have undergone sister strand union. Centric fragments that have undergone sister strand union give rise to dicentric chromatids at the next mitosis which are usually mechanically broken as they are stretched like a bridge on the spindle when the two centromeres pass to opposite poles.
Sister chromatid exchange (SCE)	A process by which sister chromatids of one chromosome exchange equivalent material and best demonstrated with techniques utilizing BUdR and other agents.
Somatic pairing	See Pairing.
S-period	The interval during which the chromosomes replicate their DNA and indispersed between the G_1- and G_2- periods.
Stalk	A thin chromatin fiber connecting the satellites to the arms of the chromosome and apparently the site of ribosomal DNA.
Stem cell	Predominent cell population with a characteristic modal number of chromosomes and a particular pattern of chromosomal aberrations that are perpetuated upon transplantation. A specific stem line is thought to characterize each tumor and at the same time to distinguish it from other tissues and from normal somatic tissues of the host.
Stickiness	One of the most common abnormalities of the chromosomes seen in tumor cells, very often resulting in clumping of chromosomes at metaphase. Sticky chromosomes also form "bridges" stretching between the poles of the mitotic spindle, thus preventing the separation of daughter chromosomes and leading to abnormal chromosomal distribution.
Submetacentric (submedian)	Describes a chromosome with the centromere located near but not at the middle of the chromosome.
Subtelocentric	Describes a chromosome with the centromere in the outer third of the chromosome. See Acrocentric.

Tandem fusion or translocation	The fusion of two acrocentrics "end-to-end" with loss of one centromere, so as to give rise to a double-length acrocentric; or, the fusion of an acrocentric, which has lost its centromere, with the end of a metacentric.
Telocentric	Chromosome with centromere at its end.
Telomere	The terminal regions of a chromosome.
Telophase	The last stage of cell division in which the cytoplasm undergoes cleavage to form two completely separate daughter cells. The chromosomes undergo in reverse the process in which they passed at prophase, i.e., they gradually elongate and become less distinct, while the nuclear membrane is reformed around them and the nucleoli reappear. At the end of this stage, cellular division is complete and the cells have truly reassumed their interphase appearance.
Terminal deletion	When deletion takes place near a terminal end.
Tetraploidy ($4n$)	Cell containing twice ($4n$) the number of chromosomes in diploid cells, i.e., in the human 92 chromosomes. See Ploidy.

Translocation
  Reciprocal (Figure 10)
  Robertsonian
  Symmetrical
  Asymmetrical
	Breakage followed by transfer of chromosome material between chromosomes:
	    Translocation may be *reciprocal*, when it is said to be balanced. If material is lost, it is unbalanced.
	    Centric fusion *(Robertsonian)* types of translocation form abnormal chromosomes by fusion of two acrocentrics in the centromeric region.
	    Symmetrical translocation is when translocated segments of the two chromosomal materials are of equal length. No change in morphology is apparent.
	    Asymmetrical translocation is characterized by unequal segments and a change in morphology.

Triploidy ($3n$)	In man, the presence of 69 ($3n$) chromosomes.
Trisomy	The presence of an extra (third) autosome in addition to a normal homologous pair. Also an extra sex-chromosome.
X-chromatin (sex-chromatin, Barr body)	Peripherally located chromatin body derived from one of the X-chromosomes in the nuclei of diploid female cells; usually characterized by the presence of one less X-chromatin body than there are X-chromosomes in the cells.
Y-chromatin (Y-body)	Fluorescent nuclear body representing the heterochromatic region of the long arm of the Y-chromosome.

References

1. For further definition of these and other terms refer to Rieger, Michaelis, et al., 1976; White, 1961.

2. Matsui, Weinfeld, et al., 1971, 1972; Matsui Yoshida et al. 1972; Sandberg, Ikeuchi et al, 1970.

2

Terminology, Classification, Symbols, and Nomenclature of Human Chromosomes

The Normal Human Karyotype

Conspectus of Human Mitotic Chromosomes
Group A (Chromosomes #1–#3)
Group B (Chromosomes #4, #5)
Group C (Chromosomes #6–#12 and the X-Chromosome)
Group D (Chromosomes #13–#15)
Group E (Chromosomes #16–#18)
Group F (Chromosomes #19, #20)
Group G (Chromosomes #21, #22)
Y-Chromosome

Chromosome Measurements

Autoradiography

Chromosome Banding Techniques
Methods and Terminology

Characterization of Chromosomes by Fluorescent Banding Techniques
Q-Banded Pattern of Individual Chromosomes

Characterization of Chromosomes by Other Banding Techniques
C-Bands
G- and R-Bands
T-Banding

Localization of Nucleolar Organizers (NOR)

Chromosome Band Nomenclature

Identification of Chromosome Landmarks and Bands
   Definitions
   Designation of arms, regions, and bands
   Diagrammatic representation of landmarks and bands
   Subdivision of an existing landmark or band

System for Designating Break Points Within Bands
   Examples

Designating Structural Chromosome Abnormalities by Breakage Points and Band Composition

Specification of Chromosome Rearrangements
   Deletions
   Translocations
   Three-break rearrangements

Specification of Break Points
Short System
   Two-break rearrangements
   Three-break rearrangements
   Rearrangements affecting two or more chromosomes

Detailed System
   Additional symbols
   Designating the band composition of a chromosome
   Examples

Four-break Rearrangements
Examples
   Terminal rearrangements
   Whole-arm translocations
   Duplication of a chromosome segment

Description of Heteromorphic Chromosomes

Code to Describe Banding Techniques

Short terminology
Complete description
Examples
Numerical aberrations
Structural alterations

Marker Chromosomes
Mosaics and Chimeras

Examples

Usage of + and − Signs

Examples

Length Changes of Secondary Constrictions

Examples

Structurally Abnormal Chromosomes

Examples

Abbreviating Lengthy Descriptions
Special Terminology

Terminology of Acquired Aberrations

Chromatid Aberrations
Chromosome Aberrations
Scoring of Aberrations
Cell Populations with Acquired Abnormalities
Tumor Cell Populations

Much of the material presented in this chapter has been taken from or based upon that of the Paris Conference (1971) and its Supplement (1975) and conferences prior or subsequent to these materials.[1]*

Many of the terms used today to describe events and cellular changes involved in mitosis were introduced years ago, and the accuracy of these descriptions is borne out by their persistence in the cytogenetic language. However, the terminology and nomenclature for human chromosomes could not be firmly established until these chromosomes were clearly visualized and described. This turned out to be more complicated than anyone had envisioned, particularly in view of the rapid developments, not only in the techniques of visualizing the structure of human chromosomes, but also in the large array of abnormal conditions accompanied by gross or subtle karyotypic changes requiring accurate notation and description. Thus, the first conference, held barely four years after the initial description of the correct number of human chromosomes, led to the Denver system of nomenclature (1960).[2] The 22 autosomal pairs were divided into 7 groups, each group containing chromosomes with similar morphology and distinct from other groups. Later modifications were introduced (London Conference, 1963), and each group was numbered by letters from A to G according to Patau's suggestion (1960, 1961 *a, b*), which is now universally accepted. The remarkable number of conditions with chromosomal abnormalities necessitated a conference, held in Chicago in 1966, at which nomenclature for these abnormalities was dealt with. The introduction of banding techniques ultimately required a conference, held in Paris in 1970, and two subsequent conferences, in 1974 and 1975, to establish standardization of the nomenclature for these subchromosomal structures. In addition, several special symposia and a number of other publications have dealt with facets of terminology of the human chromosomes.

We are fortunate in the considerable variability in the size of the human chromosomes as well as in the location of their centromere. These two characteristics afford cytogeneticists

---

*I wish to thank The National Foundation for permission to use the materials in the above publications.

a means of classifying the human chromosomes with some reliability. Hence, it was not surprising that a number of different terminologies were used by different authors to describe essentially similar and often identical chromosomes and cytogenetic aberrations. As mentioned above, this led to considerable confusion, and ultimately necessitated the holding of conferences for the express purpose of unifying the classification of chromosomes, their nomenclature, and their diseases. Even at that, the cytogenetic literature is still not characterized by a uniformity of nomenclature. It will require the special efforts and interests of reviewers, editors, and publishers to ensure that a uniform system of nomenclature is used in the description of karyotypic data published in the literature. Perhaps in this way the present confusion will eventually be ameliorated.

As mentioned above, the relative length and the location of the centromere are the most striking characteristics of the human chromosomes and serve as the most important basis for their classification (Figures 11 and 12). The centromere (kinetochore) is a unique structure of the chromosome, particularly evident during metaphase, located along the length of the chromosome, and constituting the last bond between the two chromatids of the chromosomes before their separation into daughter cells. Thus, the centromere divides the chromosomes into two segments, usually consisting of a long and a short arm, the ratio of which has been used as an important criterion for chromosome classification and identification (Figure 13).

At the time of the first conference on chromosome nomenclature and, indeed, at several of the subsequent ones, the criteria for chromosome classification, i.e., length, arm ratio, presence of satellites, and secondary constrictions, were based on findings obtained with stained material that was not always optimal (Patau, 1965). This sprang from the differences in the contraction of homologous chromosomes and occasionally between the two arms of the same chromosome and variability in the quality of the chromosomal preparations. Optimally, the metaphase set should have all the chromosomes well spread out without any overlaps, with thin and long chromatids and all chromosomes in the same state of contraction (Patau, 1960, 1961b).

Chromosomes are dark-staining structures microscopically visible in dividing cells, particularly in metaphase, and are the means by which the genetic material of a cell, contained in the nucleus, is transferred to two daughter cells in equal amounts when the cell divides. They were first described by Flemming in 1882 and, as mentioned previously, when the work of Mendel was rediscovered at the beginning of the twentieth century, it was quickly realized that they represented the structural vehicle for the segregation and assortment of the genes, the hypothetical structures then postulated as the units of genetic information.

Sometimes the chromosomes in standard preparations can be seen to show very clear helical coiling, and ways to enhance it have been described. Impressive photographs of coiling have been published,[3] and these suggest that the morphology of the coils might be an additional aid to chromosome classification. However, even though such coiling is a constant feature in some plant chromosomes, it is not sufficiently easy to induce in human chromosomes to serve as a practical method of chromosome identification, nor indeed has it been possible to show that the gyre numbers are constant for particular chromosomes; it may vary with the state of contraction. Although this coiling is of theoretical interest, inasmuch as it probably relates to the mechanism of contraction of chromosomes during metaphase, it does not appear at present that it will be useful for chromosomal identification.

A number of the chromosomes show constricted areas in their arms in certain preparations and in certain individuals, and these are called "secondary constrictions." The positions and relative frequencies of appearance of these have been described.[4] Like the centromere, the secondary constrictions appear to be true constricted regions where the chromosomal material is sparse and despiralized. Unfortunately for the cytogeneticist, the secondary constrictions are not constantly present and often can be seen on only one chromosome of a pair. Various techniques have been employed to enhance their visibility, such as modifying the fixation and spreading techniques, use of calcium-free culture medium, cold treatment, etc., but, in general, these techniques, although enhancing the appearance of the most obvious constrictions present (such as those in chromo-

Terminology, Classification, Symbols, and Nomenclature

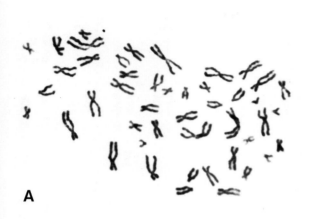

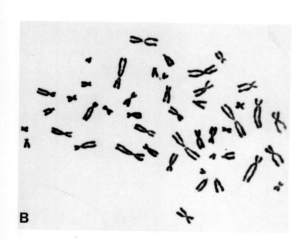

11 A. Metaphase (upper left) of a cultured blood cell from a normal female. Note variation in the size of the chromosomes, location of centromeres (kinetochores), and morphologic differences among the chromosomes. In the process of preparing the material for examination, the metaphase plate is disrupted, making it possible to spread the chromosomes and then examine them in detail. The karyotype of the metaphase (upper right) is also shown. The chromosomes have been arranged into 22 pairs of autosomes and 1 pair of sex-chromosomes (XX = female). The grouping of the chromosomes is based on their length, ratio of the arms (determined by the location of the centromere), and presence of satellites, usually present in group D and G autosomes. B. Metaphase (lower left) and karyotype (lower right) of a normal female cell.

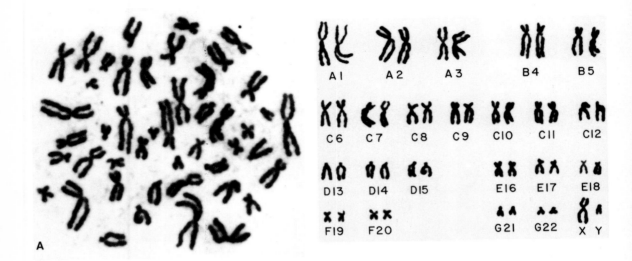
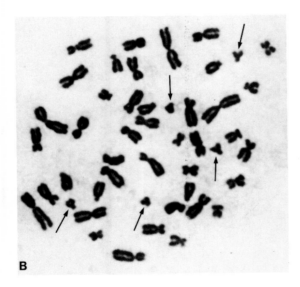

12 A. Metaphase and karyotype from a marrow cell (normal male) obtained before banding and showing the type of material that was, generally, thought at that time to be optimal for karyotypic analysis. B. Metaphase and karyotype of cells from two female cases with DS. In the metaphase (lower left) arrows point to the five acrocentrics of group G (instead of the normal four) and the karyotype (lower right) depicts trisomy of group #21, characteristic karyotypic anomaly in DS.

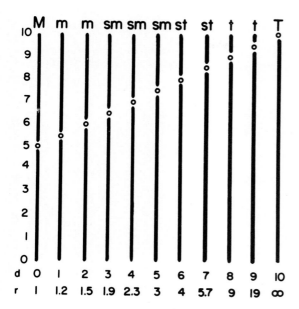

13 Nomenclature for various chromosomes based on arm ratio according to Levan, Fredga, et al. (1964a). The location of the centromere (*d*) is expressed as the difference between the length of the two arms (q−p) or as their ratio (*r*), i.e., q/p. Explanation of symbols for location of centromere: M = median point; m = median region; sm = submedian region; st = subterminal region; t = terminal region; T = terminal point.

somes #1, #3, #9, #16 and Y), do not invariably show them up as features of these chromosomes. For example, the secondary constriction most often seen is probably that on chromosome #9, on the long arm close to the centromere. Though this is often present in both members of this chromosome pair, it is also seen only in one member of the pair. This suggests that the constriction may be a structural feature of the chromosome, but one for which chromosome #9 is polymorphic. There is evidence from the constrictions associated with the chromosomal satellites that these may be inherited structural features, which segregate in Mendelian fashion. If this is so, their main importance to cytogeneticists may be as markers indicating whether a chromosome in question is paternal or maternal in origin.

In dealing with the classification, nomenclature, and some of the morphologic characteristics of the human chromosomes, the descriptions to be followed have been confined primarily to those areas relevant to chromosomes in human cancer and leukemia. Other facets of the chromosomes, particularly those related to congenital disorders and/or congenital variations in chromosomal structure and number, can be found in a number of other publications.[5]

In addition to a glossary of cytogenetic terms (presented at the end of Chapter 1), nomenclature symbols of chromosomes (Table 1), and representative figures and diagrams of the human chromosomes, the present chapter contains a collation of all pertinent descriptions, definitions, suggestions, terminology, designations, and other materials resulting from the various conferences dealing with standardization in human cytogenetics. Often these have been copied verbatim, not only because the language could not be improved upon, but also to keep the definitions and descriptions standard, uniform, and in conformity with the systems and wishes of these conferences, i.e., that all workers use a uniform nomenclature in describing cytogenetic data.

The Normal Human Karyotype[6]

Normal human somatic cells contain 46 chromosomes, though deviations from this number may be encountered in a small percentage of cells, the exact percentage depending on the tissue or origin of the cells, the techniques used in obtaining metaphases for analysis, the age and sex of the individual,[7] and the technical aspects of preparing the slides. Furthermore, it has been shown that the marrow of apparently healthy males may be missing a Y-chromosome in a few to nearly 100% of the cells.[8]

The diploid number of 46 chromosomes in man characterizes all the somatic cells in the body. The haploid number of 23 chromosomes is present only in mature geminal cells (sperm, ova). In the case of cancer or leukemia, when chromosomal changes are present they are confined to the tumor or leukemic cells only. For example, the karyotypic changes in cancer of the stomach are present only in the cancerous tissue, whereas any supporting cells or the uninvolved section of the stomach has a normal chromosome constitution. In acute leukemia, only the leukemic cells in the marrow are in-

Table 1 Nomenclature symbols

A-G	chromosome groups	
1–22	autosome numbers (Denver System)	
X, Y	sex-chromosomes	
diagonal (/)	separates cell lines in describing mosaicism	
plus sign (+) or minus sign (−)	when placed immediately before the appropriate symbol, they mean additional or missing whole chromosomes; when placed after a symbol, they mean an increase or decrease in length of chromosome	
question mark(?)	indicates questionable identification of chromosome or chromosome structure	
asterisk(*)	designates a chromosome or chromosome structure explained in text or footnote	
ace	acentric	
cen	centromere	
dic	dicentric	
end	endoreduplication	
h	secondary constriction or negatively staining region	
i	isochromosome	
inv	inversion	
inv(p+q−) or inv(p−q+)	pericentric inversion	
mar	marker chromosome	
mat	maternal origin	
p	short arm of chromosome	
pat	paternal origin	
q	long arm of chromosome	
r	ring chromosome	
s	satellite	
t	translocation	
tri	tricentric	
repeated symbols	duplication of chromosome structure	
del	deletion	
der	derivative chromosome	
dup	duplication	
ins	insertion	
inv ins	inverted insertion	
rcp	reciprocal translocation (optional, where it is desired to be more precise than provided by the use of t)	
rec	recombinant chromosome	

volved in the chromosomal changes, though when the leukemic process spreads to other tissues and organs, e.g., the spleen, aneuploid cells may be found in these locations.

Deviation from the diploid number of 46 chromosomes is observed in almost all human normal tissues examined, particularly after in vitro culture. Generally, only about 2–3% of the cells of cultured blood lymphocytes deviate from the mode of 46 chromosomes, usually consisting of metaphases with random loss of chromosomes. This is generally true of marrow cells or those observed after relatively long periods of culture. In the marrow examined by direct techniques without resort to culture, the number of aneuploid cells may run as high as 12%. This is what we observed in our laboratory some years ago.[9] It is possible that included among these numbers were males with missing Y-chromosomes, but neither our confidence in cytogenetic morphology nor the methods used at that time were sufficient to allow us an exact identification of the causes of aneuploidy. The latter consisted of both cells with loss and gain of chromosomes.

It has been demonstrated that the number of aneuploid cells found in blood cultures of normal people increases with age,[10] and this may possibly apply to the picture observed in the marrow. In both blood and marrow hypodiploid and hyperdiploid cells may be observed. Usually this change consists of a loss or gain of a single chromosome. Most commonly in *blood* it is the loss involving an X-chromosome of the female (up to 7% of the cells) and the Y-chromosome in male (1–2% of the cells). Variations in the morphology of the Y-chromosome have been established in normal individuals, varying from an unusually long or short one to a metacentric chromosome.[11]

The meaning of minor morphologic variations in the human karyotype is still obscure. The acrocentric autosomes (groups D and G) have satellites attached to their short arms, and

Table 1 — *continued*

	rob	Robertsonian translocation (optional, where it is desired to be more precise than provided by the use of t)
	tan	tandem translocation (optional, where it is desired to be more precise than provided by the use of t)
	ter	terminal or end (pter = end of short arm; qter = end of long arm)
	:	break (no reunion, as in a terminal deletion)
	::	break and join
	→	from–to
	b	break
	cs	chromosome
	ct	chromatid
	cx	complex
	dir	direct
	dis	distal
	dmin	double minute (DMS has been used also)
	e	exchange
	f	fragment
	g	gap
	med	median
	min	minute
	mn	modal number
	mos	mosaic
parentheses ( )		used to surround structurally altered chromosome(s)
	prx	proximal
	psu	pseudo
	qr	quadriradial
	rea	rearrangement
	s	satellite (also used by some authors in the past for subline)
	sce	sister chromatid exchange
	sdl	side-line, subline
	sl	stem-line (S used by some authors in the past)
semicolon (;)		separates structural rearrangements involving more than one chromosome
	tr	triradial
double underline (=)		used to distinguish homologous chromosomes (e.g., 9, 9)
	var	variable chromosome region
	xma	chiasma

these show considerable variability in size and shape, yet their exact significance regarding cellular function and integrity is unknown (Figures 36 and 37). The remarkably attenuated regions of the acrocentrics (group D and G chromosomes) connecting the short arm with the satellites, i.e., the stalks, also show variation in length and size, as evidenced by N-banding and other techniques specific for staining rDNA, and are thought to be involved in the organization of the nucleolus. Other chromosomes have secondary constrictions whose importance in cellular metabolism is unknown; and even less is known regarding the significance of the absence of or variation in such secondary constrictions.

The 46 human chromosomes can be arranged in 22 pairs of autosomes and 1 pair of sex-chromosomes (XY in the male, XX in the female; Figures 11 and 12). The chromosomes are usually classified according to their length, location of the centromere (kinetochore), the ratio of long to short arm, and presence or absence of satellites. In the past, considerable uncertainty existed in the classification and pairing of chromosomes, particularly autosomes of groups C, E, and F. Furthermore, due either to a selective effect of colcemide, to culture ecology, or to other still unknown factors, some homologues were different in size and/or staining intensity, making their exact classification uncertain. Thus, in this book the term "chromosomal changes" usually refers to relatively *gross* karyotypic changes in the morphology and/or number of chromosomes. The new chromosome techniques of banding have considerably obviated some of these difficulties and afford an approach to evaluating some of the *fine* changes in the chromosomes, these fine changes still being at a visual level and involving relatively large segments of a chromosome.

The introduction of fluorescent staining of chromosomes and other methods for "banding" patterns made it possible to identify homologues with considerable certainty. Thus, in the past, doubt was always attached to the exact pairing of chromosomes, particularly those in group C (#6–#12), and the methods mentioned above have added a new dimension of relative certainty to the identification and classification of the autosomes in the human set. Thus, by means of these methods, it has been shown that the Ph^1-chromosome is a #22 not a #21, whereas the extra autosome in Down's syndrome is a #21. Furthermore, fluorescent staining results in remarkable fluorescence of the Y-chromosome, particularly of its long arms, and makes for a very facile identification of this chromosome. In interphase cells, the Y can be identified as a fluorescent body (Srivastava, Miles, et al., 1974; Sellyei, Vass, et al., 1975). The latter would appear to have the same value as the sex-chromatin body in interphase nuclei due to the so-called inactive X-chromosome in female cells or in male cells with more than one X-chromosome. Usually, when a cell contains more than one X-chromosome, these will be late replicating and easily identified by autoradiography. These same late-replicating X-chromosomes apparently constitute the sex-chromatin bodies seen in interphase nuclei. It should be pointed out, however, that *each* Y-chromosome shows outstanding fluorescence in interphase nuclei or in metaphases.

"Marker" chromosomes are morphologically different from any of those of the human karyotype. They are often present in cancerous cells, but their exact origin has remained uncertain in most cases. Again, the fluorescent staining and banding techniques have afforded a new approach in deciphering the genesis of "markers." More importantly, in some cases *normal looking* chromosomes have been shown to be, in fact, marker chromosomes as a result of fusion of whole or part of two chromosomes. Thus, the application of these new cytogenetic techniques may, in the future, yield information on the nature of marker chromosomes, particularly regarding those of similar morphology in the same or divergent types of tumors.

Studies have shown that the Y-chromosome may be missing from marrow cells of normal males, particularly elderly males. Apparently, the incidence of the missing Y increases with the age of the males and reaches a very high percentage in very elderly males (more than 75 yr old). The number of cells in the marrow with a missing Y may vary from a few to nearly 100% of the cells. No other tissue in the body of the human has been shown to have this abnormality, though only a few tissues have been examined. The marrow cells in elderly females do not contain a comparable anomaly.

Some changes in the chromosome complement occur with aging (Court Brown, Buckton, et al., 1966; Neurath, Remer, et al., 1970), particularly in blood leukocytes. The question has been raised whether aging of tumor cells is related to the alteration of the karyotype (T. H. Yosida, 1972).

Chromosomes have been shown to replicate their DNA asynchronously in the cell cycle, with one of the X-chromosomes in female cells (or when more than one X-chromosome is present in male cells) being the latest to finish its DNA replication. Because much of the latter occurs over a relatively short period of time at the end of the S-phase, particularly in such chromosomes as the X, autoradiography after incorporation of tritiated thymidine into the DNA of chromosomes, when the cells are pulsed with labeled thymidine at the end of culture, reveals a very heavily labeled X-chromosome. This reliable identification of the X-chromosome has been of great help in some hereditary and congenital disorders and has also afforded reliable description of the morphology of the X-chromosomes.

Conspectus of Human Mitotic Chromosomes

The chromosome characteristics described below are based on observations with standard stains (Giemsa or Feulgen reactions) before the introduction of banding techniques and have taken into account the length of the chromosome, centromere location, the site of secondary constrictions (if present), and the presence of satellites. For a more detailed characterization of the various chromosomes, see the section on banding in this chapter.

Group A (Chromosomes #1–#3)

Group A is comprised of large chromosomes with approximately median centromeres (metacentric). The three chromosomes are readily distinguished from each other by size and centromere position. In many cells a secondary constriction is observed in #1 in the proximal region of the long arm. Chromosome #2 is of similar length as #1, but its centromeric position is less median than that of #1 or #3; #3 is the smallest metacentric of this group.

Group B (Chromosomes #4, #5)

Large chromosomes with submedium centromeres (submetacentric). The two chromosomes are difficult to distinguish, but #4 is slightly longer.

Group C (Chromosomes #6–#12 and the X-Chromosome)

Medium-sized chromosomes with submedium centromeres (submetacentric). Of the autosomes in this group, four are comparatively metacentric, i.e., #6, #7, #8, and #11; three are submetacentric, i.e., #9, #10, and #12. A secondary constriction is found in the proximal part of the long arm (near the centromere) of at least one chromosome of the #9 pair. The X-chromosome belongs to the subgroup with a comparatively metacentric centromere and resembles especially chromosome #6, from which it is difficult to distinguish. In normal female cells, one X-chromosome characteristically incorporates isotopically labeled thymidine over most of its length later than the others in the group.

Group D (Chromosomes #13–#15)

Medium-sized chromosomes with nearly terminal centromeres ("acrocentric" chromosomes). Satellites have been detected in all three chromosome pairs. However, all the satellites are seen only when the preparation is optimal. The three pairs are difficult to separate on morphological grounds.

Group E (Chromosomes #16–#18)

Rather short chromosomes with approximately median (in chromosome #16) or submedium centromeres. A secondary constriction is seen frequently in the proximal part of the long arm of #16. Pairs #17 and #18 are submetacentric and cannot be separated in every cell. However, chromosome #17 is longer and less submetacentric than #18.

Group F (Chromosomes #19, #20)

Short chromosomes with approximately median centromeres (metacentric). Difficult to distinguish from each other on morphological grounds.

Group G (Chromosomes #21, #22)

Very short, acrocentric chromosomes, smallest in the complement. Satellites have been detected on both #21 and #22. The Y can be separated out easily in most cases because of the following morphological characteristics: (a) it is longer than #21 and #22, (b) the long-arm chromatids diverge less than those of #21 and #22, and (c) the Y does not have satellites on its short arm.

Y-Chromosome

The most commonly observed Y is larger than either #21 or #22, its centric constriction is often indistinct, and a secondary constriction is frequently seen in the long arm; the terminal region of the long arm may be poorly defined. Typically, the two long-arm chromatids appear to diverge less than those of other chromosomes. Y-chromosomes that are unusually long (sometimes as long as D-group chromosomes), unusually short (less than half the length of G-group chromosomes), or metacentric have all been described in males who are normal and fertile. Such unusual Y-chromosomes are inherited at constant length.

Chromosome Measurements

Statements about the relative length of chromosomes identified by Q-, G-, or R-bands can now be made. The X-chromosome, both in males and in females, ranks between chromosomes #7 and #8 in total length and in short-arm length (Table 2). The X-chromosome length means were comparable in male and female cells. The X and #11 are the most metacentric chromosomes in group C. Chromosome #12 has the smallest short arm and centromere index and is demonstrably different from #10. Chromosome #15 has a higher centromere

Table 2 Measurements of relative length (in percentage of the total haploid autosome length) and centromere index (length of short arm divided by total chromosome length × 100)

Chromosome no.	Relative length A	Relative length B	Relative length C	Relative length D	Centromere index A	Centromere index B	Centromere index C	Centromere index D
1	9.08	9.08 ± 0.611	9.11 ± 0.53	8.44 ± 0.433	48.0	49.4 ± 3.04	48.6 ± 2.6	48.36 ± 1.166
2	8.45	8.17 ± 0.250	8.61 ± 0.41	8.02 ± 0.397	38.1	39.4 ± 2.05	38.9 ± 2.6	39.23 ± 1.824
3	7.06	6.96 ± 0.352	6.97 ± 0.36	6.83 ± 0.315	45.9	47.6 ± 2.10	47.3 ± 2.1	46.95 ± 1.557
4	6.55	6.62 ± 0.403	6.49 ± 0.32	6.30 ± 0.284	27.6	29.2 ± 2.97	27.8 ± 3.3	29.07 ± 1.867
5	6.13	6.34 ± 0.366	6.21 ± 0.50	6.08 ± 0.305	27.4	29.2 ± 3.03	26.8 ± 2.6	29.25 ± 1.739
6	5.84	6.19 ± 0.516	6.07 ± 0.44	5.90 ± 0.264	37.7	39.1 ± 2.63	37.9 ± 2.5	39.05 ± 1.665
7	5.28	5.60 ± 0.435	5.43 ± 0.47	5.36 ± 0.271	37.3	35.3 ± 2.90	37.0 ± 4.2	39.05 ± 1.771
X	5.80	5.45 ± 0.377	5.16 ± 0.24	5.12 ± 0.261	36.9	41.4 ± 6.16	37.5 ± 2.7	40.12 ± 2.117
8	4.96	5.13 ± 0.307	4.94 ± 0.28	4.93 ± 0.261	35.9	32.7 ± 2.80	32.8 ± 2.8	34.08 ± 1.975
9	4.83	4.81 ± 0.194	4.78 ± 0.39	4.80 ± 0.244	33.3	37.0 ± 3.04	32.7 ± 4.1	35.43 ± 2.559
10	4.68	4.66 ± 0.512	4.80 ± 0.58	4.59 ± 0.221	31.2	35.4 ± 3.81	32.3 ± 2.9	33.95 ± 2.243
11	4.63	4.70 ± 0.289	4.82 ± 0.30	4.61 ± 0.227	35.6	40.7 ± 3.07	40.5 ± 3.3	40.14 ± 2.328
12	4.46	4.66 ± 0.410	4.50 ± 0.26	4.66 ± 0.212	30.9	30.5 ± 3.64	27.4 ± 4.0	30.16 ± 2.339
13	3.64	3.22 ± 0.310	3.87 ± 0.26	3.74 ± 0.236	14.8	—	16.6 ± 3.6	17.08 ± 3.227
14	3.55	3.09 ± 0.212	3.74 ± 0.23	3.56 ± 0.229	15.5	—	18.4 ± 3.9	18.74 ± 3.596
15	3.36	2.83 ± 0.262	3.30 ± 0.25	3.46 ± 0.214	14.9	—	17.6 ± 4.6	20.30 ± 3.702
16	3.23	3.46 ± 0.353	3.14 ± 0.55	3.36 ± 0.183	40.6	42.2 ± 3.57	42.5 ± 5.6	41.33 ± 2.74
17	3.15	3.06 ± 0.377	2.97 ± 0.30	3.25 ± 0.189	31.4	36.6 ± 5.86	31.9 ± 3.3	33.86 ± 2.771
18	2.76	2.98 ± 0.316	2.78 ± 0.18	2.93 ± 0.164	26.1	31.5 ± 4.15	26.6 ± 4.2	30.93 ± 3.044
19	2.52	2.55 ± 0.269	2.46 ± 0.31	2.67 ± 0.174	42.9	48.1 ± 2.48	44.9 ± 4.0	46.54 ± 2.299
20	2.33	2.61 ± 0.144	2.25 ± 0.24	2.56 ± 0.165	44.6	46.5 ± 3.59	45.6 ± 2.5	45.45 ± 2.526
21	1.83	1.34 ± 0.189	1.70 ± 0.32	1.90 ± 0.170	25.7	—	28.6 ± 5.0	30.89 ± 5.002
22	1.68	1.53 ± 0.178	1.80 ± 0.26	2.04 ± 0.182	25.0	—	28.2 ± 6.5	30.48 ± 4.932
Y	1.96	1.82 ± 0.353	2.21 ± 0.30	2.15 ± 0.137	16.3	—	23.1 ± 5.1	27.17 ± 3.182

Chromosomes stained with orcein or the Giemsa 9 method and preidentified by Q-band patterns.

A, Denver-London data (not preidentified by Q-staining method).
B, data from 20 cells stained with orcein. The short arms in groups D and G and in the Y were excluded.
C, data from 10 cells stained with orcein.
D, Data from 95 cells from 11 normal subjects (6–10 cells per person). Average total length of chromosomes per cell: 176 μm. Cells stained with orcein or Giemsa 9 technique.
Cells in B, C, and D were measured from projected negatives of metaphase cells. Standard deviations in samples B and C are based on the total sample of measurements. Standard deviations in sample D are an average of the standard deviations found in each of 11 subjects (6–10 cells per subject).
Table taken from the Paris Conference (1972).

index than that described in the past, and #19 is longer than #20. However, because the total length of both #19 and #20 are quite similar, discrepancy occasionally reported may be due to the relatively small sample sizes. These results have been based exclusively on measurements of length. Preliminary data on a small number of cells suggest that a centromere index based on integrated optical density will generally be smaller than that based on length measurements. This difference was particularly marked in groups D and G. In summary, the measurements presented in Table 2, columns B, C, and D, provide guidelines for construction of a karyotype when homologue identification cannot be carried out.

Most of the variations in autosomal morphology are the result of changes at sites of secondary constrictions, perhaps because these are sites of repeating or redundant DNA, which either has unusual coiling properties or may contain deletions or duplications that produce no obvious phenotypic effects because of the nature of the DNA involved. The more carefully the human metaphase karyotype is examined, the more evident it seems that most of these variations are inherited polymorphisms of the chromosomes, which may be expected to vary in incidence both in individuals and in different population groups, so that many individuals can be distinguished from each other by means of these structural variations of their chromosomes. The utilization of banding techniques, however, usually leads to the identification of the chromosome involved by such variations, thus affording a means of differentiating chromosomes when their relative length and morphology have been modified by a varying amount of DNA at the secondary constriction areas.

Autoradiography

A combination of detailed morphological studies of the chromosomes with autoradiography has made possible unambiguous identification of chromosomes #4, #5, #13, #14, #15, #17, and #18. Furthermore, autoradiography is the best method for identifying the late-replicating X-chromosome(s) in cells having more than one X. It also has some value in the identification of the Y and in characterization of the commonly observed inherited autosomal variants.

The use of autoradiography together with the newer and simpler banding techniques, which permit identification of each human chromosome pair, will provide recognition of the patterns of DNA synthesis of previously unidentifiable chromosomes, such as members of the group 6-X-12, and will make possible investigation of the chromosomal features revealed by other techniques. Their combined use may be particularly advantageous in the study of structurally abnormal chromosomes, e.g., markers in leukemia and cancer and X-autosome translocations.

As a means of identifying individual chromosomes and abnormal (marker) chromosomes, autoradiography with labeled DNA precursors (especially tritiated thymidine) is of limited value when compared to the more recent observations obtained with banding techniques. Nevertheless, autoradiography does shed considerable light upon the chronology of DNA replication in the chromosomes and the fact that chromosomes replicate their DNA at different rates through the S-period, with one of the X-chromosomes in female cells (or in any cell containing more than one X-chromosome) replicating the bulk of its DNA at the end of the S-period. This behavior on the part of the X-chromosome makes possible its identification in metaphase plates after autoradiography and has helped considerably in the definition of the morphologic features of the X-chromosome.[12] Even though a few other chromosomes or parts of chromosomes are characterized by unique behavior during the S-period upon autoradiography, most of the chromosomes show a pattern too indistinct to allow these patterns to be reliable indices of chromosome identification. This is further compounded by the fact that homologues often did not show similar autoradiographic behavior.[13]

Chromosome Banding Techniques

The introduction of banding techniques has not only made possible the rigorous identification of individual chromosomes but has also afforded an opportunity to characterize and identify subchromosomal morphology. Each technique produces bands characteristic for the

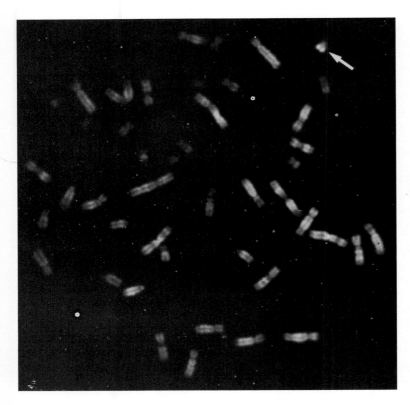

14 Q-banded metaphase of a cultured blood cell (normal male). Note prominent fluorescence of the long arms of the Y (arrow).

methodology being employed. Refinements in techniques have also allowed the subclassification of various bands. Even though standardization of the banded and unbanded regions has been introduced through several conferences and a number of publications, sufficient variability exists among laboratories and materials so that there are some differences in the banding patterns obtained. Generally, these differences are based on nuances, and undoubtedly, in time, consistent and universally accepted definitions of the various bands will be reached.

Methods and Terminology

Several different technical procedures have been reported which produce banding patterns along the metaphase chromosomes. Even though the biochemical basis of the various staining reactions remains obscure, most of these procedures give similar cytologic results. In this section, the different staining patterns have been assigned provisional names based either on the operational procedures used to obtain the patterns or on previously employed designations that have since come into general use.

Methods that demonstrate "constitutive heterochromatin" are designated as *C-staining methods*, and the term *C-band* is used to describe a unit of chromatin stained by these methods (Figure 16). The methods first published for demonstrating bands along the chromosomes were those that used quinacrine mustard or quinacrine dihydrochloride to produce a fluorescent banding pattern. These methods are named *Q-staining methods* and the resulting bands, *Q-bands* (Figure 14). Other techniques which demonstrate bands along the chromosomes use the Giemsa dye mixture as the staining agent; they are generally termed *G-staining methods* and the resulting bands, *G-bands* (Figure 17). One of the techniques using the Giemsa reagent, however, gives patterns that are opposite in staining intensity to those obtained by the G-staining methods. This technique is called the *reverse-staining Giemsa method (R-staining method)* and the resulting bands, *R-bands*.

A *band* is defined as a part of a chromosome clearly distinguishable from its adjacent segments by appearing darker or lighter with the Q-, G-, R-, or C-staining methods. Bands that stain darkly with one method may stain lightly with other methods. The chromosomes are visualized as consisting of a continuous series of light and dark bands, so that by definition there are no "interbands."

Characterization of Chromosomes by Fluorescent Banding Techniques

The description to be given of the human somatic karyotype is based on the fluorescent staining pattern. Because most laboratories are not equipped for densitometry, the description has been confined to visually recognizable patterns; these have been confirmed, however, by comparison with the densitometric results of Caspersson, Lomakka, et al. (1971*a, b*). Identification of chromosomes on the basis of length, centromeric index, autoradiographic characteristics, and location of secondary constrictions, as outlined in the Chicago Conference (1966) report, is retained in the present description. This applies to chromosomes #1 – #5, #9, #13 – #18, and the Y (X-chromosomes in numbers >1 can be identified by their late-replicating behavior). The numbers assigned to the remaining autosomes are based on their fluorescent banding patterns as given by Caspersson, Lomakka, et al. (1971 *a, b*). The designation of the additional chromosome associated with Down's syndrome has been retained as #21, although it is now known to be smaller than the #22.

In the description that follows only major fluorescent bands will be referred to, even though in some cells these may appear to consist of several smaller bands. Faintly fluorescing bands are not referred to except when they are of special significance; generally, it may be assumed that they separate the major fluorescent bands or are located at the ends of the chromosome arms.

Diagnostic features indicated by "A" are those seen in fluorescent metaphases of fair technical quality; those indicated by "B" are usually visible only in cells of good quality.

When these details are not included in the text, the banding pattern is identical to that described under "A." Features that may vary in fluorescent intensity or length or both between individuals and between homologues are indicated by "C." The terms "distal" and "proximal" refer to the position of a band in respect to the centromere; "centric" means the area occupied by the centromere.

Some mitoses show considerable nonuniformity in that the homologous chromosomes may differ greatly in overall fluorescence and relative length. Identification must be based, therefore, on the fluorescent banding patterns of the individual chromosome rather than on its overall intensity. However, intensity may serve as a secondary criterion, if due allowance is made for nonuniformity. The following terms indicate the approximate intensity of fluorescence:

Negative — no or almost no fluorescence
Pale   — as on distal 1p
Medium — as the two broad bands on 9q
Intense — as the distal half of 13q
Brilliant — as on distal Yq

Q-Banded Pattern of Individual Chromosomes

Chromosome #1 (see Figures 14 and 15)

The long arm is that previously defined as the arm with a proximal secondary constriction.

  A p: Distal, pale segment grading to a proximal, medium fluorescent segment.
      q: Central, intense band. Proximal, negative secondary constriction.
  B p: Proximal, medium fluorescent segment, divisible into two bands.
      q: Five medium fluorescent bands; central one most prominent.
  C q: Negative secondary constriction variable in length.

Chromosome #2

  A Medium fluorescence along the whole length.
  B p: Four medium fluorescent bands; two central ones often appear as a single segment.
      q: Two central bands, sometimes accompanied by another two, all of me-

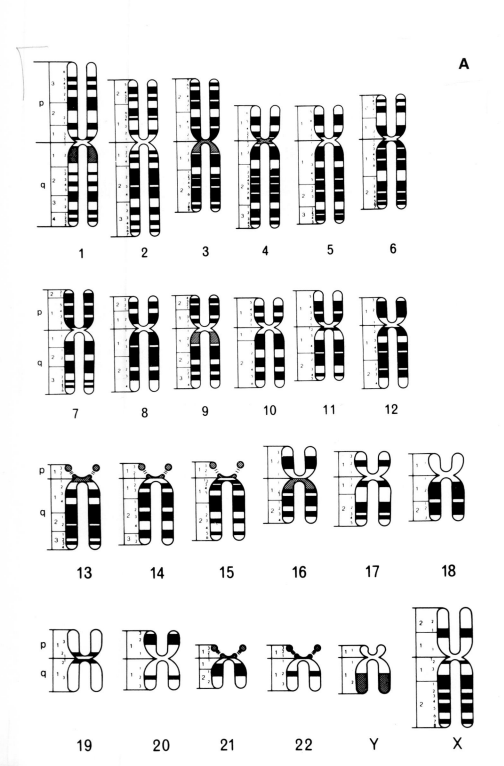

15 A. Banding pattern of the human chromosomes according to the Paris Conference (1972). Band 3p27 is now thought not to exist and the terminal band on 3p is 3p26; though not shown here, bands 8q21, 19q13, Xp11, and Xp22 have been subdivided into subbands (ISCN, 1978).

15

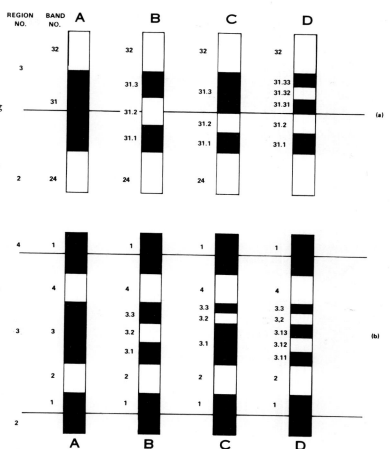

B. (a) Example illustrating the convention for numbering the subdivisions of a landmark bridging two regions: (A) the original landmark (band 31); (B) the subdivision of band 31 into 3 equal bands 31.1, 31.2, and 31.3; (C) alternatively, the subdivision of band 31 into three unequal bands; (D) further subdivision of band 31.3 into three equal bands 33.31, 33.21, and 33.33. (b) Example illustrating the convention for numbering the subdivisions of a band within a region: (A) the original band 33; (B) 3 equal bands 33.1, 33.2 and 33.3; (C) alternatively, the subdivision of band 33 into 3 equal bands; (D) further subdivision of band 33.3 into three equal bands 33.11, 33.12, and 33.13.
(Paris Conference, 1971.)

dium fluorescence. Additional bands can be seen sometimes.

Chromosome #3
A Single pale band in center of each arm separating medium fluorescent segments. Distal, medium fluorescent segment; longer in q than in p.
B Single pale band at end of each arm; longer in p than in q.
C q: Proximal band of variable fluorescence.

Chromosome #4
A Medium fluorescence along the whole length.
B p: Single central, medium fluorescent band.
   q: Proximal, intense band. Distal, pale band.
C Intense centric band.

Chromosome #5
A q: Central, long, medium fluorescent segment. Distal, pale segment.
B p: Single medium fluorescent band; shorter and brighter than on 4p.
   q: Distal, pale segment; divisible into a proximal, pale band and a distal, medium one.

Chromosome #6
A p: Central, pale band separating medium fluorescent segments.
   q: Medium fluorescence along entire length.
B q: Four medium fluorescent bands.

Chromosome #7
A p: Distal, short, medium fluorescent band.
   q: Two central, intense bands. Distal, medium fluorescent band.

B p: Proximal, medium fluorescent band.
Chromosome #8
   A Medium fluorescence along the whole length; q brighter than p.
   B p: two evenly spaced, medium fluorescent bands.
      q: Two medium fluorescent bands in distal half; brighter than those on p.
Chromosome #9
   A q: Proximal, negative segment corresponding to the secondary constriction. Two evenly spaced, medium fluorescent bands distal to the negative segment.
   B p: Central, medium fluorescent band.
   C q: Proximal, negative band (secondary constriction) variable in length.
Chromosome #10
   A p: Medium fluorescence.
      q: Three evenly spaced bands; the most proximal one intense and the others medium in fluorescence.
Chromosome #11
   A p: Medium fluorescence; longer than 12p.
      q: Short, medium fluorescent band adjacent to the centromere; separated by a negative band from a more distal, medium fluorescent segment.
Chromosome #12
   A p: Medium fluorescence; shorter than 11p.
      q: Medium fluorescent band adjacent to the centromere; separated by a short, negative band from a more distal, medium fluorescent segment. Distal segment longer than that of 11q.
Chromosome #13
   A q: Distal half intense.
   B q: Distal half intense; divisible into two bands.
   C p: Satellites and/or short arms with variable fluorescence.
      q: Proximal, intense band.
Chromosome #14
   A q: Proximal half intense. Distal half pale; medium fluorescent band close to the distal end.
   C p: Satellites and/or short arms with variable fluorescence.
Chromosome #15
   A q: Proximal half medium in fluorescence. Distal half pale; less fluorescent than either 13q or 14q.

   C p: Satellites and/or short arms with variable fluorescence.
Chromosome #16
   A p: Medium fluorescence, less fluorescent than q.
      q: Proximal, negative segment corresponding to the secondary constriction. Distal to it, a medium fluorescent segment.
   C q: Negative secondary constriction variable in length.
Chromosome #17
   A p: Overall pale fluorescence.
      q: Two segments of similar length; proximal one pale and distal one medium in fluorescence.
   B q: Narrow negative band separating proximal and distal segments.
Chromosome #18
   A p: Overall medium fluorescence.
      q: Medium fluorescence; brighter than p.
   B q: Two bands of medium intensity; proximal one longer and brighter than distal one.
Chromosome #19
   A Most weakly fluorescent chromosome in the karyotype. Short, proximal fluorescent bands on both arms; pale when compared to the whole karyotype.
   B Fluorescent band longer and brighter on p than on q.
Chromosome #20
   A Overall pale fluorescence; p medium and q pale in fluorescence.
Chromosome #21
   A q: Proximal, intense segment. Distal, pale segment.
   C p: Satellites and/or short arms with variable fluorescence.
Chromosome #22
   A Overall pale fluorescence.
   B q: Narrow, pale band in center of arm.
   C p: Satellites and/or short arms with variable fluorescence.
X-Chromosome
   A p: Proximal, pale segment. Central, medium fluorescent band.
      q: Proximal, pale segment. Distal to it, a medium fluorescent band.
   B q: Three evenly spaced, medium fluorescent bands; most proximal one brightest.

Y-Chromosome
A p: Overall pale fluorescence
   q: Proximal segment pale. Distal segment brilliant.
C q: The brilliant fluorescent segment on the end of q may vary in length and may be subdivided into two or more bands. The normal variation in length of the chromosome is associated with variation in length of the brilliant segment.

Characterization of Chromosomes by Other Banding Techniques

C-bands (Table 2A)

The banding patterns obtained with the various C-staining methods[14] (Figure 16) do not permit individual identification of each chromosome of the human somatic-cell complement. In this sense, C-bands are not strictly comparable to those obtained with the Q-[15], G-[16], or R-[17] staining methods. Used in conjunction with these techniques, however, the C-staining methods provide much useful information on the type and localization of chromatin throughout the complement.

Chromosome	C-band size
#1	Large, extends from centromere into q
#2	Small
#3 – #8	Medium
#9	Large, extends from centromere into q
#10	Medium
#11	Medium, but larger than on #10 or #12
#12	Medium
#13	Medium, but sometimes bipartite
#14 – #15	Medium
#16	Large, extends from centromere into q
#17	Medium
#18	Medium, but larger than on #17
#19 – #22	Medium
X-chromosome	Medium
Y-chromosome	Very small band at centromere; large band on distal end of q

In the above description, the C-bands are defined by their length and position along the

Table 2A Variation in staining of specific bands with various techniques*

Technique												Chromosome number												
	1	2	3	4	5	6	7	8	9	10	11	12	13	14	15	16	17	18	19	20	21	22	X	Y
C	c	c	c	c	c	c	c	c	qhv	c	c	c	c	c	c	c	c	c	c	c	c	c	c	c
G11	qhv	qhv	–	cv	–	–	p11	–	qhv	qh	–	–	pv	pv	pv	qh	–	–	–	–	pv	pv	–	q12v
R or T	p36	q37	–	p16	p15	–	p22	p24	q34	q11 q26	p15	–	p12v	p12v	p p12v	– p13	q11 q25	– p13	– p13	q11 q13	p12v	p12v q11	–	q12v –
					q35									q32		q24			q13		q22	q13		
NOR	–	–	–	–	–	–	–	–	–	–	–	–	p11v p13v cv	p11v p13v	p11v p13v	–	–	–	–	–	p11v p13v	p11v p13v	–	p12v
Q†	–	–	cv	cv	·	–	–	–	–	–	–	–	–	–	–	–	–	–	–	–	–	–	–	–

*c = cen = centromere, h = secondary constriction, v = var = variable, p = short arm, q = long arm.
†Only the brilliant and variable Q-bands have been considered.
Table taken from ISCN (1978).

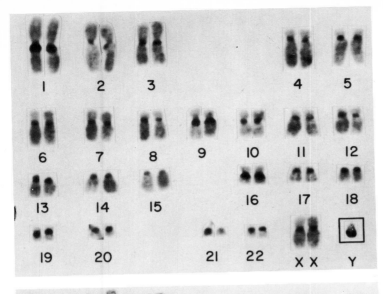

16 C-banded karyotype of a normal female cell (from marrow). The C-banding of the Y from a male cell is shown in brackets. Note some variability in C-staining between homologues constituting the so-called polymorphism of chromosomes, apparently an inherited characteristic for each chromosome.

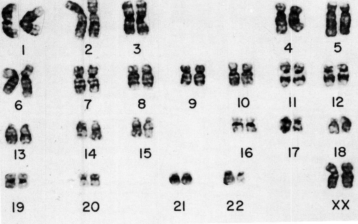

17 G-banded karyotype of a normal female marrow cell.

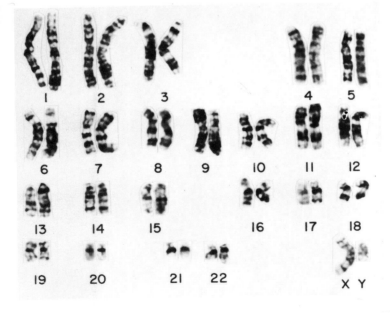

18 G-banded karyotype of a normal male marrow cell.

chromosomes. The depth, or intensity, of staining is not taken into consideration. Unless otherwise indicated, the C-band corresponds in position to the centromeric region.

The C-bands on #1, #9, and #16 and the large distal band on the Yq are all associated with obvious morphologic variability.

G- and R-bands

The banding patterns obtained with the G- and R-staining methods (Figures 17 and 18) correspond to those obtained by Q-staining, except for the following chromosome segments (h = secondary constriction):

	Q-band	G-band	R-band	C-band
1qh	Negative	+	−	+
9qh	Negative	−	−	+
16qh	Negative	+	−	+
distal Yq	Brilliant	Variable	Variable	+

Bands that appear light or unstained with G-staining in general stain darkly with the R-band technique. The only exception is the 9qh, which appears lightly stained with both methods. As a rule, neither the G- nor R-staining methods clearly demonstrate those Q-bands that vary in length or intensity and appear near the centromeres of #3, #4, #13–#15, #21, and #22.

Morphologic variability in satellite size or density is reflected by variation in the size and staining intensity of the Q-, G-, R-, and C-bands. However, none of these banding methods can distinguish late- from early-replicating X-chromosomes.

T-Banding

Under controlled thermic denaturation conditions and staining with acridine orange, it is possible to develop bands in some of the terminal regions of the chromosomes with rather characteristic location and morphology (see Table 3). The most evident banding obtained with this technique is observed in the terminal areas of the short arms of #1, #4, #7, #11, and #19 and in the terminal areas of the long arms of #8, #9, and #17. Lesser fluorescent intensity is observed on the long arms of #10, #14, #16, #20, #21, and #22 and on the short arm of #16. Apparently, the T-banding technique offers considerable advantages in demonstrating some translocations, not readily analyzable by other banding techniques, because it allows a precise location of juxtatelomeric break points. Utilizing Giemsa, instead of acridine orange, a similar distribution of banding can be obtained including evident bands in the long arms of #11 and #12 near their centromeric areas.

Table 3 Representation of the intensity of the fluorescence of the telomeric areas in the various chromosomes as demonstrated by T-banding

Chromosome pair	Short arm	Long arm	Chromosome pair	Short arm	Long arm
1	+++	0	13	var?	+
2	0	+	14	var?	++
3	0	0	15	var?	0
4	+++	0	16	++	++
5	+	+	17	0	+++
6	0	0	18	0	0
7	+++	0	19	+++	0
8	0	+++	20	0	++
9	0	+++	21	0	++
10	0	++	22	0	++
11	+++	0	X	0	0
12	0	+	Y	0	0

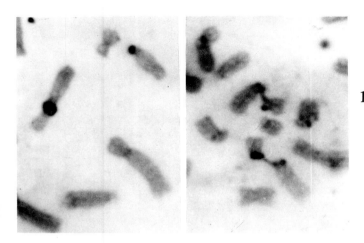

19 N-banding of acrocentric chromosomes as revealed by the dark staining areas. Connections between the acrocentric chromosomes are evident by the continuity of the staining areas. *(Hayata, Oshimura, et al., 1977.)*

Localization of Nucleolar Organizers (NOR)

The acrocentric chromosomes in the human karyotype possess terminally achromatic secondary constrictions on their short arms connected distally to chromatin pieces called "satellites." The secondary constriction regions of the human acrocentric chromosomes are known as "stalks." Various methods have been utilized to identify this region, the most common being N-banding and that based on silver-staining (see Figures 19–21). These methods have revealed that the stalks, sites of ribosomal DNA (rDNA), are probably the areas involved in nucleolar organization rather than the satellites.[18] Some variability in the number of stained stalks is observed from cell to cell and from individual to individual. Occasionally, N-banding and methods utilizing silver staining

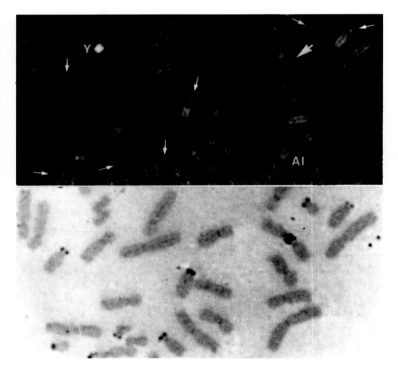

20 Partial male metaphase stained with quinacrine mustard (top), destained in 70% ethanol and restained with N-banding (bottom). A satellite association (large arrow) is clearly demonstrated. Note that N-bands are restricted to the stalk regions of D- and G-group chromosomes and are not seen in other groups.

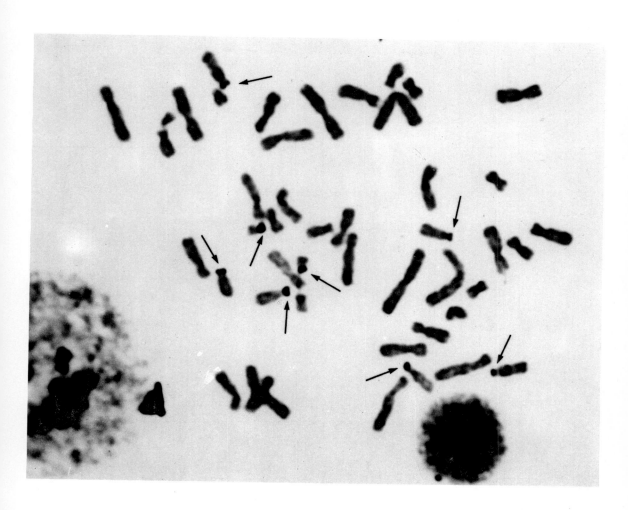

21 Silver staining of nuclear organizing regions (NOR). Arrows point to staining of eight acrocentric chromosomes, i.e., six of the D group and only two of the G group.

extend beyond the stalk and cover the satellites, especially in chromosomes with short stalks or minute satellites.

Methods have also been devised, relying primarily on ammoniacal silver staining, with apparent specificity for satellite III DNA regions on human chromosomes[19] (#1, #9, and #16), the regions usually being associated with secondary constrictions on these metacentric and submetacentric chromosomes.

Chromosome Band Nomenclature

Identification of Chromosome Landmarks and Bands

Each chromosome in the human somatic-cell complement is considered to consist of a continuous series of bands, with no unbanded areas. The bands are allocated to various regions along the chromosome arms and delimited by specific chromosome landmarks. The bands and the regions they belong to are identified by numbers, with the centromere serving as the point of reference for the numbering scheme (see Table 4).

Definitions. The definition of a *band* has been given earlier as a part of a chromosome clearly distinguishable from adjacent parts by virtue of its lighter or darker staining intensity.

A *chromosome landmark* is defined as a consistent and distinct morphologic feature that is an

Table 4 Bands serving as landmarks that divide the chromosomes into cytologically defined regions

Chromosome no.	Arm	Number of regions	Landmarks*
1	p	3	Proximal band of medium intensity (21), median band of medium intensity (31)
	q	4	Proximal negative band (21) distal to variable region, median intense band (31), distal medium band (41)
2	p	2	Median negative band (21)
	q	3	Proximal negative band (21), distal negative band (31)
3	p	2	Median negative band (21)
	q	2	Median negative band (21)
4	q	3	Proximal negative band (21), distal negative band (31)
5	q	3	Median band of medium intensity (21), distal negative band (31)
6	p	2	Median negative band (21)
	q	2	Median negative band (21)
7	p	2	Distal medium band (21)
	q	3	Proximal medium band (21), median band of medium intensity (31)
8	p	2	Median negative band (21)
	q	2	Median band of medium intensity (21)
9	p	2	Median intense band (21)
	q	3	Median band of medium intensity (21), distal band of medium intensity (31)
10	q	2	Proximal intense band (21)
11	q	2	Median negative band (21)
12	q	2	Median band of medium intensity (21)
13	q	3	Median intense band (21), distal intense band (31)
14	q	3	Proximal intense band (21), distal medium band (31)
15	q	2	Median intense band (21)
16	q	2	Median band of medium intensity (21)
17	q	2	Proximal negative band (21)
18	q	2	Median negative band (21)
21	q	2	Median intense band (21)
X	p	2	Proximal medium band (21)
	q	2	Proximal medium band (21)

The omission of an entire chromosome or a chromosome arm indicates that either both arms or the arm in question consists of only one region, delimited by the centromere and the end of the chromosome arm.

*The numbers in parentheses are the region and band numbers as shown in Figure 15.

Table taken from the Paris Conference (1972).

important diagnostic aid in identifying a chromosome. Landmarks include the ends of the chromosome arms, the centromere, and certain bands.

A *region* is defined as any area of a chromosome lying between two adjacent landmarks.

Designation of arms, regions, and bands. The symbols p and q designate the short and long arms of each chromosome, respectively. Regions and bands are numbered consecutively from the centromere outward along each chromosome arm. Thus, the two regions adjacent to the centromere are labeled "1" in each arm, the next, more distal regions, "2," and so on. A band used as a landmark is considered as belonging entirely to the region distal to the landmark and is accorded the band number of "1" in that region. A band bisected by the centromere is considered as two bands, each being labeled as band 1, in region 1, of the appropriate chromosome arm.

In designating a particular band, four items are required: the chromosome number, the arm symbol, the region number, and the band number within that region. These items are given in order without spacing or punctuation. For example, 1p33 indicates chromosome #1, short arm, region 3, band 3.

Diagrammatic representation of landmarks and bands. The chromosome banding diagram shown in Figure 15 is based on the patterns observed in different cells stained with either the Q-, G-, or R-band technique. As indicated earlier, the banding patterns obtained with these staining methods agree sufficiently to allow the construction of a single diagram representative of all three techniques, although the position of the centromere has been indicated on the basis of the Q-band technique only. The diagram is not based on measurements of the length and position of the chromosome bands; however, the relative band sizes and distributions can be taken to be approximately correct. The bands are designated on the basis of their midpoints and not by their margins. No attempt has been made to indicate the intensity of fluorescence or staining, because this will vary with different techniques. Intensity has been taken into consideration, however, in determining which bands should serve as landmarks on each chromosome, apart from the centromere and chromatid ends, in order to divide the chromosome into natural, easily recognizable morphologic regions. (A list of those bands used in constructing Figure 15 is provided in Table 4.) The C-staining method has not been taken into consideration in the preparation of this diagram.

Subdivision of an existing landmark or band. In the event that a band serving as a landmark requires subdivision, all sub-bands derived from it should retain the original region and band number of that landmark. This rule is to be followed even if subdivision should cause one or more sub-bands to lie in an adjacent region.

Whenever an existing band is to be subdivided, a decimal point should be placed after the original band designation followed by the number assigned to each sub-band. The sub-bands are numbered sequentially from the centromere outward. For example, if the original band 1p33 were subdivided into three equal or unequal sub-bands, the sub-bands would be labeled 1p33.1, 1p33.2, and 1p33.3, sub-band 33.1 being proximal and 33.3 distal to the centromere. Where the designation of the original band is in doubt, the decimal point should be followed by a question mark(?) and then the proposed sub-band number, e.g., 1p33.?1. Finally, if a sub-band is to be subdivided, additional digits but no further punctuation should be used; e.g., sub-band 1p33.1 might be further subdivided into 1p22.12, etc.

System for Designating Break Points Within Bands

A numbering system has been proposed to designate the location of break points within bands on the basis of the relative distance of the break point from the proximal margin of the band concerned. The proximal edge of a band, say, 1p22, is denoted as 1p2200. A point six-tenths of the distance from the proximal edge to the distal edge of this band would be denoted 1p2206.

Examples

1p2200
  This designation would denote the proximal edge of band 1p22.

1p2206
  This denotes a point six-tenths of the distance from the proximal edge to the distal edge of band 1p22.

Designating Structural Chromosome Abnormalities by Breakage Points and Band Composition

Two systems for designating structural abnormalities have been presented. One is a *short system* in which the nature of the rearrangement and the break point or points is identified by the bands (or regions) in which the breaks occur. Because of the conventions built into this system, the band composition of the abnormal chromosomes present can be readily inferred from the information provided in the symbolic description. The other is a *detailed system* which, besides identifying the type of rearrangement, defines each abnormal chromosome present in terms of its band composition. The two systems are not mutually exclusive and can be used to complement each other. The notation used to identify the rearrangement and the method of specifying the break points are common to both systems and will be presented first.

Specification of Chromosome Rearrangements

Single- and three-letter designations, as adopted at the Chicago Conference (1966), are used to specify rearranged (i.e., structurally altered) chromosomes. Immediately following the symbol identifying the type of rearrangement, the number of the chromosome involved in the change is specified within parentheses, e.g., r(18); inv(2). If two or more chromosomes have been altered, a semicolon (;) is used to separate their designations. If one of the rearranged chromosomes is a *sex-chromosome, then it should be listed first; otherwise the chromosome having the lowest chromosome number is always specified first*, e.g., t(X; 3), t(2; 5).

The only exception to this rule involves certain three-break rearrangements in which part of one chromosome is inserted at a point of breakage in another chromosome. In this event, the *receptor* chromosome is specified first, regardless of whether it is a sex-chromosome or whether its number is higher or lower than that of the donor chromosome.

For translocations involving three separate chromosomes, the sex-chromosome or the autosome with the lowest number is specified first. The chromosome listed next is the one that receives a segment from the first chromosome, and the chromosome specified last is the one that donates a segment to the first listed chromosome.

Some additional designations are required in the present nomenclature to identify rearrangements. These are listed in Table 1 and explained below.

Deletions. The abbreviation *del* is used to designate a chromosome deletion.

Translocations.[20] The use of the semicolon for differentiating balanced from unbalanced translocations is abandoned in the present nomenclature. All translocations are specified by the symbol *t*. If the type of translocation, i.e., Robertsonian, reciprocal, or tandem, is to be emphasized, *t* may be replaced by *rob*, *rcp*, or *tan*, respectively. (The symbol *rcp* is used for reciprocal translocations to avoid confusion with *rec*, which is used to designate a recombinant chromosome). Translocations resulting in a dicentric chromosome are designated by *tdic*. However, a dicentric generated by an internal rearrangement within the chromosome is indicated simply by *dic*.

Three-break rearrangements. These may involve one, two, or three chromosomes. Rearrangements involving three or more chromosomes will be referred to as "complex translocations." Several terms have been employed in the cytogenetic literature for three-break rearrangements involving one or two chromosomes: these include "shift," "insertion," and "transposition." All three-break rearrangements involving one or two chromosomes are referred to as "insertions" because they result from the excision of a segment after two breaks in one chromosome arm and its insertion at a point of breakage in either the same arm, the opposite arm of the same chromosome, or in another chromosome. The order of the bands on the inserted segment in relation to the centromere at the new site may be the same as the original site *(direct insertion)* or may be reversed *(inverted insertion)*. The abbreviation *ins* is used to indicate a direct insertion and *inv ins* to indicate an inverted insertion.

Specification of Break Points

The location of any given break is specified by the band in which that break has occurred (Yu, Borgaonkar, et al., 1978). Because it is not possible at present to define band interfaces accurately, a break suspected at an interface between two bands is identified arbitrarily by the higher of the two band numbers, i.e., the number of the band more distal to the centromere.

A given break may sometimes appear to be located in either of two consecutive bands. A similar situation may occur when breaks at or near an interface between two bands are studied with two or more techniques. In this event, the break can be specified by both band numbers separated by the word *or*; e.g., 1q23or24, indicating a break in either band 1q23 or 1q24. If a break can be localized to a region but not to a particular band, only the region number should be specified; e.g., 1p1, instead of 1p11or12or13. If the break point can be assigned only to two adjacent regions, both suspected regions should be specified, e.g., 1q2or3.

Short System

In this system, structurally altered chromosomes are defined only by their break points. As described earlier, the break points are specified within parentheses immediately following the designation of the type of rearrangement and the chromosome(s) involved. The break points are identified by band designations as just outlined but without specifying the chromosome number. For example, del (1) (q21) defines a terminal deletion in the long arm of chromosome #1 resulting from a break at band 1q21.

Two-break rearrangements. When both arms of a single chromosome are involved in a two-break rearrangement, *the break point in the short arm is always specified before the break point in the long arm*; e.g., inv (2) (p21q31) defines a pericentric inversion in chromosome #2 with break points in bands 2p21 and 2q31. When the two breaks occur within the same arm, *the break point more proximal to the centromere is specified first*; e.g., inv(2) (p13p23) defines a paracentric inversion in the short arm of chromosome #2 with break points in bands 2p13 and 2p23.

Three-break rearrangements. When an insertion within a single chromosome occurs, *the break point at which the chromosome segment is inserted is always specified first*. The remaining break points are specified in the same way as in a two-break rearrangement, i.e., the more proximal break point of the inserted segment is specified next and the more distal one last. "Proximal" and "distal" refer here to the positions of the break points after the rearrangement and not necessarily their original positions. For example, inv ins (2) (q13p23p13) defines an inverted insertion in chromosome #2 of the short-arm segment lying between bands 2p13 and 2p23 into the long arm at band 2q13. Because the insertion is inverted, band 2p23 is now proximal and 2p13 distal to the centromere.

Rearrangements affecting two or more chromosomes. The break points are specified in the same order as the chromosomes involved are specified, and a semicolon is used to separate the break points (punctuation is never used to separate break points in the same chromosome). For example, rcp (2;5) (q21;q31) defines a reciprocal translocation between the long arms of chromosomes #2 and #5, with break points at bands 2q21 and 5q31.

Detailed System

In this system, structurally altered chromosomes are defined by their band composition. The conventions used in the short system are retained in the present system, except that an abbreviated description of the band composition of the rearranged chromosome or chromosomes is specified within the final pair of parentheses, instead of only the break points.

Additional symbols. A single colon (:) is used to indicate a chromosome break and a double colon (: :) to indicate "break and join." To avoid an unwieldy description, an arrow (→), meaning "from-to," is employed. The end of a chromosome arm may be designated either by its band designation or by the symbol *ter*, meaning "terminal," preceded by the arm designation; e.g., *pter* indicates end of short arm and *qter*, end of long arm. When it is necessary to indicate the centromere, the abbreviation *cen* should be used.

Designating the band composition of a chromosome. The description starts at the end of the short arm and proceeds through to the end of the long arm, with the bands being identified in the order in which they occur in the rearranged chromosome. If the rearrangement is confined to a single chromosome, the chromosome number is not repeated in the band description. If more than one chromosome is involved, however, the bands and chromatid ends are identified with the appropriate chromosome number.

If, owing to a rearrangement, no short-arm segment is present at the end of either arm, the description of the structurally rearranged chromosome starts at the end of the long-arm segment with the lowest chromosome number.

When more than one chromosome is involved, the chromosome descriptions are presented in the same numerical order as the chromosomes involved in the rearrangement. In the special case of an unbalanced reciprocal translocation between the long arm of one chromosome and the short arm of another, the derivative chromosome carrying the centromere belonging to the chromosome with the lower chromosome number is described first.

Examples

In all the examples presented in this section, the short system designation is shown first and the detailed system second, followed by a brief explanation of the latter.

Isochromosomes
46,X,i(Xq)
46,X,i(X)(qter→cen→qter)

    Break points in this type of rearrangement are at or close to the centromere and cannot be specified. The designation indicates that both entire long arms of the X-chromosome are present and separated by the centromere.

Terminal deletions
46,XX,del(1)(q21)
46,XX,del(1)(pter→q21:)

    The single colon (:) indicates a break at band 1q21 and deletion of the long-arm segment distal to it. The remaining chromosome consists of the entire short arm of chromosome #1 and part of the long arm lying between the centromere and band 1q21.

Interstitial deletions
46,XX,del(1)(q21q31)
46,XX,del(1)(pter→q21::q31→qter)

    The double colon (::) indicates breakage and union of bands 1q21 and 1q31 in the long arm of chromosome #1. The segment lying between these bands has been deleted.

Paracentric Inversions
46,XY,inv(2)(p13p24)
46,XY,inv(2)(pter→p24::p13→p24::p13→qter)

    Breakage and union have occurred at bands 2p13 and 2p24 in the short arm of chromosome #2. The segment lying between these bands is still present but inverted, as indicated by the reverse order of the bands with respect to the centromere in this segment of the rearranged chromosome.

Pericentric Inversions
46,XY,inv(2)(p21q31)
46,XY,inv(2)(pter→p21::q31→p21::q31→qter)

    Breakage and union have occurred at band 2p21 in the short arm and 2q31 in the long arm of chromosome #2. The segment lying between these bands is inverted.

Ring Chromosomes
46,XY,r(2)(p21q31)
46,XY,r(2)(p21→q31)

    Breakage has occurred at band 2p21 in the short arm and 2q31 in the long arm of chromosome #2. With deletion of the segments distal to these bands, the broken ends have joined to form a ring chromosome. Note the omission of the colon or double colon.

Dicentric Chromosomes
46,X,dic(Y)(q12)
46,X,dic(Y)(pter→q12::q12→pter)

    Breakage and union have occurred at band Yq12 on sister chromatids to form a dicentric Y-chromosome.

Reciprocal Translocations
46,XY,t(2;5)(q21;q31)
46,XY,t(2;5)(2pter→2q21::5q31→5qter;5pter→5q31::2q21→2qter)

    Breakage and union have occurred at bands 2q21 and 5q31 in the long arms of chromosomes #2 and #5, respectively. The segments distal to these bands have been exchanged between the two chromosomes. Note that the derivative chromosome with the lowest number (i.e., #2) is designated first.

46,XY,t(2;5)(p12;q31)
46,XY,t(2;5)(2qter→2p12::5q31→5qter;5pter→5q31::2p12→2pter)

    Breakage and union have occurred at band 2p12 in the short arm and band 5q31 in the long arm of chromosomes #2 and #5, respectively. The segments distal to these bands have been exchanged between the two chromosomes. Note that the derivative chromosome bearing the #2 centromere has no terminal short-arm segment and, therefore, its description starts with the long-arm end having the lowest number (i.e., 2qter).

Robertsonian Translocations
45,XX,t(13;14)(p11;q11)
45,XX,t(13;14)(13qter→13p11::14q11→14qter)

Breakage and union have occurred at band 13p11 in the short arm and band 14q11 in the long arm of chromosomes #13 and #14, respectively. The segment distal to band 14q11 has been translocated onto chromosome #13 at band 14p11. The rest of chromosome #14, with its centromere, has been lost, along with the original segment distal to 13p11, i.e., 13pter→13p11.

45,XX,t(13q14q)
45,XX,t(13;14)(13qter→cen→14qter)

Breakage has occurred at or near the centromere in chromosomes #13 and #14. The rearranged chromosome has the long arms of both chromosomes separated by a centromere whose origin might have been either chromosome. Both short arms have been lost.

45,XX,tdic(13;14)(p11;p11)
45,XX,tdic(13;14)(13qter→13p11::14p11→14qter)

Breakage and union have occurred at bands 13p11 and 14p11 in the short arms of chromosomes #13 and #14, respectively. The segments distal to these bands have been deleted, and the remaining segments have joined at the break points in the short arms to form a dicentric translocation chromosome.

Direct Insertions Within a Chromosome
46,XY,ins(2)(p13q21q31)
46,XY,ins(2)(pter→p13::q31→q21::p13→q21::q31→pter)

Breakage and union have occurred at band 2p13 in the short arm and bands 2q21 and 2q31 in the long arm of chromosome #2. The long-arm segment between 2q21 and 2q31 has been inserted into the short arm at band 2p13. The original orientation of the inserted segment has been maintained in its new position, i.e., 2q21 remains more proximal to the centromere than 2q31.

Inverted Insertions Within a Chromosome
46,XY,inv ins (2) (p12q31q21)
46,XY,inv ins (2) (pter→p13::q21→q31::p13→q21::q31→qter)

Breakage and union have occurred at the same bands as in the previous example, and the insertion is the same except that the inserted segment has been inverted; i.e., 2q21 in the inserted segment is now more distal to the centromere than 2q31. The orientation of the bands within the segment has thus been reversed with respect to the centromere.

Direct Insertions Between Two Chromosomes
46,XY,ins (5;2) (p14;q22q32)
46,XY,ins (5;2) (5pter→5p14::2q32→2q22::5p14→5qter;2pter→2q22::2q32→2qter)

Breakage and union have occurred at band 5p14 in the short arm and bands 2q22 and 2q32 in the long arm of chromosomes #5 and #2, respectively. The segment between 2q22 and 2q32 has been inserted into the short arm of chromosome #5 at band 5p14. The original orientation of the inserted segment has been maintained in its new position; i.e., 2q22 remains more proximal to the centromere than 2q32. Note that the receptor chromosome is specified first.

Inverted Insertions Between Two Chromosomes
46,XY,inv ins (5;2) (p14;q32q22)
46,XY,inv ins (5;2) (5pter→5p14::2q22→2q32::5p14→5qter;2pter→2q22::2q32→2qter)

Breakage and union have occurred at the same bands as in the previous example, and the insertion is the same except that the inserted segment has been inverted; i.e., 2q22 is now more distal to the centromere of the recipient chromosome than 2q32.

Complex Translocations
46,XX,t(2;5;7)(p21;q23;q22)
46,XX,t (2;5;7) (2qter→2p21::7q22→7qter;5pter→5q23::2p21→2pter;7pter→7q22::5q23→5qter).

Breakage and union have occurred at band 2p21 in the short arm of chromosome #2 and at bands 5q23 and 7q22 in the long arms of chromosomes #5 and #7, respectively. The segment of chromosome #2 distal to 2p21 has been translocated into chromosome #5 at 5q23; the segment of chromosome #5 distal to 5q23 has been translocated onto chromosome #7 at 7q22; and the segment of chromosome #7 distal to 7q22 has been translocated onto chromosome #2 at 2p21.

Note that the chromosome specified first is the one with the lowest number; the chromosome specified next is the one receiving a segment from the first one listed, and the chromosome specified last is the one donating a segment to the first chromosome listed.

Four-break Rearrangements

There are a very large number of possible four-break rearrangements. These can be described using the conventions outlined here. A single example is illustrated here to indicate how such rearrangements can be handled:

Double Reciprocal Translocation Involving Three Chromosomes
46,XX,t (1;3) (3;9) (p12;p13q25q22)
46,XX,t (1;3) (3;9) (3pter→3p13::
1p12→1qter;1pter→1p12::3p13→3q25::
9q22→9qter;9pter→9q22::3q25→3qter)

    Breakage and union have occurred at bands 1p12 and 3p13 in the short arms of chromosomes #1 and #3, respectively, and at bands 3q25 and 9q22 in the long arms of chromosomes #3 and #9, respectively. The segments distal to 1p12 and 3p13 have been exchanged, as have the segments distal to 3q25 and 9q22.

Aided by the increasing precision of chromosome banding techniques, a special class of chromosome translocation has been discovered, termed *terminal rearrangement*, in which two chromosomes are joined end to end. Although both centromeres are apparently present, only one appears as the primary constriction.

The following abbreviations are recommended for designating whole-arm translocations between nonacrocentric chromosomes and both direct and inverted (reverse) duplications of chromosome segments: *ter rea,* terminal rearrangement; *dir dup,* direct duplication; *inv dup,* inverted duplication.

  Examples

    Terminal rearrangements. The double triplet *ter rea* can be used to describe this type of rearrangement. The expanded version of the Paris nomenclature and the triplet *cen* can be used to indicate the position of the primary constriction. For example:

45,XX,ter rea (12;14) (p13;p13)
45,XX,ter rea (12;14) (12qter→cen→12p13::
14p13→14qter)

    Breakage and union have occurred in band p13 of both chromosomes #12 and #14, and the centromere of chromosome #12 is one that appears as the primary constriction.

    In this type of rearrangement *the chromosome carrying the primary constriction is always written first.*

    Whole-arm translocations. Whole-arm exchanges that involve nonacrocentric chromosomes, where the position of the break points relative to the centromere is not known, can be described, for example, as:

46,XY,t (2;3) (2p3p;2q3q)
46,XY,t (2;3) (2pter→cen→3pter;
2qter→cen→3qter)

    This exchange involves the whole arms of chromosomes #2 and #3 with the exchange of the respective short arms and long arms of both chromosomes. The derivative chromosomes would be der(2p3p) and der(2q3q).

The alternative arrangement would be:

46,XY,t (2;3) (2p3q;2q3p)
46,XY,t (2;3) (2pter→cen→3qter;
2qter→cen→3pter)

    The two derivative chromosomes would then be der(2p3q) and der(2q3p).

In both cases the derivative chromosomes are designated according to their arms, not their centromeres. If it becomes possible to designate the centromere specifically as being derived from one or the other of the chromosomes involved, this could be indicated by preceding the symbol *cen* by the chromosome number, e.g., 2cen, if it is known that the centromere was derived from chromosome #2. For example:

46,XY,t (2;3) (2pter→2cen→3pter;
2qter→3cen→3qter)

    This indicates that the origin of the centromere in both derivative chromosomes is known.

Duplication of a chromosome segment. The symbol *dup* remains valid. It can be supplemented with the triplets *dir* or *inv* to indicate if the duplication is direct or inverted. For example:

46,XX,inv dup (2p) (p23→p14)
46,XX,inv dup (2p) (pter→p23::p14→p23::p23→qter)

Breakage and reunion have occurred at band p23 of chromosome #2, with segment p23→p14 being inverted and duplicated.

Description of Heteromorphic Chromosomes

Code to Describe Banding Techniques

In this three-letter code, the first letter denotes the type of banding, the second letter the general technique, and the third letter the stain. For example:

Q-	Q-bands
QF-	Q-bands by fluorescence
QFQ	Q-bands by fluorescence using quinacrine
QFH	Q-bands by fluorescence using Hoechst 33258
G-	G-bands
GT-	G-bands by trypsin
GTG	G-bands by trypsin using Giemsa
GTL	G-bands by trypsin using Leishman
GAG	G-bands by acetic saline using Giemsa
C-	C-bands
CB-	C-bands by barium hydroxide
CBG	C-bands by barium hydroxide using Giemsa
R-	R-bands
RF-	R-bands by fluorescence
RFA	R-bands by fluorescence using acridine orange
RH-	R-bands by heating
RHG	R-bands by heating using Giemsa
RB-	R-bands by BUdR
RBG	R-bands by BUdR using Giemsa
RBA	R-bands by BUdR using acridine orange
T-	T-bands
TH-	T-bands by heating
THG	T-bands by heating with Giemsa
THA	T-bands by heating with acridine orange

Any new triplet should be defined in the text of the publication in which it is first used.

Short terminology. The previously used short forms, such as lqh+, may still be used, but where appropriate could include a description of the technique used, e.g., lqh+(CBG).

Complete description. Heteromorphic chromosomes can be described if the term *variable*, abbreviated to *var*, is used before the chromosome number, e.g., var(13). Additional information regarding the variable region can then be conveyed by means of symbols set within brackets in the following order: 1. the location of the variable structure on the chromosome with either band numbers or code letters such as *cen, h, s*, etc. This is followed by a comma. 2. The banding technique used, whereby the triplet code given above is utilized. 3. A numerical designation for the size and staining intensity of the variable region, whereby higher numbers indicate greater size or staining intensity. Such numerical designations must be clearly defined.

A zero indicates that size or intensity was not quantitated. *The number of digits used to describe size must equal the number of digits used to describe intensity.*

When several techniques are used, each description should be separated by a comma. Their order is arbitrary. If more than one variable structure is present on the same homologue, each should be described in the same way in separate parentheses without separation by a comma.

If the same variant appears on more than one homologue, an asterisk followed by a number, e.g., *2, can be used to designate the number of chromosomes that conform to the initial *var* description. The parental origin of a chromosome can be indicated by inserting *pat* or *mat* after the last parentheses but before the asterisk indicating the number of copies.

If more than one variable chromosome of a complement is to be described, these should be listed in descending order of chromosome size, the terms relating to each chromosome being separated by a comma. Bands on a given chromosome should be listed sequentially from the centromere outward, with those bands in the short arm listed first and those in the long arm last.

Examples

46,XY,var(3)(cen,QFQ35)
Chromosome #3 with a centromeric region which, when Q-banded, is of intermediate sizes and fluoresces brilliantly.

46,XY,var(13)(p13,QFQ35)*2
Two chromosomes #13 with satellites (p13) which, when Q-banded, are of intermediate size and fluoresce brilliantly.

46,XY,var(13)(p13,QFQ55,CBG50)(q11,QFQ35,CBG30)
One chromosome #13 with very large satellites (p13) seen after both Q- and C-banding. These are brilliant after Q-banding, but C-banding intensity was not determined. In addition, band q11, when Q-banded, is of intermediate size and fluoresces brilliantly, and when C-banded, it is likewise intermediate in size. C-banding intensity was not determined.

46,XY,var(13)(p11,QFQ35,CBG30)(p13,QFQ45,CBG45)(q11,QFQ45,CBG40)
One chromosome #13 where band p11, when Q-banded, is one of intermediate size and fluoresces brilliantly. When C-banded, this band is also of intermediate size; however, its C-staining intensity was not determined. The satellites (p13) are very large, fluoresce brilliantly, and stain very darkly after C-banding. In addition, band q11, when Q-banded, is large and fluoresces brilliantly. When C-banded, this band is also large; however, its C-staining intensity was not determined.

46,XY,var(13)(q11,QFQ1205)
Chromosome #13 with band q11 in which the size of the band has been estimated quantitatively to be 12 U on a 01 to 99 scale. (Units of measurement should be defined in individual reports). Band q11 is brilliantly fluorescing. Note that *the same number of digits must be used to describe both the size and intensity.*

46,XY,var(13)(q11,QFQ55),var(21)(p13,QFQ44),var(22)(13,QFQ35)
Male with three variant chromosomes after Q-banding.

47,XY,+21,var(21)(p13,QFQ12),var(21)(p13,QFQ54)mat*2
Male with 47 chromosomes and trisomy #21. One chromosome #21 has very small satellites of pale intensity after Q-banding; the two remaining chromosomes #21 are identical, with very large and intensely fluorescent satellites derived from the mother.

Numerical aberrations. In a description of a karyotype finding, the first item to be recorded is the total number of chromosomes, including the sex-chromosomes, followed by a comma (,). The sex-chromosome constitution is given next, e.g.,

45,X
45 chromosomes, 1 X-chromosome

47,XXY
47 chromosomes, XXY sex-chromosomes

49,XXXXY
49 chromosomes, XXXXY sex-chromosomes

The autosomes are specified only when there is an abnormality present. Thus, if there is a numerical aberration of the autosomes, the group letter of the extra or missing autosome, preceded by a plus (+) or minus (−) sign, succeeds the sex-chromosome designation, e.g.,

45,XX,−C
45 chromosomes, XX sex-chromosomes, a missing C-group chromosome.

48,XXY,+G
48 chromosomes, XXY sex-chromosomes, an additional G-group chromosome.

The plus or minus sign before a chromosome letter or number indicates that the *entire* autosome is extra or missing. When the extra or missing chromosome or chromosomes have been identified with certainty, the chromosome number may be used, e.g.:

45,XX,−16
45 chromosomes, 2 X-chromosomes, a missing #16 chromosome.

47,XY,+21
: 47 chromosomes, XY sex-chromosomes, an additional #21 chromosome.

46,XY,+18,−21
: 46 chromosomes, XY sex-chromosomes, an extra #18, and a missing #21.

A question mark (?) may be used in the normal way to indicate uncertainty. If it is suspected, but uncertain, that a missing or extra chromosome belongs to a particular group, the question mark may *precede* the group designation or in some cases the chromosome number, e.g.,

45,XX,?−C
: 45 chromosomes, XX sex-chromosomes, a missing chromosome which probably belongs in group C.

Another example would describe the karyotype of a sex-chromatin positive female with an additional small acrocentric chromosome; this could be written, depending on the amount of available information, as: 47,XX,?+G; 47,XX,+G; 47,XX,?+21 or 47,XX,+21.

A *triploid* or *polyploid* cell should be evident from the chromosome number and from the further designations, e.g., 69,XXY;70,XXY,+G. An *endoreduplicated* metaphase can be indicated by preceding the karyotype designation with the abbreviation *end*, e.g., end 46,XX. If multiplicity of endoreduplications is to be indicated, an arabic numeral can be used before *end* to indicate this, e.g., 2end46,XX; 4end46,XX.

The method for describing chromosome complements containing structurally rearranged chromosomes, such as dicentrics, acentric fragments, and markers, has been previously described.

Chromosome mosaics. The chromosome constitution of the different cell lines is listed in numerical or alphabetical order, *irrespective of the frequencies of the cell types in the individual studied.* The karyotype designations are separated by a slash (/), e.g.,

45,X/46,XY
: A chromosome mosaic with two cell types, one with 45 chromosomes and a single X, the other with 46 chromosomes and XY sex-chromosomes.

46,XX/46,XY
: A chromosome mosaic with both XX and XY cell lines.

46,XY/47,XY,+G
: A chromosome mosaic with a normal male cell line and a cell line with an extra G-group chromosome.

45,X/46,XX/47,XXX
: A triple cell-line mosaic.

Structural alterations. The *short arm* of a chromosome is designated by the lower-case letter *p*, the *long arm* by the letter *q*, a *satellite* by the letter *s*, a secondary constriction by the letter *h* and the *centromere* by the abbreviation *cen*.

Increase in length of a chromosome arm is indicated by placing a plus sign (+) and *decrease in length* by placing a minus sign (−) *after* the arm designation, e.g., 2p+; Bp−; Gq−. When one arm of a mediocentric chromosome, viz., #1, #3, #19, and #20, is changed, this is indicated by placing a question mark *between* the chromosome designation and the plus or minus sign. For example, #3 chromosome with an elongated arm would be designated as 3?+.

The result of a *pericentric inversion* is indicated by p+q− or p−q+, which is enclosed in parentheses and preceded by the abbreviation *inv*, e.g., inv(Dp+q−).

A translocation is indicated by the letter *t* followed by parentheses which include the chromosomes involved, e.g.,

46,XY,t(Bp−;Dq+), or 46,XY,t(Bp+;Dq−)
: A balanced reciprocal translocation between the short arm of a B- and the long arm of a D-group chromosome.

Translocations involving a sex-chromosome and an autosome would be designated as, e.g.,

46,X,t(Xq+;16p−)
: A reciprocal translocation between the long arm of an X and the short arm of a #16 in a female.

46,X,t(Xq+;16p−)
: The same translocation in a male.

46,X,t(Yp+;16p−)
: A reciprocal translocation between the short arm of a Y and short arm of a #16.

The remaining normal sex-chromosome is written in its usual position after the chromosome number, and the other sex-chromosome involved in the translocation is included in parentheses preceding the autosome concerned.

The separation of the chromosomes within the parentheses by a semicolon (;) indicates that two structurally altered chromosomes are present and that the translocation is balanced. In a "centric fusion" type of translocation, in which only one translocation chromosome is present, the semicolon is omitted, e.g.,

45,XX,−D,−G,+t(DqGq)
: 45 chromosomes, XX sex-chromosomes, one chromosome missing from the D-group and one from the G-group, their long arms having united to form a DG translocation chromosome.

If, as rarely happens, a small centric fragment is present as well, implying a reciprocal translocation, it could be written as 46,XX,−D,−G,+t(DpGp),+t(DqGq).

Where a centric fusion type of translocation results in duplication of part of one of the chromosomes involved, this could be written as,

46,XX,−D,+t(DqGq)
: 46 chromosomes, XX sex-chromosomes, one chromosome missing from the D-group; the long arm of this chromosome is united with the long arm of a G-group chromosome. Because there are four normal G-group chromosomes, part of a G is present in triplicate.

When family studies clearly show that a particular chromosome has been inherited from the mother or the father, this may be indicated by the abbreviations *mat* or *pat*. For instance, in a family in which a father is carrying a balanced reciprocal translocation, 46,XY,t(Bp−;Dq+), and in which his malformed son has inherited only one of the two abnormal chromosomes, the son's complement would be written as 46,XY,Bp−pat or 46,XY,Dq+pat, depending on which abnormal chromosome had been transmitted. If the son had inherited both chromosomes involved in the translocation, his complement would be expressed as 46,XY,t(Bp−;Dq+)pat.

Duplicated chromosome structures are indicated by repeating the appropriate designation. Thus, 46,XX,Gpss would describe the karyotype of a female in which one of the G-group chromosomes has double satellites on the short arm. If satellites appear on a chromosome arm where they are not usually found, this arm should be designated, e.g., 46,XY,18ps, indicating a #18 with satellited short arms; 46,XX,Gpsqs, indicating a G-group chromosome with both long arms and short arms satellited. Enlarged satellites are indicated, for instance, by Gs+.

Isochromosomes are designated by the lowercase letter *i* placed after the chromosome arm involved, e.g., 46,XXqi; or if this is presumptive, then 46,XXq?i. This would indicate an isochromosome or a presumptive isochromosome of the long arm of one X-chromosome.*

Ring chromosomes are indicated by the letter *r* placed after the chromosome involved; e.g., 46,XXr would indicate a ring X, and a ring B would be written 46,XY,Br.

In describing cells damaged by ionizing radiation, chemicals, viruses, etc., the system of nomenclature that has been described should be used where applicable. This may not be possible where the cell contains a grossly unbalanced chromosome complement. In such instances it is suggested that the chromosome count in a given cell should include all centric chromosome structures present in that cell regardless of the number of centromeres.

Unidentified chromosomes are indicated by *mar* (marker). *Acentric* fragments are not included in the count but may be indicated by *ace*. *Dicentric* and *tricentric* chromosomes are counted as one body and indicated by *dic* and *tri*. As an example, a cell derived from a normal female with a total of 48 centric chromosome structures, one missing F-group chromosome, a dicentric and acentric fragment, as well as two unidentified marker chromosomes would be written as: 48,?X?X,F−,dic+,mar1+,mar2+, ace. If necessary, an asterisk (*) following a chromosome designation may be used to draw attention to an explanation in the text.

If other designations are needed for special conditions, they should, whenever possible, be

---

*An alternate designation frequently used is to place the letter *i* before the chromosome number, followed by the arm designation, e.g., i(17q), indicating an isochromosome of the long arm of #17.

taken from the first three letters of the word required, used in lower case, clearly defined, and placed immediately before or after the chromosome symbol or the bracketed chromosome designation to which they refer. If single letters are to be used for special designations, they should be in lower case and should not duplicate the capital letters A to G, X and Y, or those lower-case letters representing other structures. Lower-case letters easily confused with numerals, such as *l* or *o,* should not be used. Subscripts and superscripts should be avoided because they are easily written incorrectly, not readily handled by computers, and difficult to set in type.

In regard to abnormal chromosomes described for the first time, it was suggested in the Denver Report (1960) that such chromosomes be named after the laboratory of origin. Instead, it is now proposed that the chromosome be described according to its morphology, by use of the shorthand nomenclature recommended. The only exception to this, for historical reasons, is to be the Philadelphia (Ph[1])-chromosome first described by Nowell and Hungerford (1960 *b*).

The terms describing abnormalities of chromosome number, such as "aneuploid," "heteroploid," etc., are often used in a variety of ways. A. Levan and Müntzing (1963) have restated the original definitions of these and other terms and commented on their usage.

A further problem that often arises in chromosome nomenclature is that of defining the position of the centromere. Such terms as "telocentric" and "submetacentric" are sometimes misused. A. Levan, Fredga, et al. (1964) have proposed a standardized nomenclature defining the centromeric position in terms of the arm ratio.

Marker Chromosomes

A marker chromosome of completely unknown origin should be designated by the original Chicago Conference (1966) symbol *mar*. If part of the chromosome can be identified with one of the banding techniques, a question mark (?) and the plus (+) and minus (−) signs may be used with the short system to designate the karyotype. For example, 46 XX,t(12;?)(q15;?) defines a karyotype that includes a rearranged chromosome #12 in which the segment of the long arm distal to band 12q15 could not be identified. If such a marker happened to be longer or shorter than the chromosome from which it had been derived, this could be recorded by specifying the arm and the direction of the change in length. For example, 46,XX,t(12q+;?)(q15;?) defines a karyotype that includes a rearranged chromosome #12 with a longer-than-normal long arm owing to attachment of an unknown segment distal to band 12q15.

Mosaics and Chimeras

Individuals or tissues containing two or more types of cells that differ in their chromosome complements are said to have mixed karyotypes. When the different cell types are derived from a single zygote, the individual is a mosaic. If the two or more cell types are derived from several zygotes or zygote cell lineages, the individual is a chimera. At the Chicago Conference (1966), it was recommended that the descriptions of the several karyotypes of chromosome mosaics or chimeras be separated by a single slash. To distinguish between chromosome mosaics and chimeras, it is now recommended that the triplets *mos* and *chi* be used.

Examples

mos45,X/46,XY

A chromosome mosaic with two cell types, one with 45 chromosomes and a single X, the other with 46 chromosomes and XY sex chromosomes.

chi45,XX/46,XY

A chimera with both XX and XY cell lines.

In most instances the triplets *mos* and *chi* will be needed only for the initial description in any report; subsequently, the simple karyotype designation may be used.

Usage of + and − Signs

The + or − signs should be placed *before* the appropriate symbol where they mean additional or missing whole chromosomes. They should be placed *after* a symbol where an increase or decrease in length is meant.

Examples

47,XY,+G
  Male karyotype with 47 chromosomes, including an additional G-group chromosome.

45,XY,−21
  Male karyotype with 45 chromosomes and missing 1 chromosome #21.

46,XY,1q+
  Male karyotype with 46 chromosomes, showing an increase in the length of the long arm of 1 chromosome #1.

47,XY,+14p+
  Male karyotype with 47 chromosomes, including an additional chromosome #14 which has an increase in the length of its short arm.

45,XX,−D,−G,+t(DqGq)
  Female karyotype with a balanced Robertsonian translocation between a D- and a G-group chromosome.

46,XY,−5,−12,+t(5p12p),+t(5q12q)
  Male karyotype with two translocations involving interchange of both whole arms of chromosomes #5 and #12. The breaks have occurred at or very near the centromere, and no information is available as to which centromere is included in either product.

46,XX,−13,+t(13q21q)
  Female karyotype with an unbalanced Robertsonian translocation between chromosomes #13 and #21; the long arm of chromosome #21 is present in triplicate.

If desired, balanced Robertsonian translocations, as well as whole-arm translocations, may be recorded in briefer form, for example, 45,XX,−D,−G,+t(DqGq) may be shortened to 45,XX,t(DqGq). Unbalanced karyotypes, however, should be written out completely, as in the last example above.

Length Changes of Secondary Constrictions

Increases or decreases in the length of secondary constrictions, or negatively staining regions, should be distinguished from increases or decreases in arm length owing to other structural alterations by placing the symbol *h* between the symbol for the arm and the + or − sign.

Examples

46,XY,16qh+
  Male karyotype with 46 chromosomes, showing an increase in length of the secondary constriction on the long arms of chromosome #16.

46,XY,13ph−
  Male karyotype with 46 chromosomes, showing a decrease in the length of the negatively staining region on the short arm of chromosome #13.

Structurally Abnormal Chromosomes

All symbols for rearrangements are to be placed before the designation of the chromosome or chromosomes involved, and the rearranged chromosome or chromosomes always should be placed in parentheses.

Examples

46,XX,r(18)
  Female karyotype with 46 chromosomes, including a ring chromosome #18.

46,X,i(Xq)
  Female karyotype with 46 chromosomes, including 1 normal X-chromosome and 1 chromosome represented by an isochromosome for the long arm of the X.

46,X,dic(Y)
  Karyotype with 46 chromosomes, 1 X-chromosome, and a dicentric Y-chromosome.

Abbreviating Lengthy Descriptions

In the interests of clarity, complex rearrangements necessitating lengthy descriptions in the Chicago Conference (1966) nomenclature should be written out in full and in accord with that system the first time they are used in a report. At the discretion of the authors, an abbreviated version of the nomenclature might then be used subsequently, providing it is clearly defined immediately after the complete notation.

Special Terminology

In studies of interphase chromatin morphology, the terms *X-chromatin* (= X-body, Barr body) and *Y-chromatin* (= Y-body) should be used. The terms (chromosome) *variant* and *inherited* (chromosome) *variant* are recommended for use in situations where deviations from the norm of chromosome morphology are observed.

Terminology of Acquired Aberrations

The purpose of this section is to provide a nomenclature for those acquired aberrations not already adequately described by the existing nomenclature for constitutional aberrations. Because many induced aberrations are scored on unbanded material, recommendations are given first for unbanded preparations material. The nomenclature presented below is essentially that recommended by and based upon the ISCN (1978). (See Table 4A.)

Chromatid Aberrations

Unbanded preparations

ct — A *chromatid aberration* involves one chromatid at a single locus.

ctg — A *chromatid gap* is a nonstaining region or discontinuity of a single chromatid region where there is no misalignment of the chromatids.

ctb — A *chromatid break* is a discontinuity of a single chromatid where there is a clear misalignment of one of the chromatids.

cte — A *chromatid exchange* is the result of two or more chromatid lesions (break points) and the subsequent transposition of chromatid material. Exchanges may be interchromatid or intrachromatid. For most purposes, it will be sufficient to indicate whether the configuration is *triradial, tr,* where there are three arms to the

---

Table 4A Sample descriptions of sequential observations in a single case of chronic myeloid leukemia*

Observation 1	sl	46,XX,t(9;22)
	sdl 1	47,XX,t(9;22),+22q−
	sdl 2	47,XX,t(9;22),+8
Observation 2	sl	46,XX,t(9;22)
	sdl 1	47,XX,t(9;22), +22q−
	sdl 2	47,XX,t(9;22),+8
	sdl 3	48,XX,t(9;22),+9,+10
Observation 3	sl	47,XX,t(9;22),+8
	sdl 1	47,XX,t(9;22),+22q−
	sdl 3	48,XX,t(9;22),+9,+10
	sdl 4	46,XX,t(9;22)
Observation 4	sl	48,XX,t(9;22),+8,+8
	sdl 2	47,XX,t(9;22),+8
	sdl 3	48,XX,t(9;22),+9,+10
Observation 5	sl	49,XX,t(9;22),+8,+8,−17,+i(17q),+22q−
	sdl 2	47,XX,t(9;22),+8
	sdl 5	48,XX,t(9;22),+8,+8

*The stem-line in observation 1 is a line with a simple translocation resulting in the Ph¹ chromosome. This line is superceded by another stem-line in observation 3, becomes side-line 4, and is not seen in subsequent observations. The line with 47 chromosomes and an extra chromosome 8 is present throughout as side-line 2, except for observation 3, where it has become the stem-line. The side-lines 1 and 3 apparently disappear without further evolution, but side-line 5, and the new stem-line in observation 5, could have been derived from side-line 2.

Table taken from ISCN (1978)

pattern, quadriradial, *qr*, where there are four arms to the pattern, or complex, *cx*, where there are more than four arms to the pattern. If it is necessary to classify exchanges in more detail, the exchange may be described as *complete*, where all the broken ends are rejoined or *incomplete*, where they are not. The incompleteness may be proximal, *prx*, where the broken ends nearest to the centromere are not rejoined or distal, *dis*, where the ends furthest from the centromere are not rejoined.

sce—A *sister chromatid exchange* is the result of a lesion of the two chromatids of a single chromosome at the same locus and the subsequent exchange of the chromatids distal to these lesions.

Banded preparations—Some chromatid aberrations can only be recognized with certainty in banded preparations. For example, a chromatid deletion, *ct del*, is the absence of a banded sequence from only one of the two chromatids of a single chromosome. A chromatid inversion, *ctd inv*, is the reversal of a banded sequence of only one of the two chromatids of a single chromosome. Both are subclasses of *ctde*.

Where it is desirable to specify the location of a chromatid aberration, the appropriate symbol can be followed by the band designation: For example,

ctg(4)(q25)—chromatid gap in #4 at band 4q25
ctb(4)(q25)—chromatid break in #4 at band 4q25
cte(4;10)(q25;q22)—chromatid exchange involving #4 and #10 at bands 4q25 and 10q22, respectively
sce(4)(q25q33)—SCE in #4 at bands 4q25 and 4q33

Chromosome Aberrations

Unbanded preparations

cs—A *chromosome aberration* involves both chromatids of a single chromosome at the same locus.

csg—A *chromosome gap* is a nonstaining region or discontinuity at the same locus in both chromatids of a single chromosome where there is no misalignment of the chromatids. Chromosome gap is synonymous with *isolocus gap* and *isochromatid gap*.

csb—A *chromosome break* is a discontinuity at the same locus in both chromatids of a single chromosome giving rise to an acentric fragment and an abnormal monocentric chromosome.[21] This fragment is therefore a particular type of acentric fragment, *ace*, and *csb* should only be used where banding pattern or morphology indicates that the *ace* is the result of a single event. Chromosome break is synonymous with *isolocus break* and *isochromatid break*.

min—A *minute* is a very small acentric fragment whose length is less than the diameter of a single chromatid. It is synonymous with double minute.

Pulverization (pvz)—This notation indicates a situation where there are present in a cell chromatid gaps and breaks and/or chromosome gaps and breaks in such numbers that they cannot be accurately enumerated. Normally exchanges are not seen in these cells. Occasionally one or more chromosomes in a cell are pulverized whereas the remaining chromosomes are of normal morphology. A special class of pulverization, where the presence of multiple gaps and breaks is due to the action of some influence which causes a partially replicated chromosome or chromosomes to condense as if in preparation for mitosis, is termed "premature chromosome condensation," pcc or *prophasing*,[22] e.g., in a heterokaryon where one of the nuclei is in S-phase and the other in metaphase.

cse—A *chromosome exchange* is the result of two or more chromosome lesions and the subsequent transposition of both chromatids of a single chromosome to a new location on the same or on another chromosome. In unbanded preparations, the terminology is exactly the same as for constitutional aberrations.

Banded preparations—In general the terminology is as in the standard nomenclature but in some banded preparations further definition, particularly of fragments, is possible. An acentric fragment with band sequences from two chromosomes, in a cell with no dicentric, may be the result of a proximally incomplete dicentric, *cs dic inc prx*. Such terminology should only be used where the complementary parts of the incomplete exchange can be identified in the same cell—in this case the two monocentric chromosomes with terminal deletions.

Similarly, in a cell with an acentric fragment

and a single abnormal monocentric chromosome banding may reveal many different situations:

1. An intrachromosome exchange in which an inversion is incomplete proximally may be recognized because a compound fragment with some of the bands in reverse sequence will be present in a cell in which there is a terminally deleted chromosome complementary to the fragment, cse inv inc prx.
2. An intrachromatid exchange, in which a ring chromosome has failed to form because of proximal incompleteness, may be recognized as a compound acentric fragment in which the terminal sequences of two chromosomes are in the normal order and which is present in the same cell as the complementary abnormal monocentric chromosome with terminal deletions of both long and short arm, cse, r inc prx.

Marker chromosomes—In the special case of acquired abnormalities, a marker chromosome, *mar*, is a structurally abnormal chromosome, banded or unbanded. Where the banding pattern can be recognized, it can be adequately described by the standard nomenclature. Where two identical markers are present in the same cell, this should be indicated *mar* 1, *mar* 2.

A number before *mar* indicates the number of markers, e.g., +2mar1 indicates two markers 1, whereas +13mar indicates 13 different unidentified or unique markers.

Complex Chromosome Rearrangements

The short system of the standard nomenclature may give rise to ambiguities in describing chromosomes derived from some complex rearrangements, especially when two homologous chromosomes are involved. However, the detailed system will clarify about all situations. The position of the breaks will in most cases distinguish between the homologues. If this is not sufficient, 9mat and 9pat, for example, may be used if this is known from a study of heteromorphisms, but if the homologues cannot be distinguished, one of the numerals should be doubly underlined, i.e., 9 and 9. In some cases it will be necessary to illustrate the rearrangement and describe it in words to ensure complete clarity.

Chromosome number—The chromosome count in a given cell should include all *centric* structures present in that cell regardless of the number of centromeres. Acentric fragments, *ace*, are not included in the count. Note also that when a triradial or other complex chromatid rearrangement is present it counts only as one structure. In this way the number of centromeres in that structure is indicated. For example, 45,XX,*tr*(2cen), means a dicentric triradial; 46,XX,*tr*(1cen), means a monocentric triradial.

Where a sex-chromosome is missing or is additional as an acquired aberration, this should be indicated by adding a minus (−) or plus(+)sign followed by X or Y, e.g., (i) 45,X,−Y indicates an acquired aberration in an individual with a 46,XY chromosome consitution where the cell in question lacks the Y-chromosome; (ii) 47,XX,+X indicates an acquired aberration where the chromosome constitution is 46,XX, and where the cell in question has an additional X.

Description of single abnormal cells is often not required but a few examples may help to make the terminology clearer. Normally, put numerical abnormalities first, followed by chromosome aberrations, followed by chromatid aberrations, e.g.,

45,XY,−B,−C,+D,−G dic(B;C),+ace,5ctg

is derived from a male cell that has an additional D-group chromosome and lacks one G-group chromosome; it has a dicentric composed of the missing B- and C-chromosomes, an acentric fragment, and five chromatid gaps.

44,XX,−B−D,+mar1, csg, tr(2cen), ctb

is derived from a female cell that lacks one B-chromosome and one D-chromosome; it has a marker, a chromosome gap, a dicentric triradial which is incomplete, and a chromatid break.

Scoring of Aberrations

In the scoring of aberrations, the main types are *ctg, ctb, cte, cg, cb; ace, min, r, dic, tri, der,* and *mar,* and reports should, where possible, give the data under these headings. It is recognized, however, that frequently aberrations are grouped to give adequate numbers for statistical analysis or for some other reasons. It should be indicated how the groupings relate to the aberrations listed above. For example,

Chromatid aberrations	(ctg, ctb, cte)
Fragments (= deletions)	(csb, ace)
Asymmetric aberrations	(ace, dic, r)

Data should be presented in such a manner that it is possible to calculate "lesions per cell" where a *lesion* is an event capable of giving rise to a gap, break, or exchange point in a chromatid or chromosome. For example, give the number of each type of aberration and the total cells scored so that, say 25 dicentrics in 100 cells scored is 0.25 lesion per cell.

Cell Populations with Acquired Abnormalities

A *clone* is a population of cells derived from a single progenitor cell. It is common practice to infer a *clonal origin* where a number of cells have the same or related abnormal chromosome complement. A clone is not necessarily homogenous. Where the term *clone* is used in this sense, it should be defined by the author because the precise definition will depend both on the number of cells examined and on the nature of the aberration involved. It will always mean at least two cells (preferably three cells in hypodiploidy) with the same aberration, e.g., two cells with the same additional complex marker in 25 cells analyzed would be an acceptable clone, but two cells 45,X in 100 normal female cells would not. Thus, in some cases, especially with hypodiploid cells, three or even more cells will be necessary to constitute a clone, depending on the number of cells analyzed.

Tumor Cell Populations

The individual chromosomes of tumor cell populations can all be described by the standard nomenclature. There is a need, however, to clarify the terminology of populations of tumor cells.

Modal number (mn) is the most common chromosome number in a tumor cell population at the time it is observed in a direct preparation. Where cultured cells are studied, this should be indicated by adding the words *in vitro* in front of the term modal number.

The *modal number* may be described as $2n\pm$ (near diploid) where it is approximately diploid but where there is not a sharp mode. The modal number will be *hypodiploid* ($2n-$) where there is a mode less than 46 chromosomes but where there is not a sharp mode at a particular chromosome number, and *hyperdiploid* ($2n+$) where there is a mode more than 46 chromosomes but where there is not a sharp mode at a particular chromosome number.

Modal numbers near triploid, tetraploid, or any other multiple of the haploid number and which cannot be given as a precise number of chromosomes can be expressed as $3n\pm$ (near triploid), $3n-$ (hypotriploid), $3n+$ (hypertriploid), $4n\pm$ (near tetraploid), $4n-$ (hypotetraploid), $4n+$ (hypertetraploid), and so on.

The distinction between $2n+$ and $3n-$ would be that in the former case the majority of the counts lie below $2n+(n/2)$ and in the latter the majority of the counts lie above $2n+(n/2)$. Similarly for $3n+$ and $4n-$, and so on.

Where a line of cells varies little from triploid or tetraploid, the additional or missing chromosomes may be indicated as variations from the polyploid mode by placing the term $3n$, $4n$, etc, in front of the count, e.g., $3n$, 70, XXY, +18 is a triploid cell with an XXY sex-chromosome complement and an additional chromosome #18.

Pseudodiploid is used to describe a cell that has the diploid number of chromosomes but is abnormal because of the presence of acquired numerical and/or structural aberrations.

Stem line (S, sl) indicates the most frequent chromosome constitution of a tumor cell population at the time it is observed in a direct preparation. Again the term in vitro should be used to indicate that cultured cells were studied. All other lines are termed *side lines* or *sub lines* (s, sdl). Where there are a number of lines that appear to be related, these could be referred to as a clone.

If more than one side line is present these should be referred to as s_1 or sdl 1 and s_2 or sdl 2 and so on. The *s* number does not indicate the frequency of the side line. The same designation for a side line should be used for the same side line in sequential observations. Where a side line is found to have split into two separate side lines at a second or subsequent observation, this should be indicated by retaining the original number for the unchanged side line and giving a new *s* or *sdl*

(Table 4A) number to the new side line. The symbol S has been used in the past to denote a stem line and s a side line. The latter should not be confused with the new meaning of the symbol s = satellite (ISCN, 1978).

In writing down the chromosome constitution of a tumor, both the number of cells counted and the number of cells analyzed should be given. The chromosome constitution of a tumor will consist of two series of symbols separated by a period (.). The first series will give the numbers of cells in each category with a different number of chromosomes. The second series will give the numbers of cells in each category with a different chromosome constitution. Thus, 46=5/49=10/52=4. 46,XY= 2/49,XY,+1,+8,+12=8/49,XY,+X,+1,+7= 1/52,XY,+X,+X,+1,+7,+22,+22=4. indicates a tumor in which a total of 19 cells were counted. The number of cells with 46 chromosomes is 5, the number with 49 chromosomes is 10, and the number with 52 chromosomes is 4. Of the cells with 46 chromosomes, 2 have been analyzed and both have a normal karyotype. Of the cells with 49 chromosomes, 9 have been analyzed and 8 of these have an additional #1, an additional #8, and an additional #12, whereas 1 of them has an additional X, an additional #1, and an additional #7. Of the cells with 52 chromosomes, all 4 have been analyzed, and all have 2 additional Xs, one additional #7, and 2 additional #22s (c.f. nomenclature for mosaics).

Determination of the stem line requires consideration of the number of cells counted and analyzed as well as the number of cells with a given karyotype. Thus, for the tumor

46=5/47=35/48=10.   46,XY=3/47,XY,+8=
          6/47,XY,+X=
4/48,XY,+8,+9=1/48,XY,+21,+21=1

the calculation is as follows:

Out of the 50 cells counted 5 had 46 chromosomes and of these each of the 3 cells analyzed had a 46,XY karyotype. Assuming that the sampling is random

46,XY,	5/50 × 3/3 × 100 =	10%
47,XY,+8,	35/50 × 6/10 × 100 =	42%
47,XY,+X,	35/50 × 4/10 × 100 =	28%
48,XY,+8,+9,	10/50 × 1/2 × 100 =	10%
48,XY,+21,+21,	10/50 × 1/2 × 100 =	10%

Therefore the stem line is 47,XY,+8.

It is not feasible to describe the analysis of all side lines and unique cells in the description of a tumor because in some cases the number of separate karyotypes may be very large. It will be implicit in the description of a tumor, however, that those karyotypes not included are less numerous than those that are included.

Where variability is very great, the chromosome consitution of the tumor could be reported simply as, e.g.,

$2n\pm = 37/4n- = 3/8n-= 1$ (sl or S 47,XX,+8 = 13)

meaning that 37 near diploid, 3 hypotetraploid, and 1 hypooctaploid cells have been counted, and the stem line is represented by 13 cells with an extra chromosome #8. There will be some situations in which it will be simpler to describe the chromosome constitution of a tumor in words than to express it as a complex formula, and this should be done.

References

1. Chicago Conference, 1966; Denver Conference, 1960; London Conference, 1963.
2. Lancet, 1960, 1961b.
3. Ohnuki, 1968.
4. Ferguson-Smith, Ferguson-Smith, et al., 1962; Palmer and Funderburk, 1965; Sasaki and Makino, 1963.
5. Buhler, Jurik, et al., 1977; Carr, 1971; Dygin, 1976; Makino, 1975; J.J. Yunis, 1974; J.J. Yunis and Chandler, 1977a.
6. Even though generally accepted criteria for the human karyotype have been established, some of the persisting problems, such as the effects of age on the karyotype and the polymorphism of certain chromosomes in the human, can be found in the following references: Curnow and Franklin, 1973; Geraedts, Pearson, et al., 1975; T. C. Hsu, Pathak, et al., 1975; D. T. Hughes, 1968; Jacobs and Court Brown, 1961, 1966; Jacobs, Brunton, et al., 1963, 1964; Jarvik, Yen, et al., 1974; Kowalski, Nasjleti, et al., 1976; Lancet, 1960, 1961b; A. Levan and Hsu, 1959; Littlefield and Goh, 1973; Lüers, 1964; Simpson and Martin, 1977; J. S. S. Steward, 1960; Zankl and Zang, 1971a.
7. Golloway and Buckton, 1978; Harnden et al., 1976; Jacobs and Court Brown, 1961, 1966; Jacobs, Brunton, et al., 1963, 1964; Jarvik, Yen, et al., 1974; Sandberg, Cohen, et al., 1967.
8. Pierre and Hoagland, 1971, 1972; Golloway and Buckton, 1978; Harnden et al., 1976; Sakurai and Sandberg, 1976c.

9. Sandberg, Koepf, et al., 1960, 1961; Sandberg, Ishihara, et al., 1962a, b.
10. Jacobs, Brunton, et al., 1963, 1964; Jarvik, Yen, et al., 1974; Sandberg, Cohen, et al., 1967.
11. Geraedts, Pearson, et al., 1975; Lancet, 1971b; Makino and Muramoto, 1964; Reitalu, Bergman, et al., 1972; Schwinger, 1973; Soudek and Laraya, 1976; Soudek, Langmuir, et al., 1973.
12. Takagi and Sandberg, 1968b.
13. Sandberg, Takagi, et al., 1968a; Takagi and Sandberg, 1968a, b.
14. Arrighi and Hsu, 1971; Chen and Ruddle, 1971; Pardue and Gall, 1970.
15. Caspersson, Lomakka, et al., 1971a, b.
16. Drets and Shaw, 1971; Dutrillaux, Grouchy, et al., 1971; Finaz and Grouchy, 1971, Patil, Merrick, et al., 1971; Schnedl, 1971a, b; Seabright, 1972; Sumner, Evans, et al., 1971; Wang and Fedoroff, 1972.
17. Dutrillaux and Lejeune, 1971.
18. S. E. Bloom and Goodpasture, 1976; Funaki, Matsui, et al., 1975; Goodpasture, Bloom, et al., 1976; Hayata, Oshimura, et al., 1977; Lau and Arrighi, 1977; Matsui and Sasaki, 1973.
19. Howell and Denton, 1974, 1976; Howell, Denton, et al., 1975.
20. C. E. Ford and Clegg, 1969.
21. Aula and Koskull, 1976; Aymé, Mattei, et al., 1976; Brogger, 1971; Comings, 1974; Jacobs, Buckton, et al., 1974; J. H. Taylor, 1963.
22. H. Kato and Sandberg, 1967, 1968; Matsui, Weinfeld, et al., 1971, 1972; Matsui, Yoshida et al., 1972; Rieger, Michaelis, et al., 1976; Sandberg, Aya, et al., 1970.

3

Chromosome Structure

The Ultrastructure of a Chromosome
*Morphology of a Chromosome Observed
    with Electron Microscopy
The Fibrils
DNA Packing Ratio
Models of Chromosome Structure
Recent Considerations of Chromosome
    Structure
Heterochromatin*

The Ultrastructure of a Chromosome[1]

There is little doubt that the future of cytogenetics in cancer and leukemia resides in even further resolution of chromosomal detail and structure. It is with this thought in mind that this chapter presents a succinct overview of the structure and related facets of the human chromosome with the hope that it will serve as a fillip for further investigation of the ultrastructure of chromosomes in neoplasia and as a heuristic introduction to those embarking on the study of chromosomes in human cancer and leukemia.

For a more comprehensive treatment of chromosome structure, the reader should consult other sources (for example, Bostock and Sumner, 1978; Schwarzacher, 1976; J. J. Yunis, 1977; Sparkes, Comings, et al., 1977). The ultrastructure of chromosomes is best studied by electron microscopy of whole mount preparations.[2] Thin sectioning does not seem to be too practical for this purpose, because only very short segments appear in a given section and, furthermore, there is considerable difficulty in obtaining many serial sections for electron microscopic study. Besides, the fibrillar network of a chromosome is so dense that it is almost impossible to distinguish one fibril from another.

The methodology for the study of chromosomes by electron microscopy was greatly improved by (a) the introduction of the spreading technique of Kleinschmidt and Lang (1962), originally used for bacterial and viral fibrils, and (b) the critical-point drying method of Anderson (1951), as used by Gall (1963) and DuPraw (1966). With Kleinschmidt and Langs' method, the dividing cells are spread on an air-water interface, which bursts the cellular membrane and releases the chromosomes. The intact chromosomes are then picked up on electron microscope grids, washed, or treated with analytical reagent and the wet grid immersed in carbon dioxide under pressure. This is then passed through the critical temperature for carbon dioxide, i.e., instant passage from the liquid to the gaseous state. This technique adequately preserves the morphology of chromosomes. The collection of an adequate number of chromosomes by the spreading technique of Kleinschmidt and Lang (1962) is, however, difficult and, occasionally, structural distortion of chromosomes takes place. C. Wolff, Gilly, et al. (1974) devised a slide-centrifuging method which allows accumulation of suitable chromosomes for observation.

Morphology of a Chromosome Observed with Electron Microscopy

A well-preserved whole mount human chromosome seems to be composed of one or several "spaghetti-like fibers" of fairly uniform diameter. Generally, these fibers entangle and form loops in all directions. But, in superoptimal preparations, it is evident that very few or no free fiber ends are present, even at the telomeric ends (Figure 22). Several side-by-side sections of chromatin fibers which pass from one arm of the chromatid to the other are seen in the centromeric region (DuPraw, 1970), as

22 Electron micrograph of a water spread and critical point dried unstained human chromosome (group B) (A) (× 15,000) and its fiber (B) (× 204,000).
(Lampert, 1971.)

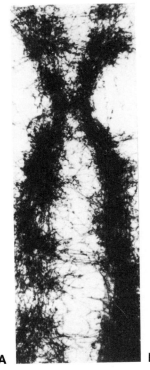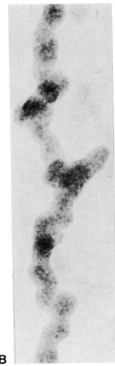

A B

well as between sister chromatids (Comings and Okada, 1970). Ruzicka (1974a, b), on the other hand, found longitudinally oriented parallel layers of fibril bundles in the centromeric region, which are particularly apparent after G-banding (Ruzicka, 1974a, b). He did not, however, rule out the possibility of these being artifacts.

The Fibrils

The diameter of the chromosomal fibers has been found to vary from 30 to 200 Å in size (Figures 23–25). The wide variation may be due to technical variability during the preparation of the chromosomes for electron microscopic study. For example, S. L. Wolfe (1965a, b) and Ris (1966) observed that with the Gall technique (1963) the size of a fibril estimated to be 250 Å in a solution of low ionic strength was reduced to 20–30 Å in one of high ionic strength. In ethylenediaminetetraacetate (EDTA)-treated chicken erythrocyte chromatin freed of hemoglobin, water spread, and then examined without exposure to ethanol, the

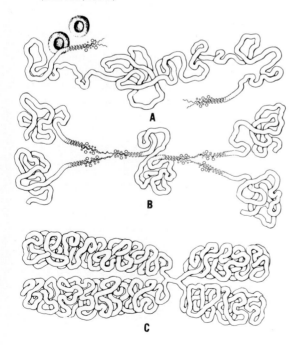

23 Diagram illustrating a folded fiber model of chromosome structure. (A) Each unreplicated chromosome (unit chromatid) is essentially a single 200–500 Å fiber, which contains a DNA double helix in supercoiled configuration; (B) replication of the chromosome occurs at several sites along the length of the fiber, where DNA polymerase catalyzes DNA synthesis at fork configurations; (C) the late-replicating segments of the fiber at the centromere and elsewhere serve to hold together the sister chromatids.
(DuPraw, 1970.)

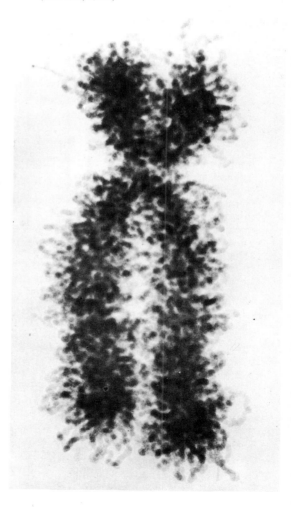

24 Human chromosome #12 in a standard configuration. The total dry mass for the two chromatids equals 13.2×10^{-13}g, of which 74% is in the long arm.
(DuPraw, 1970.)

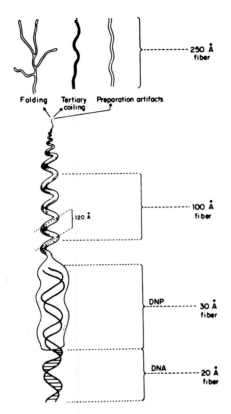

25 Structure of the DNA protein (DNP) fiber. At the bottom is the 20 Å DNA fiber. When covered with histone and nonhistone proteins it forms the basic 30 Å DNP fiber of genetically active chromatin. The scale of the diagram is then changed to illustrate the formation of the 100 Å fiber by supercoiling of the 30 Å fiber with a 120 Å pitch distance. The scale is then changed again to illustrate the three ways in which the 250 Å fiber might be formed: (a) by folding, (b) by coiling again, or (c) by adsorbing proteins or other artifacts.

(Comings, 1972.)

fiber width was about 37 Å. But, the mean fiber diameter rose to about 138Å when treated with ethanol and critical-point dried in amylacetate (Solari, 1971). If nuclei are floated on a surface containing hemoglobin, the mean fiber diameter is about 313Å. Lampert (1971) described a colcemide-blocked and hypotonically treated human chromatid composed of fibrils with a 250 Å diameter by the critical-point drying method. Ruzicka (1974a) observed 300–600 Å fibrils with the G-banding technique. According to Comings (1972), the 30 Å size of fibrils is the true width, and the high estimates are either artifacts of preparation or represent the physiologic state of increased coiling of the 30 Å fibers. However, there is evidence that a 200–300 Å fiber was formed by tertiary coiling of the 70–80 Å fibers, which in turn formed the basic secondary coiling of the deoxyribonucleoprotein (DNP) (Lampert and Lampert, 1970). Such successive coiling and supercoiled structures have also been shown by X-ray diffraction techniques (Pardon, Wilkins, et al., 1967; Richards and Pardon, 1970).

DNA Packing Ratio

The arrangement of DNA and protein in the chromatin fibrils is still an enigma. In mammals 95% of the nuclear proteins are histones and only 5% nonhistones. Older chemical models proposed that the histones were situated "outside" the DNA, with the latter possibly folded and coiled to form an inside core covered by a sheath of protein (histone). Kornberg (1974), however, proposed a model in which the DNA molecules wind around globular units of histones formed by four subunits. The subunits are connected by DNA fractions and thus stabilized and *spaced* by histone fraction 1, possibly on the outside of the chains of the globular subunits. The diameter of the subunits is about 100 Å (Fig. 25). Olins and Olins (1974) supported this concept on the basis of electron microscopic studies of the ν-bodies, called nucleosomes, isolated from chromatin fibers. These fibers are of a dimension similar to that of the model proposed by Kornberg (1974).

The DNA of a typical mammalian cell if extended would be about 1 meter long. But, this DNA is associated with protein, packed and folded in such a way that it is accommodated in a nucleus the diameter of which is about 10^{-3} cm. The ratio of the length of a segment of DNA to the length of a chromatin fiber constitutes the DNA packing ratio.

Results with quantitative electron microscopy by DuPraw and Bahr (1969) suggested a packing ratio of 50:1 for a 250 Å fiber. Bahr and Golomb (1974) determined the DNA packing ratio to be 28.3:1. Their data obtained support from X-ray diffraction studies of Par-

The Ultrastructure of a Chromosome 67

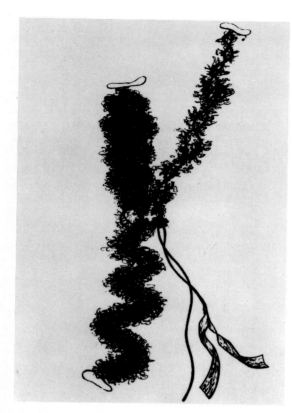

26 Model of the Chinese hamster mitotic chromosome with the arms dissociated to varying extents. The left chromatid is relatively intact with the lower arm pulled out to reveal its coiled organization. The upper right arm is pulled out further to reveal the two strands of the chromonema with attached loops of epichromatin. The chromonemata terminate in fragments of the nuclear membrane at the telomeres. The epichromatin has been removed from the lower right arm and one strand of the chromonema unwound to demonstrate its ribbon architecture. If depicted full length, it would be 6–10 times the normal chromatid length.

(*Stubblefield, 1973.*)

don, Wilkins, et al. (1967; Pardon and Wilkins, 1972). An 80 Å primary coil, wound into a 200 Å coil satisfies the 28.3:1 packing ratio.

Models of Chromosome Structure

Evidence indicates that chromosomes are composed of fibrils that contain supercoiled DNA molecules (Figures 26–28). But it is still not

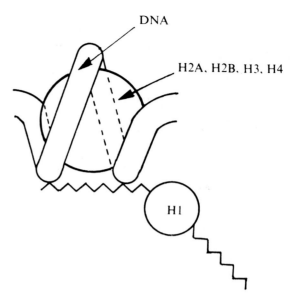

27 Schematic representation of a possible model for the chromatin subunit structure. The protein core is a complex of the apolar segment of the four histones indicated (H1–H4), the basic segments of the histones being complexed with DNA on the outside of the unit. Histone H1, possibly on the outside of the chain of globular subunits, may have a cross-linking role either between subunits in the same chain or between subunits in different chains. The pitch of the DNA, which need not be uniformly coiled, is 5.5 nm with a mean diameter of 10 nm.

(*Baldwin, Boseley, et al., 1975.*)

clear how these fibrils are arranged within a chromosome. In recent years, several reviews have been published regarding the organization of chromatin and chromosomes.[3] There are two main schools of thought. The first group believes that each chromatid consists of a single strand of DNA material that is folded in a complex manner, whereas the other group proposes that there are two or more strands of DNA per chromatid. There are indications[4] that a chromosome consists of a bundle of longitudinally arranged parallel strands, the number of which might range from 32 to 64. Stubblefield and Wray (1971) proposed a binemic model of the Chinese hamster metaphase chromosome (Figure 26). According to them, each half-chromatid contains two DNP ribbons

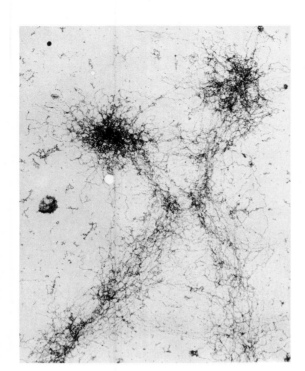

28 A well-dispersed whole-mount preparation of a human submetacentric chromosome (× 10,300).
(Comings, 1972.)

wound into a single fiber (the core) with many loops of chromatin (the epichromatin) attached along its length (Figure 26). Comings and Okada (1973), however, believe that no such chromosome exists at all, because in a well-dispersed mammalian or avian chromosome only multiple folded chromatin fibers are observed. In thin sections of the mitotic chromosomes no central cores are seen. Moreover, in partially dispersed chromosomes the superficial fibers form loops laterally and the central ones remain compacted. When these chromosomes are exposed to any additional manipulation, the central fibers tend to clump together (like wet spaghetti) and produce an artificial illusion of cores.

The other concept is that the chromosome is a tandem linkage of single DNA duplex molecules (Schwartz, 1955). Freese (1958) elaborated this idea by adding hypothetical weak bonds between protein joiners which connect the tandem strands of DNA. Later, J. H. Taylor (1966) and DuPraw (1966) eliminated the non-DNA linkers. According to these models the strand would have to be folded, looped, or coiled in a very compact way to accommodate the dimensions of a metaphase chromosome.

The model suggested by DuPraw (1966), known as the folded fiber model, has direct support from the visible structure of whole chromosomes prepared by modern techniques. The main features of the model are as follows:

1. Before interphase replication, each chromosome is thought to consist of a single long DNP molecule in which the double helix of DNA forms the main structural axis. This DNP fiber is repeatedly and randomly folded back on itself (both longitudinally and transversely) to make up the body of the chromatid.
2. This unit chromatid replicates during the S-period at two or more replication forks, giving rise to two sister-unit chromatids held together by unreplicated regions.
3. At prophase the pairs of unit chromatids fold up tightly to form visible chromosomes, but the sister chromatids continue to be held together by minute unreplicated DNA segments, especially at the centromeric region.
4. During prophase and metaphase these two daughter fibers fold up in a way that is reproducible from one generation to the next.
5. At telophase the compact anaphase chromatids unfold but retain their relative positions in the interphase nucleus, possibly by means of attachments to the newly formed nuclear envelope.
6. Finally, during the transcription and replication events of interphase, each chromatin fiber is thought to function as an independent unit, with specific parts of the DNA code being "read" at characteristic sites and rates.

The main support for this model came from the visible structure of whole chromosomes prepared by a recent method. One objection to this model is that it suggests that anaphase is initiated by replication of the centromere. However, data as to whether the centromere is actually late replicating are inconclusive. Moreover, it is difficult to reconcile this with the phenomenon of breakage and reunion with exchange of arms, which can occur between sister chromatids during mitosis. Besides, part of this model suggests that the fiber is folded in an entirely random manner; yet, recent banding techniques (where transverse bands are induced by different treatments) suggest that the structure of metaphase chromosomes is not random but highly ordered. These banding techniques will have a considerable impact on the theories of metaphase chromosome structure. Huberman (1973) suggests that in considering models of the structure of metaphase chromosomes, the orderly bands of chromosomes and the fact that there is a single chromosome fiber that runs the length of each chromosome must be given serious consideration.

Recent Considerations of Chromosome Structure

Comings (1977) in a recent evaluation of the structure of mammalian chromosomes (Figures 29–32) indicated that uninemy is no longer a subject of significant contention regarding the basic structure of mammalian chromosmes and that polynemy has been definitely proven not to play a role in such structure. In addition, the large amount of work and the voluminous explosion of literature on the ν-body and nucleosomes (Kornberg, 1974, 1977) has solved all but a few details about the structure of the chromatin in its extended "nucleosomes on a string" configuration or its compact configuration forming the 250 Å fiber. Comings summarized the numerous aspects of mammalian chromosome structure by indicating that it begins with a 20 Å DNA fiber, which is complexed with nucleosomes to form the extended "nucleosomes on a string" configuration; the latter then condense, resulting in a 100- to 300-Å thick fiber, forming the classic 250 Å chromatin fiber of metaphase and interphase chromatin. He also stressed the fact that the nuclear matrix binds to adenine–thymine (AT)-rich segments of DNA present in G-band chromatin to form chromomere loops and interchromomere DNA. These condense to form G-bands and R-bands. The fact that interband (R-bands) DNA is early replicating and guanine-cytosine (GC)-rich whereas G-band DNA is late replicating and AT-rich has long been strong evidence in favor of R-bands containing active euchromatin and G-bands containing inactive heterochromatin. Comings also suggested that the increased amount of nonhistone

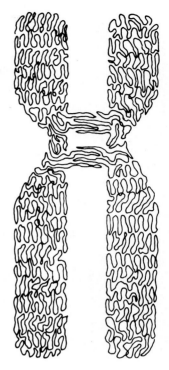

29 Single-stranded model of chromosome structure. This suggests that a single DNP fiber beginning at one telomere folds upon itself to build up the width of the chromatid and eventually progresses to the opposite telomere without lengthy longitudinal fibers, with no central core and no half- or quarter-chromatids. The centromere region in this metacentric chromosome is depicted as the result of fusion of two telocentric chromosomes with retention of the individual centromere regions. The fibers at the point of chromatid association briefly interdigitate.
(Comings, 1972.)

protein associated with euchromatin interferes with the binding of Giemsa dye to R-band DNA and is partly responsible for the poor staining of R-bands seen in G-banding. When the chromosmes are partially uncoiled they are characterized by spiralization with the dimensions of the spiral fiber ranging from 0.2 to 0.5 μm; when this chromosome is further despiralized and depending upon the conditions, the resulting strand can be 0.35–0.43 μm and as small as 0.1 μm. When this spiralized chromosome is fully condensed, it forms the classic

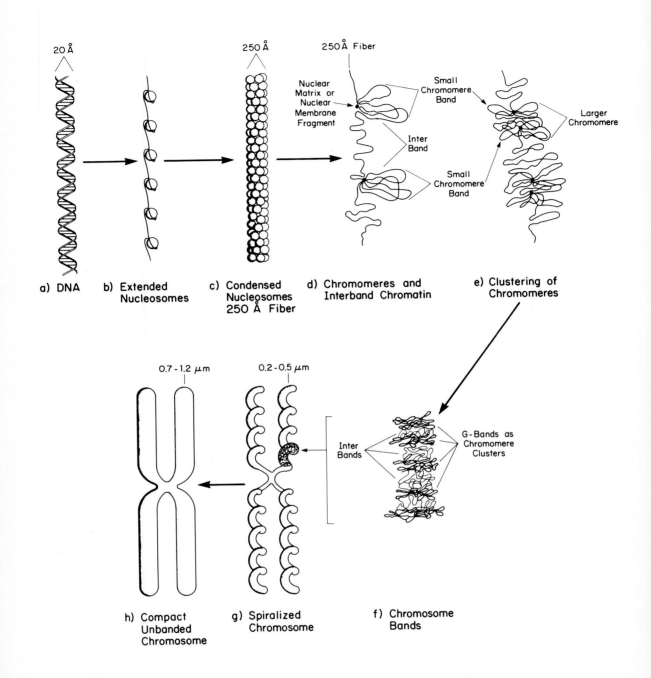

30 Schematic representation of the structure of a human chromosome starting with the basic substance DNA and carried through the formation of nucleosomes, chromomeres, and ultimately the spiralized chromosome.
(According to Comings, 1977.)

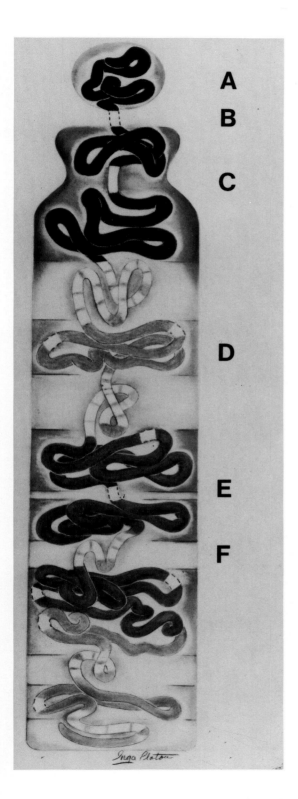

31 Schematic presentation of the internal structure of a human chromosome and its possible relation to banding. Chromosome model in which the fine structure of chromosomes, the G-banding pattern, and the localization of repetitive DNA are combined to illustrate various structural elements of the chromosome. A, perinucleolar and pericentromeric constitutive heterochromatin, rich in the highly repetitive satellite DNA; B, secondary constriction of nucleolar organizer, site of 18 and 28 S rRNA cistrons; C, primary constriction or centromere; D,E, minor and major dark G-bands, rich in intermediate repetitive DNA, spacing vital genes such as tandem gene duplicates for 5 S rRNA, 4 S tRNA, and 9 S histone IV; F, light G-bands, where the bulk of the structural genes spaced by intergenic segments are believed to be localized.
(Courtesy of Dr. J. Yunis.)

appearing chromosome seen in visual microscopy.[5]

As proposed by Kornberg (1974, 1977), chromatin consists of repeating units, termed "nucleosomes," containing a fairly well-defined length of DNA (e.g., 250 base pairs in rat liver) associated with an octomere aggregate of the histones containing pairs of each of the four main types. Enzyme digestion studies on nuclei have shown that although the DNA content of nucleosomes is often about 200 base pairs, quite large variations are found according to the species or tissue investigated. However, it has been found that further nucleus digestion produces a "core" particle containing the same number, about 140 base pairs, in all cell types so far investigated. The rest of the DNA in the repeat length not included in the core is thought of as a linker between core particles (Finch, Lutter, et al., 1977).

Heterochromatin

The term "heterochromatin" is used to denote chromosomes or chromosome regions that are condensed at interphase and prophase and do not unravel in telophase like the rest of the chromosome.[6] In mammals two main types of heterochromatin are recognized: constitutive heterochromatin, or the heterochromatin that is present in homologous chromosomes and

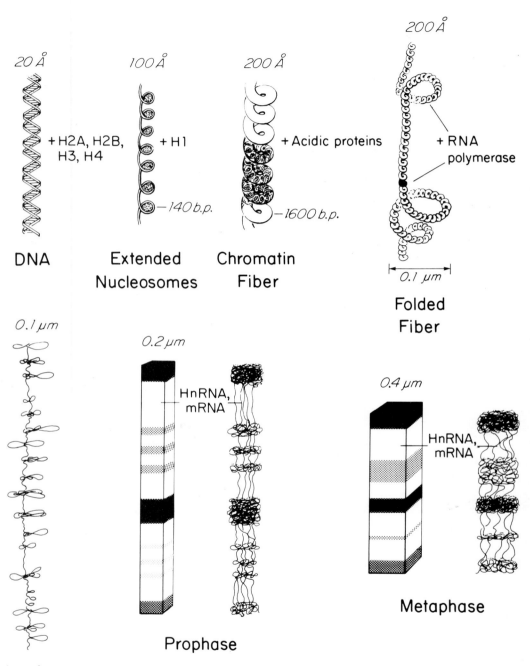

32 Schematic representation of the organization of a human chromosome and its relation to banding according to Yunis. Compare with the scheme of Comings (1977) shown in Figure 30. HnRNA = heterodisperse nuclear RNA; mRNA = messenger RNA; b.p. = base pairs; H1, H2A, H2B, H3, H4 = various histones.
(Courtesy of Dr. J. Yunis.)

facultative heterochromatin, or the heterochromatin that results from the inactivation of one of the two X-chromosomes in females. This inactivation is an effective mechanism to reduce the number of functional X-chromosomes to one in both sexes. Apparently, the DNA of constitutive heterochromatin is composed to a large extent of short, repeated polynucleotide sequences, termed "satellite" DNA. Thus, constitutive heterochromatin may be considered a special type of chromatin which contains most of the satellite DNA, i.e., highly repetitive DNA sequences of the genome not transcribed into RNA for protein synthesis. Blocks of these sequences are usually located in the regions of the nucleolar organizers, centromeres, telomeres, and sometimes intercalated within other regions of the chromosomes. An evaluation of the knowledge about constitutive heterochromatin and satellite DNA[7] suggests that the roles of constitutive heterochromatin are structural in nature and reflected in the special banding techniques for the centromeric areas.

References

1. The following references cover areas of chromosome structure not taken up in detail in this chapter: Bahr, 1978; Bahr and Golomb, 1971; Bajer, 1965; Bak, Zeuthen, et al., 1977; Baldwin, Boseley, et al., 1975; Bram, 1972; Cleaver, 1977b; Comings, 1968; Comings and Okada, 1971, 1976; Crick, 1971; Crick and Klug, 1975; Dutrillaux, 1977; Evans, 1973; Filip, Gilly, et al., 1975a, b; Geneix, Jaffray, et al., 1974; Griffith, 1975; Harrisson, 1971; Mendelsohn, Mayall, et al., 1973; Ohnuki, 1968; Schwarzacher, 1976; Schwarzacher, Bielek, et al., 1977; Silver and Elgin, 1976; Ris, 1957; J. J. Yunis, 1977.
2. Bahr and Engler, 1977; Christenhuss, Büchner, et al., 1967; Comings and Okada, 1970; Golomb and Bahr, 1971 a,b, Mouriquand, Gilly et al., 1975; Rattner, Branch, et al., 1975; Schwarzacher, 1976; Schwarzacher, Ruzicka, et al., 1976; Schwarzacher, Bielek, et al., 1977; J. J. Yunis, 1976, 1977.
3. Bahr, 1976; Comings, 1972; Huberman, 1973; Prescott, 1970; Ris and Kubai, 1970; Ruzicka, 1974a, b; Stubblefield, 1973; S. L. Wolfe, 1965a,b.
4. Huskins, 1937; Kaufmann, Gay, et al., 1960; Nebel, 1939; Ris, 1957; Steffensen, 1959.
5. Some other aspects of DNA and chromosomes may be found in the following references: Bosman, van der Ploeg, et al., 1977a; Dutrillaux, 1977; Latt, 1976; Weintraub and Groudine, 1976; Weisswichert, Schroeder, et al., 1976.
6. Balícek, Žizka, et al., 1977; P. K. Ghosh and Singh, 1976; J. J. Yunis and Yasmineh, 1971.
7. Atkin, 1974a; Chen and Ruddle, 1971; Evans, Gosden, et al., 1974; Howell and Denton, 1974; K. W. Jones, 1973, 1974; K. W. Jones and Corneo, 1971; K. W. Jones and Prosser, 1973; My. A. Kim, 1975; Olinici, Dobáy, et al., 1976.

4

Some Facets of the Cell Cycle and Unique Aspects of Chromosome Structure

Mitosis

The Morphology of Human Chromosomes

The Chromosome Complement
Ploidy
Karyotype
Characteristic Features Used for Chromosome Identification
Secondary Constrictions
Satellites
Visibility of the Satellites
Size of the Satellites
Satellite Association
Identification of Chromosomes with Autoradiographic Studies and DNA Replication
  Technique
Labeling Pattern of Specific Chromosomes
  Group A (chromosomes #1 – #3)
  Group B (chromosomes #4, #5)
  Group C (chromosomes #6 – #12)
  Group D (chromosomes #13 – #15)
  Group E (chromosomes #16 – #18)
  Group F (chromosomes #19, #20)
  Group G (chromosomes #21, #22)
  Y-chromosome
  X-chromosome

Identification of Chromosomes by Transverse Bands
Quinacrine Fluorescence or Q-Bands
Mechanism of Q-Banding
Centromeric Bands or C-Bands
  Mechanism
Giemsa or G-Bands
Comparison of Q- and G-Banding Techniques
Reverse Bands or R-Bands (Including T-Bands)
Differential Staining of the Nucleolus Organizers in Human Chromosomes

Some Comments on Polymorphism

Lateral Asymmetry

Sister Chromatid Exchange

The areas discussed in this section cover some facets of cell division and chromosomes basic to the understanding and meaning of chromosomal changes in human cancer and leukemia.

Mitosis

Mitosis (Figure 33) is a process of cell division in which a somatic cell divides into two daughter cells with karyotypes identical to that of the parent cell. The gonadal or reproductive cells undergo a specialized division, i.e., meiosis, in which the chromosome complement is reduced to half so that each sperm and egg receives half the number (haploid) of chromosomes of that of a body (somatic) cell. During *interphase* the chromosomes are active in RNA synthesis and, thus, replicate their chromatin material. This DNA synthesis period is known as the S-period. The period preceding S and succeeding the 1st mitosis is designated as G_1, and the period succeeding S and preceding the 2nd mitosis is designated as G_2. Generally, for each type of cell the S- and G_2-periods are constant.

The whole mitotic process is divided into four stages according to the function and change in morphology of the cell. They are: (1) prophase, (2) metaphase, (3) anaphase, and (4) telophase. A clear-cut demarcation, however, cannot be made from one stage to the other at the transition periods. The following are the main events that occur during different stages.

Prophase:
1. The chromosomes become visible and progressively shorter and thicker.
2. The two centrioles, which lie adjacent to each other, start to move to opposite poles and form a spindle while moving.

Metaphase:
1. The nuclear membrane and nucleoli disappear.
2. The chromosomes arrange themselves at the equator of the cell, become maximally condensed and attach to the spindle at the centromere (kinetchore).

Anaphase:
1. The centromere of each chromosome divides longitudinally, resulting in two chromatids, which become completely separated.
2. Each set of chromatids then starts moving under the influence of the spindle until they reach the opposite poles.

Telophase:
1. Cleavage of the cytoplasm takes place by which the parent cell is divided into two daughter cells.
2. The chromosomes start to elongate and gradually become less distinct.
3. The nuclear membrane reforms.
4. Nucleoli reappear.

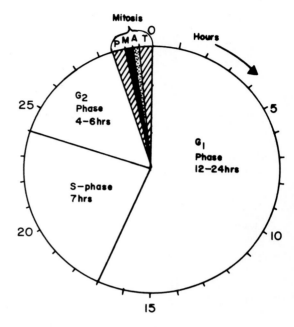

Interphase = G_1 + S + G_2

33 Division of the cell cycle into its various phases. The cycle of about 30 hr applies to most normal human cells, but may not be applicable to abnormal cells. P = prophase; M = metaphase; A = anaphase; T = telophase. Replication of the chromosomes (DNA and associated proteins) occurs during the S-phase, synthesis of enzymes and "luxury proteins" in G_1, and synthesis of proteins for the spindle and mitotic apparatus in G_2.

The Morphology of Human Chromosomes

The morphology of a chromosome can best be studied at the metaphase stage, when the chromosomes become relatively condensed and easily seen with the help of a light microscope. Human metaphases are shown in Figure 1.

A metaphase chromosome consists of two equal threadlike structures, the *chromatids*, which lie side-by-side longitudinally and are joined at a particular constricted point called the *centromere*, which in earlier days was known as the *primary constriction*. The centromere normally stains paler than the rest of the chromosome.

The location of the centromere varies from chromosome to chromosome and allows for a descriptive terminology for the chromosomes, according to the position of a centromere. If the centromere lies near the middle of a chromosome, it is called *metacentric* (median, submedian); it is *submetacentric* when the centromere is somewhat to one side of the middle; and if near one end of a chromosome, it is *acrocentric* (subterminal), and *telocentric* (terminal) when the centromere is located at the terminal end without a short arm being present (Figures 13 and 34). There are no telocentric chromosomes in the human complement.

The Chromosome Complement

Using a lung fibroblast culture from human abortuses, Tjio and Levan (1956) first discovered that human cells contain 46 chromosomes. Their finding was immediately confirmed by C. E. Ford and Hamerton (1956). They observed 46 chromosomes in spermatogonial metaphases and 23 bivalents in the great majority of primary spermatocyte at metaphase I. It is now well documented that normal human cells contain 46 chromosomes, which can be arranged into 23 pairs, according to their size and centromeric position. In a male cell, two chromosomes do not match; the longer one is submetacentric and is called the X and the shorter one is acrocentric and called Y. In a female cell there are two Xs instead of an XY. The XY chromosomes in male and XX chromosomes in female cells are called sex-chromosomes (gonosomes) and the remaining 22 pairs *autosomes* (Ferguson-Smith, 1962).

Ploidy

The 46 chromosomes (23 pairs) in the human complement constitute the *diploid* number and often are designated as $2n$. The gametes, which are produced after reduction division at meiosis, contain half the number ($22+X$ or $22+Y$), i.e., the haploid number of chromosomes (n).

The term *polyploid* is used when a cell contains a multiple of the haploid number of chromosomes (other than diploid), e.g., triploid ($3n$) cell contains 69 chromosomes, a tetraploid ($4n$) cell contains 92 chromosomes, and so on. The latter number usually results from nondisjunction or endoreduplication.[1] *Aneuploidy* is a term for a cell that contains more or less chromosomes than the exact multiple of a haploid set. The prefixes hyper- and hypo- are used before ploidy to indicate that the cells contain more or less chromosomes than a certain ploidy. Thus, hyper- and hypoploids belong to aneuploidy. A hyperdiploid cell contains more than 46 and a hypodiploid one less than 46. *Pseudodiploidy* describes a cell with 46 chromosomes but with an abnormal distribution and/or morphologic chromosomal changes.

Karyotype

For a detailed study of chromosomes, each is cut out from a photograph of a metaphase plate and arranged in homologous pairs (Figure 11). This arrangement is called a karyotype (Figure 12). A schematic presentation of a karyotype, which may be based on measurements of chromosomes from several cells, is called an *idiogram*.

Characteristic Features Used for Chromosome Identification

Besides the centromere position and chromosome length, which have already been discussed, some other morphological, structural, and physiological features can be taken into consideration in identifying chromosomes. The morphological features include secondary constrictions and satellites. The structural or physiological features include autoradiographic behavior and specific staining techniques which induce bands along the length of the chromosome. The latter are known as C-bands,

Q-bands, G-bands, R-bands, T-bands, and so forth.

Secondary Constrictions

Certain chromosomes possess constricted areas on their arms, besides the centromere or primary constriction, which are known as secondary constrictions. They represent unspiralized segments of a chromosome (Darlington, 1937). In the past, attempts have been made to use the secondary constrictions as criteria to identify individual chromosomes. Ferguson-Smith, Ferguson-Smith, et al. (1962) provided evidence for the existence of at least 15 sites where constrictions were visible with sufficient frequency at a fixed position in a chromosome. Besides the secondary constrictions (stalks) of the D- and G-group chromosomes, which separate the satellites from the short arms, obvious secondary constrictions occur in some chromosomes.

Ordinarily, secondary constrictions are not observed with consistency and when visible, they are often present in only one of the homologues. Therefore, many authors have applied special techniques to enhance their appearance. Saksela and Moorhead (1962) observed that with the flame-drying technique and 1:1 acetic acid and methyl alcohol fixation, secondary constrictions were more readily visible than with the air-drying fechnique, and 1:3 acetic acid and methyl alcohol fixation. Sasaki and Makino (1963) incubated culture material in calcium-free medium for the last few hours of culture and observed that the secondary constrictions were, generally, seen more clearly after a *longer* time in the calcium-free culture, though a prolonged stay in this medium decreased the mitotic rate. Hampel and Levan (1964) used cold treatment, and Kaback, Saksela, et al. (1964) and Palmer and Funderburk (1965) treated the cultures with 5-bromodeoxyuridine (BUdR) to optimize the visibility of secondary constrictions. Unfortunately, none of these techniques was particularly successful in the identification of every chromosome. In general, well-defined secondary constrictions are observed in chromosomes #1, #9, #16, and the Y.[2] On #1, #9, and #16 the secondary constrictions are present on the long arm, adjacent to the centromere. On the Y a constriction is seen in the medial region of the long arm. The frequency of this constriction, however, is lower than in #1, #9, and #16. Other chromosomes that show secondary constrictions but with a very low frequency are #2, #3, #4, #6, the X, #13, #14, #15, #17, #19, and #21.

Satellites

The short arms of the chromosomes of the D- and G-groups (i.e., chromosomes #13, #14, #15, #21, and #22) in human cells possess small masses of chromatin material known as satellites (Figures 34 and 35), which jut out from the main bodies of the short arms at their distal ends and are connected to and separated from them by distinct secondary constrictions (stalks) (Figure 35). During the late 1950s and early 1960s cytogeneticists tried to use these satellites as landmarks in identifying the various chromosomes of groups D and G but without much success.

Visibility of the Satellites

Until the time of the Denver Conference (1960), when cytogeneticists agreed upon the nomenclature and standardization of the human karyotype, it was thought that only pairs #13, #14, and #21 had satellites. Ferguson-Smith and Handmaker (1961 *a*, *b*) demonstrated that all five pairs of acrocentrics bear a pair of satellites on their short arms. However, in no single cell did they demonstrate satellites on all of the acrocentrics. Only 68% of group D and 70% of group G chromosomes had visible satellites. Subsequently, Gromults and Hirschhorn (1962) demonstrated satellites on all 10 acrocentrics in a metaphase plate.

The visibility of satellites varies between individuals and is related to culture conditions and, less definitely, to the stage of cell division (O. J. Miller, Mukherjee, et al., 1962; E. H. R. Ford and Woollman, 1967).

Size of the Satellites

The size of the satellites of groups D and G chromosomes shows considerable variation. Giant satellites were first described by Tjio, Puck, et al. (1960) in two of their three patients with Marfan's syndrome; the authors tried to draw a correlation between the enlarged satellites and some features of Marfan's syndrome. McKusick (1960) and Handmaker (1963),

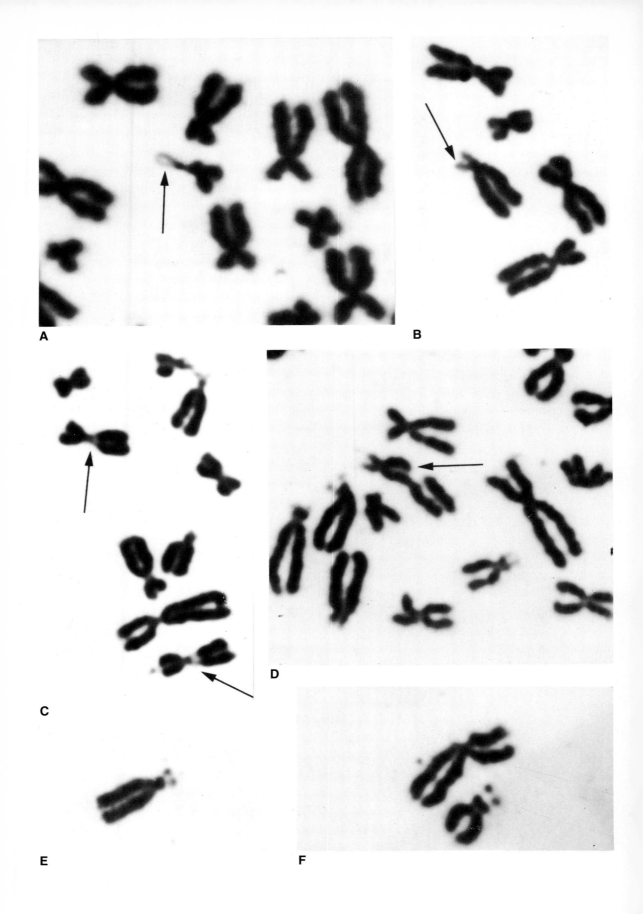

34 This figure demonstrates some of the finer morphological features of certain chromosomes in cultured blood cells from normal human subjects. The arrow in **A** points to an unusually shaped satellite attached to one of the small subterminal chromosomes of group G. In **B** an arrow points to satellites of group D chromosomes. Note that the satellites are stained more lightly than the rest of the chromosomes and differ from the antenna-like satellites of a similar chromosome shown in **E**. In **C** arrows point to the two group C (#9) chromosomes with secondary constrictions (location of satellite DNA), features characteristic of these two autosomes, which may be used as criteria for their identification. Note the satellited subterminal chromosomes of groups D and G in the upper right-hand corner. In **D** an arrow points to a gap in a chromatid of a subterminal chromosome. Note the satellites on several chromosomes of groups D and G. In **E** are shown antenna-like satellites of a group D chromosome with prominent bulbs. In **F** satellites with prominent bulbs of a chromosome belonging to group G are depicted. Compare the structure of this autosome with that in **A**.

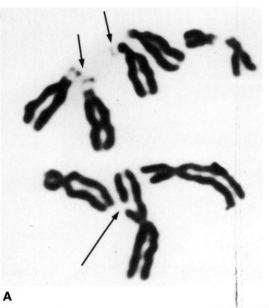

35 Satellite association among acrocentric chromosomes. A. Partial metaphase of a cultured blood cell with satellite association among four acrocentric chromosomes of group D (short arrows pointing to the satellites of these autosomes) and a gap (long arrow) in a metacentric chromosome. B and C. Further examples of satellite associations among acrocentric chromosomes. Arrows in (B) and (C) point to the general area of satellite association.

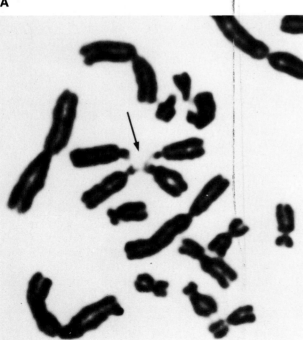

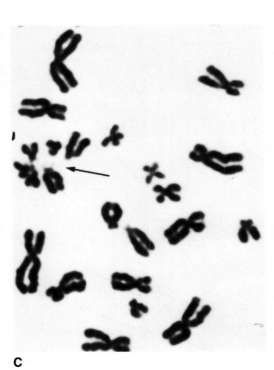

however, questioned the reliability of the diagnosis of Marfan's syndrome by enlarged satellites, because patients with this syndrome were found without enlarged satellites and, sometimes, unaffected individuals in families with Marfan's syndrome had enlarged satellites. Enlarged satellites were also seen in patients having some stigmata of Down's syndrome, in subjects with deformities of the central nervous system, patients with sexual abnormalities, and uneffected relatives of these patients.[3]

Ray-Chandhuri, Kakati, et al. (1968) described a normal individual with enlarged satellites which, however, were not consistent in all the cultures of cells from the same individual. Thus, it appears that: 1. no particular trait is associated with enlarged satellites, 2. enlarged satellites are compatible with normal development, and 3. enlarged satellites may occur sporadically, probably due to some unknown culture conditions.

Satellite Association

The satellited regions of the human acrocentric chromosomes are thought to be associated with the nucleolus.[4] A theory has been advanced that the stalks of the satellites possess the "nucleolar organizers." This assumption was based on the fact that in most plants and animals the regularly occurring secondary constrictions are found to be associated with the nucleolus.[5] Berns and Cheng (1971) in several experiments, however, observed that if the region immediately adjacent to a secondary constriction of chromosomes of the salamander *(Tricha granulosa)* is irradiated with an argon laser microbeam, there is loss of nucleolar organizing capacity. But, irradiation at 2 µm down the chromosome from the secondary constriction did not affect the capacity to organize a nucleolus. Direct irradiation on the nucleolus did not affect the ability to organize a nucleolus in 50% of the cases. Therefore, the authors postulated that there is a region adjacent to the secondary constriction that in some way is involved in nucleolar formation. The best cytological evidence for nucleolar association of the acrocentric chromosomes was given by Ferguson-Smith (1964), who showed that only five acrocentric bivalents of human meiotic chromosomes at the pachytene stage were associated with a nucleolus by their satellited ends. Henderson, Warburton, et al. (1972, 1973) demonstrated the presence of ribosomal DNA at these regions and, also, ribosomal DNA connectives between acrocentric chromosomes (at their satellites ends), even when they were quite apart. This rDNA has recently been shown to be associated with the stalks of the acrocentric chromosomes (Hayata, Oshimura, et al., 1977).

Because of nucleolar association, all the acrocentric chromosomes show a tendency to associate together with their satellited ends. This property is known as "satellite association" (Figure 35).

Satellite association of the human acrocentric chromosomes has received much attention, probably because these chromosomes are prone to nondisjunction and translocation. A review of the literature of this area is outside the scope of this book, but the following conclusions can be drawn:

1. The frequency of satellite association varies from culture to culture, depending on the techniques used, and in the same culture when different methods are used for slide preparation.
2. The frequency of satellite association also varies from person to person, which is probably influenced by age, sex, and chromosome abnormalities.
3. The participation of different acrocentric chromosomes in satellite association is random in some cases and nonrandom in others. (This discrepancy can probably be avoided if G- and Q-banding are used. Patil and Lubs [1971] demonstrated nonrandom association in a group of 20 individuals by using a Q-banding technique.)

Identification of Chromosomes with Autoradiographic Studies and DNA Replication

Living cells have the capacity to take up and then incorporate into DNA certain precursor substances which can be labeled with radioisotopes. The radioisotopes undergo a process of decaying while emitting α-, β-, or γ-rays, events which can be recorded by applying photographic emulsion on the slides.

Tritium-labeled thymidine ([³H]-TdR) is the most commonly labeled compound used for the study of the chromosomal DNA replication pattern. This method was first introduced by J. H. Taylor, Woods, et al. (1957) with *Vicia faba* chromosomes. This compound has three main advantages: 1. Its half-life is very long, 12.26 yr. 2. It emits β-rays having a short path distance and, therefore, permits high resolution. (In a photographic emulsion 99% of tritium β-particles penetrate no further than 0.8 μm and rarely beyond the first layer of silver bromide crystals in AR-10 (Kodak) stripping film [Pelc, Appleton, et al., 1965]). 3. It is readily incorporated into chromosomes.

Technique. If tritiated thymidine is added to a culture or injected into an animal, the cells in DNA synthesis, i.e., S-period, become labeled because new DNA is formed at that time.

In culture, cells can be labeled continuously or for a short period of time (called "pulse labeling"). During continuous labeling a cell is exposed to the [^{3}H]-TdR from the beginning of the S-period until the time of slide preparation. In pulse labeling, one can expose the cells (especially in a synchronized culture) to [^{3}H]-TdR for short periods of time; the medium containing the labeled thymidine is removed after the desired period of time and unlabeled (cold) thymidine is added to the new medium. The cell culture is then allowed to grow in the new medium. Slides are prepared as usual and the photographic stripping film applied on the slides; slides can also be coated with emulsion by dipping them into a jar containing emulsion. This thin film of emulsion is actually a gelatin containing very fine particles of silver bromide. The coated slides are placed in boxes in complete darkness and exposed to the β-rays for 3–4 days or as desired by the experimenter. The slides are then developed as any other photographic film. Fine grains are visible on the chromosomes which take up labeled thymidine during the process The grain density on particular chromosomes or their regions indicates the *relative* DNA synthetic activity during the time of exposure to the [^{3}H]-TdR, e.g., high grain density is compatible with active DNA synthesis and vice versa. (For detailed methodology, see reference 6.)

Labeling Pattern of Specific Chromosomes

Even though banding techniques have afforded a much more useful, facile, and reliable approach to the identification of normal and abnormal chromosomes, autoradiography can be useful in the identification of chromosomes #4 and #5, chromosomes of the D-, E-, and G-groups and the late-labeling X-chromosome. It can also be used to identify the Y-chromosomes, though here other methodologies are probably more reliable. It is of little help with chromosomes of groups A, C, and F. Autoradiography is probably too laborious a procedure to use in the routine identification of chromosomes, except for the late-labeling X, and is likely to be superceded for this purpose by the staining methods relying on banding. This can also be said for translocations, particularly involving chromosomes in groups D and G, but here again, the newer banding methods promise to be generally more convenient and reliable.

Group A (chromosomes #1 to #3). Generally, the three chromosomes of this group can be easily identified on a morphological basis and, certainly, with banding techniques. The DNA synthesis pattern of these chromosomes are pretty similar and are not of much help for further identification. They terminate DNA synthesis fairly synchronously and relatively early (Takagi and Sandberg, 1968*a*). The distal half of the long arm of #1 and #2 are first to terminate synthesis, whereas the centromeric region is last. In #3 late replication takes place in the centromeric as well as in the telomeric areas, but the grain density on the telomeric ends is comparatively sparser than in the centromeric region. Both arms of this chromosome show a similar DNA synthetic pattern (W. Schmid, 1963).

Group B (chromosomes #4 and #5). The long arms of two of the B-group chromosomes are late replicating in comparison to the other two and are thus more heavily labeled in terminally labeled preparations; in contrast, the short arms of the former two chromosomes replicate much earlier than those of the other two chromosomes. From careful measurement of B-group chromosome from (*a*) normal individuals, (*b*) individuals with short-arm deletion of a B-group chromosomes, and (*c*) patients with a 4;5 translocation, Warburton, Miller, et al. (1967) came to the conclusion that the chromosome with an early replicating long arm is 5% shorter than the late-replicating one and, therefore, is #5. This chromosome is involved in short-arm deletion in the *cri-du-chat* syndrome and has consistently been found to be one of those whose long arms replicate early. Similar results were also observed by O. J. Miller, Breg, et al. (1966). Marked asynchronous replication of pair #5 was observed by O. J. Miller, Allerdice, et al. (1968).

Group C (chromosomes #6 to #12). Autoradiographic study does not help much in the identification of the chromosomes of this group. It is only one of the X-chromosomes in normal female cells that is always out of phase and is late replicating and, thus, can be identified easily. In a normal male with an XY chromosome constitution, no late-replicating X is visible, nor is such a chromosome visible in the cells of patients with Turner's syndrome (45,XO) or those having an XY complement. On the other hand, when additional X-chromosomes are present, they are always late replicating. Thus, a complement of 47,XXX has two late-labeling X-chromosomes, a complement of 47,XXY has one late-labeling X, and so on. This late-labeling X-chromosome corresponds well with the heterochromatic sex-chromatin body.*

Group D (chromosomes #13 to #15). The three pairs of the D-group chromosome are identifiable by their autoradiographic patterns. Giannelli and Howlett (1966) observed that the longest pair of the D-group (#13) is the last to complete DNA synthesis and has a late-replicating region in the distal part and sometimes in the middle third of the long arm. The next longest pair (#14) is the second to complete DNA synthesis and shows a region of late DNA synthesis in the short arm and proximal part of the long arm near the centromere. The shortest pair (#15) is the first to complete DNA synthesis and, therefore, has few grains during the terminal period of DNA synthesis, but may have a late-replicating segment in the centromeric area. A similar pattern of DNA synthesis was found by J. J. Yunis, Hook, et al. (1964) in a patient with a D-trisomy syndrome. Differences in grain location on the D-group chromosomes are not always seen in every cell and, therefore, study of a large number of cells is necessary.

Group E (chromosomes #16 to #18). Chromosomes of this group can be identified to some extent on morphological grounds. Chromosome #16, being metacentric, is always distinctive. Confusion arises sometimes regarding #17 and #18. A combination of morphology and labeling pattern is quite helpful in the identification of these chromosomes. Chromosome #16 has a late-replicating region on the proximal end of the long arm near the centromere; probably this is the region of secondary constriction. Chromosome #17 is one of the earliest pair to finish DNA replication, whereas #18 shows heavy labeling over its entire length, because it is generally late replicating.

Group F (chromosomes #19 and #20). These chromosomes are not distinguishable by their autoradiographic patterns.

Group G (chromosomes #21 and #22). These chromosomes are all very small and appear to terminate DNA synthesis almost synchronously, but it has been consistently found that one pair is more heavily labeled than the other. Most workers have found that it is a member of this pair that is trisomic in Down's syndrome (mongolism) and that this chromosome pair is in fact the shorter of the two. In other words, Down's syndrome (DS) should really be described as trisomy #22; however, the traditional description of the condition as trisomy #21 is felt to be so deeply imprinted in the minds of cytogeneticists and clinicians that it would be very confusing to alter it at this late stage. The replication pattern of the Philadelphia (Ph[1]) chromosome has been studied and found to be late replicating in some cases and indeterminate in others. The chromosome is so small it is perhaps difficult to compare it to a normal chromosome of this

---

Sex-chromatin: In 1949 Barr and Bertram noticed that the interphase nuclei from female cats usually contain a well-defined mass of chromatin, which is absent in the male. This chromatin body is known as the Barr body (Figure 2). It was found that persons with a single X, whether it is a normal male (XY), an XYY male, or an abnormal female (XO), do not bear sex-chromatin. A normal female (XX) and an abnormal male (XXY) is sex-chromatin positive. With the addition of each extra X-chromosome, there appears an additional sex-chromatin body. Thus, a person with XX or XXY has one; those with XXX or XXXY have two; those with XXXX or XXXXY have three sex-chromatin bodies; and so on.

group by autoradiography. High grain densities over these chromosomes are generally not present during the terminal end of DNA synthesis.[7] In the early stages, however, quite a few grains can be obtained, and differences in grain densities are observed between pairs #21 and #22. In DS of both sexes, three heavily labeled and two lightly labeled G-group chromosomes are seen (W. Schmid 1963; J. J. Yunis, Hook, et al., 1965). The size of the chromosomes involved in this is smaller than that of the other two and is considered as #21.

Y-chromosome. Usually, the Y-chromosome terminates DNA synthesis later than #21 and #22. However, from pulse labeling, Cave (1966) could not find late replication of the Y-chromosome. Kikuchi and Sandberg (1965), on the other hand, found the Y to be among the last chromosomes in the complement to terminate complete DNA synthesis. Craig and Shaw (1971) by doing continuous labeling for the terminal period of DNA synthesis did not find the Y to be the last chromosome to complete DNA synthesis. Instead, they observed it to be relatively early (usually among the first 10 chromosomes) to terminate DNA synthesis. They also concluded that the Y-chromosome cannot be distinquished on the basis of its labeling pattern.

X-chromosome.[8] One chromosome in female cells is found to synthesize DNA very late and to be active autoradiographically; in fact, after most of the other chromosomes have finished replicating. No such late-replicating chromosome has been seen in normal male cells or in females with a chromosome constitution consisting of only one X. Conversely, most individuals believed to have additional X-chromosomes (47,XXX and 48,XXXY, etc.) have additional late-labeling X-chromosomes, one less in number than the presumed complement of X-chromosomes. There seems to be no doubt that these additional late-labeling chromosomes are X-chromosomes and that they correspond to the heterochromatic X-chromosomes which form the sex-chromatin bodies.

Comment. Autoradiography is useful in identifying only a few chromosomes, especially the late-labeling X-chromosome or chromosomes (Figure 36). This technique is laborious, time consuming, expensive, and probably of limited value in the investigation of neoplasia.[9] At present, other staining methods are routinely used in most laboratories, e.g., the fluorescent staining method (Q-banding), the Giemsa banding method (G-banding), the centromere banding method (C-banding), R-banding, and T-banding methods, with which the various chromosomes can be identified more readily, accurately, and reproducibly. Each one will be described in detail.

Identification of Chromosomes by Transverse Bands

In the past, the identification of individual chromosomes in the human karyotype, especially the C-group chromosomes, posed a serious problem for cytogeneticists. This problem was practically solved through the introduction of techniques with which each chromosome and chromosome segment could be differentiated by inducing transverse bands along the length of the chromosomes (Figure 37). Four banding patterns were recognized at the Paris Conference (1971) and are known as Q-bands (fluorescence banding), C-bands (centromeric banding), G-bands (Giemsa banding), and R-bands (reverse banding).

Quinacrine Fluorescence or Q-Bands

Caspersson, Farber, et al. (1968) were the first to induce fluorescing bands on chromosomes. Slides were stained with a fluorescent dye, i.e., quinacrine mustard, quinacrine dihydrochloride, and other related acridine-like acroflavins, and when the slides were viewed with ultraviolet optics, characteristic fluorescing transverse bands of variable intensities were observed along the length of the chromosomes. Caspersson and associates (Caspersson, Lindston, et al., 1970; Caspersson, Zech, et al., 1970a–c) were able to identify each homologous pair in the human metaphase on the basis of the banding pattern. Much emphasis was given to the position, width, and brightness of the bands in each chromosome. A human Q-banded karyotype is shown in Figure 4. Caspersson and associates (Caspersson, Lindston, et al., 1970; Caspersson, Zech, et al., 1970 a–c; Caspersson, Lomakka, et al., 1971a,b) by using densitometer tracings,

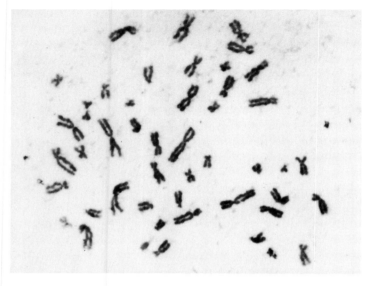

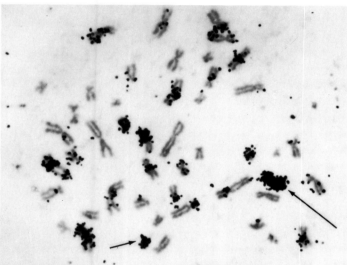

36 Autoradiographic behavior of X- and Y-chromosomes in the cell of a patient with Klinefelter's syndrome (47,XXY). Shown in the top figure is the metaphase before autoradiographic development and in the bottom figure the distribution of grains over various chromosomes following autoradiography. The late-replicating X (long arrow) is shown to be heavily labeled with tritiated thymidine, a behavior characteristic of the late-replicating X in female cells or in those cells in which more than one X-chromosome is present. The short arrow points to the rather heavy labeling of the Y-chromosome, which is also thought to be rather late replicating.

produced graphic profiles of fluorescing bands of each pair of chromosomes and felt that this method is more reliable than visual inspection (Figure 38). Zech (1969) reported that the distal part of the long arm of the human Y-chromosome fluoresces very brightly and, thus, can be easily identified. Due to this characteristic of the Y-chromosome, the Y-element can be identified even in interphase nuclei (Figures 5 and 180) (Pearson, Bobrow, et al., 1970), including those of spermatozoa (Sumner, Evans, et al., 1971; Sumner, Robinson, et al., 1971). Shettles (1971) applied this method to the prenatal sex diagnosis on smears taken from midcervical mucus in the second and third trimesters of pregnancy. Rook, Hsu, et al. (1971), on the other hand, had some reservations regarding the prenatal sex diagnosis by the presence or absence of the Y-body, because damaged or nonviable cells give unsatisfactory results and a small Y-chromosome may fail to fluoresce.

Size variation of the human Y-chromosome in the general population is not uncommon. Bobrow, Pearson, et al. (1971) observed that it is due to variation of the fluorescent area, with

continued

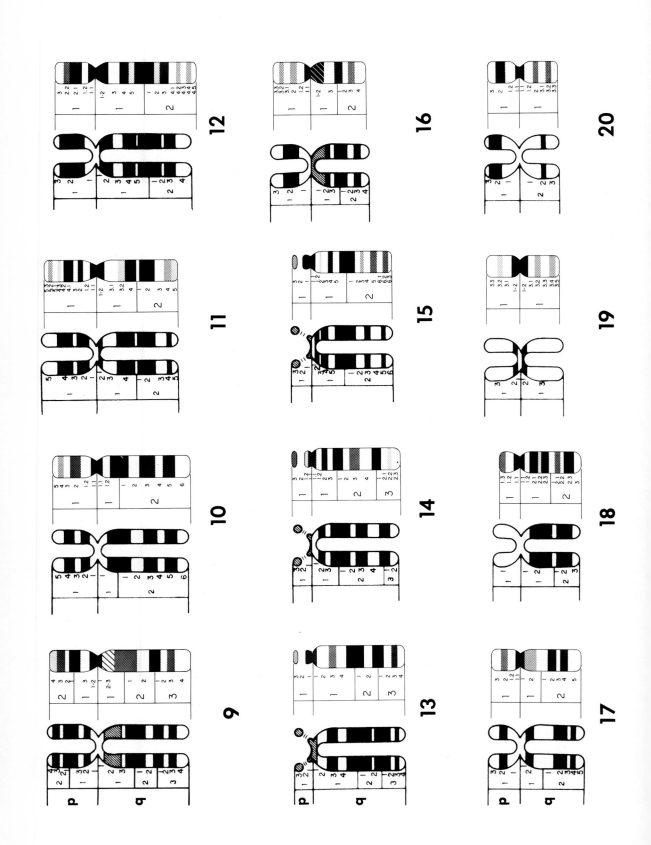

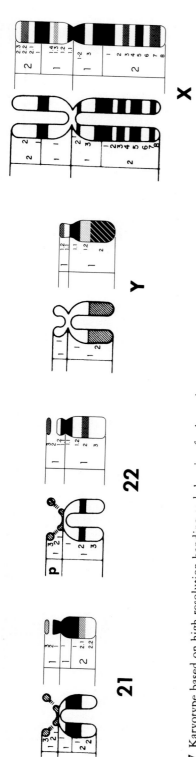

37 Karyotype based on high resolution banding and showing further substructures of many of the bands based on the Paris Conference (1971) descriptions. The metaphase chromosome shown on the left of each set represents the Paris nomenclature and the single-armed chromosomes contain the various band subdivisions. The large numbers identify each chromosome. Note that band 3p26 and 3p27 have been combined into 3p26–7. A comparison between the two systems is shown for each chromosome. Note various levels of shading in the high resolution banding; the latter results in >1,000 recognizable bands, in comparison to <300 bands with the "old" system. (*Courtesy of Dr. J. Yunis.*)

the nonfluorescent area apparently being stable. Borgaonkar and Hollander (1971) and Kakati and Sinha (1973) observed small Y-chromosomes devoid of the fluorescing region. Thus, it appears that the fluorescent analysis of the Y-element is a useful tool for sex diagnosis, though one must be cautious before making a final decision, because a small Y without the fluorescing region will remain undetected in interphase nuclei[10].

Mechanism of Q-Banding

Caspersson and associates (Caspersson, Farber, et al., 1968; Caspersson, Lomakka, et al., 1973) proposed the hypothesis that Q-bands were due to preferential binding of the mustard moiety, which is an alkylating agent, to the N-7 position of the GC residues of the DNA. Later, Caspersson, Zech, et al. (1969a,b) found, however, that agents which do not possess the alkylating group also produce a banding pattern similar to that of quinacrine mustard. Moreover, there are several recent pieces of evidence that show that polynucleotides rich in AT base pairs fluoresce more brightly with dyes like quinacrine mustard, quinacrine, proflavine, etc., than the GC base pairs.[11]

Immunofluorescent studies gave similar results. Fluorochrome molecules conjugated with anti-A immunoglobulin gave similar patterns as Q-banding excepting the Y-chromosome (Dev, Warburton, et al., 1972; O. J. Miller, Schreck, et al., 1973), whereas anti-G immunoglobulin gave the reverse of the Q-banding pattern (Schreck, Warburton, et al., 1973; Schreck, Erlander, et al., 1974). Comings, Kovacs, et al. (1975) came to the conclusion that (1) quinacrine binds to chromatin by interaction of the three planar rings with the large group at position 9 lying in the small groove of DNA, (2) most pale staining regions are due to a relative failure to bind quinacrine, and (3) this inhibition of binding is predominantly due to nonhistone proteins. Cytological evidence for fluorescence of AT-rich regions was shown by Ellison and Barr (1972) in the giant chromosomes of a fruit fly, *Samoaia leonesis*. However, there is cytological evidence in disagreement with the above hypothesis. The centromeric regions of the autosomes and the X-chromosomes in *Mus musculus* have satellite DNA high in AT base pairs (Bond, Flamm, et

38 Densitometric tracing presentations of UV absorption shown by some human chromosomes. Fluorescence patterns of chromosomes #1, #2 and #3 are depicted. The patterns in column 3A were obtained with slit-measuring technique and those in 3B with a chromatid-measuring technique.

(Caspersson, Lamakka, et al., 1971a). Also see paper by Distèche and Bontemps (1976).

al., 1967; Rowley and Bodmer, 1971) but do not show bright fluorescence with quinacrine or quinacrine mustard. Weisblum (1973) suggested that the AT/GC ratio may not be strictly related to fluorescence, whereas the relationship between base pairs to the secondary structure of DNA is more important. Distèche and Bontemps (1974), on the other hand, suggested that besides DNA base sequences and repetitiveness, the protein distribution and packing of chromatin fibers play a role in the induction of Q-banding. Correlations between G- and Q-banding and structural arrangements of the chromosome fibers observed with the electron microscope (EM) have been reported.[12]

Centromeric Bands or C-Bands

While experimenting with mouse satellite DNA by *in situ* hybridization, Pardue and Gall (1970) observed that the regions that possess satellite DNA are located at the centromeric regions of mouse chromosomes and stain more densely than other areas (see Figure 39), which they considered as heterochromatic*. This observation was later pursued by J. J. Yunis and Yasmineh (1971) and Arrighi and Hsu (1971), independently, by eliminating the hybridization step with DNA or RNA and is known as the C-banding technique (Paris Conference, 1971). Later, these techniques were further modified and simplified (Stefos and Arrighi, 1971; Sumner, 1972) (see Chapter 5, Methods). It was observed that areas near and including the centromeres of all human chromosomes stain more intensely than the rest of the chromosomes,[13] excepting the Y, where the distal end of the long arm stains more darkly than its other parts (Figure 16). However, the C-band regions are not the same in all chromosomes. The paracentromeric areas, namely, the secondary constrictions near the centromeres of #1, #9, and #16 and the distal end of the long arm of the Y-chromosome, have larger C-banding areas than the rest of the chromosomes (Figure 16). The size of the C-bands sometimes varies from individual to individual and from one homologue to another (Craig-Holmes and Shaw, 1971, Craig-Holmes, Moore, et al., 1973). Even though the C-bands demonstrate polymorphism from individual to individual, the size and shape of these bands are consistent and characteristic for each case.

Mechanism. Many interpretations have been made regarding the mechanism of C-banding (Sumner, 1978). Among them the most popular is the quick reassociation of repeated sequences of DNA. This view suggests that when the metaphase cells are treated with acid, alkali, or high temperature in a saline ci-

Heterochromatin and euchromatin: Different regions of a chromosome may remain in condensed and decondensed states. The condensed regions stain dark and are called heterochromatin and the decondensed regions stain lighter and are known as euchromatin. From a cytological point of view, the most conspicuous feature of heterochromatin is the different state of condensation of euchromatin. Heterochromatin can be divided into two categories: facultative and constitutive. Facultative heterochromatin is that which appears during physiological and developmental processes and is a temporary state of chromatin, whereas constitutive heterochromatin is the permanent part of a chromosome present in all human chromosomes in variable quantity.

39 C-banded hyperploid metaphase of a tumor cell.

trate solution, the cellular DNA is denatured and then renatured. Thus, it is presumed that regions containing highly repetitious DNA renature more quickly than the less repetitious ones. It has been shown by *in situ* hybridization techniques that the centromeric regions of all the autosomes and the X-chromosome and the distal end of the long arm of the Y-chromosome in a human complement contain repeated sequences of DNA (K. W. Jones and Corneo, 1971; Saunders, Hsu, et al., 1972; Saunders, Shirakawa, et al., 1972). Pardue and Gall (1970) and K. W. Jones (1970) observed similar situations in the centromeric regions of mouse chromosomes. Further support came from the work of Mace, Tevethia, et al. (1972) who observed that patients with systemic lupus erythematosus (SLE) develop antibodies against single- and double-stranded DNA and RNA, nucleoproteins, or some combinations of these. Fluorescent-labeled antibody against single-stranded DNA obtained from the serum of these patients gave excellent fluorescence of the chromosome only when cytologic preparations were denatured. However, when the preparations were allowed to reassociate, as in the C-banding procedure, the C-band regions failed to fluoresce.

Other evidence in support of the denaturation and reassociation hypothesis has come from the work of Stockert and Lisanti (1972) and Chapelle, Schröder, et al. (1973b, c). When stained with acridine orange, double-stranded DNA fluoresces green, whereas single-stranded DNA or RNA fluoresces red. Based on such color changes it was concluded that the C-banded technique of Arrighi and Hsu (1971) does break down double-stranded DNA to single-stranded DNA and that the C-banded regions anneal faster than the rest of the chromosomes. McKay (1973), however, interpreted the techniques of Arrighi and Hsu (1971) and Sumner (1972) as severely disruptive of all but the most tightly packed chromatin, with the latter staining dark. Comings, Avelino, et al. (1973) and T. C. Hsu, Pathak, et al. (1973) observed that there is loss of DNA during C-banding and that relatively larger amounts of DNA remain at the region of C-bands (Pathak and Arrighi, 1973; Schnedl, 1973). Kurnit (1974), on the other hand, isolated DNA from chromosomes or nuclei scraped off from the slides after each step of the C-banding procedure and observed that 80% or more of the DNA remained double-stranded after fixation of slides and 70% or more was single-stranded after treatment with alkali, whether the slides were annealed with salt or not. According to

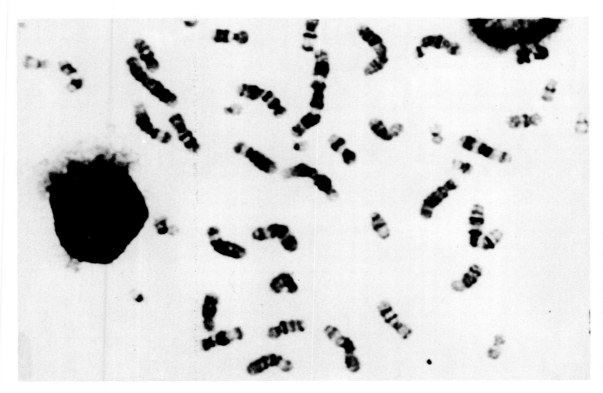

40 G-banded metaphase of a cancer cell with bands of good quality. The latter is difficult to achieve in most human cancers.

him C-banding is apparently the result of a DNA-protein interaction, rather than preferential strand reassociation of repeated DNAs.

Giemsa or G-Bands

The discovery of the C-banding technique led to the development of G-banding, where dark and light transverse bands are produced along the length of the chromosomes (Figures 17, 18 and 40) (Sumner, Evans, et al., 1971; Schnedl, 1971a, b; Gange, Tanguay, et al., 1971; Drets and Shaw, 1971). The G-banding technique is a slight modification of the one developed by Arrighi and Hsu (1971) for C-banding. Subsequently, and almost at the same time, several other workers independently developed techniques to induce G-bands by using various pretreatment procedures.[14]

Because of its simplicity, the method of Seabright (1971, 1972) became popular. In this technique, air-dried slide preparations are treated only with trypsin before staining (see Chapter V, Methods, for further information). Later, it was found that G-bands can be produced by various agents, e.g., enzymes, strong bases, etc., and a large number of approaches and modifications were introduced.[15] It was also observed that some agents, such as actinomycin D, when added to the culture for a few hours before harvesting, produce G-banding (T. C. Hsu, Pathak, et al., 1973).

Mechanism. The exact mechanism of G-banding is not yet known.[16] Whatever method is followed to induce G-banding, the pattern is the same with all the techniques. At the beginning, it was thought that denaturation and reassociation of repeated sequences of DNA were responsible for the G-bands in chromosomes. This, however, is not true, because many agents that do not react with DNA directly can induce G-bands, and some of the C-band positive regions, e.g., the centromere area of chromosome #9 and the distal end of the long arm of Y, are negatively or very lightly stained by G-banding. Some authors[17] have suggested an important role of acidic proteins in G-band formation. Comings, Avelino, et al. (1973) and Sumner, Evans, et al. (1973) believe that histone does not play an important role in G-band formation because when histones are extracted from the metaphase spreads on slides, there is no effect on band visualization. Sumner, Evans, et al. (1973) suggested that in G-banding a magenta component is formed *in situ* which was bound to the DNA by hydrogen bonding. The authors proposed that the dye molecules bridge longitudinally separated sites brought into close proximity by folding of the DNA and that the spatial arrangement of such sites in the chromosome is influenced by nonhistone proteins. McKay's (1973) observation supports the above hypothesis. He obtained G-bands in mitotic chromosomes fixed in acetic alcohol without any further treatment. He holds to the opinion that the cause of the banding is the differential distribution of chromatin. J. J. Yunis and Sanchez (1973), Sanchez and Yunis (1974), and Rodman (1974) supported this view. According to them, the physical characteristics of DNA (repetition, base composition, and secondary structure) may determine differential binding of proteins and other substances to DNA. Thus, the similarity of the banding patterns obtained by various treatments supports the concept that ultrastructural conformation is the basis of such similarity of the banding patterns. Golomb and Bahr (1974), on the basis of EM studies came to the conclusion that banding patterns appear to form an interaction of three distinct factors: 1. structural discontinuity along chromatids; 2. chemical differences in the base distribution of DNA; and 3. DNA-nonhistone protein interactions.

Comparison of Q- and G-Banding Techniques

The banding patterns produced by these techniques are very similar[18]; however, there are some minor differences. Centromeric heterochromatin generally fluoresces negatively, whereas with the G-banding technique centromeres of every chromosome stain dark. The secondary constrictions of #1, #9, and #16 stain negatively with quinacrine but positively with G-banding techniques. On the other hand, the distal end of the long arm of the Y fluoresces brightly with Q-banding, but appears very pale with G-banding.

Reverse Bands or R-Bands (Including T-Bands)

The technique for R-banding was first developed by Dutrillaux and Lejeune (1971). The pattern in R-bands is the reverse of G-bands, i.e., darkly stained G-banding areas are lightly stained and vice versa. (Dutrillaux, de Grouchy, et al., 1971; Dutrillaux, 1973*a, b*). R-banding has not been applied widely to routine chromosomal analysis. However, this technique is very helpful in detecting telomeric deletions or reciprocal translocations involving telomeres, because the telomeres of most human chromosomes stain brightly with this method. Thus, this method is particularly useful in determining the Ph^1-translocation.

The mechanism of R-banding is also not very clear. According to Schreck, Warburton, et al. (1973), R-bands are localized in chromosomal regions with high GC content.

T-banding, as termed by Dutrillaux (1973*b*), is a modification of R-banding. When stained with acridine orange or Giemsa, the telomeres fluoresce strongly. Thus, the method is useful for precise localization of juxtatelomeric break points.

Differential Staining of the Nucleolus Organizers in Human Chromosomes

Cytologists have shown a close association between the satellited areas (short arms with stalks and satellites) of the human acrocentric chromosomes of the D- and G-groups and the nucleolus organizer regions (NOR).[19] With *in situ* DNA/RNA hybridization experiments, using [^{3}H]labeled ribosomal RNA, Henderson, Warburton, et al. (1972) were able to locate the ribosomal cistrons on the short arms of the acrocentric chromosomes, but could not define their exact locations, i.e., whether the stalks or the satellites are the NOR sites. Thus, controversy existed as to whether the human NORs are on the stalk or satellite. Evans, Buckland, et al. (1974) located the site for 18S and 28S ribosomal cistrons at the constrictions (stalks) of the acrocentric chromosomes. Matsui and Sasaki (1973), using an N-banding

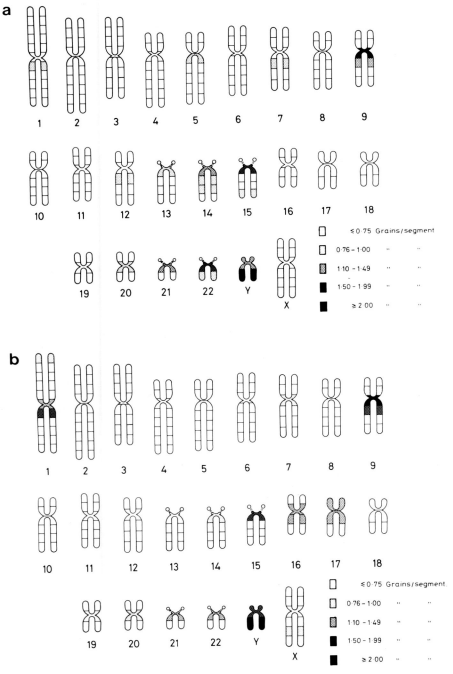

41 Distribution of Satellite I–IV DNA in the human chromosome set. Chromosome #9 appears to contain relatively large amounts of each satellite DNA with chromosome #15 and the Y also being prominent in that respect. Mean levels of hybridization of

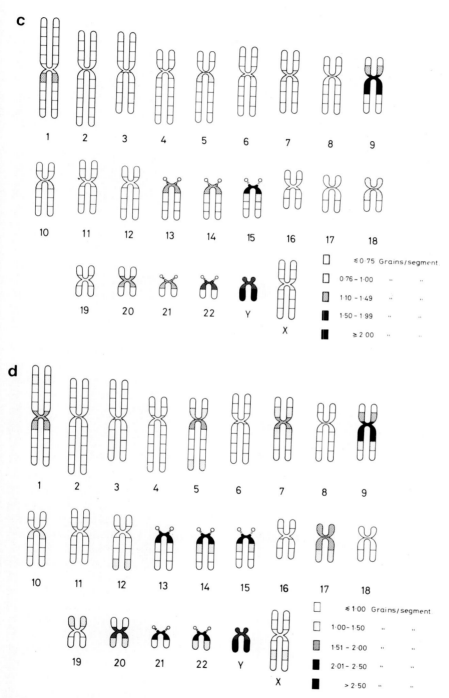

satellite complementary RNAs (cRNAs) are shown schematically: a = satellite I; b = satellite II; c = satellite III; d = satellite IV. Symbols indicate grains per segment.
(Gosden, Mitchell, et al., 1975.)

technique in which they extracted the chromosomal DNA, histones, and proteins with trichloroacetic acid (TCA) and HCl, claimed that the satellites but neither the stalks nor the short arms were the site of the NOR. Howell, Denton, et al. (1975), Howell and Denton (1976), and Denton, Howell, et al. (1976) came to a similar conclusion with a silver staining method. Utilizing the improved technique of Funaki, Matsui, et al. (1975), Hayata, Oshimura, et al. (1977) observed that if there is an increase of the stalk length of a human acrocentric chromosome, the size of the N-band is also larger. Therefore, we have postulated that it is the stalk that is the site of NOR and not the satellites. A conclusion similar to that of Hayata, Oshimura, et al. (1977) was reached by Goodpasture, Bloom, et al. (1976) using a silver staining method.

Hubbell and Hsu (1977) indicated that in humans the silver staining of NORs may not identify all areas that contain structural ribosomal genes, this applying to normal and neoplastic cells.

D. A. Miller, Dev, et al. (1978) found that the number of Ag-stained NORs is essentially proportional to that of the chromosomes and that hybridization of human and rodent cells does not inactivate the human NORs in the hybrid cells; although the nucleolus plays an important role in the origin of Robertsonian translocations in the mouse, nucleolar fusion is relatively unimportant (O. J. Miller, et al., 1978).

Some Comments on Polymorphism

The variety and frequency of chromosomal variants, i.e., polymorphisms (Schnedl, Roscher, et al., 1977), which are without apparent phenotypic effect, have been defined with the aid of banding techniques, particularly C-banding. Polymorphism is common for the juxtacentromeric heterochromatin regions (Figure 41, pages 92–93) of chromosomes #1, #9, and #16 and that on the long arm of the Y. These regions tend to vary in extent and may be the subject of pericentric inversions. They are well revealed by the C-staining technique; the Q-staining fluorescence technique shows the heterochromatic region of the Y and bright juxtacentromeric regions on chromosomes #3, #4, and the acrocentrics (#13–#15, #21 and #22), which are also subject to variations. Some individuals are subject to breakage of some of these or other sites (P. K. Ghosh and Singh, 1976; Hoehn, Au, et al., 1977).

The relation of chromosomal proteins to polymorphism is unknown, though a method for determining their *in situ* distribution has been published (Silver and Elgin, 1976). Flow microfluorometric analysis of isolated chromosomes has been described (Stubblefield, Cram, et al., 1975).

On the basis of Q-banding, Schnedl (1974a) came to the conclusion that at least 10 of the 23 chromosome pairs showed polymorphism at a high frequency and that these features are inherited. The frequency of these variabilites is high enough to allow exclusion of paternity at a likelihood of at least 70%.

Human acrocentric chromosomes were shown to have more polymorphism with the RFA than with QFQ method (Verma, Dosik, et al., 1977c). Furthermore, it was found that there was no consistent relationship between negative or brilliant Q-variants and the various colors observed with RFA; nor was there a significant difference in the overall frequencies of polymorphisms between sexes with either method.

Lateral Asymmetry

Heteromorphic segments in some chromosomes, known before banding as "secondary constrictions," have been shown with newer techniques for staining constitutive heterochromatin (repetitious DNA) in man (Arrighi and Hsu, 1971) as being large blocks of deeply staining material, particularly in chromosomes #1, #9, and #16. *In situ* hybridization studies of the four major human satellite DNAs have shown that two are localized predominantly in the heterochromatin areas of #1 (satellite DNA II) and #9 (satellite DNA III) (K. W. Jones and Prosser, 1973). Utilizing a technique to detect DNA synthesis fluorometrically (Latt, 1973), the occurrence of lateral asymmetry (LA) (see Figure 42) was reported in certain human chromosomes (Latt, Stetten, et al., 1975; My. A. Kim, 1975), i.e., the long arm of the Y, #1, and #16. Angell and Jacobs (1978) showed that the pattern of LA is simple in #15 (centromeric region), #16 (secondary constric-

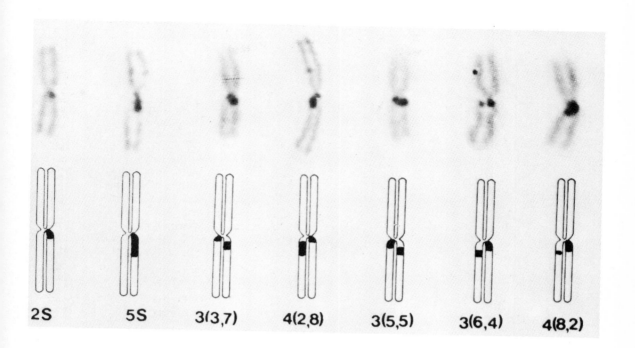

42 Seven #1 chromosomes illustrating different patterns of simple and compound lateral asymmetry. The description below each chromosome starts with the size of the C-band using a scale of 1–5 (very small, small, average, large, and very large); this is followed in parenthesis by a description of the proportion of darkly staining material on each arm expressed as a fraction of 10 irrespective of the C-band, the block proximal to the centromere being described first, e.g., if in an average sized C-band the size of the part proximal to the centromere is estimated as 2/10 of the total, then the pattern would be 3(2,8); if in a large C-band most of the darkly staining block is proximal to the centromere with only a small portion being distal, this would be 4(9,1), and if the pattern is simple rather than compound, it would be 4S.

(Angell and Jacobs, 1978.)

tion), and the Y, but was compound in #1 (secondary constriction) and possibly in #9 (secondary constriction), and postulated that the pattern of compound LA is a stable heteromorphism inherited in a simple Mendelian way and is an efficient morphological discriminator between the members of the #1 pair.

Galloway and Evans (1975) also found various types of LA and reported, in addition, LA in the C-banded region of #17, #20, #21, and #22. Emanuel (1978) demonstrated that region 6q12 → 6q14 shows compound LA. Comings (1978) has speculated on the possible mechanism of LA and, on the basis of Emanuel's (1978) findings, the possibility that the sensitivity of the BRdU techniques may be great enough to detect the location of long segments of moderately repetitious DNA that have been missed by the *in situ* hybridization techniques.

Variations in LA of #1, based on BRdU quenching of 4′,6-diamidino-2-phenylindole fluorescence, suggested a high frequency of variability in the size of the heterochromatic region of #1 (1qh) in the population. A simple system of nomenclature was proposed by the authors (Lin and Alfi, 1978) for naming the variations in the heterochromatic regions.

Sister Chromatid Exchange

The demonstration of spontaneous sister chromatid exchanges (SCE) (Figure 43) in human cells (for a general review, see H. Kato, 1977*a, b*), their possible relationship to the develop-

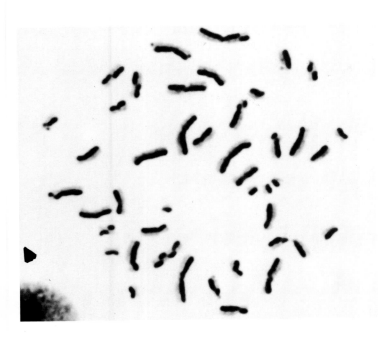

43 Sister chromatid exchanges (SCE) in a normal human cell showing the so-called harlequin chromosomes resulting from incorporation of the BUdR present in the medium in which the cells were cultured. Note exchange of segments or whole of chromatid arms in several of the chromosomes.
(Courtesy of Dr. S. Latt.)

ment of chromosomal aberrations,[20] their high incidence in some disease states (e.g., Bloom's syndrome),[21] including leukemia,[22] and the use of the technique in testing various substances for their mutagenic or carcinogenic effects, in addition to other cytogenetic events,[23] are all areas that most likely will receive much attention in the future.

A number of methods for demonstrating SCE have been published[24] and the effects of certain agents on SCE established. The techniques utilize a halogenated base analogue, 5 bromo-2'-deoxyuridine (BUdR) to label the chromosomes (Latt, 1973). A remarkable characteristic of the SCE phenomenon is its high sensitivity to some physical and chemical agents[25] and, hence, much interest has been focused on this characteristic of SCE in its use as an indicator of DNA damage caused by environmental mutagens and carcinogens.[26]

The exact mechanism of SCE production, its relationship to chromosome structure and events in the cell cycle, and to the development of chromosomal abnormalities are unclear at present.[27] While deciphering the exact mechanism for SCE remains a problem to be solved in the future, determinations of SCE have already found wide application in evaluating potential noxic, carcinogenic and mutagenic cellular effects of a host of biological (including viral), environmental, physical and therapeutic agents, as well as in many clinical states, including leukemia. In the latter disease the potential application of SCE results diagnostically and therapeutically appears promising, but needs to be explored more intensely.

References

1. Bishun, Morton, et al., 1964; Bishun and Morton, 1965; Fitzgerald, Pickerring, et al., 1975; Friedman, Saenger, et al., 1964; Obe, 1965; Powsner, 1966; Sutou and Arai, 1975.
2. Aula and Saksela, 1972; Bobrow, Madan, et al., 1972; Patil and Lubs, 1977.
3. Chapelle, Anla, et al., 1963; Cooper and Hirschhorn, 1962a; Ellis and Penrose, 1960; O. J. Miller, Mukherjee, et al., 1962.
4. Ferguson-Smith and Handmaker, 1961a, b, 1963; Ohno, Trujillo, et al., 1961; Rosenkranz and Holzer, 1972; Sele, Jalbert, et al., 1977.
5. Capoa, Ferraro, et al., 1976; Cooke, 1974; P. A. Jacobs, Mayer, et al., 1976; Mattei, Ayme, et al., 1976; Nilsson, Hansson, et al., 1975; Orye, 1974; Osztovics, Bühler, et al., 1973; Rosenkranz and Holzer, 1972; M. W. Shaw, Craig, et al., 1969; Zang and Back, 1968; Zankl and Zang, 1974; Zankl, Stengel-Rutkowski, et al., 1973.
6. Berghoff and Passarge, 1974; German, 1964;

Gilbert, Lajtha, et al., 1966; Kikuchi and Sandberg, 1964; Sandberg, Sofuni, et al., 1966; W. Schmid, 1965; A. M. R. Taylor, Harnden, et al., 1973; J. J. Yunis, 1974.
7. Fitzgerald, 1971; German, 1962.
8. Gilgenkrantz, Alexandre, et al., 1975; Hagemeijer, Smit, et al., 1977; Shiraishi, 1972; Simpson, Falk, et al., 1974; Sofuni and Sandberg, 1967; Taft and Brooks, 1963; Takagi and Sandberg, 1968b; Therman, Sarto, et al., 1976.
9. Farber and Davidson, 1977; Gavosto, Pileri, et al., 1963b; Gavosto, Pegoraro, et al., 1964; Goh and Joiner, 1968; Hand, 1977; M. W. Steele, 1969; Tiepolo, Zara, et al., 1967.
10. Dallapiccola and Ricci, 1975; Geraedts, Pearson, et al., 1975; Hellriegel, Borberg, et al., 1975; Hellweg-Fründ, Koske-Westphal, et al., 1972; Hollander and Borgaonkar, 1971; Jalal, Pfeiffer, et al., 1974; Kegel and Conen, 1972; M. Kovacs, Vass, et al., 1973; Lancet, 1971b; Latt, Davidson, et al., 1974; Luciani, 1973; Makino and Muramoto, 1964; Mukerjee, Bowen, et al., 1972; Reitalu, Bergman, et al., 1972; Retief and Rüchel, 1977; Schwinger, 1973; Sellyei and Vass, 1975; Soudek, Langmuir, et al., 1973; Spence, Francke, et al., 1973; Sulica, Borgaonkar, et al., 1974; Wyandt and Hecht, 1973a, b; Wyandt and Iorio, 1973.
11. Comings, 1973b, 1975; Comings and Avelino, 1975; Comings and Drets, 1976; Distèche and Bontemps, 1974; Pachmann and Rigler, 1972; Retief and Rüchel, 1977; Selander, 1973; Selander and de la Chapelle, 1973; Weisblum, 1973; Weisblum and de Haseth, 1972, 1973.
12. Bahr, 1970; Bahr and Larsen, 1974; Bath, 1976; Burkholder, 1975; Golomb, 1976; Golomb and Bahr, 1971a, b, 1974.
13. Abe, Morita, et al., 1975; Alfi and Menon, 1973; Cervenka, Thorn, et al., 1973; Chamla and Ruffié 1976; Galloway and Evans, 1975; Scheres, 1976a, b, 1977; Schmiady and Sperling, 1976.
14. Dutrillaux, Grouchy, et al., 1971; Finaz and Grouchy, 1972; Patil, Merrick, et al., 1971; Seabright, 1971, 1972; Wang and Fedoroff, 1972.
15. Barnett, Mac Kinnon, et al., 1973; J. P. Chaudhuri, Vogel, et al., 1971; Finaz and Grouchy, 1972; Grace and Bain, 1972; Kanda, 1976; H. Kato and Moriwaki, 1972; H. Kato and Yosida, 1972; My. A. Kim, Johannsmann, et al., 1975; Kitchin and Loudenslager, 1976; Korf, Schuh, et al., 1976; Meisner, Chuprevich, et al., 1973; W. Müller and Rosenkranz, 1972a, b; Nilsson, 1973; Schuh, Korf, et al., 1975; Shiraishi and Yosida, 1972a,b; Utakoji, 1972; Vass and Sellyei, 1973b; Walther, Stengel-Rutkowski, et al., 1974.

16. Burkholder and Weaver, 1977; Comings, Avelino, et al., 1973; Lundsteen, Kristoffersen, et al., 1974; Utakoji, 1972; Vogel, Faust, et al., 1973; Wyandt, Wysham, et al., 1976.
17. Comings, Avelino, et al., 1973; T. A. Okada and Comings, 1974; Utakoji, 1972; Vogel, Faust, et al., 1973.
18. Bosman, van der Ploeg, et al., 1977a; Dutrillaux, 1975; Gormley and Ross, 1976; Verma, Peakman, et al., 1976.
19. Ferguson-Smith and Handmaker, 1961a, b; Faust and Vogel, 1974; P. A. Jacobs, Mayer, et al., 1976; Marković, Worton, et al., 1978; Ohno, Trujillo, et al., 1961; M. Schmid, Krone, et al., 1974; Schwarzacher, Mikelsarr, et al., 1978.
20. T. C. Hsu and Somers, 1961; H. Kato, 1977a; H. Kato and Sandberg, 1977a, b.
21. Chaganti, Schonberg, et al., 1974; Shiraishi, Minowada, et al., 1978; Shiraishi and Sandberg, 1976a, b, 1977, 1978a–c.
22. Kakati, Abe, et al., 1978; Michalová, Málková, et al., 1977.
23. Carrano, Thompson, et al., 1978; Evans and O'Riordan, 1975; Frank, Trzos, et al., 1978; T. C. Hsu, Collie, et al., 1977; Nichols, 1972a, b, 1973, 1975; Savage, 1975; Stetka and Wolff, 1976a, b; Wiener, Fenyo, et al., 1972; Wiener, Klein, et al., 1973, 1974.
24. Allen, Schuler, et al., 1977; Brøgger, 1975; Crossen, Drets, et al., 1977; Dutrillaux, Aurias, et al., 1976; Goyanes, 1978; H. Kato, 1974a–c; Lambert, Hansson, et al., 1976b; Latt, 1973; Lin and Alfi, 1976; R. C. Miller, Aronson, et al., 1976; Morgan and Crossen, 1977a, b; P. Perry and Wolff, 1974; Sugiyama, Goto, et al., 1976; Vogel, Schempp, et al., 1978; Zack, Rogers, et al., 1977; Zack, Spriet, et al., 1976.
25. Abe and Sasaki, 1977a, b; Abramovsky, Vorsanger, et al., 1978; Allen and Latt, 1976; Burgdorf, Kurvink, et al., 1977; H. Kato and Sandberg, 1977a, b; Kurvink, Bloomfield, et al., 1978; Lambert, Ringborg, et al., 1979; Pant, Kamada, et al., 1976; P. Perry and Evans, 1975; Raposa, 1978; Shiraishi, Minowada, et al., 1978; Shiraishi and Sandberg, 1978b; Stoll, Borgaonkar, et al., 1976.
26. Craig-Holmes and Shaw, 1977; Faed and Mourelatos, 1978; Funes-Cravioto, Kolmondin-Hedman, et al., 1977; Hirschhorn, 1975; H. Kato, 1974a; Popescu, Turnbull, et al., 1977; Stetka and Wolff, 1976a, b; Vogel and Bauknecht, 1976.
27. Absatz and Borgaonkar, 1977; Ahnström and Natarajan, 1965; Crossen and Morgan, 1977a, b; T. C. Hsu and Pathak, 1976; Ikushima, 1977; H. Kato and Stich, 1976; Latt, 1975; Latt and Loveday, 1978; Shiraishi, Minowada, et al., 1978; Ueda, Uenaka, et al., 1976.

5

Methods

Techniques Used in My Laboratory: Details of Specific Methods in Cytogenetics

Chromosome Preparations
Leukocyte Culture — Microtechnique
Leukocyte Culture — Macrotechnique
Chromosome Preparation from Bone Marrow
Harvesting of Bone Marrow or Blood Culture Cells
Chromosome Preparation from Effusions
Chromsome Preparation from Lymph Nodes and Spleen
Chromosome Preparation from Solid Tumors (Primary and Metastatic)

Banding Methods

Q-Banding
   Staining procedure
C-Banding
G-Banding
   Giemsa staining solution
R-Banding
T-Banding
   Method 1
   Method 2
Staining Technique of Nucleolus Organizer Regions
Silver Staining Method for NOR

Method for Sister Chromatid Exchange

BRdU-Dye Methods for Detecting DNA Synthesis
   Sister chromatid exchanges
   Modifications for detection of late-replicating regions
   Modifications for detection of regions containing DNA with thymine asymmetry

Concluding Remarks
Autoradiography

High-Resolution Banding of Chromosomes

High Resolution of G-Banded Chromosomes

Amethopterin-Synchronized Lymphocyte Culture Technique
G-Banding of Chromosome Preparations with Wright Stain
   Wright's stock solution
   Phosphate buffer
   Procedure

Actinomycin D — Treated Chromosomes
   Mitogens

ABBREVIATIONS RELATED TO METHODOLOGIES

BUdR	=	Bromodeoxyuridine
Cd	=	Centromeric staining (modified C-banding) (Eiberg, 1973; Abe, Morita, et al., 1975)
RFA	=	Reverse fluorescent bands with acridine orange (Verma and Lubs, 1975a, b; 1976a, b)
BSG	=	Barium hydroxide, saline, and Giemsa method (Sumner, 1972)
ASG	=	Atabrine, saline, and Giemsa method (Sumner, Evans, et al., 1971)
LBA	=	Late-replicating bands by BUdR using acridine orange (Nakagome, Oka et al., 1977; Oka, Nakagome, et al., 1977)
FPG	=	Fluorescence plus Giemsa
CT	=	Technique for C-banding which stains the centromeric heterochromatin and probably telomeric regions (Scheres, 1976a, b)
Q	=	Quinacrine fluorescence banding
G	=	Giemsa stain banding
AO	=	Acridine orange
QFQ	=	Q-bands by fluorescence using quinacrine

Techniques Used in My Laboratory: Details of Specific Methods in Cytogenetics

A number of books[1] deal in great detail with methods used in cytogenetic studies, and the reader may find particular points regarding some techniques in these books. This chapter presents the methods used in my laboratory and which we have found reliable and yielding of good results.

Chromosome analysis requires cells in metaphase. Cells in early metaphase, when the chromatids are still closely associated, are the most suitable for counting and classification of the chromosomes. The need of cells in metaphase for chromosome analysis requires tissue with a sufficient number of dividing cells. Marrow, testes, lymph nodes, and some tumors usually contain a significant proportion of proliferating cells. Peripheral blood lymphocytes, fibroblasts, amniotic cells, and some tumors have to be subjected to in vitro conditions to obtain a sufficient number of dividing cells. Methods of chromosome preparations utilizing cultured cells are called "indirect"; the term "direct preparation" or "direct method" refers to chromosome preparations in which the cells had not been cultured. The method of choice in tumor cytogenetics is the direct method, which may guarantee a true karyotypic picture of the in vivo situation.[2] Long-term culture conditions have a tendency to favor selective cell growth, which may result in an over- or underrepresentation of normal or abnormal cells. Usually, cells with chromosome anomalies (tumor cells) have a growth disadvantage which leads to false negative results. Depending on the type of tissue, it may be necessary to incubate the cells for a brief period of time (4–12 hr) to obtain a sufficient number of metaphases for chromosome analysis. We have demonstrated that the chromosome picture of marrow or lymph node cells was not different, whether they were directly prepared or after an incubation time up to 30 hr.[3]

It could be argued that in leukemia and tumors only a small proportion of the cells belongs to the actively proliferative pool and that the great majority of the cells is either in G_0- or G_1-phase of the cell cycle and, thus, not available for chromosome analysis. However, as most neoplasias seem to be of unicellular origin, it can be anticipated that the direct method affords a representative picture of the in vitro cytogenetic situation.

Only a few reports have appeared on the chromosome findings in normal marrow.[4] Generally, over 90% of the metaphases prepared by a direct technique have a diploid chromosome

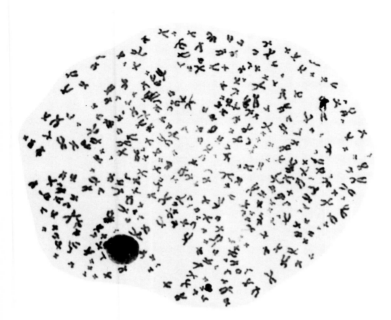

number ($2n$), with 3–4% hypodiploid (due to random loss of chromosomes as a result of preparation techniques or a missing Y) and 0–2% hyperdiploid cells being encountered (Figure 44). Polyploidy ($4n$, $8n$, or more) is not unusual in the marrow and is probably due to the presence of multinucleated cells (megakaryocytes, multinucleated normoblasts) or a result of endoreduplication. In normal subjects the frequency of polyploidy is about 0–2% (Knuutila, 1976), but may be much higher in patients with various hematological disorders, particularly leukemia (10–100%).[5] Somatic pairing (Figure 44) has been described (Sandberg, Koepf, et al., 1960), but appears to be extremely rare in the marrow. The finding of aneuploidy in the marrow should always be followed by the examination of other cells (lymphocytes, fibroblasts) in order to rule out a constitutional chromosome anomaly.

Examination of fibroblasts in culture originating from marrow cells, though on rare occasion reflective of an abnormal karyotypic state in the marrow,[6] is generally not advocated as a means of determining the cytogenetic condition in abnormal states,[7] akin to the discrepancy which may be observed between the marrow and blood.[8] Furthermore, abnormal clones may appear in fibroblast cultures even when they are known to have originated in normal subjects (Littlefield and Mailhes, 1975).

44 Three unusual metaphases. The large one on the left contains 360 chromosomes and is from the marrow of a patient with Turner's syndrome (45,XO). This metaphase contains $16n$ (or $2n \times 8$) number of chromosomes ($45 \times 8 = 360$ chromosomes) and possibly represents a multinucleated megakaryocyte. The two marrow metaphases on the right are due to somatic pairing, in which homologous chromosomes have paired in the cells of a normal female (upper) and a male with Klinefelter's syndrome. The thickness of the chromosome is particularly striking when one compares it to that of chromosomes in normal diploid metaphases, which are about half as thick.
(Sandberg, Koepf, et al., 1960.)

Chromosome Preparations

The basic steps in the chromosome preparation of all tissues are the disruption of the spindle apparatus by colchicine or colcemide or related agents, the swelling of the cells by a hypotonic solution, and the spreading of the chromosomes on a slide by a sudden change in temperature. Some workers do not consider the colchicine (or colcemide) treatment an essential step; however, these substances certainly facilitate chromosome analysis by inducing a more efficient spreading of the metaphases (through disruption of the spindle) and by contracting the chromosomes, thus facilitating their identification and examination. The cytostatic effect is not an important factor, because within the usual exposure time of 2 hr or so no accumulation of cells in metaphase can occur. The suggested concentrations of colchicine vary greatly (0.1 µg to 0.1 mg/ml of culture medium); to avoid overcondensation of chromosomes we prefer a low colchicine concentration of 0.2 µg/ml or of colcemide 0.015 µg/ml of culture medium. Some workers have found vinblastine (0.5 µg/ml) to be an excellent antimitotic agent. A large variety of different hypotonic solution has been proposed, e.g., Na-citrate 1%, distilled water, Hanks' solution, or fetal calf serum diluted 1:3 or 1:4 with distilled water, and KCl (0.075 M). We use the latter. The cells are exposed to hypotonic treatment for various periods of time, depending on the type of cells; thus, PHA-stimulated blood lymphocytes are treated for seconds only; cells derived from lymph nodes need briefer treatment (10 min) than marrow or solid tumor cells (20-30 min). There are also many ways to prepare chromosome slides; some workers use dry slides, others use wet slides, whereas some use cold and wet slides; some prefer the rapid flame-drying of the slides after the cell suspension had been put on the slide, others consider gentle drying important. It can be seen that many roads lead to Rome and no one appears to be better than another. Almost every chromosome worker has his own method, which in the hands of the beginner usually does not work unless he invents some modifications of his own. This is particularly true for special methods of chromosome identification, e.g., autoradiography, fluorescence-banding, and Giemsa banding. I, therefore, wish to summarize briefly the methods which are in use in my laboratory, rather than confuse the reader with an unlimited number of modifications described in the literature.

The buccal mucosa smear as a test for sex-chromatin is probably of little use to the clinician investigating a reproductive or other problem. Whether the results are appropriate or inappropriate to the provisional diagnosis or the sex of the patient, it is still necessary to perform full chromosome analysis. The sex-chromatin test is of use only to the cytogeneticist to complement his laboratory studies. The presence and frequency of Barr bodies are helpful in deciding whether the possibility of sex-chromatin mosaicism is great enough to warrant cytogenetic analysis of more cells or examination of other tissues. The identification of the Y-body in interphase nuclei,[9] though it may be indicative of the presence of a Y (or several)-chromosome(s), cannot supplant cytogenetic examination for establishing the presence and morphology of the Y present in the cells. In a significant number of cases, the Y-body does not show up,[10] even though a Y of normal or abnormal morphology is present in the cells.

Leukocyte Culture—Microtechnique

With this method[11] satisfactory chromosome preparation can be made with samples of whole blood as small as 0.05 ml.

1. Wet a 1-ml hypodermic syringe or a prothrombin pipette with heparin (1,000 U/ml) using a 27-gauge needle.
2. Clean the fingertips of the patient with 70% ethyl alcohol several times.
3. With the 27-gauge needle make a finger prick on the index or the second fingertip and ooze out blood. Draw 0.5-0.2 ml of blood with the heparinized syringe or the prothrombin pipette.
4. Inoculate the heparinized blood into a tissue culture flask containing:
   a. 5 ml of TC 199 or RPMI 1640 medium with 15-20% fetal calf serum and penicillin (50 U/ml) and streptomycin (50 µg/ml).
   b. 0.1 ml of phytohemagglutinin M (PHA) (Burroughs Wellcome & Co., Triangle Park, N.C.).
5. Incubate at 37°C.

Maximum number of mitosis is available at 72 hr at the 2nd or 3rd mitotic cycle. Metaphases at 1st mitosis are available at 48 hr.

Leukocyte Culture—Macrotechnique

1. Place 0.1 ml of heparin (1,000 U/ml) in a 10-ml plastic disposable syringe.
2. Aseptically draw 5-10 ml of venous blood

and mix with the heparin thoroughly by inverting the syringe.
3. Wipe the needle with alcohol, cover the needle, and allow the syringe to stand with needle up at room temperature for about 45 min to 1 hr, until 30–40% of the volume of blood is visibly clear.
4. Bend the needle at right angle and extrude the plasma into a sterile centrifuge tube if storing of the leukocyte with plasma is necessary or add 0.5–1.0 ml to a sterile culture vessel containing:
   a. 6 ml of TC 199 or RPMI 1640 medium with 50 U/ml penicillin, and 50 μg/ml streptomycin.
   b. 0.5 ml PHA M (Burroughs Wellcome).
   c. 1 ml of autologous plasma which can be obtained by centrifuging the remaining blood.
5. Incubate at 37° C.

Chromosome Preparation from Bone Marrow

Chromosome preparations reveal the in vivo composition of the marrow most reliably when prepared by a direct technique, i.e., without incubation and culture.[12] However, when the bone marrow (BM) is very hypocellular or an inadequate number of mitoses is available, short-term culture (12–24 hr) often yields most satisfactory results. The culture method is similar to that of blood cells. A syringe can be heparinized to draw BM or 0.5 ml heparin (10 U/ml) can be added to a tube containing 5 ml of culture media. For BM culture no PHA is used. Add 1–2 ml of BM to 10 ml of culture media (TC 199 or RPMI 1640 containing 20% fetal calf serum). For the direct method of subsequent culture, the procedures described next are followed.

Harvesting of Bone Marrow or Blood Culture Cells

1. Add 0.015 μg/ml colcemide (or roughly 2–3 drops of a 0.2 μg/ml stock solution) to 10 ml of the cell suspension in culture and incubate for 2 hr.
2. Transfer the culture to a 15-ml centrifuge tube.
3. Centrifuge for 5 min at 1,000–1,200 rpm.
4. Decant the supernate.
5. Slowly disturb the sediment by tapping slowly with your finger on the centrifuge tube.
6. Add 4–5 drops of 0.075 M KCl (hypotonic solution) prewarmed at 37° C and disturb the pellet as in step 5.
7. Add more 0.075 M KCl slowly and make the volume up to 10 ml.
8. Mix with a Pasteur pipette.
9. Addition of colcemide is optional at this step; we usually obtain better preparations by not adding it.
10. Incubate for 20 min in a water bath at 37° C, while disturbing the cell suspension intermittently with a Pasteur pipette.
11. Add 4–5 drops of fixative (1 vol of glacial acetic acid + 3 vol of 100% methyl alcohol).
12. Mix with a Pasteur pipette.
13. Centrifuge as in step 3.
14. Decant the supernate.
15. Disturb the pellet as in step 5.
16. Add 4–5 drops of fixative and disturb the pellet as in step 5.
17. Add fixative slowly to make the volume 2 ml.
18. Mix with a Pasteur pipette.
19. Make the volume 10 ml with fixative.
20. Repeat step 18.
21. Leave the tube at room temperature for a minimum of 30 min. (At this point the cells can be preserved indefinitely at $-15°$ C to $-20°$ C. We have experience of storing cells for 4 yr at least.)
22. Centrifuge as in step 3.
23. Repeat steps 16–20.
24. Add fixative drop by drop to make the volume about four times that of the sediment (e.g., to 0.5 ml vol of pellet, add 1.5 ml of fixative).
25. Mix with a Pasteur pipette.
26. Prepare three slides: one each for the following techniques:
    a. Flame dry (particularly useful when chromosomes are clumped).
    b. Air dry in atmospheric condition (best method when chromosomes are widely spread).
    c. Air dry in humidity.
    Put 2–3 drops of the cell suspension and:
    a. *Flame dry:* quickly ignite the slide over a flame.
    b. *Air dry in room temperature:* leave the slide to dry spontaneously at room temperature.

c. *Air dry in humidity:* leave the slide in a humid place to dry. Our procedure is to put the slide on a rack over a water bath (about 45° C) or on the sink containing water of about 45° C.
27. Check the slide under a phase contrast microscope and select the best method for slide preparation.

For routine analysis of chromosomes, the air-dried slides are stained in 5% Giemsa in water for 10 min. Chromosome analysis of PHA-stimulated lymphocytes may be necessary in some cancer or leukemia patients if a cytogenetic constitutional anomaly has to be excluded.

Chromosome Preparation from Effusions

About 20 ml of freshly aspirated, heparinized fluid is centrifuged for 7 min at 1,000 rpm. The supernate is discarded and replaced by 10 ml of KCl (0.075 M) containing 0.2 µg/ml colcemide. The suspension is incubated at 37°C for 20 min. Further steps are identical to those described for preparation of marrow cells. (See reference 13.)

Chromosome Preparation from Lymph Nodes and Spleen

As soon as possible the tissue is brought from the operating room to the laboratory. Necrotic and connective tissues are removed as much as possible. Using surgical knives (scalpels), a cell suspension is prepared in a Petri dish containing a balanced salt solution. The cell suspension is drawn off and placed in a centrifuge tube. After centrifugation for 7 min at 1,000 rpm either hypotonic treatment (with colcemide) follows or the cells are incubated at 37° C for 10–30 hr in a balanced salt solution with 10–20% fetal calf serum. In the latter case, the cell concentration should not exceed 10^{-7}/ml of culture medium, and no PHA should be added. Incubation of lymph node or splenic cells usually results in an increase of Ph^1-negative metaphases. After the hypotonic treatment with KCl (0.075 M), which should last 10 min for lymph node cells, 20 min for granulocytic cells, and 30 min for cancer cells, the respective cells are processed similarly to marrow cells.

Chromosome Preparation from Solid Tumors (Primary and Metastatic)

When a large piece of tissue is available (ca. 25 mm³), the best procedure[14] is to trim the tissue aseptically, discarding the peripheral layers so as to yield a core of tissue about 7–10 mm³. When only a small piece of tissue is available, there is obviously little choice but to proceed as outlined below.

An important step is thorough washing of the specimen with a rinsing solution; we use Hanks' balanced salt solution (HBSS) without Ca^{++} and Mg^{++}, but with 100 U/ml of penicillin and 100 µg/ml of streptomycin. Holding the piece of tissue with fine forceps, it is rinsed successively in the solution held in three Petri dishes. The specimen is then cut into very fine pieces with a surgical *blade* (scalpel) and incubated 24–48 hr in 10 ml of RPMI 1640 medium and harvested (as described for bone marrow preparations). We have found that other methods, e.g., cell isolation by trypsin digestion, are not very successful and usually affect the morphology of the chromosomes.

Banding Methods

Refer to reference 15 for further information on banding methods.

Q-Banding

Quinacrine mustard (QM) stock solution = 1 µg/ml in McIlvaine or Sorensen's buffer (pH, 6.8–7).

QM working stock solution = 3 ml QM + 50 ml of buffer (pH, 6.8–7).

See reference 16.

Staining procedure

1. Immerse the slide in buffer (1 min).
2. Stain in QM (10 min).
3. Wash in tap water.
4. Rinse in three changes of buffer (pH, 6.8–7).
5. Mount in buffer (pH, 6.8–7).

The coverslip may be sealed with nail polish. We obtain the best results when the slides are examined on the day they are prepared.

C-Banding

Treat the air-dried slides for C-banding[17] within 2–3 days after preparation. Old slides generally do not give good results.

Arrange eight Coplin jars as follows and carry the slides through the following procedures:

1. In 0.1 N HC1 (60 min); wash *vigorously* in running tap water, shaking the slide in a big jar.
2. Rinse with distilled water.
3. In saturated barium hydroxide, filtered before use, prewarmed at *50° C* for 10–15 min; *thoroughly* wash in tap water (50° C) as in step 1.
4. Rinse with distilled water.
5. In 10 × standard saline citrate solution (SSC) prewarmed at 60° C for 60 min.
6. Wash in tap water (50° C) as in step 1.
7. Rinse with distilled water.
8. Stain in Giemsa solution (20–30 min).*

To make 10 × SSC add:
    43.83 g sodium chloride (NaCl)
    22.50 g sodium citrate ($C_6H_5Na_3O_7 \cdot 2H_2O$ to 1000 ml of distilled water)

G-Banding

1. Aging of the slides is not necessary.
2. Put the slides overnight in an incubator (dry air) at 60° C.
3. Arrange four Coplin jars with the following solutions:
   a. Trypsin solution: add 5 ml of 0.25% trypsin (Grand Island Biological Co., Grand Island, N.Y.) in 45 ml of Hanks' BSS (calcium and magnesium free).
   b. 50 ml of Hanks' BBS (Ca^{++} and Mg^{++} free).
   c. 70% ethyl alcohol.
   d. 70% ethyl alcohol.
   Immerse the slides in:
       Coplin jar *(a)* about 1 min (blood culture)
       about 3.5 min (bone marrow)
       (b,c,d) just rinse.
4. Air dry.
5. Stain in Giemsa—6–8 min and air dry.

*See Giemsa Staining Solution, next column.

6. Immerse briefly in xylene (optional).
7. Mount in Permount (optional).
   See reference 18.

Giemsa staining solution. This staining solution contains 50 ml distilled water, 1.5 ml Sorensen's or McIlvaine's buffer (pH 6.8–7.0), 1.5 ml methanol, 1.5 ml 2% Giemsa. The whole mixture should have a pH of about 6.8–7.2; if not, adjust with its ingredients.

Note: Duration of trypsin treatment is not necessarily the same for all preparations. Generally, older slides take a longer time. It is necessary to check the slides after completing steps (1–5) under high power in the microscope. If undertreated by trypsin, repeat steps 3–5.

R-Banding

In R-banding[19] the air-dried slides (about 8 days old) are treated with Eagle's minimal essential (Earl's base) medium (pH 6.5) for 10–20 min at 87° C, rinsed with tap water, and stained with Giemsa (Dutrillaux, 1973a). Fluorescent reverse bands are also obtained in a similar way. Sldes are stained with acridine orange at pH 6.5 after treating the slides in phosphate buffer (pH 6.5) at 85° C for 5–20 min. The R-bands appear greenish-yellow on a red background.[20]

T-Banding

The two methods of Dutrillaux (1973*a, b*) are as follows:

Method 1

1. Heat treatment. Add 3 ml of phosphate buffer to 94 ml of distilled water. Bring temperature to 87° C. Add 3 ml of commercial Giemsa to the above solution. Immerse the slide for 5 min 30 sec in the above solution.

2. Acridine orange stain. Rehydrate the slides in alcohol grads to distilled water and then stain the slides with acridine orange, which is made in phosphate buffer, pH 6.7 (5 mg of acridine orange +100 ml of phosphate buffer). Mount the slide in phosphate buffer with coverslips and observe under UV optics.

Method 2

Denature for 20–60 min at 87° C in Earl's solution, phosphate-buffered saline (PBS) or phosphate buffer, the pH of which should be about 5.1. Stain in Giemsa as in Method 1. Acridine orange stain can be used following Giemsa in the previous method.

Chamla and Ruffié (1976) developed a method in which both C- and T-banding can be induced in the same chromosome. Their method is as follows: air-dried slides are incubated in Hanks' BSS maintained at 91° C for 30 min, cooled under tap water, and stained in 4% Giemsa for 15 min.

Staining Technique of Nucleolus Organizer Regions

N-banding technique of Matsui and Sasaki (1973) and improved N-banding technique of Funaki, Matsui, et al. (1975).

1. Incubate the slides in 5% TCA for 30 min at 85–90° C.
2. Briefly rinse in tap water.
3. Reincubate in 0.1 N HCl for 30–45 min at 60° C.
4. Thoroughly rinse in tap water and stain for 60 min in phosphate-buffered Giemsa (diluted 1:10), at pH 7.0.

The above method can be substituted by chromosomal digestion with DNase (100 μg/ml for 60 min at 37° C at pH 6.6) and pancreatic RNase A (100 μg/ml for 60 min at 37° C at pH 7.0).

The above N-banding method has been modified by Funaki, Matsui, et al. (1975) as follows:

1. Incubate the slides in 1 M NaH_2PO_4 solution (pH 4.2 ± 0.2 adjusted with 1 N NaOH) at $96° \pm 1°$ C for 15 min.
2. Wash thoroughly in distilled water.
3. Stain with Giemsa (diluted 1:25 in $\frac{1}{15}$ M phosphate buffer, pH 7.0).

Silver Staining Method for NOR

Goodpasture and Bloom (1975) described a silver (Ag) staining method[22] by which nucleolus organizers can be stained distinctly with ammoniacal silver and is known as the Ag-As technique. Later this technique was further modified and improved (S. E. Bloom and Goodpasture, 1976; Goodpasture, Bloom, et al., 1976). The preferable method, known as the Ag-1 method, is as follows:

Chromosome preparations, air- or flame-dried, are made from human blood or marrow cells, previously cultured in RPMI 1640 medium containing 0.1 μg/ml of colcemide (GIBCO) for 1–2 hr at 37° C. Inasmuch as the aging of fixed cells often results in nonreaction with Ag or the nucleolar organizer as well as in a nonspecific reaction of nonnucleolar organizer regions, the chromosome slides are processed, as described below, within several hours after cell harvest. According to the recommendation of Lau and Arrighi (1977), the chromosome slides are pretreated with a borate buffer (0.1 M Na_2SO_4 – 0.005 M $Na_2B_4O_7$, pH 9.1) for 20–30 min at room temperature; this step was empirically found to be necessary, though not proven biochemically, to eliminate the nonspecific staining of centromeres that appeared to be inherent in the original Ag-As method (Goodpasture and Bloom, 1975). The slides are then incubated in a 50% (wt/vol) $AgNO_3$ solution (made in deionized distilled water at pH 4.5–5.0) in a humid chamber for 15–18 hr at 60° C (S. E. Bloom and Goodpasture, 1976), rinsed with distilled (deionized) water, and counter-stained for 60 sec in 1–2% Giemsa.

Humidity may also be obtained by putting the slides in an air-tight plastic container and adding deionized water to the bottom of the plastic box, well away from the slides. When the chromosomes reach desired intensity, the slides are rinsed with several changes of distilled (deionized) water, air-dried, and viewed through phase contrast optics or counter-stained with 1–2% Giemsa stain or acetic orcein.

If prestaining of the chromosomes by a conventional Giemsa staining method is desired, the slides can be stained with 1–2% Giemsa in 0.01 M phosphate buffer, pH 7.0, and selected cells photographed using bright field illumination. The slides are then placed in xylene, xylene-ethanol, and ethanol series to remove immersion oil, transferred to 95% ethanol for 10–20 min to remove the Giemsa stain, and air-dried again.

Within the last few years a tremendous amount of work has been done regarding banding patterns of chromosomes. A number of review articles on this area have been published.[21]

Method for Sister Chromatid Exchange

The method is usually related to the incorporation of bromodeoxyuridine (BUdR) into the DNA of the chromosomes, i.e., into the newly synthesized DNA of each chromatid, which is an analogue of thymidine and is readily substituted by the cell (Nakagome, 1977). The incorporation of BUdR lowers remarkably the level of fluorescence of that part of the chromosome into which it has been incorporated, thus revealing the newly synthesized chromatid. Of course, all this occurs if the BUdR is present for a full mitotic cycle (in fact, the total length of the S-period) of the cell. If the BUdR is present only during the early S-phases, then the early-replicating chromatin will be "dull" compared to the high fluorescence of the late-replicating portions of the chromosomes. In contrast, if BUdR is present during the late stages of the S-period, then the late-replicating segments of the chromosomes will not fluoresce and the early-replicating ones will (Burkholder, 1978). The effects of BUdR can be negated at any stage of the cycle either by removing the cells from the medium containing it or by adding a large excess of thymidine. Fluorescence of the chromosomes is usually brought out by such stains as acridine orange (AO) or Hoechst 33258 (Misawa, Takino, et al., 1977), though other techniques (based on Giemsa or attributable to BUdR) can be utilized.

The Giemsa stainability of late-replicating chromosomal regions can be suppressed by the incorporation of BUdR as a terminal pulse at the end of the S-period. Incorporated BUdR causes a marked condensation of late-replicating regions. In normal cells, decondensation of the female inactive X-chromosome and the heterochromatic region in Yq12 acts as an aid in their identification. The replication of the male X and female active X are similar to one another, but are clearly distinguishable from that of the female inactive X.

Short-term cultures are made in RPMI 1640 culture medium with 20% fetal calf serum. BUdR at a concentration of 1 μg/ml is added within 20–24 hr of culture initiation. Cultures are allowed to grow in the dark at 37° C for another 50–52 hr including 2 hr of colcemide treatment at the end of the culture period.

Air-dried slides are made at least 1 day before the staining and studying of the slides. 0.01% AO staining solutions are made in Sorensen's buffer (pH, 6.8–7). The slides are stained for about 6 min, then rinsed thoroughly in the same buffer for 6 min and mounted using the same buffer solution. Scanning is done under UV optics (Carl Zeiss Photomicroscope III). Instead of AO, another fluorochrome often used is Hoechst 33258 (Latt, 1973).

The methods used by Latt (1974, 1978) will be presented in detail below, so that they may serve as a basis for those interested in some of the finer points of the techniques.[23]

BRdU-Dye Methods for Detecting DNA Synthesis

Sister chromatid exchanges

Peripheral lymphocyte cultures. Blood is collected in preservative-free heparin (e.g., Abbott's "Panheparin"), 10 U/ml, and allowed to settle 90 min before removal of the buffy coat. Aliquots of plasma containing $5-8 \times 10^6$ leukocytes are added to 16×125-mm culture tubes (e.g., Falcon Plastics, #3033) together with the following:

1. Medium: Eagle's MEM with Earle's BSS and 2 mM L-glutamine (e.g., Microbiological Associates #12–611), 100 U/ml penicillin, 100 μg/ml streptomycin, and 20% fetal bovine serum). Total volume used is 5 ml.
2. PHA: Either 0.2 ml of a crude saline extract of red kidney beans or 0.1 ml of Burroughs Wellcome PHA (HA-15, reagent grade).
3. Nucleosides: We routinely employ BRdU (Sigma Chemical Co.), fluorodeoxyuridine (FdU) and uridine (U) (Calbiochem) at final concentrations of 10^{-5} M, 4×10^{-6} M, and 6×10^{-7} M.

Other workers have omitted the FdU and U. Addition of deoxycytidine (dC) (10^{-4} M) improves cell growth but also results in higher base-line SCE frequencies. It is convenient to add the BRdU together with the cells, although an 18–24 hr delay in BRdU addition has been reported to improve the yield of mitotic cells.

Cells are cultured at 37° C in the absence of

light. Tubes are sealed tightly to prevent CO_2 escape and inclined at the angle of approximately 45°. Good yields of second division metaphases are generally obtained after 72–74 hr of total culture time. This time can probably be reduced to 50–54 hr if blood samples from newborn infants are being tested. Conversely, retardation of cell growth by various clastogens may necessitate the use of a longer culture time before cell harvest. All manipulations of cells after addition to BRdU to the cultures should be done in subdued light and preferably in the absence of light with wavelength <500 nm.

Lymphocyte harvest consists of the following steps: (1) addition of 0.1 µg/ml of colcemide (Calbiochem) for 2 hr; (2) removal of medium and resuspension in 0.075 M KCl for 12 min; (3) fixation in at least two changes of 3:1 mixture of methanol and acetic acid (for 15 and 10 min). Minor variations in mitotic arrest or hypotonic treatment might be appropriate in individual circumstances. For example, metaphase preparations of lymphoblasts appear to be improved by reducing the colcemide concentration to 0.023 µg/ml and the time in KCl to 8 min. At each step, the cells are centrifuged, all but about 0.2 ml liquid is removed, and the cell pellet is gently resuspended before adding the next solution. Careful cell resuspension appears to increase the probability of obtaining good metaphase spreads.

Cultured cell lines. Analogous procedures are employed when examining cells other than lymphocytes. The type of medium and container (e.g., Petri dish vs. T-flask) varies with the cell type. BRdU is generally added a few hours after cell transfer, and cell harvest is timed to occur at the second metaphase after BRdU addition. As with lymphocyte cultures, it is generally convenient to allow cells to incorporate BRdU for two replication cycles, although sister chromatid differentiation can also be effected if BRdU is present for the first but not the second cycle. Cell harvest begins with proteolytic release of cells (0.05% trypsin plus 0.02% EDTA, or 0.1% pronase) followed by the steps described for lymphocytes.

Slide staining

1. 33258 Hoechst: Slides are successively dipped in 0.14 M NaCl, 0.004 M KCl, 0.01 M phosphate, pH 7.0 (PBS) (5 min), 0.5 µg/ml 33258 Hoechst in PBS (10 min), PBS (1 min) and PBS (10 min). Stock solutions of dye (50 µg/ml) in H_2O can be stored at 4° C in the dark for at least 2 wk. Samples of 33258 Hoechst were originally obtained from Dr. H. Loewe, Hoechst AG, Frankfurt, Germany, although the dye can now be purchased from American Hoechst Co., Chemical Department, Route 202-206N, Somerville, N.J. 08876.

2. Giemsa staining: Following fluorescence photomicrography of 33258 Hoechst-stained slides (mounted in pH 7 or pH 7.5 buffer), the slides can be washed, incubated in 2 × SSC at 65° C for 15–30 min, and restained in Giemsa (e.g., 4% Gurr's R-66 in 5 mM pH 6.8 phosphate buffer). Sister chromatid differentiation previously apparent only with fluorescence can then be observed with transmitted light, as initially described by My. A. Kim (1975) and by P. Perry and Wolff (1974). In this procedure, the illumination used to excite 33258 Hoechst fluorescence also presumably promotes selective destruction of the chromatid which has been highly substituted with BRdU.

Direct photosensitization of chromosomes can be achieved by mounting slides previously stained with 33258 Hoechst in a 10^{-4} M solution of the dye in one-third strength PBS (or a comparable buffer with a pH near 7). A coverslip is applied and the slides are then placed in a Petri dish and exposed to light for a period of time which depends on the illuminating conditions. For example, approximately 4 hr exposure is sufficient after positioning the slides 6 cm from a Sylvania 20 W cool white bulb. The slides are then rinsed in H_2O, incubated at 65° C in 2 × SSC (0.30 M NaCl, 0.03 M Na citrate, pH 7) for 15 min, rinsed with H_2O, and stained with Giemsa as before. Contrast can be increased by increasing the time during which slides are exposed to light. The 2 × SSC step can in principle promote G-banding of chromosomes. However, various investigators have stated that this banding can apparently be minimized by "aging" the slides for a few days before use.

Photography

1. Fluorescence: Slides are mounted at pH 7.0 or pH 7.5 (e.g., in McIlvaine's buffer, a mixture of 0.1 M citric acid and 0.2M Na_2PO_4), and fluorescence is recorded in Tri-X film. Since success in photographing fluorescence patterns

depends on the efficiency of detection of the fluorescence, the use of an automatic camera system which removes a large fraction of light before it reaches the film is not recommended. Use of a buffer with appreciably lower ionic strength or pH reduces contrast due to BRdU-dependent dye quenching.

Fluorescence excitation of the dye-DNA complex peaks at approximately 360 nm, whereas emission peaks at approximately 475 nm. Excitation of 33258 Hoechst fluorescence is thus carried out with illumination from a high pressure mercury lamp that traverses filters transmitting maximally in the range of 360–400 nm. For example, this can be accomplished with a Leitz Orthoplan microscope with incident illumination, a UG-1 filter, and a TK 400 dichroic mirror. Emission is best observed through filter combinations excluding light below 460 nm.

If incident illumination is used, the lenses should not have a large number of correcting elements, because many of these attenuate light intensity below 400 nm and thus reduce the excitation of illumination in the range absorbed by the Hoechst dye. Achromat objectives have proved to be satisfactory for fluorescence microscopy of 33258 Hoechst-stained chromosomes.

The fluorescence of 33258 Hoechst fades rapidly under the conditions in which it is sensitive to the incorporation of BRdU. Thus, for routine detection of SCE, BRdU-Giemsa techniques are more convenient than fluorescent methods. Fluorescence analysis can often help, however, to locate the source of problems which might arise in setting up Giemsa methods.

2. Giemsa: Slides can be mounted in a standard embedding medium, or immersion can be applied directly to the slide without a coverslip. A 544-nm interference filter can be used to enhance contrast, and the optical image is recorded on high contrast copy film. Highly corrected microscope objectives can be employed.

Modifications for detection of late-replicating regions. The procedures below work for PHA-stimulated human peripheral lymphocytes. For other cell types, the details will change, but the logic (part of one cycle in BRdU, the rest in deoxythymidine [dT]) still holds.

1. Late replication = bright fluorescence.
   a. Additives at initiation of culture (per 5 ml).
      i. PHA as before.
      ii. BUdR, 0.2 ml of a 2.5×10^{-3} M solution, in Hanks' BSS.
      iii. FUdR, ~0.1 ml (2×10^{-5} M FUdR, 3×10^{-4} M uridine [U]).
   b. Growth. Cells are cultured not quite one replication cycle, after which the medium is changed (described below) and dT is added. A total culture time of approximately 50 hr (44 hr with BRdU followed by a 6-hr terminal dT-pulse) is generally used for adults. A shorter overall time, e.g., 40–42 hr, is probably better for newborns. Colcemide is added 2 hr before cell harvest as earlier.
   c. Composition of fresh medium for terminal dT-pulse.
      i. F-10 + 20% FCS.
      ii. dT, add 9×10^{-6} M to F-10 ∴ total 1.2×10^{-5} M dT.
2. Late replication = dull fluorescence.
   a. Additives at initiation of culture (per 5 ml).
      PHA 0.2 ml.
   b. Growth. Cells are cultured at 37° C for approximately 3 days (two replication cycles). If a total culture time of 72 hr is used and a pulse of x hr is required, additions (see below) are made at 72-x hr, colcemide is added at 70 hr, and the cells are collected at 72 hr.
   c. Additions (terminal BRdU-pulse).
      i. BRdU, 0.2 ml of a 2.5×10^{-3} M solution in Hanks' BSS.
      ii. 0.1 ml (2×10^{-5} M FUdR, 3×10^{-4} M U).

Modifications for detection of regions containing DNA with thymine asymmetry. Culture cells to allow BRdU incorporation for one complete cycle and then stain, e.g., with 33258 Hoechst ± light and Giemsa as described above. High ratios of thymidylate to adenylate (in a polynucleotide chain) will manifest as bright fluorescence (or dark Giemsa staining) in the chromatid containing this chain.

Concluding Remarks

With the different banding techniques just described, the identification of every single chromosome of the human chromosome set is possible. This represents immense progress when compared to chromosome identification by means of length and centromere index. Much of the chromosomal data reviewed in this book is based on the latter method, and it may well be that with the application of newer techniques a number of conclusions drawn will have to be modified or discarded. It could be that a significant part of the work performed in tumor cytogenetics since 1958, when C. E. Ford, Jacobs, et al. for the first time described an abnormal chromosome set in AL, will have to be done all over again. However, tumor cytogenetics faces one tremendous problem, i.e., the scarcity and poor quality of metaphases, when compared with the abundance and beauty of metaphases derived from PHA-stimulated lymphocytes and from cell lines and fibroblasts grown in culture. It must be realized that the newer techniques for chromosome identification were elaborated on latter cells. The application of these techniques to tumor cytogenetics is hampered by the poor spreading, fuzziness, and tendency to overcontraction of the neoplastic chromosomes. It seems to me that such problems can only be solved by improving the chromosome preparation techniques of tumor cells.

Autoradiography

After recording the location of appropriate metaphases, the Giemsa-stained slides are destained in 20% acetic acid for 20 min and then rinsed in running tap water for 30 min. We use Nuclear Track Emulsion NTB 2 (Kodak) diluted 1:2 with distilled water, in which air-dried slides are briefly dipped. To remove excessive emulsion, the slides are placed in a vertical position and allowed to dry, which takes about 30 min. The slides are then put in a special box and kept in a refrigerator for 5–7 days. For development, we use Kodak developer D-19, diluted 1:10 with distilled water. The developing time is 4 min. After a brief bath in distilled water the emulsion is fixed in Kodak's fixer for 2 min and rinsed in running tap water for 20 min. Finally, the slides are restained in 5% Giemsa for 10 min.

High-Resolution Banding of Chromosomes

Application of new approaches, particularly those based on the study of chromosomes in late prophase or early prometaphase with or without exposure to colchicine or colcemide for very short periods of time (10 min), will undoubtedly result in the identification of a much larger number of bands in human chromosomes than presently known, primarily due to resolution of existing bands into manifold components (Figure 45). Obviously, this raises the possibility that abnormalities connected with such sub-bands, and presently not detectable with methods commonly used for banding, may characterize certain human diseases, and a search for such anomalies will naturally have to be undertaken once the sub-bands are rigorously defined and characterized.

Even though automated and computerized analyses of chromosomes and their bands have been described,[24] reliable and time-saving examinations and recording of the sub-bands will require even further refinements of existing techniques.

Because high-resolution banding will undoubtedly be more widely used and may find application in human neoplasia, directly or indirectly, I have included a description of the techniques outlined by J. J. Yunis, Sawyer, et al. (1978).

High Resolution of G-Banded Chromosomes

Below is presented in detail the method for high resolution of G-banded prometaphase or late-prophase chromosomes according to J. J. Yunis and co-workers.[25, 26]

A cell synchronization technique has been developed that consistently yields with normal cells a high number (12–15%) in the various early stages of cell division, namely, late prophase, prometaphase, early metaphase, and mid-metaphase.[25] When the procedure is followed carefully, it is possible to routinely ob-

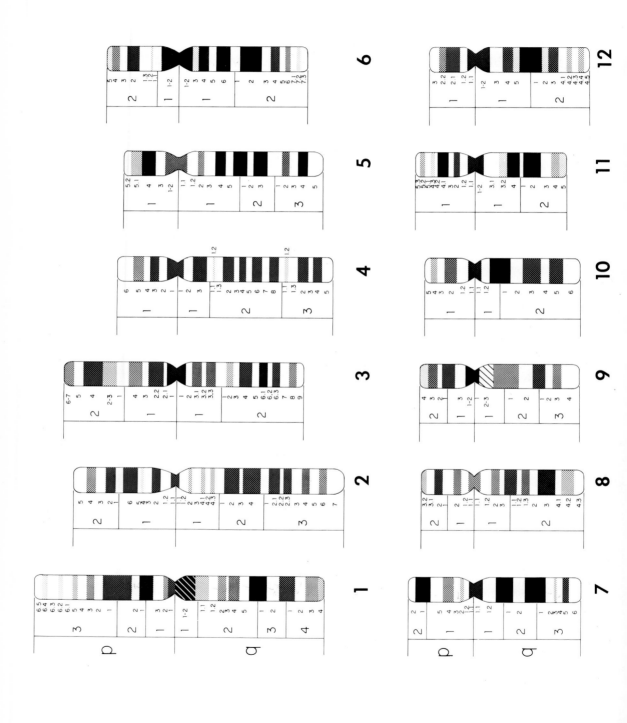

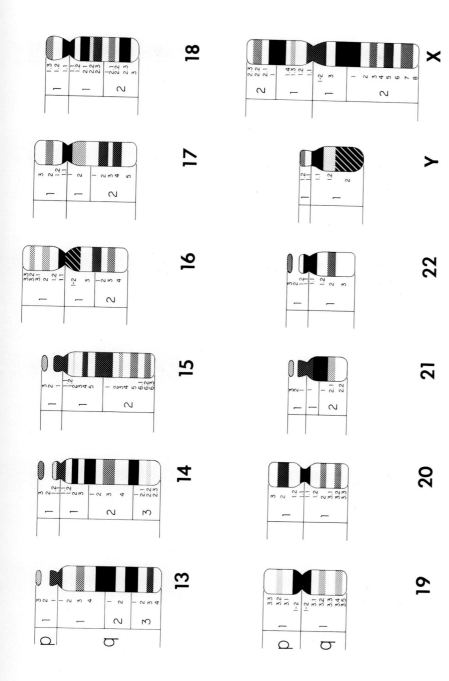

45 A. Schematic representation of sub-bands in human chromosomes obtained with a high-resolution technique in elongated (prometaphase) chromosomes. *(Courtesy of Dr. J. Yunis.)*

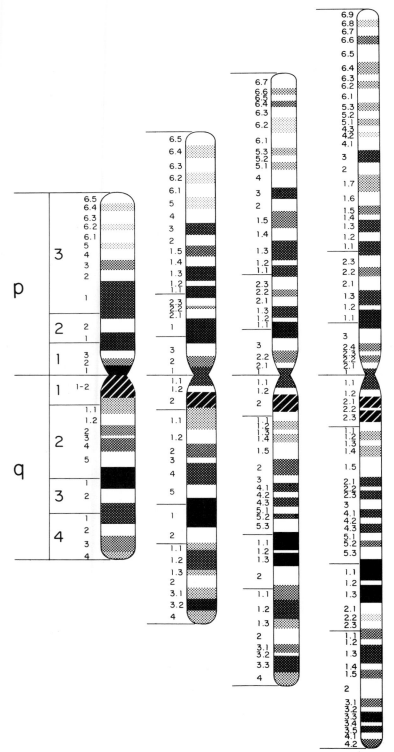

45 B. High-resolution banding of chromosome #1 showing the various sub-bands with elongation of the chromosome.
(Courtesy of Dr. J. Yunis.)

High Resolution of G-Banded Chromosomes 113

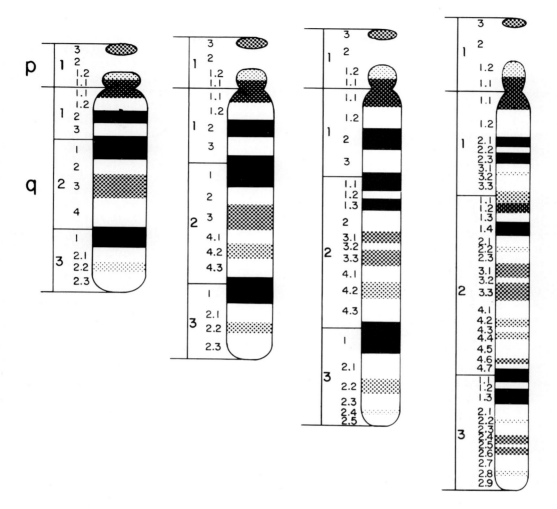

45 C. High resolution banding of chromosome #14 showing the various sub-bands with elongation of the chromosome. As in the case of #1 shown in B, so in the case of #14 more and more subbands appear as higher resolution of the bands is achieved with elongated chromosomes in late prophase and/or prometaphase.
(Courtesy of Dr. J. Yunis.)

tain 1–3% of cells in both late prophase and prometaphase and 3–5% of cells in both early metaphase and mid-metaphase.

Using a large number of straight and well-banded chromosomes (10 per stage), the characteristic position, thickness, and staining intensity of every band at each stage have been determined quantitatively for each chromosome of the haploid set. From these studies, it has been observed that metaphase bands result as a coalescence of finer sub-bands of earlier stages and that each band and its corresponding sub-bands maintain a precise location throughout the process of chromosomal contraction.

In contrast to the widely used and generally accepted idiogram provided by the Paris Conference (1971), diagrammatic representations of mitotic chromosomes with the new technique have the following features:

1. 395 bands in mid-metaphase instead of the 320 bands of the Paris 1971 scheme, are visualized, which may in part be due to the

use of a short exposure to a low dose of colcemide and a G-banding staining technique that does not require pretreatment.
2. Even though the chromosomes represent a continuum, it has been found that most chromosomes can be readily classified by relative length and typical banding patterns into four categories: mid-metaphase, early metaphase, prometaphase, and late prophase. The banding patterns in the various stages were found to be indistinguishable from those observed from nonsynchronized cell cultures.[26]
3. In describing the banding patterns, six main shades of black, gray, and white are distinguished. The schemes at each stage represent the number of bands discernible in over 50% of the chromosomes. When classified in this manner, there are approximately 400 bands per haploid set in mid-metaphase, 550 in early metaphase, 750 in prometaphase, and 100 in late prophase. An additional large number of bands (>200) are seen in the more elongated prophases.

To accomplish high resolution chromosomes, at present it is best to maintain the Paris nomenclature in which any sub-band, either in mid-metaphase or in an earlier stage, will be classified as a sub-band and assigned an arbitrary number. Eventually, however, this system will produce ambiguities such as the description of the precise localization of a particular gene.

Amethopterin-Synchronized Lymphocyte Culture Technique

Peripheral blood lymphocytes are cultured using RMPI 1603 medium (GIBCO), supplemented with 20% fetal calf serum (GIBCO), 5,000 U/ml penicillin and 5,000 μg/ml streptomycin (Microbiological Associates), and 0.25% mycostatin (E. B. Squibb & Sons). Four drops of heparinized blood and 0.2 ml PHA M (Difco Laboratories) are added to 5 ml of complete media in a 30-ml culture flask (Falcon Plastics) and tightly capped. After 72-hr growth at 37° C, amethopterin [methotrexate (Lederle Laboratories)] is added at a final concentration of 10^{-7} M (e.g., 50μl of 10^{-5} M solution to each 5-ml culture) to induce synchrony. After 17 hr additional incubation at 37° C, the cells are released from the amenthopterin block. To this end, the contents of two flasks are combined into one sterile 16 × 125-mm tube (Falcon) and centrifuged at 200 g (1,100 rpm in clinical centrifuge) for 8 min. The supernate is aspirated off leaving approximately 0.5 ml above the pellet. Ten ml of unsupplemented RPMI 1603 medium at room temperature is then added, the pellet is resuspended by inversion, and centrifugation is carried out again. After completion of a second wash, the cells are resuspended in 10 ml of complete RPMI 1603 medium containing 10^{-5} M thymidine (Sigma) (e.g., 1% of a 10^{-3} M solution), then placed in a clean sterile culture flask and returned to the incubator. After a period of 5 hr 5 min, colcemide (GIBCO) is added at a final concentration of 0.06 μg/ml (60 μl of a 10 μg/ml solution to each 10-ml culture) for 10 additional min at room temperature or 37° C.

To harvest the cells, the contents of each flask are poured into a siliconized 15-ml centrifuge tube and centrifuged at 200 g for 8 min. The supernate is aspirated off, leaving approximately 0.5 ml above the pellet, which is then gently resuspended by tapping the tube with a finger. Eight ml of 0.075 M KC1 at 37° C is gently added in 2- to 3-ml amounts with a pipette and then mixed gently but thoroughly. Cells are placed in an incubator or water bath at 37° C for 10 min. The tubes are then centrifuged at 200 g for 5 min after which the supernate is aspirated off and the cells resuspended in fresh 3:1 methanol–acetic acid fixative. To this end, the pellet is first partially resuspended by tapping the tube and the fixative is added drop by drop with a Pasteur pipette, while gently shaking the tube to keep the cells in suspension while fixation is taking place. Approximately 1–2 ml of fixative is added in this manner, followed by thorough mixing with the pipette; an additional 4–5 ml is added and mixed thoroughly. The tubes are then tightly sealed with Parafilm (American Can Co., Greenwich, CT.). After a period of 20 min, the tubes are centrifuged at 200 g for 5 min and the fixative changed. This should be repeated at least 4–6 times to ensure that the cells are clean of debris, allowing satisfactory spreading of the chromosomes. If possible, chromosome preparations should be made the same day; otherwise, cells may be stored in the refrigerator overnight in tubes tightly covered with double Parafilm, but must be allowed to come to room temperature and rinsed 2–3 times with fresh fixative before spreading onto slides.

To make slide preparations, cells are resuspended in 6:1 methanol:acetic acid at a very dilute concentration and dropped from a distance of 2.5–3 ft onto 70% ethanol-cleaned slides placed at a 30° angle. Two to three drops down the slide usually produce excellent spreading. Both wet cold and dry room temperature slides should be tested, because one method may be more successful than the other with a given pellet.

G-Banding of Chromosome Preparations with Wright Stain

Wright's stock solution. 0.25% in anhydrous acetone-free methanol. Prepare 1 liter, using powered Wright Stain (Manufacturing Chemists, Norwood, Ohio 45292). Stir with electric stirrer for 30–60 min, filter with double no. 1 filter paper into clean, dark 0.473-liter bottles, and store air-tight for at least 1 mo. Stain should not be diluted until use and then only in quantities to be used immediately.

Phosphate buffer. 409 ml of 0.06 M Na_2HPO_4 and 510 ml of 0.06 M KH_2PO_4. Adjust to pH 6.8 and store in an air-tight bottle.

Procedure. Place 3 ml phosphate buffer in clean tube, add 1 ml Wright's stock solution with a 1-ml pipette, pipette once or twice to mix, and pour immediately onto slide. After exact timing with stop watch, rinse quickly with tap water, and dry with air jet.

Important!

Exact timing is important to obtain high quality banding; therefore, it is best to stain one slide at a time, with a maximum of two. Optimum time appears to be 2.5–4 min, and the Wright's stock solution should be diluted with methanol, if necessary, to achieve this range. For each bottle of stock solution, several trials are necessary to determine the approximate staining time, which will vary somewhat with the source of cells, age of slides, type of preparation, and temperature. In our hands, the ideal temperature and humidity for banding appear to be 72–75° C and 40%, respectively. (Prophases require a few seconds longer than metaphases; actinomycin D-treated chromosomes require 20–30 sec additional time.)

Slides may be stained before 7–14 days aging. If bands are not contrasted and sharp, age slides longer.

Inadequately stained slides can sometimes be improved by further rinsing and restaining. For example, a 5- to 30-sec rinse in running tap water may be employed to minimize cytoplasmic background or improve overstained slides. Rinsing may even be as long as several minutes, followed by restaining for 30–60 sec, and possibly another rinse. Understained slides may also be rinsed and restained. Progress of staining is most easily monitored with the use of an Epiplan 80X (Carl Zeiss, Inc.) microscope objective which does not require the application of oil to the slide.

Actinomycin D–Treated Chromosomes

Actinomycin D, when administered to synchronized lymphocytes in culture before harvest, partially inhibits the normal process of chromosomal condensation, so that chromosomes of a given stage exhibit an increase in length of 25–50% as well as a 15% increase in the number of bands. This effect is more pronounced in the lightly staining G-negative bands because actinomycin D binds to GC base pairs, preponderant in G-negative bands. With this technique, the most condensed mitotic chromosomes observed, those of mid-metaphase, appear to possess the length and resolution of early metaphase chromosomes. In addition, actinomycin D-treated late prophases have a length similar to that of untreated midprophases and exhibit approximately 1,400 bands per haploid set.

Lymphocytes are cultured, synchronized with amethopterin, and released with excess thymidine for 5.25 hr, as previously described. Forty min before harvest, actinomycin D (Cosmegen; Merck Sharp & Dohme) is added to the cell cultures at a final concentration of 5 μg/ml (0.1 ml of a 0.5 mg/ml solution to each 10-ml culture). Incubation is continued at 37° C for the duration of the treatment. During the

Table 5 Lymphocyte mitogens (26)

Agent	Cells stimulated	Concentration	Reference
Phytohemagglutinin (PHA)	T-lymphocytes (human)	200 µg/ml	Nowell, 1960
	T-lymphocytes (human)	15–80 µg/ml	Punnett, Punnett, et al., 1962
	T-lymphocytes (human)	100 µg/ml	Greaves and Roitt, 1968
	Lymph node cells (mouse)	100 µg/ml	McClain, Wang, et al., 1975
	Thymocytes (mouse)	100 µg/ml	McClain, Wang, et al., 1975
Pokeweed mitogen (PWM)	B + T lymphocytes (human)	150 µg/ml	Greaves and Roitt, 1968
	B + T lymphocytes (mouse)	150 µg/ml	Greaves and Bauminger, 1972
	Spleen cells (mouse)	1–100 µg/ml	Waxdale and Basham, 1974
	Spleen cells (mouse)	100 µg/ml	McClain, Wang, et al., 1975
Concanavalin A (Con A)	T-lymphocytes (mouse)	5 µg/ml	Andersson, Edelman, et al., 1972
	Spleen cells (mouse)	1 µg/ml	Stobo, Rosenthal, et al., 1972
	Spleen cells (mouse)	1–3 µg/ml	McClain, Wang, et al., 1975
	Thymocytes (mouse)	1 µg/ml	Stobo, Rosenthal, et al., 1972
	Lymph node cells (mouse)	1 µg/ml	McClain, Wang, et al., 1975
Calcium ionophore-A23187	B + T lymphocytes (human)	5×10^{-7}–10^{-6}M	Luckasen, White, et al., 1974
	B + T lymphocytes (pig)	0.8 µg/ml	Maino, Green, et al., 1974
	B + T lymphocytes (human)	5×10^{-7}–10^{-6}M	Kersey, Sabad, et al., 1975
Sodium metaperiodate (NaIO$_4$)	T lymphocytes (human)	2×10^{-3}M	Parker, O'Brien, et al., 1972
	T lymphocytes (human)	10^{-3}–4×10^{-3}M	Parker, O'Brien, et al., 1973
	T lymphocytes (mouse)	10^{-3}M	Novogrodsky, 1974
	T lymphocytes (mouse)	2×10^{-3}M	McClain, Wang, et al., 1975
EB-virus	B lymphocytes (human)	—	Kamada, Kuramoto, et al., 1979

last 10 min of actinomycin-D exposure, colcemide is added at a final concentration of 0.06 µg/ml. Harvesting is carried out according to the procedure described above.

The chromosome preparations are G-banded according to the Wright staining procedure described above, with two modifications. First, the slides are treated before staining with 0.2 N HCl for 20 min at room temperature and dehydrated in alcohol (70% and absolute) to remove the actinomycin D. Second, the chromosomes must be stained slightly longer (20–30 sec) than those from untreated cultures.

Mitogens. A number of mitogens[27] for various types of lymphocytes is available, with some of them showing specificity toward these lymphocytes (Table 5). Unfortunately, no known mitogens have been uncovered for other types of cells, particularly various leukemic and cancer cells.

Comment. Some reports have indicated that a structural chromosome anomaly, e.g., iso(17q), apparently detected with one banding technique, may upon examination with another be revealed to be an anomaly of a different nature. However, in some of these cases the interpretation must be accepted with caution, because there is a distinct possibility for translocated material not only to affect functionally the chromatin to which it becomes newly contiguous, but may, in fact, also change its physical properties, including the staining characteristics, of that chromosomal region.

References

1. Makino, 1975; Schwarzacher and Wolf, 1974; A. K. Sharma and Sharma, 1972; J. J. Yunis, 1974.
2. Sandberg, Ishihara, et al., 1962a, b; Takagi and Sandberg, 1969.
3. Hossfeld and Sandberg, 1970b.
4. Knuutila, 1976; Knuutila, Simell, et al. 1976; O'Riordan, Berry, et al. 1970; Sandberg, Ishihara et al., 1961; Secker Walker, 1971; Sugiyama and Ishihara, 1975.
5. Borgström, Vupio, et al., 1976; Trujillo, Cork, et al., 1971.
6. Hentel and Hirschhorn, 1971.

7. Abramson, Miller, et al., 1977; Chapelle, Vupio, et al., 1973.
8. Leon, Reichhardt Epps, et al., 1961; Sandberg, Ishihara, et al., 1962*a, b*.
9. Golob, Klearchou, et al., 1973; Iinuma and Nakagome, 1972, 1973; My. A. Kim and Bier, 1972; P. L. Pearson, Bobrow, et al., 1971; Shoemaker, 1977; Soudek and Laraya, 1976; P. K. Srivastava, Miles, et al., 1974; Tishler, Rosner, et al., 1974.
10. Manolov, Manolova, et al., 1971*a, b*; Schwinger, 1973; Winters, Benirschke, et al., 1975.
11. Cardillo, Krause, et al., 1978; J. P. Chaudhuri, Ludwig, et al., 1977; H. L. Cooper and Hirschhorn, 1962*b*; Hastings, Freedman, et al., 1961; Moorhead, Nowell, et al., 1960; Rønne, 1977.
12. Kiossoglou, Mitus, et al., 1964; Lam-poo-tang, 1968; Meighan and Stich, 1961; Morse, Humbert, et al., 1977; Tjio and Whang, 1962, 1965.
13. Benedict and Porter, 1972; Ishihara, Kikuchi, et al., 1963; M. Jung, Blatnik, et al., 1964*a, b*; Kakati, Hayata, et al., 1975; Vincent, Vandenburg, et al., 1964.
14. Berger, 1969*a,b*; Carpentier and Lejeune, 1973; Grouchy, Vallée, et al., 1963; M. W. Shaw and Chen, 1974; Scherz and Louro, 1963.
15. A large number of different techniques and modifications for Q- and G-banding methods has been and will continue to be published, proving that no ideal method, particularly for leukemic and cancer cells, has been established. Included in the references under the various techniques are most of the methods described and some of the papers dealing with mechanisms of banding and factors affecting such banding. An extensive compilation of various banding and fluorescent methodologies has been published in Mammalian Chromosome Newsletter 13:21–47 (January), 1972.
16. Barnett, MacKinnon, et al., 1974; Breg, 1972; Buckton, O'Riordan, et al., 1976; Bühler, Tsuchimoto, et al., 1976; Caspersson, Zech, et al 1969*a, b*; Caspersson, Lomakka, et al., 1973; Castoldi, Grusovin, et al., 1972; Chernay, Kardon, et al., 1972; Comings, Kovacs, et al., 1975; H. J. Evans, Buckton, et al., 1971; Jarvik and Yen, 1974; My. A. Kim, 1974; My. A. Kim, Bier, et al., 1971; Latt, 1975; Lin, van de Sande, et al., 1975; Manolov, Manolova, et al., 1971*a, b*; D. A. Miller, Allerdice, et al., 1971; Nakagome, Oka, et al., 1977, 1978; Nilsson, 1973; Oka, Nakagome, et al., 1971; P. L. Pearson, Bobrow, et al., 1971; Raposa and Natarajan, 1974; Rowley, Potter, et al., 1971; Salkinder and Gear, 1962; Schiffer, Vaharu, et al., 1961; Schnedl, 1974*a*, 1978; Schnedl, Mikelsaar, et al., 1977; Schwinger, Sperling, et al., 1974; Selles, Marimuthu, et al., 1974; Verma and Dosik, 1976; Verma and Lubs, 1975*a. b*. 1976*a. b*; Verma, Dosik, et al., 1977*a*.
17. Benyush, Luckash, et al., 1977; Chen, 1974; Hoehn, Au, et al., 1977; Rubenstein, Verma, et al., 1978; Salamanca and Armendares, 1974; Scheres, 1976*a. b*.
18. Bigger, Savage, et al., 1972; Bosman, van der Ploeg, et al., 1977*a*; Comings, 1975; Crossen, 1974; Daniel and Lam-poo-tang, 1973; N. R. Davidson, 1973; Drets and Shaw, 1971; Dutrillaux, 1975; Eiberg, 1973; France, Biljlsma, et al., 1974; Goradia and Davis, 1977; Gormley and Ross, 1976; Lundsteen, Lind, et al., 1976; Rodman, 1974; Rønne and Sanderman, 1977; Rønne, Bøye, et al., 1977; Savage, Bigger, et al., 1976; Schmiady, Wegner, et al., 1975; Seabright, Cooke, et al., 1975; Skovby 1975; Sun, Chu, et al., 1974; Verma and Lubs, 1975*a, b*; J. J. Yunis and Sanchez, 1975; J. J. Yunis, Kuo, et al., 1977.
19. Ayraud, 1975*a, b*; H. F. L. Mark and Mendoza, 1976; Nakagome, Oka, et al., 1978; Sande, Lin, et al., 1976; Sehested, 1974; Verma, Peakman, et al., 1976; Verma, Dosik, et al., 1977*a, b*; Verma and Lubs, 1975*a, b*, 1976*a, b*.
20. Bobrow, Collacott, et al., 1972; Bobrow, Mandan et al., 1972; Chapelle, Schröder, et al., 1973*b*.
21. T. C. Hsu, 1974; Lubs, McKenzie, et al., 1973; Schnedl, 1973, 1974*a*, 1978.
22. Archidiacono, Capoa, et al., 1977; S. E. Bloom and Goodpasture, 1976; Faust and Vogel, 1974; Henderson, Warburton, et al., 1973; Lau and Arrighi, 1977; Lau, Pfeiffer, et al., 1978; Mikelsaar, Schmid, et al., 1977; Vogel, Schempp, et al., 1978; Zankl and Berhardt, 1977.
23. Latt, 1973; Latt, Davidson, et al., 1974; P. Perry and Wolff, 1974.
24. Bishop and Young, 1977; K. R. Castleman, Melnyk, et al., 1976; Fleischmann, Gustafsson, et al., 1971; Granlund, Zack, et al., 1976; Gray, Carrano, et al., 1975; Ladda, Atkins, et al., 1974; Marimuthu, Selles, et al., 1974; Rutovitz, 1967; Wald, Fatora, et al., 1976.
25. J. J. Yunis, 1976; J. J. Yunis and Chandler, 1977*b*; J. J. Yunis, Kuo, et al., 1977.
26. J. J. Yunis, Kuo, et al., 1977; J. J. Yunis and Sanchez, 1975.
27. Andersson, Edelman, et al., 1972; Greaves and Bauminger, 1972; Greaves and Roitt, 1968; Kersey, Sabad, et al., 1975; Luckansen, White, et al., 1974; Maino, Green, et al., 1974; McClain, Wang, et al., 1975; Novogrodsky, 1974; Nowell, 1960; Parker, O'Brien, et al., 1972, 1973; Punnett, Punnett, et al., 1962; Stobo, Rosenthal, et al., 1972; Waxdale and Basham, 1974.

6

Congenital Chromosome Anomalies and Neoplasia

Autosomal Anomalies

Down's Syndrome (Mongolism)
Other Autosomal Trisomies
Other Autosomal Anomalies
Retinoblastoma

Sex-Chromosome Anomalies

*XO Karyotype and Its Variations
  (Gonadal Dysgenesis, Including
  Turner's Syndrome)*
*XXY and Its Variants (Klinefelter's
  Syndrome)*
XXX Syndrome
*Gonadal Dysgenesis Without Chromosomal
  Changes*
XYY Males

Relation of Congenital and Familial Chromosome Abnormalities to Cancer and Leukemia

The salient features of this group of subjects with cytogenetic abnormalities is the involvement of almost all the cells in the body by the chromosomal aberrations and their presence at the earliest embryonic stage.

Autosomal Anomalies

Most of the patients with these anomalies do not live sufficiently long enough to develop cancer or leukemia. In all probability the preponderant number of cases with autosomal aberration die *in utero* or shortly after birth, with the exception of a few trisomies and, very rarely, total or partial monosomies.

Down's Syndrome (Mongolism)

This congenital disorder is due to trisomy of chromosome #21. This is the most common autosomal anomaly in man and is compatible with a relatively long survival.[1] In about 95–98% of the cases the anomaly is manifested by trisomy of #21 and in the remaining cases by translocation of the extra autosome onto a chromosome in group G or D.

The high incidence of acute leukemia (AL) in subjects with Down's Syndrome (DS) is well established and is of particular interest.[2] The role played by the extra autosome in the development of the leukemic state is uncertain, since only a small percentage of patients with DS develops AL, even though the incidence among these subjects is higher than in a comparable population with a diploid chromosome constitution. The AL in DS may be either myeloblastic or lymphoblastic, possibly depending on the age of the patient at the time of the development of leukemia (Table 6). The leukemic chromosomal changes, when present, are similar to those found in the cells of patients with AL but without trisomy. It should be mentioned that the #21 trisomy persists in the marrow cells, even when aneuploidy related to the various forms of leukemia appears.

The association of Ph[1]-positive chronic myelocytic leukemia (CML) with DS has not been convincingly described, to my knowledge, particularly since the #9–#22 translocation has not been established in such cases. In a review of leukemia and DS, it was felt that in none of the eight reported cases of DS with either a Ph[1]-chromosome or morphologic evidence of CML could the diagnosis of this form of leukemia be substantiated.[3] The presence or lack of such an association would be of great interest because it would supply considerable information on the effects of the #21 trisomy on the formation of the Ph[1]-chromosome. Even though the incidence of AL is relatively high in DS, it is probably not as high as reported by some, because leukemoid myeloblastosis may occur in these subjects and be mistaken for one or another form of leukemia.[4] The incidence of leukemia in DS with translocation trisomy (G-G or G-D) has not been established satisfactorily.

The relationship of the extra #21 autosome to the control of leukopoiesis, in general, and to the development of AL in particular, remains unclear. For example, for some years it was thought that the Ph[1] in CML was an abbreviated #21 and, thus, more emphasis was placed on this chromosome in relation to leukemogenesis than it probably deserved. However, it has been shown with banding techniques that the Ph[1] is a #22 and not a #21.[5] Nevertheless, the G-group of chromosomes continues to be implicated in leukemogenesis, e.g., the Ch[1]-chromosome in chronic lymphocytic leukemia (CLL)[6] and a statistically frequent involvement of the G-group chromosomes in AL observed by some workers.[7] In all probability this one group of chromosomes will continue to receive special attention in the future and particularly when still more refined

Table 6 Summary of morphologic types of acute leukemia in Down's syndrome*

	Acute myelo-blastic leukemia	Acute lympho-blastic leukemia
FROM THE LITERATURE	%	%
Children with DS	30.9	69.1
Newborns with DS	57.9	42.1
RECENT EXPERIENCE IN U.S.A.		
Children with DS	30.2	69.8
Newborns with DS	80.0	20.0

*All patients with "transient" leukemia have been eliminated from this analysis. This table is modified from that of Rosner and Lee (1972).

methods become available for chromosomal analysis.

Inasmuch as the development of the stigmata of DS requires the invariable presence of trisomy of chromosome #21, there must be other factors, possibly including genetic (chromosomal?), aside from trisomy, that play an important role in the genesis of the AL in this disease. So far, these factors are totally unknown.

Except for the definitely higher incidence of AL in DS,[8] a convincingly higher frequency of any other neoplastic process has not been reported in these patients with trisomy of chromosome #21.[9] However, Hodgkin's disease (HD) has been described in at least two patients with DS (McCormick, Amman, et al., 1971), and there may be an increased frequency of some neoplasms in this disease as well (R. W. Miller, 1970), but to date no reports have appeared on the karyotypic picture in these diseases complicating DS and whether the cells did or did not retain the extra #21 chromosome.

The #21 trisomy in DS leads to a definitely increased level of leukocyte alkaline phosphatase (LAP) and in the days before the use of banding techniques this particular observation made a sensible story when taken in conjunction with the deleted G-group chromosome (Ph¹) and greatly decreased LAP. However, it now appears that the relationship between chromosome #21 and CML does not exist and that the increased LAP activity in DS is based on the trisomy of #21. Thus, the control of LAP is under genetic factors residing in at least several chromosomes including #21 and #22.

The preponderant number of cases of DS and AL were reported between 1962–1967, with most of the reports dealing with single cases. Up to that time chromosomal analysis was performed in 37 cases, of which 16 contained only the trisomic anomaly in their cells, 15 had other karyotypic changes in addition to the trisomy #21, consisting primarily of extra chromosomes in groups C and G, 1 case involved a translocation of the extra #21, and rare hypodiploid cases have been reported. In a few cases the exact chromosomal picture is not given and in a small number it is difficult to ascertain whether the trisomy was established on peripheral blood or marrow cells.

An unusual case of double trisomy, i.e., XXX and trisomy #21 has been described in a female patient who had typical phenotypic signs of DS, no breast development, but normal external genitals. She had a retinoblastoma of the right eye (Day, Wright, et al., 1963). A higher sensitivity to chromosome damage by X-ray, drugs, and viruses has been demonstrated for DS cells, as well as difficulties in repairing damaged DNA resulting from these agents.[10]

No structural genes have been demonstrated with certainty on #21. It was thought that the chromosome involved in CML (Ph¹) was the same as the one involved in DS, a hypothesis that has been disproved with banding techniques. In the past, there was general agreement that the Ph¹ was a #21. Various arguments were used to support this contention. For example, it is known that the risk of developing AL in patients with trisomy #21 is increased 20- to 30-fold compared with individuals of the same age not exhibiting DS. Furthermore, lobulation of polymorphonuclear leukocytes in #21 trisomy patients is less than normal. These findings have been interpreted as evidence for "leukopoietic genes" located on #21. An excess of genetic material, as in DS, could alter the normal metabolism of leukocytes, favoring in some undetermined way the onset of a leukemic process. Other arguments were supported by enzymology, i.e., LAP levels of patients with trisomy #21 are elevated, whereas they are reduced in patients with CML. However, a simple gene enzyme relationship is not likely. In trisomy #21 the LAP level is about 50% higher than in normal subjects, whereas in CML it is usually zero. On the other hand, the LAP level is elevated in some patients with Ph¹-positive CML and also in patients with ulcerative colitis. Of course, this complicated problem was dissipated when it was shown that the chromosome involved in CML was #22, whereas the one causing trisomy in DS was #21.

Other Autosomal Trisomies

The other trisomic conditions primarily involve chromosomes in groups D, E, and F. Even though usually resulting in severe multiple congenital anomalies, they are compatible with survival after birth, though generally for less than 1 yr. Hence, these subjects do not live sufficiently long enough to develop cancer or leukemia, and the number of cases in which these abnormal states develop must be extremely small, judging by the paucity of publications dealing with the association of the above trisomies and neoplasia. Even though sporadic cases of trisomy with leukemia or cancer may be reported,[11] the exact incidence of neoplasia in these cytogenetically abnormal conditions is difficult to ascertain, because no reports have appeared on a large group of cases.

Other Autosomal Anomalies

It appears that total monosomy is not compatible with survival in human subjects, though partial monosomy may not be lethal. Included in this group is a host of chromosomal anomalies ranging from such conditions as partial deletions of group-B chromosomes (*cri-du-chat* syndrome), transformations of normal autosomes into ring chromosomes, extra chromosomes in group C, partial deletion of a number of other chromosomes besides group B, and many other abnormal chromosomal conditions.[12] The number of cases with each of these syndromes is rather small and, hence, we do not know the exact role the karyotypic changes would play in neoplasia were the affected subjects to live long enough to develop cancer or leukemia.

Only rare cases of leukemia have been reported in autosomal syndromes, e.g., AL in a patient with trisomy-D and another case in trisomy-F and one case of CML in a patient with 45 chromosomes and a D/D translocation. A lymphocytic lymphosarcoma (LS) and cancer of the breast have been reported in two patients with the latter syndrome and a carcinoma of the lung in a patient with inversion of a chromosome in group C (Harnden, Langlands, et al., 1969). A case with neonatal bilateral Wilms' tumors with a B-C translocation (?8q—; 5q+) has been described (Giangiacomo, Penchansky, et al., 1975).

An interesting autosomal anomaly related to neoplasia is that of the 13q— syndrome, as it is related to the development of retinoblastoma.[13] Apparently 20% of patients with the 13q— (absence of all or part of the 13q14 band) syndrome reported in the literature have developed retinoblastoma, with a majority of these patients having bilateral tumors diagnosed in the first year of life, which is characteristic of inheritable retinoblastoma as defined by Knudson (1971). The 13q— syndrome is characterized by a number of phenotypic anomalies including facial asymmetry, coloboma of the iris, microphthalmus, ptosis, absent or hypoplastic thumbs, and imperforate anus, in addition to retinoblastoma.[14] The genetics of the inheritable form of retinoblastoma is rather complicated, as has been discussed by Knudson (1971), who indicated that in the predominantly inherited form one mutation is inherited via the germinal cells and the second occurs in somatic cells. In the nonhereditary form, both mutations occur in somatic cells. This is of importance, because not all cases of retinoblastoma are accompanied by constitutional autosomal changes and not all patients with the 13q— syndrome develop retinoblastomas. Examination of the tumors per se has revealed a 13q— picture in one primary retinoblastoma (Lele, Penrose, et al., 1963) and analysis with banding techniques of a metastatic retinoblastoma in the marrow and spinal fluid in a 2-yr-old boy revealed the cells to contain 47 or 48 chromosomes including a marker[15] which resembled one previously described in another case of retinoblastoma.[16] It should be pointed out that no abnormalities of chromosomes in groups #13–15 were found in this metastatic tumor. Thus, the development of retinoblastoma in the 13q— syndrome obviously is related to other genetic events necessary for the development of this malignancy (Table 7). The exact nature of these events is at present unknown.

Fibroblasts derived from patients with the hereditary retinoblastoma appear to be more sensitive to the lethal effects of X-ray than do fibroblasts from patients with sporadic retinoblastoma or normal controls (Weichselbaum, Nove, et al., 1978). A defect in DNA repair was postulated to account for the high incidence of second tumors in these patients.

In 1963 Lele, Penrose, et al. found a chromosome abnormality, i.e., deletion of a D-group chromosome, in cultures of conjunctivas, skin, blood cells, and tumor tissue from a patient with bilateral retinoblastomas. It was uncertain whether the dividing cells from the tumor were neoplastic or stromal. Subsequently, other patients with retinoblastoma (about 50% being bilateral) were found to have a deleted D-group chromosome either alone or in

Table 7 Retinoblastoma and various congenital chromosomal anomalies

Karyotype	Reference
48,XXX,+G	Day, Wright, et al., 1963
47,+21 (DS)	R. D. Jensen and Miller, 1971;
47,+21	Bentley, 1975; Taktikos, 1964
48,XXY,+21	Rethoré, Saraux, et al., 1972
46,XX/47,XX,+G	Cernea, Teodorescu, et al., 1973
46,X, pericentric inversion of Y	Czeizel, Csösz, et al., 1974
47,XXY	M. G. Wilson, Ebbin, et al., 1977

46 Diagrammatic #13 on the left is shown with bands as seen in metaphase and the one on the right with sub-bands as seen in late prophase. Note that in the latter band q14 in metaphase has been resolved into sub-bands q14.1, q14.2, and q14.3 in prophase. Also shown are stained chromosomes #13 in metaphase, prometaphase, and late prophase of a case (top row) and those of another (bottom row) with retinoblastoma, with arrows indicating the break point. The latter allowed identification of sub-bands deleted in two patients with retinoblastoma, the broken and solid arrows pointing to the minute deletions observed in these two cases.

(J. Yunis and Ramsay, 1978.)

association with other cytogenetic anomalies. However, the deletions were not found in all cases of retinoblastoma and in some series contributed to only a minority or none of the cases.[17] Some patients have various congenital anomalies, usually minor in extent. M. G. Wilson, Melnyk, et al. (1969) originally felt, on the basis of autoradiographic studies, that the involved D-group autosome was #14 (14q−). However, reexamination of their material (M. G. Wilson, Towner, et al., 1973), following the suggestion of A. I. Taylor (1970) that the chromosome involved is #13, using banding techniques showed that the deleted chromosome was, in fact, #13. It appears that an interstitial q21 deletion is associated with retinoblastomas but not with severe malformations; in those patients with retinoblastoma and severe malformations, the deletion may have involved q31−q34 or q12−q14 regions.

Even though a minority of cases with retinoblastoma have a 13q−, a significantly high incidence of numerical changes, chromatid aberrations, and stable abnormal chromosomes in lymphocytes has been reported (Czeizel, Csösz, et al., 1974), as well as increased sensitivity of the fibroblasts in some cases to X-ray (J. J. Yunis, Kuo, et al., 1977; Weichselbaum, Nove, et al., 1977).

A high incidence of primary tumors in sur-

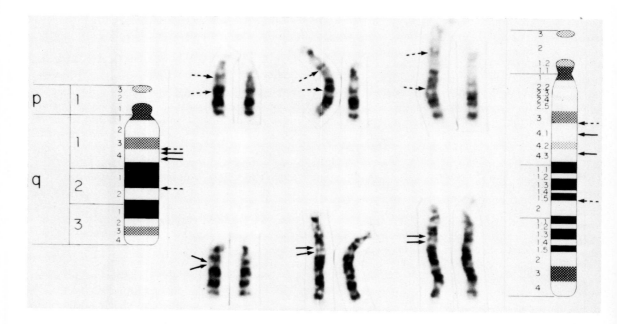

vivors of the bilateral (but not the unilateral) form of the retinoblastoma who had received radiotherapy has been reported (Kitchin, 1976). The second primaries were mainly osteogenic sarcomas in long bones away from the radiation field. The findings indicate "a pleiotrophic effect of the retinoblastoma gene which may act as initiator in more than one form of neoplasia."

Retinoblastoma

Retinoblastoma (Figures 46 and 47) is a tumor appearing in childhood with an incidence of 1 in 20,000. It occurs at a high frequency among two groups of genetically predisposed individuals: 1) subjects who inherit a tumor predisposition in an autosomal dominant fashion, and 2) persons born with a partial deletion of the long arm of chromosome #13.[18] The latter group is of particular importance because it is the only instance known in man in which a specific chromosomal defect can occur prezygotically in subjects consistently predisposed to a specific tumor. Uncertainty has existed as to the exact location of the karyotypic anomaly on #13, the uncertainty largely residing in the difficulty of establishing precise break points in that chromosome where the small, light bands q12 and q14 and the thick, dark bands q21 and q31 make up approximately 80% of the total length of the long arm and are not clearly distinguishable among themselves. Studies utilizing mid-metaphase chromosome #13, in which phase the establishment of break points is difficult, are subject to more than one interpretation of the deleted segment. With the analysis of prometaphase and late prophase chromosomes, J. J. Yunis and Ramsay (1978) was able to identify with certainty the bands involved. In these more elongated chromosomes most of the bands observed in metaphase are seen to consist of discrete sub-bands that become of critical importance in the analysis of break points. They came to the conclusion that band q14 is the specific lesion in retinoblastoma patients and suggested that a specific relationship between these two defects exists. It is possible that a partial loss of band q14 results in faulty development of the retina, which in turn predisposes to retinoblastoma. A deletion of band 13q14 is similar to an autosomal dominant mutation, in that either a mutation or deletion of a presumptive retinoblastoma gene would produce the same disorder. That the tumor has only been identified in 14% of patients with 13q— may be related to 1. more than one-half of all patients reported thus far with partial 13q— had ring chromosomes and were studied before the advent of the banding techniques, making it impossible to determine whether the same pathogenic segment was involved, 2. many patients with 13q— die during the 1st year of life, whereas the tumor usually does not manifest itself until the 2nd year of life, and 3. some patients with 13q— do not have involvement of band 13q14. About 20 patients with retinoblastoma and interstitial deletion of the long arm of chromosome #13 have been described (Table 8).

In general, the reported cases are compatible with a common deletion of band q14 (see Table 8), though unusual cases appear to exist, such as that of Ikeuchi, Sonta, et al. (1974) (without retinoblastoma) and one case reported by J. J. Yunis and Ramsay (1978) in whom only part of band q14, probably involving the total sub-band 2 and part of sub-bands 1 and 3, has been reported to be missing.

It is intriguing that even though the cytogenetic anomalies discussed in this and the fol-

47 Schematic representation of #13 and its bands according to the Paris Conference (1972). The brackets on the right depict deletions observed in retinoblastoma cases published in the literature.
(J. Yunis and Ramsay, 1978.)

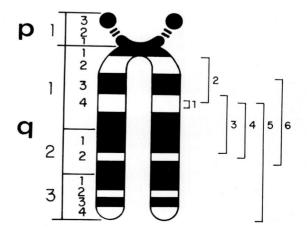

Table 8 Retinoblastomas and 13q−

Reference	Chromosome segment involved
M. G. Wilson, Melnyk, et al., 1969; M. G. Wilson, Towner, et al., 1973	q14→q22 or q22→q32
Orye, Delbeke, et al., 1971, 1974	q14→q22
Howard, Breg, et al., 1974	del prox seg. of 13q
U. Francke, 1976; U. Francke and Kung, 1976	del prox seg. of 13q (q12q14)
Knudson, Meadows, et al., 1976	q14→q22
Noël, Quack, et al., 1976	q14→q22
Noël, Quack, et al., 1976	q14→qter (q13→q14)
Welch, Fiander, et al., 1977	pter→q12::q14→qter
M. G. Wilson, Ebbin, et al., 1977	del prox. seg. of 13q (q12→q14)
J. J. Yunis and Ramsay, 1978	q14→q21
J. J. Yunis and Ramsay, 1978	part of q14 (q14.4) q 14.1 and q14.3? partially?
Cross, Hansen, et al., 1977	t(13q−;Xp+)
Knight, Gardner, et al., 1978	complex 13p inheritance
Walbaum, François, et al., 1978	13q−(q12→q14)

CASES WITH RETINOBLASTOMA IN WHICH THE EXACT SEGMENT OF #13 INVOLVED WAS NOT REPORTED

Lele, Penrose, et al., 1963
H. Thompson and Lyons, 1965
Van Kempen, 1966 [also t(2;C)]
A. I. Taylor, 1970
Gey, 1970
Grace, Drennan, et al., 1971
O'Grady, Routhstein, et al., 1974

For discussion of the significance of chromosomal changes in retinoblastoma see Knudson (1978).

lowing section are present from the time of fertilization of the ovum or very shortly thereafter, the presence of these chromosomal aberrations throughout the period of gestation, i.e., during the most crucial embryologic changes and growth, does not lead, as far as is known, to a higher incidence of neoplasia in these subjects at birth.

Sex-Chromosome Anomalies

Almost all of the following syndromes are characterized by abnormal gonadal anatomy and function, though they may be accompanied by other phenotypic abnormalities unrelated to the sex organs. Neoplasia in these subjects tends to involve the gonads and related organs. However, it is not certain whether the development of such neoplasia is the result primarily of genetic factors or of a radically modified hormonal environment in such subjects, as a result of abnormal gonadal development and physiology.

The exact incidence of neoplasia in sex-chromosome anomalies is difficult to establish with certainty, because the subjects with unusual phenotypic manifestations tend to be seen more frequently by physicians, examined more carefully, and reported on more readily than other subjects of a similar age. Nevertheless, the distinct possibility exists that certain malignancies may be seen more frequently in patients with sex-chromosome anomalies than in a comparable control group.

XO Karyotype and Its Variations (Gonadal Dysgenesis, Including Turner's Syndrome)

This abnormality[19] has had an incidence of about 1 in 1,000–1,500 births. Germ-cell tumors are extremely rare in pure Turner's syndrome (XO gonadal dysgenesis), the development of such tumors apparently requiring the presence of the Y-chromosome in the involved tissues. Hence, the phenotypic picture of these subjects appears to have little relationship to the possible development of gonadal

Table 9 Extragonadal neoplasia in Turner syndrome (gonadal dysgenesis) with 45,XO or 45,XO/46,XX

Diagnosis	Age of patient, yr	Chromosome constitution	Reference
Pituitary eosinophilic adenoma	18	45,XO	Willemse, 1962
Cancer of endometrium	22	45,XO	Dowsett, 1963
Pituitary chromophobe adenoma	38	45,XO	Milcu, Ionescu, et al., 1964
Hilus-cell adenocarcinoma	35	45,XO/46,XX	J. C. Warren, Erkman, et al., 1964
Cancer of stomach	63	45,XO	E. Engel and Forbes, 1965
Cancer of endometrium	62	45,XO/46,XX	Dumars, Kitzmiller, et al., 1967
CLL	65	45,XO/45,XX	Dumars, Kitzmiller, et al., 1967
Papillary pseudomucinous cystadenocarcinoma of ovary	21	45,XO	M. B. Goldberg and Scully, 1967
Malignant brown-fat tumor (hibernoma)	24	45,XO/46,XX	Lowry and Halmos, 1967
Cancer of thyroid	22	46,XXqi	Sparkes and Motulsky, 1967
Medulloblastoma	16	45,XO	Brun and Sköld, 1968
Retroperitoneal mesenchymoma	2	45,XO/46,XX	M. B. Goldberg, Scully, et al., 1968
Squamous cell carcinoma of vulva	45	45,XO/46,XX	M. B. Goldberg, Scully, et al., 1968
Ganglioneuroma (adrenal)	4 mo	45,XO	R. W. Miller, Fraumeni, et al., 1968
Cancer of uterus	31	45,XO	Scott, 1968
Anaplastic lung tumor	23	45,XO	Li and Fraumeni, 1969
AML	13	45,XO	Wertelecki, Fraumeni, et al., 1970
Granular cell myoblastoma (Schwannoma)	12	46,XXqi?/46,XX	Wertelecki, Fraumeni, et al., 1970
Hygroma of neck	Abortuses	45,XO	R. P. Singh, 1970
Epithelioid sarcoma (right wrist)	16	45,XO/46,XX	Males and Lain, 1972
Cancer of stomach	34	45,XO	Siegler, 1975

germ-cell tumors, which tend to occur at a much higher frequency in subjects with gonadal dysgenesis associated with the presence of a Y-chromosome (XO/XY, XO/XYY, XY),[20] though it has been reported in subjects with XO and XO/XX fragments or an iso-X-chromosome constitution. Gonadoblastomas tend to occur more frequently in patients with female phenotypes showing primary amenorrhea, signs of masculinization, and usually negative sex-chromatin pattern. It is of interest that the pure XO syndrome has been found very rarely among patients with cancer or leukemia,[21] though some patients with XO/XX mosaicism and these diseases have been described, including a case of CLL (Dumars, Kitzmiller, et al., 1967) and one of an epithelioid sarcoma of the wrist (Males and Lain, 1972) (Table 9).

Gonadoblastoma (dysgenetic gonadoma or gonocytoma type III), an infrequent gonadal tumor, was recognized as a distinct pathological entity about 25 yr ago.[22] The tumor has a distinctive histological appearance and is usually found in the abdomen or the inguinal canal. However, its external appearance can be confusing and may vary from patient to patient. Gonadoblastoma develops exclusively in patients with gonadal dysgenesis, predominantly in those with a female phenotype, though it

Table 10 Gonadoblastoma cases reported in the literature

Sex chromosome constitution	Age (yr)	Clinical aspects	Reference
XY Monozygotic	6	Normal ♀	Frasier, Bashore, et al., 1961
XY Twins	6	Normal ♀	Frasier, Bashore, et al., 1964
XY	13.5	Androidal	Siebenmann, 1961
XO/X, minute marker			R. W. Miller, 1964
XX/XY	16	Hermaphrodite male habitus	Overzier, 1964
XO/XY**	17.5	PA*	A. Robinson, Priest, et al., 1964
XY**	19	Eunuchoidal	Teter, Philip, et al., 1964
XY	19	Androidal; eunuchoidal	Teter, Philip, et al., 1964
XY	10(♂)	Gynecomastia	Warkany, Weinstein, et al., 1964
XO/XY**	11	TS‡	Borghi, Moutali, et al., 1965
XY	20	TS, small breasts	M. M. Cohen and Shaw, 1965
XO/XY	29	PA	Milcu, Ionescu, et al., 1965
XY	12	Normal, breasts beginning to develop	Strumpf, 1965
XO/XY	19	TS	Teter and Boczkowski, 1967
XY (2 cases)	19	Androidal, eunuchoidal	Teter and Boczkowski, 1967
XY	24	PA, eunuchism	Guinet and Eyraud, 1968
XY**	17	GD§	Freeman and Miller, 1969
XO/XY	13	Stunted growth, lack of puberty	Josso, Nezelof, et al., 1969
XO/XY	15	Growth failure and virilism	Josso, Nezelof, et al., 1969
XY	54	TE ‖	Breckenridge, Nash, et al., 1970
XO/XXY	21	PA, short stature	Suñé, Centeno, et al., 1970
XO/XY + minute metacentric	25	GD	Ferenczy, Richart, et al., 1971
XY	15	Mixed GD (with virilization)	Kariminejad, Movlavi, et al., 1972
XO/XY	19	Mixed GD (with virilization)	Goldner, Hale, et al., 1973
XO/XY	8	Short stature	Segall, Shapiro, et al., 1973
XO/XY/XYY	18	PA, no puberty	Osztovics, Ivady, et al., 1974
XO/XY	10	TS	Serra, Moneta, et al., 1974
XY	10	Slight masculinization	Similä, Jukarainen, et al., 1974
46,X, Dicentric Yp**	19	PA, sexual infantilism	Málková, Michalová, et al., 1975
XY	22		Muller, Clavert, et al., 1970
XY	18	PA, small breasts slight virilization	J. Philip, 1975 J. Philip, Hansen, et al., 1975
XY**	25(♂)	Bilateral hernias	J. Philip, 1975 J. Philip, Hansen, et al., 1975
XO/XY	9(♂)	Many congenital defects, male retarded growth, abnormal genitalia, testicular dysgenesis, (tumor in testis)	Boczkowski, Teter, et al., 1967
46,XY/45,Y with partial deletion of X**	31	PA	Garvin, Pratt-Thomas, et al., 1976 Williamson, Underwood, et al., 1976
46,XY/45,X	22	PA	Garvin, Pratt-Thomas, et al., 1976 Williamson, Underwood, et al., 1976

Table 10—continued

Sex chromosome constitution	Age (yr)	Clinical aspects	Reference
46,XY**	19	PA, sterility	Garvin, Pratt-Thomas, et al., 1976
46,XX/45,X	22	Ruptured ectopic pregnancy	Williamson, Underwood, et al., 1976
			Garvin, Pratt-Thomas, et al., 1976
X,X + minute	21	PA, slight hirsutism, facial acne	Williamson, Underwood, et al., 1976
			McDonough, Byrd, et al., 1976
XX	13	Hermaphrodite (♂) hypospadias, gynecomastia	McDonough, Byrd, et al., 1976
XY	19	TS	
XO/XY	14	PA, poorly developed breasts, no virilization	Govan, Woodcock, et al., 1977
			Govan, Woodcock, et al., 1977
XO/XY**	26	PA, normal breast development, no virilization	Govan, Woodcock, et al., 1977
XY	24	Hermaphrodite	Szokol, Kondrai, et al., 1977
XY	7.5	Hypercletorism	Desbois, Coicadan, et al., 1977
45,XO/46,X-ring Y	17	Secondary amenorrhea, virilization	Ishida, Tagatz, et al., 1976

*PA, primary amenorrhea.
**, Bilateral tumors.
‡TS, Turner syndrome.
§GD, gonadal dysgenesis.
‖TF, testicular feminization.

may be rarely observed in patients with a male appearance (J. Philip, Hansen, et al., 1975). Even though all dysgenetic gonads are threatened by tumor development, a higher risk is incurred by those patients with a 46,XY stem line with or without an associated 45,X cell line (Tables 10 and 11). Mosaic patterns (45,X/46,X, fragments; 46,XY/46,XX mosaic, etc.) have also been described. In some cases chromosome analysis was not performed, but when examined the sex-chromatin was found to be negative, except for rare cases (Perrin and Landing, 1961). A patient with gonadal dysgenesis and bilaterial gonadoblastomas with 46X,dic(Y) as the major line in blood and skin and a minor line with a partially deleted Y-chromosome has been described (Málková, Michalová, et al., 1975). Another patient, a 19-yr-old woman with an XO/XY mosaicism and signs of Turner's syndrome, was found to have a tumor arising from an intraabdominal dysgenetic testis and shown to be an interstitial cell carcinoma (Goldner, Hale, et al., 1973). Cells cultured from the tumor revealed a heteroploid line, clustered around the triploid and tetraploid modes with the presence of 1 or 2 Y-chromosomes in the cells. Cytogenetic studies of cultured lymphocytes and fascial and fallopian cells revealed the preponderance of XO/XY cells, accounting for the Turner's features. A case of an 18-yr-old patient with 45,X/46,XY/47,XYY mosaicism with a gonadoblastoma was described by Osztovics, Ivady et al. (1974). The triple mosaicism was also found in the cultured cells of the gonadoblastoma; in the blood the 45,X line made up 62% of the cells and the most common cells (56%) in the tumor had a 46,XY constitution. Several of the authors have stressed the presence of a Y in the gonadoblastomas, even when other cells, e.g., lymphocytes or fibroblasts, did not reveal its presence.[23]

Dysgerminoma (gonocytoma type I) has been described in two patients with gonadal dysgenesis of the 45,X/46,XY constitution (Psaroudakis, Oettinger, et al., 1976). The au-

Table 11 Mixed tumors (mainly gonadoblastoma and dysgerminoma)

Sex chromosome constitution	Age (yr)	Clinical aspects	Reference
XY	19	GD*	Schellhas, Trujillo, et al., 1971
XY	18	GD	Schellhas, Trujillo, et al., 1971
XY	23	Induced menses	Schellhas, Trujillo, et al., 1971
XY	16	GD	Zárate, Karchmer, et al., 1971
XY		GD (variant?)	Williamson, Underwood, et al., 1976
XY		GD	Hill and McKenna, 1974
45,X,−16, +t(16;Yp−)/ 46,XYq−,−16, +t(16;4p)		Normal	Park, Heller, et al., 1974
46,XY	17	Mass palpated on precollege exam	Garvin, Pratt-Thomas, et al., 1976 Williamson, Underwood, et al., 1976
XY	18	PA‡, infertility	Amarose, Kyriazis, et al., 1977 Dorus, Amarose, et al., 1977
XY	13	PA, some breast development; no virilization	Govan, Woodcock, et al., 1977
XO/XY	19	TS§	Govan, Woodcock, et al., 1977
XO/XY	17	PA, poor breast development; no virilization	Govan, Woodcock, et al., 1977
XY	24	PA, very poor breast development; severe virilization	Govan, Woodcock, et al., 1977
XO/XY	21	PA, normal breasts; no virilization	Govan, Woodcock, et al., 1977

*GD, gonadal dysgenesis. ‡PA, primary amenorrhea. §TS, Turner syndrome.

Table 12 Dysgerminomas in patients with sex-chromosome abnormalities

Sex chromosome constitution	Age (yr)	Clinical aspects	Reference
Sex chromatin negative	15	TS*	Dominguez and Greenblatt, 1962
XY + minute metacentric	17	GD‡	Fettig, Schröter, et al., 1968
XY	57	TF§ (seminoma)	Volpe, Knowlton, et al., 1968
XY	17	GD	P. Fischer, Golob, et al., 1969
XY	18	Induced menses, small breasts	Schellhas, Trujillo, et al., 1971
XY	15	Amenorrhea, no breast development	Schellhas, Trujillo, et al., 1971
45,XO/46,XY	15	TS	Psaroudakis, Oettinger, et al., 1976
XO/XY (seminoma)	12(♂)	Testicular tumor	Doll, Kandzari, et al., 1978
XY sisters	17	PA‖ with breast development	Boczkowski, 1976
XY	22		Boczkowski, 1976
XY (5 cases)	19–26	Clitoral enlargement, eunuchoidal	Teter and Boczkowski, 1967

*TS, Turner's syndrome. ‡GD, gonadal dysgenesis. §TF, testicular feminization. ‖PA, primary amenovihea.

thors concluded that there is an association between the male karyotype and dysgerminoma, though the XY occurrence may be minimal. Hence, it is important to count large numbers of cells in cytogenetic studies in order to establish the XO/XY mosaicism. A hilus-cell adenoma has been described in a 33-yr-old woman with an XX/XO mosaicism and virilization (Warren, Erkman, et al., 1964). Thus, in cases with gonadal dysgenesis it behooves the physician to search for a Y-chromosome cell line, because of its implications regarding gonadal malignancy (Table 12). The presence of any cell line containing the Y-chromosome should alert the responsible persons to look-out for any changes in the patient's condition, particularly those that might be indicative of the development of a gonadoblastoma or other tumors.

XXY and Its Variants (Klinefelter's Syndrome)

This syndrome[24] appears to occur in about 1 out of 500–700 births. Even though patients with Klinefelter's syndrome have been described to develop various forms of leukemia and lymphoma, the exact incidence of these neoplastic diseases in this syndrome is still unknown.[25] On firmer ground is the distinct possibility that cancer of the breast is much more common in these male subjects than in other males (Table 13).[26] To establish this contention more securely,[27] it is imperative that a survey be made of as many patients with Klinefelter's syndrome as possible; and of the exact incidence of cancer of the breast in these subjects and in other males. Only through an evaluation of a large group of such subjects will the exact incidence of cancer of the breast in Klinefelter's syndrome be established.

A case of Klinefelter's syndrome with XXY and lung carcinoma has been presented, in which a large dicentric marker and a ring chromosome were present in the cancer cells of a pleural effusion, these markers probably being related to the missing chromosomes in groups X, C, D, and E (Cervenka and Koulischer, 1973).

A 32-yr-old man with Klinefelter's syndrome and a karyotype consisting of 47,XXY,15S+ with enlarged and fluorescent satellites on #15 developed a seminoma of the right testis which had been subjected to orchiopexy 14 yr earlier for cryptorchidism. The seminoma was not examined cytogenetically (Isurugi, Imao, et al., 1977).

An infant with incontinentia pigmenti, Bloch-Sulzberger, and an XXY constitution has been reported (Kunze, Frenzel, et al., 1977).

XXX Syndrome

As in the case of the XXY syndrome, occasional cases of XXX with cancer or leukemia are encountered (Table 14), but the exact incidence of neoplastic diseases in this condition is unknown.[28] Even though this chromosomal abnormality is not uncommon, occurring in about 1 out of 1,000–1,500 births, these patients are not easily detected because of a lack of characteristic phenotypic manifestations. A survey of these cases may yield appropriate information regarding cancer and leukemia in these females.

Gonadal Dysgenesis Without Chromosomal Changes

This group of patients has been discussed under the section of Turner's syndrome and includes those female subjects with an XY, or rarely XX, chromosome constitution and gonadal dysgenesis. The high incidence of gonadoblastoma in these subjects has been established.[29] For some reason the cells of cancer patients with XY-gonadal dysgenesis have an increased susceptibility to transformation by SV40 virus (Mukerjee, Bowen, et al., 1972) (Table 15). Thus, it is incumbent upon the physician to establish the cytogenetic nature of the gonadal dysgenesis in these patients, particularly those with a female phenotype and an XY sex-chromatin complement, and probably submit these patients to surgical exploration and the removal of the abnormal gonads.

XYY Males

Almost nothing is known about the incidence of neoplasia in this not uncommon sex-chromosome anomaly,[30] first described from my laboratory.[31] A case of XYY with medulloblastoma has been described. Unfortunately, the

Table 13 Cases of Klinefelter's syndrome, their chromosome anomalies and neoplasia reported in the literature

Diagnosis	Sex-chromosome constitution	Age (yr)	Reference
	BREAST CANCER		
	XXY	75	Dodge, Jackson, et al., 1969 (also benign interstitial tumor of testis)
	XXY	59	A. W. Jackson, Muldal, et al., 1965
	XXY	52	A. W. Jackson, Muldal, et al., 1965
	XY/XXY/XXXY	55	A. W. Jackson, Muldal, et al., 1965
	XXY	63	Cuenca and Becker, 1968
	XXY	69	Robson, Santiago, et al., 1968
	XXY/XY	34	Coley, Otis, et al., 1971 (bilateral)
	XXY	33 and 73	Harnden, Maclean, et al., 1971 (two cases)
	XX/XXY	58	Harnden, Maclean, et al., 1971
	XXY	66	Berger, May-Lewin, et al., 1973a (also cancer of tongue)
	XXY	72	Scheike, Visfeldt, et al., 1973 (bilateral)
	XXY	55	Lynch, Kaplan, et al., 1974
	LEUKEMIA		
AL	XXY	73	N. H. Kemp, Stafford, et al., 1961
AL	XXY	36	Mamunes, Lapidus, et al., 1961
CML	XY/XXY	59	Tough, Court Brown, et al., 1961
ALL	XXY	32	Bousser and Tanzer, 1963
RCS	XY/XXY/XXXY	53	MacSween, 1965
ALL	XXY	?	Ruffié, Ducos, et al., 1966b
AL	XXY	2.5	Borges, Nicklas, et al., 1967
CML	XY/XXY	69	Fitzgerald, Pickering, et al., 1971
ALL	XXXY	19	Sohn and Boggs, 1974
Lymphoma	XXY/XY	46	Tsung and Heckman, 1974
	OTHER TUMORS		
Oat cell cancer of bronchus	XY/XXY	74	Lubs, 1962
Cancer esophagus	XXY	71	Atkin and Baker, 1965b
Cancer prostate	XY/XXY	67	Arduino, 1967
Cancer bronchus	XY/XXY	38	Mukerjee, Bowen, et al., 1970

Table 13 — *continued*

Diagnosis	Sex-chromosome constitution	Age (yr)	Reference
	OTHER TUMORS		
Extragenital seminoma	XXY	43	Doll, Weiss, et al., 1976
Cancer bladder	XY/XXY	41	Fujita and Fujita, 1976
Seminoma	XXY, 15s+	32	Isurugi, Imao, et al., 1977
Incontinentia pigmenti	XXY	Newborn	Kunze, Frenzel, et al., 1977
Retinoblastoma	XXY	2	M. G. Wilson, Ebbin, et al., 1977

Table 14 Females with sex chromosome anomalies and neoplasia

Sex-chromosome constitution	Age (yr)	Phenotype	Type of disease	Reference
X/XXX	64	Normal	AL	F. J. W. Lewis, Poulding, et al., 1963c
X/XX/XXX	60	Normal? small breasts	Meningioma	Ayraud, Duplay, et al., 1972
X/XX/XXX	9	Normal	Astrocytoma	Ayraud, Duplay, et al., 1972
XXpi	53	Normal	Pituitary tumor (adenoma?)	Keogh, Kretser, et al., 1973
XXX	18	Normal? (slight hypoplastic uterus)	HD	Lech, Polaniecka, et al., 1974
X/XX/XXX	75	Normal	Cancer of endometrium	Atkin, 1976c

Table 15 Transformation frequency of fibroblasts by SV40 in controls and patient with Klinefelter's syndrome*

Cell strains	Karyotypes	Transformation frequency per 10^4 cells	Number of foci per 8×15^5 cells
Control males	44 ± XY	2.7 ± 0.10	199
Control females	44 ± XX	3.0 ± 0.15	244
Cell strains in patients	44 ± XY	9.7 ± 0.30	780
	44 ± XXY	28.5 ± 4.30	2,282

*Based on the data of Mukerjee, Trujillo, et al. (1971).

tumor was not examined for its chromosome constitution; this 7-yr-old child was shown to have a XYY karyotype probably on the basis of blood culture (Rosano, Delellis, et al., 1970).

Relation of Congenital and Familial Chromosome Abnormalities to Cancer and Leukemia

Congenital abnormalities in several members of one family associated with the presence of cancer in other members has been described. The following families will serve as examples: (1) the parents were normal, one child died from acute lymphoblastic leukemia (ALL), one from acute myeloblastic leukemia (AML), one from pneumonia, and one was found to have a mosaic XY/XXY;[32] (2) the mother was a D/G translocation carrier and had four children of whom two were normal, one was trisomic for chromosome #21, and one had AL;[33] (3) the father had CLL, a son an XXXXY karyotype, and an aunt and cousin trisomy #21;[34] (4) the mother was found to be mosaic of XX/trisomy #21, her son to have trisomy #21, and the father and two paternal uncles had CML;[35] (5) five cases of AL and one of trisomy #21 in three generations in one family;[36] and (6) three siblings, two with trisomy #21 and one with AL.[37] Based on such observations it has been suggested that "prezygotic determinants in acute leukemias" exist.

The familial association of congenital chromosome abnormalities and disorders with frequent malignant transformation has been described. Thus, a mother with XXX, who had a child with Fanconi's anemia (FA), showed chromosomal breaks in a higher frequency than normal.[38] A mother with a W-chromosome marker and Waldenström's macroglobulinemia (WM) had a boy with a D/D translocation,[39] and another patient with a similar marker had an increased level of gammaglobulins and a child with trisomy #21.[40] It has been suggested that autoimmune diseases, especially of the thyroid, seem to be linked either in the patients or in their parents with congenital chromosome anomalies. It has been also suggested that impairment of immunologic mechanisms predisposing to certain congenital chromosome disorders and some types of cancer may be linked.[41] However, much more conclusive data are necessary to ascertain the relationship between these cytogenetic abnormalities and the development of leukemia or cancer.

If a monozygotic twin develops leukemia there is a 20% probability that the other twin will also develop the disease. No such risk appears to exist for dizygotic twins. Moreover, it has been stated that "it seems likely that the true incidence of concordant leukemia in identical twins is greater than can be explained by chance, and considerably less than could be expected if strong genetic factors were operating."[42] Bone marrow chromosomes of at least four pairs of identical twins, discordant for CML, have been studied (Table 33). In each pair, the Ph¹-chromosome was found in the leukemic twin but not in the other. This clearly demonstrates that both CML and the Ph¹-chromosome were acquired and not inherited. A similar situation appears to exist in WM, in which only the affected twin was shown to have a marker chromosome in blood lymphocytes.[43] A similar observation in dizygotic twins, one with AL and the other normal, has also been reported, i.e., the twin with AL was the only one to show an abnormal stem line.[44] The chromosomes of two pairs of monozygotic twins, both with AL have been studied (Table 33). In one pair,[45] it was shown that the chromosome abnormality in the bone marrow cells was identical in both twins; in the other study,[46] one child was in relapse and the other in remission. In the first child an abnormal cell line with 65 chromosomes was observed, whereas in the second, even though the preponderant number of mitoses were diploid, three cells with 65 chromosomes were observed. These results seem to indicate that a similar cell line was present in both twins, that both patients reacted the same way to the leukemogenic agent, and that chromosome evolution patterns could be genetically predetermined. These findings stand in contrast to that found in dizygotic twins with AML.[47] One had a stem line with 45 and the other one with 52 chromosomes with the karyotypic pictures being totally divergent (Figure 48). Each twin appears then to have reacted to the leukemogenic agent in a genetically different manner, showing different stem line karyotypes yet displaying the same clinical disease. It would appear that chromosome abnormalities in leukemia and perhaps in cancer in general are acquired and not inherited. Fur-

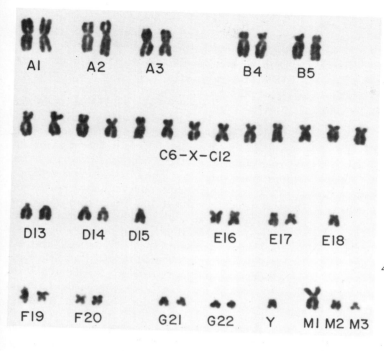
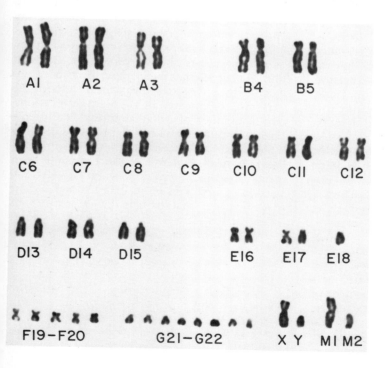

48 Karyotypes of twins (paternal) with AML (probably congenital), one (upper) with hypodiploid and the other (lower) with a hyperdiploid modal chromosome number in the leukemic cells of the marrow. Markers, but of different morphology, are present in each karyotype. Unless one is willing to impute two different causes for the disease in each twin, it would appear that the chromosome constitution in the leukemic cells is probably related more directly to the genome of the host than to the cause of the AL. Incidentally, the blood lymphocytes of these AML cases were diploid.

(Sandberg, Cortner, et al., 1966.)

Table 16 Some Mendelian inherited diseases frequently associated with cancer or leukemia

Neurofibromatosis	—Sarcoma
Enchondromatosis	—Chondrosarcoma
Gardner's syndrome* (familial intestinal polyposis)	—Adenocarcinoma of colon
Tuberous sclerosis	—Glioma
BS*	—Leukemia, lymphoma
FA*	—Leukemia, cancer
Bruton's agammaglobulinemia	—CLL
AT*	—Lymphoma
Xeroderma pigmentosum*	—Skin cancers
Chediak-Higashi syndrome	—Lymphoma

*Chromosome aberrations observed in these conditions.

Not included in this table are a number of conditions associated with chromosomal changes and development of neoplasia (benign and malignant), e.g., 15 p+ in multiple endocrine adenomatosis (Janson, Roberts, et al., 1978).

thermore, the chromosome evolution patterns are influenced by the genome of the patient.

"Cancer families" do exist but little is known about the factors that lead to a higher incidence of disease.[48] Certain Mendelian inherited diseases are not infrequently associated with neoplasia and these are shown in Table 16. Occasionally, an increase of a specific chromosome anomaly such as breaks, gaps, rings, or other chromosome rearrangements can be observed in unaffected relatives of patients with various forms of cancer and leukemia. At other times, more "specific" abnormalities are noticed in apparently healthy relatives. Thus, a peculiar G-chromosome with deleted short arms has been reported in many healthy members of a family as well as in those relatives who developed CLL.[49] A "Ph¹-chromosome" (not based on banding) has been observed in a daughter and two of three grandchildren of a patient with Ph¹-positive CML.[50] At the time of observation, all were healthy. A characteristic marker in the blood cells of a patient with WM and in four healthy relatives has been described;[51] the sister of the patient, after encephalitis, showed an increase in serum globulin level with a corresponding increase (8%) of cells having the marker. A "Ph¹-like" chromosome in two brothers with polycythemia vera (PV) has been reported,[52] but no follow-up (particularly, banding analysis) on these patients has appeared to date. Furthermore, in the case of the Ph¹, the presence of a congenital chromosome abnormality possibly resembling the Ph¹ and having a very different meaning than an acquired karyotypic abnormality has not been ruled out. The LAP, which is generally low in CML, has been found to be normal in the subjects with the Ph¹-like chromosome. In addition, a small deleted Ph¹-like chromosome (Gq−) has been reported in several congenital malformation syndromes. At present, the familial incidence of a Ph¹ is still a subject of discussion, and more research is needed before any conclusions can be drawn, particularly as present studies are not based on banding analyses.

It is beyond the scope of this book to review or even summarize the extensive literature on the remarkable array of congenital chromosomal abnormalities in the human described to date. The list is very lengthy, the phenotypic manifestations are extremely variable and complicated, and the effects on development, survival, and appearance of other complicating diseases have not been established with certainty for the bulk of these conditions. It appears that to date almost every chromosome in the human set has been described to be involved by either trisomy, partial deletion or trisomy, complicated or simple translocations, and a number of other abnormal karyotypic manifestations. Certainly, the relation of these congenital karyotypic abnormalities to the development of leukemia or cancer has not been established and it will require a relatively long period of time and more detailed and cogent data in order to establish the importance of these abnormalities in oncogenesis (Koller, 1973; R. W. Miller, 1966).

References

1. R. W. Miller, 1968, 1970; Rosner and Lee, 1972.
2. Baikie, Court Brown, et al., 1960c; Blattner, 1961; Castel, Riviere, et al., 1971; Conen and

Erkman, 1966; R. W. Miller, 1968, 1970; Rosner and Lee, 1972.
3. Rosner and Lee, 1972.
4. R. R., Engel, Hammond, et al., 1964.
5. Caspersson, Gahrton, et al., 1970; O'Riordan, Langlands, et al., 1972.
6. Fitzgerald, 1965; Gunz, Fitzgerald, et al., 1962.
7. Gunz, Bach, et al., 1973.
8. Schuler, Dobos, et al., 1972.
9. R. W. Miller, 1970.
10. Banerjee, Jung, et al., 1977; Countryman, Heddle, et al., 1977; Higurashi and Conen, 1972; Higurashi, Tamura, et al., 1973; Huang, Banerjee, et al., 1977; Lambert, Hansson, et al., 1976a; O'Brien, Poon, et al., 1971; Sandberg and Sakurai, 1973a, 1976; Sasaki and Tonomura, 1969; Sasaki, Tonomura, et al., 1970; D. Young, 1971b.
11. Borges, Nicklas, et al., 1966; Dische and Gardner, 1978; Forteza-Bover, Baguena-Candela, et al., 1963; Nevin, Dodge, et al., 1972; Schade, Schoeller, et al., 1962; Zuelzer, Thompson, et al., 1968.
12. Berger, 1971; Goh, 1968c.
13. Bishop and Madson, 1975; Falls and Neel 1951; Francois, Matton, et al., 1975; Hashem and Khalifa, 1975; Jensen and Miller, 1971; Kitchin, 1976; Knudson 1971, 1975, 1976, 1977; Knudson and Strong, 1972; Lancet, 1971a; Schappert-Kimmijser, Hammes, et al., 1966; Shimada, Oda, et al., 1960.
14. Allderdice, Davis, et al., 1969.
15. Inoue, Ravindranath, et al., 1974.
16. J. Mark, 1970c.
17. Czeizel, Csösz, et al., 1974; Ladda, Atkins, et al., 1973; Pruett and Atkins, 1969; Welch, Fiander, et al., 1977; S. Wiener, Reese, et al., 1963.
18. Allderdice, Davis, et al., 1969; Grace, Drennan, et al., 1971; Ikeuchi, Sonta et al., 1974; Ladda, Atkins, et al., 1973.
19. Caspersson, Lindston, et al., 1970; R. W. Miller, 1964.
20. Ferrier, Ferrier, et al., 1967; Fournier, Saint-Aubert, et al., 1976; J. Philip and Teter, 1964; Schellhas, Trujillo, et al., 1971; Suñé, Centeno, et al., 1970; Teter and Boczkowski, 1967.
21. Pawliger, Barrow, et al., 1970; Say, Balci, et al., 1971; R. P. Singh, 1970; Wertelecki and Shapiro, 1970.
22. Ishida, Tagatz, et al., 1976; Josso, Nezelof, et al., 1969.
23. Garvin, Pratt-Thomas, et al., 1976; Mulvihill, Wade, et al., 1975; Williamson, Underwood, et al., 1976.
24. Lubs, 1962.
25. O. J. Miller, Breg, et al., 1961.
26. Cuenca and Becker, 1968; Harnden, Maclean, et al., 1971; A. W. Jackson, Muldal, et al., 1965; Robson, Santiago, et al., 1968.
27. Nadel and Koss, 1967.
28. Harnden, Langlands, et al., 1969.
29. Dewurst, Ferreira, et al., 1971; Teter and Boczkowski, 1967.
30. Borgaonkar, 1968.
31. Sandberg, Koepf, et al., 1961.
32. Baikie, Jacobs, et al., 1961.
33. Buckton, Harnden, et al., 1961.
34. Hungerford, 1961.
35. Vereensen, van den Berghe, et al., 1964.
36. Heath and Moloney, 1965a.
37. Conen, Erkman, et al., 1966.
38. G. E. Bloom, Warner, et al., 1966.
38. Lustman, Stoffes-de Saint Georges, et al., 1968.
40. Elves and Israëls, 1963.
41. Fialkow, 1976b.
42. M. A. Pearson, Grello, et al., 1963.
43. Spengler, Siebner, et al., 1966.
44. Kiossoglou, Rosenbaum, et al., 1964.
45. M. A. Pearson, Grello, et al., 1963.
46. Hilton, Lewis, et al., 1970.
47. Sandberg, Cortner, et al., 1966.
48. Baikie, Jacobs, et al., 1961; Bouton, Phillips, et al., 1961; Cervenka, Anderson, et al., 1977; O.J. Miller, Breg, et al., 1961.
49. Gunz, Fitzgerald, et al., 1962.
50. Weiner, 1965.
51. A. K. Brown, Elves, et al., 1967.
52. Levin, Houston, et al., 1967.

7

Effects of Noxious Agents and Viruses on the Human Karyotype

Effects of X-Ray and Other Forms of Irradiation

Effects of Chemical Agents
Benzene Poisoning and Leukemia

Effects of Viruses

DNA Viruses
Adenoviruses
Herpes Viruses

    HSV type 1
    HSV type 2
    Cytomegalovirus
    Herpes zoster virus
    EB virus
    Papovaviruses
    Polyoma virus
    SV40
    Poxviruses

RNA Viruses
Paramyxoviruses

    Measles virus
    Mump virus
    Sendai virus
    Myxoviruses, arboviruses, picornaviruses, and unclassified viruses

Oncornaviruses

    Avian leukemia-sarcoma complex
    Murine leukemia-sarcoma complex

Chromosomal Changes upon Exposure to X-Ray or to Transformation by Certain Viruses

Survey of Cancer Patients for Chromosomal Anomalities

Inasmuch as a large number of biologic, physical, and chemical agents has been implicated in the possible causation of certain human cancers and leukemia, it is imperative that we review briefly some of the effects of these agents on human chromosomes. The number of physical and chemical agents shown to produce deleterious effects on the human chromosomal set in vitro is literally legion (Table 17). It appears that almost any substance added to cultures of human cells leads to the production of chromosomal abnormalities (breaks, deletions, gaps, multiradials, pulverization, etc.). In this section we shall concern ourselves with those noxious agents that produce chromosomal damage primarily in vivo and, thus, have been considered as possible causes of cancer or leukemia. The mechanisms by which chromosomal changes, induced by biologic, physical, or chemical agents, lead to neoplasia is poorly understood (Allison and Paton, 1965; Meisner, Chuprevich, et al., 1977), for the karyotypic changes in human neoplasia very seldom resemble the type of disruptive changes in the chromosomes caused by these noxious agents (Evans, 1977).

Heterochromatically altered regions in chromosomes, generated during the induction of the damage to the genome, may be responsible for nondisjunction of chromosomes, thus leading to nonrandom rearrangement of the chromosomal complement with newly acquired neoplastic characteristics. The importance of gaps, deletions, and translocations in chromosomes induced by these agents, followed by changes in ploidy due to nondisjunction and/or duplication of chromosome sets, have already received attention.[1]

Effects of X-Ray and Other Forms of Irradiation

The immediate effect of ionizing radiation is the production of a number of different types of damage to the chromosomal set. These consist of deletions, rearrangements, formation of ring chromosomes, dicentrics, acentric fragments and many other morphologic abnormalities (e.g., premature chromosome condensation [PCC])[2] of the chromosomes. The incidence, persistence, and nature of the chromosomal changes induced by ionizing radiation are dependent on the type of tissue being examined, the nature of the irradiation, source, dosage, and duration of exposure,[3] the length of time between the exposure and the cytogenetic examination,[4] and the method utilized for karyotypic observation.

Apparently, after exposure to irradiation, the cytogenetic abnormalities may persist for a very long time (more than 20 yr) in the lymphocytes of blood, when these are cultured in vitro with PHA. On the other hand, the changes in marrow may be rapidly eliminated or may be encountered after an interval of days or weeks after the exposure to radiation. However, the persistence of some karyotypic changes of a stable type in the marrow has been described in individuals exposed during an H-bomb explosion.[5] Because the most common neoplastic process resulting from ionizing irradiation is one form of leukemia or another,[6] it would seem that the modification of chromosomal function leading to neoplasia occurs at a level below the resolution of visual microscopy and may, in fact, have little to do with the gross chromosomal changes observed in cultured lymphocytes. Nevertheless, there is the distinct possibility that the chromosomal damage and changes produced by the initial ionizing insult may lead to permanent (usually structural rather than numerical) alterations in the karyotypes of some hematopoietic cells. It should be pointed out, again, that the types of chromosomal changes induced by ionizing irradiation are quite different from those observed in established forms of leukemia. The crucial chromosomal changes induced by irradiation may be dormant and submicroscopic and lead to the genesis of leukemia many years after exposure to the agent. Although so far based on a rather small number of cases, it is interesting to note that all subjects who have developed leukemia after exposure to irradiation have had karyotypic changes in their leukemic cells.[7] This contrasts with the 50% incidence of karyotypic abnormalities in other patients with so-called spontaneous AL[8] and indicates that more extensive studies and follow-ups should be performed on subjects who develop leukemia after irradiation.

The relation of ionizing irradiation to the development of leukemia is further illustrated by a recent rash of publications on leukemia-complicating HD[9] and other lymphomas

Table 17a Effects of biological, physical, and chemical agents on human chromosomes in vivo

Infectious agents	Reference
Viral hepatitis	Vormittag, 1972
Viral infections and vaccinations	Grouchy, Tudela, et al., 1967
Rubella, chickenpox	Knuutila, Laasonen, et al., 1977
Measles	Nichols, Levan, et al., 1962; Knuutila, Laasonen, et al., 1977
Measles (no changes)	Tanzer, Stoitchkov, et al., 1963
Congenital rubella	Nusbacker, Hirschhorn, et al., 1967
Smallpox vaccine	Frovlov, Slysuarev, et al., 1975; Knuttila, Mäki-Paakkanen, et al., 1978 (increased SCE)
Viral infections	Gripenberg, 1965
Viral infections	Nichols, 1963, 1970, 1972a, 1973
Measles	Koskull and Aula, 1977
Rickettsia	Halkka, Meynadier, et al., 1970
SV40 virus, in vitro, SCE ↑	Nichols, Bradt, et al., 1978

Drugs and chemicals	Reference
Anti-neoplastic agents	Sieber and Adamson, 1975
Chlorambucil	Reeves, Pickup, et al., 1974; Reeves and Margoles, 1974
Bleomycin	Dresp, Schmid, et al., 1978
6-Azauridine	Elves, Buttoo, et al., 1963
Benzene	Forni, 1966
Organic phosphate insecticides	Van Bao, Szabó, et al., 1974
Busulfan (myleran)	Gebhardt, 1974; Gebhardt, Schwanitz, et al., 1974; Krajinćanić, Lazarov, et al., 1976
Vinyl chloride	Ducatman, Hirschhorn, et al., 1975; Hansteen, Hillestad, et al., 1978; Purchase, Richardson, et al., 1978; Fleig and Thiess, 1978
Spray adhesives (no effect)	Lubs, Verma, et al., 1976
Colchicine	M. M. Cohen, Levy, et al., 1977
22 cytostatic drugs (no significant changes)	Schinzel and Schmid, 1976
Sulphonyl ureas (Rx in diabetes)	Watson, Petrie, et al., 1976
Arsenic therapy	Petres, Schmid-Ullrich, et al., 1970
Benzene	Tough and Court Brown, 1965
Vincristine	Gebhardt, Schwanitz, et al., 1969
Metronidazole	Mitelman, Hartley-Asp, et al., 1976
Hair dyes	Kirkland, Lawler, et al., 1978
Nitrogen mustard, 6-mercaptopurine, A-649 (antibiotics)	Nasjleti and Spencer, 1966
Folic acid antagonists	Ryan, Boddington, et al., 1965; Krogh Jensen, 1967a
Oral contraceptives (probably no changes)	Klinger, Glasser, et al., 1976
Adriamycin (SCE ↑)	Nevstad, 1978
Cytoxan	Arrighi, Hsu, et al., 1962
Imuran	Krogh Jensen and Soborg, 1966
Lead exposure	Schwanitz, Lehnert, et al., 1970; Schwanitz, Gebhart, et al., 1975; Bauchinger, Dresp, et al., 1977; Nordenson, Beckman, et al., 1978
Ara-C	Bell, Whang, et al., 1966
Vinyl chloride	Funes-Cravioto, Lambert, et al., 1975
Chloramphenicol	Hampel, Lohr, et al., 1969
Oral contraceptives	Littlefield, Lever, et al., 1975
Drugs	Nichols, 1972b
Heavy smokers	Obe and Herha, 1978
Chemotherapy and gynecological conditions	Kawasaki, 1968a
Chlorambucil (in PV)	Westin, 1976

Table 17a—continued

Drugs and chemicals	Reference
Mineral oil exposure	Benn, Hardnen, et al., 1978
8-Methoxypsoralen and UV (SCE ↑)	Mourelatos, Faed, et al., 1977
Chemicals	Bishun, Williams, et al., 1973
Melphalan (in rat marrow)	Wantzin and Jensen, 1973
Drugs in cancer patients	Amarose, 1964
Stilbestrol(diethyl) (no changes)	Bishun, Eddie, et al., 1978
Industrial monitory	Kilian, Picciano, et al., 1975
Environmental monitory	Kamarov, 1973
Azathioprine	Krogh Jensen and Hüttel, 1976
Trichlorethylene	Konietzko, Haberlandt, et al., 1978

Radiation	Reference
Clinical X-ray	Slowiskowska and Grzymala, 1972
Clinical X-ray	Takumura, Sakurai, et al., 1970
Thorotrast	Visfeldt, Jensen, et al., 1975; Ishihara, Kohno, et al., 1978
Atomic bomb survivors	Watanabe, 1964
Plutonium workers	Brandom, Bloom, et al., 1976
Radiation exposure	Ishihara and Kumatori, 1965
^{131}I therapy	Lloyd, Puttott, et al., 1976
H-bomb exposure (15 yrs later)	Ishihara and Kumatori, 1969
^{192}Ir (γ-rays) exposure	Ishihara, Kohno, et al., 1973
^{131}I therapy (no changes)	Blein, Garnier, et al., 1972
Thorotrast (19–27 yrs later)	P. Fischer, Golob, et al., 1966c
^{131}I therapy	E. Boyd, Buchanan, et al., 1961
Diagnostic ultrasound (no changes)	E. Boyd, Abdulla, et al., 1971
Diagnostic ultrasound (no changes)	Abdulla, Campbell, et al., 1971
Radioactive gold	Stevenson, Bedford, et al., 1971a
Diagnostic X-ray	A. D. Bloom and Tjio, 1964
X-ray therapy	Amarose and Baxter, 1965
^{131}I therapy	Speight, Smith, et al., 1968
X-ray therapy	Millard, 1965
X-rays and neutrons	Bender and Gooch, 1963
X-ray for spondylitis	Tough, Buckton, et al., 1960
X-ray	Court Brown, Buckton et al., 1965, 1967
Irradiation	H. J. Evans, 1974
Total body irradiation	Goh, 1968b, 1971b, 1973
Thorotrast	Haim, Dudley, et al., 1966
10 yr after fallout	Kumatori, Ishihara, et al., 1965
X-ray and ^{131}I	G. E. Moore and Sandberg, 1961; G. E. Moore, Ishihara, et al., 1963
X-ray (accidental exposure)	Popescu and Stephanescu, 1971
Natural radioactivity (in Brazil)	Barcinski, Abreu, et al., 1975
Atomic bomb	A. D. Bloom, Neriishi, et al., 1966, 1967
X-ray of gynecological conditions	Kawasaki, 1968c
^{131}I	Haglund, Lundell, et al., 1977
γ-Irradiation	Takumura, Sakurai, et al., 1970

This table lists only those publications that deal with the in vivo effects, for a tremendous number of such agents have been shown to affect chromosomes in vitro without necessarily having an effect in vivo. In choosing the agents, I have paid particular attention to those related to therapy or having an association with malignancy in the human. In my opinion, the in vivo results are more germane to neoplasia, though in vitro some agents can raise havoc cytogenetically in a laboratory, for example, mycoplasma infection (Romano, Comes, et al., 1970).

Table 17b Increased SCE frequency resulting from in vivo effects

Agent	Reference
Virus infections (Herpes, common cold)	Kurvink, Bloomfield, et al., 1978a
Small-pox vaccination	Knuutila, Mäki-Paakkanen, et al., 1978
Workers in chemical laboratories	Funes-Cravioto, Kolmodin-Hedman, et al., 1977
Smoking	Lambert, Lindblad, et al., 1978
Cytoxan, vincristine, and other antileukemic agents	Raposa, 1978
Adriamycin	Nevstad, 1978
8-Methoxypsoralen and UV	Mourelatos, Faed, et al., 1977
Chemotherapy for lung cancer (i.v. cytoxan, vincristine, methotrexate, BCNU)	Hollander, Tockman, et al., 1978
Therapy for CML?	Kakati, Abe, et al., 1978

(Zarrabhi, Rosner, et al., 1978) in patients with and without cancer (Bloomfield and Bruning, 1976; Bukowski, Weick, et al., 1977). In all cases so far described, X-ray and/or chemotherapy was given a few years before the appearance of leukemia. In our experience the application of radiotherapy to the abdomen appeared to be essential for the development of leukemia, and an analysis of cases published in the literature revealed that almost all of them had received abdominal irradiation for one condition or another. The chromosomal picture present in the cases of leukemia-complicating HD or other lymphomas does not differ qualitatively from that observed in other leukemias, including the presence of a Ph¹-chromosome in CML-complicating HD.[10] It will be of interest to establish whether patients with diseases other than HD and lymphoma develop leukemia when given the same amount of X-ray therapy to the abdomen and other areas of the body. Of course, there is always a possibility that the ionizing irradiation and/or chemotherapy further modify an agent or activate one in the patients with HD or lymphoma leading to the development of leukemia. Even though rare exceptions exist, it has been demonstrated that those patients with PV who develop the 20q− abnormality have received irradiation through ^{32}P administration in the past. In this case, the specificity and location of the chromosome involved are intriguing.

That ionizing irradiation leads to the development of leukemia and other neoplastic processes in a significant portion of an exposed population is most cogently illustrated by the high incidence of such diseases in individuals exposed to the A-bomb explosions in Hiroshima and Nagasaki.[11] For many years it was known that exposure to ionizing irradiation by radiologists,[12] and patients receiving X-ray therapy for ankylosing spondylitis or other diseases, was associated with a high incidence of leukemia in these subjects.[13] Because at least 50% of the patients with AL have no demonstrable cytogenetic changes in their leukemic cells, it will be of interest to establish with time the incidence of chromosomal changes in patients developing leukemia after irradiation, since almost all of them to date, though this is based on a small number of cases, have had karyotypic changes in their leukemic cells.

Noxious effects on chromosomes have been shown to be produced by "mild" ionizing irradiation, such as fluorescein applied by luminous dial painters,[14] in whom the incidence of karyotypic abnormalities appears to be related to the body content of radium, in patients receiving isotope therapy for thyroid disease, and even after diagnostic X-ray.[15]

Irradiation-induced neoplasia remains an area in which chromosomal analysis could be of considerable help in heralding or possibly predicting the susceptibility of affected individuals to the development of cancer or leukemia. Unfortunately, the cytogenetic criteria remain to be clearly and more definitely delineated.

Effects of Chemical Agents

A host of chemical substances, ranging from inorganic ions to organic drugs, to which man is exposed either accidentally or through daily use, has been shown to produce chromosomal damage in vitro and in vivo.[16] Most of the time the results are based on effects on cultured lymphocytes, but some of these agents produce chromosomal damage in marrow cells, when the latter are examined without any in vitro procedures. A number of these substances has been shown to be carcinogenic or leukemogenic in man, and others have been considered as potentially carcinogenic hazards.

Benzene Poisoning and Leukemia

Among the definitely carcinogenic substances shown to cause chromosomal damage is benzene.[17] In some of the subjects exposed to this chemical, chromosomal aberrations in cultured lymphocytes can be found some time before the development of leukemia, particularly during the period of severe anemia preceding the leukemia. When leukemia develops, the cytogenetic changes are not different than those observed either in AML or ALL, though AML and related leukemias tend to be much more frequent in these subjects than ALL (ca. 7:1). Again, as in the case of irradiation-induced chromosomal changes, we do not know the exact significance of the karyotypic changes in the cultured lymphocytes, for not all subjects who have these changes in their lymphocytes develop leukemia and neither do all the patients who succumb to AL necessarily have karyotypic aberrations in their cultured blood cells.

Cytogenetic findings were obtained in 10 leukemia patients (shoemakers in Turkey using adhesives containing benzene) with chronic exposure to benzene (Aksoy, 1977). Numerical changes were found in five (polyploidy in two, trisomy C in two, and monosomy C in one). Chromosome aberrations (acentric fragments, "achromatic lesions," dicentrics, breaks) were observed in nine of the cases. The author also mentioned a Ph^1-positive case of leukopenia with benzene exposure without signs of leukemia and stated that the chromosomal changes seen in the subjects with leukemia can also be observed in cases of chronic benzene poisoning without evidence of leukemia. The author further stated that in view of the great variation of chromosome aberrations, it is difficult to incriminate karyotypic defects as a cause of leukemia in benzene poisoning.

Forni and Moreo (1967) made serial studies on a 38-yr-old female who had been exposed occupationally to benzene vapor for over 20 yr. At the beginning of 1966 when the patient had an aplastic anemia, the cultured blood cells showed a high incidence of stable and unstable chromosome changes. A few months later cells with 47 chromosomes appeared in the marrow. At the end of 1966, AML supervened accompanied by the emergence of a 47,XX,+C clone in the marrow and blood.

Sellyei and Kelemen (1971) found a 47,XY,+D clone in a 32-yr-old male with subacute myeloid leukemia which developed 7 yr after benzene-induced pancytopenia. A pseudodiploid karyotype including a ring and a minute were observed in a similar patient with erythroleukemia (EL) (Forni and Moreo, 1969). A 35-yr-old patient who had been exposed to benzene since the age of 18 and ultimately developed CML with a possible Ph^1 has been reported (Liaudet and Combaz, 1973). Thus, benzene-induced leukemia has been shown to be associated with aneuploid clones which may be preceded by the presence of stable or unstable chromosome changes in the marrow. The possible value of chromosome studies on blood cells in the evaluation of marrow damage in subjects exposed to benzene has been discussed.[18] Age and genetic factors may play a role both in the chromosomal changes observed and in the nature of the leukemia, e.g., ALL in a man and AML in his paternal uncle, both of whom had been exposed to benzene.[19]

Two cases of interest have been reported by Verwilghen and Van Den Berghe (1977). Two males developed leukocytosis (>100.00 leukocytes/mm^3) and a marked shift to the left after exposure to organic solvents, apparently containing benzene. The first patient developed a 46,XY,t(9p+;10p−) pattern in 100% of the marrow and unstimulated blood cells. The leukemic blood and marrow picture gradually disappeared and normal cells partially replaced the abnormal clone in the marrow. After a spontaneous remission of 5 mo, the abnormal clone with additional karyotypic changes reappeared and ultimately replaced all the normal cells. The patient developed frank AML and died within a few months. In the second patient, a 45,XY,−7 pattern was consistently found in 100% of the marrow and unstimulated blood cells. These abnormal cells disappeared spontaneously from blood and marrow. The patient died from an infectious complication during this regenerative period. The authors postulated that the spontaneous rejection of abnormal leukemia-like cell clones with temporary regeneration of normal hematopoiesis in these two patients with benzene poisoning suggests an immunological mechanism. Marked eosinophilia in one patient and plasmacytosis in the other tend to support this contention.

Only the cases reported by Verwilghen and Van Den Berghe (1977) have been studied with banding (R and Q). However, a number of cases of benzene-associated hematological disorders and leukemia have been studied cytogenetically, including cases of EL, CML with ?Ph[1], and aplastic anemias.[20] The clear demonstration of a Ph[1] with a 9;22 translocation or other structural, nonrandom chromosome anomalies in benzene-induced leukemias would be of crucial importance to our better understanding of the significance of these cytogenetic anomalies in these malignant processes.

Other chemical substances, such as arsenic, lead, and many others, have been shown to produce changes similar to those observed after benzene poisoning[21] and some have been implicated in the genesis of cancer and leukemia; they deserve the special attention of cytogeneticists and pathologists. These changes, which have usually been described in cultured lymphocytes, consist of breaks, gaps, deletions, rearrangements, the production of acentric fragments, and a number of other morphologic changes in the chromosomes.

Generally, toxic agents, be they physical or chemical, that lead to the development of leukemia cause initial and severe depression of marrow function, following which the proliferative stage sets in with a possibility of a leukemic picture. There is little doubt that our natural environment has undergone and is continuing to undergo rapid changes, including the exposure of populations to a remarkable variety of chemical substances, drugs, industrial products, food additives, and radiation of one type or another, all of which bring new problems in their wake. Not infrequently, it has been found that such substances are linked with specific or nonspecific dangers to health, including cytogenetic effects. Karyotypic analysis of cells for chromosomal changes could

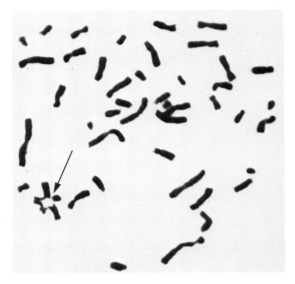

49 Metaphase containing abnormal chromosomes from the cultured blood cell of a patient with AL. The chromosomal changes were due to chemotherapy, which affected only the lymphocytes of the blood and not the cells in the marrow. The arrow points to a very abnormal unit formed by exchanges between chromosomes.

possibly indicate the carcinogenic potentiality of these substances.

Many, if not all, chemotherapeutic agents used in cancer or leukemia have been shown to produce chromosomal damage either in vitro or in vivo (Figures 49 and 50).[22] It is probable that some of the clinical effects of these drugs are mediated through such chromosomal effects and, hence, their significance in this regard is of a different quality than those karyotypic changes taken up in the previous paragraphs. The untoward effect of chemotherapeutic agents are best determined in cultured lymphocytes, in which cells such aberrations may persist for a long period of time (years). I have seen similar changes in marrow cells which, however, are short lived and disappear after a relatively short period of time (weeks?). I am of the opinion that chemotherapy almost never leads to changes in the established leukemic karyotype, and I have not seen karyotypic alterations of a relatively permanent nature that could be ascribed to chemotherapy. It is possible that in the future we will have to become

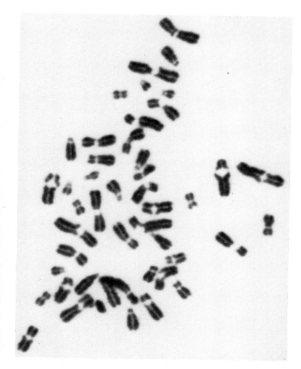

50 Metaphase of an uncultured marrow cell from a patient with AML with "ballooning" of the centromeric regions apparently due to Ara-C therapy. It should be pointed out that the chromosomal changes shown in this figure can be produced by a number of physical and chemical agents and by certain vitamin (B_{12}, folic acid) deficiencies.

diotherapy. Nevertheless, the development of AL in patients treated with chemotherapeutic agents for malignant and nonmalignant diseases indicates the serious potential of this problem.

Effects of Viruses

In writing the following section on the chromosomal effects of viruses,[23] I have relied heavily on the critical review by Harnden (1974a), in whose publication the reader will find the bulk of appropriate references. Space and other limitations of this book preclude doing justice to the large field of cytogenetics and viruses and the relation of viruses to the induction of leukemia and cancer in human subjects and animals. The studies in the latter have been numerous, and the reader may wish to consult the more comprehensive and extensive publications in this area.

There is a possibility that viruses may play a role in the genesis of human neoplasia.[24] Viruses are capable of producing a number of chromosomal changes, including breakage of chromosomes, pulverization, possibly numerical changes and, more importantly, specific alterations in certain chromosomes (Table 18). An example of the latter is the relation of a heterochromatic region in a group C autosome (possibly #10) to the presence of a virus associated with Burkitt lymphoma (BL).[25] It should be pointed out, however, that not all of the materials examined have been found to contain this karyotypic anomaly in BL[26] and, thus, it does not appear to be an essential prerequisite for the genesis of this disease. As a matter of fact, most of the cases with BL are accompanied either by a diploid chromosomal picture or abnormalities of #14[27] in the tumors, and no

more concerned with the long-term effects of chemotherapeutic agents on normal chromosomal integrity, but at the moment there are more pressing facets connected with these agents then their effects on chromosomes. Possibly, this also applies to various forms of ra-

Table 18 Reported chromosome changes in infectious and other conditions

Condition	Reference
Hepatitis	Aya and Makino, 1966
	Aya, Makino, et al., 1966
	Matsaniotis, Kiossoglou, et al., 1966b
Ulcerative colitis	Konstantinowa and Bratanowa, 1969
	Emerit, Emerit, et al., 1972
	Xavier, Prolla, et al., 1974
Infectious mononucleosis	J. Philip, Hansen, et al., 1975; Watt, Hamilton, et al., 1977
Aseptic meningitis	Makino, Yamada, et al., 1965

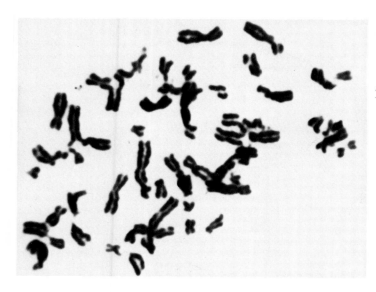

51 Marrow metaphase from a male patient with ALL who developed measles during his disease and went into a prolonged remission of the leukemia. For a year or more abnormalities shown in the figure were commonly observed both in marrow and blood cells. Note the complex exchanges, broken chromosomes, and some ascentric units.

consistent karyotypic anomaly has been established for *all* cases with BL. An intriguing finding has been the demonstration of SV40-related antigens in human meningiomas with monosomy #22 (A. F. Weiss, Portmann, et al., 1975).

Infections with measles (Figure 51), chicken pox, infectious mononucleosis, and other viruses have been shown to result in the presence and persistence of chromosomal anomalies, i.e., primarily breaks, gaps, deletions, and pulverizations in cultured lymphocytes. These changes may persist for a relatively long period of time. Because there is evidence that some oncogenic viruses may, in fact, become incorporated into the genome of the host and produce relatively specific chromosomal changes, future studies will have to demonstrate the reproducibility and specificity of such changes.

Some of the changes observed in cultured lymphocytes as a result of exposure to viruses may be seen in cervical lesions of the uterus. Because no definitive evidence exists as to whether a virus resident within the cervical tissue plays a role in the development of cervical neoplasia, it is difficult, at present, to evaluate the significance of the association, even if the virus were responsible for some of the cytogenetic changes observed in cervical cells.

Until the exact relationship between viruses and the genesis of human cancer and leukemia is established, it is premature to ascribe any basic significance to chromosomal changes produced by these agents, however specific they may be. Even when such a relationship will be ascertained, the various karyotypic changes produced by viruses will have to be more specifically and clearly defined and the changes resulting from immunologic parameters separated from those due to the virion per se. In addition, the cytogenetic changes induced by viruses leading to neoplasia will have to be clearly separated from those due to viruses not implicated in cancer and leukemia.

All the substances and conditions discussed in this section of the book, be they directly or inferentially accepted as possible causes of human neoplasia, have not been shown to cause the type of chromosomal changes usually associated with either cancer or leukemia. It is possible that if these agents do cause neoplasia, they do so by changing the genetic material at the submicroscopic level, the chromosomal changes preceding, and especially those succeeding, the appearance of neoplasia being merely an expression of altered metabolic conditions resulting in a "phenotypic" change of the karyotype.

The chromosomal effects of some of the viruses implicated in possible oncogenesis in the human and animals will be briefly summarized.

DNA Viruses

Adenoviruses

All members of the adenovirus group, and there are more than 30 different serotypes of human adenoviruses and many other of nonhuman origin, have the potential for transformation of some cell types, though attempts to demonstrate that adenoviruses are involved in naturally occurring neoplasms have so far proved unsuccessful. Some of the human adenoviruses have been shown to have a low to high degree of oncogenecity in hamsters. The effects of adenovirus type 12 on human cells have been demonstrated in PHA-stimulated lymphocytes and cultured fibroblasts, in which cells the infection led to breaks and gaps of apparent random nature.[28] However, more careful studies have revealed a frequent involvement of chromosomes #1[29] and #17, in addition to the random changes observed in a human amnion line (AV3), human embryo kidney (HEK) cells, and normal and abnormal human dermal fibroblasts. All the aberrations produced by adenovirus 12 were either chromatid or chromosome gaps and breaks, with a few chromatid interchanges. No dicentrics or abnormal chromosomes have been observed.

Cells from patients with Fanconi's anemia (FA) survived for several weeks after infection with adenovirus 12; some specific chromosome damage could be demonstrated throughout this period of survival. However, there is no evidence, at present, to suggest that either the random changes or the specific damage associated with the adenovirus 12 infection of human cells are in any way connected with cellular transformation, tumor production, or even with the integration of the virus genome into the cell. Adenovirus 18 has been shown to be highly oncogenic and adenovirus 7 weakly oncogenic in hamsters; adenoviruses 2 and 4 are incapable of producing tumors in these animals. It appears that the capacity to induce chromosome damage early after infection is not associated with the oncogenic potential of these viruses in hamsters. It has also been shown that, whereas human adenovirus types 2, 7, 18, and 31 all induce random chromosome damage in HEK cells at a low multiplicity of infection, adenovirus 18 and 31 also cause a specific lesion on chromosome #17 similar to that induced by adenovirus 12. These three viruses (adenovirus 12, 18, and 31) constitute the highly oncogenic group. It is probable that induction of chromosome damage is associated with an early event after infection, as ultraviolet (UV)-impaired virus can still cause chromosome damage.

Herpes Viruses

The herpes group of viruses have been associated with the occurrence of neoplasms, even though in some instances the association is tenuous. A relationship has been suggested between herpes simplex virus (HSV) and lip carcinoma in man but this association has not been confirmed; there is at least one report of cellular transformation by HSV types 1 and 2.[30] A seroepidemiological relationship has been demonstrated between HSV type 2 and carcinoma of the cervix. It is accepted by many workers that the association between Epstein-Barr (EB) virus and BL may be a causal relationship, and there seems to be little doubt that the herpes virus is associated with induction of tumors in a number of different animals. Hence, there is good justification for considering as relevant to tumor genesis the chromosome damage caused by all members of the herpes virus group, even though some, such as herpes zoster virus, have not been associated with malignancy.

HSV type 1. The first demonstration that viruses could cause damage to chromosomes was related to HSV type 1 in which gaps, breaks, chromatid exchanges, and possibly dicentric chromosomes with a tendency for breakage to occur at the centromeric and secondary constriction areas of chromosomes #1 and #2 in an established train of Chinese hamster cells were demonstrated. Similar changes were described for the established human cell lines HeLa and HEP-2, the latter originating from a cancer of the larynx. In contrast to the results with adenoviruses, inactivated HSV type 1 did not cause chromosomal damage. The effects on other human cell lines have varied from a high frequency of chromatid gaps, breaks, and pulverizations to only a small increase in these chromosomal abnormalities;

however, a striking accentuation of the secondary constrictions of chromosomes #1, #9, and #16 in HEK cells and a large excess of chromatid aberrations 12 hr after HSV infection have been described. In addition, a few dicentrics were found and there was said to be some specificity of breakage at the midpoints of group D and B chromosomes, at the middle and proximal segments of the long arm of #2 and near the middle of the long arm of a group C chromosome.

HSV type 2. Relatively few observations have been made with this virus, except for some increase in breaks and accentuation of the secondary constrictions in HEL and HEK cells, cell lines originating from human embryonic lung and kidney, respectively, and cultured blood leukocytes. A small number of pulverized cells were observed in carcinoma of the cervix; it is surprising that no observations have been made with this virus in cultures of cervical cells or other human cells.

Cytomegalovirus. No chromosomal abnormalities in human embryo fibroblasts or in blood cultures infected with cytomegalovirus have been described.

Herpes zoster virus. Though the results with chicken pox virus have not been consistent, there appears to be little doubt that the virus can cause chromosome breaks and gaps in cultured HEL cells. Overcontraction of the chromosomes occurred and this was sometimes accompanied by apparent fragmentation.

EB virus. The changes induced by this virus in established cell cultures or those seen in cells originating from patients subsequent to their establishment in vitro have not been consistent. Thus, in cell cultures obtained from lymphoid tissue chromosome exchanges leading to the formation of marker chromosomes, pseudodiploidy, and an occasional dicentric chromosome are the most common, but gaps and breaks are occasionally seen. Later in the evolution of the cultures, the cells tend to become polyploid and at this stage marker chromosomes become more common. Marker chromosomes similar in appearance have been seen in several different cell lines, but most recent evidence suggests that this is a morphological similarity only, in that similar markers are not derived from identical rearrangements. The description of an extra band at the end of the long arm of chromosome #14 in cultured cells and biopsies from BL cases does give the strong impression that the exchange points are not random. Even though workers have differed on the occurrence and nonoccurrence of certain chromosomal changes in established lines, a near terminal constriction on the long arm of a group C chromosome in some of the established lines from BL, probably chromosome #7, has been described. Similar findings have been described in cell lines obtained from patients with leukemia or infectious mononucleosis. The induction of such a change in normal cells by co-cultivation with irradiated virus-containing cells has been described, though it has not been accepted by all workers.

In a study utilizing hybrid clones derived from fused human lymphoblastoid cells and unique mouse fibroblastic cells grown under special conditions, the findings suggested that the resident EB virus genome is closely associated with chromosome #14 and the presence of this particular chromosome is sufficient for the maintenance and expression of the EB virus genetic infection in human lymphoblastoid cells (Yamamoto, Mizuno, et al., 1978).

Papovaviruses. This group includes the papilloma viruses, polyoma virus, and simian vacuolating virus 40 (SV40). The papilloma viruses are associated with the production of benign tumors in several species, including man. Unfortunately, no cytogenetic studies on cells infected with human wart virus or with papovaviruses have appeared. Apparently the papilloma virus does produce chromosomal changes in rabbit tumors, which are considered to be benign papillomas.

Polyoma virus. The tumors induced in various animals with this virus have generally been shown to contain chromosomal changes both in morphology and chromosome number. However, a number of tumors have been described in which the established tumors were shown to be diploid, though the majority were aneuploid, containing marker chromosomes.

SV40. The effects of this virus on established human cell lines has not been consistent, depending on the origin and age of the line when exposed to the virus. However, the occurrence of gaps, breaks, dicentric chromosomes, and chromatid interchanges appeared to be relatively common with enhancement of secondary constrictions and loss of chromosomes. Apparently, there was no correlation between the chromosomal changes and the transformation of the cell lines by the virus. It was found, however, that cultures of human fibroblasts have become aneuploid, and cells from patients with genetically determined syndromes with chromosomal instability are more readily transformed by SV40 virus.

When fully transformed human cells were studied, chromosome changes were found in almost all the cells, with a definite tendency toward polyploidy. Some chromosome and chromatid breakage occurred, and abnormal chromosomes including dicentrics were common. Such instability of the karyotype did not usually lead to a selection of stable clones of karyotypically marked cells, though transient clones were sometimes found; the heteroploidy and chromosomal instability may persist. Some evidence of specificity for the chromosome aberration has been reported, particularly loss of chromosomes in groups B, C, and D. An association of break points and secondary constriction regions in chromosomes #9 and #16 has been noted.

Assignment of a gene(s) for cell transformation and malignancy to #7 in one human cell line and to #17 as the integration site for the SV40 virus in another line has been reported by Croce (1976, 1977). The presence of SV40-like antigens in human meningiomas with monosomy #22 has been alluded to previously (A. F. Weiss, Portmann, et al., 1975).

Poxviruses. This group includes several members that produce benign tumors, including the molluscum contagiosum virus of man. Observations on human cells or tumors infected with these viruses have been very scanty. Thus, a high incidence of breakage in lymphocytes has been shown in two out of eight children after vaccination with vaccinia virus. A report has appeared indicating the chromosomes to be normal in the marrow of four boys and eight girls between the 9th and 24th day after successful vaccination with vaccinia virus.[31] In vitro infection of lymphocytes, however, resulted in complete disintegration of the chromosomes, presumably something akin to the "pulverization" seen with many RNA viruses.

RNA Viruses

Paramyxoviruses

Measles virus. There have been many reports of chromosome breakage by measle virus in vitro and in vivo,[32] the damage consisting primarily of simple chromatid and chromosome breakage and multiple fragmentation of the chromosomes, a phenomenon that has come to be known as "pulverization." There is no evidence of either unstable or stable chromosome rearrangements after infection with measle virus. Even though the biological *phenomena* of chromosome damage induced by measle virus are undoubtedly of interest, it is difficult to envision any role for such damage in the induction and/or progression of neoplastic disease, as it seems probable that the affected cells are not viable.

Mump virus. Changes similar to those observed in measles have been described in lymphocytes of patients with mumps.

Sendai virus. The changes induced by this virus are probably related more to its ability to induce cell fusion than to direct effects on the chromosomes.

Myxoviruses, arboviruses, picornaviruses, and unclassified viruses. This group of viruses, which includes yellow fever, polio, hepatitis, and rubella viruses, have been shown to produce chromosome breakage in lymphocytes of patients with these infections. "Pulverization" in cultured lymphocytes has also been observed. The results are usually inconsistent from one patient to another.

Oncornaviruses

Avian leukemia-sarcoma complex. This large group of morphologically identical (C-type) RNA viruses can be subdivided according to biological and serological characteristics; however, a distinction between oncogenic and nononcogenic members cannot be made in any meaningful manner, because even viruses of proven oncogenicity may behave in a nononcogenic manner in certain cell systems. Furthermore, there is no clear-cut evidence that any RNA viruses are involved in the transformation of cells or the production of malignant tumors in any species, with the exception of the C-type viruses of the leukemia-sarcoma complex and the mammary tumor viruses. The tumors induced by the avian leukemia-sarcoma complex have been described to be both chromosomally normal and aneuploid.

Murine leukemia-sarcoma complex. No evidence of cytogenetic changes in the preleukemic phase of virus-inoculated mice has been found, and the abnormalities in chromosome number appearing in the leukemic phase may be a consequence of the neoplastic transformation. Chromosomally abnormal cells have been found in transplantable Moloney virus-induced leukemia and in Friend virus and Rauscher leukemias.

Chromosomal Changes upon Exposure to X-Ray or to Transformation by Certain Viruses

Certain cells undergo transformation into neoplastic ones upon exposure to oncogenic viruses[33] or develop chromosomal abnormalities in excess of those shown by normal cells after exposure to X-ray in vitro.[34] Thus, it has been shown that cells from patients and their heterozygous relatives with some conditions in which the development of neoplasia is more frequent than in the general populations, e.g., DS and FA, show a much higher frequency of cell transformation in vitro when exposed to SV40 virus. In addition, cells from subjects with an extra X-chromosome (Klinefelter's syndrome with an XXY consitution) were shown to undergo a marked malignant transformation by SV40 virus when compared to normal XY cells or to the XY cells present in the tissues of the affected individuals.

The findings with SV40 transformation in cells of Bloom's syndrome (BS) or FA (Table 19) suggest that chromosomal anomalies observed in vitro may be one of the heralding

Table 19 Frequency of cell transformation in vitro by SV40*

Source of cells	Number of transformed colonies‡
Controls (7)	1.6 – 5.1
Franconi's anemia	
Homozygotes	
Patient 1	79.7 ± 18.1
Patient 2	41.1 ± 12.1
Heterozygotes	
Patient 3	20.1 ± 3.2
Patient 4	28.2 ± 8.7

*Based on the data of Todaro, Green, et al. (1966).
‡10,000 cells plated.

signals of abnormal cellular behavior which can predispose to malignant transformation. Hence, it may be possible to utilize a karyotypic aberration which occurs in the cells of persons from higher susceptibility families or those exposed to environmental agents as indicators of the carcinogenic hazards to these individuals.

A simple, though possibly less reliable approach, in ascertaining the susceptibility of individuals to neoplasia is to study certain chromosomal aberrations after exposure of cells to X-ray in vitro (Table 20). A number of studies cited above have appeared indicating that what was found with SV40 virus transformation is pretty much paralleled by the findings in cells after exposure to X-ray. Thus, the number of chromosomal aberrations in the cells of subjects with DS, FA, BS, and other diseases is much higher than that in control cells. We have investigated the frequency of such chromosomal changes in the irradiated cells of subjects with sex-chromosome anomalies, because the data with SV40 virus transformation indicated that an unbalanced chromosome constitution increases the cellular susceptibility to transformation. The results of our studies are shown in Table 20.

A comparative study has been made of fibroblasts obtained from patients with differing susceptibilities to malignant disease (FA, ataxia telangiectasia [AT] and BS) and a family with neurofibromatosis with respect to their transformation with SV40 virus. Despite confirmation of earlier findings that FA fibroblasts have an increased transformation rate with the

Table 20 Chromosomal changes (dicentrics, rings, acentrics) induced by X-ray in cultured cells

Chromosome constitution of subjects	Amount of change*	Reference
XX or XY	+	Higurashi and Conen, 1972; Sandberg and Sakurai, 1973a; M. S. Sasaki, Tonomura, et al., 1970
XXY	+++	Sandberg and Sakurai, 1973a; M. S. Sasaki, Tonomura, et al., 1970
XXX	+++	Sandberg and Sakurai, 1973a
XYY	++	Sandberg and Sakurai, 1973a
G-trisomy	++++	Higurashi and Conen, 1972; Sandberg and Sakurai, 1973a; M. S. Sasaki, Tonomura, et al., 1970
D-trisomy	++++	Higurashi and Conen, 1972; M. S. Sasaki, Tonomura, et al., 1970
E-trisomy	++++	M. S. Sasaki, Tonomura, et al., 1970
XO	+	Sandberg and Sakurai, 1973a; M. S. Sasaki, Tonomura, et al., 1970
Deletions (B,E)	+	M. S. Sasaki, Tonomura, et al., 1970
	+++	Higurashi and Conen, 1972

*The changes shown represent deviations from the findings with control (nonirradiated) cells.

virus, fibroblasts from two BS patients were found not to have a raised transformation rate, and no correlation was found between chromosome abnormalities per se and transformation (Webb and Harding, 1977). Of two cell types with greatly increased rates of transformation, one was derived from a neurofibromatosis patient and the other from an AT heterozygote.

The results of Kaplan, Zamansky, et al. (1978) indicate that for SV40-transformed cells there is some correlation between induction of SCE and that of an infectious virus, which could suggest a special relationship between SV40 virus and some chromosomal event. In another paper (S. Wolff, Rodin, et al., 1977) dealing with an increased rate of SCE in a xeroderma pigmentosum (XP) cell line transformed by SV40 virus, caused by UV-like and X-ray-like chemicals, the role played by the virus in increasing the SCE rate cannot be ignored, even though the authors did so.

Survey of Cancer Patients for Chromosomal Anomalies

Even though the approaches in establishing constitutional aberrations in cells, usually blood lymphocytes, of patients with cancer, lymphoma, and leukemia vary from laboratory to laboratory and involve facets which may affect the interpretation or even the results themselves, it was thought advisable to pool data on large groups of patients[35] in order to obtain an inkling regarding this parameter, i.e., the incidence of constitutional chromosomal anomalies with cancer or leukemia.

The autosomal and gonosomal anomalies in patients with neoplastic disease are summarized in Tables 21–23. The frequency of these anomalies is in some respects slightly higher than would be expected, but the figures are so small and variations between individuals so great that it would not be wise to attach too much significance to these observations at present. But they do suggest that a continuation of these surveys is necessary. Thus, in 228 women with breast cancer only 4 abnormal chromosomal constitutions were found,[36] only 1 XXY in 50 patients with testicular tumors, and 9 in 730 patients with leukemia or lymphoma. The frequency of variants of normal chromosomes (elongated short arms of chromosomes in groups G and D, long Y, large #1 or #16) does not appear to be much different in patients with cancer or leukemia than that found in surveys of general populations.

The reporting of single cases of patients with constitutional chromosomal defects with either cancer or leukemia tends to distort the

Table 21 Incidence of autosomal abnormalities in control and cancer subjects

Groups studied	Incidence of abnormalities (%)	Total number of subjects studied
Incidence of autosomal aberrations (exclusive of trisomies) in control population	0.4	3,045
Incidence of autosomal aberrations (exclusive of trisomies) in cancer patients	0.6	2,624
D-D translocation in control population	0.13	3,045
D-D translocation in cancer patients	0.11	2,624

Table 22 Type of constitutional chromosomal anomalies in 4,543 cancer patients

Sex chromosome aberrations	10 (?12)
Autosomal translocations (6 D-D)	15 (9 D-D)
Marker chromosomes	3
Autosomal deletion	1
Autosomal inversion	1
Total	30

Table 23 Incidence of constitutional chromosomal anomalies in cancer and leukemia patients

34 anomalies in 4,543 patients — 0.75%	Reference
1,149 cancer patients — 8 anomalies	Harnden, Langlands, et al., 1969
356 cancer and leukemia patients — 6 anomalies	Berger, 1971
100 leukemia and lymphoma patients — 2 anomalies	Prigogina, Stavroskaja, et al., 1970
1,919 cancer patients — 12 anomalies	Harnden, quoted by Koller, 1972
1,019 cancer patients — 6 anomalies	Sandberg and Sakurai, 1973a

actual incidence of these conditions, as borne out by the surveys cited above. Nevertheless, more data are necessary on the possible protective or enhancing effect autosomal or sex-chromosomal anomalies have on the development or progression of neoplasia in man. The presence of certain cytogenetic abnormalities in families (translocations) with a high incidence of cancer or leukemia[37] may or may not play a role in the genesis of the neoplasia, as familial predisposition to cancer occurs without any familial karyotypic abnormalities; but, they do afford an opportunity to evaluate the role of chromosomal changes in the development of cancer and leukemia.

There are some conditions known to be associated with neoplasia in which chromosome studies have not been extensively performed or have given inconsistent results. Among these is the Chediak-Higashi syndrome, which is apparently inherited as an autosomal recessive trait and is lethal in the homozygous state. It is characterized by partial albinism, recurrent infections, a distinctive leukocyte anomaly (giant cytoplasmic granules that are probably lysosomal in nature) and hepatosplenomegaly. Patients tend to develop lymphoreticular malignancy. In one patient, a 3-yr-old boy, a mosaic condition was found in cultured lymphocytes, about half the metaphases being diploid while a G-group chromosome (apparently not the Y) was either missing or replaced by a ring chromosome in most of the remainder. In another study, chromatid and chromosome breaks were found. The same authors later suggested (Rozenszajn, Radnai, et al., 1969; Rosenszajn and Radnay, 1970; Say, Tuncbilek, et al., 1970) that a defective lysosomal membrane might allow DNase to diffuse into the cytoplasm and from there into the nucleus where it might cause the observed chromosomal damage and perhaps neoplastic transformation.

The relation of other syndromes with chromosomal changes, e.g., Blackfan-Diamond syndrome (Amarose, Tartaglia, et al., 1965), to neoplasia is at present unknown. A case of AML has been reported in a 31-yr-old female with this disease (Wasser, Yolken, et al., 1978).

References

1. Koller, 1972.
2. Witkowski and Anger, 1976.
3. Amarose, Plotz, et al., 1967; Awa and Bloom, 1967; Barnes, Holmes, et al., 1969; Bender, 1964; A. D. Bloom, Nakagome, et al., 1970; J. T. Boyd, Court Brown, et al., 1966; J. K. Brown and McNeill, 1969; Buckton, Jacobs, et al., 1962a; Hennekeuser, Citoler, et al., 1970; Horvat, 1975; Tomonaga, Ichimaru, et al., 1967.
4. Cantolino, Schmickel, et al., 1966; Doida, Hoke, et al., 1971; E. Engel, Flexner, et al., 1964; Ishihara and Kumatori, 1967; Kamada, 1969a,b; Macdiarmid, 1965; R. Moore, 1965; S. Warren and Meisner, 1965.
5. Ishihara and Kumatori, 1967.
6. Conen, Erkman, et al., 1966; E. B. Lewis, 1963; A. Stewart, Pennypacker, et al., 1962; Uchino, 1968.
7. A. D. Bloom, Nakagome, et al., 1970; Ezdinli, Sokal, et al., 1969.
8. Sandberg and Hossfeld, 1970.
9. Canellos, Devita, et al., 1975; Durant and Tassoni, 1967; Ezdinli, Sokal, et al., 1969; Raich, Carr, et al., 1975; Rowley, Golomb, et al., 1977b; Steinberg, Geary, et al., 1970.
10. Ezdinli, Sokal, et al., 1969.
11. Brile, Tomanaga, et al., 1962; Tomonaga, Ichimaru, et al., 1967; Watanabe, 1964.
12. E. B. Lewis, 1963.
13. Sandberg, 1966b, 1974; Sandberg and Hossfeld, 1973.
14. J. T. Boyd, Court Brown, et al., 1966.
15. F. H. Adams, Norman, et al., 1978; Sandberg and Hossfeld, 1973.
16. Bishun, 1971; Nakamuro, Yoshikawa, et al., 1976; Newsome and Singh, 1977; M. W. Shaw, 1970; Stahl and Luciani, 1970; Stebbings, 1976.
17. Forni, 1966; Forni and Moreo 1967; Forni, Baroni, et al., 1971; Forni, Pacifico, et al., 1971; Haberlandt, 1971; Hartwich and Schricker, 1969; Hartwich and Schwanitz, 1972; Hartwich, Schwanitz, et al., 1969; H. Khan and Khan, 1973; Kissling and Speck, 1972; Sellyei and Kelemen, 1971; Tough, Smith, et al., 1970; Vigliani and Forni, 1969.
18. Forni, Baroni, et al., 1971; Forni, Pacifico, et al., 1971; Hartwich and Schwanitz, 1972; H. Khan and Khan, 1973; Tough and Court Brown, 1965.
19. Aksoy, Erdem, et al., 1974.
20. Erdogan and Aksoy, 1973; Forni and Moreo, 1967, 1969; Hartwich, Schwanitz, et al., 1969; Liaudet and Combaz, 1973; Pollini and Colombi, 1964a, b.
21. Nordenson, Beckman, et al., 1978; Schwanitz, Lehnert, et al., 1970; M. W. Shaw, 1970; Sherfving, Hansson, et al., 1970; Verschaeve, Kirsch-Volders, et al., 1978.
22. Bishun, 1971; Conen and Lansky, 1961; Elves, Buttoo, et al., 1963; Kaung and Swartzendruber, 1966; Krogh Jensen, 1967a; Pedersen, 1964b; Stevenson, Bedford, et al., 1971a, b; Zara, Fraccaro, et al., 1966.
23. Harnden, 1974a; Montaldo and Zucca, 1970; Nichols, 1963, 1970, 1972a, 1973.
24. Awano, Toshima, et al., 1961; Gallo, 1977.
25. M. A. Epstein, Achong, et al., 1964; Kohn, Mellmann, et al., 1967; Nichols, 1969.
26. Huang, Minowada, et al., 1970; Toshima, Takagi, et al., 1967.
27. Kaiser-McCaw, Epstein, et al., 1977a; Zech, Haglund, et al., 1976.
28. H. F. Stich, 1973.
29. McDougall, 1971; Steffensen, Szabo, et al., 1976.
30. Darai, Braun, et al., 1977.
31. Frolov, Slysuarev, et al., 1975.
32. Nichols, 1963, 1973.
33. R. W. Miller and Todaro, 1969; Mukerjee, Trujillo, et al., 1971; Todaro, Green, et al., 1966; Todaro and Martin, 1967; Webb and Harding, 1977; Webb, Harnden, et al., 1977.
34. Higuraishi and Conen 1971, 1972; Sandberg and Sakurai, 1973a, 1976; Sasaki and Tonomura, 1969; Sasaki, Tonomura, et al., 1970.
35. Berger, 1971; Dumars, Kitzmiller, et al., 1967; Fraumeni, 1969; Harnden, Langlands, et al., 1969; O'Riordan, Langlands, et al., 1972; Prigogina, Stavrovskaja, et al., 1970; Sandberg and Sakurai, 1973a, 1976.
36. Harnden, Langlands, et al., 1969; Harnden, 1970.
37. Bottomley, Trainer, et al., 1971; R. W. Miller, 1968, 1970.

8

Chromosome Breakage Syndromes

Fanconi's Anemia (Congenital Pancytopenia)
Early Chromosome Studies of Fanconi's Anemia
Recent Studies of Fanconi's Anemia
Acute Leukemia in Fanconi's Anemia

Bloom's Syndrome
Studies of Bloom's Syndrome Cells in My Laboratory
DNA Replication in Bloom's Syndrome

Louis-Barr Syndrome (Ataxia Telangiectasia Syndrome)

Xeroderma Pigmentosum
Studies of DNA Defects in Xeroderma Pigmentosum Cells

New Chromosome Instability Syndromes
Incontinentia Pigmenti (Bloch-Sulzberger Syndrome)
Scleroderma (Progressive Systemic Sclerosis)
Porokeratosis of Mibelli
Glutathione Reductase Deficiency Anemia
Kostmann's Agranulocytosis
Basal Cell Nevus Syndrome

According to German (1969b) the term "chromosomal breakage syndrome" was introduced by McKusick and comprises a number of rare but distinct clinical entities, i.e., Fanconi's anemia (FA), Bloom's syndrome (BS), Louis-Barr syndrome (ataxia telangiectasia, AT), xeroderma pigmentosum (XP), Kostmann's agranulocytosis, and glutathione reductase deficiency anemia. These cases have in common an autosomal recessive or dominant transmission, an increased tendency to develop malignancies, and a more or less pronounced disposition to spontaneous chromosomal breakage of cells in vitro while retaining a diploid set of chromosomes (Tables 24 and 25). The blood cells of subjects with AT, BS, or FA are significantly more radiosensitive than those of controls, particularly the occurrence of dicentrics and rings (Higurashi and Conen, 1973). Reviews dealing with these syndromes, particularly regarding their implication in neoplasia, have been written by Schroeder and Kurth (1971) and German (1972a, b, 1973). I shall discuss FA, BS, XP, and AT in some detail, but other diseases belonging to this syndrome, e.g., Kostmann's agranulocytosis, only briefly because they are too poorly documented (Matsaniotis, Kiossoglou, et al., 1966a) for more extensive comment.

The significance of the chromosomal changes to be discussed in this section, particularly as they relate to the development of cancer, lymphoma, or leukemia[1] in some of these disorders, is difficult to evaluate convincingly and reliably as a result of some disturbing and confusing observations. Thus, not all clinical conditions in which "breakage" of chromosomes occurs have a propensity to develop neoplasia at a higher rate than would normally be expected. Furthermore, chromosome breakage has been observed in diseases due to infectious agents, toxic substances, and in congenital disorders in which neither cancer nor leukemia is considered to be a significant risk. In most of the conditions to be discussed in this section, the cytogenetic changes have been demonstrated in cultured blood lymphocytes and in the few cases in which the vivo status has been evaluated by examining the chromosomes in the bone marrow, little deviation from normal has been found in most cases. It is possible that the chromosomal changes in "breakage syndromes" and in other conditions in which cytogenetically similar findings have been seen are but a limited expression of a number of different disorders, some of which may show the full clinical and metabolic syndrome without the cytogenetic abnormalties, e.g., the "variant" type of AT.

The human genome is a fragile thing, tending to reflect this fragility in chromosome or chromatid gaps, breaks, and rearrangements[2] known as "breakage," due to a large variety of external and internal agents and conditions. The tendency in some of these conditions to develop cancer or leukemia, though it may be related to chromosomal changes resulting from "breakage," is probably more likely related to gene mutation resulting from a basic hereditary defect. Thus, the chromosomal changes observed (gaps, breaks, etc.) may be but another phenotypic expression of the basic genetic defect manifesting itself in altered and unstable chromosome abnormalities.

The chromosomal breakage syndromes, though rare in incidence, are of direct interest to the aims of this book, because some of them have a distinctly higher predisposition to cancer or leukemia development than observed in other conditions or the general population. This higher incidence of neoplasia apparently applies also to family members of affected individuals. These conditions include BS, FA, AT, and XP. Other conditions developing chromosomal breakage, e.g., Kostmann's agranulocytosis or glutathione reductase deficiency anemia, have not been shown to be definitely predisposing to a higher incidence of neoplasia. More importantly, more and more conditions with chromosomal changes are being described, particulary in cultured blood lymphocytes, very similar to those observed in the syndromes enumerated above. In some of these conditions, though, the number of cases reported and studied cytogenetically has been small and, hence, the conclusions to be drawn from these cases should be very guarded. Furthermore, the in vivo status of the chromosomes, as revealed by direct examination of the bone marrow cells, is also lacking, as is more detailed information on the clinical picture (e.g., recent viral or other infections, drug administration, etc.).

Almost all of the conditions to be considered are transmitted as autosomal recessive traits. It is possible that one of the "phenotypic" manifestations of these syndromes is the

Table 24 Chromosomal breakage syndromes

Syndrome	Major phenotypic manifestations	Complicating neoplasia	Outstanding chromosomal aberrations
BS	1. Small body size 2. Sun-sensitive telangiectatic skin lesions over face	AL	Quadriradial formation
FA	1. Pancytopenia 2. Anatomic defects—multiple	AL	Chromosomal breaks and gaps
AT (the Louis-Bar syndrome)	1. Cerebellar ataxia 2. Telangiectasia 3. Stunted growth and hypogonadism	Cancer and lymphoma	Chromosomal breaks and rearrangements
Xeroderma pigmentosum	1. Severe skin lesions 2. Other defects	Skin cancer	Chromosomal rearrangements

instability of the chromosomal complement manifested by increased chromosomal disruption and rearrangement, referred to as "chromosomal breakage." The latter is particulary striking in cultured blood lymphocytes or skin fibroblasts, possibly indicating that the genetic defect is manifold, but under the "stressful" conditions of in vitro culture the chromosomal aberrations are more clearly brought out. However, inasmuch as some of these karyotypic abnormalities have been observed in uncultured bone marrow cells, the chromosomal manifestations of these syndromes are also operative in vivo. The exact nature of the bone marrow cells involed by the breakage has not been established.

It is possible that the genetic defects in these diseases, responsible for the development of neoplasia and the breakage manifestations exhibited by the chromosomes, have little to do with the development of malignancy. The genes that, in addition to producing characteristic clinical symptoms, increase chromosomal instability in somatic cells and predispose to cancer appear to characterize certain cellular systems (German, 1973, 1974a, c). Even though the degree of chromosomal disruption and rearrangement in such systems may be impressive, it is probable that the instability alone has nothing to do directly with cancer other than to provide a predisposing background. Thus, the incidence of chromosomal

Table 25 Major cytogenetic features of the four main chromosome instability syndromes

	SCE*	ctb‡	cte§	Homologous exchanges	
BS	++	+	+	++	Cells have slow rate of fork motion during DNA replication; SCE incidence can be further increased, e.g., by MMC
FA	N‖	++	++	−	Cells have defect in repairing DNA cross links; SCE incidence difficult to increase
AT	N	?+	?+	−	Cells have difficulty in repairing X-ray damage; difficult to induce increased SCE incidence
XP	N	−	−	−	Cells have difficulty in repairing UV damage of DNA; incidence of SCE increased by UV (Cheng et al., 1978).

*SCE, sister chromatid exchange.
‡ctb, chromatid breakage.
§cte, chromatid interchange.
‖N = normal.

aberrations may remain the same in lymphocytes after leukemia or other malignancy has appeared in BS or FA. However, it is also possible to visualize considerable loss or malfunction of chromosomes as a result of the breakage, leading to the neoplastic transformation of a cell or cells, with ultimate proliferation of such a cell into an overt cancer or leukemia. It should be pointed out, however, that the cancerous or leukemic cells have not been shown to have any of the characteristics of the chromosomes known as breakage. In other words, the karyotypic changes observed in the leukemic cells of subjects with breakage syndromes are similar to those seen in other patients with the same type of leukemia but without any of the manifestations of the breakage syndrome.

Inasmuch as the main aim of the present review is an evaluation of *visible* chromosome anomalies and their bearing on the pathogenesis of human neoplasia, we shall not be concerned with the invisible but very important chromosome anomalies, such as point mutations, gene deletions, and gene duplications, which may result from chromosomal breakage and subsequent repair.[3] The significance of these submicroscopic cytogenetic changes deserves close attention and has been documented by others.[4] Furthermore, it is my opinion that for the discussion only the results obtained in vivo are meaningful and applicable. Thus, the discussion of the chromosomal breakage syndromes will be focused to some extent on the karyotypic aspects of bone marrow cells.

Fanconi's Anemia (Congenital Pancytopenia)

This disease is probably inherited as an autosomal recessive trait and its outstanding clinical manifestations are generalized failure of the bone marrow leading to pancytopenia, major anatomical defects especially of the radius, thumb, and kidney, mild mental and growth retardation, and patchy brown pigmentation of the skin.[5] Symptoms related to the bone marrow failure (e.g., pancytopenia) usually appear between the ages of 4–12 and are progressive. Some of these manifestations have been controlled somewhat by the administration of androgens and corticosteroids. Probably a higher than normal proportion of patients with Fanconi's anemia (FA) develops leukemia and possibly other forms of cancer. If current therapy prolongs the life of patients with FA signficantly, it will be of interest to follow the incidence of these neoplastic states concomitantly with the course of the disease. Apparently, the heterozygote may also have an increased risk of developing malignancy, and the chromosomal defects have been described in both parents of a patient with FA. In the homozygote patient, the typical chromosomal changes are already present long before the anemia or symptoms related to it appear (Schroeder, Drings, et al., 1976; Schroeder, Tilgen, et al., 1976).

Fanconi's anemia was the first disease in which spontaneous chromosomal breakage was detected in 1964, both in vitro and in vivo (Schroeder, Anschütz, et al. 1964). More than 40% of the analyzed metaphases from blood cultures of two affected brothers had chromosomal breakage and rearrangements. In most laboratories the incidence of these chromosomal anomalies in the blood cells of normal individuals is < 5%. The percentage of cells with breakage and rearrangements in the marrow has been found to be much lower than in the blood or, frequently, not to be present at all. It is hardly justifiable to assume that the pancytopenia in FA is due to chromosome breakage in marrow cells. Of course, the bone marrow studies are often hampered by a paucity of cells due to the hypoplastic state of the marrow. However, the incidence of the chromosomal aberrations is high in cultured skin fibroblasts. The chromosomal changes may be present before the anemia becomes manifest.

Early Chromosome Studies of Fanconi's Anemia

Results on chromosome studies of marrow cells of only 24 patients are available.[6] Considering that fewer than 200 (5–6 with neoplasia) cases with FA have been reported, this figure is surprisingly low. This may be due to the nature of the disease, i.e., that no material for analysis was available because of the marrow aplasia. Though exceptions occur,[7] in 17 patients the percentage of marrow metaphases with chromosome anomalies was usually not higher than 10%; in 5 patients the percentage was about 5% or less,[8] and in 4 patients no anomalies could be detected.[9] The lymphocytes of these 24 patients, however, showed a chromosomal breakage incidence varying between 15 and 74% of the metaphases; whereas, the respective figure for fibroblasts of 6 patients varied between 10 and 49%. The kind of chromosome aberrations, i.e., chromatid and chromosome breaks and less frequently reunion figures, were identical in the different tissues. From the figures given, it is clear that chromosomal breakage in FA occurs in marrow cells distinctly less frequently than in those cells subjected to in vitro conditions. In the majority of marrow studies in this anemia the chromosome aberration rate does not appear to be higher than that in normal subjects.[11]

Some observations are worthy of special note. In two patients, lymphocyte studies were done before and after leukemia became apparent and in both instances the breakage rate remained unchanged.[12] One patient's marrow was examined while he had leukemia, and no significant structural or numerical chromosome anomalies were encountered.[13] In six patients stable chromosome anomalies were found consisting of a marker chromosome in 100% of the marrow cells in one case and a trisomy C in 10% of the cells in another case.[14] In two patients the majority of the cells in the marrow were abnormal: one patient having 50 chromosomes in over 80% of the cells, possibly related to terminal leukemia;[15] and another patient having an abnormal D-chromosome (Dq+) in 60% of the marrow cells.[16] Anomalies seen in the marrow of the four patients were not present in the cultured lymphocytes. These particularly remarkable cases suggest that on rare occasions chromosomal breakage may lead to stable chromosome anomalies, which may endow the affected cells with a growth advantage, a phenomenon not observed in megaloblastic anemia. In three of the patients with stable chromosome anomalies, no signs indicative of leukemia were detected.

In cultured blood lymphocytes, chromatid and isochromatid gaps (achromatic regions), breaks (usually not located at the centromere), fragments (often acentric), dicentric, ring and endoreduplicated chromosomes are present in a high proportion of PHA-stimulated cells and constitute the most frequent breakage abnormalities in the cells of about 50 cases of FA studied to-date,[17] including four and possibly five with leukemia.[18] Apparently, both T and B lymphocytes are involved equally by breakage (Bushkell, Kersey, et al., 1976). Loss of chromosomal material is often evident. This incidence of gaps and breaks is higher than observed in BS. Even though definitely increased above normal, the incidence of chromatid interchanges and rearrangements of chromosomes leading to abnormal chromosomes or polycentrics is lower in FA than in BS.[19] Apparently, the appearance of quadriradial (Qr) formation originating from homologous chromosomes is rare in FA as compared with BS; Qrs affecting nonhomologous chromosomal regions as well as other types of complex rearrangements make up the majority of rearrangements in FA.

Another important difference between FA and BS is the normal frequency of SCE in the former vs. the high incidence in the latter.[20] Furthermore, the SCE frequency in FA apparently cannot be raised significantly by mitomycin C as compared with normal cells.

Recent Studies of Fanconi's Anemia

In a 9-year-old girl with Fanconi's anemia (FA) 60% of the marrow cells had a Dq+ chromosome. The authors suggested that an in vivo arrangement had taken place in the past producing the abnormal D-chromosome, evidently conferring selective advantage to the cells with this cytogenetic abnormality. The blood lymphocytes of the patient showed a clearly increased incidence of chromosome breakage; none of obligate or possible heterozygotes in the family had over 10% abnormalities in their lymphocytes. At the time of the report, the patient had not developed any malignant state (Lisker and Gutierrez, 1974).

In another study (based on G-banding) two sibships, including four patients with FA, were studied cytogenetically. Interfamilial variations of chromosome breakage and rearrangements were found, suggesting heterogeneity of the disease. An abnormal clone with 47 chromosomes (+21) was found in the marrow of one patient (Berger and Aubert, 1975; Berger, Bussel, et al., 1977).

The finding of "premature chromosome condensation" (PCC) in a case of FA (Obe, Lüdeke, et al., 1975) indicates the presence of micronuclei which are subject to prophasing when the major nucleus is in metaphase.

In FA more chromosomal aberrations per abnormal cell occur than in the cells of BS, but whether these are or are not located preferentially in any particular chromosomes or their segments to form a pattern as in BS is still unsettled. Thus, chromosome loci 3q27 and 13q32 were reported to be susceptible to breaks, with the distal regions of the A-group chromosomes being particularly involved by an excess of breaks and the centromeric area to a lesser extent (Koskull and Aula, 1973, 1977). The sex-chromosomes had a deficit of breaks. In another study, on the other hand, #3, #17–#22, and the Y were affected less often than others. Apparently, in at least five cases with this disease no chromosomal breakage in cultured blood cells was observed, though the results are based on only one examination and no data were given on the marrow. No increases in SCE in the lymphocytes of cases with FA have been observed, including the heterozygous cells.

There is little doubt that chromosomal breakage in somatic cells observed in vitro and in vivo in FA is a secondary consequence to a primary genetically determined defect. Chromosomal breakage in vivo, however, is not a prerequisite for the spontaneous development of malignancy; chromosomal breakage in vitro evidently represents the results of certain cell properties due to an unknown primary genetic defect, which if observed exclusively in vitro must be provoked by the in vitro ecology. However, it is conceivable that in vivo the same defect may lead to point mutation, gene deletions, and duplications instead of breaks. This is one way of linking the frequent coincidence of leukemia and other cancers with this disease. On the other hand, the appearance of chromosomally abnormal clones, mainly pseudodiploid, in FA lymphocytes kept in culture or in the marrow of patients with FA may be indicative of the potential of these cells to undergo malignant transformation.

Because chromosomal breakage in the somatic cells of FA, both in vivo and in vitro, is related to a metabolic defect reflecting a basic genetic derangement, the tendency to malignancy in such cells should be either provoked or enhanced by some external factors. Hence, the tissues of patients with FA might be more sensitive to chromosome breaking agents and viruses,[21] akin to the findings observed with X-ray or thymidine-analogues damaged cells. Thus, it was shown that fibroblasts from patients with FA and their heterozygous relatives have a greatly increased percentage of transformed cells with SV_{40} virus over that found with normal cells (Table 19). Cells from the homozygotes were more susceptible to transformation than those from heterozygotes. The sensitivity to breakage was further demonstrated by the much higher number of breaks per cell after irradiation with 10 and 100 rads in cells of patients with FA than shown by normal cells. It should be pointed out, though, that a similar sensitivity has been demonstrated not only for cells from other breakage syndromes, but also for those from other congenital anomalies accompanied by aneuploidy, particularly hyperdiploidy. An increased sensitivity of blood lymphocytes of FA to DNA cross-linking agents (nitrogen mustard, mitomycin C, and special UV light) has been reported. The interpretation was that these cells are defective in the repair mechanism necessary for the repair of cross-links produced in their DNA,[22] which may have some relevance to the increased risk of malignancy in this hereditary disease. Some evidence has recently been presented that the cells in FA have an increased susceptibility to chromosome damage from mutagens and carcinogens. The persistence of the chromosome aberrations after exposure to a carcinogen further suggested a mechanism for cellular evolution to malignancy (Auerbach and Wolman, 1976).

In skin fibroblasts from individuals with FA, the autoradiographic patterns of DNA replication were found to be normal, indicating that S-phase DNA synthesis is not abnormal in FA (Hand, 1977).

Acute Leukemia in Fanconi's Anemia

Berger (1977) reported a case of FA diagnosed in 1965 and in whom chromosome anomalies similar to those described in FA were found in cultured blood cells. In 1973 a clone of 47,XX,+21 cells was found in the marrow (40% of the metaphases). It was never seen again. After 1974 a second type of anomaly was found in the marrow: 46,XX,−3,−12,+2mar. A few cells with 46,XX,−12,+mar were encountered. No normal cells were seen. In 1976 the marrow cells consisted of 46,XX,−1,−3,−7,−12,+4mar and 46,XX,−7,−12,+2mar. The patient had developed AL at that time.

Bloom's Syndrome

The major clinical features of Bloom's syndrome (BS), as conditioned by an autosomal recessive gene, are small body size of well-proportioned minuteness and sun-sensitive telangiectatic skin lesions in a butterfly distribution over the face. Not uncommonly dolichocephly and facial narrowness are present. Possibly related to a severe disturbance of immunity[23] is the predisposition to infections, particularly upper respiratory and intestinal.

The incidence of leukemia and cancer is much higher among patients with BS than in the general population.[24] Thus, in a close follow-up of 49 patients with this syndrome, 8 have already died of malignancy (4 of AL), all but one (age 13) 25 yr or older.

The cytogenetic findings in BS were first described in 1964[25] and are present in the blood cells of all affected individuals. These chromosomal changes persist and are found upon repeated examinations over a period of years. The chromosomal aberrations are rarer in long-term cultures of dividing skin fibroblasts,[26] but clearly more frequent than in control cultures. In contrast to FA, in which chromatid abnormalities are numerous, those in BS are of the chromosome variety. The most characteristic of these aberrations (Figure 52B) is in 0.5–14% of all dividing PHA-stimulated lymphocytes and consists of a quadriradial configuration (Qr) of symmetrical shape with centromeres in opposite arms of the figure. This rearrangement occurs before mitosis and is a consequence of an equal exchange of chromatid segments of two homologous chromosomes, usually at the centromere (Comings, 1975b). In the absence of Qr the diagnosis of BS should be seriously held in abeyance. Chromosomes of groups C, F, and #1 appear to be involved more often than others. The Qr is rarely seen in other chromosomal instability syndromes. These quadriradials have been described in cultured fibroblasts of BS; we have found chromosome instability, including chromatid and chromosome breaks, but no homologous chromatid interchanges in BS marrow cells incubated in vitro for either 1.5 or 46 hr. This observation points to the existence of chromosome instability in vivo (Shiraishi, Freeman, et al., 1976).

There are more than 50 known individuals with BS.[27] Bone marrow chromosome studies of only a few cases have been reported; in one case, among 364 metaphases, 7–8 were found to have decentrics, triradials, and other structural anomalies, and in another case, 4 out of 50 metaphases showed breaks.[28] In BS 5–20% of skin-and/or lymphocyte-derived metaphases had structural anomalies, and reunion figures (tri- or quadri-radials) were particularly characteristic.[29] Therefore, in spite of the low percentage of marrow cells with abnormal chromosomes, the type of aberrations found in German's case, as well as in ours mentioned above (Shiraishi, Freeman, et al., 1976), indicate that the findings may be significant, which may not be true for the other case. Obviously, much more data are necessary to establish with certainty the exact in vivo karyotypic picture in BS as manifested in the cells of the marrow.

Other chromosomal changes are seen in BS;[30] quadriradial from nonhomologous sites, breakage near the centromere or elsewhere, with or without sister chromatid reunion, acentric fragments of paired chromatids, dicentric

52 A. Greatly increased incidence of SCE in a cell from a patient with BS, to date the only condition in the human in which such an incidence occurs "spontaneously." Compare with the normal SCE incidence shown in Figure 43.

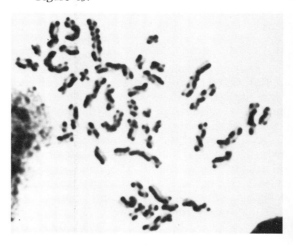

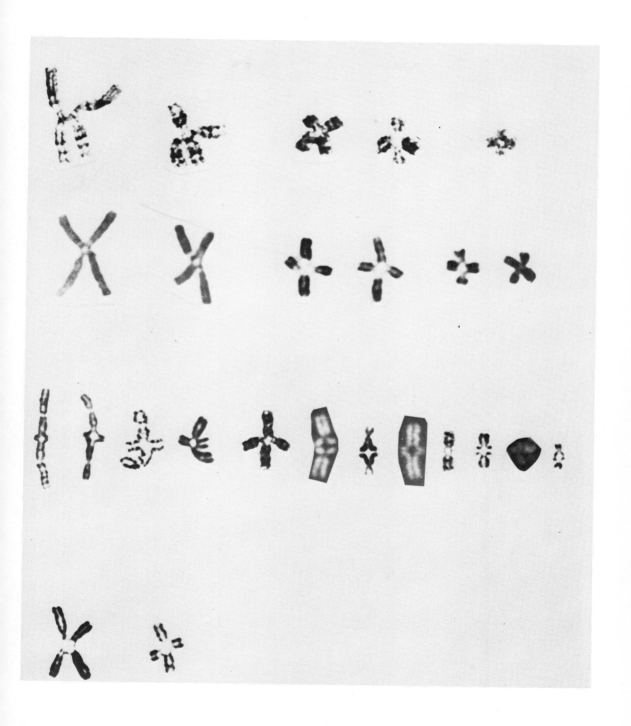

52 B. Group of *homologous* quadriradial chromosomes observed in the cells of a patient with BS. G and Q-banding were applied to these preparations.

chromosomes, and abnormal monocentric chromosomes. Chromatid and isochromatid gaps and breaks, common in FA, also occur in BS but at a lower frequency and inconsistently. At anaphase and telophase the various chromosomal aberrations result in increased numbers of chromatin bridges or of lagging or displaced chromatin fragments resulting in the not infrequent visualization of cells with distorted nuclei and micronuclei in interphase. More recently, a manifold increase in SCE was observed in lymphocytes of BS (Chaganti, Schonberg, et al., 1974). The heterozygous cells of BS and the homo- and heterozygous cells of FA and AT did not appear to have a higher frequency of SCE, when compared to that in control cells. In one report the incidence of SCE was higher in lymphocytes than in the skin fibroblasts of an 8-yr-old girl with BS, i.e., 133 exchanges vs. 49, respectively (Sperling, Goll, et al., 1976). In our studies (Shiraishi, Freeman, et al., 1976) we found the incidence of SCE and chromosome abnormalities in the marrow to be similar to that in lymphocytes (Figure 52). A recent report by German, Schonberg, et al. (1977) indicates that a population of lymphocytes which exhibit a normal amount of SCE can be detected in the blood of some individuals with BS. This coexistence of cells with a normal and a greatly increased SCE frequency results in a phenotypic dimorphism, in apparent contradiction to the autosomal recessive mode of inheritance of the syndrome. In a study using Q-banding, the distribution of mitotic chiasmata in lymphocytes of three patients with BS was determined. It was found that mitotic chiasmata, similar to chromosome breakage of various types and to SCE, take place preferentially in Q-dark bands ("interband" regions) or at borders between bands and interbands (Kuhn, 1976, 1978).

The sensitivity to UV light in BS might point to the chromosomal changes as being a cumulative effect of such light, these changes possibly being responsible for the exacerbation of skin lesions. An increased susceptibility of cells from BS to X-ray has been reported.[31]

A pseudodiploid subclone in growing cultures of diploid skin fibroblasts of a patient with BS has been described.[32] During life the proportion of quadriradial and other chromatid exchanges remains constant in BS, whereas there is a gradual rise in the frequency of abnormal monocentric and dicentric chromosomes, leading to a higher incidence of micronuclei. The latter may be reflected in the increased incidence of PCC (prophasing) in BS (Hecht and McCaw, 1977; Sperling, Goll, et al., 1976).

The question arises as to whether progeny cells of a parental cell with Qr (or other chromosomal changes) from BS have genomes different from one another and different from all other cells in the body, possibly developed by a process equivalent to crossing-over.[33] Under such conditions, a cell could become homozygous for all genes on the affected chromosomes distal to the point of exchange.

Studies of Bloom's Syndrome Cells in My Laboratory[34]

Marrow cells from a patient with BS, originally misdiagnosed,[35] cultured for 48 hr in the presence of BUdR exhibited a striking increase in the number of SCE, in comparison with that in the marrow cells of a patient with treated PV, i.e., 80 vs. 8 exchanges per cell, respectively. Thus, it appears that an increased incidence of SCE in BS occurs in various differentiated types of cells, not just blood lymphocytes, and constitutes the syndrome's most characteristic cytogenetic feature. In contrast, the incidence of SCE was not increased in marrow cells and lymphocytes of the PV patient studied, whose cells did exhibit increased numbers of chromatid and chromosome gaps and breaks, presumably as a result of earlier treatment.

An increased frequency of SCE was demonstrated in BS lymphocytes using both a technique based on BUdR incorporation and one based on labeling with tritiated deoxycytidine. This observation constitutes evidence against the increase of SCE being due to an unusual reaction to BUdR.

The distribution of the break points of SCE was compared with that of chromosome aberrations in BS by using differential sister chromatid staining and banding techniques. A comparison was made of the distribution in #1, #2, and #3, because the exact identification of other chromosomes is difficult with the technique used. It was shown that SCE and chromosome breaks do not necessarily correlate as to location. Some chromosome break points (e.g., 1q21, 1p36, 2q31, 3q12, and 3p13) were common with those of SCE, whereas others (at 1p13, 2p11, 2q11, and 3q11) showed little or no SCE. SCE breaks were not observed in the centromeric regions. In addition, the SCE frequency was examined in BS cells with and without chromosome aberrations, and no significant differences of SCE frequency were observed between cells with chromatid- or chromosome-type of aberrations and those with normal complements. Banding analyses indicated a nonrandom distribution of chromosome breaks in the lymphocytes and marrow cells of the BS patient.

BS lymphocytes, which are characterized by a high incidence of SCE (80 per cell), were treated with mitomycin C (MMC) and the effect of the chemical on SCE frequency compared with that in normal cells. Raising the concentration of MMC from 1×10^{-9} to 1×10^{-7} g/ml led to about a 10-fold increase (61.7 SCE per cell) in the SCE frequency over the base line in normal lymphocytes (6.4 SCE per cell), though chromosome aberrations remained at a relatively low frequency. MMC caused about a twofold rise in SCE in cells of BS (128.8 SCE at 10^{-9} g/ml; 139.3 SCE at 10^{-8} g/ml). The frequency of chromosome aberrations in BS cells at concentrations of MMC of 1×10^{-9} and 1×10^{-8} g/ml was 0.350 and 0.825 per cell, respectively, and low when compared to the increased number of SCE. The increased frequency of SCE in normal and BS cells caused by MMC is in contrast to the reported findings with cells from FA and XP. The distribution of SCE in MMC-treated normal cell correlates with that of spontaneous SCE in cells of BS.

The pattern of sister chromatid labeling and incidence of SCE was examined with several techniques, including the effects of caffeine on SCE, in normal human cells and in those of a patient with BS after 1–3 mitotic cycles. The results indicate, in confirmation of those obtained by others with different cells, that SCE in the cells studied is not due to a postreplication repair process, but more likely involves exchanges between the double strands of chromatids.

DNA Replication in Bloom's Syndrome

Analysis of DNA fiber autoradiograms from BS skin fibroblasts and blood lymphocytes has revealed a retarded rate of replication fork movement compared to normal adult controls. Other measurements from the autoradiograms were normal and, hence, the results suggest that a slow rate of fork movement is a specific manifestation of defective DNA synthesis in all BS cells (Hand and German, 1975, 1977).

Louis-Bar Syndrome (Ataxia Telangiectasia Syndrome)

This rare syndrome is transmitted as an autosomal recessive trait. Much interest has centered on the immunological deficiency found in this syndrome and its possible relationship to an apparently increased incidence of malignant diseases. The affected individuals usually appear normal at birth but exhibit cerebellar ataxia after a few months of life. Usually, between the 3rd and 6th yr of life telangiectasia appears in the bulbar conjunctivae and skin. The disease is progressive and when neoplasia occurs, it is usually of the lymphoid type, though other types of cancer have been encountered and with a much higher frequency than would be expected in such a small group of patients.[36] Many of the patients die of respiratory tract infections, this possibly being related to a deficiency in circulating immunoglobulins.

The blood lymphocytes do not respond adequately to PHA in culture and, hence, cytogenetic studies have been hampered by too few metaphases for examination.[37] As in some of the other chromosome instability syndromes, e.g., BS, the lymphocytes in ataxia telangiectasia (AT) grow sluggishly and fibroblasts have a prolonged doubling time and dally unduly in the S-phase of the cell cycle, these aspects possibly contributing to the susceptibility of the chromosome to breakage (Ros, 1975; Hecht and McCaw, 1977). A suggestion has been made that both the immunologic deficiency and the small number of lymphocytes sensitive to PHA stimulation may be the result of thymic deficiency leading to greatly reduced numbers of circulating T cells. This phenomenon may explain why early studies, usually performed on single cases, indicated that the karyotype was normal, but no details were available. More recently, detailed analysis of the cytogenetic findings in cultured lymphocytes has become available. A total of 30 patients appear to have had their blood lymphocytes examined cytogenetically after stimulation with PHA. In eight apparently no chromosomal or chromatid abnormalities were found. In only six cases was the percentage of cells with gaps, breaks, and rearrangements high, and in nine cases abnormal clones have been described. Gaps, breaks, and fragments when present had a frequency that ranged from 0 to nearly 40%; however, these did not appear to be regular findings in AT. In addition, the appearance of karyotypically abnormal clones has been reported and has ranged from 11–80% of the lymphocytes in culture. There is a striking absence of quadriradial configurations (Qr), as seen in BS, whereas clones of abnormal lymphocytes occur much more frequently in AT than in either FA or BS. Unstable and stable chromosomal aberrations appear to occur with a higher frequency (20–85%) in patients with AT than in the cells of their relatives or in normal cells. The unstable aberrations are those that will tend to undergo further changes at the next mitosis, e.g., dicentrics, fragments, and rings; whereas the stable aberrations are usually abnormal monocentric chromosomes. Often, the clones contained abnormal chromosomes originating from those of group D (particularly 14q−) exchanges and rearrangements. It is possible that gaps and breaks occur more often in younger patients and rearrangements in older ones. Of more interest is the finding of clones primarily in older patients. So far, the patients in whom abnormal clones have been observed have been hematologically normal.

About 250 patients with AT have been described. Bone marrow chromosome studies of three cases have been published to-date. The karyotypes were normal in one case[38] and abnormal with chromatid breaks in 33% of the metaphases in the second case.[39] The third patient had leukemia at the time of chromosome analysis; 40% of the metaphases had an abnormally small G-chromosome.[40] It is not certain whether this anomaly was associated with AT or due to the leukemic process. The latter is more likely. The chromosomal behavior after culture of lymphocytes is not completely clear, because the findings in AT are far from being as evident as in FA or BS. In the 50 or more patients with AT in whom *lymphocyte* chromosome studies have been done,[41] chromosomal aberrations, mostly chromatid breaks and a few triradials, were found in over 30 cases, with 20–25% of the metaphases being affected. A dozen cases had normal chromosomes.

As in FA, patients with stable chromosomal anomalies may be of particular interest. Such stable clones were observed in the cultured lymphocytes of at least four patients;[42] in one case a gradual increase of the abnormal clones from 1–3% in the initial study to 46% 32 mo later was demonstrated in the blood cells.[43] The chromosomes of the bone marrow cells of this latter patient, however, were normal.

Examination of skin *fibroblasts* has only been performed in a small number of cases of AT,[44] possibly because of the slow growth of these cells in vitro. In one case an increased incidence of fragments and dicentrics was found and in another a translocation involving groups D and F chromosomes in all cells examined, accompanied by numerous other abnormalities (frequent chromosome exchanges, some gaps, breaks, and one chromatid exchange).[45] In the mother of one patient with AT, 1 cell out of 50 had a ring chromosome, and 13 had an extra group F chromosome. The father's fibroblasts were normal. In another case, where the fibroblasts of the father were normal cytogenetically, the mother's cells had a high incidence of chromosomal abnormalities, most of which were chromosome exchanges. Thus, it appears that as in FA and BS, the chromosomal instability in AT is not confined to the lymphoid cells.

Hayashi and Schmid (1975) studied cultured lymphocytes in five patients with AT. Four of these patients had an increased incidence of chromosome-type aberrations. In one patient over 95% of the metaphases contained a tandem duplication of almost the entire long arm of #14; some cells also contained various translocations, e.g., 1;2, and dicentrics of varying origin. The 31 dicentrics seen in 724 metaphases from four patients were all of a unique variety, i.e., no chromosome material was lost, and they seem to have arisen by end-to-end fusions. The incidence of chromatid-type aberrations and SCE in AT were found to be in the normal range (Bartram, Koske-Westphal, et al., 1976).

The modal karyotype in AT is generally normal, which is the case in all of the chromosome instability syndromes. In many patients, increased chromosome breakage is evident in lymphocytes and, less strikingly, in fibroblasts. Breakage is of the chromosomal type, involving both chromatids at homologous sites and apparently random in location. This is in contrast to BS where the breaks are

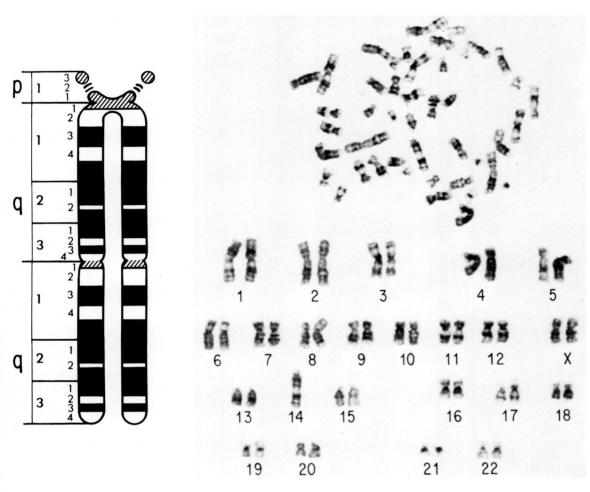

53 A. (right) Metaphase and karyotype from a patient with AT and a 14;14 tandem translocation.
B. (left) Schematic presentation of the translocation: t(14q11:14q34).
(Courtesy of Dr. R. V. Pierre.)

predominantly located near the centromere and are nonrandom with regard to the chromosomes involved.

The level of chromosome breakage often fluctuates. At birth and in early infancy there may be no detectable increase in breakage. It may rise perceptibly and then decline to normal levels in later years. The reasons for these fluctuations in chromosome breakage are not well understood.

Pseudodiploid clones are usually common in AT. For example, a young man with AT was followed clinically and cytogenetically for almost 5 yr until his death of progressive pulmonary insufficiency. When first studied (at 18 yr) a small percentage (1-2%) of his lymphocytes had a D/D translocation of the t(14q−;14q+) type. Over the 5 yr, the translocation clone of lymphocytes increased to become the predominant type of peripheral lymphocytes, eventually constituting over 70% of sampled lymphocytes. There was a gradual decline in chromosome breakage and a concomitant rise in mitotic activity. Dicentric chromosomes, as observed in BS and FA, also became more frequent. Significantly, the patient never developed leukemia or other malignancy, although two of his affected siblings had died of ALL (Hecht and McCaw, 1977).

Most patients with AT have clones marked by a translocation involving #14 (Figure 53). The break in #14 is always in the long arm at the q11−q12 region. In the translocation,

Table 26 Lymphocyte clones containing abnormal D-group chromosomes in patients affected with AT

Case no.	Karyotype of clone	Reference
1	45,XX,−D,t(D;D)	Goodman, Cooper, et al., 1969
2	45,XY,−D,+Dq+	Pfeiffer, 1970
3	46,XY,Dq+	Pfeiffer, 1970
4	46,XY,Dq+	W. Schmid and Jerusalem, 1972
5	46,XY,t(14q−;14q+)	Hecht, McCaw, et al., 1973
6	46,XX,t(14q−;7q+)	Hecht, McCaw, et al., 1973
7	46,XY,t(14q−;7p+)	Hecht, McCaw, et al., 1973
8	46,XX,t(14q−;Xq+)	Harnden, 1974b
9	46,XX,t(14q−;14q+)	Harnden, 1974b
10	46,XX,t(Dq−;Cq+)	Bochkov, Lopukhin, et al., 1974
11	46,XX,Dq−	Bochkov, Lopukhin, et al., 1974
12	46,XY,t(Dq−;Cq+)	Bochkov, Lopukhin, et al., 1974
13	46,XX,t(Dq−;Cq+)	Bochkov, Lopukhin, et al., 1974
14	46,XX,t(Dq−;Dq+)	Bochkov, Lopukhin, et al., 1974
15	46,XX,t(Dq−;Dq+)	Bochkov, Lopukhin, et al., 1974
16	45,XY,−14,−14,+tan(14q14q)	Rary, Bender, et al., 1974
17	D-group chromosome	Hatcher, Pollara, et al., 1974
18	D-group chromosome	Hatcher, Pollara, et al., 1974
19	46,XY,t(X,14)(q28q13)	Nelson-Rees, Flandermeyer, et al., 1975
20	46,XY,t(14q−;6q+)	McCaw, Hecht, et al., 1975
21	46,XX,r14	McCaw, Hecht, et al., 1975
22	46,XY,14q+	McCaw, Hecht, et al., 1975
23	46,XX,t(14q−;14q+)	McCaw, Hecht, et al., 1975
24	46,XY,14q+	M. M. Cohen, Shaham, et al., 1975
25	46,XY,14q+	M. M. Cohen, Shaham, et al., 1975
26	46,XY,14q+	M. M. Cohen, Shaham, et al., 1975
27	46,XY,14q+	M. M. Cohen, Shaham, et al., 1975
28	46,XY,14q+	M. M. Cohen, Shaham, et al., 1975
29	46,XX,14q+	M. M. Cohen, Shaham, et al., 1975
30	46,XX,14q+	M. M. Cohen, Shaham, et al., 1975
31	t(13;14)	Van Hemel 1976
32	tan(14q;14q)	Levitt, Pierre, et al., 1978
33	tan(14q;14q)	Hayashi and Schmid, 1975

the other involved chromosomes have been #7, #8, #14, and X. There is no apparent loss or gain of genetic material. The only AT patient reported with a nontranslocation had a ring #14 marker (McCaw, Hecht, et al., 1975). In a study by M. M. Cohen, Shaham, et al. (1975), a 14q+ was found in varying proportions (2–5% in lymphocytes; 1–9% in fibroblasts) in the seven patients with AT studied.

A common denominator among AT pseudodiploid clones appears, therefore, to be either rearrangement of material on the long arm (q) of #14 or, more likely, position effect (i.e., a change in the genetic activity of the distal portion of 14q due to repositioning of that segment in the genome) (Table 26).

A higher radiosensitivity of blood lymphocytes of patients with AT to damage by X-ray has been reported,[46] though another report indicated the fibroblasts from two such patients not to be unusually sensitive to transformation by SV_{40} virus.

A. M. R. Taylor (1978) suggested that the unusually high level of radiation-induced chromosome- and chromatid-type aberrations in cells of AT is due to a significantly increased fraction of unrepaired double and single-strand breaks. M. C. Paterson, Smith, et al. (1976) indicated that AT cells lack the full complement of functional endonuclease and thus fail to initiate normal excision repair of γ-induced base defects.

Obviously, much more data are necessary,

not only on the in vitro chromosomal picture as it appears in cultured lymphocytes and fibroblasts, but particularly in marrow cells reflective of the in vivo status; and of the leukemic or cancer cells of those patients with AT who developed these diseases. The normal incidence of SCE in cultured lymphocytes of AT has been mentioned previously, in contrast to the high frequency in BS.

In vitro studies with cell strains from AT patients indicate that they are evidently defective in excision repair of radiation-induced DNA damage,[47] comparable to the defect in repairing UV-induced DNA damage in XP. This may explain the radiosensitivity of AT patients and the chromosomal changes observed (R. D. A. Peterson, Kelly, et al., 1964; R. D. A. Peterson, Cooper, et al., 1966; A. M. R. Taylor, Metcalfe, et al., 1976a,b). Incidentally, no significant differences were found between fibroblasts from normal subjects and patients with AT in either the production or repair of double-strand breaks induced by X-ray (Lehmann and Stevens, 1977).

Xeroderma Pigmentosum

This rare disorder is transmitted as an autosomal recessive trait and is characterized by an increased sensitivity to sunlight and a high incidence of neoplasia, predominantly but not exclusively of the skin. Basal cell and squamous cell epitheliomas may appear in large number over the course of years.

The number of cases of xeroderma pigmentosum (XP) in which the chromosomes have been examined is very small and those reports that have appeared have relied primarily on cultured lymphocytes or fibroblasts. We know next to nothing about the cytogenetic picture in the marrow or tumors observed in this disease.

Even though the chromosomal picture is normal in XP[48] (as in FA, BS, and AT), cytogenetic changes have been described.[49] Pseudodiploid clones of skin fibroblasts or lymphocytes with apparently balanced translocations, e.g., 46,XY,t(Cp+,Cq−) and 46,XY,t(Cp−; Dq+), found among primarily diploid cells growing in vitro have been reported. The latter clone was subsequently shown to be 46,XY, t(1q−;Dq+) in the skin fibroblasts (German, Gilleran, et al., 1973). The significance of "an abnormally asynchronous chromosome F" described in a case of XP remains to be established and confirmed (Grouchy, Nava, et al., 1967b). Chromosomal instability, i.e., chromosomal breakage, translocation figures, or dicentrics have not been described to occur with a high frequency in XP. XP cells do, however, have a defect in DNA repair after UV irradiation, possibly leading to chromosomal changes.[50] It is possible that exposure to UV light may lead to such changes in vivo. Such exposure may lead to unexcised pyrimidine dimers causing increased numbers of chromosomal rearrangements, some of which could be stable and become incorporated as marker chromosomes in the complement of a cell and, thus, start a pseudodiploid clone. It is unknown whether such a cell with a chromosomal mutation can become the progenitor of a cancer stem line. In other words, does the mutation with its concomitant duplication or deletion of chromosomal material itself constitute the conversion to cancer or is it only a part of the process of conversion?

A normal chiasma count was found in the first spermatocytes from one patient with XP.[51] Blood lymphocytes and skin fibroblasts from this patient and his younger brother were diploid, with no definite increased frequency of chromosomal anomalies. These two patients are representative of XP "variant" subjects, who have the clinical manifestations of the disease but whose cells lack the repair defects. However, utilizing a host-cell reactivation technique to estimate the repair of adenovirus type 2 (anuclear-replicating, double-stranded DNA virus), it was shown that the cells of five known variant kindreds of XP had defects in the repair of DNA damaged by UV irradiation when compared with normal cells.[52]

It has been demonstrated that "repair replication" of DNA occurs at a reduced rate or not at all in XP cells after UV irradiation and that the amount of unscheduled synthesis is negligible when compared with that of normal human fibroblasts. Furthermore, it has been shown that XP cells are defective in the endonuclease-mediated chain breakage constituting the initial step in dimer excision after UV irradiation and subsequent repair of the affected DNA strand. It is possible that the genetic expression of XP varies with the cell source. For example when four different fibroblastic lines from patients with XP were studied in vitro, it was found that in early passages all lines had normal chromosome constitutions, with rates of polyploidy and chromosome aberrations not different from those of normal cells.[53] In the latter passages, two of the lines developed polyploidy (up to 50%), and chromosome aberrations (up to 79%), whereas the other two lines had normal levels of these changes. Incidentally, the most common type of aberration in the two abnormal lines was dicentrics. It is possible that the changes in SCE observed in XP cells when exposed to certain carcinogens is related to this impaired repair of DNA. SCE in XP had been shown to be normal.[54]

In a study of 50 patients with XP (22 of whom developed skin cancer) in Japan, Takebe and associates (Takebe, Nii, et al., 1974; Takebe, 1976; Takebe, Miki, et al., 1977) showed that host-cell reactivation, i.e., the degree of survival of herpes simplex virus after UV irradiation, was low and clone-forming ability very low in XP cells when compared with normal cells. The authors suggested that the age distribution of the cancer-bearing patients and their DNA repair characteristics indicate that almost all XP patients will develop skin cancers unless their cells have relatively normal DNA repair capacity.

In the study of a patient with XP and a squamous cell carcinoma of the skin, Bloch-Shtacher, Goodman, et al. (1972) showed that 84% and 57% of the unexposed and exposed skin were diploid, respectively, and only 10% of the squamous cell carcinoma.

Studies of DNA Defects in Xeroderma Pigmentosum Cells

It has also been shown that double-strand breaks induced by X-ray in XP fibroblasts show no differences either in production or repair when compared to normal cells (Lehmann and Stevens, 1977). Furthermore, decreased amounts of UV-induced, unscheduled DNA synthesis was observed in XP fibroblasts; XP variants were unable to repair the UV-irradiated adenovirus as well as normal cells. This indicated that XP variants may have abnormal postreplication DNA repair (R. S. Day, 1975; D'Ambrosio and Setlow, 1978). The UV-induced breaks did not appear in XP fibroblasts known to have defects in DNA repair synthesis; the appearance of such breaks required a short post-UV incubation, consistent with the expected action of an endonuclease. Cells of the variant form of XP, characterized by normal DNA repair synthesis, exhibited normal production of breaks after UV, but were slower than normal cells in resealing these breaks. It was concluded that these cells have a defect in postreplication repair (Fornace, Kohn, et al., 1976). Unscheduled DNA synthesis characterizes XP lymphocytes, these cells having apparently normal levels of excision repair but high sensitivity to UV. A new variant type of XP cells was described, pointing to genetic heterogeneity of the cells in this condition (Sandhofer, Tuschl, et al., 1976). XP and variant XP fibroblasts in culture can carry out single-strand excision of DNA normally; however, the breaks so induced appear to be slowly rejoined in the XP variant and not at all by XP cells. Caffeine aggravated these effects (Dingman & Kakunaga 1976). Both homozygous and heterozygous XP cells show a definite increased susceptibility to transformation by murine and feline sarcoma viruses (Chang, 1976). A low DNA repair capacity has been demonstrated in XP cells by Takebe and associates (Takebe, 1976; Takebe, Nii, et al., 1974; Takebe, Miki, et al., 1977) and Weichselbaum, Nove, et al., 1978.

The XP cells have been shown to have a greatly increased sensitivity to the chromosome damaging effect of 4NQO (H. F. Stich, Stich, et al., 1973). Some XP lines show a definitely increased level of SCE after exposure to UV (Weerd-Kastelein, Keijzer, et al., 1977) or mutagenic carcinogens (S. Wolff Rodin, et al., 1977). The authors indicated that SCE is the most sensitive indicator yet found of determining chromosomal effect of potential mutagens and carcinogens in mammalian cells.

Decreased incorporation of thymidine into the DNA of XP lymphocytes after UV-irradiation in vitro has been demonstrated (Burk, Lutzner, et al., 1969). An increased susceptibility of XP cells to chromosome damage by adenovirus type 12 has been described by H. F. Stich, Stich, et al. (1974) and unscheduled DNA synthesis, UV-induced chromosome aberrations, and SV40 virus transformation by Parrington, Delhanty, et al. (1971). Some of these observations, particularly those following UV-induced changes, may reflect defective excision of UV-induced pyrimidine dimers of DNA (Cleaver, 1969, 1970, 1972, 1973, 1977; Robbins, 1978).

New Chromosome Instability Syndromes

A number of diseases has been described in which the cytogenetic changes are similar to those in FA, BS, AT, or XP. Unquestionably, more conditions with chromosomal changes similar to those present in the "breakage syndromes" will be described. Some of these appear to have no higher incidence of neoplasia than does the general population and others are of an infectious or external toxic nature with no putative relevancy to the development of cancer or leukemia.

In Itai-Itai disease, due to toxic concentrations of cadmium, definite chromosomal aberrations in cultured lymphocytes of seven patients have been described and consisted of quadriradials and triradials, chromatid breaks and gaps, dicentrics and acentrics with occasional ring chromosomes being present.[55] These findings could be duplicated by the addition of cadmium to cultures of normal leukocytes.

In a family with a high incidence of spinocerebellar ataxia, four of the affected subjects were studied cytogenetically, and an unusually high incidence of chromosomal breakage (15–100%) was found in the cultured lymphocytes.[56] These changes consisted of chromatid type breaks, rare chromosomal rearrangements, and fragmentations. Basically, the karyotypes were normal.

Incontinentia Pigmenti (Bloch-Sulzberger Syndrome)

This disease is characterized by skin pigmentation anomalies in combination with a variety of malformations, suggesting features of FA and AT (Hecht and McCaw, 1977). The skin lesions, usually evident at birth or shortly thereafter, may first resemble tattooing with inflammation and in its full form has the swirling appearance of multicolored marble. Histologically, the basal layer of the epidermis is "incontinent" of melanin. Malformations may involve the eyes, dentitia, heart, and skeleton. Mental retardation is still a debatable feature.

In incontinentia pigmenti, a disease due to a dominant gene probably located on the X-chromosome and fatal to males, chromosomal changes in blood leukocytes have been described in affected females and their carrier mothers;[57] these changes have consisted of gaps and a greatly increased number of abnormal mitosis in the cells of two families examined. The disease is not known to have a higher incidence of malignancy, though cases with AML and pheochromocytoma have been described (see Hecht and McCaw, 1977).

Scleroderma (Progressive Systemic Sclerosis)

Scleroderma is a generalized disorder of connective tissue and perhaps an autoimmune disease. The genetics of the disease are unknown, though some familial clustering is seen and, hence, it may have in part a genetic basis. Over 25 patients with scleroderma developing alveolar cell adenocarcinoma of the lung have been described (see Hecht and McCaw, 1977). However, the demonstration of increased chromosome breakage in a large group of apparently healthy relatives of patients with scleroderma may have special importance with regard to the concept of scleroderma being a familial autoimmune disease (Emerit, Housset, et al., 1976). Increased breakage is also observed in related disorders such as lupus erythematosus, dermatomyositis, periarteritis nodosa, and rheumatoid arthritis (Emerit, 1976).

An increased frequency of structural chromosomal anomalies in blood lymphocytes has been reported in relatively large numbers of patients with scleroderma (systemic sclerosis).[58] These consisted of chromatid aberrations (> 20%), including telomeric and centromeric breaks, chromatid exchanges, and selective endoreduplication. Chromosomal abnormalities were noted in only 12% of mitoses. In addition to acentric fragments, dicentrics, and ring chromosomes, there were many other marker chromosomes. Even though the incidence of hypermodal cells was increased, the total number of aneuploid cells did not differ from that in controls. Structural chromosome abnormalities have been described in the marrow of such patients, consisting of over 25% incidence of abnormalities as compared with a little over 7% in controls. Besides numerous gaps and breaks of one or both chromatids, acentric fragments and structurally abnormal chromosomes were described. These cytogenetic findings in patients with scleroderma appear to be at variance with the relatively normal karyotypic findings obtained earlier.

Porokeratosis of Mibelli[59]

Skin biopsies were obtained from four patients with porokeratosis of Mibelli, a rare inherited (autosomal dominant trait) skin disease (genodermatosis) associated with malignancy. Two patients had squamous cell carcinoma in some of the lesions. Fibroblasts from affected and unaffected areas of the patients' skin were cultured, and clones of cytogenetically abnormal cells were found in 8 of 16 lines of fibroblasts from *affected* areas of skin of two patients. A third patient had no chromosomal abnormalities, and a further patient showed one line of cells from an apparently normal area of skin with two abnormal clones. No specific chromosomal abnormalities were common to all cell lines, and no unstable rearrangements were found. Ten lines of cells from 7 healthy individuals, age-matched to the porokeratosis patients, had no abnormal clones of cells. The evidence suggests a chromosomal instability in cultured cells from patients with this syndrome, which may have relevance to the high incidence of skin cancer in this condition. No chromosomal changes have been observed in unaffected skin (fibroblast cultures of normal areas of skin) or in lymphocytes.

Glutathione Reductase Deficiency Anemia

This autosomally dominant state is very rare and has been reported to be associated with a high risk of leukemia. However, the disorder has not been well defined and much more information will be required before glutathione reductase (GR) can be put in proper perspective, clinically, biochemically, and cytogenetically. Unbalanced nuclear division, with chromatin bridges, fragments, and unbalanced nuclei, have been described in the marrow cells, similar to the findings in the marrow of patients with FA, Kostmann's agranulocytosis, and in some cases of pernicious anemia (PA). Chromosome breakage in blood lymphocytes in some and not in others cases of GR has been reported.[60]

Kostmann's Agranulocytosis

Kostmann's agranulocytosis (KA; infantile genetic agranulocytosis) is inherited as an autosomal recessive trait and is characterized by severe neutropenia, accompanied by frequent infections, primarily furuncles and carbuncles. Patients rarely survive more than 1 yr. One patient is reported to have died of leukemia. In the bone marrow 20% of the cells were found to be cytogenetically abnormal in the only patient with KA on whom such an examination has been performed.[61]

Basal Cell Nevus Syndrome[62]

This syndrome is a complex hereditary disorder including multiple nevoid basal cell carcinomas, jaw cysts, skeletal malformations, pits in the palms and soles, and various other anomalies. The syndrome is an autosomal dominant trait; there may also be a much rarer autosomal recessive form.

The blood lymphocytes of several patients with the syndrome have been shown to contain chromatid and isochromatid gaps and breaks, acentric fragments, dicentric and quadriradial chromosomes, and markers in a small percentage of the total metaphases (Happle, Mehrle, et al., 1971; Happle and Kupferschmid, 1972).

Crucial to the understanding of chromosome breakage and instability syndromes is an understanding of the nature and mechanisms of chromosome or chromatid breaks. Unfortunately, our knowledge in this area is incomplete, though some aspects of breaks have been studied.[63] Thus, autoimmunity, chromosomal vacuolization, and sensitivity of certain regions to breakage have received attention. More knowledge will have to be gained in order to ascertain the mechanisms underlying those conditions (FA, BS, AT) in which the chromosome changes occur in vivo and, hence, are primary in nature as compared to those states (XP, basal cell nevus, scleroderma) in which these changes are secondary in nature and can only be observed in vitro.

References

1. Lehmann and Arlett, 1977; Passarge, 1972; Pfeiffer and Kim, 1971; Setlow, 1978.
2. Brøgger, 1971; P. A. Jacobs, Buckton, et al., 1974; Purchase, Richardson, et al., 1976; Schuler, Dobos, et al., 1975.
3. Schroeder and Kurth, 1971; Tartaglia, Propp, et al., 1966.
4. German, 1972a,b; Schroeder and Kurth, 1971; Schuler, Kiss, et al., 1969a,b.
5. Beard, 1976; German, 1972a,b; R. D. A. Peterson, Cooper, et al., 1966; Polani, 1976; Schroeder and Kurth, 1971; Schroeder, Drings, et al., 1976; Schroeder, Tilgen, et al., 1976; Swift, 1971, 1976.
6. Berger, Bussel, et al., 1977; G. E. Bloom, Warner, et al., 1966; Coutinho, Falcao, et al., 1971; Crossen, Mellor, et al., 1972; Dosik, Hsu, et al., 1970; Gmyrek, Witkowski, et al., 1968; Hirschman, Shulman, et al., 1969; Hoefnagel, Sullivan, et al., 1966; Lieber, Hsu, et al., 1972; Lisker and Gutiérrez, 1974; Macciotta, Cao, et al., 1965; W. Schmid, Schärer, et al., 1965; Shahid, Khouri, et al., 1972; Swift and Hirschhorn, 1966; Varela and Sternberg, 1967; Wolman and Swift, 1972.
7. Berger, Bussel, et al., 1977; Lisker and Gutiérrez, 1974; Shahid, Khouri, et al., 1972.
8. Coutinho, Falcao, et al., 1971; Dosik, Hsu, et al., 1970; Varela and Sternberg, 1967.
9. G. E. Bloom, Warner, et al., 1966; Lieber, Hsu, et al., 1972; Macciotta, Cao, et al., 1965; Rozman, Sans-Sabrafen, et al., 1963; W. Schmid, Schärer, et al., 1965.
10. Hirschman, Shulman, et al., 1969; W. Schmid, Schärer, et al., 1965; Swift and Hirschhorn, 1966; Visfeldt and Mortensen, 1970.
11. Hirschman, Shulman, et al., 1969; W. Schmid, 1967; Schroeder, 1966a,b.
12. Crossen, Mellor, et al., 1972; Hirschman, Shulman, et al., 1969.
13. Dosik, Hsu, et al., 1970; Gmyrek, Witkowski, et al., 1968; Palade, Postelmieu, et al., 1970; Schuler, Kiss, et al., 1969a,b.
14. Dosik, Hsu, et al., 1970.
15. Motomura and Yamamoto, 1971.
16. Lisker and Gutiérrez, 1974.
17. Bersi and Gasparini, 1973; Bushkell, Kersey, et al., 1976; Dosik, Verma, et al., 1977; Dutrillaux, Couturier, et al., 1977; Gastearena, Erice, Lasa Doria, et al., 1972; Germain, Requin, et al., 1968; Grouchy, Nava, et al., 1972; Guanti, Petrinelli, et al., 1971; Hand, 1977; Koskull and Aula, 1973, 1977; Meme, Oduori, et al., 1975; Nordenson, 1977; Ortega-Aramburu and Garcia, 1972; Perkins, Timson, et al., 1969; Perona and Testolin, 1966; Sarna, Tomasulo, et al., 1975; W. Schmid and Fanconi, 1978; Schroeder and Drings, 1973; Schroeder, Drings, et al., 1976; Schroeder, Tilgen, et al., 1976; Schuler, Kiss, et al., 1969a; Sperling, Wegner, et al., 1975; Therman and Kuhn, 1976.
18. Berger, Bussel, et al., 1977; Bourgeois and Hill, 1977; Dosik and Verma, 1977; Motomura and Yamamoto, 1971.
19. Beard, 1976; German, 1973; German and Crippa, 1966; Hecht and McCaw, 1977; Polani, 1976; Schroeder and German, 1971, 1974; Schroeder and Passarge, 1973; Schroeder and Stahl Mauge, 1976.
20. Bartram, Koske-Westphal, et al., 1976; Latt, Stetten, et al., 1975; Shafer, 1977; Sperling, Wegner, et al., 1975.
21. Auerbach and Wolman, 1976; Higurashi and Conen, 1973; Lubiniecki, Blattner, et al., 1977.
22. Poon, O'Brien, et al., 1974; Sasaki, 1975; Sasaki and Tonomura, 1973.
23. Baccaredda, 1939; Hütteroth, Litwin, et al., 1975; Schoen and Shearn, 1967.
24. Braun-Falco and Marhgescu, 1969; Brunel, Donnadio, et al., 1977; Festa, Meadows, et al., 1979; German, 1973, 1974a,b; German, Schonberg, et al., 1977; Schroeder, 1972, 1975a; Schroeder and Stahl Mauge, 1976.
25. German, 1969a; German, Archibald, et al., 1965.
26. German and Crippa, 1966.
27. German, 1972a,b, 1973; Keutel, 1969.
28. Landau, Sasaki, et al., 1966.
29. Bourgeois, Calverley, et al., 1975; German, 1969b, 1972a,b, 1974b; German, Crippa, et al., 1974; Keutel 1969; Keutel, Maigheseu, et al., 1967; Rauh and Soukup, 1968; Schroeder and German, 1974; Schroeder and Stahl Mauge, 1976; Shiraishi, Freeman, et al., 1976; Sperling, Goll et al., 1976.
30. Chaganti, Schonberg, et al., 1974; Comings, 1974; German, Schonberg, et al., 1977; Schroeder, 1975a,b; Shafer, 1977; Shiraishi, Freeman, et al., 1976; Shiraishi and Sandberg, 1976a,b, 1977, 1978a–c.
31. Higurashi and Conen, 1973.
32. Rauh and Soukup, 1968.
33. Therman and Kuhn, 1976.
34. Shiraishi, Freeman, et al., 1976; Shiraishi and Sandberg, 1976a, 1977, 1978a–c.
35. A. I. Freeman, Sinks, et al., 1970; A. I. Freeman, Edwards, et al., 1972.
36. Daly, 1976; German, Passarge, et al., 1971; Goldsmith and Hart, 1975; Harris and Seeler, 1973; R. D. A. Peterson, Kelly, et al., 1964; Schuster, Hart, et al., 1966.
37. German, 1973; Hecht and McCaw, 1977.

38. Hecht and Case, 1969.
39. Lisker and Cobo, 1970.
40. Lampert, 1969; for comparison see Sawitsky, Bloom, et al., 1966.
41. Bochkov, Lopukhin, et al., 1974; M. M. Cohen, Kohn, et al., 1973; M. M. Cohen, Shaham, et al., 1975; German, 1972a,b; Goodman, Cooper, et al., 1969; Gropp and Flatz, 1967; Haerer, Jackson, et al., 1969; Harnden, 1974b; Hatcher, Pollara, et al., 1974; Hayashi and Schmid, 1975; Hecht and Case, 1969; Hecht, Koler, et al., 1966; Higurashi and Conen, 1973; Hook, Hatcher, et al., 1975; Levitt, Pierre, et al., 1977; Lisker and Cobo, 1970; Nelson, Blom, et al., 1975; Oxford, Harnden, et al., 1975; Pfeiffer, 1970; Rary, Bender, et al., 1975; Ros, 1975; I. R. Rosenthal, Makowitz, et al., 1965; W. Schmid and Jerusalem, 1972; Schuler, Schöngut, et al., 1971; A. M. R. Taylor, Metcalfe, et al., 1976a,b; Utian and Plit, 1964; Van Hemel, 1976; R. R. Young, Austen, et al., 1964; Zelweger and Khalifeh, 1963.
42. Hayashi and Schmid, 1975; Hecht and Case, 1969; Pfeiffer, 1970.
43. Hecht and Case, 1969; Hecht and McCaw, 1977.
44. M. M. Cohen, Shaham, et al., 1975; Harnden 1974b; Hecht, Koler, et al., 1966; Oxford, Harnden, et al., 1975; Van Hemel, 1976; Webb, Harnden, et al., 1977.
45. Harnden, 1974a.
46. Higurashi and Conen, 1973; A. M. R. Taylor, Harnden, et al., 1975; A. M. R. Taylor, Metcalfe, et al., 1976a.
47. A. M. R. Taylor, Metcalfe, et al., 1976b; M. C. Paterson, Smith, et al., 1976.
48. German, 1974b.
49. Reed, Landing, et al., 1969.
50. Burk, Lutzner, et al., 1969; Cleaver, 1969, 1970, 1972, 1973; Parrington, 1971; Parrington, Delhanty, et al., 1971; Parrington, Casey, et al., 1977; Robbins, Kraemer, et al., 1974; Sasaki, 1973; Setlow, Regan, et al., 1969; Tates, 1976.
51. Hultén, Weerd-Kastelein, et al., 1974.
52. Fornace, Kohn, et al., 1976.
53. Huang, Banerjee, et al., 1975.
54. S. Wolff, Rodin, et al., 1977.
55. Shiraishi, 1975; Shiraishi and Yosida, 1972b.
56. Ikeuchi and Kawasaki, 1973.
57. Cantu, Castillo, et al., 1973; Grouchy, Bonnette, et al., 1972.
58. Dutrillaux, Croquette, et al., 1978; Emerit, 1976; Emerit, Housset, et al., 1971, 1976; Emerit, Emerit, et al., 1972; Emerit, Levy, et al., 1973; Emerit and Housset, 1973; Emerit and Marteau, 1971; Housset, Emerit, et al., 1969; Khondkarian, Burak, et al., 1967; Pan, Rodnan, et al., 1975.
59. Goerttler and Jung, 1975; A. M. R. Taylor, Harnden, et al., 1973.
60. Hampel, Lohr, et al., 1969.
61. Matsaniotis, Kiossoglou, et al., 1966a.
62. Happle and Hoehn, 1973; Happle and Kupferschmid, 1972; Scully, Galdabini, et al., 1976.
63. Aula and Koskull, 1976; Aymé, Mattei, et al., 1976; Castoldi and Mitus, 1968; Fialkow, 1964; Kihlman, 1971; Reeves and Lawler, 1970.

9

Preleukemia

Idiopathic Aplastic Anemia
Pure Red Cell Aplasia
Refractory Anemias
Sideroblastic Anemia

Cytogenetic studies are assuming an ever increasing diagnostic and prognostic significance in many hematologic disorders. In no groups of disorders are these studies of more potential significance than in those presented in this chapter. Yet, no other hematologic disorders are more in need of further extensive and complete chromosomal (and other) data than this group of potentially preleukemic conditions. This raises the question as to whether the finding of abnormal stem lines or aneuploid cells in the marrow of patients with hematologic disorders is synonymous with a leukemic state. Even though the available cytogenetic data on some of the conditions to be discussed are meager, they do indicate that the demonstration of chromosomally abnormal clones of cells is associated with a high risk of developing AL. However, in some cases the presence of karyotypically abnormal cells has not always heralded the development of leukemia.[1] My opinion is that when chromosomal changes are observed a leukemic condition already exists regardless of the clinical manifestations. The fact that some abnormal laboratory findings and/or symptomatology are invariably associated with the karyotypic changes is sufficient evidence that a malignant state is associated with the chromosomal changes (Heller and Gross, 1977), the "preleukemic" picture being merely an extreme variant of the biologic behavior of the state, just as an overwhelming and fulminating AL is at the other end of the spectrum.

The relatively long survival of some patients with hematologic disorders and karyotypically abnormal cells in their marrow has been interpreted severally. Nowell (1965b, 1971b,c) is of the opinion that when aneuploidy is present in the marrow of subjects who have not been exposed to radiation, the risk of developing leukemia within a few months of the chromosomal findings is high and that patients who have such leukemic progression probably have subclinical leukemia at the time of the initial cytogenetic studies. On the other hand, if frank leukemia does not develop within 3 mo of the chromosomal findings, patients with these karyotypic changes have no greater risk of developing leukemia than patients without such changes. Krogh Jensen and Philip (1973) and I take a somewhat different view than that of Nowell (1971b,c), particularly in regard to those patients who have survived for years with cytogenetically abnormal clones in the marrow. The number of such patients with "silent" clones in the marrow is small, and they have not been followed long enough for one to be categorical about their nonleukemic states. It is possible that these abnormal cells represent leukemic ones but of low "virulence," leading to a balance between the host and the leukemic cells. Stress has been put on the appearance of the chromosomes in relation to the development of leukemia, i.e., fuzzy and ill-defined chromosomes being observed in those patients who developed leukemia rapidly and well-defined and thin chromatids in those patients who have survived for a relatively long period of time and have not developed leukemia.

My opinions in relation to the chromosomal changes in the conditions to be discussed in this chapter may be summarized as follows:

1. Most of the patients with so-called preleukemia will be found to have a diploid chromosomal picture in the marrow when examined initially and very often over a period of time. However, the presence of a diploid chromosomal picture does not preclude the development and/or the presence of leukemia because 50% of patients with AL have no demonstrable chromosomal changes.
2. The presence of chromosomally abnormal cells or stem lines is synonymous with a neoplastic clone of cells (Table 27), in the case of the conditions to be discussed below, usually AML. However, it is possible for the host to eliminate such abnormal clones, possibly through immunologic means.
3. The fuzzy and ill-defined appearance of the chromatids, whether in diploid or aneuploid metaphases, is a prognostically "bad" sign and actually indicates the presence of a leukemic process in the marrow (Gahrton, Lindsten, et al., 1974c).
4. Patients who survive for a relatively long period of time with demonstrable chromosomal changes in their marrow without the development of leukemia are, in my opinion, to be considered leukemic, even though clinically the disease may not appear to be very malignant. These patients very often have well-defined chromatids in the

Table 27 Karyotypic findings in various preleukemic conditions

Karyotype	Reference
45,−B,−C,+mar(abnormal E)	Nowell, 1965b
44,45,88,90 chromosomes (+min,mar) } pancytopenia	Nowell, 1965b
45,−2C,+E (abnormal D)	Nowell, 1965b
45,XX,−C	Humbert, Hathaway, et al., 1971
47,XY,+C/48,XY,+C,+C	M. H. Khan, 1972a
45,XX,−7	Kaufmann, Löffler, et al., 1974
(preleukemia → AML)	
45,−C (5 patients)	Pierre, 1974, 1975
47,+C (2 patients)	Pierre, 1974, 1975
47,+D (1 patient)	Pierre, 1974, 1975
45,X,−Y/44,X,−Y,−A (1 patient)	Pierre, 1974, 1975
44,XX,−C,−F/48,XX,+2C (→ Ph¹−negative CML)	Kirchner and Hofmann, 1976
−6,+8,+t(3;6),+3p−	Panani, Papayannis, et al., 1977
11q−	Panani, Papayannis, et al., 1977
+8 and normal	Panani, Papayannis, et al., 1977
+3,+9,+12	Panani, Papayannis, et al., 1977
+3,+9/N	Panani, Papayannis, et al., 1977
+9/N	Panani, Papayannis, et al., 1977
45,XY,−7,del(20(qll) → AL	Ruutu, Ruutu, et al., 1977b
45,XY,−7	Ruutu, Ruutu, et al., 1977b
46,XY,−17,i(17q)	Ruutu, Ruutu, et al., 1977b
47,XX,+8	Ruutu, Ruutu, et al., 1977b
47,XY,+8	Ruutu, Ruutu, et al., 1977b
46,XX,t(1;3;11)(pl;q2;q?)	Ruutu, Ruutu, et al., 1977b
46,XX,del(5)(q13or14),del(22)(q11or12)	Ruutu, Ruutu, et al., 1977b
46,XX→46,XX,20q− →47,XX,+21,20q− →48,XX,+9,20q−	Testa, Kinnealey, et al., 1977
(with no changes in clinical picture)	
46,XY,12p−,20q−(preleukemia)	Mitelman, personal communication
47,XY,+8	Koeffler and Golde, 1978
45,XX,−C	Koeffler and Golde, 1978
47,XY,+9	Koeffler and Golde, 1978
46,XY,−11,+t(1;11)(q11or12;q25)(→ AMMoL)	Najfeld, Singer, et al., 1978
47,XY,−7→EL	Shiloh, Naparstek, et al., 1979
46,XY,t(9;10)(p24?;p12) } benzene	Van Den Berghe, Louwagie, et al.,
45,XY,−7 } exposure	1979a
45,X,−Y (3 cases)	Pierre, 1978a
45,XY,−7 (3 cases)	Pierre, 1978a
47,XY,+8	Pierre, 1978a
47,XY,+19	Pierre, 1978a
46,XY,20q−	Pierre, 1978a
47,XX,+mar	Pierre, 1978a
45,XY,−5,−17,−22,+2mar	Pierre, 1978a
46,XY,5q−,22q+	Pierre, 1978a
46,XY,−2,13q+,+ace	Pierre, 1978a
46,XY,del(7)(p13),r(18)	Pierre, 1978a

metaphases of their marrow. This is consonant with the view of Killmann (1976) "that preleukemic states are AMLs that present in partial and sometimes long-lasting remission, which only after months or years lose their differentiation ability and then are classified as AML." The description of a Ph¹-positive preleukemic state in several patients deserves comment. In two cases definite leukocytosis (ca 30,000/mm³) was already present when the Ph¹ was observed and, thus, the condition could hardly be considered preleukemic, even though progression of the disease required a number of

Table 28 Some criteria for the possible diagnosis of so-called preleukemia

1. Refractory anemia with normo- or hypercellular erythropoiesis in the BM. Megablastoid changes usually seen with ring sideroblasts
2. Neutropenia and/or abnormal neutrophil morphology (pseudo-Pelger anomaly and/or defective granulation)
3. Thrombocytopenia and/or abnormal megakaryocyte morphology (abnormally small forms, abnormal arrangement of nuclear lobes)
4. Normal percentage of blast cells (myeloblasts) in the BM
5. No other hematological disease can be discovered upon careful investigation, especially vitamin B_{12} or folate deficiency

Ruutu, Ruutu, et al. (1977a).

years to appear. In a third case a Ph^1 was seen during the course of PV. Five years later no Ph^1 could be seen; neither could it be established subsequently with banding techniques. In my opinion, the čase cannot be classified as a preleukemia.

In terms of chromosomal changes, the interpretation of the so-called preleukemic states or preleukemia is complicated by the lack of a terminological and clinical definition (Table 28) of these states and the variety of views presented by authors as to what are considered to be preleukemic conditions.[2] This discussion is not concerned with those conditions associated with cytogenetic aberrations, e.g., trisomy of chromosome #21 in DS, chromosomal breakage syndromes (e.g., FA), and a higher risk of leukemia (each of these conditions is discussed in separate sections of this book), but with those unusual hematologic conditions that often (?) terminate in AML or, rarely, in other forms of leukemia. The question arises not so much as to what the chromosomal alterations are in these conditions, but what is the meaning of the karyotypic changes when these are found in the marrow.[3] Do these indicate that leukemia already exists, even in the absence of cytologic and clinical evidence? In some cases the cytogenetic changes may be present for a relatively long period without any overt transformation occurring in the disease, and this leads to the perplexing problem in the interpretation of the findings.

Idiopathic Aplastic Anemia

Idiopathic aplastic anemia (IAA), characterized by pancytopenia and hypoplasia or aplasia of the marrow without demonstrable causes, is considered by some to be a preleukemic condition, because not infrequently it precedes the development of AML. A rare case of ALL after the development of IAA has been observed. In most patients with IAA the possibility of exposure to drugs or chemicals is difficult to establish. Only a few chromosomal studies have been reported in IAA, primarily due to the difficulty in obtaining sufficient material from the marrow for cytogenetic analysis. In a recently reported series of 18 patients with IAA, only 4 were found to have some cytogenetic changes in the marrow consisting of 3% polyploid cells. However, in none of the 18 patients has the disease progressed to AL (Cobo, Lisker, et al., 1970; Lisker, Cobo de Gutierrez, et al., 1973). There is the possibility that were patients with IAA to survive longer, more would then develop leukemia and, consequently, more remarkable chromosomal changes might be encountered (Table 29). A 24-yr-old white male with trisomy #6 associated with IAA (Geraedts and Haak, 1976) and the development of IAA followed by AL in a subject with congenital trisomy #8 mosaicism (Gafter, Shabtai, et al., 1976) have been reported. Because of a report, Hashimoto, Takaku, et al., 1975, that the lymphocytes of patients with IAA may contain significantly more single-strand breaks in their DNA, the authors (Geraedts and Haak, 1976) examined the SCE frequency in the lymphocytes of their patient and found it normal. A male patient (33 yr old) with pancytopenia was found to have 49 chromosomes in the marrow (46,XY,+3C) 50 mo before he developed leukemia, at which time the chromosome number was also 49 (Krogh Jensen and Philip, 1973); another male patient with pancytopenia was found to have a 45,XY,−7 karyotype (Rowley, 1973d). To date, fewer than 100 cases of leukemia superseding IAA have been reported in the literature. The karyotypic changes so far described in IAA are very similar to those seen in AML.

Table 29 Chromosome findings in aplastic anemia and other refractory anemias and pancytopenias

Karyotype	Reference
45,XY,−C	Rowley, Blaisdell, et al., 1966
47,XY,+C (pancytopenia associated with viral hepatitis)	Albahary, Grouchy, et al., 1971
49,XY,+C,+C,+C (pancytopenia → AML)	Krogh Jensen and Philip, 1973
45,X,−Y/46,XY→45,XY,−7	Rowley, 1973d
47,XYq−,+1/46,XYq−	Warburton and Bluming, 1973
46,XX,−B,+mar/45,XX,−B,−E,+mar	Maldonado and Pierre, 1975
46,XX,−B,+E/45,XX,−B,−C,+E	Maldonado and Pierre, 1975
42−43chromosomes,−B,Bp+,−C,−E,−F,+dicC	Moretti, Broustet, et al., 1975
48,+8,+C	Moretti, Broustet, et al., 1975
46,22p+	Moretti, Broustet, et al., 1975
46,Fp−	Moretti, Broustet, et al., 1975
48,XX,+1,−5,+11,+mar (anemia, pancytopenia → AML)	Rowley, 1975b
46,XX (RA → AMoL)	Kamada and Uchino, 1976
45,XY,−C (IAA → EL)	Kamada and Uchino, 1976
46,XX,−B,−B,−D,+3mar (leukopenia → AL)	Kamada and Uchino, 1976
46,XX,Cp−,Gq− (leukopenia → EL)	Kamada and Uchino, 1976
45,XX,−A,Cp− (monocytosis → AMoL)	Kamada and Uchino, 1976
48,XX,+C,+C	Kamada and Uchino, 1976
46,XY,Dq+/46,XY	Kamada and Uchino, 1976
46,XY,Cq−,Gq−/46,XY	Kamada and Uchino, 1976
+C (2 cases)	Prieto, Badia, et al., 1976
+C,Ph¹+ (1 case)	Prieto, Badia, et al., 1976
46,XX,5q−(q15)	Prieto, Badia, et al., 1976
46/47 with i(17q)	Prieto, Badia, et al., 1976
47,XX,+8	Yamada and Furasawa, 1976
47,XY,+8 (pancytopenia → AML)	Yamada and Furasawa, 1976
47,XY,+8 (IAA → EL)	Yamada and Furasawa, 1976
N → 45,XY,−7 (no leukemia)	Boetius, Hustinx, et al., 1977
+8	Hagemeijer, Smit, et al., 1977
+8	Hagemeijer, personal communication
45,XX,5q−	Mitelman, personal communication
46,XX,i(17q)	Nowell and Finan, 1978a
45,XY,−C(?7 or 8)	Nowell and Finan, 1978a
47,XX,11q−,?21q−	Nowell and Finan, 1978a
47,XX,+?14	Nowell and Finan, 1978a
46,XX,−6,−16,+i(6p−;?16p;?16q),+min	Nowell and Finan, 1978a
44,XY,−2C,−D,+E,6p+	Nowell and Finan, 1978a
46,XX,−4,−5,−6,−7,−11,+4C,−13,−16,+19, +20,+t(7q;13q)/43,XX,−2,−3,−5,−C(?7), +C,−16,15q+,+19	Nowell and Finan, 1978a
46,XX,−7,+8,−12,−16,−17,+E,+t(6q;12q), 5q−,6q−,+min	Nowell and Finan, 1978a

Pure Red Cell Aplasia

Pure red cell aplasia of the chronic variety is a relatively rare disease, which may be acquired or consitutional. In the former, immunologic abnormalities have been implicated in the pathogenesis of the disease. Acute leukemia has been a terminal event in some of the patients. However, a case has been reported in whom repeated marrow studies over a period of 5 yr showed an abnormal clone (46,XX,Bq−) replacing the marrow but not involving the blood lymphocytes or skin fibroblasts (Fitzgerald and Hamer, 1971). The pa-

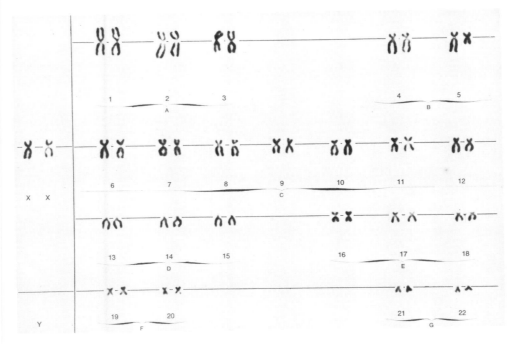

54 Karyotype showing a 5q− anomaly, shown with banding to be an interstitial deletion, observed in a marrow cell a patient with refractory anemia.
(Courtesy of Dr. H. Van Den Berghe.)

tient died of infection, showing no evidence of leukemia. Obviously, it is impossible to speculate as to whether the patient would have developed frank leukemia had she lived longer. The relation of this syndrome to that described by Van Den Berghe and his associates (Van Den Berghe, Cassiman, et al., 1974) is probably very close. These authors described an interstitial deletion of #5 (pter→q12::q31→qter) associated with erythroid hypoplasia (Figure 54), megakaryocytic nuclear anomalies, and a slightly reduced number of granulocytes with moderate increase in myeloblasts (up to 15%) in the marrow of subjects with refractory anemia.[4] The chromosomal findings in cultured blood lymphocytes were normal. Because in some of the patients 90% of the metaphases examined demonstrated deletion of #5, the authors speculated that the involvement of the stem cell may take place, though the evident clinical results from such involvement appear to be only related to erythrocyte maturation and production. Nevertheless, a number of these patients with 5q− have been described recently to have developed AML,[5] and patients with various leukemias and other hematologic disorders have been reported to have the 5q− anomaly in addition to other karyotypic changes in the marrow (Table 30).

In one of the constitutional forms of pure erythrocyte aplasia (Diamond-Blackfan syndrome), chromosomal changes have been described, but this disease apparently does not progress to leukemia (Amarose, Plotz, et al., 1967; Tartaglia, Propp, et al., 1966).

Refractory Anemias

Refractory anemias of several types may transform into leukemia, particularly of the acute myeloblastic type. Thus, an anemia is encountered in patients in whom no diagnosis can be established and who develop AML within a few months or a year after the onset of the anemia. Most of the affected subjects are elderly, though the disease may occur at any age, including infants. The presaging of AML by such an anemia is observed in 10–20% of all patients with AML, and except for the perplexing phase of the unexplained anemia, the course of the leukemia is no different from that of other patients without the anemic phase.

Table 30 5q− anomaly in various conditions

Disorder	Reference
Refractory anemia (RA) with slight increase in myeloblasts (8 cases) (7 cases with only 5q−, 1 case with other changes)	Van Den Berghe, Cassiman, et al., 1974; Van Den Berghe, David, et al., 1976a; Van Den Berghe, Michaux, et al., 1977b; Kaffe, Hsu, et al., 1978
	G. Sokal, Michaux, et al., 1975; Prieto Badia, et al., 1976, Verhest, Lustman, et al., 1977
Refractory anemia → AMMoL (1 case) 46XX,5q−/47,XX,5q−,+9	Verhest, van Schoubroeck, et al., 1976b; Verhest, personal communication
ANLL (5 patients, 3 with other changes)	Van Den Berghe, Michaux, et al., 1977b
(1 case)	Hagemeijer, Smit, et al., 1977
(3 cases)	P. Philip, 1977a
(1 case)	Mitelman, personal communication
(5 cases; includes one case of multiple myeloma →AML)	Rowley, 1976c
AMMoL (1 case)	Swolin, Waldenström, et al., 1977
EL (1 case)	Swolin, Waldenström, et al., 1977
PV → MMM (1 case)	Swolin, Waldenström, et al., 1977
Refractory anemia with morphologically immature megakaryocytes and/or an increase of myeloblasts (4 cases, 1 developed AML)	Swolin, Waldenström, et al., 1977
PV (2 cases)	Westin and Weinfeld, 1977
BP of CML (1 case)	Hagemeijer, Smit, et al., 1977
Aplastic Anemia (1 case)	Mitelman, personal communication
Sideroblastic anemia (SBA) (in atomic bomb survivor, 1 case) Bq−	Kamada and Uchino, 1976
Myelofibrosis (1 case) t(3q+;5q−)	Hagemeijer, Smit, et al., 1977
Pancytopenia with increased myeloblasts, hypergammaglobulinemia, etc. (2 cases)	Van Den Berghe, Michaux, et al., 1977b
AML del(5)(q15→q23) del(5)(q22;q31→q33)	Oshimura, Hayata, et al., 1976
Blast cell leukemia t(1q+;5q−)	Soderström, 1974
Anemia and pancytopenia t(5q−;21q+)	Meisner, Inhorn, et al., 1973
AML in child	Zuelzer, Inoue, et al., 1976
Thrombocytosis	Nowell and Finan, 1977
ALL	S. Abe, Kohno, et al., 1979
Myelofibrosis	Whang-Peng, Lee, et al., 1978
Histiocytic lymphoma	Mark, Ekedahl, et al., 1978
LS	Fukuhara and Rowley, 1978
MM del(5)(q22)	Liang, Hopper, et al., 1979
Lymphoma → ANLL (2 cases)	Whang-Peng, Knutsen, et al., 1979
Preleukemia	Pierre, 1978a
Pancytopenia	Nowell and Finan, 1978a
AL del(5)(q22)	Humbert, Morse, et al., 1978
RA → AML	Cabrol and Abele, 1978

Generally, the various reports on anemias and their chromosomal changes are difficult to interpret, not only because of considerable variation in terminology, thoroughness of patient work-up, inadequate or inappropriate cytogenetic examination (e.g., only blood leukocytes examined), but also due to incomplete description of the clinical and cytogenic findings.

Unexplained refractory anemias, usually normochromic and normocytic, though they may also be hypochromic and microcytic and at times macrocytic and normochromic (sideroblastic anemia will be discussed below), often constitute a permanent premonitory sign of leukemia and not infrequently are associated with chromosomal changes, particularly of group C chromosomes. These refractory anemias are usually associated with a normal or only slightly increased percentage of myeloblasts. When chromosomal changes are observed in refractory anemias, they persist into the frank leukemic phase of the disease. In this respect, such anemias could be considered preleukemias, although I feel that when karyotypic changes are encountered in these conditions, the leukemic process is already operative. Inasmuch as about 50% of the AL are not accompanied by cytogenetic changes in the leukemic cells, the finding of diploidy in these refractory anemias should not be considered evidence against a leukemic process. In fact, the bulk of these refractory anemias is not associated with any chromosomal changes, including a substantial number that do not progress to the leukemic stage. Crucial to the diagnosis of these cases is the percent of myeloblasts (and in very rare cases of ALL, lymphoblasts) in the marrow. Opinion varies, but generally it is assumed that 10% or more of myeloblasts are indicative of an AML, particularly if accompanied by certain laboratory features (e.g., low LAP), abnormal or unusual behavior of the cells in vitro, and appropriate clinical symptoms and signs. Macrocytic anemia and the presence of cells in the marrow suggestive of myeloblasts point to the possibility of Di Guglielmo's syndrome, particularly when accompanied by chromosomal changes.

In many cases of leukemia reported in the literature, cytogenetic examination was undertaken because of the presence of unexplained anemia. For example, Whang-Peng, Henderson, et al. (1970) in reporting on their series of patients with AML, described two subjects who were examined for refractory anemia before the diagnosis of AML. Knospe and Gregory (1971) and Nowell (1971b) also reported several such patients. In general, the experience of those reporting relatively large series of patients with AL studied cytogenetically has been to include several patients who initially were investigated because of unexplained anemia. However, forms of these anemias exist that are unrelated to leukemia and it was hoped that chromosomal analysis would be of help in separating these anemias from the preleukemic state and aid in their classification. Unfortunately, even though karyotypic changes may exist in some of these anemias, generally, the cytogenetic examinations have been more confusing than helpful.

The most comprehensive study on preleukemic states reported to-date involved 223 patients in whom a bone marrow aspiration was requested as part of the clinical evaluation of hematologic findings suggestive of preleukemia (Pierre, 1974, 1975). The latter were primarily cytopenias and refractory macrocytic anemias. Also included in the study were patients presenting with selected and potentially leukemic disorders: FA, BS, DS, pure erythrocyte aplasia, and Klinefelter's syndrome. Forty of the patients have developed leukemia; all had anemia and 22 had leukopenia and thrombocytopenia. The most striking characteristics of the subjects undergoing leukemic transformation were "anemia with oval macrocytes and abnormal erythrocytic maturation." The incidence of these was somewhat lower in the 180 patients who had not developed leukemia. The chromosomal changes observed in the 40 preleukemic subjects who developed AML consisted of karyotypic changes that did not differ from those of the group with overt AML, except for the occurrence of ring chromosomes in the latter group and the higher incidence of microchromosomes in the former group. The only statistically significant difference was the 66.8% diploidy in the 180 patients with preleukemia who have not developed leukemia and the 45 and 43.9% diploidy in the 40 patients with preleukemia who did develop AL, and in the overt acute leukemic subjects, respectively.

Krough Jensen and Philip (1973) described a case of pancytopenia in a 33-yr-old male in whom the bone marrow was studied twice at an interval of more than 4 yr. AML developed approximately 50 mo after the first examination. Both aspirates had a hyperdiploid mode, the majority of metaphases containing 49 chromosomes (46,XY,+3C). In the second aspirate, the chromosomes had a blurred, fuzzy appearance often observed in the cells of AL. Apparently, five other cases with pancytopenia, six with AA, and two with sideroblastic ane-

Table 31 Karyotypic changes associated with refractory idiopathic sideroblastic anemia (SBA)

Karyotype	Reference
46,XY,Fq− (2 cases)	Grouchy, Nava, et al., 1966c
46,XY,inv(F) (3 cases)	Grouchy, Nava, et al., 1966c
47,XY,+C	Rowley, Blaisdell, et al., 1966
46,X,−Y or 46,XY,−G	Goodall and Robertson, 1970
46,XX,2q+	Goodall and Robertson, 1970
47,XY,+8	Hellström, Hagenfeldt, et al., 1971
48,XY,+C,+D	Abrahamson and Edgington, 1972
47,XY,+C	Abrahamson and Edgington, 1972
47,XY,+C	Chapelle, Schröder, et al., 1972
46,XX,+C	Chapelle, Schröder, et al., 1972
45,XY,−C	Krogh Jensen and Philip, 1973
47,XY,+C	Lisker, Cobo de Gutiérrez, et al., 1973
45,XY,−C	Lisker, Cobo de Gutiérrez, et al., 1973
46,XY with chromosome breakage (Friedreich's ataxia with ?SBA)	Samad, Engel, et al., 1973
47,XYq−,+1	Warburton and Bluming, 1973
46,XX,20q−	M. M. Cohen, Ariel, et al., 1974
46,XX/47,XX,+8	Gahrton, Zech, et al., 1975
46,XX/46,XX,+8	Jonasson, Gahrton, et al., 1974
46/47,+F	Moretti, Broustet, et al., 1975
46/47,+C,t(Bq−;Cq+)	Moretti, Broustet, et al., 1975
46,XX,−17,i(17q)	Kamada in Killmann, 1976
46,XY,Ep+ or Cq− (SBA → AML)	Kamada and Uchino, 1976
46,XY→46,XY,Bq−	Kamada and Uchino, 1976
46,XY,Cq−	Kamada and Uchino, 1976
45,X,−Y	Krogh Jensen and Mikkelsen, 1976
44,X,−Y,−21/45,XY,−21/46,XY,1q+(q12−q44)	Yamada and Furasawa, 1976
47,+8	Bitran, Golomb, et al., 1977
45,X,−Y	Mende, Fülle, et al., 1977
45,XX,−G	Mende, Fülle, et al., 1977
47,XX,+G	Mende, Fülle, et al., 1977
47,XX,+8	Knuutila, Helminen, et al., 1978
47,XY,+8/46,XY	Shiloh, Naparstek, et al., 1979

mia, had a diploid chromosome number. The authors state that 3 of the 29 diploid cases (including the 13 mentioned above) studied by them developed AL, but it is not clear whether any of the 3 had pancytopenia, AA, or sideroblastic anemia (16 of the other patients had myeloproliferative disorders). Moretti, Broustet, et al. (1972, 1975) studied a group of patients that included three with sideroblastic anemia and seven (four males and three females) with refractory anemia and some myeloblastosis in the marrow. Four patients with 10% or less of myeloblasts in the marrow had an essentially normal diploid picture, whereas of the three patients with a higher percentage of myeloblasts in the marrow, two had definite chromosomal abnormalities, one consisting of hypodiploidy and another of hyperdiploidy (48 chromosomes with 12% myeloblasts in the marrow).

Sideroblastic Anemia

Sideroblastic (sideroachrestic; iron-loading) anemia (Table 31) probably represents an entity of several heterogenous disorders due to a variety of causes, but often associated with or leading to leukemia and other malignant diseases (e.g., acute myelofibrosis) (Yeung and Trowbridge, 1977). This type of anemia is characterized by the accumulation of excessive iron within the mitochondria of normoblasts. The latter process results in the so-called ringed sideroblasts, nucleated erythrocytes in which Prussian blue-positive granules form a full or partial ring around the nucleus. The excess iron in the mitochondria is probably a consequence of defective heme synthesis. Erythroid hyperplasia of the marrow is usually present. Included in this group of sideroblastic anemias is idiopathic refractory sideroblastic anemia (IRSA), which usually affects older adults and very rarely involves patients below the age of 50. Characteristic findings in IRSA include a normocytic or macrocytic anemia, occasional heavily stippled hypochromic cells, and lack of periodic acid-Schiff (PAS)-positive material in the normoblasts (Kushner, Lee, et al., 1971). Even though this form of sideroblastic anemia may be accompanied by definite chromosomal changes, apparently it is compatible with a relatively long survival, and most of the patients do not seem to develop leukemia or other forms of hematopoietic malignancy.

In reviewing the literature, it appears that the majority of cases with *refractory anemias* that have been described have consisted of the acquired sideroblastic anemia (Dreyfus, Rochant, et al., 1971). However, difficulties in classifying the various anemias arise from a lack of uniformity in the descriptions of the disease and its classification. Thus, Dreyfus, Sultan, et al. (1969) distinguished between two types of refractory anemias, one type being identical to chronic erythremic myelosis with an excess of myeloblasts in the marrow, and the other type being acquired sideroblastic anemia characterized mainly by a significant number of ring sideroblasts. According to these authors, a transition to AL is a frequent event in the first type and rare in the second. Due to the paucity of cases reported in the literature, the chromosomal data on these two types are sometimes difficult to separate and, hence, some overlap is unavoidable.

Of a total of 62 cases from the literature in which the chromosomal findings in sideroblastic anemia have been described, 31 had a normal chromosomal constitution.[6] Among the cases with aneuploid clones, hyperdiploidy, pseudodiploidy, and hypodiploidy[7] were common. One case had a pseudotetraploid mode (H. Singh, Boyd, et al., 1972), four cases had trisomy of #8, two a missing Y, and one an abnormal Y. Grouchy's (1966c) group attached particular significance to a structurally abnormal chromosome in group #19–#20. They claim to have seen this abnormality, which consisted either in a partial deletion of one arm (resembling the F-anomaly described by Lawler, Millard, et al., in 1970 in PV) or in a paricentric inversion in five out of six cases (Nava, Zittoun, et al., 1965). We observed one case of refractory anemia, probably of an acquired sideroblastic type, in whom 95% of the metaphases had 47 chromosomes with a missing #19–#20 and two small metacentric chromosomes; the morphology and the autoradiographic patterns suggested that the anomaly was derived from an endoreduplicated #19–#20 chromosome with deletion of both arms (Hossfeld, Schmidt, et al., 1972a). This patient went on to develop what resembled AL and from which he died. With the possible exception of group #19–#20 (and #8?) no other chromosome group appears to be preferentially affected in sideroblastic anemia; extra or missing chromosomes in group #6–#12 (#7; #8?) have been described, and the significance of this finding remains to be determined.

Included in the total of 62 cases with refractory anemia were some that progressed to AL after chromosomal anomalies had been detected.[8] Several cases already had leukemia at the time of chromosome analysis; one of them was chromosomally normal (Catovsky, Shaw, et al., 1971) and one had a prominent clone of hypodiploid cells in the marrow (Silberman and Krmpotic, 1969). One case has been reported to have terminated in acute myelofibrosis (Yeung and Trowbridge, 1977). It would be interesting to know whether the more malignant variant of refractory anemia (chronic erythremic myelosis) is associated more frequently with karyotypic changes than is sideroblastic anemia; at present, no conclusions can be drawn.

A Ph¹-like chromosome in a 76-yr-old male, thought to have refractory dysplastic anemia (erythremic myelosis) but also showing some evidence of early AML, was shown with Q-banding to be a short Y-chromosome in the marrow. The latter was of normal morphology in blood lymphocytes. Trisomy of #1 was also present in the marrow cells (Warburton and Bluming, 1973).

References

1. Lisker, Cobo de Gutierrez, et al., 1973; Nowell, 1965b, 1971a; Nowell, Jensen, et al., 1976a.
2. Furusawa, Kawada, et al., 1971; Heimpel, 1972. Linman and Bagby, 1976; Milner, Testa, et al., 1976; Nowell, Jensen, et al., 1976a; Saarni and Linman, 1973.
3. Bauke, Hoerler, et al., 1973; Baserga, 1970; G. E. Bloom and Diamond, 1968; Bottura and Ferrari, 1962b, Hellriegel, 1975; Knospe and Gregory, 1971.
4. Sokal, Michaux, et al., 1975; Van Den Berghe, Cassiman, et al., 1974; Van Den Berghe, Michaux et al., 1977b; Verhest, Van Schoubroeck, et al., 1976b.
5. Rowley, 1976c; Swolin, Waldenström, et al., 1977; Van Den Berghe, David, et al., 1976b.
6. Bitran, Golomb, et al., 1977; Goodall and Robertson, 1970; Grouchy, Nava, et al., 1966c; Heath and Moloney, 1965a; Krogh Jensen and Mikkelsen, 1976; Moretti, Broustet, et al., 1975; Nowell, 1971b, c; Visfeldt, Franzén, et al., 1970.
7. Freireich, Whang, et al., 1964; Heath, Bennett, et al., 1969; Mende, Fülle et al., 1977; Reimann, Endogan, et al., 1965; Silberman and Krmpotic, 1969.
8. Freireich, Whang, et al., 1964; Heath, Bennett, et al., 1969; Kamada and Uchino, 1976.

ABBREVIATIONS USED FOR CLINICAL TERMS

A	=	Leukemia characterized by an abnormal karyotype
AA	=	Leukemia characterized by *only* (cytogenetically) abnormal cells in the marrow
AIL	=	Angioimmunoblastic lymphadenopathy
AL	=	Acute leukemia
ALL	=	Acute lymphoblastic leukemia
AML	=	Acute myeloblastic leukemia
AMMoL	=	Acute myelomonocytic leukemia
AMoL	=	Acute monocytic leukemia
AN	=	Leukemia characterized by the presence of (cytogenetically) normal and abnormal cells in the marrow
ANLL	=	Acute nonlymphocytic leukemia
APL	=	Acute promyelocytic leukemia
BL	=	Burkitt lymphoma
BM	=	Bone marrow
BP	=	Blastic phase (of CML)
CLL	=	Chronic lymphocytic leukemia
CML	=	Chronic myelocytic leukemia
CMMoL	=	Chronic myelomonocytic leukemia
CP	=	Chronic phase (of CML)
DMS	=	Double-minute chromosomes
EL	=	Erythroleukemia
EM	=	Electron microscopy
ET	=	Essential (primary or idiopathic) thrombocythemia
FL	=	Follicular lymphoma
HD	=	Hodgkin's disease
IAA	=	Idiopathic aplastic anemia
LAP	=	Leukocyte alkaline phosphatase
LS	=	Lymphosarcoma
MA	=	Megaloblastic anemia
MAKA	=	Major karyotypic abnormalities
MIKA	=	Minor karyotypic abnormalities
MF	=	Myelofibrosis
MM	=	Multiple myeloma
MMM	=	Myeloid metaplasia with myelofibrosis
MS	=	Myeloproliferative syndrome
N	=	Leukemia characterized by a normal (diploid) karyotype
PB	=	Peripheral blood
PCC	=	Premature chromosome condensation (prophasing)
PNH	=	Paroxysmal nocturnal hemoglobinemia
Ph^1	=	Philadelphia chromosome
PV	=	Polycythemia vera
RA	=	Refractory anemia
RC	=	Reticulum cell sarcoma
TdT	=	Terminal deoxynucleotidyl transferase

10

The Leukemias
Chronic Granulocytic Leukemia

Introduction

Chronic Myelocytic Leukemia (CML)
Definition of the Ph^1
Nature of the Ph^1
Possible Mechanism for the Genesis of the Ph^1-Chromosome
The Ph^1 and the Genesis of CML
Some Novel Views of the Ph^1
Pathogenetic Aspects of the Ph^1 in CML and Related Disorders
Polyploid Cells with Ph^1-Chromosomes
Nature of Precursor Cells of CML and the Ph^1
Extramedullary Origin of Ph^1-Positive Cells, Including Those in the Blastic Phase
Therapy and the Ph^1
Ph^1-Negative CML
Value of Detailed Chromosome Studies on Large Numbers of Cells in CML
Clinical Implications of Chromosomal Findings in CML
Unusual and Complex Ph^1-Translocations
Ph^1 Without Evidence of Translocation
Ph^1-Positive CML and the Missing Y-Chromosome
Ph^1-Negative CML with a Missing Y-Chromosome
Further Aspects of CML and the Missing Y-Chromosome
Additional Chromosomal Changes in the Chronic Phase of CML

CML in Children
Eosinophilic Leukemia
Chromosome Findings in Eosinophilic Leukemia
Basophilic Leukemia
Leukocyte Alkaline Phosphatase and the Ph^1
Muramidase (Lysozyme) and the Ph^1
Peroxidase Activity, Periodic Acid–Schiff, and Sudan Black B Reactions of Blasts in CML
Cell Surface and Other Markers in the Blastic Phase of Ph^1-Positive CML
Ph^1 in Hematopoietic Disorders Other Than CML
Chromosomal Findings in the Blastic Phase of CML
Multiple Ph^1-Chromosomes in the Blastic Phase
Double-Ph^1, Blastic Phase, and Cell Source in CML
Ph^1-Negative, Diploid Cells in CML
Karyotypic "Staging" of CML and Its Prognostic Implications
Clonal Origin and Evolution of CML
Extramedullary Aspects of Ph^1-Positive CML
Miscellaneous Aspects of Ph^1-Positive CML

Chronic Myelomonocytic Leukemia

Introduction

The chromosomal changes in human leukemia are almost invariably confined to the leukemic cells.[1] Thus, the karyotypic picture is most reliably established by examining the cells in the marrow, preferably with a "direct technique" (Shiloh and Cohen, 1978). This does not mean that resort should not be made to short-term culture of marrow and/or blood cells, especially when the former does not contain sufficient metaphases for chromosome analysis. However, there is a distinct tendency under these circumstances for the normal diploid cells to overgrow the leukemic ones, particularly when PHA is used in the culture and, thus, yield a spuriously normal karyotypic picture in leukemia (Sandberg, Ishihara, et al., 1962a; Leon, Reichhardt, Epps, et al., 1961). This differs from the findings in marrow cells, where diploidy does not militate against the diagnosis of AL; however, the finding of aneuploidy of any type strongly indicates the presence of a malignant (usually leukemic) process. In the case of Ph[1]-positive chronic myelocytic leukemia (CML), we have obtained satisfactory results in the absence of PHA, as we have with short-term (12–48 hr) culture of marrow cells of CML and AL. In some cases of leukemia with high leukocyte counts, the blood cells can be processed by a direct technique, similar to that of the marrow and, not infrequently, a large number of metaphases is available for analysis.

Because human leukemias can be considered, for all practical purposes, to be essentially diseases of the marrow, direct examination of marrow cells for their chromosome constitution would appear putatively to be the most reliable means of establishing the karyotypic picture in leukemia by a number of different methods.[2] Not only is the marrow, generally, the source of the leukemic cells and, hence, the cytogenetic findings in these cells should represent the most reliable karyotypic picture of the disease, but there are also several other cogent reasons for examining the marrow cells by a direct method. As stated above, the chromosomal changes in leukemia are confined to the leukemic cells and, hence, examination of cells other than those of the marrow (or of the leukemic cells in other tissues, e.g., spleen), such as the blood lymphocytes (Goh, 1964), marrow (Greenberg, Wilson, et al., 1978a) or skin fibroblasts, etc., will usually reveal the karyotype of the individual, which in the preponderant number of patients with leukemia is diploid. The number of mitotic cells is much higher in the marrow than in any other readily available cell source, and the marrow usually affords a sufficient number of metaphases for establishing the karyotypic picture with reliability. The marrow also contains dividing cells of different origin and characteristics (erythroid, granulocytic, megakaryocytic) and, thus, the chromosomal picture of all the cells present in the marrow can be established. The importance of chromosomally normal cells in the *marrow* in AL and CML is another cogent and practical reason for examining the marrow in addition to or rather than blood cells, though examination of the latter can be useful for diagnostic purposes (e.g., is the CML Ph[1]-positive or negative?) or to possibly ascertain blastic transformation at extramedullary sites, in which case aneuploid cells may appear in the blood before the marrow. Even though dividing leukemic cells may be present in the circulating blood, and in a small number of cases their number may, indeed, be sufficient for chromosome analysis (Fitzgerald, 1961a), in most cases of leukemia the karyotypic profile cannot be established by examining the blood. When resort is made to culture of blood cells (with or without PHA), the normal lymphocytes have a tendency to overgrow the leukemic cells (if, indeed, the latter grow at all) and a spurious diploid picture is often obtained (Sandberg, Koepf, et al., 1961; Sandberg, Ishihara et al., 1962a,b). This does not mean that in rare cases one cannot obtain a large number of dividing leukemic cells, particularly in the absence of PHA, upon submitting blood to culture conditions for 24–72 hr (Dieška, Izaković, et al., 1967b). The reservations stated above can also be made regarding culture of marrow, but in the case of marrow we have found that short-term culture for 12–24 hr often yields metaphases in a marrow essentially devoid of these originally and that these are representative of the true chromosomal picture and similar to the findings obtained with a direct method.

In all probability the addition of colchicine or colcemide to marrow material aids not only in arresting cells in mitosis and, consequently, increasing the number of metaphases for analysis, but also in causing morphologic changes in the chromosomes leading to their clearer delineation and chromatid visualization. However, excessive exposure to these agents may obscure chromosomal structure and actually make the metaphases unsuitable for detailed analysis.

A few words are in order regarding long-term cultures of human leukemic cells, particularly those of AL. A sine qua non of such cultures should be the demonstration that the original karyotypic abnormality present in the leukemic cells prior to culture has persisted.[3] Even though modification of an aneuploid karyotype may take place in vitro, the basic chromosomal motif of the cells should persist. Except possibly for some Ph^1-positive cell lines (C. B. Lozzio and Lozzio, 1975; C. B. Lozzio, Lozzio, et al., 1976; B. B. Lozzio, Machado, et al., 1976, 1978; Sonta, Minowada, et al., 1977) and a few of acute leukemic origin (Minowada, Oshimura, et al., 1977; Minowada, Tsubota, et al., 1977), no convincing evidence exists that human leukemic cells have been maintained in long-term culture. This statement does not preclude the possibility that a virus related to the leukemia (Hall and Schidlovsky, 1976) or other factors associated with the leukemic process may not have been propagated in vitro, but it does say that, generally, the persistence of an abnormal chromosomal picture in leukemic cells cultured in vitro for lengthy periods of time has not been convincingly established.*

When examination of the marrow, which contains the usual types of cells, yields results that characterize 100% of the cells, we may assume that the karyotypic picture reflects that of the three major cell lines, i.e., erythroid, myeloid, and megakaryocytic. However, in cases of AL or in marrows having mixed chromosomal populations this may not be the case. In AL the marrow may be crowded overwhelmingly with abnormal leukemic cells, so that either normal cells are extremely rare or not in division and, hence, the cytogenetic analysis will reveal a totally aneuploid chromosomal constitution without reflecting the diploid cells in the marrow. This does not mean that totally aneuploid marrows do not exist in leukemia, but caution must be exercised in making such a statement until a sufficient number of cells and marrows have been examined. A mixed karyotypic population of cells, i.e., aneuploid and diploid, coexisting in the same marrow, offers difficulties regarding the cell source of the chromosomally different metaphases. One series of cells may have a much higher mitotic index than another and, thus, the preponderant number of metaphases, either aneuploid or diploid, may originate from this series only. On the other hand, more than one cell series can have such a chromosomally mixed population, and the interpretation under these conditions is extremely difficult, unless specific aspects of the cells (presence of hemoglobin, isoenzymes) can be demonstrated.

The effects of therapy on the marrow picture in leukemia have led to some confusion in the literature. Therapy may affect the chromosomal picture in leukemia, particularly in AL, by affecting the cellular composition of the marrow. Not only can therapy lead to a remarkable reduction in leukemic cells, and, thus, to a greatly reduced number of aneuploid cells, but it may also result in the disappearance of the abnormal metaphases and lead to the spurious diagnosis of a diploid leukemia, when, in fact, the chromosomally normal cells are not of leukemic origin but of normal cells repopulating the marrow as a result of the beneficial effects of the antileukemic therapy.

Chronic Myelocytic Leukemia (CML)

The chromosomal findings in chronic myelocytic leukemia (CML) constitute a most exciting, interesting, and, yet, provocative chapter in the history of human cytogenetics and a milestone in medicine. Chronic myelocytic leukemia is a disease characterized by a greatly increased number of granulocytic cells (primarily of the myelocytic variety), in the marrow, blood, spleen, and other organs, often leading to splenomegaly, hepatomegaly, lymph node enlargement, and extramedullary hematopoiesis (Table 32). The disease may occasionally be preceded or its development accompanied

---

*Apparently, Ph^1-positive cells have survived and proliferated when injected into newborn hamsters (Miyoshi, Kubonishi, et al., 1976) or hereditary asplenic–athymic mice (B. B. Lozzio, Machado, et al., 1976).

Table 32 Some salient cytogenetic, clinical, and survival features of CML

Percent Ph¹-positive	85% (range in literature 74–100%)
Percent Ph¹-negative	10–15% (range in literature 0–26%)
Percent Ph¹-positive with missing Y	Ca 8–10% of male patients (range in literature 0–14%), most often in patients < 60 yr old
Percent with unusual or complex Ph¹-translocations	Probably < 10% of all cases with Ph¹-positive CML
Percent with only Ph¹-positive cells	67–84%
Most common karyotypic changes during the CP (in addition to the Ph¹)	+Ph¹,+8,−Y (incidence for *each* is about 3% of the cases)
Percent of cases developing further chromosomal changes in BP	Ca. 70%
Most common karyotypic changes in the BP of CML	Hyperdiploidy → (over 60% of all cases) Hypodiploidy ←, exclusive of −Y cases, (< 10% of all cases) +Ph¹,+8,iso(17q),+19 Chromosome number ranges from 44–58 (most common 46–49, peak at 47) Loss of chromosomes is infrequent—most commonly autosomes in group C or E
Two Ph¹-chromosomes in CP	May be encountered in as much as 10% of all patients
Normal (karyotypically) cells in CP and BP	In CP: Exact % unknown, probably much less than 30% In BP: Rare, < 5%
Median survival Ph¹+ cases Ph¹− cases −Y,Ph¹+ cases	>4 yr <1 yr >6 yr
Survival after BP (all cases)	Depending on type of therapy; usually about 2–3 mo; with newer therapeutic approaches 3–6 mo and occasionally longer
Length of CP in Ph¹-positive CML	3–4 yr
Signs of impending BP	*Change* in chromosome picture, particularly the appearance of an i(17q) Increase in spleen size Fever and general deterioration of health Increase in leukocytes Anemia Increased % of myeloblasts (>20%) in BM

by concomitant thrombocythemia (Bauters and Goudemand, 1975; Hossfeld, Bremer, et al., 1975) and, more rarely, erythremia. Often, CML terminates in a blastic phase (BP) reminiscent of AML (Wintrobe, 1974; Moloney, 1977). To date, CML is the only human malignant disease in which a characteristic and consistent karyotypic finding, i.e., the Philadelphia chromosome (Ph¹), has been established. Since its description by Nowell and Hungerford in 1960, the Ph¹ remains the most significant and interesting, but also the most puzzling and perplexing chromosomal finding in human and experimental oncology. It is the only chromosome anomaly closely related to a specific type of neoplastic disease, i.e., CML,[4] and, hence, by far the strongest argument in favor of the theory that chromosomal changes are causally related to neoplasia. With techniques currently available, it is doubtful whether a cytogenetic finding comparable in significance and consistency to those of the Ph¹ is likely to be established in the near future.* As will be apparent from the content of this and other chapters, the continued surveillance of the cytogenetic picture in CML or, for that matter, in other leukemias and myeloproliferative disorders may ultimately have an important bearing on the therapeutic approaches to be taken in this dis-

*The 14q+ anomaly in lymphoproliferative disorders (e.g., Burkitt lymphoma, lymphosarcoma, etc.) is slowly emerging as a candidate for a specific karyotypic anomaly comparable to that of the Ph¹ in CML.

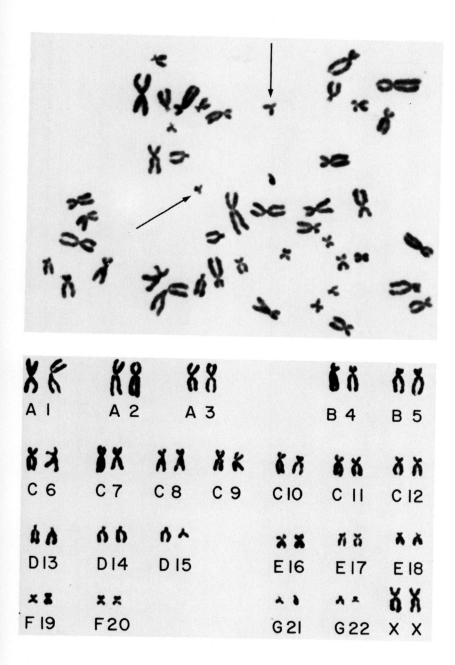

ease and, in particular, on the prediction of the onset of the BP of CML and the treatment of this phase of the disease.

Definition of the Ph^1

The Ph^1 is defined (Figure 55) as a postzygotic anomaly associated most often with myeloproliferatives diseases, preponderantly CML, and

55 A. Metaphase and karyotype (not banded) of a marrow cell from a patient with Ph^1-positive CML published in 1962 (Sandberg, Ishihara, et al., 1962b). Arrows point to the Ph^1 and a D-group autosome (#15) with deleted long arms. In the karyotype the Ph^1 is shown as belonging to group #22. Deletions of the long arms of #13 and #15 are not unusual in myeloproliferative disorders (Kohno, Van Den Berghe, et al., 1978).

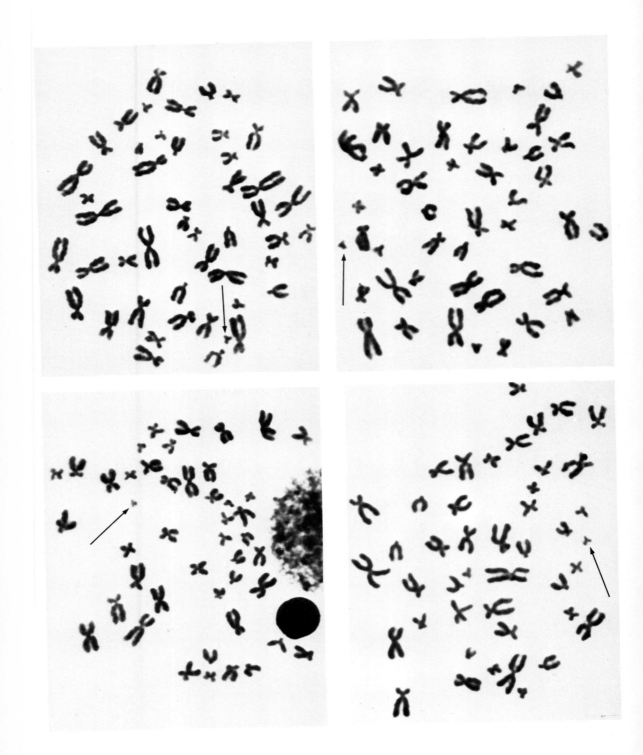

55 B. Four marrow metaphases each containing a Ph¹ (arrows) from cases of CML.

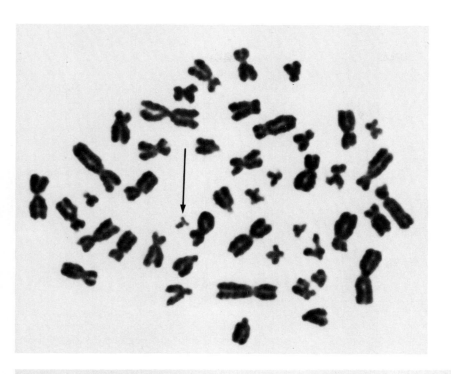

55 C. Metaphase and its karyotype of a Ph[1]-positive cell from a patient with CML (Sandberg, Ishihara, et al., 1962b) and showing the Ph[1] as a #22.

Table 33 Reports on chromosome studies in twins with leukemia

Reference	Diagnosis
Goh, 1965	CML
Goh and Swisher, 1965	CML
Dougan, Scott, et al., 1966	CML
E. M. Jacobs, Luce, et al., 1966	CML
Woodliff, Dougan, et al., 1966	CML
Woodliff and Onesti, 1967	CML
Goh, Swisher, et al., 1967	CML
Tokuhata, Neely, et al., 1968*	CML
Holton and Johnson, 1968	CML
E. W. Jackson, Norris, et al., 1969	CML
Kosenow and Pfeiffer, 1969	CML
Bauke, 1969; 1970b‡	CML
Hilton, Lewis, et al., 1970	AML
Sandberg, Cortner, et al., 1966	AML
M. A. Pearson, Grello, et al., 1963	AL
Svarch and de la Torre, 1977‡	AMMoL (?)
Kiossoglou, Rosenbaum, et al., 1964a	AL

*Ph¹-positive CML in twins and their brother.
‡Ph¹-positive in one of identical twins.

occurring in cells of the hematopoietic system only. The Ph¹ is due to deletion of the long arm of a #22, most often involving a reciprocal translocation with a #9, i.e., t(9;22)(q34;q11) or, rarely, other chromosomes. Even though uncertainty existed as to whether band q11 or q12 was involved in the Ph¹ formation, and there are those who still are uncertain, the general opinion holds that it is band q11 that is affected in #22, leading to the formation of the Ph¹.[5] The Ph¹ behaves like an autosomal dominant gene; the loss of an entire #22 does not produce CML. The postzygotic, acquired nature of the Ph¹ is borne out by these findings: (1) in identical twins, of whom one has CML, only the hematopoietic cells of the affected twin were Ph¹-positive (Table 33); (2) examination of the marrow cells of either the parents or siblings of individuals affected with Ph¹-positive CML fails to reveal such cells in the relatives (Bauke, 1970b); and (3) in patients with CML, the Ph¹ has never been observed in nonhematopoietic cells, e.g., lymphocytes or fibroblasts (Baikie, Court Brown, et al., 1960b; Baikie, Garson, et al., 1969; Maniatis, Amsel, et al., 1969). The pathogenetic implications of an inherited[6] or a radiation-induced Ph¹-*like* chromosome,[7] which occurs in other tissues besides the hematopoietic ones, are totally different from those related to the Ph¹ in CML. With one exception (Berger, 1965), none of these persons, even after an observation time of several years, has developed a myeloproliferative disease. Furthermore, banding techniques, developed subsequent to these publications, should shed considerable light upon the nature of the Ph¹-like chromosome in these cases, i.e., whether the chromosomal abnormality is the result of a characteristic translocation between #9 and #22 or is an abnormal #21. The presence of a Ph¹ in hematopoietic cells is invariably associated with a myeloproliferative disease, no matter how high the percentage of Ph¹-positive cells. Not a single case with Ph¹-positive marrow cells and a normal hematological status has been described to-date.

In interpreting past reports (before banding) dealing with a Ph¹ in unusual cases of CML or other diseases, caution must be exercised in accepting this cytogenetic finding.[8] Unless a deletion of #22 (22q−) and translocation between the #22 and another chromosome have been demonstrated [the rare occurrence of a Ph¹ (22q−) without evidence of translocation will be discussed below], the possibility exists that a deletion of a #22 (or #21), leading to a Ph¹-like chromosome, may have taken place with a loss of the deleted material; whether this can be considered a bona fide Ph¹ remains to be determined. Both events, i.e., specific deletion of the long arm of #22 and translocation to another chromosome, are essential for accepting Ph¹-positivity. Such caution applies in particular to conditions other than CML.

Nature of the Ph¹

In their original description of an "abnormally small" chromosome found in the leukemic cells of two males with CML, Nowell and Hungerford (1960a) considered the chromosome involved to be the Y. Having noted the same anomaly in the leukemic cells of females with CML, Baikie, Court Brown, et al. (1960b, d) established that the abnormality involves one of the small acrocentric autosomes and applied the term "Philadelphia-chromosome" to this abnormality (Tough, Court Brown, et al., 1961), according to the location of the laboratory in which the karyotypic abnormality was first recognized. When it became clear that the Ph¹ is probably derived from one of the small acro-

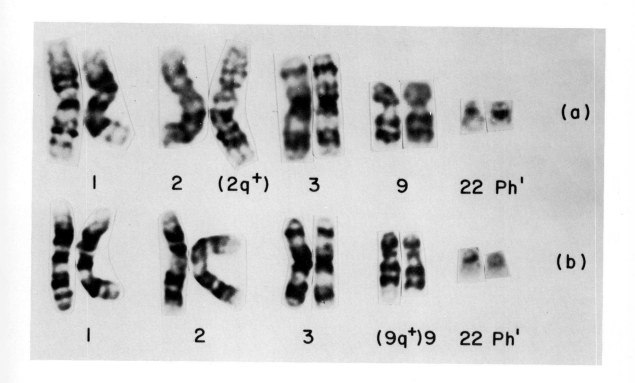

56 Partial karyotypes depicting the usual (standard) Ph¹-translocation (b), i.e., that between the long arm of chromosome #22 and the long arm of one of the chromosomes of group #9, and an unusual translocation (a) between chromosomes #2 and #22.

centric chromosomes, which has lost a portion of its long arms, the questions arose as to whether the missing chromosomal material was deleted or translocated, whether the amount of missing chromosomal material was always identical, and whether the Ph¹ was a member of group #21 or #22 (Chandra, 1968; Woodliff, Dougan, et al., 1965). The last question will be dealt with first, but each can be definitely answered now. Until 1970 the Ph¹ was generally assumed to be derived from a chromosome of group #21. This was based on several observations: (1) the interpretation that the Ph¹ appears to replace one of the larger G-group chromosomes;[9] (2) extrapolation from the theory with regard to the behavior of the enzyme leukocyte alkaline phosphatase (LAP); and (3) the higher incidence of leukemia in patients with mongolism (DS), in which it was assumed that the same chromosome pair was affected by trisomy as the partial monosomy in CML. However, in the past, no unequivocal morphological differences existed between groups #21 and #22, and we know now that the behavior of LAP is unrelated to visible chromosome anomalies. Parenthetically, we showed in 1962 in our first publication on this subject (Sandberg, Ishihara, et al., 1962b) that the Ph¹ was one of the #22 autosomes. In subsequent publications, both the prevailing cytogenetic opinion and the insistence of some editors led us to classify the Ph¹ as a #21. Autoradiography did not help to clarify the question as to the origin of the Ph¹, mainly because of the absence of clear-cut morphological differences and autoradiographic behavior between the G-group chromosomes. Accordingly, the results of such studies were contradictory, some indicating the Ph¹ to be derived from pair #21, others from pair #22.[10]

Evidence for the identity of the Ph¹ supplied before the introduction of banding techniques was the publication of Prieto, Egozcue, et al. (1970) in which these authors combined morphological and autoradiographic information to establish a #22 as the source of the abbreviated Ph¹. These authors studied a patient with CML, one of whose G-chromosomes had remarkably enlarged satellites, both in the normal lymphocytes and on the Ph¹. In the normal lymphocytes of the patient, the chromosome with pronounced satellites was a member of the pair that replicated early and, thus, it could be deduced that the Ph¹ was one of an early-replicating pair and in this way could be distinguished from the extra G-group chromosome that is found in patients with DS, which is late labeling. Prieto, Egozcue, et al. (1970) suggested from their data that, inasmuch as trisomy #21 was the accepted nomenclature for DS, the Ph¹ should be considered a deleted #22. The priority of the Spanish workers in identifying the Ph¹ has not been sufficiently appreciated.

Once the fluorescence banding technique became available, the problem was readily solved (Figure 56). It was demonstrated by Caspersson, Gahrton, et al. (1970) and soon confirmed by O'Riordan, Robinson, et al. (1971) that, firstly, different G-group chromosomes are involved in DS (mongolism) and the Ph¹ anomaly; and, secondly, that patients with DS are trisomic for the smaller pair of acrocentric chromosomes, whereas Ph¹-positive cells are monosomic for the larger pair. Similar results were revealed by the Giemsa banding technique (W. Müller and Rosenkranz, 1972b). Only for convenience, but in contrast to the Denver nomenclature according to which #1 is the largest and #22 the smallest, the chromosome that is trisomic in mongolism is #21, and consequently the Ph¹ is a #22.

The clarification of the nature and genesis of the Ph¹ and the fate of the deleted part of #22, now easily identified with rather simple techniques (Sehested, 1974; Buckton and O'Riordan, 1976), constitutes an excellent example of the dependence of cytogenetic advances on methodologic evolution.

Whether the formation of the Ph¹ is the consequence of a terminal or an intercalary deletion was a matter of considerable debate in the past. The simplest and most plausible mechanism seemed to be terminal deletion of a G-group chromosome, which is in good agreement with the finding of Caspersson, Gahrton, et al. (1970) that the breakage apparently occurs in the fluorescent region of the long arm of #22. The broken ends must have some telomeric properties that prevent them from fusing. The fate of the chromatin lost subsequent to the formation of the Ph¹ was unknown until 1973. The question existed: does it, in fact, get lost (deleted) or is it translocated onto another chromosome? Even with optimally prepared material, past techniques for chromosome identification offered no convincing evidence of translocation in Ph¹-positive cells. The Ph¹-positive case of CML described by Bottura and Coutinho (1971a) as a presumed 21/21 translocation product has been shown by these authors (1974) to be a t(21q+,22q−) with the Ph¹ being due to the usual deletion of #22. The introduction of banding techniques afforded the opportunity for visualization and identification of small segments of chromosomes. Thus, Rowley (1973e) showed that the Ph¹ is due to deletion of the long arms of chromosome #22, the deleted portion being translocated onto the long arms of a #9, i.e., t(9;22)(q34;q11). There is little doubt that in the preponderant number of cases with CML, the Ph¹ is the result of such a translocation,[11] but an array of other sites other than 9q for this translocation has been described (see below and Table 35).

The deleted chromatin from #22 is not late-labeling, Giemsa-positive, or highly fluorescent and, hence, it must contain genetically active euchromatin. A Giemsa-positive band (also shown with R-banding, the reverse of G-banding) is present on the intact #22 at the site of the postulated break point. If this band does not have telomeric (end point) properties, i.e., lack of the tendency to fuse with other telomeres or ends of broken chromosomes, the deletion must be intercalary (and require two breaks) or must be the result of translocation. The morphology of the Ph¹-chromosome is, generally, remarkably uniform in the early stages of CML, even though unevenness in coiling may be seen in individual cells. The chromatin missing from the #22 represents only a small amount of the total DNA (<0.5%) present in the cell and the Ph¹ behaves like an autosomal dominant gene (Borgaonkar, 1973).

Another question with regard to the nature of the Ph¹ is whether the same amount of chromatin is translocated in each case. A number of authors have pointed to differences in size and shape in the Ph¹ among different cases and, rarely, within the same case.¹² An interesting variability in morphology is not infrequently observed among supernumerary Ph¹-chromosomes often seen in association with the BP of CML; this is particularly perplexing because the extra Ph¹(s) are thought to arise through a process of endoreduplication or nondisjunction. In some a dicentric Ph¹ is encountered, probably having the same significance as two Ph¹-chromosomes. More difficult to explain is the morphologic variation not related to dicentric formation; even though some of the heterogeneity of morphology of the Ph¹ may be due to minor rearrangements, inversions, or translocations, there is the possibility that this morphologic heterogeneity may be the result of differences in the amounts of chromatin deleted from and/or translocated to the #22. The Ph¹ can be either very small or relatively large (Dosik and Verma, 1977), acrocentric or metacentric. In the case of atypical Ph¹-chromosomes, particularly metacentric ones, the possibility of a dicentric Ph¹ must be ruled out with various banding techniques before ascribing the unusual morphology of the Ph¹ to such mechanism as pericentric inversion or others (Schneider, Stecher, et al., 1967). The most reasonable explanation for these morphological differences in the variability in the amount of chromatin deleted from #22, though no cogent evidence exists for this concept. According to some, it seems best not to consider the Ph¹ as an abnormality that originates by the loss from #22 of a well-defined and definite amount of DNA, an opinion based mainly on cytophotometric results on a single case (Rudkin, Hungerford, et al., 1964). Provided the deletion of the #22 involves loss of genetically active DNA, the important question remaining to be clarified is whether differences in the degree of deletion have any bearing on the course of the disease.

The short arm of the Ph¹ remains intact, though it is subject to coiling variations probably related to the shortened long arm. A normal frequency of association of the Ph¹ with other acrocentrics, though lower than that of normal lymphocytes, has been seen. The Ph¹ has a slower recovery rate after colcemide when compared to normal lymphocytes. In these studies, however, the behavior of the Ph¹ should have been compared to that of normal myeloid cells in the marrow, because leukemic cells may have abnormal centriole structure.

The description of dicentric Ph¹-chromosomes, established with banding techniques, indicates another variant of its morphology. In one study, such a dicentric Ph¹-chromosome was observed in five patients (Figure 57), in one of whom two such dicentric chromosomes were seen (Whang-Peng, Knutsen, et al., 1973). The most likely explanation for the formation of the dicentric Ph¹, which with regular staining appears metacentric but with banding has been shown to have two centromeres, is that it represents two Ph¹-chromosomes joined at the satellite region or the short arms by either fusion or translocation. At this region the bond between the two is weak, and either the hypotonic treatment or other forces during the preparation of the slides may frequently break them into two regular Ph¹-chromosomes. The dicentric chromosome itself can replicate abnormally, as evidence by the high incidence of two dicentric Ph¹-chromosomes in the patient already mentioned. The clinical significance of the dicentric chromosome is probably similar to that of two Ph¹-chromosomes, in that it is often an indication of the onset of the BP and is usually associated with a relatively short survival, once the dicentric Ph¹ appears.

In summary, then, the Ph¹-chromosome results from the deletion of the long arm of #22 at q11 (Pravtcheva and Manolov, 1975) with translocation of the material to the long arm of a #9 (q34), the translocated material apparently being equal to that missing from the long arm of the #22. In other words, the amount of additional material on #9 is approximately equal in size to the amount of chromosomal material missing from the Ph¹ (22q−). These findings strongly suggest that a translocation had occurred between the #9 and the Ph¹, producing the 22q− and 9q+. Though it cannot be visualized with presently available techniques, the studies of Mayall, Carrano, et al. (1974, 1977), based on the measurement of the DNA content of human metaphase chromosomes by means of a quantitative cytochemistry, scanning cytophotometry, and image-processing techniques, and the studies of Watt and Page (1978) indicate that the telomeric area of the long arm of #9 (q34) has been translocated to the deleted arms of the #22, resulting in a *balanced* reciprocal translocation. Evidence that

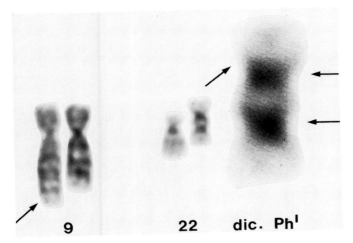

57 Dicentric Ph¹-chromosome which without banding appeared as a metacentric chromosome. Translocation to the long arm of one #9 is evident and the other #9 appears to be normal. The signficance of the dicentric Ph¹ is equivalent to that of two Ph¹-chromosomes, a situation usually present in the BP of CML.

(*Whang-Peng, Knutsen, et al., 1973.*)

band 22q11 is related to CML is supplied by the study of an interesting family in which the father and two of his three children carried an abnormally short #22, resembling the Ph¹ but due to t(11;22)(q25;q13), i.e., the break point on #22 was at the q12/13 band interface instead of q11/12 interface of the Ph¹. The absence of leukemia or other hematological disorders in members of this family suggests that the genetic site on #22 concerned with abnormal cell proliferation in CML is located in the 22q11 band (Fitzgerald, 1976). This interpretation has been challenged (Hecht and Kaiser McCaw, 1977*a*), though not without generating some controversy (Fitzgerald, 1977). Furthermore, opinion exists that the mechanism producing the Ph¹ is not necessarily universal (Mitelman, 1974*c*). Abe, Morita, et al. (1978) reported a 3-yr-old girl with multiple congenital malformations whose cells, as well as those of her normal mother, had a 45,X,−22,+der(X), rcp(X;22)(p22;q11)*mat* karyotype. No hematological changes were observed either in the child or her mother, leading the authors to doubt "whether the chromosomal rearrangement which causes the Ph¹, especially the break at the band 22q11 prior to the translocation, is relevant to the genesis of CGL."

Possible Mechanism for the Genesis of the Ph¹-Chromosome

The results of a study in my laboratory on the effects of mitomycin C (MMC) on human chromosomes and the formation and behavior of quadriradials, in particular, may shed some light on the possible mechanism by which the Ph¹ is formed (Figure 58) (Oshimura, Sonta, et al., 1976). It is possible that a nonhomologous exchange between #9 and #22 may be responsible for the Ph¹. Through segregation at anaphase, daughter cells with a Ph¹ and others with normal karyotypes could be produced through alternate-segregation, or two daughter cells with abnormal #9 and #22 could be produced through adjacent-segregation. If the former mechanism prevailed, Ph¹-negative CML may represent overproliferation of the karyotypically normal (biologically leukemic) cells over the Ph¹-positive cell resulting from alternate-segregation. In the case of adjacent-segregation, Ph¹-positive cells without translocations could be produced. Even though cells with only a translocation on #9 but not containing a Ph¹ have not been reported, detailed analysis of Ph¹-negative cases might reveal such cells. It is also possible that the disease associated with 9q+;22 may manifest itself not in the usual form of CML or any other leukemia; hence, it becomes important to analyze all conditions in search of this particular karyotypic picture. Possibly, the variable incidence of each karyotype, i.e., 9q+;22q−;9;22q−;9q+;22 and 9;22, can be explained by the growth advantages characterizing the cells with the standard type of Ph¹-translocation in CML, which tend to overgrow the cells with the other karyotypes; a case of Ph¹-positive CML (5-yr-old male) changing to a normal karyotypic picture in the BP has been reported (Lawler, 1977*a*). Variant

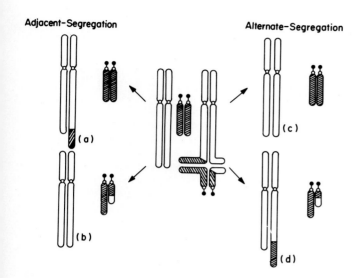

58 Genesis of the Ph¹-chromosome through different processes of segregation: (a) 9q+ and normal #22; (b) Normal #9 and 22q−. (c) Normal #9 and normal #22. (d) 9q+ and 22q−. Types b, c, and d have been seen in Ph¹-positive CML. See Figure 119 for further aspects of exchanges.

cases of 9;22q− and 9q;22 appear to exist, though admittedly the former are rare. Moreover, an interesting case in whom Ph¹-positive normal lymphocytes were observed in AL associated with a normal karyotype has been reported (Secker Walker and Hardy, 1978). Thus, under special conditions, leukemic Ph¹-positive and karyotypically normal cells, i.e., complementary cells, may exist at the same time.

The Ph¹ and the Genesis of CML

In no known neoplastic condition (either human or animal) is there a stronger case for the direct involvement of a chromosomal change in the causation of the disease than in CML (Ishihara, Kohno, et al., 1975). This is based on the consistency with which the Ph¹ is found in the involved cells very *early* in the disease and, generally, its persistence throughout the course of the CML, including its BP. In very rare cases of CML, the Ph¹ may be lost from the leukemic cells (Michaux, Van Den Berghe, et al., 1975), usually in association with a changing cytogenetic picture (Miyamoto and Takita, 1973) and, of course, when therapy leads to normalization of the marrow with the resulting appearance of normal cells without the Ph¹ (Jung, Blatnik, et al., 1963). However, it is still unknown as to whether the Ph¹ is directly responsible for CML or is the result of the factor(s) leading to the disease. This statement is based on the fact that the Ph¹ has been invariably found in patients who have some demonstrable anomalous laboratory or clinical findings (Baccarani, Zaccaria, et al., 1973; Spiers, Bain, et al., 1977). Thus, for example, even in subjects without any symptoms or signs of CML, the Ph¹ has been looked for either because of an unexplained leukocytosis or discovered accidentally during some routine procedure and, consequently, found to be accomplished by abnormal blood findings. The Ph¹, when seen early in CML, may be found in a small percentage of the marrow cells, and this percentage increases with advancing disease. It is not known whether in this early phase of CML the diploid cells are already leukemic or not. Of interest in this connection have been two cases of CML (both male) observed by us who originally were apparently Ph¹-negative and developed a Ph¹ when subsequently seen in the BP (Sonta and Sandberg, 1978a).

There are those who view the chronic phase (CP) of Ph¹-positive CML merely as a preleukemic condition.[13] This is based on the relatively "benign" course taken by CML until the BP develops, the latter often being accompanied by additional cytogenetic changes and, thus, can be considered as a true leukemic state. Such a concept of CML is supported, it can be supposed, by the rigorous demonstration that no DNA is lost as a result of the deletion and translocation involving #22 and #9, indicative of a balanced genome in the CP of CML. On the other hand, were some genetic material from #22 to be lost or functionally modified, it would be difficult to reconcile the karyotypic abnormality with a "benign" state of CML.

The fact that the preponderant majority (>80%) of CMLs is Ph^1-positive strongly indicates a possible causal relationship between this chromosome anomaly and CML. But, what is the role of the Ph^1 in the pathogenesis of CML? In vivo and in vitro maturation and differentiation of Ph^1-positive marrow cells are not grossly disturbed[14] and this may possibly reflect the concept that there is no loss of DNA in Ph^1-positive cells (Mayall, Carrano, et al., 1974, 1977), even though very small amounts of genetically critical DNA may be lost from the cell without being detected by the best of methods. A gradual increase in the granulocytic pool and the propensity of precursor cells to develop predominantly into granulocytic cells appear to be pathognomonic for CML. However, these characteristics can be found in Ph^1-positive as well as in Ph^1-negative CML. Even though Ph^1-positive and Ph^1-negative CML generally present distinct clinical, hematologic, and prognostic entities, which will be discussed in more detail in following sections, there are cases of CML that fulfill all the criteria of typical (Ph^1-positive) CML, but do not have the Ph^1. Thus, it can be argued that the Ph^1 as such may not be responsible for the leukemic process. Similarities between Ph^1-positive and Ph^1-negative CML allow the postulate that the basic defect could be a microscopically invisible chromosomal deletion of an interstitial segment in the long arms of a #22.

Some Novel Views of the Ph^1

The studies and views of Muldal, Mir, et al. (1975) may be worthy of mention at this juncture. These authors started with the observation that in CML the extra material on the standard recipient chromosome (9q+) appears to be constant, even though the size of the Ph^1 could vary. Because in their view the translocation appears to be nonreciprocal, i.e., the deleted #22 does not apparently receive material from the recipient chromosome (for another view, see Mayall, Carrano, et al., 1974, 1977), the deleted chromatin, situated usually on the 9q arm, must equal the size of the original deletion. By comparing the sizes of the normal and recipient homologues, together with the size of the Ph^1, these authors gained the impression that there is further loss from the Ph^1 during the CP. The initial deletion of the #22 comprises typically all chromatin distal to the edge of the Giemsa-positive band q11 (the authors state q12). Even though this band is capable of serving as an end point, it appears to be unstable and prone to further loss. It is, of course, difficult to measure the Ph^1 in such unfavorable material as leukemic cells, but the disappearance of terminal stainability and reduction in size coincide apparently with the onset of the BP and often with endoreduplication or nondisjunction of the Ph^1, leading to the appearance of several Ph^1-chromosomes.

Variability of the size of Ph^1 has been tacitly assumed to be due to variation in the size of the original deletion. Apparently, this is not always so. The above mentioned authors (Muldal, Mir, et al., 1975) quote cases that show that a small Ph^1 can result from additional, larger than usual, translocations. Such nonreciprocal translocations, even though allegedly uncommon, are by no means unprecedented; the most exacting chromosomal analysis known, i.e., of the giant salivary chromosomes of *Drosophilia*, clearly showed the presence of noncompensated terminal deletions.

The possibility exists for translocated material to affect the physical (staining) and functional aspects of the chromatin with which it becomes newly contiguous, and this may account for some of the unusual banding pictures obtained in some cases of the Ph^1. Hence, it is advisable to perform several different banding analyses if the Ph^1-translocation is not clear or appears to be complicated (for an example see Verma and Dosik, 1977).

Because the Ph^1 clone evidently is abnormal, whereas the karyotype appears to be a balanced heterozygote, the abnormal genetic constitution may be due to "position effect" associated with the normally intercalary q11 band becoming terminal.

G-positive bands apparently contain inactive genetic material, more or less in repetitious DNA sequences. Broken chromatid ends differ from normal ends in their diffuseness and lack of sharp definition. The morphology of the deleted arm of the Ph^1 is consistent with the absence of a normal end point (telomere)

(Muldal, Mir, et al., 1975). The terminal q11 band is also less condensed at its distal edge. This may possibly indicate that transcription can take place here. Position effects are usually thought to be associated with the presence or absence of heterochromatin in the immediate neighborhood of a "euchromatic" gene. In this case, therefore, Muldal, Mir, et al. further postulate (1975) a position effect may involve the expression of heterochromatin itself. The terminal part of the q11 band, which is normally tightly coiled when intercalary, may therefore not only be exposed to transcription but also be vulnerable to lesions in successive mitoses. The presence of the q11 band on the Ph^1 appears to be a feature associated with the CP of CML. It seems, therefore, that when this band is "used up," the Ph^1 becomes prone to nondisjunction, hence the association with the BP and the acquisition of a second Ph^1 derived from the first one. At this time, the previous stability of the cell is often lost and signs of karyotypic evolution may be observed. This rather sudden onset of instability may be associated with the disappearance of the q11 band with the Ph^1 now having an euchromatic end. Nondisjunction could therefore be associated with exchanges in this now euchromatic region.

Even though in rare cases of CML, radiation (E. Engel, 1965b; Schneider, Stecher, et al., 1967) or drugs (Sandberg and Hossfeld, 1974) can be imputed as the likely causes of the disease (and the Ph^1), in the bulk of the cases with CML the direct cause of the disease is unknown and until such cause or causes are established, we are not in a position to say categorically whether the Ph^1 is or is not directly responsible for CML. To use the argument that CML may be seen without a Ph^1 or any other chromosomal change is, in my opinion, not a very convincing argument, because in all probability Ph^1-negative CML is a different disease than the Ph^1-positive one. There are those who wish to consider the possibility that some loss or modification of #22, possibly in addition to the Ph^1-translocation, does occur in Ph^1-positive CML and may be directly responsible for the disease and that in Ph^1-negative CML the same phenomenon occurs but without the #9–#22 translocation. The latter, to carry this argument further, actually plays a somewhat ameliorating role in CML in delaying the onset of the BP. This phase is known to take place much earlier in Ph^1-negative than in Ph^1-positive CML (J. E. Sokal, 1976; Sandberg, 1978a).

Even though the development of additional Ph^1-chromosomes and/or other chromosomal abnormalities may herald the onset of the BP and often accompanies that stage, a substantial number of CML cases (ca 10–30%) develops the BP without any additional chromosomal aberrations, except for the persistence of the Ph^1 (Rowley, 1977a). Thus, even in the BP it is difficult to implicate chromosomal changes as playing a direct role in its development. Those CML cases in the BP that are associated with chromosomal changes might be due to causes other than those producing the BP in Ph^1-positive CML but without other chromosomal changes during that stage of the disease. Only when the exact causation of CML and its BP is established will we be in a position to evaluate the role of the karyotypic changes in the genesis of this syndrome. Though some chromosomes are involved in the BP more often than others (Lawler, 1977a; Rowley, 1977a), the fact that only a portion of the cases with the BP have a similar (often not identical) karyotypic picture, points to the possibility of several different or even a variety of causes leading to the BP of CML.

Susceptibility of the q11 region of #22 to breakage may be an epiphenomenon accompanying the causation of CML in individuals who are metabolically or genetically predisposed to such breakage. Of course, there is always a possibility that whatever causes the CML produces the break in #22. Even though much has been written about the cellular kinetics, life cycle, and proliferation of the cells in the marrow and blood of patients with CML,[15] studies dealing with these parameters have not been particularly useful in ascertaining the cause of CML. Some of these approaches can be utilized in the diagnosis of the disease, particularly in cases which do not present a characteristic hematological or clinical picture, but, in general, they do not offer the type of basic information that would help us understand the genesis of the disease and the mechanisms by which the chromosomal changes are produced. Still, a close correlation, based on rigorous data and their analysis, between some of these cellular parameters and the chromosomal changes observed during the course of CML may be of considerable value in throwing light upon the relationship between these two parameters.

Pathogenetic Aspects of the Ph^1 in CML and Related Disorders

As far as is known, the Ph^1 has never been observed in the marrow cells of a hematologically normal person. It has been demonstrated, however, in marrow cells of rare patients who were in an early and clinically noncharacteristic phase of CML (Kemp, Stafford, et al., 1964; Hossfeld and Sandberg, 1970a). A remarkable example has been published by Hellriegel and Gross (1969): This patient had a leukocytosis of $17,000/mm^3$, a normal hemoglobin and platelet count, no hepatosplenomegaly, and a decreased LAP. The demonstration of a Ph^1 in marrow cells revealed the patient to have CML. A young asymptomatic man in one of the laboratories of our Institute was diagnosed as having CML in 1969 after a Ph^1 was found in his marrow cells, which were examined when he noticed a somewhat enlarged buffy coat in his blood which he was preparing as a source of cells for tissue culture. Except for an elevated leukocyte count (about $17,000$ mm^3), he had no other demonstrable clinical, laboratory, or cytologic abnormalities. Only a portion of his marrow cells were Ph^1-positive at the time of diagnosis, but this has changed since then, though the patient is still alive (in 1977) and his disease under control.

Of great theoretical interest is the finding that 90–100% of the marrow metaphases may already be Ph^1-positive in the very early stages of the disease. It is reasonable to expect that at the onset of CML only a portion of the metaphases will be Ph^1-positive, if one assumes that the induction of the Ph^1 and the initiation of CML originate in a single altered precursor cell. However, among some 300 cases studied, we have observed only a handful of cases with an increasing number of marrow metaphases becoming Ph^1-positive during the progression of CML; one patient originally had only 2% Ph^1-positive metaphases in his marrow and within 1 yr these increased to 70%. The failure to find more frequently patients with early CML and a double marrow population of normal and abnormal (Ph^1-positive) cells may have several reasons:

1. The most important consideration is that CML is a disease in which clinical manifestations become evident only when the disease has progressed (cellularly) to a stage where the preponderant percentage of the cells is Ph^1-positive.
2. Too few metaphases are analyzed; rarely are > 50 metaphases studied.
3. A very low percentage of Ph^1-positive metaphases is difficult to interpret, because differences in sizes among the small acrocentric chromosomes may occur normally, and hence, the significance of these findings may not be appreciated.
4. The possibility exists that the Ph^1 is not induced in a single precursor cell but may originate in several or many precursor cells (Pedersen, 1975b); hence, the process by which all marrow cells become Ph^1-positive would appear to proceed very rapidly. The last possibility is not very likely, but it confronts us with the question of how, on the basis of the single cell origin of the Ph^1, the original Ph^1-positive precursor cell manages to outgrow many billions of normal marrow precursor cells? Ph^1-positive cells neither grow faster nor live significantly longer than normal cells. The original Ph^1-positive precursor cell must acquire other unidentified properties in order to dominate ultimately the whole of the hematopoietic system, including sites of potential hematopoiesis (spleen, liver, lymph nodes, etc.).

Polyploid Cells with Ph^1-Chromosomes

Polyploid cells observed in the bone marrow specimens constituted <10% (average 4.3%, range 1.7–7.2%) of the metaphases during the CP of the CML. However, polyploid cells occurred in higher frequency in the BP (9.5–31.3%, average 18.8%) than in the CP. In some cases, the specimens with a high frequency of polyploid cells were accompanied by hyperdiploidy which appeared in the terminal period of the BP (Sonta and Sandberg, 1978a).

Nature of Precursor Cells of CML and the Ph^1

Whether or not CML and the Ph^1 are of single cell origin has not been demonstrated conclusively.* The main argument for the hypothesis

---

*See also section on Clonal Origin and Evolution of Chronic Myelocytic Leukemia.

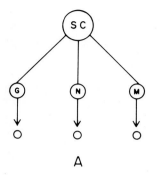
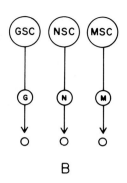
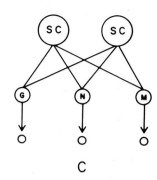

of the single cell origin is the finding that in female patients with Ph¹-positive CML, who were heterozygous for the sex-linked enzyme glucose-6-phosphate dehydrogenase (G6PD), granulocytic and erythrocytic cells carried only one enzyme allele (type A or B) (Fialkow, Gartler, et al., 1967; R. D. Barr and Fialkow, 1973). However, before considering this "convincing evidence in favor of single cell origin," the possibility must be excluded that type B marrow precursor cells are resistant to the leukemic producing agent. It is remarkable that in all the cases studied by Fialkow, Gartler, et al. (1967), the marrow-derived cells had only type A of the enzyme alleles. What was thought to be additional evidence for the single cell origin of CML was reported by Fitzgerald, Pickering, et al. (1971): in a male with sex-chromosome mosaicism, the marrow cells with a chromosome constitution of 47,XXY were found to be Ph¹-negative, whereas the cell line 46,XY and its presumptive derivatives were Ph¹-positive. The findings of this case are in contrast to those in the only other case of a patient with 46,XY/47,XXY mosaicism and CML (Tough, Court Brown, et al., 1961); in this case 100% of the marrow cells, both those with 46,XY and 47,XXY were Ph¹-positive, which points to the origin of the Ph¹ in at least two precursor cells. No conclusions as to single or multiple cell origin of CML and the Ph¹ could be reached by studying the distribution of the erythrocyte antigen Xg in female patients with CML, because the Xg locus on the so-called inactivated X-chromosome is probably not inactivated (Lawler and Sanger, 1970). We used the somewhat different, though rarely observed, morphology of the Ph¹ in the same patient as an argument for the multiple cell origin in CML (Hossfeld and Sandberg, 1970a), but most chromosome workers insist on an identical morphology of the Ph¹ in the same patient. Thus, no conclusive answer is available as yet with regard to the single or multiple cell origin of CML and the Ph¹ (Figure 59).*

General agreement exists that in the majority of cases of CML the Ph¹ originates in an early precursor cell, which under appropriate

59 Schematic representations of three different theories regarding the origin of marrow cells on the basis of observations with the Ph¹-chromosome in CML. In scheme A, one stem cell (SC) gives rise to granulocytic (G), normoblastic (N), and megakaryocytic (M) precursors, and the causes leading to CML and/or the Ph¹ may affect the SC, thus resulting in 100% Ph¹-positive cells. In scheme B, each cell line has its own precursor and the causative agent of CML and/or the Ph¹ may affect all the stem cells (GSC, NSC, and MSC) or only one or two, thus resulting in only a portion of the marrow cells being Ph¹-positive, with the percentage varying as to the type of SC affected and its contribution to the metaphase population. Scheme C indicates that different pools of stem cells (SC) may exist, with each contributing precursors to the various types of marrow cells. In this scheme, only one pool may be affected in CML and, thus, lead to only a partial population of Ph¹-positive cells. The causative agent of CML and/or the Ph¹-chromosome may affect the precursor cells beyond the stem-cell stage.

*For more recent findings and views see section on Clonal Origin and Evaluation of Chronic Myelocytic Leukemia below.

stimuli may differentiate into myelocytic, erythrocytic, and megakaryocytic cells. Evidence for the existence of such a common precursor cell is derived from several observations:

1. The finding that the Ph[1] occurs not only in myelocytic but also in erythrocytic and megakaryocytic cells; this was demonstrated indirectly by a correlation of cytological with cytogenetic data,[16] and, directly, by the detection of hemoglobin-derivatives in Ph[1]-positive metaphases (Clein and Flemans, 1966; Rastrick, 1969);
2. The finding that erythrocytic and granulocytic cells share the same G6PD locus in females with Ph[1]-positive CML and heterozygous for this enzyme (Fialkow, Gartler, et al., 1967). These observations also suggest that lymphocytes, which have not been observed to be Ph[1]-positive, originate from precursor cells different from other hematopoietic cells. However, it must be realized that the evidence is only circumstantial, because it is possible that lymphocytic cells and other hematopoietic cells may have a common precursor cell, but that those with the Ph[1] originate only in the marrow-committed precursor cell(s).

We have theorized that the Ph[1] may not always originate in an early precursor cell (Hossfeld, Han, et al., 1971; Sandberg, Hossfeld, et al., 1971), i.e., the Ph[1] may be induced in more or less differentiated precursor cell categories (Pedersen, 1975b; Hitzeroth, Bender, et al., 1977). Rare cases of ALL have been reported to have Ph[1]-positive leukemic cells (Table 48). Whether or not all of these cases were actually ALLs and not undifferentiated (so-called stem cell) leukemias or the BP of CML and whether or not a bona fide Ph[1] was present is difficult to say. However, a number of recent reports based on banding analyses, including one from our laboratory, leave little doubt that the Ph[1], at least in some cases of ALL, is due to a standard type of Ph[1]-translocation, i.e., t(9;22)(q34;q11). In the light of some findings, i.e., that the earliest precursor cell (stem cell) in the marrow is one resembling lymphocytes,[17] these cases of Ph[1]-positive ALL could conceivably have their origin from such an undifferentiated cell category. In cases in which the Ph[1] appeared to be restricted to one cell series (myelocytic or erythrocytic) (Hossfeld, Han, et al., 1971; Sandberg, Hossfeld, et al., 1971), it might have originated in a more differentiated (committed) precursor cell category. However, in explaining the cytogenetic and cytologic findings of the last mentioned cases, the possibility should be considered that a common precursor cell may not exist, i.e., each series has its own precursor cell.

A report that the blood lymphocytes of two patients in partial remission of CML contained the Ph[1] indicated to the authors (Barr and Watt, 1978) a common origin of these cells and other blood cells in the human. However, rigorous identification of the Ph[1]-positive cells was lacking.

Extramedullary Origin of Ph[1]-Positive Cells, Including Those in the Blastic Phase

The demonstration of Ph[1]-positive metaphases in cells derived from spleen, liver, lymph nodes, and skin,[18] and in macrophages derived from the marrow (Golde, Burgaleta, et al., 1977) raises the question as to whether the Ph[1] originates not only in marrow but also in other tissues capable of extramedullary hematopoiesis. Most of the cases on which such studies have been performed were in the blastic phase (BP) of CML, and the simplest explanation for the presence of Ph[1] positive cells in the spleen and other organs is that it is the result of invasion of these tissues by Ph[1]-positive marrow cells. However, there is evidence indicating that the spleen may be capable of producing leukemic cells in CML,[19] which could also be true for other sites of extramedullary hematopoiesis. In fact, there are those who believe that the spleen may be the site of genesis of cells in the BP, such cells apparently invading the marrow subsequent to their appearance in the spleen (or other extramedullary sites). The rationale of this theory is based on kinetic and cytogenetic data, which will be discussed later. At this point I only wish to indicate the possibility that the Ph[1] is not necessarily an anomaly originating exclusively in marrow cells, but possibly also in cells of CML of extramedullary origin.

Therapy and the Ph[1]

Even though new approaches to the therapy of CML have led in some cases to the apparently total disappearance of the Ph[1]-positive cells from the marrow,[20] generally, these cells persist in the marrow in most cases (Pedersen, 1968b). In our series of patients with CML (Sonta and Sandberg, 1978a), we found that in a number of cases the chromosomal anomalies, in addition to the Ph[1], disappeared when the patients were treated. Thus, one case developed +8 but reverted to a purely Ph[1]-positive picture when treated for the BP. Another developed a translocation between #1 and #9, which disappeared when the patient underwent therapy for the BP, and one patient, who did not develop the BP even though a +8,+8 karyotypic picture developed, returned to a purely Ph[1]-positive status upon therapy.

What must be considered a unique situation in oncology is the fact that despite the persistence of the chromosome anomaly, i.e., the Ph[1], the induction of which most probably is related to the genesis of the neoplastic process (i.e., CML) and which persists throughout the course of the disease, chemotherapy may achieve a reversion to normal differentiation and maturation of the marrow cells. In almost all CMLs remissions can be obtained with various agents, in which not only the bone marrow morphology returns to normal, but also such aspects as kinetic parameters, LAP activity, acid phosphatase, glycogen content (Gahrton, Brandt, et al., 1969), and phagocytic and migration indices. And, yet, the percentage of Ph[1]-positive marrow cells remains unchanged. It is a generally accepted concept in oncology that chemotherapy and radiation therapy can only be effective by eradication of malignant cells. In CML, however, presently available therapy often only reduces the leukemic cell population, and the remaining leukemic cells (Ph[1]-positive cells) may be associated with remission of the CML. Thus, unless one denies the causative role of the Ph[1] in the genesis of CML, remissions in this disease confront us with the perplexing situation in which, without concurrent normalization of the visibly recognizable genome, formerly leukemic cells behave like normal cells.

The cytogenetic conversion of Ph[1]-positive CML to a Ph[1]-negative state in the marrow was recently attempted in 16 patients in the CP of the disease. Twelve patients had an adequate trial of drug treatment (cytosine arabinoside 100 mg/sq m/day × 5 and thioguanine 100 mg/m²/day × 5 every 21 days), and 10 of these had adequate chromosome examinations. In only one patient was a conversion maintained (>5 mo), possibly as a result of the presence of 20% normal cells in the marrow of this patient before treatment, again indicative of the importance of such cells in attempts to "normalize" the marrow in Ph[1]-positive CML (Smalley, Vogel, et al., 1977).

Ph[1]-Negative CML

The patients with this variant of CML tend to be older than those with the Ph[1]-positive disease (Figure 60), consist primarily of males, have a relatively small increase of leukocytes and more pronounced thrombocytopenia (Ezdinli, Sandberg, et al., 1969; Ezdinli, Sokal, et al., 1970). The urinary and serum muramidase levels are greatly increased (Perillie and

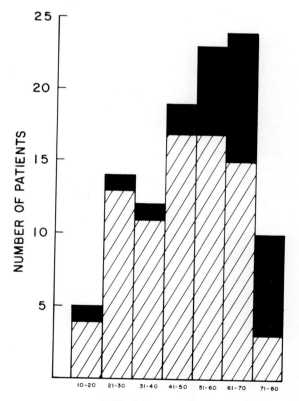

60 Distribution of Ph[1]-positive (hatched bars) and Ph[1]-negative (dark bars) patients with CML according to age. The preponderant number of Ph[1]-negative patients were over the age of 50.

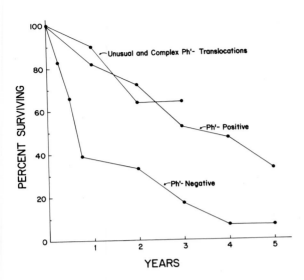

61 Survival of Ph¹-positive and Ph¹-negative CML patients based on data of Ezdinli, Sokal, et al. (1979). Survival of patients with unusual or complex Ph¹-translocations is shown for the initial 3 yr after diagnosis, because an insufficient number had been diagnosed 5 yr before the time this figure was published (Sonta and Sandberg, 1977b). However, an analysis since then has revealed that the survival continues to parallel that of the Ph¹-positive cases.

Finch, 1970; Mintz, Pinkhas, et al., 1973). In general, the Ph¹-negative patients respond poorly to therapy, resulting in a shorter survival (Figure 61) than observed for the Ph¹-positive cases (ca 15 vs. over 40 mo, respectively) (Krauss, Sokal, et al., 1964; Canellos, Whang-Peng, et al., 1976; Jacquillat, Chastang, et al., 1975). It appears that once the BP sets in the survival is not much different for the Ph¹-negative cases than for the Ph¹-positive ones (Vallejos, Trujillo, et al., 1974).

In children the Ph¹-negative disease appears to be associated with marked lymph node enlargement and thrombocytopenia at the onset of the disease, whereas the Ph¹-positive disease resembles Ph¹-positive CML in adults.

Of the >1600 cases of CML, published before banding techniques were introduced, 85% were Ph¹-positive, with 84% of these having the Ph¹ as the only abnormality at the time of the initial examination during the CP of the disease. Our own experience during the last 10 yr with nearly 300 cases of CML indicates a somewhat <15% incidence of Ph¹-negative CML, with males predominating 3-5:1. The incidence of Ph¹-negative CML reported since the introduction of banding and the exact identification of the Ph¹ has ranged from as low as 0% (Grouchy and Turleau, 1974, were only Ph¹-positive cases considered to have CML in this study?) to nearly 30%,[21] including an incidence of nearly 29% in a group of 35 patients with the BP of CML (Vallejos, Trujillo, et al., 1974).

Karyotypic progression in Ph¹-negative CML has not been established on as firm a basis as in the Ph¹-positive disease. This is possibly related to the rather short duration of the disease, though in most cases the karyotype is diploid at the time of diagnosis. Karyotypic changes observed in Ph¹-negative CML are shown in Table 30. In 40 of our patients with this disease, 10 were found to have abnormal clones. Most of these involved abnormalities of groups C and E. Banding analysis has revealed normal karyotypes in the BP of Ph¹-negative CML (Hossfeld, 1975b), though abnormal karyotypes have been observed in about 30-50% of the adult patients studied (Hossfeld, 1975b; Vallejos, Trujillo, et al., 1974; Canellos, Whang-Peng, et al., 1976). In those cases of Ph¹-negative CML in which chromosome changes are observed, both in the CP and BP, #8, #9 and #19 have been reported to be commonly affected (Gahrton, Lindsten, et al., 1974c); other karyotypic changes have also been reported (Table 34). A Ph¹-negative CML case (35-yr-old male) was shown to have a normal karyotype in the marrow at the time of diagnosis in 1970, a 46,XY,t(11;18)(q23;q12) karyotype in 1975 and in 1976, 3 months prior to the BP, an extra copy of the marker #18 (18p11→18q12). The authors (Bagby, Kaiser-McCaw, et al., 1978) expressed the view that this pattern is analogous to that of the extra Ph¹ in the BP of Ph¹-positive CML.

An interesting observation is that of L. Brandt, Mitelman, et al. (1974) who showed a much lower mitotic activity of granulocytic precursor cells in Ph¹-negative CML as compared with the Ph¹-positive disease in the CP, with both diseases being much lower than normal. The results support the view that an accumulation of precursor cells with a low mitot-

Table 34a Ph[1]-negative CML cases with karyotypic abnormalities (exclusive of juvenile type)

Chromosomal changes	Reference
46,XY,−B,+mar	Tanzer, Harel, et al., 1964
21p− (familial)	Obara, 1968; Obara, Makino, et al., 1968
47,XY,+C	Sandberg and Hossfeld, 1974
45,XY,−C	Sandberg and Hossfeld, 1974
47,XX,+D(?13)	L. Y. F. Hsu, Alter, et al., 1974; L. Y. F. Hsu, Papenhausen,
47,XY,+8	et al., 1974
45,X,−Y	Hossfeld, 1975b
47,XY,20q−,+21	Hossfeld, 1975b
46,XX,21q+,Xq−	Hossfeld, 1975b
45,−E	M. T. Shaw, Bottomley, et al., 1975
46,t(1q−;1q+)	M. T. Shaw, Bottomley, et al., 1975
47,+2C,−E	M. T. Shaw, Bottomley, et al., 1975
47,+C (4 cases)	Canellos, Whang-Peng, et al., 1976
1 pseudodiploid	Canellos, Whang-Peng, et al., 1976
2 hypodiploid (44 chromosomes)	Canellos, Whang-Peng, et al., 1976
45,X,−Y	Whittaker, Davies, et al., 1975
45,X,−Y	Sonta, Oshimura, et al., 1976
47,XY,+C (2 cases) (BP)	Vallejos, Trujillo, et al., 1974
47,XY,+C,(Cp−;Ep−) (BP)	Vallejos, Trujillo, et al., 1974
45,XX,−B or −C (BP)	Vallejos, Trujillo, et al., 1974
46,XY/44−47 chromosomes (BP)	Vallejos, Trujillo, et al., 1974
51−60 chromosomes (BP)	Vallejos, Trujillo, et al., 1974
47,+D	Conroy, Wieczoreck, et al., 1975
47,XY,+21	Wurster-Hill, 1975
46,8q− (in CP)	Lawler, Lobb, et al., 1974
48,XY,+2C,+F,−E	Meisner, Inhorn, et al., 1970
46,XX/46,XX,del(9)(q22)	Geraedts, personal communication
45,−7	Hagemeijer, Smit, et al., 1977
46,XY/46,XY,t(6;14)(p21;q32)	Mintz, Pinkhas, et al., 1973
46,XY/47,XY,+8	Mintz, Pinkhas, et al., 1973
47,XY,+8/48,XY,+8,+8	Mintz, Pinkhas, et al., 1973
46,XX,+16q+,−17/47,XX,+16q+,−17,+C	Sandberg, unpublished data
46,XY,−C,+16	Sandberg, unpublished data
47,XY,+C	Sandberg, unpublished data
45,X,−Y	Sandberg, unpublished data
47,XX,+17	Sandberg, unpublished data
49,XY,+C,+D,+16	Sandberg, unpublished data
45,XY,−C	Sandberg, unpublished data
48,XY,+E,+C	Khare, Bhisey, et al., 1978
47,XY,+C/45,XY,−E	Khare, Bhisey, et al., 1978
47,XY,+G	Khare, Bhisey, et al., 1978
45,XY,−E	Khare, Bhisey, et al., 1978
46,XY,−16,+18 (myelofibrosis → CML)	Khare, Bhisey, et al., 1978
49,XY,+3C (myelofibrosis)	Whang-Peng, Lee, et al., 1978
47,XX,+8	Lindquist, Gahrton, et al., 1978

ic activity has a serious prognostic implication in CML.

Over the years the percentage of Ph[1]-negative CML cases has decreased, so that now very few workers report > 5–10% of such patients, with the preponderant percentage (80% or more) of the CML cases being Ph[1]-positive. In fact, there are those who advocate that only Ph[1]-positive cases be considered as characteristic CML. However, it is difficult to escape the fact that Ph[1]-negative cases with CML whose clinical, laboratory, and cytologic parameters are very similar to those of Ph[1]-positive CML are encountered (Mitelman, Brandt, et al.,

Table 34b Some further cytogenetic changes observed in the chronic phase of CML (Ph¹-positive)

Chromosomal changes	Reference
(4;17)	Lawler, O'Malley, et al., 1976
t(8;22)	Lawler, O'Malley, et al., 1976
t(16;17)	Lawler, O'Malley, et al., 1976
21q+	Sharp, O'Malley, et al., 1976a,b
t(15;19)	Sonta and Sandberg, 1978a
t(2;4)	Sonta and Sandberg, 1978a
t(14;21)	Velazquez, Arechavala, et al., 1976

1974a; Rowley, 1974c). It is possible that the decline in the percentage of Ph¹-negative CML may be spurious, because there is a tendency nowadays for clinicians and cytogeneticists to exclude the diagnosis of CML in the absence of the Ph¹. At the same time, there is little doubt that cytogenetics has supplied the clinician with a most reliable and specific tool in the diagnosis of leukemia, even in the most complicated or earliest of cases (Kovary, Lonauer, et al., 1977). In my opinion, CML appears to consist of two entities: a preponderant number of cases with Ph¹-positive cells in the marrow and a small proportion (about 10%) of Ph¹-negative cases.

Value of Detailed Chromosome Studies on Large Numbers of Cells in CML

Clones with additional chromosomal abnormalities besides the Ph¹ are rarely observed in the early, CP of Ph¹-positive CML upon routine cytogenetic examination, usually based on a relatively small number of cells, i.e., < 50. Furthermore, when a few cells with additional chromosomal abnormalities are observed on the initial cytogenetic examination, the question arises as to the significance of such a finding, e.g., does the presence of 1–2 abnormal cells out of 30 metaphases examined constitute a clone? On the other hand, when 10–20 such cells are found among hundreds (e.g., 300 metaphases or more) examined, the existence of a definite clone is on a much firmer basis.

Additional clones were demonstrated by us to exist in the uncomplicated, CP of CML when a large number of cells was examined and which were not evident upon routine cytogenetic study (Sonta and Sandberg, 1978c). Furthermore, clones, observed when a large number of cells was examined during the CP, became the dominant ones during the BP. Even though examination and karyotyping of a large number of cells is an arduous and time-consuming task and predicated on the presence of a sufficiently large number of analyzable metaphases, which often is not the case with marrow material in CML, the results have rather important (karyotypically) theoretical implications.

Chronic myelocytic leukemia is characterized by variability in the rate of its development and progression and is generally assumed to be often accompanied by progressive karyotypic changes, so-called clonal evolution, particularly preceding or associated with the BP of CML. Furthermore, the clonal evolution theory holds that a step-wise progression of chromosomal changes is characteristic of the BP. The findings upon examination of a large number of cells are not at variance with the operational concepts of clonal evolution, but they do point to the strong possibility that, in some cases of CML, cells with the karyotypic changes observed in the BP may be present in small numbers during the early or uncomplicated CP. In some cases with only a Ph¹ even the examination of a large number of cells may fail to reveal a cytogenetic picture different from that seen on routine examination and, yet, in these cases karyotypically abnormal cells can appear during the BP. This implies that either an insufficient number of cells was examined in marrows which probably contained very few of the abnormal cells or that karyotypic progression can occur at any stage of CML. The crucial point to be made, though, is that in some cases of CML a variety of karyotypically abnormal cells are already present in the marrow during CP and that the quantitative shift in favor of these abnormal cells as a result of their increased proliferative capacity constitutes the chromosomal aspects of the BP, rather than the de novo genesis of cells with chromosomal changes previously not seen in the marrow.

A clinical significance of the findings is yet to be established on much larger groups of cases with CML. It is possible that cases with abnormal cells early in the disease may have a propensity to develop the BP earlier (or later) than those without such findings in the marrow

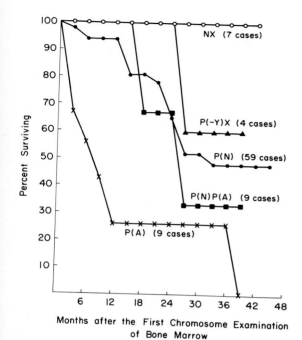

62 Life table survival curves from Ph¹-positive CML patients classified according to the chromosomal findings in the first marrow examination. NX = presence in the marrow of cells with a normal chromosome constitution, regardless of the karyotypic picture of the Ph¹-positive cells; P(−Y) = missing Y-chromosome in Ph¹-positive cells with or without other karyotypic changes; P(N) = marrow consisting exclusively of Ph¹-positive cells without any other chromosomal changes; P(N) P(A) = marrow consisting of Ph¹-positive cells with or without other karyotypic abnormalities; P(A) = marrow consisting exclusively of abnormal Ph¹-positive cells.

(Sakurai, Hayata, et al., 1976.)

Clinical Implications of Chromosomal Findings in CML

Not until chromosome analysis had been introduced into hematological diagnosis was it apparent that CML is a syndrome rather than a clinical entity. There appear to be two main subgroups, i.e., Ph¹-positive and Ph¹-negative CML.[22] There are good reasons to consider Ph¹-positive CML with the missing Y in the hematopoietic cells as a special subentity, and there are some indications that among the Ph¹-positive CML cases two more subentities may exist, i.e., cases with additional karyotypic anomalies and those with mixed Ph¹-positive and Ph¹-negative cell populations.

Cases of CML in which 100% of the metaphases derived from marrow cells are Ph¹-positive correspond to the "classic" form of CML. In such cases at the time of diagnosis the leukocyte count is usually higher than 100,000/mm³, the platelet count is normal or elevated, slight anemia is present, the spleen is markedly enlarged, and there are only occasional myeloblasts in the peripheral blood. The response to radiation or chemotherapy is good to excellent. The median survival time from the time of diagnosis is about 40 mo. About 85% of all CMLs belong to this "classic" category of 100% Ph¹-positive type of CML.

Chronic myelocytic leukemias with 100% of the marrow-derived metaphases being Ph¹-negative made up about 15% of all CMLs. Patients with this variant of CML are generally characterized by a leukocytosis below 100,000/mm³, severe thrombocytopenia, pronounced anemia, a lesser degree of splenomegaly, and a higher percentage of myeloblasts in the marrow and peripheral blood than observed in Ph¹-positive CML. Most of these patients respond poorly to therapy, develop the BP sooner than the Ph¹-positive patients, and their median survival time is only 10–20 mo.[23] Older males (more than 60 yr of age) are predominantly affected by Ph¹-negative CML.

Caution is indicated in labeling a CML Ph¹-negative before a sufficient number of marrow-derived metaphases has been analyzed. Furthermore, blood cells incubated in the presence of PHA and harvested more than 24 hr later are not suitable in establishing the cytogenetic constitution of marrow cells, because Ph¹-negative metaphases stemming from lymphocytes

and, hence, the chromosomal data could be of help to clinicians in devising effective therapy for the BP. Furthermore, the demonstration that karyotypically *normal* cells were present in some marrows upon examination of a large number of metaphases could be of considerable value to clinicians treating these patients, because the presence of such cells carries with it a better prognosis and response to therapy than when chromosomally abnormal cells are present exclusively (Figure 62).

may yield false-negative results (Lüers, Struck, et al., 1962). The analysis of a sufficient number of metaphases is imperative in order to exclude the possibility that a minor Ph^1-positive cell line has not been overlooked. Because of the unfavorable course of Ph^1-negative CML, it may represent a clinical entity different and apart from Ph^1-positive CML; in fact, a suggestion has been proffered to substitute the term "subacute" for "chronic" in Ph^1-negative CML (Ezdinli, Sokal, et al., 1970). Other workers claim that every typical CML is Ph^1-positive and that a CML without the Ph^1 is not CML (Whang, Frei, et al., 1963; Goh and Swisher, 1964a, b). On the basis of our experience we still maintain that there are CMLs that fulfill all the criteria mentioned above for the diagnosis of "typical" CML and, yet, in whom the marrow cells are 100% Ph^1-negative. Admittedly, Ph^1-negative CML seems to be a less homogeneous group than the Ph^1-positive one. From this it follows that on the basis of clinical and hematological data alone no parameter can be used to predict reliably the prognosis of a given patient (Hossfeld, Ezdinli, et al., 1971).

In typical 100% Ph^1-positive CMLs the percentage of Ph^1-positive metaphases remains fixed throughout the course of the disease, no matter whether the patient is in an excellent remission or in frank relapse. If the leukemic process in CML is, in fact, causally related to the Ph^1, and taking into account the observations that in CML in remission a number of hematological, kinetic, and cytochemical parameters may greatly improve or even return to normal, it must be assumed that therapy in CML is effective but not curative. There are, however, some rare cases in which after iatrogenic induction of bone marrow hypoplasia the majority or all of the marrow-derived metaphases were reported to be Ph^1-negative.[24] It is probable that with newer approaches to the therapy of CML more and more such cases will be described. In all of these cases the emergence of subsequent marrow hyperplasia and the unusually long survival time were thought to be causally related to the emergence of Ph^1-negative cells. Hence, overtreatment was considered a possible curative therapeutic approach to CML. Unfortunately, none of these cases is sufficiently documented karyotypically or otherwise to draw such important conclusions. Regretfully, in almost all of these cases no chromosome studies were performed before the initiation of therapy and, more importantly, no consideration was given to the possibility that these cases belonged to the subgroup of CMLs characterized from the beginning of the disease by a double population of Ph^1-positive and Ph^1-negative marrow cells. The existence of such a subgroup has been documented by a number of workers.[25] The percentage of Ph^1-positive metaphases can be as low as 20%, but differs from case to case. In remarkable contrast to 100% Ph^1-positive CML is the observation that within an individual case the percentage of Ph^1-positive metaphases may fluctuate during the course of the disease (Whang-Peng, Canellos, et al., 1968). Thus, in one of our patients with clinically typical CML, on initial examination the marrow had 10% Ph^1-positive metaphases; chromosome analyses were repeated several times within a 3-yr period, and the percentage of Ph^1-positive cells varied from 10–30%. No increase in the number of Ph^1-positive metaphases with duration of the disease was noted in our case, but may have occurred in other cases.[26] The median survival time of this subgroup of Ph^1-positive CML, but with some Ph^1-negative cells, was calculated in one study to be 31 mo (Whang-Peng, Canellos, et al., 1968). This short median survival time contrasts with long remissions lasting from several to over 10 yr of predominantly or totally Ph^1-negative, overtreated patients with CML.[27] From this it follows that Ph^1-negativity could have different meanings. In some cases, both the Ph^1-positive and the Ph^1-negative cell populations may be leukemic, and it can be expected that the median survival time would be somewhere between that of 100% Ph^1-positive CML (40 mo) and 100% Ph^1-negative CML (10 mo). In other cases a normal marrow cell population besides a leukemic, Ph^1-positive one may exist. Successful, intensive therapy with resultant marrow hypoplasia erradicates the leukemic cells, and the marrow will then be repopulated by normal cells. If this concept is correct, it must be concluded that no cure of 100% Ph^1-positive CML is possible, because no normal precursor cells are available in the marrow.

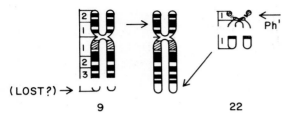

63 Schematic presentation of the "standard" (usual) type of Ph¹-translocation between #9 and #22 shown in Fig. 56(b), i.e., t(9;22)(q34;q11). Some evidence exists that the translocation may be reciprocal and balanced (Mayall, Carrano, et al., 1977) and, hence, the part shown in this figure as "(lost?)" may have been translocated to the "open" ends of the deleted #22.

Unusual and Complex Ph¹-Translocations

Shortly after the description of the Ph¹ as being due to a reciprocal translocation between #9 and #22 (Figure 63), a case was encountered in my laboratory in which the translocation occurred onto #2 rather than #9 (Hayata, Kakati, et al., 1973) (Figure 64). Since then, a relatively large number (>50) of Ph¹-translocations differing from the usual one have been described (Table 35).

Most cases of CML are associated with a Ph¹ due to a translocation, possibly balanced, between #9 and #22, i.e., t(9;22)(q34;q11). This will be referred to as the "standard" Ph¹-translocation. However, unusual and complex Ph¹-translocations have been described in a small number (probably not exceeding 10%) of the patients with CML (Figure 65). An "unusual" (simple) translocation is one in which the deleted part of #22 is translocated onto chromosomes or chromosomal areas other than the long arm of #9. In the "complex" Ph¹-translocation a rearrangement involving at least three chromosomes, but invariably #22 and usually #9, takes place (Figures 66 and 67). Even though the number of patients with unusual and complex Ph¹-translocations available for analysis is still relatively small, the data collected indicate that the clinical, prognostic, and hematologic features of these cases are not significantly different from those of CML cases with the standard type of Ph¹-translocation (Sonta and Sandberg, 1977a, b). Except for the long arms of #9 and #22, no other chromosome appeared to be involved more often than others in unusual and complex translocations. It would appear that the crucial event in CML is the genesis of the Ph¹ from the #22, and that the site and complexity of the translocation of the deleted part of the #22 may be of only minor significance in determining the course of the disease. To-date, #18, #20, and the Y have not been described to be involved in either an unusual or complex Ph¹-translocation; in addition, #1, #4, #5, #7, and #8, have not been involved in the former nor #12 and #16 in the latter type of Ph¹-translocation.

We have encountered a translocation from #22 to #9 with the latter showing an unusual banding pattern of the long arm (Hayata, Kakati, et al., 1974) (Figures 68 and 69). A negative

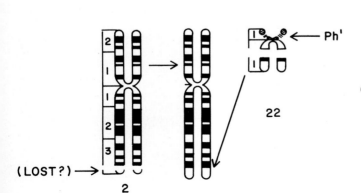

64 Schematic presentation of an unusual Ph¹-translocation between #2 and #22 (shown in Figure 56(a)), i.e., t(2;22)(q37;q11).

Table 35 Simple and complex Ph¹-translocations in CML involving chromosome #22 and those listed in the table

Chromosome involved	Type of translocation	Reference
	SIMPLE TRANSLOCATION	
#2	t(2;22)(q37;q11or12)	Hayata, Kakati, et al., 1973
#2	t(2q+;22q−)	Van Den Berghe, personal communication
#3	t(3;22)(p21;q11)	Pravtcheva, Andreeva, et al., 1976
#6	t(6;22)(p25;q12)	Mammon, Grinblat, et al., 1976
#6	t(6;22)(p26;q11)	Berger, Gyger, et al., 1976
#7	t(7p+;22q−)	Adler, Lempert, et al., 1978
#9	t(9p+q+;11p−;22q−)	Gahrton, Friberg, et al., 1977
#9	t(9;22)(p24;q12)	Ross, Wiernik, et al., 1979
#10	t(10;22)(q26;q11)	E. W. Fleischman, Prigogina, et al., 1977
#11	t(11;22)(p15;q11?)	Muldal, Mir, et al., 1975
#11	t(11p+;22q−)	E. Engel, personal communication
#12	t(12;22)(p13;q13)	Geraedts, Mol, et al., 1977
#12	t(12;22)(p13;q11)	Blij-Philipsen, Breed, et al., 1977
#12	t(12;22)(p13;q11)	E. Engel, McGee, et al., 1977
#13	t(13;22)(p13;q11or12)	Hayata, Kakati, et al., 1975
#14	t(14;22)(q23;q11?)	Ishihara, Kohno, et al., 1975
#15	t(15;22)(q26;q11)	Hossfeld and Köhler, 1979
#15	t(15p+;22q−)	Jotterand-Bellomo, 1978
#16	t(16p+;22q−)(p?;q11)	E. Engel, McGee, et al., 1975a
#16	t(16;22)(p13;q11)	Lyall, Brodie, et al., 1978
#17	t(17;22)(q25;q11)	Matsunaga, Sadamori, et al., 1976b
#17	t(17q+;22q−)	Hagemeijer, Smit, et al., 1977
#17	t(17p+;22q−)	Hagemeijer, Smit, et al., 1977
#17	t(17;22)(p13;q11)†	Rowley, personal communication
#19	t(19q+;22q−)	Gahrton, Zech, et al., 1974
#19	t(19;22)(q13;q11)	Lawler, O'Malley, et al., 1976
#19	t(19q+;22q−)	Van Den Berghe, personal communication
#21	t(21q+;22q−)	Bottura and Coutinho, 1974
#22	t(22;22)(q13;q12)***	Foerster, Medau, et al., 1974
X	t(X;9)(q21;p21);t(7;9)(p11;q11), −9,22q−,22q+	Geraedts, Mol, et al., 1977
X	t(X;22)?	Geraedts, Mol, et al., 1977
X	t(X;22)(p22;q12)	Hossfeld and Köhler, 1979
	COMPLEX TRANSLOCATION	
#1,4	t(1;4;22)¶	Geraedts, Mol, et al., 1977
#1,9	t(1;9;22)(q23→qter)	Verma and Dosik, 1977
#2,9	t(2;9;22)(q24or31;q33or34;q11)	Tanzer, Najean, et al., 1977
#2,9	t(2q−;9q+;22q−)	Alimena, Annino, et al., 1977
#3,9	t(3;9;22)(p14or21;q34;q?)	Nowell, Jensen, et al., 1975
#3,9	t(3;9;22)(p22;q34;q12)	Rozýnkowa, Stepién, et al., 1977
#3,9	t(3;9;22)(p21;q34;q11)	C. Anderson, Mohándas, et al., 1978
#3,9	t(3;9;22)(q11;q34;q11)	Tanzer, personal communication
#3,4,9	t(3;4;9;22)(q21;q31;q34;q11)	Chessells, Janossy, et al., 1979
#4,9	t(4;9;22)(q23;q34;q11)	Rowley, Wolman, et al., 1976¶¶
#5,9	t(5;9;22)(q12;q34;q?)	Nowell, Jensen, et al., 1975
#6,9 (2 cases)	t(6;9;22)(p21;q34;q11)	Potter, Sharp, et al., 1975
#6,9	t(6;9;22)(q21;q34;q11)	Chessells, Janossy, et al., 1979
#7,9	t(7;9;22)	Nowell, personal communication
#7,9	t(7;9;22)(q22;q34;q11)	Francesconi and Pasquali, 1978a
#7,9,11	t(7;9;11;22)	Ishihara, personal communication

Table 35—*continued*

Chromosome involved	Type of translocation	Reference
	COMPLEX TRANSLOCATION	
#8,9	t(8;22;9)(q22;q11;q34)	Lawler, Lobb, et al., 1974
#9	Abnormal 9q+;22q—	Hayata, Kakati, et al., 1975
#9,10	t(9;10;22)(q34;q11;q11or12)	Hayata, Kakati, et al., 1975
#9,10	t(9;10;22)(q34;q22;q11)	Geraedts, Mol, et al., 1977
#9,10,15,19	t(9q+;10p+q—;15q—;19q+; 22q—)(q34;p15q22;q22;q12; q12)	Hayata and Sasaki, 1976
#9,11	t(9;22;11)(q34;q11;q13)	Lawler, O'Malley, et al., 1976
#9,11	t(9;11;22)(q34;q13;q11)	Ishihara, personal communication
#9,11	t(9;11;22)(22qter→22q12::9pter→9qter::11pter→11p13::11p13→11 qter; 22pter→22q11)	Gahrton, Friberg, et al., 1977
#9,13	9q—q+,ins(13;9)(,22q—)*	Hayata, Kakati, et al., 1975
#9,13,15	t(9q+;13q—q+;15q—;22q—) (q34;q12;q22;q11)	Potter, Sharp, et al., 1975
#9,14	t(9;14;22)(q34;q24;q11)	Potter, Sharp, et al., 1975
#9,17	t(9;17;22)(q34;q21;q11)	Sonta and Sandberg, 1977b
#17,22	ins(Ph¹;17)(q11;p1q24)**	E. Engel, McGee, et al., 1974
#17,22	t(Ph¹;17)††	E. Engel, McGee, et al., 1975a,b
#21,22	t(21p+;22p+;22q—)	Ishihara, Kohno, et al., 1974b
X,9	t(X;9;22)	Alimena, Annino, et al., 1977; Dallapiccola and Alimena, 1979

*t(9;13;22)(9pter→9q21::22q11or12;13pter→13q13::9q21→9q34::13q13→13qter;22pter→22q11or12::?).

**No Ph¹ present as such.

***The translocation was also seen in normal lymphocytes.

†Also interpreted as t(X;q;22;17)(q13;q34;q11;q11).

††No Ph¹ present as such. In all cells a 9q+, similar to the usual Ph¹-translocation, and a 9p+, material apparently derived from a missing X, were observed in all cells. In some cells an i(17q) replaced the t(Ph¹;17) with the result that no evidence of a Ph¹ was observed.

¶(1qter→1p1; 4pter→4q2::22q1→22qter; 22pter→22q1::1pter→1p1::4q2→4qter).

¶¶Originally published by Horland, Wolman, et al. (1976).

G-band or an intermediate pale Q-band between evenly spaced, darkly stained G-bands or medium fluorescing Q-bands at the middle part (9q22) of the long arm of #9 was missing in every cell with the Ph¹. Blood cells without the Ph¹ showed a normal banding pattern of the #9 pair. Therefore, it can be assumed that band 9q22 was probably lost during the chromosomal rearrangement involved in the Ph¹-translocation. In another complicated case the segment between 9q13 or 21 and 9q34 was inserted into the long arm of #13 at band 13q13 with the original orientation of the inserted segment being maintained in its new position. The segment of #22 distal to band 22q11 was probably translocated onto #9 at band 9q13 or 21. As a result of these translocations, an acrocentric chromosome of large C-group size, a metacentric chromosome of D-group size, and a Ph¹ were produced. These abnormal chromosomes were detected in all of the 100 Ph¹-positive cells observed. Cultured blood lymphocytes (with PHA) had a normal male karyotype.

In another complicated case (Hayata, Kakati, et al., 1974), the segment of #22 distal to 22q11 was probably translocated onto #10 at band 10q11 and the segment of #10 distal to band 10q11 translocated onto #9 at band 9q34. As a result of these translocations, a Ph¹ and a large subtelocentric marker of A-group size and a submetacentric marker of the E-group size were created. In the marrow two normal cells without the Ph¹ were present without any of the anomalies mentioned above. In addition, no "standard" type of Ph¹-positive cell was observed in over 200 cells examined. Therefore, there is a strong possibility that the chromosomal rearrangements among the three chromosomes described occurred simultaneously. Others have also described complex translocations related to the Ph¹ including three interesting cases. In one the translocation between chromosome #22 and #9 were readily apparent, but the Ph¹ was masked by a translocation involving #2 (2q−) and the shortened arms of the Ph¹ (Tanzer, Najean, et al., 1977). Another case involved a 38-yr-old male with sickle cell anemia and CML (Nowell, Jensen, et al., 1975). In this case the complex translocation involved deletion of the long arm of one #5 with translocation to the long arm of #9 and a translocation between #22 and the abbreviated arm of the #5 (Figure 70). Thus, in this case the genesis of the Ph¹ was related to a translocation between #22 and #5. In another case involving a 55-yr-old female (Nowell, Jensen, et al., 1975), the complex translocation related to the Ph¹ appeared to involve a translocation of the short arm of #3 to the long arm of #9 and the genesis of the Ph¹ through a translocation between 22q− and the abbreviated arm of the #3.

In a patient with sickle cell anemia and CML associated with a complicated Ph¹-translocation, a constitutional chromosome variation, apparently without phenotypic effects, involving a pericentric inversion of #9 and satellite polymorphism in the G-group chro-

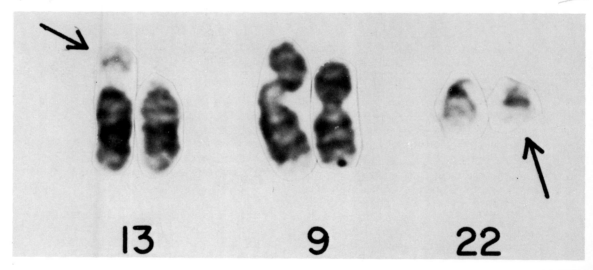

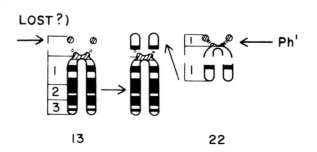

65 Banded partial or complete karyotypes from four patients with unusual Ph¹-translocations.
A. (above) Partial G-banded karyotype from a male with CML showing an unusual Ph¹-translocation between #13 and #22, i.e., t(13p+;22q−). B. (left) Schematic presentation of the unusual Ph¹-translocation between #13 and #22 (shown in Figure 65A).
C. (opposite) R- and G-banded karyotypes from a patient with an unusual Ph¹-translocation, i.e., t(X;22) (p22;q12).
(Hossfeld and Köhler, 1979.)

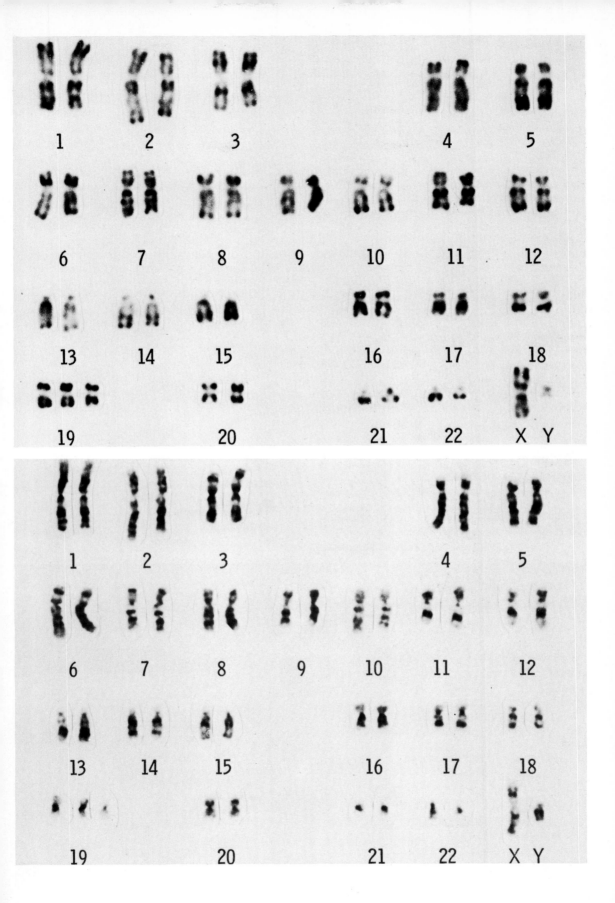

65 D. (left) Partial karyotypes (G-banded) showing a 22;17 translocation. E. (below) Q-banded partial karyotype with a 22q−; 12p+ in a patient with CML.
(Courtesy of Dr. A. Hagemeijer.)

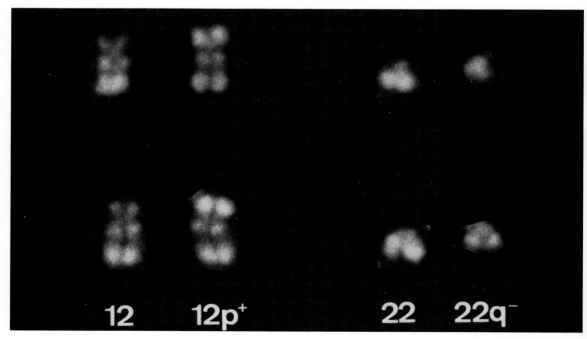

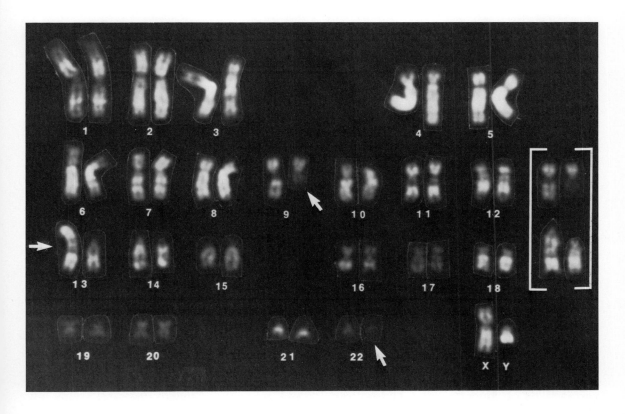

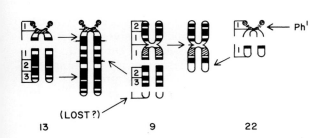

66 A. (above) Q-banded and partial karyotype of a male with CML including a complex Ph¹-translocation among chromosomes #9, #13, and #22, leading to an abbreviated #9 (arrow), an elongated #13, and the Ph¹ (arrow). The rearrangements involved insertion of material from 9q into 13q and translocation of the 22q material on the residual portion of 9q. Chromosome pairs #9 and #13 (in brackets without numbers) from another cell are shown to indicate consistency of the abnormalities in these pairs. B. (left) Schematic presentation of the complex Ph¹-translocation, shown in Figure 66A, involving #9, #13, and #22.

67 Schematic presentation of a complex Ph¹-translocation involving #9, #10, and #22.

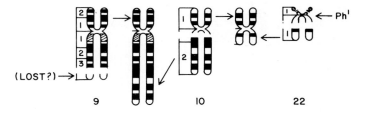

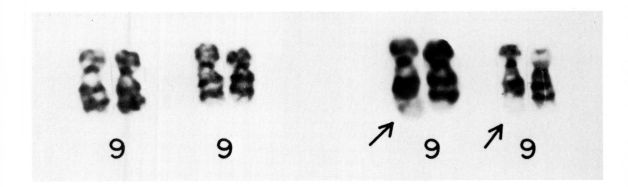

mosomes was observed (Nowell, Jensen, et al., 1975). These constitutional variants have been recognized in a number of different population studies, but there is no evidence that these cytogenetic alterations or the inborn genetic defect of sickle cell anemia are related to either the development of leukemia or the unusual chromosome rearrangement in leukemic cells. However, involvement of #5 in the three-way Ph¹-translocation may be significant. It has been observed that a deletion without translocation of approximately two-thirds of 5q in marrow cells of patients with a hematologic disorder characterized by refractory anemia and minor abnormalities of the myeloid and megakaryocytic cells may occur. The authors (Nowell, Jensen, et al., 1975) suggested an association between the specific marrow chromosome anomaly (5q−) and the hematologic disorder, citing the possible location of a structural gene for hemoglobin on the long arm of a group B chromosome. If 5q− contributes, indeed, to marrow dysfunction, it could be a factor in some patients with hematological problems already complicated by the combination of leukemia and other anemias.

68 G-banded #9 chromosomes in a case of CML; two sets of #9 chromosomes from normal cells are shown on the left and two sets, including an abnormal #9, from Ph¹-positive cells on the right (arrows). The Ph¹-translocation invariably involved the abnormal #9 (missing one of the bands in its long arm), thus pointing to the unicellular origin of Ph¹-positive CML.
(Hayata, Kakati, et al., 1974.)

Dor, Mattéi, et al. (1977b) described a case of CML with a standard Ph¹-translocation but in which the other #9 was involved in a translocation with #12, i.e., t(9;12)(q33;q13).

On the basis of polymorphism of the C chromatin on #9, Watt and Page (1978) indicated that the material translocated from #22 onto #9 was located exclusively on the #9 with the largest block of C chromatin. However, Berger and Bernheim (1978) found the translocated material was as often on the #9 with shorter C chromatin as on the #9 with the longer C chromatin.

Ph¹ Without Evidence of Translocation

In the preponderant number of cases of Ph¹-positive CML, the genesis of the Ph¹ has been shown to occur through deletion of the distal part of the long arm of #22 and translocation of this part to the distal end of the long arm of #9, i.e., t(9;22)(q34;q11). In <10% of the cases of Ph¹-positive CML, the translocation occurs to chromosomes other than #9 (see above section). These findings appear to indicate that the deletion of #22 is the crucial event related to CML, rather than the site of the translocation.

69 Schematic presentation of an unusual type of Ph¹-translocation shown in Figure 68 in which a band in the recipient #9 was invariably lost.

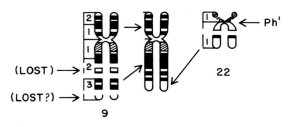

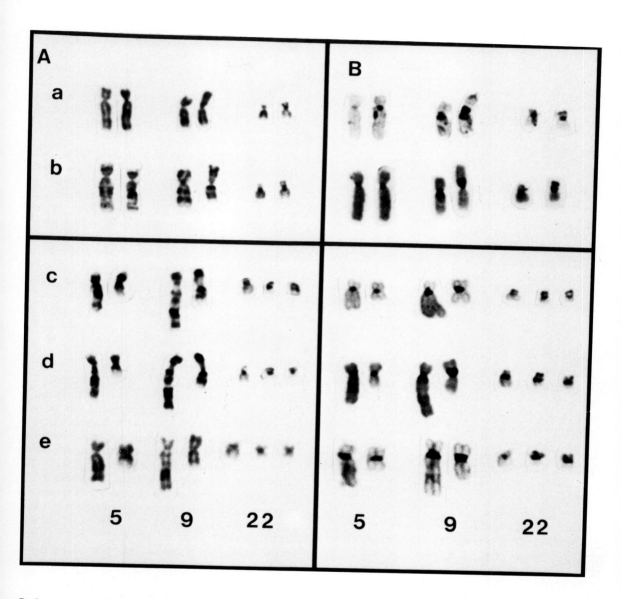

70 Complex Ph¹-translocation involving chromosomes #5, #9, and #22. Apparently, only one translocation from #22 to #9 is evident even though two Ph¹-chromosomes are present. The partial karyotypes shown were obtained with G- and C-banding in A and B, respectively.

(*Nowell, Jensen, et al., 1975.*)

Only two cases of Ph¹-positive CML, that of a 55-yr-old female "with typical CML" and a prominent basophilia (Mitelman, 1974c), and another studied by us (Sonta and Sandberg, 1977b; Sonta, Oshimura, et al., 1976) in whom the translocation could not be found, have been published. A 45-yr-old female with Ph¹-positive CML but without evidence of a Ph¹-translocation has been seen by Van Den Berghe (personal communication), whose cells also contained a t(10;16)(p11;p11). The patient died 15 mo after the onset of symptoms. Our case, the second to be reported, was that of a

54-yr-old male who was considered before admission to have either AML or CML and was karyotypically and clinically complicated. Our patient was definitely in the BP of CML and his clinical features, course, and the presence of substantial splenomegaly pointed to the strong possibility that he was not a patient with AML. All the cells in his marrow and spleen were Ph¹-positive but no evidence of a translocation was found with banding. The additional chromosomal changes observed in the marrow and spleen of this patient, i.e., +8,+11,+13,+21, are not infrequent accompaniments of the BP of CML. Obviously, because the number of cases of Ph¹-positive CML without evidence of translocation is extremely small, it is impossible to ascertain the role this finding plays in the course of the disease. As stated above, our patient was in the BP of his disease when diagnosed, whereas the patient described by Mitelman (1974c) was not. However, the latter patient developed the BP in June 1974 (28 mo after diagnosis), and died in August 1974. Thus, the rather short survival of these two patients with CML and the early appearance of the BP may possibly be related to the missing DNA.

An instructive case is that of Hossfeld and Köhler (1978), a patient who was considered to be a Ph¹-positive CML without evidence of translocation until careful analysis of the banding pattern revealed the deleted part of the #22 to have been translocated to the short arms of an X-chromosome (Xp+22q−). This case points to the necessity of carefully analyzing the karyotypes (based on several different banding techniques) before deciding that a Ph¹ is present without evidence of translocation.

Ph¹-Positive CML and the Missing Y-Chromosome

Before discussing CML with a missing Y (Table 36), the background for the missing Y in the marrow will be presented. Early in our studies on chromosomes in the human marrow we reported a not insignificant percentage of cells in nonleukemic males as having 45 chromosomes (Sandberg, Koepf, et al., 1960). At that time, not only were reliable methods for the identification of the Y, such as the newer fluorescent banding technique, unavailable, but also our own lack of expertise, diffidence in the reliable identification of the Y, and other methodological shortcomings precluded commitment as to the nature of the cells with 45 chromosomes in male marrows. With time, the experience and observations of other workers led to the conclusion that marrows of males without hematological disease, particularly those over the age of 70, may not contain a Y.[28] The incidence of the missing Y increases with the age of males and is very high after the age of 80 (about 75% of the subjects), but may be seen in 10% of males over the age of 60 (Figures 71 and 72). Apparently, the incidence of a missing Y is rare below the age of 40, though it may occur in occasional males. The percentage of marrow cells with a missing Y may range from 1 to over 90%, but rarely reaches 100%.

Table 36 List of papers that include cases of Ph¹-positive CML with a missing Y

Atkin and Taylor 1962	Hossfeld and Sandberg, 1970a
	Meisner, Inhorn, et al., 1970, 1973
Tough, Court Brown, et al., 1962	
Tough, Jacobs, et al., 1963	
Reisman and Trujillo, 1963	Serra, Sargentini, et al., 1970
Speed and Lawler, 1964	Garson and Milligan, 1972
	Motomura, Ogi, et al., 1973
E. Engel, McGee, et al., 1965	Lawler, Lobb, et al., 1974
Grouchy, Nava, et al., 1966a	
Lawler and Galton, 1966	Shiffman, Stecker, et al., 1974
Elves and Israels, 1967	Hayata, Sakurai, et al., 1975
Pedersen, 1968b	Whittaker, Davies, et al., 1975
	Nigam and Dosik, 1976
Tanzer, Jacquillat, et al., 1969	Mende, 1976
Bauters, Croquette, et al., 1970	
Berger and Bernheim, 1979	Lilleyman, Potter, et al., 1978
Alimena, Brandt, et al., 1979	Stoll, Oberling, et al., 1978

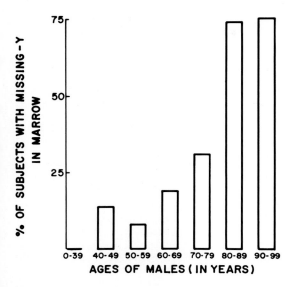

71 Percent of male subjects of various ages with missing Y in the marrow, based on data obtained on subjects without hematological disorders and normal males.

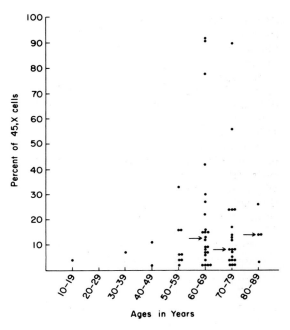

72 Percent of metaphases with a missing Y (45,X) in the marrow of male control subjects by age groups. Arrows indicate median values. The group consisted of cancer patients with normal marrows.

Whether there is a progressive increase in the percentage of cells with a missing Y with advancing age in males is not known at present. In our experience, in which we have accepted a patient as having a missing Y in the marrow if at least 3% of his cells had this karyotypic anomaly, confirmed by fluorescent staining and karyotypes, we have found about 15% of nonleukemic males over the age of 60 with a missing Y in the marrow cells (Sakurai, Hayata, et al., 1976).

The missing Y has to date only been shown in the marrow cells, and there is no reason to believe that it affects many other tissues of these males. In fact, the blood lymphocytes have been shown to have a normal XY complement in the patients with a missing Y. We have not been able to find a missing Y in the urinary bladder epithelium, and there is indirect evidence that the frequency of this karyotypic anomaly is relatively low in the cells of the buccal mucosa of males (Iinuma and Nakagome, 1972). Subjects with a missing Y are endocrinologically normal, and no evidence for sterility has been presented. Female subjects of an age comparable to that of the males do not show a missing X in the marrow cells. As the missing Y has been observed in nearly 100% of the metaphases in the marrow of some males, it appears that this karyotypic anomaly may involve all the cellular elements of the marrow.

The subgroup of Ph^1-positive CML and a missing Y deserves particular attention. Since its original description by Atkin and Taylor (1962), about 8–10% of the male cases with CML have been described to have a chromosome constitution $45,X,-Y,1Ph^1$ in marrow cells.[29] With one exception (Hossfeld and Sandberg, 1970a), in none of the cases reported up to 1970 had the missing Y been unequivocally identified by fluorescent or autoradiographic studies; nevertheless, morphologically there is little doubt that in the majority of the cases it was the Y that was lacking. As mentioned previously, the loss of the Y is restricted to hematopoietic cells; blood lymphocytes, fibroblasts, and seminal cells

(Bauters, Croquette, et al., 1970) were shown to have normal chromosomes. Loss of the Y is a phenomenon that appears to be secondary to the induction of the Ph¹ but closely related to it. Bauke (1971) reported a case in which loss of the Y occurred during the course of the disease. The percentage of hematopoietic cells lacking the Y varied between 30 and 100%.

The missing Y in marrow cells may be a normal aging phenomenon of hematopoietic cells (Pierre and Hoagland, 1971, 1972), which, nevertheless, may have significance in the pathogenesis and prognosis of CML and some other hematopoietic disorders (Table 37). Thus, several considerations support this interpretation (for a further consideration of the loss of the Y and leukemia, see below):

1. The mean age of patients with Ph¹-positive CML and missing Y is only 49 yr.
2. In 50% of such patients the disease lasted more than 4 yr, in >⅓ of them between 7 and 14 yr; and most of them required minimal treatment. Thus, CML with a missing Y seems to be remarkably benign, whereas Ph¹-negative CML without a missing Y, usually affecting older males, has a poor prognosis. However, reports challenging the more favorable prognosis of Ph¹-positive CML and a missing Y have been published (Lawler, Lobb, et al., 1974; Shiffman, Stecker, et al., 1974).
3. Loss of the Y has been reported much more frequently in older males with some chronic hematopoietic diseases than in comparable normal control groups; in addition to CML the chromosome consitution 45,XO of marrow cells had been observed in patients with PV (Kay, Lawler, et al., 1966; Lawler, Millard, et al., 1970), chronic EL (Heath, Bennett, et al., 1969), megaloblastic and other anemias,[29] chronic myelomonocytic leukemia (CMMoL) (Hurdle, Garson, et al., 1972; Mende, Fülle, et al., 1977), and other ill-defined hematopoietic disorders.[31] This high incidence may be spurious, however, because control groups should consist of male patients of ages similar to those of the hematopoietically involved group. Furthermore, hematologic patients are more likely to have their marrows examined, and the results should be compared to those obtained on a large group of patients without hematopoietic disease. Nevertheless, it is possible that the missing Y is a reflection of a basic defect occurring in elderly males in whom the marrow is involved by a hematologic disorder.
4. We have seen one case with CML and a mixed Ph¹-positive and Ph¹-negative marrow cell population; only the Ph¹-positive metaphases lacked the Y, indicating that the

Table 37a Hematological conditions associated with a missing Y (conditions other than CML)

Diagnosis	Reference
PV	Kay, Lawler, et al., 1966
Chronic EL	Heath, Bennett, et al., 1969
Megaloblastic anemia	Lawler, Roberts, et al., 1971; Pierre and Hoagland, 1971, 1972
CMMoL	Hurdle, Garson, et al., 1972
Myelofibrosis	Di Leo, Müller, et al., 1977
SBA	Krogh Jensen and Mikkelsen, 1976; Mende, Fülle, et al., 1977
AML	Sandberg et al., unpublished data
Other	E. Boyd, Pinkerton, et al., 1965; Kajii, Neu, et al., 1968; Rowley, 1971
ALL	Padre-Mendoza, Farnes, et al., 1979
PV	Lawler, Millard, et al., 1970; Shiraishi, Holdsworth, et al., 1975; Wurster-Hill, Whang-Peng, et al., 1976; Zech, Gahrton, et al., 1976; Berger and Bernheim, 1979
MA	Kiossoglou, Mitus, et al., 1965c
Myelofibrosis	Nowell, 1971b; Berger and Bernheim, 1979
SBA	Goodall and Robertson, 1970
AML	Cardini, Clemente, et al., 1968; Heath, Bennett, et al., 1969 (EL); Yamada, Shinohara, et al., 1966; Abe, Golomb, et al., 1979
ALL	Schmidt, Dar, et al., 1975; Berger and Bernheim, 1979
AMMoL	Berger and Bernheim, 1979
Pancytopenia	Sakurai, 1970a
Thrombocythemia	E. Engel, McKee, et al., 1968
Leukocytosis	Nowell, 1971b
Preleukemia	Pierre, 1978a

Table 37b Age, karyotypic, and prognostic features of the ANLL patients with a missing Y in the literature

Age, yr	Cells with 45,X only, %	Other anomaly in addition to −Y	Survival after diagnosis, mo	Reference
39	4–84	No	0.5	Yamada, Shinohara, et al., 1966
72	55	Yes		Cardini, Clemente, et al., 1968
64	100	No	6.5	Heath, Bennett, et al., 1969
49	0	t(8;21)	6.4	Sakurai, Oshimura, et al., 1974
	4 AML cases with presumable −Y and t(8;21)			Trujillo, Cork, et al., 1974
21	0	Yes	7	Golomb, Vardiman, et al., 1976
	3 AML cases with −Y and t(8;21)			Kamada, Okada, et al., 1976
57	0	Yes		Mitelman, Nilsson, et al., 1976
44	0	t(8;21)	4.9	Sakurai and Sandberg, 1976a
64	0	t(8;21) and others	14.2	Sakurai and Sandberg, 1976a
26	0	t(8;21)	7.1	Sakurai and Sandberg, 1976a
60	0	t(8;21) and others	1.5	Sakurai and Sandberg, 1976a
60	0	Yes	1.7	Sakurai and Sandberg, 1976a
59	0	Yes	0.4	Sakurai and Sandberg, 1976a
76	0–24	No	6	Sakurai and Sandberg, 1976c
71	0–23	Yes	0.5	Sakurai and Sandberg, 1976c
76	0	Yes		Oshimura, Hayata, et al., 1976
24		t(8;21)		Yamada and Furusawa, 1976
?	90	Yes	15	Alimena, Annino, et al., 1977
73	100	No		Muir, Occomore, et al., 1977
14	0	t(8;21)		Padre-Mendoza, Farnes, et al., 1979
29	0	t(8;21)		Philip, Jensen, et al., 1978
37	0	Yes		Philip, Jensen, et al., 1978
77	67	No		Philip, Jensen, et al., 1978
32	0	Yes	0.5	Abe and Sandberg, unpublished data
77	100	No	18	Abe, Golomb, et al., 1979
74	100	No		Abe, Golomb, et al., 1979
76	78–100	No		Abe, Golomb, et al., 1979
62	82–100	No	1.8	Abe, Golomb, et al., 1979
58	78–100	No	60	Abe, Golomb, et al., 1979
29	9–82	Yes* +5,13q+	7.3	Abe, Golomb, et al., 1979

Table taken from S. Abe, Golomb, et al. (1979).
*Originally 45,X/46,XY

Table 37c Survival in months (mo) of ANLL patients with missing Y

	All patients with −Y	Patients with 45,X/46,XY cell lines only	Patients with −Y and other chromosome abnormalities
Survival (in mo)	6.5	7.5	3.4
Number of patients	21	10	11

loss of the Y is a process closely linked with the leukemic state (Sakurai, Hayata, et al., 1976).
5. Finally, the XO-marrow cell population disappeared or diminished with therapy in some cases of CML in which serial chromosome studies were done (Kay, Lawler, et al., 1966; Rowley, 1971).

All these different points make me believe that loss of the Y in the marrow cells of males with CML or other hematological disorders (Furusawa, Adachi, et al., 1973; Mende, Fülle, et al., 1977) does have a significance related to these disorders and may not, necessarily, merely represent a normal aging phenomenon. I believe that loss of the Y in marrow cells is not a consequence of the absolute biological age of the body, but the consequence of the relative age of hematopoietic cells, which may be due to a higher turnover rate in diseased than in normal marrows.

The chromosome constitution XO has not been reported in female patients with CML, even though loss of chromosomes has been found in some cases. Autoradiography is the method of choice in deciding whether in such cases the missing chromosome is an X.

We have drawn attention to the strong possibility that a missing Y as the *sole* karyotypic anomaly in marrow cells may actually be a protection against such cells ceveloping AL (Sandberg and Sakurai, 1973b). Even though this subject will be discussed further in the chapter on AL, the possible relevance of the missing Y to the development of the BP in CML deserves attention.

A case of Ph^1-positive CML characterized by the presence of $45,XO,Ph^1$ cells and $46,XY,Ph^1$ cells in a ratio of 3:1 in the marrow has been described (Motomura, Ogi, et al., 1973). When the patient developed the BP, the cells with a missing Y gradually decreased in the marrow and were ultimately replaced by *polyploid* all containing the Y. The authors postulated that this change gave evidence for the monoclonal origin of the BP of CML; it can be further interpreted, as indicated in other sections of this book, that the cells with the missing Y were, apparently, resistant to blastic transformation.

Ph^1-Negative CML with a Missing Y-Chromosome

Of the four reported cases of Ph^1-negative CML (Table 38) with a missing Y, the first was that of a 69-yr-old male patient first diagnosed in April 1972, and in whom no 46,XY cells were found in the marrow on the two occasions (Hossfeld and Wendehorst, 1974). The patient

Table 38 Missing Y-chromosome in Ph^1-negative CML*

Age of patient, yr	Source of cells and %−Y	Survival, mo	Complications	Reference
84	BM,Bl‡	ca 12	Sepsis from genitourinary complications	E. Engel, McGee, et al., 1975a and personal communication
78	BM,Bl (without PHA) 70−90% Bl with PHA <5%	>30	No BP, living	Sonta, Oshimura, et al., 1976
61	BM,Bl 100%	24	BP (7 mo)	Whittaker, Davies, et al., 1975
69	BM 100% Bl 12%	>18	No BP, living	Hossfeld and Wendehorst, 1974
3	BM 96% Bl 53%	12.5	BP (3 mo)	Hays, Humbert, et al., 1975
74	BM 100%	24	Myocardial infarction	Di Leo, Müller, et al., 1977
79 (? CML)	BM 100% Bl 40%	>12	No BP, living	Di Leo, Müller, et al., 1977

*A case of CMMoL with 45,X,−Y has been reported (M. A. S. Moore and Metcalf, 1973).
‡Missing −Y present in 12 of 15 blood cells (Bl) after 72-hr culture with PHA.

is still alive. The second case was that of a 61-yr-old male patient who died 2 yr after the diagnosis (Whittaker, Davies, et al., 1975). At that time, the marrow contained some 46,XY cells, but the preponderant number was 45,X. On three subsequent occasions, including one just before death, the marrow contained exclusively 45,X cells. The third case described was of the juvenile form of CML (Hays, Humbert, et al., 1975), and whether or not the same criteria of evaluation and significance can be applied to this case in comparison with those of the adult form is uncertain. This 3.5-yr-old boy died 12.5 mo after being diagnosed. The marrow was examined on three occasions; on one, a few 46,XY cells were seen ($^2/_{13}$), but in the others only 45,X cells were observed. In all reported patients, cultured blood lymphocytes and/or fibroblasts revealed a 46,XY constitution, indicating that the 45,X picture was neither an acquired nor a congenital one. We have been observing a 78-yr-old male with Ph^1-negative CML and a missing Y (Sonta and Sandberg, 1977b) for nearly 4 yr. The patient is alive and responding to therapy.

Even though the authors of one case (Hays, Humbert, et al., 1975) state that the short course of their patient's disease is in contrast to the hypothesis that absence of the Y in CML is compatible with long survival, it should be pointed out that their patient lived 2 yr after diagnosis, a figure significantly higher than the median survival of 10 mo for patients with Ph^1-negative CML. The course of the 3.5-yr-old patient with the juvenile form of CML is difficult to interpret because no comparative data are available for this form of the disease.

We have postulated that the missing Y as the *sole* karyotypic anomaly may prevent the cells from becoming involved in an acute leukemic process, citing the extremely rare occurrence (if ever) of male patients with AL and a missing Y as the only karyotypic anomaly in the leukemic cells. It is interesting to note that the case of Ph^1-negative CML with a missing Y described by us and the ones previously reported had features that were closer to those of "classic" Ph^1-positive CML than of the Ph^1-negative variety. At this writing, two of the patients are alive, and it will be of interest to establish whether their therapeutic and clinical courses will be relatively favorable, as would be predicted on the basis of our postulate regarding the protective role of the missing Y. So far, neither patient has entered the BP, and both patients are doing well clinically. Obviously, it is impossible to generalize on the missing Y and the course of Ph^1-negative CML on the basis of a few rare cases. Only through the reporting of other such cases will it be possible to collect a number sufficient for critical evaluation of the missing Y in this type of CML.

Further Aspects of CML and the Missing Y-Chromosome

Ph^1-positive marrow cell lines with a missing Y (45,X,—Y,Ph^1) have been reported in from 0 to 14% of patients with Ph^1-positive CML. The proportion of patients with 45,X,—Y,Ph^1 cell line in one of our studies (4 out of 59) (Sakurai, Hayata, et al., 1976) is in keeping with these reports. In summarizing the data on patients with Ph^1-positive CML with a missing Y in the literature, including those on our patients, we found that only 6 out of the 24 patients were over the age of 60 yr. Thus, the bulk of these patients were of ages in which nonleukemic individuals seldom acquire a missing Y in their marrow cells. Eight patients also exhibited metaphases with a 46,XY,Ph^1, in addition to those with 45,X,Ph^1. In contrast, 45,X metaphases were extremely rare in the marrows of patients with 45,X,Ph^1 cells. These findings indicate that the missing Y in Ph^1-positive cells of CML is a phenomenon closely associated with the presence of the Ph^1 or its translocation. It might well be that 45,X cells cannot acquire a Ph^1 and become CML cells. However, the patient's age may also play a role. None of the eight patients who exhibited 46,XY,Ph^1 in addition to 45,X,Ph^1 metaphases in the marrow was over 60 yr old, whereas 6 of the 16 patients who did not exhibit 46,XY,Ph^1 metaphases in the marrow were over that age. The difference between the age distributions of the two groups was not statistically significant, but was highly suggestive (P=0.06; according to a direct calculation of probability). In our 55 male patients with Ph^1-positive CML not exhibiting a missing Y, 25 were under 40 yr of age. Combining these patients with those in the literature, 50 of 116 such patients were under 40 yr of age. In contrast, only 3 of the above 24 patients with a Ph^1 and a missing Y were younger than that age. The difference in the age distribution between both groups was statistically significant at the 0.01 level according to a χ^2 test. The median age for the former group was 52.5 yr, whereas that for the latter was 45 yr.

None of our patients developed chromosome changes in their 45,X,Ph¹ cells in association with the BP. In one patient, additional abnormalities involved the 46,XY,Ph¹ rather than the 45,X,Ph¹ cell lines. There is a report of a patient in whom additional abnormalities occurred in the 46,XY,Ph¹ but not in the 45,X,Ph¹ cells (Motomura, Ogi, et al., 1973).

The long survival of some patients with 45,X,Ph¹ cell lines may have been due to the fact that the cells with this karyotype are relatively resistant to the development of additional karyotypic abnormalities, as compared with 46,XY,Ph¹ cells; this, in our opinion, reflects the resistance of these cells to blastic transformation, despite occasional reports to the contrary.[32]

It is my view that the survival in these patients is longer than in male patients with a Y; an analysis of publications[33] and our data dealing with groups of patients with Ph¹-positive CML and a missing Y reveals that about 50% of these live more than 7 yr after diagnosis, a figure significantly higher than that for other patients with CML. Reports dealing with shorter survival times often present a complicated karyotypic picture, e.g., the presence of "several aneuploid cell lines" containing the Y-chromosome in a case with a survival of 17 mo (Whittaker, Davies, et al., 1975). Obviously, in such a case a mosaicism of 45,XO/46,XY cells existed, with the latter clone being the one involved by additional chromosomal changes compatible with the BP.

Thus, in the future, studies must be undertaken to investigate the following:

1. Whether or not 45,X,Ph¹ cells in the BP of CML have other chromosomal abnormalities that could only be detected with detailed analysis by banding techniques.
2. Whether or not the clinical manifestations of the BP of CML with 45,X,Ph¹ karyotypes are different from or milder than those of the blastic CML with other karyotypes.
3. Whether or not 45,X cells in CML can acquire additional chromosomal abnormalities as often as 46,XY or 46,XY,Ph¹ cells.

Other questions to be answered regarding the missing Y include the following: Why is it that in none of the control subjects do 45,X cells constitute 100% of the bone marrow cells, whereas in patients with CML, 45,X,Ph¹ cells often replace entirely the 46,XY,Ph¹ cells? Conceding that the presence of the Ph¹ causes the premature appearance of a missing Y, why is it that only 10% or so of the Ph¹-positive patients acquire a missing Y? Banding analysis by quinacrine fluorescence of the cells of the latter patients revealed that their karyotypes had a simple translocation between #9 and #22 in addition to the missing Y, findings that deny the possible association of a missing Y with a Ph¹ resulting from another translocation.

Additional Chromosomal Changes in the Chronic Phase of CML

The incidence of the various karyotypic abnormalities, except possibly the presence of a Ph¹, varies considerably from one laboratory to another (Table 39). This applies particularly to the chronic phase (CP) in which, unless a large number of cells is examined, the exact incidence of aneuploid (besides the Ph¹) and karyotypically normal cells may not be apparent. In our experience (Sonta and Sandberg, 1978a), the larger the number of metaphases examined, particularly when 300 cells are observed, the higher the percentage of patients in whom karyotypically abnormal and/or normal cells will be established. The diligence with which observers examine cytogenetic material,

Table 39 Cytogenetic features of the CP of CML

	Percent	
Ph¹ only	67*	70‡
Variant translocations of the Ph¹	8	8
−Y	8	<1
+8	3	2
+22q−	3	5
+8,+22q−	6	0
Other stable rearrangements	8	}12
Ph¹ and other clone	6	

*Data from Lawler (1977a).

‡Data from First International Workshop on Chromosomes in Leukemia (1978b).

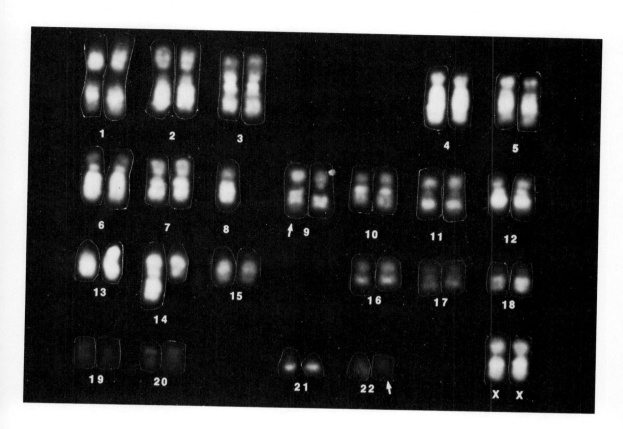

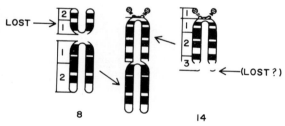

73 A. (above) Q-banded karyotype of a female with CML showing the standard Ph¹-translocation of 9;22 (arrows) and, in addition, an 8;14 translocation resulting in a missing #8 and an abnormally large marker shown in pair #14. B. (left) Schematic presentation of the marker resulting from t(8;14)(q11;q32) shown in Figure 73A.

some basic differences in the biology and background of the patients with CML, and factors that as yet remain undetermined, may account for the variability of the chromosomal picture in CML observed by different investigators, including the incidence of unusual Ph¹-translocations, the presence of cells with two Ph¹-chromosomes, the incidence of trisomy #8, iso-17, etc.

In about 5–10% of the Ph¹-positive CMLs in the CP additional chromosome anomalies are present in the hematopoietic cells (Figure 73). Figures of 16% and higher are given by Rowley (1977a) and Lawler (1977a, b), respec-

tively.* These may be present at the time of diagnosis or may develop during the course of the disease, still without any evidence of the BP. The karyotypic changes are, like the Ph¹, restricted to the leukemic cells, and they may be structural[34] and/or numerical[35] in nature. All or part of the hematopoietic cells may be

*At the First International Workshop on Chromosomes in Leukemia (1978) held at Helsinki, Finland, in September 1977, general experience regarding karyotypic changes and their implications in the CP and BP of CML and AML was discussed and analyzed and the summaries of the results were published [some refer to the publication as Van Den Berghe et al., Cancer Res. 38, 867–868 (1978)].

affected by these additional cytogenetic anomalies, which we found in some cases before treatment was started and in others after long-term radiation or chemotherapy. For the majority of cases of CML, therapy is unlikely to be the direct cause of additional chromosome anomalies, even though the emergence of Ph^1-positive mutants resistant to radiation or chemotherapy may play a role. An open question is which event came first, i.e., the induction of the Ph^1 or of the additional anomalies? Cases with a double marrow cell population of Ph^1-negative and Ph^1-positive cells plus additional cytogenetic anomalies suggest that the Ph^1 is the primary event. However, variations in the degree of susceptibility of different precursors to the same causative agent acting at the same time cannot be excluded. The most common additional anomalies (exclusive of a missing Y) are an acquisition of a second Ph^1 and an extra #8, each of which has been reported in about 3% of the cases as the only additional anomaly, combination of the two anomalies in 6% of the cases and in 10–15% of the cases in conjunction with other numerical and/or structural changes (Lawler, 1977a, b). Besides these changes, no other common karyotypic alterations appear to exist in the CP. Certain karyotypic changes that appear to be associated with or characterize the BP, particularly the presence of an i(17q) and +19, are seldom observed in the CP, though they have been reported in that phase of the disease (Sonta and Sandberg, 1978a; Lawler, O'Malley, et al., 1976). The fact that the emergence of additional chromosome anomalies is a frequent event in the BP of CML raises the important question as to whether the observation of the changes just described in the CP heralds the appearance of the BP, which could necessitate more aggressive treatment. In agreement with other workers,[36] we believe that the observation of additional anomalies may have no important bearing upon the evolution of the disease (Hossfeld and Sandberg, 1970a). The time at which additional chromosomal anomalies are observed is significant, e.g., the detection of additional anomalies at an early stage of CML (usually at the time of diagnosis) does not impair prognosis, whereas the appearance of such anomalies after many months or years of the disease (during which at least one chromosome analysis was performed in order to realize the chromosomal change) does herald the BP (Michaux, Van Den Berghe, et al., 1975).

Of interest is the observation in a patient with CML (Ph^1-positive) who had a population of cells with trisomy #8 in the marrow and a similar proportion of such aneuploid cells in the spleen. Ten mo after splenectomy the trisomic population could no longer be found in the marrow, and the situation remained the same after a further interval of 10 mo (Gomez, Hossfeld, et al., 1975). The patient had been treated after splenectomy by multiple drug chemotherapy. This case illustrates the point that follow-up studies may reveal that the chromosomal picture is by no means stable and that the proportion of cells with different karyotypes in the marrow may vary from time-to-time. On the other hand, in another case (L. Brandt, Mitelman et al., 1975a), the karyotypes in the marrow and spleen were divergent, with the marrow cells, but not those in the spleen, showing chromosomal changes associated with the BP. This case demonstrates that splenectomy cannot be expected to prolong the CP in all cases. In tissues obtained from elective splenectomies at the Royal Marsden Hospital in London, no gross difference in the karyotypes, as compared with those of marrow cells, has yet been found, although the chance of finding Ph^1-negative myeloid metaphases is greater in spleen than in marrow cells (Lawler, 1977a, b).

The occurrence of two Ph^1-chromosomes has been reported in as many as 10% of patients in the CP and is not accompanied by duplication of the 9q+ chromosome. In one such patient, a number of clonal variants was observed including cells with two monocentric Ph^1-chromosomes, some with one dicentric Ph^1, and others with two such dicentrics; only a single 9q+ was ever seen in any variant among >50 cells analyzed. Whang-Peng, Knutsen, et al. (1973) have observed the appearance of a dicentric Ph^1 during the CP and have commented on abnormal replication resulting in their presence in duplicate.

In the CP of CML stable rearrangements in the form of additional translocations (Table 34a) occur in about 8–10% of the cases. These additional translocations appear to be very stable and, thus, it seems likely that the arrangements arise simultaneously with the Ph^1. Other stable rearrangements may be present in the CP in minor clones of cells, and these do not

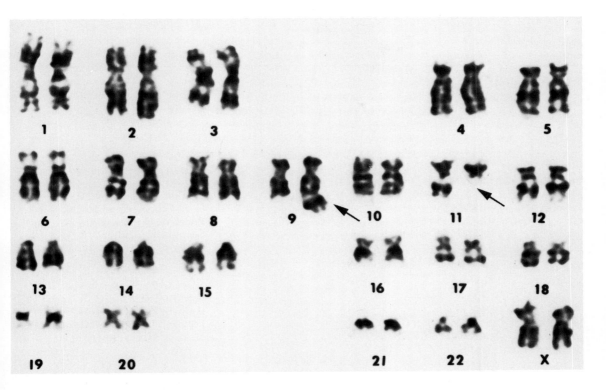

74 A translocation between #9 and #11 (9q+;11q−) in a child with Ph¹-negative CML. *(Warburton and Shah, 1976.)*

seem to be associated with any particular prognostic features. The only chromosomal rearrangement that appears to be pathognomonic of the BP is the appearance of an isochromosome for the long arm of #17, even if it is found only in a few cells.

CML in Children

Even though CML occurs very rarely in children (only 2% of the cases occur in patients younger than 15 yr), it has been observed in all age groups, even at the age of <1 yr (Figure 74). Reported cases in the literature are shown in Table 40. Wahrman, Voss, et al. in 1967 collected 34 cases of CML in children whose chromosomes (marrow and blood) had been studied, with the distribution being nearly equal between Ph¹-positive and Ph¹-negative cases. As is true for adults, the Ph¹-positive and Ph¹-negative forms in children present two distinct disease entities; the Ph¹-positive form with a typical CML picture, clinically and hematologically, and a good prognosis and the Ph¹-negative form with an atypical picture and a poor prognosis.[37] The Ph¹-negative form occurs predominantly in very young children (ages 1–3 yr) and was, thus, designated as the "juvenile type" of CML (Hardisty, Speed, et al., 1964). In contrast to the Ph¹-positive (adult) type of CML in children and also the Ph¹-negative type in adults, increased values of hemoglobin-F were found in the juvenile type (Hardisty, Speed, et al., 1964), which, however, does not appear to be a regular feature (Wahrman, Voss, et al., 1967).

Even though Ph¹-positive CML is very rare in infants (only five cases have been described chromosomally), the course appears to resemble that of the adult disease. An interesting report (Sekine and Alva, 1976) is that of a male infant with Ph¹-positive CML whose symptoms could be traced as early as 1 mo of age accompanied by hepatosplenomegaly, a leukocyte count of 28,700/mm³ and a myeloid–erythroid ratio (M:E) ratio of 10:1 in the marrow.

Table 40 Chromosome findings in childhood leukemia of the CML type

Age in yr, and sex	Karyotype	Reference
	PH¹-POSITIVE CASES (ADULT TYPE)	
8 ♀		Reisman and Trujillo, 1963
9 ♀		Reisman and Trujillo, 1963
9 ♂		Reisman and Trujillo, 1963
10 ♂		Reisman and Trujillo, 1963
10 mo ♂		Reisman and Trujillo, 1963
7 ♂	46,XX,Ph¹+	Hardisty, Speed, et al., 1964
11 ♀	46,XX,Ph¹+	Hardisty, Speed, et al., 1964
11.5 ♂	46,XY,Ph¹+	Hardisty, Speed, et al., 1964
11 ♀	46,XX,Ph¹+	G. E. Bloom, Gerald, et al., 1966
8 mo ♀	46,XX,Ph¹+	Ponzone, Rovera, et al., 1967
5 ♀	46,XX,Ph¹+	Sekine and Alva, 1976
1 mo ♂	46,XY,Ph¹+	
3 ♀	46,XX,Ph¹+	Schettini, Guanti, et al., 1969
6 ♀	46,XX,Ph¹+/52,XX,Ph¹+,+6C (some are iso(17q) in my view)	
4 ♀	46,XX,Ph¹+	Forman, Padre-Mendoza, et al., 1977
	ALL→CML→BP	Forman, Padre-Mendoza, et al., 1977
9 ♂	BP of CML	Forman, Padre-Mendoza, et al., 1977
	46,XY,Ph¹+,t(4p−;15p+)	Forman, Padre-Mendoza, et al., 1977
7.5 ♂	46,XY/47,XYY/46,XY,Ph¹+ in BP 46,XYY,Ph¹+,−C	M. A. S. Moore, Ekert, et al., 1974b
	Long course may have been due to presence of 47,XYY cells without a Ph¹ in the BM	
11 ♀	48,XX,Ph¹+,+C,+F/46,XX,Ph¹+	Wahrman, Voss, et al., 1967
3 ♀	46,XX,Ph¹+ and chromosome aberrations	Wahrman, Voss, et al., 1967
9 ♀	97% of cells Ph¹+	Whang-Peng, Canellos, et al., 1968
8 ♂	100% Ph¹+	Tjio, Carbone, et al., 1966
14 ♂	36% Ph¹+ in blood	Tjio, Carbone, et al., 1966
8 mo ♀	?Abnormal	Nowell and Hungerford, 1961, 1962
	?CML	Nowell and Hungerford, 1961, 1962
9 ♀	Ph¹-positive CML	Nowell and Hungerford, 1961, 1962
10 ♀	Ph¹-positive CML	Nowell and Hungerford, 1961, 1962
14 ♀	Ph¹-positive CML	Tough, Court Brown, et al., 1961
2.5 ♀	Ph¹-positive CML	Fortune, Lewis, et al., 1962
2 ♂	Ph¹-positive CML	Saffhill, Dexter, et al., 1976
	PH¹-NEGATIVE CASES (JUVENILE TYPE)	
14 mo ♂	48,XY,+2M	Reisman and Trujillo, 1963
15 mo ♀	47,XX,+M	Reisman and Trujillo, 1963
3 ♂	47,XY,+M	Reisman and Trujillo, 1963
18 mo ♂	46,XY	Reisman and Trujillo, 1963
3 ♂		Hardisty, Speed, et al., 1964
19 mo ♀	46,XX	Hardisty, Speed, et al., 1964
16 mo ♀	46,XX,?21q−	Hardisty, Speed, et al., 1964
22 mo ♂	46,XY,?21q−	Hardisty, Speed, et al., 1964
10 ♀	46,XX,t(9q+;11q−)	Warburton and Shah, 1976
3.5 ♂	45,X,−Y	Hays, Humbert, et al., 1975
16 mo ♂	46,XY,3q+,7q−,t(3p;7p)	Altman, Palmer, et al., 1974
5 ♂	46,XY	Altman, Palmer, et al., 1974
2.5 ♀	46,XX/46,XX,3P−,11p+,t(11p+;3p−)	Inoue, Ravindranath, et al., 1977
3 ♂	45,XY,−E(18?)/46,XY	Inoue, Ravindranath, et al., 1977
3 ♀	46,XX/45,X,−X(?)/47,XX,+G	Inoue, Ravindranath, et al., 1977

Table 40—continued

Age in yr, and sex	Karyotype	Reference
	PH¹-NEGATIVE CASES (JUVENILE TYPE)	
4 ♂	46,XY	Prigogina and Fleischman, personal communication
2 ♂	46–50 chromosomes	Whang-Peng, Canellos, et al., 1968
4 ♂	46,XY	Tjio, Carbone, et al., 1966
8 ♀	46,XX	Tjio, Carbone, et al., 1966
7 ♀	46,XX	Tjio, Carbone, et al., 1966
4 ♀	46,XX	Nowell and Hungerford, 1962
18 mo ♂	Agammaglobulinemia	Reisman, Mitani, et al., 1964
	CMMoL	Reisman, Mitani, et al., 1964
	47 chromosomes (minute marker)	Reisman, Mitani, et al., 1964
3 ♂	XY,46	Reisman, Mitani, et al., 1964
	Ph¹+ CML in child	Cao, Leone, et al., 1969
"Infant boy"	Ph¹+ CML with normal cells also (possibly CMMoL)	Biscatti and Vaccaro, 1965 Wehinger, Niederhoff, et al., 1975
3.5 ♂	Juvenile type 47,XY,+21 (no symptom of DS)	Cáp, Izakovič, et al., 1977

A report (Forman, Padre-Mendoza, et al., 1977) of two rather complicated cases of Ph¹-positive leukemia in children points to the complexity of the disease in this age group. One child (a 4-yr-old white female) initially presented a picture of ALL, followed by development of CML 2 yr later. A second child (9-yr-old white male) presented in the BP. Both patients showed blast cells possessing both lymphoid and myeloid abnormalities, as evidenced by histochemical, biochemical, and surface receptor properties.

Three patients (2.5–3 yr old) with Ph¹-negative CML (one originally presented like an ALL, but an overt clinical picture of juvenile CML emerged 2.5 yr later) were each characterized by a different karyotype: 46,XX, 3p−,11p+,t(3;11p); 45,XY,−E and 45,X(X), −C/47,XX,+G (Inoue, Ravindranath, et al., 1977).

The presentation of "lymphoid" BP of Ph¹-positive CML as ALL is discussed in another section. However, it appears that the Ph¹-negative form of juvenile CML may occasionally present initially as ALL (Inoue, Ravindranath, et al., 1977; Forman, Padre-Mendoza, et al., 1977). An unresolved question is whether the chemotherapy and/or irradiation given to these patients caused what appears to be a second malignancy, i.e., CML.

A child with Ph¹-negative juvenile CML and a missing Y as the only abnormality has been reported by Hays, Humbert, et al. (1975).

A case of juvenile CML (myelomonocytic type) in an infant boy with half of the marrow cells being Ph¹-positive has been described (Wehinger, Niederhoff, et al., 1975). During the course of the disease the Ph¹-positive clone was replaced by cells with different chromosome aberrations. In another case (Boque and Wilson, 1977), the course of the Ph¹-positive CML in a 5-mo-old boy was rather rapid (the patient died at the age of 16 mo) and was accompanied by a very high leukocyte count, gross splenomegaly, facial rash, and purpura.

Eosinophilic Leukemia

Cytogenetic studies have greatly aided in establishing eosinophilic leukemia of several varieties as true entities (Table 41). From such general experience it would appear best to classify the eosinophilic leukemias along the same lines as other leukemias, i.e., acute and chronic forms. The chronic form, similar to CML, can be further subdivided into the Ph¹-positive and Ph¹-negative varieties. The subjects with the chronic form have a very striking eosinophilia

Table 41 Summary of chromosome findings in peripheral blood or bone marrow cells from cases of eosinophilic leukemia from the literature

Patient	Reference	Type of chromosome analysis	Type of abnormality
1	Sandberg, Ishihara, et al., 1962b	BM direct	Pseudodiploid(?) with a G-group abnormality; no Ph^1 seen
2	Kauer and Engle, 1964	Peripheral blood culture	Ph^1-chromosome in 4/6 cells
3	Gruenwald, Kiossoglou, et al., 1965	BM direct	15/17 cells and 43/47 cells on two occasions Ph^1-positive stem lines, one with an additional C-group chromosome
4	Goh, Swisher, et al., 1965	Peripheral blood culture	4/43 metaphases contained a large abnormal acrocentric chromosome + random losses. One cell with 48 chromosomes (3 very large abnormal chromosomes + loss of Y)
5	Goh, Swisher, et al., 1965	BM direct	6/30 cells studied contained a very large acrocentric chromosome + random losses
6	Kiossoglou, Mitus, et al., 1966a	BM direct	95% Ph^1-positive cells. Occasional cells with C-trisomy and E-monosomy
7	Elves and Israëls, 1967	BM direct	3/15 cells Ph^1-positive + other abnormalities
8	Elves and Israëls, 1967	BM direct	9/16 cells Ph^1-positive
9	Dallapiccola, 1970	BM direct	46,XY,−A,+C
10	Flannery, Dillon, et al., 1972	BM direct and peripheral blood culture	Short Y-chromosome in all metaphases; (I interpret the BM to contain a Ph^1)
11	Chusid, Dale, et al., 1975	Cultured BM	All cells Ph^1-positive
12	Chusid, Dale, et al., 1975	Cultured BM	1% of cells Ph^1-positive at first study. Repeat study normal
13	Chusid, Dale, et al., 1975	Cultured BM	36% of metaphases lacking a C-group chromosome

and an otherwise fairly typical picture of CML. In most of these patients increased numbers of mature and immature neutrophils are found in the blood in addition to mature and immature eosinophils (Wintrobe, 1974). In the past it was thought that such findings merely represented extreme instances of the less striking increase in the eosinophils present in most patients with CML, but recent karyotypic evidence indicates that chronic eosinophilic leukemia is a definite form in itself. Those cases in whom the Ph^1 is found have a somewhat longer course than those in whom it is absent. The development of a BP, akin to that seen in CML, has been observed in these patients (Huang, Gomez, et al., 1979). It is my opinion that in all probability the Ph^1-positive cases represent variants of CML, however high the percentage of eosinophils in the blood and/or marrow. A case in point is one of CML (Ph^1-positive) in whom we observed over 80% eosinophils in the marrow and blood, but in whom wide variations in eosinophil numbers were observed during the course of the disease.

Numerous reports of similar cases have appeared under diverse headings such as "eosinophilic leukemia," "disseminated eosinophilic collagen disease," Loffler's endocarditis with eosinophilia; however, chromosomal studies have differentiated those cases with true leukemia from the severe eosinophilia that may accompany a number of infections, certain malignancies (HD, PV), parasitic infections, and allergic disorders such as periarteri-

Table 41 — *continued*

Patient	Reference	Type of chromosome analysis	Type of abnormality
14	Chusid, Dale, et al., 1975	Cultured BM	20% of metaphases lacking a C-group chromosome
15	Chusid, Dale, et al., 1975	Cultured BM	5% aneuploid cells (random losses)
16	Chusid, Dale, et al., 1975	Cultured BM	10% aneuploid cells (random losses)
17	Chusid, Dale, et al., 1975	Cultured BM	30% aneuploid cells (random losses)
18	Mitelman, Panani, et al., 1975	BM direct G-banding	All cells containing an iso-chromosome #17
19	Goldman, Najfeld, et al., 1975	BM direct G-banding	All cells (?) containing an extra chromosome #10
20	Chusid and Dale, 1975	BM culture	30% aneuploid cells; possible +C
21	L. Brandt, Mitelman et al., 1977	Blood cells	Extra C-chromosome
22	Weinfeld, Westin, et al., 1977b	BM direct G- and Q-banding	All cells containing an extra chromosome #8
23	Bitran, Golomb, et al., 1977	BM (G-banding)	49,XYY,t(3:5),+8,+mar/46,XY
24	Lisker, Cobo de Gutiérrez, et al., 1973	BM direct	46,XY,Dq+
25	Ruzicka, Pawlowsky, et al., 1976	BM and blood culture	Hypodiploidy
26	Ruzicka, Pawlowsky, et al., 1976	BM and blood culture	45 chromosomes; endoreduplication
27	Calculli, Donisi, et al., 1977	Blood culture	15% 45,XX,−C,mar D(?13q+)
28	Huang, Gomez, et al., 1979	Chloroma	49,XY,+10,+15,+19,3q−

Cases in whom no karyotypic abnormalities were observed include those described by: Dallapiccolla (1970), three cases; Marcovitch, Cain, et al. (1973), one case; Polliack and Douglas (1975), one case; Yam, Ki, et al. (1972), two cases; Benvenisti and Ultman (1969), two cases; Chusid, Dale, et al. (1975), six cases; Grant, Horowitz, et al. (1974), one case; Ruzicka, Pawlowsky, et al. (1976), one case.

tis nodosa, bronchial asthma, and various skin diseases and the hypereosinophilic syndrome (HES), possibly secondary to a profound antigenic stimulus.

A distinction has to be made also between the eosinophilic leukemias and other syndromes associated with severe eosinophilia, such as the "PIE syndrome" (pulmonary infiltration with eosinophilia), a disease compatible with a long survival and in which most of the eosinophils are mature, from several closely related disease entities that make up the idiopathic HES. The syndrome is manifest by persistent and prolonged eosinophilia with organ damage and hematologic, cardiac, and neurologic abnormalities attributable to this disease (Chusid, Dale, et al., 1975). However, a substantial proportion of patients with this syndrome has been lately shown to constitute, in fact, an eosinophilic leukemia as revealed by the presence of definite chromosomal abnormalities. Furthermore, the patients with an increased number of blasts in the blood or marrow survive only a few months and their symptoms more closely approximate those due to leukemia than other, less benign conditions. Generally, the eosinophils make up >50% of the cells in the blood and/or marrow, with eosinophilic promyelocytes and myelocytes predominating in the latter. A total number of eosinophils of >40,000/mm^3 is not unusual in the eosinophilic leukemias. In the leukemia pulmonary symptoms associated with shifting infiltrates, sterile endocarditis with congestive heart failure, and migratory thrombophlebitis with embolization characterize the course of the disease and constitute the usual causes of death. It is this form that in all likelihood should be called acute eosinophilic leukemia.

With coworkers, I have reported a case of eosinophilic leukemia in a 55-yr-old male who for nearly 5 yr had a rather benign chronic course with prominent eosinophilia but who in the later stages of the disease developed an acute phase characterized by an extramedullary (chest wall) mass diagnosed as a chloroma and CNS involvement by the leukemic cells. Interestingly, the marrow cells were diploid throughout the course of the disease, whereas those of the chloroma had a 49,XY,+10,+15,+19,3q− karyotype. I think that this case is an example of extramedullary blastic transformation of a Ph^1-negative chronic eosinophilic leukemia with terminal CNS involvement. I put stress on the chromosome analysis playing a crucial role in deciphering the complicated phases of this leukemia and differentiating it from other HES (Huang, Gomez, et al., 1979).

In some of the patients with eosinophilic leukemia in whom the Ph^1 has been absent, the course of the disease has resembled that described as acute eosinophilic leukemia, a condition very reminiscent of the Ph^1-negative form of CML.

Chromosome Findings in Eosinophilic Leukemia

The demonstration of a Ph^1 in the leukemic cells in at least some cases of eosinophilic leukemia may indicate that the disease is a variant form of CML. However, most (20 of 28) of the reported cases of eosinophilic leukemia with chromosomal changes have been Ph^1-negative and 17 of the 45 reported cases did not show any chromosomal aberrations. Thus, Mitelman, Panani, et al. (1975) found a marker chromosome in all marrow metaphases identified with G-banding as an isochromosome 17, Goldman, Najfeld, et al. (1975) demonstrated an extra C-group chromosome (probably #10) in their case, and Weinfeld, Westin, et al. (1977b) showed an extra C-group chromosome identified with G- and Q-banding as #8 in all the metaphases in the marrow of their case. In none of the above mentioned cases was a Ph^1 present. It is on the basis of such cytogenetic findings that it seems appropriate to classify eosinophilic leukemia like CML as Ph^1-negative and Ph^1-positive varieties. In both groups clonal chromosome aberrations may occur during the disease. Sequential cytogenetic and other studies in eosinophilic leukemia, as well as in the other syndromes with eosinophilia such as HES of unknown etiology, will undoubtedly give us more understanding and easier classification of these conditions.

Treatment with vincristine and high doses of prednisone has given a prompt clinical and laboratory response in the eosinophilic leukemias with a phase comparable to that of the BP of CML (Canellos, DeVita, et al., 1971). Good results with such therapy have been reported in two cases of eosinophilic leukemia (Weinfeld, Westin, et al., 1977b), and in one of them a complete and long-standing remission was obtained. Prednisone, which is known for its eosinophilytic effect, has been found not to be able to control the disease when given alone (Chusid, Dale, et al., 1975). However, in some patients the therapeutic response is of short duration and the course of the disease rather rapid, comparable to that seen with unresponsive AL.

A recent report (L. Brandt, Mitelman, et al., 1977) has dealt with the composition of the eosinophilic cell series in the marrow of patients with pronounced reactive eosinophilia (leukemoid reaction), including patients with HD, periarteritis nodosa, and others, and in patients with eosinophilic leukemia. An impaired differentiation of the eosinophils was found in the leukemic patients when compared to the reactive group. Thus, the ratio of the eosinophilic promyelocytes plus myelocytes to that of segmented eosinophils was over 9 in the patients with leukemia and averaged 1.3 in the reactive patients. It is suggested that eosinophilic leukemia is characterized by an impaired differentiation of the eosinophilic marrow cells and that the recognition of this abnormality is of value in the differential diagnosis of leukemia and reactive eosinophilia.

Basophilic Leukemia

So-called basophilic CML may or may not be associated with a Ph^1, though the number of cases reported is very small.[38] Some of these patients must be differentiated from those with

Ph[1]-positive CML who develop considerable basophilia after therapy and those in whom such basophilia can be transient and fluctuating. We have observed one such case who developed >70% basophilia in the marrow, though originally when diagnosed the number of basophils was relatively normal and became normal again when the patient was treated for his Ph[1]-positive CML. An interesting case of basophilic Ph[1]-positive CML has been reported (McKenzie and Perrotta, 1975). In this 49-yr-old white female, pruritus, hepatosplenomegaly, and an erythematous rash of the lower extremities were prominent findings. The absolute basophil count was 7×10^3 of the 12×10^3 leukocytes in the blood, accompanied by a highly elevated blood histamine but not serotonin, a decreased LAP score, and elevated B_{12} levels. Basophilic precursors comprised 33% of the cells in the marrow. About 20 mo after the diagnosis, splenectomy was performed, at which time about half of the cells in the marrow in addition to the Ph[1] contained two extra chromosomes in group C. Blastic transformation occurred 4 mo later, and the patient died about 2 yr after the initial diagnosis. In our experience, basophilia is almost always an integral part of Ph[1]-positive CML, the percentage of mature and immature basophilic cells being remarkably striking in some cases of CML.

The observations of in vitro growth and behavior of blood basophils from a patient in the BP of Ph[1]-positive CML led the authors (Denegri, Naiman et al., 1977) to conclude that the case was one of acute basophilic leukemia and that the Ph[1]-positive basophils arose from "the pluripotential hematopoietic stem cell." Elevated blood histamine levels have been encountered in this disease (Debray, Cheymol, et al., 1975).

Leukocyte Alkaline Phosphatase and the Ph[1]

The leukocyte alkaline phosphatase (LAP) reaction of granulocytic leukocytes, primarily neutrophils, has been used as a clinical index for many years and, more recently, as a marker in genetic studies.

In typical CML the activity of LAP is usually either zero or very low, whereas in DS it is usually 50% higher than normal.[39] On the assumption that in both diseases the same chromosome (#21) was affected, an assumption which we now know is incorrect, the theory was advanced that LAP is controlled by a gene located on that chromosomal segment deleted in CML and trisomic in DS.[40] Subsequently, it was realized that no such simple gene-dose relationship could exist.[41] A number of findings were not in accord with this theory. The decisive finding, of course, was that in CML and DS different chromosome groups are involved, i.e., #22 in the former and #21 in the latter. With regard to the relationship between the deleted chromosomal material in CML and LAP, we are confronted with findings indicating that LAP activity may be normal in Ph[1]-positive CML,[42] low or absent in Ph[1]-negative CML,[43] increased or becoming normal upon remission or in response to an inflammatory process,[44] and increased in the BP of CML.[45] Thus, there is hardly a basis for the assumption that regulatory genes for LAP are located on the distal part of chromosome #22.

Decreased or absent activity of LAP in CML is probably not due to an absolute lack of the enzyme, but may be a reflection of a profoundly altered isozyme pattern characterized by the regression of normal components and the emergence of others (J. C. Robinson, Pierce, et al., 1965); it is possible that these other components cannot be demonstrated cytochemically. Pedersen and Hayhoe (1971a,b) showed that LAP and phagocytic activity are closely correlated. Ford Bainton and Farquhar (1968) described two types of granules in the granulocytic cells, i.e., azurophilic and specific granules, which differ in their enzyme content and are formed in different stages of granulocyte development. Specific granules contain LAP, and these are greatly reduced in the leukocytes of CML (Bessis, 1968). Because specific granules are only formed at a later stage of cellular differentiation, this finding indicates cytoplasmic immaturity of polymorphic leukocytes in CML (Pedersen and Hayhoe, 1971a,b). In the light of these complex findings, decreased LAP activity in CML appears to be but one expression of a disturbed maturation of granulocytic cells, and not the consequence of a simple gene loss. Also, in accord with what was said earlier, LAP activity is probably unrelated to the Ph[1] as such; absent or low LAP in Ph[1]-negative CML should be due to similar mechanisms as in Ph[1]-positive CML (Hossfeld and Sandberg, 1970a).

Muramidase (Lysozyme) and the Ph¹

In 1970 Perillie and Finch reported that marked muramidasuria and an increased serum muramidase value can be demonstrated in patients with Ph¹-negative CML, whereas serum and urine of patients with Ph¹-positive CML contain normal amounts of the enzyme. These results were confirmed by other groups.[46] Tischendorf, Ledderose, et al. (1972) found additionally that by irradiation of the enlarged spleen heavy muramidasuria occurs also in Ph¹-positive CML. This finding makes the assumption that Ph¹-negative granulocytic cells contain more enzyme than Ph¹-positive cells a tenuous one. It remains to be seen whether spontaneous muramidasuria in Ph¹-negative CML is a reflection of a high turnover and death rate of leukocytes and what kind of relationship exists between the Ph¹ and this enzyme.

Peroxidase Activity, Periodic Acid–Schiff, and Sudan Black B Reactions of Blasts in CML

Generally, the cells in the CP and BP of CML are cytochemically similar (Castoldi, Grusovin, et al., 1975). Like normal myeloblasts, those observed in the CP are generally devoid of periodic acid–Schiff (PAS) positive substances, lipid staining with Sudan Black B, and peroxidase activity; however, PAS-positive myeloblasts are frequently observed in the BP (Pedersen, 1973a,d, 1975a,b) and, similarly, Sudan Black B positivity and peroxidase activity have been demonstrated in such cells. The findings have not been consistent from case to case, and the positivity of the reactions has varied among cases (M. T. Shaw, Bottomley, et al., 1975). Attempts at correlating the morphological, cytochemical, and cytogenetic features with the response to therapy in the BP of CML have revealed considerable heterogeneity and no definite conclusions. Some of the features of the cells just described may be related to the fact that differentiation of the cells in the BP, including a differential between the cytoplasm and nucleus, has not advanced as far as in the CP (Pedersen, 1973a). A correlation between the above-mentioned activities and the karyotypic picture has not been extensively performed, though one report correlating PAS-positive myeloblasts with several karyotypic parameters indicated that the frequency of circulating PAS-positive myeloblasts correlated positively with the relative size of the myeloblastic compartment and with the incidence of metaphases containing additional chromosomes, in particular those showing extra chromosomes in group C (Pedersen, 1973d). The observations suggested that premature development of PAS-positivity, like the accumulation in the BP of poorly differentiated cells, reflected defective granulocyte precursor differentiation possibly due to evolution of certain types of aneuploid clones (Pedersen, 1973a).

Cell Surface and Other Markers in the Blastic Phase of Ph¹-Positive CML

The clinical and morphological picture of the BP of CML may be highly variable, though a progressive change in blood and marrow cytology resembling that in AL is most often seen. In a minority of cases, the blastic change appears to occur in extramedullary sites, such as the spleen and lymph nodes. Morphologically, the BP may resemble AML but in a minority of cases the blast cells have a lymphoid appearance (Boggs, 1974, 1976) and can be shown to react cytochemically and immunologically in a manner atypical for AML (Hammouda, Quaglino, et al., 1964). These morphological appearances have led to the speculation that in some cases the blastic transformation may actually be "lymphoid" in nature (Canellos, DeVita, et al., 1971; Boggs, 1974); recent studies have reinforced this relationship by demonstrating that in some cases the blast cells contain terminal deoxynucleotidyl transferase (TdT) activity (Marks, Baltimore, et al., 1978), which is also found predominantly in normal thymus and ALL cells.[47] The recent introduction of surface marker analysis has considerably improved the characterization of nonmyelogenous AL and has shown that most cases of ALL can be positively identified by cell surface antigen analysis (Janossy, Greaves, et al., 1976; Janossy, Roberts, et al., 1976, 1977). A large proportion (about 70%) of cases are undifferentiated and react with antisera specific for ALL (ALL+) and are unreactive with T- and B-cell markers (null cells = non-T, non-B ALL). About 20–25% are positive for T-lymphocyte markers (T-ALL) and rare cases (about 1%) have been found to have a B-cell surface phenotype. Both T and B ALLs are unreactive with anti-ALL sera that have been produced against non-T, non-B ALL. A most important observation, in relation to the BP of CML, was that in a number of cases blast cells of this condition were strongly ALL+ (Janossy, Greaves, et al., 1976, 1977; Janossy, Roberts, et al., 1976). These blast cells were, in addition, negative for surface markers which identify human thymocytes, T and B lymphocytes. These observations excluded the possibility that the "lymphoid" BP in CML may derive from thymocytes or T and B lymphocytes, but did not exclude the possibility that these leukemias originate from prelymphoid (prethymocyte) cells or stem cells. Leukemic cells that had lymphoid appearances but were unreactive with conventional lymphocyte and thymocyte markers are designated as "lymphoid." The results are in good correlation between the surface marker analysis and the morphological data. ALL+ blast cells were observed exclusively in cases originally classified as "lymphoid," though not all lymphoid cases of the BP of CML are characterized by ALL+ blasts. This finding is in agreement with observations that suggest that in a proportion of cases lymphoid blast cells observed during the BP of CML are indistinguishable by a number of criteria from typical ALL blast cells. These include the morphological appearances, high nuclear cytoplasmic ratio, rather coarse nuclear chromatin, agranular cytoplasm, negative Sudan Black reaction, and lack of organelles (Hammouda, Quaglino, et al., 1964; M. T. Shaw, Bottomley, et al., 1975). Other criteria include demonstration of TdT and adenosine deaminase enzyme activities, specific and characteristic for ALL cells, as well as the characteristics shown by in vitro culture of leukemic cells (M. A. S. Moore, 1975; M. A. S. Moore, Spitzer, et al., 1974). Of interest is the fact that a favorable response to prednisolone and vincristine therapy may occur more often in lymphoid cases with the BP than in others, characteristics which apply to the response in typical ALL (Canellos, DeVita, et al., 1971; Boggs, 1974).

Theories have been advanced to explain such findings, one favoring the involvement of different stem cells (prelymphoid and myeloid) by the Ph^1, thus leading to lymphoid and myeloid forms of the BP. Alternatively, the genesis of the Ph^1 aberration in an ALL-like BP may take place in an undifferentiated pluripotential stem-cell precursor, which is also the target cell in typical non-T, non-B ALL or possibly in a precursor cell with many of the morphologic, enzyme, and surface marker characteristics of lymphoid cells but not necessarily a parental lymphoid cell. It is likely that this stem cell is the common precursor of prelymphoid and myeloid stem cells. In that case the early development of the BP is likely to occur occasionally before the CP has had an opportunity to produce symptoms. This possibility is supported by cases who develop ALL+ with a double Ph^1 in CML and who, after therapy, have a typical Ph^1-positive CML phase of the disease. The two possibilities just discussed are not exclusive in the sense that perhaps in different CML patients the Ph^1 aberration takes place at different levels of stem-cell development.

The results discussed above suggest that the BP of CML may involve different cellular derivatives of a pluripotential stem cell in which the primary malignant changes reside. The BP of CML can, therefore, be heterogeneous with the cellular expression in a significant proportion of patients involving a cell which, by membrane markers and morphological criteria, is indistinguishable from that seen in the common form of ALL. In these cases the Ph^1 may be the only distinguishing cellular characteristic.[46] In any case, this field is currently developing rapidly and many of the questions that remain unanswered should find a solution in the near future (Janossy, Greaves, et al., 1976, 1978; Janossy, Roberts, et al., 1976; Janossy, Woodruff, et al., 1978). The role played by present-day therapy in the appearance, if not the genesis, of the lymphoid form of the BP of CML is yet to be evaluated. Some relationship between membrane markers (anti-ALL and anti-Ia antisera) and TdT activity and response to therapy and survival in Ph^1-positive CML has been reported (Janossy, Greaves, et al., 1978). In another report, the same group (Janossy, Greaves, et al., 1978; Janossy, Woodruff, et al., 1978) studied five lymphoid cases with the BP of Ph^1-positive CML for the ALL antigen and TdT activity and indicated their important role in the diagnosis of this state.

The appearance of lymphoblastlike cells in

the BP of CML has led to the occasional Ph¹-positive BP masquerading as an ALL in children. In one report (Secker Walker, Summersgill, et al., 1976), the experience suggested that above 2% of all children diagnosed as ALL without the benefit of chromosome analysis may actually have Ph¹-positive CML presenting in the BP. Thus, chromosome analysis should be performed particularly in ALL patients in whom the blasts defy precise classification on morphological, cytochemical, and immunological grounds and on patients with ALL who fail to remit on standard therapy, such as vincristine, prednisolone, and L-asparaginase, in <2 mo. These observations raised some doubt regarding the actual occurrence of Ph¹-positive ALL, though these authors indicated that one of their cases probably had such a condition because the Ph¹ was found predominantly in lymphocytes, and the course of the case indicated it to be one of ALL (see section on Ph¹-positive ALL). As stated in several other places in this book, in my mind there is little doubt that Ph¹-positive ALL is an existing entity.

Not all cases of CML terminate in the BP. A proportion of patients (ca 25%) is characterized by an increased resistance of the disease to all standard therapeutic agents without an overt elevation of the percentage of blasts in the hematopoietic tissues, whereas the leukocyte count remains high and a thrombocytopenic stage develops. Another group of CML patients develops progressive myelofibrosis (Buyssens and Bourgeois, 1977) accompanied by anemia, thrombocytopenia, and a high leukocyte count. In very rare cases the BP is characterized by megakaryoblastic transformation of the CML (Bain, Catovsky, et al., 1977), by megaloblastosis (V. A. E. Klein, 1969; Srodes, Hyde, et al., 1973), or neutrophilic predominance (Gingold, Oproiu, et al., 1964). However, at least 50% of patients with CML develop a terminal stage compatible with the BP of the disease. The prognosis of this group is poor with an overall remission rate of only about 10%.

The number of colony-forming cells and the presence of colony-stimulating factors in the blood and marrow of patients with CML have been investigated in several laboratories, and these approaches appear to hold some promise of being helpful in following the course of the disease, e.g., the relative proportion of normal cells present in the blood and marrow. However, to my knowledge no comprehensive correlation between the chromosomal findings in CML and the number or character of colony-forming cells in semisolid media has been published. Such an approach might be informative and of particular value.

Ph^1 in Hematopoietic Disorders Other than CML

The Ph^1 is not an anomaly absolutely specific for CML. A chromosome anomaly morphologically indistinguishable from the Ph^1 has been found in AML,[49] ALL (see Table 48), EL,[50] PV,[51] thrombocythemia (Woodliff, Onesti, et al., 1967a), and osteomyelofibrosis and sclerosis.[52] However, in only a small number of these cases was the nature of the Ph^1 ascertained with banding techniques. Until the Ph^1 in these cases is confirmed with banding methods, it may be best to accept the Ph^1-positivity in most of these cases (published before banding) with some reservations. Most remarkable and provocative is the demonstration of the Ph^1 in ALL, because this disorder cannot be considered as belonging to the myeloproliferative syndrome as defined by Dameshek (1951). All other disorders could be regarded as variants of CML. CML may be preceded, for instance, by PV or thrombocythemia, states in which the typical CML picture may supervene later. An interesting report is that of Van Den Berghe, Louwagie, et al. (1979b) regarding 4 patients with multiple myeloma (MM) in whom a Ph^1 was found: one case with a 9;22 translocation, two without demonstrable evidence of a translocation, and one with a complex one (3;8;22). Ph^1-positive AML could be a BP of CML with the CP having been extremely short or easily tolerated by the patient. However, the Ph^1-positive AML cases we studied (Hossfeld, Han, et al., 1971; Sonta, Oshimura, et al., 1976) had no clinical symptoms suggestive of the BP of CML; these patients had no splenomegaly, basophilia or eosinophilia, and comparatively low leukocyte counts. Also, the distribution of the Ph^1-positive cells in AML and other disorders is different from typical CML, i.e., the percentage of Ph^1-positive metaphases is in most cases below 90%. To explain the occurrence of the Ph^1 in diseases other than CML several theories can be advanced.

1. The induction of the Ph[1] may occur at different cell maturation stages; the most immature cell may be the lymphocytelike stem cell, and this may explain the finding of the Ph[1] in ALL. The induction of the Ph[1] in an erythrocytic precursor cell (Rastrick, Fitzgerald, et al., 1968) may be associated with EL.
2. The causes involved in the induction of the Ph[1] may be individually different.
3. The response of the human genome to the Ph[1] or its inducing factors may be individually different.
4. Differences in the degree of deleted material may have different clinical implications. Banding analyses have shown this to be an unlikely cause.

Chromosomal Findings in the Blastic Phase of CML

No universally accepted definition of the BP of CML has been established (Figures 75–77). Furthermore, different terms, e.g., blast crisis, acute or blastic transformation, acutization, blast or acute phase, acute presentation, acute metamorphosis, acute, terminal, or accelerated stage and others, have been used to describe the BP. A small proportion of patients with CML first appear in the BP without a preceding period of symptomatic CML and may be confused with AML[53] (and rarely ALL).[54] Cytologic criteria for the BP vary, e.g., the number of blasts and promyelocytes exceeding 20% of the blood leukocytes in a previously untreated patient, in untreated CML 30% or more blasts and promyelocytes in the marrow in patients with $<100 \times 10^9$ blood leukocytes/liter, etc. In my laboratory we have generally felt that the BP exists when the percentage of myeloblasts exceeds 20% of the cells in the marrow, taking into consideration, at the same time, the clinical picture of the patient (Table 42).

Not all patients with CML, whether Ph[1]-positive or Ph[1]-negative, develop further kary-

75 Q-banded karyotype of a cell from a female patient in the BP of CML and containing the most common karyotypic abnormalities observed in that condition, i.e., extra #8, #19, and Ph[1]-chromosomes. A standard type of Ph[1]-translocation was present.

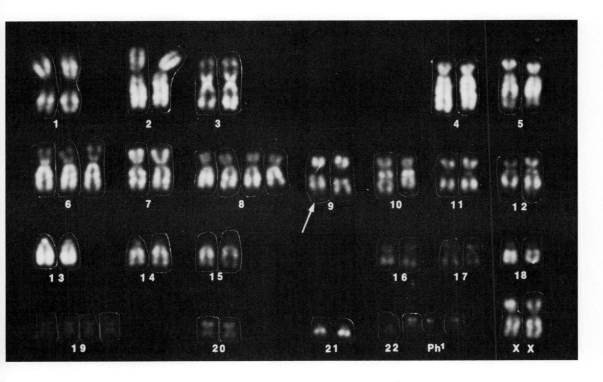

236   The Leukemias

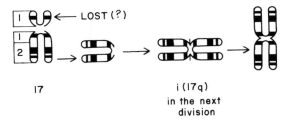

76 A. (above) Q-banded karyotype of a male patient with BP of CML. The long arrow points to a G-banded iso-17q. Other complex deletions and losses in the karyotype were del(10)(q24), t(3;17)(q21;p1?) and the presence of a Ph¹ due to a 9;22 translocation (short arrows). B. (left) Schematic presentation of the formation of an iso-17q chromosome.

otypic changes when they enter the BP (Pedersen, 1973b,c). Depending on the origin of the reports, up to 30% of Ph¹-positive patients may not develop additional chromosomal abnormalities in the BP (Rowley, 1976d, 1977b,c) and in the case of Ph¹-negative CML, even though the number of reports is meager, most of the patients do not seem to have chromosomal changes in the BP that did not exist previously. Furthermore, most of the patients with CML develop changes that are different from case to case, though admittedly changes such as an additional #8 and/or Ph¹, the appearance of an iso-17, etc., tend to be the most common and involve more than one case. The question naturally arises as to why some Ph¹-positive CML cases develop nonrandom chromosomal changes whereas others do not. Is it related to the type of therapy? Is it genetically determined? Clinically, there appear to be no outstanding differences between the CML cases with the nonrandom karyotypic changes and those without such aberrations (Sonta and Sandberg, 1978a).

Even though the number of such cases is relatively small, apparently cells cytogenetically normal have been observed in the marrow of patients in the BP of Ph¹-positive CML. Whether these diploid cells are leukemic or not is uncertain. However, were they to be

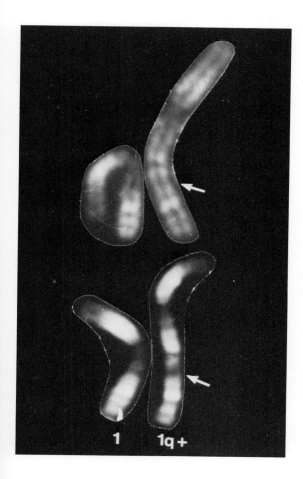

77 Q-banded normal and abnormal #1 chromosomes from the marrow of a Ph¹-positive case of CML in the BP. The abnormal #1 (1q+) is due to trisomy of the long arm, i.e., segment q23→qter of a #1 has been translocated to the long arm of another #1 at q42.

(Sonta and Sandberg, 1978a.)

shown to be leukemic, it would indicate that the leukemic cells in Ph¹-positive CML probably may start out devoid of the Ph¹, which may develop in the later stage of the disease. On the other hand, if the diploid cells in the BP are nonleukemic, they may assume a significance in the evolution, course, and therapy of the BP of CML, akin to their role in AML. The problem as to why in very rare patients with Ph¹-positive CML cytogenetically normal cells survive into the BP of the disease, whereas in other patients they do not, has yet to be solved.

Transformation of CML into the BP is frequently accompanied by striking chromosomal changes of the leukemic cells (Scully, Galdabini, et al., 1976; B. Castelman, Scully, et al., 1973; Das, Nayak, et al., 1978). In about 10–30% of the cases of CML the chromosomal make-up of the CP remains unchanged

Table 42 Some possible differentiating points between Ph¹-positive AML and the BP of CML

CML	AML
1. Usually 100% Ph¹-positive	Usually a mixture of Ph¹-positive and normal cells
2. More than one Ph¹ are not rare	More than one Ph¹ rare
3. Auer bodies very rare	Auer bodies common
4. Other karyotypic changes rare (in <10% of CP); common (>70% in BP)	Other karyotypic changes not rare
5. Return of the BM to a normal karyotypic picture after therapy rare	Return of BM to normal cytogenetically not uncommon
6. Response to therapy rather poor; remission short	Response to therapy fairly good; remission relatively long
7. Some basophilia and/or eosinophilia in BM common	Basophilia and/or eosinophilia relatively rare in BM
8. iso-17[i(17q)] common in BP of CML	iso-17 rare in AML
9. +19 not uncommon in BP of CML	+19 very rare in AML
10. −7,5q− rare	−7,5q− common
11. Hypodiploidy rare (<5%)	Hypodiploidy common
12. Organomegaly common	Organomegaly rare
13. Cells markers often indicate "lymphoid" component; however, such cells may be Ph¹-negative (Foadi, Slater, et al., 1977)	No "lymphoid" component present

during the BP, indicating that the development of chromosomal changes is not the causative mechanism of the BP (Sandberg and Sonta, 1978). However, the significance of newly arising chromosome anomalies after years of chromosomal stability has been pointed out already; in these cases the appearance of cells with additional anomalies usually heralds the BP, sometimes many months before corresponding clinical and cytological changes are apparent (R. T. Silver, Schleider, et al., 1977). Chemotherapy or radiation are in all probability not causally related to the evolution of the chromosomal changes in the BP because a number of such cases with extensive chromosome anomalies but without preceding therapy has been described,[55] and cases with intensive therapy before the BP, but without additional chromosome anomalies in this phase, militate also against such an assumption. A recent study in my laboratory (Sonta and Sandberg, 1978a) indicated that when a large number of cells (300 or more) is examined during the CP, chromosomal changes in a small number of cells not observed upon examination of 30–50 cells (routine in most laboratories) became evident. It is possible that an accelerated proliferation of such abnormal cells is associated with the BP, rather than a de novo development of cells with karyotypic alterations.

An unusual case of CML in a 55-yr-old man with Ph^1-positive CML whose BM contained predominately cells with 24 chromosomes during the BP has been reported (Daniel, Francis, et al., 1978). Only a haploid set, including the t(9;22), of 22 autosomes and the X and Y were present in the cells with 24 chromosomes.

The time interval between the appearance of chromosomal changes during the CP and the development of the BP varies considerably among patients. In some cases the development of such chromosome abnormalities occurs after the development of the BP, but in most cases the karyotypic changes antedate that phase from periods of less than a few months to as long as several years. Thus, once a karyotypic change develops during the CP, it is difficult to predict with certainty the time at which the BP will appear, although it can be assumed that it will do so. There does not appear to be any apparent correlation between the type of karyotypic changes developing in the CP (except possibly for the missing Y) and the appearance of the BP, though there is a strong suggestion that genesis of an iso-17 chromosome may possibly be associated with a shorter interval between its development and that of the BP; however, much more data are necessary to establish this assumption on a firmer basis.

Occasionally, the chromosomal changes besides the Ph^1, both in the CP and BP of CML, may be observed in the blood rather than in the marrow cells, particularly when the former is incubated without PHA. The exact reasons for such an observation are still unclear, though there is a distinct possibility that the genesis of cells with chromosomal abnormalities not seen in the marrow may take place at extramedullary sites (spleen, lymph nodes, etc.) and may possibly be related to the metamorphosis of the disease at such sites rather than in the marrow. Were such cells to be generated in the spleen, they may disappear from the circulating blood after splenectomy and, in fact, such observations have been made by others and us.[56] Generally, however, it is assumed that the abnormal cells in the BP of CML are generated in the marrow, from which they migrate to other organs, e.g., the spleen, from which further spread may occur (Verhest, van Schoubroeck, et al., 1976a).

The Ph^1 as such cannot be regarded as predisposing to transformation of the CP into the BP, because patients with Ph^1-negative CML develop the BP more readily than the Ph^1-positive cases of CML. Several cases of CML have been published in whose hypodiploid leukemic cells the Ph^1 disappeared during the BP (M. H. Khan, 1973b; Miyamoto and Takita, 1973), though the development of additional chromosomal changes including markers may have masked a translocated Ph^1 (these cases were studied before banding) in two of these cases. The one case studied with banding (Van Den Berghe, personal communication), with a 45,XY9q+,−22 karyotype in 75% of the marrow cells, apparently had no such loss in the blood cells several years before the BP.

It is reasonable to expect that in the BP aneuploid metaphases are derived from the greatly increased number of immature cells. Clein and Flemans (1966), however, showed that erythrocytic cells carry the same type of anomalies as the granulocytic cells. Thus, additional chromosomal anomalies characterizing the BP must originate in the same cell category in which the Ph^1 is induced.

We have expressed the view that the development of the BP in CML may not necessarily be due to the genesis of two or more Ph^1-chromosomes and/or the appearance of aneuploidy in the leukemic cells, because the percentage of cells with two Ph^1-chromosomes in the BP is often much less than the percentage of the immature cells (myeloblasts, promyelocytes) in the marrow. Furthermore, as stated previously, not all cases of CML in the BP are accompanied by more than one Ph^1 or aneuploidy and cases of CML with two or more such chromosomes are encountered who do not develop the BP (Nowell, Jensen, et al., 1975). Incidentally, the appearance of more than one Ph^1 is not accompanied by any further evidence of translocation, i.e., only one 9q+ chromosome is seen (for an exception, see the case described by Rowley, 1973e), indicating that the extra Ph^1 (and possibly other) chromosomes are generated by a process of selective nondisjunction or endoreduplication (Hampel, 1963).

The hypodiploid cases with the BP, constitute <10% (in our experience <5%) of the aneuploid cases, save those with a missing Y. The hypodiploid cases appear not to be characterized by nonrandom karyotypic changes, except possibly for loss of a C-group chromosome, though more cases will have to be studied to be certain about this. It is of interest to note that three patients with hypodiploid cells studied by us (Sonta and Sandberg, 1978a) had −17. Even though loss of the Y is the most common cause of hypodiploidy in CML, we have the impression that #17 is frequently involved in the relatively small number of other hypodiploid CML cases.

With regard to the incidence of aneuploidy in Ph^1-negative CML in the BP, the number of such cases studied cytogenetically and published is rather small. I am aware of 15 cases, 4 of which showed hyperdiploidy, 1 hypodiploidy, 1 44–47 chromosomes, and 5 remained diploid.[57] The median survival time after the onset of the BP was demonstrated not to be much different in Ph^1-positive and Ph^1-negative CML (Ezdinli, Sokal, et al., 1970).

Upon complete remission of the BP a pseudonormalization of the hematopoietic cells may be found, which means that the marrow cells revert to the chromosome constitution that characterized the CP.[58] The Ph^1 remains always; it does not disappear in the BP or after complete remission of the BP (rare exceptions occur, e.g., see Fefer, Cheever, et al., 1977, and Greenberg, Ikeda, et al., 1978). Based on their material, Canellos, DeVita, et al. (1971, 1976) believe that patients whose BP was characterized by hypodiploidy of the leukemic cells respond more favorably to a certain type of combination chemotherapy than other CML patients. It is my belief that it is not the chromosome constitution of the leukemic cells that influences prognosis, but the presence or

Table 43 Chromosome changes in CML in BP

	−Y	+C	+Ph^1	i(17q)	+F	+G	Other	No change
1. Acute and chronic phases (123 patients)*	10	63	54	18	27	4	3	
2. Acute phase (72 patients)*		28	28	18	10	14	14	31

	+8	+C	+Ph^1	i(17q)	+19	+G	−C	
3. Acute phase (40 patients with banding)‡	38	58	58	25	20	12	10	
4. 223 patients (acute and chronic phases)‡	20		17	8				
5. First International Workshop on Chromosomes in Leukemia(1977§)	37		31	18				

*Prior to banding.
‡Mainly based on banding.
§Published in (1978) Br. J. Haematol., 39, 305–309.

Table 44 Summarized information on the patients with Ph¹-positive CML and with unusual karyotypes*

Case no.	Patient	Age in yr and sex	Survival, mo	Initial karyotype in CP	Blastic crisis when karyotyped	Further chromosomal changes‡
1	FP	54F	106§	Standard‖	Yes	+8,+12,+21,+Ph¹,+X,+X/+8,+12,+21,+Ph¹,+X,iso-17q
2	JP	68M	86§	Standard/iso-17q‖	Yes	
3	MS	44M	86§	Standard/+8‖	Yes	+8;+8→+1,+3,+8,+8,+10,+11,+13,+13,+13,+14,+19,+19,+19,+20,+21,+21,+Ph¹,+Ph¹,+X,+X
4	AG	54F	85§	Standard/iso-17q‖	Yes	iso-17q/iso-17q,+iso-17q
5	HC	42M	77§	Standard‖	Yes	Standard/+8; standard/+8→standard
6	DW	46M	41§	Standard	Yes	+6,+9q+,+10,+12,+19,+21,+Ph¹/+6,+10,+12,+19,+21,+Ph¹/standard
7	RW	40M	39§	Standard	Yes	+8,+21,+Ph¹,iso-17q/+Ph¹,iso-17q/standard
8	AK	48M	38§	Standard/−Y	Yes	−Y;−16,−17,−Y/−Y
9	DA	24M	34§	Standard	Yes	+8,+17,+21,+Ph¹/+8,+8,+17,+21,+Ph¹/+8,+8,+8,+17,+21,+Ph¹/standard
10	HW	43M	33§	Standard/+8,+10,+11,+13,+14,+19,+21,+21	Yes	+8,+10,+11,+13,+14,+19,+21,+21,+8,+10,+11,+13,+14,+19,+21,+21/+8,+8,+10,+11,+13,+14,+19,+21,+21,+21
11	AF	45M	32§	t(9;10;22)	Yes	Ph¹-translocations(9q;10q;22q); t(9;10;22)/t(9;10;22),+8,+12,+14,+17,+Ph¹
12	MR	42F	31§	Standard	Yes	Standard/−7,−17,+13,+14
13	TLR	33M	28§	Standard	Yes	t(1;9)(9;22)→standard
14	DR	54M	3§	+8,+11,+21,+21/+8,+11,+13,+21,+21 (no Ph−translocation)¶	—	
15	BL	61F	84	Standard‖	Yes	Standard/t(15;19)
16	SM	40F	80	Standard‖	No	Standard/−17
17	CM	25F	70	Standard	No	t(2;4)/standard/t(2;4),+19,+Ph¹
18	MC	29F	70	Standard‖	No	Standard/+Ph¹
19	EW	53F	62	Standard	No	Standard/+8,+8;standard/+8,+8→standard
20	WA	26M	59	Standard/+8,+8	Yes	Standard
21	LS	28M	53	t(13;22)	Yes	Ph¹-translocation (13p;22q);t(13;22),+Ph¹

22	IR	53F	50	Standard	Yes	1q+/standard
23	RR	24M	47	Standard	No	Standard/−17
24	OT	65M	38	Standard	No	Standard/+Ph¹
25	AE	36M	34	Standard‖	Yes	Standard/iso-17q
26	DL	45F	30	Standard	No	Standard/+8,+19
27	WK	65M	28	Standard/iso-17q	No	
28	DM	20F	19	Standard‖	No	Standard/+8,+X
29	HA	41M	12	Standard	No	Standard/+Ph¹
30	SB	43M	57§	Standard	Yes**	Ph¹-translocations(Abnormal 9q;22q)
31	HS	38M	55§	Standard	Yes	iso-17q,+8/iso-17q,+4,+6,+8,+8/t(3;17)(q21;p1?),10q−,iso-17q
32	MTS	57F	48§	Standard	Yes	+6,+8,+8,+19,+19,+Ph¹,+Ph¹/+6,+8,+8,+19,+19,+Ph¹/+6,+8,+8,+19,+19,+Ph¹,+Ph¹,+X
33	AS	49F	39§	Standard	No	Standard/iso-17q
34	JD	18M	86	Standard	Yes	Standard/+Ph¹
35	VS	58F	83	Standard	Yes	Standard/1q+,17p+
36	DS	50M	70	(9;13;22)	No	Ph¹-translocation (9q;22q;13q)
37	HK	26M	52	Standard	No	Standard/iso-17q
38	PD	51M	42	Standard	No	−Y
39	CH	28F	40	Standard	No	Standard/iso-17q
40	DD	32M	39	Standard	No	iso-17q
41	FP	58M	26	Standard	Yes	Standard/iso-17q
42	GE	53M	20	(2;22)	No	Ph¹-translocation(2q;22q)
43	JH	33F	5	Standard	Yes	Standard/t(8;14)(q11;q32)

*In each case, except for those with unusual Ph¹-translocations, each cell line listed contained a Ph¹ in all cells in addition to the karyotypic changes shown. Thus, when a Ph¹ is listed in the table, it indicates an additional such chromosome. "Standard" karyotype refers to a chromosome constitution of cells in which the only abnormality is a Ph¹ resulting from deletion of chromosome 22 and translocation to 9, i.e., t(9;22)(q34; q11). Table based on data from Hayata, Sakurai, et al. (1975); Sonta and Sandberg (1978a).

‡Chromosomal mosaics in marrow in order of frequency.

§Expired.

‖Presence of karyotypically normal cells.

¶Patient in BP when first seen.

**Apparently Ph¹-negative in initial exams.

241

Table 45 Karyotypic patterns in Ph¹-positive CML in the BP—determined with banding techniques

Karyotype	Reference
46,XX,t(C;C)	Rowley, 1977a
46,XX,Cq−	Rowley, 1977a
46,XX,t(3;11)(q21;p15),i(17q)	Rowley, 1977a
46,XX,i(17q)/47,XX,+C,i(17q)	Rowley, 1977a
47,XX,+8,i(17q)	Rowley, 1977a
47,XX,+C	Rowley, 1977a
47,XY,+22q−	Rowley, 1977a
48,XY,+8,i(17q)?,+19	Rowley, 1977a
48,XY,+8,+22q−	Rowley, 1977a
49,XY,+8,+10,+22q−	Rowley, 1977a
49,XXY,+8,+22q−	Rowley, 1977a
50,XY,+8,+8,+mar,+22q−	Rowley, 1977a
52,XY,+C,+C,+F,+F,+G,+22q−	Rowley, 1977a
54,XY,+3,−7,+8,+11,+13,+14,+19,+2mar,+22q−	Rowley, 1977a
46,XY,i(17q)/47,XY,+C,i(17q)	Lobb, Reeves, et al., 1972
46,XX,+C,−16,i(17q)/47,XX,+C,+C,−16,i(17q)	Lobb, Reeves, et al., 1972
46,XX,+C,−16,i(17q)/47,XX,+C,i(17q)	Lobb, Reeves, et al., 1972
49,XX,+10,+21,+22q−	Beck and Chesney, 1973
47,XX,+8	Van Den Berghe, 1973
46,XY,dic(22q−)	Whang-Peng, Knutsen, et al., 1973; Whang-Peng, Gralnick, et al., 1974
46,XY,dic(22q−)	Whang-Peng, Knutsen, et al., 1973; Whang-Peng, Gralnick, et al., 1974
46,XY,8q−(q23)	Whang-Peng, Broder, et al., 1976
48,XY,+7,+12	Kaffe, Hsu, et al., 1974
49,XY,+8,+19,+22q−	Mitelman, Brandt, et al., 1974b, Mitelman, Nilsson, et al., 1975b
48,XY,7p−,+8,+22q−	Gahrton, Lindsten, et al., 1974c
53,XX,+8,+11,+12,+16,+17,+19,+22q−	Gahrton, Lindsten, et al., 1974c
51,XY,+10,+14,+19,+22q−,+mar	Hossfeld, 1974a
51,XY,+4,+17,+19,+22q−,+mar	Hossfeld, 1974a
49,XX,+19,+22q−,+mar	Hossfeld, 1974a
46,XX,i(17q)/47,XX,+8,+i(17q)	McCaw, personal communication to Rowley, 1977
45,X(−X),8p−,+minute	E. Engel, McGee, et al., 1975b
46,XX,6q−,17q+	E. Engel, McGee, et al., 1975b
46,XY,8p+/47,XY,8p+,+22q−	E. Engel, McGee, et al., 1975b
47,XY,+19	E. Engel, McGee, et al., 1975b
47,XY,+22q−	E. Engel, McGee, et al., 1975b
49,XY,+16,17p+,+18,+21/50,same karyotype, +22q−	E. Engel, McGee, et al., 1975b
51,XX,+7,+9,+12,+18,+22q−	E. Engel, McGee, et al., 1975b
51,XX,+9,+12,+22,+22,+22q−	E. Engel, McGee, et al., 1975b
57,XY,+5,+6,+9,+10,+14,+18,+21,+22,+22q−*	E. Engel, McGee, et al., 1975b
49,XY,+10,−17,+20,+22q−	Hossfeld, 1975b
49,XY,+10,+14,+17	Hossfeld, 1975b
47,XX,−17,22q+,+2M	Hossfeld, 1975b
46,XY,−18,+22q−	Hossfeld, 1975b
46,XX,−21,+M	Hossfeld, 1975b
46,XX,−20,+M	Hossfeld, 1975b
46,XY	Hossfeld, 1975b
47,XY,+8,−17,−18,t(17q;18q),+22q−/	
48,XY,+8,−17,−18,t(17q;18q),t(17q;18q),+22q−	
47,XX,+8,−17,i(17q)	Sharp, Potter, et al., 1975

Table 45 — *continued*

Karyotype	Reference
48,XY,+8,−9,−11,−12,−14,−19,+22q−,+5M	Sharp, Potter, et al., 1975
50,XY,+8,+15,+17,+22q−	Sharp, Potter, et al., 1976a,b
48,XY,+8,−17,i(17q),+22q−	Sharp, Potter, et al., 1976a,b
47,XY,+8,−17,i(17q)	Sharp, Potter, et al., 1976a,b
46,X,−Y,21q+,+M	Sharp, Potter, et al., 1976a,b
46,XX,del(4)(q31)	Sharp, Potter, et al., 1976a,b
50,XX,t(2;12)(p13;q24),+8,+8,+12q+,+22q−	Sharp, Potter, et al., 1976a,b
48,XX,+11,+22q−	Philip, 1975b
45,XO/46,XY	Mitelman, personal communication
47,XX,+8	Mitelman, personal communication
46,XX,dic(Ph¹)/47,XX,dic(Ph¹),+8	Mitelman, personal communication
49,XY,+Ph¹,+8,+13	Mitelman, personal communication
47,XY,+Ph¹ (3 cases)	Mitelman, personal communication
47,XY,+Ph¹ (2 cases)	Mitelman, Levan, et al., 1976b
46,XX,dicPh¹	Mitelman, Levan, et al., 1976b
49,XY,+3,+8,+12/50,XY,+3,+8,+11,+12	Mitelman, Levan, et al., 1976b
46,XY,iso(17q)	Mitelman, Levan, et al., 1976b
48,XX,+8,+Ph¹/49,XX,+8,+8,+Ph¹	Mitelman, Levan, et al., 1976b
49,XX,+8,+17,+Ph¹	Mitelman, Levan, et al., 1976b
48,XY,+8,i(17q),+Ph¹	Mitelman, Levan, et al., 1976b
48,XX,+17,+Ph¹	Mitelman, Levan, et al., 1976b
47,XY,+Ph¹/48,XY,+19,+Ph¹/49,XY,+8,+19,+Ph¹	Mitelman, Levan, et al., 1976b
46,XY accelerated phase	Prigogina and Fleischman, 1975a, and Prigogina, Fleischman, et al., 1978
46,XY,11p−/47,11p−,+22q−/48,+Ph;11p−,+8/46,+dicPh¹ accelerated phase	Prigogina and Fleischman, 1975a, and Prigogina, Fleischman, et al., 1978
46,XY,i(17q),20q−	Prigogina and Fleischman, 1975a, and Prigogina, Fleischman, et al., 1978
46,XY,i(17q)	Prigogina and Fleischman, 1975a, and Prigogina, Fleischman, et al., 1978
46,XX,i(17q)	
47,XY,i(17q),+8 (3 cases)	Prigogina and Fleischman, 1975a, and Prigogina, Fleischman, et al., 1978
47,XX,i(17q),+8,6q−	Prigogina and Fleischman, 1975a, and Prigogina, Fleischman, et al., 1978
46,XX,i(17q)/48,XX,i(17q),+3,+8	Prigogina and Fleischman, 1975a, and Prigogina, Fleischman, et al., 1978
46,XY,i(17q)/47,XY,i(17q),+8	Prigogina and Fleischman, 1975a, and Prigogina, Fleischman, et al., 1978
47,XX,i(17q),+8	Prigogina and Fleischman, 1975a, and Prigogina, Fleischman, et al., 1978
48,XX,i(17q),+8,+19/49,i(17q),+8,+19,+22q−	Prigogina and Fleischman, 1975a, and Prigogina, Fleischman, et al., 1978
50,XX,+5,+6,+9,+22q−	Prigogina and Fleischman, 1975a, and Prigogina, Fleischman, et al., 1978
50,XY,+6,+11,+10,+22q−	Prigogina and Fleischman, 1975a, and Prigogina, Fleischman, et al., 1978
54,XY,+6,+8,+10,+11,+19,+21,+Y,+Ph¹	Prigogina and Fleischman, 1975a, and Prigogina, Fleischman, et al., 1978
51,XY,+7,+8,+12,+19,+21	Prigogina and Fleischman, 1975a, and Prigogina, Fleischman, et al., 1978
46,XY,r(7)	Prigogina and Fleischman, 1975a, and Prigogina, Fleischman, et al., 1978

continued

Table 45 Karyotypic patterns in Ph¹-positive CML in the BP—determined with banding techniques—*continued*

Karyotype	Reference
46,XX/46,X,−X	Prigogina and Fleischman, 1975a, and Prigogina, Fleischman, et al., 1978
46,XX,t(1q;17q)	Prigogina and Fleischman, 1975a, and Prigogina, Fleischman, et al., 1978
46,XY,i(9q+),−22,+M	Prigogina and Fleischman, 1975a, and Prigogina, Fleischman, et al., 1978
47,XY,+22q−	Fleischman and Prigogina, 1975, and Prigogina, Fleischman, et al., 1978
46,XX,13q+	Fleischman and Prigogina, 1975, and Prigogina, Fleischman, et al., 1978
46,XX,19q+	Lawler, O'Malley, et al., 1976
46 (2 cases)	Lawler, O'Malley, et al., 1976
46,XY	Lawler, O'Malley, et al., 1976
46,XY	Lawler, O'Malley, et al., 1976
46/47,i(17q)	Lawler, O'Malley, et al., 1976
47,−16,17p+,+22q−	Lawler, O'Malley, et al., 1976
46/48,+2C	Lawler, O'Malley, et al., 1976
48,+19,+22q−	Lawler, O'Malley, et al., 1976
46,t(2:2)(q21;p25)	Lawler, O'Malley, et al., 1976
46,15q−	Hagemeijer, Smit, et al., 1977
46 (3 cases)	Hagemeijer, Smit, et al., 1977
−X,−7	Hagemeijer, Smit, et al., 1977
−17,−18,17p+	Hagemeijer, Smit, et al., 1977
Xq−,5q−,6p+,inv(11),16q+,i(7q),20q+	Hagemeijer, Smit, et al., 1977
i(17q)/i(17q),+8/i(17q),+8,Xp+	Hagemeijer, Smit, et al., 1977
+8/+8,+19/i(17q)	Hagemeijer, Smit, et al., 1977
47,XY,−12,−13,t(12;13),+Ph¹+mar (?+8)	Rożynkowa, Stepién, et al., 1977
49,XY,+8,+10,+Ph¹	Rożynkowa, Stepién, et al., 1977
+8,+8,+19,+mar	Lilleyman, Potter, et al., 1978
+Ph¹ (4 cases)	Lilleyman, Potter, et al., 1978
+12,+21	Lilleyman, Potter, et al., 1978
−Y,21q+,+mar	Lilleyman, Potter, et al., 1978
+8,+19,−17,i(17q)	Lilleyman, Potter, et al., 1978
t(7;8),1q+	Lilleyman, Potter, et al., 1978
17p−	Lilleyman, Potter, et al., 1978
−17,i(17q) (2 cases)	Lilleyman, Potter, et al., 1978
+8,+8,−17,i(17q)	Lilleyman, Potter, et al., 1978
+8 (3 cases)	Lilleyman, Potter, et al., 1978
+Ph¹,+8,−17,i(17q)	Lilleyman, Potter, et al., 1978
+8,−17,i(17q)	Lilleyman, Potter, et al., 1978
+8,+19,+22	Lilleyman, Potter, et al., 1978
+8,−17,i(17q) (2 cases)	Lilleyman, Potter, et al., 1978
12p−,−17,i(17q)	Lilleyman, Potter, et al., 1978
Inv(11)	Lilleyman, Potter, et al., 1978
10q−	Lilleyman, Potter, et al., 1978
+Ph¹,+3,+8,+12,+14,+15,+19,+20,+22	Lilleyman, Potter, et al., 1978
+Ph¹,+8	Lilleyman, Potter, et al., 1978
46,XY,+Ph¹	Lyall and Garson, 1978
46,XY,i(17q)	Lyall and Garson, 1978
46,XX,i(17q)	Lyall and Garson, 1978
47,XY,+8,i(17q)	Lyall and Garson, 1978
47,XX,+8,i(17q)	Lyall and Garson, 1978
48,XY,+8,+Ph¹	Lyall and Garson, 1978

Table 45—*continued*

Karyotype	Reference
50,XY,+4C,−E,+Ph¹(BM)/47,XY,+8 i(17q)(Blood)	Lyall and Garson, 1978
49,XY,+9,+14,+Ph¹	Lyall and Garson, 1978
53,XY,+8,+8,+8,+9,+9,+19,+Ph¹,+Ph¹/45,X,−Y	Lyall and Garson, 1978
51,XX,+6der(8),t(8;Ph¹)(p12;p11)/51,XX,−8, +6der(8)(p12),+19,−Ph¹	Lyall and Garson, 1978
46,XX,+2der(8),t(8;13)(p12;q34),−20/46,XX	Lyall and Garson, 1978
46,XX/46,XX,i(Xq) or i(9q)	Lyall and Garson, 1978
45,XY,+del(1)(q12),−3,−7,del(11)(q21),−19, +mar	Lyall and Garson, 1978
47,XX,+8	Alimena, Brandt, et al., 1979
50,XY,+6,+8,+13,+Ph¹	Alimena, Brandt, et al., 1979
47,XY,+8,i(17q)	Alimena, Brandt, et al., 1979
47,X,−Y,+8,+19	Alimena, Brandt, et al., 1979
47,XY,+8 (2 cases)	Alimena, Brandt, et al., 1979
50,XX,+X,+8,+9,+19	Alimena, Brandt, et al., 1979
48,XY,+Y,+21	Alimena, Brandt, et al., 1979
49,XY,+8,+14,+19	Alimena, Brandt, et al., 1979
51,XY,+Y,+8,+19,+Ph¹,i(17q),t(6;8)(q34;q11)	Alimena, Brandt, et al., 1979
47,XY,+8,i(17q),del(6)(q13)	Alimena, Brandt, et al., 1979
47,XX,+Ph¹,5p+,t(7;8)(q36;q22)	Alimena, Brandt, et al., 1979
47,XY,+Ph¹	Alimena, Brandt, et al., 1979
48,XY,+8,+Ph¹	Alimena, Brandt, et al., 1979
45,XY,+8,−18,−19	Alimena, Brandt, et al., 1979
47,XX,+8,i(17q)	Alimena, Brandt, et al., 1979
48,XX,+8,+Ph¹	Alimena, Brandt, et al., 1979
52,XY,+8,+9,i(17q),+19,+20,+21,+Ph¹,inv(7)(p15q22),1p+	Alimena, Brandt, et al., 1979
46,XY,+8,+9,−14,−16,−21,+Ph¹,del(1)(q21), dup(1)(q21)	Alimena, Brandt, et al., 1979
46,XX,1p+	Alimena, Brandt, et al., 1979
46,XX,t(1;8)(q25;qter),5q+,17p+	Alimena, Brandt, et al., 1979
47,XX,+del(1)(p11)	Alimena, Brandt, et al., 1979
48,XX,+1,+4,+8,+9,−14,i(17q),−18,−19,+Ph¹	Dallapiccola and Alimena, 1979

Unless indicated otherwise, the cases shown had a t(9;22) leading to a Ph¹, hence only extra Ph¹-chromosomes are shown in the "Karyotype" column and indicated as either +Ph¹ or +22q−.
*Plus 2 unidentified chromosomes.

absence of a pseudodiploid (Ph¹-positive) cell population in the marrow. This opinion had been expressed by Garson, Burgess, et al. (1969) and is supported by the fact that in all the patients, including those studied by us, in whom complete remissions could be obtained, a number of pseudodiploid metaphases were found besides the aneuploid ones. No complete remission could be achieved in a patient whose marrow was devoid of pseudodiploid metaphases upon initial examination, despite a remarkable response to chemotherapy (Pedersen, 1971).

The chromosomal findings in the BP of CML can be summarized as follows (Tables 43–45). In a third or fewer of the cases no additional changes can be detected[59] whereas in two-thirds or more of the cases aneuploid cell clones can be seen.[60] Aneuploidy may be hypo- or hyperdiploid in nature; the great majority of the cases (>90%), however, is associated with a hyperdiploid pattern. The chromosome number per metaphase varies between 44–58 (Berger, 1970b). The karyotypic changes in the BP (as well as in the CP) have been considerably clarified with banding techniques. These changes have been found to be nonrandom in character and, generally, to consist primarily of those changes seen as additional ones (to the Ph[1]) in the CP of the disease. Thus, the most frequently observed karyotypic alterations seen in the BP are: +Ph[1] (about 30–50% of the cases),[61] +8 (about 40%),[62] iso-17q (about 25%),[63] +19 (about 15%), +21 (about 10%), extra chromosomes in group C(#9, #10, and #11), etc.[64] These anomalies may occur separately or in association with others, e.g., +Ph[1] plus iso-17q or +19; +8 plus +Ph[1] or +19. However, iso-17q almost never occurs with +19, possibly being one of the postulated "forbidden combinations."[65]

The story of the iso-17[i(17q)] anomaly in CML is interesting, for it points to the crucial role played by banding in deciphering karyotypic anomalies (Table 46). Pedersen (1964a) first drew attention to the −E,+C picture in CML. E. Engel, McKee, et al. (1967) proposed that the formation of an iso-17 may be responsible for the −E,+C. This hypothesis was confirmed with banding by Lobb, Reeves, et al. (1972). The importance of this chromosome anomaly in the BP of CML was already pointed to by Grouchy, Nava, et al. in 1968. The observation that patients with Ph[1]-positive CML who developed additional cytogenetic changes, i.e., i(17q), +8,−7, etc., had a survival that did not differ from that of patients without such changes, the fact that the BP may appear in patients without further associated karyotypic aberrations, and the fact that the chromosomal picture did not seem to affect the clinical course or survival of these patients, point to factors other than the chromosomal changes as playing a crucial role in the progression and prognosis of Ph[1]-positive CML (Sonta and Sandberg, 1978a).

Even though we (Sonta and Sandberg, 1978) could not find any correlation between the development of additional chromosomal abnormalities in the BP and survival, the latter being similar to that of patients without further changes, Prigogina, Fleischman, et al.

Table 46 Iso-17 [i(17q)] chromosome in conditions other than CML

Diagnosis	Reference
AML	Nowell and Hungerford, 1962; E. Engel, McGee, et al., 1967; Mitelman, Brandt, et al., 1973; L. Brandt, Levan, et al., 1974b; E. Engel, McKee, et al., 1975; Barlogie, Hittelman, et al., 1977
AL, primarily EL	Fitzgerald and Adams, 1965; E. Engel, McKee, et al., 1967; Seif and Spriggs, 1967; Millard, 1968; Sakurai and Sandberg, 1976a; Najfeld, 1976 (chronic EL); Hagemeijer, Smit, et al., 1977
Lymphoma and HD	Castoldi, 1970, 1973a; Ricci, Punturieri, et al., 1962b; Fitzgerald and Adams, 1965; Sasaki, Sofuni, et al., 1965; Toshima, Takagi, et al., 1967 (Burkitt cell line); Spiers and Baikie, 1968a; Fleischman and Prigogina, 1975, 1977; Zech, Haglund, et al., 1976 (Burkitt cell line); Mark, 1977b; Hossfeld, 1978
Eosinophilic leukemia	Mitelman, Panani, et al., 1975
ALL	Fitzgerald and Adams, 1965; Yamada and Furasawa, 1976
SBA	Kamada in Killmann, 1976
Refractory anemia with excess myeloblasts	Prieto, Badia, et al., 1976; Van Den Berghe, Michaux, et al., 1977a
Plasma cell leukemia	Fitzgerald, Rastrick, et al., 1973
Preleukemia	Ruutu, Ruutu, et al., 1977b
Myeloproliferative syndromes	E. Engel, McKee, et al., 1967; Hagemeijer, Smit, et al., 1977; Nowell and Finan, 1977; Whang-Peng, Lee, et al., 1978; McDermott, Romain, et al., 1978
Cancer of prostate	Oshimura and Sandberg, 1976
Sézary syndrome (CLL)	Nowell and Finan, 1977
Sézary syndrome	Bosman, van Vloten, et al., 1976 (2 cases)
Myelofibrosis→AMMOL	Whang-Peng, Lee, et al., 1978
RA	Nowell and Finan, 1978a

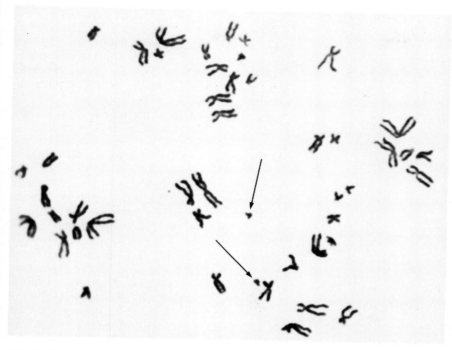

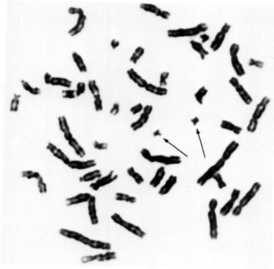

78 Metaphases showing the presence of two Ph¹-chromosomes (arrows), a condition usually associated with or heralding the BP of CML.

(1978) reported a correlation of the chromosomal changes in the acute phase with the morphology of the blast cells and with the clinical course. Thus, in 11 of 24 cases, the survival exceeded 6 mo (in 6 of the 11 patients only a 9q+;22q− was present), whereas the 13 patients surviving 6 mo had additional chromosomal changes. Remission was obtained in 6 of 7 patients in the former group and in only 50% of the latter. This type of correlation must be extended to larger groups of patients in various clinics.

Multiple Ph¹-Chromosomes in the Blastic Phase

Particular attention has been paid to the duplication of the Ph¹, which has been described in nearly 50% of the cases reported in the literature (Figure 78). Cases with three or more Ph¹-chromosomes have been described (see reference 27); those with two such chromosomes often have a number of cells with three or more Ph¹-chromosomes. Duplication of the Ph¹ may be the only additional anomaly in the BP,[66] but is usually associated with other anomalies. Duplication of the Ph¹ has been observed in splenic cells but not in the marrow cells of a patient in the BP (Oberling, et al., Stoll, 1975a). The most common chromosome combination of group #21–22 in the BP is three #21–22 and two Ph¹; next in frequency is the combination of four #21–22 and two Ph¹.[67] Rarely, the combination of two #21–22 and two Ph¹ was seen.[68] The last combination could be due to loss of a #21–22; the other explanation would be that the second Ph¹ is derived from the missing #22. Duplication of the Ph¹ and other chromosomes (including the 9q+) is commonly considered to be the consequence of endoreduplication or nondisjunction (Hampel, 1963; Cabrol, Peytremann, et al., 1976). Both mechanisms result in two chromosomes of identical morphology. When extra Ph¹-chromosomes develop, no additional translocation of the deleted material from #22 is seen and the G-group maintains its previous number of autosomes. This points to endoreduplication (or nondisjunction) as the most plausible method for the genesis of the extra Ph¹-chromosomes. A number of authors, however, have pointed to morphological differences of the two Ph¹ in the same metaphase.[69] We saw a patient (Sandberg and Hossfeld, 1974) with the chromosome constitution 47,XY,−E,+C,2Ph¹ in his marrow cells; in a third of the metaphases two morphlogically different Ph¹ˢ were present, one being metacentric, the other acrocentric. By means of autoradiography the replication pattern of the Ph¹ˢ and the small acrocentric chromosomes was studied. It was found that the median grain count of the more metacentric Ph¹ was significantly higher than that of the acrocentric Ph¹. The results indicate that the two Ph¹-chromosomes may have originated from two different chromosomes or by dissimilar processes.

Even though AML and the BP of CML have some aspects in common (including very rare cases with Auer rods in CML; Kundel, Tanaka, et al., 1963), the chromosomal behavior is quite different. Only about 50% of AMLs have chromosomal anomalies, whereas >70% of CMLs in the BP are aneuploid, irrespective of the Ph¹. Only rarely are chromosome numbers higher than 48 encountered in AML, and half of the chromosomally abnormal AMLs are hypodiploid. Hypodiploidy in the BP of CML is rare, whereas hyperdiploidy or >50 chromosomes is frequent. The most remarkable differences are the chromosomal heterogeneity and instability of the leukemic cells in the BP of CML, when compared with AML. Apart from the pseudodiploid, Ph¹-positive cell population, which can be detected in 70% of the cases in 5–95% of all metaphases, several other cell clones can be seen in a significant number of cases.[70] Up to nine chromosomally different clones were found in a case reported by Crossen, Mellor, et al. (1971b). With duration of the disease, whether or not favorably influenced by therapy, the modal chromosome constitution may fluctuate. This fluctuation is usually characterized by an increase of the chromosome number,[71] but a decrease may also occur.[72] A common pathway in a step-by-step evolution of complex chromosome anomalies has been proposed by many workers, but actually demonstrated in rare cases only (Pedersen and Videbaek, 1964). There can be no doubt that most or all of the chromosomally different cell clones in a given case are related to each other. It is debatable, however, whether clonal evolution follows a certain pathway, as suggested by Grouchy, Nava, et al. (1966b) and echoed by others.[73] Chromosomal instability and heterogeneity are very uncommon events in AML, where a given chromosome constitution remains usually stable from the start to the end of the disease (see subsequent section).

Double-Ph¹, Blastic Phase, and Cell Source in CML

The transformation of CML into the blastic phase (BP) is a poorly understood phenomenon (Pedersen, 1977). The onset of this phase usually leads to a downhill course of the disease (Pedersen-Bjergaard, Worm, et al., 1977) with the survival time after the development of the BP apparently not differing in Ph¹-positive or Ph¹-negative cases. A substantial number of cases with the BP of CML is associated with chromosomal changes which may precede and herald the onset of this transformation. In Ph¹-negative CML, little information is available on the cytogenetic changes in marrow cells that precede and/or accompany the BP of the disease; besides, it is often difficult to clearly delineate the onset of the BP in Ph¹-negative CML.

Even though a significant number (ca 25%) of Ph¹-positive CML cases is not associated with any further karyotypic changes when transformation into the BP occurs, most cases develop aneuploidy (usually hyperdiploidy) and/or extra Ph¹-chromosomes before or in conjunction with the blastic transformation. Usually the appearance of extra Ph^1-chromosomes is related to the genesis of the BP. Two Ph^1-chromosomes may appear in CML a considerable period of time before the BP develops and, in fact, in occasional cases such a phase did not appear because the patients died of other causes. These observations indicate that the development of the BP, a phase very similar to AML, is not due to the extra Ph^1 and that, in all probability, the genesis of the chromosome abnormalities is a result of the BP rather than the cause of it. This promulgation is further substantiated by the finding of metaphases with two Ph^1-chromosomes in cells at a time of almost complete remission of the leukemic picture in the marrow. Whether the persistence of the two Ph^1-chromosomes is evidence that complete remission of the BP has not been achieved or whether the presence of two Ph^1-chromosomes, once having taken place, becomes a permanent feature of the cytogenetic make-up of the marrow regardless of the cytologic and clinical status, is a problem that cannot be answered without more cogent and direct information than is presently available. The number of cells containing the extra Ph^1 may be variable from case to case, and, in the same case from time to time, constituting less than the preponderant percentage of the total cell population, and not bear any relationship to the number of myeloblasts and other immature granulocytic cells in the marrow. In fact, in my experience, in some marrows in which leukemic blasts predominated, the percentage of metaphases with two Ph^1-chromosomes may be impressively low.

Until accurate methods are introduced that will identify with certainty the type of cells from which all metaphases of the marrow originate, considerable doubt will continue to exist regarding the type of cell involved by two Ph^1-chromosomes. Nevertheless, on the basis of the marrow cell differential, it is possible to infer, on occasion, the nature of the cells involved by the two Ph^1-chromosomes.

The persistence of Ph^1-positive pseudodiploid cells in consortium with aneuploid cells with two Ph^1-chromosomes suggest that either the latter cells must have originated from the former or that they have arisen *de novo* out of a precursor cell different from that responsible for the pseudodiploid cells. Inasmuch as such a picture may be encountered in marrows essentially devoid of normoblasts, it appears that the granulocytic cell population consists of both diploid and aneuploid elements (with one or two Ph^1-chromosomes). In other patients, all diploid elements disappear from the marrow with progression of the BP and in these it must be assumed that all precursors had become aneuploid, including those of normoblasts and other cells. On the other hand, in one of our patients most of the metaphases were diploid (with one Ph^1) when the marrow was blastic with an M:E ratio of 24:1 and only 4% nucleated erythrocytes. When the patient responded to therapy and the marrow contained 58% nucleated erythrocytes, no metaphases with 46 chromosomes were present; most of the cells had two Ph^1-chromosomes. Therefore, it is possible that the aneuploid cells with two Ph^1-chromosomes may have belonged to only one series of cells, e.g., normoblasts, whereas the aneuploid cells with one Ph^1-chromosome were confined to the myleoblastic series. However, the possibility also exists that some of the normoblastic and myeloblastic precursors contained two Ph^1-chromsomes and the remaining precursors only one. Unfortunately, neither our data nor those in the literature have indicated consistent reversion of aneuploid marrows in the BP of CML to a truly diploid state upon remission, because this, combined with the differential picture, may give us an inkling as to the possible source of aneuploid cells in the BP of CML, with one or two Ph^1-chromosomes. It should be pointed out again, though, that the BP of CML may take place in the presence of only 46 chromosomes with one Ph^1 and that the genesis of aneuploidy and supernumerary Ph^1-chromosomes is not a necessary prerequisite for the development of the BP. Furthermore, the karyotypic changes vary from one case to another and, thus, resemble the cytogenetic variability found in AML or ALL, with no consistent or truly characteristic picture being present. Hence, the chromosomal changes are more likely to be secondary or epiphenomena of the leukemic process rather than causally related to it.

Ph^1-Negative, Diploid Cells in CML

The enigma as to which cells are involved by two Ph^1-chromosomes can be carried over into the area related to the type of cells involved by one Ph^1 in CML. The presence of a significant number of cells without the Ph¹, so-called Ph¹-negative cells, in the marrows of some patients with CML gives rise to the promulgation of several theories regarding the origin of marrow cells. In a recent study (Sonta and Sandberg, 1978a), normal cells in the marrow were encountered in 10 of the 41 patients studied in the early CP of the disease, and ranged from 1.8% to 61.0% of the cells counted. These normal cells gradually decreased in percentage during the course of the disease. In the BP of CML we did not observe any normal cells in the specimens examined directly or cultured for 1 day without PHA. Some caution must be exercised in committing oneself to the presence of normal cells in the BP of Ph¹-positive CML in patients without a prior history or evidence of the CP of CML. In a significant number of patients with AML both karyotypically normal and abnormal cells exist in the marrow and the possibility always exists that the patients with the Ph¹-positive disease described above are, in fact, cases of Ph¹-positive AML. Hence, the presence of any chromosomally normal cells in the marrow of a patient with Ph¹-positive CML should alert one to the possibility of AML or other forms of AL.

The unicellular stem cell theory of marrow cell origin implies that all three types of cells are derived from one precursor and the presence of the Ph¹ in 100% of marrow metaphases has been used as a strong argument in favor of such a theory. The significance of this theory lies in its implication that whatever is the cause of CML and/or the Ph¹ produces its effects originally at the level of the one precursor for all the marrow cells. However, the finding of Ph¹-negative cells is perplexing in the context of the theory of the unicellular origin of marrow cells and requires explanation. At least three plausible explanations can be advanced:

1. The presence of Ph¹-negative cells could be explained within the framework of this theory by assuming that some precursors escape the effects leading to the genesis of the Ph¹, and the percentage and presence of such cells may vary throughout the course of CML. The genesis of the BP, usually affecting only the myeloblastic series, could be due to causative factors affecting the already identifiable cells of the granulocytic series at a step beyond the stem cell stage.

2. It is also possible to explain the Ph¹-negative cells in other terms and, yet, be compatible with a unicellular stem cell origin of all marrow cells. It is possible that the factors leading to Ph¹-positive CML work at the level of already differentiated cells just beyond the stem cell stage (Abramson, Miller, et al., 1977). In this fashion, the factors would affect precursors of the myeloblastic, normoblastic, and megakaryocytic series of cells. In the 100% Ph¹-positive cases, all series are totally involved, whereas in some cases one or another of the cell series may escape involvement by the factors leading to the genesis of the Ph¹. This explanation would imply that the Ph¹ may possibly predispose the granulocytic series of cells to the risk of becoming blastic.

3. Another hypothesis postulates that each series of cells has its own precursor cell and that not all three have to be affected by the process leading to the development of CML and/or the Ph¹. Thus, even though in most cases of CML all precursors are involved, it is possible for the normoblastic or megakaryocytic stem cells not to be affected and reflected in perfectly diploid cells.

The various theories advanced above are all compatible with the cytogenetic data obtained in CML and its varied stages, and my contention is that it may be premature and inappropriate to utilize the findings with the Ph¹ as support for the unicellular stem cell origin of marrow cells. Until all the variegated facets of the karyotypic picture in CML are explained and put on a firm scientific basis, it may be more prudent not to rely on the Ph¹ as an index of cellular genesis in marrow. Whether one theory or another is correct will depend largely on rigorous demonstrations of the cellular level at which the Ph¹ and CML are produced and the behavior of the various cells during the progression of CML and its BP. Until such data are secured, it is best to hold any single theory in abeyance.

Karyotypic "Staging" of CML and Its Prognostic Implications

An analysis of the chromosomal data obtained in a series of patients with Ph¹-positive CML at our Institute (Sakurai and Sandberg, 1976c; Sakurai, Hayata, et al., 1976) has not only revealed the probable importance of normal cells in the marrow to the course of this disease, but has also strongly indicated that the clinical progression of CML is usually accompanied by concomitant progression of the karyotypic picture; and that the latter may reflect reliably the prognostic aspects of the disease.

The "purely" Ph[1]-positive condition is an essentially early manifestation of CML. The development of aneuploidy in addition to the Ph[1] is a manifestation of CML which tends to occur at least 2 yr after the diagnosis is made and, once developed, is associated with a generally poor prognosis and a death rate significantly higher than that for patients who only have a Ph[1] anomaly. Clinical progression of the CP of Ph[1]-positive CML is accompanied by gradual progression of the cytogenetic picture, especially in the BP (Sonta and Sandberg, 1978a). Because a missing Y may affect the prognosis of CML, I have separated these patients from the other male patients with CML. No female patients had a missing sex-chromosome in the marrow cells.

Karyotypically normal cells are encountered early in the course of CML, as indicated by the short interval between the diagnosis and the chromosomal examination. Furthermore, the presence of these normal cells appeared to endow this group of patients with a much longer survival than that seen for patients without normal cells in their marrow. It is possible that the Ph[1]-positive cells may be more sensitive to therapy than the karyotypically normal cells and, hence, repopulation of the marrow with normal cells, when the leukemic cells decrease in number or totally disappear from the marrow, takes place. That this may, indeed, occur is pointed to by the emergence of cytogenetically normal marrow in four of our CML patients with some normal cells in their marrows before initiation of the chemotherapy for the disease.

As an extension of the above, and in confirmation of the findings of others (Prigogina, Fleischman, et al., 1978), additional chromosomal changes besides the Ph[1] were usually seen in advanced stages of CML, commonly in the BP. The latter was particularly frequent in those patients whose marrow contained not only Ph[1]-positive cells but also additional cytogenetic abnormalities. Thus, the study points to the need for an earlier diagnosis of CML than is presently available in the preponderant number of cases with this disease, because, apparently, at a very early stage, chromosomally normal cells will usually be encountered; and the survival of these patients, when appropriately treated, appears to be significantly longer than when the disease is diagnosed at a relatively late stage. At present, normal metaphases are encountered in the marrow of only a small percentage of CML patients, whether the disease is in the CP or BP; or, when present, they are rather scanty. In the later stages of Ph[1]-positive CML, normal cells are only rarely encountered in the marrow, and the presence of additional cytogenetic abnormalities is indicative of progression of the CML, usually into the BP.

The intervals from the date of diagnosis to the date of the first examination were significantly shorter for the patients with some normal cells in their marrow than for those with Ph[1]-positive and otherwise chromosomally normal cells. This fact indicates that normal metaphases are usually present in the marrow much more frequently at an early stage of the disease than at a later one and that they will disappear from the marrow unless the CML is successfully treated; a long survival after appropriate therapy, in fact, has been reported in patients with normal metaphases among the Ph[1]-positive cells. However, reports indicating that the normal cells may persist for long periods without treatment of the CML have also been published.

The median survival after the first examination was significantly shorter for the patients with only abnormal Ph[1]-positive cells as compared with that of the patients with only Ph[1]-positive cells with or without some other karyotypic anomalies. This fact is reminiscent of our findings in AML, in which condition any normal cells in the marrow were vitally important to the success of chemotherapy and, more importantly, to the survival of the patients. The Ph[1]-positive, but otherwise karyotypically normal cells in CML appear to have a role similar to that of the normal cells in AML, even though the former cells must themselves become the target for therapy when their number is too high and especially when the patient has normal metaphases besides the Ph[1]-positive cells in the marrow.

The distribution of the percentages of myeloblasts plus promyelocytes among Ph[1]-positive but otherwise chromosomally normal marrows seemed to follow a normal pattern up to the 30% point; this observation, and the influence of the percentage of such cells on the patients' prognosis, indicated that from a practical aspect, a marrow with >30% of such cells should be considered in the BP. A Ph¹-positive but otherwise chromosomally normal marrow was more likely to be seen when there were <30% myeloblasts plus promyelocytes; when there were >30% such cells, the chances for the marrow to be only Ph¹-positive or contain some or all cells with other karyotypic changes were almost equal. About 10% of the marrows with only Ph¹-positive cells were accompanied by >30% myeloblasts plus promyelocytes, and more than one-half of the marrows with only chromosomally abnormal Ph¹-positive cells were accompanied by <30% of such cells. These findings indicate that the former cells can undergo a blastic transformation without further chromosomal changes. Thus, these Ph¹-positive cells can be compared to normal cells in the marrow of normal individuals before developing AL. Both types of cells behave like normal cells, and without acquiring (further) chromosomal changes, such subjects can develop AML or the BP of CML.

In addition to the chromosome constitution, the cell differential of the marrow seemed to correlate with the patient's prognosis. Out of the nine patients with only karyotypically abnormal Ph¹-positive cells in the marrow, five died within 6.3 mo, four of whom had >30% myeloblasts plus promyelocytes at their first examination. The other four survived more than 6.4 mo, all of whom had <30% such cells. There were also nine patients whose first bone marrow examinations exhibited >30% myeloblasts plus promyelocytes. Among them, five died within 6.3 mo, four of them being patients with only chromosomally abnormal cells in the marrow (the same patients as the above four). None of the other four patients, who had survived 1.6, 3.0, 7.4 and 15.8 mo, respectively, and were living at the time of the study or, when last seen, had only cytogenetically abnormal cells in the marrow. Thus, I believe that, for a more complete appraisal of the prognosis of Ph¹-positive CML, an evaluation of both the chromosomal findings and the marrow cell differential is indicated.

Even though in the preponderant number of patients with Ph¹-negative CML the initial and subsequent cytogenetic studies reveal a diploid picture, including those performed with banding techniques, the significance of the diploidy in the marrow is of a different nature than in Ph¹-positive CML. Hence, the proposed hypothesis regarding the significance of cytogenetically normal cells in the bone marrow of CML applies only to the Ph¹-positive disease.

Based on the findings of the study (Sakurai, Hayata, et al., 1976), we believe that every effort should be directed toward the early detection and diagnosis of Ph¹-positive CML for the following reasons: (1) marrows with normal cells are mostly found in the very early stages of the disease; (2) improvement in the cellular karyotypic picture is very difficult to achieve; and (3) the hope for the permanent cure of the disease must reside in the eradication of the Ph¹-positive cells and the restoration of a normal marrow.

Clonal Origin and Evolution of CML

Clonal evolution is defined as a step-wise rearrangement of the karyotype occurring in an apparently orderly fashion; when many cell populations are present, one type may predominate, a process corresponding to the "stemline" concept of Makino (Figure 79).[74] Clonal evolution appears to proceed according to certain patterns or models involving numerical changes, structural rearrangements, or a combination of both. Acquisition or loss of normal and/or development of marker chromosomes are essential to the progression of clonal evolution. However, as many cases have been analyzed, it has become apparent that several of these particular evolutionary models may be associated with specific patterns of progressive clinical symptoms. Future studies must, therefore, attempt to characterize completely such models and to discover still unrevealed clinical features associated with each.

Selective endoreduplication, defined by Lejeune, Berger, et al. (1966) as the result of double replication of part of the genome (a whole or part of a chromosome), may explain the acquisition of supernumerary chromosomes in the absence of complementary cells. The latter rules out nondisjunction, when the competing cells are nonviable. Loss of chromosomes can also be explained by nondisjunction. But here again complementary cells are not

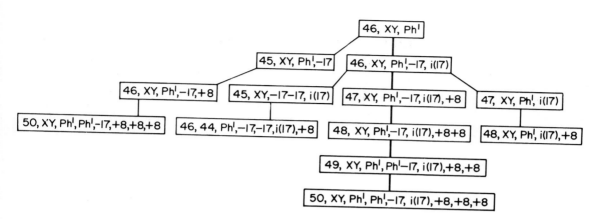

79 Schematic presentation of clonal evolution in CML. Note that the accepted scheme involves progressive development of karyotypic changes, in this case those common to the BP of CML. The major clone proceeds in a vertical fashion, whereas the subclones are shown on both sides of the major pathway leading to a number of cell clones with differing karyotypic pictures.

found. This would be in favor of another possible mechanism, e.g., chromosome lag at anaphase, which results in chromosome loss without production of complementary cells. In any case, clonal evolution may represent no more than a process of selective endoreduplication of certain chromosomes (Ph¹, #8, and #19) in CML.

Some evidence for the monoclonal origin of CML has recently been obtained with chromosomal findings based on banding analyses. We described a male patient with CML who had an unusual #9 with a typical Ph¹-translocation (Hayata, Kakati, et al., 1974). This unusual #9 showed heteromorphism of the 9qh region and a missing band in the middle of its long arm. No translocation was seen on the normal #9. These characteristic features of the leukemic cells strongly support the monoclonal origin of CML. However, the possible multiclonal origin of the CML in our case, though extremely unlikely, cannot be absolutely ruled out. Thus, it is possible to envision that the modified #9 preferentially attracted the translocation from #22 and, hence, the karyotypic changes consistently involved the modified #9. The additional modification of the substructure of the #9, however, may be due to the leukemic process or induced by the translocation. Even if there is a predilection for the chromosomal changes to exist in one of the chromosomes of pair #9, it is difficult to explain with the multicellular origin concept of CML the fact that in the case studied by us the #9 with the translocation always had a missing band. Hence, the case strongly favors the monoclonal origin of CML. Other chromosomal evidence for the monoclonal origin of CML was presented by Gahrton, Lindsten, et al. (1973, 1974b), who reported a case of CML with a Ph¹ derived from a maternal #22 that had a characteristic Q-band at the satellite region. Hossfeld (1975b) reported a female patient with CML in whom the 9;22 translocation was constantly found to involve only a variant #9 with an unusually long secondary constriction. He concluded that the findings indicated a unicellular origin of CML. M. A. S. Moore, Ekert, et al. (1974a) found a Ph¹ only in a 46,XY line in a case of 46,XY/47,XYY mosaicism with CML and suggested that the presence of the 46,XY,Ph¹ line in the mosaicism reported by Fitzgerald, Pickering, et al. (1971) might be the result of transformation occuring after the appearance of the Ph¹ because the skin cells were carrying only 47,XXY.

Some doubt on the clonal evolution of CML has been cast by the subsequent observations of M. A. S. Moore, Ekert, et al. (1974b) on the case mentioned above (M. A. S. Moore, Ekert, et al., 1974a). These authors followed a case of CML in a 7.5-yr-old boy with a constitutional mosaicism (46,XY/47,XYY) in skin, blood, and marrow. At diagnosis of CML a third 46,XY, Ph¹ cell line predominated in the marrow. Subsequent to the development and therapy of the BP, 46,XYY,-C,Ph¹ was present in 85% of the marrow cells and the remaining were 47,XYY. Upon relapse of the BP only 46,XYY,-C,Ph¹ cells were present. A significant fraction of the metaphases in the marrow were Ph¹-negative (predominantly 47,XYY) for a prolonged period of the clinical course. All of these facets led the authors to reevaluate their conclusions regarding the clonal origin of CML without coming to any definite conclusions. Lawler, O'Malley, et al. (1976) described a patient in whom both a #9 and #22 could be distinguished from their homologues; the same member of each pair was involved in the Ph¹. These authors further indicated that the presence of another translocation (in about 8% of the patients) in every cell as a stable rearrangement in addition to the Ph¹ provides further evidence for the monoclonal origin of CML.

The question of clonal evolution must also be considered in connection with the three-way rearrangements observed in some of the complicated cases of Ph¹-translocation. Inasmuch as no cells with intermediate alterations have been observed, it would appear that in each case the translocations involving chromosomes #22, #9, and a third chromosome result from three simultaneous breaks in a single cell and not from a step-wise process.

Ph¹-positive CML involving only one of the cell lines in the marrow of patients with mosaic congenital chromosome abnormalities, e.g., XY/XXY or XY/XYY, may be explained on the basis of involvement of only one type of precursor cells, which, at the same time, carries only one of the mosaic karyotypes, e.g., the myeloid cells being XY and the erythroid, XXY. As the disease advances and involves other or all cellular elements of the marrow, the Ph¹ will appear in the previously uninvolved cell precursors. This may explain the interesting and perplexing observations of several authors, who originally tried to attribute Ph¹-positive CML to a monoclonal evolution in only one cell line of the mosaic. It is also possible that a similar mechanism may apply to some cases of AML or erythroleukemia (EL). Thus, it may be visualized that only the myeloid cells would have a 46,XY constitution and, thus, be reflected in the cytogenetic profile of the leukemic cells in the initial stages of the disease, with the 47,XXY cells characterizing the uninvolved erythroid elements. As the leukemic process progresses, the latter cells become involved in the process and concomitantly in the chromosomal changes previously seen only in the 46,XY cells.

Though there appears to be considerable evidence in favor of the monoclonal (unicellular) origin of CML (Adamson and Fialkow; 1978, Nowell, 1978), some questions still remain. Even though the argument for the monoclonal, if not unicellular, origin of CML is based on some cogent cytogenetic and enzymatic data, with the original leukemic insult involving only one type of cell (usually granulocytic), the question still remains as to how the leukemic process ultimately spreads to involve other (normoblastic, megakaryocytic) elements of the marrow?

Extramedullary Aspects of Ph¹-Positive CML

The transformation of a pseudodiploid to an aneuploid state has been correlated with a more immature morphology of the leukemic cells, more autonomous growth, and more aggressive and invasive behavior of such cells.[75] Leukemic cells with two Ph¹'s, particularly, have been thought to be able to infiltrate skin, lymphatic tissue, mammary glands, ovaries, pleural and abdominal cavities. A number of these tumors have been erroneously classified as lympho- or reticulosarcoma; the demonstration of the Ph¹ disclosed the real nature of some of these tumors.[76] It is probable, however, that CML is a systemic disease, and the presence of hemopoietic cells in various tissues may not be due to infiltration of these tissues by leukemic marrow cells but to the production of hemopoietic cells *in situ*. This theory is favored by chromosomal findings in lymph nodes and spleen. For instance, a report (Canellos, DeVita, et al., 1971) has appeared in which a case was shown to have double Ph¹-positive aneuploid cells in an enlarged lymph node before the subsequent infiltration of the marrow; no such cells could be detected in the spleen. We compared the chromosome constitution of leukemic cells derived from marrow and spleen and found that aneuploidy in spleen cells was higher than in marrow cells; in one case all the splenic cells but only 50% of the marrow cells had an identical marker chromosome, suggesting that cells with a marker chromosome originated in the spleen and infiltrated the marrow. The same case was 100% Ph^1-positive before splenectomy; 6 wk after splenectomy only 30% of the marrow metaphases revealed a Ph^1. Another case was observed in our laboratory in whom the spleen cells had the chromosome constitution 45,XY and a missing #22, whereas the marrow cells were 100% Ph^1-positive. All these findings indicate an autonomous production of leukemic cells by the spleen, lymph nodes, and possibly other organs and tissues.

In some patients with CML, particularly those with marked splenomegaly, in whom the spleen may be the dominant blood-producing organ rather than the marrow, the development of an extra Ph^1 and/or aneuploidy indicates that the BP is capable of being generated in that organ. Subsequent to the development of these additional karyotypic changes in the spleen, such abnormal cells may be found in the marrow.[77] The possible genesis of the BP of CML in the spleen has led some to advocate the use of spleen biopsies for early detection of the BP (L. Brandt and Mitelman, 1978). Whether such cells get to the marrow from the spleen or are the result of subsequent development of the BP in the marrow per se is at present unknown.

Two patients with extramedullary manifestations of the BP of CML were described by Hossfeld, Bremer, et al. (1975) and shown to have 100% Ph^1-positive aneuploid metaphases. Despite the absence of blastic changes in the marrow and blood cells, a significant proportion of the metaphases derived from other tissues also showed aneuploidy. The authors postulated "that maturation and differentiation of aneuploid myeloblasts are influenced by their environment." Even though one finds little to argue with in this paper, the possibility exists that aneuploid cells were generated in extramedullary sites and then migrated to the marrow, where they may or may not have undergone further maturation and differentiation. This may have been particularly true in the second case described, in whom the lymph nodes contained granulocytic cells at various stages of maturation, though admittedly the specimen was shown to contain 50% estrase-negative blasts. At the same time the possibility exists that aneuploidy in the marrow may precede by a considerable interval the cytologic evidence of blastic transformation, with a much shorter interval applying to the cells in extramedullary environments. Nevertheless, evidence does exist that aneuploidy associated with myeloblastosis and/or myeloblastomas can be observed whereas the nonblastic marrow has a chromosomal pattern typical for the CP.

Evidence is accumulating that the appearance of aneuploidy with or without blastic transformation of CML may be related to the generation of such cells in extramedullary sites, particularly the spleen. Olinici, Petrov, et al. (1978) have described a patient with Ph^1-positive CML whose bone marrow (BM) contained cells with +8, whereas those in the spleen showed a +19, $2Ph^1$ pattern. The authors interpreted the findings to indicate that the splenic cells were responsible for the BP. The postulate by some authors (Motomura, Ogi, et al., 1976) that patients with Ph^1-positive CML entering the BP can be divided hematologically into two different groups, one characterized by predominance of myeloblasts in the marrow and cytogenetically by either diploidy or hypodiploidy, and another with low myeloblast counts in the marrow without significant difference between the peripheral blood and the marrow and with hyperdiploidy, suggesting that an extramedullary acute transformation in the spleen occurs primarily in most of the cases of the latter group, was not, in fact, supported by chromosomal analysis of splenic material, and this conclusion cannot be accepted until proven by such data. However, the demonstration of abnormal clones in the spleen preceding the BP in the marrow has been recorded by several authors and indicates that transformation to the BP may be recognized earlier in the spleen than in the marrow. An interesting case has been published where blastic transformation was observed in scar tissue at a time when there were no such signs of a transformation in the marrow (Killmann, 1972). It is possible that blastic transformation can be generated in several sites including the marrow, spleen, lymph nodes, and other tissues, with the appearance of aneuploidy followed by such blastic transformation requiring different latent periods in different tissues (Sjögren and Brandt, 1976). It is possible that the location of such transformation may account for the fact that some patients with the BP of CML are benefited by splenectomy, whereas others are not, i.e., the former group benefiting from removal of the spleen due to the fact that the blastic cells are being generated in that organ rather than in the marrow.

On the assumption that the spleen may be an important site for generating the BP of CML, splenectomy has been advocated as a preventative and/or therapeutic approach (Fuscaldo, Brodsky, et al., 1977; Hester, Kwaan, et al., 1977). In one report, for example, splenectomy in patients with 100% Ph^1-positive cells resulted in the presence of about 20% normal cells even 1 yr after removal of the spleen (Rosenfeld and Venuat, 1976).

Miscellaneous Aspects of Ph^1-Positive CML

The characteristic course of CML includes a relatively long (chronic) phase where, after nonintensive treatment, the patients remain in a clinical and hematological condition which barely resembles a malignant state. After the development of the BP, however, the condition becomes highly reminiscent of AML, clinically as well as hematologically. The BP is more resistant to therapy than AML and consequently shows a very poor prognosis. The current strategy is to administer busulphan and a variety of other drugs used in the CP, either singly or in combination, after development of the BP. The fact that such treatment may change, but rarely succeeds in normalizing the cytogenetic picture of the BP cell population, suggests that some of the aneuploid clones are drug resistant and consequently responsible for the poor therapeutic results. The probability of complete remission may thus be reversely related to the cytogenetic heterogeneity of the hemopoietic cells. If, on the other hand, karyotype studies are performed when the patient shows signs of decreased sensitivity to busulphan and intensive treatment is administered, provided more than sporadic aneuploidy is observed, then a radical elimination of the abnormal cells might be obtained because of their relatively limited numbers and because relatively few cytogenetically different variants occur at this stage. Such a strategy might improve the life expectancy in CML by a prolongation of the period during which the patients are still able to live a normal life. In fact, recent experience at our Institute indicates that less aggressive therapy (e.g., colcemide or its trimethyl derivative and 6-MP) may lead to a normalization of the marrow in occasional cases and produce more beneficial results (sometimes combined with splenectomy) than the more aggressive therapeutic approach to the BP with chemotherapy similar to that used in AML (e.g., Ara-C and thioguanine) (Gomez, Sokal, et al., 1977).

Even though during the BP aneuploidy may not be observed in 100% of the leukemic cells, the presence of a Ph^1 is invariable. A few exceptions have been published, but as the analyses in these cases were performed either before or without resort to banding, it is difficult to ascertain whether the Ph^1 had been translocated onto another chromosome and was, in fact, absent. Of interest is the fact that a case has been published in which apparently karyotypically normal cells appeared in the marrow of a patient who had been 100% Ph^1-positive during the BP of his CML. Subsequently, the Ph^1 reappeared in the marrow in 100% of the cells.

The presence of a second Ph^1 is a very common additional finding in CML, usually in the later stages of the disease. The absence of a second 9q+ in such cases, as well as the enlarged short arms and satellites on both Ph^1-chromosomes in some cases, argues strongly for the production of the second Ph^1 from the original one by selective endoreduplication or nondisjunction. However, morphological differences between the Ph^1-chromosomes in the same metaphase raise some questions regarding this hypotheses.[78] The presence of any number of cells with a single Ph^1 in such cases suggests either a very early development of the double Ph^1 clone or a selective growth advantage for it.

The reports of cases of Ph^1-positive CML without involvement of or the presence of a translocation to #9 appear to indicate that it is the rearrangement of one or more gene loci on 22q that is of crucial importance in the pathogenesis of CML, and the specific site of the translocation is of lesser importance. The high frequency of involvement of 9q, however, in the usual Ph^1-translocation, as well as the apparent simultaneous breakage of #22 and another chromosome in some of the cases, clearly suggest a nonrandom phenomenon. Perhaps some type of nonhomologous somatic association commonly exists between the relevant segments of #22 and #9 during interphase, greatly increasing the probability of simultaneous breakage and mutual rearrangement. Reports by several workers of the association at metaphase between satellited chromosomes and the secondary constrictions of chromosomes #1, #9, and #16 support this hypothesis, as may some indications of #9 being involved with the satellited acrocentrics in nucleolar organization. However, no evidence specifically relating #9 and #22 has yet emerged from such studies or from other investigations of nonrandom arrangements of human chromatin during interphase. It is also possible that an entirely different mechanism, such as the proliferation of an oncogenic virus or other mechanism leading to the genesis of CML with involvement of specific sites in nonhomologous chromosomes, could explain the high frequency of simultaneous breakage of #22 and #9. The mechanism by which chromosomes other than #9 are involved in the translocation with #22 leading to the genesis of the Ph[1], is very much unknown.

The development of Ph[1]-positive CML in two males (62 and 74 yr old) with CLL is of interest. The 62-yr-old patient had received radiation for the CLL and in his case the development of the leukemia may or may not be related to the radiotherapy; however, the second patient had not received any radiation therapy. The latter patient also had a significant number of marrow cells that were Ph[1]-positive but lacked the Y. Some of the cytogenetic changes observed in the metaphases in the spleen of the irradiated patient may be related to the considerable radiation therapy he had received. It is of interest that these metaphases did not contain the Ph[1].

There are several aspects of CML, as they relate to the Ph[1], that are worthy of note. Apparently, CML is being diagnosed earlier during the course of the disease, possibly because of the more frequent checking of the leukocyte count in the population and, of course, by the presence of the Ph[1]. This had resulted not infrequently in the institution of early therapy in CML and, very importantly, at a time in the disease when normal cells may still be present in the marrow. When Ph[1]-positive cells respond to therapy, the normal cells are then in a position to repopulate the marrow. Newer approaches to the therapy of CML have led to more frequent responses in such patients with concomitant eradication, however temporary, of the Ph[1]-positive leukemic cells.

Chronic Myelomonocytic Leukemia

There is considerable debate regarding the existence of a separate form of leukemia labeled "chronic (and subacute) myelomonocytic" (CMMoL). Apparently, this disease is predominantly found in elderly males, starts with weakness and long-standing anemia and, either in the beginning or at some time during the course of the disease, a persistent monocytosis and often thrombocytopenia, granulocytopenia, or granulocytosis may develop (Zittoun, 1976). Death from infection or a terminal transition to AL is common. Six patients with CMMoL have been studied cytogenetically by Hurdle, Garson, et al. (1972) of whom three were shown to have a normal karyotype in the marrow and three a 45,XO,−Y karyotype, the latter confirmed by fluorescent banding. The Y was missing in all the cells in the three cases; this is in contrast to the varying percentage of cells (usually < 100%) with Y loss found in the marrow cells of normal males over the age of 60. In a few cases, Zittoun, Bernadou, et al. (1972) found −C,+F or −Y. Mende, Fülle, et al. (1977) studied the marrow chromosomes in seven patients with subacute MMoL: two were normal, three had structural anomalies (isochromosome of group G, break in #2, etc.), one female patient had a deletion of the short arm of a C-chromosome and one 81-yr-old male had a missing Y. Six patients with CMMoL were studied cytogenetically, and all were essentially normal (Mende, Fülle, et al., 1977). The Ph[1] has not been observed in patients with MMoL of either the chronic or subacute variety (Miescher and Farquet, 1974; Garson, Milligan, et al., 1971). The relationship between the loss of the sex-chromosome in this form of leukemia, as well as findings in female patients with CMMoL, and detailed studies of cases who have developed AL will be of interest and should go a long way toward elucidating the nature of this unusual leukemia.

References

1. In the following references are included articles on the general cytogenetic experience in leukemia, some correlative studies between the chromosome findings and enzymatic, morphologic, genetic, and other aspects of the leukemias, and review papers on the karyotypic results in these diseases. For additional general papers, see the section on acute leukemia. Astaldi, 1963; Baikie, 1966a, b; Baikie, Garson, et al., 1969; Baikie, Court Brown, et al., 1961; Barlogie, Hittleman, et al., 1977; Berger, 1968a, b, 1972b; Block, 1965; Bottomley, 1968, 1976; Conen and Erkman, 1968; Elejalde and Restrepo, 1972; E. W. Fleischman and Prigogina, 1975; C. E. Ford, 1960, 1962; Forteza-Bover, Baguena-Candela, et al., 1964a, b; Gavosto, 1965a; Grouchy, 1961; Gutierrez, Cobo, et al., 1974; Hampel and Palme, 1964; Hecht and McCaw, 1974, Higurashi, Nakagome, et al., 1970; Hossfeld and Cohnen, in press; S. J. Jackson, 1961; Kamada, Oguma, et al., 1976; Kurita, Kamei, et al., 1974; Lancet (editorials) 1959, 1961b, 1964, 1977; Lawler, 1969, 1973, 1977a, b; Lawler and Millard, 1967; Lawler and Reeves, 1976; Mandelli, Amadori, et al., 1977; Meisner, Inhorn, et al., 1973; O. J. Miller, 1964; Mitelman, 1975a; Motoiu-Raileanu, 1971; Hj Müller and Stalder, 1976; Hj Müller, Buhler, et al., 1976; V. D. Müller, Orywall, et al., 1971; Nava, 1969; Nowell and Hungerford, 1974; Nowell, Hungerford, et al., 1958; Padeh, Bianu, et al., 1965b; Pogosianz, Prigogina, et al., 1976; Queisser, Queisser, et al., 1974; Ritzman, Stoufflet, et al., 1966; Rowley, 1969, 1975b, 1976d, 1977a–c, 1978b; Rożynkowa, Stepién, et al., 1977; Ruffié, 1962; Ruffié, Marquès, et al., 1964; Sandberg, 1966a, 1968a, b, 1973; Sandberg and Hossfeld, 1974; Sandberg, Kikuchi, et al., 1964; I. P. Singh and Ghosh, 1976; Spiers, 1972; Spiers, Bain, et al., 1977; Spitzer, Schwartz, et al., 1976; Takino and Nakazima, 1966; Tanzer, 1967; Teplitz, 1968; Tough, 1965; Trujillo, 1976; Trujillo, Cork, et al., 1971; Wahrman, Schaap, et al., 1963; Whang-Peng, 1977; Woodliff, 1971, 1975.

2. Borch, Stanley, et al., 1969; Bottura, 1960; Bottura and Ferrari, 1961; Chaudhuri and Roy, 1967; Christensen, 1965; Hellriegel, Heit, et al., 1977; Kiossoglou, Mitus, et al., 1964; Lam-Poo-Tang, 1968; Meighan, and Stich, 1961; Morse, Humbert, et al., 1977; Rothfels and Siminovitch, 1958; Tjio and Whang, 1962, 1965.

3. Chervenick, Ellis, et al., 1971; Huang, Imamura, et al., 1969; Huang, Hou, et al., 1974; E. Klein, Ben-Bassat, et al. 1976; C. B. Lozzio, Lozzio, et al., 1976; C. B. Lozzio and Lozzio, 1975; Minowada, Oshimura, et al., 1977; Minowada, Tsubota, et al., 1977; Pedersen, 1966c; Rosenfeld, Goutner, et al., 1977; Sonta, Minowada, et al., 1977; Ueyama, Morita, et al., 1977; Zur Hausen, 1967.

4. Baikie, 1974; Baserga and Castoldi, 1972a, b, 1973; Berger, 1973; Block, 1965; Brit. Med. J. 1961, 1966; W. M. C. Brown and Tough, 1963; Erkman, Crookston, et al., 1966; Fitzgerald, 1961b; Forteza-Bover and Baguena Candela, 1962; Kamada and Uchino, 1978; Kardinal, Bateman, et al., 1976; Lawler, 1976; M. R. Leibowitz, Derman, et al., 1976; Lubs and Lurie-Blitman, 1965; Nicoara, Butoianu, et al., 1967; Pisano, 1976; Stryckmans, 1974.

5. Fitzgerald, 1976, 1977; Hecht and Kaiser McCaw, 1977a; Lawler, 1977a, b; Lawler, O'Malley, et al., 1976; Misawa, Takino, et al., 1975; Muldal, Mir, et al., 1975; Pravtcheva and Manolov, 1975; Rowley, 1977a–c; Watt, Hamilton, et al., 1977; Whang-Peng, Lee, et al., 1974.

6. Court Brown, 1967; Endo, Yamamoto, et al., 1969; G. E. Moore, Fjelde, et al., 1969.

7. E. Engel, 1965b; Gavosto, Pileri, et al., 1965; Goh, 1966; Ishihara and Kumatori, 1967; Kamada, Okada, et al., 1967.

8. Flannery and Corder, 1974; B. Hall, 1963; Kontras, Robbins, et al., 1966; Ricci, Dallapiccola, et al., 1970; Weiner, 1965.

9. Bauke, 1971; Lima-de-Faria, Bianchi, et al., 1967; Nowell and Hungerford, 1961.

10. Büchner and Gatermann, 1971; Goh, 1967c, d; Haines, 1965; Higurashi, Nakagome, et al., 1968; Lima-de-Faria, Bianchi, et al., 1967; Muldal, Lajtha, et al., 1965; Prieto, Egozcue, et al., 1970; W. Schmid, 1963; Sofuni, Kikuchi, 1967; W. Stich, Back, et al., 1966.

11. Buckton and O'Riordan, 1976; Dinauer and Pierre, 1973; Gahrton, Lindsten, et al., 1973; Ishihara, Kohno, et al., 1974a, b; Petit and Cauchie, 1973; Raposa, Natarajan, et al., 1974; Van Den Berghe, 1973.

12. Baikie, 1964, 1966a, b; Bosi, Castoldi, et al., 1962; Court Brown and Tough, 1963; Duvall, Carbone, et al., 1967; Hecht and McCaw, 1975; Hossfeld and Sandberg, 1970a; Nowell, 1962; Pedersen, 1969; Watt, Hamilton, et al., 1977; Whang, Frei, et al., 1963.

13. Canellos and Whang-Peng, 1972; Killmann, 1972, 1976; Moretti, Broustet, et al., 1972; Pedersen, 1973a.

14. Athens, 1969; Boll and Mersch, 1969; Chervenick, Ellis, et al., 1971; Galbraith, 1969; M. A. S. Moore and Metcalf, 1973; Muldal and Lajtha, 1974; S. Perry, 1968; Rhodes, Robinson, et al., 1978.

15. Golde, Byers, et al., 1974; Pedersen and Killmann, 1971; Sultan, Marquet, et al., 1975.

16. Hossfeld, Han, et al., 1971; Sandberg, Ishihara,

et al., 1962b; Trujillo and Ohno, 1963; Tsessarskaya, Osechenskaya, et al., 1964a, b.
17. Bekkum, Noord, et al., 1971; Crosby, 1972; Fliedner, Thomas, et al., 1964.
18. Armenta, Cadotte, et al., 1976; L. Brandt and Mitelman, 1978; Hossfeld and Cohnen, in press; Hossfeld and Hirche, 1976; Hossfeld and Schmidt, 1973; Hossfeld, Schmidt, et al., 1972b; Hossfeld, Bremer, et al., 1975; Long, Whang-Peng, et al., 1978 (Ph[1]+ CML as primary tumor of bone); Mitelman, 1975b; Spiers and Baikie, 1965a, 1968b; Spiers, Baikie, et al., 1975a; Trujillo, Fernandez, et al., 1970; Zaccaria, Baccarani, et al., 1974.
19. L. Brandt, 1969; R. Fischer, Hennekeuser, et al., 1970; Hossfeld, Schmidt, et al., 1972b; Ogawa, Fried, et al., 1970; S. Perry. Moxley, et al., 1966.
20. Clarkson, Dowling, et al., 1974; Fefer, Cheever, et al., 1977; Smalley, Vogel et al., 1977.
21. Canellos, DeVita, et al., 1976; Hagemeijer, Smit, et al., 1977; Hossfeld, 1975a, b; Lawler 1977a; Rowley 1977a−c.
22. Ezdinli, Sandberg, et al., 1969; Ezdinli, Sokal, et al., 1970; Fitzgerald, 1962; Frei, Tjio, et al., 1964; Hardisty, Speed, et al., 1964; Krauss, Sokal, 1964; Reisman and Trujillo, 1963.
23. Crawfurd and Pegrum, 1964; E. Engel, McKee, et al., 1968; Ezdinli, Sokal, et al., 1970; Hayhoe and Hammouda, 1965; Sandberg, Ishihara, et al., 1962b; Sjögren, Brandt, et al., 1974; Speed and Lawler, 1964; Tanzer, Jacquillat, et al., 1969; Tough, Jacobs, et al., 1963; Whang-Peng, Canellos, et al., 1968.
24. Clarkson, Dowling, et al., 1974; Fefer, Cheever, et al., 1977; Finney, McDonald, et al., 1972; Maurice, Alberto, et al., 1971; Smalley, Vogel, et al., 1977; Speed and Lawler, 1964.
25. L. Brandt, Mitelman, et al., 1976; Golde, Bersch, et al., 1976; Kamada, Okada, et al., 1967; Kenis and Koulischer, 1967; Sandberg, Hossfeld, et al., 1971; Vagner-Capodano, 1972; Whang-Peng, Canellos, et al., 1968.
26. Fitzgerald, Adams, et al., 1963; Grouchy, Nava, et al., 1966b; Kamada, Okada, et al., 1967; Kenis and Koulischer, 1967.
27. Arlin, Chaganti, et al., 1977; L. Brandt, Mitelman, et al., 1976; DiLeo, Müller, et al., 1977; Djaldetti, Padeh, et al., 1966; Finney, McDonald, et al., 1972; Maurice, Alberto, et al., 1971; Speed and Lawler, 1964.
28. O'Riordan, Berry, et al., 1970; Pierre and Hoagland, 1971; Sandberg and Sakurai, 1973b; Secker Walker, 1971.
29. Atkin and Taylor, 1962; Elves and Israëls, 1967; E. Engel, McGee, et al., 1965; Garson and Milligan, 1972; Grouchy, Nava, et al., 1966a; Lawler and Galton, 1966; Lawler, Lobb, et al., 1974; Nigam and Dosik, 1976; Pedersen, 1968a; Pierre and Hoagland, 1973; Reisman and Trujillo, 1963; Sakurai and Sandberg, 1976c; Sakurai, Hayata, et al., 1976; Shiffman, Stecker, et al., 1974; Speed and Lawler, 1964; Tanzer, Jacquillat, et al., 1969.
30. Lawler, Roberts, et al., 1971; Mende, 1976; Pierre and Hoagland, 1971, 1972.
31. E. Boyd, Pinkerton, et al., 1965; DiLeo, Müller, et al., 1977; Hurdle, Garson, et al., 1972; Kajii, Neu, et al., 1968; Rowley, 1971.
32. Garson and Milligan, 1972; Lawler, Lobb, et al., 1974; Shiffman, Stecker, et al., 1974.
33. Grozdea, Kessous, et al., 1973; Pedersen, 1968b, c; Shiffman, Stecker, et al., 1974; Speed and Lawler, 1964.
34. Adams, Fitzgerald, et al., 1961; Cascos and Barreiro, 1964; E. Engel, McKee, et al., 1968; C. E. Ford, and Clarke, 1963; Forteza-Bover, Baguena-Candela, et al., 1963; Goh, Swisher, et al., 1964; Houston, Levin, et al., 1964b; Makino and Awa, 1964; Merker, Schneider, et al., 1968; Tough, Jacobs, et al., 1963; Whang-Peng, Canellos, et al., 1968.
35. E. Engel, McKee, et al., 1975; E. W. Fleishman and Wolkowa, 1971; Hayhoe and Hammouda, 1965; Hungerford, 1964; Lawler, 1977a, b; A. Levan, Nichols, et al., 1963; Meisner, Inhorn, et al., 1970; Mitelman, Nilsson, et al., 1975; Miyamoto, 1974; Pedersen, 1964a, b, 1967a−c; Pedersen and Killmann, 1971; Pedersen and Videbaek, 1964; Reisman and Trujillo, 1963; Rowley, 1977a−c; Sharp, 1976; Tanaka, 1974.
36. Crist, Ragab, et al., 1978; E. W. Fleishman and Wolkowa, 1971; Meisner, Inhorn, et al., 1970; Nicolau, Popescu, et al., 1964; Padeh, Bainu, et al., 1965a; Pedersen, 1966a−c; Watson-Williams, Lewis, et al., 1977; Whang-Peng, Canellos, et al., 1968.
37. Biscatti and Vaccaro, 1965; G. E. Bloom, Gerald et al., 1966; Davis-Lawas and Lawas, 1965; Fortune, Lewis, et al., 1962; Hardisty, Speed, et al., 1964; Hays, Morse, et al., 1978; Kosenow and Pfeiffer, 1969; Nowell and Hungerford, 1962; Reisman and Trujillo, 1963; Rosen and Nishiyama, 1968; K. L. Smith and Johnson, 1974; Tanzer, Jacquillat, et al., 1969; Tjio, Carbone, et al., 1966; Whang-Peng, Canellos, et al., 1968.
38. Kyle and Pease, 1966; McKenzie and Perrotta, 1975; Nau and Hoagland, 1971; S. Rosenthal, Canellos, et al., 1977; Shohet and Blum, 1968.
39. Grinblat, Mammon, et al., 1977; Grozdea, Colombiès, et al., 1975; Kaplow, 1968; Masera, Pogoraro, et al., 1965; Sewell, 1972; Teplitz, 1966; Teplitz, Rosen, et al., 1964; Winkelstein, Goldberg, et al., 1967; Wintrobe, 1974.
40. Alter, Lee, et al., 1963; Baikie, 1962; Hook and Engel, 1964; Kamada, Okada, et al., 1968; King, Gillis, et al., 1962; Masera, Pogorara, et al., 1965; Tough, Jacobs et al., 1963.

41. R. P. Cox, 1969; Gropp, Fischer, et al., 1968; Krauss, 1968; Merker, 1965, 1971; Mitus and Kiossoglou, 1968; J. C. Robinson, Pierce, et al., 1965; Rosen and Nishiyama, 1968; W. Stich, Back, et al., 1966.
42. Ezdinli, Sokal, et al., 1970; Kamada, Okada, et al., 1967; Mitus and Kiossoglou, 1968; Tjio, Carbone et al., 1966.
43. Ezdinli, Sokal, et al., 1970; Perillie and Finch, 1970; Rosen and Nishiyama, 1968.
44. Bauters, Croquette, et al., 1970; Carbone, Tjio, et al., 1963; Kamada, Okada, et al., 1967; Kenis and Koulischer, 1967; Krauss, 1968; Merker, 1965; Rosen and Teplitz, 1965; Rosner, Schreiber, et al., 1972.
45. Hammouda, Quaglino, et al., 1964; Kaplow, 1968; Kenis and Koulischer, 1967; Marks, McCaffrey, et al., 1978; Rosen and Nishiyama, 1968; Rosner, Schreiber, et al., 1972.
46. Asamer, Schmalzl, et al., 1971; Mintz, Pinkhas, et al., 1973; Senn and Rhomberg, 1970; Tischendorf, Ledderose, et al., 1972.
47. McCaffrey, Harrison, et al., 1975; Sarin, Anderson, et al., 1976; Srivastava, Khan, et al., 1977.
48. Crist, Ragab, et al., 1978; Harousseau, Smadja, et al., 1977; Mauri, Torelli, et al., 1977.
49. Baguena Candela and Forteza-Bover, 1964; Bloomfield, Peterson, et al., 1977; Fortune, Lewis, et al., 1962; Grossbard, Rosen, et al., 1968; Hossfeld, Han, et al., 1971; M. H. Khan and Martin, 1967a; Kiossoglou, Mitus, et al., 1965b; Kohn, Manny, et al., 1975; Mastrangelo, Zuelzer, et al., 1967; Nagao, Yonemitsu, et al., 1977; Obara, Sasaki, et al., 1971a; Oshimura and Sandberg, 1977; Sasaki, Muramoto, et al., 1975; Sonta, Oshimura, et al., 1976; Whang-Peng, Henderson, et al., 1970.
50. Castoldi, Yam, et al., 1968; Hossfeld, Han, et al., 1971; Kiossoglou, Mitus, et al., 1965a; Oshimura and Sandberg, 1977.
51. Koulischer, Frühling, et al., 1967; Levin, Houston, et al., 1967; Modan, Padeh, et al., 1970.
52. Bowen and Lee, 1963; S. M. Cohen, 1967; Forrester and Louro, 1966; Frey and Siebner, 1968; Heath and Moloney, 1965b; M. H. Khan and Martin, 1968a; Kiossoglou, Mitus, et al., 1966a; Krauss, 1966; D. Müller and Haberlandt, 1970.
53. Bornstein, Nesbit, et al., 1972; Briere, Castro-Malaspina, et al., 1975; Killmann, Philip, et al., 1976; Misset, Venuat, et al., 1977; Neerhout, 1968; L. C. Peterson, Bloomfield, et al., 1976a, b; S. Rosenthal, Canellos, et al., 1977; Venuat, Rosenfeld, et al., 1976; Worm and Pedersen-Bjergarrd, 1977.
54. Beard, Durrant, et al., 1976; Gall, Boggs, et al., 1976; Gallo, Bhattacharyya, et al., 1974; Sebahoun, Gratecos, et al., 1976; Secker Walker, Summersgill, et al., 1976.

Grouchy, et al., 1969a; Rowley, 1978c; Rozynkowa and Stepién, 1975a, b; Sandberg, Hossfeld, et al., 1971; Spiers, Bain, et al., 1977; Spiers, Janis, et al., 1977; Turleau, Trebuchet, et al., 1971.
56. Armenta, Cadotte, et al., 1976; Gomez, Hossfeld et al., 1975; Rosenfeld and Venuat, 1976; Schwarze, Schwalbe, et al., 1975; Spiers, Baikie, et al., 1975b.
57. Bauke and Bach, 1972; Canellos, Whang-Peng, et al., 1976; E. Engel, McKee, et al., 1968; Sandberg, Hossfeld, et al., 1971; Vagner-Capodano, 1972; Vallejos, Trujillo, et al., 1974.
58. Canellos, DeVita, et al., 1971; Canellos, Whang-Peng, et al., 1972; Court Brown and Tough, 1963; Garson, Burgess, et al., 1969; Jung, Blatnik, et al., 1963; Khouri, Shahid, et al., 1969; Srodes, Hyde, et al., 1973; Tough, Jacobs, et al., 1963.
59. Canellos, DeVita, et al., 1971; Court Brown and Tough, 1963; Dougan, Onesti, et al., 1967; Garson and Gruchy, 1975; Hayhoe and Hammouda, 1965; Hellriegel and Koebke, 1971; Kamada, Okada, et al., 1967; Kenis and Koulischer, 1967; Kiossoglou, Mitus, et al., 1965b; Knospe, Klatt, et al., 1967; Kundel, Tanaka, et al., 1963; Speed and Lawler, 1964; Tjio, Carbone, et al., 1966.
60. Anderson, Pearson, et al., 1968; Canellos, DeVita, et al., 1970, 1971; Castro-Sierra, Gorman, et al., 1967; Court Brown and Tough, 1963; E. Engel, McKee, et al., 1968; Erkman, Hazlett, et al., 1967; C. E. Ford and Clarke, 1963; Goh, 1967a; Grouchy, 1967b; Hammouda, Quaglino, et al., 1964; Hayhoe and Hammouda, 1965; Hellriegel and Koebke, 1971; N. H. Kemp, Stafford, et al., 1964; M. H. Khan and Martin, 1967b; Khouri, Shahid, et al., 1969; Kiossoglou, Mitus, et al., 1965b; Krompotic, Lewis, et al., 1968; Lawler and Galton, 1966; A. Levan, 1967; Muldal, Taylor, et al., 1967; Nava, Grouchy, et al., 1969a; Pegoraro, Pileri, et al., 1967; Rigo, Stannard, et al., 1966; Sakurai, 1970c; Sandberg, Hossfeld, et al., 1971; Schroeder and Bock, 1965; Spiers and Baikie, 1968b; W. Stich, Back, et al., 1966; Streiff, Peters, et al., 1966a; Tjio, Carbone, et al., 1966.
61. Dougan and Woodliff, 1965; Duvall, Carbone, et al., 1967; E. W. Fleishman and Wolkowa, 1971; Goh, 1974; Kenis and Koulischer, 1967; Meisner, Inhorn, et al., 1970; Pedersen, 1968c; Smalley, 1966, 1968; Whang-Peng, Henderson, et al., 1970.
62. Bauke, 1971; Berger, 1970b; Grouchy, Nava, et al., 1966a, b; Lobb, Reeves, et al., 1972; Meisner, Inhorn, et al., 1970; Reeves and Lawler, 1972; Spiers and Baikie, 1972; Whang-Peng, Canellos, et al., 1968.

63. Beck and Chesney, 1973; Berger, 1970a; Crossen, Mellor, et al., 1971b; Hossfeld, 1974a; Lawler, 1977a, b; P. Philip, 1975b; Rovera and Peggoraro, 1968; Rowley, 1977 a–d; Ruffié, Ducos, et al., 1965; Sandberg and Hossfeld, 1974; Sharp, Potter, et al., 1975; Volkova and Fleischman, 1973.
64. Cabrol, Peytremann, et al., 1976; Diéska, Izaković, et al., 1967a, b; Kahn, 1974.
65. Grouchy and Turleau, 1974; Lejeune, Berger, et al., 1965; Rowley, 1977a–d.
66. Duvall, Carbone, et al., 1967; E. Engel and McKee, 1966; Hammouda, Quaglino, et al., 1964; Hampel and Palme, 1964; Kamada and Uchino, 1967; Kiossoglou, Mitus, et al., 1966b; Rolović, Markovic, et al., 1970; Sandberg, Hossfeld, et al., 1971; Streiff, Peters, et al., 1966a; Tjio, Carbone, et al., 1966; Vagner-Capodano, 1972; Woodliff and Dougan, 1966.
67. Bauke, Neubauer, et al., 1967; E. Engel, McKee, et al., 1968; Erkman, Hazlett, et al., 1967; Grouchy, Nava, et al., 1968; Hammouda, Quaglino, et al., 1964; N. H. Kemp, Stafford, et al., 1964; Rigo, Stannard, et al., 1966; Schroeder and Bock, 1965; Spiers and Baikie, 1972.
68. Anderson, Pearson, et al., 1968; M. H. Khan and Martin, 1969; Knospe, Klatt, et al., 1967; Nowell and Hungerford, 1962; Spiers and Baikie, 1972.
69. Duvall, Carbone, et al., 1967; E. Engel and McKee, 1966; Hossfeld and Sandberg, 1970a; Pedersen, 1969; Schoyer, 1964; Watt, Hamilton, et al., 1977.
70. Bauke, Neubauer, et al., 1967; Crossen, Mellor, et al., 1971b; Erkman, Hazlett, et al., 1967; Fitzgerald, 1966; Grouchy, Nava, et al., 1966b; Hammouda, Quaglino, et al., 1964; Hellriegel and Koebke, 1971; Kamada, Okada, et al., 1967; Knospe, Klatt, et al., 1967; Krompotic, Lewis, et al., 1968; Sakurai 1970c; Sandberg, Hossfeld, et al., 1971; Schroeder and Bock, 1965; Vagner-Capodano, 1972.
71. Canellos, DeVita, et al., 1971; Crossen, Mellor, et al., 1971b; E. W. Fleishman and Wolkowa, 1971.
72. Bauke and Bach, 1972; M. H. Khan, 1973b; Miyamoto and Takita, 1973.
73. Bauke, Neubauer, et al., 1967; Crossen, Mellor, et al., 1971b; Erkman, Hazlett, et al., 1967; Hellriegel and Koebke, 1971; Krompotic, Lewis, et al., 1968; Vagner-Capodano, 1972.
74. Bauke, 1973; Fialkow, Gartler, et al., 1967; Fitzgerald, Pickering, et al., 1971; Gahrton, Lindsten, et al., 1973, 1974a, b; Grouchy and Turleau, 1974; Hayata, Sakurai, et al., 1975; Hossfeld, 1975a, b; Lederlin, Puchelle, et al., 1975; Lejeune, Berger, et al., 1965; M. A. S. Moore, Ekert, et al., 1974a, b; Motomura, Ogi, et al., 1973, 1976; Pedersen, 1975a; Whang-Peng, Broder, et al., 1976.
75. Bauke, 1971; Duvall, Carbone, et al., 1967; Ilberry and Louer, 1966; Joseph, Zarafonetis, et al., 1966; Knopse, Klatt, et al., 1967; Kwan, Singh, et al., 1977; Nava, Grouchy, et al., 1969a, b; Pascoe, 1970; Pedersen, 1971.
76. Duvall, Carbone, et al., 1967; Ellman and McChesney, 1973; Gall, Boggs, et al., 1976; Garfinkel and Bennet, 1969; Joseph, Zarafonetis, et al., 1966; Knospe, Klatt, et al., 1967; Libre and McFarland, 1967; Naman, Cadotte, et al., 1972; Oberling, Stoll, et al., 1975a, b.
77. Hossfeld and Schmidt, 1973; Hossfeld, Bremer, et al., 1975; Mitelman, Brandt, et al., 1974b; Mitelman, Nilsson, et al., 1975; Oberling, Stoll, et al., 1975a, b; Verhest, van Schoubroeck, et al., 1976a; Zaccaria, Baccarani, et al., 1974.
78. Duvall, Carbone, et al., 1967; E. Engel and McKee, 1966; Hossfeld and Sandberg, 1970a; Rowley, 1973a, b.

# 11

The Leukemias
Acute Leukemias

Acute Leukemia (AL)
Significance of the Ph¹ in Various Acute Leukemias: Personal Experience
Ph¹-Positive Acute Lymphoblastic Leukemia
Ph¹-Positive Acute Myeloblastic Leukemia
Ph¹-Positive Erythroleukemia
General Remarks Regarding the Ph¹ in AL
Marrow Transplantation in AL and Chromosome Markers
Nuclear Blebs and Aneuploidy in AL
Cellular Markers in Differentiating Various Leukemia

Acute Lymphoblastic Leukemia (ALL)
Studies Before Banding
Frequency of Abnormalities and Number of Chromosomes
Common Abnormalities in ALL, Including the 6q−
Karyotypic Evolution
Cell Surface Markers and Chromosomal Abnormalities
General Comments
Cytogenetic Experience in ALL
Congenital Chromosome Abnormalities and ALL

Acute Myeloblastic Leukemia (AML)
Common Chromosomal Changes in AML
Chromosome Changes and Prognosis of AML
The Ph¹ in AML
Chromosomes and the Genesis of AML
Chromosome Patterns in AML Revealed by Banding
Classification of AML by the Karyotypic Patterns and Their Relation to Prognosis
AML Complicating Other Conditions
The Missing Y-Chromosome in AML
Monosomy-7 and Trisomy-8 in AML

Congenital Leukemia

Erythroleukemia (EL)
Chromosome Changes in EL
MIKA and MAKA Acute Leukemias
The MAKA Group—A Type of Erythroleukemia?
Polyloidy—A Function of Abnormal Erythroid Cells and Karyotypic Instability
A Survey—Three Chromosomal Types of EL?
Further Aspects of MAKA and EL
Prophasing and EL

Down's Syndrome and Leukemia
Acute Promyelocytic Leukemia
Acute Myelomonocytic Leukemia
Acute Megakaryocytic Leukemia
Acute Monocytic Leukemia
"Hairy Cell" Leukemia
Near-Haploidy in Acute Leukemia

Acute Leukemia (AL)

In discussing the chromosomal findings in AL,[1] this group of diseases has been divided into three major categories: acute myeloblastic leukemia (AML), acute lymphoblastic leukemia (ALL), and erythroleukemia (EL). Only minor attention will be given to other forms of AL, primarily because of scarcity of cytogenetic information in most of these conditions and the existing confusion regarding their exact status. Even though it is possible to classify AL in several ways (Bennett, Catovsky, et al., 1976) and the one I have chosen can be, in all probability, further subclassified or radically modified, I think that a meaningful presentation of the cytogenetic data in AL is best served by the classification given. In this section I shall discuss separately the karyotypic aspects of ALL, AML, and EL and some of the rarer forms of AL.

The first description of chromosomal changes in AL appeared in 1958 (C. E. Ford, Jacobs, et al., 1958; C. E. Ford and Mole, 1959) and indicated that the cytogenetic findings varied from case to case. Furthermore, it soon became clear that many cases of AL were not associated with any *visibly recognizable* chromosomal changes in the marrow.[2] Since then it has been established that about 50% of the patients with AL do not have chromosomal changes in the leukemic cells.[3] Thus, even with various banding techniques the incidence of karyotypic changes in AL continues to be approximately 50%, though the percentage has varied from laboratory to laboratory, i.e., ranging from 30% to over 65%.[4] The existence of diploid karyotypes in *all* forms of AL has also been established with various banding techniques, even though such techniques have revealed that some cases (one out of five to six cases at the most), considered to be diploid with routine staining, upon closer karyotypic analysis were revealed with banding not to be perfectly diploid (Milligan and Garson, 1974; Rowley, 1977a–d).

Even though banding techniques have unquestionably unraveled a number of nonrandom karyotypic changes in AL (Mitelman and Levan, 1979), there is considerable variability in the cytogenetic profiles among the cases with AL, this being particularly prominent in ALL. Such a state was evident before banding,[5] even though some chromosome groups were thought to be involved more often than others (Steenis, 1966; A. Levan, 1966), without general agreement (Sandberg, Bross, et al., 1968; Hirschhorn, 1968) existing on the consistency of such findings.

A unique morphologic feature of AL chromosomes, first described by us (Sandberg, Ishihara, et al., 1961, 1964b), is the frequently ill-defined and fuzzy appearance of the chromatids, which often seem to be poorly separated, as compared with the well-defined, clear and thin chromatids in normal cells.[6] This fuzzy appearance of the chromatids characterizes a substantial percentage of AL cases, both diploid and aneuploid: in about 30% of AML and as much as 70% of ALL and EL. This is particularly evident when normal cells coexist with leukemic cells in the same marrow. In the case of diploid AL, the appearance of metaphases with fuzzy chromatids probably has the same significance as that in aneuploid metaphases. Because the metaphases with the fuzzy chromosomes may disappear from the marrow upon successful therapy, the reappearance of cells with such chromosomes invariably heralded relapse of the AL. These morphological features of leukemic chromosomes (and in some other malignant conditions) have been confirmed by others (Krogh Jensen, 1971), but the reasons for these unique features of leukemic chromosomes are unknown. Whether this fuzzy appearance is due to such causes as modified DNA structure, changes of protein and/or RNA associated with the DNA, or substances extraneous to the chromosomes (e.g., cytoplasmic) affecting the staining characteristics of the chromosomes, remains unknown. The fuzziness of the leukemic chromosomes is evident regardless of the stain used or the method for preparing the chromosomes. Certainly, the elucidation of these reasons may shed light on the nature of AL, as would studies on the chromosome-associated proteins. The fuzzy appearance (or its cause) of the chromosomes in AL, particularly in ALL, may be responsible for the frequent failure, partial or complete, of the chromosomes in these conditions to show the characteristic banding pattern of normal chromosomes (Lawler, Secker Walker, et al., 1975; Oshimura, Freeman, et al., 1977b).

The distribution of the modal chromosome numbers in AML (including the cases of EL)

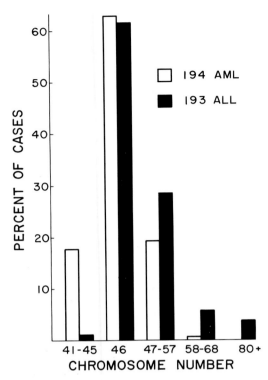

80 Modal chromosome number distribution in patients with AML and ALL. Note that hypodiploidy is extremely rare in ALL and that very high counts rarely appear in AML.

and ALL is shown in Figure 80 (also see Table 53). As can be seen, >50% of the cases with either form of leukemia have no *visibly recognizable* karyotypic changes in the leukemic cells. As first reported by us some years ago (Sandberg, Ishihara, et al., 1964b), and subsequently corroborated by further observations in our laboratory and by the data of others,[7] the aneuploid cases with ALL almost invariably have chromosomal modes of 46 or higher, whereas AML may be associated with hypo-, pseudo-, or hyperdiploidy, but seldom with a chromosome number exceeding 55. On the other hand, the chromosome number may be extremely high in ALL. Very low counts may be encountered in rare cases of AML. Cases of AL with a tetraploid number of chromosomes have been described.[8] The chromosome mode in AL tends to be rather sharp, as contrasted to the flat modes in most cancers.

In contrast to the usual findings in Ph¹-positive CML, not necessarily all the marrow cells in AL are involved by cytogenetic changes, when these are present.[9] Thus, the percentage of aneuploid cells may vary from relatively low numbers, 10% or less in some cases of AML, to as high as 100% in both ALL and AML.

The karyotypic changes are confined to the leukemic cells only, though more than one series of cells may be involved by the AL, e.g., myeloid and erythroid elements in EL (Krogh Jensen and Killmann, 1971). Thus, in almost all cases of ALL with aneuploid cells in the marrow, some normal (diploid) metaphases are found, and this is also true for most of the cases of AML (for further discussion of the significance of these findings, see a following section). During successful therapy of AL the marrow may contain only normal metaphases, the aneuploid ones having disappeared. During relapse, the previously observed karyotypically abnormal cells invariably reappear in the marrow. Only minor deviations may occur in the abnormal karyotype of AL cells throughout the course of the disease, including various courses of chemotherapy (Blatnik, Jung, et al., 1962), several remissions and relapses, and anatomic spread of the leukemia (spleen, liver, etc.). These karyotypic changes during the course of AL are usually minor variations on the original karyotypic theme observed in the cells. Imminent relapse of AL may be heralded by the reappearance of aneuploid cells in the marrow, from which they had completely disappeared after therapy. The chromosomally abnormal cells usually disappear totally from the marrow when the AL responds to chemotherapy. This is particularly true of ALL and somewhat less frequently of AML. However, newer chemotherapeutic approaches have successfully resulted in the not infrequent appearance of cytologically and chromosomally normal marrows in AML. The persistence of karyotypically abnormal cells would indicate that the leukemic cell population has not been totally eradicated from the marrow. The reappearance of chromosomally anomalous cells, invariably of the same karyotypic makeup as that before therapy, as already indicated, heralds or is associated with relapse of the AL. Thus, the cytogenetic findings may prove to be of considerable value in planning the therapy of AL, though little use of this approach has been taken to date. In previous studies, we have shown (Sandberg, Ishihara, et al., 1962a) that examining the chromosome constitution of the marrow in AL is best accomplished by a direct technique without resort to in vitro culture. The latter leads to overgrowth of the aneuploid leukemic cells by the normal elements of the marrow with the result that a spuriously normal, diploid picture is obtained for AL.

The same mechanism and results appear to apply to culture of peripheral blood cells in AL. Thus, I continue to recommend that direct examination of marrow cells, whenever possible and successful, should be the approach of choice in establishing the chromosomal picture in AL. Furthermore, except for a few reports, successful long term in vitro culture of aneuploid leukemic cells has not been, to my satisfaction, established, in spite of reports to the contrary. In other words, I am not aware of any leukemic cell lines that have maintained their original karyotype in vitro. A rare exception was the establishment of a cell line from a patient with AL and characterized by the same chromosomal changes observed in vivo in the marrow and reported from our laboratory (Minowada, Oshimura, et al., 1977). Nevertheless, short-term incubation (12–18 hr) of marrow cells from AML yields metaphases that represent a relatively faithful karyotypic picture of that in the marrow (Hossfeld and Sandberg, 1970b). We have not observed a case of AML in which the marrow was diploid and the short-term culture aneuploid. The findings of Obara, Sasaki, et al., (1971b) differ and are of special interest, i.e., the cultured cells were aneuploid, and those examined by a direct method were diploid. Though it would be tempting to state that the metaphases observed by the direct method belonged to nonleukemic cells of the marrow, the possibility exists that their case of AL may have contained two cell populations in the marrow, a diploid one and an aneuploid one. The metaphases of the latter predominated in culture, whereas the former cells were the only ones in mitosis in the marrow at the time of examination.

Generally, once a diploid or aneuploid picture has been established in ALL or AML, the chromosomal constitution does not appear to change with the clinical phases of the disease or therapy. Thus, a diploid ALL or AML has not been generally observed by us to become aneuploid, even after years of the disease and through a number of cycles of remission and relapse. Some exceptions have been reported (Gunz, Ravich, et al., 1970; Cimino, Rowley, et al., 1979). The aneuploid leukemias may show some variations in the modal number of chromosomes during the course of the disease, but these are usually a reflection of the variability of the chromosome number about the mode in aneuploid ALL and AML.

In both ALL and AML rare, but theoretically important, cases have either progressed from karyotypic normality (including that based on banding) to definite cytogenetic abnormalities (Mitelman and Brandt, 1974; Cimino, Kinnealey, et al., 1977; Cimino, Rowley, et al., 1979) or have shown remarkable progression of the chromosome changes present initially (Oshimura, Hayata, et al., 1976). There is always the possibility that therapy played a role in these changes, though I do not subscribe to this hypothesis.

Marker chromosomes occur at a much lower (20–30%) frequency in AL than in cancers (40–60%). The morphology of the markers does not differ basically among the various leukemias or from those in cancer (the Ph^1 is an exception) and aside from specific exceptions that will be described, there is considerable variability in the structure of the markers from case to case and occasionally in the same case.[10] As in cancer cells, the markers may be present in only a few to as many as 100% of the leukemic cells. The origin of the few markers can usually be ascertained in AL, whereas frequently this is not the case in cancerous tissues.

In all conditions that predispose to AL, be they congenital (DS, FA), (Table 47), acquired (benzene poisoning, radiation), or other myeloproliferative disorders (myelofibrosis, IAA, PV), when the acute leukemic phase develops the cytogenetic changes are similar to those encountered in other ALs, particularly AML. I do have the impression, however, that aneuploidy occurs much more frequently in these syndromes than in "spontaneous" AL without an apparent predisposing cause. This is particularly true of AL after radiation and/or chemotherapy.

Analysis with banding techniques affords a rigorous recognition of chromosomes in AL and other neoplastic states which, though ap-

Table 47 Leukemia and other proliferative disorders in patients with congenital chromosome abnormalities (exclusive of DS)

Type of leukemia	Chromosome disorder	Reference
AL	47,XXY	N. H. Kemp, Stafford, et al., 1961; Mamunes, Lapidus, et al., 1961; Bousser and Tanzer, 1963; Borges, Nicklas, et al., 1966; Ruffié Ducos, et al., 1966; Lawler, Secker Walker, et al., 1975
AL	49,XXXY	Sohn and Boggs, 1974
CML	47,XXY/46,XY	Tough, Court Brown, et al., 1961; Fitzgerald, Pickering, et al., 1971
AL	45,XO?	Wertelecki and Shapiro, 1970; Pridie and Dimitrescu-Purvu, 1961
CLL	45,XO/46,XX	Dumars, Kitzmiller, et al., 1967
AL	45,XO/47,XXX	F.J.W. Lewis, Poulding, et al., 1963
AL	46,XX/46,X,i(Xq)	Castoldi, Grusovin, et al., 1971
CML (Ph[1]+)	46,XY/47,XYY	M.A.S. Moore, Ekert, et al., 1974a,b
Lymphoma	XY/XXY/XXXY	MacSween, 1965*
CML	45,D/D	E. Engel, McGee, et al., 1965; Kohno and Sandberg, 1979
AL	45,D/D	Prigogina, Stavrovskaja, et al., 1968
EL	45,D/D	Dallapiccola and Malacarne, 1971
AML	Partial D-trisomy	Zuelzer, Thompson, et al., 1968
AML	D-trisomy	Schade, Schoeller, et al., 1962
ALL	B/D translocation, t(4q+;14q−) familial	Garson and Milligan, 1974 (leukemic cells 47,+C)
AL	C/G translocation	Hinkes, Crandall, et al., 1973
AL	F+ trisomy	Borges, Nicklas, et al., 1966
AL	D/G translocation t(13q+;21q−)	Whang-Peng, Knutsen, et al., 1976b
AL	Ring −#1	Bobrow, Emerson, et al., 1973
HD	XXX	Lech, Polaniecka, et al., 1974
CML	t(13;14),Ph[1]+	Kohno and Sandberg, 1979
AML	XY/XO	Cardini, Clemente, et al., 1968 (leukemic cells −G)
ALL	XYY	Kessous, Corberand, et al., 1975b
CML	45,XX,t(13q;14q), Ph[1]+	Wennström and Schröder, 1973
Retinoblastoma	47,XXY	M. G. Wilson, Ebbin, et al., 1977
Retinoblastoma	48,XXX,+21	Day, Wright, et al., 1963
Medulloblastoma	XYY	Rosano, Delellis, et al., 1970
OTHER CONSTITUTIONAL CHROMOSOME ANOMALIES		
Neurofibroma	Bq+;15q−(4q+;15q−)	Ganner, 1969a
AML	(2;5)(q23;q35)	Lawler, Secker Walker, et al., 1975
PV	45,−D,−G,t(Dq;Gq)	Berger, Parmentier, et al., 1974
PV	46,Dq+/Dq−	Berger, Parmentier, et al., 1974
CLL	(Gp−) Ch[1]-anomaly	Fitzgerald and Hamer, 1969
Acute plasma cell leukemia	(GP−) Ch[1]-anomaly	Fitzgerald, Rastrick, et al., 1973
MM → AML	t(13q;14q)	Rowley, 1976b
Ph[1] + CML	t(14/21)	Velazquez, Arechavala, et al., 1976
ALL	t(12p−;22p+)	Hinkes, Crandall, et al., 1973
AML	Ring-1-chromosome	Bobrow, Emerson, et al., 1973
AL	Mosaicism of trisomy 8	Riccardi, 1976; Gafter, Shabtai, et al. 1976

Table 47—continued

Type of leukemia	Chromosome disorder	Reference
	OTHER CONSTITUTIONAL CHROMOSOME ANOMALIES	
AML	5p−, +21 (Cri-du-chat and DS)	Petit, Maurus, et al., 1968
Adrenocortical cancer	13 trisomy	Nevin, Dodge, et al., 1972
Microscopic neuroblastoma	13 trisomy	Nevin, Dodge, et al., 1972
Wilms' tumor	18 trisomy	Geiser and Schindler, 1969
Neurofibromatosis	46,XX,t(2p−;22p+)	Alvarez and Villalpando, 1971
ALL	t(Gp−;Dp+)	Goh, 1968d
Neuroblastoma (familial)	21p−q−; 11q21 wide 11q23 narrow	Pegelow, Ebbin, et al., 1975
Embryonal sarcoma	ring-D (?13)	Ayraud, Rey, et al., 1971
Wilms' tumor and aniridia	t(8p+;11p−) with some 8p−	Ladda, Atkins, et al., 1974
AML	Familial t(7;20) (p13;p12)	Riccardi, Svjansky, et al., 1978
Myeloproliferative disorder	t(13q14q)	Goh, Bauman, et al., 1979

*The reports of the cases of lymphosarcoma and Klinefelter's by Augustine and Jaworski (1968) and Becker et al. (1966) do not contain the age of the patients and their chromosome constitution.

pearing normal upon staining with standard Giemsa or other dyes, are, in fact, abnormal chromosomes due either to reciprocal translocations or subtle modifications in the structure of chromosomes. Thus, it is essential that all metaphases be analyzed with banding methods in order to ascertain the exact karyotypic picture of leukemic (or cancer) cells.

Significance of the Ph¹ in Various Acute Leukemias: Personal Experience

It has been generally accepted that the Ph¹ exists in the granulocytic, erythrocytic, and megakaryocytic series of the marrow in "classic" CML, but not in the lymphocytic series (Sandberg and Hossfeld, 1974). This evidence has been used as a strong argument for the unicellular origin of the three cell series and as an indicator of the divergence of their origin from that of the lymphoid elements. However, the occurrence of an abnormal G-group chromosome indistinguishable from the Ph¹ in AL and states other than CML, i.e., AML, EL, myeloid metaplasia with myelofibrosis, PV, thrombocythemia, and particularly ALL, raises a number of questions requiring clarification to establish the etiologic and clinical significance of the Ph¹.

Many investigators reported a "Ph¹-like" chromosome in conditions other than CML before the development of banding techniques. Recently, a substantial number of cases have been studied with banding techniques and in these the Ph¹-like chromosome was shown to be due to a standard translocation between chromosome #9 and #22, i.e., t(9;22) (q34;q11), thus indicating that the chromosome was in fact a bona fide Ph¹. Our experience in this area may be appropriately presented at this juncture and related to the Ph¹-positive cases of AL published in the literature. In seven of our Ph¹-positive cases of AL[11] the origin of the Ph¹ was shown to be due to a deletion of the long arm of a #22 at q11 and translocation of the deleted part to other chromosomes, including two unusual translocations. Thus, the data show that not only the standard but also unusual Ph¹-translocations may exist in conditions other than CML.

The salient features of the cytogenetic

Table 48 Published case reports of Ph¹-positive ALL*

Age/sex yr	BM karyotypes	Chromosomal findings in remission	Banding	Ph¹-translocation
53M	46,XY/46,XY,Ph¹	No complete remission achieved	No	?
34M	46,XY/46,XY,Ph¹/ 46,XY,+14,−15,Ph¹	Normal	Yes	14;22
5M	46,XY/46,XY,Ph¹ then only 46,XY,Ph¹	No complete remission achieved	Yes	9;22
7F	46,XX/46,XX,Ph¹	Normal	No	?
2M	46,XY/46,XY,Ph¹	Normal	Yes	9;22
56F	45,XX,−7,Ph¹/45,XX, −7/46,XX,Ph¹/46,XX, 7q−,Ph¹	Normal	Yes	9;22
4M	45,XY,−20,−17,Ph¹, marker/?46,XY	Normal	?	?
34F	43,X,−X,−7,−8,Ph¹/ 46,XX,Ph¹	Normal	Yes	9;22
45M	46,XY,Ph¹/46,XY	Unknown	Yes	9;22
42M	46,XY,Ph¹/46,XY	Normal	Yes	21;22
20F	43,X,−X,−7,−15,Ph¹	Normal	Yes	9;22

*Does not include the six Ph¹-positive cases of adults with ALL described by Bloomfield, Peterson, et al. (1977), and Bloomfield, Linquist, et al. (1978). Two cases described by Sakurai (1970b) and one by Cerny, Šimánková, et al. (1962) may be Ph¹-positive ALL.

studies in these adults with Ph¹-positive AL were:

1. The presence of Ph¹-positive hypodiploidy in a case of ALL followed by a normal cytogenetic constitution during remission and then by Ph¹-positive pseudodiploidy during relapse of the ALL.
2. The demonstration of two new and unusual Ph¹-translocations in conditions other than CML, i.e., AML (19;22) and EL (4;22).
3. The first description (to my knowledge) of a Ph¹ in EL.
4. The first description of a 4;22 Ph¹-translocation in any disease.

The significance of these findings will be discussed separately for each disease entity.

Ph¹-Positive Acute Lymphoblastic Leukemia

On the basis of the cytologic findings, cell surface and antigen markers, and many of the clinical aspects of our case, the diagnosis of ALL appeared to be most likely, even though prominent hepatosplenomegaly, which the patient had, is not characteristic of the disease. However, it has been reported (Bloomfield, Peterson, et al., 1977) that substantial splenomegaly is more likely to be encountered in Ph¹-positive ALL (Table 48) than in its negative variant. The initial BM cytology, the positive ALL antigen (cALL) reaction, the cell surface markers, and the cellular TdT activity speak against our case being confused with one of AML, although the cytogenetic findings and their changes were compatible with the latter. Even though several features of our case may be seen in the BP of CML (hepatosplenomegaly, some of the cytogenetic findings, and cellular enzyme activities), the BM morphology, the response to particular antileukemic therapy, the clinical course, and, above all, the return of the marrow to total diploidy point against the case being one of CML in the BP. It is possible that in exceptional cases of AL (and our case may be

Remarks	Reference
Survived 4 mo	Propp and Lizzi, 1970
11 mo	Ayraud, Dujardin, et al., 1975
Survived 14 mo	Schmidt, Dar, et al., 1975
Survived 15 mo	Biervliet, Hemel, et al., 1975 Secker Walker, Summersgill, et al., 1976
Still alive	Mandel, Shabtai, et al., 1977
16 mo	Rausen, Kim, et al., 1977
Survived 22 mo	Oshimura and Sandberg, 1977
2⅓ yr	Philip, Müller-Berat, et al., 1976
½ yr	Cimino, Kinnealey, et al., 1977
Still alive	Gibbs, Wheeler, et al., 1977 S. Walker, Wheeler, et al., 1977

one of them), the cytogenetic findings may be a more reliable index as to the origin of the leukemic cells and the nature of the disease than morphologic and other criteria. Were it not for the presence of the Ph^1 and some of the other karyotypic data on our case (e.g., hypodiploidy, a missing #7), there is doubt whether the diagnosis of ALL would have been challenged. This means that either the cytologic, laboratory, and/or clinical criteria for the diagnosis of ALL are not always reliable, particularly when ALL is confused with the so-called lymphoid BP of CML (Gall, Boggs, et al., 1976), or that the unusual cytogenetic aspects encountered in our case, which in some ways are more typical of Ph^1-positive CML and AML than ALL, may be generated in rare cases of the latter disease. There is no reason to believe that the precursor cells in ALL cannot be involved in the genesis of a Ph^1 indistinguishable from that in CML. The total disappearance of the Ph^1-positive aneuploid cells from the marrow during remission of the ALL is rarely observed in the BP of CML, but is often seen during remission in ALL (or AML) (Sandberg, Hossfeld, et al., 1971).

On the basis of presently used and acceptable cytologic and clinical criteria, supported in some of the cases by electromicroscopic morphology and immunologic and surface cell markers, it would appear that rare cases of ALL can be Ph^1-positive.

The Ph^1 in ALL has been shown with banding techniques to be a deleted #22, thus dissipating doubt as to the authenticity and origin of the Ph^1. However, a report relative to Ph^1-positive ALL by Secker Walker, Summersgill, et al. (1976) indicates that the cells in BP of CML may, in rare cases, have many of the characteristics of lymphoblasts. When such a BP is not preceded by the CP, ALL may be diagnosed, and the true diagnosis rests on the presence or absence of the Ph^1.

A summary of all Ph^1-positive ALL case reports published to-date is presented in Table 48. With the reservations already expressed earlier regarding Ph^1-positive ALL, in at least eight of the patients in whom a complete remission was induced, the marrow assumed a normal diploid picture karyotypically. In each patient some normal cells were present in the marrow before therapy. The patient of Schmidt, Dar, et al., (1975), although initially having normal cells in the marrow, failed to achieve full remission and, in fact, on two occasions, i.e., 5 mo after diagnosis and 2 mo before the patient died, the marrow consisted of only Ph^1-positive cells. Neither was a complete remission achieved in the patient of Propp and Lizzi (1970). In eight case reports banding analysis was performed: a standard type of Ph^1-translocation was present in six and unusual translocations involving #14 and #21, respectively, in 2 cases. The latter translocations are to my knowledge, the only unusual translocations to be described in ALL, although in the brief publications no pictorial evidence is shown for these unusual translocations.

In six adult patients with Ph^1-positive ALL reported by Bloomfield, Peterson, et al. (1977), Q-banding revealed the Ph^1 to be due to a 22q− in four patients and in two of these to involve a translocation with chromosome #9 (9q+). In the two other cases the material was not of sufficiently good quality to establish the exact site of the translocation. In three of the six patients karyotypically normal cells were present either at diagnosis or during relapse, and in one patient a completely diploid picture was observed during remission. Additionally, in three of the patients two or more Ph¹-chromosomes were present, accompanying in each case a hypodiploid picture.

Ph¹-Positive Acute Myeloblastic Leukemia

Ph¹-positive cases of acute myeloblastic leukemia (AML) were described in dozens of cases before the demonstration of the Ph¹ source and its genesis. To date, at least two dozen more cases have been studied with banding, and a standard Ph¹-translocation, i.e., t(9;22) (q34;q11), has been shown to exist. The first two cases studied with banding were described by Sasaki, Muramoto, et al. (1975) in two young women without any other karyotypic changes. The case reported by us (Oshimura and Sandberg, 1977) was the first in which an unusual Ph¹-translocation (19;22) was described in AML. Such a translocation has been observed in CML and appears to involve the same segments in #19 (q13) and #22 (q11) as in the case of Ph¹-positive AML described by us. In addition to the Ph¹, other chromosomal changes were found in the marrow of our case, including a missing #7 and #15, and several chromosomes with deleted or extra material. Some karyotypically normal cells were found in the BM on both examinations. This in itself would mitigate against the case being a BP of CML, because in the latter condition only very rare cases with normal cells have been described to date (Sandberg, Hossfeld, et al., 1971). Two more cases with unusual Ph¹-translocations in AML have been observed involving #3 and #12 and, of course, #22 (Table 49).

Karyotypically normal cells have been observed in the initial marrow examination in almost all the Ph¹-positive patients with AML (as well as those with ALL) and in some of these cases the occurrence of remission was accompanied by a cytogenetically normal picture in the marrow. However, upon relapse the Ph¹-positive cells reappear in the marrow in spite of the preceding karyotypic normality.

The chromosomal picture of our case of Ph¹-positive AML would place him in the major karyotypic abnormalities (MAKA) group of AML as defined by Sakurai, Hayata, et al. (1976). The prognosis in these patients is usually much poorer than that of the minor karyotypic abnormalities (MIKA) group. The patient described by us lived < 1 yr after developing symptoms, though he did have several episodes of remission. Several cases of AML in whom more than one Ph¹ were present have been described. In one case (M. H. Khan, 1973a) karyotypically normal cells appeared in the marrow after therapy and made up 100% of the metaphases for substantial periods of time.

Ph¹-Positive Erythroleukemia

Even though half-a-dozen Ph¹-positive cases of erythroleukemia (EL) were described before the introduction of banding techniques (Sandberg and Hossfeld, 1974), the case described by us (Oshimura and Sandberg, 1977) is the first in which 1. a deletion of #22 was shown to be the source of the Ph¹ in EL, and 2. the Ph¹-translocation not only involved a chromosome other than #9, but appears to be the first description of a Ph¹-translocation involving #4 (4p) (see Table 50). A few cells with two Ph¹-chromosomes were present, but only one translocation to 4p was evident in the karyotype, indicating that the extra Ph¹ was probably generated by nondisjunction or endoreduplication. The cell population, all Ph¹-positive, in the patient's marrow when first examined consisted of two cell types: one with 45 chromosomes due to a missing #7, i.e., 45,XX,−7, t(4;22) (p16;q11) and another line with a 46,XX,t(4;22) (p16) (q11) karyotype. The latter cell line was not seen in the marrow on subsequent examinations. Except for a few Ph¹-positive cells, the blood cells examined after culture with PHA had a normal female karyotype, indicating no constitutional chromosomal anomaly in the patient. Reexamination of the BM revealed no normal cells among the additional 100 metaphases observed.

The general cytogenetic picture encountered in this patient with EL is not unusual for this disease, i.e., the absence of normal cells with an AA picture as defined by Sakurai and Sandberg (1974, 1976b) and a missing #7. Because of the AA picture, the patient has not been treated with chemotherapy and only given blood, platelets, and other supportive therapy whenever necessary. The patient is still

Table 49 Ph¹-positive cases of ANLL (AML, EL, and AMMoL) described in the literature

Case no.	Age in yr and sex	Material	Cells Ph¹+,%	Karyotypic changes in addition to Ph¹	Blasts,%	Reference
1	?	BM and Bl*	75	Dq+	?	Borges, Wald, et al., 1962
2	2½ ♀	Bl	58	46,XX	12	Fortune, Lewis, et al., 1962
3‡	47 ♂	Bl	32	48,XY,+C,+mar	64	Lüers, Struck, et al., 1962
4	71 ♀	BM	43	48,XX,+A1,+E18/48, XX,+C6,+E18	84	Baguena-Candela and Forteza-Bover, 1964
5§	65 ♂	BM	100	44,XY,2q−,5q−,−D,−E, 18q−,−G,+Ph¹	10	Kiossoglou, Mitus, et al., 1965a
6	49 ♂	BM	20	46,XY	−	Kiossoglou, Mitus, et al., 1965b
7	47 ♂	BM	29−68	47,XX,+9	−	Kiossoglou, Mitus, et al., 1965b
8	54 ♀	BM	86	46,XX	−	Kiossoglou, Mitus, et al., 1965b
9	62 ♀	BM	100	46,XX/47,XX,+D	−	Kiossoglou, Mitus, et al., 1965b
10	55 ♀	BM	85−100	46,XX	−	Kiossoglou, Mitus, et al., 1965b
11 ‖	62 ♀	BM	100	44,XX,+2,−4,−4,−C,−D, +E/45,XX,+2,−4,−4, −2C,+E,+F/46,XX,+2,−4 −4,−4,−2C,+E,+2G	−	Kiossoglou, Mitus, et al., 1965b
12	23 ♀	Bl	−	46,XX/45,XX,−G	−	Bayle and Linhard, 1966
13¶	5 ♂	BM	20−100	46,XY,Fq+	0.4−82.2	Mastrangelo, Zuelzer, et al., 1967
14 ‖	73 ♂	BM	100	46,XY	−	Castoldi, Yam, et al., 1968
15 ‖	66 ♂	BM	17	46,XY	−	Castoldi, Yam, et al., 1968
16**	19 ♂	Bl	100	46,XY	38	Grossbard, Rosen, et al., 1968
17 18	2 cases diagnosed as Ph¹-positive AML; No further information given					Ezdinli, Sokal, et al., 1970
19‡	35 ♂	BM	87	45,XY,−C/46,XY	50	Onesti and Woodliff, 1970
20**	9 ♂	BM	26−100	46,XY/65−70 chromosomes	−	Whang-Peng, Henderson, et al., 1970
21‡	68 ♀	BM	100	46,XX/48,XX,+C,+D	−	Whang-Peng, Henderson, et al., 1970
22**	3 ♂	BM	83	46,XY	−	Whang-Peng, Henderson, et al., 1970
23¶	46 ♀	BM	74	46,XX,−B,−D,+Dq+, +Dq+	−	Whang-Peng, Henderson, et al., 1970
24‡	42 ♂	BM	80	46,XY	45−69	Hossfeld, Han, et al., 1971

continued

Table 49 Ph¹-positive cases of ANLL (AML, EL, and AMMoL) described in the literature — *continued*

Case no.	Age in yr and sex	Material	Cells Ph¹+, %	Karyotypic changes in addition to Ph¹	Blasts, %	Reference
25¶	43♂	BM	80–100	46,XY	5–89	Hossfeld, Han, et al., 1971
26‡§ ‖	42♂	BM	60–100	46,XY/48,XY,+?,+Ph¹	23–68	Hossfeld, Han, et al., 1971
27‡	63♀	Bl	40–64	47,XX,+mar/ 47,XX,−C,+2mar/ 46,XX,−2C,−D,+3mar/ 47,XX,−B,−D,+3mar/ 48,XX,−B,−C,+4mar	3–86	Motoiu-Raileanu and Bercaneanu, 1971
28	?	BM	54–65	46 chromosomes	98	V.D. Müller, Orywall, et al., 1971
29§	?♂	BM	80	46,XY/ 53,XY,+3C,+2F,+G,+Ph	38	V.D. Müller, Orywall, et al., 1971
30	27♀	BM and Bl	7–94	45,XX,−C/	57 (BM)	Obara, Sasaki, et al., 1971a
				46,XX	11 (Bl)	Obara, Sasaki, et al., 1971a
31¶	43♂	BM	10–100	46,XY/47,XXY	2–85	Canellos and Whang-Peng, 1972
32‡	61♂	BM	11–53	46,XY,−3,−3,−B,−G,+D,+E,+2mar occ. Bp−,Dq+ in Bl.	70	Khan, 1972b
33§**	59♂	BM and Bl	3–100	46,XY/46,XY,−G,+Ph¹/46,XY,−C,+Ph¹	75–97	Khan, 1973a
34‖	73♀	BM	—	46,XX	7–11	Pacheco, Gabuzda, et al., 1973
35	55♂	BM	86	47,XY,+?C	95	Moraine, Brémond, et al., 1974
36‡‡‡	38♀	BM and Bl	29–100	45,XX,−6	68 (Bl) 85 (BM)	Kohn, Manny, et al., 1975
37¶‡‡	17♀	BM	85	46,XX	17–26	Sasaki, Muramoto, et al., 1975
38¶‡‡	25♀	BM	100	46,XX	93	Sasaki, Muramoto, et al., 1975
39‡¶‡‡	21♀	BM	100	48,XX,+2,+16/ 66,+2,+2,+2,+3,+2B,+5C,+D,+16,+3E,+F,+G,+Ph¹,+Ph¹	84.2	Sakurai and Sandberg, 1976a
40‡‡	24♂	BM	70	46,XY,t(7;10)(q34;q22)	72 (Bl)	Sharp, Potter, et al., 1976a
41‡‡‡	50♂	BM and Bl	0–100	46,XY→	0–28 (BM)	Sonta, Oshimura, et al., 1976
				46,XY,21q+	0–62 (Bl)	Abe and Sandberg, 1979
42	29♂	BM and Bl	100	46,XY	78 (BM) 63 (Bl)	Bloomfield, Peterson, et al., 1977
43	74♂	BM and Bl	100	46,XY	27 BM	Bloomfield, Peterson, et al., 1977
					7 (Bl)	Bloomfield, Peterson, et al., 1977
44§	66♂	BM and Bl	97	46,XY	33 (BM) 17 (Bl)	Bloomfield, Peterson, et al., 1977

Table 49—*continued*

Case no.	Age in yr and sex	Material	Cells Ph¹+, %	Karyotypic changes in addition to Ph¹	Blasts, %	Reference
45§‡‡	29♂	BM and Bl	100	46,XY	36 (BM) 43 (Bl)	Bloomfield, Peterson, et al., 1977
46**‡‡	26♂	BM	0–100	46,XY	—	Gustavsson, Mitelman, et al., 1977
47‡‡	29♀	BM and Bl	60	46,XX	24–34	Misset, Venuat, et al., 1977
48‡‡	40♂	BM and Bl	25	46,XY	80–89	Misset, Venuat, et al., 1977
49‡‡	53♀	BM and Bl	25	46,XX	10–65	Misset, Venuat, et al., 1977
50‡‡	13♂	BM and Bl	60	46,XY	15–85	Misset, Venuat, et al., 1977
51§¶‡‡	81♂	BM and Bl	97–100	47,XY,+Ph¹/46,XY/ 47,XY,+17	88–91	Nagao, Yonemitsu, et al., 1977
52**‡‡	39♂	BM	84–96	45,XY,7q+,−7,9p+,8p−, −15t(19;22)	39–90	Oshimura and Sandberg, 1977
53**‡‡	66♀	BM and Bl	100	45,XX,−7,t(4;22)/ 46,XX,t(4;22)	4–36	Oshimura and Sandberg, 1977
54‡‡	?♀	BM	95–100	45,XX,−7	—	Ruutu, Ruutu, et al., 1977a
55‡‡	?♀	BM	50–70	46,XX,del(5)	—	Ruutu, Ruutu, et al., 1977b
56‡‡ 57‡‡ 58‡‡	3 cases: 2 with a portion of cells +Ph¹ and 1 case 100% +Ph¹					Sharp, Wayne, et al., 1977
59‡‡	?	BM and Bl	10	10q−; t(9q+;22q−)	?	Rożynkowa, Stepień, 1977
60‡‡	24♀	BM and Bl	100	47,XX,1q+,t(3;22),+7, −8,der(8),14q+,−18, +mar	67–80	Van Den Berghe, David, et al., 1978a
61‡‡	?	BM	?	46,XY	?	Mitelman, personal communication
62‡‡	15♂	BM and Bl	38	48,+B,+G,Yq−	64	Padre-Mendoza, Forman, et al., 1978
63‡‡	57♂	BM and Bl	100	46,XY	43 (Bl)	Bloomfield, Lindquist, et al., 1978
64‡‡	74♂	BM and Bl	82	46,XY	96 (BM) 96 (Bl)	Bloomfield, Lindquist, et al., 1978
65§‡‡	42♀	BM and Bl	59	46,XX/47,XX,+Ph¹/ 48,XX,+21,+Ph¹	49	Abe and Sandberg, 1979
66§‡‡	50♂	BM and Bl	33–56	46,XY/46,XY,21q+	62	Abe and Sandberg, 1979
67‡‡	33♀	BM and Bl	88–100	46,XX	61	Abe and Sandberg, 1979
68§‡‡	59♀	BM and Bl	100	50,XX,+6,+8,del(9) (p13),+10,+Ph¹	89	Abe and Sandberg, 1979
69‡‡	40♂	BM and Bl	71	46,XY	90	Abe and Sandberg, 1979

*BM, bone marrow; Bl, blood cells.
‡Failure of remission.
§=Presence of double Ph¹.
‖Erythroleukemia.
¶Incomplete remission achieved.
**Complete remission achieved.
‡‡Banding performed.

Table 50 Ph¹-positive cases with EL

Age in yr and sex	Karyotype	Reference
62♀	44,XX,+2,−4,−4,−C,−D,+E,Ph¹,etc.	Kiossoglou, Mitus, et al., 1965b
73♂	46,XY,Ph¹	Castoldi, Yam, et al., 1968
66♂	46,XY,Ph¹	Castoldi, Yam, et al., 1968
42♂	46,XY,Ph¹	Hossfeld, Han, et al., 1971
73♀	46,XX,Ph¹	Pacheco, Gabuzda, et al., 1973
66♀	45,XX,−7,t(4;22)	Oshimura and Sandberg, 1977

alive and her disease is relatively stable. We have indicated that chemotherapy given to AA patients often leads to severe depletion of the marrow elements, which in the absence of any normal cells cannot repopulate the marrow and leads to a very short survival of such patients.

General Remarks Regarding the Ph¹ in AL

Bloomfield, Peterson, et al. (1977) have correlated the clinical, survival, and laboratory data of Ph¹-positive cases of *adult* ALL and AML with those of the patients with these diseases but without the Ph¹; a significantly longer survival was observed in the Ph¹-negative cases than in the Ph¹-positive cases (see Table 51). Thus, none of the six Ph¹-positive ALL cases survived > 2 yr (five lived < 1 yr) and none of those with Ph¹-positive AML lived > 3 mo. The latter is similar to the survival observed in the BP of CML. The presence of the Ph¹ in AL apparently carries with it a dire prognosis (L. C. Peterson, Bloomfield, et al., 1976a). The high incidence of Ph¹-positive cases in the group reported by Bloomfield, Peterson, et al. (1977), i.e., 6 out of 15 with ALL and 4 out of 55 with AML or acute myelomonocytic leukemia (AMMoL), the difference in survival, and some of the laboratory and clinical features, when compared with the experience of other clinics and laboratories, are difficult to explain. Obviously, a much larger number of patients with Ph¹-positive AL will have to be studied before a definite conclusion can be reached. Differences in diagnostic criteria may account, to some extent, for the variability in the incidence of Ph¹-positive AL and, hence, there is a great need for standardization of cytologic criteria in AL, akin to that already accomplished in chromosomology. However, there is also the possibility that Ph¹-positive AL may be increasing in incidence, because it is unlikely that an abnormal chromosome of the morphology of the Ph¹ would have been missed, even before the introduction of banding techniques, i.e., prior to 1970.

The role played by antileukemic therapy in the chromosomal changes observed in our Ph¹-positive cases of AL is difficult to evaluate. It is doubtful whether it played a role in the karyotypic observations of most cases, whereas it undoubtedly led to a shift in the chromosome pattern of one of our cases (ALL), i.e., from a Ph¹-positive hypodiploidy with 43 chromosomes to diploidy and then to Ph¹-positive pseudodiploidy and hypodiploidy of 44–45 chromosomes, during various remission and relapse phases of the disease.

A missing #7 was encountered in all three cases with Ph¹-positive AL studied by us. This is not an unusual karyotypic finding in AL (Sandberg and Hossfeld, 1974), particularly in

Table 51 Unusual Ph¹-translocations in AL

Chromosome involved	Type of translocation	Diagnosis	Reference
#4	t(4;22)(p16;q11)	EL	Oshimura and Sandberg, 1977
#19	t(19;22)(q13;q11)	AML	Oshimura and Sandberg, 1977
#14	t(14;22)	ALL	Ayraud, Dujardin, et al., 1975
#3	t(3q+;22q−)	AML	Van Den Berghe, David, et al., 1978a
#21	t(21;22)(q22;q11)	ALL	Cimino, Kinnealey, et al., 1977
#12	t(12p+;22q−)	AML	Hagemeijer, Smit, et al., 1977
#3,#8	t(3;8;22)	MM	Van Den Berghe, Louwagie, et al., 1979c

AML and EL, and although it may be seen on occasion in the BP of Ph¹-positive CML (Rowley, 1975b), it is rarely present without other more common changes, i.e., +8, iso(17), +19, etc. It is doubtful whether the presence of the Ph¹ per se had any relation to the missing #7, because the latter has been observed in AL without a Ph¹ (Rowley and Potter, 1976; Zech, Lindsten, et al., 1975). A missing #7 is very rare in ALL.

The role of the Ph¹ in the course of AL is uncertain because the number of such patients is entirely too small for evaluation, though there are indications that the Ph¹-positive AML and ALL do not fare as well as their counterparts without the Ph¹, yet much better than the patients with the BP of CML. Even less is known about the effects of unusual Ph¹-translocations in AL, although in CML we have shown that unusual and complex Ph¹-translocations do not affect the course of the disease, when compared with that of patients with the standard type of Ph¹-translocation (i.e., #9–#22) (Sonta and Sandberg, 1977b).

The presence of a significant proportion of karyotypically normal cells in the marrow of a Ph¹-positive AL strongly argues against it being the BP of CML, because normal cells have been only very rarely reported in the BP. Thus, the presence of chromosomally normal cells in a Ph¹-positive marrow associated with conditions resembling AL is indicative of an acute process per se rather than the BP of CML. Thus, in my opinion, several of the 10 cases reported by Misset, Venaut, et al. (1977) and 1 of the 3 cases by Sharp, Potter, et al. (1976a, b) as the BP of CML are cases of AML. This is primarily based on the presence of a large proportion of normal cells.

The experience of Secker Walker, Summersgill, et al. (1976) at the Royal Marsden Hospital and the Hospital for Sick Children in London is pertinent to the preceding discussion. These authors have indicated that the blood cells in the BP of CML may, in a few cases, have many of the characteristics of lymphoblasts and when such a phase is not preceded by the classic CP, ALL may be diagnosed; the true diagnosis then rests on the presence or absence of the Ph¹. In the authors' experience about 2% of all children diagnosed as ALL without the benefit of chromosome analysis may actually have Ph¹-positive CML presenting in the BP. So far, they have found 2 cases of CML in the BP out of 72 patients with ALL. In one case (13-yr-old girl) with poorly differentiated blasts, which were Sudan black negative and did not react with anti-ALL serum, the Ph¹ was found in all the marrow cells examined, and the diagnosis was changed to CML in BP. The authors are continuing to examine the chromosomes of BM or unstimulated blood cells in selected cases of presumed ALL according to the following criteria: (1) any patient in whom the blasts defy precise classification on morphological, cytochemical, and immunological grounds; and (2) any patient with ALL who fails to remit on standard therapy with vincristine, prednisolone, and L-asparaginase in < 2 mo.

The authors indicate that cytogenetic analysis of ALL is important because treatment and prognosis are quite different in the two conditions, i.e., BP of CML vs. ALL. The experience of Secker Walker, Summersgill, et al. (1976) raises a number of questions regarding some unusual cases published in the past,[12] e.g., change from ALL to Ph¹-positive CML (Tanzer, Jacquillat, et al., 1975). The latter case is that of a patient with ALL who developed typical CML with a Ph¹ during a long drug-induced remission of the ALL; the patient died 6 yr later in the BP.

Marrow Transplantation in AL and Chromosome Markers

The utilization of cytogenetic studies in following events in patients who have received marrow transplants is an important application of such a technique.[13] In patients with AL and aneuploidy, the success of the transplantation can be monitored cytogenetically (Littlefield, 1972). In other subjects with diploid AL, events following the marrow transplantation, e.g., involvement of the donor cells by the leukemic process, can be established if the sex of the donor and host differ or when the donor cells are characterized by karyotypic markers (Ph¹, XXY, etc.). The case of AML in a 4-yr-old girl engrafted with the cytogenetically normal marrow cells of her brother is a good example (Goh and Klemperer, 1977). The leukemic cells of the patient before the graft had a 46,XX/47,XX,+F constitution; at autopsy all hyperdiploid cells with XY pattern and 43% of those with XX had an additional F-chromosome, pointing to leukemic transformation of the donor male cells. In another case of AML in a 22-yr-old man, whose leukemic cells were apparently shown to be diploid and who had received a marrow transplant from his genotypically HLA-identical sister, relapse of the disease appeared to involve the donor cells, which developed a new consistent chromosome abnormality [45,X,−X,t(8;21) (q22;q22)] (Elfenbein, Borgaonkar, et al., 1978). There is little doubt that cytogenetic studies will play an important role in marrow transplantation as this approach to treating AL and other disorders is more widely and successfully used.

Nuclear Blebs and Aneuploidy in AL

Ahearn, Trujillo, et al. (1974) reported on the apparent association between aneuploidy and a specific ultrastructural cellular abnormality (nuclear blebs). Blebs were not found in the BM cells of normal individuals. In AL during the active phase (AML and ALL) the mean number of blebs (59.4/100 cell sections) was much higher than in diploid cases (2.5/100 cell sections). Successful response to therapy was accompanied by a marked reduction or total disappearance of the chromosomally and ultrastructurally abnormal cells. Conversely, impending relapse was indicated by their reappearance (Trujillo, Ahearn, et al., 1974, 1975; Trujillo, Cork, et al., 1974). No substantial difference in the mitotic activity of diploid and aneuploid cells was observed in AML, in which disease such activity is very low (L. Brandt, Mitelman, et al., 1975b, c).

Cellular Markers in Differentiating Various Leukemias

The recognition of the nature of the leukemic cells, particularly when the morphologic criteria are inconclusive, has led to the development of a number of approaches for their characterization (Greaves, 1975; Janossy, Greaves, et al., 1976; Janossy, Roberts, et al., 1976). These approaches have been particularly useful in differentiating among the various ALs, though they also may be useful in differentiating the "lymphoid" BP of CML from the myeloid one (Pedersen, 1973a) and possibly the blasts in AL from those in CML, and include a number of cellular phenotypic markers based on: 1. the morphologic criteria as seen in Romanovky preparations (Bennett, Catovsky, et al., 1976); 2. cytochemical features[14]: peroxidase (Hennekeuser and Mobius, 1974), PAS reaction (Feldges, Aur, et al., 1974), Sudan black B, various esterases, LAP (Kamada, Okada, et al., 1968, 1976), and acid phosphatase, arylsulfatase, oil red O (M. T. Shaw, 1976), etc.; 3. ultrastructural aspects; 4. immunocytological properties: HLA antigens,[15] cell surface markers,[16] etc.; 5. functional characteristics; 6. enzymatic and other biochemical markers (Abbrederis, 1977): terminal deoxynucleotidyl transferase (TdT),[17] adenosine deaminase (G. E. Bloom, 1972), and β-glucuronidase (Wasastjerna, Vuorinen, et al., 1975) activities, DNA-binding serum proteins (Hoch, Longmire, et al., 1975); and 7. in vitro culture behavior,[18] particularly on soft agar[19] (see Table 52).

At the time of this writing it was still premature to correlate the chromosome findings in leukemia and other states with some of the surface characteristics and other markers (immunologic, enzymatic) of the cells involved. Such a correlation may yield important information and I hope that results will be forthcoming in the near future on the cytogenetic aspects of cells tested for HLA antigens, B-, T-, and null-cell characteristics, TdT, PAS, esterases, and so on, because some of these have been shown to be related to the prognostic aspects of the leukemias (L. J. Wolff, Richardson, et al., 1976) and found useful in predicting relapse of the disease.

Acute Lymphoblastic Leukemia (ALL)

Studies Before Banding

The distribution of the modal chromosome numbers in the aneuploid cases of acute lymphoblastic leukemia (ALL) is shown in Table 53 and is characterized by a very low incidence of hypodiploidy and the occurrence of modes near and beyond the tetraploid range (Figures 81–85). It is possible that the hypodiploid cases of ALL published in the literature may represent AL of stem cell origin, possibly more akin to AML than ALL. On the other hand, the extreme rarity of hypodiploid cases of ALL

Table 52 Markers in leukemia

| | Surface markers | | | | | | | Others | |
| | Rosette assay | | | Membrane immunofluorescence | | | | | |
Leukemia type	E	EA	EAC	SmIg	HTLA (T-Ag)	Ia-like	cALL	TdT	ADA
ALL 75% null-cell ALL	−	−	−	−	−	+	+	+	−
20% T-cell ALL	+/−	−	−/+	−	+	−	−/+	+	+
5% B-cell ALL	−	−	−/+	+	−	+	−/+	−/+	−
Undifferentiated leukemia	−	−	−	−	−	+	+/−	?	−
AML	−	+/−	+/−	−	−	+/−	−	−/+	−
CLL 95 B-cell CLL	−	−	+	+	−	+	−	−	−
5% T-cell CLL	+	−	−	−	−/+	−	−	−	−
HCL	−	+/−	+/−	+/−	−	+	−	−	−
CML (CP)	−	−	−	−	−	−	−	−	−
BP									
"lymphoid" (50%)	−	−	−	−	−	+	+	+	−
"myeloid" (50%)	−	−	−	−	−	+/−	−	−/+	−

Abbreviations in and explanations of Table 52:
- E: Sheep red blood cell rosettes; nonimmune, human T-cell marker.
  Normal blood T-cells, thymocytes, T-cell leukemia, T-cell lymphoma.
- EA: Fc receptor, bovine erythrocyte-IgG antibovine erythrocyte antibody complex.
  Some normal B and T cell subsets, monocytes, macrophages, some Burkitt lymphomas (B-cell lymphomas), some AML, hairy cell leukemia (HCL), and some lymphomas.
- EAC: C_3 receptor, bovine erythrocyte-IgM antibovine erythrocyte antibody-complement complex.
  Normal B-cells, some fetal thymocytes, monocytes, CLL, some AML, and rare ALL (both T-cell and B-cell types).
- SmIg: Surface membrane immunoglobulins, human B-cell marker.
  Normal B-cells in blood, CLL, HCL, lymphomas (most of adults).
- HTLA: Human thymus-leukemia antigens.
  Normal thymocytes; ALL of T-cell type; T-cell lymphoma.
- T-Ag: Normal T-cell antigens.
  Normal T-cells in peripheral lymphoid tissues and thymocytes.
  ALL of T-cell type, CLL of T-cell type, T-cell lymphomas, Sézary cells.
- Ia-like: Ia-like human B-cell associated antigens.
  Normal B-cells in peripheral lymphoid organs, monocytes(?)—"null" cells.
  ALL of non-T, non-B (null) cell type, undifferentiated leukemia, majority of AML and some of CML, majority of CML in blast phase ("lymphoid" and "myeloid" type).
- cALL: Antigen specific/or associated with common form of ALL (null-cell ALL).
  Normal stem cells (?).
  ALL of non-T, non-B-cell type, some of ALL of T-cell type, some ALL or lymphoma of B-cell type, CML in blast phase with "lymphoid" features.
- TdT: Terminal deoxynucleotidyl transferase.
  Normal thymocytes, some cells in normal bone marrow(?).
  T-cell ALL and lymphoma, non-T, non-B ALL, "lymphoid" type CML in BP, some AML, and ALL of B-cell type.
- ADA: Adenosine deaminase.
  Normal T-cells and thymocytes.
  T-cell ALL and lymphomas.

may be related to diagnostic propensities peculiar to our clinics. In any case, hypodiploidy is certainly a rare occurrence in ALL and, at least in our laboratory, we tend to reconsider the diagnosis when it is observed in cases of AL.[20]

The already mentioned fuzziness of the chromosomes, including those of diploid cases, occurs much more frequently in ALL than AML (50–70% in ALL vs. 30% in AML) (Figure 86). In contrast to AML, the karyotypic findings in ALL before the banding era did not reveal cytogenetic findings of any consistency

Table 53 Distribution of modal chromosome number in 716 cases of AL

	Modal chromosome number							
	41–45	46*	46‡	47–50	51–55	56–62	>63	Total
			AML					
No. of cases	71	29	231§	74	13	8		425
Percent of cases	(16.7)	(6.8)	(54.3)	(17.4)	(3.0)	(1.9)		
			ALL					
No. of cases	2	11	155‖	55	34	22	12	291
Percent of cases	(0.7)	(3.8)	(53.3)	(18.9)	(11.7)	(7.6)	(4.1)	

(The difference between the number of patients with AML and ALL is solely due to special studies conducted by us).
*Pseudodiploid cells.
‡Diploid (normal) cells.
§Includes 51 cases with some aneuploid cells in a preponderantly diploid population.
‖Includes 29 cases with some aneuploid cells in a preponderantly diploid population.

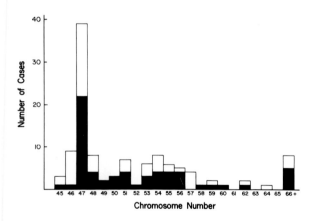

81 Distribution of aneuploid and pseudodiploid modal chromosome numbers of ALL cases. The cases shown with 46 chromosomes were all pseudodiploid. The black columns indicate cases reported during the first 10 yr of our experience with ALL (Sandberg, Tagaki, et al., 1968a), and the open columns the subsequent 10 yr (Oshimura, Freeman, et al., 1977b). Similar distribution of the modal chromosome numbers in the cases is apparent.

82 Metaphase from the marrow cell of a patient with ALL containing a large number of chromosomes, including a definite marker (arrow).

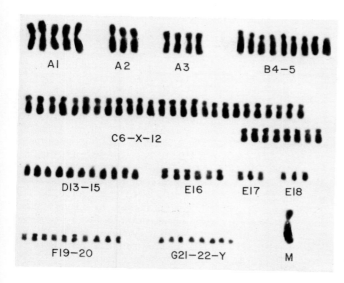

83 Karyotype of a marrow cell from a patient with ALL including a large marker (M). Obviously, arbitrariness had to be used in assigning the chromosomes to the various groups, as banding was not performed. Even with banding, the chromosomes of AL, particularly ALL, are often very fuzzy and ill defined and defy identification with any banding technique. The important point being made with this figure is the large number of chromosomes that may be observed in ALL, >100 chromosomes not being unusual.

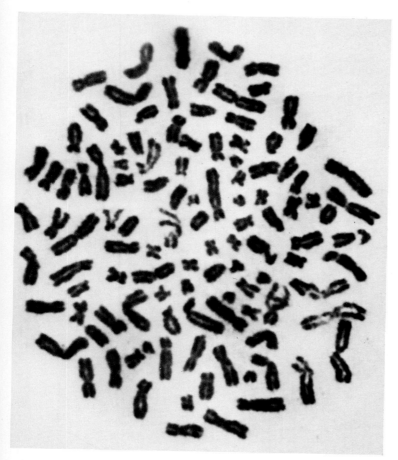

84 Metaphase from a marrow cell of a patient with ALL containing a large number of chromosomes, including some abnormal ones. Some broken chromosomes and several with definite breaks are seen in the metaphase.

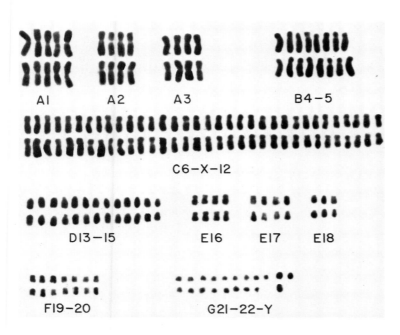

85 Karyotype of a leukemic cell from the marrow of a patient with ALL with an extremely high modal chromosome number. The cell shown contained 192 chromosomes. Even with banding it would have been difficult to classify all the chromosomes and, hence, some arbitrariness was used by us in grouping the large number of chromosomes in this case. Note two ringlike chromosomes shown with the G group and the very fuzzy appearance of the chromosomes, a finding not unusual in ALL.

86 A. Two contiguous metaphases in the marrow of a patient with partial remission of ALL. The metaphase on the right is diploid (normal) and the one on the left hyperdiploid (leukemic). Note the fuzzy and ill-definded appearance of the leukemic chromosomes compared with the thin and well-defined chromatids of the normal cell. This fuzzy appearance of the leukemic chromosomes is not uncommon in AL (and other neoplastic disorders) and is as characteristic of the disease as the karyotypic changes. The cells were obtained without resort to culture or exposure to colchicine or colcemide. B–F. (pages 281 and 282) Three marrow metaphases demonstrating the fuzzy and ill-defined appearance of the chromatids in the cells of AL.

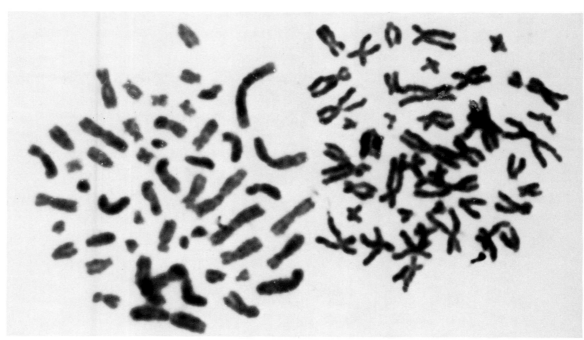

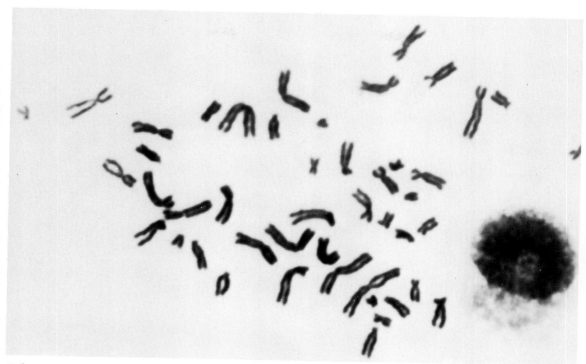

86 B. Thin and well defined metaphase of a normal diploid cell in the marrow of a patient with partial remission of AL.

86 C. A diploid and probably leukemic cell of above patient showing the fuzzy and ill defined appearance of the chromatids. No definitely abnormal chromosomes can be ascertained.

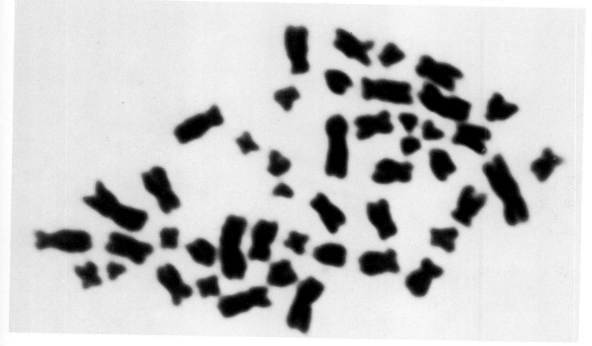

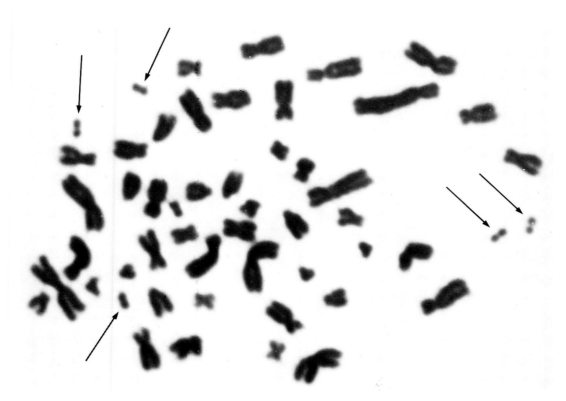

86 D. A leukemia metaphase with fuzzy chromosomes from a case of AL, containing markers and DMS (arrows).

86 E and F. Metaphases with abnormal chromosomes (markers) (arrows) and fuzzy appearance of the chromatids, a finding characteristic of the chromosomes in AL.

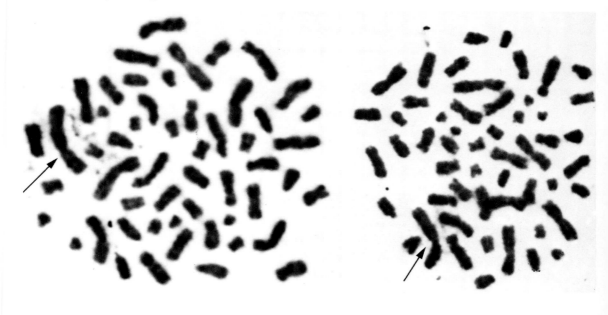

87 A. (above) Karyotype (unbanded) of a marrow cell from a patient with ALL containing a somewhat abbreviated #1 (1p−), +9, +G, and a small marker chromosome (M). B. (middle) Karyotype of a BM cell from a patient with ALL containing 92 chromosomes. Apparently, the only abnormality consisted of a marker chromosome (M), which appeared to replace one of the G21 group autosomes. C. (below) Metaphase from the marrow of a patient with ALL showing a large number of chromosomes, a finding more common in ALL than in AML.

or uniqueness, except possibly the more frequent involvement of group C and G chromosomes than those of any other group (Sandberg, Tagaki, et al., 1968c). The karyotypic picture appeared to differ from one case to another, with very few cases having similar chromosomal findings, even when based on careful analysis. Thus, in spite of attempts to implicate this or that chromosome as being more frequently involved in AL, careful scrutiny of cytogenetic data in this disease before banding analysis failed to reveal such a consistency (Figure 87). Studies with banding on large series of patients with ALL have been scarce, probably because of the fuzzy and ill-defined appearance of the leukemic chromosomes and the indistinct features of the chromatids, which is not rare in AL, making banding studies in ALL very difficult, and in some cases impossible. In a recently completed study by us of ALL cases since 1968, (Oshimura, Freeman, et al., 1977b), banding was performed successfully on a substantial number of cases, and the results indicate some nonrandomness of the karyotypic pattern in ALL. Thus, of 50 recently unselected cases of ALL in whom Q- and G-banded karyotypes were attempted, 31 were successfully analyzed with these techniques.

Frequency of Abnormalities and Number of Chromosomes

The data show that > 50% (54/101) of the patients with ALL had aneuploid clones constituting varying percentages of the total cell population in their marrow (Figure 80). Two of the 54 cases were hypodiploid. The frequency of karyotypic abnormalities and the distribution of the chromosomal number were very similar to those of 106 cases studied in our laboratory prior to 1968 (Sandberg, Tagaki, et al., 1968c). Thus, the previous report contained 63 cases with aneuploid clones in their marrow cells: only one of these cases had a clone with a hypodiploid number of chromosomes in a preponderantly hyperdiploid cell population, whereas the remaining 62 cases had clones with a pseudodiploid or hyperdiploid number of chromosomes. The findings are not incompatible with the results of others.[21] The experience of Zuelzer, Inoue, et al. (1976) is somewhat at variance with the results of Whang-Peng, Lee, et al. (1976) and those obtained by us, in that only 21% of the children with ALL were found to have a diploid chromosome constitution, whereas 79% of the 53 patients studied had chromosome abnormalities. About 50% of our patients were studied when they were still untreated with chemotherapy or radiation, and the distribution of aneuploidy and diploidy in this group of subjects was similar to that of the total group. Therefore, we believe that karyotypic abnormalities are usually not induced by the therapy of AL. In addition, the consistency of the karyotypic pictures, in spite of changes in treatment, points to therapy not playing an important role in the induction of chromosomal changes in ALL.

Common Abnormalities in ALL, Including the 6q—

Before analysis with banding techniques, the chromosomal changes appeared to be variable from case to case both in ALL and AML, even though the frequent involvement of C-and G-group chromosomes was reported. Several recent studies of the karyotypic picture in AML with banding techniques (Table 54) have afforded some evidence regarding the possible nonrandom distribution of chromosomal abnormalities in this disease (Figure 88), i.e., an #8–#21 translocation, and trisomy for all or part of the long arm of #1[22]. The data obtained by us on 16 ALL cases show that some chromosomal changes are also nonrandom in ALL (Oshimura and Sandberg, 1976), e.g., the frequent occurrence of deletion of the long arm of #6 (4 cases), although their break points were not the same. One case of ALL with the deletion of the long arm of #6, del(6)(q13), was reported by Lawler, Secker Walker, et al. (1975). Also, even though the majority of the patients were not studied with banding techniques, Zuelzer, Inoue, et al. (1976) did describe four cases with 6q— and five or more cases with a deletion of a Cq, some of which may possibly be #6, among their 53 patients with ALL (42 aneuploid). Recently, we had the opportunity to perform chromosomal studies on cell lines originating from thymus-derived lymphocytes of the peripheral blood from seven patients with ALL. Four of the seven cell lines had an abnormality similar to the one

Table 54 Published chromosome anomalies in ALL subsequent to the introduction of banding techniques

Case no.	Karyotype
1	47,XX,iso17q,+21
2	45,XX,−21/52,undetermined
3	47,XX,+16
4	46,XY,1q+
5	45,X(−X)
6	45,−C,Cq−/46,Dq−46,1q−,Cq−,mar/46,mar
7	46,+C,−F/46,Cq−/46,mar1mar2/>2N,−−−5q+,Dq−
8	46,del(6)q13/46,del(7)q22/46,19p+
9	46,XX,4q+
10	46,XY,2p−,19p+
11	27,haploid, except for X,10,18,21/54
12	45,XY,−8
13	46,XY,13q+,21q−
14	47,XY,4q−,+17q−
15	46,XX,2q+,4q−,4q−,12p+,13q+,14q+,−20,+21
16	46,XX,21q−
17	46,t(1q−,?+),−−−
18	46,inv(4)(p−,q+),−−−
19	46,del(1)q,−−−
20	46,t1h+(p−,?+),−−−
21	46,del(1)p−/46,del(11)q−,−−−
22	46,tB(q−,?+),del(C)q−,−−−
23	46,XY,Ph¹,del(2)q−/t2(2q−;?+) of which 2 in combination with del(2)q−,−−−
24	46,t1(p−,q−,?q+)(?AML)
25	45,t12(12p−;D+)(AL)
26	46,t4(4q−,15q+),del(15)q−,del(17)q−,−−−
27	46,XX,Cp−,Cq−,−−−
28	46,XX,t1(q−,Cq+),−−−
29	46,XX,−−−del(2)q−,−−−
30	46,XY,del(C)p−,del(C)q−,−−−
31	46,XX,−C,+t?inv(q−,q+),−−−
32	46,XY,Bp+,del(C)(p−,q−)tC(q−,?+),Dq−,−−−
33	52,XX,t2(q−,?+)−−−
34	51,t2(q−,?+),CP−−−
35	50,del(3)q11,del(Y)q11,del(1)p31,t12(12p−,Yq+),−−−
36	52,del(3)p24−−−
37	45,6q−,−−−
38	52−54,tC(p−,?1q)Cq°
39	55−60,XY,dic?(2p−,?+)Gq−,−−−
40	48−55,XY,t1(p−,?+)−−−
41	49−55,XX,t1(q−,?2q)−−−
42	50−62,XX,t2(q−,?+)−−−
43	55−58,XY,inv(2),tD(q−,?+),−−−
44	46−56,XY,dic(1)(p−,1q+),del(3)(p−,q−),del(6)q°,t2(q−,Cp+),−−−
45	47,XY,+18,22q−(?)→45,XY,−15
46	47,XX,+8
47	N→t(2;2)(q12;q22),+X
48	N→t(12;17)(p13;q12),9p−
49	46,XX,t(4;11)(q13,q22)
50	46,XY,t(4;11)(q13,q22)
51	46,XX,t(4;11)(q13,q22)
52	46,XXp+,del(1)(q21),del(6)(q21),i(17q)
53	47,X,−X,del(6)(p23),+17p+,+18p−

continued

Table 54 — continued

Case no.	Karyotype
54	45,XY,1q+,−6
55	46,XY,t(4;11)(q21;q23)
56	55,XX,+2,+4,+9,+14,+15,+19,+20,+21,+21
57	53,XX,+2,+5,i(7q),+8,+13,+21,+21,+22
58	47,XX,1q+,+13
59	61,XX,+X,+1,+2,+4,+5,+6,+6,del(6)(q21),+13,+14,+15,+16,+18,+21,+21,+22
60	55,XY,+X,+3,+6,+10,+13,+14,+16,+21,+21
61	57,XY,+X,+5,+6,i(7q),+10,+12,+13,+15,+18,+21,+21,+22
62	53,XY,+X,+4,+8,+14,+15,+17,+21
63	46,XY,+2,6p+,+7,−12,−13,14q+,17p+
64	47,XX,+12
65	54,XX,7p+,7p+,+10,+10,+14,+14,+18,+18,+21,+21
66	46,XY,del(6)(q23−25)
67	46,XY,del(6)(q23−25)
68	N→47,XX,+19
69	46,XY,t(21;22)(q22;q11)
70	47,XY,−10,−12,+17,+D(14or15),mar(T-cell leukemia)
71	58,XX,+4,+6,+7,+10,+14,+14,+21,del(18)(q22),+C,+3mar
72	48,XY,+18,+mar
73	48,XX,+13,+19
74	49,XY,+7,+12,−13,+9p+,t(1;13),t(6;18),t(11;14)
75	46,XX,del(9)(p21)
76	49,XY,+7,+12,−13,t(1;13)(q12.2;p13),t(6;18)(p25;q21),+9p+,t(11;14)(q23;q32)

Cases 1−5 (Yamada and Furasawa, 1976); cases 6−8 (Lawler, Secker Walker, et al., 1975); cases 9 and 10 (Fleischman and Prigogina, 1975); case 11 (Kessous, Corberand, et al., 1975b); cases 12−16 (Whang-Peng, Knutsen, et al., 1976b); cases 17−44 (Zuelzer, Inoue, et al., 1976); cases 45−48 (Cimino, Kinnealey, et al., 1977); cases 49−51 (Van Den Berghe, David, et al., 1978a); cases 52−67 (Oshimura, Freeman, et al., 1977b), cases 68−75 (Cimino, Rowley, et al., 1979); case 76 (Roth, Cimino, et al., 1979).

———Additional anomalies not readily identified.

described above, i.e., 6q−. The abnormality of #6 was not observed in > 40 AML patients with aneuploid karyotypes. Thus, the incidence of the 6q− abnormality in ALL is at least as common as some of the so-called nonrandom changes described for AML[23]. Furthermore, five cases had two extra #21 chromosomes and one case had one extra such chromosome, and three cases had one extra #22. We did not observe a missing Y in any of our male patients, whereas two such patients were observed by Zuelzer, Inoue, et al. (1976) and one by Whang-Peng, Lee, et al. (1976), these cases being accompanied by other karyotypic abnormalities. This rather low incidence of a missing Y contrasts with the much higher incidence observed in AML and CML. Even though some chromosomal changes seem to be nonrandom in ALL (Figures 89−92), many more cases will have to be studied with banding techniques to establish a more precise karyotypic picture, especially as ALL is often associated with a substantially increased number of chromosomes, thus complicating cytogenetic interpretation.

The report of Zuelzer, Inoue, et al. (1976), based predominantly on results obtained before the introduction of banding analyses, deals with cytogenetic studies in 71 children with AL. Of these patients, 52 had the diagnosis of acute "stem cell" (lymphatic) leukemia (ALL), with 79% of the patients having chromosomal abnormalities. As mentioned previously, this percentage is much higher than that observed by most workers. Furthermore, only 1 patient of the 11 with AML and 1 of the 5 with AMMoL had a diploid chromosome constitution. Whether this is a reflection of the age of the patients (all below 16 yr) is difficult to ascertain, because generally only about 50% of

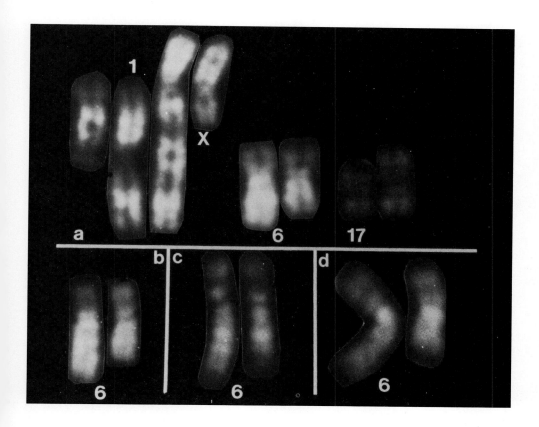

the patients with these leukemias have been found to have chromosomal abnormalities. It should be mentioned that out of the 71 patients studied, 20 had marker chromosomes in the leukemic cells. The distribution of the modal counts among the 52 patients with ALL was interesting in that the largest number of patients had a chromosome mode of 50 or more, with pseudodiploid karyotypes being next in frequency. Among the 52 patients with ALL, there were apparently 2 patients with hypodiploid clones. Besides the 6q− anomaly, already alluded to, one patient with an alleged Ph[1] and several with a missing Y (in addition to other chromosomal anomalies) were observed. Based on their experience, the authors stated that in the overwhelming majority of cases the cytogenetic findings suggested clonal identity of the leukemic cell population in relapse with that studied at the onset of the disease, notwithstanding considerable karyotypic instability in almost half of the patients. In a small minority an independent origin of the

88 Q-banded partial karyotypes of four ALL cases with partial deletion of the long arm of chromosome #6: a. (from left to right). del(1)(q21), a normal #1, Xp+ originating through a translocation of the long arm of #1 (q21) and an unknown chromosome to the short arm of the X-chromosome (p22), a normal X, a normal #6, deletion of chromosome #6 (q21), a normal #17, and an isochromosome of #17. b. A normal #6 and a deleted #6 (q21). c. A normal #6 and a deleted #6 (q23−25). d. A normal #6 and a deleted #6 (q23−24).

relapse clone could not be excluded on cytogenetic grounds but was considered unlikely because mechanisms capable of accounting for the changes observed in these patients could be demonstrated in other cases. The persistence of diploid leukemic cells and the presence of an aneuploid subclone were demonstrated in the relapsed BM and/or spinal fluid in all ac-

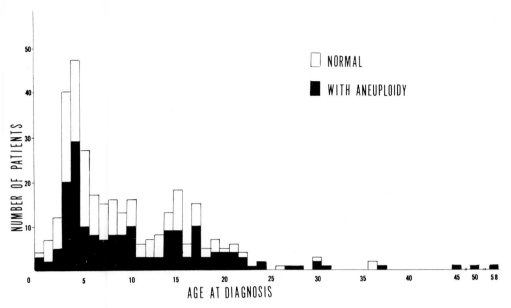

89 Distribution of patients with ALL according to the age at diagnosis. It is apparent that the bulk of the patients are diagnosed below the age of 10 yr. As has been the general experience, about 50% of the patients with ALL are associated with aneuploidy in the leukemic cells.
(Whang-Peng, Knutsen, et al., 1976b)

90 Q-banded karyotype of a leukemic cell from a patient with ALL showing an interstitial deletion of #5 (5q−) and the short arm of #9 (9p−). Such an interstitial deletion of #5 occurs characteristically in some refractory anemias and AML and rarely in ALL.

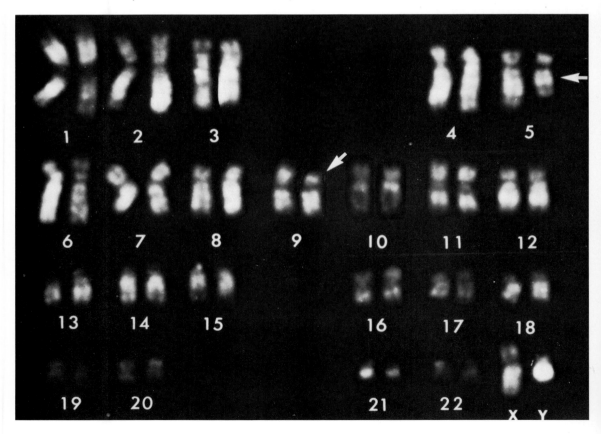

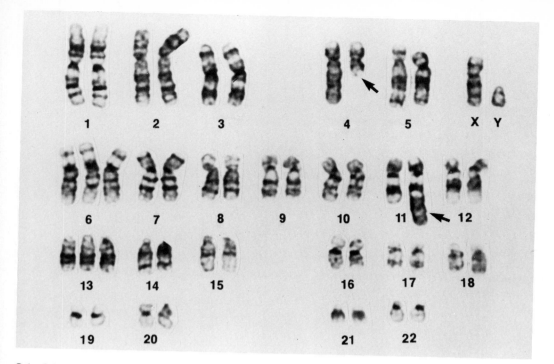

91 G-banded karyotype of an ALL case showing 48 chromosomes, including an extra #6 and #13 and a reciprocal translocation between #4 and #11, i.e., t(4;11)(q21;q23).

92 Q- and G-banded partial karyotypes of four ALL cases showing structurally changed chromosomes: a. 6p−, +17p+, and +18p−. b. 1q−, normal #1, and 1q+. c. 17p+ composed of most of the long arm of #12, most of #17 and an unknown chromosome and a 14q+. d. An isochromosome of #7.

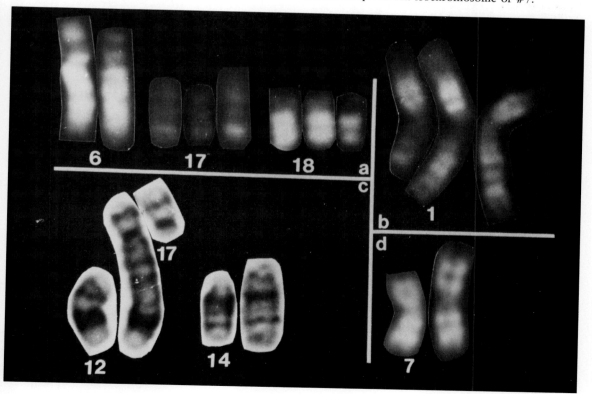

tive phases of the disease. On this basis the conversion from an aneuploid to a predominantly or exclusively diploid karyotype could be visualized and a new model of clonal evolution involving repetitive formation of abnormal karyotypes from a surviving diploid clone could be suggested. The last hypothesis of the authors would seem to indicate to me that the development of chromosomal changes during the course of ALL is a reflection of factors other than causative ones, the authors' implication being that the diploid leukemic cells can become aneuploid, as part of the progressive process of the disease, and not be directly related to the initiation of ALL.

Karyotypic Evolution

Almost half of the ALL cases examined by us (Oshimura, Freeman, et al., 1977b) had some degree of karyotypic instability. In 50% of the cases examined by banding techniques, more than two clones, most of which could be traced karyotypically, were observed in the same marrow aspirate or in the following one. Banding studies of cases with karyotypic instability revealed that whenever karyotypic evolution could be observed, the chromosome number always had a tendency to increase gradually. In almost all cases, however, the karyotypic findings seemed to represent cytogenetically advanced changes and not the pristine karyotypic picture, especially in the cases with large numbers of chromosomes.

We failed to find any specific chromosome engaged in increasing the number of chromosomes. Karyotypic instability, selection, and/or evolution during the disease studied in large groups of patients with ALL have been reported by Zuelzer, Inoue, et al. (1976), as well as by Whang-Peng, Lee, et al. (1976). Frequent and careful chromosome analysis based on banding analysis in long-term studies of ALL will undoubtedly reveal a clearer picture than is presently apparent, although from the previously mentioned studies, it would appear that selection of different karyotypes may occur throughout the disease, ranging in rare cases from a shift of diploidy to aneuploidy to the development of a host of clones with variable karyotypic pictures.

Cell Surface Markers and Chromosomal Abnormalities

Acute lymphoblastic leukemia can be subgrouped phenotypically into three categories; null-cell (non-T/non-B), T-cell, and B-cell types. The great majority of the 42 ALL cases studied by us were neither T- nor B-cell types (null-cell), one case was the B-cell type and four of the T-cell type. Eighteen cases out of 37 with null-cell, 1 case out of 4 with T-cell, and the 1 case with B-cell ALL had chromosome abnormalities in their leukemic cells. Study of the surface markers in two cases with the 6q− anomaly revealed one to be typically T cell and the other to be of the null-cell variety of ALL. As mentioned previously, four of the seven cell lines derived from the T cells of patients with ALL had a similar abnormality, i.e., 6q−; three of the cell lines had other abnormalities in addition to the 6q−. Inasmuch as only T-type leukemic cells have been successfully cultured as cell lines, it is safe to say that only further studies will shed light on the significance of the findings.

Abnormalities of #14 have been noted often in a variety of lymphoid malignancies, such as lymphoma, multiple myeloma, and plasma cell leukemia (see Table 74). In these cases, including one of ours with a B-cell ALL, the #14 abnormality seemed to be restricted only to those lymphoid neoplasias having B-cell characteristics. Thus, the significance, implications, and clinical importance of the cytogenetic abnormalities in ALL, as well as of the cell surface markers, remain to be determined in future studies.

Further characterization of the cell type involved in ALL may be forthcoming from such approaches as the β-glucuronidase reaction of the cells (Wasastjerna, Vuorinen, et al., 1975) (apparently more specific for ALL than PAS), the acid phosphatase reaction of the lymphoblasts (only positive in T cells) (Catovsky, 1975), and ultrastructural studies of the leukemic cells (Trujillo, Cork, et al., 1974; Trujillo, Ahearn, et al., 1975).

General Comments

No attempt can be made, at present, to perform an in-depth correlation between specific cytogenetic findings in ALL and some of the clinical, laboratory, and survival parameters. This awaits a larger body of karyotypic data based on banding analyses. Nevertheless, some comparison of the data published by us can be made with those obtained on large series of patients with ALL, particularly by Whang-Peng, Lee, et al (1976). These authors found that aneuploid ALL (hyperdiploidy being most prominent) is usually accompanied by some chromosomally normal cells in the marrow. In general, this has also been our experience. Whang-Peng, Lee, et al. (1976) also indicated that the persistence of aneuploidy and, particularly, the development of total aneuploidy augur a poor prognosis. Again, not only recent results, but also our past experience, support this concept. In agreement with the findings of Whang-Peng, Lee, et al. (1976) and Zuelzer, Inoue, et al. (1976), we have found that aneuploidy disappears after successful therapy and remission of the disease and that new clones, occasionally unrelated to the preexisting aneuploid karyotype, may develop in a significant proportion of the cases. Because aneuploidy was observed in half of the cases before antileukemic therapy, it can be assumed that such therapy per se does not play a crucial role in the genesis and/or evolution of the chromosomal changes. Whang-Peng, Lee, et al. (1976) in their large series found a high incidence of aneuploidy in patients under the age of 1 yr or over the age of 20 yr, and in those with low or markedly elevated leukocyte counts at diagnosis.

Cytogenetic Experience in ALL

In a study of 49 patients with ALL utilizing a G-banding technique, Lawler, Secker Walker, et al. (1975) found that banding patterns were difficult to visualize in the marrow chromosomes in AL because of their fuzzy appearance, particularly in ALL. Hyperdiploidy was the most common numerical abnormality in ALL with a large proportion of the patients having > 30% of hyperdiploid cells. Hypodiploidy was very uncommon. Abnormal clones included: —C, Cq—; Dq—; 1q—, Cq—, mar; tC,—F; Cq—; 5q+, Dq—; XXY, mar; 6q—(q13), 7q—(q22); 19p+ and cases with unidentified markers. The authors also stated that the high proportion of randomly distributed chromosomal breakage found in a high proportion of the patients may be related to the disease process.

The experience of Whang-Peng, Lee, et al. (1976) covered 331 patients with ALL (Figure 93) and a period of 15 yr (1961–1976), including some banding studies. Aneuploidy was observed in pretreatment marrows in 42.6% and at some stage of the disease in 54.1% of the patients. Aneuploid cells usually coexisted with normal cells; hyperdiploidy was predominant with the most common groups involved being C, G, and B (the latter two statistically significant). Aneuploidy disappeared after successful remission, and new clones developed in 12 patients during relapse. The appearance of aneuploid cells in the marrow at the onset or later in the disease was of no prognostic significance, but persistence of these lines and the development of total aneuploidy signal a poor prognosis. Eradication of these aneuploid cells is, therefore, essential for the achieve-

93 Gain and loss of chromosomes in various groups in ALL showing an essentially common involvement of most of the chromosomes, though groups C and G appear to be more commonly involved than the others. For comparison see the findings in AML (Figure 94).

(Whang-Peng, Knutsen, et al., 1976b.)

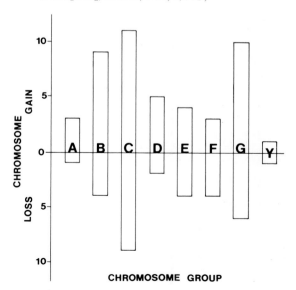

ment of a long remission and progress to a permanent cure. In two patients of this group a suspected Ph¹ was shown to be 21q− with banding; in two other cases with a suspected Ph¹ apparently no banding was done. In the five aneuploid cases in whom G-banding was successful the karyotypes were: 45,XY,−8; 47,XY,4q−,+17q−; 46,XX,2q+,12p+,13q+,14q+,4q−,4q−,+21,−20; 46,XX,21q−; and 46,XY,t(13q+,21q−).

Karyotypes were analyzed with banding in 16 cases of ALL (10 adults, 6 children) by Cimino, Rowley, et al. (1979). In 11 patients karyotypic abnormalities were found, in 3 cases the abnormalities appearing in a marrow apparently originally diploid. No hypodiploid cases were encountered (modal number ranged from 46–58). Eight of the 11 patients had different abnormalities involving the bulk of the chromosome groups. One patient had a Ph¹ due to an unusual translocation, i.e., t(21;22) (q22;q11). One patient with a B-cell ALL had a 14q+ marker. The median survival of patients with initially normal karyotypes may be longer than that of patients whose karyotypes are abnormal initially. In each of the patients with aneuploidy, normal (diploid) cells were present in the marrow, either before therapy or following relapse, though the percentage of the normal cells varied from case to case and during different stages of the disease.

A 30-yr-old woman with ALL and a 46,XX,21q− karyotype in the leukemic cells, after a remission period of 5 yr, developed Ph¹-positive CML with a 46,XX,t(9;22)(q34;q12) karyotype (Tosato, Whang-Peng, et al., 1978). In contrast to AML, no clear-cut prognostic and clinical differences have been established, as yet, between the aneuploid and the diploid forms of ALL. No correlation could be established between a cytological subclassification of ALL and the chromosome abnormalities in this disease in children (Kessous, Corberand, et al., 1975b). A comparison of survival in 29 ALL patients with (9 patients) and without chromosomal changes revealed the former to have a mean survival 7 mo longer than the latter, but both groups showed wide survival ranges (including mean and median survivals) with no statistically significant differences. Patients over the age of 20 yr had a significantly shorter survival than those 20 yr old or less at diagnosis.

Two short reports have dealt with the relation of the cytogenetic findings to the prognosis of ALL. Cimino, Kinnealey, et al. (1977) in a study of 10 cases found that if the chromosome constitution is initially normal, the median survival will be longer than if it is abnormal, i.e., 12 vs. 6 mo, respectively. Lange, Alfi, et al. (1977) found values of 7.1 and 16 mo. The presence of chromosomally abnormal clones in children under 2 yr of age carried with it a poor prognosis. The results of Alimena, Annino, et al. (1977) appear to be in general agreement with those given above.

Patients with childhood ALL in the hyperdiploid category had significantly longer first remissions than those in all other categories (hypodiploid, pseudodiploid, diploid, and mixed); those in the pseudodiploid category had the shortest remissions (Secker-Walker, Lawler, et al., 1978). The authors suggested that the proportion of hyperdiploid cells, determined by conventional chromosomal staining techniques, may be used as an additional prognostic feature in childhood ALL. The complete disappearance of the aneuploid cells from the marrow of ALL patients is not unusual, because complete remission is more readily obtained in this disease than in AML. Furthermore, in almost all cases of ALL some diploid cells are found in the marrow at diagnosis or during frank relapse of the disease; whereas in AML a substantial number of the cases are 100% aneuploid and, hence, no normal cells are available to repopulate the marrow when the leukemic cells disappear as a result of chemotherapy.

Congenital Chromosome Abnormalities and ALL

Acute lymphoblastic leukemia has been reported to occur in patients with congenital chromosome abnormalities (Table 47) and includes males with Klinefelter's syndrome (XXY or XXXY),[24] a case (female) in a family with a (4p;14q) translocation (Garson and Milligan, 1974), a child with XYY (Kessous, Corberand, et al., 1975b), and a case in a family with C-G translocation (Hinkes, Crandall, et al., 1973). In some of these cases additional karyotypic changes were associated with the ALL, but in each case the congenital chromosome anomaly persisted in the leukemic cells. Thus, in the patient with the 4;14 transloca-

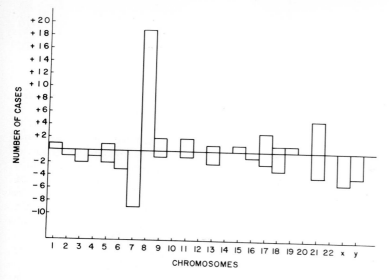

94 A. Gains and losses of various chromosomes in AML, with loss of #7 and gain of #8 being the most common. Very similar results were obtained at the First International Workshop on Chromosomes in Leukemia, 1977 (Br. J. Haematol. 39, 311–316, 1978).

94 B. Breakpoints of chromosomes participating in in the structural rearrangements leading to the formation of 50 markers in the leukemic cells of 19 patients with ANLL (Philip, Krogh Jensen, et al., 1978). The lines connecting some of the breakpoints indicate points identified within the same case.

(Courtesy of Dr. P. Philip.)

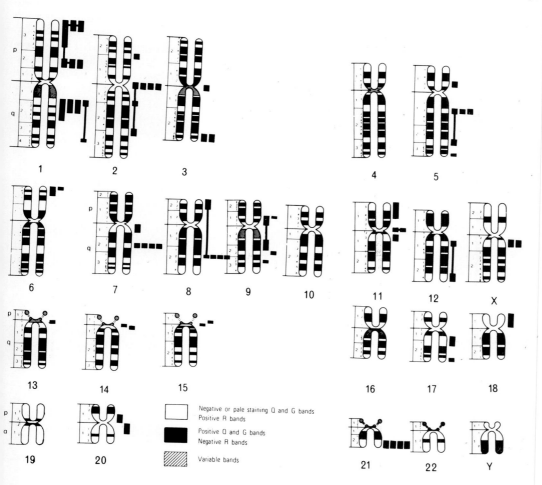

tion, the leukemic cells contained 47–51 chromosomes on account of +C,+E,+F and +G, the +C being most common. In the XYY patient −C and a marker were seen. Another case of ALL allegedly occurred in the family with the C-G (12;22) translocation. A case of D-G translocation has been reported (Whang-Peng, Lee, et al., 1976) in whom a Ph¹ was thought to exist, whereas banding revealed it to be a 21q−, the result of a 13;21 translocation.

A new characteristic chromosome anomaly in ALL, consisting of a reciprocal translocation (4;11) (q13;q22) has been described in four patients (Van Den Berghe, David, et al., 1978b) and appeared in one of our cases with ALL (Oshimura, Freeman, et al., 1977b). The authors indicated that the presence of the 4;11 and/or 6q− may help the clinician in establishing the diagnosis of ALL.

Acute Myeloblastic Leukemia (AML)
Common Chromosomal Changes in AML

Banding techniques have revealed certain cytogenetic patterns in AML which not only occur with more consistency than chance alone would allow,[23] but may also be used as indices for therapy and classification of the disease (Sakurai and Sandberg, 1973, 1974, 1976a). These developments are in sharp contrast to the relative scarcity to-date of such fundings in ALL. However, it is possible that AML may constitute a somewhat heterogenous group of diseases and the cytogenetic findings would, thus, only reflect that state. Even though reexamination with banding techniques of cells in AML previously stained with standard Giemsa has revealed that in 10–20% of such cases subtle karyotypic changes could only be found with the newer techniques (Rowley and Potter, 1976), it does appear, nevertheless, that about 50% of the patients with AML will continue to be characterized by diploid karyotypes (Mitelman, Levan, et al., 1975), and not until methods revealing chromosome substructure below the level of banding become available will the diploidy of such cases be challenged.

The nonrandom chromosomal changes in AML (Table 55) most commonly involve #8, #7, and #21; the most common change is +8, followed by −7 and then either gain or loss of #21 (Figure 94). The occurrence of an iso-17 [i(17q)], though very rare in AML, has been described (see Table 89). An identical translocation of 6p;9q in two female patients with AML and del(17) in another two have been reported (Rowley and Potter, 1976). Six vulnerable chromosome points in AML have been

Table 55 Chromosome numbers in acute myeloblastic leukemia and EL from major papers published since 1968

	Hart, Trujillo, et al. (1971)	Trujillo, Ahearn, et al. (1974); Trujillo, Cork, et al. (1974)	Sandberg, Takagi, et al. (1968c)	Sakurai and Sandberg (1976a,b)	Gunz, Bach, et al. (1973)	Zuelzer, Inoue, et al. (1976)	Rowley (1976d)	Sasaki, Okada et al. (1976)
No. of cases	95	160	113	69	79	19	50	27
Hypodiploid	14%	10%	21%	16%	13%	0%	6%	7.4%
Pseudodiploid	9%	11%	7%	10%	7.5%	21%	20%	22.2%
Diploid	59%	79%	51%	58%	69.6%	26.3%	50%	55.6%
Hyperdiploid (46–50) chromosomes	18%		18%	14.5%	7.5%	42.1%	24%	11.1%
Hyperdiploid (51–55) chromosomes			3%	1.5%	2.5%	15.8%		
Total abnormal percent	41	20	49	42	30.4	73.7	50	44.4

emphasized by P. Philip (1977a): 1p22, 1q2(ql—3), 2 cen, 7q22, 8q22, and 21q22. It should be emphasized, though, that in most patients with AML either chromosomal changes are not present (about 50% of all AML cases are normal diploid) or are of a heterogenous variety defying classification at present (Table 56). The findings of J. H. Ford, Pittman, et al. (1974), which have not been confirmed by others, are somewhat at variance with the general experience. These authors suggested that patients who are in remission and whose cells have 46 chromosomes do not have a normal karyotype but are −8,+9. Rowley (1977a-d) has suggested that, whereas it is very easy to differentiate between #8 and #9 with Q-banding, these two chromosomes may be confused when they are studied with G-banding.

The most typical karyotypic anomaly in AML appears to be a probable reciprocal translocation between #8 and #21, i.e., t(8;21) (q22;q22),[25] which we have called the "prototypic karyotype" in AML (Figure 95). This anomaly is probably present in about 10–15% of the patients with AML and not infrequently is accompanied by loss of a sex-chromosome, particularly in males (Rowley, 1974b; Sakurai and Sandberg, 1974). That the loss of the sex-chromosomes is probably consequential to the development of the prototypic karyotype is pointed to by the following: (1) in some patients with the prototypic karyotype the missing sex-chromosome appears subsequent to the development of the #8–#21 translocation; (2) a mixture of cells with the prototypic karyotype but with and without missing sex-chromosomes has been observed in some patients; and (3) with extremely rare exceptions, no cases with a missing sex-chromosome in the leukemic cells have been observed who did not have other karyotypic abnormalities.

The existence of the #8–#21 translocation was revealed with banding techniques.[26] In the past, studies with standard Giemsa staining had indicated the not uncommon (15–20% of AML cases) occurrence of karyotypes in AML represented by 46,−C,+D,+E,−G.[27] That this was a spurious picture was due to the fact that deletion of #8 leads to a chromosome morphologically similar to autosomes in group E; and the #21 with the translocated portion from #8 resembles a group D chromosome, thus accounting for the placing of an extra chromosome in that group. Banding techniques revealed a pseudodiploid picture as the result of a relatively simple translocation between #8 and #21 (Table 57). It should be pointed out, however, that Makino, Obara, et al. (1969), in describing a 16-yr-old girl with AML and a C-G translocation, recognized the nature of the

Table 55—continued

Kamada, Oguma, et al. (1976)	Krogh Jensen (1971)	Whang-Peng, Henderson, et al. (1970)	Pierre (1974)	Mittelman and Brandt (1974)	Lawler, Millard, et al. (1970)	Barlogie, Hittelman, et al. (1977)	Rowley (1977a)*	Philip, Krogh Jensen, et al. (1978)
84	50	103	114	30	38	68	313	88
11.9%	18%		19.3%	10%	7%	16.2%	10.9%	11%
34.5%	10%		3.5%	10%	8%	16.1%	6.2%	8%
40%	56%	70.9%	43.9%	43%	55%	53%		51%
9.5%	16%		33.3%	37%	29%	11.8%	65%	16%
					5.9%		17.9%	2%
60	44	29.1	56.1	57	45	47	35	42

*Summarized data available in Cervenka and Koulischer (1973).

Table 56 Chromosome abnormalities observed in AML without banding

Karyotype	Reference
46,XY,+12,−21	Ruffie, Biermé, et al., 1964
46,XY,Ep+orCq−	Kamada and Uchino, 1976
43,−B,−C,−E with ↓ fluorescence of Y	Sofuni and Okada, 1975
50,XY,+2C,+2D,+E,−G	Nicolau, Nicoară, et al., 1966
47,XY,+D	Pedersen, 1964b
43,XX,−C,−E(17),−G→51,XX,+C,+2D,+E,+G	Emberger, Taib, et al., 1975
46,−C,+mar	Zuelzer, Inoue, et al., 1976
46,−1,−2,+2mar/47,−1,5q−,+F,+mar	Zuelzer, Inoue, et al., 1976
52,+6,+9,+11,+18,+21,+21	Zuelzer, Inoue, et al., 1976
66,+9C,+D,+3E,+3F,+4G	Zuelzer, Inoue, et al., 1976
51,−2B,+C,+D,+E,+F,−G,+3mar	Zuelzer, Inoue, et al., 1976
53,XYY,−2,+22,+6mar	Zuelzer, Inoue, et al., 1976
47,−1,4q−,+19,+mar	Zuelzer, Inoue, et al., 1976
47,XX,+14	Zuelzer, Inoue, et al., 1976
50,XX,+F,+3G	Meisner, Inhorn, et al., 1973
47,XX,+C	Meisner, Inhorn, et al., 1973
46,XX,t(8;21)	Meisner, Inhorn, et al., 1973
45,X,?−Y,+D,−G probably t(8;21) & −Y	Littlefield, 1972
45,XX,−7 (2 cases)	Larsen and Schimke, 1976
−21	Catti, Cork, et al., 1973
+21	M. H. Khan and Martin, 1967c
47,XXqi,+C	Castoldi, Grusovin, et al., 1971
47,XY,+C(?9)	M. H. Khan, 1972a
+C,−21	Ruffie, Biermé, et al., 1964
46,−C,+D,+16,−G/45,−2C,+D,+16,−G	Sakurai and Sandberg, 1976a
46,−C,+D,+16,−G	Sakurai and Sandberg, 1976a
45,−C,+D,+16,−G,−Y (2 cases)	Sakurai and Sandberg, 1976a
45,−C,+D,+16,−G,−Y/45,2p−,Bq+,−C,+D,+16,−G,−Y/ 44,2p−Bq+,−C,+D,+16,−F,−G,−Y	Sakurai and Sandberg, 1976a
45,−2C,+D,+16,−G	Sakurai and Sandberg, 1976a
46,+3,−C,+D,−G/46,+3,Bp−q+,−C,+D,−G/ 45,+3,−C,−G	Sakurai and Sandberg, 1976a
47,+C (9 cases)	Sakurai and Sandberg, 1976a
46,+C,−D,?17p−,F?p−,Gq−(?Ph¹)	Sakurai and Sandberg, 1976a
45,−C (3 cases)	Sakurai and Sandberg, 1976a
46,Bq+,Cq− (2 cases)	Sakurai and Sandberg, 1976a
47,Bq+,Cq−,+r(C)/46,Bq+,Cq−	Sakurai and Sandberg, 1976a
48,+2C	Sakurai and Sandberg, 1976a
46,16q+,Fq+/46,−C,16q+,+?G	Sakurai and Sandberg, 1976a
46,+3,−C	Sakurai and Sandberg, 1976a
47,Cq+,+G	Sakurai and Sandberg, 1976a
47,+?G	Sakurai and Sandberg, 1976a
45,−G (2 cases)	Sakurai and Sandberg, 1976a
46,17/18q+/46,+1,−3,Bq−,17/18q+	Sakurai and Sandberg, 1976a
47,+C,F?p−	Sakurai and Sandberg, 1976a
49,+D,+F,+mar	Sakurai and Sandberg, 1976a
46,−C,−F,+2mar	Sakurai and Sandberg, 1976a
47,−F,+2mar	Sakurai and Sandberg, 1976a
45,+1,−3,−B,+C,+Dq−,+16,−2F,+F?q−,−G,−Y/ 45,+1,−3,−B,+C,Dq−,+16,−2F,+F?q−,−G	Sakurai and Sandberg, 1976a
45,−2B,Bq+,+2C,+16,−2G,−Y,+mar	Sakurai and Sandberg, 1976a
44,+3,−4C,Dq+,+16,−17/18,−2G,+3mar	Sakurai and Sandberg, 1976a
49,+3,−D,+F,+2G	Sakurai and Sandberg, 1976a
46,+2,−2B,+16	Sakurai and Sandberg, 1976a
66,+2,+2,+2,+3,+2B,+5C,+D,+16,+3E,+F,+G.+2Gq− (?2Ph¹)/48,+2,+16,Gq−(?Ph¹)	Sakurai and Sandberg, 1976a

Table 56—*continued*

Karyotype	Reference
45,—C	M.A.S. Moore and Metcalf, 1973
45,—C	M.A.S. Moore and Metcalf, 1973
46,XY/47,XY,+C/48–51 chromosomes	M.A.S. Moore and Metcalf, 1973
48,+C,+G	M.A.S. Moore and Metcalf, 1973
46,XY,t(8;21)/46,XY,—C,t(8;21),+ring	Obara, Makino, et al., 1969; Obara, Sasaki, et al., 1971b
46,XX,t(8;21)	Obara, Makino, et al., 1969; Obara, Sasaki, et al., 1971b
52,XX,+4C,+2G	Obara, Makino, et al., 1969; Obara, Sasaki, et al., 1971b
46,XX,—C,+ring	Obara, Makino, et al., 1969; Obara, Sasaki, et al., 1971b
47,XXorXY,+C (12 cases)	Trujillo, Cork, et al., 1974
47,XX,+C/48,XX,+2C	Trujillo, Cork, et al., 1974
47,XY,+C/49,XY,+2C,+D	Trujillo, Cork, et al., 1974
47,XX,+C/47,XX,+G	Trujillo, Cork, et al., 1974
45,XXorXY—C (4 cases)	Trujillo, Cork, et al., 1974
45,XX,—C/45,XX,—C,—D,+G	Trujillo, Cork, et al., 1974
45,XY,—D,+E,—2G (4 cases)	Trujillo, Cork, et al., 1974
46,XX,—D,+E,—G (3 cases)	Trujillo, Cork, et al., 1974
46,XX,—C,+D,+E,—G/46,XX,—C,+D	Trujillo, Cork, et al., 1974
46,XY,—C,+D,+E,—G/46,XY,—2C,+D,+E,—G,+mar	Trujillo, Cork, et al., 1974
46,XY,—C,+D,+E,—G/47,XY,—C,+2D,+E,—G	Trujillo, Cork, et al., 1974
47,XY,—C,+2D,+E,—G	Trujillo, Cork, et al., 1974
46,XY,+C,+E,—2G	Trujillo, Cork, et al., 1974
46,XY,Bq+	Trujillo, Cork, et al., 1974
46,XX,18q— (2 cases)	Trujillo, Cork, et al., 1974
46,XY,17q—	Trujillo, Cork, et al., 1974
46,XX,11p—q—,18q—	Trujillo, Cork, et al., 1974
46,XY,Cq—	Trujillo, Cork, et al., 1974
46,XX,Gq—	Trujillo, Cork, et al., 1974
47,XY,+C,18q—	Trujillo, Cork, et al., 1974
47,XX,+C,8p—,17q—	Trujillo, Cork, et al., 1974
46,XY,+C,Dq+,+E,—G	Trujillo, Cork, et al., 1974
46,XY,+C,—G	Trujillo, Cork, et al., 1974
45,XY,+C,—B,—D	Trujillo, Cork, et al., 1974
45,XX,—E	Trujillo, Cork, et al., 1974
46,XY,—G,+mar	Trujillo, Cork, et al., 1974
47,XX,+B,—C,Dq+,+E	Trujillo, Cork, et al., 1974
45,X,—Y,+D,—G/44,X,—Y,—B,+D,—G	Trujillo, Cork, et al., 1974
44,XY,—B,+E,—2G	Trujillo, Cork, et al., 1974
47,XY,+G (2 cases)	Trujillo, Cork, et al., 1974
48,XY,+E,+G	Trujillo, Cork, et al., 1974
92,XXYY	Trujillo, Cork, et al., 1974
46,XX,invl p—q+	Trujillo, Cork, et al., 1974
48,XY,+2mar	Trujillo, Cork, et al., 1974
43,XY,—A,—B,+C,—3E,—G,+min	Trujillo, Cork, et al., 1974
49,XY,2q+,+C,+D,+F/50,XY,2q+,+C,+D,+2F/51,XY,2q+,+3C,+D,+F	Trujillo, Cork, et al., 1974
55,XY,+2A,+D,Dq+,+4E,+G	Trujillo, Cork, et al., 1974
46,XY,Gq—	Barlogie, Hittelman, et al., 1977
46,XY,Fq—	Barlogie, Hittelman, et al., 1977
46,XX,Fq+	Barlogie, Hittelman, et al., 1977
46,XX,t(8;21)	Barlogie, Hittelman, et al., 1977
46,XY,t(12;17)	Barlogie, Hittelman, et al., 1977
46,X,—Y,+8	Barlogie, Hittelman, et al., 1977

continued

Table 56 Chromosome abnormalities observed in AML without banding—*continued*

Karyotype	Reference
46,Cq−/47,XX,+A,Cq−	Barlogie, Hittelman, et al., 1977
46,XY,t(1;3),+6,−13,t(16;?)	Barlogie, Hittelman, et al., 1977
46,XY,i(17q)	Barlogie, Hittelman, et al., 1977
45,XY,−C (2 cases)	Barlogie, Hittelman, et al., 1977
45,X,−Y (3 cases)	Barlogie, Hittelman, et al., 1977
45,XX,−E	Barlogie, Hittelman, et al., 1977
45,XY,−7	Barlogie, Hittelman, et al., 1977
45,X,−X,t(8;11),12q−	Barlogie, Hittelman, et al., 1977
47,XX,+C	Barlogie, Hittelman, et al., 1977
47,XX,t(8;21),+frag	Barlogie, Hittelman, et al., 1977
47,XX,+8/48,XX,+E,+E	Barlogie, Hittelman, et al., 1977
47,XY,+D/94,XXYY,+2D	Barlogie, Hittelman, et al., 1977
53,XX,+A,+4C,+F,+G	Barlogie, Hittelman, et al., 1977
53,XY,+A,+B,+2C,+D,+E,+G	Barlogie, Hittelman, et al., 1977
92,XXYY	Barlogie, Hittelman, et al., 1977

This table does not include those cases associated with DS, lymphosarcoma, or multiple myeloma. These are shown in other tables in the book.

abnormality by depicting the #8−#21 translocation in two of the karyotypes in their publication. The possibly favorable relation of this prototypic karyotype to the prognosis of AML will be discussed below.

Because complete remissions have been less readily obtained in AML than in ALL, it was rare in the past to observe complete disappearance of aneuploid cells (when these are present) from the marrow. Nevertheless, we have recently observed more and more cases with AML in which the BM picture, cytologic and cytogenetic, became normal after newer combinations of chemotherapy, the marrows containing a normal number of myeloblasts, promyelocytes, and myelocytes (granulocytic cells supplying the metaphases in the marrow) and normoblasts. These findings indicate that the myeloblastic cells in AML may be of a different origin than the normal diploid cells, including the normal myeloblasts, during remission.

As indicated, in most ALs with an aneuploid chromosome constitution in the marrow, some diploid cells are also present. These numbers tend to be rather small in ALL, particularly in those cases in which the marrow consists almost exclusively of lymphoblasts. With eradication of the leukemic cells as a result of therapy, which is more readily achieved in ALL than in AML, the marrow may have a totally diploid picture. Inasmuch as lymphoblasts are neoplastic cells "foreign" to the marrow, the results with therapy are readily understandable. For some years it was not clear whether in aneuploid AML a diploid picture would be present in the marrow, were it to return to normal (including a normal number of myeloblasts). This problem has been essentially clarified recently, and it appears that the normal myeloblasts, present in the marrow in full remission of AML, are diploid and that the leukemic, aneuploid myeloblasts seen originally or in relapse appear to have different precursor cells than the normal, diploid ones. This is a basic aspect in the understanding of the leukemogenic process of AML.

We have encountered a karyotypic picture in AML, i.e., trisomy of part or of the whole long arm of #1 in four patients with leukemia (Oshimura, Sonta, et al., 1976): three with AML and one with CML in the BP (Figure 96). Studies with banding techniques revealed in two of the former patients and in the latter one that this karyotypic abnormality was not an initial one and was apparently associated with clinical progression of the diseases. This chromosomal change seems to be a rather common karyotypic abnormality in blood disorders, especially in AML, and possibly bears a relationship to selective growth advantages of the leukemic cells. Similar findings have been reported in other patients with myeloproliferative disorders in which marrow cells were shown to be trisomic either for the entire #1 or for its long arm. In two of these patients the

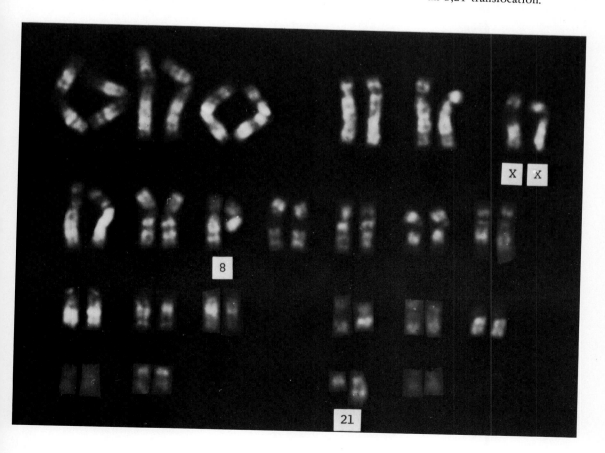

95 A. (upper) Karyotype (not banded) of a leukemic cell not infrequently observed in AML and containing a number of abnormalities which were clarified by banding. Thus, the 8;21 translocation of the "prototypic karyotype" leads to a 8q− (shown as an E16) and an acrocentric (D?) larger than #21. Note missing Y. B. (lower) Karyotype (Q-banded) from a female patient with AML representing the "prototypic karyotype" with an 8;21 translocation.

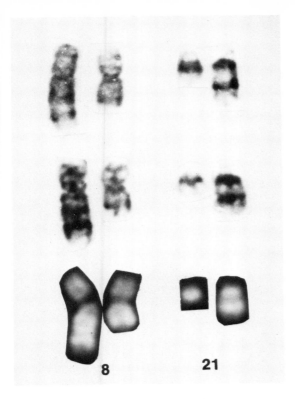

95 C. (left) Partial Q- and C-banded karyotypes of AML cells showing the prototypic 8;21 translocation. D. (below) Partial karyotypes from two patients with AML obtained with G- (upper) and Q- (lower) banding. In each case it can be seen that a reciprocal translocation between #8 and #21 had occurred. Thus, #8 had lost part of its long arm (8q−), and the material had been translocated to the lower arm of #21 (21q+). In the case of the male patient (lower partial karyotype) the Y-chromosome had been lost, a frequent occurrence in AML with this prototypic karyotype [t(8;21)].

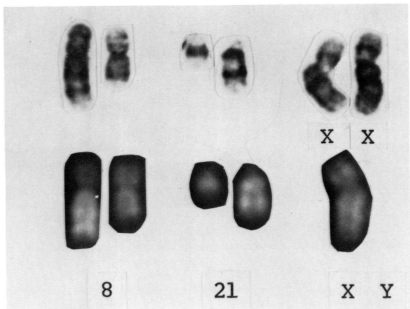

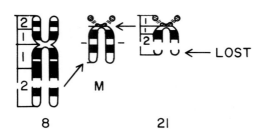

95 E. Schematic presentation of the prototypic translocation in AML consisting of a translocation between chromosomes #8 and #21, i.e., [t(8;21)(q22;q22)]. Whether or not the translocation is reciprocal is uncertain, because the very small band of the deleted part of long arm of #21 has not been identified on the #8 or any other chromosome. Hence, I have indicated in the figure that it may possibly have been lost.

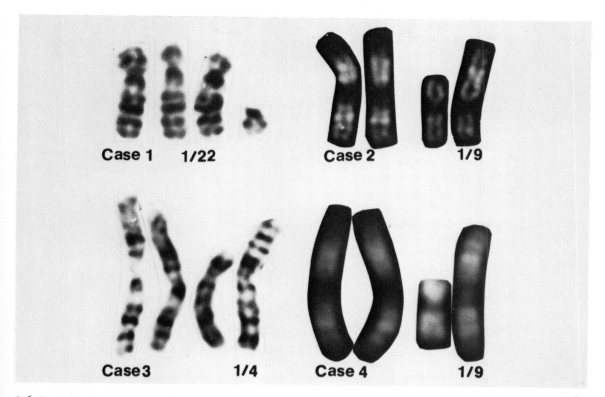

96 Partial G-banded (cases #1 and #3) and Q-banded (cases #2 and #4) karyotypes of four cases with trisomy of the long arm of #1: A. Case 1 (AML) (G-banding). The two chromosomes in the middle consist of the translocated long arm of chromosome #1 to the short arm of #22 (1;22). On the left is shown a #1 with its short arm deleted (1p−) (translocated to a #8) and on the right a normal #22. B. Case 2 (AML) (Q-banding). On the right is an abnormal chromosome due to translocation of most of the long arm of #1 to the short arm of #9 (1;9). The other chromosomes shown are a pair of normal chromosomes #1 and a normal #9. C. Case 3 (AML) (G-banding). On the right is an abnormal chromosome due to translocation of most of the long arm of #1 to the short arm of #4 (1;4). The other chromosomes shown are a pair of normal #1 chromosomes and a normal #4. D. Case 4 (BP of CML) (Q-banding). On the right is an abnormal chromosome due to translocation of the long arm of a #1 to the proximal end of the abnormal short arm of a #9 (1;9). The long arm of this #9 had the standard type of Ph¹-translocation at its proximal end. The other chromosomes shown consist of a pair of normal #1 chromosomes and a normal #9.

chromosomal abnormality served as an important factor in establishing the diagnosis of AL developing in one patient with thrombocytopenia and anemia and in another with PV. An interesting case of trisomy of #1 involved a 76-yr-old male with refractory dysplastic anemia thought to represent erythremic myelosis. On the first marrow analysis, the cells contained 46 chromosomes including a small G-chromosome (?Ph¹) which was later shown to be a deleted Y. A second analysis of marrow cells performed 4 mo later showed an aneuploid karyotype with most of the cells having an extra #1 (Warburton and Bluming, 1973). The patient was given transfusions and received no chemotherapy for his disease but died shortly after the second marrow aspiration of a pulmonary embolism. A giant #1, possibly due to a translocation between 1;5 (1q+;5q−) in a 67-yr-old woman with a blast cell leukemia has

Table 57 Published cases of AML with probable and proved (with banding) 8;21 translocations

Karyotype published	Karyotype reinterpreted	Reference
	PROBABLE	
45 Chromosomes; −C,−G,+D t(Cq−;Gq+) (4 cases)	45,XX,t(1q−,Gq+),−C t(Cq−;Gq+)	Krogh Jensen, 1967b Kamada, Okada, et al., 1968
46 Chromosomes; very short #11,−G,+D	46,XX,t(Cq−;Gq+)	Khouri, Shahid, et al., 1968
+D,+E16,−C,+G	46,XX,t(Cq−;Gq+)	Sandberg, Takagi, et al., 1968c Makino Obara, et al., 1969
C/G translocation shown as 8;21	46,XX,t(8;21)	Ezdinli, Sokal, et al., 1969
−G,+D,+E,−C	46,XX,t(Cq−;Gq+)	Obara, Sasaki, et al., 1971a
C/G translocation; 45,−C/46,C/G translocation,−C,+ring	46,XX,−C,t(Cq−;Gq+),+ring	Obara, Sasaki, et al., 1971a
46,XY,C/G translocation	45,XY,t(Cq−;Gq+)	Littlefield, 1972
Two chromosomes missing in the G+Y group, one extra in group D (1 case)	45,X,−Y,t(Cq−;Gq+)	
46,+D,+16,−C,−E	46,t(Cq−;Gq+)	Gunz, Bach, et al., 1973
46,XX,8q−,21q+	46,XX,t(8;21)	Meisner, Inhorn, et al., 1973
45,X,−X,t(8q;21q+)	45,X,−X,t(8q−;21q+)	Rowley, 1973b
46,−C,+D,+E,−G (2 cases)	46,XY,t(Cq−;Gq+)	Sakurai, Oshimura, et al., 1974
45,X,−Y−C,+D,+16,−G (3 cases)	45,X,−Y,t(Cq−;Gq+)	Sakurai, Oshimura, et al., 1974
45,XX,−2C,+D,+16,−G	45,X,(−X),t(Cq−;Gq+),(−C)	Sakurai, Oshimura, et al., 1974
45,XY,−C,+D,+E,−2G (4 cases)	45,X,(−Y),t(Cq−;Gq+),(−C)	Trujillo, Ahearn, et al., 1974; Trujillo, Cork, et al., 1974
46,XY or XX,−C,+D,+E,−G (6 cases)	46,XX or XY,t(Cq−;Gq+)	Trujillo, Ahearn, et al., 1974; Trujillo, Cork, et al., 1974
46,XY or XX,t(Cq−;Gq+) (9 cases)	46,XY or XX,t(Cq−;Gq+)	Kamada, Okada, et al., 1976
45,X,−Y,t(Cq−;Gq+) (3 cases)	45,X,−Y,t(Cq−;Gq+)	Kamada, Okada, et al., 1976
t(Cq−;Gq+) C group aberration (3 cases)	t(Cq−;Gq+)	Kamada, Okada, et al., 1976
	BANDED	
46,XX,t(8;21)(q22;q22)		Sasaki, Okada, et al., 1976
46,XY,t(8;21)(q22;q22)		Sasaki, Okada, et al., 1976
45,X,−X,t(8;21)(q22;q22)		Sasaki, Okada, et al., 1976
45,X,−Y,t(8;21)(q22;q22)		Sasaki, Okada, et al., 1976
46,XY/45,X,−Y,t(8;21)(q22;q22)		Sasaki, Okada, et al., 1976
45,X,−Y,t(8;21)(q22;q22) (1 case)		Yamada and Furasawa, 1976
47,XX,+8, t(8;11;21)(q22;p15;q22)		Lindgren and Rowley, 1977
45,X,−X,t(8;17;21)(q22;q23?;q22) Corrected kryotype of one of the cases of Rowley (1977) thought to be only (8;21)		Lindgren and Rowley, 1977
45,X,−X,t(8;21)(q22;q22) (2 cases)		Rowley, 1977a
46,XX,9q−,t(8q−;21q+)		Rożynkowa, Stepień, et al., 1977
45,X,−Y,t(8;21)(q22;q22)		Kaneko, Sakurai, et al., 1978a
t(8;21) (3 cases)		Hagemeijer, personal communication
45,X,−X,t(8;21)/46,XX		Mitelman, personal communication
46,XX,t(8;21)(8pter−8q22::21q22−21qter;21pter−21q22::8q22−8qter)		P. Philip, 1977a, and personal communication
46,XX,t(8;21)		Philip, 1977a, and personal communication
45,X,−Y,(8;21)		Philip, 1977a, and personal communication

Table 57—*continued*

Karyotype published	Karyotype reinterpreted	Reference
	BANDED	
t(8;21)		Barlogie, Hittelman, et al., 1977
45,X,−Y,t(8;21)(q22;q22)t(9;22) (q34;q11)		Francesconi and Pasquali, 1978b
t(8;21)(2 cases)		T. Abe and Misawa, 1978
45,X,−Y,t(8;21)		T. Abe and Misawa, 1978
45,X,−Y,t(8;21)		Golomb, Vardiman, et al., 1978
45,X,−X,t(8;21)(q22;q22) (In donor cells)		Elfenbein, Borgaonkar, et al., 1978
45,X,−X,t(8;21)(q22;q22)/46,XX, t(8;21)(q22;q22)/46,XX		Kaneko, Sakurai, et al., 1978b
45,X,−Y,t(8;21)(q22;q23)/46,X,−Y, t(8;21),+M(11)t(11;12)(p11;q13)/ 46,XY		Kaneko, Sakurai, et al., 1978b
46,XY,t(1;8;21)(q32;q22;q22)		Kondo, Sasaki, et al., 1978
45,X,−Y,del(7)(q32),t(8;21)(q22;q22) AMMoL		Hustinx, Burghouts, et al., 1979
45,X,−X,t(8;21)(q22;q22)		G. Levan and Mitelman, 1979

Trujillo, Ahearn, et al. (1979) have described a group of 32 patients (out of a total of 546 cases with AL, 42.9% of whom were aneuploid) with a t(8;21) in at least 15 of which the translocation was established with banding. All of these cases had Auer rods in the leukemic cells. The latter were all peroxidase-positive. Of the 32 cases, 8 of 19 males also had a missing Y and 2 of the 13 females a missing X. In 8 of the 32 patients other chromosomal changes were also present.

been described (Soderström, 1974). An elongated #1 (1q+?) was observed in the presumed leukemic cells of a 9-yr-old girl with microcephalic dwarfism and a ring #1 (Bobrow, Emerson, et al., 1973). Other karyotypic changes in AML include, in addition to the common ones already mentioned, 18q−, an abnormal #6, 7q− (q22) (in two cases), +E, del(20) (q12), 7p−, and the presence of markers of unknown origin (Table 58). Of special interest is the description of monosomy−7 as the sole abnormality in two adult patients with AML (Zech, Lindsten, et al., 1975) and an iso-17 [i(17q)] in one (Mitelman, Brandt, et al., 1973); similar findings have been reported in EL (see Table 61). A family has been described in which five members died of AML and two siblings had monosomy C (#7?) (Larsen and Schimke, 1976).

Chromosome Changes and Prognosis of AML

Of considerable interest has been the demonstration in my laboratory that chromosomal findings may have an important relationship to the prognosis of AML.[28] It was shown that immediate survival, either from time of diagnosis or beginning of treatment, of patients with *only* normal karyotypes in their marrow (N patients, about 50% of the cases) throughout their disease was somewhat longer than, but not significantly different from, that of patients having abnormal karyotypes in their marrow (A patients) in at least one sample of marrow. When the latter patients (A patients) were further subdivided into those with some normal karyotypes in their marrow (AN patients, about 35% of AML cases), during the course of the AML and those with none (AA patients, about 15% of AML cases), the immediate survival of the latter group was significantly shorter than that for the AN or N group (Figure 97). It is possible, of course, for N patients to have a totally leukemic marrow cell population regardless of the diploid karyotype; such cases would tend to reduce the median survival of the N group, because they would behave clinically in a manner similar to that of the AA patients (Sandberg, 1978b). The implications of these chromosomal findings in the therapy of AML should be realized by all who treat AL.

Table 58a Chromosome abnormalities in AML as shown by banding

Karyotype	Reference
47,XY,+8 (3 cases)	Yamada and Furasawa, 1976
45,X,−Y,t(8;21)(q22;q22)	Yamada and Furasawa, 1976
45,XY,−21,18p+	Yamada and Furasawa, 1976
45,XY,−21,t(4;9)(p16;q22)	Yamada and Furasawa, 1976
46,XY,t(1;6;11)(q12;q23;p15)	Yamada and Furasawa, 1976
45,XY,−7	Yamada and Furasawa, 1976
45,XX,−2,5q+	Yamada and Furasawa, 1976
46,XY,del(11)(q23),del(12)(p11)	Yamada and Furasawa, 1976
47,XX,+8	Sasaki, Okada, et al., 1976
48,XX,+8,+?22	Sasaki, Okada, et al., 1976
47,XY,+12	Sasaki, Okada, et al., 1976
46,XY,del(3)(pter→q21)	Sasaki, Okada, et al., 1976
46,XY,del(9)(pter→q22)	Sasaki, Okada, et al., 1976
46,XX,t(8;21)(q22;q22) (2 cases)	Sasaki, Okada, et al., 1976
45,X,−X,t(8;21)(q22;q22)	Sasaki, Okada, et al., 1976
45,X,−Y,t(8;21)(q22;q22)	Sasaki, Okada, et al., 1976
46,XY/45,X,−Y,t(8;21)(q22;q22)	Sasaki, Okada, et al., 1976
46,XY,t(9;Y)(p11;p11)	Sasaki, Okada, et al., 1976
46,XY,−C,+D	Fleischmann and Bodor, 1975
46,XX,abnormal #16	Bodor and Fleischmann, 1975
45,XX,−7/46,XX (3 cases)	Mitelman, Nilsson, et al., 1976, Mitelman, Brandt, et al., 1978
46,XY,+8,−17/47,XY,+8/46,XY (AMMoL)	Mitelman, Nilsson, et al., 1976, Mitelman, Brandt, et al., 1978
47,XY,+8/46,XY	Mitelman, Nilsson, et al., 1976, Mitelman, Brandt, et al., 1978
47,XY,+9/46,XY	Mitelman, Nilsson, et al., 1976, Mitelman, Brandt, et al., 1978
47,XY,+21/46,XY (AMMoL)	Mitelman, Nilsson, et al., 1976, Mitelman, Brandt, et al., 1978
47,XY,+8/48,XY,+8+8/46,XY	Mitelman, Nilsson, et al., 1976, Mitelman, Brandt, et al., 1978
45,XY,−7/44,X,−Y,−7	Mitelman, Nilsson, et al., 1976, Mitelman, Brandt, et al., 1978
47,XY,+8 (3 cases)	Mitelman, Nilsson, et al., 1976, Mitelman, Brandt, et al., 1978
47,XY,−7,r(7),+19	Mitelman, Nilsson, et al., 1976, Mitelman, Brandt, et al., 1978
47,XY,+8/47,XY,+8,−11,+21/47,XY,−7,+8,+21	Mitelman, Nilsson, et al., 1976, Mitelman, Brandt, et al., 1978
47,XY,+14,−17,+18/48,XY,+8,+14,−17,+18	Mitelman, Nilsson, et al., 1976, Mitelman, Brandt, et al., 1978
47,XY,+8/48,XY,+8,+9	Mitelman, Nilsson, et al., 1976, Mitelman, Brandt, et al., 1978
46,XY,del(7)(q22),del(11)(q14),del(12)(p12)	Mitelman, Nilsson, et al., 1976 Mitelman, Brandt, et al., 1978
46,XY,+2,−7/46,XY,−7,+21	Mitelman, Nilsson, et al., 1976, Mitelman, Brandt, et al., 1978
44,X,−X,−21	Mitelman, Nilsson, et al., 1976, Mitelman, Brandt, et al., 1978
46,X,del(Xq),del(5q),t(5;21)	Mitelman, Nilsson, et al., 1976, Mitelman, Brandt, et al., 1978
45,XO/46,XY (AMMoL)	Mitelman, Nilsson, et al., 1976, Mitelman, Brandt, et al., 1978

Table 58a—*continued*

Karyotype	Reference
45,X,−X,t(8;21)/46,XX	Mitelman, Nilsson, et al., 1976, Mitelman, Brandt, et al., 1978
46,XX,17q−/46,XX	Mitelman, Nilsson, et al., 1976, Mitelman, Brandt, et al., 1978
47,XX,+mar(21q−?)/46,XX	Mitelman, Nilsson, et al., 1976, Mitelman, Brandt, et al., 1978
47,XX,+21/46,XX	Mitelman, Nilsson, et al., 1976, Mitelman, Brandt, et al., 1978
46,XY,22q−(Ph¹)	Mitelman, Nillson, et al., 1976, Mitelman, Brandt, et al., 1978
48,XX,+3,+8,i(17q)	Mitelman, Nilsson, et al., 1976, Mitelman, Brandt, et al., 1978
45,XY,3p−,−12,−13,−18,t(3;12),t(13;18)	Mitelman, Nilsson, et al., 1976, Mitelman, Brandt, et al., 1978
46,XY/48,XY,+9,+21	Mitelman and Brandt, 1974
46,X,−X,+8	Mitelman and Brandt, 1974
46,XX,i(17),−17	Mitelman and Brandt, 1974
47,XX,+8	Mitelman and Brandt, 1974
44,XX,−3,−21	Mitelman and Brandt, 1974
45,XY,−5,−12,t(5;12),−18,i(18q),−19	Mitelman and Brandt, 1974
46,XX,t(16p;16q)	Mitelman and Brandt, 1974
46,XY,21q+	Mitelman and Brandt, 1974
48,XX,+1,del(5)(q13−15),+11	Rowley and Potter, 1976
47,XX,+8	Rowley, 1977a,b
47,XX,+C,t(C;C)	Rowley, 1977a,b
46,XX,t(6;9)(p23;q34) (2 cases)	Rowley, 1977a,b
46,XY,−7,+mar	Rowley, 1977a,b
45,XY,−7	Rowley, 1977a,b
45,X,−X,t(8;21)(q22;q22) (2 cases)	Rowley, 1977a,b
43,XY,−3,−5,−21	Rowley, 1977a,b
43,XX,−4,−5,−6,−7,−13,−17,1p+,del(18)(q21),+3mar	Rowley, 1977a,b
47,XX,+8→47,XX,+8,t(2;18)(p13;q23),t(9;15)(p11;q11)	Jonasson, Gahrton, et al., 1974
46,XY,i(17q)	E. Engel, McKee, et al., 1975
46,+8 (6 cases)	P. Philip, Wantzin, et al., 1977; P. Philip, 1975a
45,XX,−7	Zech, Lindsten, et al., 1975
46,XX/45,XX,−7	Zech, Lindsten, et al., 1975
45,XX,−7	U. Kaufmann, Löffler, et al., 1974
47,XX,+9	Lampert, Phebus, et al., 1972
46,XX,−C,−17,+2mar	Littlefield and Vodopick, 1975
(16 yr. ♀, lived 7.5 yrs.)	Littlefield and Vodopick, 1975
−7 (following Rx of ALL)	Secker Walker and Sandler, 1978
Includes cases with:	
+8	Lawler, Secker Walker, et al., 1975
18q−	Lawler, Secker Walker, et al., 1975
mar(6)	Lawler, Secker Walker, et al., 1975
2q+,+mar	Lawler, Secker Walker, et al., 1975
7q−(q22)	Lawler, Secker Walker, et al., 1975
Abnormal C-group chromosome (2 cases)	Milligan and Garson, 1974
46,XX,t(8;21)(q22;q22)	Oshimura, Hayata, et al., 1976
45,X,−Y,t(8;21)(q22;q22)	Oshimura, Hayata, et al., 1976
44,XX,−4,−5,−21,t(8;21)(q22;q22),+frag/	Oshimura, Hayata, et al., 1976
43,XX,−4,−5,−21,t(8;21)(q22;q22)	Oshimura, Hayata, et al., 1976
45,XX,−21 (2 cases)	Oshimura, Hayata, et al., 1976

continued

Table 58a Chromosome abnormalities in AML as shown by banding—*continued*

Karyotype	Reference
51,XY,+8,+9,+13,+14,+21	Oshimura, Hayata, et al., 1976
47,XX,del(5)(q15→23),+8	Oshimura, Hayata, et al., 1976
47,XY,+22	Oshimura, Hayata, et al., 1976
44,XX,−9,−10,−17,del(14)(q22),+mar(9)t(9;10)(q34;q22)	Oshimura, Hayata, et al., 1976
46,XY,−16,del(10)(p11 or 12),+ring	Oshimura, Hayata, et al., 1976
47,XX,−4,+13,+mar(4)t(1;4)(q22→25;p14→16)	Oshimura, Hayata, et al., 1976
43,X,−Y,t(1;8),−1,−11,−11,−13,−16,−17,−22,+2mar(22), t(1;22),+mar(11)t(11;13),+mar(17p+)+mar	Oshimura, Hayata, et al., 1976
47,XX,t(3;5)(q21;q31),−1,+8,−13,+22,+mar(1)(1;13)(p36;q14)	Oshimura, Hayata, et al., 1976
47,XX,+1q+,+13,+20,−16,−16	Prieto, Badia, et al., 1978*a*
47,XY,+C(?8)	T. Abe and Misawa, 1978
46,XY,−6,der(6),t(1;6)(q23;p21)	Gahrton, Friberg, et al., 1978*a,b*
47,XY,+13	Golomb, Vardiman, et al., 1978
47,XX,+8,t(8;11;21)	Golomb, Vardiman, et al., 1978
47,X,−X,+2−4mar	Golomb, Vardiman, et al., 1978
46,XX,6p−	Golomb, Vardiman, et al., 1978
46,XX,20q−/48,XX,+9,20q−,+21	Golomb, Vardiman, et al., 1978
45,XX,ins(3;3),−7	Golomb, Vardiman, et al., 1978
46,XY,complex;44,XX,complex;43,XX,complex (3 cases)	Golomb, Vardiman, et al., 1978
46,X,−X?,complex;44,XY,complex/46,XY,complex (2 cases)	Golomb, Vardiman, et al., 1978
42,XY,complex;55,XX,complex (2 cases)	Golomb, Vardiman, et al., 1978
58,XX,complex; 65,XXY,complex (2 cases)	Golomb, Vardiman, et al., 1978
46,XX,12p− (following radiation for cancer of colon)	Wiggans, Jacobson, et al., 1978
47,XX,+del(9)(q13)/46,XX	Mitelman, Nilsson, et al., 1978
46,XX,t(8;16)(p11;p13)	Mitelman, Nilsson, et al., 1978
44,XX,+8,−12,−13,−14,t(2;5)(q12;q35)/46,XX	Mitelman, Nilsson, et al., 1978
48,XX,+3,+8/46,XX	Mitelman, Nilsson, et al., 1978
46,XX,del(5)(q14)	Mitelman, Nilsson, et al., 1978
47,X,−X,+del(1)(p11),del(5)(q13),t(5;12)(q21;q24),t(11;12)(q25;q13),t(13;14)(q11;p11),+16,+21	Mitelman, Nilsson, et al., 1978
48,XX,t(1;3)(q44;p14),del(3)(q21;p14),−5,+10,+11,+21	Mitelman, Nilsson, et al., 1978
45,XY,−7	Mitelman, Nilsson, et al., 1978
45,XY,+2,−5,−7/46,XY	Mitelman, Nilsson, et al., 1978
47,XY,del(7)(q22),+del(8)(q22) (AMoL)	Mitelman, Nilsson, et al., 1978

This table does not include those cases associated with DS, lymphosarcoma, or multiple myeloma. These are shown in other tables in the book.

Table 58b Abnormal karyotypes in bone marrow cells from 37 ANLL cases with clonal chromosomal aberrations

Case no.	Years of age	Treated or untreated	No. of mitoses Abnormal Clonal	No. of mitoses Abnormal Nonclonal	Normal	Abnormal stemlines
1	15	U	25	0	0	47,XY,+8
2	44	U	11	0	0	47,XX,+8
3	69	U	8	0	0	47,XX,+8
4	56	U	10	1	1	47,XX,+8
5	78	U	3	0	0	47,XX,+8
6	79	T	6	0	10	47,XY,+8

Table 58b—*continued*

Case no.	Years of age	Treated or untreated	No. of mitoses Abnormal Clonal	No. of mitoses Abnormal Nonclonal	No. of mitoses Normal	Abnormal stemlines
7	21	U	10	1	0	47,XX,+8
8	78	U	10	0	0	47,XX,+8
9	84	U	9	1	1	47,XX,+8
10	61	U	6	5	0	45,XY,−7
11	60	T	2	4	37	45,XY,−7
12	69	U	15	1	2	45,XX,−7
13	10	T	5 3	2	1	46,XX,9qh−,t(8;21) 46,XX,−9,t(8;21),+mar
14	58	U	12	0	0	46,XX,t(8;21)
15	29	U	20	2	0	45,X,−Y,t(8;21)
16	65	U	2 2	20	1	44,XX,−8,−11,−20,+t(8;11) 46,XX,7q−,−20,−21,−22,t(1;3),+3mar
17	8/12	U	2 17	1	19	46,XY,2q−,−2,−7,9p−,−21,+t(2;21),+t(2;?) 46,XY,−1,2q−,−2,−7,9p−,−17,−21,+t(1;?),+t(2;21),+t(2;?),+t(17;?)
18	82	U	2	1	17	??,XY,+6,+10,+11,+13,+?
19	62	U	4	0	1	46,XX,t(1;17)
20	61	U	7	0	0	45,XX,3p−,−5,−7,−12,+t(5;?),+t(5,7,12)
21	51	T	8 2	1	0	46,XY,inv(9),−15,+t(11;15) 46,XY,inv(9),−13,−15,+t(11;13),+t(11;15)
22	72	U	4 2 4 5	0	1	45,XX,−5,−6,+t(5;?) 45,XX,−5,−6,−18,+t(5;?) 44,XX,−5,−6,−19,+t(5;?) 44,XX,−3,−5,−6,−19,+t(1;2),+t(5;?)
23	77	U	13	1	2	47,XY,+15
24	71	U	7	0	0	48,XX,+8,−9,+t(X;9),+t(X;9)
25	69	U	6 3	2	0	48,XX,−5,15q−,+19,+t(5;?),+mar 49,XX,−5,15q−,+19,+t(5;?),+2mar
26	81	U	3 6	0	0	47,XY,+21 46,XY,−7,+21
27	11	T	2	2	29	47,XY,+6
28	83	U	6 5	2	0	46,X,−X,+t(X;X) 45,X,−X,−7,+t(X;X)
29	26	U	4	2	2	46,XX,−1,+t(1;1),+t(3;7)
30	41	U	3	1	0	47,XX,+mar
31	53	U	3	5	0	51,XY,+3,+5,−9,+19,+20,+21,t(6;14)+t(9;?)
32	79	U	11	0	1	45,XY,−18,−20,+t(18;20)
33	83	T	15	3	0	49,XY,−5,+6,+13,+14,+t(5;?)
34	37	U	8	0	0	45,X,−Y,−6,8q−,−11,+t(6;11),+t(8;11)
35	71	U	3 7	0	0	46,XY,20q− 46,XY,7q−,20q−
36	72	T	8	3	2	45,XY,−1,−2,5p−,−14,−15,−15,−17,+t(1;14),+t(1;15),+t(1,?),+t(2;?),+mar
37	77	U	1*		1*	45,X,−Y

*Case No. 37 was supplemented with 18 mitoses in conventional Giemsa staining; 12:45,X,−Y and 6:46, XY.
Based on a table of P. Philip, Krogh Jensen, et al., 1978.

Table 58c Median survival time and cytogenetic pattern in 241 patients with ANLL*

Cytogenetic pattern	No. of patients	Median survival—all, mo	Median survival—CR, mo	Alive after 1 yr, no./%
NN	102	6	13	21/(21)
AN	80	5	12	17/(21)
AA	59	4	10	3/(5)

*Results based on an analysis by the First International Workshop on Chromosomes in Leukemia (1978).
CR = Survival of patients with complete remission.

With successful antileukemic therapy, AN patients (as well as N patients) are capable of repopulating of their marrow with the normal elements present in their marrows, when the leukemic cells disappear. Not so with AA patients. When the leukemic cells totally disappear from their marrows, these patients do not have normal elements available for repopulating their marrow and these subjects with AML usually die of complications related to marrow failure (infection, bleeding). Thus, it may be wise in some of these patients not to give any therapy, if the clinical picture allows it, or only sufficient therapy for partial elimination of the leukemic cells from the marrow. Of course, when marrow transplantation becomes a routinely succesful procedure, the AA patients would be prime candidates for such an approach, thus allowing intensive therapy of the AML. It is interesting to note that with very rare exceptions, all the ALL patients examined by us have been of the AN or the N variety, and this may explain, to some extent, the more favorable prognosis of these patients as compared to subjects with AML.

The Ph^1 in AML

The Ph^1-chromosome may be seen in rare cases of AML (Table 49); it is usually accompanied by karyotypically normal cells (S. Abe and Sandberg, 1978). This contrasts with the findings in the BP of CML in which the cells are usually 100% Ph^1-positive. It is possible that some cases of Ph^1-positive AML represent a *fulminant* type of CML without a CP, though it appears more likely that the Ph^1 may be associated with some cases of AML. More than one Ph^1 has rarely been observed in AML (M. H. Khan, 1973a), such a karyotypic picture being much more common in the BP of CML. The return of the marrow to cytogenetic normality after anti-AML chemotherapy definitely points to the fact that two Ph^1-chromosomes may occur in AML, as this is very rarely observed in the BP of CML. In the case reported by M. H. Khan (1973a) the two to three Ph^1-chromosomes were accompanied by three chromosomes in group G + Y, i.e., 46,XY, $-2G,+2Ph^1/46,XY,-C,-2G,+3Ph^1$.

I believe that the Ph^1 may only be an expression of a causation of AML similar to that of CML or a morphologically similar karyotypic expression of the same autosome (#22) to different factors causing leukemia. Of course, it is possible that there may exist a variety of forms of AML, the presence or absence of the Ph^1 being but one parameter of difference. Nevertheless, further information of a more basic character will have to be obtained before we can be sure of this hypothesis. Thus, some years

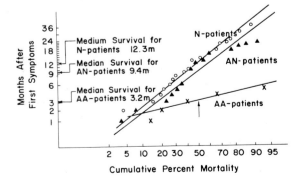

97 The cumulative percent of mortality from first symptoms among the various group of patients with AML, plotted on log probability paper. For a definition of N, AN, and AA patients see text.

ago (Sandberg, Ishihara, et al., 1964b), we promulgated the possibility that hypodiploid AML may differ from the hyperdiploid form of this disease. A subsequent observation reported by us (Sandberg, Cortner, et al., 1966) essentially destroyed this possibility. We observed a pair of paternal twins with congenital AML, one with hypodiploidy and the other with hyperdiploidy (see Figure 48). Unless one wishes to impute two different causes for the development of AML in utero, the findings definitely indicated that the chromosomal changes in AML may not necessarily be indicative of different forms of this disease.

Chromosomes and the Genesis of AML

In the genesis of AML, initially the leukemic process probably involves a small portion of the myeloblastic precursors, so that only a portion of the metaphases is aneuploid (Sandberg, Sakurai, et al., 1973). With progression of the disease more of the myeloblastic elements and/or erythroblastic cells may become involved, the latter resulting in EL. With further progression of the AL all the precursor elements of both the myeloblastic and normoblastic cells may be affected, leading to a marrow totally aneuploid (AA patients). When first seen, the AML may be characterized by any of the karyotypic pictures hypothesized (and some not, e.g., the involvement initially of the normoblastic elements only), depending on the biologic behavior of the AL in a particular patient, the rate of progression of the disease, and the time elapsed between the beginning of the AML and cytogenetic examinations.

The possibility exists that a dual population of leukemic cells may exist in AML, one diploid and the other aneuploid;[29] for if the chromosome changes are a result of the AL process and not the direct cause of it, a view held to by me, such a situation in the marrow would not be surprising. The diploid leukemic cells would only represent a more recently affected precursor population. Convincing evidence for the transition from diploidy to aneuploidy is supplied by some interesting cases published.[30] Of course, at least 50% of all cases of AML and ALL never develop aneuploidy and remain diploid throughout the course of the disease.

The problem as to whether AML, akin to CML, is a clonal disease has not been settled, particularly in view of the reports of karyotypic instability occurring in this disease. Even though chromosomal abnormalities detected in patients before treatment have shown remarkable uniformity in all or almost all of the abnormal cells obtained in the initial marrow samples, pointing to the possibility of clonal evolution of AML (Motomura, Misutake, et al., 1972), others have reported karyotypic instability in about 10% of patients with AL (Trujillo, Ahearn, et al., 1974; Trujillo, Cork, et al, 1974; Sakurai and Sandberg, 1976a). Even though such instability may occur more frequently in EL, the presence in AML of structurally abnormal chromosomes, such as rings, dicentrics, and fragments, and the degree of karyotypic variability was such that no consistent pattern could be discerned. However, it is possible that future studies with banding techniques will permit the accurate identification of the structurally altered chromosomes and, thus, the recognition of stable clones even in patients with multiple rearrangements and karyotypic instability.

When patients with AML enter remission, the chromosomally abnormal cell lines decrease or disappear. When relapse occurs, the same alterations which were observed initially usually recur. Even though chemotherapy and/or progression of the AML may produce chromosomes with some morphologic changes and lead to an increase in chromosome breaks or fragments, there is no evidence that antileukemic treatment induces a clone of stable aneuploid cells (Sandberg and Hossfeld, 1974). Furthermore, when karyotypic evolution occurs, it involves the original chromosomally abnormal clone of cells. Even though such evolution frequently takes place in the BP of CML, changes in the karyotype have been observed in a small proportion of patients with AL, though the exact incidence appears to vary from one laboratory to another. Of interest is the fact that when the leukemic cells of AML are analyzed with banding techniques, nonrandom patterns is evident in the karyotypic evolution of a substantial number of cases, which may not have been apparent in the initial karyotype with standard staining (Rowley and Potter, 1976). Thus, the acquisition of an additional #8 appears to be the most frequent finding. This is similar to the addition of a #8 in patients with CML who undergo blastic transformation.[31] These observations seem paradoxical, because these relatively consistent changes have been noted in patients who have received occasional radiotherapy as well as a variety of chemotherapeutic and potentially mutagenic drugs, administered in some individuals over a long period of time. One would expect, therefore, to observe bizarre, random chromosomal alterations. Instead, the majority of patients whose karyotype has evolved is found to have an extra #8 and/or a missing #7, regardless of whether the original disease was AML or CML.

Chromosome Patterns in AML Revealed by Banding

Even though chromosomal variability exists in AML, as has been emphasized repeatedly in previously mentioned reports, the introduction of banding techniques has revealed the existence of some nonrandom cytogenetic patterns in this disease (Table 54). This was due to the fact that all previous reports, published before analysis with banding techniques, had important deficiencies related to the impossibility of detecting cells which may have had structural rearrangements that did not alter the apparent morphological appearance of the chromosomes; furthermore, when a chromosome was missing or when an additional chromosome was observed, it was often impossible to determine the precise identity of these chromosomes. Thus, the exact percentage of AML patients with chromosomal abnormalities and the nature of these abnormalities when present were uncertain. Banding techniques permit the precise identification of each human chromosome and provide a means for reexamining these areas. Even though such studies continue to reveal that chromosome abnormalities are only present in about 50% of patients with AML, the emergence of some nonrandom karyotypic patterns in this disease appears to have taken place. These nonrandom patterns must have some significance, even though we do not fully appreciate it at the present time. More importantly, when these karyotypic changes occur at the level detectable with the light microscope, they are stable and are a very important clinical and biological feature of the leukemia; clinicians will have to become more cognizant of their significance in the future than they have been in the past.

The presence of nonrandom patterns in AML becomes much more apparent if published data on chromosomal abnormalities, determined with banding techniques, are analyzed on the basis of cases published in the literature. The results are shown in Table 58. Not included in this table is a number of cases with AML in whom the chromosomal abnormalities either have not been identified rigorously or present a problem in interpretation. The latter applies in particular to the report of patients in remission of AML and whose cells had 46 chromosomes but with an abnormal karyotype, the cytogenetic abnormality apparently being related to the loss of a #8 and the presence of an extra #9 (J. H. Ford, Pittman, et al., 1974; J. H. Ford and Pittman, 1974). Because a confirmation of this unusual finding has not been forthcoming from any laboratory, the data have not been included in the table. It is interesting to note in Table 58 that the modal chromosome number fell within a rather narrow range with most of the cases having either 45, 46, or 47 chromosomes. Cases with < 45 (Figure 98) or > 47 chromosomes constituted a minority of the cases analyzed.

Prior to banding, a number of authors had noted a frequent involvement of chromosomes in groups C and G[32] and the absence of chromosomal abnormalities of group F in AML (Trujillo, Ahearn, et al, 1974; Trujillo, Cork, et al., 1974). The use of banding confirmed these observations and revealed that there is a nonrandom pattern in both the C and G groups as to the particular chromosomes involved (Figures 99–101). Thus, in cases of AML in which banding was performed (Table 58), the following were the most common numerical changes: +8 in 26 cases, −21 in 15 cases, −Y in 7 cases, −7 or −17 in 6 cases each, −5,−9,+21 or −X in 5 cases each, and −14,−11,−13, or +13 in 4 cases each. The loss of an X− or Y− chromosome was frequently associated with a balanced translocation involving #8 and often #21. Thus, it appears that the #8−#21 translocation is an abnormality noted with some frequency in AML, followed by gain of #8, loss of #7 and #17, and abnormalities of #21. The incidence of the prototypic karyotype, involving an #8−#21 translocation, may be higher than indicated by the data of the table, as we

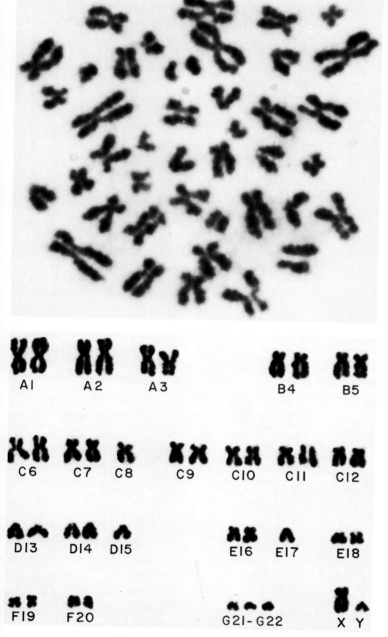

98 Metaphase (top) and karyotype (bottom) of a hypodiploid marrow cell from a patient with AML, consisting of only 42 chromosomes without any apparent markers (unbanded preparation).

have found this abnormality in 11 of 51 patients, and another report, though based on standard Giemsa staining, indicates the possibility that such a translocation was present in 12 out of 69 patients (Trujillo, Ahearn, et al., 1974; Trujillo, Cork, et al., 1974). The relationship of the tendency for #8 and #21 to occur in an aneuploid state in AML and their frequent involvement in the reciprocal translocation is yet unresolved. In a cytogenetic evaluation of 65 adults with AML (Kamada, Okada, et al., 1976), 15 were found to have a consistent

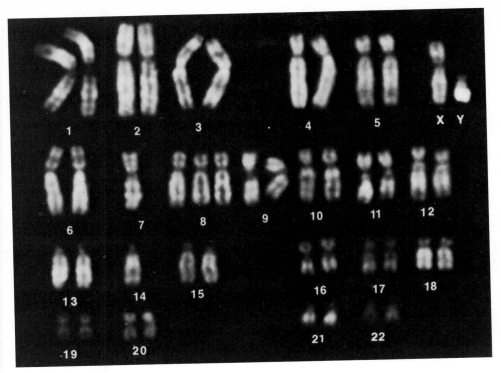

99 Q-banded karyotype of an AML cell with 45 chromosomes and containing two of the commonest changes in AML, i.e., −7 and +8. A −14 is also shown.

100 Q-banded karyotype of an AML cell containing a number of abnormal chromosomes. The origin of M_1 appears to have involved translocation of the long arm of #1 to #22, and M_2 probably originated by translocation of the distal part of #13 (q12→14) to the distal part of #11 (q23→25). M_3 was probably composed of the long arm of #17 (p11) and part of an unknown chromosome. The origin of M_4 could not be determined.

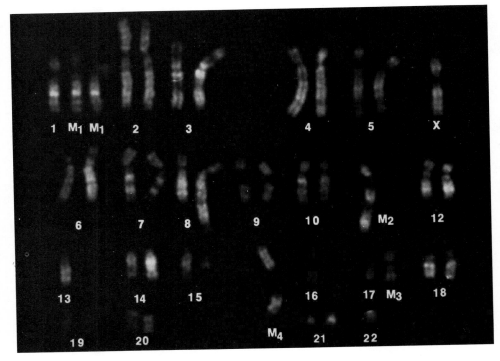

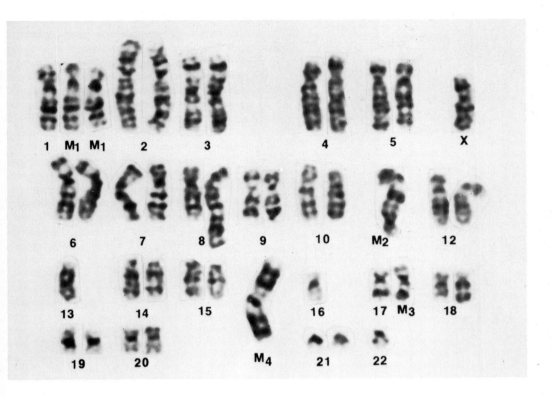

101 G-banded karyotype of a cell of the case with AML shown in the previous figure with a very similar karyotype.

chromosome abnormality consisting of a C-G translocation [t(8;21)(q22;q22)] including 3 of 6 males with a missing Y, in whom the neutrophils had a low LAP activity and specific morphologic abnormalities. The marrow in these patients showed a definite tendency for maturation of the leukemic cells. The median survival of these patients was significantly longer than that of patients with normal or high LAP activity.

At the First International Workshop on Chromosomes in Leukemia held in Helsinki in 1977 (1978c) the pooled data from various laboratories revealed the leukemic cells of 50% of the ANLL patients (140 out of 279) to be diploid. The most common chromosome aberrations were +8 (17%) and −7 (15%). These anomalies appeared to be less frequent in AMMoL and AMoL than in AML. Next in frequency were −5, −17, and ±21. Structural changes affected all chromosomes except #10 and the sex chromosomes; #3, #5, #8, #9, #15, #17, and #21 were most frequently involved. Only three structural rearrangements were seen in more than two patients: t(8q−; 21q+), the so-called prototypic karyotype, in 12 patients, including two cases with a three-way translocation [t(8;11;21) and t(8;17;21)], t(15q+;17q−) in 9 patients with APL and a Ph[1] due to t(9q+;22q−) in 4 cases. One patient had a t(3q+;22q−).

Patients with −7 had an incidence of complete remission (CR) of only 13% and a median survival of 4 mo, compared with 50% and 6 mo for the cases with t(8;21) and 33% and 1 mo for the APL patients with t(15;17).

Table 58a shows the overall median survival, that for the patients who achieved CR and the percentage of patients alive after 1 yr according to the cytogenetic pattern and based on 241 cases of ANLL with documented clinical information. Noteworthy is the poor prognosis for the AA patients. These differences did not apply to patients with AMMoL or AMoL.

Only 4 cases with initial normal karyotypes developed a chromosome aberration during the course of the disease: one developed +8, a second a t(8q−;21q+), a third +8 and other abnormalities, and a fourth a marker chromosome. No evidence was obtained that chemotherapy produces new stable abnormal clones.

The role of #8 in malignancy is one of the most perplexing problems facing cytogeneticists investigating hematological diseases (Rowley and Potter, 1976). On the one hand, an additional #8 is the single most frequent abnormality seen in leukemic patients initially, as well as in the further evolution of this disease. On the other hand, an additional #8 is also the chromosomal abnormality seen most commonly in nonmalignant hematologic disorders such as refractory anemia (Rowley, 1975b). Whereas some of the latter patients may develop leukemia in the future, others have had no change in their hematologic status for at least 10 yr (Rowley, 1973c). It has not yet been determined whether these frequent additions of #8 are related to genes carried on this chromosome, for example, glutathione reductase (Chapelle, Vuopio, et al., 1976), or whether #8 is the site of viral incorporation or genes related to malignancy, the expression of which is independent of the tissue involved.

Classification of AML by the Karyotypic Patterns and Their Relation to Prognosis

Even though a correlation between the chromosomal findings in AML and the prognostic aspects of the disease has not revealed consistency in different laboratories[33], it is our contention that analysis of the karyotypic picture *throughout* the course of AML does lead to the possibility of characterizing the disease and the chromosomal changes (Figures 102 and 103) (Sakurai and Sandberg, 1973, 1974). Thus, we have found that the median survival after initiation of antileukemic therapy of patients with AML was surprisingly short for those patients who never had normal metaphases in their marrow *during the course* of the disease (AA patients, 1.2 mo), particularly when compared to the survival of patients with both abnormal and normal metaphases (AN patients, 7.2 mo) or those with only normal metaphases (N patients, 9.1 mo). Among the latter two groups, the majority of patients

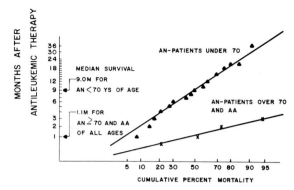

102 Cumulative percent mortality after the initiation of chemotherapy in patients with AML. AN patients over the age of 70 and AA patients of all ages, generally, had a much shorter survival time than AN patients <70 yr old.

over 70 yr of age had a very poor survival after therapy (0.7 mo). The remarkable shortness of the life-span of these patients after therapy, as compared to that from the onset of symptoms, indicates that current therapies are of little help to AA patients and most patients over the age of 70 yr[34]. Some of the AA patients and a few of the AN patients constitute a unique group of AML patients in whom the erythroid and myeloid series are involved by the leukemic process[35] and who are, thus, actually affected by EL. Their survival, however, seems to depend on whether or not they have any normal metaphases in their marrow eligible to repopulate the marrow with normal cells when the leukemic ones have responded to therapy. A study has shown that the mitotic indices of the granulocytic precursors in AML patients were generally low, and that there were no indications that the cells with a normal karyotype (N patients) had a different mitotic activity than those with 100% abnormal karyotypes (AA patients) (L. Brandt, Mitelman, et al., 1975b). Thus, this kinetic abnormality is not related to an abnormal chromosome constitution of these cells.

Acute myeloblastic leukemia patients with chromosomal abnormalities in their marrow can be further subdivided into those with minor karyotypic abnormalities MIKA and those with major karyotypic abnormalities MAKA

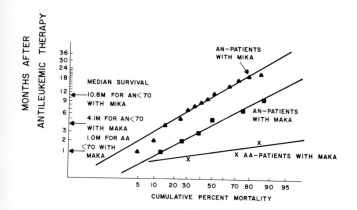

103 Cumulative percent mortality after the initiation of chemotherapy in patients of various karyotypic groups 69 yr old or less. AN patients with MAKA had a significantly shorter median survival than their counterparts with MIKA, even though the former group still had a significantly longer median survival time than AA patients of the same karyotypic group (MAKA).

(Sakurai and Sandberg, 1974). One of the largest subgroups in the MIKA group was shown to have cytological characteristics typical of so-called classic AML and a prototypic karyotype, which has been shown to be the result of the translocation between #8 and #21. A missing Y-chromosome in AML was mostly associated with this karyotype. Patients of the MAKA group were usually erythroleukemic, had no or very few normal metaphases among the abnormal ones in the marrow, and almost invariably showed karyotypic instability. Karyotypic differences did not affect the survival of the patients after the initiation of chemotherapy as much as the presence or absence of normal metaphases. However, the karyotypes do appear to be relevant to the decision as to whether or not a patient should be administered chemotherapy when no normal metaphases are found on the initial marrow examination.

All patients with a prototypic karyotype in our studies (Sakurai and Sandberg, 1973, 1974) and those reported in the literature[36] had typical AML. As in our patients, Auer bodies have been reported in many of the AML patients with the prototypic karyotype on whom detailed hematological data are given. The above facts would indicate that a prototypic karyotype may be relatively specific for "classical" AML. Just as the Ph^1-chromosome is associated with and possibly determines the clinical fea-

104 Karyotype with MIKA from the marrow of a patient with AML. No more than one or two independent events are present in MIKA, e.g., in this karyotype −17 and ? marker.

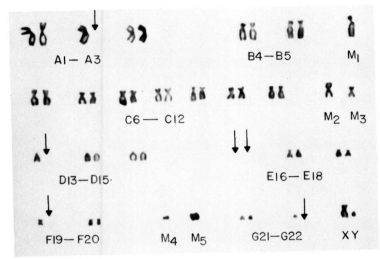

a

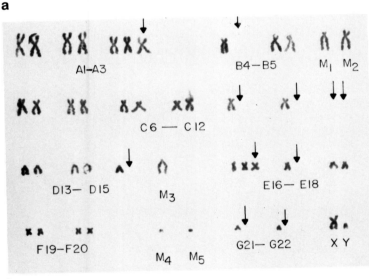

b

105 Karyotypes (a–d) from marrow cells of patients with EL showing chromosome constitutions considered to be MAKA in character, i.e., three or more independent events have occurred in each karyotype.

tures of CML, the prototypic karyotype consisting of the #8–#21 translocation in AML may also be a determining factor in the apparently homogenous clinical features of these patients.

Among the patients with minor karyotypic abnormalities (MIKA group), the most common abnormality was −C, +D, +15, −G. This abnormality has been amply reported in the literature and confirmed as being the result of presumably balanced translocation between chromosomes #8 and #21 (Rowley, 1974b; Sakurai, Oshimura, et al., 1974). A very common occurrence with this karyotype was a missing sex-chromosome. The prototypic karyotype is the only one in AML that often leads to or is associated with the missing Y-chromosome. The incidence of a missing Y in the marrow cells in four of six male patients of our series is surprisingly high, as compared with that in Ph^1-positive CML. CML, to our knowledge, has been the only leukemic condition that can acquire a missing Y, in which, however, the incidence is < 10%. In the AML patients a missing Y was invariably accompanied by the prototypic karyotype. This would mean that the missing Y (or missing X) phenomenon occurred almost

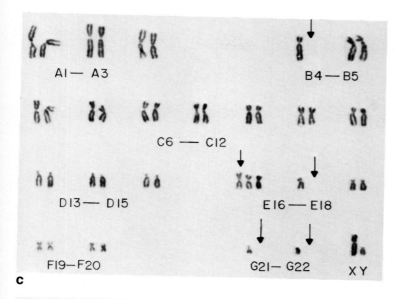

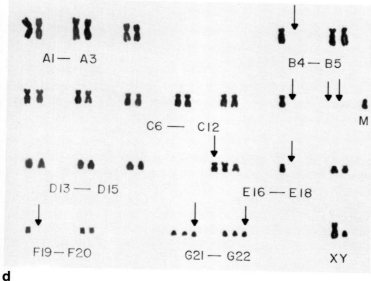

at the same time or subsequent to the #8–#21 translocation. Inasmuch as 45,X (missing Y and otherwise normal) cells, which are often present in the marrow, especially in elderly males, were not observed in the marrows of these patients with the missing Y prototypic karyotype (except for a sporadic and rare occurrence) and inasmuch as 45,X cells are thought not to be involved in leukemogenesis (Sandberg and Sakurai, 1973a), it does not seem likely that the missing Y prototypic cells were derived from these 45,X cells. Another possibility might be that the cells with missing sex-chromosomes had a selective advantage over those not missing such chromosomes. Of interest also was the observation that the presence of the prototypic karyotype seemed to be related to the long survival of these patients with AML (Sakurai and Sandberg, 1974). Others have also reported long and complete remission in patients with the prototypic karyotype.[37] Many of the cases in the literature had no missing sex-chromosomes, indicating a lack of its relation to the long survival, in contrast to the possible longer survival observed in patients with a missing Y in Ph^1-positive CML.

Another important finding in our series was that all our patients with the prototypic karyotype (except for a rare case with its variant) had similar cytological features in the marrow, consisting of typical myeloblasts with round or oval, but very often indented, nuclei with finely dispersed chromatin, one to two nucleoli, and a moderate amount of blue cytoplasm with few or no granules. Very often, but not invariably, a significant percentage of promyelocytes and sometimes also myelocytes was found in the same marrow before therapy, even when no normal metaphases were detected, indicating a certain tendency toward maturation. Auer rods were found in most cases, even though the frequency differed from case to case. Peroxidase and Sudan black staining were clearly positive in the leukemic cells. Erythroblasts and megakaryocytes were markedly decreased. Such cytologic homogeneity was almost never encountered in patients with AML whose karyotypes were characterized either by missing or extra C-group chromosomes or MAKA.

Next to the prototypic karyotype the most common abnormality encountered in our series consisted of extra or missing group C chromosomes (Sakurai and Sandberg, 1974). Patients of this variety did not show homogenous clinical and cytologic features as was observed for the patients with the prototypic karyotype. Thus, their myeloblasts were of a dissimilar appearance even within the same karyotypic group, peroxidase and Sudan black staining of the leukemic blast cells was positive in some and negative in others; and when positive, they differed in intensity and incidence among the cases. We have also observed a few patients whose leukemic cells contained a Bq+, Cq− abnormality, but whether this is a cytogenetically unique group has yet to be established, even though other cases have been reported in the literature (Van Den Berghe, Fryns, et al., 1972; Trujillo, Ahearn, et al., 1974; Trujillo, Cork, et al., 1974).

In some of our patients chromosomal abnormalities developed for the first time late in the course of the disease, despite a high percentage of myeloblasts in their earlier marrows. The later development of chromosomal abnormalities in these patients would indicate that at least some of the chromosomal changes in AML are merely secondary phenomena and have little to do with the original leukemogenesis. The genesis of chromosomal changes in a primarily diploid marrow as well as the karyotypic progression of new clones in the late stages of AML would indicate, however, a certain relationship of the karyotypes with the clinical facets of the disease.

The survival of patients with AML was shown to be dependent on whether or not the patient had normal metaphases in the marrow and also on whether he or she was under or over the age of 70 (Sakurai and Sandberg, 1976a; Nilsson, Brandt, et al., 1977, 1978). These principles held true regardless of karyotypic differences. Those with no normal metaphases (AA patients) and those over the age of 70 yr, even with some normal metaphases in their marrows, had a very short survival as compared to that of patients with AML under 70 yr and with normal metaphases among the abnormal ones. Among the latter, however, those with MAKA had a relatively shorter survival than those with MIKA. Most of the AN patients with MAKA had very few normal metaphases (near−AA patients). Some of this group responded to chemotherapy just as poorly as did the AA patients. This findings suggests a possible adverse effect on the normal cells in the marrow either as a result of karyotypic instability, complicated abnormalities, or the involvement of both the erythroid and myeloid cells.

Further information in this area was provided by Nilsson, Brandt, et al. (1978) who examined the karyotypic profiles of 30 adult patients with AML and came to the conclusion that in older patients (>50 yr old) the incidence of AA is lower (16%) than in the younger group (42%). These authors indicated that AML in elderly patients is preceded by a rather long period of smoldering development of the AML during which the patients retain an AN pattern. When the disease ultimately advances into the AA stage, the prognosis becomes poor.

The correlation of the karyotypic findings with the clinical and prognostic aspects of AML have been dealt with in several publications. In some of these no correlation between these parameters could be established (Fitzgerald and Hamer, 1976; Gunz, Bach, et al., 1973), whereas in others, including the papers published from our laboratory, a definite correlation appeared to be present.[38] The differences for this correlation cannot be established readily, though they may be related partially to the fact, for example, that in our laboratory the karyotypic findings utilized were based on those obtained throughout the course of the disease rather than on a single examination, even at the initial phases of the disease. Several studies reflecting agreement with the results obtained by us deserve special attention.

A relationship between chromosome patterns and prognosis in AML was already pointed to by Hart, Trujillo, et al. (1971); Trujillo, Ahearn, et al. (1974); and Trujillo, Cork et al. (1974), who found that certain cytogenetic profiles carry prognostic implications, i.e., patients with diploid or hyperdiploid karyotypes survived for 49 wk (mean survival time), whereas those with pseudodiploidy or hypodiploidy survived only 36 and 13 wk, respectively. Of the 170 patients studied, 69 (40%) had chromosomal changes in their leukemic cells; of these 69 patients, 15 had an extra C-group chromosome (?mostly +8) and a mean survival of 68 wk and 12 patients a −C,+D,+E,−G karyotype (prototypic consisting of 8;21) and a mean survival of 80 wk. Patients with unstable karyotypes including various markers (e.g., rings, dicentrics) and acentric fragments had a mean survival of only 8 wk. Early disappearance of abnormal clones usually indicated a favorable response to therapy.

A study by Golomb, Vardiman, et al. (1976) led the authors to conclude that AML patients with chromosome abnormalities had a median survival time markedly different from the patients without such abnormalities, i.e., 2 vs. 18 mo, respectively. Furthermore, when these authors subclassified their patients according to criteria suggested by us (Sakurai and Sandberg, 1976a,b), the following median survivals were obtained (our values in parentheses): AA patients, 2 mo (1.2 mo), AN patients, 9 mo (7.2 mo), and N patients, 10 mo (9.1 mo). A more recent analysis by the same group (Golomb, Vardiman, et al., 1978) yielded results which did not differ materially from the previous ones. Golomb, Vardiman, et al. (1978) further showed that none of the AA cases achieved a complete remission of the AML vs. 38 and 69% of the AN and NN cases, respectively. In the study of Nilsson, Brandt, et al. (1977) on a series of 30 adult patients with acute nonlymphocytic leukemia the values were 1.0, 2.5, and 8.0 mo, respectively. The results indicated that the karyotype analysis by banding techniques is an important prognostic indicator in AML. The authors suggested that in patients with cytogenetically abnormal marrow cells the disease has reached a more advanced stage than in those with exclusively diploid cells (Golomb, 1978).

In 34 children with AML, 15 of whom had chromosome abnormalities before treatment and 2 of whom developed such abnormalities after therapy, the average length of survival from the detection of a clone was 7.1 mo, which was < 50% of that for the group with normal karyotypes (Lange, Alfi, et al., 1977).

AML Complicating Other Conditions

In all conditions that predispose to AL, be they congenital (DS, FA), acquired (benzene poisoning, radiation), or other myeloproliferative disorders (myelofibrosis, IAA), when AL (usually AML) develops the cytogenetic changes are similar to those usually encountered in these states. I do have the impression, however, that aneuploidy occurs much more frequently in these syndromes than in ALL and AML without an *apparent* predisposing cause. This is particularly true of AL after radiation or developing in patients with lymphoma, multiple myeloma, HD, and various forms of cancer. In this group of patients with AML, diploidy has not been reported to-date. In one patient with AML and HD a +14,−17,+18 pattern was established with banding (Lundh, Mitelman, et al., 1975), in two other such cases abnormalities of group D and +E were found (Ezdinli, Sokal, et al., 1969; Weiden, Lerner, et al., 1973), the latter anomaly was also seen in one patient with EL complicating HD (Ezdinli, Sokal, et al., 1969).

It would be of great interest and value to know the evolution of the chromosomal changes in AL superimposed on lymphosarcoma or multiple myeloma, i.e., to ascertain the karyotypic changes in such conditions before the onset of leukemia and to follow the changes in the leukemic cells in order to find out any possible relationship between the karyotypes of the two neoplasms and whether they do or do not resemble, are identical to or reflect, those of the initial malignancy. A few cases in which we have been able to follow the progression from lymphoma or multiple myeloma to AML, the chromosomal changes in the initial disease were quite different from those in the subsequently developed AML, indicating that the karyotypic changes are unique to each disease. However, many more cases are required before we can be certain of this observation.

For more details regarding the karyotypic finding in AML complicating lymphoma or multiple myeloma, see Tables 72 and 78. Rowley (1977c) has stressed the loss of a #5 in eight of the ten cases studied, loss of #7 in five cases, and rearrangements of #17, #6, or #8. Some of the differences between these karyotypic changes and those in "spontaneous" AML were pointed out (Rowley, Golomb, et al., 1977b).

Acute myeloblastic leukemia in patients with or without neoplastic diseases treated with various chemotherapeutic agents is being described with increasing frequency.[39] Except for two cases for which scanty cytogenetic information is supplied (Greenspan and Tung, 1974), others have had chromosomal changes, e.g., a marker chromosome in a leukemia associated with chloramphenicol therapy, hyperdiploidy (H. J. Cohen and Huang, 1973), ring chromosomes, and abnormalities of groups A-E in two cases with structural abnormalities treated with cytoxan and azathioprine (Seidenfeld, Smythe, et al., 1976).

We have found cytogenetic monitoring of patients to be useful in those cases suspected of developing AL while receiving radio- and/or chemotherapy. The appearance of any aneuploidy should alert the clinician to the possibility of AL. Except for discontinuing the therapy, no definite criteria as to the course to be taken in such cases are available at present. However, there is little doubt that this area will receive more attention in the future due to the increasing use of chemotherapy in human cancer and in some noncancerous conditions.

The Missing Y-Chromosome in AML

Pierre and Hoagland (1972) found that the number of males exhibiting loss of the Y in the BM cells correlated with age, but that the degree of the loss did not. Our findings are generally in accordance with theirs, although the degree of Y loss in any one individual appears to have been generally higher in their report than in our series (Sakurai and Sandberg, 1976c; Sakurai, Hayata, et al., 1976). Since in our study the incidence of individuals in whose BM one or two 45,X metaphases were detected (probable cases) also increased with age, it is likely that the missing Y in such a small proportion of the BM cells was, in many cases, not due to loss of such a chromosome during the cell preparation, but to a phenomenon occurring in vivo.

Only an extremely rare number of our patients with AML had a missing Y as the *only* karyotypic anomaly in the leukemic cells. The 45,X cells detected in two elderly patients with AML were considered to be nonleukemic on the basis of the following findings: 1. the substantial number of 45,X cells present in the early stages of the disease and during remission; 2. failure of the 45,X cells to acquire additional karyotypic changes; and 3. the advanced age of the patients in whom 45,X cells normally would be expected to be present. Based on such observations, we speculated that in these two patients the cells with a missing Y were somehow protected from being involved in the AL.

It should be pointed out, however, that a missing Y can occur rather frequently in a leukemic cell line when certain autosomal abnormalities are present. There were 6 such patients among our 35 males with AML with chromosomal abnormalities (Sakurai, Hayata, et al., 1976). Four of the 6 "prototypic" patients and 2 of the 16 MAKA patients had a missing Y in their abnormal karyotypes (Sakurai and Sandberg, 1976c). In none of them were metaphases with 45 chromosomes with only a missing Y detected. These findings, along with the relatively young ages of many of the patients, indicated that the aneuploid leukemic lines were probably derived from 46,XY and not from 45,X cells. The association of a missing Y with a "prototypic" karyotype (#8-#21 translocation) has now been well established.[40]

Based on the data from the literature[41] and those of our study (Sakurai and Sandberg, 1976c), out of a total of 64 male patients with AML over the age of 60 yr, there were 23 who had only normal diploid metaphases. Forty-one patients had leukemic cell lines with chromosomal abnormalities. Only one patient among them, however, aged 64 yr, and with a diagnosis of EL, had a leukemic 45,X cell line (Heath, Bennett, et al., 1969). If we assume, on the basis of the data in control subjects, that roughly 10%–20% of the patients with AML over the age of 60 yr should have had a missing Y cell line (or up to 40% including those with sporadic missing Y cells) before the development of leukemia and that such cells had been preferably involved in leukemogenesis, a minimum of 10 (or up to 25) patients should have had a leukemic 45,X cell line or cells lines derived from 45,X cells. Thus, it is safe to state that even though the missing Y phenomenon in marrow cells is an age-associated condition, 45,X (missing Y) cells are much less readily involved in leukemogenesis than 46,XY cells. In addition to the above-mentioned cases, there have been two single case reports of AML patients of 39 and 72 yr of age with 46,XY and 45,X (Yamada, Takagi, et al., 1966), and 46,XY, 45,X and 44,X,−G cell lines (Cardini, Clemente, et al., 1968), respectively.

Our hypothesis that 45,X (or 45,X,Ph^1 cells) are generally resistant to the development of further chromosomal abnormalities and, thus, reflect their resistance to being involved in an acute leukemic process, appears, on the surface, to be contradicted, for example, by six AML patients with missing Y cells containing additional karyotypic changes observed by us (Sakurai and Sandberg, 1976c), in particular the four cases with the prototypic karyotype. However, that the loss of the Y-chromosome is probably consequential to the development of the prototypic karyotype is pointed to by observations in our group of patients with AML: (1) in some patients with the prototypic karyotype the missing Y appeared subsequent to the development of the #8–#21 translocation; (2) a mixture of cells with the prototypic karyotype, but with and without a missing Y in the leukemic cells has been observed; and (3) with very rare exceptions, no cases of AML have been observed who did not have other karyotypic abnormalities in addition to the missing Y in the leukemic cells.

The marrow chromosomes in 7 ANLL patients with a missing Y were evaluated (Abe, Golomb, et al., 1979). One patient presented complex karyotypic abnormalities in addition to the missing Y, i.e., 45,X,t(1q+;8q−), 21q−, whereas the remaining six cases exhibited the missing Y as the *only* abnormality at diagnosis, which persisted through the clinical course of the disease. One patient had a shift from the 45,X to a 47,XY,+5,13q+ karyotype in the terminal phase of the disease. The findings obtained in the 7 patients indicate that no case with a missing Y as the *only* chromosomal abnormality in the marrow cells developed further karyotypic changes in the 45,X cells; the leukemic process seemed to involve the 46,XY cells rather than the 45,X cells. Based on the data of the study and those in the literature, Abe, Golomb, et al. (1979) concluded that the presence of a missing Y in ANLL may afford a rather benign clinical course to such patients when compared to that of ANLL patients without a missing Y. It was suggested that the age of the patient, the karyotypic differences, and the degree of a missing Y in the marrow cells possibly affect the clinical course of ANLL.

Berger and Bernheim (1979) have presented two cases of AL in whom a missing Y appeared to characterize the leukemic cells: one case was that of a 73-yr-old male with AMMoL, and the other that of a 32-yr-old patient with ALL. In addition, these authors reported a 65-yr-old male with PV whose marrow contained a 45,X,−Y cell line for 2 yr before transformation into AL. A few months before the acute phase, the cell line with the −Y predominated in the BM, with other chromosome aberrations being superimposed during the leukemic phase. In another case, a 45,X,−Y,Ph^1-positive karyotype was found in the majority of the BM metaphases in a 51-yr-old patient with CML in the BP. When diagnosed 1.5 yr before the BP, this patient's leukemic cells had a 46,XY,t(9;22) karyotype. The 45,X,−Y karyotype predominated during remission of the BP. These observations would appear to be exceptions to the suggested protective role of a missing Y in leukemic transformation.

Care should be taken in stating that AL or any other condition has involved the marrow cells of an individual with a missing Y, but who also had some normal, diploid cells in the marrow. There is a distinct possibility that only the diploid cells were involved by the malignant process and not the cells with the missing Y. This is exemplified by several of our patients with AML who during complete remission of the disease apparently had only cells with a missing Y, whereas XY cells appeared during relapse of the AL.

Of course, in patients who have 100% cells with a missing Y both during remission and relapse of the AL (Heath, Bennett, et al., 1969; Abe, Golomb, et al., 1979), the findings would point to the strong possibility of these cells being affected by the leukemia process. Thus, in reporting on the findings of cases with a missing Y in the marrow it behooves the authors to pay particular attention to the presence or absence of diploid cells and the percentage of each type of cell present in the marrow during various phases of the leukemic process. Only through careful reporting and analysis will we be able to establish the susceptibility of the missing Y cells to the development of AL and the effects on the course of the disease.

An interesting case has been reported by Francesconi and Pasquali (1978) of a 13-yr-old boy with AML, a prototypic karyotype, −Y, and a Ph¹. The leukemic cells contained Auer rods. The authors implied that their case might have been one of the BP of CML with the rather unusual t(8;21) in such disease. I believe that their case is a Ph¹-positive AML with such a translocation.

Monosomy-7 and Trisomy-8 in AML

Before banding some authors stressed the importance of certain C-group chromosomes in AL (Rowley, Blaisdell, et al., 1966), following a report from my laboratory in which C-group trisomy was the only karyotypic aberration found in a case of AL developing in a patient with myeloid metaplasia (Sandberg, Ishihara, et al., 1964a). Thus, C-trisomy as the sole karyotypic abnormality had been described in a number of cases with AL, including both ALL and AML.[42] C-group trisomy was shown to occur with a relatively high frequency in association with other cytogenetic changes in AL. Monosomy of group C had also been described before banding. It is now apparent that most of the cases of group C trisomy were due to +8 and those of monosomy to −7 (Gahrton, Zech, et al., 1975).

There is a tendency for some authors to attribute a leukopoietic function to group C autosomes (#8 and #9) and the chromosomal changes in that group as being causally related to AL (J. H. Ford and Pittman, 1974; J. H. Ford, Pittman, et al., 1975). Even though this possibility exists, group C trisomy occurs only in a portion of the cases with AL. Nevertheless, this observation certainly deserves continuing study and correlation with cytologic and clinical aspects of AL in order to establish its exact significance.

Monosomy-7 in the leukemic cells of patients with AML (or EL) is a condition that may be accompanied by defective chemotaxis and abnormalities of the Colton blood groups (Chapelle, Vuopio, et al., 1975; Ruutu, Ruutu, et al., 1977a, b). It is possible that these parameters may have some bearing on the relatively poor prognosis of the cases of AML with monosomy-7. Further discussion on monosomy-7 can be found in the section on myeloproliferative disorders.

Trisomy-8, like monosomy-7, is not an unusual finding in AML or EL and may be accompanied by abnormalities of the normoblastic cells (e.g., changes in glutathione reductase in the erythrocytes) (Chapelle, Erickson, et al., 1971; Chapelle, Schröder, et al., 1972, 1973a; Chapelle, Vuopio, et al., 1976), possibly faulty chemotaxis (Ruutu, Ruutu, et al., 1977a,b), and changes in coagulation (Grouchy, Josso, et al., 1974). Studies indicate that AML cases with trisomy-8 do not fare worse than those without chromosomal changes or those that are karyotypically abnormal but are without trisomy 8.[43]

Leukemia occurring in patients with congenital +8 has been described (Gafter, Shabtai, et al., 1976; Riccardi, 1976), though the two conditions share little, if any, of the phenotypic manifestations (Cassidy, McGee, et al., 1975).

Congenital Leukemia

Congenital leukemia is a very rare condition, with evidence for the disease being present at birth (connatal type) or becoming apparent within a month of birth (neonatal type). In contrast to childhood leukemia, AML is the predominant type in congenital leukemia and that developing in the first year of life (Nakagome, Kudo, et al., 1964). Occasionally, the leukemia accompanies a congenital chromosome anomaly, e.g., Turner's syndrome (Pridie and Dimi-

Table 59 Congenital leukemia

Type of leukemia	Karyotype	Source of cells	Reference
ALL	N*	Bl	Bouton, Phillips, et al., 1961
AML	47,+21 and tetraploidy	?	C. D. Brown and Propp, 1962
AML	D-trisomy constitution	skin	Schade, Schoeller, et al., 1962
AML	48,XY,+C,+21 (also DS)	BM	Honda, Punnett, et al., 1964
AML	59,XY,+4C,+3D, +E,+2F,+2G,+21 (DS)	?	Conen, Erkman, et al., 1966
AML	3q−	Bl	Zussman, Khan, et al., 1967
ALL	25% pseudo- and hyperdiploid	Bl	H. P. Wagner, Tönz, et al., 1968
AML	N	Bl	O. J. Miller, Allerdice, et al., 1968
AMMoL	46,XX,+D,−E	BM,Bl	Bauke, Cremer, et al., 1970
AML	Bq+;Dq−(4q+;15q−)	BM,Bl	Ponzone, de Sanctis, et al., 1972
AML?	46,XX,t(Bq+;Cq−)	Bl	Van Den Berghe, Fryns, et al., 1972
ALL	45,XY,+C,−E,−2G,+mar	Bl,BM?	Sharp, Potter, et al., 1973
AML	G-trisomy (not DS)	Bl	Björness, Bühler, et al., 1974
	46/?47	BM	Björness, Bühler, et al., 1974
AML	Abnormalities of #17 and #18	?	Larripa, Brieux de Salum, et al., 1975
ALL	46,XX,t(4;11)(q13;q22)	BM,Bl	Van Den Berghe, David, et al., 1978b (2 cases)

*N = normal karyotype.

trescu-Purvu, 1961) and trisomy D (Schade, Scholler, et al., 1962). As shown in Table 59, some of the congenital leukemias are normal cytogenetically (J. Y. Chu, O'Connor, et al., 1977), whereas others show a variety of karyotypic changes. The two cases described by Van Den Berghe, David, et al. (1978b) had a 4;11 translocation in their leukemic cells.

Erythroleukemia (EL)

Erythroleukemia (EL, acute erythremic myelosis) is defined as a condition characterized by the neoplastic proliferation of erythrocytic cells leading to ineffective erythropoiesis, megaloblastoid changes, and multinucleated cells, and is usually associated with more or less prominent neoplastic proliferation of the granulocytic system (myeloblasts). However, uncertainty continues to exist as to whether erythroleukemia (DiGuglielmo's syndrome) constitutes a well-defined entity or is merely a variant form of AML. Because EL tends to show more variability in its numerous aspects than does AML, attempts have been made at the karyotypic characterization of EL, with the hope that the chromosomal findings will result in a more reliable definition and diagnosis of this disease entity.

Chromosome Changes in EL

Evaluation of the exact incidence of karyotypic changes in EL (Tables 60 and 61) is made difficult by the inclusion of such cases in the general category of acute nonlymphocytic leukemia (ANLL) in published reports, without a differentiation being made between AML and EL. I am aware of 125 cases of EL reported in the literature. A summary of the numerical

Table 60 Chromosome number in 125 cases of EL reported in the literature

	No. of cases	%
Hypodiploid (35−45 chromosomes)	36	29
Pseudodiploid	10	8
Diploid (normal)	61	49
Hyperdiploid (46−50 chromosomes)	17	14
Hypotetraploid	1	<1

Table 61 Karyotypes in EL

Karyotype	Reference
39,X,−Y,+2,+C,−2D,−16,−18,−F,−3G	Heath and Moloney, 1965a
45,XY,−C	McClure, Thaler, et al., 1965 (chronic erythroleukemia)
44,XX,−4,−18	Pawelski, Topolska, et al., 1965
45,XX,−18	Pawelski, Topolska, et al., 1965
47,XX,+6?	Weatherall and Walker, 1965
High ploidy with basic karyotype 44,XX,−2C,+D,−2E,+mar	Nicolau, Nicoară, et al., 1966
45,XY,+B,−2C,+2E,−2G	Nicolau, Nicoară, et al., 1966
44,Cp−,−D,−G/45,−2C,17−18p+,+mar	E. Engel, McKee, et al., 1967
45,XY,−G	Castoldi, Yam, et al., 1968
47,XY,+C,+D,−G	Castoldi, Yam, et al., 1968
45,XY,−C,(Ph¹+)	Castoldi, Yam, et al., 1968
45,XY,−E,(Ph¹+)	Castoldi, Yam, et al., 1968
46,XX,(Ph¹+)	Castoldi, Yam, et al., 1968
44,XX,−2B,−2C,+E,+ring	Crossen, Fitzgerald, et al., 1969
46,XY/47,XY,+ring/47,XY,−C,+2 rings	Diekmann, Rickers, et al., 1969
47,XY,+C,−16,+mar	Ezdinli, Sokal, et al., 1969
48,XY,1q+,+D,+F/48,XY,1q+,+C,+D,+F,−G	Kroll and Schlesinger, 1970
49,XY,+C,+C,+D	Meisner, Inhorn, et al., 1973
45,XY,−7	Petit, Alexander, et al., 1973
45,XX,−7	MacDougall, Brown, et al., 1974
47,XX,+12	Inoue, Ravindranath, et al., 1975
47,XX,+22	Inoue, Ravindranath, et al., 1975
47,+C	Moretti, Broustet, et al., 1975
47,+C/48,+C,+F,Dp−	Moretti, Broustet, et al., 1975
46,XX,del(7),t(1;3)	P. Philip, 1975d, 1976
45,XX,−7	Zech, Lindsten, et al., 1975
47,XX,−5,i(17q),−19,+3mar	Najfeld, 1976
46,XX,del(7)(q22)/46,XX,del(7)(q22),del(5)(q22;q31→33)	Oshimura, Hayata, et al., 1976
47,XX,t(3;5),−1,+8,−13,+22,t(1;13)	Oshimura, Hayata, et al., 1976
45,XY,del(#4−#8),−6/45,XY,del(#4−#8),−4,−6,+mar	Rowley and Potter, 1976
49,XY,+8,+12,+17,t(17;21)(p11;q22)(from PNH)	Yamada and Furasawa, 1976
44,XY,−3,−18	Yamada and Furasawa, 1976
45,XX,−18	Yamada and Furasawa, 1976
47,XY,+8 (from AA)	Yamada and Furasawa, 1976
46,XX,+F,−G/47,XX,+C	Meytes, Seligsohn, et al., 1976 (after MM)
45,XX,−7,t(4;22)(p16;q11)/46,XX,t(4;22)(p16;q11)	Oshimura and Sandberg, 1977
43?,XX,−B,del(6),−7,−12,−?,+mar	Rowley, Golomb, et al., 1977b (after HD)
47,XY,−F,+2mar	Sakurai and Sandberg, 1976b
45,XY,−B,−C,Dq+,+16,+16,−18	Sakurai and Sandberg, 1976b
43,XY,−B,+16,−?17,−2G	Sakurai and Sandberg, 1976b
45,XY,+3,−3C,+mar/44,1q+,−2C	Sakurai and Sandberg, 1976b
45,XY,−B,+C,−17,Fp−?	Sakurai and Sandberg, 1976b
46,XY,+2,−2B,+16/43,−2,−2B,+D,+16,−F,−2G,+mar	Whang-Peng, Gralnick, et al., 1977
46,XX,19q−p−,11q+/47,XX,19q−p−,+21	Whang-Peng, Gralnick, et al., 1977
46,XY,20q−p−/47,XY,20q−p,+C	Anomalies observed in a number of cases, for which no complete karyotypes were given
Dq−	
−C	
+C	
+F	
Gq+	
Ph¹	
51,XY,+C,+F,+3G (Down's syndrome) (3Gp− : Ch¹ chromosome)	Juberg and Jones, 1970

Table 61 — *continued*

Karyotype	Reference
Ring chromosome in 90% of cells	DiGrado, Mendes, et al., 1964
47,XY,+9	Smalley and Bouroncle, 1967
45,XY,−C (AA→EL)	Kamada and Uchino, 1976
46,XX,−B,−B,−D,+3mar (leukopenia→EL)	Kamada and Uchino, 1976
46,XX,Cp−,Gp−	Kamada and Uchino, 1976
45,XY,−C	Dyment, Melnyk, et al., 1968
45,XX,−C	Pegoraro and Rovera, 1964
46,XX,−E,−E,+2mar (benzene)	Forni and Moreo, 1969
44,XY,−3,−B,+C,−E,−F,+M	Heath, Bennett, et al., 1969
44,XY,−2B	Heath, Bennett, et al., 1969
45,X,−Y	Heath, Bennett, et al., 1969
44,XX,−C,−G and ↑ polyploidy	Heath, Bennett, et al., 1969
46,XX	Heath, Bennett, et al., 1969
45,XY,−A,−B,+2C,−D,−E,+mar	Heath, Bennett, et al., 1969
47,XY,−B,+3C,−D	Heath, Bennett, et al., 1969
45,XY,−D	Heath, Bennett, et al., 1969
44,XY,−A,−B,−G,+mar	Heath, Bennett, et al., 1969
45,XY, with complex karyotype	M. H. Khan and Martin, 1970a,b
45,XX,−11,2q+	Golomb, Vardiman, et al. 1976b
47,XX,+C/48,XY,+C,+C } siblings	Izakovič, Steruská, et al., 1977
47,XX,+C/48,XX,+C,+C } siblings	R. J. L. Davidson, Walker, et al., 1978
45,XY,−7	Shiloh, Naparstek, et al., 1979

changes found in these subjects is shown in Table 60.

In 18 cases the percentage of polyploid metaphases (10-50%) was thought to be significantly increased. In 16 of these 18 patients the chromosome mode was in the hypodiploid region;[44] in only 1 case did the polyploid metaphase represent an exact duplication of the chromosome mode (Heath and Moloney, 1965a), whereas in the others various chromosomes had been lost after duplication of the mode. A multimodal or nonmodal chromosome distribution was observed in 31 of 64 cases with chromosome anomalies.[45]

In some of the cases with aneuploid chromosome modes, apparently normal diploid cells were encountered. Structural anomalies were found in at least 26 cases and consisted of rings derived from different chromosomes,[46] abnormally large or shortened acrocentric chromosomes,[47] and an abnormal #18 (Fitzgerald, Adams, et al., 1964). Minutes, breaks, dicentrics, and enhanced secondary constrictions were seen in 5-33% of the metaphases in 12 instances.[48] The application of banding techniques has led to the description of a number of different karyotypic findings in EL (Tables 61 and 62). Most interesting are an i(17q) and −7, the former in a 57-yr-old woman with 47,XX,−5, i(17q), −19,+3mar (Najfeld, 1976) and the latter in a 4.5-yr-old boy (Petit, Alexander, et al., 1973). Monosomy #7 has been described in other patients with typical and atypical EL and the relation of the chromosomal findings to the −7 syndrome in children is still unknown, though there is a strong suggestion that the association is more than chance would allow. Other interesting karyotypic findings in EL include: Dq−; 19p−q−, 11q+ or +21; Fp−q− with and without +C; del(20)(q12), 7p−,7q− or +E; del of chromosome 4q, 5q, 6q, 7p, 8p, 9q, −6, 16p+ and/or −4 and markers; −3,−18; −18; +8,+12,+17,t(17;21)(p11;q22); and 1q+,+C,+D,+F. Erythroleukemia in a 62-yr-old male of a family with a t(Dq;Dq) translocation has been described (Dallapiccola and Malacarne, 1971). Except for the translocation no other chromosomal changes were seen. Data gathered from the literature do not support the notion (M. H. Khan and Martin, 1970b) that the G-group is more often involved in EL than other chromosome groups.

Table 62 Ring chromosome in lympho- and myeloproliferative disorders

Disorder	Reference
AML	Sandberg, Ishihara, et al., 1962a (untreated); Nava, 1969; Obara, Sasaki, et al., 1971a (2 cases); Oshimura, Hayata, et al., 1976; Mitelman, Nilsson, et al., 1976; Sakurai and Sandberg, 1976a; Williams, Scott, et al., 1976 (2 cases); Hagemeijer, Smit, et al., 1977
EL	Di Grado, Mendes, et al., 1964 (90% of cells); Krogh Jensen, 1966; Durant and Tassoni, 1967; Crossen, Fitzgerald, et al., 1969; Diekmann, Rickers, et al., 1969; Heath, Bennett, et al., 1969; M. H. Khan and Martin, 1970a,b; Sakurai and Sandberg, 1974; Alimena, Annino, et al., 1975; Sakurai and Sandberg, 1976b (2 cases)
ALL	Sandberg, 1966a
AMMoL	Pierre, Hoagland, et al., 1971; Alimena, Annino, et al., 1975; Rowley and Potter, 1976; Olinici, Marinca, et al., 1978
PV	Lawler, Millard, et al., 1970 (with AL); Visfeldt, Franzén, et al., 1973; Manoharan and Garson, 1976; Zech, Gahrton, et al., 1976 (with AML)
CML (Ph^1+)	Nava, 1969; Meisner, Inhorn, et al., 1973; Prigogina, Fleischman, et al., 1978
Lymphoma	Sandberg, Ishihara, et al., 1964c; Fleischman and Prigogina, 1975; Mark, 1977b; Mark, Ekedahl, et al., 1978
Sézary syndrome	Lutzner, Edelson, et al., 1973; Lutzner, Emerit, et al., 1973; Whang-Peng, Lutzner, et al., 1976
Acute myeloproliferative syndrome	Berry and Desforges, 1969
Multiple myeloma	Dartnall, Mundy, et al., 1973
BP of CML	Srodes, Hyde, et al., 1973; Knuutila, Helminen, et al., 1978
AML (after X-ray)	Motoiu-Raileanu, Bercaneanu, et al., 1970
Myelofibrosis	Goodman, Bouroncle, et al., 1968
RCS	Obara, Sasaki, et al., 1971b
HD	Reeves, 1973
Gamma globulin abnormality	K. Jensen, Christensen, et al., 1969
EL after benzene	Forni and Moreo, 1969
Preleukemia	Pierre, 1978a
Lymphoma→AML	Goh, Bakemeier, et al., 1978

A comparison of the karyotypic pictures of EL and AML revealed no basic differences between these two groups of patients, except for those to be discussed below. As already emphasized by others (Diekmann, Rickers, et al., 1969; Naman, Cadotte, et al., 1971), it is likely that the cases of EL reported in the literature do not reflect the actual incidence of chromosomal changes in EL, because there is a tendency not to communicate normal findings, particularly in a single case. Exceptions to this finding are the completely normal cases reported by Sakurai (1970b), Nichols, Nordén, et al. (1970), and Hayhoe and Hammouda (1965). Also, in several publications on chromosomal anomalies in AML, some cases of EL have been included. Nevertheless, even with these reservations in mind, the published data indicate three interesting features of EL vs. AML: (1) hypodiploidy seems to be more common in EL than in AML; (2) structural chromosome anomalies with subsequent rearrangements giving origin to ring and dicentric chromosomes, karyotypic instability and polyploidy occur more frequently in EL than in AML; and (3) MAKA type of karyotypic changes and an AA chromosomal picture occurs much more often in EL than in AML (Sakurai and Sandberg, 1976b).

Among more than 500 cases of AML, a ring chromosome has been described in a half-dozen patients (Table 62). Reference to a case of Baikie and co-workers (Baikie, Court Brown, et al., 1961; Baikie, Jacobs, et al., 1961) is wrong because the cells were lymphocytes of an irradiated patient who developed leukemia. Even though Krogh Jensen (1969) observed breaks in 44% of his untreated AML patients, these breaks were "almost exclusively chromatid breaks" and no fragments, minutes, or dicentrics were mentioned. Considering the fact that megaloblastosis is a common finding in EL, and that megaloblastic anemias (MA) are almost invariably associated with structural chromosome changes, it has been proposed that in EL the structural chromosome anomalies are the result of an impaired DNA metabolism (Crossen, Fitzgerald, et al., 1969). Inasmuch as the B_{12} and folic acids levels in EL are usually normal, the pathogenesis of megaloblastosis and chromosome anomalies must be different in EL from that in MA.

Even though evidence has been presented that the erythroblastic series is involved by the leukemic process in EL, Inoue, Ravindranath, et al. (1975) indicated that this may not have been the case in two children with EL. Using several cytogenetic studies, the authors stated that erythropoiesis is not involved in the malignant process, but represents a response caused by unknown stimuli. This was based on a correlation of the number of aneuploid cells (47,XX,+C and 47,XX,+G in the two cases) with that of blasts in mitosis, peripheral normoblast count, and hemoglobin levels. L. Brandt, Mitelman, et al. (1975c) investigated the relation of aneuploidy to megaloblastosis in AML with 100% normal, diploid, or 100% abnormal cells. The findings indicated that chromosomal aberrations are not a prerequisite for the development of megaloblasts in AML and that abnormalities in the DNA synthesis responsible for megaloblastosis may occur without affecting the karyotype.

Usually EL transforms into AML during the latter course of the disease. The cytological change, however, is not accompanied by a cytogenetic one; the original karyotypic pattern persists throughout the course of the disease. From this it must follow that the chromosomal abnormalities are induced in precursor cells which give origin to erythroblasts and myeloblasts (Krogh Jensen, 1966; Krogh Jensen and Killmann, 1971), or that the leukemic process affects predominantly the myeloblastic series during the later stages of EL. The involvement of such precursor cells, however, does not appear to be restricted to EL. We have obtained indirect evidence from cytogenetic findings (Hossfeld, Han, et al., 1971) and others have more direct evidence from kinetic studies (Huber, Huber, et al., 1971), that in other varieties of DiGuglielmo's syndrome and in some cases of AML, the pristine precursor cell is the target cell for the leukemic process, which is certainly true for the great majority of cases with CML. As has been discussed for AML, the involvement of the precursor cell compartment by the leukemic process need not be a complete one. The presence of some normal diploid metaphases, which were noted in some cases of EL, suggests that some normal precursor cells are still resident within the marrow, a state that may be crucial in the prognosis and treatment of EL, as indicated by us (Sakurai and Sandberg, 1973). Consequently, after a complete remission had been obtained, the karyotypic picture of the marrow may become totally normal (Dyment, Melnyk, et al., 1968).

MIKA and MAKA Acute Leukemias

Because in our previous studies it appeared that the MAKA (major karyotypic abnormalities) group contained most of our patients with EL (Sakurai and Sandberg, 1976a), we undertook a detailed reexamination of MAKA patients to ascertain the possible existence of cytogenetic features characteristic for or at least indicative of EL (Sakurai and Sandberg, 1976b). We classified patients with AML into MIKA (minor karyotypic abnormalities) and MAKA groups. Patients of the former group had karyotypes that were the result of one, or at the most, two translocations or nondisjunctions. Patients of the latter group, on the other hand, had karyotypes that were the result of three or more such cellular events. Whereas the MIKA group consisted of patients with "classic" AML, and a small number of patients with AMMoL *(Naegeli)*, AMoL *(Schilling)* or EL, the MAKA group included many patients with a diagnosis of EL or possible EL and the remaining patients also had either cytologic and/or clinical features suggestive of EL. Thus, of our series of 113 patients with AML we analyzed more closely the karyotypes of patients of the MAKA group. Among the 19 patients who had been classified as belonging to the

MAKA group, 2 were interpreted as extreme cases of the MIKA group because they did share many of the specific features possessed by other patients with MIKA. The 17 patients with MAKA, who were included in the study, accounted for 1/3 of the 51 AML patients with chromosomal abnormalities studied by us; interestingly, nearly 50% of the patients of the MAKA group had a diagnosis of EL or possible EL. This proportion is surprisingly high considering that there was only 1 patient with EL among 34 with MIKA and only 4 with EL or possible EL among the 62 with no chromosomal abnormalities (N patients). The problem, then, is whether or not the remaining 50% of the patients of the MAKA group had EL.

The MAKA Group—A Type of Erythroleukemia?

In this section, the karyotypic aspects and their application to the clinical conditions of the MAKA group of patients will be discussed, because it was in this group of patients that the preponderant number of EL cases existed. Furthermore, those cases not diagnosed as having typical EL, nevertheless had a number of clinical and cytologic facets which resembled those present in EL. Thus, it is felt that the discussion of the MAKA group of patients bears primarily on the chromosomal interpretation of EL.

When representative karyotypes of patients of this group were compared, most of them had a certain degree of similarity to one another. The most common abnormalities were missing chromosomes in group B, and/or G, and the gain of a #16 and/or marker chromosomes. Even though the incidence of missing #17 or #18 or F group chromosomes was not proven to be significantly higher, the possibility existed that the same member within these groups was always being lost. Even though there were no markers in all or many of the cases of this group, markers such as 2q+, Dq+ or those consisting primarily of chromosome #2 or #12, minute metacentrics, and acentrics were found in more than one patient of the MAKA group. It is possible that at least some of the markers found in our series of patients will probably be established in the future, by utilizing banding techniques, as rather specific for this group, because most of the markers were never observed in patients with MIKA. The possibility of an isochromosome of the long arm of #17 occurring more frequently than chance alone would allow appears to be real. Besides these cytogenetic findings, karyotypic instability and high polyploidy were also features of this group of patients. All of the above features were equally observed in patients with a diagnosis of EL or possible EL, as well as in those patients in the MAKA group without such a diagnosis. Moreover, even though in the latter patients' marrows, the erythroid precursor cells generally constituted a small percentage of the total cell population, they were often morphologically abnormal, mostly megaloblastoid in nature and exhibiting PAS positivity. The total or nearly total lack of normal metaphases in the marrow (AA patients), whether or not the patient had a diagnosis of EL, was another important feature of this group, and strongly points to the involvement of both the erythroid and myeloid cell systems by chromosomal abnormalities. All of the above findings suggest that patients with MAKA constitute a unique group of AML, with all the patients in this group apparently having EL, whether or not they were so diagnosed.

High polyploidy and karyotypic instability were highly specific to the patients of the MAKA group. Because practically no polyploid mitotic cells were found in the BM of AA patients with MIKA, in whom all metaphases examined had chromosomal abnormalities, as in most patients with MAKA, and because karyotypic instability was almost never observed in patients with MIKA, the presence of karyotypic instability and any polyploid metaphases with chromosomal abnormalities in a patient with AML should strongly suggest that the patient belongs to the MAKA group, and hence, has EL. As discussed below, an increased polyploidy may be observed also in patients with EL who have MIKA or no chromosomal abnormalities. Hence, polyploid leukemic cells should never or seldom be encountered in patients with AML who have no suggestion of EL.

"Typical" MAKA should satisfy at least three of the following five criteria: (1) karyotypic instability, (2) polyploidy of abnormal cells, (3) markers, such as Dq+, rings, minute metacentrics, or acentrics, etc., (4) a mode of 47 chromosomes or less, and (5) a suggestion of three or more cellular events of nondisjunction or translocation (or deletion). The last two criteria, which originally were used to sort out the MAKA patients, may not always be satisfied.

Polyploidy—A Function of Abnormal Erythroid Cells and Karyotypic Instability

As stated earlier, high polyploidy is a characteristic feature of the MAKA group. What are the possible causes of this polyploidy? What is the meaning of polyploidy to the MAKA group? Even though dicentrics and rings were observed more often in polyploid cells than in cells in the diploid range, the incidence of such abnormalities generally was not high enough to fully explain the increased polyploidy. The fact that there was no difference in the incidence of dicentrics or rings among the cells with chromosome numbers in the diploid range in patients with a high incidence of polyploidy vs. those with a low incidence also suggests that these abnormalities did not play a critical role in the causation of the high polyploidy. However, exceptions may occur. In one of our patients who showed the highest polyploidy, dicentrics of various sizes were observed in most of the polyploid mitotic cells, whereas none was found in cells in the diploid range. The dicentrics possibly had some relation to the increase of polyploidy in this particular patient. The incidence of dicentrics or rings is too low to explain the karyotypic instability. Megakaryocytes were only scarcely seen in the marrow of patients with this type of AL, indicating that they did not account for the presence of high polyploidy, as reported in CML.

A positive correlation between polyploidy and the percent of proerythroblasts and erythroblasts indicates that the increased polyploidy is related to the higher percentage of erythyroid cells. However, even though generally the marrows of patients with high polyploidy had more binucleate or giant-nucleated abnormal cells, compared with those with a low polyploidy, and the majority of these cells consisted of proerythro- and erythroblasts, nevertheless, many of these cells were myeloblasts, promyelocytes, or myelocytes, and in a few cases the latter classes of cells outnumbered the former. It might well be, therefore, that common unknown factors may be operating in this group of patients to stimulate proliferations of the leukemic erythroid cells and at the same time to stimulate the polyploidization of abnormal erythroid as well as myeloid cells.

The positive correlation between the percent polyploid mitotic cells and the percent of karyotypic instability suggests that the latter may also have some influence on the increase of polyploidy. An interesting finding was that three of our patients with 100% karyotypic instability with low polyploidy had a low percentage of erythroid cells, whereas the three patients with 100% karyotypic instability and high polyploidy, had a high percentage of erythroid cells in their marrow (Sakurai and Sandberg, 1976b). The probability for the occurrence by chance alone of three patients with < 2% polyploidy having < 20% erythroid cells and three patients with > 5% polyploidy having > 20% erythroid cells is 0.05 by direct calculation of probability. Thus, these findings indicate that the presence of > 20% erythroid cells is essential for an increase of polyploid cells, and that only when this prerequisite is fulfilled, the presence of high karyotypic instability could further increase the polyploidy beyond the 2% threshhold.

Polyploid mitotic cells often seem incapable of completing cell division. Double-minute chromosomes (DMS) and pulverization, together with protracted mitosis, as suggested by frequently observed parted sister chromatids in the polyploid mitotic cells, indicate that many such cells might commit a mitotic death rather than complete mitosis (Urasinski, 1976). It would be tempting to speculate that this mechanism contributes to the maintainance of hypo- or normocellularity of the BM in most cases with MAKA. As a matter of fact, we have observed a number of patients of the MAKA group who did well for months with only conservative treatment before requiring any chemotherapy, an experience described by others. In contrast, almost all of the published cases with MIKA or with no chromosomal abnormalities[49] received chemotherapy, because of the high proliferative activity of the marrow and the consequences resulting therefrom.

A Survey—Three Chromosomal Types of EL?

Of the 30 patients with MAKA published in the literature,[50] 15 had a diagnosis of EL. The others had the following diagnoses: refractory anemia, AML, monocytic leukemia, stem cell leukemia, AL, and myeloproliferative disorders. Even in the patients with these disorders, the clinical aspects, when given in detail, often suggest that they had an erythroleukemic condition. As in our patients, their karyotypes differed from case to case, including karyotypic instability in most cases. Yet, when these patients were looked at as a whole, they were seen to have the same cytogenetic tendency as the patients we studied. Thus, most of these patients had a hypodiploid mode, karyotypic instability, and increased polyploidy and appeared to have no or very few normal dividing cells (i.e., AA or near-AA patients). Their karyotypes often lack B-, E-, or G-group chromosomes and have rings, minutes markers, dicentrics, or breaks.

The other reported chromosomal abnormalities in EL have been shown to include cases with simple loss[51] or gain[52] of chromosomes, translocations or partial deletions (Fitzgerald, Adams, et al., 1964; Kiossoglou, Mitus, et al., 1965a), and cases without any chromosomal abnormalities (N patients).[53] Therefore, from the cytogenetic viewpoint there seem to be three groups in EL, i.e., one group with MAKA, another with MIKA, and a third without chromosomal abnormalities.

Increased polyploidy has been reported in the literature in 10 of the 15 EL patients with MAKA, in 3 of 17 patients with MIKA, and in 1 of 20 patients without chromosomal abnormalities. The above figures probably reflect a generally lower proportion of polyploid cells in patients with MIKA and in those without chromosomal abnormalities, than in those with MAKA. Not only patients with MAKA, but many patients of EL with MIKA and those with no chromosomal abnormalities have been reported to have multinucleated or bizarre erythroid cells.

Even though in our group of patients with EL or possible EL we have observed only one with MIKA and four with no chromosomal abnormalities (Sakurai and Sandberg, 1976a), the findings in these patients do not seem to contradict those of the cases reported in the literature. Of our five patients, only a patient with MIKA showed polyploidy and was the only one with some binucleate erythroid cells and giant myeloblasts. Thus, our speculation concerning polyploidy in patients of the MAKA group could be generalized for patients with EL. There is a tendency in patients with EL for the marrow cells to fail in nuclear division or very often in cytoplasmic division. This may be the mechanism responsible for the multinucleated or giant nucleated abnormal cells and the polyploid cells in EL. This mechanism could operate in patients with EL regardless of the presence or absence of chromosomal abnormalities or of their karyotypic status. However, the tendency might be the strongest in the MAKA group, possibly being enhanced by the karyotypic instability.

Because the majority of the patients with MAKA have no normal dividing cells among the normal cells in their marrow (AA patients), and the others have only a few normal cells (near-AA patients), their response to antileukemic therapy is very poor (for AA) or at the best fair (for near-AA), as far as survival is concerned. Patients with MIKA, on the other hand, usually have substantial numbers of dividing normal cells (AN patients) and, hence, show a much better response to therapy—as do patients without chromosomal abnormalities. These principles seem to hold true also for EL in the literature. Summarizing the patients in the literature, including our own series (Sakuari and Sandberg, 1974, 1976a, b), in whom the survival time after the initiation of chemotherapy was available, it appeared that 5 out of 10 patients whom we have specified at MAKA died within a month,[54] whereas only 2 of the 18 with MIKA or no chromosomal changes[55] died within the same period of time.

Cytogenetic results presented for EL, taken in conjunction with certain clinical and cytologic features of the disease, e.g., the fact that we have not observed Auer bodies but PAS positivity in the cells of our patients with MIKA, may lead to a more precise classification and diagnosis of this disease. For example, in our series of 51 patients with AML, the EL patients were particularly characterized by MAKA, and #8–#21 translocation that is relatively common in AML was not seen, at least as an isolated chromosomal abnormality, in the cells of any of our patients with EL. Of interest is the fact that three of our patients with MAKA had multiple myeloma before they developed AL (Sakurai and Sandberg, 1976a, b). Even though there has been no description of such a history among the patients with MAKA in the literature, multiple myeloma or the therapy for this disease may have some relation to the development of the MAKA type of EL, because none of our AML patients with MIKA or without chromosomal abnormalities had such a history. It is hoped that further analysis with banding techniques of the abnormal chromosomes in EL may reveal karyotypic features for this disease more thoroughly and when combined with careful evaluation of its clinical aspects will ultimately lead to a better understanding of this form of AL.

Further Aspects of MAKA and EL

As stated above, about 50% of patients of the MAKA group have a diagnosis of EL or possible EL. This finding contrasts remarkably with that in the MIKA group, in which only 1 out of 32 patients had such a diagnosis (Sakurai and Sandberg, 1976a). In the MAKA group, besides an overwhelming majority of AA or near AA-patients, predominant hypodiploidy and karyotypic instability, the karyotypes were highly complicated, most with markers, including minutes and rings, which were rarely seen in other karyotypic groups of AML. These findings strongly indicate a homogeneity of the patients of the MAKA group. In the literature, these chromosomal features have been reported in many patients with EL.[56]

Despite our presumption (Sakurai and Sandberg, 1973, 1974) of the presence of chromosomal abnormalities both in the erythroid and myeloid cells in EL of the MAKA group, the same does not hold true for patients of the MIKA group. In fact, very few erythroblasts were encountered in the marrow of these patients, even when chromosomal analysis showed 100% abnormal metaphases, and when present these erythroblasts did not have any morphologic abnormalities. Therefore, even though some patients of this group had megaloblastoid erythropoiesis in their marrow before therapy, most of the patients of the MIKA group did not. Hence, both on the basis of the morphologic aspects of the marrow and from graphanalysis (Sakurai and Sandberg, 1974), we cannot agree at this time with the generalized concept that in all patients with AML both the myeloid and erythroid elements are invariably involved (Krogh Jensen and Killmann, 1971; Sandberg, Sakurai, et al., 1973).

Although a missing Y is chiefly associated with the prototypic karyotype in AML, it may also occur in some patients of the MAKA group. However, whether or not the original chromosomal abnormalities of these patients resided in a prototypic karyotype (Sakurai and Sandberg, 1974) is unknown. The highly complicated karyotype of these MAKA patients had certain similarities with the prototypic one, but also many features in common with other MAKA group patients. Even though banding studies have not been applied extensively to the cells of these patients, before long such studies should elucidate whether or not some MAKA patients really have a missing Y without the involvement of a #8-#21 translocation. Some published karyotypes suggest that a missing Y may have occurred in a MAKA type patient with EL (Heath and Moloney, 1965a).

Prophasing and EL

Prophasing ("chromosome pulveration";PCC) (Matsui, Weinfeld, et al., 1971, 1972) is a term applied to the induction by a metaphase cell of premature prophasing in an interphase cell in a fused cell containing the nuclei of two such cells in a common cytoplasm (Figures 106 and 107; Takagi, Aya, et al., 1969; Sandberg, Aya, et al., 1970). The presence of a substance in the metaphase cell necessary for prophasing has been established in my laboratory (H. Kato and Sandberg, 1967, 1968). Under normal circumstances prophasing is not seen in the marrow. The finding of multinucleate cells due to fusion, usually indicated by the asynchrony of the nuclei and, in particular, by prophasing, indicates that cell fusion has occurred in the marrow. In all probability only certain viruses (adenoviruses) are capable of inducing cell fusion in vivo and the finding of prophasing merely indicates the presence of such a virus. The latter is not necessary for prophasing, once cell fusion between a metaphase cell and an interphase cell has occurred. In other words, prophasing is prima facie evidence for the presence of certain viruses in the marrow, though the possibility that it is due to viral or other products cannot be excluded. The use of prophasing for predicting the response and progression of leukemia has been advocated (Hittelman and Rao, 1978).

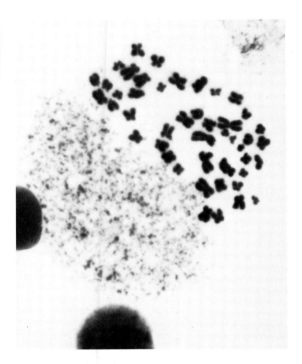

In only one condition has prophasing been encountered in the marrow of a significant number of cases, i.e., EL. Even though prophasing may be seen rarely in some tumors (e.g., neuroblastoma), we have observed it in a significant portion of our patients with EL. This finding definitely points to the presence of a virus or its products in the BM of EL. The exact relationship of this virus to the disease is unknown at present. We have not seen prophasing in the marrow of patients with ALL or AML without abnormalities of the erythroid precursor cells. In this connection, the findings of Williams, Scott, et al. (1976) are cogent, because these authors observed a very high incidence of prophasing in seven patients, six of whom had AML/AMMoL "with abnormal erythroid precursor morphology" compatible with EL. In three similar cases prophasing was not noted.

Obviously, more appropriate studies are needed in order to explore the implications of these findings.

106 Prophasing resulting from fusion of a human metaphase with an interphase cell (upper figure). The interphase cell appears to have been in the S-phase, as shown by the uptake of tritiated thymidine (lower figure).

Down's Syndrome and Leukemia

Much has been written and heard about the relation of DS (mongolism) to the development of leukemia, particularly AL (Table 63). Recent evaluations of this relationship point to the very strong possibility that in a substantial number of cases of DS with alleged leukemia, the hematologic picture was more properly related to a leukemoid reaction, which was reversible and not leukemic (R. R. Engel, Hammond, et al., 1964). Nevertheless, the incidence of leukemia does appear to be increased in subjects with DS over that in the general population and deserves special attention.[57] Furthermore, a higher frequency of leukemia appears to be present in families with a high incidence of Down's syndrome, the latter not necessarily involving the subjects affected by leukemia.[58] A higher frequency of other cancers has been reported in DS (R. W. Miller, 1963, 1970). The exact origin of the extra #21 (Mutton, 1973) and some of the enzymic changes (c.f., N. J. Brandt, Froland, et al., 1963) apparently associated with it remain to be clarified. The incidence of AL in DS due to translocation (G-G or G-D translocation) is not known because of the low number of cases involved (German, Demayo, et al., 1962). The

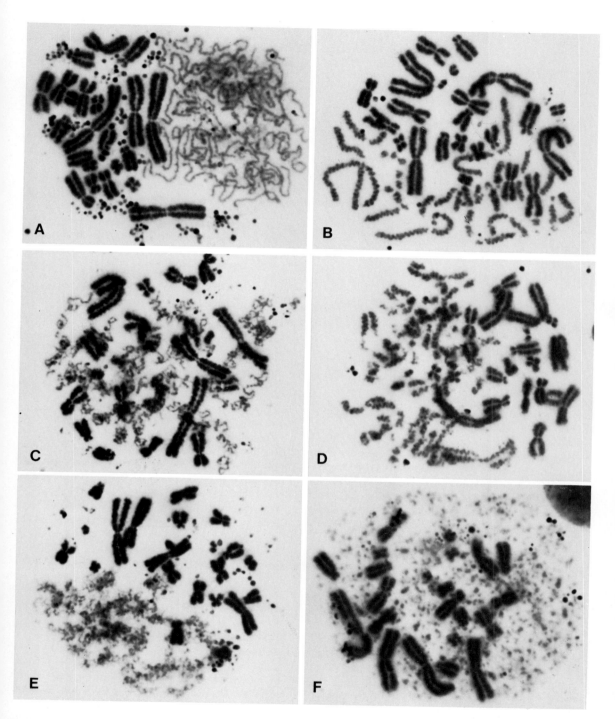

107 Various stages of "prophasing" (premature chromosome condensation, PCC) observed in Chinese hamster cells when a metaphase is fused with an interphase cell. A substance in the metaphase cell leads to the prophasing of the *interphase* nucleus, as long as the latter is in either the G_1, S, or G_2 period. A and B represent prophasing of nuclei in the G_1 period; C, D, and E in the S-phase, and F in the G_2 phase, as witnessed by the presence of two chromatids.

Table 63 DS and AL

Karyotype	Type of AL	Reference
47,XY,+21/48,XY,+C,+21,(+frag)	AML	Johnston, 1961
47,XX,+21,+tetraploid cells		Propp, Brown, et al., 1966
49,XX,+21,+2G (iso-X)		Ross and Atkins, 1962
51,XY,+13,+19,+21,+22,+22	AML	Kiossoglou, Rosenbaum, et al., 1963
54,XX,+2C,+2D,+F,+2G,+21	AML	Lejeune, Berger, et al., 1963
47,XX,+21/48,XX,+C,+21,Bq+	AML	Mercer, Keller, et al., 1963
48,XX,+C(8),+21	AML	Warkany, Schubert, et al., 1963
47,XY,+21/48,XY,+C,+21	AML	Honda, Punnett, et al., 1964
59,XY,+4C,+3D,+E,+2F,+2G,+21	AML	Conen, Erkman, et al., 1966
47,XX,+21/ca. 8% tetraploid with 94 chromosomes	AML	Propp, Brown, et al., 1966
47,XY,+21/50,XY,+D,+20,+21,+22	AMMoL	DeMayo, Kiossoglou, et al., 1967
46,XX,−C(7),+21/47,XX,+21		M. H. Khan and Martin, 1967c
48,XX,Bq+,+C,+21	?AML	Schleiermacher and Kroll, 1967
48,XY,+F,+G/49,XY,+C,+F,+G/51,XY,+C,+D,+F,+2G	AML	Petit Maurus, et al., 1968
49,XX,+C,+G,+G	AML	Buchanan and Becroft, 1970
51,XY,+C,+F,+3G(3Gp−;3Ch¹)	EL	Juberg and Jones, 1970
50,XY,+21,+2C,+G	AML	Rethoré, Prieur-Lecuyer, et al., 1971
46,XX,+21,−C/45,XX,+21,−C,−D/45,XX,−D/ 47,XX,+21,−C,+F	ALL	Crisalli, Monteverde, et al., 1971
52,XX,1p+,+10,+13,+19,+21,+22,+22	AL	Berger, Weisgerber, et al., 1973b

role played by the extra #21 in the development of the AL is uncertain, because only a small percentage of patients with Down's syndrome develop AL, even though admittedly the incidence of AL in these patients is much higher than in a comparable population with diploid chromosome constitutions. The AL in DS may be either myeloblastic or lymphoblastic, the exact type apparently depending on the time of the development of the leukemic state (Rosner and Lee, 1972; Boisseau and Le Menn, 1974). However, it does appear that AML is more frequent than ALL, when compared to the incidence in a similar group of subjects without DS. The chromosomal changes in AL with DS resemble those of other forms of AL,[59] i.e., about 50% of the cases developed karyotypic changes, in addition to the #21 trisomy. The latter persists in the marrow cells, even when aneuploidy appears. With successful therapy the aneuploid cells may disappear from the marrow, but the #21 trisomy is retained.

The susceptibility of the cells in DS to damage by X-ray and UV and their impaired ability to repair the DNA lesions has been proposed as a possible cause for the higher rate of leukemia in these subjects (Lambert, Hansson, et al., 1976a). Whether the sensitivity of the cells in DS to other agents (SV40 virus, dimethylbenz(a)anthracene (DMBA), mitomycin C)[60] is also indicative of this higher risk of leukemia remains unknown.

An interesting case of an infant with DS and AML in whom clonal evolution was thought to occur was described by Berger, Weisgerber, et al. (1973). In the leukemic cells of this patient the chromosomal number ranged from 48–53 in the marrow; all cells contained 1p+,+C, +F (#19), +G(#22), and +21. In some cells tetrasomy of #21 apparently was observed. Trisomy #13 was also present in cells with the higher number of chromosomes.

In another case of DS and AML (3-yr-old female) the chromosome count ranged from 42–49 in the marrow with loss of a G-group chromosome, either alone or with loss of a D-group or gain of an F-group autosome, being most common. Loss of a D-group chromosome coupled with loss of a G-group chromosome was also observed. Of interest was the presence of cells with a 45,XX,D− pattern, not includ-

Table 64 Chromosome findings in APL

Karyotype	Reference
46,XX,17q− or 18q−	E. Engel, McKee, et al., 1967
46,XX,t(15;17)	Golomb, Vardiman, et al., 1976; 1978
46,XY,t(15;17)	Golomb, Vardiman, et al., 1976; 1978
46,XY,t(15;17)/46,XY,7q−,9q−,t(15;17)	Golomb, Vardiman, et al., 1978
46,XY,t(15;17)(q22;q21)	Kaneko and Sakurai, 1977
?Ph¹-positive case	Misset, Venaut, et al., 1977
46,XX,t(15;17)(q22;q21)	M. Okada, Miyazaki, et al., 1977
46,XY,t(15;17)(q22?;q21),del(7)(q22),del(9)(q22)	Rowley, Golomb, et al., 1977a
48,XY,+8,+13,t(15;17)/47,XY,+8,t(15;17)/46,XY,t(15;17)/ 49,XY,+8,+13,+18,t(15;17)	Testa, Golomb, et al., 1977; 1978
t(15;17)(q24;q21)	Scheres, Hustinx, et al., 1978
46,XY,del(17)(q12)/45,X,−Y→46,XY→46,XY/46,XY,del(12) (q11)→46,XY/45,XY,−12→46,XY/46,XY,del(12)(q11)	Van Den Berghe, Louwagie, et al., 1979b
46,XX,t(15;17)(q26;q22)	Van Den Berghe, Louwagie, et al., 1979b
46,XX/46,XX,t(15;17)(q26;q22)	Van Den Berghe, Louwagie, et al., 1979b
47,XY,+8	Van Den Berghe, Louwagie, et al., 1979b
46,XX,t(15;17)(q26;q22)/46,XX,del(7)(q33or35)t(15;17) (q26;q22)→46,XX→46,XX,t(15;17)(q26;q22)/46,XX	Van Den Berghe, Louwagie, et al., 1979b
47,XX,+8/47,XX,+8,t(15;17)(q26;q22)/46,XX,−7,+8, t(15;17)(q26;q22)	Van Den Berghe, Louwagie, et al., 1979b
46,XY,t(15;17)(q21?:q25?)/47,XY,+8,t(15;17)q21?:q25?)/ 47,XY,+9,t(15;17)(q21?;q25?)	Van Den Berghe, Louwagie, et al., 1979b
46,XY/46,XY,t(15;17)(q26;q22)/47,XY,+8/47,XY,+8,t(15;17)	Van Den Berghe, Louwagie, et al., 1979b
46,XX,t(15;17)(q26;q22)	Van Den Berghe, Louwagie, et al., 1979b
46,XX	Van Den Berghe, Louwagie, et al., 1979b
46,XX,t(15;17)(q26;q22)/47,XX,−7,+8,−10,t(15;17),+2mar/ 46,XX,−7,−9,t(15;17),+2mar/45,XX,−10,t(15;17)	Van Den Berghe, Louwagie, et al., 1979b
46,XY/46,XY,t(15;17)(q26;q22)/47,XY,+8,t(15;17)	Van Den Berghe, Louwagie, et al., 1979b
46,XY,t(15;17)(q26;q22)	Van Den Berghe, Louwagie, et al., 1979b
46,XY,t(15;17)(q26;q22)	Van Den Berghe, Louwagie, et al., 1979b
47,XY,del(12)(p11),del(20)(q11),+9	Mitelman, Nilsson, et al., 1978

ing the #21 trisomy (Crisalli, Monteverde, et al., 1971). A 2-yr-old boy trisomic for #21 with AML fatal within 5 mo was shown to have a major clone of 50 chromosomes (50,XX,+10, +11,+21,+22) with apparent clonal evolution (Rethoré, Prieur, et al., 1971).

The association of CML with DS has not been acceptably described, to my satisfaction or that of others (Rosner and Lee, 1972), because in the few reports on this subject the patients appear not to meet the criteria for CML. Such an association would be of great interest, because it would supply cogent information on the susceptibility of patients with group-G trisomy to CML and Ph¹ formation. The demonstration that the trisomy in DS is due to chromosome #21 (J. J. Yunis, Hook, et al., 1965) and the Ph¹ in CML is a #22 autosome (Cas-

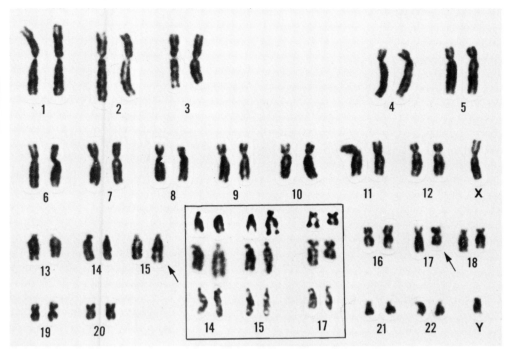

108 **A.** (above) Two karyotypes (not banded) stained with Giemsa and quinacrine, respectively, and partial metaphases from a patient with APL showing the 15;17 translocation observed in a significant number of the patients with this disease. *(Golomb, Rowley, et al., 1976.)* **B.** (below) Karyotype (Q-banded) from one of our patients with APL and demonstrating the 15;17 (arrows) translocation. This is the only patient with APL with this anomaly observed by us to date among a group of at least 10 patients with this form of AL. Note the fuzzy and ill-defined appearance of the chromosomes, a finding characteristic for leukemia chromosomes.

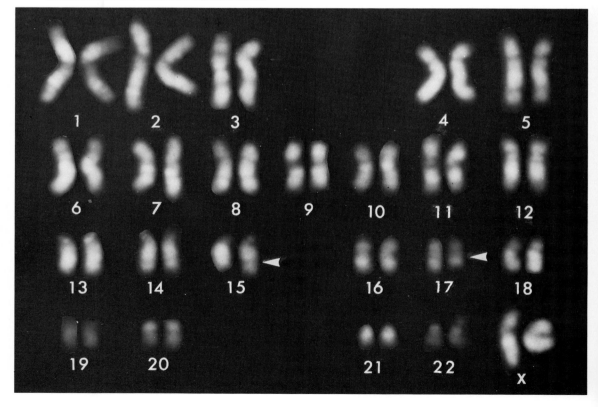

persson, Gahrton, et al., 1970; O'Riordan, Robinson, et al., 1971) further enhances the interest in the incidence of CML in DS. Furthermore, the chromosomal picture in DS and leukemia is complicated by demonstrations that indicate that not an inconsiderable number of subjects thought to have leukemia are, in fact, involved by a leukemoid reaction rather than a leukemic process (Rosner and Lee, 1972; Séé, Dayras, et al., 1974). Thus, much remains to be clarified in the area of leukemia associated with DS.

Acute Promyelocytic Leukemia

Acute promyelocytic leukemia (APL) (Table 64) is thought by some to be a unique form of AML (Bennett, Catovsky, et al., 1976) characterized by the predominance of leukemic promyelocytes filled with coarse azurophilic granules and by the presence of diffuse intravascular coagulation (DIC) (Goldman, 1974). An unusual case of DIC in a patient with leukemia and trisomy #21 has also been published (Cosson, Depres, et al., 1974).

Banding analysis has revealed an apparently unique cytogenetic pattern in APL, i.e., (15;17)(q22;q21) (Figure 108). Originally, Golomb, Rowley, et al. (1976) described a partial deletion of the long arm of #17 in the only two patients with APL in their group of 50 patients with ANLL. A third patient with APL was reported by Rowley, Golomb, et al. (1977d) in whom the exact nature of the karyotypic changes was established. Since then two other cases of APL with a 15/17 translocation have been described (M. Okada, Mikazaki, et al., 1977; Kaneko and Sakurai, 1977). In 16 patients with APL, 14 (87%) had abnormal karyotypes, of whom 11 showed a t(15;17)(q26;q22), 1 patient a 47,+8, 1 a rearrangement of #15 and #17, and 1 a deletion of #17 without apparent translocation (Van Den Berghe, David, et al., 1978a). Trisomy #8 was found in five and −7 in two of these cases. It is still uncertain, though, whether the break occurs at 17q12 instead of 17q21 because band q12 of the abnormal #17 did not seem to be as broad as that of the normal #17. We have observed one 15/17 translocation in 5 patients with APL; Nava (1969) published two karyotypically normal cases and one with +C. In 1967 E. Engel, McKee, et al. described a deleted long arm in a chromosome of group #17−#18 in a 6-yr-old girl with APL. Zuelzer, Inoue, et al. (1976) described a boy with APL with a normal karyotype. Other cytogenetic changes have been observed in some of the patients with the 15/17 translocation, e.g., 46,XY,t(15;17),del(7)(q22),del(9)(q22) (Testa, Golomb, et al., 1977).

Teerenhovi, Borgström, et al. (1978) found only 1 case of APL with chromosome abnormalities out of a group of 12 such cases in Finland and Sweden. No 15;17 translocation was observed. The authors proposed an uneven geographical distribution of 15;17 translocations of APL in view of the high incidence of this translocation in APL in Chicago and Belgium.

Acute Myelomonocytic Leukemia

The exact definition of acute myelomonocytic leukemia (AMMoL) (Table 65) has not been established nor, in particular, has its relation to AML and the subacute and chronic forms of myelomonocytic leukemia. Abnormalities of leukocyte adenosine deaminase (ADA) phenotypes were observed in seven patients with AMMoL and rarely in other types of AL; it has been suggested that electrophoresis of ADA in cells from patients with AL may be a useful adjunct to the diagnosis of AMMoL. Abnormalities of the platelets have also been described in this leukemia.

Saarni and Linman (1971) have emphasized the multicellular nature of AMMoL and its rather common occurrence. An admixture of atypical monocytic, granulocytic, erythrocytic, and megakaryocytic precursors is readily recognized in AMMoL, according to the above authors. Others have considered AMMoL to be part of a spectrum of AML, a view subscribed to by me.

There is little doubt that the putative incidence of AMMoL differs from laboratory to laboratory as a result of differences in diagnostic criteria. Thus, in our experience and apparently in that of Mitelman, Nilsson, and colleagues (1976), AMMoL is rather rare, whereas a high incidence among cases of AL has been reported by Golomb, Vardiman, et al. (1976). These authors also indicated that on the basis of their experience with 24 cases of AMMoL

Table 65 Chromosome findings in AMMoL

Karyotype	Reference
44,X(Y),–B,–C,–E,–G,+D,+mar/43,X(Y),–B,–C,–E,–G,+D/44,XY, –B,–E,–G,+D	Nowell and Hungerford, 1962
47,XX,+C/48,XX,+C,+mar	Nowell and Hungerford, 1962
46,XY/50,XY,+4G	Atkins and Goulian, 1965
46,XY,+2,–3,–3,+mar	Ragen, McGuire, et al., 1968
45,XY,–F/46,XY	Pierre, Hoagland, et al., 1971
50,XY,+4min/50,XY,–C,+5min	Pierre, Hoagland, et al., 1971
52,XX,–A,+mar,+6min/51,XX,–A,+mar,+5min	Pierre, Hoagland, et al., 1971
44,XY,–C,–D/45,XY,–C/44,XY,–C,–D,–G,+min	Pierre, Hoagland, et al., 1971
48,XX,–A,+E,+mar,+ring	Pierre, Hoagland, et al., 1971
47,XY,+min/46,XY	Pierre, Hoagland, et al., 1971
46,XY,–C,+E	Pierre, Hoagland, et al., 1971
45,XX,–B,–D,+ring/46,XX	Pierre, Hoagland, et al., 1971
48,XX,–C,–2D,+5G,+min/49,XX,–C,–2D,+5G,+4min	Pierre, Hoagland, et al., 1971
47,+C	Berger, 1972a
47,1q+,+C	Berger, 1972a
45,XX,–9 (with cutaneous xanthoma)	Lampert, Phebus, et al., 1972
45,–G	M. A. S. Moore and Metcalf, 1973
46,XX	M. A. S. Moore and Metcalf, 1973
47,XY,+21	L. Brandt, Levan, et al., 1974a
45,XX,–7 (myeloproliferative disease → AMMoL)	MacDougall, Brown, et al., 1974
47,XY,+9	Rutten, Hustinx, et al., 1974
46,XY/45,XY,–8,+9,–21/47,XY,+9	J. H. Ford, Pittman, et al., 1975
46,XX,+9,–14/46,XX	J. H. Ford, Pittman, et al., 1975
45,XX,–13	Golomb, Vardiman, et al., 1976
42,XY, with complex karyotype	Golomb, Vardiman, et al., 1976
46,XY,t(1;9)(p34?;q34),inv(8)(p11q12)	Rowley and Potter, 1976
46,XX,ins(3;3)(q21;q21q26)	Rowley and Potter, 1976
46,XY,t(3;5)(q25;q33?)	Rowley and Potter, 1976

the presence of an abnormal karyotype did not influence complete response to therapy or median survival, compared to the presence of a normal karyotype. Their series included 14 N patients, 9 AN patients and 1 AA patient with AMMoL. The AN group included two male patients (58 and 61 yr old) with a missing Y apparently as the only karyotypic abnormality.

The karyotypic abnormalities in AMMoL have been variable and have included gain or loss of C- and G-group chromosomes (L. S. Goldberg, Winkelstein, et al., 1968), loss of the Y and marker chromosomes. Trisomy #21 was revealed with G-banding in a 57-yr-old male with AMMoL (L. Brandt, Levan, et al., 1974a) and multiple chromosomal associations and paracentromeric region instability in another (Castoldi, Gruvosin, et al., 1975). In the patient of L. Brandt, Levan, et al. (1974a) almost all marrow cells were trisomic for #21. No evidence of DS was present. The trisomy was evidently present in both granulocytic and monocytic cells, because dividing cells of both types were present in the marrow in almost equal proportions. The authors refer to a dozen similar cases showing loss of one G-group autosome and suggest the possible specific involvement of #21 in AMMoL. Other anomalies obtained with banding included: ins(3;3)(q21;q21 q26); t(1;9)(p34?;34), inv(8)(p11q12); t(3;5) (q25;q33?); t(?;11)(?;q23); t(8;17)(q22;q25?); ring consisting of 2 #21 chromosomes; del (2) (q31?), del(7)(p13),–13; and other more com-

Table 65 — *continued*

Karyotype	Reference
46,XX,t(?;11)(?;q23)	Rowley and Potter, 1976
45,X,−X,t(8;17)(q22;q25?)	Rowley and Potter, 1976
45,X,−Y (2 cases)	Rowley and Potter, 1976
45,X,−Y,r(21)	Rowley and Potter, 1976
45,XX,del(2)(q31?),del(7)(p13),−13	Rowley and Potter, 1976
42,XY,−7,i(8q),−16,−17,−18,t(21;21)(q12;q11),+mar,+fragment	Rowley and Potter, 1976
46,XX/47,XX,2q+,−5,+3,+mar/48,XX,−5,+14,+22,+mar	Zuelzer, Inoue, et al., 1976
46,XX,Ph¹+/47,XX,+C,Ph¹+	Zuelzer, Inoue, et al., 1976
46,XX/47,XX,+C/47,XX,+22	Zuelzer, Inoue, et al., 1976
Two cases of Ph¹+	Bloomfield, Peterson, et al., 1977
46,XY/46,XY,−C,+mar (rare cell with 45,X,−Y)	Castoldi, Grusovin, et al., 1977
45,X,−Y,del(7)(q32),t(8;21)(q22;q22)	Castoldi, Grusovin, et al., 1977
47,+19,20p+	Hustinx, Scheres, et al., 1977
?Ph¹+	Morse, Humbert, et al., 1977
46,XY/46,XY,19p−q−/47,XY,+C or +G	Svarch and de la Torre, 1977
46,XYq−,Ph¹/48,XYq−,+B,+G,Ph¹/46,XYq−	Whang-Peng, Gralnick, et al., 1977
46,XX,16p+	Padre-Mendoza, Forman, et al., 1978
46,XY,20q−	Golomb, Vardiman, et al., 1978
46,XY,t(5;19),16p+,19p+	Golomb, Vardiman, et al., 1978
47,XY,+F(20)/45,X,−Y,−C(or E18),+F(20)	Golomb, Vardiman, et al., 1978
47,XX,+8/46,XX	Karpas, Khalid, et al., 1978
47,XY,+22	Shiloh, Naparstek, et al., 1979
46,XY,−11,+t(1;11)(q11 or 12; q25) → 48,XY,+8,−11,+t(1;11)(q11 or 12;q25),+19p−(Preleukemia → AMMoL)	Shiloh, Naparstek, et al., 1979
46,XY,+8,−17/47,XY,+8/46,XY	Najfeld, Singer, et al., 1978
47,XY,+21/46,XY	Mitelman, Nilsson, et al., 1976, 1978
45,X,−Y/46,XY	Mitelman, Nilsson, et al., 1976, 1978
46,XX,i(17q)	Mitelman, Nilsson, et al., 1976, 1978
45,X,−X,r(21)	Mitelman, Nilsson, et al., 1978
	Olinici, Marinea, et al., 1978

plicated karyotypes. Trisomy #9 has been seen in a 35-yr-old male with AMMoL 4 yr after removal of a Wilms' tumor for which he received radiation and chemotherapy (Rutten, Hustinx, et al., 1974). A case of AMMoL (57-yr-old male) with Auer rods in some of the cells was shown to have a 45,X,−Y,del(7)(q32), t(8;21)(q22;q22) karyotype. The latter translocation is the "prototypic" one described for AML, with which −Y is not uncommon. The chromosome picture became totally diploid during full remission (Hustinx, Scheres, et al., 1977). Such a case would probably have been diagnosed by me as AML and, in fact, the various findings and clinical course are consistent with this assumption. A case of AMMoL in a 76-yr-old woman containing a ring chromosome and a mode of 47 chromosomes has been described (Alimena, Annino, et al., 1975).

A 2-yr-old girl with an incomplete form of neurofibromatosis who developed leukemic xanthomatosis (AMMoL with cutaneous xanthomas) was found by Lampert, Phebus, et al. (1972) to have 45 chromosomes in the marrow and blood cells in which #9 was missing.

Trujillo, Cork, et al. (1971) described a unique finding in a male aged 69 with AL and a normal tetraploid line in the marrow (87% of the cells), with 8% aneuploid cells in the tetraploid range and 5% diploid cells. Even though the initial diagnosis was AMMoL, full investigation failed to confirm the cell type of the leukemic cells. The tetraploidy, however, suggested the possibility that they were neoplastic megakaryoblasts. Cases with diploid karyotypes have also been seen, constituting > 50% of the patients in one series (Maldonado and Pierre, 1975).

In 16 patients with AMMoL in whom major morphologic alterations of the platelets were found, no correlation with the karyotypic picture could be established (Maldonado and Pierre, 1975). As stated above, patients with AMMoL demonstrated no difference in median survival times when subgrouped according to the presence or absence of chromosome abnormalities (Golomb, Vardiman, et al., 1976). This contrasts with the findings in AML.

Padre-Mendoza, Forman, et al. (1978) described the case of a 15-yr-old boy with AMMoL with aneuploidy and a Ph¹. In some cells two Ph¹-chromosomes were thought to exist, but it was shown with banding that one of them was a short Y.

A 26-yr-old woman with AMMoL was shown to have a 15q+ clone in the BM and went into complete remission lasting 4 yr (to date) after cytosine arabinoside therapy (Rouesse, Berger, et al., 1978).

Acute Megakaryocytic Leukemia

Acute megakaryocytic leukemia is a rare disease. The authors of a recent article found only 15 acceptable cases following scrutiny of the literature (Balducci, Weitzner, et al., 1978), and presented a case of 52-yr-old man with the disease in whose cells a hypodiploid picture with loss of C and D group autosomes was observed, accompanied by the presence of markers.

Acute Monocytic Leukemia

The relation of acute monocytic leukemia (AMoL) to AMMoL and AML is unclear though there are those who believe that the former is merely a variant of the latter two. The existence of an AL consisting exclusively of malignant monocytes (AMoL) was suggested by Schilling, whereas others, including Naegeli, recognized cases of AL in which immature monocytic as well as granulocytic cells were present and regarded these cases as AMMoL, which they considered to be a variant of AML.

In most morphologic discussions of AMoL, the preponderantly monocytic nature of the marrow, cellular infiltrates, and/or peripheral blood has been emphasized. Because of the rarity of "pure" AMoL, efforts to define the disease further have been infrequent. The application of EM and possibly scanning studies and newly described cytochemical parameters, particularly the α-naphtyl butyrate technique for monocytic identification, may be of great help. The utilization of cellular phenotypic markers may also prove of value (Koziner, McKenzie, et al., 1977).

Only one case of AMoL has been studied with banding. Brynes, Golomb, et al. (1976) described a case of a 56-yr-old man with the disease with characteristic cyto- and histopathologic, cytochemical, ultrastructural, and clinical patterns for AMoL and in whose leukemic cells the chromosome number ranged from 45–51 without any normal cells being seen. The cells with 49 chromosomes contained only one normal #8 and #9; in addition three cells with 8p− (p21) and one with 9p+ were seen. Studies before banding include a case (7-yr-old girl) with 47 chromosomes due to a small marker and endoreduplication (Reisman, Zuelzer, et al., 1963), and an extra small acrocentric (#21) in a man (Ilbery and Ahmad, 1965). An apparently diploid case of AMoL studied without banding was reported by V. D. Müller, Orywall, et al. (1971).

"Hairy Cell" Leukemia

"Hairy cell" leukemia (leukemic reticuloendotheliosis) is a somewhat ill-defined clinical condition the outstanding characteristic of which is the histology of the cells present in the marrow. These cells have been referred to as "hairy cells" because of the irregular cytoplasmic villi that give the cell a flagellated appearance in stained, living, or EM preparations. The cell is large, usually ranging from 15–30 μm, with a round or oval nucleus and fairly abundant gray-blue cytoplasm. The exact origin of the "hairy cell" is unknown; it has been shown that the acid phosphatase activity in these cells is prominent and resistant to degradation by tartrate, whereas cells in CLL or LS have minimal acid phosphatase activity and are degradeable by tartrate (Nanba, Jaffe, et al., 1977). Hairy cell leukemia appears to affect predominantly middle-aged males, with only 20% of the patients being females. The patients usually have pancytopenia, moderate to massive splenomegaly, and moderate to severe thrombocytopenia (Hahner and Burkhardt, 1977). The abnormal hairy cells are usually found in only small numbers in the circulating blood, but the marrow is usually infiltrated by these cells. A "dry tap" is not uncommon in this type of disease, necessitating a biopsy of the marrow for establishing the diagnosis. Biopsy of the spleen usually shows diffuse invasion of the pulp cords with the hairy cells.

Cytogenetic studies of hairy cell leukemia have been meager. It has been noted that the hairy cells have a decreased response to mitogen stimulation and the low mitotic index of the cells precludes the study of large numbers of metaphases. This low mitotic activity of hairy cells could result in an inability to identify abnormalities from marrow aspirates, a normal karyotype merely indicating that it may be reflecting proliferative cells normally present in the marrow and not including an abnormality in the hairy cells. A recent study of 20 patients with hairy cell leukemia revealed most of them to be karyotypically normal (Golomb, Rowley, et al., 1978; Golomb, Lindgren, et al., 1978). In two patients, however, consistent chromosomal abnormalities were evident in cells from unstimulated peripheral blood samples. One of the patients lacked the Y and was trisomic for #12. The second patient had a similar karyotype, also missing the Y, with addition of a C-sized marker, the long arm of which closely resembled #12 from band q14 to q-terminal. A third patient had in one sample an abnormal cell with extra #3 and #12 and in a later sample a second cell which also may have been trisomic for #12. Obviously, further data are needed on hairy cell leukemia before the chromosomal abnormalities can be interpreted in relation to the clinical, histologic, and therapeutic aspects of the disease (Golomb, Rowley, et al., 1978).

There is uncertainty regarding the exact origin of the hairy cells, some favoring a lymphocytic malignancy (Golde, Bersch, et al., 1976), others indicating that the cells are incompletely developed monocytes (Seshadri, Brown, et al., 1976). However, there is evidence that the hairy cell is of B-lymphocyte origin (W. I. Smith, Zidar, et al., 1977; Matre, Talstad, et al., 1977; Utsinger, Yount, et al., 1977; Zidar, Winkelstein, et al., 1977). One study indicated (Seshadri, Brown, et al., 1976) that in hairy cell leukemia the properties of the cell are consistent with those of incompletely developed monocytes, though denial for that has also been published (Golde, Bersch, et al., 1976; Keusch, Rüttner, et al., 1976). Others have disagreed with all these concepts (Burns, Cawley, et al., 1977).

In a study of 10 patients with hairy cell leukemia, Burns, Cawley, et al. (1978), utilizing surface membrane immunoglobulin and other markers, cytochemical, ultrastructural, and surological data, came to the conclusion that in this disease many more cells were involved in the B-cell neoplastic proliferation than was apparent by simple morphological examination and that a severe deficiency of circulating normal B lymphocytes existed.

Near-Haploidy in Acute Leukemia

Cytogenetically unusual cases of AL can be very instructive regarding the chromosomal changes and their significance. The observation of near-haploidy in AL is extremely rare and only two such cases have been described to-date. It is known that leukemic or cancerous cells may contain a significantly reduced number of chromosomes, e.g., < 40, and continue to function as neoplastic cells in spite of the severe hypodiploidy. However, a modal chromosome number of < 40 is extremely rare in AL. It is unknown how much of the chromosomal DNA is or is not functional in these hypodiploid cells. The encounter of leukemic cells with < 30 chromosomes or a near-haploid number is a very rare phenomenon. The two cases described in the literature are that of Kessous, Corberand, et al. (1975a) and one observed by us (Oshimura, Freeman, et al., 1977a). The patient studied by us was a 12-yr-old white girl who was shown to have ALL and a significant number of cells with 27 chromosomes (Figure 109).

The salient features of the cytogenetic and cytologic studies consisted of the following.

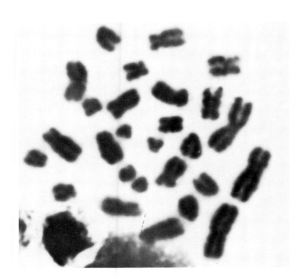

109 Metaphase containing 27 chromosomes, from the marrow of a patient with ALL. Note fuzzy appearance of the chromosomes.

(Oshimura, Freeman, et al., 1977a.)

(1) The presence of a dual cytogenetic population of cells in the marrow; cells with 27 and 54 chromosomes. (2) The cells with 54 chromosomes contained an exact duplicate set of those present in cells with 27 chromosomes. (3) The near-haploidy of 27 chromosomes appears to be the lowest ever described for human cancerous or leukemic cells. (4) The cytogenetic duality seems to have been accompanied by a cytologic one, in which lymphoblasts of large and small size (15 and 7 μm) existed in the marrow and paralleled (?) the chromosomal picture. (5) The marked similarity in the karyotypic picture in the case described by us with that of Kessous, Corberand, et al. (1975a) points to a nonrandom evolution of the cytogenetic near-haploid picture.

A comparison of the karyotypes, based on banding techniques, in the cells with 27 chromosomes of our case and the one of Kessous, Corberand, et al. (1975a) shows remarkable similarity. The case reported by Kessous, Corberand, et al. (1975a) was that of a girl (about 5 yr old) with ALL who, on initial examination, was found to have 22 metaphases with 27 chromosomes and 16 diploid cells out of the 38 marrow cells examined, at a time when she had 91% lymphoblasts and 4% erythroblasts in the marrow. No metaphases with 54 chromosomes were encountered at that time. Apparently, the lymphoblast population in the marrow consisted of a morphologically dual population: microlymphoblasts (40%) and blasts with azurophilic granulation in the cytoplasm (60%). The patient responded to therapy for ALL and did well for about 6 mo, when relapse occurred and in spite of treatment she succumbed to her disease a little over 7 mo after diagnosis. During the remission, two marrow examinations revealed an essentially diploid picture, though some "aneuploid" cells were encountered and one cell with 54 but none with 27 chromosomes were encountered. At the time of relapse, when the marrow contained 93% lymphoblasts and 2% erythroblasts, out of the 98 metaphases examined, 45 had 54 chromosomes, 47 had 27 chromosomes, 1 was diploid; and 5 were "aneuploid." The cells with 54 chromosomes were shown to contain an exact duplicate number of the chromosomes in the cells with 27. Even though not specifically stated by the authors, this would mean that the cells with 54 chromosomes contained four X-chromosomes.

A comparison of the karyotypic findings in our case (Oshimura, Freeman, et al., 1977a) with those of Kessous, Corberand, et al. (1975a) reveals that chromosomes of groups #10, #17, and #21 were present in diploid number with all other groups having a haploid number of chromosomes. The exception in the two cases is related to the presence of two X-chromosomes in the case of Kessous, Corberand, et al. (1975a) and only one X-chromosome in our case in cells with 27 chromosomes; and the presence of a diploid number of chromosomes in group #14 in our case and only a haploid number in the case of Kessous, Corberand, et al. (1975a). Thus, there appears to be a remarkable similarity (with the exceptions just mentioned) in the karyotypic picture in the two cases of ALL in whom cells with 27 chromosomes made up a significant, and on occasion predominant, proportion of the cells in the marrow.

The exact mechanism by which the near-haploid cells were generated is uncertain. The spontaneous occurrence of near-haploidy in a mammalian (male opossum) peripheral blood (Sinha, 1967) has been described, and it was postulated that in the BM diploid nuclei fused to form tetraploid "hybrid cell-types" which subsequently gave rise to near-haploid daughter cells by a process of double reduction division. Inasmuch as no tetraploid cells were encountered in the marrow of our case nor, apparently, in that of Kessous, Corberand, et al. (1975a), the process just described probably did not play a role in the production of near-haploid cells in ALL. Somatic pairing, only very rarely seen in human marrow, in cells with 54 chromosomes could possibly have led to those with 27, but it was not seen in the marrow of the ALL cases described.

It is possible that any cell type in the marrow (e.g., diploid, tetraploid, or that with 54 chromosomes) could through a process of multipolar mitosis yield cells with 27 chromosomes. Evidence for the phenomenon of multipolar mitosis, which may give rise to haploid cells, has been described in cultured cells by a number of investigators. Multipolar mitosis has also been observed as a common mitotic event leading to chromosomal anomalies in various cancers. Thus, multipolar mitosis in diploid, tetraploid, or 2S cells could possibly have led to the genesis of cells with a near-haploid number of chromosomes, as observed in the two known cases of ALL with 27 chromosomes. However, the most likely possibility is that the cells with 27 chromosomes constituted the stem line (S) of the leukemia and that the cells with 54 chromosomes represented 2S cells, for the following reasons: 1. the cells with 54 chromosomes contained an *exact* duplication of the chromosome set found in the cells with 27 chromosomes; 2. the cells with 54 chromosomes contained two markers (#7) at the time when the cells with 27 chromosomes had only one abnormal #7; 3. the cells with 54 chromosomes had only two normal #7 chromosomes when the patient relapsed after a full remission and the cells with 27 chromosomes contained only one normal #7; and 4. a diploid complement in the cells with 27 chromosomes was *consistently* present in groups #10, #14, #18, and #21.

The relation of the chromosome number in the lymphoblasts to their response to therapy and the course of the disease are difficult to ascertain. In the case studied by us, cells with 54 and 27 chromosomes were present upon the initial diagnosis and both types of cells disappeared from the marrow when the patient responded to therapy. During relapse most of the cells contained 27 chromosomes, indicating, at least during this episode, that these cells were capable of behaving like leukemic cells in spite of their near-haploidy.

We have been able to establish a cell line from the blood cells of this patient in which the number of cells with 27 chromosomes has varied from over 95% to less than 5% (Kohno, Minowada, et al., 1978). Interestingly, cells with 26 chromosomes were also observed, this apparently being the lowest chromosome number observed in human somatic cells in vitro.

A case of ALL with a single clone of cells with 26 chromosomes has been described by Prieto, Badia, et al. (1978).

A clone of Ph^1-positive cells with only 24 chromosomes constituting the preponderant number of metaphases in the marrow in a male patient with the BP of CML has been described by Daniel, Francis, et al. (1978). These cells contained the haploid number for each autosome and the X and Y. The Ph^1 was due to t(9;22). Before the BP, the marrow was revealed to contain only cells with a 46,XY, t(9;22)(q34;q11) karyotype. The marrow of the patient contained predominately miniature blast cells which in all probability contained 24 chromosomes. This number appears to be the lowest described for human somatic cells in vivo. Much has been written about the role of chromosomal changes in carcinogenesis and leukemogenesis. In this regard, the importance of numerical changes and/or abnormal (marker) chromosomes has had its proponents, particularly those who champion the concept that chromosomal imbalance, as opposed to a single gene locus mutation, plays a crucial role in the development of neoplasia. Because about 50% of human ALs are not accompanied by any visible chromosomal changes, the imbalance concept is not universally applicable. Furthermore, in the cases of ALL with 27 chromosomes, no marker chromosomes were present and obvious AL was associated with cells, though chromosomally imbalanced, containing either a much reduced or increased number of chromosomes, i.e., cells with 27 or 54 chromosomes, respectively. Thus, there must be much more to the causation of neoplasia in the human than chromosomal imbalance or the presence of abnormal chromosomes. This does not take away from the importance of nonrandom karyotypic changes which may be encountered in human leukemia and cancer, as evidenced by the remarkable similarity in the karyotypes of cells with 27 chromosomes of the two such cases of ALL described; but, in all probability, the changes are a reflection of similar mechanisms operating in carcinogenesis and leukemogenesis.

References

1. The following references include general papers dealing with chromosomes in the AL, particularly early findings, studies in familial AL, observations on cellular DNA, and autoradiography in leukemic cells, effects and relation of radiation and drugs in AL: Ahearn and Trujillo, 1972; Awano, 1961; Awano, Tsuda, et al., 1965; Bagby, Richert-Boe, et al., 1978; Borges, Nicklas, et al., 1966, 1967; Bottomley, 1976; Bottura and Coutinho, 1971b; Castoldi, Spanedda, et al., 1975; Conrad, 1975; Dameshek and Gunz, 1964; Fitzgerald, Rastrick, et al., 1973; C. E. Ford, 1960, 1961; Gavosto, 1965a,b; Gavosto, Pileri, et al., 1963; Gavosto, Ghemi, et al., 1964; Gavosto, Pegoraro, et al., 1963, 1964; Goh, 1971a; Hayhoe, 1960; Heath and Moloney, 1965a; Kaur, Catovsky, et al., 1972; N. H. Kemp, 1961; Khouri, Shahid, et al., 1968; Kiossoglou and Mitus, 1963, 1964; Lancet, 1959, 1964; F. J. W. Lewis, MacTaggart, et al., 1964; Lüers, Struck, et al., 1963; Mamaev, Gerchak, et al., 1976; McBride, 1964; McCormack and Sheline, 1963; Mitani and Okochi, 1967; Mitelman, Brandt, et al., 1979; Nicolau, Nicoară, et al., 1966; Orywall and Muller, 1971; Sandberg, 1978b; Schuler and Kiss, 1963; R. N. P. Sutton, Bishun, et al., 1969; Tessarscaia, Osencencaia, et al., 1964; Twomey, Levin, et al., 1967.

   Brieux de Salum, 1968; Brieux de Salum and Pavlovsky, 1968; Ducos, Ruffié, et al., 1964; E. Engel, 1977; E. Engel, Flexner, et al., 1964; B. L. Fischer, Lyons, et al., 1977; German, 1978; Gilman, Jackson, et al., 1970; N. H. Kemp, Stafford, et al., 1963; Kessous and Colombies, 1975; Kiossoglou, Rosenbaum, et al., 1964a; Mastrangelo, Zuelzer, et al., 1970; Morse, Ducore, et al., 1978; Muir, Occomore, et al., 1977; Nowell and Hungerford, 1960a; Ruffié, 1963; Ruffié, Ducos, et al., 1970; Zakharova, Korenevskaya, et al., 1971; Zuelzer, 1977, 1978.

2. Bernard and Tanzer, 1964; Chatterjea, 1965; Sandberg, 1965; Sandberg, Takagi, et al., 1968c.
3. Fitzgerald, Rastrick, et al., 1973; Gunz, Bach, et al., 1973; Kamada, Oguma, et al., 1976; Sakurai and Sandberg, 1976a,b; Sandberg, Takagi, et al., 1968c; Sandberg and Hossfeld, 1970, 1974; Trujillo, Ahearn, et al., 1974; Trujillo, Cork, et al., 1974; Whang-Peng, Lee, et al., 1976.
4. Alimena, Annino, et al., 1977; Golomb, Vardiman, et al., 1976; Kisliak, 1976; Lawler, Secker Walker, et al., 1975; Mitelman and Brandt, 1974; Mitelman, Brandt, et al., 1974b; Mitelman, Nilsson, et al., 1976; Oshimura, Freeman, et al., 1977b; Rowley, 1977a–d; Sasaki, Okada, et al., 1976; Zuelzer, Inoue, et al., 1976.
5. Atkin and Taylor, 1962; Conen, 1967; Davidenkova and Kolosova, 1964; E. Engel, 1965a; Fitzgerald, Rastrick, et al., 1973; C. E. Ford, 1961; Gavosto, Ghemi, et al., 1964; Gavosto, Pegoraro, et al., 1964; Gunz and Fitzgerald, 1964; Lampert, 1967a; Lampert and Gauger, 1968; Ruffié, Ducos, et al., 1970; Sandberg, 1968a; Sandberg, Takagi, et al., 1968c; Schleiermacher and Kroll, 1967; E. D. Weinstein and Weinstein, 1963.
6. Krogh Jensen, 1967b,c, 1971; Krogh Jensen and Killmann, 1967.
7. Cimino, Rowley, et al., 1979; Kiossoglou, Mitus, et al., 1965a; Krogh Jensen, 1967b, c, 1971; Krogh Jensen and Killmann, 1967; Obara, Makino, et al., 1969; Onesti and Woodliff, 1968; Pegoraro and Rovera, 1964; Sandberg, Takagi, et al., 1968c; Whang-Peng, Freireich, et al., 1969.
8. Bottura and Ferrari, 1963; Foadi, 1977; Trujillo, Cork, et al., 1971.
9. Borges, Wald, et al., 1962; Brieux de Salum, 1968; Hungerford and Nowell, 1962; Sandberg and Hossfeld, 1970.
10. Bottura, Ferrari, et al., 1961a; Grouchy and Lamy, 1962; Ilbery and Ahmad, 1965; Pierre, Hoagland, et al., 1971; Reisman, Zuelzer, et al., 1963; Ricci, Punturieri, et al., 1962a; Rowley, 1977a–d; Sandberg, Cortner, et al., 1966; Wahrman, Schaap, et al., 1962; Whang-Peng, Lee, et al., 1976.
11. S. Abe and Sandberg, 1978; Oshimura and Sandberg, 1977; Sonta, Oshimura, et al., 1976.

12. Jenkins, Rivera, et al., 1972; Ravindranath, Inoue, et al., 1977; Todd, Wood, et al., 1969.
13. Burgess and Garson, 1969; Fefer, Thomas, et al., 1974; Feig, Falk, et al., 1977; Fialkow, Thomas, et al., 1971; Gahrton, Zech, et al., 1976; Goh, 1975; Johnson, Hartmann, et al., 1976; Kolb, Wündisch, et al., 1975; Thomas, Bryant, et al., 1972; Thomas, Storb, et al., 1975; Thomas, Clift, et al., 1976.
14. Abbrederis, Schmalzl, et al., 1969; Bennett and Reed, 1975; Grusovin and Castoldi, 1976; Kurz and Haas, 1974.
15. Kamada, Okada, et al., 1968; Kamada, Oguma, et al., 1976; Lawler, Lobb, et al., 1974; Oliver, Pillai, et al., 1977; L. J. Wolff, Richardson, et al., 1976.
16. Brouet, Preud'Homme, et al., 1975b; Chechik, Pyke, et al., 1976; Hollinshead, 1976; Huhn, Rodt, et al., 1976; Mendes, Musatti, et al., 1974; Tsubota, Minowada, et al., 1977; L. J. Wolff, Richardson, et al., 1976.
17. McCaffrey, Harrison, et al., 1975; Mertelsmann, Mertelsmann, et al., 1978; Roberts, Greaves, et al., 1978; Sarin, Anderson, et al., 1976; Srivastava, Khan, et al., 1977; Sutherland, Smart, et al., 1978.
18. Duttera, Bull, et al., 1972; Golde and Cline, 1973; Hiraki, Miyoshi, et al., 1977; Knudtzon, 1977; Vincent, Sutherland, et al., 1977.
19. Altman and Baehner, 1975; Buick, Till, et al., 1977; Bull, 1975; Hoffbrand, Ganeshaguru, et al., 1977; McLaren, Tebbi, et al., 1976; M. A. S. Moore, 1975, 1976; M. A. S. Moore, Spitzer, et al., 1974; Spitzer, Schwarz, et al., 1976; Srivastava, Khan, et al., 1978.
20. Oshimura, Freeman, et al., 1977b; Sandberg, 1965, 1966a, b, 1968a, b, 1973; Sandberg, Ishihara, et al., 1964b; Sandberg, Takagi, et al., 1968c; Sandberg and Sakurai, 1976.
21. Kessous, Corberand, et al., 1975b; Lawler, Secker Walker, et al., 1975; E. D. Weinstein and Weinstein 1963; Whang-Peng, Freireich, et al., 1969; Whang-Peng, Knutsen, et al., 1976b; Zuelzer, Inoue, et al., 1976.
22. Oshimura, Hayata, et al., 1976; Oshimura, Sonta, et al., 1976; Rowley, 1975b, 1976d; Sakurai and Sandberg, 1976a.
23. Kamada, Oguma, et al., 1976; Mitelman, Levan, et al., 1976b; Mitelman, Nilsson, et al., 1976; Morse, Ducore, et al., 1979; Oshimura, Hayata, et al., 1976; Rowley, 1976d,e, 1977a–c, Rowley and Potter, 1976; Sandberg, 1978b; Sasaki, Okada, et al., 1976; Yamada and Furasawa, 1976.
24. Bousser and Tanzer, 1963; Ruffié, Colombiès, et al., 1966; Sohn and Boggs, 1974.
25. Rowley, 1973b; Sakurai, Oshimura, et al., 1974; Sakurai and Sandberg, 1974; Sakurai, Oshimura, et al., 1974.
26. Kamada, Okada, et al., 1976; Lindgren and Rowley, 1977; Oshimura, Hayata, et al., 1976; Rowley, 1973b, 1974b; Sakurai, Oshimura, et al., 1974; Sakurai and Sandberg, 1974; Sasaki, Okada, et al., 1976; Yamada and Furasawa, 1976.
27. Gunz, Bach, et al., 1973; Heath, Bennett, et al., 1969; Littlefield, 1972; Makino, Obara, et al., 1969; Meisner, Inhorn, et al., 1973; Sandberg, Takagi, et al., 1968c; Trujillo, Cork, et al., 1974; Whang-Peng, Henderson, et al., 1970.
28. Sakurai and Sandberg, 1973, 1974, 1976a; Sandberg, 1978b.
29. Atkins and Goulian, 1965; Awa, Mitani, et al., 1961; Baserga, Bosi, et al., 1962; Bersi, Gasparini, et al., 1971a,b; Černy, Šimánková, et al., 1962; Hammouda, 1963; Hart, Trujillo, et al., 1971; Kamada, 1969a; Makino and Sasaki, 1964; Pegoraro, Pileri, et al., 1963; Reisman, Mitani, et al., 1964; Reisman, Zuelzer, et al., 1964; Remy, 1964; Ruffié and Lejeune, 1962; Stowens, 1963.
30. Gunz, Ravich, et al., 1970; Houston, Levin, et al., 1964a; Koulischer, 1966.
31. Hayata, Sakurai, et al., 1975; Rowley, 1973b, 1975b; Rowley and Potter, 1976.
32. Fitzgerald, Rastrick, et al., 1973; Trujillo, Cork, et al., 1974; Whang-Peng, Hendersen, et al., 1970.
33. Alimena, Annino, et al., 1977; Fitzgerald and Hamer, 1976; Gunz, Bach, et al., 1973; Golomb, Vardiman, et al., 1976; Hart, Trujillo, et al., 1971; Kamada, Oguma, et al., 1976; Lange, Alfi, et al., 1977; Nilsson, Brandt, et al., 1977; Trujillo, Ahearn, et al., 1974.
34. Bloomfield and Theologides, 1973; Nilsson, Brandt, et al., 1978; Sakurai and Sandberg, 1974.
35. Blackstock and Garson, 1974; L. Brandt, Mitelman, et al., 1975b; Sakurai and Sandberg, 1976a.
36. Ezdinli, Sokal, et al., 1969; Gunz, Bach, et al., 1973; Hart, Trujillo, et al., 1971; Littlefield, 1972; Makino, Obara, et al., 1969; Meisner, Inhorn, et al., 1973; Rowley, 1973b; Trujillo, Ahearn, et al., 1974; Trujillo, Cork, et al., 1974; Whang-Peng, Henderson, et al., 1970.
37. Gunz, Bach, et al., 1973; Hart, Trujillo, et al., 1971; Makino, Obara, et al., 1969; Trujillo, Ahearn, et al., 1974; Trujillo, Cork, et al., 1974.
38. Alimena, Annino, et al., 1977; Golomb, Vardiman, et al., 1976; Hart, Trujillo, et al., 1971; Lange, Alfi, et al., 1977; Nilsson, Brandt, et al., 1978; Sandberg, 1978b; Trujillo, Ahearn, et al., 1974; Trujillo, Cork, et al., 1974.
39. H. J. Cohen and Huang, 1973; Hersh, Whitecar, et al., 1971; Seidenfeld, Smythe, et al., 1976;

Silvergleid and Schrier, 1974; Tchernia, Mielot, et al., 1976.
40. Rowley, 1974a,b; Sakurai, Oshimura, et al., 1974; Sakurai and Sandberg, 1976b.
41. Castoldi, Yam, et al., 1968; Fitzgerald, Adams, et al., 1964; Hayhoe and Hammouda, 1965; Heath, Bennett, et al., 1969; Hungerford and Nowell, 1962; Krogh Jensen, 1967b; Sakurai, 1970b; Sandberg, Ishihara, et al., 1962a, 1964b.
42. Kiossoglou, Mitus, et al., 1965a; Krogh Jensen, 1967a–c; Rowley, Blaisdell, et al., 1966; Whang-Peng, Freireich, et al., 1969.
43. Hagemeijer, Smit, et al., 1977; Jonasson, Gahrton, et al., 1974; P. Philip, Wantzin, et al., 1977.
44. Baikie, Court Brown, et al., 1961; Durant and Tassoni, 1967; Fitzgerald, Adams, et al., 1964; Heath and Moloney, 1965a; Heath, Bennett, et al., 1969; Oshimura and Sandberg, 1977; Pawelski, Topolska, et al., 1965; Petit, Alexander, et al., 1973; Sakurai and Sandberg, 1976b; Stahl, Papy, et al., 1965; Yamada and Furasawa, 1976.
45. Baikie, Court Brown, et al., 1961; Castoldi, Yam, et al., 1968; Diekmann, Rickers, et al., 1969; Di Grado, Mendes, et al., 1964; Durant and Tassoni, 1967; Fitzgerald, Adams, et al., 1964; Heath, Bennett, et al., 1969; M. H. Khan and Martin, 1970a,b; Kiossoglou, Mitus, et al., 1965a; Kroll and Schleisinger, 1970; Krompotic, Silberman, et al., 1968; Naman, Cadotte, et al., 1971; Oshimura, Sonta, et al., 1976; Oshimura and Sandberg, 1977; Pajares and Espinos, 1970; Sakurai and Sandberg, 1976b; Stahl, Papy, et al., 1965; Strosselli and Bernadelli, 1964; Whang-Peng, Gralnick, et al., 1977.
46. Alimena, Annino, et al., 1975; Crossen, Fitzgerald, et al., 1969; Diekmann, Rickers, et al., 1969; Di Grado, Mendes, et al., 1964; Durant and Tassoni, 1967; Heath, Bennett, et al., 1969; M. H. Khan and Martin, 1970a,b; Krogh Jensen, 1966, 1969; Sakurai and Sandberg, 1974, 1976b.
47. Baserga and Ricci, 1964; Becak, Becak, et al., 1967; Ceppelini, Celada, et al., 1964; Introzzi and Buscarini, 1966; Kroll and Schlesinger, 1970; Lawler, Secker Walker, et al., 1975; Ricci, 1965; Sakurai and Sandberg, 1976b; Stahl, Papy, et al., 1965.
48. Crossen, Fitzgerald, et al., 1969; Forni and Moreo, 1969; Heath and Moloney, 1965a; Heath, Bennett, et al., 1969; M. H. Khan and Martin, 1970a,b; Rowley, Golomb, et al., 1977b; Smalley and Bouroncle, 1967; Steruska, Hrubisko, et al., 1977; Weatherall and Walker, 1965.
49. Berry and Desforges, 1969; Crossen, Fitzgerald, et al., 1969; Durant and Tassoni, 1967; Erkman, Crookston, et al., 1967; Fitzgerald, Adams, et al., 1964; Forni and Moreo, 1969; Heath, Bennett, et al., 1969; Heath and Moloney, 1965a; Khan and Martin, 1970a,b; Krogh Jensen, 1966; Krogh Jensen and Philip, 1970; Sandberg, Ishihara, et al., 1964b.
50. Berry and Desforges, 1969; Castoldi, Yam, et al., 1968; Crossen, Fitzgerald, et al., 1969; Diekmann, Rickers, et al., 1969; Di Grado, Mendes, et al., 1964; Durant and Tassoni, 1967; Erkman, Crookston, et al., 1967; Fitzgerald, Adams, et al., 1964; Forni and Moreo, 1969; Heath, Bennett, et al., 1969; Heath and Moloney, 1965a; M. H. Khan and Martin, 1970a,b; Kiossoglou, Mitus, et al., 1965a; Krogh Jensen, 1966; Krogh Jensen and Philip, 1970; Sakurai, 1970b; Sandberg, Ishihara, et al., 1964b; Trujillo, Ahearn, et al., 1974; Trujillo, Cork, et al., 1974; Whang-Peng, Henderson, et al., 1970.
51. Dyment, Melnyk, et al., 1968; Fitzgerald, Adams, et al., 1964; Heath, Bennett, et al., 1969; Kiossoglou, Mitus, et al., 1965a; Petit, Alexander, et al., 1973.
52. Kiossoglou, Mitus, et al., 1965a; Krompotic, Silberman, et al., 1968; Sakurai, 1970b; Smalley and Bouroncle, 1967; Weatherall and Walker, 1965.
53. Dyment, Melnyk, et al., 1968; Fitzgerald, Adams, et al., 1964; Heath, Bennett, et al., 1969; Kiossoglou, Mitus, et al., 1965a; Krogh Jensen, 1966; Nichols, Nordén, et al., 1970; Sakurai, 1970b.
54. Durant and Tassoni, 1967; Forni and Moreo, 1969; M. H. Khan and Martin, 1970a,b; Sakurai and Sandberg, 1976a,b.
55. Dyment, Melnyk, et al., 1968; Heath, Bennett, et al., 1969; Krompotic, Silberman, et al., 1968; Nichols, Nordén, et al., 1970; Petit, Alexander, et al., 1973; Sakurai and Sandberg, 1974, 1976a,b; Smalley and Bouroncle, 1967.
56. Castoldi, Yam, et al., 1968; Crossen, Fitzgerald, et al., 1969; Diekmann, Rickers, et al., 1969; Forni and Moreo, 1969; Heath, Bennett, et al., 1969; Heath and Moloney, 1965a; M. H. Khan and Martin, 1970a,b; Kiossoglou, Mitus, et al., 1965a; Krogh Jensen, 1966; Kroll and Schlesinger, 1970.
57. Buckton, Harnden, et al., 1961; Goh, Lee, et al., 1978; Hellriegel, Pfeiffer, et al., 1969; Mikkelsen, Petersen, et al., 1964; R. W. Miller, 1963; H. Okada, Lin, et al., 1970; M. W. Thompson Bell, et al., 1963.
58. Buckton, Harnden, et al., 1961; Conen, Erkman, et al., 1966; Ebbin, Heather, et al., 1968; Heath and Moloney, 1965a; Lahey, Beier, et al., 1963; O. J. Miller, Breg, et al., 1961.
59. Borges, Wald, et al., 1962; Cawein and Lappat, 1964; Conen and Erkman, 1966; De Mayo,

Kiossoglou, et al., 1967; Honda, Punnett, et al., 1964; M. H. Khan and Martin, 1967c; Kiossoglou, Rosenbaum, et al., 1963; Kiossoglou, Mitus, et al., 1964; Lampert, 1967b; Lejeune, Berger, et al., 1963; Mercer, Keller, et al., 1963; Reisman, Mitani, et al., 1964; Reisman, Zuelzer, et al., 1964; Ross and Atkins, 1962; Rowley, 1964; Sandberg, Cortner, et al., 1966; Tough, Court Brown, et al., 1961; Vincent, Sinha, et al., 1963; Wahrman, Schaap, et al., 1963; Warkany, Schubert, et al., 1963.

60. Banerjee, Jung, et al., 1977; Higurashi, Tamura, et al., 1973; Huang, Banjeree, et al., 1977; O'Brien, Poon, et al., 1971; Sasaki and Tonomura, 1969; D. Young, 1971b.

12

Chronic Lymphocytic Leukemia

The chromosomal pattern in chronic lymphocytic leukemia (CLL) is poorly understood, even though considerable efforts have been made to elucidate this aspect of leukemic cytogenetics. Inadequate knowledge of the chromosomal pattern in CLL is intimately related to the pathophysiology of the disease, which among other features, is characterized by a slow cellular proliferation rate and a delayed and diminished response to various mitogens of the leukemic lymphocytes.[1] Direct chromosome preparation, the method of choice for revealing the chromosome constitution of any neoplastic tissue, is not practicable in CLL because of the very small number of spontaneously dividing cells, whether they are recovered from infiltrated lymph nodes, the BM, or leukemic blood.[2] To overcome this problem PHA-stimulated peripheral blood cells have been used for chromosome studies. However, it was soon realized that an inverse relation existed between PHA responsiveness and the height of the peripheral blood lymphocyte count.[3] This indicated that two (or more) lymphocyte populations are present in CLL, i.e., a normal one that responds to PHA and an abnormal leukemic one that does not. The relative proportion of normal lymphocytes decreases with increasing expansion of the leukemic population.[4] Thus, chromosome analysis of blood lymphocytes cultured for 72 hr in the presence of PHA either did not yield any results when the patient had a very high lymphocyte count or disclosed a normal chromosomal situation when the relative proportion of normal lymphocytes was high enough to be detected.[5] Later, it was demonstrated that leukemic blood lymphocytes in CLL are not totally unresponsive to PHA and other mitogens; they do respond, but this response is delayed and reduced with a maximal transformation rate between the 5th and 9th day of incubation and a [^{3}H]-thymidine incorporation rate of about 50% compared to normal cells.[6] Hence, prolongation of the culture time up to 6 or more days appeared to be optimal for obtaining chromosome preparations actually derived from leukemic lymphocytes. It seems remarkable that leukemic lymphocytes derived from lymph nodes appear to respond normally to PHA (Spiers and Baikie, 1968a). But, recently, the situation became even more complicated by immunologic findings suggesting that CLL is preponderantly a B-cell disease.[7] Because only T lymphocytes are thought to respond to PHA, the question must be raised again whether any of the chromosomal data based on cytogenetic analysis of PHA-stimulated lymphocytes can be ascribed to leukemic B lymphocytes. Until the exact nature and behavior of the leukemic lymphocyte is known, no definite statements regarding the chromosome constitution of these cells should be made.

This chapter does not include a discussion of chromosomal changes observed in the blood of patients with various types of lymphoma infiltrating the blood and simulating or being responsible for a leukemic condition. Such changes will be discussed in Chapter 14. The occurrence of Ph¹-positive CML in patients with established CLL, though admittedly very rare, should be mentioned and constitutes an intriguing aspect of double malignancies, and the relation of the two leukemias, in particular.[8] A 70-yr-old patient with CLL of 10 yr duration was found to have a 47,XX,+C, Gq−(22q−) karyotype without evidence of CML. The Ph¹-like chromosome persisted until death, due to "the final stage of CLL," still without evidence of CML (Rolović and Ćirić, 1974). A patient with AT with a 46,XX, t(14q−;14q+) constitutional karyotype was reported to have developed CLL (McCaw, Hecht, et al., 1975).

For reasons discussed above, only a few reports are available on the chromosomes of marrow-derived lymphocytes in CLL which were not subjected to in vitro conditions (Table 66).[9] In these reports all or the overwhelming majority of metaphases were normal. A pseudodiploid chromosome constitution with a subtelocentric marker in one case and an acrocentric marker in another in < 5% of the metaphases was observed by Baserga, Castoldi, et al. (1966). A small number of hyperdiploid metaphases (47–48 chromosomes) was found in 4 out of 32 cases studied.[10] In all these marrow studies the possibility cannot be excluded that even in the presence of > 95% lymphocytes, the metaphases observed were not derived from marrow cells, particularly nucleated erythrocytes. To avoid this possibility, direct chromosome prepara-

Table 66 Chromosome constitutions of karyotypically abnormal clones in CLL

Karyotype	Reference
−G	Brieux de Salum, 1968
E-group abnormality	Różynkowa, Marczak, et al., 1968
47,XX,+C,22q−	Rolović and Ćirić, 1974
48,XY,+C,+F/45,XY,−C,−E,+mar (CLL → CML)	Whang-Peng, Gralnick, et al., 1974
45,X,−Y,Ph¹+ (CLL → CML)	Whang-Peng, Gralnick, et al., 1974
47,XY,+C (T−CLL)	Brody, Burningham, et al., 1975
45,XX,1p+,−8,+9,−D,−22,+mar (T−CLL)	Nowell, Jensen, et al., 1976b
46,XY,2p+,del(11)(q22)	Fleischman and Prigogina, 1977
46,XY,t(11;17)(q21;qter)	Fleischman and Prigogina, 1977
46,XY,t(8;14)(q24;q32),19p−	Fleischman and Prigogina, 1977
46,XY,t(1;10)(q22;qter)	Fleischman and Prigogina, 1977
(all with tumorous lymph nodes)	

tions from lymph nodes were attempted, but failed to most instances. Short-term incubation without PHA did not improve the yield of metaphases. Published findings are hardly conclusive because of the small number of mataphases analyzed; for example, among eight cases in the literature was one with a (radiation-induced?) fragment in some cells (Millard, 1968) and another pseudodiploid case with a subacrocentric marker in 20% of the metaphases (Baserga, Castoldi et al., 1966); all other metaphases were diploid.[11] With regard to PHA-stimulated blood lymphocytes, chromosomal data of a total of 173 cases were reviewed. No chromosomal anomalies were described in the early papers.[12] Goh (1967c) analyzed chromosomes of lymphocytes cultured for 5–6 days instead of 3 days, and found pseudodiploidy in 40% (range 20–80%) of the metaphases; a consistent anomaly was not seen in any of the six patients studied who were either treated or untreated. These results were generally confirmed by Ducos and Colombiès (1968), who found some pseudodiploid cells with marker chromosomes in 6-day but not in 3-day cultures. Goh (1967c) and Ducos and Colombiès (1968) concluded that the abnormal metaphases came from leukemic lymphocytes. The results of Goh (1967c) were not corroborated by two other studies, however; though Lawler, Pentycross, et al. (1968) observed an increased number of pseudodiploid cells (8% versus 2% in normal controls), the percentage was considerably smaller when compared to Goh's (1967c) figure, and no great difference was shown to exist between 3- and 5-day cultures. Also, an increase of pseudodiploidy was only apparent in treated and not in untreated patients; the last group was found to have a higher percentage of aneuploid cells.

Berger and Parmentier (1971), too, did not see differences with regard to ploidy or pseudodiploidy between 2-, 3- and 6-day cultures. In this material (10 cases) the proportion of hyperploid metaphases was elevated in untreated patients, whereas the proportion of metaphases with structural anomalies was increased in treated patients only, results that are quite the opposite to the findings of Lawler, Pentycross, et al. (1968).

No valid conclusions as to the effects of different culture times on CLL lymphocytes can be drawn (Woodliff and Cohen, 1972). In this paper a significantly increased percentage of aneuploid, but not pseudodiploid cells (14.2 vs. 5% in normal controls), is reported; aneuploidy was mainly characterized by hypodiploidy, as in the Lawler, Pentycross, et al. (1968) study. Treated and untreated patients were not different chromosome-wise. The authors paid special attention to the morphology of chromosomes #21–#22, because Fitzgerald (1965) had found that the small acrocentric chromosomes were significantly shorter than in normal men. Fitzgerald's (1965) findings were not confirmed (Woodliff, Leong, et al., 1972). Psuedodiploidy and hyperdiploidy are mentioned as an occasional finding in some cells by a number of other workers,[13] the significance of which, however, is difficult to establish as no clinical data or control studies are given. Considering the low frequency of chromosomal anomalies found in CLL, careful control studies are necessary to endow these findings with any meaning at all.

One case of CLL shall be mentioned separately, because it provides, on the basis of chromosomal findings, evidence that two spontaneously dividing lymphocyte populations and a PHA-dependent lymphocyte population were present; in cultures grown without PHA the great majority of metaphases was characterized by a translocation, the rest of the metaphases were normal; PHA-dependent lymphocytes and fibroblasts were also normal (Obara, Makino, et al., 1970a).

Finally, the Christchurch (Ch¹) anomaly (Gunz, Fitzgerald, et al., 1962; Fitzgerald and Gunz, 1964) shall be discussed. The Ch¹ is an autosome #21–#22 that had lost all or the greater part of its short arms. In contrast to the Ph¹, the Ch¹ is a constitutional, inherited anomaly that is present in all tissues, not only in leukemic cells, and can be transmitted to the offspring. From this it follows that the pathogenetic implications of the Ch¹ are totally different from those of the Ph¹. In a sibship of seven sisters and brothers the Ch¹ was found in four, and three of the carriers developed CLL (Fitzgerald and Hamer, 1969). In this particular family the Ch¹ may predispose the carriers to the development of CLL. Apart from this, no other implication should be attached to this anomaly, because neither was it found in other instances of familial CLL,[14] nor did CLL occur in other families with a very similar anomaly,[15] nor was it demonstrated in any other cases of CLL.

In a more recent paper (Crossen, 1975), an analysis of blood cell chromosomes of 20 subjects with CLL was undertaken on 3-day cultures of defibrinated, gelatin-sedimented blood, probably in the presence of PHA. It was shown that 97 of 100 metaphases examined with Giemsa banding had a normal pattern. The three remaining metaphases, all from one patient, had a pattern similar to that seen after aging of slides. However, the karyotype of one patient shown in the publication is definitely abnormal, containing 43 chromosomes including +5, −16, and possible loss of the Y. It was Crossen's conclusion that the chromosomes in CLL have normal banding patterns. The author was careful to point out, however, that if PHA proves to be a specific T-cell stimulant and CLL a B-cell leukemia, then the majority of chromosomes studied in CLL have been made on normal lymphocytes and the absence of chromosomal abnormalities is not surprising. If, on the other hand, PHA is not a specific T-cell stimulant, but capable of stimulating both T and B cells, then it is possible that chromosome studies have been made on both T and B cells. However, it is not possible to assess whether these studies have been made on leukemic T or B cells. The problem will remain unsolved until methods of isolating and differentiating leukemic cells are established and means to stimulate their division found. Even though it is generally accepted that CML is a B-cell leukemia, recent work indicates that there may be a minor number of T-cell leukemias as well.[16] Experimental studies with mouse lymphocytes have shown that PHA selectively stimulates T cells to divide, but there is as yet no unequivocal evidence that this is true for human cells. In another study (Lawler and Lele, 1972) the finding of chromosomally rearranged cells in patients with CLL treated with chlorambucil can be interpreted as surviving cells with drug-induced changes; both PHA and pokeweed mitogen (PWM) were used because of the suggestion that these two mitogens stimulate different classes of lymphocytes.

A hypothesis has been advanced that a high T-cell count may occur in stable CLL as an expression of successful antileukemic therapy.[17] If that is so, it is not surprising that the blood or marrow of some patients with CLL responds to mitogenic agents, leading to the appearance of normal metaphases. However, whether this is a common occurrence and, in fact, accounts for the karyotypically normal cells observed in the preponderant number of patients with CLL remains to be established.

Recent advances in determining cell surface and other markers in leukemia have indicated that the preponderant number of cases with CLL are of the B-cell variety with the T-cell type being rather uncommon, though the exact proportion in clinically typical CLL has not been established, as yet (Brouet, Flandrin, et al., 1973, Brouet, Labaume, et al., 1975). A good number of T-cell CLL cases apparently are not classic CLL, and the largest series reported (11 patients) suggests a relatively characteristic picture of splenomegaly with only moderate marrow infiltration, a high content of lysozomal enzymes and cytoplasmic granules in the neoplastic lymphocytes, increased frequency of skin lesions, and severe neutropenia. Other workers have also noted the common occurrence of skin involvement, sometimes suggesting either the Sézary syndrome or mycosis fungoides (Brouet, Labaume, et al., 1975).

Because human B cells respond poorly to mitogens in culture, it has been frustrating to evaluate the proliferative capacity of the chromosomal pattern of the leukemic lymphocytes in most cases of CLL.[18] In normal blood the T lymphocytes are stimulated by mitogens in culture, e.g., PHA, concanavalin A (Con A), and the calcium ionophore A23187, and constitute the metaphases available for analysis. Thus, variations in the number and proportion of T cells during various phases of CLL, the effects of therapy and the leukemic environment may play an important role in the type of cytogenetic observations encountered in this disease.

Of considerable interest in this regard are several publications dealing with the T-cell variant of CLL (Thiel, Bauchinger, et al., 1977; Finan, Daniele, et al., 1978) accompanied by defective response of the cells to mitogens. In one such patient (80-yr-old female) a clone of karyotypically abnormal cells was present before therapy and during the course of the CLL and consisted of numerical and morphologic changes (Nowell, Jensen, et al., 1976b). Only 8 of the 136 cultured cells examined had a normal diploid karyotype, with 95 having 45 chromosomes. The modal cells contained a large marker resulting from elongation of the short arm of #1, an extra #2, probable −8,+9,−14(?), −15(?)−22, and a very small metacentric marker of unknown origin. In a similar case of T-cell CLL, chromosome abnormalities were also found. These observations and the finding of cytogenetic changes in some cases of the Sézary syndrome seem clearly to indicate that at least in the chronic T-cell neoplasms karyotypic alterations are present, as in nearly all other tumors. The frequency of such abnormalities and their specificity, if any, must await additional studies and wider application of banding techniques; but, it is intriguing that #1 marker chromosomes, resembling the abnormal #1 described by Nowell, Jensen, et al. (1976b),

have been previously reported in a number of lymphoproliferative disorders and other tumors.

The above group of workers (Finan, Daniele, et al., 1978a, b) reported on the cytogenetic findings in seven patients with T-cell dyscrasias (three with CLL, two with the "small cell variant" of Sézary syndrome, and two unclassified). The cells of these patients all responded "moderately" to one or more T-cell mitogens (PHA, Con A, and the calcium ionophore A23187) after 3–5 days of culture. There was no consistent karyotypic change, but alterations of #2 were noted in four of the seven cases. The authors speculated that the necessity of using more than one mitogen in order to obtain abnormal metaphases possibly suggests that different neoplasms may arise from different T-cell subpopulations.

To reiterate, in the common B-cell form of CLL dividing cells for study are very difficult to obtain, either in vivo or in vitro. There have been numerous reports of normal chromosome patterns from PHA-stimulated CLL cultures, but, as noted above, dividing cells in these instances are probably residual normal T cells rather than the neoplastic B cells. It appears likely that questions concerning cytogenetic changes in B-cell CLL will remain unresolved until a satisfactory human B-cell mitogen is available.

References

1. Bernard, Geraldes, et al., 1964a,b; Bouroncle, Clausen, et al., 1969; Catovsky, Holt, et al., 1972; Hayhoe, Sinks, et al., 1966; Perebra and Pegrum, 1974; Rubin, Havemann, et al., 1969; Rundles and Moore, 1978; Schiffer, 1968; Schrek, 1967; Sharman, Crossen, et al., 1966; Winter, Osmond, et al., 1964.
2. Fitzgerald, 1967; Oppenheim, Whang, et al., 1965; Ruffié, Ducos, et al., 1966.
3. Bernard, Geraldes, et al., 1964a,b; Hayhoe and Hammouda, 1965; Quaglino and Cowling, 1964; Schrek and Rabinowitz, 1963.
4. Macavei and Halmos, 1975; Oppenheim, Whang, et al., 1965; Thomson, Robinson, et al., 1966.
5. Colombiès, Ducos, et al., 1965; Fitzgerald and Adams, 1965; Macavei and Halmos, 1975; Oppenheim, Whang, et al., 1965.

6. Goh, 1967c, 1968a; König, Cohnen, et al., 1972; Rubin, Havemann, et al., 1969.
7. McLaughlin, Wetherly-Mein, et al., 1973; Nowell, Daniele, et al., 1975; Nowell, Jensen, et al., 1976b; Papamichail, Brown, et al., 1971; Shevach, Edelson, et al., 1974; Stein, Lennert, et al., 1972; J. D. Wilson and Nossal, 1971.
8. B. L. Fisher, Lyons, et al., 1977; Whang-Peng, Gralnick, et al., 1974.
9. Baserga and Castoldi, 1965; Baserga, Castoldi, et al., 1966; Chitham and McIver, 1964; Court Brown, 1964; Fitzgerald and Adams, 1965; Hayhoe and Hammouda, 1965; Kinlough and Robson, 1961; Lawler, Pentycross, et al., 1968; Sandberg, Ishihara, et al., 1961; Woodliff and Cohen, 1972.
10. Fitzgerald and Adams, 1965; Sandberg, Ishihara, et al., 1961; Woodliff and Cohen, 1972.
11. Baker and Atkin, 1965; Baserga, Castoldi, et al., 1966; Lawler, Pentycross, et al., 1968; Millard, 1968; Spiers and Baikie, 1968b.
12. Fitzgerald and Adams, 1965; Hayhoe and Hammouda, 1965; Nowell and Hungerford, 1964; Oppenheim, Whang, et al., 1965.
13. Columbiès, Ducos, et al., 1965; Fitzgerald and Adams, 1965; Hayhoe and Hammouda, 1965; Heni and Siebner, 1964b; Rozynkowa and Marczak, 1970.
14. Court Brown, 1964; Fitzgerald, Crossen, et al., 1966; Fraumeni, Vogel, et al., 1969; Heni and Siebner, 1964b; Millard, 1968; Sandberg, Koepf, et al., 1960.
15. Abbott, 1966; M. W. Shaw, 1962.
16. Brody, Burningham, et al., 1975; Huang, Hou, et al., 1974; Nowell, Daniele, et al., 1975; Nowell, Jensen, et al., 1976b; Pierseus, Shur, et al., 1973; Sumiya, Mizoguchi, et al., 1973.
17. Daguillard, Fontaine, et al., 1974; Han, Moayeri, et al., 1976; Macavei and Halmos, 1975.
18. Bouroncle, Clausen, et al., 1969; Daguillard, Fontaine, et al., 1974; Pegoraro, Jakšić, et al., 1973.

13

Myeloproliferative Disorders

Myeloid Metaplasia with Myelofibrosis and/or Osteomyelosclerosis

Acute Myelofibrosis

Myeloproliferative Syndrome in Children with #7 Monosomy

Essential (Primary or Idiopathic) Thrombocythemia

Megaloblastic Anemia

Polycythemia Vera

Experience Before Banding
Chromosome Changes in PV
Chromosomal Changes and Progression of Polycythemia Vera
The Ph^1 in Polycythemia Vera
Banding Studies in Polycythemia Vera

Comments on Chromosomal Changes in Myeloproliferative Disease

Most of the myeloproliferative disorders (MD) described in this section (Figure 110) may become frankly leukemic, though the incidence and the type of leukemia developed differ from one condition to another. It is probably a misnomer to call any of the conditions "preleukemic," because when first seen these patients have already some clinical, histologic, or hematologic evidence of disease. Furthermore, some of these conditions are considered by some to be, in fact, just one facet of the total picture of leukemia. For example, idiopathic aplastic anemia (IAA) or the severe marrow hypoplasia after benzene exposure is considered to be initial manifestations of AL, which usually appears in these patients if they survive the aplastic phase. Hence, karyotypic changes may be present at any stage of these diseases and probably should not be considered as preleukemic karyotypic evidence, when, in essence, the leukemic phase already exists. When the characteristic leukemic picture does develop, the cytogenetic changes do not differ from those observed in other leukemias not preceded by the so-called preleukemia phase.

Included in this section are the hematopoietic disorders polycythemia vera (PV), myoproliferative syndrome with monosomy of #7, myeloid metaplasia with myelofibrosis (MMM) or osteomyelosclerosis, panmyelosis, paroxysmal nocturnal hemoglobinuria, pernicious anemia, and idiopathic thrombocythemia. The interpretation of the chromosomal data in these disorders is a very delicate and difficult task, because none of these disorders, with the possible exception of PV, represents a well-defined entity (Table 67); in addition, all of them share a number of features and most of them can merge or change into another, including CML.[1] Thus, primary or secondary myelofibrosis (MF) may develop early or late during the course of CML and PV;[2] thrombocythemia may be the initial event in CML; and refractory anemia may precede AL. A tendency to progress to frank leukemia is one of the most significant features of this heterogenous group of diseases (D. S. Rosenthal and Moloney, 1977), constituting the bulk of the so-called preleukemias. It is important to realize that such terms as polycythemia, megaloblastic anemia, or thrombocythemia are but descriptive in character, pointing to the most prominent hematologic features. What actually happens in most of the diseases of the myeloproliferative syndrome is an involvement of a variable proportion of the precursor cell compartment, resulting in a more or less pronounced maturation defect of the erythroid, myeloid, or megakaryocytic series. It is this situation which, among others, accounts for the difficulties in classifying myeloproliferative disorders.

At this juncture it may be appropriate to say a few words regarding leukocyte alkaline phosphatase (LAP) activity, because this activity is often elevated in MD and has been used as a point of differentiation from CML. Much has been written on LAP activity in myeloid metaplasia, DS and leukemia (particularly CML).[3] Because in the former two conditions the LAP is invariably very much elevated and in the latter very much decreased, it was thought that these changes were related to the excessive chromosomal material in group #21 in DS and to missing chromosomal material in the case of the Ph^1 in CML.[4] Of course, it has been established that these two states are associated with two different autosomal changes: #21 trisomy in DS and deletion of #22 in CML. Even before this demonstration, there existed findings which complicated the previous interpretations of the LAP data, e.g., the normal LAP levels in CML cells during remission of the leukemia or severe infection, the occurrence of occasional subjects with CML and normal LAP, high levels of LAP in patients with MD and a Ph^1, and the very low levels of LAP in CML without a Ph^1.[5] More importantly, most cases of MD and PV are associated with high levels of LAP and, yet, are not accompanied by chromosomal changes. It is possible that both chromosomes (#21 and #22), or for that matter other chromosomes, play a role in the LAP levels, and there is little doubt that an interpretation of LAP activity in relation to karyotypic changes is much more complicated than previously thought.

As a group of ill-defined and heterogeneous diseases, the MD to be discussed are not accompanied by any specific or characteristic karyocytic changes. The exact incidence of chromosomal abnormalities in this group remains uncertain (Figure 111), because the preponderant number of cases is not accompanied by karyotypic abnormalities, and there is a tendency for workers to report only those cases that are karyotypically abnormal or to report a

	1	2	3	4	5	6	7	8	9	10	11	12	13	14	15	16	17	18	19	20	21	22	X	Y
	ALL	CLL	C		AML	ALL	AML	AML	ALL		ML		S	BET			ALL				ALL	ALL		
	C	ML	ML		C	BL	BL	BET	AML		S			BL			AML			PV	AML	CML		
	ML	MM	MM		MD	C	C	BL	CLL					CLL			CLL				CLL	M		
	MM					MD	MD	C	CML					ML			CML				S			
	PV						MM	CML	ML					MM										
								MM	MM															
								PV	MD															
									PV															

AdV12						EBV SV40		HV						EBV	HV	AdV5 AdV12 SV40						PV	
ENO1	MDHs	ACON	PGM2	HexB	PGM3	MDHm	GSR	AK3	GoTs	ACP2	LDHB	RNr	RNr	RNr	APRT	TK	PepA	GPI	ADA	RNr	RNr	95	TDF
PGD	ACP1	GALT	Hb	If2	GLO	HACDH	F-7r	ABO	PP	LDHA	PepB	Rb1	TrpRS	MPI	Hpα	GALK	hCG	PepD	DCE	SODs		loci	H-Y
AK2	Gal-Act			DTS	HLA	Co		AK1	HK1	EsA4	TPI	ESD		HexA	LCAT				ITP	AVP		incl.	
Rh	ICDs				C2	Jk		NPa	ADK	AL	SHMT			PK3						PRGAS		PGK	
E11	If1				C4	HaF		Ws1	GSAS	SA1	ENO2			β2m						Ag		αGAL	
UMPK					C8			XP1			GAPD			IDHm						Gpx		HGPRT	
PGM1					Bf															Down		SAX	
Amy1					Ch																	G6PD	
Amy2					Rg																		
Cae					MEs																		
Fy					SODs																		
UGPP					Pg																		
PepC																							
RN5S																							
FH1																							
FH2																							
GUK1																							
GUK2																							
αFUC																							

110 Representation of the 22 autosomes and the 2 sex-chromosomes and their involvement in various myeloproliferative disorders. Also shown are some of the gene and viral interaction loci in the various chromosomes. Clustering of aberrations to specific chromosomes in human neoplasia. At the top are represented the 24 human chromosome types and below them the various classes of neoplasias with which each chromosome has been most frequently involved in aberrations:

C = carcinoma   ML = malignant lymphoma   BET = benign epithelial.tumors   S = sarcoma   M = meningioma

Table 67 Cytogenetic changes observed in various myeloproliferative syndromes

Karyotype	Reference
46,XY,−C,+mar(?Cq−)	Solari, Sverdlick, et al., 1962
46,XY,Dq−	W. Engel, Merker, et al., 1968
46,XX,−C,+ring,?Fq−(PV−Rx with P^{32}→myelofibrosis)	Goodman, Bouroncle, et al., 1968
45,XY,−C,−F,+ring→58,XY,+7C,+D,+E,+3G,(+frag)	Berry and Desforges, 1969
47,XY,+C (myelofibrosis→AML)	Krogh Jensen and Philip, 1970, 1973
46,XX/48,XX,+C,+C	Krogh Jensen and Philip, 1970, 1973
46,XX/46,XX,+2acentric fragments (MM→AML)	Krogh Jensen and Philip, 1970, 1973
46,XX/44,XX,−C,−D (myelofibrosis→AML)	Krogh Jensen and Philip, 1970, 1973
46,XX,t(1;3)	Slyck, Weiss, et al., 1970
45,XX,−C	Humbert, Hathaway, et al., 1971
45,XX,−C	Woodliff, Chipper, et al., 1972
50,XX,+C,+C,+C,+mar	Woodliff, Chipper, et al., 1972
47,XY,+9	W. M. Davidson and Knight, 1973
47,XX,+9	W. M. Davidson and Knight, 1973
47,+C	Krogh Jensen and Philip, 1973
48,+C,+C/46	Krogh Jensen and Philip, 1973
44,−C,−D/46	Krogh Jensen and Philip, 1973
47,XY,+E	Lisker, Cobo de Gutiérrez, et al., 1973
46,XX,−C,+mar/45,XX,−C	Lisker, Cobo de Gutiérrez, et al., 1973
47,XX,+mar	Lisker, Cobo de Gutiérrez, et al., 1973
47,XY,+mar	Vormittag, Kuhbock, et al., 1973
46,XX,Dq−	Ganner-Millonig, 1974
45,XX,−7 (myeloproliferative disease→AMMoL)	MacDougall, Brown, et al., 1974
48,X,−Y,+C,+2mar(#17or#18)/47,X,−Y,+C,−F,+2mar(#17or#18)/ 49,XY,+C,+2mar(#17or#18)	Moake, Lebos, et al., 1974
47,XX,+9	Knight, Davidson, et al., 1974
47,+mar	L. Brandt, Mitelman, et al., 1975a
46,XY,−12,+mar	Fleischmann and Krizsa, 1975
49,XX,+10,+19,+2Ph¹,1q+,−G	Cehreli, Ezdinli, et al., 1976
45,XY,−1,−7,+t(1p+;4q−;7q−),4q−	Nowell, Jensen, et al., 1976a
45,XY,−C(?7−9)	Nowell, Jensen, et al., 1976a
47,XX,+?20q−,18p+	Nowell, Jensen, et al., 1976a
46,XX,1q+,1q−	Nowell, Jensen, et al., 1976a
44,XY,−5,−7,−7,−11,−12,+t(7p−;12q+),+t(17;18),17p+,+min (cancer of prostate; X-ray; MF→PV−P^{32})	Nowell, Jensen, et al., 1976a
47,XX,−1,−9,+2t(1q;9p),+t(1p;9q),7q−,20q−	Nowell, Jensen, et al., 1976a
47,XX,+8 (2 cases)	Nowell, Jensen, et al., 1976a
44,XY,−18,−21	Yamada and Furasawa, 1976
45,X,−Y	DiLeo, Müller, et al., 1977
3N	Hagemeijer, Smit, et al., 1977 and personal communication
t(3q+;5q−)	Hagemeijer, Smit, et al., 1977 and personal communication
−7	Hagemeijer, Smit, et al., 1977 and personal communication
−7/+8	Hagemeijer, Smit, et al., 1977 and personal communication
i(17q)	Hagemeijer, Smit, et al., 1977 and personal communication
46,XX,−6,+t(1;6)(q25;p22)	L. Y. F. Hsu, Pinchiaroli, et al., 1977
46,XX,i(17q)	Nowell and Finan, 1977
20q− (2 patients with atypical myeloproliferative syndrome, 1 with AMM)	Testa, Kinnealey, et al., 1977
47,XX,+8	Jacobson, Salo, et al., 1978

continued

Table 67 Cytogenetic changes observed in various myeloproliferative syndromes—*continued*

Karyotype		Reference
51,XX,+1,2q−(q33),+6,+9,+11,−19,+20q,+mar (myelofibrosis following P^{32} Rx of erythrocytosis)		Najfeld, Price, et al., 1978
47,XX,+21 (→ acute leukemia)		Whang-Peng, Lee, et al., 1978
47,XX,+G		Whang-Peng, Lee, et al., 1978
47,XX,+16		Whang-Peng, Lee, et al., 1978
47,XY,+21		Whang-Peng, Lee, et al., 1978
46,XX,8q−		Whang-Peng, Lee, et al., 1978
46,XXp−q−,5q−,13q+,15q−		Whang-Peng, Lee, et al., 1978
46,XY,+t(1q2q),−8,inv(13)(pter→q12::q31→q13::q32→qter),−17,+i(17q)		Whang-Peng, Lee, et al., 1978
46,XX,+t(5;12)(5qter→5q14::12q13→12pter),−12/45,XX,−16/47,XX,+2		Whang-Peng, Lee, et al., 1978
45,XY,−C,−D,+F/46,XY,−C,+F		Whang-Peng, Lee, et al., 1978
49,XY,+3C (→ CML, Ph^1-negative)		Whang-Peng, Lee, et al., 1978
45,XY,5p−,−7,−10,11q−,+i(17q)/46,XY,−C,−E,+?i(17q),min (→ AMMoL)		Whang-Peng, Lee, et al., 1978
Ph^1 + basophilic leukemia		Goh and Anderson, 1979
45,XY,−2C,+E,t(3q+;3q−)	Acute MF	Nowell and Finan, 1978*b*
47,XX,−21,+2,i(21q)	Acute MF	Nowell and Finan, 1978*b*
50,XY,−1,−B,+2C,+1q,+1q,+D,+19	Acute MF	Nowell and Finan, 1978*b*

series of cases in which those with chromosomal deviations constitute only a small percentage of the total series. It is a safe statement to make that the bulk of the cases with myeloproliferative disorders have a normal, diploid chromosome picture.

Another aspect of this group of MD that bears a cogent relation to the cytogenetic findings is the development of a leukemia-like picture, usually resembling AML or CML, in a number of these cases. Hence, the chromosomal changes, when they occur, tend to be of the same nature as those observed in the leukemias, i.e., ranging from hypodiploidy to hyperdiploidy but with an inconsistent and variable karyotypic picture in those cases who develop AML and the presence of a Ph^1 in those with CML. At present, we do not know whether this group of diseases represents variants of AML or CML, this concept deserving some consideration because of the occurrence of the Ph^1 in occasional cases of PV, thrombocythemia, and other related disorders.

With these reservations in mind, an attempt will be made to describe separately the chromosomal findings in PV, megaloblastic anemia, MMM, thrombocythemia, and their related disorders.

Myeloid Metaplasia with Myelofibrosis and/or Osteomyelosclerosis

The essential features of this ill-defined, *heterogenous*, and probably etiologically different group of diseases are splenohepatomegaly, the presence of nucleated erythrocytic cells, and a greatly increased number of immature granulocytic cells, with or without an acquired Pelger-Huët anomaly in the blood, and generalized or focal MF or osteomyelosclerosis. We know that CML and PV may be associated with all the features considered essential for MMM in the initial stages or during progression of these diseases. Therefore, parameters are necessary which would permit drawing a distinction between agnogenic MMM and diseases simulating it. The most valuable parameters are the LAP activity and the Ph^1. Even with the application of these two criteria, a number of cases remains unclassifiable. Most hematologists rely on the presence of absence of the Ph^1 in order to classify such border-line cases as CML or MMM.[6] On the other hand, there are at least 20 cases with MMM or PV described in the literature in which a variable proportion of the BM cells (between 10–100%) were revealed to contain the Ph^1.[7]

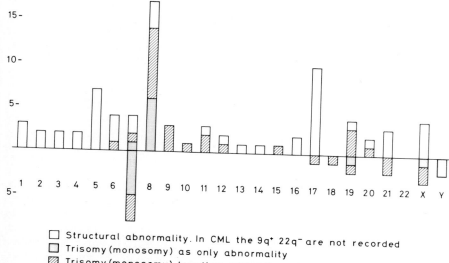

111 Distribution of chromosome abnormalities in 45 cases of myeloproliferative disorders seen at the clinics of Erasmus University at Rotterdam.

(Courtesy of Dr. A. Hagemeijer.)

Myeloid metaphasia is a somewhat heterogenous group of diseases of unknown etiology which are probably related to one form of leukemia or another. It appears, that if patients with this disease live long enough, most of them would develop a leukemic state. It is at this stage that chromosomal anomalies may become evident in the marrow or blood cells.

Some workers have attached significance to the occurrence of group C trisomy in patients with MMM or other conditions considered to be preleukemic. Even though a C-trisomy of one group or another may occur, evaluation of this observation with banding techniques and establishing the exact chromosome or chromosomes involved in such trisomy should shed considerable light on the karyotypic significance of this anomaly.

In the light of the demonstration that about 20% of the patients with Ph[1]-positive CML have myelofibrosis (Gralnick, Harbor, et al., 1971), it is possible that the Ph[1]-positive MMM cases may have to be reclassified. Because there is a strong possibility, however, that the Ph[1] is not specific solely for CML, no single parameter can be considered absolutely pathognomonic for any variant of the myeloproliferative syndrome.

The chromosomal data of the published cases with Ph[1]-negative MMM including diseases described as myelofibrosis, panmyelosis, and marrow dysplasia have been combined and reviewed. The data indicate that over 50% of the cases had no chromosome anomalies. Analysis of the data showed that most of the karyotypically normal cases were contributed by a small number of laboratories, which together reported over half of the cases (20-25%).[8] Thus, the percentage of chromosomally abnormal MMM is probably too high, since there is a tendency to report only cytogenetically abnormal cases in the form of single·case reports. Of the abnormal cases, about 25% were reported to be trisomic[9] and nearly 25% monosomic[10] for a group #6-12 chromosome. It is my impression that the exact significance of trisomy or monosomy of group C chromosomes in MMM remains to be elucidated in a larger series of patients by the application of banding techniques in order to identify the extra or missing chromosomes and the relation to the course of the disease, particularly the development of leukemia.

Of the more than 80 cases with MMM published to date, 50% were not accompanied by any demonstrable chromosomal changes. Definite karyotypic abnormalities were shown to exist in 41 of the 80 cases, including 2 cases with a Ph¹ (Kiossoglou, Mitus, et al., 1966a; Forrester and Louro, 1966), one with a t(1;3) (Slyck, Weiss, et al., 1970), more than a dozen with group C changes consisting either of missing or extra autosomes, one case with a 5q⁻ (Rowley, 1976c) and a number with complex cytogenetic abnormalities (Mitus, Coleman, et al., 1969; Moake, Lebos, et al., 1974; Nowell, Jensen, et al., 1976a; Whang-Peng, Henderson, et al., 1970).

In contrast to CML, for which in the great majority of cases 90–100% of the metaphases derived from BM cells contain the Ph¹, in MMM a chromosomally normal cell population successfully competes with the abnormal one, e.g., 90–100% abnormal metaphases were encountered in only 15 cases with MMM. However, published data suggest (Nowell 1971b, c) that the percentage of a chromosomally normal cell population is not related to prognosis, e.g., two cases are included who had only 20–40% abnormal metaphases and who died of AL within 1–3 mo after the detection of the chromosome anomalies. These data raise the possibility that at least a portion of the chromosomally normal cells may be leukemic, similar to the situation in some cases with AL. A case of childhood MMM characterized by two clones with 48 and 49 chromosomes, respectively, has been described (Teasdale, Worth, et al., 1970); these clones disappeared during remission of the disease and did not reappear when the patient died of AL. This case illustrates impressively the difficulties involved when one tries to correlate a normal chromosome constitution with normal cells.

Generally, the chromosomal anomalies in MMM with or without myelofibrosis or myelosclerosis have varied in their picture. However, two patients, one male and one female, with deletion of a group D chromosome have been described (W. Engel, Merker, et al., 1968; Ganner-Millonig, 1974), and in another male patient a chromosome in group D was missing and an extra chromosome, possibly #3, was found; and a young man who had apparently a myelofibrotic disease for over 10 yr was shown to have two small metacentric marker chromosomes in all the marrow cells and to be accompanied by an extra chromosome in group C, missing chromosomes in group F and loss of the Y (Moake, Lebos, et al., 1974). The latter case is rather complicated and the two metacentric markers bear some resemblance to those of a case observed by us with sideroblastic anemia (Hossfeld, Schmidt, et al., 1972a).

A 20q⁻ anomaly has been described in five patients: two with an atypical myeloproliferative syndrome, one with agnogenic myeloid metaplasia, one with "smoldering" leukemia, and one with AMMoL (Testa, Kinnealey, et al., 1977).

In some reports the exact nature of the myeloproliferative syndrome has not been stated; in all probability a substantial percentage of such cases represent early stages of one or another form of leukemia. Many of these cases were characterized by the presence of an aneuploid cell line, particularly when the disease transformed into an acute or subacute leukemia with rapid development of the abnormal chromosome picture. However, exceptions do exist and a case in point is one with 48 chromosomes (+F,+G) in whom no further karyotypic changes or transformation into AL were observed after 43 mo.

Even though banding techniques have somewhat clarified the nature of the chromosomes usually involved in aneuploid MMM, they have not brought any further elucidation of the types of clinical syndromes belonging to this group of diseases. Whether those entities which have similar or identical karyotypic characteristics should be considered so clinically remains a moot point, requiring much more data, chromosomal and clinical, before some order in the classification of these diseases can be introduced.

The most common identifiable (with banding) karyotypic change appears to be trisomy of #8, followed by that of #9 and either a partial or full monosomy of #7 in a rough ratio of 6:3:2, respectively. A host of other changes have been observed in MMM, but these do not occur with sufficient regularity to characterize any specific segments of the cases. Thus, a patient (82-yr-old female) with a history of anemia for about 1 yr with morphologic changes of an acquired Pelger-Huët anomaly, normal LAP, disordered erythrocyte maturation, and some increase in marrow myeloblasts was thought to represent an atypical myeloproliferative disorder and to have a 47,XX,−17, i(17q) karyotype in all 25 cells from the marrow

(Nowell and Finan, 1977). So far, no evidence of leukemia has been found, though it is too early (1 yr) to be sure that a leukemic process does not exist.

Another interesting case was described by Cehreli, Ezdinli, et al. (1976) involving a 55-yr-old black woman with agnogenic MMM in blastic transformation with unusually prominent lymphadenopathy simulating lymphoma. The initial evidence of myeloblastic transformation was present in lymph nodes but not in the marrow and serial BM biopsies showed fibrosis throughout the course of the disease, in spite of a gradual increase in the percentage of blood myeloblasts. The abnormal cells were demonstrated to have a karyotype of 48,XX,C+,F+, and the presence of two Ph^1-like chromosomes as well as markers. Such aneuploid cells were present in lymph nodes, BM, and blood.

In another study, cytogenetic analysis was done on 18 patients with myelofibrosis or the closely related syndrome of undifferentiated myeloproliferative disorder (Nowell, Jensen, et al., 1976a). Clones (referring to a population of cells with the same or related abnormal karyotype presumably descendant from the same abnormal precursors) of cells with chromosome abnormalities were demonstrated in the blood of eight patients, including two with a history of radiation therapy and two with "acute myelofibrosis." Trisomy #8 was present in the latter two patients, but otherwise there was no consistent cytogenetic pattern or correlation with specific hematologic or clinical findings. None of the patients had progressed to leukemia. The authors concluded that chromosome abnormalities are relatively common in myelofibrosis but (as with PV) the aberrant clones do not appear to indicate that clinical leukemia is imminent.

In a series of 20 patients with MMM studied cytogenetically Whang-Peng, Lee, et al. (1978) found 64% to be aneuploid at diagnosis; ultimately 75% of the patients had aneuploidy. In six patients the disease terminated in leukemia (AML, CML, AMMoL and unclassified type): two had normal karyotypes, one a normal karyotype initially and aneuploidy 3 yr later, and three had varying degrees of aneuploidy throughout the disease. Of the 14 patients without evidence of leukemia, only three had completely normal karyotypes. These studies demonstrate that extensive chromosomal abnormalities of a complex and nonspecific nature are common in myelofibrosis. Cytogenetic data reflect the relationship that exists between leukemia and the various myeloproliferative disorders and also indicate that myelofibrosis should, in fact, be considered an expression of another underlying myeloproliferative disorder rather than a separate entity. A change in karyotype may herald the appearance of a malignant clone and subsequent development of leukemia. Among patients without leukemia several have had the karyotypic anomalies for prolonged period of time (5–10 yr). No consistent karyotypic picture emerged from the analysis, including the findings obtained with G banding. The two patients who developed CML, one with and one without chromosomal changes prior to the development of the leukemia, were apparently both Ph^1-negative.

Divergence of cytogenetic findings between the BM and spleen in a case of agnogenic MMM developing a blast crisis was described by L. Brandt, Mitelman, et al. (1975a). Chromosome studies were performed in a patient (62 yr-old-male) who developed a blast crisis after a chronic phase lasting over 6 years. The proportion of blast cells were about equal in the marrow and spleen. However, in the former 43% of the metaphases were abnormal including a marker, whereas all the spleen metaphases were karyotypically normal. The authors indicated that chromosomal changes associated with blasts transformation in MMM may occur in the marrow prior to such changes in the spleen and that the data supported the concept that marrow and spleen may constitute relatively separate pools of hematopoietic tissue in chronic myeloproliferative diseases. To me, the observations hold further significance in that they indicate that chromosomal changes are not necessarily the basis for transformation, since in this case the percentage of myeloblasts was the same in the marrow and spleen and, yet, chromosomal changes were only observed in the former tissue. This would speak strongly against the chromosomes playing a directly causal role in blast transformation in MMM.

The cytogenetic results obtained in MMM must be viewed, however, in the same light as those in leukemia, particularly since the diseases making up MMM appear to be just as malignant as leukemia but manifesting themselves in a variety of ways. As in the leukemias, though on a more extended scale, the manifestations in MMM may assume a wide spectrum of signs, symptoms, hematological findings, and pathology, and the presence or absence of diploidy or aneuploidy does not speak for or against a leukemic process or different nature of the disease, just as diploidy does not speak against the presence of acute or chronic leukemia.

Acute Myelofibrosis

Various anomalies, including a Ph¹-like chromosome and dicentrics (M. H. Khan and Martin, 1968a), and #8 trisomy (Nowell, Jensen, et al., 1976a; Nowell and Finan, 1978), have been described in acute myelofibrosis (Mitus, Coleman, et al., 1969). Slyck, Weiss, et al. (1970) found an apparent t(1q+;3p−) in a case of acute myelofibrosis, but failed to demonstrate this anomaly in cultured marrow fibroblasts and blood lymphocytes, suggesting that they were not neoplastic and that the marrow fibrosis was secondary to a neoplastic change in the hematopoietic cells (Bergsman and Slyck, 1971). A case of acute myelofibrosis apparently induced by procarbazine therapy was accompanied by nearly 30% aneuploid cells and 7% with structural abnormalities in the blood cells (Pinedo, Van Hemel, et al., 1974). Only a few karyotypic studies have been reported in other unusual myeloproliferative conditions, e.g., megakaryocytic myelosis, coexistence of a lympho- and myeloproliferative disorder (Dougan, Onesti, et al., 1967; Louwagie, Desmet, et al., 1973).

Myeloproliferative Syndrome in Children with #7 Monosomy

A myeloproliferative disorder associated with an absent C-group chromosome was first described in 1964 (Freireich, Whang, et al., 1964). These children have shown leukocytosis with immature granulocytes and erythrocytes, hepatosplenomegaly, anemia and thrombocytopenia. In all cases, a C-group chromosome (#7) has been missing in BM cells; cells from the spleen, skin, and peripheral blood have been shown to be normal karyotypically. Most children affected have developed AML, AMMoL, or EL. In some of these cases, the identity of the missing #7 has been established with banding techniques. To date, apparently 14 such cases have been reported, the disease affecting both male and female children.[11] In one somewhat complicated case a ring chromosome was also observed in the marrow cells (Crossen, Fitzgerald, et al., 1969). Even though a similar karyotypic picture, i.e., 45,−7 has been observed in adults and children with AML or EL without the preleukemic findings described above, the occurrence of a 45,C− karyotype in such a relatively large series of children, with the leukemic picture developing subsequent to the cytogenetic observations in the marrow, may indicate that the chromosomal changes may have a direct bearing on the development of leukemia in these patients.

Paroxysmal nocturnal hemoglobinuria (PNH) is an acquired disease, thought by some to fall into the general category of autoimmune conditions, though its transformation into AL has been described.[12] The literature contains at least 20 bona fide cases of PNH studied cytogenetically.[13] In eight cases only blood cells were studied and no chromosomal changes were found. This is not surprising, because the cytogenetic studies of the blood cells represent an analysis of the chromosomes of lymphocytes in the majority of individuals, rather than of the myeloproliferative cells involved in PNH. Undetected marrow abnormalities may have existed in these patients and, in fact, in two other patients with PNH no lymphocyte abnormalities were encountered in face of marked karyotypic changes in the marrow. Two patients with PNH studied by Zaccaria, Ricci, et al. (1973) were shown to have significant aneuploidy (35−40% of the cells) in the marrow, without any specific clones being observed, and similar changes in the PHA-stimulated lymphocytes of the blood.

Eight cases of PNH without leukemia had chromosome studies performed on marrow; in three no anomalies were found. In three cases clones of abnormal cells were present in the marrow: −C and/or −G, Gp− in 76−95% of the cells; 45,XY,−C in 50% of the cells; and 45,XO in 84%−92% of the cells. Inasmuch as in the latter case (Whang-Peng, Knutsen, et al., 1976a) a missing Y was the sole abnormality in a 77-yr-old male with PNH, the significance of the chromosomal findings in relation to the disease can be questioned. Loss of the Y in marrow cells of elderly but otherwise normal males is not unusual. In the two patients studied by Zaccaria, Ricci, et al. (1973) the chromosomal studies, although suggesting PNH as a myeloproliferative disorder, did not yield direct evidence either for a clonal origin of PNH or for the participation of all marrow cells. However, Oni, Osunkoya, et al. (1970) studied a patient with PNH who was heterozygous for G6PD, and it appeared that the PNH was produced by the overgrowth of a single clone of cells as defined by G6PD typing. Unfortunately, no cytogenetic data on the case were presented. Yamada and Furasawa (1976) reported a 49,XY,+8,+12,+17,t(17;21) (p11; q22) karyotype based on banding in a 40-yr-old patient with PNH who developed EL. A 46,XY,−C,+mar(minute) was observed by Kamada and Uchino (1976) in an atomic bomb survivor who developed PNH without evidence of leukemia.

Essential (Primary or Idiopathic) Thrombocythemia

This rare disorder is characterized by a permanent increase in the number of platelets above 1,000,000/mm³, by a hyperplastic marrow and splenomegaly. Chronic myelocytic leukemia may rarely present itself initially with identical features. Accordingly, it is not surprising that cases of essential thrombocythemia (ET) were found to be Ph¹-positive.[14] However, one of these Ph¹-positive cases later developed typical CML (Woodliff, Onesti, et al., 1967a). Among the nearly 30 cases reported to date in the literature, most were cytogenetically normal, and only a few showed prominent hyperdiploid modes with extra chromosomes in group #6−12, possibly involving #8, #9, or #10.[15]

In one case with primary hemorrhagic thrombocythemia (Wiik, Paulson, et al., 1971), the marrow cells were karyotypically normal, whereas in the blood a small proportion of the cells was found to contain dicentric chromosomes, acentric fragments, and stable chromosome aberrations with translocation. These anomalies were attributed to busulfan therapy given 6 yr before the study. Another case of interest is that of a 45-yr-old white housewife with megakaryoblastic leukemia observed at our Institute and described by Hossfeld, Tormey, et al. (1975). The case was characterized by megakaryoblastic and erythrocytic hyperplasia of the BM, thrombocythemia, and hepatosplenomegaly. All the marrow-derived metaphases were Ph¹-positive and the additional karyotypic changes included a marker replacing one of the C-group chromosomes in the majority of the cells. A case associated with a karyotype 48,XX,+9,+8 or +10, established with banding, has been described by Rowley (1973c). Nicoara, Butoianu, et al. (1967) encountered aneuploidy in one case and a Ph¹ with a missing chromosome in the G group in another case with possible transition to CML. The importance of a 21q− anomaly observed in a high proportion of cases with ET (and some with PV or CML) remains to be established.

A 21q− has been described in the marrow cells of patients with untreated primary thrombocythemia (Zaccaria and Tura, 1978), as well as in myeloproliferative disorders presenting with thrombocythemia (Fuscaldo, Erlick, et al., 1978). The 21q− when seen in the latter group of patients was present in 25−30% of the dividing cells in the BM or unstimulated blood cells, with the size of the clone increasing with disease progression.

Megaloblastic Anemia

Even though megaloblastic anemia (MA) is neither a neoplastic disease nor a condition that predisposes to cancer or leukemia,[16] it is included in this chapter because of the striking chromosomal anomalies that may accompany this disease and because of some cytological and cytogenetic similarities between MA and EL. Particularly striking may be the megaloblastic or megaloblastoid changes observed in some treated and untreated cases of EL or AML. Treatment with some of the antileukemic agents may also lead to megaloblastosis. Morphologically the chromosome aberrations observed in MA resemble those seen in breakage syndromes; MA, however, is unique in that the cytogenetic changes are observed in uncultured marrow cells, in addition to those which may be seen in the cultured blood lymphocytes. No distinction will be made between vitamin B_{12} and folic acid deficiency forms of MA because no chromosomal differences have been described.

Even though sporadic cases were described before 1965, it was in that year that conclusive evidence was first provided on a relatively large group of patients that megaloblastic anemias are accompanied by chromosomal changes in BM cells (Kiossoglou, Mitus, et al., 1965c). Apparently, nearly 100 cases of MA have been described in the literature and chromosomal anomalies of various kinds were found in nearly 90%. It is surprising that this phenomenon had not been substantiated earlier. This most probably was due to three previous conflicting reports. In 1960 one report (Court Brown, Jacobs, et al., 1960) stated that patients with MA undoubtedly have a normal chromosome constitution in the marrow cells; retrospectively, one must assume that the authors had concentrated on the chromosome number and that either no attention was paid to structural anomalies or that culture conditions (the marrows were cultured for 24 hr) had caused a disappearance of such anomalies. Two years later (Astaldi, Strosseli, et al., 1962), striking hypodiploidy in 81% of the BM metaphases in one case of MA was reported; these authors also observed structural chromosome anomalies. However, since the BM cells were cultured with PHA for 48 hr, it is very probable that the metaphases were derived from lymphocytes. In 1963 (Chapelle and Gräsbeck, 1963) completely normal chromosome findings in PHA-stimulated lymphocytes were described in four patients with untreated MA.

Since 1965, chromosomal abnormalities have been described in the BM cells of 90% of the patients with MA. One study (Lawler, Roberts, et al. 1971) substantiated the previous findings of others,[17] i.e., that the circulating lymphocytes can also be shown to have chromosomal anomalies. Chromosomes in cultured fibroblasts were investigated only once in MA and found to be normal.

The most prominent and common chromosome anomaly in BM cells of patients with MA is the chromatid break (Krogh Jensen, 1977). The breakage frequency varies from 5–86% of the BM metaphases, with the breaks being randomly distributed among the chromosome groups.[18] Whereas chromosome breaks in the hands of most workers are a very frequent finding in MA, one group stated in 1966 (Menzies, Crossen, et al., 1966) that the absence of chromosome breaks in the material from their 10 patients indicates chromosomal changes in MA to be "distinctively different from those induced by radiation or radiometric agents." Next in frequency to chromosomal breakage in MA is incomplete (reduced) condensation or contraction of some of the chromosomes leading to "giant chromosome" (Powsner and Berman, 1961) formation. Exaggerated centromeric constrictions and centromere spreading were observed less frequently.[19] Structural rearrangements, giving rise to dicentric chromosomes, interchange figures, ring chromosomes, etc, are a point of controversy. Their occurrence is expressively denied by some (Menzies, Crossen, et al., 1966), but it had been observed by others in some metaphases of about a third of their patients.[20] With regard to numerical changes, no clear picture emerges from a review of the literature. Some workers[21] claim that no numerical changes have been found; whereas other groups state that hypodiploidy is a prominent feature of BM cells in MA.[22] Preferential loss of G- and Y-chromosomes, described by one group (Kiossoglou, Mitus, et al., 1965c), has not been observed by other workers (Lawler, Roberts, et al., 1971).

Hypodiploidy, involving the small acrocentric chromosomes, was also seen in metaphases derived from PHA-stimulated lymphocytes of MA patients. Even though B_{12} or folic acid deficiency may affect not only BM cells but also other proliferating tissues, relatively little attention has been paid to the chromosome behavior of blood lymphocytes. According to one report (Lawler, Roberts, et al., 1971), the frequency of hypodiploidy (27–30%) in lymphocytes was as high as in the BM cells. This figure corresponds well to the percentage given by another group (Keller and Nordén, 1967). Structural chromosome anomalies, basically of the same type as described in BM cells, were found less frequently in lymphocytes by some workers (Krogh Jensen and Friis-Møller, 1967; Lawler, Roberts, et al., 1971) and as frequently as in BM cells by others (Keller, Lindstrand, et al., 1970). An interesting case of a 58-yr-old male with persistent macrocytic anemia, refractory to therapy, leukopenia, and a megaloblastic marrow of 10-yr duration, and in whose marrow a tetraploid line predominated has been described (H. Singh, Boyd, et al. 1972). In addition, a submetacentric marker chromosome (slightly larger than #6) replacing one #16, was found in varying percentages among the tetraploid clones and, thus, the patient's marrow karyotypes consisted predominantly of 92,XXYY,−16,+mar and 92,XXYY.

After vitamin B_{12} and folic acid therapy the numerical and structural chromosome anomalies will usually disappear completely.[23] Whether or not a complete disappearance of the cytogenetic aberrations takes place, depends on the therapeutic dose of vitamin given to the patients, the length of time between the treatment and the chromosome study, and the types of tissue utilized for analysis. The higher the vitamin dose, the more rapid the improvement of the cytogenetic changes (Lawler, Roberts, et al., 1971; Keller, Lindstrand, et al., 1970). Normalization of the chromosomal picture most probably does not result from repair of the broken DNA-strands; it is much more likely that the chromosomally damaged cells are eliminated and that the normal cells originate from "cured" precursors. This recovery requires time and depends on the proliferative behavior of different tissues. Thus, normalization of chromosomal changes of the slowly proliferating lymphocytic population may take years, (Lawler, Roberts, et al., 1971), whereas the BM recovers within weeks.

Those workers who did not observe structural rearrangements and breakage of chromosomes suggested that chromosomal changes occurring in MA are "distinctively different" from those caused by X-ray and some mutagenic compounds (Menzies, Crossen, et al., 1966; Keller and Nordén, 1967). The chromosomal damage was felt to be comparable to that produced by agents which are active in G_1- or S-phase (cytosine arabinoside, 5-fluorouracil, 5-bromodeoxyuridine, viruses) of the cell cycle (Menzies, Crossen, et al., 1966; Krogh Jensen and Friis-Møller, 1967). Inasmuch as structural rearrangements were found by several other workers, however, the mechanism underlying chromosomal damage in B_{12} and folic acid deficiency remains obscure.

Polycythemia Vera

Experience Before Banding

A total of 150 cases of polycythemia vera (PV) were reported in the literature before banding, more than half of which were communicated by one group (Lawler, Millard, et al., 1970). It is still a subject of controversy as to whether uncomplicated PV is a benign or neoplastic disorder,[24] and one wonders whether chromosomal analyses may help to elucidate this question. This problem can be approached only by separating chromosomal findings of untreated and treated patients, because of the possibility that the chromosomal changes observed may be related to radio- and/or chemotherapy rather than to the disease as such. Apart from the question as to whether uncomplicated PV is a benign or malignant disease, stands the undeniable fact that the incidence of AL (usually AML) is 20–40 times greater in patients with PV than in the normal population. Transition to AL is in most cases associated with MMM.[25] Whether the chromosomal changes in the leukemic phase are related to those in the polycythemic phase, and whether such changes may herald the transition to leukemia when observed in the polycythemic phase are questions that have not been fully answered.

Of the total of the 150 unbanded cases of PV, there were 51 who had not received any kind of radio- or chemotherapy.[26] Among these 51 untreated cases there were 7 (13.7%) in which BM-derived metaphases were totally or partially aneuploid;[27] in six cases the aneuploidy was due to extra chromosomes in groups #6–#12, in one case the Y was missing, and in one case 90% of the marrow cells were characterized by a deleted chromosome of group #19–#20, the so-called F-abnormality. Among the 99 cases who had received treatment, there were 36 cases (ca 37%) with a varying proportion of aneuploid BM-derived metaphases. Thus, chromosome anomalies were encountered in treated patients about three times more frequently than in untreated cases. In most patients with PV, the disease can be controlled for a relatively long time by venesection; it is unfortunate that in the group of untreated patients with chromosome anomalies the time interval between diagnosis and the development of the chromosome anomaly is not known. Thus, it remains to be seen whether in some cases the onset of PV is related to the karyotypic anomalies (involving chromosomes of groups #6–#12 or #19–#20), or whether there is actually no chromosomal basis for PV. It is possible that the chromosomal changes are associated only with progression of PV rather than with its onset. The chro-

Table 68 Cytogenetically abnormal clones in the BM of untreated patients with PV

Karyotype	Reference
48,XX,+C,+C	Kay, Lawler, et al., 1966 (33 cases)
45,XO	Kay, Lawler, et al., 1966
47,XX,+C/48,XX,+C,+C	Kay, Lawler, et al., 1966
47,XX,+C (2 cases)	Kay, Lawler, et al., 1966
46,XX,Gq−(Ph¹)	Koulischer, Frühling, et al., 1967 (4 cases)
46,XY,t(12;17)/47,X,mar(Y:1),+9	Rowley, 1973d (1 case)
47,XY,+C	Visfeldt, Franzén, et al., 1973 (17 cases)
46,XY,F?−	Visfeldt, Franzén, et al., 1973
48,XX,+C,+C/48,XX,+B,+C	Gras, Cowling, et al., 1975 (13 cases)
48,XX,+B,+C	Gras, Cowling, et al., 1975
47,XX,+A	Wurster-Hill, Whang-Peng, et al., 1976 (131 cases)*
47,XX,+1p−	Wurster-Hill, Whang-Peng, et al., 1976
47,XX,+C (3 cases)	Wurster-Hill, Whang-Peng, et al., 1976
47,XX,+9	Wurster-Hill, Whang-Peng, et al., 1976
47,XY,+12	Wurster-Hill, Whang-Peng, et al., 1976
47,XY,+C	Wurster-Hill, Whang-Peng, et al., 1976
48,XY,+C,+C	Wurster-Hill, Whang-Peng, et al., 1976
47,XX,+8	Wurster-Hill, Whang-Peng, et al., 1976
47,XX,+mar	Wurster-Hill, Whang-Peng, et al., 1976
46,XX,13q−	Wurster-Hill, Whang-Peng, et al., 1976
47,XX,+G/45,XX,−A,−C,+G	Wurster-Hill, Whang-Peng, et al., 1976
47,XY,+D	Wurster-Hill, Whang-Peng, et al., 1976
46,XY,+D,−G	Wurster-Hill, Whang-Peng, et al., 1976
46,XY,−C,+E/45,XY,−C,+E,−G	Wurster-Hill, Whang-Peng, et al., 1976
46,XY/48,XY,+8,+9	Westin, Wahlström, et al., 1976 (50 cases)
48,XY,+8,+9/46,XY	Westin, Wahlström, et al., 1976
46,XY/45,XY,−16	Westin, Wahlström, et al., 1976
47,XY,+del(1)(?p21),del(20)(q11)	Westin, Wahlström, et al., 1976
47,XX,+t(1;9)(q22;q13)	Westin, Wahlström, et al., 1976
47,XY,+mar/46,XY	Westin, Wahlström, et al., 1976
46,XX/48,XX,+8,+9	Westin, Wahlström, et al., 1976
47,XX,+C/48,XX,+C,+C/48,XX,+B,+C/49,XX,+C,+C,+mar	Lawler, Millard, et al., 1970
46,XY/45,X,−Y	Lawler, Millard, et al., 1970
47,XY,+8	L. Y. F. Hsu, Alter, et al., 1974 (1 case)
47,XY,+9/45,X,−Y/?Ph¹	Shabtai, Weiss, et al., 1978 (15 cases)
47,XX,+8	Shabtai, Weiss, et al., 1978
47,XX,+8 (→ myelofibrosis)	Shabtai, Weiss, et al., 1978
47,XX,+20	Shabtai, Weiss, et al., 1978

Total number of patients, 305‡
Total number with chromosome anomalies, 41
Abnormalities involved 25–100% of the marrow cells

Total cases studied shown in parentheses.
*Does not include the cases of Westin, Wahlström, et al. (1976) shown in this table.
‡Includes 40 cases in whom no changes were found (Nagy and Yurgutis, 1968 [5 cases]; Barnes, Holmes, et al., 1969 [5 cases]; Berger, Parmentier, et al., 1974 [27 cases]; Shiraishi, Hayata, et al., 1975 [3 cases]).

mosome constitution of the abnormal clones in treated PV patients was pseudodiploid in 60% of the cases and about 20% (each) were hyper- and hypodiploid, with no indication for a preferential loss or gain in certain chromosomal groups, except group #19–#20.

It was Lawler's group[28] that brought attention to an anomaly involving one of the chromosomes of group #19–#20, and that was characterized by a deletion of what was thought to be a variable portion of one arm. They observed this anomaly in 10 out of 21 cases of PV with chromosome aberrations; because all the cases bearing this anomaly were treated, the possibility could not be excluded that it was radiation induced. A similar anomaly was reported in three more cases, one of which was untreated (Visfeldt, Franzén, et al., 1970). In 11 out of 13 cases the "F-anomaly" was the only aberration, and it was present in 32–100% of the metaphases. The significance of this observation was underlined by the ultimate demonstration with banding that the chromosome involved is always a #20 (20q−) (Reeves, Lobb, et al., 1972).

Looking into the chromosomal data of PVs that progressed to AL, two interesting facets emerge; firstly, all but one of the cases had chromosome anomalies when they were in the polycythemic phase;[29] secondly, all but one case (Perlin, Granatir, et al., 1973) exhibited striking additional chromosomal anomalies in the leukemic phase.[30] Considering that uncomplicated ALs have chromosome anomalies only in 40–50% of the cases, this appears to be a very remarkable finding and resembles the chromosomal behavior in the BP of CML.

Caution, however, is indicated in generalizing about these statements. It is by no means justified to label all PVs that have aneuploid clones preleukemic states, and to initiate prophylactic chemotherapy.[31] From the data available it can be deduced that only a third of the patients transform into frank leukemia. It has been calculated (Nowell, 1971 b, c) that the critical period of time between the detection of chromosome anomalies in the marrow and development of leukemia is about 3 mo; no such relation appears to exist for treated patients with PV.

Chromosome Changes in PV

The most common chromosomal abnormalities in *untreated* PV appear to involve chromosomes in group C, with an extra #8 or #9 apparently being the most frequent abnormality (Tables 68 and 69).[32] The occurrence of PV cases in whose cells the Ph¹ is present has already been referred to in the chapter on CML.[33] However, this finding awaits confirmation with banding techniques to establish whether the Ph¹ observed in PV is due to a #9−#22 translocation. In treated PV, particularly with ³²P and/or radiation therapy, the most common abnormality appears to be a

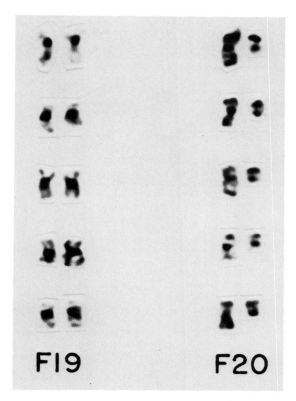

112 Five sets of group F chromosomes from the marrow cells of PV. It is to be noted that five deleted #20 chromosomes have a very similar appearance by G-banding.

deletion of the long arm of one of a #20 (Figure 112) and this karyotypic abnormality has been studied and confirmed with various banding techniques.[34] It should be noted that rare cases of 20q− in untreated patients with PV have been encountered (Westin, Wahlström, et al., 1976). It appears that the long arm of one #20 is susceptible to a break followed by loss of the involved segment, because the missing part of the #20 has not been found in the karyotype with any technique (an exception is the case of Berger, described in 1972, with an 11;20 translocation). The 20q− may be related to the possible genesis of PV, the #20 being particularly susceptible to the effects of radiation. In the untreated subject with PV and the 20q−, the loss of the genetic material may be a result of extreme susceptibility to natural radiation or the small amounts of radiation to

Table 69 Chromosome findings in treated patients with PV

Karyotype	Reference
46,XY,Ph¹-positive	N. H. Kemp, Stafford, et al., 1964
46,XY,Dq−	Kay, Lawler, et al., 1966; Millard, Lawler, et al., 1968; Lawler, Millard, et al., 1970
46,XY,Cq−	Kay, Lawler, et al., 1966; Millard, Lawler, et al., 1968; Lawler, Millard, et al., 1970
46,XX,Dq−	Kay, Lawler, et al., 1966; Millard, Lawler, et al., 1968; Lawler, Millard, et al., 1970
46,XY,1q+/46,XY,Dq−/46,XY,+2,−C	Kay, Lawler, et al., 1966; Millard, Lawler, et al., 1968; Lawler, Millard, et al., 1970
46,XX,+C,−16−18,+F,t(16?q+;18q−)	Kay, Lawler, et al., 1966; Millard, Lawler, et al., 1968; Lawler, Millard, et al., 1970
46,XX,Fq−	Kay, Lawler, et al., 1966; Millard, Lawler, et al., 1968; Lawler, Millard, et al., 1970
47,XX,+C,+F?− ⎫	Kay, Lawler, et al., 1966; Millard, Lawler, et al., 1968; Lawler, Millard, et al., 1970
44,XY,−3,−B/45,XY,−B	Kay, Lawler, et al., 1966; Millard, Lawler, et al., 1968; Lawler, Millard, et al., 1970
46,XY,−C,+ring(C) with	Kay, Lawler, et al., 1966; Millard, Lawler, et al., 1968; Lawler, Millard, et al., 1970
46,XY,Cq−,F?− leukemia	Kay, Lawler, et al., 1966; Millard, Lawler, et al., 1968; Lawler, Millard, et al., 1970
46,XY,Dq−/47,XY,+2,Dq−/47,XY,+2, −C, Dq−,+18	Kay, Lawler, et al., 1966; Millard, Lawler, et al., 1968; Lawler, Millard, et al., 1970
45,X,−X,−17,+mar ⎭	Kay, Lawler, et al., 1966; Millard, Lawler, et al., 1968; Lawler, Millard, et al., 1970
46,XY,Ph¹+ (PV → CML)	Koulischer, Frühling, et al., 1967
44,−A,−C(X) (PV → AL)	Koulischer, Frühling, et al., 1967
51,XY,+E,+4mar (no ³²P)	Modan, Padeh, et al., 1970
47,XX,−2B,+mar,+2rings	Modan, Padeh, et al., 1970
44,XX,−B,−C,−D,+mar(dic)	Modan, Padeh, et al., 1970
46,XY,−7,+mar → Ph¹+	Modan, Padeh, et al., 1970
Ph¹+ PV and Ph¹+ benign erythrocytosis after ³²P	Modan, Padeh, et al., 1970
46,XY,−B,+16	Visfeldt, 1971; Visfeldt, Franzén, et al., 1973
46,XY,F?− (4 cases)	Visfeldt, 1971; Visfeldt, Franzén, et al., 1973
46,XX,1?+,−B,−2C,+16,+16,+16	Visfeldt, 1971; Visfeldt, Franzén, et al., 1973
46,XX,−2,−3,Bq+	Visfeldt, 1971; Visfeldt, Franzén, et al., 1973
46,XX,F?− (2 cases)	Visfeldt, 1971; Visfeldt, Franzén, et al., 1973
47,XY,+C	Visfeldt, 1971; Visfeldt, Franzén, et al., 1973
46,XX,−C,+D,Dq−,F?−	Visfeldt, 1971; Visfeldt, Franzén, et al., 1973
46,XX,+2,−C	Visfeldt, 1971; Visfeldt, Franzén, et al., 1973
46,XY,Dq−	Visfeldt, 1971; Visfeldt, Franzén, et al., 1973
46,XY,+B,−2C,+D	Visfeldt, 1971; Visfeldt, Franzén, et al., 1973
46,XX,1?+/44,XX,2q−,−3,Bq+,−E(16?)	Visfeldt, 1971; Visfeldt, Franzén, et al., 1973
46,XX,−C,+D,Gs+	Visfeldt, 1971; Visfeldt, Franzén, et al., 1973
46,XX,Dq−,Ep+	Visfeldt, 1971; Visfeldt, Franzén, et al., 1973
45,XY,−3,+ring(C)	Visfeldt, 1971; Visfeldt, Franzén, et al., 1973
47,XY,+C,F?−	Visfeldt, 1971; Visfeldt, Franzén, et al., 1973
46,XX,−E,+G	Visfeldt, 1971; Visfeldt, Franzén, et al., 1973
46,XX,Gq−	Visfeldt, 1971; Visfeldt, Franzén, et al., 1973
46,XX,Dq−	Visfeldt, 1971; Visfeldt, Franzén, et al., 1973
47,XX,+C	Visfeldt, 1971; Visfeldt, Franzén, et al., 1973
44,XY,−B,−G,Eq+	Visfeldt, 1971; Visfeldt, Franzén, et al., 1973
46,XY,7q−(q22)	Rowley, 1973d
46,XY,t(12;17)(q13;p11)/47,X,+9,t(1;Y)(q21;q12)	Rowley, 1973d
46,XY,7q−(q22)	Tsuchimoto, Bühler, et al., 1974
48,XX,+C,+C	Gras, Cowling, et al., 1975

Table 69—continued

Karyotype	Reference
51,XX,+3,+8?,+8?,+C,−15,16p−,+19,t(1;15)(P1?;q1?),t(2;11)(p13;q21) (PV → MM → ?AL)	Rowley, 1975b
46,XY,20q−? (2 cases)	Shiraishi, Hayata, et al., 1975
46,XY,11q−,13q−/46,XY,11q−/46,XY,13q−	Shiraishi, Hayata, et al., 1975
46,XY,del(20)(q11)	Shiraishi, Hayata, et al., 1975
46,XY/45,X,−Y	Shiraishi, Hayata, et al., 1975
46,XX,t(1;8)(q11;q12),del(12)	Stavem, Hagen, et al., 1975
46,XX/46,XX,−E,ring(E)/47,XX,+C	Manoharan and Garson, 1976
46,XX,−A,+mar(?1q+)	Manoharan and Garson, 1976
46,XX → 46,XX,del(12)(p11)/46,XX/46,XX, del(12)(P11),del(20)(q11)	Westin, 1976
46,XY → 46,XY,t(1;5)(p36;q31)/46,XY (no leukemia)	Westin, 1976
46,XX,t(1;15)	Wurster-Hill, Whang-Peng, et al., 1976
47,+B	Wurster-Hill, Whang-Peng, et al., 1976
47,XX,+9	Wurster-Hill, Whang-Peng, et al., 1976
48,XY,+8,+9	Wurster-Hill, Whang-Peng, et al., 1976
48,XX,+C,+E/47,XX,+E	Wurster-Hill, Whang-Peng, et al., 1976
45,XY,−16,−17,+21/48,XYY,+11,+16,−17, −17,+22	Wurster-Hill, Whang-Peng, et al., 1976
44,XX,−18,−18/44,XX,−18,−22/43,XX,−9,−22,−22	Wurster-Hill, Whang-Peng, et al., 1976
45,XY,−16/45,XY,−17/44,XY,−17,−20/44,XY, −10,−13	Wurster-Hill, Whang-Peng, et al., 1976
46,XY,tan(14;20),+fragment	Wurster-Hill, Whang-Peng, et al., 1976
46,XY → 46,XY,t(1;13)(q12;p12)/47,XY,+t(1;13)	L. Y. F. Hsu, Pinchiaroli, et al., 1977
46,XX → 49−51 chromosomes,XX,−1,+8,+?15, +1−3mar/46,XX(AL)	Weinfeld, Westin, et al., 1977a
46,XX → 46,XX,del(20)(q11)/46,X,−X,+mar/ 46,XX(AL)	Weinfeld, Westin, et al., 1977a
46,XY → 46,XY (AL)	Weinfeld, Westin, et al., 1977a
48−50chromosomes,+8,+9,del(12)(p11),del(12)(p11)+2fragments(AL)	Weinfeld, Westin, et al., 1977a
46,XY/46,XY,−3,+mar(2q+?),+mar,−(B,C or D chromosomes)/44,XY,t(2;13)(q37;q12?),t(13;?)(p11;?),−13,−15 (AL)	Weinfeld, Westin, et al., 1977a
48,XX,+8,+9,+9,−11,del(4)(q23)/46,XX (AL)	Weinfeld, Westin, et al., 1977a
47,XX,+8/46,XX,−21,+i(21q)	Shabtai, Weiss, et al., 1978
45,XX,−F/47,XX,+20	Shabtai, Weiss, et al., 1978
46,XX,Ph¹ (→ CML)	Shabtai, Weiss, et al., 1978
47,XX,−1,−9,+2t(1q;9p),+t(1p;9q),7q−,20q−	Nowell and Finan, 1978b
48,XY,+3,−5,+C,−13,−14,+t(13q;14q),+2min	Nowell and Finan, 1978b
44,XY,−3,−5,+D,−E(?17−18)	Nowell and Finan, 1978b
47,XY,+1p−,+C,−20	Nowell and Finan, 1978b

which these patients have been exposed during various diagnostic X-ray procedures (Table 70).

The development of CML in a patient with PV treated with ^{32}P and Myleran after a splenectomy has been reported in a female patient (Stavem, van der Hagen, et al., 1975). However, the clinical course of this patient appears to be more compatible with some form of AL; the fact that no Ph¹ was found either in the blood or marrow cells places the diagnosis of CML in this patient in doubt.

Chromosomal Changes and Progression of Polycythemia Vera

Because PV may turn into CML, EL, AL, or MMM, the chromosomal abnormalities encountered in this disease are often difficult to

Table 70 20q− anomaly in conditions other than PV

Diagnosis	Reference
CML(Ph¹+)	Prigogina and Fleischman, 1975a
CML(Ph¹−)	Hossfeld, 1974b
AML	P. Philip, 1977a; Barlogie, Hittelman, et al., 1977; Golomb, Vardiman, et al., 1978
AMMoL	Testa, Kinnealey, et al., 1977; Golomb, Vardiman, et al., 1978
Subacute myeloid leukemia	Grouchy, Nava, et al., 1966c
Smoldering leukemia	Testa, Kinnealey, et al., 1977
Preleukemia	Ruutu, Ruutu, et al., 1977b; Mitelman, personal communication; Pierre, 1978a
SBA	Grouchy, Nava, et al., 1966c (6 cases); M. M. Cohen, Ariel, et al., 1974; Bitran, Golomb, et al., 1977
Myeloproliferative syndrome	Testa, Kinnealey, et al., 1977 (3 cases); Nowell, Jensen, et al., 1976a (2 cases)
Refractory anemia with excess myeloblasts	Van Den Berghe, Michaux, et al., 1977a
LS → AML	Whang-Peng, Gralnick, et al., 1977; Whang-Peng, Knutsen, et al., 1979

interpret in relation to their role in the disease proper. Because almost all the PV cases reported with karyotypic changes were accompanied by a leukocytosis, it is possible that the abnormal metaphases observed belonged to leukocytes during their early evolution into a leukemic state, particularly if the disease has been treated with ^{32}P. However, some of the observations militate against such an interpretation and require explanation. In some untreated patients with PV, the marrow cells were all found to be aneuploid and their number to decrease significantly when the patients were treated. The question arises as to the origin of these cells before therapy. Were the aneuploid metaphases of erythroblastic or leukocytic origin or both? What is the origin of the diploid cells after therapy? Does treatment affect the precursors of only one series of cells, e.g., leukocytes, and not those of others? When marrow metaphases are examined, it is conceivable that those originating from one series of cells may not be spread sufficiently in the cytologic preparation to be optimally or reliably examined. It is possible, for example, to visualize only the leukocytic elements as yielding technically suitable metaphases for examination and the erythroblastic ones as being inadequate or *vice versa*. Thus, if a series of cells is accompanied by diploidy or aneuploidy, this series may spuriously contribute to the interpretation of a condition as being overwhelmingly diploid or aneuploid. In addition, few chronologic in-depth studies of PV, i.e., from the original phase of the disease through its evolution into a leukemic state, have been described; hence, it is difficult to be certain of the significance and genesis of the chromosomal changes in PV.

A missing Y has been described in patients with treated and untreated PV.[35] One of our patients had a normal marrow picture for many months after busulfan (Myeleran) therapy. The karyotypic findings in the marrow consisted of 50–70% of the metaphases containing 45 chromosomes, with apparently a missing Y being the only visible abnormality (Shiraishi, Hayata, et al., 1975). The patient had not received ^{32}P, and there was no clinical or histological evidence that he was developing a leukemic state. The granulocytes had a normal LAP activity, such activity having been higher than normal in the past. Because a missing Y has been shown to exist in the marrow cells of normal elderly males, this case could be interpreted as indicating that the cells with a normal chromosome constitution were the ones involved by the polycythemic process; whereas after response to therapy and return of the BM to normal cellularity, the 45,X (missing Y) cells, which may have been of normal origin, repopulated the marrow.

The chromosomal findings in another reported case of PV are difficult to interpret, because the essentially diploid picture was obtained both in the blood and marrow cells following 72-hr culture in the presence of PHA (Nava, 1969). Under these circumstances one would expect the normal lymphocytes to undergo cellular division with the probability that the cells involved in the polycythemic process were not being examined under these circumstances. The same author also observed a 20q− marker in an acquired idiopathic sideroblastic anemia but could not find a common link between this disease and PV.

Of cogent significance is an evaluation of the chromosomal changes in PV as they relate to the progression of the disease into several directions. Eventually, about a quarter of the patients with PV develop myelofibrosis and progressive reduction in erythrocyte population; extra medullary hematopoiesis can then be observed in the spleen, liver, and other tissues. A rising leukocytosis often with increased immature myeloid forms accompanies these changes, as well as enlargement of the spleen. About one-third of such patients ultimately develop a picture simulating that of AML. The presence of chromosomal changes in these patients when such were not observed during the polycythemic stage is of great help in establishing the diagnosis of AML. It is of further importance to know whether the PV patients with karyotypic changes, prior to the development of the leukemic stage, are the ones who develop this complication; and whether those patients who do not develop a leukemic picture are the ones with diploid karyotypes. In other words, does the presence of chromosomal abnormalities before the progression of PV into a stage resembling AL indicate that such patients are more likely to develop such a complication than those who do not have chromosomal changes? The presence of chromosomal changes may also serve to differentiate patients with PV developing a leukemoid reaction from those who have a bona fide AL. Even though the number of PV patients developing AL has varied from one series to another, the consensus appears to be that as many as 15% may develop an AL-like picture without preceding evidence of MMM or myelofibrosis. Whether these changes are the result of irradiation therapy (Court Brown, Jacobs, et al., 1962) or reflect a natural evolution of the disease, when early death from thrombosis or hemorrhage is avoided, is still an unsettled question. Again, cytogenetic studies of the disease from its earliest stages to the development of the leukemia-like picture would aid greatly in deciphering this enigma. There are those who believe that, because about 10% of untreated patients with PV have chromosomal changes in the marrow, a subgroup of patients with PV possess a propensity for cytogenetic accidents and perhaps these are the patients who develop leukemia. In any case, much more has to be elucidated in the cytogenetics of PV in order to ascertain the significance of the chromosomal changes during the various stages of the disease, particularly as they relate to the development of complications, the most serious of which is AL.

The Ph^1 in Polycythemia Vera

Rare cases of patients with PV and a Ph^1 in the marrow cells have been described. The picture of PV and the characteristic morphology of the Ph^1 in one case are quite convincing (Koulischer, Frühling, et al., 1967). In the bulk of the cases published to-date, however, no banding analysis has been performed in the cases of PV with a Ph^1 and, hence, it is unknown whether the genesis of the Ph^1 is due to a #9−#22 translocation. The patient mentioned above did not go on to develop CML and was still untreated when the cytogenetic analysis was performed and a Ph^1 observed in 10% of the mitoses analyzed. Later the patient had still not developed CML (Cervenka and Koulischer, 1973). Another 71-yr-old patient in whom PV had been diagnosed in 1960 and who was treated with ^{32}P, transformed into CML in 1965; in 1967 over 70% of the BM cells were Ph^1-positive. A Ph^1-positive PV reported in 1961 transformed into CML in 1964 (N. H. Kemp, Stafford, et al., 1961, 1964). A Ph^1 was observed in the marrows of two brothers with PV in whom the LAP level was moderately increased (Levin, Houston, et al., 1967). However, the authors agreed (Levin, Ritzmann, et al., 1968) with Summitt's (1968) interpretation that the karyotypic anomaly was a shortened Y, transmitted familially, rather than a Ph^1.

The presence of a Ph^1 in untreated patients with uncomplicated PV raises an important question regarding the development of CML in this disease. Even though the number of patients developing CML during the course of the PV is rather small, cases in which the Ph[1] has been established with certainty with banding techniques have been reported in recent years (Pris, Launais, et al., 1978). Even though in most of these cases the Ph[1] was not found in the preleukemic stage, rare cases have been reported in which a Ph[1] was present during the nonleukemic stage of the PV. Subsequently, CML developed in these patients and, thus, there appears to be little doubt that, at least in these patients, PV with a Ph[1] progressed to CML. The case described by Hoppin and Lewis (1975) not only had a Ph[1] but also a missing Y in all the *marrow* cells. *Peripheral blood* cells, cultured with PHA, showed 15 of 18 cells to be 46,XY,Ph[1]-negative and three metaphases to have 45,X,Ph[1]-positive, 45,X,Ph[1]-negative, and 45,XY,—C karyotypes, respectively. Banding revealed the 9q+ abnormality. After therapy with busulfan and normalization of the peripheral blood count, the marrow remained 100% 45,XO,Ph[1]-positive.

A recent report (Hoppin and Lewis, 1975) has summarized the recent literature on PV progressing to Ph[1]-positive CML. In the cases reported, the patients were exposed to radiation in the form of ^{32}P, the dose varying from 4 to 30 mCi; one patient has been reported who had not received any form of radiotherapy. Thus, it would appear that patients with PV do develop on rare occasions Ph[1]-positive CML; and that it is possible for PV patients with the Ph[1] to progress to CML and retain the Ph[1]. In some of these cases, the presence of a #9–#22 translocation has been demonstrated and, thus, there is little doubt that the genesis of the Ph[1] is identical to that in CML.

An unusual case has been described in an 81-yr-old female who was diagnosed as having PV in 1966, at which time about 10% of her marrow cells were Ph[1]-positive (Verhest and Schoubroeck, 1973). In subsequent years, the patient received ^{32}P therapy leading to a normalization of her blood count in 1973. In 1971 and 1972 her LAP reaction appeared to be increased and repeat marrow cytogenetic studies failed to reveal the Ph[1]. It was suggested that an abortive attempt to establish a clone of Ph[1]-positive cells occurred in this patient with PV. However, the finding of three and possibly four patients who initially began with PV and later progressed to CML makes the likelihood of this occurring by chance alone exceedingly small.

Thus, the Ph[1] provides a clearer picture of the relation between these two myeloproliferative disorders, i.e., PV and CML, than afforded by other aspects of these states. In some of the PV patients the Ph[1]-positive cells constituted only a partial percentage of the total cell population, with karyotypically normal cells also being present in the marrow. Still uncertain is whether the PV patients without the Ph[1] ultimately develop CML with the Ph[1]. It is hoped that serial studies in PV will shed light on this important aspect of cytogenetics as related to PV and CML and to the MD, in general.

Banding Studies in Polycythemia Vera

A number of studies dealing with relatively large series of patients with PV and including observations with banding techniques have appeared during the last few years. Although these studies have not established any characteristic or specific karyotypic change characterizing PV (except possibly for the 20q−) these have gone a long way toward clarifying the nature of the chromosomes involved in the aneuploid cases with this disease and revealed a somewhat nonrandom chromosomal picture characterizing a significant number of the cases with PV. Thus, some of the confusion existing in regard to the cytogenetic changes observed in PV has been clarified with banding, though the bulk of the chromosomal changes continue to be of a variable and inconsistent nature.

The first observations with banding methods in PV were performed by Reeves, Lobb, et al. (1972) in four cases with the disease, in two of whom these authors were able to show a 20q− anomaly, thus confirming the nature of the chromosome frequently involved in PV and reported as a partial deletion of an F-chromosome in the past.

A report by Berger (1972a) in which a 20q− anomaly in a case of PV was shown with G-banding to be due to a translocation between the short arm of #11 and the long arm of #20, i.e., t(11;20)(p15;q11), remains to be confirmed. Were this anomaly shown to be consistent in PV, it might assume a significance similar to that of the Ph[1] in CML and 14q+ in lymphoma. Berger, Parmentier, et al., (1974) reported the cytogenetic studies in 35 cases of

PV: 27 in the initial stages of the disease, 8 with myelofibrosis, and 7 in the AL stage. Two constitutionally balanced translocations (between chromosomes of groups D and G) were found in the group. In 13 of the cases cytogenetically abnormal cells were observed but in only 1 was evidence of clonal evolution found. Aneuploid cells were most common in the myelosclerotic stage.

A chromosome study performed by a cooperative group (Wurster-Hill, Whang-Peng, et al., 1976) on a total of 149 patients with PV revealed chromosomal changes of one nature or another in 48 (32% incidence). Nearly 70% of the 48 patients with chromosomal changes were characterized by hyperdiploidy or polyploidy due to a variety of chromosomal changes, both numerical and morphologic. Only four cases of bona fide hypodiploidy were observed. To me the most striking aspect of the study published by Wurster-Hill, Whang-Peng, et al. (1976) was the absence of a single patient with 20q−, though apparently one case was thought to have such a karyotypic anomaly but to be involved in a complex translocation with #14, i.e., 46,XY,tan(14;20), plus an acentric fragment. The most common anomaly observed in the aneuploid group was trisomy C, most frequently involving either #8 or #9 (four cases). Loss of a C-group chromosome, possibly #7, was observed in a number of patients, but because banding studies were not apparently fruitful, it is difficult to ascertain whether, in fact, the monosomy C observed was due to loss of #7. Partial deletion of the long arm of #7 (7q−) has been reported in PV by Tsuchimoto, Bühler, et al., (1974) and by Rowley (1973b). Two of the cases studied by Wurster-Hill, Whang-Peng, et al., (1976) were shown to have extra material from #1, which according to the authors brings the total of such cases reported in the literature to six. Special emphasis was put by the authors on the occurrence of trisomy 1q, particularly that involved in translocation with #15 [t(1q;15q)], as opposed to trisomy of the whole #1 observed in patients with AL, refractory displastic anemia, and other conditions. No patients with a Ph^1 were observed in this group of 149 subjects with PV.

In another study (Zech, Gahrton, et al., 1976) based on Q−, G−, and C-banding techniques, ten patients with PV were investigated: four had received no treatment with cytotoxic drugs, three had been given ^{32}P only, and the other three had received ^{32}P, busulfan, or busulfan and procarbazine. One 73-yr-old male, treated with venesection only for 4 yr, lacked the Y and had a deletion of the long arm of #20 (20q−) in all cells investigated. One of the other three patients who had received no drugs had a chromosome abnormality in only 1 of 19 identifiable metaphases, consisting, however, of trisomy #9, the most common abnormality observed in treated patients. In the group of treated patients, trisomy #9 was found in three patients, trisomy #8 in one, and a deletion of the long arm of one #20 (20q−) in one patient. Multiple aberrations in addition to the extra #9 were found in one patient in whom the disease had transformed into AML. The finding of identical chromosomal aberrations, i.e., 20q− and trisomy #9, in two patients who had received no drugs and in four who had received ^{32}P, busulfan, or procarbazine, favors the view that these aberrations are specifically associated with the disease and not induced by the drugs. With the exception of the patient with AML, all the other patients were alive 1–11 mo after the chromosomes were examined and 1–230 mo after diagnosis. The authors speculated that whatever the mechanism of the appearance of chromosomal aberrations in PV may be, these abnormalities do not have the same prognostic implications as similar aberrations in other MD, such as CML. They stressed in particular the trisomy of #8, which usually indicates a poor prognosis in CML after its acquisition, whereas in the patients studied this aberration did not apparently accelerate the progression of the PV. Three of the patients had no chromosomal abnormalities and the incidence of karyotypic changes observed by Zech, Gahrton, et al., (1976) is higher than is the general experience. However, it is unknown whether these were selected patients or patients seen at random.

In contrast to the above is a study by Westin, Wahlström, et al., (1976), who found only seven patients with crhomosome abnormalities in a series of 50 consecutive, unselected cases of untreated PV (an incidence of only 14%). The essential karyotypic findings in the seven cases consisted of the following: three cases of trisomy #8 and #9, two with a similar but not identical marker including material from the long arm of #1 (one of these patients also had a

20q— deletion), one patient with an extra unidentified isochromosome, and one patient with monosomy #16. In five of the cases with PV the abnormal clones were accompanied by cells with a normal karyotype. No hypodiploid cases were encountered. As in other series of PV, except for the 20q— deletion, no abnormality typical for the disease has been detected though some chromosomes (#1, #8, #9, and #20) seem to be involved more often than others. The patients with the chromosome aberrations did not show any clinical, laboratory, or morphologic features that would separate them from cases with normal karyotypes. Two of the patients with initial normal chromosomal findings later developed structural chromosome abnormalities after therapy with chlorambucil. In one case after 25 mo of myelosuppressive therapy a del(12)(p11) in 60% of the cells was observed and 10 mo later all the metaphases in the marrow contained this deletion. In addition, 20q— was found in three cells. In the second case, after 40 mo on chlorambucil, the presence of a balanced translocation, t(1;5)(p36;p31), in 90% of the cells was established. In neither one of the patients were any definite clinical, laboratory, or morphological signs of an approaching leukemic transformation present.

Our experience with PV, though rather limited, is based on a study of 13 patients, 4 of whom were untreated (Shiraishi, Hayata, et al., 1975). Eight of the 13 patients, including the untreated ones, had a normal diploid chromosome constitution in the marrow. Three of the patients appeared to have a 20q— anomaly, one had 11q— and 13q—, and one patient (70-yr-old male) had a mixture of normal and cells with a missing Y. The latter cytogenetic finding is thought not to be related to the PV. The four patients with partial deletions had been treated with ^{32}P and three of them with chemotherapy also. As has been the experience of others, the missing portion of the deleted #20 could not be found on any other chromosome. It appears that most patients with PV do not develop chromosomal abnormalities through the course of their disease; when such abnormalities appear (20q— in particular), these seem to result possibly from radiation or chemotherapy. Thus, cytogenetic changes do not appear to play a crucial role in the genesis of PV.

Cytogenetic findings in a group of 15 patients with PV reported by Shabtai, Weiss, et al. (1978) were based on G and C banding of marrow and blood preparations. Major chromosomal aberrations were found in 4 of the 15 patients and minor ones in 3 more. The most interesting finding concerned the chromosomal polymorphism of pair #19; the authors indicated a possible relation of this karyotypic aberration to the etiology of the disease. The 20q— anomaly was not found in any of the patients. In four patients chromosomal breaks, usually of the chromatid type, were found in more than 15% of the marrow cells.

The relationship of the chromosomal changes in PV to the development of AL is not totally understood. A comprehensive study has been published by Weinfeld, Westin, et al. (1977a) in which the authors evaluated the clinical, morphologic, and cytogenetic findings in eight patients with PV terminating in AL. These cases were derived from a total of 120 patients with well-defined PV observed over a period of 5 yr. AL constituted 24% of the total deaths in the 120 patients. No distinctive initial clinical and morphological characteristics made it possible to predict the development of AL. In one case MMM was already present at the time of PV diagnosis and in another MMM developed several years before the AL. Chemotherapy with alkylating drugs was the main treatment for all patients. Four of the patients were treated with chlorambucil or busulfan alone, without radiation or ^{32}P. The onset of AL in some of the cases was very rapid and had a character of an unexpected progressive pancytopenia, appearing during chemotherapy. One patient had a leukemia of undifferentiated blastic type, three had AML, one had a promyelocytic type, and three had acute EL. Of the seven patients studied cytogenetically all but one had multiple and complex abnormalities indicating the presence of several superimposed cell lines. In one patient all marrow cells were hypertetraploid. In three of the patients pretreatment chromosome analyses were performed and were found to be normal. Three of the patients had trisomy or tetrasomy of #8 or #9, one patient had a 20q— anomaly developed during therapy, one patient had two complex translocations, one involving #2 and #13 and the other #13 and another unknown chromosome, in addition to —15. In two of the patients karyotypically normal cells were present in the marrow at the time aneuploidy developed.

The presence of diploidy in most patients

with PV, the inconsistency of the chromosomal changes, and the frequent presence of normal diploid cells in aneuploid marrows are difficult to reconcile with a view advanced that PV is a disease of clonal origin (Adamson, Fialkow, et al., 1976). The authors felt that results based on G6PD determinations in two women with PV and heterozygosity (Gd^b/Gd^a) at the X-chromosome linked locus for the enzyme provided direct evidence for the stem-cell nature of PV and strongly imply a clonal origin for this disease. The chromosomal changes are somewhat difficult to reconcile with such a view, though the possibility that the enzymatic environment favors involvement of such cells, if not their survival, in PV, as in CML, is to be considered.

Comments on Chromosomal Changes in Myeloproliferative Disease

In my opinion, what the spectrum of chromosomal findings has demonstrated, particularly in myeloproliferative disorders, is the malignant nature of these diseases, ranging from certain anemias through myelofibrosis to frank leukemia. The frequent and somewhat characteristic involvement of #1 and the vulnerability of some points on #3 in various myeloproliferative disorders (including the leukemias) have been discussed and stressed by several authors.[36] Even though in some cases, particularly those of PV who had received ^{32}P and/or chemotherapy (e.g., chlorambucil), the chromosomal changes may have been induced, there is little doubt that so-called spontaneous cytogenetic aberrations may develop in some of these cases with various MD. The concepts of Nicoara, Butoianu, et al. (1967), for example, regarding the specificity of the Ph^1 and other chromosomal changes in the progression of these disorders (e.g., PV to CML or thrombocythemia to CML), appear to indicate an intimate relationship between the karyotypic changes. Nevertheless, the possibility remains that in some of these conditions radiation and/or chemotherapy may play a role in accelerating the induction and/or unmasking of chromosomal changes related to the appearance of a new, and certainly more malignant, phase of the disease.

The findings in the series of studies by Westin, Wahlström, et al. (1976) and Weinfeld, Westin, et al. (1977a) indicate that even though abnormal clones are already present in certain patients with PV at the time of diagnosis, they may also develop during the course of the disease, and do so irrespective of the type of therapy given. The presence of a chromosomally abnormal clone in PV does not seem to denote a subgroup of patients more susceptible to the development of AL.

In a recent review (Meytes, Akstein, et al., 1977) of the chromosome changes during the course of PV, the authors pointed out methodological pitfalls related to data collection, presentation and interpretation. Furthermore, the karyotypic derangements are not specific to PV and clone formation is infrequent, with therapy increasing the quantity of aberrations rather than any specific aberration.

References

1. Baserga, 1970; Cadiou, Ruff, et al., 1975; Dameshek, 1951; W. Fischer and Fölsch, 1975; Goh, 1965a; Nau and Hoagland, 1971; D. S. Rosenthal and Moloney, 1969.
2. Burkhardt, Pabst, et al., 1969; Catti, 1971; Ferreyra and Dondo Lascano, 1969; Gralnick, Harbor, et al., 1971; Prokofjeva-Belgovskaya, Kosmachevskaya, et al., 1964; Slyck, Weiss, et al., 1970.
3. Kaplow, 1968; Sewell, 1972; Teplitz, 1966; Teplitz, Rosen, et al., 1964.
4. Alter, Lee, et al., 1963; Tough, Jacobs, et al., 1963.
5. Ezdinli, Sokal, et al., 1970; Kamada, Okada, et al., 1967; Kenis and Koulischer, 1967; Krauss, 1968; Merker, 1965.
6. Kiossoglou, Mitus, et al., 1966a; Mitus, Coleman, et al., 1969; D. S. Rosenthal and Moloney, 1969; Sandberg, Ishihara, et al., 1962b.
7. Bowen and Lee, 1963; Cehreli, Ezdinli, et al., 1976; S. M. Cohen, 1967; Forrester and Louro, 1966; Frey and Siebner, 1968; Heath and Moloney, 1965b; N. H. Kemp, Stafford, et al., 1963; M. H. Khan and Martin, 1968a; Kiossoglou and Mitus, 1965; Kiossoglou, Mitus, et al., 1966a; Koulischer, Frühling, et al., 1967; Krauss, 1966; Rudders and Kilcoyne, 1974; Wolf, Merker, et al., 1966.
8. Better, Brandstaetter, et al., 1964; Borgaonkar, 1972; Hellriegel, 1968; Kay, Millard, et al., 1970; Krogh Jensen and Philip, 1970, 1973; Nowell, 1971b,c; 1977; Nowell, Jensen, et al., 1976a; Pileri, Masera, et al., 1966; Sandberg, Ishihara, et al., 1962b; Whang-Peng, Broder, et al., 1976.
9. Borgaonkar, 1972; Chapelle, Wenström, et al.,

1970; Kiossoglou, Mitus, et al., 1966a; Krogh Jensen, 1968; Lawler, Kay, et al., 1966; Leeksma, Friden-Kill, et al., 1965; Pawelski, Maj, et al., 1967; Sandberg, Ishihara, et al., 1964a; Winkelstein, Sparkes, et al., 1966.
10. Bock, Haberlandt, et al., 1970; Holden, Garcia, et al., 1971; J. F. Jackson and Higgins, 1967; Krogh Jensen and Philip, 1970; Polák and Žižka, 1970; Teasdale, Worth, et al., 1970.
11. Boetius, Hustinx, et al., 1977 (adults); Chapelle, Vuopio, et al., 1975; Freireich, Whang, et al., 1964; J. D. Holden, Garcia, et al., 1971; Humbert, Hathaway, et al., 1971; J. F. Jackson and Higgins, 1967 (adult); Kamiyama, Shibata, et al., 1973; U. Kaufmann, Löffler, et al., 1974; MacDougall, Brown, et al., 1974; McClure, Thaler, et al., 1965; Polák and Žižka, 1970; Rowley, Blaisdell, et al., 1966 (adult); Teasdale, Worth, et al., 1970.
12. Beutler, Goldenburg, et al., 1964; Bottura and Ferrari, 1962b; Fitzgerald and Hamer, 1971; Fleischmann and Bodor, 1970; Goh and Swisher, 1963; Hartmann and Jenkins, 1965; Kamada and Uchino, 1976; R. W. Kaufmann, Schechter, et al., 1969; Ross and Rosenbaum, 1964; Tsuchimoto, Ishii, et al., 1970; Whang-Peng, Knutsen, et al., 1976a; Zaccaria, Ricci, et al., 1973.
13. Dameshek, 1969; D. Holden and Lichtman, 1969; Jenkins and Hartmann, 1969.
14. D. A. G. Galton, 1965; M. L. Ghosh, 1972; Tough, Jacobs, et al., 1963; Woodliff, Onesti, et al., 1967a.
15. Frick, 1969; Rowley and Blaisdell, 1966; Rowley, Blaisdell, et al., 1966; Woodliff, Onesti, et al., 1967a.
16. Blackburn, Callender, et al., 1968; Nielsen and Krogh Jensen, 1970.
17. Forteza-Bover and Baguena-Candela, 1963; Keller and Nordén, 1967; Kiossoglou and Mitus, 1964; Krogh Jensen and Friis-Møller, 1967.
18. Heath, 1966; Keller and Nordén, 1967; Krogh Jensen, 1977; Krogh Jensen and Friis-Møller, 1967; Menzies, Crossen, et al., 1966.
19. Castoldi, Scapoli, et al., 1969; Heath, 1966; Keller and Nordén, 1967; M. H. Khan and Martin, 1968b; Powsner and Berman, 1965.
20. Bottura and Coutinho, 1968; Castoldi, Scapoli, et al., 1969; Heath, 1966; M. H. Khan and Martin, 1968b; Krogh Jensen and Friis-Møller, 1967.
21. Bottura and Coutinho, 1968; Heath, 1966; Keller, Lindstrand, et al., 1970.
22. Castoldi, Scapoli, et al., 1969; Keller and Nordén, 1967; M. H. Khan and Martin, 1968b; Kiossoglou, Mitus, et al., 1965c; Lawler, Roberts, et al., 1971; Matsaniotis, Kiossoglou, et al., 1968.
23. Astaldi, Strosseli, et al., 1962; Bottura and Coutinho, 1968; Heath, 1966; Keller and Nordén, 1967; Keller, Lindstrand, et al., 1970; Kiossoglou, Mitus, et al., 1965c; Krogh Jensen and Friis-Møller, 1967; Lawler, Roberts, et al., 1971; Matsaniotis, Kiossoglou, et al., 1968.
24. Adamson, Fialkow, et al., 1976; Burkhardt, Pabst, et al., 1969; Modan and Lilienfeld, 1965.
25. Burkhardt, Pabst, et al., 1969; Landaw, 1976; Lawrence, Winchell, et al., 1969; Weinfeld, Westin, et al., 1977a.
26. Callender, Kay, et al., 1971; Hossfeld, Schmidt, et al., 1972a; Kiossoglou, Mitus, et al., 1966a; Krogh Jensen, 1968; Lawler, Millard, et al., 1970; A. Levan, Nichols, et al., 1964; Sakurai, 1970a; Scharfman, Amarose, et al., 1969 (erythrocytosis of childhood); Visfeldt, Franzén, et al., 1970.
27. Hossfeld, Schmidt, et al., 1972a; Lawler, Millard, et al., 1970; Nowell, 1971b,c.
28. Kay, Lawler, et al., 1966; Lawler, Millard, et al., 1970; Millard, Lawler, et al., 1968.
29. Lawler, Millard, et al., 1970; Visfeldt, Franzén, et al., 1970; Weinfeld, Westin, et al., 1977a.
30. Erkman, Hazlett, et al., 1967; Koulischer, Frühling, et al., 1967; Lawler, Millard, et al., 1970; Solari, Sverdlick, et al., 1962; Visfeldt, Franzén, et al., 1970.
31. Nowell, 1971b, c; Visfeldt, 1971; Visfeldt, Franzén, et al., 1971; Wurster-Hill and McIntyre, 1978.
32. Berger, Parmentier, et al., 1974; McIntyre, 1970; Motoiu-Raileanu, Berceanu, et al., 1970; Nagy and Yurgutis, 1968; Shiraishi, Hayata, et al., 1975; Traczyk, 1963.
33. Berger, Parmentier, et al., 1974; Bousser, Zittoun, et al., 1977; Hoppin and Lewis, 1975; Koulischer, Frühling, et al., 1967; Krmpotic, Vykoupil, et al., 1977; Levin, Houston, et al., 1967; Modan, Padeh, et al., 1970; Tanzer, Levy, et al., 1973; Verhest and Schoubroeck, 1973.
34. Reeves, Lobb, et al., 1972; Shiraishi, Hayata, et al., 1975; Westin, Wahlström, et al., 1976; Zech, Gahrton, et al., 1976.
35. Hoppin and Lewis, 1975; Shiraishi, Hayata, et al., 1975; Wurster-Hill, Whang-Peng, et al., 1976; Zech, Gahrton, et al., 1976.
36. Gahrton, Friberg, et al., 1978a; P. Philip, 1975d; Rowley, 1977a, b; Van Den Berghe, David, et al., 1976a.

14

The Lymphomas

Hodgkin's Disease (HD)
Some Cytogenetic Findings in Hodgkin's Disease
Acute Leukemia in Hodgkin's Disease

Reticulum-Cell Sarcoma, Lymphocytic Lymphosarcoma, and Follicular Lymphoma
Chromosome Studies Before Banding
Banding Studies in Lymphoma
Chromosome Findings in Follicular Lymphoma
The 14q+ Anomaly in Lymphoma

Acute Leukemia in Patients with Lymphoma

Burkitt Lymphoma (BL)
Nonendemic Burkitt Lymphoma

Comparison of Karyotypic Findings in Various Lymphomas
Studies in Established Cell Lines
The 14q+ Anomaly in Burkitt Lymphoma

Chromosome Changes in Nonendemic Burkitt Lymphoma
Some In Vitro Observations
Sézary Syndrome and Mycosis Fungoides

Augioimmunoblastic Lymphadenopathy

Histiocytic Medullary Reticulosis
General Comments

Karyotypic findings in this group of diseases will be presented in several separate sections. Even though the division into sections may appear to be somewhat arbitrary, the genomic characteristics of the various diseases considered in the lymphoma group find, surprisingly, correlatives with their known clinical aspects and assumed etiologies. A common characteristic of these diseases is their presence in tissues in which they may be surrounded by or admixed with normal cells. Hence, the presence of diploid cells is difficult to interpret as to their origin, i.e., normal vs. malignant. Because most of these diseases probably originate in lymphoid tissues, e.g., lymph nodes, spleen, thymus (?), the best source for metaphases of lymphomas are these tissues. When the marrow is involved by lymphoma, which, in my experience, is encountered in only about 10–15% of the cases, karyotypically abnormal cells may be found in the marrow. When lymphoma cells circulate in the blood (Koziner, McKenzie, et al., 1977), their abnormal metaphases may be examined either by a direct method or after culture, with or without PHA.

Even though chromosomal changes in the marrow have been described in lymphoma and HD (Sandberg, Ishihara, et al., 1964c), this usually turns out to be a not very fruitful approach to establishing the karyotypic picture in lymphoma. Not only does the marrow have to be infiltrated by abnormal cells and such cells be in mitosis at the time of their examination,[1] but often a totally diploid picture is obtained in a marrow that may have abnormal cells in it. Hence, a spurious diploid picture may be interpreted as indicative of lack of infiltration of the marrow by such cells. This applies, in particular, to cases with HD or BL.

Human and animal lymphoid cells appear to be comprised of two major populations; the bursa-derived or BM derived (B cells) and the thymus-dependent (T cells) lymphocytes. These B and T populations can be distinguished by cytoplasmic membrane markers. Membrane-associated immunoglobulins (Ig) detectable by rather sensitive methods such as direct immunofluorescence, the receptor for antigen-antibody complement complexes, and the ability to bind heat-aggregated human IgG, have been shown to be B-cell markers in man. On the other hand, the spontaneous formation of rosettes with sheep erythrocytes under specific conditions appears to be a characteristic property of human T lymphocytes. The use of T-cell specific heteroantiserums may turn out to be a sensitive way of detecting T lymphocytes. These markers have been used to classify lymphoproliferative diseases on the basis of the type of cells involved.[2] This classification may be of considerable clinical and etiologic interest. Thus, reports from several laboratories have indicated that CLL is usually identifiable as a B-cell neoplasia and, moreover, the study of surface Ig has led to the concept that this B-cell proliferation is presumably of a monoclonal nature. On the other hand, other syndromes, including Sézary syndrome and mycosis fungoides, are thought to be derived from T cells. Thus, it is possible that when the cellular origin of the various lymphomas is rigorously established, the chromosomal picture may be found to be related to the cellular origin of the malignancy.

Hodgkin's Disease

Discussion on the nature of Hodgkin's disease (HD) ranges between two extreme concepts, one advocating inflammation and the other advocating neoplasia. On the basis that inflammation is not usually associated with chromosomal anomalies whereas neoplasia is, cytogenetic studies of Hodgkin-diseased tissue appear to be of utmost importance in settling the problem (Castoldi, 1973). Obviously, the tissue in question must be the lymph node, cells from which have to be subjected to chromosome analysis either directly or after short-term incubation (Baker and Atkin, 1963). Chromosomal data on lymph node-derived cells, which were incubated for 48 hr or more in the presence of PHA will not be considered in the following discussion.

I was able to collect data on 118 cases in which the chromosome constitution of lymph node-derived cells was investigated. This is a surprisingly small number, if one remembers that in adults HD is not only the most common type of malignant lymphoma, but also as common as leukemia. An analytical review of chromosome results in HD, however, is not so much hampered by scarcity of data, as by the pathogenesis of the disease itself. No meaningful interpretation is possible without a correlation of the chromosomal data with an exact histological staging of the tissue under

Table 71 Summary of chromosomal findings in Hodgkin's disease

Total number of cases	135
Number of cases with no anomalies	32
Number of cases with anomalies in the triploid range	55
Number of cases with anomalies in the diploid range	23
Number of cases with anomalies in the tetraploid range	25

(Based only on those cases in which a definite mode was established.)

investigation (Hamann, Oehlert, et al., 1970). In the literature available to me a number of cases is included in which no histological classification is given.[3] In other papers, histological changes were classified, but either according to H. Jackson and Parker (1944a, b), Rappaport (1966), Rappaport and Braylan (1975) or Lukes and associates (Lukes, 1972; Lukes, Butler, et al., 1966); hence, the results are difficult to compare. Above all, histological classification in many cases of HD is a matter of bias, largely depending on the personal experience of the pathologist.

The most important diagnostic features of HD are Sternberg-Reed cells and Hodgkin's cells, both cell types being considered to be intimately related to the pathogenesis of the disease. Consequently, the main interest of chromosome workers has been focused on the chromosome constitution of these cells.

The most valuable finding of chromosome studies in HD is the demonstration that in the majority of cases a proportion of the metaphases does have numerical and/or structural anomalies (Table 71). This indicates very strongly that HD is a neoplastic and not an inflammatory disease (Spriggs, 1971). The other remarkable observation is that the percentage of abnormal metaphases is small in most cases, with about 70–90% of all metaphases being apparently normal. In 11 out of 118 cases reported in the literature, no chromosome anomalies were detectable at all;[4] a high proportion of chromosomally normal cases was communicated by Whitelaw (1969), who interpreted his findings as evidence for HD not being a neoplasia. Fleischmann, Håkansson, et al. (1976) found four diploid cases of HD among nine examined. A characteristic anomaly, be it structural or numerical in nature, does not exist in HD. Even though the 14q+ anomaly appears to be rather characteristic for lymphoma and has been reported in HD (Fukuhara, Shirakawa, et al., 1976; Zech, Haglund, et al., 1976), the exact incidence in HD has to be established. Among the 118 cases with chromosomally normal and abnormal cell populations were 43 cases in which the chromosome mode of the abnormal population was in the triploid (57–82 chromosomes), 24 cases in the tetraploid (83–115 chromosomes), and 24 cases in the diploid range (46–56 chromosomes). Generally, the modes were very flat, in some cases barely recognizable, with chromosome numbers equally distributed between 70–85. A wide variety of different markers was found in about two-thirds of the cases with an abnormal cell population. Deletions of chromosomes of group #17–#18 (Melbourne-chromosome) (Spiers and Baikie, 1970) and accentuated secondary constrictions of a C-chromosome (Miles, Geller, et al., 1966) do not appear to me to have particular significance in HD. In a number of cases the presence of an identical marker chromosome in metaphases with different chromosome numbers pointed to a possible common precursor cell.[5] In other cases the presence of several abnormal clones was suggested by morphologically different markers.[6] Almost all of the cases with structural and/or numerical chromosome anomalies were untreated at the time of chromosome analysis; thus, iatrogenic events can be excluded as responsible factors. Instead, chromosomally abnormal metaphases must be considered to bear witness to the neoplastic properties of a proportion of the cells, which is usually quite small in HD. It is tempting to conclude that the abnormal metaphases were derived from cells that have been identified with HD, i.e., Hodgkin's and Sternberg-Reed cells. Typically, Sternberg-Reed cells are not only multilobulated but also multinucleated, and their chromosome number can be expected to be beyond the tetraploid range. The majority of abnormal metaphases in HD, however, was shown to have chromosome numbers in the triploid range. It therefore seems reasonable to speculate that the abnormal metaphases belong to Hodgkin's cells or, more generally speaking, as indicated by Seif and Spriggs (1967) as well as by Coutinho, Bottura, et al. (1971) and Peckham and Cooper (1969), to malignant reticulum rather than Sternberg-Reed cells. This notion is also corroborated by kinetic cell findings indicating that Hodgkin's cells do synthesize DNA, whereas Sternberg-Reed cells do not.[6]

The significant number of cases with no apparent chromosome anomaly does not disprove the concept of HD as a neoplasia. Even though in most of the cases a substantial number of metaphases was analyzed, the possibility cannot be excluded that the abnormal metaphases were missed. It is known that the number of Hodgkin's and Sternberg-Reed cells is inversely related to the number of lymphocytes, being small during early stages of the disease (lymphocytic predominance) and large during later stages (mixed cellularity, lymphocyte depletion). Indeed, in analyzing the chromosome data in HD, published by Coutinho, Bottura, et al. (1971), it can be noticed that of the three cases with the histological type of lymphocytic predominance, two were completely cytogenetically normal, and the third case had one hypertetraploid metaphase among 47 normal ones. The only other case with a completely normal chromosomal complement was in the group of cases with nodular sclerosis. The rest of the cases with nodular sclerosis and all cases with mixed cellularity were demonstrated to have chromosomally abnormal populations, usually in the triploid range. A similar impression can be gained from the data of Peckham and Cooper (1969). No additional support in favor of this concept can be found in the literature, because staging of the lymph nodes was not done or they were classified according to the older and less accurate system of H. Jackson and Parker (1944a, b).

Fleischman, Prigogina, et al. (1974) examined the cytogenetic findings in 17 cases with HD and correlated cytological and chromosomal findings and came to somewhat different conclusions.

The phenomenon that all or most of the metaphases derived from Hodgkin's lymph nodes have a normal chromosome constitution is perplexing. Agreement exists that the normal metaphases are derived from the cell population consisting predominantly of lymphocytes, but also of plasma cells.[7] The fact that such lymphocytes show high labeling indices (Peckham and Cooper, 1969) and divide without being stimulated by mitogens is remarkable, and, hence, was considered to indicate a malignant state (Spiers and Baikie, 1968a). The latter authors suggested that these cells "may carry a fine specific chromosome anomaly." The possibility exists that this is true; however, it seems much more likely that the high proliferative activity of the lymphocytic population reflects an immunological reaction,[8] comparable to the situation in lymph nodes with benign, reactive hyperplasia.[9] A computer display based on the fluorescent pattern of cells with a normal chromosome constitution in biopsies of HD and malignant lymphoma has revealed a perfectly normal pattern (Fleischmann, Gustafsson, et al., 1972).

As in the case of AL, the hypothesis has been advanced that HD may start out in diploid cells at a stage which cannot be identified histologically and that this stage then progresses to a predominantly lymphocytic stage, still accompanied by diploidy or near diploidy, with subsequent progression to nodular sclerosis, lymphocyte depletion, and ultimately the development of Hodgkin's (disease) sarcoma.

Some Cytogenetic Findings in Hodgkin's Disease

In a study of 30 cases of HD, five were found to be diploid and 25 (20 untreated) to have definite chromosomal abnormalities (Baker *in* Atkin, 1976a). Except for three cases in whom splenic tissue was examined, all others were based on lymph node material examined either directly or after short-term culture (Baker and Atkin, 1965). In cases with karyotypic abnormalities diploid cells were nearly always present. Eight of the cases had a chromosome number in the diploid range, eight in the triploid range, and nine in the tetraploid range. Thirteen cases had markers, in five of these the markers were subtelocentric in nature. Except for one case, the cases with lymphocytic and histiocytic HD were in the diploid range and most of those with mixed cellularity or nodular sclerosis in the hypotriploid-tetraploid range. Karyotype analysis showed the consistency within each case characteristic of a neoplastic cell population. Two nodes from one untreated case removed at an interval of 3 mo showed the same markers and other karyotypic features, even though the first was hyperdiploid and the second hypertetraploid.

In a case described by Olinici (1972) the

ascitic fluid containing Hodgkin cells had a bimodal chromosome number of 66–68 and 46. Some of the cells with 46 chromosomes were diploid, whereas others had a 46,XX, −C,+D karyotype, the latter possibly representing a clone of lymphocytes which arose as a consequence of previous radiotherapy. In another case (Fleischmann, Håkansson, et al., 1972) three medium-sized markers were seen, one possibly being an isochromosome. Reeves (1973) found possible iso(18q) and iso(18p) and 6p− in three cases with HD.

In one of four established cell lines originating from HD, Zech, Haglund, et al. (1976) found 16q−, t(9q+;22q−) and four markers. The authors do not state whether the 9;22 translocation resulted in a Ph[1].

Studies of the in vitro behavior of cells from patients with HD,[10] the ratio of T and B lymphocytes in the spleen,[11] the response of lymphocytes to PHA (Havemann, 1969), or determination of TdT activity (Donlon, Jaffe, et al., 1977) have not been correlated with chromosome studies.

A 14q+ anomaly has been reported in the cells of two out of five patients with HD and in those of a cell line derived from the pleural effusion of a patient with Hodgkin's sarcoma (Fukuhara, Shirakawa, et al., 1976). This anomaly is being observed in a wide variety of lymphomas and other diseases (e.g., multiple myeloma).

Hossfeld and Schmidt (1978) studied effusion cells from six patients with advanced HD and even though no Hodgkin or Reed-Sternberg cells could be identified cytogenetically, in four of the six cases chromosome anomalies were demonstrated. Not only was cytogenetic analysis in these cases productive in terms of diagnosis, but the authors also concluded that the effusion cells with abnormal karyotypes were intimately related to the pathogenesis of the HD.

A 14q+ anomaly has been described in a number of instances in HD,[12] including two cell lines established from tissues affected by HD.[13]

Acute Leukemia in Hodgkin's Disease

Leukemia-complicating HD is being reported more frequently, usually as a result of chemo- or radiation therapy, though this has not been rigorously proven.[14] However, the possibility exists that the two malignancies (and this also applies to lymphosarcoma, multiple myeloma, and others) are pathogenetically and/or etiologically related (Weiss, Brunning, et al., 1972). Both chronic and AL have been described to develop in HD, AML being the most frequent, with the chromosomal findings in the marrow being comparable to those one would expect in the various types of leukemia. Thus, CML with a Ph[1] has been described by us as a complication of HD (Ezdinli, Sokal, et al., 1969). We feel that the development of leukemia in HD is particularly related to radiation applied to the abdomen, though it has been observed after extensive radiation to other parts of the body. It is possible that, with wider application of radiation and/or chemotherapy in HD, and in other diseases, and with concomitant prolongation of life in these patients as a result of more effective therapy, more cases of leukemia complicating these diseases will be encountered (Zaccaria, Barbieri, et al., 1978). We must ascertain whether these leukemias will be associated with the same karyotypic picture as in other patients with leukemias of less known cause.

Judging from the karyotypic findings in cases of leukemia-complicating HD (Table 72), it would appear that the chromosomal changes do not reflect and certainly are not similar to those present in the original HD. However, findings in more cases are necessary to be sure of this generalization. Also of interest would be the demonstration of the chromosomal picture of other tumors developing in patients with HD, such as cancer of the breast, lung, or pancreas, to ascertain whether the cytogenetic pictures have anything in common.

Of the cases with AL-complicating HD on whom chromosome data are available, all were cytogenetically abnormal. The case described by Lundh, Mitelman, et al. (1975) is of interest because all the cells in the marrow had +14, −17 and +18, as shown with banding analysis. Some of the cells were shown at a later date to have either +8 or −8 and −21. Because the latter chromosomes are two of the most commonly involved chromosomes in spontaneous AML, the authors stated that the karyotypic pattern might imply similar oncogens in both types of AML, the leukemogenic effect being favored indirectly by chemotherapy or radio-

Table 72 Leukemias complicating lymphomas

Initial disease	Complicating leukemia	Karyotype	Reference
HD	AML	44,XX,−5,−7,−16(17?),−18,+2mar/46,XX	Rowley, Golomb, et al., 1977b
HD	AML	48,XX,−5,t(8;?)(p23?;?),t(15;?)(p11;?),−16?,−21,+5mar	Rowley, Golomb, et al., 1977b
HD	AML	46,XY,t(1;17)(p36;q21)	Rowley, Golomb, et al., 1977b
HD	Preleukemia	45,XX,−7/46,XX→45,XX,−5,−7,+mar	Rowley, Golomb, et al., 1977b
HD	APL	45,XX,−4,−5,−14,t(21;?)(q22;?),+2mar/46,XX	Rowley, Golomb, et al., 1977b
HD	EL	43?,XX,−B,del(6)(q15or11),−7,−12,−?,+min	Rowley, Golomb, et al., 1977b
HD	AML	45,XY,−5,−7,−9,+2mar/46,XY	Rowley, Golomb, et al., 1977b
HD	Preleukemia	45,XX,t(1;?)(p36?;?),−5,−7,−12,t(13;?)(q34?;?),del(14−15)(q24?;?),t(14?;?)(p1?;?),−17,−22,+mar,+3min	Rowley, Golomb, et al., 1977b
Lymphoma	Preleukemia	45,XX,t(1;?)(p36?;?),−2,−3,−5,−7,−12,−13?,−14?,−17,−17,−22,+8mar,+min	Rowley, Golomb, et al., 1977b
Lymphoma	Preleukemia	46,XX,−5,del(6)(q13),−7,+8,−17,+2mar	Rowley, Golomb, et al., 1977b
Lymphoma	Preleukemia	43,XX,−5,−13,−18/44,XX,−5,t(13;?)(p11;?),−18/46,XX	Rowley, Golomb, et al., 1977b
HD	AML	"Minor degree of aneuploidy"	Raich, Carr, et al., 1975
HD	AML	"Primarily a normal karyotype"	Raich, Carr, et al., 1975
HD	AML	Hypodiploidy with a stem line of 42 or 43 chromosomes. Aneuploidy with breaks, gaps, and deletions	Steinberg, Geary, et al., 1970
HD	EL	34−45 chromosomes with mode of 43 with polyploid multiples	Durant and Tassoni, 1967
HD	AML	46/45,−C	Canellos, DeVita, et al., 1975
HD	AML	45,−C/46	Canellos, DeVita, et al., 1975
HD	AML	44 chromosomes (100%)	Canellos, 1975
HD	CML	Ph¹−positive	Canellos, 1975
HD	AML	46,XY,21q+	Cavallin-Ståhl, Landberg, et al., 1977; Lundh, Mitelman, et al., 1975
HD	AML	47,XY,+8,+14,−17,+18,−21	
HD	AML	15%−89% Dq−	Weiden, Lerner, et al., 1973
PDL−N	ANLL	45 chromosomes	Collins, Bloomfield, et al., 1977
PDL−N	ANLL	50,+mar	Collins, Bloomfield, et al., 1977
PDL−N	ANLL	80−85 chromosomes	Collins, Bloomfield, et al., 1977
HD	AML	46,XX,−C,+D,+16,−G	Ezdinli, Sokal, et al., 1969
HD	EL	47,XY,+C,−16,+mar	Ezdinli, Sokal, et al., 1969
HD	CML	46,XY,Ph¹−positive	Ezdinli, Sokal, et al., 1969
Lymphoma	AML	46,XY,7p−,7q−,del(20)(q12)/47,XY,del(20)(q12),+E	Whang-Peng, Gralnick, et al., 1977

Table 72—continued

Initial disease	Complicating leukemia	Karyotype	Reference
Lymphoma	AML	42,XX,−5,−12,−14,−16,−18,+mar	Oshimura, Hayata, et al., 1976
	EL	46,XX,del(7)(q22)/46,XX,del(7)(q22),−9,+M(9)t(1;9)(q22or23;p24)	Oshimura, Hayata, et al., 1976
Lymphoma and CLL	AL	43,XY,−3,−4C,+F,+min/44,XY,−3,−3C,+F,+min	Whang-Peng, Knutsen, et al., 1979
Lymphoma and CLL	AL	45,XY,+marker (complicated karyotype)	Whang-Peng, Knutsen, et al., 1979
Lymphoma and CLL	AL	46,XY,7p−,7q−,20q12−/47,XY,+E,20q12−	Whang-Peng, Knutsen, et al., 1979
Lymphoma and CLL	AL	44,XY,5q−,−7,12p−,−22	Whang-Peng, Knutsen, et al., 1979
Lymphoma and CLL	AL	46,XX,7q−/45,XX,−7/46,X,−X,+7q−,−8,+22/47,X,−X,+7q−,+22/47,X,−X,−2,7q−,+15,+2markers	Whang-Peng, Knutsen, et al., 1979
Lymphoma and CLL	AL	44,XY,−4C,+E,+marker/45,XY,−4C,+E,+2mar	Whang-Peng, Knutsen, et al., 1979
Lymphoma and CLL	AL	45,XX,−7/43,X,−Y,−5,5q−,−7/44,XY,−7,−19/45,XY,−7,−7,+mar	Whang-Peng, Knutsen, et al., 1979
Lymphoma and CLL	AL	43,XX,−1,+3p+q−,4q+,−5,−5,−7,+9,+9,10p+,+10+10,+10p−q−,+11p−,−13,−15,−16,−17,−18,−21	Whang-Peng, Knutsen, et al., 1979

therapy, e.g., by suppressing host resistance and/or activating a preexisting oncogenic agent or virus. On the other hand, the main changes observed in the case (+14,−17,+18) are uncommon in AML but common in various lymphoproliferative disorders, e.g., #14 in Burkitt lymphoma, #17 and #18 in lymphoma. Whether these changes in the case described are epiphenomena in leukemic progression or indicate a real pathogenetic relationship to lymphoma, remains to be clarified. So far, the cytogenetic data neither support nor rule out such a relationship.

Rowley and co-workers (Rowley, Blaisdell, et al., 1966; Rowley, Golomb, et al., 1977a, b) have reported eight cases of leukemia or preleukemia in HD and found that the incidence of hypodiploidy was much higher than observed in spontaneous AL and that −5 and −7 were rather common occurrences. Other published cases tend to support these conclusions.

Rowley, Golomb, et al. (1977b, c) advised that any patient who is at risk of AL and who shows cytopenia associated with morphologic changes should have a cytogenetic analysis (with banding) of marrow. The presence of a consistent karyotypic abnormality is definite evidence that the patient is in a preleukemic stage. The fact that almost all cases of AL complicating HD (as well as lymphosarcoma and multiple myeloma) have been accompanied by karyotypic changes has been used as an argument for a different causation than in spontaneous AML, because the latter is just as often diploid as aneuploid.

Reticulum-Cell Sarcoma, Lymphocytic Lymphosarcoma, and Follicular Lymphoma

Reticulum-cell sarcoma (RCS; includes histiocytic type) is characterized by the predominance of histiocytic cells of varying degrees of maturity, which may or may not have reticulin fibers. The predominant cell in lymphocytic lymphosarcoma (LS) is a more or less differentiated lymphocytic cell. The typical feature of giant follicular lymphoma (FL) is an increase in the number and size of folliclelike nodes consisting of lymphocytic cells. In this chapter, FL will be considered a variant of LS; hence, some aspects of the chromosomal changes in FL and LS will be pooled. An attempt has been made to compare the chromosomal changes in LS with those in RCS. The validity of such a comparison is jeopardized by the heterogeneity of both groups of diseases; subgroups of LS and RCS will not be considered separately.

The method of choice for the analysis and establishment of the chromosome constitution in lymphoma (and solid tumor generally) is the direct preparation of tumor tissue, i.e., lymph nodes. Fortunately, in the great majority of cases with RCS and LS cytogenetic analyses are based on results with this technique.

Even though the cytogenetic picture in lymphoma was first successfully obtained on the basis of direct chromosome observations in cells from lymph nodes, after somewhat unsuccessful experiments with blood cultures and marrow preparations, even this method had certain drawbacks, including low numbers of dividing cells and poor quality of the metaphases. The introduction of short-term (ca 72-hr) culture of cells from tumorous lymph nodes represented a step forward and resulted in significantly larger numbers of mitoses and chromosomes of good quality (Baker and Atkin, 1965). With this technique it became possible to investigate the chromosomes of malignant lymphomas more or less routinely.

One of the methods described utilizes fresh surgical biopsy material from lymph nodes set up in Earle-Eagle's medium with 20% fetal calf serum. Chromosome preparations were made after an incubation time of 24-96 hr, usually 48-72 hr. Vinblastine may yield better results than colchicine as a mitotic inhibitor (Fleischmann, Hakansson, et al., 1976).

Chromosome Studies Before Banding

I was able to collect and review the chromosomal data of a total of 80 cases of RCS; lymph nodes were used for chromosome analysis in most instances (ca 70%)[15]; in some 15 cases the chromosome constitution was ascertained in the skin, blood cells,[16] tumors from involved organs (e.g., brain, intestine, stomach),[17] or cells obtained from effusions.[18] The chromosome constitution of marrow invaded by tumor cells and of PHA-stimulated blood cells has been reported in several papers.[19] As a result of contamination with cells not involved by the sarcomatous process in studies based on blood cells (Lawler, Pentycross, et al., 1968), these cases have to be considered separately. Data of all other cases were pooled.

A normal chromosome constitution has been described in a small number of cases, i.e., < 10.[20] Exclusively abnormal metaphases were observed in 28 cases; a mixed population of abnormal and apparently normal metaphases, with the majority (60-95%) of the metaphases being abnormal, was found in 35 cases. The modal chromosome number varied between 46 and 57 in 49 cases; modal numbers of 46 (pseudodiploid) and 47 prevailed. Chromosome modes in the tetraploid range (85-100 chromosomes) were seen in 9 cases,[21] and hypodiploid modes in 8.[22] With the exception of a small number of cases,[23] multimodal chromosome constitution was encountered, the different clones being always more or less closely related to each other. Fleischman, Prigogina, et al. (1974), however, stressed the sharp mode of the chromosome numbers in RCS, as compared to HD, and similar to that in AL. Markers were frequently encountered in RCS (Figure 113); in the 80 cases one or more markers were seen in at least 50%. The most striking example of the complexity of chromosome anomalies and rearrangements is the case published by Obara, Sasaki et al. (1971b): four different markers and four chromosome modes were detected in three different tissues (lymph node, marrow, and peripheral blood); one marker was common to all metaphases of all tissues, and, hence, a monoclonal origin of the neoplastic process may be entertained. A typical marker, characteristic for RCS, does not exist. However, recent studies utilizing various banding techniques have revealed the existence of some common chromosomes in lymphoma, and this will be discussed below. The morphological similarity of acro- or submetacentric markers observed in some cases has no significance so long as their origin is totally unclear. However,

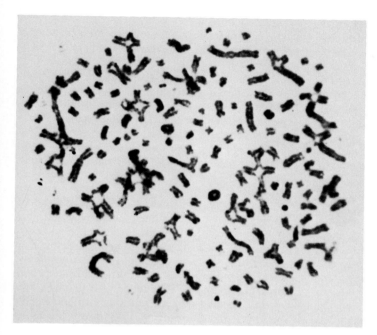

113 Metaphase from a lymphosarcoma cell in the marrow showing a multitude of heterologous exchanges, rings, and other chromosome abnormalities.

it is hoped that work with banding techniques will resolve this problem. It remains to be determined whether our observation (Sandberg and Hossfeld, 1974) that chromosome losses of group #1–3 occur with unexpected frequency (in 9 of 24 cases) has a relationship to the genesis of such markers. The chromosome group most frequently involved in numerical and/or structural anomalies appears to be group #17–18, a finding first described by Spiers and Baikie (1966, 1970) and confirmed by others.[24] The relevance of this finding will be discussed below.

Chromosome studies of marrow infiltrated by RCS revealed, as would be expected, a normal marrow-derived and an abnormal sarcoma-derived cell population in two cases.[25] Both cases were described as leukemic reticulosis. Chromosome-wise both cases do not provide additional information, which is also true for the only case in which peripheral blood cells were studied (Lawler, Pentycross, et al., 1968). In contrast, the one case described by Fleischman, Prigogina, et al., (1974) had hundreds of aneuploid cells in a marrow aspiration.

In discussing the chromosomal changes in LS, cognizance should be taken of the fact that no clear-cut division, either at the clinical or histologic level, may exist in some cases between RCS and LS; hence, the chromosomal findings in the two conditions not only may overlap but also may apply equally to both. The problem is further complicated by the propensity of workers in different locations to utilize diverse systems of classifying LS, e.g., one group classified LS as lymphoblastic and lymphocytic, whereas another classified it as lymphocytic, mixed, histiocytic, or undifferentiated. Thus, in summarizing the data available in the literature on the karyotypic changes in malignant lymphoma, some arbitrariness had to be utilized in dividing those cases thought to be RCS from others which apparently belong more rightfully to LS.

To ascertain the chromosomal picture of LS, I collected data on a total of 92 cases. In 79 instances chromosome analysis was performed on cells derived from lymph nodes or other tumor tissue[26]; in 2 cases cells were obtained from effusions[27]; in 11 cases heavily infiltrated BM was studied[28]; PHA-stimulated peripheral blood cells were used for chromosome analysis in 6 cases[29]; and cultured cell lines in 4.[30] Again, marrow- and blood-derived results will be dealt with separately, whereas all other data will be pooled.

Of the 92 cases only 6 had a normal chromosome constitution.[31] A mixed cell population with most of the metaphases being abnormal and with a varying percentage of normal metaphases was found in most (65%) of the cases, whereas exclusively abnormal metaphases were observed in about 25%. The chromosome modal numbers ranged from 46 to 51 in most cases; modes of 46 to 48 predominated. Modal numbers in the triploid or tetraploid range (69–98) were found in eight[32] and hypodiploid numbers (44–45) in six cases.[33] No chromosome counts are given for some of the cases (Millard, 1968; Lawler, Pentycross, et al., 1968). Inasmuch as the preponderant number of LS is characterized by several modes, it should be emphasized that the figures given above reflect only the prevailing chromosome number, which in several cases was not very distinct.

Apart from a higher percentage of diploid metaphases (most probably marrow derived), the chromosome constitution of LS cells in marrow closely resembled the pattern of lymph nodes.[34] In some cases a striking discrepancy between percentages of LS cells and abnormal metaphases was seen (Sandberg, Ishihara, et al., 1964c). PHA-stimulated peripheral blood cells yielded some abnormal metaphases with 47–48 chromosomes in only two out of six cases with leukemic phase of LS, indicating the inadequacy of this kind of approach (Hayhoe and Hammouda, 1965). One or several markers were detected in over 50% of the cases in which lymph nodes or other tissues were studied. The occurrence of markers in malignant lymphoma has been documented under various conditions but controversy continues to remain as to whether some markers are specific of one form of lymphoma or another. In the past, uncertainty existed as to the exact origin of markers, but banding techniques have helped greatly in deciphering the genesis of such chromosomes.

In a most detailed and extensively investigated case reported by Fleischmann, Håkansson, et al. (1972) seven markers were identified. These authors applied the fluorescence technique and chromosome measurement in order to characterize and identify the abnormal chromosomes, and, yet, the origin of five of the seven markers could not be traced. They provided circumstantial evidence that three of the markers represented an isochromosome in triplicate. In the discussion the authors refer to the presence of the same type of medium-sized metacentric marker in other, unpublished cases of lymphoma and to those published by others. They then made reference to the findings of Spiers and Baikie (1970), who believe that chromosomes of group #17–18 are more often involved in numerical and structural changes of the chromosomes in LS than might be expected, but are careful not to identify the markers as isochromosomes of group #17–18.

In screening the literature on RCS, LS, and HD, I found only the 14q+ marker chromosome to be possibly characteristic for the lymphomas. This does not hold true for the so-called Melbourne chromosome which is a #18 deleted in its short arms,[35] and which was found in only 9 out of a total of 118 cases of RCS, LS, and HD. On the other hand, we (Sandberg and Hossfeld, 1974), too, have found some evidence that numerical anomalies of group #17–#18 appear to occur more frequently than expected randomly in LS. In our hands, RCS predominantly showed losses (8 out of 12 cases with numerical anomalies) and LS predominantly gains of chromosomes of groups #17–#18 (7 out of 8 cases with numerical anomalies). Nevertheless, not until a solidly based statistical analysis of this problem has been accomplished in a large group of patients should etiological and evolutionary significance of group #17–18 chromosomes in reticuloendothelial neoplasia, as claimed by Spiers and Baikie (1970), be accepted.

In comparing the chromosomal behavior of RCS and LS, it appears that no basic difference possibly of diagnostic value exists. In both diseases a normal chromosome constitution is exceptional and is in striking contrast to ALL. In 95% of RCS and LS, all or the majority of the metaphases are abnormal. A near-diploid chromosome mode (45–48) occurs in about 80% of both diseases; usually several closely related modes are present, the distribution of the chromosome number about the modal number being rather flat. The proportion of apparently normal, diploid metaphases in cases with a mixed population amounts to 30–35% in both diseases. Markers can be seen in RCS more often than in LS (72 vs. 58%). Whether it is true that loss of chromosomes of group #17–#18 occurs more often in RCS, and gains more often in LS, remains to be confirmed.

There are a number of diseases in which lymphoma occurs at a much higher frequency than in the general population, and this includes some of the chromosome breakage syndromes, particularly BS. Malignant lymphoma seems also to be related to several genetic diseases such as Chediak-Higashi syndrome, Wiskott-Aldridge syndrome, and, in fact, many other severe congenital immunologic deficiencies. When lymphoma develops in these states, the karyotypic changes in the lymphoma appear to be not much different from those observed in patients without the underlying congenital anomalies.

Banding Studies in Lymphoma

Banding studies[36] in various lymphomas (Table 73) in nearly 100 cases have led to several interesting points. Outstanding among these has been the demonstration of the 14q+ in several different lymphoma types, e.g., HD (8 cases), LS (28 cases), RCS (24 cases), and FL (2 cases). The presence of a 14q+ in two cases of CLL has been described (Fleischman and Prigogina, 1977), though it is difficult to ascertain whether the cases might not be lymphomas of small cell type associated with a lymphocytic leukocytosis. Banding has been done in 23 cases of HD, 41 cases of LS, 36 cases of RCS, and 6 cases of FL (Brill-Symmer's syndrome). In one report (Fleischman and Prigogina, 1977), 12 cases of CLL are reported, of whom 8 were diploid and 4 with tumorous lymph nodes had definite karyotypic changes. It is possible that these four cases represent some form of LS. In addition to the 14q+ anomaly mentioned above, 6q− occurred in a number of cases (2 of HD, 13 of LS, see chapter on ALL). Patients with diploidy occurred in all categories of lymphoma (9/22 in HD, 8/30 in LS, 1/26 in RCS, and 1/5 in FL). The preponderant number of cases (> 80%) contained markers. In aneuploid lymphomas certain chromosomes, i.e., #14,#11,#3, and #18, were thought to be preferentially involved with the incidence indicated by the order of the chromosomes shown (Reeves, 1973). Fleischmann, Håkansson, et al. (1976) found that C- and especially E-group chromosomes are most frequently involved in LS. These authors stressed the presence of isomarkers (isochromosomes) in lymphoma, suggesting that the development of the malignant stem line is stimulated by the duplication of the genes located on the isochromosomes. Deletion of the short and long arms of #17 and #18 in lymphomas has been stressed by Reeves (1973).

Two lymphosarcomatous lymph nodes removed from one case on different occasions had the same karyotype when analyzed with Q-banding (Fleischmann, Håkansson, et al., 1972). The modal number was 48 with 1q−,+3,3p−q−, −4,−6,−9,−17,−18,−19, and −X and several markers which could not be identified completely, including three medium-sized metacentrics which had the appearance of isochromosomes. Similar markers had previously been seen in other malignant lymphomas including HD (Fleischmann, Håkansson, et al., 1971; Mark, 1977b).

Reeves (1973) after studying 10 cases of lymphoma with banding stated that all the tumors contained, in addition to numerical abnormalities, structural chromosome changes as a result of simple or complex deletions, involving translocations and isochromosome formation. Almost without exception, the break points appeared to be in the light-staining bands and at the centromeric region. Involvement of chromosomes #3,#1,#9 (9q−), #6 (6q−), #12 (12q+), 14q+, and #15 were most common.

The value of chromosomal studies in the diagnosis of lymphoma in complicated cases was pointed to by Mark, Ekedahl, et al. (1976). On the basis of chromosomal results obtained in direct preparations of suspected tissues, these authors unequivocally showed with G-banding a malignant process in two suspected malignant lymphomas, thereby contributing to a firm and rapid diagnosis. The karyotypic changes were similar to those observed in other banded non-Burkitt lymphomas, i.e., 46,XX,1q+,2q+,−8,+12,11q+, and 50,XX,3q+,+4,6p−,9p−q−,+11,+12, and +21. As is often seen in lymphomas, the tumor tissues contained a significant number of diploid cells (nearly 50%). The presence of such cells is a perplexing problem considering the long duration of lymphoma in the two patients and the striking predominance of mitoses in areas wholly dominated by tumor cells. The authors postulated that both lymphomas originated from cells without any gross chromosomal deviations and the rapid rate of growth of the tumors was concomitant with the development of a new clone with karyotypic changes.

Table 73a Karyotypes of published cases with various types of lymphoma, based primarily on banding studies

Case no.	Type of lymphoma	D/C	Banding	Karyotypes
1	RS	D	+	47–48,XX,1q–,+i(1q–),+5,12p–,14q+
2	RS	D	+	47–48,XX,13q–,13q–,14q+,+mar
3	RS	C	+	46–47,XX,+C,+E
4	RS	C	+	47,XY,der11[t(11;11)(pter23::q13qter)],+18 or r(18),der 19
5	RS	C	+	47–48,XY,+X,–13,–15,+t(11q;6p),+t(?;13q;14q),+mar
6	RS	C	+	47–50,XY,t(1q;11p);1q–,1p–,3q+,11p+,t(17;11),+mars
7	HD	C	–	46,XY,–1,+mar
8	HD	C	–	46,XY,–16,+mar
9	HD	C	–	47,XY,–2,–5,+16,–17,+3mar
10	HD	C	–	46–48,XY,+2mar
11	HD/AML	D	+	46,XY,21q+
12	HD/AML	D	+	47,XY,(±8),+14,–17,+18,(–21)
13	HD	STCL	+	1q–,6q–,–13,+19
14	HD	C/D	+	a) 82,XX,+2X,+1,+2,+3,+3,+4,+5,+5,+6,+6,+7,+8,+9,+10,+14,+14,+16,+16,+17,+18,+19,+20,+20,+21,+21,+22,i(2q),[del(9)(q11)×2],del(9)(q31),t(14q+;?)(q32;?),i(18p)i(18q),2mar b) 82,XX,+1,+2,+2,+3,+3,+4,+5,+5,+6,+6,+7,+7,+8,+8,+8,+9,+9,+10,+12,–13,+14,+14,+16,+16,+16,+18,+20,+20,+20,+21,+21,+21,del(6)(q21),t(9q;15p)(q11;p12),t(14q+;?)(q32;?)
15	HD	STCL	+	46,XY t(3;13)–11,+mar,+mar
16	HD	C/D	+	102,XY,+X,+1,+1,+5,+6,+6,+6,+8,+8,+8,+9,+9,+10,+10,+11,+11,+12,+12,+13,+13,+14(x7),+15,+15,+16,+17,+19(x4),+20(x6),+21,+21,+22,+22,[t(3p+;?)(p21 or 23;?)×3],del(12)(q22),1 ring, 5mar
17	HD	C/D	+	numerous anomalies among which: del(1)(q32),i(1q),del(2)(p13),del(3)(p13),del(5)(q31),del(6)(q13 or q15),del(6)(q23 or q25),del(7)(q11),del(10)(q26),12q+,15q+
18	HD	LTCL	+	36–49, numerous anomalies among which: +7,t(9q+,22q–),14q+,16q–,+mars
19	SEZ	C	+	43–44,XX,–1,–1,–6,–17,–19,–22,+3–7mars
20	LS	C	–	47,XX,+C
21	LS	C	+	46,XY,–4,–15,–16,+3mar
22	LS	C	+	48,X(–Y),+3,–4,–6,+7,–9,–17,–18,–19,+6mar
23	LS	C	+	50,XY–––+C
24	LS	C	+	47–48,XX,+3,+18
25	LS	C	+	47–49,XX,+X,1p+,+r(2),–14,[1q;i(17q)]
26	LS	C	+	42–43,XX,t(11;14)(q13;qter)
27	LS	C	+	45,X(–Y),t(11;14)(q13;qter),t(11;19)(q13;qter)
28	LS	C	+	45,XX,t[i(17q);21q]
29	LS	C	+	46,XY,D(14?)q+,del 11(q12)
30	LS	C	+	46,XX,+3,t(13q;22q)
31	LS	C	+	47–48,XY,4q+,+7,9p–,–11,–14,D(14?)q+,–15
32	LS	C	+	47–50,XX,+3,7q–,+18,+mars
33	LS	C	+	50,XY,+3,+3,+12q–,14q+,+18
34	LS	LTCL	+	47,+7,8q–,14q+
35	LS	LTCL	+	47,+7,inv(14),+mars
36	LS	LTCL	+	50,1q+,+7,+8,+8,+12,14q+,+mar
37	LS	C	+	46,XX,–14,–18,+mar$_1$[i(14q)?],+mar$_2$[i(18q)?]
38	ML	C	+	43,XX,–4,–5,–9,–16,+1mar

Table 73a—continued

Case no.	Type of lymphoma	D/C	Banding	Karyotypes
39	ML	D	+	46,XX,1q+,2q+,−8,+12,16q+
40	ML	D	+	50,XX,3q+,+4,6p−,9p−,q−,+11,+12,+21
41	ML	C/D	+	46,XX,t(2p+;?)(p2;?),+8,−15,del(16)(q13)
42	ML	C/D	+	45,XY,+del(6)(q13),−8,t(15q+;?)(q24 or q26;?), +19,−20,−21
43	ML	C/D	+	a) 77,XY,+Y,+2,+4,+5(x5),+6,+8,+9,+10,−11,+13,+14, +14,+15,+16,+16,+17,+17,+17,+18,+18,+19,+20, −21,[del(3)(p21)×2],[del(9)q22)×2],[t(11q+;?)(q23;?) x2][t(12q+;?)(q22;?)×2] b) 45,XY,−3,−6,−8,−9,−11,−12,−19,−22,[del(3)(p21) x2],del(9)(q22),t(11q+;?)(q23;?),t(12q+;?) (q22;?),t(15q+;?)(q22;?),1mar
44	ML	C/D	+	46,XY,−1,−3,−7,−8,−9,+14,+17,−18,+22,t(3q+;?) (q27 or 29;?),del(7)(q32),del(9)(q22)
45	ML	C/D	+	46,XX,abnormalities include del(1)(p34 and q32), 14q+,del(18)(q21)
46	ML	C/D	+	92,XY,+X,+Y,+1,+1,+2,+2,+3,+6,+6,−9,+10,+10, +10,+11,+11,+12,+12,+13,+13,+14,+14,+14,+15, +16,+16,+18,+18,+19,+19,+20,+20,+21,+21,+22, +22,[del(1)(p22)×2],[t(2;9)(q11;p11)×2][t(5p+; ?)(p15;?)×2],[del(11)(q23)×2],t(17q+;?)(q25;?), 17mar, 2 rings
47	ML	C/D	+	43,Y,−X,−7,−13,−13,−14,−17,−18,del(9)(q22), 2−4mars
48	ML	C	−	47,XY,+C
49	CLL	C	+	46,XY,2p+,del(11)(q22)
50	CLL	C	+	46,XY,t(11;17)(q21;qter)
51	CLL	C	+	46,XY,t(8;14)(q24;q32),19p−
52	CLL	C	+	46,XY,t(1;10)(q22;qter)
53	AIL	D	+	46,XX,t(1;2)(p?,p?)
54	AIL	C	+	47,XY,+3/47,XY,+3,−9,+21/48,XY,+3,+8,+9,−20
55	AIL	D	−	46,XY,Bq+/47,XY,Bq+,+C/47,XY,+C
56	AIL	D	+	47,XX,+minute marker
57	AIL	D	+	45,X(−Y)t(1q+,10q−)/46,XY,t(1q+,10q−)
58	NABL	LTCL	+	46,XY,t(4;5;7)(4qter→q13::5p13→5qter;4q13→4pter; 7qter→7p22::5p13→5qter),t(8;14)(8pter→8q23; 14pter→14q32::8q23→8qter/47,idem+(4q13→4pter)
59	NABL	LTCL	+	46,XY,t(8;14)(8pter→8q23;14pter→14q32::8q23→8qter)
60	ABL	LTCL	+	47,XY,+2derC,2Cqh+,13q+,14q+,−G(?21)
61	ABL	C	+	46,8q−,14q+
62	ABL	C	+	46,8q−,14q+
63	ABL	C	+	46,8q−,14q+
64	ABL	C	+	46,8q−,14q+
65	ABL	C	+	46,8q−,14q+,+ unidentified anomalies
66	ABL	C	+	48,+7,8q−,14q+,+D
67	ABL	C	+	46−52,+X,+1,+3,+5,+7,8q−,+8,14q+,i(17q)
68	ABL	C	+	48−53,+X,+2,+3,8q−,+9,+10,+11,+12,14q+,+20
69	ABL	C	+	46,(8q−?),14q+
70	ABL	C	+	46,(8q−?),14q+
71	ABL	LTCL	+	(46)?,(8q−?),14q+ (1 cell)
72	ABL	LTCL	+	46−48,+6,+7,8q−,+13,14q+,15q−
73	ABL	LTCL	+	45,t(10p;11),t(10q;13)

continued

Table 73a Karyotypes of published cases with various types of lymphoma, based primarily on banding studies — *continued*

Case no.	Type of lymphoma	D/C	Banding	Karyotypes
74	ABL	LTCL	+	45,t(8q−,14q+),t(7q;16),t(6q;17)
75	ABL	LTCL	+	46,i(3q),t(3p:13),t(8q+,22q−),t(10p:17)
76	HL	C (Blood)	+	50,XY,1q+,+2,4p+,+4p+,t(8;14)(q24;q32),+17p+,18q+,+20
77	PDLS	C/D	+	48,XY,−C,−14,14q+,21q+?,+4mar
78	PDLS	C/D	+	48,XY,−6,−11,+12,+1p−,3q+,t(14;18)(q32;q21),+3mar
79	PDLS	C/D	+	44,XX,−3,−9,−13,−18,1p+,t(1;9?)(p32;q11?),t(11;14)(q13;q32),+2 mar
80	PDLS	C/D	+	45,X,−X,Dq+/Dq−
81	PDLS	C/D	+	44,X,−Y,−9,−10,+12,−14,−17,−21,t(1;14)(p21;q13),t(11;21)(q23;q22),del(6)(q13),del(8)(p21),5p+,17p+,13q+,22q+,del(9)(q22)(p+),19p− or q−,+3mar
82	HL	C/D	+	50,XY,+2,+20,t(8;14)(q24;q32),1q+,4p+,+4p+,+17p+,18q+
83	HL	C/D	+	45,X,−Y,−15,−15,t(14;15)(q32;q21orq22),+18,−22,1q+,3p−,5p−,7p−,9p+,12p+,16q+,17p−,+2mar
84	HL	C/D	+	47,X,+X,−Y,−4,t(4q;14q),−18,1q+,2p+,6q−,7q+,9q−,12q+,+3mar
85	HL	C/D	+	84,XXY,−Y,tetraploid(−5,−C,−3D,−3E,−20,−21,11p+,+11p+,+2mar),t(1;14)(q2;q24 or q32)
86	LS	C/D	+	53,X,+X,−Y,+20,1q−,+1p−,2q−,4q+,5q−,5p+,6q−,+7q+,t(8;14)(q24;q32) and t(14;18)(q32;q21),10q+,11q+,11q−,13p+,+4mar
87	PDLS	C/D	+	48,XX,1p+,12q+,t(14;18)(q32;q21),+mar
88	PDLS	C/D	+	47,XY,−9,−10,+12,−16,8p−,13q+,t(14;18)(q32;q21),+3mar
89	PDLS	C/D	+	46,XY,−1,−17,−22,t(14;18)(q32;q21) or t(10;14)(q24;q32),+3mar
90	PDLS	C/D	+	45,XY,+2,−5,−6,−11,t(14;18)(q32;q21),+mar
91	PDLS	C/D	+	47,XY,+7,−9,−11,1p+,6q−,10p−,t(14;?;14q),15q−,22q−,+2mar
92	MF	C/D	+	48,XX,−9,1p−,5q−,+5q−,+5q−,9q−,10p−,18q+,t(2;8;14)(q3;q24;q24) lymph node
93	MF	C/D	+	46,XX,t(1;14)(q32;q32) blood cell (case 92)
94	HD	C/D	+	66,t(1;14)(q2;q32) with large number of rearranged chromosomes
95	HL	C/D	+	45,XYq−(q12),t(4;10)(4pter→4q12::10q24→10qter),t(6;7)(6pter→6q21::7q32→7qter),del(8)(p21),del(Y)(q12),−6
96	HL	C/D	+	45,X,−X,del(1)(q22),1p−,t(1q17q),del(2)(q33),del(3)(p21p25),ins(3;3?)(3qter→3p14::3p21→3p25::3p14→3qter),+5,−6,−14,+22,del(6)(q14),t(1;7)(7pter→7q32::1p22→1pter),t(7;?)(7qter→7p22::?),t(7q8q),del(8)(p12),t(9;?)(9pter→9q34::?),del(13)(q14),t(13;14)(14pter→14q31::13q14→13qter),t(1;21)(21pter→21q22::1q22→1qter)
97	HL	C/D	+	47,XX,+7,t(1;18)(1qter→1p36::18q21→18qter),t(8;12)(8pter→8q11::12q15→12qter),t(8;12;13)(12pter→12q15::8q22→8qter),del(13)(q14),t(8;14)(14pter→14q32::8q22→8qter)
98	HL	C/D	+	48,XY,t(3.8)(3pter→3q22::8q11→8qter),t(5;10)(5pter→5q33::10q22→10qter),del(7)(q21),t(4;9)(9qter→9p24::4q11→4qter),t(10;13)(10pter→10q22::13q12→13qter),r(12),del(19)(p13),del(22)(q11),+7,+7q−

Table 73a—continued

Case no.	Type of lymphoma	D/C	Banding	Karyotypes
99	HL	C/D	+	50,XY,t(2;?)(2qter→2p25::?),del(2)(p13),i(3p),del(6)(p21),t(6;?)(6qter→6p21::?),del(11)(q13),t(11;14)(14qter→14q31::11q13→11qter),i(15q),+7,+8,+9,+9,−11,−13,+14,−15,+20,+21,−22
100	HL	C/D	+	82,XXX,t(3;12)(3pter→3q13::2q15→12qter),t(3;8)(3pter→3q22::8q11→8qter),del(3)(q11),del(5)(q31),i(7q),del(10)(q24),t(10;10)(10pter→10q26::10q24→10qter), del of above at p13, inv(11)(p15q21),del(12)(q13),del(13)(q14),+1,+1,+2,+3q+,+3q−,+3q−,+4,+4,+5q−,+5q−,+6,+7,+7,+7p+,+7p+,+9,+10p−q+,+10p−q+,+11,+11p+q−,+12,+12q−,+12q−,+12q−,+13q−,+14,+14,+16,+17,+18,+18,+20,+20,+21,+21
101	HL	C/D	+	47,XY,−6,−14,1q−,+1p−,2q+,2q+,3q+,9p+,10p−,10p+,12p−,18q+,19(p+orq+),+22p+,+mar
102	HL	C/D	+	45,XY,−6,−9,−17,2q−,2q−,3(p− and q−),4q−,5q−,6p+,7p+,7q+,+7q−,9p−,12q+,13p+,13p+,20p−?,+mar
103	HL	C/D	+	47,XY,+Y,2q−,3q+
104	LS	D	+	46,XY,+3,−8
105	LS	D	+	46,XY,t(6;14),−8,−9,+16,+20/46,XY
106	LS	D	+	47,XY,+3
107	LS	D	+	50,XX,+3,+4,+10,+18/51,XX,+3,+4,+9,+10,+18/46,XX
108	LS	D	+	87,XXYY,1q−,+1q−,+2,+2,+3,+3p−,+3p−,+3p−,+4,+4,+5,+6,+6q−,7p−,+8,+8,+8,+9,+9,+10,+11q−,+11q−,+12,+13,+13,+14,+14,+15,+16,+17,+17p+,+18,+20,+20,+21,+21,+22,+4mar
109	LS	D	+	88,XXX,+1,+1p−,+2xt(3q9q),t(3q9p),+4,+5,+5x6q−,+7,+8,+8,−9,+10,+11,+11,+12,+12,+13,+13,+14,+14,+15,+16,+17,+17,+17,+18,+18,+19,+19,+20,+20,+21,+21,+22,+2xt(9;22)
110	Lennert's lymphoma (lymphoepithelioid)	D	+	46,XY/49,XY,+2q−,+3,t(10;17),+17,+17p+,−19
111	HL	D	+	45,X,Xq−,1q−,+1p−,2q−,3q−,6p−,6q−,t(7;9),t(8;10?),9q−,−10,10q−,t(2?;11),−13,14q−,15q−,16q−,+17,−19
112	HL	D	+	50,XX,t(1;14),+3,+7,+12,+18/46,XX
113	HL	D	+	82,XX,Xq−,t(1;14),i(1q),+2xi(1qh+),+2xi(1p),+2,+3,+4,+4,+5,+6,+6,+2x7p−,+9,+i(9p),+10,+11,+11,+12,+13,14q−,4x14q+,i(15q),−16,−17,i(17q),+18,+18,+20,+21,+21,+22,+22,+2mar
114	HL	D	+	89,XXYY,+1p−,+1p−,+2,+2,+3,+3,+4,+5,+5,+6,+6,+7,+7,+8,+8,+9,+9,+10,+10,+11,+11,+12,+12,t(5?;13),t(14;20),+14q+,+15,+16,+16,+21,+21,+22,+22,t(1;17),t(1;19),+20q−,+19p+

ABL, African Burkitt lymphoma; AIL, angioimmunoblastic lymphadenopathy; C, culture; D, direct investigation; HL, histiocytic lymphoma; LTCL, long-term cell line; MF, mycosis fungoides; ML, malignant lymphoma; NABL, North American Burkitt lymphoma; PDLS, poorly differentiated lymphosarcoma; RS, reticulosarcoma; SEZ, Sézary syndrome; STCL, short-term cell line.

Cases 1 and 2—Mark (1975b); cases 3, 20–23, 38—Fleischmann, Håkansson, et al. (1976); cases 4, 5, 24–33, 49–52—Fleischman and Prigogina (1977); cases 39–40—Mark, Ekedahl, et al. (1976); cases 11–12—Cavallin-Ståhl, Landberg, et al. (1977); cases 58–59—Kaiser-McCaw, Epstein, et al. (1977a,b); cases 13, 15—Boecker, Hossfeld, et al. (1975); case 60—Petit, Verhest, et al. (1972); cases 19—Dewald, Spurbeck, et al. (1974) case nos. 14, 16–17, 41–47—Reeves (1973); cases 18, 34–36, 61–75—Zech, Haglund, et al. (1976); case 48—Gariepy and Cadotte (1970); case no. 53—Goh and Bakemeier (1978); cases 54–55—Hossfeld, Höffken, et al. (1976); case 37—Siegal, Voss, et al. (1976); cases 56–57—Castoldi, Scapoli, et al. (1976); case 76—Brynes, Golomb, et al. (1978); cases 77–81—Fukuhara, Rowley, et al. (1979); cases 82–94—Fukuhara and Rowley (1978); cases 95–100—Mark, Ekedahl, et al. (1978); cases 101–103—Fukuhara, Rowley, et al. (1978b); cases 104–114—Mark, Dahlenfors, et al. (1979).

Chromosome studies were performed on malignant cells obtained from 27 patients with non-Burkitt lymphomas by Fukuhara and Rowley (1978). A 14q+ was the single most frequent abnormality and was noted in 17 of these patients. The frequency of the 14q+ varied with the type of lymphoma, i.e., five of eight histiocytic, one of three mixed cell type, eight of eight poorly differentiated lymphocytic, 0 of three well-differentiated lymphocytic, none of one lymphoblastic, two of three HD, and one of one mycosis fungoides. The donor chromosome involved in the 14q translocation was identified in 12 cases; certain chromosomes appear to be affected more frequently than others. Even though the breakpoint was band 14q32 in most cases, the exact location of the receptor site on the long arm of #14 was not always consistent. The distal part of 14q24 was also involved as a receptor site in at least one translocation. These findings suggest that in some types of lymphoid malignancy cells with a 14q+ have a proliferative advantage over cells with other karyotypic derangements. The presence of the 14q+ may be important in the future for the distinction among morphologically different but functionally comparable subgroups of lymphoid disorders.

In another study (Fukuhara, Rowley, et al., 1979) banding studies were performed on cells from tissues primarily involved by poorly differentiated lymphocytic type of malignant lymphoma. Chromosomally abnormal cells were observed in each of the nine patients and in each a consistent clonal abnormality could be detected. An abnormality of #14 was seen in every patient and in nine it was 14q+. These findings were correlated with the histopathology based on the Lukes and Collins (1977) classification and indicated that the lymphoma cells from five of eight patients with small cleaved cell type had a 14q marker that originated as a result of t(14;18) (q32;q21). Leukemic cells in one patient had a very complex rearrangement of chromosomes including two 14q+ chromosomes, i.e., one t(14;18) and the other a t(8;14) (q24;q32). The cells from this patient were of the null-type. In another patient the donor chromosome for the 14q+ was undetermined and the leukemic cells had B markers. In the eight patients two cells from PHA stimulated blood cells had a 14q+ and a Dq+ marker respectively. In one of two patients whose tumors were classified as small noncleaved cell lymphoma, 2 hypodiploid cells from the lymph node had a 14q marker which could have resulted from a t(11;14) (q13;q32). Leukemic cells from the second patient had a #14 translocated to 1p[t(1;14) (p22;q13)], and the cells had B markers. The data suggested to the authors that the 14;18 translocation could be preferentially associated with the small cleaved cell of lymphoma. No clear correlation between the cytogenetic and the immunologic markers was observed and in some patients failure to express surface markers may have been related to increasing aneuploidy.

The banding patterns in six histiocytic lymphomas and those of seven cases reported previously were surveyed together with data (often incomplete) from 32 lymphomas in the literature (Mark, Ekedahl, et al., 1978). The following features characterize these types of lymphomas: #14 was commonly affected, usually by structural deviations involving its long arm, preferentially at band q32. This as a rule resulted in the formation of a 14q+ found in 17 of the 45 cases, the extra material on #14 showing inconsistent derivation, a difference in comparison with BL. Autosomes #3, #7, #8 and #11 were next in frequency of involvement, structural deviations predominating for #3 and #11, and some recurrent marker types were seen. Chromosomes #7 and #8 were mostly affected by numerical deviations, usually gains of the former and losses of the latter. Structural deviations often affected #1, #6, #9 and #13 and recurrent marker types related to #6 and #13 were observed. The centromeric and the light stained regions were preferentially affected by the breakpoints.

Noteworthy features of the study by Mark, Ekedahl, et al. (1978) were the demonstration of a remarkable number of translocations in the six lymphomas examined and the identification with banding of a large number of abnormal chromosomes present in the lymphomatous cells (Table 73).

Chromosome Findings in Follicular Lymphoma

Even though the cytogenetic data on follicular lymphoma (FL) have been incorporated into those of LS, some statements regarding the

Table 73b Karyotypic findings obtained with banding in nonendemic Burkitt lymphoma

Source of cells	Karyotype	Reference
BM (direct)	46,XY/46,XY,14q+/45,XY,−22 and various types of markers	Philip, Krogh Jensen, et al., 1977
Cell line (4–12 mo old)	t(8;14),t(7q;16),t(6q;17),−16,−17	Zech, Haglund, et al., 1976
Established cell line	46,XY,t(4;5;7)(4qter→q13::5p13→5qter;4q13→4pter;7qter→7p22::5p13→5pter),t(8;14)(8pter→8q23;14pter→14q32::8q23→8qter)	
Established cell line	46,XX,t(8;14)(8pter→8q23;14pter→14q32::8q23→8qter)	Kaiser-McCaw, Epstein, et al., 1977a
Established cell line (from patient with AT)	46,XY,14q+	Kaiser-McCaw, Epstein, et al., 1977b
Established NAB line (EBV-genome positive)	t(1;14)(q1;q32) 4 cells out of 30; rcp (1;5)(p1;q33),t(3;7)(p25;q11) in all cells; markers involved #2,#8,#13,#15 and #22	Temple, Baumiller, et al., 1976
Ascitic fluid (direct)	47,XY,+7,+7,−10,−11,t(8;14)(q23;q32),+t(X;1)(p21;q21)	Kakati, Barcos, et al., 1979
Culture of splenic and LN tissues	t(2;8)(p12;q23)	Van Den Berghe, Parloir, et al., 1979b

former are in order. Follicular lymphoma is a term applied to well-differentiated lymph node tumors of lymphoreticular origin in which progression is slow, though anaplastic changes eventually occur. About 40 such tumors have been studied cytogenetically.[37] Spiers and Baikie (1968a) studied the chromosomes in lymph node material from six cases after culture for 16–20 hr of which three were hyperdiploid [(48,+C,+G; 48,+C,+C and 47,Ep−(17−18)]. Two tumors were near-tetraploid and one consisted of a mixed population of cells with near-diploid and near-tetraploid chromosome numbers.

Nine follicular lymphomas (seven untreated; 2 treated) studied by Baker (Atkin, 1976a) were all in the diploid-hyperdiploid range (46–52 chromosomes), with most of them containing markers. In one case studied by Fleischmann, Håkansson, et al. (1976) with a pseudodiploid karyotype, a large submetacentric marker was present. Three cases studied by Reeves (1973) had near-diploid modes (46, 45–47, and 92 chromosomes) and very abnormal karyotypes. A 14q+ has been described in one case (Reeves, 1973).

Studies of the lymph nodes of six patients with FL (Spiers and Baikie, 1968a) revealed all to be aneuploid with the cases being 48,XY,+Y,−C; 47,XY,+mar, hypotetraploid, 47,XY,+2C/47,XY,+C,+F, possible hyperdiploidy and tetraploidy, and 91 chromosomes (including loss of chromosomes in groups B, E, and F and gain in C), respectively.

A case of FL in transformation was shown to have a 14q+ in the ascitic cells (Catovsky, Pittman, et al., 1977); a substantial number of cells also contained a 16q−,−17,18q−,13q−,+7,+8,8q− and some an additional 14q+. Minute chromosomes were present in 83% of the cells. The origin of the "donor" chromosome to the 14q could not be identified, but it possibly could have been #8, #13, or #18.

Acute Leukemia in Patients with Lymphoma

Seven cases of ANLL and one with a malignant myeloproliferative syndrome were identified from a pool of 189 cases of non-Hodgkin's lymphoma and CLL treated primarily with extensive radiotherapy (Whang-Peng, Knutsen, et al., 1979). The median time interval from the diagnosis of the primary malignancy to the development of leukemia was 61 mo (range 33–98 mo) and the median survival after the diagnosis of leukemia was 2 mo (range 0–9 mo). All eight patients were cytogenetically abnormal; hypodiploidy was the most commonly observed abnormality (in seven out of the eight patients). Abnormalities of #7, the chromosome most frequently involved in aneuploidy, were seen in all five patients analyzed with banding; four of them had monosomy of #7. Two patients had deletions of the long arm of this chromosome (7q−), both with the breakpoint at the q22 region; one of these 2 patients had an extra #7 that was also deleted (+7q−). The high incidence of hypodiploidy and the frequent involvement of #7 are in keeping with the reports of others (Lundh, Mitelman, et al., 1975; Rowley, Golomb, et al., 1977b). Chromosome #5 was the next most frequently involved in abnormalities; this usually consisted of loss of one or both #5 chromosomes and/or the presence of a 5q− anomaly. The deletion of the 5q was interstitial in one case and terminal in another. All eight patients in this series showed chromosome abnormalities at some point in their disease, in one case the appearance of monosomy of #7 heralding the onset of AL a considerable time period prior to the appearance of evidence of AL. The complexity and extensive nature of the cytogenetic abnormalities seen in these patients may be characteristic of secondary leukemia in radiation-treated lymphoma and the presence of such anomalies may predict leukemic transformation.

In another study (Collins, Bloomfield, et al., 1977) ANLL developed in four patients 66 to 157 mo following the diagnosis of nodular poorly differentiated lymphocytic lymphoma. Marrow cytogenetic studies demonstrated abnormalities in all three patients studied: one patient had a hypodiploid mode of 45, another patient a mode of 50 chromosomes, including a marker, and the third patient had 80–85 chromosomes.

The 14q+ Anomaly in Lymphoma

Some special attention must be given to #14 in lymphoma (Table 74). This anomaly (14q+) was first described by Manolov and Manolova in 1972 in Burkitt lymphoma (BL) and since then found not only in cases of BL but also in non-BL and other diseases. On the basis of initial banding studies it was thought that the 14q+ is due exclusively to t(8q;14q). However, subsequent banding studies revealed that although an exchange between #8 and #14 may be operative in BL, other chromosomes (1q,4q,5q,10q,11q, and 18q) may be involved in the contribution to the 14q+ through translocations in conditions other than BL, i.e., histiocytic lymphoma, lymphosarcoma, RCS, ALL, etc. There is little doubt that band q32 of #14 constitutes a very vulnerable point subject to breakage and a "receptor" site for material from other chromosomes. It is interesting to note that when #14 serves as a "donor" chromosome, it is band 14q12 that is involved (Kaiser-McCaw, Epstein, et al., 1977b). The exact frequency of 14q+ has not been established, though it appears to occur in a significant percentage of all lymphomas.

The relation of #14 in certain hereditary disorders to the development of lymphoma is worthy of some discussion. Clones of aberrant lymphocytes have been seen in patients with AT due primarily to rearrangement of chromosome(s) #14. A gradual rise in the aberrant clone during the course of the disease and development of CLL in the clone with tandem translocation 14q;14q in the case of McCaw, Hecht, et al. (1975) speak in favor of selective advantages of lymphocytes in patients with such rearrangements. McCaw, Hecht, et al. (1975) noted that the rearrangements of #14 in AT are mostly associated with breaks near the centromeric region of the long arm, whereas the changes in lymphoma are in the terminal region of #14. These observations raise some doubt, in my mind, as to the relation of the cytogenetic findings in AT to lymphoma and related disorders.

A new characteristic chromosome anomaly in lymphoproliferative disorders, i.e., t(11;14)(q14;q32?), has been described by Van Den Berghe, Parloir, et al. (1979) in four cases (three lymphomas and one ALL) leading to the formation of a 14q+. This extra material belonged to the long arm of a #11 in three cases and to the long arm of a #14 in the other case. The authors indicated that these findings confirm the postulate that the distal end of chromosome 14q may function as a receptor site and that chromosome #14 may not be unique in showing so-called donor and receptor sites in that other chromosomes such as #11 may behave similarly.

Burkitt Lymphoma

Among all human neoplastic diseases, Burkitt lymphoma (BL) is the one for which a viral etiology has been more seriously considered than for any other malignant tumor. Because viruses are known to be capable of producing structural chromosome abnormalities (Nichols, 1969), and, thereby, possibly induce somatic mutation which could result in malignant growth, the chromosomal aspects of BL have attracted much attention. Both African and American BL cells have been shown to be of B-cell origin, having IgM on their surface (Binder, Jencks, et al., 1975; E. Klein, Klein, et al., 1968).

The discussion of the results of chromosomal analyses of BL will be divided into two parts: 1. results obtained in direct preparations of tumors, and 2. results of tumor-derived lymphoblastoid cell lines maintained in vitro for various periods of time.

Detailed chromosomal data from preparations of tumor tissue before banding are available from 12 cases only[38]; from six additional cases only "rough" data were reported.[39] In five instances an apparently normal chromosome constitution was found; all of them were untreated at the time of tumor biopsy.[40] However, the distinct possibility exists that cells with only the 14q+ anomaly might have escaped detection with nonbanding methods. All the aneuploid tumors were revealed to have structural chromosomal anomalies. With regard to the chromosome number, remarkably minor deviations from the diploid number are to be noted. A pseudodiploid chromosome mode was encountered in five tumors[41]; hyperdiploidy ranging from 47 to 50 was seen in five[42]; and hypodiploidy (45 chromosomes) in two tumors.[43] One tumor had a tetraploid mode with 20% of the metaphases in the diploid range (Clifford, Gripenberg, et al., 1968). In evaluating the individual tumors it becomes apparent that, generally, the great majority of metaphases belonged to a certain modal karyotype; metaphases with additional numerical and/or structural anomalies were shown to be variants of the modal karyotype. The percentage of apparently normal metaphases in a given aneuploid tumor was either too small or zero, with one exception (Jacobs, Tough, et al., 1963); this one case may not be characteristic because the material was obtained from a tumor-infiltrated lymph node. Cytogenetic studies in two American children with BL revealed a normal, diploid picture in one and an abnormal clone in the other (Hübner and Littlefield, 1975). In both cases the marrow was infiltrated by lymphoma cells. The abnormal clone consisted of 46,XX,−1,−3C,+4M. The large submetacentric marker seen in this case was morphologically identical to a large #1 that had been seen in three African children and in one American woman with this disease, possibly indicative of this change being nonrandom in BL.

A number of different markers have been observed in BL tumors and within the same tumor. Jacob, Tough, et al. (1963) drew attention to an acrocentric marker (of the size of the long arms of a #2) which they found in 5 out of 16 tumors, and which replaced a #2 in 4 tumors. A similar situation was seen in the tumor studied by S. E. Stewart, Lovelace, et al. (1965). All the tumors with this particular marker were accompanied by additional but different markers, the morphology of which varied from submetacentric to subacrocentric, and which either replaced chromosomes of different groups or were extra. Even though the association of an acrocentric marker with a missing #2 has been observed, mainly in one laboratory, it appears more than coincidental; the number of direct chromosome studies is, however, too small to evaluate its significance. As is true for all kinds of recurrent markers, similar morphology does not necessarily mean similar origin or biological consequences. It can be safely predicted that, after a large scale application of banding techniques, the fruitless discussions on the significance of acrocentric or other markers frequently observed in tumors will be resolved positively. Along these lines, Steel (1971) reported that similar chromosome anomalies in different lymphoblastoid cell lines were derived from different chromosomes. Manolov and Manolova (1971, 1972), on the other hand, reported that a morphologically normal D-group chromosome (#14) showed an extra band of bright fluorescence (darkly stained with Giemsa) at the end of the large arms seen

Table 74a 14q+ due to various translocations in lymphoma

Translocation	Disorder and source of material	Reference
t(14;14) (2 cases)	Histiocytic lymphoma	Mark, 1975b
t(1;14)(q2;q24 or q32)	Histiocytic lymphoma	Fukuhara, Shirakawa, et al., 1976
t(1;14)(q2;q32)	HD	Fukuhara, Shirakawa, et al., 1976
t(14;14)	2 cases (1 of lymphosarcoma, 1 of HD)	Fukuhara, Shirakawa, et al., 1976
t(1;14)(q1;q32)	American BL cell line	Temple, Baumiller, et al., 1976
t(8q−;14q+)	8 Burkitt lymphoma (possibly 10) biopsies; 3 established BL lines; 1 lymphosarcoma biopsy; possibly 1 lymphosarcoma cell line and 1 HD cell line	Zech, Haglund, et al., 1976
t(11;14)(q;3;qter)	Lymphosarcoma	Fleischman and Prigogina, 1977
t(8;14)(q24;q32)	(2 cases of CLL with tumorous lymphosarcoma)	
t(8;14)(q24;q32)	Histiocytic lymphoma	Fukuhara and Rowley, 1977,1978; Fukuhara, Rowley, et al., 1979
t(4q;14q)	Histiocytic lymphoma	Fukuhara and Rowley, 1977,1978; Fukuhara, Rowley, et al., 1979
t(1q;14q)	Histiocytic lymphoma	Fukuhara and Rowley, 1977,1978; Fukuhara, Rowley, et al., 1979
t(14;15)(q32;q21 or 22)	Histiocytic lymphoma	Fukuhara and Rowley, 1977,1978; Fukuhara, Rowley, et al., 1979
t(11;14)(q13;q32)	Lymphosarcoma	Fukuhara and Rowley, 1977,1978; Fukuhara, Rowley, et al., 1979
t(8;14)(q24;q32) and	Lymphosarcoma	Fukuhara and Rowley, 1977,1978; Fukuhara, Rowley, et al., 1979
t(14;18)(q32;q21)	Lymphosarcoma	Fukuhara and Rowley, 1977,1978; Fukuhara, Rowley, et al., 1979
t(14;18)(q32;q21) (6 cases)	Lymphosarcoma	Fukuhara and Rowley, 1977,1978; Fukuhara, Rowley, et al., 1979
t(8;14)(8pter→8q23; 14pter→14q32::8q23→8qter)	American BL cell lines (2 lines)	Kaiser-McCaw, Epstein, et al., 1977a
t(8;14)	(3 lines from African BL)	Kaiser-McCaw, Epstein, et al., 1977a
t(8;14)	American BL + AT	Kaiser-McCaw, Epstein, et al., 1977b
t(11q−;14q+)(q13;q32)	Histiocytic lymphoma	Mark, Ekedahl, et al., 1977
t(10;14)(q21;q22)	Histiocytic lymphoma	Mark, Ekedahl, et al., 1977
t(1;14)(q23;q32)	RCS	Yamada, Yoshioka, et al., 1977
t(2;8;14)(q3;q24;q24) and t(1;14)(q32;q32)	Mycosis fungoides	Fukuhara, Rowley, et al., 1978a
t(11;14)(q23;q32)	Plasma cell leukemia	Liang and Rowley, 1978; Liang, Hopper, et al., 1979
t(13;14)(13q14→13qter→::14pter→14q31)	Histiocytic lymphoma	Mark, Ekedahl, et al., 1978
t(8;14)(8q22→8qter::14pter→14q32)	Histiocytic lymphoma	Mark, Ekedahl, et al., 1978
t(11;14)(11q13→11qter::14pter→14q31)	Histiocytic lymphoma	Mark, Ekedahl, et al., 1978
t(11;14)(q14;q32?)	3 cases of lymphosarcoma 1 case of ALL	Van Den Berghe, David, et al., 1978b

Table 74a—*continued*

Translocation	Disorder and source of material	Reference
t(11;14)(q13;q32)	Plasma cell leukemia	Liang, Hopper, et al., 1979
t(11;14)(q23;q32)	ALL (B-cell)	Roth, Cimino, et al., 1979; Cimino, Roth, et al., 1978
t(14;14)(q21?;q32)	Lymphosarcoma	Van Den Berghe, personal communication
t(8q−;14q+)	American BL (ascites)	Kakati, Barcos, et al., 1979
t(6;14)(q21;q32)	Lymphocytic lymphoma	Mark, Dahlenfors, et al., 1979
t(1;14)(p22;q32)	Lymphocytic lymphoma	Mark, Dahlenfors, et al., 1979
t(1;14)(q23;q32)	Histiocytic lymphoma	Mark, Dahlenfors, et al., 1979
t(14;20)(q32;q12)	Histiocytic lymphoma	Mark, Dahlenfors, et al., 1979

Table 74b Donor chromosomes involved in the 14q+ translocation in various lymphomas and/or their cell lines

Type of lymphoma	Chromosome involved	References
Nonendemic BL	#1, #8	Temple, Baumiller, et al., 1976; Kaiser-McCaw, Epstein, et al., 1977a,b; Kakati, Barcos, et al., 1979
African BL	#8	Zech, Haglund, et al., 1976; Kaiser-McCaw, Epstein, et al, 1977a
HD	#1, #8, #14	Fukuhara, Shirakawa, et al., 1976; Zech, Haglund, et al., 1976
Lymphosarcoma	#1, #6, #8, #11, #14, #18	Fukuhara, Shirakawa, et al., 1976; Fukuhara, Rowley, et al., 1979; Zech, Haglund, et al., 1976; Fleischman and Prigogina, 1977; Fukuhara and Rowley, 1977, 1978; Van Den Berghe, David, et al., 1978b; Mark, Dahlenfors, et al., 1979
Histiocytic lymphoma	#1, #4, #8, #10, #11, #13, #14, #15, #20	Fukuhara and Rowley, 1977, 1978; Mark, Ekedahl, et al., 1977; Brynes, Golomb, et al., 1978; Mark, Ekedahl, et al., 1978
RCS	#1	Yamada, Yoshioka, et al., 1977
ALL	#11	Van Den Berghe, David, et al., 1978b; Roth, Cimino, et al., 1979
Plasma cell leukemia	#11	Liang and Rowley, 1978; Liang, Hopper, et al., 1979
Mycosis fungoides	#1, #2, #8	Fukuhara, Rowley, et al., 1978a

by means of the fluorescence and Giemsa techniques. This marker band was seen in all metaphases of five out of six BL tumors and in seven out of nine BL-derived lymphoblastoid cell lines. This extremely interesting finding has been amply confirmed by a number of investigators (see sections on 14q+).

Burkitt lymphoma is usually highly susceptible to chemotherapy, and long lasting remissions can be obtained in many cases. Only one incompletely documented case is available to answer the question of what the chromosome constitution of the recurrent tumors is when compared to the primary one; in this case the pretreatment mode was 49, and after recurrence 47 (Clifford, Gripenberg, et al., 1968). The demonstration of an apparently identical marker in both preparations strongly indicates that the new tumor, after a disease-free interval of years, was due to the reemergence of the original tumor cells and not to the induction of a second tumor (Fialkow, Klein, et al., 1970). Tumors that did not respond to therapy were shown to have either remarkably stable anomalies (Manolov, Levan, et al., 1970; Manolov, Manolova, et al., 1971c, d) or to acquire additional anomalies (Gripenberg, Levan, et al., 1969). The last case offered another striking feature, i.e., tumors of the left side of the face and neck had a normal chromosome constitution, whereas that of the tumors of the right side of the face and neck was abnormal. It is possible that this situation reflects an independent tumor induction at different sites of the body. It is more likely that the same cause led to the genesis of all the tumors, some being diploid and others aneuploid. Also, it is possible that the right-side tumors could be a metastasis of the original diploid left-side tumor. In light of Fialkow, Klein, et al. (1970) finding the same 6GPD enzyme type in four differently located tumors in one patient who was heterozygous at the 6GPD-locus, the last assumptions seem more likely.

An unexpected and unexplained finding by Manolov, Levan, et al. (1970) in a tumor from a 9-yr-old boy was the presence of a female diploid karyotype in the neoplastic cells of BL, with an allocyclic X shown by the late labeling with autoradiography and which formed a Barr body in interphase. The presence of female cells in a male patient suggested that the cells had been transferred from another host. Histocompatibility tests indicated that the tumor was not of maternal origin and that transfer of tumor cells by a mosquito from a female patient with BL remained a possibility.

Nonendemic Burkitt Lymphoma

Burkitt lymphoma occurs with characteristic manifestations in tropical Africa and New Guinea. Approximately 50% of the patients, especially the young ones, suffer from typical jaw lesions, but the clinical picture is variable and practically every organ may be affected. The neoplasm may develop in the thoracic cavity or retroperitoneally, affect the central nervous system (CNS), marrow, and endocrine organs. In regions where it is epidemic, this multifocal and often bilateral malignant neoplasm is recognized and confirmed without difficulty. However, in areas where BL is rare, the variability of the presenting signs and symptoms may be misleading and delay the diagnosis. In this respect, the finding of 14q+ and possibly other chromosomal changes has helped greatly in establishing the diagnosis of nonendemic BL. Cases of nonendemic BL have been described both in the United States and other countries where they tend to be labeled by the geographic location of the patients, e.g., North American BL, etc. Very few reports have appeared on the cytogenetic findings in nonendemic BL (Table 74a). Lymphoma cells in ascitic fluid from an American patient described as a "poorly differentiated lymphocytic lymphoma" were studied in direct preparations by E. W. Chu, Whang, et al. (1966) and of the 125 metaphases, only 2 were diploid, whereas the rest had 46 chromosomes including a marker that was larger than the A-group chromosomes and apparently replacing one #1 in an otherwise normal karyotype. Nearly half of the cells had 47 chromosomes, due to an extra D-group autosome. A cell line derived from the ascitic fluid was studied 7–9 mo after initiation of the cultures, and the cells were found to be predominantly diploid, possibly indicating that they had been derived from contaminating normal cells present in the fluid. Two cases of American BL with massive infiltration of lymphoma cells into the marrow have been examined cytogenetically (Hübner and Littlefield, 1975). In one patient all the cells were found to be cytogenetically normal, whereas in the other definite karyotypic changes existed. Upon initial examination 87% of the cells were pseudodiploid (46,XX,−1,−C,−C,−C,+4 mar) with the possibility that one of the markers may have been a 1q+ and another a 14q+. The number of these abnormal cells decreased and apparently was replaced by Ph¹-positive cells (the authors do not discuss this parameter, which is only shown in the table of the paper), but the pseudodiploid cells with the markers returned again several weeks later only to be replaced by the Ph¹-positive cells before death. The authors indicate that the large marker chromosome observed by them is similar to that seen in direct preparations from three African BL, one American BL, and in some of the established cell lines from BL.

In another case of nonendemic BL the chromosome aberrations in tumor cells obtained by marrow aspiration and determined with banding revealed a karyotype that was somewhat variable but predominantly 46,XY,14q+. Some of the cells contained −22, deletions of #3,−9, and markers (P. Philip, Krogh-Jensen, et al., 1977). The origin of the extra segment on the 14q+ could not be determined, but apparently did not involve the long arm of #8.

Two other cases (Slater, Philip, et al., 1979) with leukemia nonendemic BL were found to have a 14q+ in the abnormal cells, but the origin of the extra material on #14 could not be detected. A number of other karyotypic abnormalities were present in these cases, for example, t(1;6), 6q+, 13q+, 1q+ del(6)(pter→q13:) and del(8)(qter→p21:).

Comparison of Karyotypic Findings in Various Lymphomas

A comparison of the chromosome behavior of RCS and LS, on the one hand, and HD and BL, on the other, may be of value. It appears that all groups have some rather typical chromosomal features. In both BL and HD tumors without detectable major chromosome anomalies can be observed, which is exceptional for RCS and LS. In HD a large percentage of apparently normal, diploid metaphases can be encountered and totally normal tumors are practically nonexistent; chromosomally abnormal clones in HD are almost equally distributed between the near-diploid, near triploid, and near-tetraploid ranges with most of the cases belonging to the two latter high ploidy groups. Burkitt lymphoma, RCS, and LS usually have near-diploid modes. Burkitt lymphoma is characterized by sharp, prominent modes, whereas all other lymphomas have flat modes with a multimodal chromosome distribution. (For a different opinion, see the report of Fleischmann, Håkansson, et al., 1976.) Markers occur most frequently in RCS (72%), followed by LS (58%) and BL (55%); they are least common in HD (40%).

In contrast to those of other cancers and leukemias, cells from BL readily divide under in vitro conditions[44] and, most importantly, can be established as permanent cell lines retaining much of the cytogenetic and other characteristics of the parent tumor. The success with growing the lymphoblastoid cells of BL in vitro has been of great help in establishing many features of these cells and the effects on the chromosome constitution and possible propagation of the virus (Epstein-Barr [EB] virus) associated with BL.[45]

Studies in Established Cell Lines

It is beyond the scope of this chapter to give a detailed description of chromosomal findings in the numerous lymphoblastoid cell lines derived from BL. However, for the sake of completeness some data shall be given. It is important to know that BL cells can be grown in vitro. This was proven by the demonstration of the continuous presence of markers, which characterized the original tumor, in some cell lines.[46] However, in the majority of BL cell lines the original tumor chromosome constitution is unknown, and it should be noted that in such cell lines neither the demonstration of virus particles, nor the EB-specific DNA,[47] nor of certain chromosomal changes gives evidence as to the origin of the cell lines, i.e., whether they are derived from tumor cells or normal lymphocytes. With the possible exception of the D-marker band (14q+) described by Manolov and Manolova (1972), no typical karyotypic changes exist in BL cell lines. The so-called C-marker which had some specificity for BL,[48] could possibly be related to the viral genome incorporated into the DNA of the host, but it is doubtful whether the C-marker is specific for BL. Identical (?) markers, characterized by a subterminal constriction, were seen in other (non-Burkitt) lymphoblastoid cell lines,[49] and viral DNA could be demonstrated in other (non-Burkitt) lymphoblastoid cell lines also (Zur Hausen, Diehl, et al., 1972). Additionally, the presence of the C-marker appears to depend on culture conditions; a cell line reported to be positive for the C-marker when examined in Philadelphia, was negative when examined in Edinburgh (Tough, Harnden, et al., 1968). When the chromosome constitutions of a tumor and its established cell line were compared, markers characteristic for the tumor were demonstrable in the cell line. The percentage of marker-positive metaphases was similar in both tissues in two instances,[50] and lower in the cell lines in another case (Gripenberg, Levan, et al., 1969). Additional structural changes such as breaks, acentric fragments, and exchange figures were observed in two cases.[51] The chromosome number remained in the diploid range. Burkitt lymphoma established lines have been also shown to be predominantly diploid, with some of these undergoing karyotypic changes when maintained in continuous culture (S. E. Stewart, Lovelace, et al., 1965; E. H. Cooper, Hughes, et al., 1966).

A cell line from a 29-yr-old Japanese woman with BL was shown to have a submetacentric marker involving #1, but lacking the #14 translocation (Kishimoto, 1978).

Q-banding of an established BL cell line harboring the EB virus revealed a hyperdiploid number, −21,13q+,14q+,+2M, and secondary constrictions of #7 and #9 (Petit, Verhest, et al., 1972). In several other BL cell lines the 14q+ abnormality was detected in all seven lines and present in 100% of the cells. In contrast, the abnormality was not observed in 775 cells from 31 infectious mononucleosis-derived lines nor in 450 cells from 18 lines obtained from cord blood lymphocytes experimentally transformed by EB virus in vitro (Jarvis, Ball, et al., 1974). The results indicate that the cells transformed by the virus in vivo are characterized by the chromosomal abnormality and may somehow be related to the genesis of the lymphoma. The results may also indicate that EB virus DNA is present in malignant and nonmalignant cells at two different chromosomal sites leading to two different types of infections. Certainly, the observations point to a clear-cut cytogenetic difference between BL-derived lines, on the one hand, and all other lines, on the other, findings which add to the immunochemical differences already reported between these two groups of cells. These authors (Jarvis, Ball, et al., 1974) felt that the demonstration of two kinds of nonproductive EB virus infections, the one involving malignant cells and the other normal lymphocytes, fits well with the concept of EB virus as an etiological agent in BL.

The 14q+ Anomaly in Burkitt Lymphoma

Using banding techniques, Manolov and Manolova (1971, 1972) detected an anomaly of one #14 in five out of six biopsy specimens and seven out of nine cell lines from a total of 12 patients with African BL (Figures 114 and 115). There was an extra band at the end of the long arm of one #14. Various other markers were present, though none so consistently. The quinacrine banding patterns of some of these markers had previously been described (Manolov, Manolova, et al., 1971c, d). It has now been established that in every case of BL in which a translocation can be demonstrated the 14q+ is due to t(8;14) (q24;q32). Even though this type of translocation may be observed in other forms of lymphoma, in which other types of translocations leading to a 14q+ appear to be more common, the 8;14 rearrangement appears to be characteristic, if not diagnostic, for BL. Even though cases of North American BL have been described with the 8;14 transloca-

114 The 14q+ anomaly in cells of BL. **A.** Complete karyotype with the arrows indicating translocation from the long arm of #8 to the long arm of #14. No other evident anomalies are present. B and C. Two partial karyotypes showing the 14q+ in North American and African BL cells and the translocation from the #8 to the #14, respectively. D. Schematic presentation of the translocation. Obviously, it is still unknown whether the translocation is reciprocal, i.e., whether there is a balanced translocation with part of #14 being translocated to the long arm of #8.
(Kaiser-McCaw, Epstein, et al., 1977).

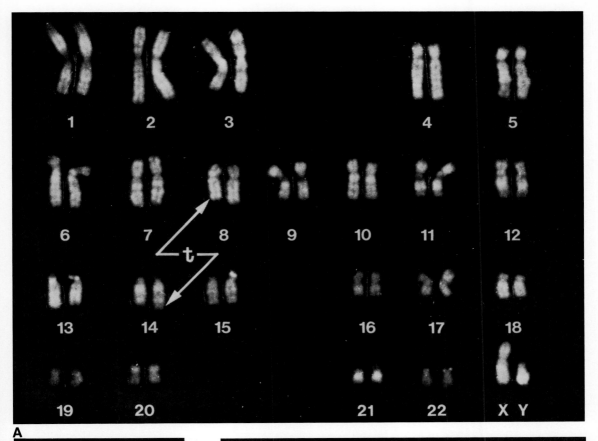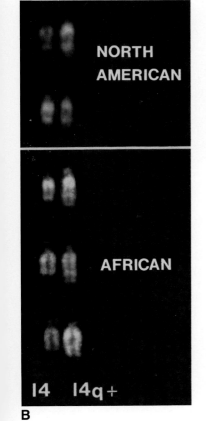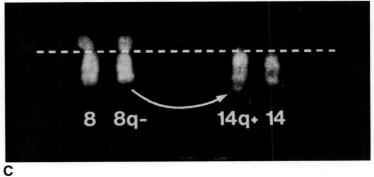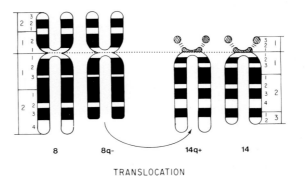

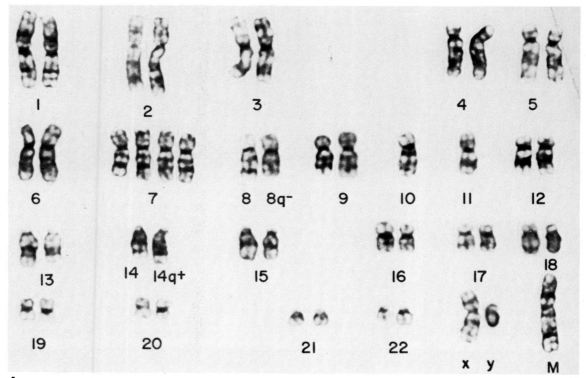

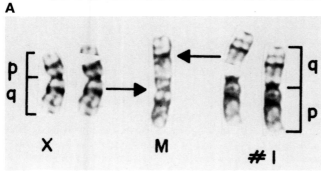

115 A. G-banded karyotype of an ascites cell from a patient with American Burkitt lymphoma containing a 14q+ due to t(8;14) (q23;q32). Other anomalies consisted of +7,+7,−10,−11 and a marker (M). B. Partial karyotype (G-banded) showing the origin of the marker as being due to a translocation between the X and #1 [(t(X;1)(p21;q21)].

tion, a cell line derived from one such patient (Temple, Baumiller, et al., 1976) has been shown to a few cells with 14q+ due to a 1;14 translocation (Table 74).

An indication of the authenticity of the various lymphoblastoid lines established in vitro as originating from BL tumor tissue is the demonstration in almost all of these lines of the presence of a 14q+ anomaly.[52] This applies both to African and North American BL. Even though in some of these cases banding analysis has not been performed, examination of the published karyotypes leaves little doubt as to the nature of the anomaly, even in those cases in which the authors have not drawn attention to the possibility of the presence of a 14q+.

Some recent developments in connection with the 14q+ anomaly in lymphoma are worthy of review. Even though 14q+ apparently occurs in a very high percentage of BL tumors, i.e., in nearly 85% of the 36 cases published to-date, it has been observed in other types of lymphoma and indicates that the 14q+ marker is lymphoma-associated rather than BL-specific. The question as to whether the clinical course of BL (or other lymphomas) differs in patients

Table 75 Rearrangements of chromosome #14

Class	Number of #14 in cell	Morphology One #14	Morphology Other #14	Mechanism	Balanced rearrangement	Result
I	2	Normal	Normal	—	Yes	Normality
II	2	14q+	14q−	t(14;14)	Yes	Normality
III	2	14q+	Normal	t(14;other chromosomes)	Yes	Normality
IV	1	14q+	Lost	t(14;14)	No	del(14p) and prox(14q)
V	2	iso 14q/ t(14q;14q)	Normal	Isochromosome formation/ t(14;14)	No	del(14p) and dup(14q)
VI	2	14q+	Normal	Tandem dup 14q	No	dup(14q)

Source: Kaiser-McCaw, Epstein, et al. (1977b).

with or without 14q+ merits further investigation.

Studies with banding methods have recently revealed the possible origin of the extra band on #14. Zech, Haglund, et al. (1976) showed with Q-banding that the 14q+ was due to a translocation of material from the terminal region of the long arm from #8, i.e., t(8;14). The above authors found the 14q+ in the cells of 10 BL biopsies, in four of nine established BL cell lines, in non-BL biopsy, in the cells of a LS-derived cell line, and in a long-established cell line derived from the pleural exudate of a patient with HD. The translocation was absent in the blood lymphocytes of five BL patients and in five lymphoblastoid cell lines of non-BL origin. Incidentally, trisomy #7 was found in two of the BL biopsies, in two BL cell lines, in one non-BL biopsy, in two LS-derived cell lines, and in one cell line from a patient with HD. The chromosome number and/or the karyotypes varied from sample to sample and included a variety of translocations (other than 8;14) and isochromosomes.

The findings of Zech, Haglund, et al. (1976) have been confirmed in two established North American BL cell lines (Kaiser-McCaw, Epstein, et al., 1977a, b). The t(8;14) was detected in both EB-positive and negative cell lines, indicating that the chromosomal abnormality correlates with lymphoid neoplasia rather than with EB virus.

Even though some chromosome rearrangements are observed in the lymphoproliferative disorders that do not involve 14q (Spiers, Janis, et al. 1977), it is clear that 14q+ commonly occurs in lymphoma. An analysis of the published data reveals that most neoplastic rearrangements of #14 can be placed in one of the categories shown in Table 75 (Kaiser-McCaw, Epstein, et al., 1977b). The first three categories result in a balanced karyotype with category II being a balanced rearrangement between the two homologues and category III between a #14 with another chromosome. The 8;14 translocation so characteristic of BL and the 11;14 translocation observed in lymphosarcoma are examples of the latter. Categories IV-VI result in an unbalanced karyotype with only a 14q+; category IV probably arises through the same mechanism as the 14;14 translocation of category II with the subsequent loss of the smaller deleted #14. Such an event has apparently been observed during clonal evolution in AT (McCaw, Hecht, et al., 1975). Thus, the short arm and part of the long arm of one #14 are lost. In category V one normal #14 and a metacentric chromosome composed of two #14 chromosomes may represent an iso-14. A possible case with this translocation has been described by Siegal, Voss, et al. (1976). Category VI includes a normal #14 and a 14q+ which is a tandem repeat of the long arm of #14. A ring composed of two #14 chromosomes can be included in category VI.

All these rearrangements of #14 could arise from either an interchromosomal event, such as somatic pairing, misalignment or chromatid breakage at 14q12 and 14q31, followed by a refusion or an intrachromosomal event if there is breakage between sister chromatids, crossing over, followed by refusion, and perhaps promoted by a loop. An intrachromosomal scheme must be invoked to explain a normal #14 remaining in such a cell.

Heteromorphisms, such as satellites, can

sometimes be used to designate each #14. For example, one #14 may be marked by large or bright satellites and the other by very dull or small ones. In a benign lymphocyte clone marked by a 14;14 translocation (category II), the donor 14 and the receptor 14 could be correlated with a #14 found in the patients fibroblasts.

Cells with a rearrangement of #14 may have a proliferative advantage. However, there may also be selection against cells not carrying a rearrangement of #14. For example, spontaneous chromosome breakage in cultured lymphocytes from normal people occur randomly throughout the karyotype. Only about 1% of all these random breaks are destined to be involved in balanced chromosome rearrangements; all others remain a single chromatid or isochromatid break. By contrast, over 60% of breaks in 14q result in a balanced chromosome rearrangement, usually with the other #14 or #7. The frequency of 7;14 balanced translocations in cultured lymphocytes is relatively high.[53] Thus, cells with random chromosome breakage may not survive because this breakage results in monosomy lethal to the cell. Breaks in #14 usually result in a balanced translocation, and as genetic material is not lost, the cells survive.

The proliferative advantage of neoplastic cells with a balanced rearrangement of #14 may also be due to "position effect," i.e., the effect of a gene or genes may be dependent upon its location and position in respect to neighboring genes. The chromosome material translocated on the end of #14 in BL frequently shows altered fluorochromatic staining with acridine orange (Kaiser-McCaw, Epstein, et al., 1977b). If this can be verified, it would constitute cytochemical evidence for "position effect" in malignant lymphoid cells with a balanced translocation.

The specificity of rearrangements of #14 in lymphoproliferative disorders has been compared to the specificity of the 9;22 translocation most frequently seen in CML. Rearrangements of #14 may yet prove to be of diagnostic value (Fukuhara, 1978). Chromosome #14 rearrangements consist of a variety of alternative rearrangements which may be either balanced or unbalanced. The instability of #14 and its lability to various rearrangements must ultimately be explored and explained at the molecular level. It is possible that these rearrangements can be correlated with an arrest of the normal differentiation pathway for lymphocytes, resulting in abnormal proliferation and disease.

Two consistent types of translocations between #7 and #14 have been found in lymphocytes stimulated with PHA (Zech and Haglund, 1978), and a correlation between the breakage point on #14 and the attachment of the 14q segment to either the short or long arm of #7 was found. A possible role of PHA in producing these translocations was advanced by the authors.

Not only in lymphoma, but also in other human cancers, it will be imperative to ascertain whether certain karyotypic changes, e.g., 8;14 translocation or the nature of the translocation (e.g., the donor chromosome to the #14), characterize one form of the disease akin to the Ph[1]-chromosome in the various CML.

The specificity of the 14q+ anomaly in BL has recently been questioned by Van Den Berghe, Parloir, et al. (1979), who demonstrated a t(2;8)(p12;q23) in a child with typical European BL and pointed to the possibility that chromosome #8 (8q−) may be more specific for the disease than the 14q+.

Chromosome Changes in Nonendemic Burkitt Lymphoma

All nonendemic Burkitt lymphomas (NEBL) have been shown to have chromosome anomalies, including a 14q+, whether the materials were examined directly or following their establishment as cell lines in culture. In each case in which the source of the translocation onto the long arm of the #14 could be identified, it was #8, with the break occurring at band 14q32 and acting as a receptor site. Table 74a indicates that the chromosomal changes (other than 14q+) differed from one case to another. In the case of a 10-yr-old boy with American Burkitt lymphoma observed by us (Kakati, Barcos, et al., 1979), the 14q+ was due to a t(8;14)(q23;q32), shown to exist in the ascitic lymphoma cells examined directly. Each cell had a different karyotype, with the 14q+ anomaly being present in most of the cells. Markers, some of which could not be identified with banding, were observed in most of the cells. Furthermore, a hitherto undescribed translocation in Burkitt tumors, i.e., involving the X chromosome, was also seen. The case of Temple, Baumiller, et al. (1976), in which 4 out of 30 cells of an established cell line from an American Burkitt lymphoma were found to have a t(1;14) leading to a 14q+, is difficult to interpret, as the change may have occured in vitro. The low incidence of the 14q+ in this established line in culture suggested to the authors that it arose subsequent to the establishment of the cell line and that the lability of the 14q3 region may be an important feature in some lymphoid diseases, possibly reflecting an in vivo predisposition toward abnormal growth. Thus, it appears that on the basis of the general experience that BL of any type can be ruled out if the donor chromosome for the 14q+ is an autosome other than #8. The application of this observation may help in a more concise definition of lymphomas (Lukes and Collins 1974, 1977), particularly BL and NEBL (Berard, O'Conor, et al., 1969), with the hope that ultimately all variations of lymphoma can be defined or abetted in their definition by their chromosomal pattern.

Some In Vitro Observations

Ten established cell lines from patients with histiocytic lymphoma were examined with banding techniques (Kaiser-McCaw, Epstein, et al., 1977b). In three of these a 6q− was present and in three other lines 14q+ was observed. Trisomy of #9, #11, and #13 was present in two of the lines (both also had 6q−) and +8 and 3p+ was observed in two of the lines with 14q+. The identity in the chromosome number, sex-chromosome constitution, and other markers, as well as the chromosomal picture in some of the lines, may indicate cross-contamination in the laboratory, an occurrence that may take place even under the strictest control of laborarory conditions. The authors also observed a patient with AT who developed North American BL. In his established cell line, a t(8;14) was found with no other chromosomal abnormalities. A repeat study of one of the cell lines with 14q+, +8, and 3p+ after a 6-mo interval demonstrated that the line was now monosomic for #8 and #14 and had several new large marker chromosomes. The latter may indicate a general chromosomal instability of these lines under in vitro conditions. The authors stressed the importance of the 14q+ finding, particularly as it related to its presence in other types of lymphoma and pointed out that there appear to be only two sites on chromosome #14 that are involved in all of these rearrangements: 14q12 is a "donor" site contributing material either to its homologue or, as in some AT clones, to another autosome, usually #7. The other site, 14q31, acts as a "receptor" site, receiving material either from its homologue or from another chromosome, usually #8 in BL or #11 in lymphosarcoma. The possibility exists that there may be homology between these two regions in neoplastic cells promoting inter- and, particularly, intrachromosomal exchanges at these two points. These regions may be polymorphic in the general population with more than one alternative structure at the two sites in each #14. The latter is the only chromosome in the karyotype that behaves in this fashion and the resultant rearrangements often appear to be balanced. There is selection for cells carrying a rearrangement of 14q and this selection or the proliferative advantages of the cell may be due to a position effect.

Sézary Syndrome and Mycosis Fungoides

The Sézary syndrome (Table 76), first described by Sézary and Bouvrain in 1938, is characterized by an intense pruritic erythroderma, leonine facies, lymphadenopathy, and the constant presence of abnormal cells in the blood. Apparently, skin tumors and involvement of the marrow do not occur in Sézary syndrome. This abnormal cell, termed the "Sézary cell," is characterized by unique cytoplasmic vacuoles and their content of PAS-positive material and cerebriform and serpentine aspects of the nucleus in electron microscopy. The Sézary cell resembles normal lymphocytes in responding to PHA (not always well [Kalden, Peter, et al., 1974; Vance, Cervenka, et al., 1977]), but differs from them by a poor reaction to PWM. There is a distinct possibility that the pathologic cells are produced in normal sites of lymphopoiesis such as the lymph nodes but may be derived as "prelymphomatous" in character, with the risk of true lymphoma or leukemia developing eventually. However, there is agreement among workers that chromosomal abnormalities, invariably found in Sézary cells, represent an additional

Table 76 Chromosome findings in Sézary syndrome and mycosis fungoides

Chromosome finding	Reference
SÉZARY SYNDROME	
65–100; mode 76; 2 markers	Crossen, Mellor, et al., 1971a
<92,>92 no mode marker (2 cases)	Brouet, Flandrin, et al., 1973
Pseudodiploid and hypodiploid, no mode, marker (Cases included in Lutzner et al., 1973?)	Brouet, Flandrin, et al., 1973
50–92, mode around 80–85,XXYY, extra chromosome in all groups except #1; 1–4 markers were present including one possible Dq+ (?14)	Lutzner, Edelson, et al., 1973; Prunieras, 1974
52–99; mode around 90; extra chromosome in all groups with markers and Gq−	Lutzner, Edelson, et al., 1973; Prunieras, 1974
47–48, mode 47, 46,XY,−9,−10,−16,−17,−20,+5mar (including 2 large acrocentrics)	Lutzner, Edelson, et al., 1973; Prunieras, 1974
Mode 47, 47,XX,−2,−17,−20,−21,+5mar (one large acrocentric),Gq−	Lutzner, Edelson, et al., 1973; Prunieras, 1974
42–45; mode 45,−1,−1,−6,−17,−19,−22, 3–7 markers (1 large acrocentric)	Dewald, Spurbeck, et al., 1974
42–47; long submetacentric, ring	Whang-Peng, Lutzner, et al., 1976
48, ring	Whang-Peng, Lutzner, et al., 1976
45, long acrocentric	Whang-Peng, Lutzner, et al., 1976
43–44, no markers	Whang-Peng, Lutzner, et al., 1976
47–48, marker, ring; 1–3 submetacentrics and minute	Whang-Peng, Lutzner, et al., 1976
45, long acrocentric	Whang-Peng, Lutzner, et al., 1976
46, min	Whang-Peng, Lutzner, et al., 1976
72, 1–2 min	Whang-Peng, Lutzner, et al., 1976
75–88, min	Whang-Peng, Lutzner, et al., 1976
47,XY,−9,−13,−17,−19,+del(9)(q11;p11),+t(9;13)(q11;p12),+t(9;19)(q11;q13),+i(17q),+(?)2	Nowell and Finan, 1977
MYCOSIS FUNGOIDES	
46,XY,2q+/48,XY,+2mar(min)	Obara, Makino, et al., 1970b
49,XY,+B,+21,+21	Erkman-Balis and Rappaport, 1974
47,XX,−1,−2,−D,+G,+3mar (2 large acrocentrics)	Erkman-Balis and Rappaport, 1974
45,XY,−3,−E,−F,+2mar (1 min)	Erkman-Balis and Rappaport, 1974
47,XY,+3,+C,−E	Erkman-Balis and Rappaport, 1974
48,XY,−1,+2C,−2D,+E,+2mar (both large acrocentrics)	Erkman-Balis and Rappaport, 1974
48,XY,−B,+E,+G,+mar	Erkman-Balis and Rappaport, 1974
44,XY,−1,−3,−E,−G,+2mar (1 minute)	Erkman-Balis and Rappaport, 1974
47,XY,Bp−,−C,+2E,−F,+mar	Erkman-Balis and Rappaport, 1974
45–47,XX,−E/47,XX,−3,+E,+mar	Erkman-Balis and Rappaport, 1974
47,XX,−2,−B,+2C,+mar	Erkman-Balis and Rappaport, 1974
49,XY,+B,+D,+marker	Fukuhara, Rowley, et al., 1978a
t(2;8;14)	Fukuhara, Rowley, et al., 1978a
t(1q−;14q+)	Goh, Reddy, et al., 1978
47,+F	Goh, Reddy, et al., 1978
5,−E (2 cases)	Goh, Reddy, et al., 1978
45,−C	
Pseudodiploidy (24% to 42% of cells) in all 4 patients; 2q+, −C,−E,+F found in some cells of the 4 cases	

crucial criterion for the characterization of the Sézary cell (Prunieras, 1973).

The abnormal circulating, mononuclear leukocytes (Smetana, Daskal, et al., 1977) are thought to be T lymphocytes[54], the disease being characterized in some patients by the presence of large-cell and/or small-cell variety. Sézary syndrome has been regarded as closely related to some of the lymphomas and the abnormal blood cells as modified lymphocytes. The pigmented erythroderma and lymphadenopathy are due to the presence of abnormal leukocytes, these being related to those in the peripheral blood.

The first chromosomal findings in Sézary syndrome were published in 1971 (Crossen, Mellor, et al., 1971a) and indicated the presence of a cytogenetically dual population of abnormal cells: one with a near-triploid chromosome number with a distinctive marker and another of 86–100 chromosomes with another characteristic marker. A near-diploid chromosome number but with an abnormal chromosome constitution may also occur in the Sézary syndrome. Whether the cells with a high chromosome number are related to the large variety of abnormal lymphocytes and those in the diploid range to the small variety of lymphocytes has been difficult to ascertain. However, Lutzner, Emerit, et al. (1973) studied four patients, two of whom had large and two small abnormal cells in blood smears. Cytogenetic studies showed that the former were in the tetraploid range and the latter two were hyperdiploid or pseudodiploid. In each, one or more markers were constantly present. Dewald, Spurbeck, et al. (1974) described a case with a mode of 44 chromosomes, including three markers in most cells. The latter consisted of a large chromosome whose banding pattern suggested an origin from 6p and 2q, a slightly smaller marker perhaps composed of 1q and 13q, and a medium-sized marker that may have been 2q−. Various other abnormal chromosomes including dicentrics were also seen (the patient had received chemotherapy). In almost all cases, abnormal markers, varying from large chromosomes to minutes, have been described. A large number of markers may not only be seen in the triploid-tetraploid cases but also in those with hypodiploid chromosome numbers. In our laboratory we have studied two cases with Sézary syndrome in whom BM material was examined and found to be karyotypically normal; the culture blood cells were aneuploid when grown with human but not with calf serum.

Of the 25 cases of Sézary syndrome in the literature on whom successful cytogenetic studies have been done, 20 contained markers with rings and minutes being very common. Diploid cells may accompany the aneuploid ones, though 100% aneuploidy has been described (Lutzner, Emerit, et al., 1973). Hypodiploidy is much less common than hyperdiploidy and very high chromosome counts are not unusual. Because of the remarkable variability in chromosome number it is difficult to state that one chromosome group is involved in the karyotypic changes more often than others. A case without chromosomal changes has been mentioned in one publication (Löffler, Meyhöfer, et al., 1974).

Bosman, van Vloten, et al. (1976) performed cytogenetic analysis on blood cultures (with PHA) obtained from three patients with Sézary syndrome before cytostatic therapy was started. In one patient two hypodiploid cell clones could be identified with 39 and 40 chromosomes, respectively. Moreover, a large number of hypotetraploid cells was found corresponding to a doubling of the S=40 clone. In all these abnormal cells multiple double-minute fragments were present and multiple translocations were identified, though they varied considerably from cell to cell. Representative changes in the cell line with 39 chromosomes included t(4q;3p),t(6q;3q),t(9q;13q) and i(q17) and in the clone with 40 chromosomes t(9q;13q),t(9q;14q),t(9q;12q) and ?i(q17). In the second patient aneuploidy and some tetraploidy were present, with numerous translocations being particularly evident in the cells with 47 chromosomes, as well as in the pseudodiploid cells. A considerable number of the cells contained 45 chromosomes. A somewhat representative picture in the pseudodiploid cells of this patient contained +7,t(9q;14q),t(9q;15q),t(11q;2q) and i(q17). In the third patient the degree of aneuploidy was less than in the two other patients, though a few polyploid cells were present; a small number of cells showed a translocation and an abnormal chromosome with a banding pattern resembling that of #2 was observed. One of the translocations was t(Xq;10q) and an isochromosome of undetermined nature was also seen. The break points, though they seemed to have been nonrandomly distributed, were almost all located adjacent to the centromere with chromosomes of group C and D being rather frequently involved.

In a patient with T-cell CLL with skin involvement, probably identical with "small cell" variant of the Sézary syndrome, 123 metaphases were counted, 53 from blood cultures and 65 from lymph node cultures. Most of the cells were diploid and apparently normal, but 17 cells from the blood and 7 from the lymph node had 47 chromosomes with multiple abnormalities and a karyotype interpreted as 47,XY,−9,−13,−17,−19,+del(9) (q11; p11),+t(9;13) (q11;p12),+t(9;19) (p11;q13),+i(17q),+(?2). Analysis of diploid metaphases from blood and lymph node cultures showed no abnormality. It is interesting to note that Lutzner et al. (1973) found an apparent isochromosome 17 in two individuals with "small cell" type variant of Sézary syndrome, clinically similar to the patient of Nowell and Finan (1977), but not in two other cases with typical large Sézary cells. Obviously, further reports are required to determine whether an isochromosome 17 is more common in T-cell diseases such as CLL with skin involvement, in Sézary syndrome, and in other lymphocytic neoplasms.

Mycosis fungoides is a chronic, slowly progressive skin disease characterized by a sustained proliferation of abnormal "lymphoid" cells in the dermis. Even though the neoplastic nature of this primarily cutaneous disorder is at times difficult to establish on purely histologic grounds, particularly during the early phases of the disease, most investigators consider it to belong to the general category of malignant lymphomas.

The chromosome studies in mycosis fungoides have been few and in the past primarily based on either the findings in lymph nodes or blood. However, in 1974 (Erkman-Balis and Rappaport) a study of 11 patients with mycosis fungoides was published in which the chromosomes were examined in the skin, lymph nodes, spleen, BM, and peripheral blood; all were revealed to contain abnormal karyotypes with a modal chromosome number ranging from 44 to 50. The chromosome abnormalities were consistent in each case, though they differed from case to case. Multiple tissues taken from six cases, simultaneously and/or consecutively at time intervals ranging from 2.5 wk to 1.5 yr revealed, in each case, identical chromosome abnormalities which remained essentially unchanged over periods of up to 18 mo. In three cases cytogenetic studies were of additional help in the evaluation of the involvement of lymph nodes and marrow by mycosis fungoides at a time when histologic examination was negative or doubtful. In 9 of the 11 cases markers were present and frequent involvement of E-group chromosomes were noted in 7 cases. An interesting case of mycosis fungoides was described by Obara, Makino, et al. (1970b) of a 45-yr-old male in whom two chromosomally abnormal cell lines were present in skin tumors, enlarged lymph node, marrow and blood. One line was characterized by a large submetacentric marker (2q+) and the other by two minutes.

Chromosomes from a patient with mycosis fungoides were examined with banding techniques (Fukuhara, Rowley, et al., 1978a). Hyperdiploid cells from a lymph node had common anomalies of certain chromosomes, and abnormal karyotypes were found in some cells from peripheral blood cultured with and without PHA. A structural anomaly of the long arm of #14 (14q+) was observed in all hyperdiploid lymph node cells and in one blood cell. The abnormality in the lymph node cells could be the result of t(8;14;2), whereas that in the blood resulted from t(1q−;14q+). These findings support the notion that mycosis fungoides is a type of malignant lymphoma and that rearrangement of the long arm of #14 might be an abnormality that is specific for the disease, as observed in other malignant lymphomas.

There is a controversy among different groups of investigators as to whether mycosis fungoides and Sézary syndrome are distinct entities or whether the latter represents merely an erythrodermic variant of the former with circulatory neoplastic cells. Currently, PHA stimulation of the circulating abnormal cells is thought to be very significant in Sézary syndrome. Dividing cells with abnormal karyotypes were observed only in 72-hr PHA-stimulated cultures. This PHA responsiveness has led to the view that the abnormal cells are of lymphoid origin. The behavior of the cells in mycosis fungoides is sufficiently different from that in Sézary syndrome to indicate that these two diseases may not be of common nature.

Angioimmunoblastic Lymphadenopathy

Angioimmunoblastic lymphadenopathy (AILD) is apparently a distinct clinical disorder characterized by a hyperimmune proliferation of B lymphocytes and T-cell deficiency (Kosmidis, Axelrod, et al., 1978) associated with polyclonal gammopathy and often with autoimmune hemolytic anemia. Histologically, the distinctive features of the disorder consist of proliferation of small, arborizing vessels and deposition of acidophilic-staining interstitial material and proliferation of cells compatible with transformed lymphocytes and mature plasma cells in lymph nodes and other lymphoid organs. Clinically, the patients develop progressive lymphadenopathy, usually associated with hepatosplenomegaly, itching, fever, and progressive weight loss. Usually, the course of the disease is progressive and evolution into lymphoma has been frequently observed (Frizzera, Moran, et al., 1975; Nathwani, Rappaport, et al., 1978). The chromosomal findings support the view that AILD is at the least a prelymphomatous state, if not already a lymphoma, when karyotypic changes are demonstrated (Goh and Bakemeier, 1978).

The chromosomal changes are observed in the involved lymph nodes and affected tissues; the marrow and blood cells are usually normal cytogenetically. The chromosomal changes observed in the six cases published to date are shown in Table 77 and indicate little consistency in the karyotypic aberrations observed.

A lymphoproliferative syndrome with variable phenotypic expression and apparently related to X-linked recessiveness has been described (Purtilo, DeFlorio, et al., 1977; Purtilo, Bhawan, et al., 1978; Paquin and Purtilo, 1978). Apparently, the patients develop abnormal lymphoproliferative responses upon exposure to EB virus: either a proliferative phenotype such as agammaglobulinemia, aplastic anemia or B-cell proliferation phenotypes such as fatal infectious mononucleosis, Burkitt lymphoma, or immunoblastic sarcoma of B cells. The 14q+ marker has not been seen in the established cell lines of XLRLS.

Histiocytic Medullary Reticulosis

Histiocytic medullary reticulosis (HMR) is a peculiar disorder characterized by the proliferation of morphologically atypical histiocytes

Table 77 Chromosomal changes in angioimmunoblastic lymphadenopathy

Age and sex of patient	Chromosomal findings	Source of material	Reference
62♂	Chromosome counts of 48–60 "with aneuploidy of numerous pairs"; no markers (G-banding)	Blood cells	Volk, Monteleone, et al., 1975
32♂	47,XY,+3 47,XY,+3,−9,+21 48,XY,+3,+8,+9,−20	Lymph node	Hossfeld, Höffken, et al., 1976
45♂	46,XY,Bq+ 47,XY,Bq+,+C	Lymph node	Hossfeld, Höffken, et al., 1976
64♀	47,XX,+ centric fragment	Lymph node	Castoldi, Scapoli, et al., 1976
52♂	47,X,−Y,+2,t(2q+;10q−)	Pleural sediment	Castoldi, Scapoli, et al., 1976
59♀	46,XX,t(1p−;2p+)/ 47,XX,t(1p−;2p+),+5	Lymph node	Goh and Bakemeier, 1978

through the lymphoreticular system and clinically marked by fever, progressive wasting, lymphadenopathy, hepatosplenomegaly, and often pancytopenia. Pathologically, the entity is characterized by prominent infiltration of several organs with anaplastic or bizarre histiocytic cells, often showing erythro- and plateletphagocytosis and, less frequently, ingestion of granulocytic cells. Because the course of the disease is usually of short duration, a clinical diagnosis is rarely made during the life of the patient on the basis of lymph node or liver biopsy. Even though most of the reported cases appear to be systemic proliferations of histiocytic cells, in the last few years some descriptions have suggested the possibility that the disorder is a terminal feature of leukemic processes. Castoldi, Grusovin, et al. (1977) reported a case of a 46-yr-old man with AML that terminated in HMR, of which the cytomorphological and biological features were investigated during the life of the patient. During the course of the disease, which lasted about 1.5 yr, the cytogenetic picture was studied in the marrow and blood. No stable chromosomal profile was obtained on this patient, though the presence of a very small marker and C-trisomy appeared to be the most salient cytogenetic feature.

General Comments

The theoretical and practical contributions of cytogenetics to the expansion of our knowledge and understanding regarding malignant lymphomas have been considerable.[55] The unequivocal demonstration that karyotypically HD is a true neoplasia has settled an argument that lasted many years. The findings that non-Hodgkin's lymphomas are characteristically near-diploid and half of HD in the triploid range, that in HD the karyotypes are more deranged than in other lymphomas, even when the chromosome number is in the diploid range (Lawler, Reeves, et al., 1975), the unique 14q+ anomaly present in a significant number of non-Burkitt lymphomas and its seeming specificity in BL, the frequent involvement of #18 and its possible relation to a "good" progression (Lawler, 1975), and the use of cytogenetic analysis in differentiating between reactive hyperplasia and malignancy and even in establishing the diagnosis of lymphoma,[56] represent an impressive list of contributions to our knowledge of lymphoma through cytogenetic studies. Furthermore, the 14q+ and the type of translocation leading to it, e.g., t(8;14) in BL vs. t(11;14) in LS, may ultimately be diagnostically and etiologically useful in lymphoma recognition. This 14q+ anomaly combined with 6q− and/or changes in #18 may prove to be rather characteristic for lymphoma, as the Ph¹ in CML. There is little doubt in my mind that future techniques capable of revealing more intimate substructural changes in chromosomes than available presently will further serve to differentiate the various lymphomas, at least on a cytogenetic basis.

Even though the 14q+ anomaly in lymphoma is starting to assume an importance only second to that of the Ph¹ in CML and the possibility exists that lymphomas may ultimately be defined or characterized in terms of the nature of the 14q+ translocation (Table 74b), sight must not be lost of the fact that in the past the involvement of certain chromosomes in lymphoma was pointed to by a number of workers without the benefit of banding. Thus, Spiers and Baikie (1970) indicated the frequent involvement of autosomes in group #17−#18; Reeves (1973) pointed out the frequent deletions involving the short or long arms of the same chromosomes. Because involvement of #18 in a translocation with the long arm of #14 appears to be a rather frequent occurrence in certain lymphomas, the observations of these authors should be noted in particular. The involvement of #1, #11, and #18 was frequently pointed to by other workers in the past and the recent demonstration of the nonrandom involvement of these chromosomes in the 14q+ translocation is again an indication of the importance of past observations. However, not all lymphomas or the cells in any particular lymphoma contain the 14q+ anomaly. In time it will have to be established as to the nature of lymphomas which do or do not have this interesting anomaly and whether there is any relationship between the presence of a 14q+ and the histologic type of the lymphoma, a history of previous therapy, the demonstration of a viral or a viral products in the tumors, etc. Certainly, the application of banding techniques in lymphoma has been an important development, in that it points to the possibility that a specific chromosomal change, i.e., 14q+, may yield important information regarding the tumor with the potential use of such a karyotypic anomaly in the therapy of and other clinical approaches in lymphoma.

The description (Kurvink, Bloomfield, et al., 1978) of an elevated SCE incidence in the lymphocytes from 47 patients with malignant lymphoma (12.7 ± 0.9/cell vs. 6.1 ± 0.3 in cells of 40 controls) and the further elevation of this incidence (to 14.3 ± 1.3/cell) by chemotherapy deserve further study and evaluation.

References

1. Ellman, 1976; O'Carroll, McKenna, et al., 1976.
2. Berard, Jaffe, et al., 1978; Brouet, Labaume, et al., 1975, Brouet, Preud'Homme, et al., 1975, 1976; Goldblum, 1977; Koziner, Filippa, et al., 1977.
3. Baker and Atkin, 1965; Fleischmann, Håkansson, et al., 1976; Galan, Lida, et al., 1963; Miles, 1973; Miles, Geller, et al., 1966; Millard, 1968; Ricci, Punturieri, et al., 1962b; Spriggs and Boddington, 1962.
4. Coutinho, Bottura, et al., 1971; Fleischman, Prigogina, et al., 1974, 1977; Fleischmann, Gustafsson, et al., 1972; Fleischmann, Håkansson, et al., 1976; Miles, 1973; Miles, Geller, et al., 1966; Millard, 1968; Seif and Spriggs, 1967; Spiers and Baikie, 1968a; Whitelaw, 1969.
5. Coutinho, Bottura, et al., 1971; Galan, Lida, et al., 1963; Miles, Geller, et al., 1966; Peckham and Cooper, 1969; Ricci, Punturieri, et al., 1962b; Spiers and Baikie, 1968a.
6. Coutinho, Bottura, et al., 1971; Miles, 1966.
7. Coutinho, Bottura, et al., 1971; Seif and Spriggs, 1967; Spiers and Baikie, 1968a; Whitelaw, 1969.
8. Peckham and Cooper, 1969; Sinks and Clein, 1966.
9. Coutinho, Bottura, et al., 1971; Seif and Spriggs, 1967; Whitelaw, 1969.
10. Lawler, Pentycross, et al., 1967; Long, Zamecnik, et al., 1977; Olinici, 1972; Trujillo, Butler, et al., 1967.
11. Joseph and Belpomme, 1975; Kaur, Spiers, et al., 1975.
12. Fukuhara and Rowley, 1978a; Fukuhara, Shirakawa, et al., 1976; Hossfeld, 1978; Reeves, 1973.
13. Fukuhara, Shirakawa, et al., 1976; Lawler, Reeves, et al., 1975.
14. Canellos, 1975; Canellos, Devita, et al., 1975; Coleman, Williams, et al., 1977; Collins, Bloomfield, et al., 1977; Ezdinli, Sokal, et al., 1969; Focan, Brictieux, et al., 1974; Jacquillat, Belpomme, et al., 1973; Lundh, Mitelman, et al., 1975; Raich, Carr, et al., 1975; Rowley, Golomb, et al., 1977b; Steinberg, Geary, et al., 1970; Weiden, Lerner, et al., 1973.
15. Coutinho, Bottura, et al., 1971; Fleischman and Prigogina, 1977; Fleischman, Prigogina, et al., 1974; Goh, 1968a; Kajii, Neu, et al., 1968; Khouri and Nassar, 1974; Mark, 1975b, 1977b; Mark, Ekedahl, et al., 1976; Miles, Geller, et al., 1966; Obara, Sasaki, et al., 1971b; Reeves, 1973; Spiers and Baikie, 1965a, b, 1966, 1967, 1968a.
16. Fleischman and Prigogina, 1977; Fleischman, Prigogina, et al., 1974; Ganner, 1969b; Reeves, 1973; Yam, Castoldi, et al., 1968.
17. Fleischman and Prigogina, 1977; Mark, 1977b; Yamada, Yoshioka, et al., 1977.
18. Coutinho, Bottura, et al., 1971; Sasaki, Sofuni, et al., 1965.
19. Bauke and Schöffling, 1968; Fleischman and Prigogina, 1977; Fleischman, Pirogina, et al., 1974; Forni, Baroni, et al., 1971; Lawler, Pentycross, et al., 1968; Reeves, 1973.
20. Demin, Radjabli, et al., 1972; Fleischman, Pirogina, et al., 1974; Spiers and Baikie, 1968a.
21. Coutinho, Bottura, et al., 1971; Fleischman, Prigogina, et al., 1974; Fleischmann, Håkansson, et al., 1976; Ganner, 1969b; Reeves, 1973; Spiers and Baikie, 1968a.
22. Fleischman, Prigogina, et al., 1974; Fleischman and Prigogina, 1977; Fukuhara, Shirakawa, et al., 1976; Miles, Geller, et al., 1966; Spiers and Baikie, 1968a.
23. Fleischman, Prigogina, et al., 1974; Fleischman and Prigogina, 1977; Fleischmann, Håkansson, et al., 1976; Fukuhara, Shirakawa, et al., 1976; Ganner, 1969b.
24. Fleischmann, Gustafsson, et al., 1972; Lawler, 1975; Millard and Seif, 1967; Reeves, 1973.
25. Bauke and Schöffling, 1968; Forni, Baroni, et al., 1971.
26. Baker and Atkin, 1965; Brieux de Salum, Suárez, et al., 1971; Broustet, Meuge, et al., 1970; Coutinho, Bottura, et al., 1971; Fleischman and Prigogina, 1977; Fleischman, Prigogina, et al., 1974; Fleischmann, Gustafsson, et al., 1972; Fleischmann, Håkansson, et al., 1976; Fukuhara, Shirakawa, et al., 1976; Gariepy and Cadotte, 1970; Lawler and Pentycross, 1968; Mark, Ekedahl, et al., 1976; Millard, 1968; Miles, Geller, et al., 1966; Nayar, 1976; Reeves, 1973; Siegal, Voss, et al., 1976; Spiers and Baikie, 1968a; Suárez, Brieux de Salum, et al., 1969; Tjio, Marsh, et al., 1963; Tormey, Ellison, et al., 1975; Zech, Haglund, et al., 1976.
27. Coutinho, Bottura, et al., 1971; Siegal, Voss, et al., 1976.

28. Fitzgerald and Adams, 1965; Millard, 1968; Sandberg, Ishihara, et al., 1964c; Spiers and Baikie, 1968a; Tormey, Ellison, et al., 1975.
29. Fleischman and Prigogina, 1977; Hayhoe and Hammoud, 1965; Prigogina and Fleischman, 1975b; Siegal, Voss, et al., 1976; Tormey, Ellison, 1975.
30. Brieux de Salum, Suárez, et al., 1971; Suárez, Brieux de Salum, et al., 1969; Zech, Haglund, et al., 1976.
31. Fleischman and Prigogina, 1977; Fleischmann, Håkansson, et al., 1976; Spiers and Baikie, 1968a.
32. Coutinho, Bottura, et al., 1971; Fleischman, Prigogina, et al., 1974; Millard, 1968; Spiers and Baikie, 1968a; Suárez, Brieux de Salum, et al., 1969.
33. Coutinho, Bottura, et al., 1971; Fleischmann, Håkansson, et al., 1976; Millard, 1968; Miles, Geller, et al., 1966; Tormey, Ellison, et al., 1975.
34. Erkman-Balis and Conen, 1972; Fitzgerald and Adams, 1965; Fleischman, Prigogina, et al., 1974; Millard, 1968; Sandberg, Ishihara, et al., 1964c; Spiers and Baikie, 1968a; Wisniewski and Korsak, 1970.
35. Erkman-Balis and Conen, 1972; Fleischmann, Håkansson, et al., 1971; Spiers and Baikie, 1970.
36. T. Abe, Morita, et al., 1976; Fleischman and Prigogina, 1977; Fleischmann and Krizsa, 1976; Fleischmann, Håkansson, et al., 1976; Fukuhara, Shirakawa, et al., 1976; Mark, 1975b; 1977b; Mark, Ekedahl, et al., 1976; Pierre, 1978; Prigogina and Fleischman, 1975b; Reeves, 1973; Siegal, Voss, et al., 1976; Van Den Berghe, David, et al., 1978b; Yamada, Yoshioka, et al., 1977; Zech, Haglund, et al., 1976.
37. Baker (in Atkin, 1976c); Fleischmann, Håkansson, et al., 1976; Fukuhara, Shirakawa, et al., 1976; Reeves, 1973; Spiers and Baikie, 1968a.
38. Gripenberg, Levan, et al., 1969; Jacobs, Tough, et al., 1963; Manolov, Levan, et al., 1970; S. E. Stewart, Lovelace, et al., 1965.
39. Clifford, Gripenberg, et al., 1968; Jacobs, Tough, et al. 1963; Lampert, Bahr, et al., 1969; Sinkovics, Drewinko, et al., 1970.
40. Gripenberg, Levan, et al., 1969; Jacobs, Tough, et al., 1963.
41. Gripenberg, Levan, et al., 1969; Jacobs, Tough, et al., 1963; S. E. Stewart, Lovelace, et al. 1965.
42. Clifford, Gripenberg, et al., 1968; Jacobs, Tough, et al., 1963.
43. Jacobs, Tough, et al., 1963; Manolov, Levan, et al., 1970.
44. Amano, Shigeta, et al., 1973; Bishun and Sutton, 1967; Furukara and Huang, 1976; Gerber, Whang-Peng, et al., 1969; Huang, Imamura, et al., 1969; Kaiser-McCaw, Epstein, et al., 1977b; Lambert, Hansson, et al., 1976a; Macek, Seidel, et al., 1971; Minowada, Moore, et al., 1968; Nadkarni, Nadkarni, et al., 1969; Steel, Woodward, et al., 1977.
45. M. A. Epstein, Barr, et al., 1966; A. L. Epstein, Henle, et al., 1976; M. A. Epstein and Barr, 1964.
46. Gripenberg, Levan, et al., 1969; Manolov and Manolova, 1972; Manolov, Levan, et al., 1970; Manolov, Manolova, et al., 1971a, b.
47. Tomkins, 1968; Zur Hausen and Schulte-Holthausen, 1970.
48. E. W. Chu, Whang, et al., 1966; Henle, Diehl, et al., 1967; Ikeuchi, Minowada, et al., 1971; Kohn, Mellmann, et al., 1967; Tough, Harnden, et al., 1968; Zajac and Kohn, 1970.
49. Huang, Minowada, et al. 1970; Macek and Benyesh-Melnick, 1972; Miles and O'Neill, 1967; Whang-Peng, Gerber, et al., 1970.
50. Manolov, Levan, et al., 1970; S. E. Stewart, Lovelace, et al., 1965.
51. Gripenberg, Levan, et al., 1969; S. E. Stewart, Lovelace, et al., 1965.
52. Bishun and Sutton, 1967; Hillman, Charamella, et al., 1977; Huang, Imamura, et al., 1969; Jarvis, Ball, et al., 1974; Kaiser-McCaw, Epstein, et al., 1977a,b; Kohn, Mellmann, et al., 1967; Macek, Seidel, et al., 1971; Manolov and Manolova, 1972; Petit, Verhest, et al., 1972; Rabson, O'Conor, et al., 1966; S. E. Stewart, Lovelace, et al., 1965; Temple, Baumiller, et al., 1976; Zech, Haglund, et al., 1976.
53. Beatty-De Sana, Hoggard, et al., 1975; Hecht, McCaw, et al., 1973, 1975; Welch and Lee, 1975.
54. Broder, Edelson, et al., 1976; Brouet, Flandrin, et al., 1973; Brouet, Preud'Homme, et al., 1976; Ding, Adams, et al., 1975; Kint, de Weert, et al., 1976; Lutzner, Edelson, et al., 1973; Schuller and van de Merwe, 1976.
55. Castoldi, 1970, 1973; Lawler, 1975; Lawler, Reeves, et al., 1975; Miles, 1973.
56. Hossfeld, 1978; Lawler, 1975; Nassar and Khouri, 1974.

15

Plasma Cell Dyscrasias

Waldenstrom's Macroglobulinemia

Multiple Myeloma
Leukemia Complicating Multiple Myeloma

Plasma Cell Dyscrasias of Unknown Significance

The disorders to be discussed in this chapter are accompanied by profound changes in the immunologic milieu of the patients. Whether such changes, and those in other conditions with an altered immunologic environment, affect the chromosomes in vivo and/or in vitro has not been answered satisfactorily. Views pro and con have been presented in the literature. It is likely that chromosomal integrity can be modified by an altered immunologic environment, with the chromosomal changes depending on the nature of the disease, the type of cells examined, the method of examination, and the conditions under which the cells were obtained, e.g., in vivo material or that obtained after culture. However, it is probable that most of the cytogenetic changes observed in the conditions to be discussed are the result of their neoplastic aspects, with the changed immunologic milieu playing a secondary role in most cases.

Patients with certain infections, e.g., measles and chickenpox, may have chromosomal changes in their cultured lymphocytes. These changes may persist for a relatively long time, and it is possible that even though the initial karyotypic aberrations may be virus induced, those seen subsequently may be due to the patients' immunologic reactions resulting from the viral infection.

The term "plasma cell dyscrasia," introduced by Osserman (1968), includes disorders characterized by the more or less unbalanced proliferation of a single clone of lymphocytic and/or plasmacytic cells resulting in the synthesis of abnormal quantities of a single protein. Some of the disorders included in this syndrome have typical biochemical and clinical patterns, i.e., Waldenström's macroglobulinemia and multiple myeloma; in others, the production of M-type gamma globulin is not associated with a well-defined clinical picture, and they are referred to as "plasma cell dyscrasias of unknown significance." Waldenström's macroglobulinemia and multiple myeloma are neoplastic diseases, whereas in the "plasma cell dyscrasias of unknown significance" symptoms indicating neoplasia of the lymphocytic-plasmacytic systems may never become apparent. Being malignant diseases, Waldenström's macroglobulinemia and multiple myeloma can be expected to have chromosomal anomalies. The demonstration of chromosomal anomalies in "plasma cell dyscrasias of unknown significance" could have pathognomonic and prognostic significance.

Waldenström's Macroglobulinemia

Waldenström's macroglobulinemia (WM) is characterized by an excessive proliferation of the M-type gamma globulin-producing cell clone. Waldenström's macroglobulinemia is a rare disease, occurring primarily in the elderly and often terminating in a clinical picture indistinguishable from CLL or lymphoma. Chromosome studies, however, have been performed in a relatively large number of cases. The discrepancy between the rarity of the disease and the number of reported chromosome studies is probably due to the description of a marker in WM, found in the early years of clinical chromosome research and considered to be intimately related to the genesis of the disease.[1] This marker has been termed the "W" chromosome. In the first four papers on the chromosome constitution of WM, the marker was an extra one resulting in a chromosome count of 47 per metaphase. The percentage of metaphases with 47 chromosomes plus the marker varied between 1.5 (Pfeiffer, Kosenow, et al., 1962) and 50% (Bottura, Ferrari, et al., 1961b). The length of the marker was either equal to or longer than the longest chromosome in the human chromosome complement; the position of the centromere was exactly median,[2] submedian,[3] or subtelic.[4] In one case,[5] some cells contained a metacentric, others a submetacentric marker. The essentials of the chromosome changes in WM are remarkably represented by these early papers. These findings were confirmed by studies of about 50 additional cases; no major cytogenetic characteristics could be uncovered, even though some interesting aspects were elaborated during the following years. For unknown reasons banding studies in WM have not been forthcoming as readily as would be expected from the cytogentic contributions related to this disease prior to banding.

A review of the chromosomal data of a total of more than 70 cases of WM studied before banding reveals the following results: An aneuploid cell clone was seen in 70% of the cases and solely normal, diploid metaphases were present in 12 cases (<20%).[6] The chromosome number of aneuploid cell clones was 47 in the great majority of cases;[7] in some cases the number was 46-pseudodiploid[8] or hypodiploid.[9] A marker with the characteristics described above was found in 52 cases (ca 75%). The percentage of marker-positive metaphases varied between 1.5 and 50% (average 13%). With the exception of three cases (5%),[10] all others were pseudo- or hypodiploid due to the presence of the marker. Different groups stressed the significance of diverse additional anomalies: extra-chromosomes in groups #6–#12, #19–#20, and #21–#22;[11] numerical and structural anomalies of the smallest chromosome pair of groups #6–#12,[12] loss of chromosomes of groups #6–#12 and #16–#18,[13] and long secondary constrictions.[14]

Cells used for chromosome analysis in WM were in most instances PHA-stimulated peripheral blood cells. Typically, lymphocytosis can be encountered in WM, and the marrow is diffusely infiltrated by lymphoplasmacytic cells. By means of immunofluorescence IgM synthesis by such cells was demonstrated (Preud'Homme, Hurez, et al., 1970). Thus, theoretically both marrow and blood contain the cells in question, and these should be suitable material for the study of the chromosome constitution of cells involved in WM. Yet, one is confronted with the fact that only a small percentage of the metaphases is aneuploid. If we postulate, and there is reasonable basis for this, that the neoplastic, IgM-producing cell clone is always accompanied by an aneuploid chromosome constitution, then several possibilities could account for the cytogenetic findings; the neoplastic cell clone is indeed very small, and, because it is mainly engaged in globulin synthesis, it proliferates slowly; the neoplastic cells do not respond, or do so very poorly, to PHA, which results in a relative predominance of normal lymphocytes. A poor response of blood cells of patients with WM to PHA has been described by several workers.[15] The "smallness" of the neoplastic cell clone was indicated by immune fluorescence studies (Preud'Homme, Hurez, et al., 1970). Therefore, it is necessary that a large number of metaphases be analyzed in order to find a sufficient number of those belonging to the neoplastic clone.[16] Hence, a small percentage of aneuploid metaphases in WM does not militate against the significance of the karyotypic findings.

The morphologic variability of the markers among different cases of WM and within the same cases[17] makes it difficult to believe that the marker as such has pathognomonic significance. The use of banding techniques will be of crucial value in elucidating the question of whether the marker is composed of similar or different chromosomal material. Autoradiographic studies suggest that the marker is derived from chromosomes of group #1–#3 (Siebner, 1967).

Other hypotheses have been put forward to explain the marker. One regards the marker as a #2 produced by nondisjunction and pericentric inversion (Patau, 1961a), whereas another considers it to be an isochromosome derived from the long arm of #2 (German, Biro, et al., 1961). With banding (Contrafatto, 1977), the marker has been shown to be a del(3) (qter→p25) in two cases with WM, whereas in a third case there might be a translocation of group F to #1 and from group G to #2.

Several workers have considered the possibility that the causation of chromosome abnormalities and impaired response of the lymphocytic cells to PHA in WM are related to the abnormal protein content of the serum. By culturing cells of patients with WM in normal plasma and, conversely, normal cells in plasma of patients with WM, Siebner (1967) and Salmon and Fudenberg (1969) did not find evidence in favor of this possibility, whereas Goh and Swisher (1970) did. Goh and Swisher's findings are difficult to understand, for on the one hand there is most probably no "abnormal" protein in WM but only an increased amount of IgM globulin, and no correlation exists between the absolute amount of immunoglobulin, immunological and physicochemical types of the protein, and the frequency and type of chromosomal anomalies[18]; on the other hand, Goh and Swisher (1970) did find the same kind of chromosome anomalies when they cultured WM cells in normal plasma, indicating that the chromosome anomaly is linked to the genome of the cells rather than to their environment. Finally, the clonal nature of the chromosomal anomaly, i.e., the restriction to a certain proportion of cells of the lymphoplasmacytic system, and the fact that the chromosomal anomalies cannot be found in fibroblasts of diseased persons (Ferguson and Mackay, 1963), strongly militate against Goh and Swisher's hypothesis.

Several theoretical considerations may be worthy of note. Waldenström macroglobulinemia may be transmitted as a dominant trait, in which case the chromosomal abnormalities or the tendency to such abnormalities should also be so transmitted. In fact, an assumption has been made, based on a patient with WM and a marker and whose four healthy relatives had at least one cell with such a marker (A. K. Brown, Elves, et al., 1967), that the tendency for the chromosomal changes is an hereditary feature of the disease. However, an opposite view was offered on the basis of a study in which the presence of an abnormal cell line in one monozygotic twin with macroglobulinemia but not in his identical twin was demonstrated (Spengler, Siebner, et al., 1966). These findings strongly suggest that the abnormalities acquired represent an analogy to the cases of monozygotic twins in CML in which only the affected twin has the Ph[1], whereas the healthy one does not. In relation to this is the description of a female patient with WM and a large marker chromosome in 12% of her cells, whose healthy son did not have this chromosomal abnormality but did have an abnormal peak of γ-globulins in his blood (Lustman, Stoffes-de Saint Georges, et al., 1968). In addition, the son of a patient with WM and a large marker was shown to have a congenital abnormality consisting of 45,XY,D−,D−,t(Dq,Dq).

An intense constriction in #1 has been described in a 3-yr-old girl with recurrent infections, low IgM, and no isoagglutinins in her serum (Østergaard, 1973). The chromosome abnormality was seen in 80% of her blood cells, though in my opinion, the published karyotype appears to contain other abnormalities (4q+?).

Abnormalities in serum gamma globulins have been associated with aberrations of chromosome #18,[19] including a ring chromosome (Jensen, Christensen, et al., 1969). However, there appears to be no correlation between the protein anomaly in the serum or urine of WM (or of MM) and the chromosome anomalies (Hellriegel, 1976).

The development of AMMoL in an untreated patient with WM (Salberg, Kurtides, et al., 1977) is of interest, particularly in view of the occurrence of AL in lymphoma and MM in association with radiation and/or chemotherapy.

Multiple Myeloma

Multiple myeloma (MM) is characterized by the unlimited proliferation of either IgG, IgA, or IgD globulins producing cell clones. The most common type is IgG myeloma, followed by IgA myeloma; IgD myeloma is very rare. Multiple myeloma occurs probably as often as AL. The number of cases, however, in which chromosome studies have been performed is relatively small. The main reason for this seems to be the very slow proliferation rate of plasma cells (Astaldi, Airò, et al., 1965) and, consequently, the paucity of metaphases. It is a common experience of chromosome workers to find only rare metaphases or none in marrows heavily infiltrated with myeloma cells. Inasmuch as negative results are unlikely to be reported, the literature most probably does not provide true evidence of the amount of work invested in this field. The results of chromosome analysis in a total of 275 cases of MM were reviewed by me. In 214 instances, *direct* preparations of BM or peripheral blood were used for chromosome analysis; in 60 instances PHA-stimulated blood cultures, and in one case tumor tissue were investigated. It is interesting to note that blood cultures with PHA showed abnormalities in the metaphases of 68% of the specimens examined, whereas only about 40% of the BM examinations revealed chromosome abnormalities. Even though an increase of lymphoplasmacytic cells can be seen in the blood of many cases of MM, marrow must be considered the tissue of choice for chromosome analysis, due to the usual presence of a high percentage of myeloma cells. Exceptions to this are rare cases with plasma cell leukemia.[20] Of the total of 275 cases,[21] 57% were revealed to have varying percentages of metaphases with chromosomal anomalies. In 43% of the cases, no chromosomal anomalies could be detected; among these "normal" cases were included those in which PHA-stimulated blood cultures were studied,[22] and some in which marrow studies were done.[23] In 259 marrows from patients with MM, the frequency of abnormal stem lines in the aspirates was nearly 40% (99 out of 259). Because of the presence of normal cells, diploid (normal) metaphases were encountered in the great majority of cases; 95–100% aneuploid metaphases were found only in those rare cases in which cytology revealed only myeloma cells.[24] From cases with a 100% aneuploid number of metaphases it may be tentatively concluded that all myeloma cells have an abnormal chromosome constitution. However, this question is open to discussion, and it could also be that there are myeloma cells with no apparent chromosome anomalies, or that chromosomal changes develop only in an advanced stage of the disease. With regard to the chromosome number of the cases with aneuploid clones, it was found that one third had a pseudodiploid and one third had a hyperdiploid (47–51 chromosomes) mode; about 10% of each group had chromosome numbers in the hypodiploid or hypotetraploid range, and in 15% the chromosome number was distributed between 52 and 60. Loss and addition of chromosomes apparently follows a random pattern. Markers were detected in about 55% of the cases.[25] Markers resembling the ones frequently seen in WM were disclosed in 15% of the total cases,[26] and most of these cases were contributed to by just two laboratories.[27] Recent results indicate variation in the incidence of such a marker from laboratory to laboratory. Another group pointed to the significance of an acrocentric marker (Tassoni, Durant, et al., 1967), which was also found by others.[28] Additional structural anomalies, in some instances probably related to chemo- or radiation therapy, were occasionally observed,[29] though it may not be a universal experience.

Chromosome Studies in Multiple Myeloma Before Banding

The exact incidence of karyotypic abnormalities in MM is difficult to establish on the basis of the chromosomal studies published in the literature. Because the marrow is involved in almost 100% of the patients with MM, it was anticipated in the early days of cytogenetic studies that it would be rather easy to establish the chromosomal picture in MM. Nothing could have been further from reality. The principle difficulty is that chromosome preparations of marrow in MM seldom yield satisfactory results. With a few exceptions, most investigators have found that the chromsomes in these cells spread poorly and are difficult to analyze or identify.[30] In addition, problems of sampling and interpretation are complicated by the irregular distribution of varying percentages of myeloma cells in many marrows. This is particularly difficult in marrows in which solely diploid cells are found. However, in about 50% of the patients with MM a mild neutropenia and a relative lymphocytosis are present, with immature-appearing lymphocytes and plasma cells being encountered. In a few cases, numerous plasma cells are present in the blood, a condition often referred to as plasma cell leukemia, with a course not much different from that of therapy-resistant AL. The study of blood cells (with or without PHA) may yield sufficient metaphases for analysis, though their exact origin cannot be confidently ascertained, even when aneuploidy is present. A review of the results obtained solely with marrow examinations revealed that in about 55% of the specimens, abnormal metaphases were present as compared with nearly 70% when stimulated blood lymphocytes were examined. The results certainly suggest that MM cells are circulating in the peripheral blood of many patients with this disease.

The nature of the cells in the BM that contribute to the metaphase population cannot be established with certainty in MM, because the number of abnormal plasma cells present in the marrow may vary from a small percentage to nearly 100%. In addition, the number of mitotic myeloma cells may be very small and, hence, the preponderant population may be found to be diploid in nature, being contributed to by the other cells of the marrow. Thus, when a marrow is totally diploid the decision has to be made whether any of these diploid cells originate from the myeloma cells. Reports do exist in the literature in which BMs consisting almost exclusively of myeloma cells were found to be diploid, and it must be assumed that the diploidy truly reflected the karyotypic picture of these abnormal cells. Whether it is possible to have a double population, i.e., diploid and aneuploid cells originating from myeloma cells, is a difficult question to answer at present. A similar enigma exists in the case of blood, particularly that which has been cultured in the presence of PHA. If aneuploid cells are present it must be assumed that they have originated from myeloma cells, as part of the infiltration in the blood by myeloma cells. However, if the picture is totally diploid, the possibility still exists that myeloma cells may constitute part of the diploid population. It is of interest that ready response to PHA is accepted as a characteristic of thymus-derived lymphocytes (T cells), whereas MM is usually regarded as a disease of BM lymphocytes (B cells).[31] In man, the distinction between these two classes of cells is not clear. The findings in the literature on PHA-stimulated cells in MM may be interpreted either as evidence against the simple distinction of B cells from T cells or, equally as well, to be in accord with a relatively poor response of B cells to PHA.

If one considers IgG, IgA, and IgD myeloma separately, no chromosomal differences seem to exist among the groups. However, the number of cases studied is too small to reach a definite conclusion. In future studies attention should be paid to the question of whether or not biclonal myelomas have two different abnormal chromosome modes.

Comparing the chromosomal makeup of MM and WM it can be said that WM is characterized by cells with a near-diploid chromosome mode, with a large meta- or submetacentric marker being present in a high number of the cases, whereas in MM the chromosome number ranges from hypodiploid to hypotetraploid, and no typical markers exist. The lower percentage of chromosomally abnormal cases with MM may be related to the slower proliferation rate of the cells involved, and does not necessarily reflect the karyotypic state of such cells.

Even though the marrow is almost always involved in MM, abnormalities of the chromosomes have not been encountered with consistency in this tissue. I was able to collect 260 cases of MM in which the BM was examined for its chromosome constitution either directly or after short-term culture, and in slightly < 40% of the patients karyotypic aberrations were encountered. In some of these cases examination of tumorous tissue revealed definite chromosome abnormalities (Mancinelli, Durant, et al., 1969), as well as in the blood cells when myeloma cells were present in the circulation (Okada, Hattori, et al., 1977; Dartnall, Mundy, et al., 1973), in face of normal findings in the marrow. This would seem to indicate that in order to rule out the absence of chromosome abnormalities in MM, it may be worthwhile to examine the cells in a number of different tissues and blood. Among these 260 patients with MM were observed four males with a missing Y in part of their marrow cells.

Bauke, Kaiser, et al. (1972) examined the BM in eight patients with MM and stated that "abnormal clones were not found in any of the eight patients." However, abnormalities occurred in sporadic hypodiploid, pseudodiploid, and hyperdiploid cells particularly in groups C and G. A C-group trisomy was found in 15 cells and G-group trisomy in 8 cells from different patients. A submetacentric chromosome of the group A size was observed in two cells from one patient. There was no correlation between the chromosome findings and the paraprotein type or the clinical course in the various patients. However, the experience with marrow examination varied from laboratory to laboratory. For example, no chromosomal changes have been described by a number of authors,[32] whereas others have demonstrated a high incidence of abnormalities,[33] with those authors presenting relatively large series of patients,[34] finding abnormalities in the marrow on the average in 38% of the cases (range 19 − > 75%).

The first report on chromosomes in MM appeared in 1959 (Baikie, Court Brown, et al., 1959) in which diploidy was found in two patients with this disease. Despite the difficulties discussed above, there is now little doubt that a variety of chromosomal abnormalities does occur commonly in the BM and the blood cells of patients with MM.

Two studies on large series of patients with MM will serve to illustrate the points raised in the above paragraph. In one study (Dartnall, Mundy, et al., 1973), 30 patients (no sex breakdown given) with MM were studied by several different methods: unstimulated culture of blood, stimulated (with PHA) culture of blood, and direct examination of the BM. It was possible to analyze only a small number of cells from stimulated blood cultures. Mitoses were present in unstimulated blood in 13 cultures from 12 patients, but it was not possible to analyze any of these with certainty. Similar unanalyzable mitoses were seen in four of the eight direct BM preparations obtained. In only one of these four cases were any abnormal mitoses from marrow cells analyzable: 20 cells were examined and only 3 contained karyotypic abnormalities. Abnormal mitoses were present in 19 stimulated (PHA) cultures of blood from 17 cases. In seven patients only diploid cells were found and in four others the cultures failed. Abnormalities of group #13–#15 included an abnormal chromosome in five of the cases, the marker being about the size of the long arms of a #2. In some cells it occurred in addition to the normal set in group #13–#15. In four patients additional normal #13–#15 chromosomes were observed. All these findings are similar to those reported by others. Abnormalities of groups #6–#12 and #21–#22 were rather frequent, and a large marker the size of #2 was seen in only one case. Marked aneuploidy was not a feature of the results, and no correlation was found between the cytogenetic abnormalities and previous therapy, disease stage, further progression of the disease, or serum and urinary protein abnormalities.

In another study the story was quite different. Chromosomes were examined in the marrow of 38 patients (37 males, 1 female) with MM, and in the marrow of only 3 specimens out of 70 examined no abnormalities were found. Of these 38 patients 21 were studied once and 17 several times. Diploidy was found in the marrow of only two untreated patients. Over 3,000 metaphases were studied in 70 marrow specimens. Structural abnormalities were insignificant except for markers in 11 cases. These markers were of large size but varied in morphology from case to case. Numerical abnormalities ranged from hypodiploidy to hyperdiploidy. No specific or characteristic chromosomal changes were established. Therapy may lead to the disappearance of the abnormal clones from the marrow; an increase in abnormal metaphases and appearance of markers and clones was often associated with relapse of the MM (Anday, Fishkin, et al., 1974).

Obviously, it is impossible to determine the discrepancy in the successful analysis of marrow material in MM among different laboratories. Even though dogged persistence in examining a large number of preparations may play some role, it is difficult to pinpoint the failure of some researchers to find adequate numbers of metaphases for cytogenetic analysis in the marrow of patients with MM and the seemingly facile success of others.

Banding Studies in Multiple Myeloma

A variety of markers have been described not only among the cases of MM, but also in the cells of the same marrow or blood. For example, in two studies (P. Philip and Drivsholm, 1974; Wurster-Hill, McIntyre, et al., 1973) in which banding analysis was used to determine the nature of the markers, G-banding analysis revealed three markers, i.e., t(1;16), t(1;5), and t(3;5) in the cells of one patient; in another study it was shown with G-banding that one of the markers was an abnormal #14 (14q+) in two cases, one with plasma cell leukemia and one with MM (Wurster-Hill, McIntyre, et al., 1973). The former patient had a pseudodiploid karyotype and the latter both diploid and hypodiploid cells (42 chromosomes) with a variety of markers. The 14q+ marker contained two extra bands terminally. A similar 14q+ has been reported in a case of MM by Philip (1975c), who showed with G-banding that the 14q+ replaced one normal #14 (as in the two cases of Wurster-Hill, McIntyre, et al., 1973). The origin of the extra segment could not be ascertained. Present also was a large marker due to t(1;3). Banding analysis in two patients (P. Philip and Drivsholm, 1976) with MM revealed in one case the basic karyotype to be 52,XY,−1,+3,+5,+7,+8,+9,+11,−15,+18, −20,+t(1;15),+t(1;16) and the other with 46,XY and clones of 45,XO,−1,−3,+11,14q+, +t(1;3) and 44,XO,−1,−1,−3,+t(1;1),+t(1;1). The frequent involvement of #1 in translocations in these two cases and in others with MM is of special interest. The authors (P. Philip and Drivsholm, 1976) stated that nonrandom chromosomal participation in the translocations and the existence of specific vulnerable points on #13 and #16 are suggested by their studies.

Banding studies have revealed a number of karyotypic abnormalities in MM[35] that have a nonrandom incidence. Thus, chromosomes #1, #3, and #16 participate in translocations and have specific vulnerable points in the order shown. A 1q+ anomaly in a case of MM (Spriggs, Holt, et al., 1976) and abnormalities of #1 in six of seven patients with chromosome changes (P. Philip, 1977b) have been described. Among the latter patients two had 14q+ and breaks at 3q2(789), a picture almost identical to that observed by Wurster-Hill, McIntyre, et al. (1973) in two cases.

The finding of 14q+ in MM, first described by Wurster-Hill, McIntyre, (1973), is of great interest, for up to that time this anomaly had been established with banding almost exclusively in lymphomas. A 14q+ has been described in six more patients with MM or plasma cell leukemia.[36] In at least one patient a translocation t(11;14)(q23;q32) was demonstrated (Liang and Rowley, 1978). Chromosomes #1, #5, and #21 were most frequently involved, with #6, #7, #10, and #15 being next in frequency (Liang and McLean, 1978). The possible presence of 14q+ in studies performed without banding can be inferred for the published figures (Dartnall, Mundy, et al., 1973; Tassoni, Durant, et al., 1967). It is possible that the 14q+ in MM is indicative of common cellular origins for lymphoma and myeloma and/or a common causation for such cases.

In one study the chromosomes were examined in aspirates from 25 patients with MM and abnormal stem lines found in 7 of the cases. The authors presented evidence that the cytogenetic changes are confined to myeloma cells and do not involve the erythrocytic or granulocytic elements of the marrow. Furthermore, the view was expressed that the karyotypic changes in MM are much more uniform than in other malignant disorders with the exception of CML (Krogh Jensen, Eriksen, et al., 1975). Of the seven abnormal cases, two were hyperdiploid, three pseudodiploid, and two were associated with more than 50 chromosomes. In two cases large submetacentric markers were present, in two others large acrocentric markers, and in two further cases small acrocentric markers. The cytogenetic findings could not be correlated with the type of abnormal immunoglobulin present in the serum of the patients, nor was it possible to find any correlation between the cytogenetic findings and the course of the disease.

Karyotypic abnormalities were detected in the malignant cells of 18 patients with MM (Liang, Hopper, et al., 1979). Six patients with benign monoclonal gammapathy, one with amyloidosis of immunoglobulin origin, and two with WM had normal karyotypes. All the patients with MM with aneuploidy were in a group of ten patients in the relapse phase of their disease; four had high serum paraprotein levels (4.24–7.29 g%) when their abnormal karyotypes were detected. Aneuploidy was not observed in eight stable MM patients.

Abnormalities of #14 were present in all six patients, a 14q+ in five and loss of a #14 in one. A translocation between #11 and #14 was found in the aneuploid cells of two patients who had plasma cell leukemia (PCL). However, the breakpoint in the long arm of #11 differed in the two patients. A gain of #5, #9, and #11 was seen in three patients, a gain of #1 in two cases, and rearrangements of #9 in five MM patients, including all four who had a 14q+ initially. A deletion of #6 at band q25 was detected in two MM patients and a pericentric inversion of #6 (6p21–6q13) was seen in the patient with PCL. Three of the MM patients had a nonrandom loss of a #8.

Two other MM patients who were treated with melphalan and prednisone developed ANLL 2+ and 4+ yr after the diagnosis of MM. Marrow cells of one patient showed a 5q− and a constitutional translocation involving #13 and #14 during the preleukemic stage; during the leukemic phase, the karyotype evolved had 50 chromosomes including extra #1,#6,#8, #10, and #21 and a missing #7, in addition to the originally detected 5q− and the 13;14 translocation. The blood cells from the other patient were hypodiploid, with a missing #7 and a translocation between 3q and 9q. These patterns of chromosome changes resembled those of ANLL rather than MM and are similar to those seen in ANLL after treatment of malignant lymphoma (Liang, Hopper, et al., 1979).

Leukemia Complicating Multiple Myeloma

As in the case of lymphoma, when patients with MM develop AL, the latter is almost invariably accompanied by karyotypic changes in the marrow (Table 78). This contracts with only a 50% incidence of cytogenetic aberrations in "spontaneous" AL. A presumptive case of MM presenting as an AL with a hypodiploid cell line has been described (Lewis, MacTaggart, et al., 1966).

The development of AL in patients with MM

Table 78 Leukemias complicating multiple myeloma

Complicating leukemia	Therapy	Karyotype	Reference
AML?	Alkeran (phenylalanine mustard)	44,45,46	Nowell, 1968
AMMoL	Melphalan	44,XX,−B,−17(18?) 43,XX,−B,−17(18?),−G	Hossfeld, Holland, et al., 1975
AMMoL	X-ray, melphalan	45,XY,−3,−B,+C/44,XY, −3,−B,−16,+C?	Hossfeld, Holland, et al., 1975
AMMoL	X-ray, melphalan, cyclophosphamide	56,XX,+B,+C,+C,+C,+D, +D,+F,+F,+2mar	Hossfeld, Holland, et al., 1975
Reticulosarcoma, leukemia	Melphalan, vincristine	42,XX,−D,−D,−D,−17(18?), −F,+mar	Hossfeld, Holland, et al., 1975
EL	X-ray, melphalan	43 and 44 chromosomes, −C	Bennett, 1971
AML	Melphalan	44,XY,−D,−E	Dahlke and Nowell, 1975
AL	Melphalan	30−35/45−48/65−70	Dahlke and Nowell, 1975
EL	X-ray	Chromatid breaks	Dahlke and Nowell, 1975
AML	Melphalan	5q−	Swolin, Waldenström, et al., 1977
EL	Melphalan	46,XX,−G,+F/47,XX,+C	Meytes, Seligsohn, et al., 1976
AMMoL	Melphalan	Hypodiploidy (45)	Khaleeli, Keane, et al., 1973
AMMoL	Melphalan	N	Khaleeli, Keane, et al., 1973
AMMoL	Melphalan	Hypodiploidy (45)	Khaleeli, Keane, et al., 1973
AMMoL	Melphalan	N	Khaleeli, Keane, et al., 1973
AML	Melphalan	60% 40−45; 15% polyploid; 25% N;long D and A	Gonzalez, Trujillo, et al., 1977
AML	Melphalan	85% 43−45; 15% polyploid; (chromosome damage); long D and A	Gonzalez, Trujillo, et al., 1977
AML	Melphalan	46,XY,−22,M(22) t(11;22)(q13;p11or12)	Oshimura, Hayata, et al., 1976
AML	Melphalan, Ara-C, prednisone	45,XX,del(5)(q13),t(13;14) (cen;cen)→50,XX,plus +1, +6,+8,+10,+21,−7	Liang, Hopper, et al., 1979
AML	Cytoxan, x-ray, prednisone, melphalan	45,XX,−7,t(3;9)(q29;p21)	Liang, Hopper, et al., 1979

who had received alkylating agents, particularly phenylalanine mustard (melphalan), is being reported more and more. A possibility of a causal relationship between the development of AL and such treatment has been suggested, though the coexistence of the two diseases has been reported in rare cases without such therapy (Tursz, Flandrin, et al., 1974; Cleary, Binder, et al., 1978). The chromosome-damaging radiomimetic effect of alkylating agents has been implicated as playing a critical role in the genesis of AL. However, the immunodeficiency of the MM may predispose to the development of AL resulting from the prolonged survival of MM patients treated with alkylating agents.

In six cases of MM complicated by AL, the chromosomal picture was variable. Unfortunately, in no case were the karyotypic changes established in the myeloma phase before the development of the AL. Hossfeld, Holland, et al. (1975) studied four such cases and found one with a chromosome mode of 56 with one to three markers; the morphology and size of the markers varied in different metaphases. Two cases had a bimodal, hypodiploid distribution (43–45 chromosomes) with the karyotypes in one case being 43,XX,–B,–17(18?), –G and 44,XX,–B,–17(18) and in the other 45,XY,–3,–B,+C and 44,XY,–3,–B,–16,+C? The latter case also contained a large acrocentric marker. The fourth case, one with reticulosarcoma leukemia, was characterized by 42,XX,–3D,–17(18?),–F,+mar. The two other published cases (Nowell, 1968; Bennett, 1971) had hypodiploid modes (43–45 chromosomes), with one case having –C. A case of Ph¹-positive AML developing in a 53-yr-old patient with MM has been reported (Daneshvar-Alavi, Lutcher, et al., 1977).

A case of MM in one member of monozygotic twins (67-yr-old brothers) but with chromosomal changes in some of the blood and BM cells in both members has been described (Ogawa, Fried, et al., 1970). The major karyotypic change was the presence of a small marker chromosome (G-group size) in BM and blood cells of the affected member and only in the blood cells of the unaffected member.

An interesting report is that of Van Den Berghe, Louwagie, et al. (1979c) of four patients with MM in whom a Ph¹ was found in BM and/or blood cells. In one case a 9;22 Ph¹-translocation and in another a complex one, t(3;8;22), were established. Though one cannot be absolutely certain, apparently the existence of AL was ruled out in these cases of MM and the authors expressed the view that the findings lend considerable support to the common progenitor cell concept.

The case of Spriggs, Holt, et al. (1976)—a patient with clinical MM treated with melphalan and who developed a clone of cells in the marrow with a chromosomal abnormality interpreted as consisting of duplication of part of the long arm of #1 (segment q21 to q31)—is of interest. Observations over a 2-year period failed to reveal any evidence of leukemia, even though the abnormality of #1 was seen during remission of the MM. The authors indicated that the hemopoietic cells, rather than plasmacytoma cells, were implicated. However, because the karyotypic picture of the myeloma cells was unknown in this case, and the marrow was not examined cytogenetically before the melphalan therapy, it is difficult to interpret the chromosome findings with certainty. Furthermore, it is still possible that the patient will develop leukemia in the future.

An interesting case is that of a 68-yr-old woman with a Christchurch chromosome (Ch¹), who developed CLL and after 8.5 yr of treatment with chlorambucil presented a picture of acute plasma cell leukemia (Fitzgerald, Rastrick, et al., 1973). During the evolution of the plasma cell leukemia the patient developed a large marker, extra chromosomes in group C, and a missing chromosome in group E. The relationship between lymphocytes and plasma cells appears to be consistent with a second leukemia in this patient and the presence of the constitutional Ch¹ abnormality conferring a propensity to lymphoid forms of leukemia. The role played by the long exposure to chlorambucil is difficult to ascertain, though it may have been responsible for or a factor in the blastic transformation. It should be pointed out that the Ch¹ persisted in the cells. Four patients with MM of long duration developed AML; two of the patients were treated with melphalan, one with cyclophosphamide, and one with both drugs (Andersen and Videbaek, 1970). A case of EL in a patient with MM has been described (Meytes, Seligsohn, et al., 1976).

Allusion has already been made above to a patient with the Ch¹ and CLL (Fitzgerald, Rastrick, et al., 1973) who ultimately developed MM with additional karyotypic changes. This case seems to illustrate that the karyotypic changes in CLL are probably different from those in MM and this may be related to the difference in causation of the two diseases. In not a small number of cases recently reported, it has been shown that various forms of leukemia can develop during the course of MM, particularly after certain courses of chemotherapy. Unfortunately, no reported cases have appeared to date in which the karyotypic picture was established in the myeloma before the development of leukemia and in which the cytogenetic findings have been examined after the development of the leukemia. Such cases are of extreme interest, because they would be of great value in ascertaining whether the chromosomal changes in MM and those in AL are of a different nature and reflect different oncogenic processes in man.

There is little doubt that chromosomal studies in MM based on various banding techniques will be published in the near future and these will not only shed considerable light on the origin of the markers in this disease, but will also supply more reliable and detailed data regarding the karyotypic picture and its evolution and relation to progression of the disease, including the development of leukemia. Already, the finding of a 14q+ abnormality in a number of cases of MM would seem to indicate that there may be some specificity to the changes observed in this disease.

Comments on Chromosome Findings in Multiple Myeloma

It appears that a variety of karyotypic pictures have been observed in MM without any consistency being evident in the disease or any of its subclassifications. The progression or lack of such progression in the course of MM is difficult to evaluate on account of the small number of cases studied cytogenetically on several occasions during the course of the disease. From the few reports available, it appears that clinical relapse was frequently associated with a decrease in the number of diploid metaphases, an increase in both hypodiploid and especially hyperdiploid metaphases, the appearance or persistence of marker chromosomes, and the development of definite clones in the blood or marrow.

In cases of purely diploid marrows it has often been observed that the chromosomes in MM show the fuzziness and ill-defined nature of the chromatids observed not infrequently in the BM of patients with AL. In such cases the distinct possibility exists that this fuzziness and ill definition of the chromosomes may be related to their malignant nature and, hence, the interpretation of the diploid marrow would be compatible with that of a MM with an essentially normal chromosome picture. It is hoped that banding studies in such diploid cases, as has been the case in AML, will define the nature of these chromosomes and establish, once and for all, whether these cases are, in fact, diploid or not. Thus, not only in MM but also in WM and other plasma cell dyscrasias, it is imperative that chromosomal studies be performed with banding techniques to make the karyotypic picture clear.

Plasma Cell Dyscrasias of Unknown Significance

Of great theoretical importance are chromosome analyses in "plasma cell dyscrasias of unknown significance." Chromosome studies of PHA-stimulated lymphocytes were performed in 34 cases with this disorder.[37] No chromosome abnormalities were seen in 25 cases.[38] Pseudodiploid metaphases containing a large meta- or submetacentric marker, resembling the marker seen in WM, were found in six cases[39] and hyperdiploid metaphases with a similar marker in two others.[40] The percentage of marker-positive metaphases was 2–8%. Siebner (1969) has followed a case with plasma cell dyscrasias of unknown significance, which 2 yr after the demonstration of an increased number of structural chromosome anomalies developed frank MM. It would be extremely interesting to know whether other cases with chromosome anomalies are more likely to develop MM or macroglobulinemia than those without karyotypic changes. On the basis of chromosomal findings, we must conclude that no such disease as "benign paraproteinemia" exists; the production of abnormal amounts of gamma globulin, together with the occurrence of chromosomal anomalies, indicate a malignant state (Audebert, Krulik, et al., 1977). It is a question of time before this becomes clinically apparent.

References

1. Benirschke, Brownhill, et al., 1962; Bottura, Ferrari, et al., 1961b; German, Biro, et al., 1961; Pfeiffer, Kosenow, et al., 1962.
2. German, Biro, et al., 1961.
3. Benirschke, Brownhill, et al., 1962; Pfeiffer, Kosenow, et al., 1962.
4. Bottura, Ferrari, et al., 1961b.
5. Benirschke, Brownhill, et al., 1962.
6. Braunsteiner, Rothenbuchner, et al., 1964;

Buchanan, Scott, et al., 1967; Ferguson and Mackay, 1963; Hayhoe and Hammouda, 1965; Kanzow, Lange, et al., 1967; Moretti, Hartmann, et al., 1965; Petit, Vyrens, et al., 1968; Petite, 1964; Pfeiffer, Kosenow, et al., 1962; Ponti, Valentini, et al., 1965.

7. Baserga, 1972; Benirschke, Brownhill, et al., 1962; Blake, 1966; Bottura, Ferrari, et al., 1961b; Broustet, Hartmann, et al., 1967; A. K. Brown, Elves, et al., 1967; Elves and Israëls, 1963; Ferguson and Mackay, 1963; C. Francke and Robert, 1970; German, Biro, et al., 1961; Hellriegel, 1971, 1976; Houston, Ritzmann, et al., 1967; Kanzow, Lange, et al., 1967; Pfeiffer, Kosenow, et al., 1962; Riva, Spengler, et al., 1971; Siebner, 1967; Wallace, 1963.
8. Broustet, Hartmann, et al., 1967; Goh and Swisher, 1970; Hellriegel, 1976; Houston, Ritzmann, et al., 1967; Siebner, 1967.
9. Gött and Vitéz 1970; Hellriegel, 1976.
10. Ganner, 1967; Kanzow, Lange, et al., 1967.
11. Siebner, 1967.
12. Houston, Ritzmann, et al., 1967.
13. C. Francke and Robert, 1970.
14. Braunsteiner, Rothenbuchner, et al., 1964; Pfeiffer, Kosenow, et al., 1962.
15. Houston, Ritzmann, et al., 1967; Salmon and Fudenberg, 1969; Siebner, 1967; Siebner, Spengler, et al., 1965; Siebner, Aly, et al., 1969.
16. Hellriegel, 1971; Houston, Ritzmann, et al., 1967; Siebner, 1967.
17. Banerjee, 1964; Benirschke, Brownhill, et al., 1962; Elves and Brown, 1968; Hellriegel, 1971; Heni and Siebner, 1964; Patau, 1961a; Siebner, 1967; Spengler, Siebner, et al., 1966.
18. Broustet, Hartmann, et al., 1967; Fialkow, 1964.
19. Borgaonkar, Bias, et al., 1969; Jensen, Christensen, et al., 1969; Ruvalcaba and Thuline, 1969.
20. Cardini, Bersi, et al., 1970; Hayhoe and Hammouda, 1965; Liang and Rowley, 1978; Takatsuki 1968.
21. Anday, Fishkin, et al., 1974; Baikie, Court Brown, et al., 1959; Bauke, Kaiser, et al., 1972; Bottura, Ferrari, et al., 1961b; Bottura, 1963; Cardini, Bersi, et al., 1970; Castoldi, Ricci, et al., 1963; Dammacco, Trizio, et al., 1969; Dartnall, Mundy, et al., 1973; Das and Aikat, 1967; Dubrova, Dygin, et al., 1966; Grouchy, Nava, et al., 1967a; Guillan, Ranjini, et al., 1970; Hayhoe and Hammouda, 1965; Houston, Ritzmann, et al., 1967; Houston, Hoshino, et al., 1970; Itani, Hoshino, et al., 1970; Kanzow, Lange, et al., 1967; Krogh Jensen, Eriksen, et al., 1975; F.J.W. Lewis, Fraser, et al., 1963; F.J.W. Lewis, MacTaggart, et al., 1963; Liang and McLean, 1978; Mancinelli, Durant, et al., 1969; M. Okada, Miyazaki, et al., 1977; P. Philip, 1977b; Ponti, Valentini, et al., 1965; Ranjini, 1971; Rashad and Morton, 1964; Richmond, Ohnuki, et al., 1961; Rochon, Cadotte, et al., 1971; Siebner, Spengler, et al., 1965; Spriggs, Holt, et al., 1976; Takatsuki, 1968; Tassoni, Durant, et al., 1967; Tsuchimoto, Kamada, et al., 1968; Van Den Berghe, personal communication; Wurster-Hill, McIntyre, et al., 1973.
22. Houston, Ritzmann, et al., 1967; Houston, Hoshino, et al., 1970; Kanzow, Lange, et al., 1967; Pfeiffer, Kosenow, et al., 1962; Pfeiffer and Kosenow, 1962.
23. Baikie, Court Brown, et al., 1959; Bottura, 1963; Bottura and Ferrari, 1962a; Hayhoe and Hammouda, 1965; F.J.W. Lewis and MacTaggart, 1962; Ponti, Valentini, et al., 1965; Richmond, Ohnuki, et al., 1961; Rochon, Cadotte, et al., 1971; Tassoni, Durant, et al., 1967.
24. Cardini, Bersi, et al., 1970; Castoldi, Ricci, et al., 1963; F.J.W. Lewis, Fraser, et al., 1963; F.J.W. Lewis, MacTaggart, et al., 1963; Mancinelli, Durant, et al. 1969; Rochon, Cadotte, et al., 1971.
25. Anday, Fishkin, et al., 1974; Cardini, Bersi, et al., 1970; Castoldi, Ricci, et al., 1963; Castoldi, 1964; Coltman, 1967; Das and Aikat, 1967; Grouchy, Nava, et al., 1967a; Houston, Ritzmann, et al., 1967; Kanzow, Lange, et al., 1967; Krogh Jensen, Eriksen, et al., 1975; Mancinelli, Durant, et al., 1969; Siebner, 1967; Tassoni, Durant, et al., 1967.
26. Coltman, 1967; Houston, Ritzmann, et al., 1967; Mancinelli, Durant, et al., 1969; M. Okada, Miyazaki, et al., 1977; Siebner, 1967; Tassoni, Durant, et al., 1967.
27. Houston, Ritzmann, et al., 1967; Siebner, 1967.
28. Cardini, Bersi, et al., 1970; Castoldi, Ricci, et al., 1963; Kanzow, Lange, et al., 1967; Mancinelli, Durant, et al., 1969.
29. Bobzien, 1967; Dammacco, Trizio, et al., 1969; Dubrova, Dygin, et al., 1966; Guillan, Ranjini, et al., 1970; Houston, Ritzmann, et al., 1967; Itani, Hoshino, et al., 1970; Kanzow, Lange, et al., 1967; Ranjini, 1971; Richmond, Ohnuki, et al., 1961; Tsuchimoto, Kamada, et al., 1968; Vagner-Capodano, Detolle, et al., 1970.
30. Baikie, Court Brown, et al., 1959; Dartnall, Mundy, et al., 1973; Krogh Jensen, Eriksen, et al., 1975; P. Philip and Drivsholm, 1976.
31. A plasma cell leukemia with T-cell population has been described by Wetter, Reis, et al., 1973. No chromosomal studies were presented.
32. Hayhoe and Hammouda, 1965; Kanzow, Lange, et al., 1967; Pfeiffer and Kosenow, 1962; Ponti, Valentini, et al., 1965.
33. Das and Aikat, 1967; Ranjini, 1971; Siebner, Spengler, et al., 1965; Tassoni, Durant, et al., 1967.

34. Anday, Fishkin, et al., 1974; Houston, Ritzmann, et al., 1967; Krogh Jensen, Eriksen, et al., 1975; Liang and McLean, 1978; P. Philip, 1977b; Van Den Berghe personal communication.
35. P. Philip, 1977b; P. Philip and Drivsholm, 1974, 1976; Spriggs, Holt, et al., 1976.
36. Liang and McLean, 1978; Liang and Rowley, 1978; P. Philip, 1975c; Philip and Drivsholm, 1976; Wurster-Hill, McIntyre, et al., 1978.
37. Elves and Israëls, 1963; C. Francke and Robert, 1970; Siebner, 1967, 1969.
38. C. Francke and Robert, 1970; Siebner, 1967.
39. C. Francke and Robert, 1970.
40. Elves and Israëls, 1963.

16

Chromosomes and Cancer

Chromosomes in Cancers: Some "Practical" Aspects

Significance of Chromosome Abnormalities in Cancer

Chromosome Changes in Malignant Transformation
DNA Measurement in Tumors

Chromosomes and Causation of Human Cancer and Leukemia

The Chromosomes in Animal Tumors and Leukemia
  Spontaneous tumors and leukemia
  Tumors due to nutritional or endocrine imbalance
  Chemical carcinogens
  Oncogenic viruses
  Contagious tumors
  Minimal deviation hepatomas
  Further comments on some experimental tumors

Chromosomes and Progression of Human Neoplasia

Chromosomes in the Diagnosis, Therapy, and Prognosis of Cancer and Leukemia

Chromosomal Findings in Preneoplastic Lesions

Chromosomes in Cancers: Some "Practical" Aspects

Much material in this section is based not only on my personal experience with tumor chromosomes, but also on that of the many investigators who have published on this subject and on information obtained from personal contact with cytogeneticists working in the area of human cancer. However, this subject was best covered and most aptly stated by Atkin (1974a); hence, I have liberally "borrowed" from his statements in order to express some of the putative points relative to chromosome studies in tumors and the egregious pitfalls associated with this facet of cytogenetics.

The visualization of metaphase chromosomes in solid tumors requires that the material be processed (Figure 116). This step usually involves pretreatment of the minced tissue with hypotonic solutions, perhaps exposure to colcemide or colchicine, flattening and spreading of the chromosomes, slide preparation by air drying or flaming, and an application of staining procedures, such as in banding techniques. To obtain satisfactory spreading it may be necessary to employ fairly vigorous treatments, e.g., the material may have to be resuspended several times in fresh fixative. If such treatment is too vigorous, many disrupted metaphases may result. When samples of the same tumor are given differnet treatments, isolated and small groups of chromosomes may be found in some preparations but not in others. Unfortunately, it is impossible to be always absolutely sure whether a given metaphase is in fact complete. The presence of many broken metaphases will result in distortion of the distribution curve of chromosome numbers, even though the position of the mode or modes may not be appreciably altered. There may also be a bias favoring inclusion of broken metaphases, because they tend to exhibit better spreading and chromosome morphology.[1]

At present, it is usually impossible to obtain successful chromosome preparations from every tumor studied. Of course, the yield depends on the source of the tumor (solid vs. effusion), the site of the tumor, and a number of other parameters. In my laboratory the yield with solid tumors is about 10–20% and with effusion material 30–35%. Furthermore, adequate banding is still a difficult chore in most tumors, due to the condensed and fuzzy nature of the chromosomes in many of the samples (Figure 117). A major difficulty may be a paucity of dividing cells. Various subtle features of the tumor, including the type and degree of differentiation may influence the spreading and morphology of metaphase chromosomes. In all, therefore, there may be a bias in favor of selecting the more malignant tumors with frequent mitoses and/or optimal chromosome morphology.

By employing culture techniques it may be possible to obtain larger numbers of divisions than could be obtained with a direct method. However, long-term cultures may be suspect because the cells studied may either have originated from nonmalignant cells also present in the tumor or, if malignant, they may have undergone some change in their karyotypes in vitro. In short-term cultures, the posssibility that such changes have occurred during culture is obviously smaller, and we have used it to advantage with marrow material and such tumors as lymphomas and malignant testicular tumors. These and other tumors we have found to be amenable to short-term culture techniques (up to 96 h) and chromosome preparations made from both uncultured and cultured materials; one or the other may prove to be more productive, even though the results are generally comparable.

Another approach that may prove useful is the growth of human tumors in mice, particularly the "nude" type, an approach that has been shown to yield material for banding analysis.[2] What will have to be established first is that the karyotype of the tumor does not always change in these mice.

When tumor cells are present in the marrow (Anner and Drewinko, 1977), it should be processed as in leukemia.

Although dependent to some extent on the techniques, the morphology of the chromosomes of tumor cells tends, on the whole, to be less satisfactory than that of normal cells. This may be noticeable when comparing the chromosomes of aneuploid tumor cells and diploid normal (or presumed normal) cells coexisting in the same preparation, whether these are from solid tumors or malignant effusions; the cancer chromosomes have an indistinct outline, ill-defined primary constrictions, and sometimes are overcontracted, rendering them

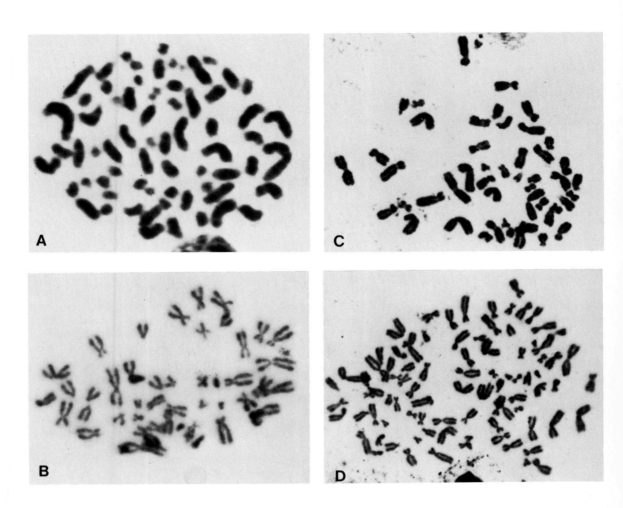

116 Metaphases from solid tumors prepared without (A and B) and with (C and D) previous incubation for short periods of time. In general, the incubated preparations yielded well-defined and more easily identifiable chromosomes. It is not uncommon in cancerous material to obtain chromosomes with the fuzzy and ill-defined morphology shown in A.

difficult to classify. A similar difference in chromosome morphology between aneuploid and diploid cells from leukemia patients was first drawn attention to in one of our early papers (Sandberg, Ishihara, et al., 1961). Varying degrees of spreading of chromosomes within a metaphase group, such as greater spreading of those lying peripherally, can present a problem; however, the varying intensity of staining of the chromosomes that results can often be compensated for by the experienced observer.

In many tumors there is a limited number of karyotypes of good quality available for study. Often, therefore, it is only possible to describe some limited features of the karyotypes, which are then taken with greater or lesser assurance to be representative of the tumor as a whole. Nevertheless, in some studies there have been large numbers of metaphases with chromosome complements which were countable, in spite of the inevitable presence of broken metaphases already mentioned. As a generality, it is found that in any given sample of tumor tissue the chromosome numbers are grouped more or less closely around a mode, with a secondary mode of metaphases with doubled chromosome complements. Upon karyotype analy-

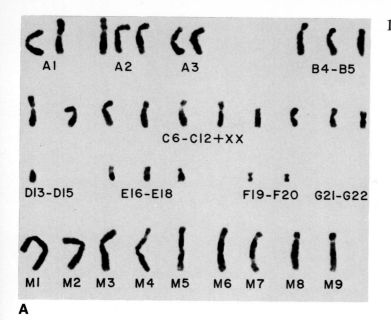

117 A karyotype and metaphase from breast cancer cells in effusion showing the presence of marker chromosomes. Such marker chromosomes are much more common in cancerous tissue than in leukemic cells and in the former they are more common in metastatic (particulary effusion) material than in primary tumors. A. Extremely hypodiploid karyotype, consisting of only 35 chromosomes, which constitute the modal chromosome number of this particular cancer and one of the lowest numbers observed by us. Note the 9 markers which may account for the large number of chromosomes missing in the other groups. B. The metaphase contains a number of abnormal chromosomes (arrows), in addition to the ring, which cannot be identified due to the fuzzy and ill-defined appearance of the chromosomes.

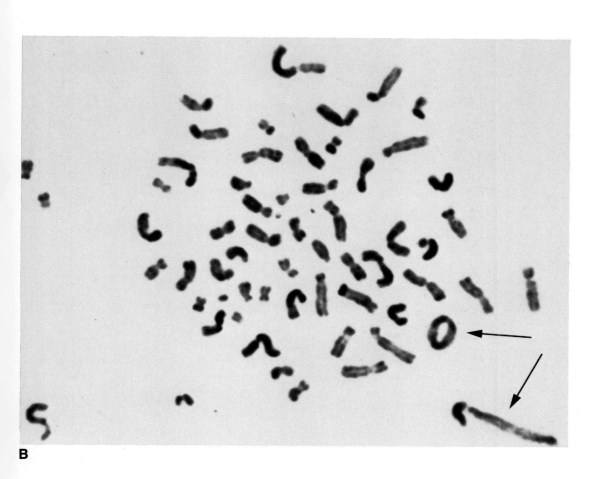

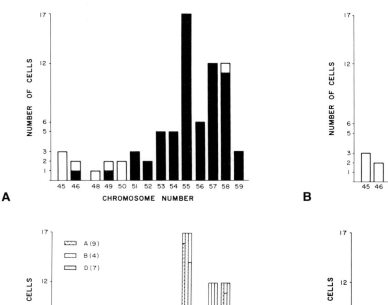

118 Karyotypic diversity in the cancerous effusion of a patient with a mucocele of the appendix and extensive pseudomyxoma peritoneii. **A.** Distribution of marker chromosomes (M_1 and M_2) in the cells with various chromosome numbers, shown as black bars. Most of the cells shown in undarkened areas contained only M_1. **B.** Cells with identical karyotypes among cells with the same chromosome number; three karyotypic groups within population had a modal number of 55 chromosomes and those with 57 also had three sets of cells with identical karyotypes. **C.** Distribution of chromosomes of groups A, B, and D among the cells with 45–59 chromosomes. Note that groups A and D have more than a diploid number of chromosomes. **D.** Dotted bars indicate the number of cells containing the following combination of chromosomes in the groups indicated: A–9; B–4; D–7; M_1 and M_2. Of the 72 cells, 32 (44.4%) had this combination.

(Sandberg and Yamada, 1966.)

sis, similarities are found which are common to most if not all metaphases within the sample. One or two diploid cells may be present and these can usually be presumed to be nonmalignant stromal cells; however, substantial numbers of diploid metaphases have been found in tumors at some sites, such as the bladder and corpus uterus.

The most noticeable feature of tumor metaphases is the presence of one or more markers (Atkin, 1960*b*), especially when these are larger than the largest normal chromosome. Identical markers are often found in virtually every metaphase. On detailed examination, it may be apparent that the number of chromosomes of each type tends to be the same in every metaphase and twice this number in those with doubled complements. Even though the similarities are usually close enough to indicate a clonal relationship, minor variations or at least uncertainties are common. Where a reasonable number of metaphases with a modal number are available, they mostly appear to have the same karyotypes; on the other hand, it may seem that everyone is different (Figure 118). Perhaps it is significant that the differences or uncertainties tend to involve chromosomes with near-median centromeres rather than those that are subtelocentric or acrocentric. However, in spite of the uncertainties, it is highly probable that the degree of variation within a tumor, such as minor variations in chromosome number and karyotypes among metaphases with the same chromosome number, or the degree of polyploidization, is itself a variable characteristic that may have a significant relationship to the behavior pattern of the tumor. For example, data obtained on carcinoma of the large bowel suggest that the greater the variation in karyotypes, the worse the prognosis.

Most tumors can be characterized by a modal chromosome number which, however, may fall anywhere within a wide range. Nevertheless, at most sites, the modal chromosome numbers tend to fall into one or another of two or more less well-defined groups, one situated around the diploid level and the other in the triploid-tetraploid range. Within the diploid range, chromosome numbers may vary from a little below 40 to 50 or so; tumors with modal chromosome numbers of 37 and 38 are seen at some sites, such as the ovary and breast, but modes below this are exceptional, even though a malignant melanoma with a modal number of 29 or 30 has been reported.[3] At the upper end of the scale, malignant cells in an effusion from a patient with carcinoma of the breast had a mode of 133 chromosomes[4] and a primary seminoma had a mode of 156 chromosomes.[5]

Certain features of the karyotypes should be noted which are to some extent independent of the modal chromosome number. First is the presence of one or more markers (Figure 119): markers are very commonly and, perhaps universally, present in tumors with hypodiploid chromosome numbers, whereas they are frequently but less regularly present in other tumors. In one study (Atkin, 1974a), markers were present in each of 11 hypodiploid ovarian carcinomas and in 18 of 21 with modal numbers of 46 or more; they were present in each of 12 hypodiploid cervical carcinomas but in only 8 of 18 tumors at the site with modes of 46 or more.

The second feature of tumors is a tendency

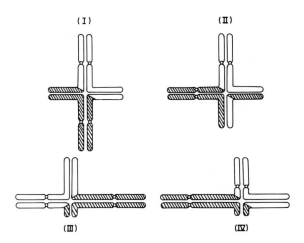

119 Schematic presentation of the four types of homologous exchanges between chromosomes: (I) alternate symmetrical; (II) adjacent symmetrical; (III) alternate asymmetrical; and (IV) adjacent asymmetrical. See text for explanation of how different abnormal chromosomes may arise through one of these processes.

toward certain changes in the relative numbers of chromosomes that can be fitted into each of the normal groups or subgroups. In particular, there are fewer than the expected number of chromosomes in groups B, D, and G; that is, the numbers of chromosomes in these groups are fewer in proportion to the total number of chromosomes present than in cells with normal karyotypes. Conversely, there are often more #3, group C, #16, and group F chromosomes than expected; the tendency, on the whole, appears to be toward an increase in chromosomes with medially placed as compared with those with distally placed centromeres.[6] Even though commonly group B, D, and G chromosomes may be missing, it is not known whether this represents actual loss of the chromosomes or transfer of the material to new chromosomes after structural rearrangements. Banding techniques should certainly shed considerable light on these changes. It should be stressed that each individual tumor varies with respect to the direction and extent of any change in the number of chromosomes of a particular type; the tendency is usually clear, however, when the average extent to which chromosomes of each type is under- or overrepresented in a series of tumors is calculated. Thus, the numbers

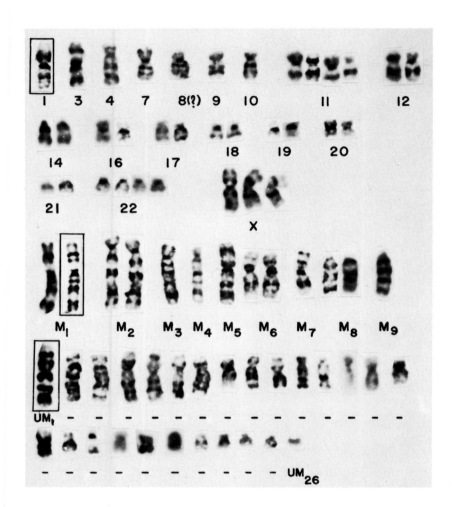

120 Karyotype (G-banded) from a breast cancer cell. Bracketed chromosomes came from another metaphase of the same case. Nine of the markers could be identified (Kakati, Hayata, et al., 1975). An extremely large number (UM_{1-26}) of unidentified markers was present in this complicated case.

of groups B, D, and G chromosomes in a series of 12 carcinomas of the cervix, 11 of the large bowel, and 20 of the ovary were on the average reduced 62–97% of the expected number.[7] The postulate that an increase in the proportion of #16 chromosomes is of special significance in human cancer was based on a semiautomatic system of chromosome analysis,[8] but the method used has since been criticized and the results not confirmed.[9]

The bimodality of the distributions of modal chromosome numbers and DNA values of the tumors suggest that chromosome doubling has occurred at some stage in the evolution of a proportion of the tumors. However, other interpretations are possible, as suggested by the finding that tumors in the higher range are commonly near-triploid rather than near-tetraploid.

Significance of Chromosome Abnormalities in Cancer

The great majority of human cancers exhibit gross (i.e., visible) chromosome changes and the pattern within each tumor indicates a clonal relationship. Gain or loss of individual chromosomes may be a consequence of nondisjunction, and the formation of new chromosomes a consequence of breakage followed by reunion.

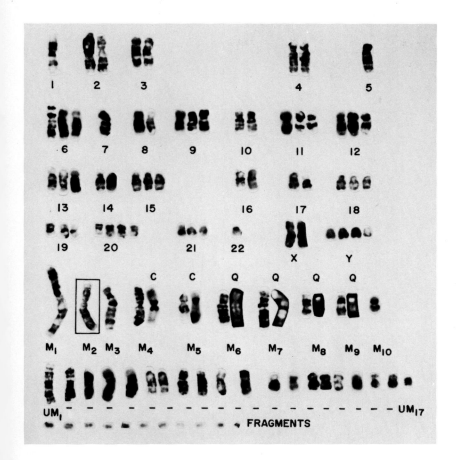

121 Karyotype (G-banded) from a lung cancer. A large number of identifiable markers was present (M_{1-10}) as well as 17 unidentifiable markers and fragments. The bracketed M_2 and those chromosomes with C- and Q-banding were from another metaphase of the same case.
(Kakati, Hayata, et al., 1975.)

Breakage may involve chromosomes or chromatids (Fraccaro, Mannini, et al., 1965); the structural changes that follow may be intra- or interchromosomal. An alternative hypothesis suggests selective endoreduplication of whole chromosomes or segments of chromosomes.[10]

Markers may be several times larger than the largest and smaller than the smallest of the normal chromosomes. Naturally, only those abnormal chromosomes that deviate significantly from normal are recognizable as markers by conventional cytogenetic staining methods. Banding studies in my laboratory have shown that it is possible to identify *all* the abnormal chromosomes in some tumors, but not in others (Figures 120–122). Furthermore, markers of identical derivations may be present in tumors of different organs and a variety of markers characterize tumors of the same organ (Figures 123–126). The differences may involve length, centromere position, and other features such as secondary constrictions. Ring chromosomes are encountered from time to time and may be present in tumors of common sites including those of the large bowel, cancer of the cervix, and others. Dicentric markers may be difficult to distinguish from chromosomes with prominent secondary constrictions;[11] it has been suggested that dicentric chromosomes with short intercentric distances, which are difficult to recognize, may occur in tumors as a result of translocations, e.g., between group G chromosomes.[12] Metacentric chromosomes may be isochromosomes; an isochromosome

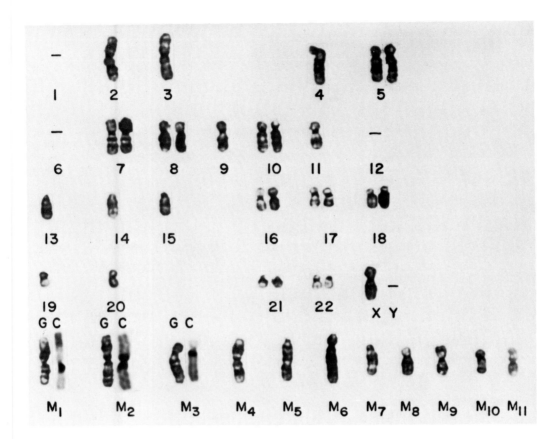

122 A G-banded karyotype of a malignant melanoma cell with missing chromosomes in most of the groups and containing 11 marker chromosomes (M_1–M_{11}), shown at the bottom of the figure (G = G-bands; C = C-bands), all of which were identified as to their origin. This case demonstrates the relative ease with which marker chromosomes can be identified in some tumors, even when the number of such markers is large. Unfortunately, this applies to a small percentage of all tumors.

for the long arm of #17, for example, may regularly appear in the course of CML, and abnormal medium-sized metacentric chromosomes with even and equal fluorescence in both arms have been observed in lymphomas after staining with quinacrine.[13] They are rare in tumors of adults, although they have been described in carcinoma of the ovary and observed in cancer of the rectum. In any case, though not always successful, banding analysis utilizing a number of techniques (G, Q, C, R, and T) offers a means of identifying the origin and nature of markers in human tumors and should obviate speculation and controversy regarding markers.

High chromosome numbers may be achieved by a doubling of the complete set through several mechanisms (e.g., nondisjunction, endoreduplication). Tumors with modal numbers in the triploid or tetraploid region often have markers that are not present in duplicate; this may include the presence of an uneven number of chromosomes in one or more of the normal groups, indicating that changes must have occurred subsequent to the doubling.[14] The commonly observed near-triploid numbers could be achieved by a series of nondisjunctions or by a combination of chromosomal loss through nondisjunction and a complete doubling. Other mechanisms might,

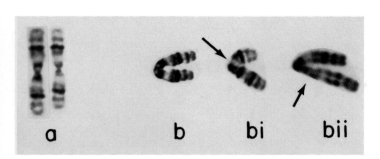

123 1q markers: **a.** Pair of i(1q) from a single metaphase spread originating from a highly undifferentiated cancer. **b.** Identical isochromosome of the long arm of #1 [i(1q)] from an ovarian tumor. **bi** and **bii.** Further changes of the marker shown in b with loss of dark-banded region q12 and one of the arms in bi and an insertion of a light banded q21 in one of the arms in bii.

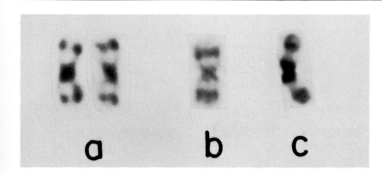

124 i(17q) markers: **a.** pair of markers i(17q) from a metastatic cancer of the prostate. **b** and **c.** One copy of the identical marker i(17q) from a highly undifferentiated cancer and an ovarian tumor, respectively.

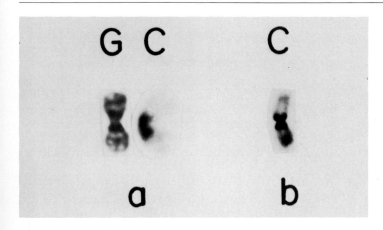

125 i(16q) markers: **a** and **b.** Markers from the ascitic fluid of two lung cancers (G = G-bands; C = C-bands).

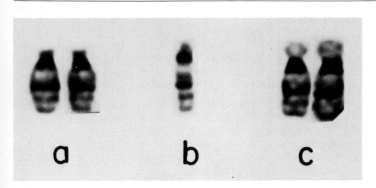

126 del (1)(p13) markers. **a.** Pair of markers [del(1)(p13)] obtained from a cancer of the breast. **b.** One marker in an ovarian cancer. **c.** t(1;3)(p22;p25) markers in a case of ovarian cancinoma.

however, account for near-triploid complements; in cultured marrow from a patient with CML and localized lymphadenopathy it was noted that among the pseudodiploid and related pseudotetraploid leukemia cells a near-triploid cell that could have been derived from a pseudodiploid cell by duplication of a single (abnormal) haploid set was present.[15] Sporadic triploid cells have been observed in blood and fibroblast cultures derived from patients with congenital abnormalities and in fibroblasts derived from placenta; their origin remains speculative, however. Fusion of tumor cells with host cells or among themselves with a consequent increase in the chromosome number may be a mechanism whereby the degree of malignancy of a tumor may be increased.[16]

Well-differentiated tumors at some sites may have minimal chromosome changes and contain cells with normal karyotypes, e.g., possible differences between seminomas and malignant teratomas of the testis with respect to the proportion of chromosomes in the various normal groups. Also, there may be differences in the proportions of tumors at the various ploidy levels according to the site (for instance, the frequency of hypodiploid tumors in the breast and ovary and their absence in the testis). Large markers have frequently been observed in tumors at sites such as the breast, ovary, and testis; despite the fact that markers of similar appearance are present in different tumors, there is no real evidence for the existence of markers specific to tumors of any given site or groups of sites. The presence of large markers may be related to the chromosome number rather than to the sites. Perhaps detailed comparison among tumors of various sites will reveal subtle differences in the karyotypes similar to those described for seminomas and teratomas.

In *hypodiploid* tumors the relative constancy of the various features characteristic of malignant cell karyotypes is impressive; markers often as large or larger than group A chromosomes and the reduction in the number of chromosomes in groups B, D, and G are notable. Such karyotypes apparently do not occur in preinvasive lesions, and they may, therefore, be indicative of malignancy.[17] Atkin (1974a) has made the interesting observation that although hypodiploid tumors of the ovary and other sites may have some spread in the chromosome number immediately around the mode, this very rarely extends above 48. *Pseudodiploid* tumors show the same characteristic features of their karyotypes, though perhaps less constantly. *Hyperdiploid* tumors (in the range of 47–57 chromosomes) sometimes show minimal changes. There may only be additional chromosomes or gains combined with losses, and markers are often absent. Frequently, though not so consistently, there are relatively few B-, D- (may be present in excess), and G-chromosomes. The distribution of chromosome numbers often has a positive tail, analysis of the hypermodal cells showing a fairly consistent pattern of additional, with perhaps some lost, chromosomes. Thus, in a carcinoma of the rectum closely related karyotypes with 50–55 chromosomes were found, but whereas the mode in the one region of the tumor was 51, in another region it was 54 or 55 chromosomes. Tumors in the *triploid* and *tetraploid* ranges frequently show the same changes in the proportions of the normal chromosomes as hypodiploid tumors, and markers are usually, although by no means constantly, present. In some tumors, particularly those with modes near or above the tetraploid level, a few cells with related karyotypes but only half the number of chromosomes may be found; such near or hyperdiploid cells may predominate in other regions of the tumor. As already mentioned, where markers are present in near diploid cells, these markers are generally present in duplicate in near-tetraploid cells from the same tumor. However, many tumors with high modes, particularly in the hypertriploid or hypotetraploid regions, lack a minor component with half the number of chromosomes; in these tumors the markers are usually not present in duplicate.

When pathological and clinical criteria are equivocal, chromosome studies may be of diagnostic value, because the presence of a clone of cells with an unequivocally altered karyotype might, in the present state of our knowledge, provide a strong indication that a malignant condition is present. Clones with less extensive changes should be interpreted more cautiously, even though they might suggest that a (pre)malignant lesion is present. These include clones in which the only abnormalities are a single trisomy or in which the karyotype is pseudodiploid with evidence of the presence of one or more balanced translocations.

Sex-chromatin may indicate a chromosome

change, when it presents an anomalous pattern, for example, triple sex-chromatin.[18] The presence of sex-chromatin in a malignant melanoma in a male patient whose metastases regressed[19] and in a male child with BL in which most of the tumor cells had a female diploid karyotype,[20] raises questions concerning the origin of the sex-chromatin and the possible relation to the unusual clinical courses of some of these tumors. A nuclear protrusion or projection in interphase has been related to large marker chromosomes present in cancer cells.[21] These protrusions are more readily apparent in preparations from pleural or ascitic fluids or in cultured materials. However, these projections may also depend on the total nuclear volume and may be seen more often in near-diploid or hypodiploid nuclei than in polyploid ones. In some cases with nuclear protrusions neither a large marker nor any other abnormal chromosomes have been demonstrated. However, a direct correspondence between extra-large markers and nuclear projections has been demonstrated with banding (Castoldi, Grusovin, et al., 1976).

The nature of the chromosome involved in nuclear protrusions was studied with C-banding and the authors demonstrated interstitial heterochromatic regions on the long arm (when present), but not the centromeric heterochromatin, which frequently gave rise to chromocenters in the protrusions (Atkin and Baker, 1979a). The four tumors studied included two fully differentiated cancers of the ovary and two transitional cell cancers of the bladder. Each of the tumors contained nuclear protrusions as well as an abnormal chromosome whose long arm approached or exceeded the length of the whole of #1. The constitution of the abnormal chromosome varied from case to case. The authors believed that the findings left little doubt that protrusions are formed by part of the long arm of the markers. The striking configuration of protrusions, projections in otherwise smoothly outlined nuclei, indicates a change in the normal pattern of the interphase nucleus, at least as far as the nuclear membrane is concerned, but whether in itself is of any significance functionally is unknown.

In general, the length of the protrusions seems to bear a relationship to the length of the marker's long arm, a factor that together with others may also influence the incidence of protrusions. Even though protrusions may sometimes be seen in over half the nuclei, their incidence is usually somewhat lower than this. Possibly, factors such as the position in the cell cycle and the metabolic activity of the cell influence the appearance of protrusions and may result in different incidences in different regions of the tumor.

Lagging of the abnormal chromosome at anaphase may be at least a contributory factor in the formation of protrusions. Prominently protruding chromosome arms can be seen in anaphase and telophase as well as in metaphase in histological sections of a variety of malignant tumors and in a carcinoma in situ of the cervix shown to have large abnormal chromosomes (Brandão and Atkin, 1968; Atkin and Brandão, 1968).

Protrusions serve as a useful indicator that an aneuploid clone having a large abnormal chromosome and, although per se not indicative of malignancy, may provide evidence of a neoplastic condition.

The significance of the presence and number of sex-chromatin bodies (both X and Y) in human tumors has been reported on by a number of investigators.[22] Correlations between the number of sex-chromatin bodies, progression of the malignancy (particularly metastases), and prognosis of the cancer have been attempted. The presence of a Y-body in the preponderant number of male cancers has been reported by Atkin (1973a), including an occasional female twin. However, the author argues, and I strongly concur with him, that the nature of the fluorescent body (thought to be the Y) must be confirmed with chromosome analysis. A missing Y as the *sole* chromosome abnormality in a human cancer has not been reported, a missing Y always being accompanied by other karyotypic changes in the tumors.[23]

Chromosome Changes in Malignant Transformation

Carcinogenic agents, whether they be viruses, ionizing radiation, or chemicals, generally cause chromosome breakage; certain inborn anomalies in which there is an excessive liability to chromosome breakage carry an increased risk of malignant disease. There is, therefore, a link between chromosome breakage and cancer; however, a key question concerns the sequence of events that occurs between the initial chromosome damage and the acquisition of malignant properties by the cell or its descendants. In man, at least, it seems that this is usually a lengthy process of sequential changes (clonal evolution).[24] There is the possibility, however, that chromosome breakage, per se, may be a "phenotypic" expression of a more fundamental genetic defect, probably at the gene level, and that the chromosomal changes observed bear little relationship to the development of malignancy in these patients.

A further question is whether and how the aneuploidy of malignant cells directly determines their abnormal behavior. A number of authors have recently discussed this and similar problems with particular reference to human cancer.[25] Even though aneuploidy can safely be regarded as a characteristic of most forms of human cancer, and degrees of aneuploidy bear some relation to the degree or state of malignancy, it is at present impossible to answer how an unbalanced chromosome set is related to the properties of malignancy or premalignancy.

In many experimental animal tumors, the presence of normal chromosome complements in the early stages of tumor development suggests that chromosome changes are only concerned with the "progression" of tumors to higher degrees of malignancy. In man, there indeed appears to be "progression" that is related to chromosome changes, particularly in the early stages of the tumor. Once established, human cancers are usually already aneuploid, with highly unbalanced karyotypes; serial studies indicate that they tend to retain the same chromosome complements and undergo only minor changes, somewhat comparable to the karyotypic differences found in different regions of the same tumor. Moreover, in terms of changes in the histological picture progression in the direction of greater neoplasia, increased growth rate, and greater liability to metastasize is the exception rather than the rule; recurrences are usually similar to the primary tumor in these respects.

Studies on experimental tumors in rats and Chinese hamsters, which were initially diploid, may be of significance: there is apparently a regular and predetermined sequence of chromosome changes in both species which, however, varies (i.e., different chromosomes are involved) according to whether a virus or a chemical carcinogen is the inducing agent.[26] This finding raises the exciting possibility that different pathways of chromosomal evolution in human tumors may be related to different causative agents.

Normal or apparently normal karyotypes found in some human tumors raise several questions. Obviously, some small change may not be discernable with present methodologies. On the other hand, at times quite gross changes may have occurred that have not altered the appearance of the karyotypes, although this may be revealed by refined banding techniques. The presence of diploid cells and those in which the only abnormality is a single trisomy in some early, usually well-differentiated tumors of the corpus uteri and bladder raises the possibility that these cells, though tumor cells, are not very invasive or subject to metastasis. The malignancy of these tumors on histological and clinical grounds is in fact often in doubt, and even when it is not in doubt the situation might be that some of the cells, e.g., those with diploid complements or other minor changes, constitute the remnants of precursors of the aneuploid malignant cell line which are still present in the tumor.

An essential feature of aneuploidy in malignant cells might be that it constitutes the attainment of an unbalanced combination of chromosomes (or parts of chromosomes) that ensure their ability to grow and divide and at the same time to be released from normal controls; however, the ability of cells or groups of cells to differentiate may be retained. The frequency of markers in the karyotypes of cancer cells suggests that the balance necessary for malignancy may require chromosome breakage and reunion rather than the gain or loss of whole chromosomes. Perhaps in the course of human evolution, selective forces have operated to ensure that dangerous combinations of genes may result only from unlikely events, e.g., particular chromosome rearrangement as opposed to the gain or loss of chromosomes.

Banding techniques have already proved useful in the study of abnormal chromosomes in solid tumors, e.g., lymphoma and meningioma.[27] When the new techniques for chromosome identification have been applied to many more tumors, including those that have so far received little or no study (e.g., basal cell and prostatic carcinoma), it is hoped that questions about the relationship between aneuploidy and cancer in its many forms, including their developmental stages, can be answered in more detail.

DNA Measurement in Tumors

As discussed elsewhere, the measurement of the DNA content of tumor cells, even with the most sophisticated techniques, although capable of revealing gross deviations from diploidy, has several drawbacks which make this approach to karyotypic analysis less than optimal. Foremost is the inability to ascertain monosomy or trisomy of the smallest chromosomes (#21, #22, and the Y) or deletion of chromosomes involving too small an amount of DNA to be reliably monitored by DNA measurement. When a cell is in the S-period, part or the whole of its DNA is replicated, and DNA measurements may under these conditions yield spuriously abnormal values, even in diploid cells. Nevertheless, in spite of these shortcomings of DNA measurement, this approach has been utilized in human neoplasia[28] and is likely to be used for many years to come.

Changes in Blood Leukocytes

A number of publications have described instances in which chromosome changes in the blood leukocytes of patients with various cancers have been noted.[29] Whether these are reflective of the malignant state and/or effects of therapy has yet to be established.

Chromosomes and Causation of Human Cancer and Leukemia

Even though general agreement seems to exist that the neoplastic process in mammalian cells is caused by alterations in the function of chromosomal chromatin (DNA), the possibility has not been ruled out that the cancerous state may also be caused and abetted by abnormalities resident outside the DNA and nucleus, i.e., in the histones and other proteins associated with the chromosomes, in the nucleolus, the elements present within the cytoplasm, or the cell membrane. Furthermore, agreement exists that

References to footnotes 1–29, pages 427–439, appear on pages 456 and 457.

minor molecular rearrangements within genes, certainly not detectable visually even with the most sophisticated microscope techniques presently available, may lead to the development of cancer and leukemia. Hence, in discussing the role of chromosomal changes in the causation of cancer and leukemia in human subjects and animals, it must be realized that one is considering only those cytogenetic abnormalities that are visibly recognizable and that consist primarily of numerical and morphological deviations of the chromosomes.

In the present chapter chromosomal changes in human neoplasia will be interpreted as they relate to the causation and progression of cancer and leukemia, though appropriate observations in animals will be presented to support (or not to support) and supplement the findings in human subjects.

This section will further develop concepts previously discussed regarding the role of chromosomal changes in cancer causation. The general thesis is that the development of cancer and leukemia in human subjects is usually not associated with or the result of karyotypic abnormalities, though it is possible that some cases may develop such cytogenetic aberrations concomitant with, or as a result of, the neoplastic state and that chromosomal changes may play a role in the progression and biologic behavior of tumors. Only a small number of relatively specific entities of human cancer and leukemia may be caused primarily by chromosomal changes.

The consistency and morphological integrity of the chromosomal set in mammalian cells are guaranteed by the remarkable process of DNA replication and cellular mitosis. As a result of these events the amount of DNA per cell, as reflected in the relative stability of the chromosome number and morphology, is kept essentially identical in all somatic cells. Thus, when the chromosome number is determined in normal cells, it is found to be constant ($2n = 46$ in the human, 40 in the mouse, 42 in the rat, etc.). Furthermore, the structure and relative length of the chromosomes are remarkably similar from cell to cell and from tissue to tissue, allowing rather facile classification of the chromosomes, at least in the human, into fairly specific groups. This has afforded the recognition of unique karyotypic abnormalities in various diseases and the deciphering of certain abnormal states on the basis of specific cytogenetic anomalies.

Inasmuch as studies in animals lend themselves to a more direct examination between cause and chromosome changes in tumors, a short review of the findings in animals is given before the data in human neoplasia are discussed.

The Chromosomes in Animal Tumors and Leukemia

Because chromosomal changes in human neoplasia are difficult and often impossible to study throughout the genesis of the cancer or leukemia, animal studies are essential to obtain cytogenetic information regarding this facet of carcino- and leukemogenesis. Spontaneous tumors and those induced by a number of different agents in animals are used to obtain data on the importance of karyotypic changes in the development and induction of cancer and leukemia in a variety of species.

Studies of chromosomes in various animal neoplasms have been performed to elucidate their role in carcinogenesis, because the chromosomal changes are studied only after they have been well established in human cancer and leukemia. There is some possibility in animal studies that such changes can be found during the process of initiation and progression of these diseases. The effect of carcinogens on chromosomes can also be studied in the latter. In the following sections a comparison is attempted of the chromosomal changes in various neoplasms, mainly primary ones, induced by various agents.

Spontaneous tumors and leukemia. There have not been many reports of chromosome studies on spontaneous tumors or leukemias of animals, particularly in recent years. This is probably because such studies, similar to those in human cancer and leukemia, cannot contribute much to the understanding of the role of chromosomal changes in the causation of these states, as the states are fairly well established when the studies are undertaken. Nevertheless, early observations indicated that mammary carcinoma and leukemia of mice were accompanied by a diploid or hyperdiploid chromosome constitution and that bovine LS was characterized by either diploidy or aneuploidy, the latter being either hypo- or hyperdiploid.

A number of leukemias and lymphomas in the dog were found to be diploid (78 chromosomes) (Whang-Peng, 1969). A RCS in the golden hamster was observed to be diploid after many passage generations, though it ultimately became aneuploid (Haemmerli, Zweidler, et al., 1966). The chromosomes were found to be essentially diploid in one study of Marek's disease of chickens (Owen, Moore, et al., 1966; Bloom, 1970). One author (Kawamura, 1965), after studying the chromosomes in spontaneous and induced leukemia of mice, stated that "the aberrant karyotypes that can be detected by the presently available technics are not essential for the development of leukemias." Friend virus-induced ascites tumors in rats and leukemias in mice have been observed to be diploid as often as aneuploid (Sofuni, Makino, et al., 1967; Sasaki, Mori, et al., 1970), as was true for a number of clinically and radiation-induced tumors in animals (Yosida, 1975; Yosida, Imai, et al., 1970). As in advanced human malignancy, established animal tumors are characterized by a variable chromosome picture (Hare, Yang, et al., 1967).

Tumors due to nutritional or endocrine imbalance. The findings presented in Table 79 indicate that the least malignant thyroid tumors were diploid, and hence, resembled in

Table 79 Chromosomal findings in thyroid tumors of rats

Type of tumor	Conditions of growth	Rate of growth	Chromosome findings
Dependent	Iodine deficiency and thyroidectomy	Slowly growing	Diploid
Transitional	Growth faster in iodine deficiency and thyroidectomy	Slowly growing	Hypodiploid; missing #15
Autonomous	No dependence on iodine deficiency and thyroidectomy	Fast growing	Hyperdiploid; missing #15

some respects normal thyroid tissue. The more undifferentiated the thyroid tissue became, the more complex were the chromosomal anomalies. However, convincing evidence was not found that the chromosomal changes could just as well have been the result of the underlying neoplastic process, the metabolic changes, or the iodine deficiency.[1]

Little information is available on chromosomal findings in early thyroid tumors in the human. Frank cancers of the thyroid have invariably been found to be aneuploid, without any consistent karyotypic picture and with each cancer having its own unique chromosomal profile. No two tumors had identical chromosomal changes. Even though the cytogenetic findings in rats with thyroid tumors presented above (Table 79) were relatively more homogeneous than the findings in human cases, they were, nevertheless, variable within the groups studied. In addition, it is probable that the relative hereditary homogeneity of the rats used led to a somewhat more stable and characteristic karyotypic picture than that obtained in human tumors, the latter possibly as a result of the extremely variable heredity of human subjects.

There is no absolute proof that nutritional or endocrine imbalance causes cancer of the thyroid in human subjects, although this area is being intensely investigated in a number of laboratories. It is possible that a number of human tumors may result from endocrine or metabolic imbalance, e.g., adrenal tumors possibly due to hypothalmic-pituitary imbalance, ovarian tumors due to pituitary malfunction, or cancer of the prostate due to hormonal factors. However, no definitive proof exists that there is a relationship between the hormonal or metabolic upheaval and development of any cancer. Nevertheless, when cancers or benign tumors develop in these tissues, the chromosomal characteristics are not specific or unique, and the malignant tumors tend to have a variable karyotypic picture. The benign tumors are invariably accompanied by chromosomal diploidy.

Chemical carcinogens. A host of chemicals has been shown to cause cancer and leukemia in a number of animals. Studies have also been performed on the role of chromosomal changes in the production of these tumors; the general opinion is that most, if not all, experimental tumors start out as diploid. Karyotypic changes become evident either with the biological progression of the tumor, transformation to a more malignant phase, or in the metastases of the tumor. Agreement also appears to exist that chromosomal abnormalities are not required for chemical carcinogens to induce tumor development or growth and that tumors with quite different karyotypes may exist at the same time in the same organ of an animal.

The mechanism for the production of a chromosomal change by chemicals, whether they are carcinogenic or not, is poorly understood. Heterochromatically altered regions in chromosomes, generated during carcinogen-induced malignant transformation, may be responsible for nondisjunction of chromosomes, leading to nonrandom rearrangement of a chromosomal complement with neoplastic, biochemical, and cytological characteristics. An example of such a heterochromatic region is that of a C1 chromosome in DMBA-induced leukemia in rats. The importance of gaps and deletions in chromosomes induced by chemicals followed by changes in ploidy due to nondisjunction and/or duplication of chromosome sets has also been stressed.[2]

Even though many chemicals play a role in the causation of human cancer and leukemia, only a few have been strongly implicated in these diseases. Benzene is probably a cause of AL in some subjects exposed to this chemical. Chromosomal changes, consisting primarily of gaps, breaks, deletions, translocations, and others, have been described in the cultured lymphocytes and marrow cells of subjects exposed to benzene, particularly during the severe anemia phase of the toxicity.[3] When leukemia supersedes, the cytogenetic findings do not differ from those observed in other human ALs. Many other chemicals have been shown to produce chromosomal changes of a nonspecific nature; whether these changes play any role in the development of neoplasia in human subjects is a moot point, because only a few chemicals have been imputed as carcinogenic.

Thus, a survey of the data (Table 80) and a review of the literature of experimentally induced tumors in animals, as well as the data pertaining to human subjects, appear to indicate that the genesis of the neoplastic state by these chemicals does not require any alteration

Table 80 Chromosomal changes in tumors induced by chemical carcinogens

Chemical	Animal	Neoplasia	Chromosome findings
3-Methylcholanthrene	Mice	Subcutaneous sarcoma	Mostly diploid, some hyperdiploid
	Wistar rats	Leukemia	Diploid
	Golden hamsters	Primary tumors (subcutaneous)	Diploid, aneuploid, heteroploid
7-12-Dimethylbenz(a)anthracene	Mice	Thymic lymphoma	Hyperdiploid
	African mice	Sarcoma	Pseudodiploid, hypo-, and hyperdiploid
	Rats	Leukemia	Trisomy of specific chromosomes (40% trisomy of C1, some of A6)
N-2-Fluorenylacetamide	Rats	Hepatic nodules	Diploid
		Hepatoma	Diploid, subtetraploid
3'-Methyl-4-dimethylaminoazobenzene	Rats	Hepatoma	Diploid, hypo-, and hyperdiploid
4-Nitroquinoline 1-oxide	Golden hamsters	Malignant transformation of embryo cells	Diploid, near tetraploid
Urethane	Mice	Thymic lymphosarcoma	Diploid, some cells with 41 chromosomes (hyperdiploid)
Freund adjuvant	Mice	Plasma cell tumor	Diploid, hypo-, and hypertetraploid*

*Yosida, Imai, et al. (1970); for all other references regarding this table, see Sakurai and Sandberg (1974).

of the chromosomal set from diploidy. It is possible that some chromosomal changes are involved in the causation of neoplasia in animals, but the overwhelming evidence appears to indicate that the development of such a state can occur just as readily in the presence of a diploid set of chromosomes, without any evident alterations.

Oncogenic viruses. Many reports exist regarding the chromosomal changes in virus-induced tumors or leukemias in animals. Generally, no outstanding abnormalities have been observed in conditions induced by RNA viruses. The abnormalities are more striking in tumors or leukemia induced by DNA viruses.

RNA viruses. Even though there are no significant alterations in the chromosome number or karyotypes in mice infected with *Friend leukemia virus*, apparently a significant correlation between the number of secondary chromosomal constrictions and progression of the disease was observed. As the weight of the spleen increased, the number of secondary constrictions per metaphase also increased. Furthermore, there was an inverse relationship between the increased weight of the spleen and the number of diploid cells with no secondary constrictions. Inasmuch as the secondary constrictions were observed early in the disease and appeared to progressively increase throughout the course of the study, it was felt that the changes had significant mutational importance in relation to the development of leukemia.

Several interesting studies have been performed on the chromosome constitution in mice infected with *Rous sarcoma virus* (RSV). The chromosomes were examined in the primary tumors induced by the Schmidt-Ruppin strain of RSV in Chinese hamsters, such an analysis being performed on 42 tumors in 29 animals.[4] Twenty-eight different karyotypes were found in 68 tumor stem lines and side lines. About 40-52% of the stem and side lines had a normal karyotype. The most common karyotypic abnormality observed was the presence of extra chromosomes. The proportion of normal cells in any given tumor correlated inversely with the age of such a tumor. The conclusion reached was that malignant transformation may be initiated without any visible chromosomal changes.

Earlier studies in vitro on the effect of RSV on cultured Chinese hamster fibroblasts indicated a higher incidence of chromosomal breakage than observed in control cells. Apparently, the spontaneous and virus-induced breakages were distributed nonrandomly among the cells, the chromosomes, and chromosomal regions. Most of the breaks were localized in three specific regions of the chromosomes. Furthermore, a correlation was demonstrated between the pattern of chromosomal breakage and secondary constrictions. The most significant difference, however, was found between the patterns of distribution of spontaneous and virus-induced chromosomal alterations.

More recent studies[5] dealt with the cytogenetic findings in 50 primary Rous sarcomas induced in the rat. About 80% of the sarcomas had a normal diploid stem line, although about half of these had some minor alterations of the chromosomal number in the cell population. These alterations most commonly consisted of hyperdiploidy, with trisomy of one group or another being the predominant karyotypic anomaly. Pseudodiploidy appeared to occur as the second most common karyotypic aberration. The karyotypic changes during the development of heteroploidy appeared to be of a nonrandom nature. One extra telocentric or subtelocentric chromosome or both were found in 76% of all heteroploid cells and in about 70% of all variant cells. There appeared to be no relationship between the latent period of tumor development and the chromosomal findings. With increase in age of the tumor, the number of diploid cells decreased, and this was associated with a progressive histological anaplasia of the tumors.

Studies of the chromosomal constitution of 16 metastatic Rous sarcomas in rats revealed the close karyotypic relationship between the primary sarcomas and their metastases.[6] The outstanding difference was an accelerated chromosomal progression in the secondary tumors, this progression being associated with a decrease in the histological maturity of the tumors. In 56% of the metaphases, heteroploid stem lines were found, as compared with 23% in primary tumors. Furthermore, about one third of the primary sarcomas with normal diploid chromosome constitutions had some cytogenetically abnormal cells, whereas about two thirds of the metastatic tumors had such abnormal cells.

DNA viruses. As indicated previously, the tumors induced by DNA viruses have chromosomal abnormalities more consistently than those induced by RNA viruses. Thus, it has been shown that the chromosomes of eight primary tumors induced by adenovirus type 12 in Syrian hamsters were characterized by a karyotypically heterogeneous cell population, consisting of aneuploid, pseudodiploid, and polyploid cells. These changes were accompanied by marker chromosomes in some of the tumors and by a relatively high incidence of chromosomal and mitotic irregularities, including chromatid and isochromatid breaks, gaps, overcontracted chromosomes, and chromosome fragmentation. Each tumor contained many clones of cells with abnormal karyotypes, but a karyotypically predominant abnormal stem line was not present. It was speculated that the chromosomal instability was a reflection of the host genetic alteration associated with the viral-induced antigenic changes.

To date, there is no cogent or definitive evidence that any human cancer or leukemia is caused by a virus. Hence, one cannot ascertain whether any of the karyotypic abnormalities observed in human neoplasia have been contributed to or are a result of viral effects. The chromosomal changes observed after infection with various viruses (some of which may in the future be shown to be oncogenic) have been observed primarily in cultured lymphocytes. These changes, which may persist for relatively long periods of time, have consisted primarily of gaps, breaks, fragmentation, and occasionally more serious alterations in chromosomal morphology. To my knowledge, no change in ploidy or the production of consistent markers has been demonstrated to occur as a result of or concomitant with such a viral infection. The chromosomal changes just described have also been observed in early lesions, not necessarily cancerous ones, of the uterine cervix in human subjects, and some workers have related these changes to the presence of virus in the tissue, with the possibility that such a virus may play a role in the oncogenic process leading to carcinoma of the cervix. No definite correlation, however, has been shown to exist between the presence of such a virus, the chromosomal changes in the cervical cells, and the subsequent development of neoplasia.

Contagious tumors. An allegedly contagious RCS of the Syrian hamster has been studied after the subcutaneous implantation of such a tumor or after fragments of tumor were fed to animals, and in tumor-free animals who were caged with tumor-bearing ones. The stock tumor, maintained by serial subcutaneous passage in male hamsters, had a sharp modal number of 51, including a small marker. Tumors developing in animals of either sex or induced by several modes of implantation were shown to have karyotypes identical to those of the parent tumors (with only one exception).

The chromosomal make-up of 17 cases of venereal tumors of dogs obtained in distantly separate localities and at different times in Japan have been investigated. All the tumors had stem lines with the chromosomal number of 59, instead of the normal number of 78, with one tumor having 58 chromosomes. The karyotypes of the various tumors were very similar and included a number of markers. In these two examples, the transmission of intact tumor cells from animal to animal is more likely than the induction of the same chromosomal changes in a succession of primary tumors.

Considerable attention has been paid to the Sticker venereal sarcoma whose existence has been known for about a century and which spreads naturally by the transfer of cellular material. The tumor, probably a lymphoma in nature, is particularly common in Japan from where many of the studies have emanated.[7] The chromosomes were first studied over 20 years ago (Oshimura, Sasaki, et al., 1973), and the subject has been extensively reviewed by Makino (1974). The most remarkable feature of the karyotype of these tumors, whether from Japan, France, the United States, or Jamaica, is their similarity, with most at or close to 59 chromosomes. The reduction in the chromosome number from the diploid number of 78 in the dog (all the autosomes being acrocentric) is accompanied by the formation of bi-armed chromosomes, perhaps through centric fusion. Estimation of the nuclear DNA content showed that there has been no significant loss of chromosomal material. The banding patterns of two primary and one transplanted tumor with modal numbers of 57–59 including 16–17 metacentrics and submetacentrics and 40–42 acrocentrics have been observed (Oshimura, Sasaki, et al., 1973). The banding patterns of the larger chromosomes in all three tumors were very similar. The second largest submetacentric contained an extensive heterochromatic region on its long arm which in C-stain preparations formed a conspicuous chromocenter in interphase adjacent to a nucleolus. The similarity of the karyotypes of the venereal sarcomas is striking although it may only be superficial. As discussed by Makino (1974), some tumors may have identical karyotypes because they have a common origin, even though the dogs may be from widely separated geographic locations.

To my knowledge, no contagious tumors, particularly warts (verruca vulgaris), have been studied in human subjects. Such studies should be performed because warts in all probability have a rather high mitotic index and should be a source of sufficient metaphase plates. To date, however, no human cancers or leukemias have been shown to be contagious.

Minimal deviation hepatomas. The most comprehensive study to date was performed on 42 transplantable rat hepatomas, including the so-called minimal deviation hepatomas.[8] The tumors were divided into three groups according to the chromosomal findings: 9 were diploid, 17 had abnormal chromosome numbers with minimal deviation of the morphology of the chromosomes, and 16 had major changes in the chromosome number and morphology. Transplanted tumors, studied in hosts of the opposite sex from which they originated, allowed the investigators to distinguish contaminating host cells from those of the original tumor. Metabolic studies indicated that the "minimal deviation" tumors had enzymic alterations which certainly were more than minimal and evidenced considerable variation from tumor to tumor. Interestingly, one of the hyperdiploid tumors (45 chromosomes) was metabolically less "deviated" than any of those with the diploid number of chromosomes. The conclusion was that a diploid chromosome complement did not guarantee that a tumor was "minimally deviated" from normal hepatic tissue either in its growth rate or metabolic profile. In addition, aneuploid tumors did not necessarily have striking morphological or metabolic deviations. Thus, "regardless of how normal these cells appear by other criteria, the term 'minimally deviated,' compared with normal liver cells, would seem inappropriate, at least from a cytogenetic standpoint" (Nowell, Morris, et al., 1967). A more recent study showed that two of the diploid hepatomas previously reported had undergone transition to an aneuploid state.

Further comments on some experimental tumors. Mark (1967b, 1969a) described a series of studies on the chromosomes of Rous sarcomas in mice (induced with BRSV-SR strain of virus). About 40% of 91 primary tumors had a diploid stem line; the others were aneuploid, mostly hyperdiploid or near-tetraploid. Chromosomal progression was investigated in 11 tumors subjected to repeated subtotal excisions. The aneuploid tumors often underwent further chromosome changes; in six tumors with diploid stem lines, aneuploid side lines, which may or may not have been present when the tumors were first examined, tended to become predominant.

Mitelman (1972b) followed the changes during in vitro passages of two Rous sarcomas. Initially, the tumors were diploid without any deviating cells; there were indications that the earliest changes were predetermined, with a specificity with regard not only to the direction of the pathway (toward hyperdiploidy), but also to the particular chromosomes involved. In the course of the heteroploid evolution, the normal diploid cells decreased by a nonlinear process with a slow initiation and a rapid fall. The chromosomal progression was related to a histological dedifferentiation.

In contrast to the studies of Rous sarcoma in rat, in those induced by DMBA in these animals and indistinguishable histologically from the Rous sarcomas, trisomy of the longest terminal chromosome (A2) or markers containing segments of A2, was the first characteristic feature of the tumors.[9] In Chinese hamsters, of six DMBA-induced sarcomas, four had mostly diploid stem lines, with one to two hypodiploid side lines or DMS. The two heteroploid sarcomas consisted of one hypodiploid and one triploid tumor. The deviating cells displayed a nonrandom karyotypic pattern and one different from that of Rous sarcoma in the same species.

Inasmuch as the DMBA-induced sarcomas in the rat had certain karyotypic features that have been previously observed in rat leukemias and epitheliomas of the ear induced with DMBA, the above authors believed that a specific pattern of chromosomal changes appears to be related to a particular class of inducing agents. Later studies showed that different hydrocarbons induced slightly different patterns of cytogenetic changes during carcinogenesis. Thus, only rarely were diploid cells seen in tumors induced by single injections of 20-methylcholanthrene or 3,4,-benzpyrene, unlike those induced by DMBA where 55% of the cells had a diploid karyotype. Also, there were many more markers and a high degree of karyotypic variability, with nonrandom involvement of certain chromosomes. It was suggested that this variability was the result of a relative lack of site specificity on the part of these chemicals in their effect on the chromosomes, in contrast to DMBA, which apparently attacks decisive gene loci directly.

These authors[10] have by extension related their findings in animals to the possibility that only a few chromosomes carry genes of prime importance for malignant transformation in the human. Thus, they have analyzed a large number of cases and have indicated that certain chromosomes (#7, #8, #9, #14, #17, #20, #21, and #22) were more frequently involved in some neoplastic diseases than any other chromosomes, these diseases consisting primarily of lympho- and myeloproliferative disorders. Based on their previous contention that the etiological factor is important in determining the chromosomal evolution in tumors, with some of the experimental tumors characterized by step-wise karyotypic evolution with predetermined sequences of chromosomal changes, they speculated that in the human various neoplasms develop distinctive karyotypic patterns under the influence of different oncogenic agents, though in man it cannot be tested experimentally. These authors concluded that the clustering of aberrations indicates that only a few chromosomes carry genes of prime importance for malignant transformation and that these genes are selectively engaged in the interaction with the inducing agent during carcinogenesis. The primary change induced may or may not be discernable microscopically, but once the crucial genes have been activated, the sequence of karyotypic changes will follow. The fact that they often take different routes, perhaps under the influence of different agents, shows that the development of malignancy is not bound to one specific chromosome constitution.

In my opinion, there are a number of problems that make the above interpretation less plausible. There is a distinct possibility that some of the chromosomes in the human, particularly in hematopoietic and lymphoid cells, represent the "Achilles' heel" of the human genome, showing changes of a rather frequent nature regardless of the oncogenic agent involved. This is supported by the fact that in many of the conditions enumerated by the above authors chromosomal changes are not observed, e.g., the ALs, and that the chromosomes that appear to be frequently involved, often do so *after* the disease has become well established. Thus, it appears difficult to reconcile changes in these chromosomes, which allegedly carry genes of importance for malignant transformation, with the fact that they are not involved in a substantial number of cases of human neoplasia. Furthermore, there is also a distinct possibility that the genes responsible for the development of neoplasia in the human are located on a number of chromosomes and not necessarily on those enumerated above, the involvement of the chromosomes enumerated being merely a "phenotypic" expression of the basic neoplastic process. Also, the fact that a variety of conditions, e.g., acute and chronic leukemias, meningioma, colonic polyps, lymphoma, etc., appear to have predisposition to show changes in the chromosomes enumerated above would seem to me to indicate that they cannot be related to a common etiologic factor, because it is inconceivable that this variety of conditions is caused by the same agent; hence, they must reflect a secondary response to the malignant condition, as do a number of other biochemical and cellular parameters. Surely, neoplasia must be initiated at a genic level that cannot be extrapolated to the gigantic, in terms of cellular parameters, and catastrophic events reflected by such changes as extra or missing chromosomes or the presence of very high aneuploidy.

Two chemically induced hepatomas, which had apparently diploid karyotypes when studied with conventional techniques, were reinvestigated with Q- and G-banding (Wolman and Horland, 1975). One tumor showed no significant variation from normal, whereas the other was characterized by an asymmetry of #2. Mori and Sasaki (1974) studied the chromosomes of 9 primary and 13 transplanted leukemias and LS induced by 9-N-nitrosobutylurea or various RNA viruses and 1 spontaneous leukemia. Eight primary and six transplanted neoplasms retained the diploid number, the banding patterns being indistinguishable from those of normal diploid cells except in one which had "minor structural changes." One chemically induced primary leukemia had a diploid stem line with hypotetraploid variants. The remaining eight transplanted neoplasms were pseudodiploid, hyperdiploid, or hypodiploid. Two transplanted Rauscher LSs showed a complete or partial trisomy for #1, but no other possibly consistent changes were discovered. The transplanted leukemia of spontaneous origin, which was the oldest tumor in the series, showed the most extensive changes in the karyotypes, even though the diploid number of 42 was retained in most of the cells.

By use of G-banding techniques, the chromosomes of five tumors induced by DMBA and three cell lines transformed in vitro by the same agent have been investigated.[11] The stem lines of three tumors and one transformed cell line were diploid, the others being aneuploid with preferential involvement of #1, #2, and #3. However, the same chromosomes were involved in two nontransformed cell lines derived from clone cells treated by the carcinogen (though aneuploid, these cell lines retained the cellular morphology of untreated controls and did not produce tumors when injected into newborn rats).

The occurrence of trisomy #15 in T-cells of mice involved by a leukemic process of either spontaneous or induced (virus, X-ray, DMBA) origin (Wiener, Ohno, et al., 1978) and in spontaneous lymphomas and leukemias of rodents (Dofoku, Biedler, et al., 1975; T. O. Chang, Biedler, et al., 1977) is an indication of nonrandom karyotypic changes in animal neoplasia. However, in both the spontaneous and induced forms of the above states diploidy may be present not only during the initial phases of the development of the diseases, but also during their progression.

Even though the chromosomes of tumors and leukemias often show a blurred outline in comparison with their clear-cut appearance of normal cells, Mitelman (1974a) found a contradictory and unexplained difference between the diploid metaphases seen in the early stages of Rous sarcoma and the aneuploid metaphases seen in the later stages: the latter, though presumably of a higher degree of malignancy, had clearly outlined chromosomes whereas those in the diploid metaphases had a blurred appearance.

Chromosomes and Progression of Human Neoplasia

Some possible pathways for the involvement of chromosomes in cancer (or leukemia) causation in human subjects are shown schematically in Figure 127.[12] Evidence for or against a particular path playing a role in the direct induction of the neoplastic state in human subjects will be given. At present, there is no absolute evidence that a *visibly recognizable* karyotypic abnormality, with the possible exception of the Ph¹ in CML, plays a *direct* role in the causation of human cancer or leukemia.[13] For a cytogenetic abnormality to be the responsible cause of neoplasia it should be characteristically found in a well-defined cancerous or leukemic entity and be present in the affected cells of all such cases. In essence, the only cytogenetic abnormality that meets these criteria is the Ph¹ in CML. Even though there are about 15–20% of the cases with CML that do not have the Ph¹ in the marrow cells, there is little doubt that > 80% of the CML cases are characterized by the presence of this karyotypic anomaly in the marrow cells.[14] It is very possible that the Ph¹-negative cases may constitute a group separate from the Ph¹-positive ones; thus, CML may have to be defined as a disease invariably associated with this chromosomal anomaly.[15] Even though the exact cause for the formation of the Ph¹ is still unknown, it is conceivable that it may be directly responsible for the leukemic state, although there is no cogent proof of this. However, one case of the de novo appearance of a Ph¹ in a previously Ph¹-negative CML has been observed by us and indicates that, at least in this case, the induction of the Ph¹ was an event secondary to the CML. Furthermore, a

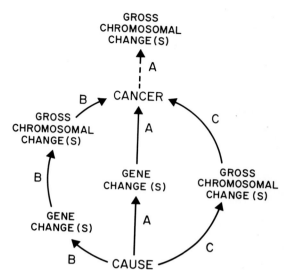

127 Three possible paths (A, B, and C) leading to cancer and the chromosomal changes in relation to these paths. In path A, the direct causation of cancer is a change in genic function, with the chromosomal changes being secondary to the neoplastic state. In path B, the chromosomal aberrations are induced by genic malfunction, but it is the chromosomal change directly responsible for the causation of cancer. In path C, the chromosomal changes are produced by an agent that affects chromosomal structure directly without intervening genic involvement. The chromosomal changes then induce the cancerous state. Note that only in paths B and C do the chromosomal changes play a direct role in the causation of cancer.

Ph¹ due to a translocation between #22 and #9 (and sometimes between other chromosomes) has been seen not infrequently in cases of AML, ALL, and other hematologic disorders. It is of interest that the Ph¹ persists in the marrow cells for the duration of the disease, including remissions, and more than one Ph¹ may be present in the cells, this latter condition usually heralding or being associated with the BP of CML.

Some observers have made much of the relation of increased breakage of chromosomes in vivo and in vitro in certain human conditions (FA, BS) to the incidence of neoplasia in affected subjects.[16] The contention in this case is that the anatomic instability of chromosomes ultimately leads to the genesis of a single cell bearing a stable mutation that induces the development of cancer. This concept has been advanced in the past by Makino (1957a, 1959) as the stem-cell origin of cancer, which is related to the theory of clonal evolution of human neoplasia advanced more recently by other workers.[17] Of course, another explanation could rely on the interpretation that whatever genetic abnormality leads to the chromosomal instability may also be responsible for the development of neoplasia in such subjects by affecting another segment of the genetic material. That the genetic picture in chromosome instability syndromes in relation to development of neoplasia is not totally clear is evident from the fact that not all patients with the chromosome breakage syndrome develop a neoplastic process; and that when the chromosomal changes do occur in neoplastic cells, they differ from one subject to another.[18] In other words, no consistent karyotypic picture characterizes the neoplastic state in such subjects nor, for that matter, is the type of malignancy the same from subject to subject.[19] Another dichotomous finding is the demonstration that the susceptibility of chromosomes to breakage in vitro may be seen in the relatives of subjects affected with FA, BS, and other breakage syndromes and not in the chromosomes of the *propositi*.[20] Inasmuch as the relatives appear not to have a higher susceptibility to neoplasia, such findings raise considerable uncertainty regarding the interpretation of chromosomal breakage in relation to the development of neoplasia.

Much has been made of the susceptibility of karyotypically abnormal cells or those from subjects with chromosomal breakage to abnormalities after exposure to irradiation or oncogenic viruses.[21] In these studies an attempt was made to ascertain whether the development of neoplasia in such a group of subjects could be based on the predictability from the findings obtained in vitro. Thus, it has been shown that the cells of patients with FA or those of their relatives appear to have a higher susceptibility to undergo "malignant" transformation when exposed to certain viruses than do normal cells.[22] These findings may merely indicate, however, that whatever is the basic genetic defect leading to chromosomal breakage in FA

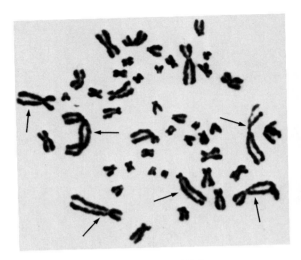

128 Metaphase (pseudodiploid) from a cancerous ascites containing marker chromosomes (arrows). In this case as well as the one shown in Figure 129 spectrophotometric DNA measurements probably would not have revealed any significant abnormalities, pointing to the necessity of performing cytogenetic analysis in order to establish karyotypic abnormalities.

and other syndromes may not only be responsible for the disease entity itself, but also for the chromosomal findings. The higher susceptibility to X-ray-induced changes of abnormal cells from patients with XXY, XYY, or an extra-G (DS) appears to be related to the presence of extra chromosomes in the set.[23] Thus, cells from patients with Klinefelter's or other syndromes in which there are extra sex-chromosomes appear to be more sensitive to radiation-induced changes than cells containing < 46 chromosomes, e.g., Turner's syndrome with an XO and 45 chromosomes. I do not think that any of these tests are, in fact, indicative of susceptibility to neoplasia in human subjects but are merely a reflection of a sensitivity of the karyotype to one agent or another in vitro.[24]

Markers morphologically distinct from any of the known chromosomes in the human set may originate through a number of different mechanisms (Figures 128–131). The presence of an identical marker in most and sometimes in all cancer cells has been used as evidence for the single-cell origin of human neoplasia.[25] It

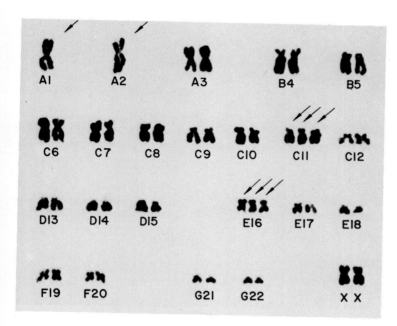

129 Pseudodiploid karyotype from a cancer of the colon without any apparent markers. Chromosomes are missing in groups A1 and A2 and extra chromosomes are present in groups C11 and E16.

should be pointed out, though, that markers resembling the morphology of normal elements of the human karyotype may be present in a cell and escape detection. Banding of chromosomes may shed light on this possibility,[26] particularly in cells with a hypodiploid chromosome number. Even though it is possible

130 Examples of G-banded isochromosomes originating from various autosomes in cancers and leukemias.

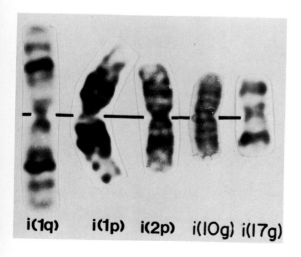

that most, if not all, human cancers and leukemias may originate from a single cell, particularly when such a contention is based on observations of cells with markers, the argument used by some investigators that these chromosomal abnormalities may play a direct role in the causation of cancer or leukemia is not as convincing as one would like it to be. A definite possibility exists that whatever the mechanism that causes development of aneuploid karyotypes in neoplastic cells may also be responsible for the development of markers. Furthermore, there is a host of human cancers and leukemias in which markers, as defined above, have not been found.[27] In addition, at least 50% of subjects with AL do not have cytogenetic changes in their leukemic cells,[28] indicating that the development of the leukemic state does not necessarily have to be associated with an abnormal karyotypic picture. In relation to the markers again, much has been made of the occurrence of similar markers in several neoplastic states. Except for the Ph^1, such markers are not present consistently in human cancer or leukemia and when these markers are present, they do show morphologic variability from case to case. Furthermore, very similar markers can be observed in cancers of different origin. It is possible that the genesis of markers has

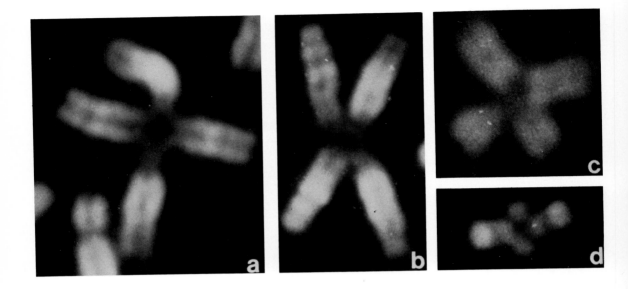

131 Q-banded *homologous* exchanges at centromeric regions induced by MMC. a = alternate symmetrical exchange of #1; b = adjacent symmetrical exchange of two #1 chromosomes; c = adjacent symmetrical exchange of #9 chromosomes; and d = alternate symmetrical exchange of #17 chromosomes.

certain morphologic limitations in human cells, based on the type of chromosomes participating in the formation of markers and the type of mechanism leading to such karyotypic abnormalities. Thus, there is no convincing evidence that, except for the Ph[1], markers can be used as criteria for the rigorous identification of one or another type of leukemia or cancer in human subjects.

Our past conclusion that in human malignant disorders different chromosome groups were involved *randomly* must be revised. Banding techniques have disclosed a *nonrandom* involvement (Rowley, 1976f) of certain chromosomes in the BP of CML, AML, meningioma, PV, etc. (Table 81a); in all probability other neoplastic conditions will be added. Furthermore, an analysis of nearly 1000 malignant human conditions has revealed the nonrandom involvement of certain chromosomes (Table 81b), and extrapolations of these observations to the causation of human neoplasia have been attempted.[29] Based on banding analyses, it has been indicated through an examination of the cytogenetic data in the literature[30] that certain chromosomes are involved more frequently than others in various hematological and other malignancies, i.e., #7, #8, #9, #14, #17, #20, #21, and #22. The importance of #1, particularly partial or complete trisomy of 1q, in human cancers was pointed to by us[31] and since confirmed by others.[32] In addition, certain regions or bands in specific chromosomes have been pointed to as being involved more frequently than others in deletions and translocations in human cancer and leukemia. Though this may be true, even the smallest bands in a human chromosome contain a large number of chromosomal subunits (e.g., nucleosomes) and certainly genes and, hence, a lesion occurring in a single band may not necessarily affect the same genetic material in each case of malignancy. Thus, caution must be exercised in assuming that band involvement always indicates identical genetic material being affected in a particular band.

Nonrandom karyotypic changes in human cancer and leukemia, however extrapolated, should not be confused with the direct causation of these states. It has been sufficiently demonstrated that certain chromosome abnormalities occur in a nonrandom fashion in human cancer and leukemia (Table 81), e.g., −22 in meningiomas, +8 in AML, and 14q+ in BL.

Table 81a Chromosomal abnormalities associated with human neoplasia, primarily hemopoietic

Disease	Chromosomal abnormality		Reference
Chronic and acute myeloproliferative disorders	Aneuploidy; trisomy 8, 9 and $-Y$		Jonasson, Gahrton, et al., 1974; Rutten, Hustinx, et al., 1974 Atkin and Taylor, 1962; Kay, Lawler, et al., 1966; Sakurai, Oshimura, et al., 1974
		Monosomy 7	Petit, Alexander, et al., 1973
OML	Structural changes	$t(Ph^1) = t(9;22)$	Rowley, 1973a
Unusual and complex Ph^1-translocations		$t(2;22); t(13;22)$, etc.	Hayata, Kakati, et al., 1973, 1975 Sonta and Sandberg, 1977b
CML blast crisis		i(17q)	Lobb, Reeves, et al., 1972
Polycythaemia rubra vera		20q–	Reeves, Lobb, et al., 1972
SBA		20q–	M.M. Cohen, Ariel, et al., 1974
AML		t(8;21)	Rowley, 1973b; Sakurai, Oshimura, et al., 1974
Refractory anemia		5q–	Van Den Berghe, Cassiman, et al., 1974
ALL		6q–	Oshimura and Sandberg, 1976
		t(4;11)(q13;q22)	Van Den Berghe, David, et al., 1978b
Myeloma		14q+	Wurster-Hill, McIntyre, et al., 1973
BL		14q+	Manolov and Manolova, 1972
BL		t(8;14)	Zech, Haglund, et al., 1976
Non-BL		14q+	Reeves, 1973
Lymphoma		t(11;14)	Van Den Berghe, Parloir., et al., 1979a; Fleischman and Prigogina, 1977
APL		t(15;17)	Rowley, Golomb, et al., 1977a; Testa, Kinnealey, et al., 1977 Van Den Berghe, Louwagie, et al., 1979b
AT	Fragility	Increased breakage	Hecht, Koler, et al., 1966
BS		Increased breakage	German, Archibald, et al., 1965
Increased SCE in BS		Many sister chromatid exchanges	Chaganti, Schonberg, et al., 1974; Shiraishi, Freeman, et al., 1976
FA		Increased breakage	Schroeder, Anschütz, et al., 1964
Meningioma		–22	Zang and Singer, 1967, 1968a,b; Mark, 1970c, 1971c
Bladder cancer		Markers in papillary cancers	Sandberg, 1977; Falor and Ward, 1978

Based on table by Lawler (1977b).

Clustering of the cytogenetic aberrations in human neoplasia to a few specific chromosome types, i.e., #7, #8, #9, #14, #17, #20, #21, and #22, has been indicated by Mitelman and Levan (1976a). Parenthetically, #1 and possibly #5 should be added to this list. These authors stated that etiologic factors of the neoplasia, at least in animals and possibly in humans, are important in determining the chromosomal evolution in tumors, i.e., similar cytogenetic changes being due to the same agents, even though the neoplastic processes may involve diverse organs and be of a different nature. Some stress was put by Mitelman and

Table 81b Chromosomes preferentially involved in various human neoplasias

Type of neoplasia	Chromosomes preferentially involved
Myeloproliferative disorders	
AML	5,7,8,21
CML	8,9,17,22
PV	1,8,9,20
Others	1,5,7,8
Lymphoproliferative disorders	
Lymphomas	1,3,9,14
BL	7,8,14
ALL	1,21,22
CML	1,14,17
Monoclonal gammopathies	1,3,14
Solid tumors	
Meningiomas	8,22
Benign epithelial tumors	8,14
Carcinomas	1,3,5,7,8
Sarcomas	13,14
Neurogenic tumors	1,22
Malignant melanomas	1,9

Based on the data of Mitelman and Levan (1979).

Levan on the role of chromosomal changes in malignancy.

The nonrandom changes do not occur in every case in any specific neoplasm, e.g., not all meningiomas have a −22, not all AML cases have a +8, and not every lymphoma has a 14q+. Furthermore, even though the same chromosomal group may be involved in diverse conditions, the nature of the involvement may vary, e.g., the amount of material translocated onto the long arm of #14 in lymphoma varies from case to case. It is also difficult to see how diseases that have similar chromosomal changes can be due to the same etiologic agents, e.g., meningioma and CML. In my opinion, the nonrandom changes merely reflect the limited and predilective *response* of the human genome to the neoplastic state, though it should be added that every chromosome in the human set has been observed at one time or another to be involved in one malignant state or another. As has been aptly stated, the chromosome changes in neoplasia are but "fingerprints of some collateral action of the carcinogenic agents." The narrow range of response, so-called nonrandomness, may merely be a reflection of the propensity of certain chromosomal groups to respond more frequently to neoplasia than others. For, if neoplasia requires involvement of only one gene, the expression of which will be evident whether it is located in a normal or abnormal chromosomal set, then the karyotypic changes may be merely a "phenotypic" expression of the consequences related to this gene and probably significantly influenced by the remarkable variability of the human genome.

At least 50% of the patients with AL do not have chromosomal changes in their leukemic cells, and this constitutes the most cogent evidence that visibly *recognizable* karyotypic changes are not necessary for the development of neoplasia in human subjects. Furthermore, when the chromosomal changes occur in such cells, they are, with some exceptions, diverse in AL; this may indicate that the cytogenetic changes are secondary to the neoplastic state. In such a state, it is possible that the original neoplasia is induced by a mutation at the gene level with a chromosomal change being a consequence of the development of the leukemic state. A similar situation probably applies to other leukemias and cancers in human subjects. Further evidence in support of this hypothesis is supplied by cytogenetic observations in CML with a Ph[1]. Almost all of these cases have a chromosome number of 46, including the Ph[1], without any other observable karyotypic abnormalities. When the BP of CML is about to appear or is established, other karyotypic changes, though some are nonrandom, may occur in the leukemic cells. These karyotypic changes have been of a diverse nature. Thus, except for the consistent and persistent chromosomal abnormality of the Ph[1], other karyotypic changes that occur during the course of CML appear to be secondary to the original causation of the leukemia.

Certain observations bear directly on the problem just discussed above. We have had the opportunity to examine twins with congenital AML in whom the chromosomal picture was very different.[33] One member of these paternal twins had hypodiploid leukemic cells and the other hyperdiploid. Unless one wishes to impute two different causes for the AML, which was induced in utero, the chromosomal changes are probably not involved in the causation of the AML, and the diversity of the karyotypic picture may be but a reflection of the genetic difference between the two individuals involved. It is of interest that identical twins with AL have identical chromosomal changes, when these are present, in their leukemic cells.[34]

In a previous section an unusual case of BL was discussed in which the tumors in the same individual were either hyperdiploid or diploid. Because it is extremely unlikely that the causes for these tumors were different in the individual involved, the findings definitely point to the fact that the induction of neoplasia in human subjects, as demonstrated by this case and by most of the cases in AL, is not necessarily accompanied by or due to chromosomal changes and that in all probability the chromosomal changes are a result of or follow the development of the neoplastic process.

The interesting observations in meningiomas have raised the general problem of the relation of chromosomal changes to neoplasia. A substantial number of meningiomas contain a relatively consistent karyotypic abnormality consisting of a missing #22. In some other meningiomas, cytogenetic observations similar to those observed in most malignant cancers have also been reported. Because generally, meningiomas are considered to be "benign" tumors, the question arises as to the significance of these chromosomal changes. Meningiomas, though considered benign, may represent an extreme stage of early low-grade but malignant cancers; though histologically benign, the chromosomal changes would point to the possibility that these tumors are, in fact, cancerous in nature at the time of observation. The fact that such tumors may become sarcomatous and show local metastatic spread in occasional cases supports this contention. Thus, the occurrence of chromosomal changes in these tumors points to the possibility that meningiomas may be cancers of low malignancy, particularly those in which definite karyotypic changes have been observed. The chromosomal changes may indicate that these meningiomas may either be in the process of becoming more malignant or show biologic progression in their behavior. The gist of this is the promulgation that occurrence of chromosomal changes in any neoplasia points to the tumor as being of a cancerous nature rather than benign.

In conclusion, I interpret the chromosomal data obtained in human neoplasia to date as indicating that cytogenetic abnormalities have not been shown to be directly involved in the causation of human cancer or leukemia (CML and the Ph^1 may be exceptions). I believe that the cytogenetic changes in neoplastic states, characterized by an extreme diversity of the chromosomal picture among different cancers and leukemias, are due to the extreme variability of the genetic make-up of the individuals involved.

Chromosomes in the Diagnosis, Therapy, and Prognosis of Cancer and Leukemia

The application of cytogenetic findings to the diagnosis of cancer and leukemia in human subjects has not been fully explored. Because it appears that even the slightest karyotypic deviation of a cell from its normal diploid chromosome constitution may indicate a definitely altered biologic behavior, compatible with that of the neoplastic state, utilization of karyotypic examination in the diagnosis of cancer or leukemia, in conjunction with other cytologic and clinical criteria, particularly when these are not clear cut, may yet become an important tool in the diagnosis of neoplasia in human subjects. Reference has already been made to the Ph^1 as being specific and characteristic for CML (and in rare instances, for allied conditions), and undoubtedly the presence of this abnormal chromosome in the cells of the marrow is diagnostic of this leukemic state. Unfortunately, no cytogenetic abnormalities similar to that of the Ph^1 (and possibly the $14q+$ in lymphoma) have been found in other human cancers or leukemias, in spite of some attempts to correlate some markers with certain tumors. In the latter cases, the occurrence of a marker has not been described in all cases, and similar markers may occur in different cancers. However, the presence of such a marker in any cell is, in my opinion, indicative of a neoplastic state. It is doubtful whether with presently available techniques any other karyotypic anomaly similar to that of the Ph^1 or $14q+$ will be found for a neoplastic state in humans. However, newly available methodologies for the banding patterns of human chromosomes have already revealed cytogenetic features for certain tumors and leukemias which may turn out to be of considerable help in the diagnosis of such conditions (Wurster-Hill and Maurer, 1978).

To date only a very rare human cancer has been found to be diploid, possibly due to the fact that human cancers tend to be seen at very late stages of their development when the karyotypic analysis is performed. In contrast to the findings in animals, almost all of the human cancers have been established to have an abnormal chromosome constitution with either numerical and/or morphological changes in the karyotype. These chromosomal changes are of extremely protean nature and it is rare for two cancers to have an identical karyotypic make-up. Furthermore, cancers of the same organ and of similar histology may have extremely divergent karyotypes, thus indicating that the chromosomal picture may not necessarily correlate with the biologic behavior of a cancer. Nevertheless, the presence of any karyotypic changes, particularly morphologic (including markers), is indicative that a cell is cancerous.

Chromosomal analysis can be of diagnostic help in cases of AL. In aneuploid cases the appearance of cells with an abnormal chromosome constitution in a previously diploid marrow may indicate imminent relapse of the AL and, thus, serve as an indicator for therapeutic purposes. The relation of chromosomal findings to the prognosis of AML has already been alluded to previously in the section on AL.

Allusion has been made to the possibility that newly developed techniques may reveal karyotypic changes in some human cancers which may prove to be of diagnostic significance. These techniques include, at the moment, the banding patterns of individual chromosomes, which allow a more critical evaluation and identification of the human chromosomal set than afforded by previous methodologies (Figure 45). Thus, it has been shown that the Ph^1 belongs to group #22, whereas the chromosome involved in trisomy of DS belongs to group #21. A similar situation exists in meningioma, as mentioned previously, in which it was shown that the missing chromosome is a #22. It is possible that future studies may reveal deviations from the normal karyotype in certain cancers or leukemia and that these new cytogenetic criteria may assist in the diagnosis of neoplastic states.

Several recent reports are of interest in that they indicate the usefulness of cytogenetic analysis in the diagnosis of certain conditions. Thus, Hossfeld and Schmidt (1978) have reported the presence of definite chromosome abnormalities in a number of effusions from patients with HD, in some of whom the diagnosis could not be made on the basis of the cytologic examination. In another report (Hansson and Korsgaard, 1974), a number of effusions from cancers of the lung or breast were shown to contain abnormal clones of cells based on cytogenetic analysis, whereas cytologic examination was negative in a significant number of these effusions. Korsgaard (1974) has stated that chromosome analysis is a valuable diagnostic tool in malignant pleural effusions, especially when cytological examinations yield no definite diagnosis. In some cases, e.g., mesotheliomas, karyotype analysis has shown itself to be superior to examination with standard cytological techniques. Chromosome analysis as a diagnostic tool is a simple one, which demands no expensive reagents, can be performed rather expeditiously and reliably and "completely without false positive diagnoses." A similar view has been expressed by Dewald, Dines, et al. (1976) on the basis of their experience with 104 unselected effusions from patients with benign and malignant conditions.

The application of chromosomal findings to the therapeutic approaches in cancer and leukemia has not been fully evaluated or realized. It is possible that when large series of particular cancers are examined, in relation to the chromosomal findings, it may be possible to arrive at a utilization of the karyotypic picture in the decision regarding therapy.

Thus, in certain tumors it may be shown that those with a high chromosome number are less susceptible to radiation therapy than those with a low chromosome number. Such a situation appears to apply to cancer of the colon and an opposite one karyotypically to cancer of the cervix. In certain conditions, there appears to be some value in utilizing the cytogenetic findings in deciding upon the exact course of therapy. It is not difficult to appreciate that with newly developed techniques and with a critical appraisal of the chromosomal data in large series of patients, it may be feasible to utilize the chromosomal observations in the therapeutic approaches to a large number of malignant states in human subjects (see section on papillary tumors of the bladder).

In the case of CML it has been shown that those cases that contain the Ph^1 in their cells have a more favorable prognosis, due primarily to the fact that the Ph^1-positive cases respond more readily to therapy than the Ph^1-negative cases. Thus, it appears that a definite approach toward the chemotherapy of Ph^1-negative CML will have to be undertaken, and this situation is, again, an example of the utilization of chromosomal findings in the therapy of a malignant state.

The importance of ascertaining the cytogenetic picture in AML has already been discussed in the section on AL and indicates a crucial understanding and utilization of the karyotypic findings in planning therapy of AML. Furthermore, the appearance of aneuploid cells in a patient with remission in AL may indicate imminent relapse of the disease, and these types of cytogenetic findings should find applicability in the therapeutic approaches to AL.

The outstanding example of a correlation between chromosomal findings and prognosis has already been mentioned, i.e., Ph^1-positive vs. Ph^1-negative CML. The much longer survival of the Ph^1-positive patients appears to be related to their more ready response to chemotherapy and significantly longer periods of time between the diagnosis and the development of the BP. Apparently, once the BP develops, the survival is very similar in the two types of CML. Thus, the absence of a Ph^1 in CML indicates a much shorter survival for such a patient, than when a Ph^1 is present. Unfortunately, no condition similar to CML in relation to the prognostic aspects of cytogenetics has been described. Some data have been presented on the invasiveness of cervical cancer in relation to the karyotypic findings, with the indications that the lower the chromosome number the more likely the tumor is to be invasive. Because invasiveness may be related to prognosis, this would appear to be a fruitful approach to be taken in future studies, not only for cancer of the cervix but for other cancers in human subjects.

As indicated, the potentialities of applying chromosomal findings to the therapy of cancer and leukemia in human subjects has probably not been fully explored, and it is hoped that newly developed techniques, the study of larger series of patients, and a competent correlation of the various clinical, cytological, and karyotypic parameters may lead to a more realistic and frequent utilization of chromosomal data in the therapy of human neoplasia.

An example of the usefulness of chromosomal monitoring (e.g., Ph^1) is in engrafting after marrow transplantation (Graze, Sprakes, et al., 1977).

Chromosomal Findings in Preneoplastic Lesions

Much has been written about chromosomes in preleukemic states and in precancerous lesions. In the case of the former, it is my contention that by the time chromosomal analysis is performed, the conditions are not preleukemic, because definite clinical and histological evidence of disease is usually present. This does not preclude the possibility of such aneuploid cells disappearing as a result of the host's "defense," e.g., immunologic. In the case of precancerous lesions, there is more cogent information available on the subject on the basis, primarily, of karyotypic studies on various lesions of the human uterine cervix.

No other tissue in the body, except possibly the skin, offers such a unique opportunity in determining the changes in the chromosome constitution of cells over the wide range from normality to invasive cancer, as does the uterine cervix. It is readily accessible and visualized, easily biopsied, and symptoms resulting from cervical disease are noticed relatively early by patients. Furthermore, established programs for testing cervical cytology and the frequency of examinations in women afford another parameter of availability of this tissue. Thus, if there is any hope of ascertaining the role of chromosomal changes in human carcinogenesis, lesions of the cervix appear to offer the best opportunity for it.

Cytogenetic examinations of cervical tissue are not without their difficulties and shortcomings. The mitotic index in normal cervical epithelium is very low, so that resort has to be made often to vitro culture, a procedure known to yield cells with considerable variation from the number of chromosomes of the tissue in vivo and, thus, to give inaccurate results. Cervical tissue is often infected, and as some viruses have been implicated in the possible causation of cervical cancer, it is difficult to interpret the potential effect such agents may have on the genome of cervical cells or, for that matter, that of nononcogenic viruses and other infective agents. Furthermore, the cervical cells are sensitive to estrogens and other steroid hormones, and we are still pretty much in the dark about the effects of these hormones on the chromosomal make-up of the cervical cells during the cyclicity of the female hormonal milieu. These parameters are of import if we are to determine the relation of karyotypic findings to precancerous lesions of the cervix.

The question must arise as to whether the findings of an abnormal chromosomal picture in a cervical cell indicates a precancerous or even a cancerous condition. It is my opinion that it does. Thus, I consider the presence of an abnormal cytogenetic picture in a cervical cell to indicate a definite cancerous transformation. This does not rule out the possibility of such cells being destroyed by the host, as, in fact, spontaneous regression of dysplasia and carcinoma *in situ* has been observed.

The statements made above regarding cervical cells apply also to other potentially precancerous tumors, e.g., polyps of the colon, in that the presence of chromosomal changes of any nature indicates malignant transformation in these tumors. In my opinion, then, chromosome analysis cannot reveal a true precancerous cell, for if karyotypic changes are present in such a cell, they indicate that the cell is already cancerous. No criteria based on chromosomal findings have established a definition of the precancerous state.

References

Pages 427–439

1. Berger, 1969a, b.
2. Bordelon and Stubblefield, 1974; Visfeldt, Povlsen, et al., 1972.
3. Berger, 1968a, b.
4. Ishihara, Moore, et al., 1961.
5. Martineau, 1969.
6. Atkin, 1970b.
7. Atkin, 1970b; A. Levan, 1966; Steenis, 1966.
8. Minkler, Gofman, et al., 1970a.
9. Bender, Kastenbaum, et al., 1972.
10. Lejeune, Berger, et al., 1966.
11. Atkin, Baker, et al., 1967b.
12. Muldal, Elejalde, et al., 1971.
13. Fleischmann, Gustafsson, et al., 1971, 1972.
14. Muldal, Elejalde, et al., 1971.
15. Nava, Grouchy, et al., 1969b.
16. Wiener, Fenyo, et al., 1972.
17. Atkin and Baker, 1969.
18. Atkin, 1967a.
19. Atkin, 1971c.
20. Manolov, Levan, et al., 1970.
21. Atkin, 1960a, 1964b, 1969b; Atkin and Baker, 1964; Atkin and Brandão, 1968; Brandão and Atkin, 1968; Castoldi, Grusovin, et al., 1976; L.Y.F. Hsu, Alter, et al., 1974; J.F. Jackson and Clement, 1974; LoCurto and Fraccaro, 1974; Nayar and Sharma, 1975; Pegoraro, Rovera, et al., 1967; Uyeda, Davis, et al. 1966.
22. Atkin, 1960a, 1967a, 1971c, 1973a; Atkin and Petrović, 1973; Ganina, 1972; Miles, 1959; Nieburgs, Herman, et al., 1962; Tavares 1962.
23. Kegel and Conen, 1972; Sellyei and Vass, 1975; Straub, Lucas, et al., 1969; Takahashi 1977; Vass and Sellyei 1973a; Vass, Sellyei, et al. 1973.
24. Dallapiccola and Tataranni, 1969; Fialkow, 1976a, b; Lejeune, 1973; Nava, Grouchy, et al., 1969b; Nowell, 1965a, 1970, 1971c, 1974b, 1976; Olinici, 1973; Šlot, 1967a, 1970.
25. Included in the following references are papers dealing with the role of chromosomes in cancer causation, discussion of the genetic conditions predisposing to cancer, peculiarities and uniqueness of the cancer cell as reflected karyotypically and cellular mechanisms which may lead to the cytogenetic changes seen in cancer: Atkin, 1970a, 1974a, 1976b; Awano, 1961; Bartalos, 1971; Benedict, Rucker, et al., 1975; Berger, 1968a,b, 1969c; Bloch-Shtacher and Sachs, 1976; Buckton, Jacobs, et al., 1962b; Codish and Paul, 1974; R. P. Cox, 1976; DiPaolo, 1975; Field, 1972; P. Fischer and Hebrard, 1974; Gartler, 1977; Goh, Reddy, et al., 1976; Grouchy, 1967a, b, 1973; Grouchy and Nava, 1968; Hamerton, 1971a, b; Harnden, 1974a, 1976a, b, 1977; Hauschka, 1963; Hecht and McCaw, 1977; Heston, 1976; Hirschhorn, 1976; Jami and Aviles, 1976; K. W. Jones, 1974; R. Kato and Levan, 1968; Klein, Bregula, et al., 1971; Knudson, Strong, et al., 1973; Kohler, Bridson, et al., 1971; Lejeune, 1967; A. Levan, 1967, 1969, 1973; A. Levan, Levan, et al., 1977; G. Levan and Levan, 1975; R.W. Miller, 1968, 1975; Mouriquand, 1968; Hj. Müller and Stald-

er, 1976; Oksala and Therman, 1974; Parmentier and Dustin, 1951; Refsum and Hansteen, 1974; Rowley, 1974a, 1976a,b; Sandberg, 1965, 1973, 1974; Sandberg and Hossfeld, 1970, 1974; Sandberg and Sakurai, 1973a, 1974, 1976; Schroeder, 1972, 1975; Shapot, Krechetova, et al., 1976; Simons, 1966; Therman and Kuhn, 1976; Turpin and Lejeune, 1965; Wakonig-Vaartaja, 1962; Weber, 1977; Wolman and Horland, 1975; Wurster-Hill, Cornwell, et al. 1974; T. Yamamoto, Rabinowitz, et al. 1973; Yoshida and Tabata, 1957.

Allard and Cadotte, 1972; Comings, 1973a; Court Brown, 1962; DiPaolo and Popescu, 1976; Erkman and Conen, 1964b; Ishihara, 1958; L.G. Jackson, 1978; Koller, 1964; A. Levan, 1956c, 1958; Merz, El-Mahdi, et al., 1968; Nowell, 1969, 1974a,b; Ohno, 1971; Sharma, Parshad, et al., 1963; Srivastava and Lucas, 1976; Yung, Blatnik, et al., 1964; Zankl and Zang, 1978a.

26. Mitelman, 1972a; Mitelman, Mark, et al., 1972a.
27. Fleischmann, Gustafsson, et al., 1971, 1972; Fleischmann, Håkansson, et al., 1972; Kakati, Hayata, et al., 1975, 1976; Kakati, Oshimura, et al., 1976; Kakati, Song, et al., 1977; Manolov and Manolova, 1972; Manolov, Manolova, et al., 1971a,b; Sonta, Oshimura, et al. 1977; Sonta and Sandberg, 1978a; Zankl and Zang, 1972a.
28. Atkin, 1969c, 1976a; Atkin, Mattinson, et al., 1966; Barlogie, Hittelman, et al., 1977; Kraemer, Peterson, et al., 1971; Paulete-Vanrell, 1970; Pfitzer and Pape, 1973; Stich and Emson, 1959; Stich and Steele, 1962.
29. Bridge and Melamed, 1972; Dutta-Choudhuri and Choudhuri, 1967.

Pages 440-453
1. Al-Saadi and Mizejewski, 1972.
2. Ikeuchi and Honda, 1971; Yosida, 1966, 1975; Yosida, Imai, et al., 1970.
3. Tough, Smith, et al., 1970.
4. R. Kato, 1968.
5. G. Levan and Mitelman, 1976; Mitelman, 1971, 1972a,b, 1973, 1974a; Mitelman and Mark, 1970.
6. Mitelman, 1972a,b.
7. Makino, 1974; Oshimura, Sasaki, et al., 1973; Sasaki, Oshimura, et al., 1974; Sellyei, Tury, et al., 1970.
8. Nowell, Morris, et al., 1967.
9. Ahlström, 1974; G. Levan, 1974, 1975; G. Levan, Ahlström, et al., 1974; G. Levan and Levan, 1975; Mitelman and Levan, 1972; Mitelman, Mark, et al., 1972a.
10. G. Levan and Mitelman, 1975; Mitelman and Levan, 1976a, b; Mitelman, Mark, et al., 1972a.
11. DiPaolo, 1977; DiPaolo, Nelson, et al., 1971; DiPaolo and Popescu, 1973; Popescu, Olinici, et al., 1974.
12. Sandberg, 1966b.
13. Nowell, 1965a; Sandberg and Hossfeld, 1970.
14. Sandberg and Hossfeld, 1973; Whang-Peng, Canellos, et al., 1968.
15. Ezdinli, Sokal, et al., 1970.
16. German, 1972a, b.
17. Fialkow, 1977; Grouchy and Nava, 1968; Grouchy, Nava, et al., 1965a, b.
18. German, 1972a, b; Swift, Zimmermann, et al., 1971.
19. German, 1969b.
20. Lieber, Hsu, et al., 1972.
21. Higurashi and Conen, 1971, 1972; R.W. Miller and Todaro, 1969; Mukerjee, Trujillo, et al., 1971; Todaro, Green, et al., 1966; Todaro and Martin, 1967; D. Young, 1971a.
22. McDougall, 1971a; Swift and Hirschhorn, 1966; D. Young, 1971a.
23. Mukerjee, Trujillo, et al. 1971; Sandberg and Sakurai, 1973a.
24. Sandberg and Sakurai, 1973a; Sasaki, Tonomura, et al., 1970.
25. Atkin and Baker, 1966; German, 1972a, b.
26. Caspersson, Zech, et al., 1970b; O'Riordan, Langlands, et al., 1972.
27. Sandberg, Takagi, et al., 1968c.
28. Sandberg and Hossfeld, 1973.
29. A. Levan, 1966; Sandberg, Takagi, et al., 1968a; Steenis, 1966. The hypothesis advanced by Gofman, Minkler, et al. (1967) and Minkler, Gofman, et al. (1970a, b) for autosomes of group E playing a major role in the genesis of human cancer has been refuted by Bender, Kastenbaum, et al. (1972).
30. G. Levan and Mitelman, 1975, 1977; Mitelman, 1973; Mitelman, Mark, et al., 1972b; Mitelman and Levan, 1976a, b, 1979.
31. Kakati, Hayata, et al., 1975, 1976; Kakati, Oshimura, et al., 1976; Kakati, Song, et al., 1977; Oshimura, Sonta, et al., 1976.
32. Atkin, 1977; Jones Cruciger, Pathak, et al., 1976; Kovacs, 1978; Rowley, 1975a, 1978a; Spriggs, Holt, et al., 1976; Yamada and Furasawa, 1976.
33. Sandberg, Cortner, et al., 1966.
34. Hilton, Lewis, et al., 1970.

17

Solid Tumors and Metastatic Cancer

Benign Tumors
Benign Tumors Without Chromosome Anomalies
Benign Tumorous Conditions with Chromosome Anomalies
Cancer: Primary and Metastatic

Tumors of the Alimentary Tract
Tumors of the Oral Cavity and Esophagus
Tumors of the Stomach
Some Recent Studies in Cancer of the Stomach
Tumors of the Colon, Cecum, Appendix, Rectum, and Anus
    Benign lesions of the colon
Cancer of the Colon, Cecum, Appendix, and Rectum
Banding Studies of Primary Tumors of the Large Bowel
Tumors of the Liver, Pancreas, and Peritoneum

Tumors of the Female
Tumors of the Breast
Tumors of the Ovary
    Early studies of cancer of the ovary
    Chromsome studies in cancer of the ovary based on banding analyses
    Ovarian teratomas
    Further comments on the origin of benign or variant teratomas
Cancer of the Corpus Uteri and Endometrium
    Studies of cancer of the uterus before banding
    Banding studies in cancer of the uterus

Chorionic Villi, Hydatidiform Mole, and Chorionepithelioma
Miscellaneous Gynecologic Tumors
    Hydatidiform moles
    Trophoblastic tumors

Tumors of the Urinary Tract
Chromosomes in Cancer of the Bladder
    Spectrophotometric DNA studies
    Cytogenetic studies: chromosome numbers and morphology
    Studies with banding
Cancer of the Kidney

Tumors of the Male
Tumors of the Testis
    Further comments on tumors of the testis
Cancer of the Prostate

Tumors of the Lung
Cancer of the Lung (Bronchus) and Larynx

Malignant Melanoma
Some Chromosome Findings in Malignant Melanoma
Chromosome #1, Malignant Melanoma, and Other Cancers

Tumors of the Thyroid and Adrenal Glands
Further Comments on Tumors of the Thyroid

Miscellaneous Tumors

Cancer of the Uterine Cervix
Pathology of Cervical Lesions

Problems Related to Cytogenetic Studies in Cancer of the Cervix
Chromosomal Findings of Cervical Lesions
    Invasive carcinoma
    Microcarcinoma (microinvasive carcinoma)
    Carcinoma in situ and dysplasia
    Mild dysplasia
    Evidence for clonal evolution in cervical lesions

Certain Aspects of Chromosomal Changes in the Uterine Cervix
    Polyploidy
    Aneuploidy
    Chromosome breakage

Summary of Chromosome Findings in Lesions of the Uterine Cervix
    Some facets of chromosomes in cervical cancer

Brain Tumors

Meningiomas
    Studies prior to banding
    Banding studies
    Summary of cytogenetic findings in meningioma

Malignant Gliomas (Astrocytic)
Oligodendrogliomas
Ependymomas
Medulloblastomas and Neuroblastomas
Retinoblastomas and Optical Gliomas
Neurinomas (Schwannomas)
Pituitary Adenomas
Metastatic Tumors to the Nervous System
Double Minute Chromosomes (DMS)
Comments on DMS Observed in Human Tumors
Homogeneously Staining Regions and Double Minute Chromosomes

The following sections will discuss the chromosomal findings in benign tumors, solid cancers, and metastatic lesions. In some respects the division between some benign lesions and malignant ones is not well defined histologically, biologically, and karyotypically. I believe that the presence of cells with an aneuploid cytogenetic picture is prima facie evidence of cancerous transformation, such as may be observed in areas of certain benign tumors, e.g., polyps of the colon, lesions of the uterine cervix. In spite of the large body of data collected to date on the chromosome constitution of human cancers, cytogeneticists have been frustrated by the disappointment of not finding a characteristic or specific karyotypic change in each human cancer. It is hoped that new methodologies will prove to be more successful in unraveling the remarkable cytogenetic picture present in the various human cancers.

Tumors considered to be benign, i.e., meningiomas, pituitary adenomas, and other CNS tumors, are discussed below. Furthermore, the bulk of the tumors to be described below was studied before banding and, hence, some of the results lack karyotypic specificity. It is hoped that findings based on banding techniques will shed further light on the significance of the absence or presence of chromosomal changes in tumors considered to be benign.

Benign Tumors

The interpretation of chromosomal findings in benign tumors is not simple. Whether a tumor is truly benign is often a question not easily answered by the pathologist, and the finding of abnormal metaplasia in a benign tumor is not easily interpreted. Generally, most benign tumors have been found to be perfectly diploid. When chromosomal changes are found in benign tumors, e.g., polyps of the colon or dysplasia of the endometrium, they most probably indicate cancerous transformation of that section of a benign lesion. This is particularly likely because most of the benign tumors in which aneuploid cells have been found do have a tendency to become clinically malignant. Thus, abnormal cytogenetic findings are indicative of such malignant transformation.

Our knowledge of the chromosome constitution of benign tumors is rather poor when compared with that of primary malignant tumors. A very low mitotic rate, which represents one of the histopathological essentials for making the diagnosis of a benign tumor, is the main limiting factor. Attempts to overcome this difficulty were made by culture of benign tumor explants. From those instances in which the culture time did not exceed 5–10 days, it can be assumed that the chromosomal findings possibly reflect the in vivo situation of the tumor cells. The results from culturing the cells for prolonged periods of time, however, cast serious doubt on the validity of such an assumption.

There is no reason not to believe that the biology of human cancer does not represent a wide spectrum, with the extremely anaplastic and invasive cancers constituting one end of the spectrum and relatively benign (histologically), slow growing tumors constituting the other end of the spectrum. The latter, although not considered from a histologic or clinical viewpoint to be malignant tumors, can be accompanied by karyotypic changes, and it would appear that these may be a more sensitive and reliable index of malignant transformation than afforded by the histologic or clinical picture (Carlevaro, Rossi, et al., 1978). Meningiomas can serve as a cogent example. Even though histologically these tumors probably constitute the extreme end of the spectrum due to their benign histologic appearance, slow growth, and failure to metastasize, there appears to be little doubt that transformation to a sarcomalike picture may occur in occasional tumors of this type. This fact correlates with the finding of frequent chromosomal changes, some of them of rather specific nature, in a high proportion of meningiomas; this would seem to indicate that these cells are truly malignant, but the level of malignancy is at the extreme end from those tumors that are extremely invasive and metastasize readily. In my opinion then, such tumors (meningiomas) are malignant.

Before discussing the chromosomal findings in benign tumorous conditions, the reader should realize that all conditions that have atypical histological changes suggesting transformation into frank cancer were excluded from this section. Thus, the chromosome constitution of dysplasia of the cervix uteri, of carcinoid bronchial adenoma, adenoma of the colon with atypia, Bowen's disease, and atypical hyperplasia of the endometrium, will be considered in their appropriate sections.

There is no firm basis for the statement that malignant tumors have an abnormal and benign tumors a normal chromosome constitution. It has been shown that the great majority of "benign tumors" of the nervous system have striking chromosome anomalies. Thus, Mark (1970 a, b, 1971a-c, 1972a) demonstrated that meningiomas, pituitary adenomas, and neurinomas have characteristic karyotypic profiles. Most polyps of the colon were revealed to have definite chromosomal anomalies, and, though less convincing, a similar situation seems to be true for cystic adenoma of the ovary, and some hydatidiform, noninvasive moles. Reports on a normal chromosome constitution of adenomas of the thyroid, giant cell tumors of the bone, and cystic mammary tumors are based on too scanty data to make a general statement that most benign tumors are chromosomally normal. We are, thus, faced with the fundamental question as to whether the generally accepted view that chromosomal anomalies indicate a malignant state is wrong or whether all benign tumors with chromosome anomalies should not be considered preneoplastic, if not already neoplastic conditions. It is Mark's (1971b) opinion that neither considerable heteroploidy nor structural chromosome aberrations are "reliable cytogenetical indicators of malignant transformation." In this context, the situation in hematological disorders must be considered. Nowell (1971a) showed that the development of aneuploid clones in a significant number of cases was followed by leukemic transformation, whereas other patients lived for years with a more benign hematological disease. In CML the appearance of clones with additional karyotypic anomalies usually heralds acute transformation. In AL the demonstration of chromosomal anomalies may precede hematological relapse for weeks. From these observations it appears that in hematological disorders cytological and karyotypical parameters do not necessarily coincide. The problem with solid benign tumors is that such tumors are usually removed in toto and, therefore, we do not know what the behavior of a benign tumor with chromosome anomalies would be if a par of it were left in situ. Such a novel situation was reported by Benedict, Porter, et al. (1970); in this case a meningioma that was chromosomally highly abnormal did not show any signs of malignancy when it was thought to have been completely removed; however, when it recurred 18 mo later the meningioma contained similar chromosome anomalies but was definitely malignant. Experiments with benign tumors in animals should be performed in order to clarify whether so-called benign tumors with chromosome anomalies have a greater tendency to progress to cancer than those without. As long as these questions remain to be answered, I continue to believe that the chromosomal changes are indicative of a malignant state. To begin with, such a malignant state may not become apparent except at the cellular level. In some tissues chromosomal anomalies may not interfere grossly with maturation and differentiation of the affected cells, as was shown for Ph^1-positive BM cells. Thus, I agree with Mark (1970a, 1971b) that chromosomal anomalies do not necessarily coincide with histological and/or clinical malignancy, but at the cellular level the development of chromosome aberrations signals a change to a malignant state. Hence, even though benign tumors with chromosome aberrations may be referred to by some as benign on a histological basis, I feel that they are malignant at the chromosomal level.

Benign Tumors Without Chromosome Anomalies

Socolow, Engel, et al. (1964) studied the chromosome constitution of five *follicular adenomas of the thyroid*. Despite the presence of some hyperdiploid metaphases, four of them were considered to be chromosomally normal. In one microfollicular adenoma with marked nuclear pleomorphism 75% of the metaphases were pseudodiploid, but no consistent karyotypic pattern was apparent. The patient had received therapy to his head and neck 30 yr before he developed an adenoma of the thyroid; thus, the authors could not decide whether the chromosomal anomalies observed were exogenous or endogenous in nature.

Ishii (1965) analyzed the chromosomes of five *giant cell tumors of the bone*. None of the tumors showed anomalies in direct preparations. An increasing percentage of polyploid and aneuploid metaphases appeared with prolonged culture time.

Toews, Katayama, et al. (1968b), performed chromosome studies on four *cystic mammary tumors*. No abnormal metaphases could be seen. However, the number of metaphases studied was so small that no definite conclusions can be drawn.

A *uterine fibromyoma* was found to be diploid.

Hauschka (1961) briefly refers to the normal chromosome constitution of one *verruca papilloma* he had studied, without providing exact data.

Benign Tumorous Conditions with Chromosome Anomalies

Adenomatous polyps of the colon without atypia have been studied by a number groups.[1] Of a total of 20 such tumors in which a sufficient number of well-spread metaphases could be obtained, 9 were described to have normal chromosomes. Six tumors had abnormal clones in the hyperdiploid (47–50 chromosomes) range. In two cases the chromosome counts ranged between 23–85 and 47–56, respectively, with no consistent karyotypic pattern. Marker chromosomes were absent in all the cases.

On the basis of microspectrophotometry, H. F. Stich, Florian, et al. (1960) attempted to estimate the DNA content of specimens from eight patients with rectal polyps and found a diploid amount in six and a predominant tetraploid DNA in the other two. Four additional samples from one patient with hereditary multiple polyposis were examined, and the DNA content was found to be in the diploid range in all of them. Because such polyps are often accompanied by chromosome changes that may not significantly affect the total amount of DNA, the method used would miss such subtle karyotypic changes.

In polyps with some atypia, the chromosomal evolution was characterized by the preferential involvement of chromosomes in groups C and D. Other chromosome groups were rarely involved. Examination of the literature before banding revealed that out of 16 polyps of the large bowel with deviating cells and in which karyotype analyses were recorded, 13 had stem lines or variant cells with involvement of C and/or D chromosomes. This selective involvement is a clear indication of a nonrandom karyotypic pattern in colonic polyps.

An interesting observation is related to *Gardner's syndrome*, a condition in which colonic polyps occur in high frequency.[2] Increased endoreduplication with tetraploidy was shown to occur in cultures established only from tissue containing epithelium (skin and colonic polyps) from patients with this syndrome, whereas those from blood (lymphocytes and lymphoid suspensions) and connective tissue from the same patients did not show increased tetraploidy.[3] The author advocated the use of this approach for the detection of the Gardner gene in families at risk.

Studies of several types of *bronchial adenomas* apparently revealed deviations in chromosome number (Falor, 1971b).

In a brief report (Khishin, El-Zawahri, et al., 1976), the karyotypes of 16 patients with *fibroadenoma of the breast* revealed 12 normal and 3 abnormal chromosome constitutions. In view of the authors' experience in cancer of the breast, the data suggest that these three cases with abnormal karyotypes are precancerous stages, and the cytogenetic analysis may be used as an early diagnostic aid.

Cystic adenomas of the ovary were subjected to chromosome analysis in two instances and abnormalities were found in both. Hyperdiploid clones with extra chromosomes in group #6–12 were seen in one case, and some polyploid metaphases with an acrocentric marker in the other. Makino, Sasaki et al. (1965) investigated the chromosome constitution of hydatidiform, noninvasive moles. In addition to some metaphases with markers they noted an increase in tetraploid and aneuploid cells in this tumor type as compared to normal chorionic villi. No consistent karyotypic anomaly, however, was apparent among the 13 moles studied.

Cancer: Primary and Metastatic

Several facets of karyotypic examination of human primary and metastatic tumor cells (Figure 132) are worthy of emphasis. When such tissues are examined they are usually well established and, hence, it has been impossible to ascertain the genesis of the chromosomal changes in relation to cancer development. In

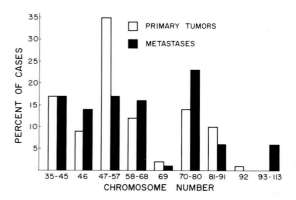

132 Distribution of the modal chromosome numbers in primary and metastatic tumors. Generally, the latter were characterized by higher chromosome modes than the primary tumors.

addition, only a small fraction of human primary cancers are suitable for chromosome analysis, because most of them do not contain sufficient metaphases for analysis. Inasmuch as, under such circumstances, resort has to be made to short- or long-term culture, some doubt may be raised regarding the exact karyotypic picture in the original tumor. The same features apply to solid metastatic tumors. Cancerous effusions, on the other hand, usually contain sufficient metaphases for analysis, and this probably accounts for the fact that most of the original publications dealing with human cancer chromosomes were established with effusion materials.[4] Furthermore, the quality of the chromosomal morphology in effusion is usually excellent and lends itself to careful and reliable analysis.[5]

The modal chromosome numbers in cancer, primary and metastatic, are shown in Figure 126 and Table 82.[6] It is obvious that remarkable variation characterizes the modal chromosome numbers of these tumors, ranging from severe hypodiploidy (Table 83) to more than 100 chromosomes.[7] The tumors shown to have 46 chromosomes were pseudodiploid. Generally, metastatic cells tended to have a higher chromosome number than primary tumors. Even when tumors have the same chromosome number, their karyotypes differ to the extent that no two human cancers have been shown to have identical karyotypes. Thus, even tumors

Table 82 Distribution of modal number of chromosomes in 209 cases of human primary cancer and in 305 cancerous effusions reported in the literature

	Range of modal chromosomal number	Percent of cases	
		Primary cancers	Cancerous effusions
Hypodiploid	32–40	3	1
Diploid	41–57	60	42
Triploid	58–80	29	44
Tetraploid	81–133	8	13

originating in the same organ or tissue have remarkably different karyotypes, and no cytogenetic findings have been found to characterize any human tumor. The exception to the statements in this paragraph has already been mentioned, i.e., meningioma.

Even though scarce observations are available on the karyotypic findings in primary tumors (Tables 84 and 85) before and after recurrence or therapy, the observations with metastatic lesions, primarily effusions of one type or another, indicate that the karyotypes of cancer cells tend to be rather stable. This stability characterizes the metastases and recurrences of tumors, and the appearance of an effusion in a previously uninvolved body cavity is usually accompanied by a chromosomal picture similar to that found before therapy or spread of the disease. In general, the metastatic lesions contain karyotypes that reflect those of the original tumor, though the metastatic cells tend to have a higher ploidy and more variability in the chromosome number.[8]

Markers occur with a higher frequency in cancer cells (ca. 50%) than in leukemic ones (< 30%) except, of course, CML in which the Ph^1 occurs regularly.[9] The morphology and number of markers differ from case to case and even

Table 83 Unusually low ploidy in neoplasia (40 chromosomes or less)

Chromosome no.	Diagnosis	Reference
35	Cancer of ovary (effusion)	Spriggs, Boddington, et al., 1962b
30–35	Metastasis to brain from cancer of lung	Spriggs, Boddington, et al., 1962b
40	Cancer of breast (effusion)	Ishihara, Kikuchi, et al., 1963
41	Cancer of breast (effusion)	Ishihara, Kikuchi, et al., 1963
39	Cancer of ovary (effusion)	Ishihara, Kikuchi, et al., 1963
40	Cancer of larynx	Yamada, Tagaki, et al., 1966
39	Metastatic cancer of breast	Yamada, Tagaki, et al., 1966
35	Cancer of breast (effusion)	Sandberg, Yamada, et al., 1967
40	Cancer of ovary (effusion)	Sandberg, Yamada, et al., 1967
40	Cancer of ovary (effusion)	Sandberg, Yamada, et al., 1967
32	Metastatic malignant melanoma	Visfeldt, Povlsen, et al., 1972
39	Erythremic myelosis	Heath and Moloney, 1965a
30–35	Multiple myeloma → AL	Dahlke and Nowell, 1975
27	ALL	Kessous, Corberand, et al., 1975a
27	ALL	Oshimura, Freeman, et al., 1977a
39–40	Sézary syndrome	Bosman, van Vloten, et al., 1976
24	BP of CML	Daniel, Francis, et al., 1978
26	ALL	Prieto, Badia, et al., 1978b
16–24	BP of CML	Kohno and Sandberg, unpublished

Table 84 Summarized information on patients with malignant solid tumors

Case no.	Patient	Age yr	Sex	Diagnosis
1	FV	43	M	Retroperitoneal histicytoma
2	DL	17	F	Sarcoma of thigh
3	RW	72	F	Carcinoma of transverse colon
4	JH	41	F	Carcinoma of breast
5	PW	21	M	Teratosarcoma of testis
6	RH	43	M	Seminoma of testis
7	DH	57	M	Liposarcoma of thigh
8	HH	61	F	Bronchogenic carcinoma
9	DS	24	M	Embryonal carcinoma of lung (metastatic testicular tumor to lungs)
10	GB	65	F	Adenocarcinoma of rectum
11	HS	54	M	Recurrent cancer of colon
12	CS	72	M	Carcinoma of rectum
13	FG	61	F	Carcinoma of sigmoid colon
14	PM	56	M	Adenocarcinoma of rectum
15	FH	63	F	Carcinoma of pancreas
16	WS	56	M	Adenocarcinoma of sigmoid colon
17	MC	76	F	Carcinoma of rectum
18	DC	63	M	Carcinoma of transverse colon
19	HW	71	F	Carcinoma of rectum
20	GM	58	F	Carcinoma of anus
21	CM	44	M	Carcinoma of transverse colon
22	MW	59	F	Carcinoma of colon
23	PC	55	M	Carcinoma of colon
24	EM	55	F	Carcinoma of cecum
25	DB	13	F	Ewing's sarcoma of right ileum
26	MK	68	M	Adenocarcinoma of rectosigmoid and anus, villous adenoma with atypia and cancer *in situ* of rectosigmoid colon
27	EC	65	F	Carcinoma of rectum
28	WE	53	M	Cecal carcinoma (glandular) of right colon arising in adenomatous polyp
29	WW	61	M	Carcinoma of rectum
30	BH	83	F	Diverticulum of colon
31	NN	58	M	Carcinoma of colon

Data from Sonta, Oshimura, et al. 1977, and Sonta and Sandberg, 1978b.
This table presents some of the clinical information for the patients shown in Table 85.

among the metaphases of the same tumor. In the latter case it is usually the number of markers that varies and not their morphology. Markers may be much larger or much smaller than any of the chromosomes in the human cell. The genesis of markers still remains obscure in most cases, though the application of banding analyses may (and already has in some cases) help to clarify the origin of markers in human cancer (and leukemia).[10]

Several reports have stressed the presence of characteristic markers in certain tumors. Thus, a large marker, with or without a secondary constriction, has been thought to be characteristic for testicular tumors.[11] In my opinion, this is unlikely because the morphology of this marker varies from case to case, and a very similar marker has been found in tumors of different histological types. In addition, similar markers have been demonstrated in other unrelated tumors and in the cells of AL.[12] Furthermore, many of the testicular tumors do not contain such a marker at all.[13] The significance of other recurrent markers, e.g., a large submetacentric chromosome replacing an A-group autosome (Api-chromosome) in cancer of the uterine cervix[14] or a medium-sized acrocentric marker in colon tumors[15] remains to be established. Characteristic or not of any tumor, the presence of a marker or markers in a cell undoubtedly indicates the malignant character of such a cell.[16] It must not be assumed, however, that cancer cells require markers for their malignant behavior, because markers may not be present in all of the cells of the same tumor, or may even vastly outnumber the cells with markers, and be just as biologically virulent as the cancer cells with markers. A significant correlation in future studies will be that between the presence or absence of markers and their morphology and the biologic and clinical behavior of tumors.

In contrast to the observations with normal and leukemic cells (cultured and/or otherwise) in which the chromosomal number is usually characterized by a sharp mode[17], the situation in cancerous cells is more complicated. The distribution of the chromosome counts in cancers is often rather "flat" with the modal cells constituting as low as 10–20% of the counts.[18] This arrangement allows for a wide variety of karyotypes; in fact, in some of the cancers the possibility of several different stem lines with karyotypes unique for each and existing side-by-side in the same tumor or its metastases cannot be ruled out.[19] It is my opinion that, when such a situation exists, the various karyotypes are the outgrowth of variations upon the original karyotype. Hence, it is not surprising that tumors and their metastases may be characterized by karyotypes that show a variation from one section to another, depending on which cell type has the upper hand in its growth potential at the time of examination. When a mode becomes established, however, it tends to remain stable and usually persists for the duration of the life of the tumor, including its recurrences.

The persistence of the karyotypic picture in cancer is well illustrated by our observation of the chromosome constitution of a cancerous peritoneal infusion in which an aneuploid karyotype was present.[20] After chemotherapy was instilled into the peritoneal cavity, the cancerous effusion disappeared. Many months later the effusion recurred, but this time in the pleural cavity, and the karyotypic findings were almost identical to those observed in the previous peritoneal effusion. When the latter also recurred later, the chromosomal findings were of the same nature as in the original peritoneal effusion.

In all probability, the occurrence of aneuploid cells in an effusion is very strongly indicative of cancer (or other neoplasia) and the opposite findings, i.e., the presence of only diploid cells, probably is sufficient to rule out a cancerous state. These statements are based on the observations in many laboratories and of a large number of cases.

In regard to characteristic karyotypes of cancer and leukemia, particularly of the former, such tumors are often characterized by a "flat" distribution of the chromosome number, so that the modal cells may not constitute $>$ 20–30% of the metaphases examined. Under such conditions, only a representative karyotype can be arrived at and cognizance must be had of the deviations that may occur from this karyotype. These deviations may only be minor variations on the representative karyotypic theme, e.g., the presence of duplicates of a chromosome or marker, or they may be extensive and in some cases variable in the extreme. Generally, the modal cells constitute the preponderant percentage of the abnormal metaphases in leukemia, whereas in the case of solid tumors, more often than not they make up $<$ 50% of the cells. Thus, in the case of solid cancers it may be difficult to arrive at a characteristic karyotype and only a representative one can often be presented, with the limitations that such an interpretation of the cytogenetic profile of a tumor offer.

Revealing great variation from tumor to tumor, the cytogenetic data in human cancers have failed to disclose any simple changes common to all cancers, which might be comparable to that giving rise to the Ph^1 in CML; either there is no such change, or it is at present undetectable. However, the improved resolution that can be achieved with the use of chromosome banding techniques has, as yet, made little impact on the field of cancer cytogenetics. Perhaps we should look for common changes, specific for a given tumor or histologic type, which may be more readily discernible in the premalignant than in the invasive phase of tumor development. An example of such a change may be the loss of a #22 in meningiomas.

The association of diploidy with some pri-

Table 85 Distribution of chromosome numbers and banding karyotypes in patients with primary tumors*

Patient	Number of cells counted	Cells with chromosome no.																	Chromosome no. of mode	Karyotypes
		≤43	44	45	46	47	48	49	50	51	52	53	54	55	56–65	66	67	≥68		
FV	29					1	4	3	19									2	50	50,XY,+5,+12,+13,+21
DL	18		1		1	2	14												48	48,XX,+4,+5
RW	55				1		1		1	5	3	10	28	2	2			2	54	54,XX,+1,+2,+8,+11,+13,+14,+21,+21
JH	46	1		3	11		1	5	17	2	2	1						3	46/50	46,XX/50,XXX,+8,+12,+21
PW	41								1		2	1	1		8	8	15	5	67	67,XY,+1,+3,+4,+5,+6,+7,+7,+8,+10,+11,+13,+13,+14,+16,+17,+18,+20,+20,+21,+21
RH	45	1		5	11		1	4		1	3	19							46/53	46,XY/53,XY,+3,+4,+6,+7,+11,+13,+19
DH	24	1				1	2	5	13	1					1				50	50,XY,+3,+6,+8,+11,+14,−22,17p+
HH	39		1	3	6	27		1										1	47	47,XX,+2
DS	37	1	2	2	31													1	46	46,XY
GB	43							1		3	5	8	14	2	6			4	54	54,XX,+5,+6,+7,+9,+11,+21,+21,+1q−
HS	55	3	2	5	43													2	46	46,XY
CS	20	1		1		4	2	11	1										49	49,XY,+4,+5,+8
FG	15			2	2	9			1									1	47	47,XX,+8,+13,−14
PM	18	3	1	4	3	6	1												47	47,XY,−1,−2D,+4mar(?)
FH	7				7														46	46,XX
WS	24						1			1	5	10	2	4	1				53	53,XY,+1,+3C,+D,+2mar(?)
MC	36	2			1	3	29	1											48	48,XX,−7,+15,+17,+mar

DC	25						1		4	54	54,XY,+1,+5,+9,+17,+19,+20,+21,+21
HW	19		1		2	2	14			49	49,XX,+8,+17,+20
GM	39			1	3	2	9	18	2	1	50	50,XXXX,+8,+8,+13
CM	30	1		1		1				1	26	73	73,XXXY,+1,+1,+2,+3,+4,+5,+6,+7,+7,+8,+9,+10,+11,+11,+12,+13,+13,+14,+15,+17,+17,+18,+18,+20,+21,+22
MW	38	1	3	34	5	9					46	46,XX
PC	73	3	1	4	19		29	1		1	1	49/46	49,XY,+3,+8,+21/46,XY
EM	41	2		4	35	2	5					46	46,XX
DB	29	2	1	1	1	4	15	1	2			49	49,XX,+3,+13,+14
MK	12	4	7			2						45	45,XY,-1,-6,-14,-16,+21,+2mar
EC	50	1			1	2	1	5	14	3	2	1	15	50/63	50,XX,+8,+14,+21,+mar/63,XXX,+2,+4,+5,+6,+8,+10,+13,+14,+15,+16,+18,+21,+21,+4mar
WE	47	1	2						1					46	46,XY
WW	50		1	44		1	1		1	2	5	38	1	55	55,XY,1p+,-10,-17,+2,+2,+4,+8,+8,+15,+18,+20,+21,+2mar
BH	18				1	3	3	10	1					50	50,XX,+1,+5,+17,+19
NN	62		1	2	3	44	6	2	2	1		1		48	48,XY,+1,-5,+8,+17

For some clinical information on these cases see Table 84.

mary cancers, though relatively rare, has been reported,[21] and the possibility exists that the metaphases observed were of cancerous origin, even though it cannot be ruled out with certainty that the diploid cells were not of noncancerous origin, i.e., host elements of a nonmalignant nature. Even though we have encountered only rare diploid primary or metastatic cancers in my laboratory, the finding of such tumors should not be surprising, however, when viewed in the light of karyotypic observations in human AL in which at least 50% of the cases have diploid karyotypes in their leukemic cells. The presence of diploidy in some tumors supports my contention that the chromosomal changes in human cancer are secondary phenomena and not causally related to the cancers.[22] The tremendous variability of the karyotypes from one case to another may be a reflection of the variability of the human genotype in its repsonse to the cancerous state.

The presence in some tumors of cells with normal or apparently normal karyotypes requires special consideration. Many tumors, largely experimental tumors of animals, have been found to have normal chromosome complements in the early stages. Often, there are questions of interpretation that may be difficult to answer. For example, are the cells with normal karyotypes stromal or inflammatory cells or are they tumor cells? Are the karyotypes really normal or have they undergone some small as yet undetectable but significant change? In my mind there is little doubt that, were human cancers to be examined in their earliest stages, diploidy would be much more common than observed when the tumors are studied cytogenetically at a rather advanced stage of the disease; for in my opinion the chromosomal changes, however nonrandom or specific they may be for a given tumor, are a rather late expression of the already existing neoplastic state with the chromosomal changes being epiphenomena to the primary causative factors.

Tumors of the Alimentary Tract

Tumors of the Oral Cavity and Esophagus

Even though the oral cavity is characterized by a relatively high frequency of tumors of varying histology, chromosome reports on these tumors have been rather scarce. Furthermore, no odontogenic tumors have been examined, and the published reports deal primarily with oral squamous cell carcinoma. Early reports dealt with culture of such carcinoma cells which were found to be characterized by marked polyploidy and fragmentation even though exact counts could not be established.[1] With microspectrophotometry, squamous cell carcinomas of the tongue and buccal mucosa were found to have considerable variation of DNA, suggesting aneuploidy.[2] Examination of three squamous cell carcinomas of the maxilla revealed chromosome modes of 71, 73, and 69 with none of the tumors having > 40% of the cells in the mode.[3] In a fourth tumor the counts ranged from 63 to 104. In another study 48-hr cultures of three carcinomas of the oral cavity were undertaken;[4] all were found to have counts with considerable spread with no notable markers or prominent stem lines. A report[5] on the chromosome picture in squamous cell carcinoma of the esophagus revealed in six cells the number to be 43 or less. Only one karyotype was constructed and contained a large acrocentric marker, as well as a very long marker similar to an A- or B-group chromosome, probably representing association of missing acrocentrics. Cells in the malignant ascites from an epidermoid esophageal cancer were found to have a mode of 50 chromosomes, including markers.

In one esophageal cancer established in long-term culture the chromosome number ranged from 54 to 60 with several marker chromosomes (K. M. Robinson, 1977). In a short report (Gripenberg, Ahlqvist, et al., 1977) the findings in nasopharyngeal carcinomas indicated that in each case a mixture of diploid and aneuploid (triploid-tetraploid region) cells was observed. (A similar mixture was seen in a biopsy of a solitary liver metastasis of a rapidly progressing renal cancer.) The authors stated that diploid cells in these malignancies represent normal host cells involved in an immune response reaction, with the neoplasm triggering the mitotic activity of the host cells.

To date, insufficient data are available to present any correlation between the chromosomal findings and either the site or histopathologic picture of oral tumors; no reports have appeared on benign or precancerous lesions of the oral cavity.

A carcinoma of the esophagus from a South African Zulu was cultured in vitro, and the

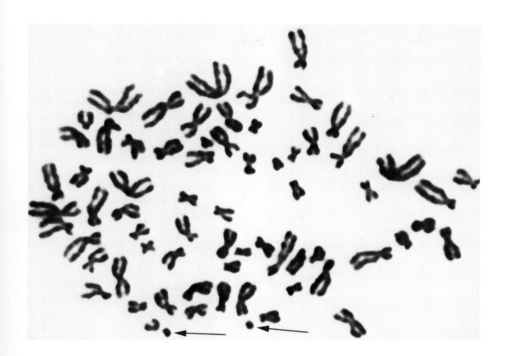

chromosome constitution was found to consist of three unusual markers and hyperdiploidy (Bey, Alexander, et al., 1976).

Tumors of the Stomach

Until recently, most of the information regarding the chromosomes of cancer of the stomach was based on an analysis of metastatic cells in ascites (Figure 133). Even though sporadic cases of cancer of the stomach analyzed cytogenetically appeared prior to 1959,[6] the first sizable series was published in that year in which 10 cases with cancer of the stomach were examined:[7] three specimens came from solid tumors and seven from peritoneal effusions. One of the tumors had a mode of 56, whereas the two others had a considerable spread in the chromosome number, i.e., 89–112 and 80–109, respectively, without a sufficient number being present in the modal area. In five effusion samples the modal numbers ranged from 42 to 89 with two of the effusions having bimodal numbers. In one case three effusion samples were obtained over a 6-wk period in a patient who had been treated with radiation for cancer of the stomach. The first sampling showed a predominant stem line of 45 chromosomes and

133 Metaphase with 69 chromosomes from a peritoneal effusion of a patient with cancer of the stomach. Arrows point to two small markers. The other chromosomes appear to be compatible with normal morphology.

the last one a predominant stem line of 42, suggesting that the cells with 42 chromosomes, which were present in the original sample, grew at the expense of the cells with 45 chromosomes. In another case a sessile adenomatous polyp of the stomach showing malignant transformation had a mode of 58 chromosomes but was also accompanied by a significant number of diploid cells. In another case the cancer cells in effusion were found to have a modal number that changed from 60 to 64 over a period of 2 mo.

In studies on the evolution of stem lines in gastric carcinoma most authors have observed a diminution in the number of cells with extreme hyperploidy.[8] This was true for cells without apparent relation to existing stem lines and for those which most likely originated from stem cells by endoreduplication. Thus, in three samples from a gastric cancer effusion taken over a period of 3 mo from a patient who had been irradiated before and during the sampling, three main cell populations were shown:[9] two near-diploid and one in the range of 80–84 chromosomes. In another case[10] cells with 45 chromosomes were of highest frequency in the first sampling (28%) and of lower frequency in the second and third samplings. Stem line cells with 42 chromosomes were lower in the first samplings and their occurrence markedly increased in the second (34%) and third (31.2%) samplings. Cells with chromosome numbers of 80–84, which probably arose by endoreduplication from stem cells containing 42 chromosomes, gradually decreased with time in the tumor.

In another investigation[11] cells from a cancerous effusion from a carcinoma of the stomach with metastases were examined. Four different samples were obtained over a period of 1 mo after the patient had been treated with Carzinophilin before and after the chromosome analyses. Each examination showed a marked peak of cells having 62 chromosomes, which persisted without being affected by treatment and even increased in number in the third and fourth samples. On the other hand, cells with hyperploidy (octoploidy) and chromosomal breaks greatly diminished in number.

An extensive series of patients with cancer of the stomach whose effusions were studied cytogenetically appeared in 1964.[12] In most of these clear stem lines could be detected, despite a rather remarkable spread of the chromosome number. Four of the tumors appeared to contain an essentially diploid karyotype. Two tumors were examined before and after treatment with chemotherapy and radioactive gold and in both therapy failed to change the chromosome constitution of the stem lines which contained 51 chromosomes in the first and 63 and 65 in the second tumor.

The presence of markers in cancer of the stomach has been known for some years, e.g., two ring chromosomes in a malignant effusion from a patient with cancer of the stomach who had a modal number of 80 chromosomes were observed during the early studies of cytogenetics of human tumors.[13] However, acrocentric, metacentric, and other markers have been described in cancer of the stomach, but no typical marker has emerged for this tumor.[14] Furthermore, the cytogenetic data from gastric carcinoma are characterized by considerable variability of the chromosome and stem line numbers (range 42–120) and the type of abnormalities encountered. Studies of possible evolution of stem lines within a single tumor have not yielded convincing results. In two studies apparently one stem line appeared to gain advantage over another. However, subsequent studies failed to confirm such observations in which it was shown[15] that stem lines with hypodiploid numbers of 42 chromosomes were replaced by stem lines with a higher number of chromosomes and in another study[16] stem lines with hyperdiploid numbers of 62, 51, 63, and 65 remained unchanged even after various forms of therapy. Thus, the question remains unanswered as to whether hyperdiploid stem lines in effusions of gastric cancer are generally more stable and adaptable than cells with a hypodiploid number. More importantly, examination of more tumors of the primary variety will have to be established to gain a clearer picture of the exact karyotypic findings in these tumors, rather than basing the cytogenetic data and conclusions on the behavior of cells in ascites in which considerable latitude is gained by such cells in the evolution of their karyotypic picture.

One specimen of an adenomatous sessile polyp of the stomach had clearly aberrant chromosomes with a mode of 58 and chromosome numbers ranging from 45 to 106.[17]

Even though several analyses of the involvement of chromosomes in gastric carcinoma have appeared, unfortunately, these were based on the exclusion of abnormal chromosomes which may have a more direct importance on the genesis of the karyotypic picture than the variability in the numerical picture. More importantly, because banding analysis of the chromosomes has not been performed on a large number of cancers of the stomach, it is difficult to be certain about the various conclusions reached, i.e., that excessive C-group chromosomes are most likely caused by a centric fusion type of translocation between acrocentrics, which according to one analysis was consistent with the finding of an underrepresentation of acrocentrics in groups D and G.[18] We have challenged this hypothesis on the basis of vector analysis of cells in cancers and in AL.[19]

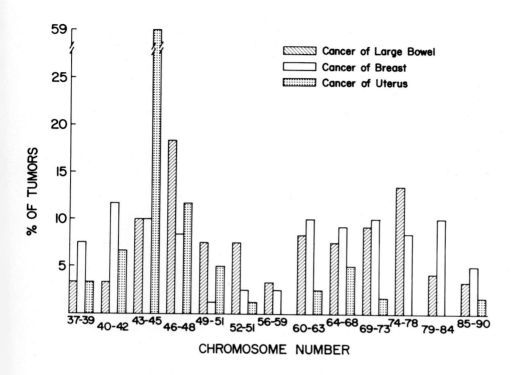

134 Modal chromosome numbers in cancer of the large bowel, breast, and uterus.

Some Recent Studies in Cancer of the Stomach

A malignant effusion with a modal number of 51 including two markers has been described (Dallapiccola and Tataranni, 1969). There was a number of metaphases with 46–50 chromosomes whose karyotypes suggested a process of clonal evolution with successive acquisition of E-group chromosomes. Šlot (1967a) followed an untreated malignant ascites for a 2-mo period; the modal number changed from 60 to 64 but the karyotypes were dissimilar and the change could not be explained simply on the basis of the addition of a few chromosomes.

Messinetti, Zelli, et al. (1968), in addition to studying a sessile adenomatous polyp, followed a primary carcinoma the modal number of which was 48, with various numerical and structural changes. A tumor maintained in organotypic culture for several years has been described (Grouchy and Wolff, 1969). The tumor originated from a hepatic metastasis in 1962, the chromosomes were studied in 1966 when the mode was 64 and again in 1968 when it had shifted to 69. On both occasions there were only three or four chromosomes in the D group and a dicentric marker.

In a pleural effusion secondary to an anaplastic cancer of the stomach (Granberg, Gupta, et al., 1973), Q-fluorescence revealed a modal number of 49 with trisomy for #3, #7, and #8 and the following structural changes: 1q+, 4q+, a possible reciprocal translocation involving the arms of #12 and an X-chromosome and 17q+.

Tumors of the Colon, Cecum, Appendix, Rectum, and Anus

Benign lesions of the colon. A number of chromosomal investigations of benign or precancerous lesions has been carried out on adenomas or polyps of the colon. These lesions are regarded as a most important "etiologic" factor in cancer of the colon and, in most cases, are of genetic origin.

Utilizing the technique of microspectrophotometry on Feulgen-stained nuclei, a study as early as 1960[20] attempted to estimate diploidy based on the DNA content of colonic polyps. Specimens from eight patients with rectal polyps yielded diploid DNA content in six and a predominant tetraploid one in the other two. Four additional samples from one patient with hereditary multiple polyposis were examined and the DNA content found to be in the diploid range in all. It should be noted, however, that microspectrophotometry is not precise enough to establish the exact number of chromosomes and, thus, estimation of the diploid value of the specimens noted above may represent near-diploidy or pseudodiploidy with one or more chromosomes being in excess or absent, or abnormal chromosomes being present.

Cancer of the large bowel may often commence in polyps, the smallest carcinomas being seen as a malignant transformation in polyps rather than arising de novo from the mucosa. Studies on polyps, therefore, have a direct relevance to cancer of the colon.[21] In one study[22] it was shown that 1. cells with normal karyotypes may be present in polyps (these may or may not be epithelial cells); 2. cells with abnormal karyotypes may be found alone or may be present together with diploid cells; 3. the chromosome abnormalities commonly, though not invariably, involve simple changes such as the addition of one or two group C chromosomes, and their pattern may indicate or suggest a clonal relationship. The chromosome numbers of cells with abnormal karyotypes are often near-diploid (e.g., 47 or 48 chromosomes), but a polyp with a mode of about 80 chromosomes has been described;[23] and 4. especially when there is histological evidence of early malignant transformation, more complex changes, including markers, may be found.

In 1963[24] the chromosomes were determined in benign adenomatous polyps removed from a 68-yr-old man. The chromosomes present were obtained by a direct technique and yielded two karyotypes, both apparently of the diploid variety. In another case a sigmoid polyp with marked villous and adenomatous components was examined, and the chromosomes of 12 cells were counted; all except two were abnormal, containing 47 chromosomes with a supernumerary chromosome in group F. A more extensive study on colonic (precancerous) lesions appeared in 1967[25] in which the authors classified their specimens into various groups. Chromosomal aberrations, mainly hyperploidy, were found with significant frequency in all groups, including six patients with typical "benign" adenoma. Two tumors in that group displayed a sharp mode of 47 chromosomes, one had a supernumerary chromosome in the C group and the other in group D. Examination of polyps with major and minor atypia revealed higher ploidy than those in benign-looking tumors. In two cases with major atypia the chromosome numbers ranged between 51 to 60 and 51 to 90, respectively. In polyps with less marked atypia the counts were 48–50. There was a suggestion of stem lines in three of the five cases, and several large markers were noted. All groups contained pseudodiploid cells. In adenocarcinoma, hyperploidy was even more marked than in the other four groups.

In seven noninvasive polyps of the large bowel in a case of Gardner's syndrome,[26] four polyps had a normal diploid stem line and showed a predominantly hyperdiploid chromosome variation and three had heteroploid stem lines (two were pseudodiploid and one was hyperdiploid). Chromosomal differences between polyps of different size were merely quantitative and the same nonrandom karyotypic pattern was found in all polyps irrespective of size and histological grading. The chromosome gains or losses primarily involved those in groups C and D. Attempts at analysis with banding failed. The chromosome pattern was studied by conventional and Giemsa banding techniques in five polyps from a patient with polyposis coli and in sporadic adenomas from four patients.[27] In both the familial and nonfamilial types the chromosomal changes preceded histologic signs of invasiveness, with particular involvement of #8 and #14, suggesting a similar chromosomal evolution in colonic adenomas regardless of whether they have a hereditary basis or not. The rate of progression, however, was faster in the familial than in the nonfamilial types of colonic polyps. These cytogenetic findings are in accord with observations made before banding.

Examination of inflammatory tissue from the colon was performed and 47 metaphases counted.[28] As expected, all were normal diploid except for one cell with 48 chromosomes. In another study an adenomatous polyp of the colon was examined but only one karyotype was successfully obtained, and that was normal.[29] A second specimen of a villous adenoma with *in situ* carcinoma contained at least one cell with 48 chromosomes.

An extensive study on the cytogenetic findings in chronic ulcerative colitis and benign and malignant lesions of the colon has been published.[30] Cytogenetic preparations were obtained by brushing the rectal mucosa during proctoscopy. Of 61 individuals examined, 33 provided reliable material for chromosome analysis: 10 normal controls, 5 with Crohn's disease involving the colon, 10 with chronic ulcerative colitis, 4 with ulcerative colitis complicated by cancer of the colon (special care was taken to obtain separate samples from both colitic and cancer tissues), and 4 with cancer of the colon alone. Chromosome counts and karyotypes were normal in all controls and in patients with Crohn's disease. In chronic ulcerative colitis, the modal chromosome number remained in the diploid range. However, aneuploid cells or possibly broken polyploid cells were seen in two patients and in the inflamed mucosa of two other patients with cancer of the colon associated with ulcerative colitis. All of these cells were in the hypotetraploid range. No chromatid breaks were found in the controls or in patients with Crohn's disease, but patients with ulcerative colitis and colon cancer demonstrated a very variable number of breaks.[31] Profound abnormalities were demonstrated in the malignant tumors, characterized by aneuploidy, chromosome breaks, and markers. This study documented significant chromosomal abnormalities in chronic ulcerative colitis, both in patients with or without cancer of the colon.

Two villous adenomas of the colon have been analyzed and aberrant chromosome numbers found in one specimen, ranging from 23 to 85. No predominant stem lines were observed and no common karyotypic aberrations were found.[32] Karyotypes indicated groups C, F, and G to be chiefly affected. The other adenoma contained no markers or an aberrant stem line, and the chromosome count ranged from 29–68. Karyotypes of 18 cells in two cases of benign adenomatous polyps of the sigmoid colon without atypia were found to be diploid.[33]

A single colonic polyp having C-trisomy as the only change and two rectal polyps with C-trisomy, one with 49 chromosomes and one having a pseudodiploid chromosome constitution, have been described.[34] The karyotype in the last case showed two markers (one being a ring chromosome), C-trisomy, and missing chromosomes in groups A, B, and E.

Assuming that an abnormal karyotype is characteristic of the majority of cases of cancer, the data cited above on polyps and adenomas of the colon and rectum would then agree with the commonly held opinion that these lesions are precursors of cancer of the colon. Malignant transformation of villous adenomas has been estimated to occur in 30–75% of the cases. It may be that even this estimation is not sufficiently high because in familial polyposis almost all affected individuals may be expected to develop carcinoma of the colon 10–15 yr after the initial appearance of polyps.

Cancer of the Colon, Cecum, Appendix, and Rectum

The frequency of tumors of the colon is, at present, second to that of the breast and lung in females and males, respectively; in the United States it is the most common among gastrointestinal neoplasms. There have been numerous studies of the karyotypic picture in cancer of the colon, before the introduction of banding techniques.[35] These findings have been established on relatively large numbers of primary tumors as well as on metastatic lesions, mainly of the ascitic form. The first report on the chromosomes in carcinoma of the rectum appeared 30 years ago[36]; although an exact chromosome count was not established, the photographs presented indicate some of the tumors to be polyploid and some hypodiploid (Figures 134 and 135). In a study of nearly 120 carcinomas,[37] including samples from two or more regions of most of the tumors, the distribution of the modal chromosome numbers accumulated around the diploid level (40–50 chromosomes) and again near the triploid level (60–75 chromosomes). Based on these fairly extensive data on carcinoma of the large intestine, some generalizations were made that are probably equally applicable to tumors of other sites: 1. generally, all the malignant cells are aneuploid; 2. their chromosome patterns indicate that they belong to a clone; 3. their karyotypes tend to show the change in distribution characteristic of the aneuploidy of the malig-

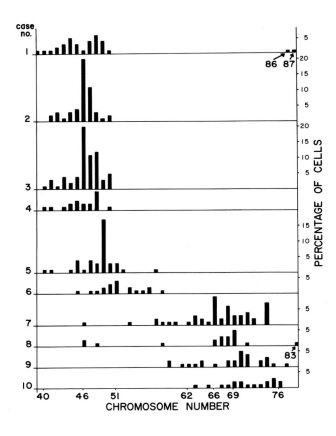

135 Distribution of chromosome numbers in the primary tumors of 10 cases with cancer of the colon. The number ranged from hypodiploidy to near triploidy, a condition not uncommon in a number of human cancers.

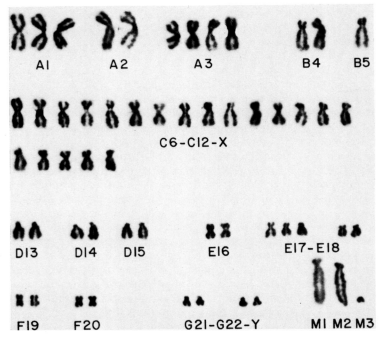

136 Karyotype from an effusion of a patient with mucocele of the appendix with extensive pseudomyxoma peritoneii. The effusion was characterized by a variety of modal chromosome numbers and karyotypes as shown in the bar figures of Figure 118. The karyotype shown contains three markers.

nant cells, and there are relatively fewer chromosomes in groups B, D, and G; 4. one or more easily recognizable marker chromosomes are frequently, but not invariably, found in tumors; and 5. with few exceptions, the chromosome patterns of the cells in samples from different regions of the primary tumor are indistinguishable or at least sufficiently similar to indicate that each belongs to the same clone. Sometimes two regions differ, e.g., one has a mode about twice that of the other, the similar chromosome pattern indicating, however, that the higher mode has been derived from the lower by chromosome doubling. The occasional exceptions, in which the chromosomes of two regions are dissimilar, may represent two tumors. Thus, two different regions of carcinoma of the cecum varied histologically and also had quite different karyotypes with modes of 40–42 and 68 chromosomes, respectively. According to these authors, different tumors from the same patient seem as likely to have different chromosome patterns as if they came from different patients. For example, in one unusual patient,[38] a carcinoma of the cecum and two carcinomas of the sigmoid colon, which developed 14 mo later, each had a distinctive chromosome pattern. In one of our cases a malignant ascites from a patient with a mucocele of the appendix and a pseudomyxoma peritonei was studied on four different occasions over a period of 3 wk (Figure 136). Modes of 44 or 45 chromosomes were found with three markers, but there was a considerable amount of variation among the karyotypes.[39]

In another one of our cases an unusually high ploidy in a peritoneal effusion from an anaplastic carcinoma of the sigmoid colon was encountered (Figures 137 and 138). In direct preparations over 50% of the cells contained > 600 chromosomes, including chromosomes of abnormal size and shape.[40] Some cells had over 1,000 chromosomes, and even as many as 2,000 chromosomes were estimated in some of the cells. A mode of 77 chromosomes was found in over 20% of the cells, suggesting that the original tumor had not been extremely hyperploid. The high numbers appeared to arise by endoreduplication, because, in some cells, the topographic proximity of morphologically identical chromosomes was still evident. After culturing the cells for 12 days the frequency of markers did not change but the hyperploid cells decreased. Only 8% of the cells had over 600 chromosomes and 77 chromosomes occurred in 42% of the cultured cells.

Lubs (1970) examined (before banding) the chromosomes in 11 primary tumors of the large bowel (1 benign adenoma, 3 villous adenomas, 3 nonmetastatic adenocarcinomas, and 4 metastatic ones). Only the benign adenoma had a diploid complement; all other tumors were abnormal. An acrocentric marker (intermediate in size between groups D and G) was present in at least three of the seven malignant tumors. All of the four metastatic cancers contained cells with > 60 chromosomes and in three it was the predominant cell line. In the three nonmetastatic cancers the counts ranged from 44 to 49. Each of the three villous adenomas was characterized by both an absence of markers and a lack of predominant cell lines. The author postulated that cancers of the colon with high chromosome numbers are more likely to metastasize than those with numbers close to 46. He also came to the conclusion that colonic tumors are complex cytogenetically, with significant chromosomal variation occurring between and within individual tumors. Nowell (1970) commented on the significance of these findings.

In cancer of the colon chromosome numbers have ranged widely, being as low as 30 and reaching extreme hyperploidy of about 1,000–2,000 chromosomes, as discussed above. Nevertheless, the tendency to eliminate very hyperploid cells in culture is to be noticed. Cells with too many chromosomes are probably not capable of dividing under artificial conditions or that reversion to a more normal mitotic division had occurred in culture.[41] As in tumors from other sites, correlation between the degree of malignancy and specific chromosomal constitutions has not been clearly demonstrated. However, hyperdiploidy and near-diploidy seem to prevail in advanced tumors of the colon. Of course, it is possible that analysis with banding techniques may reveal some consistency among the various cancers of the colon and, hence, some unique karyotypic characteristics for this malignancy.

Large markers occur rather frequently, and in some tumors they may be present in 100% of the cells (Figures 139–143). Often, these markers survive the process of evolution of stem lines and appear to be a characteristic fea-

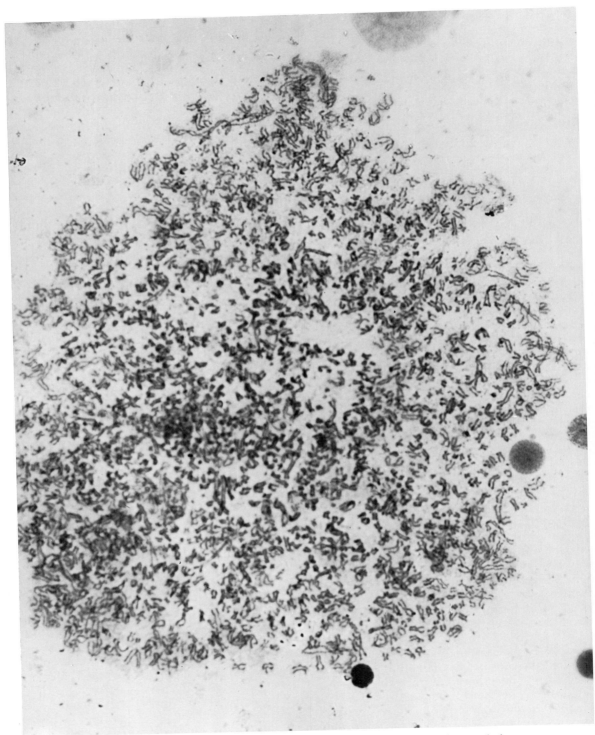

137 Metaphase with a tremendous number of chromosomes, including many abnormal chromosomes, from the ascites of a patient with cancer of the colon.

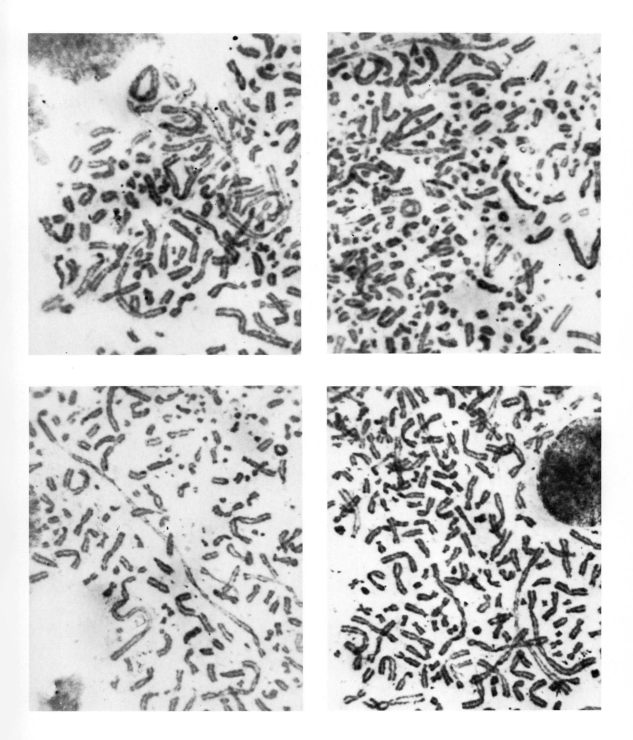

138 Partial metaphases from the ascites of a patient with cancer of the colon characterized by a large number of chromosomes. The remarkable abnormalities in this case are shown by ring chromosomes, extremely long chromosomes, many acentric fragments, and minute units.

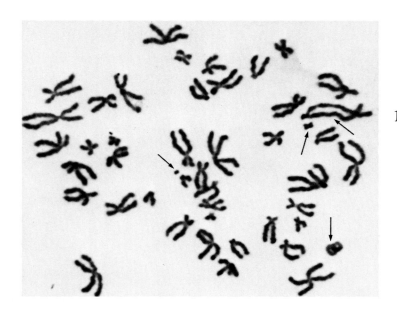

139 Metaphase from the ascites of an untreated patient with cancer of the colon. A number of marker chromosomes were present including a ring chromosome, a very large acrocentric, and a small metacentric.

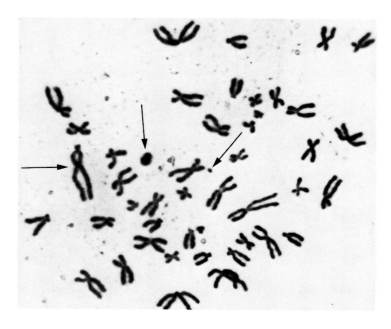

140 Pseudodiploid metaphase consisting of 46 chromosomes from the effusion of a patient with cancer of the colon. Arrows point to three markers present in the cells.

ture of most of the tumor cells despite the difference in chromosome numbers. Some authors[42] feel that the consistent perpetuation of markers strongly agrees with the hypothesis that human malignant tumors originate chiefly from a single cell in which chromosome changes have occurred. This applies in the case of cancer of the colon arising simultaneously in several polyps in familial multiple polyposis, i.e., the invasive tumor may develop in several places independently. Furthermore, in an examination of 7 cancers of the colon among 23 malignant tumors, the authors[43] found significant underrepresentation of chromosomes of groups B, D, and G in most of the tumors. When taken together, groups D and G were

Tumors of the Alimentary Tract 479

141 Karyotype of a cell with 49 chromosomes from a case of cancer of the colon. Though unbanded, no marker chromosomes were evident. Extra chromosomes are present in groups C and E.

142 Karyotype containing 87 chromosomes from a cancer of the colon. A marker (M) chromosome is present.

deficient in all but one tumor. However, these data do not appear to be in agreement with the findings of others for tumors of the alimentary tract, as well as tumors of other sites. These authors calculated the "specific common chromosomal pathway for the origin of malignancy" to be an imbalance in the #16 chromosome, i.e., an excess of #16, either absolute or relative to the other groups. However, refutation of this particular hypothesis has already appeared, and preliminary data with banding techniques in cancer of the colon would seem to refute the concept of the unique involvement of #16.

G-band analysis of an intestinal leiomyosarcoma revealed a mode of 42 chromo-

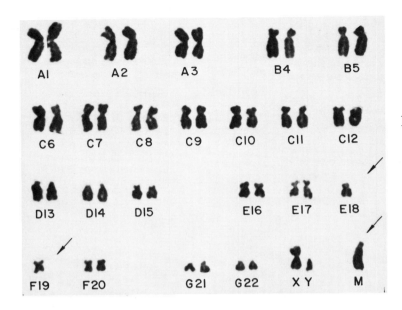

143 Karyotype containing 45 chromosomes (unbanded) from a cancer of the colon. Arrows point to missing chromosomes and to a marker (M).

144 A Q-banded karyotype of an adenocarcinoma of the sigmoid; showing the tumor cells to be hypodiploid: 45,XY,t(1;14),−16,+21. The markers (M_1 and M_2) were probably due to a translocation between chromosomes #1 and #14.

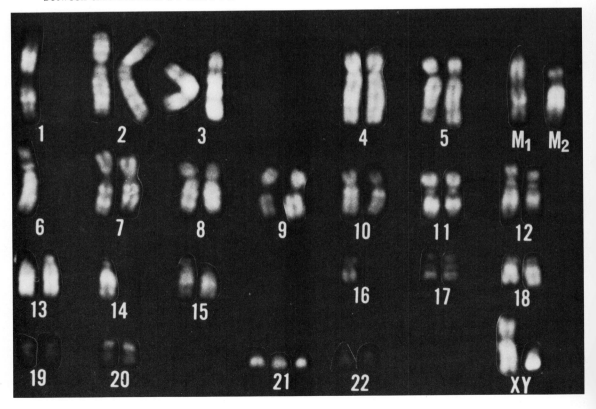

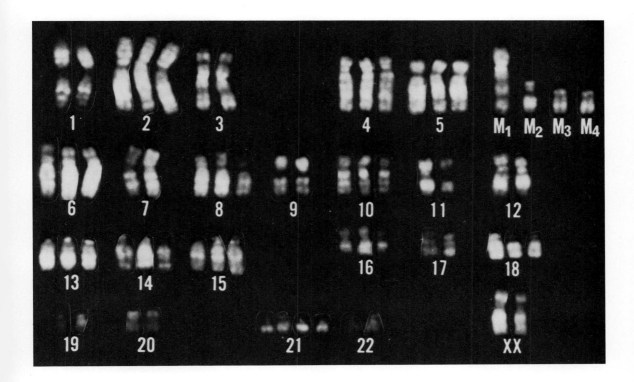

145 A Q-banded karyotype of a cancer of the rectum consisting of 63 chromosomes, including four markers. It is probable that M_1 is due to an inversion of #3 and M_2 to a deletion of the distal part of the short arm of #11. The origins of M_3 and M_4 are unknown.

somes (42,XX,+7,−13,−14,−15,−18,−22,1p−, 11q−) and a karyotypic pattern reminiscent of some meningiomas (Mark, 1976).

The significance of chromosome polymorphism in linkage studies in Gardner syndrome and familial polyposis has been reported by Borgaonkar, Trips, et al. (1977). However, the usefulness and limitations of this approach have yet to be defined rigorously. The same can be said for the chromosome findings in families with a high incidence of colonic and uterine cancers and lymphoma (Law, Hollinshead, et al., 1977).

Banding Studies of Primary Tumors of the Large Bowel

The chromosomes of 14 primary tumors of the large bowel (Sonta and Sandberg, 1978b) were analyzed with banding techniques (Figures 144–148). Of the 14 tumors, 11 had some chromosomal abnormalities (seven with numerical and four with both structural and numerical abnormalities) and in the remaining two no karyotypic abnormalities were found.

The modal chromosome number ranged from 46–73, with most of the tumors being hyperdiploid. Two tumors had a bimodal distribution; one consisted of 50 and 63 chromosomes and the other of 49 and 46 chromosomes, the latter being diploid. No common marker chromosomes were seen among the various tumors and no two tumors with identical karyotypes were encountered, though some chromosomes were involved more often than others. Excessive chromosomes in the primary tumors were usually due to extra chromosomes in the following groups: #8, #13, #15, #17, and #21. On the other hand, chromosome losses, though much less frequent, involved #5, #6, #7, #10, and #16. Most of the tumors were hyper-

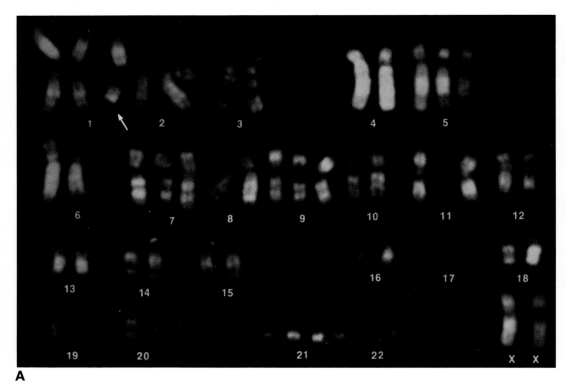

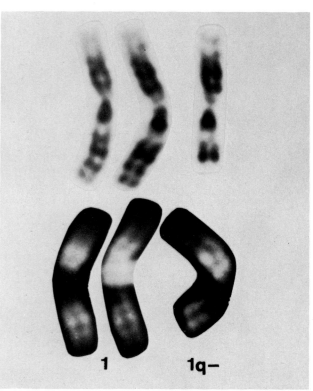

146 A Q-banded karyotype (A) and a partial Q and G-banded karyotype (B) from a cancer of the rectum. Arrow in (A) indicates deletion of a segment of the long arm of #1 (q32→qter). Therefore, the karyotype of this tumor was shown to be 54,XX,+5,+6,+9,+11,+21,+21, and +1q−. In (B) are shown two sets of #1 chromosomes containing a 1q−.

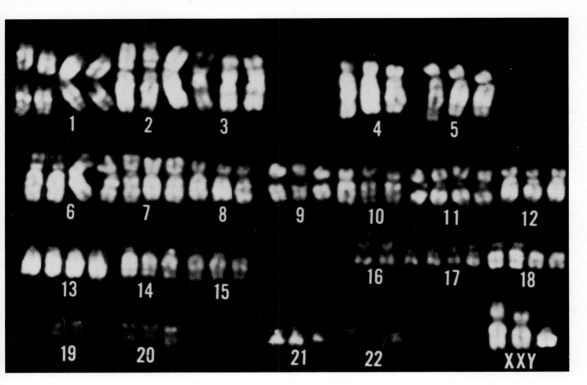

147 Q-banded karyotype of a cancer of the colon showing the tumor to have a hypertriploid karyotype with 73 chromosomes. Extra chromosomes are present in 19 of the groups, including an XXY. *(Sonta and Sandberg, 1978b.)*

148 Q-banded karyotype of the tumor from a male patient with a retroperitoneal histiocytoma, showing 50,XY,+5,+12,+13, and +21.

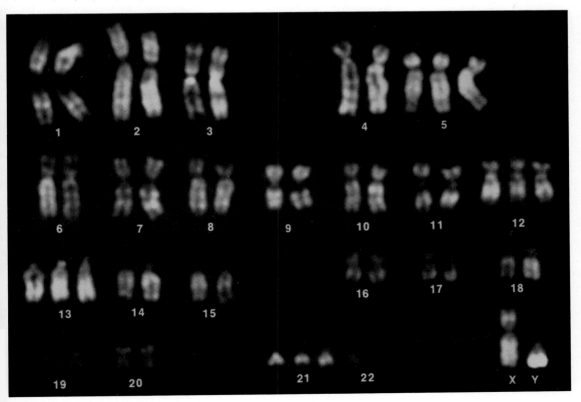

diploid and usually contained < 60 chromosomes. Only one tumor contained hypodiploid cells. The cytogenetic data obtained on these primary tumors indicate that they consist primarily of numerical changes, relative infrequency (when compared to metastases), and small numbers (1 – 4) of markers.

Martin, Levin, et al. (1979) utilizing a medium containing L-arterenol-HCl in a final concentration of 10^{-4} M, described a method which apparently yields chromosomes of a high quality in tumors of the colon. The samples were minced in such medium and the suspension was then divided into aliquots incubated at 37°C in 5% CO_2/95% O_2 for 3- to 24-hr periods. Colcemide (1 μg/10 ml of medium) was added immediately to the 3-hr sample before incubation or was present during the last 3 hr in incubations of longer duration. At the end of incubation the samples were centrifuged, the supernate was removed, and the cells were resuspended in 0.027 M sodium citrate balanced salt solution (pH 7.2), and incubated in a shaker bath for 30 min in an atmosphere of 5% CO_2/95% O_2. At the end of 30 min the larger pieces were pulsated several times through a syringe and then discarded. The resulting fine suspension was again centrifuged for 5 min, the supernate removed and 0.075 M KCl added, the sample incubated for 20 min at 37° C, centrifuged for 5 min and placed in a fixative of 3:1 methyl alcohol/glacial acetic acid for at least 1 hr in a refrigerator at 4°C. The fixative was then changed to 55% methyl alcohol/ 45% glacial acetic acid. Changing of the fixative was done at least five times before slides were made by usual techniques.

With this technique, 12 of 17 consecutive tumors of the large bowel had countable metaphases ranging from 5 to nearly 60 cells per tumor (Martin, Levin, et al., 1979). Seven of the 12 cases had over 25 countable cells. Seven tumors were karyotyped, six with banding techniques. Of these, two were hyperdiploid, one pseudodiploid, and four hyperdiploid. There were some similarities in the karyotypes, though no two were identical. At least three of the tumors contained markers whose origin could not be identified. The six tumors examined with banding were shown to have a variable karyotypic picture consisting of missing or extra chromosomes or with deleted arms. The involvement of #20 was stressed by the authors, but just as common was the involvement of #5 and some of the other chromosomes. The two hypodiploid tumors ha the following constitutions: 44,XX,–1,+5,–1 –18,2q+/44,XX,–1,+5,–6,–15,–18,2q+,10p– +marker and 44,XX,+5,–9,–13,–13,–15,–1 –18,+22,+2 markers, respectively.

Chromosome studies have been made o two human tumors of the colon maintained a xenografts in immune-deprived mice (Reeve and Houghton, 1978). In both tumors huma karyotypes were retained, although progressiv changes occurred during serial passages. Thu one line had a mode of 46 chromosomes i passage 3 and a mode of 47 in passage 1 whereas the other line was found to be hypo diploid with a mode of 41 chromosomes i passages 4 and 10. The first line had a ste line of 46,X,–Y,+12 and side lines charac terized by 46,X,–Y,–11,+12,+3 markers an another one 46,X,–Y,–10,+12,–13,+2 mark ers. The shift of the mode to 47 by passage 1 was due to a gain of #19 in the stem line. Th second line, characterized by hypodiploid had a number of translocations and deletion involving #1, #3, #6, #7, #11, #12, #1 #17, #22, and 25 markers. The similar mo phology of many of the smaller markers mad the identification difficult from cell to cel but some of them could be recognized wit banding. In both tumor lines prophasing wa present in some of the cells.

Tumors of the Liver, Pancreas, and Peritoneum

The first report on the chromosomes of a hu man liver tumor originated in my laboratory i 1963[44] in which it was shown that an effusio of liver carcinoma contained 30% cells with normal set of 46 chromosomes. Because th histologic examination did not reveal any nec plastic cells, it was felt that nonmalignant cell originating from the host had been examinec However, abnormal chromosomal finding were reported for another effusion of liver car cinoma (epithelium-like carcinoma in the as cites form);[45] the chromosome number range from 65 to 83 with a sharp mode of 75. A strik ing increase was observed in group G, includ ing as many as 16 G-group chromosomes. I another patient with an effusion of liver carci noma, the chromosomes ranged from 48 to 10 with 40% of the cells containing 46 chromo somes.[46] Some years later we described[47] th findings in the peritoneal effusion of a liver carcinoma with a mode of 79 chromosomes with three constant markers. Chromosome #1 was consistently diploid in all the karyotypes.

So far the only chromosomal findings in carcinoma of the pancreas were described from my laboratory in 1967 and by Berger in 1968 a, b,[48] in which a modal number exceeding 100 chromosomes with a rather infrequent ring marker and 61–62 chromosomes with a large subtelocentric marker were observed, respectively. The peritoneal effusion of a rare sarcoma of the mesentery was also analyzed by us on two separate occasions.[49] In the first examination all the cells contained apparently diploid karyotypes. The second sampling still showed nearly 90% of the cells to be diploid and histological examination revealed an absence of tumor cells.

Some years later we described the findings from the effusions of a cancer of the peritoneum.[50] Fluid was withdrawn on four separate occasions over a period of a month; the last three samples were obtained before and one after 5-FU therapy. Despite the fact that 65% of the cells were karyotyped, none of which was completely identical with any other, selection of one type cell over another was observed. Some characteristics were constant, such as two markers (98% of the cells) and the presence of four chromosomes in group B and G (80% of the cells). The modal number ranged from 44 to 56. These features, in our opinion, indicated maintenance of the basic chromosome pattern which might be important for survival of the cancerous cells.

Even though the number of cases reported for hepatic, pancreatic, and peritoneal cancers is small and the analyses have not been performed with more refined cytogenetic techniques, it would appear that no specific chromosomal features have emerged from the observations.

A cell line from a pancreatic cancer had 58–71 chromosomes (J. J. Yunis, Kuo, et al., 1977).

Tumors of the Female

Tumors of the Breast

Tumors of the breast (Figure 134) constitute the most frequent cancer in women of all ages. Chromosomes of > 100 cases of breast cancer have been reported. Large markers[1] and rings[2] were observed in some but without any consistency. In one study, four cases of benign cystic mammary disease were examined and all had a diploid chromosome picture.[3] The same findings were made in two cases of *in situ* lobular carcinoma. Also, specimens were examined after 2 hr incubation with colchicine. Examination by a two-wavelength microspectrophotometric method in hyperplasia and cystic hyperplasia revealed a normal diploid picture.[4] A review of 104 reports of chromosomal analysis of breast cancer containing modal cells, 54 from effusions and 50 from solid tumors, revealed the mean modal number in effusions to be 62.3 and in solid tumors 53.3, with hyperdiploidy prevailing both in the original tumor and in its metastases.[5]

A study on 77 primary breast tumors indicated that they fall into two categories:[6] one in the near-diploid group with predominance of hypodiploid tumors (as in the ovary) and another with 60–85 chromosomes. Among the published reports is one of a fairly well differentiated infiltrating ductal carcinoma in a woman of 77 which remained stationary over a 6-mo period and had a mode of 46 chromosomes, of which the 11 metaphases karyotyped were shown to have a diploid complement.[7] A metastatic breast carcinoma yielded only diploid karyotypes, but areas of lymphocytes were present and the identity of the cells in the chromosome preparation was uncertain.[8] Most carcinomas of the breast, however, are aneuploid with markers. A pseudodiploid breast carcinoma studied with conventional staining was shown to have gained one B chromosome and lost one #16. In seven out of eight effusions of breast cancer patients markers were observed, with most of the cells having several markers. The chromosome counts ranged from 35–258 with the modes ranging from 43–76.

A giant fibroadenoma in a 19-year-old patient yielded nine cells with 46 and eight with 47 chromosomes;[9] an analysis of three of the former showed a diploid karyotype, whereas four with 47 chromosomes had an extra group F chromosome. In a lobular carcinoma in situ, one of the two karyotyped cells with 47 chromosomes also had an extra group F chromosome;[10] in the same study another lobular carcinoma in situ and four benign cystic hyperplasias yielded only a few diploid karyotypes. It should be emphasized, however, that in the preponderant number of cases of cancer of the breast, both primary and metastatic, the chromosomal picture, regardless of the modal chromosome number, is extremely abnormal and the occurrence of diploid cells is a rare phenomenon. Banding (G) analysis of two poorly differentiated breast cancers revealed both to be pseudodiploid (Mark, 1975a). One of these had the constitution 46,XX,2p−,+5,+8,9p−,16q−,17p+,−20 and −21. The other tumor was shown to be 46,XX,3q−,−6,+7,−8,11q+,−15,16q+ and two different ring chromosomes. The details of the structural changes could be readily clarified with G-banding.

Attention was drawn by us, on the basis of banding analyses, to the fact that #1 is often involved in breast cancer (Kakati, Hayata, et al., 1975). Subsequently, Jones Cruciger, Pathak, et al. (1976) analyzed the chromosome constitution of seven established long-term cell lines originating from breast cancers. In all of these lines it appeared that the distal segment of the long arm of #1 was involved in a translocation, even though the translocated partner may have been different, e.g., #3, #5, #7, #11, and #12, and the amount of 1q translocated variable.

Another aspect related to chromosome abnormalities in breast cancer is the incidence of sex-chromatin in interphase cells.[11] Several authors have concluded that there are *masculine* types of breast cancer in females depending on the presence or absence of Barr bodies. Furthermore, it was suggested that those tumors with a high frequency of sex-chromatin have an unfavorable prognosis, i.e., hormonal therapy does not meet with success. This hypothesis was not confirmed by others and the evidence to date is equivocal.

A positive significant correlation between sex-chromatin in breast cancer and a 5-yr survival time or disease-free interval of distant metastasis was observed; however, no correlation was seen between sex-chromatin counts and lymph node metastasis or the size of the tumor (Ghosh and Shah, 1975). In another study, measurements were carried out on the group A chromosomes of 50 metaphases from patients with breast cancer and 50 metaphases from normal controls. The authors were unable to confirm a previous report in which a large mismatched #2 was found more frequently in patients with cancer of the breast (Bishun, Raven, et al., 1974).

Chromosomes in breast cancer were studied about the time the diploid number was established for the human;[12] aneuploid stem lines were demonstrated. Subsequent studies[13] based on effusion, metastatic, or primary tumor materials have frequently revealed large or medium-sized markers with chromosome numbers being in the near-diploid (often hypodiploid with 38−45 chromosomes) or in the hypotriploid−hypotetraploid range (60−85 chromosomes) (Figures 149−153). Based on modal DNA values in 107 tumors, Atkin (1972) concluded that those in the near-diploid group had a significantly better survival rate than those in the near-triploid−tetraploid group. The data also suggested that where the mode of the tumors was close to the center of the mode (i.e., at about 46 or 70−75 chromosomes), the prognosis tended to be better than where it deviated widely from these levels.

An adenocarcinoma of the breast examined in effusion material was shown to contain a subtelocentric marker larger than #1 and present in all the cells. The modal chromosome number was hypodiploid, i.e., 40 chromosomes. Hypotetraploid cells were rare, as compared to tumors at other sites (Grouchy, Vallée, et al., 1963).

One fairly well-differentiated ductal carcinoma had an apparently diploid chromosome constitution;[14] there have been two reports of pseudodiploid tumors: one lacked a #16 and had an additional member in the B-group,[15] whereas the other lacked a C-group chromosome and had an additional chromosome morphologically similar to a Y, thus superficially resembling the male diploid karyotype.[16] However, none of the above tumors was studied with chromosome banding techniques. A banded metaphase was illustrated in a report describing a cell line derived from a pleural effusion secondary to a mammary carcinoma;[17] the mode was at about 64−66 chromosomes, and a giant and a minute marker were seen.

In one study short-term tissue cultures of human breast explants arrested with colchicine were used to provide mitotic figures.[18] In such explants from 99 patients, including several premalignant or benign tumors plus preinvasive breast cancer, only 7 of the explants could not be cultured. Of the 48 premalignant and benign lesions (adenosis, simple cystic disease, proliferating cystic disease, cystic disease associated with cancer, fibroadenoma), 19 of 48

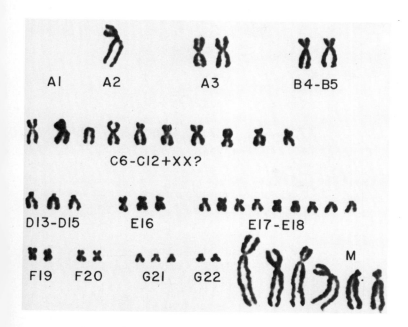

149 Karyotype of a cell from the pleural effusion of a patient with cancer of the breast containing six marker chromosomes (M) and missing chromosomes in groups A1, A2, B, C, D, and extra chromosomes in groups E and G. In all probability the markers were composed of some of the missing units and had banding analysis been performed (this case was studied prior to the introduction of banding methods) undoubtedly the origin of some of the markers could have been identified.

150 Metaphase from the pleural effusion of a patient with cancer of the breast containing two ring chromosomes and several other large markers (arrows). This patient had not been treated and, hence, the formation of the abnormal chromosomes cannot be ascribed either to radiation or chemotherapy.

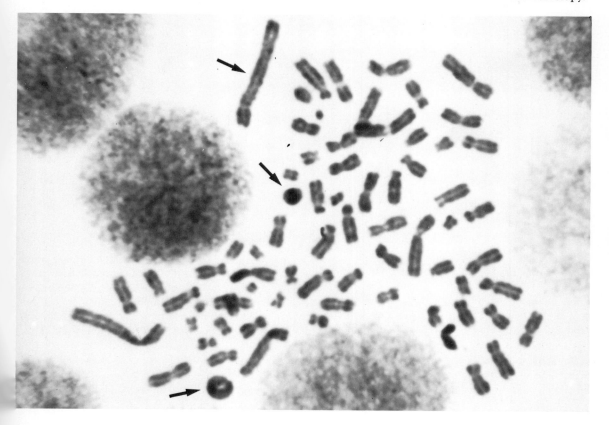

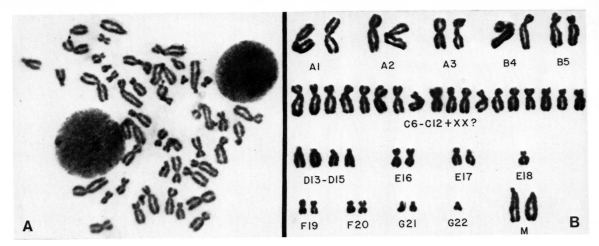

151 Pseudodiploid metaphase (A) and karyotype (B) of effusion cell of a patient with ovarian cancer. Even though 46 chromosomes are present, their distribution is very abnormal and the set includes 2 acrocentric markers (M). Two extra chromosomes are present in group C, and chromosomes are missing in groups D, E18, and C22.

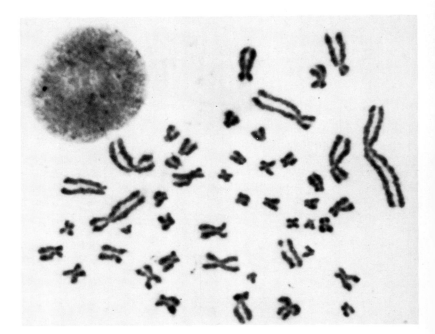

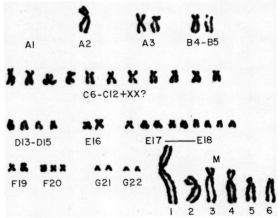

152 Above: Metaphase from the pleural effusion of a patient with breast cancer. Left: Pseudodiploid karyotype of the metaphase shown in A, containing six abnormal chromosomes (M_{1-6}) with missing chromosomes in groups A, C, D, and extra chromosomes in groups E and F.

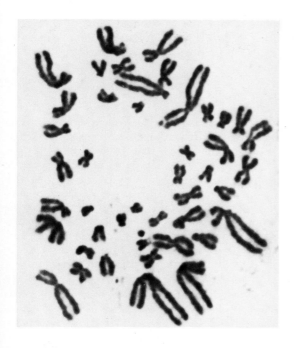

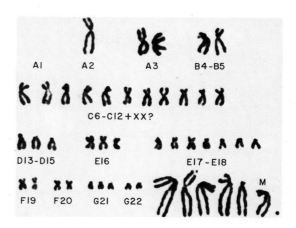

153 Metaphase and karyotype of a cancer cell from the pleural effusion of a patient with carcinoma of the breast. Note the markers of differing morphology and size.

were hyperdiploid, including 3 preinvasive breast cancers. From karyotype analysis nearly half of the 40 benign and premalignant explants had chromosome anomalies; the preinvasive breast cancers were karyotypically abnormal. The most frequent anomaly was a large subtelocentric chromosome. The author concluded that these studies support the view that cystic lesions and possibly fibroadenoma with epithelial proliferation share chromosome anomalies with mammary cancer.

Comparative studies of chromosome and mitotic anomalies were conducted on specimens from 50 patients with proliferating fibroadenomatosis, 50 more patients with similar lesions undergoing malignant transformation, and another 50 patients with breast cancer. Chromosome analyses of 81 cells from four patients with malignant breast fibroadenomatosis and in 69 cells from five patients with breast cancer showed that aneuploidy was present in 74 and 81% of the specimens, respectively. Analysis of 33 metaphases from four patients with proliferating fibroadenomatosis revealed only a few abnormal cells. The author postulated that malfunctioning of the mitotic system may be the cause of cellular aneuploidy, malignant tumors, and hyperplastic processes; also, aneuploidy in the cells of proliferating fibroadenomatosis is evidence for its precancerous nature.

Chromosome studies[19] were made of a group of 118 patients with breast tumors. Of these patients, 16 had fibroadenomas and 102 carcinoma of the breast. Mitotic figures were obtained on only 78 patients. The cells of all patients with carcinoma of the breast showed aneuploidy with many in the tetraploid and hypertriploid range. The modal numbers varied from 36 to > 200 chromosomes. A high percentage of endoreduplication was found in many cases. Markers of various sizes were found. The authors postulated that the primary tumor and its metastases generally contained the same chromosomal abnormalities. In eight patients with recurrent breast cancer, five showed the same cytogenetic abnormalities in both primary and recurrent tumors. In 16 patients with fibroadenomas, 12 had normal chromosome constitutions in the tumors, and 3 had abnormal ones. This suggested that the three cases with abnormal karyotypes were in a precancerous stage and that the cytogenetic analysis may be used as an early diagnostic aid.

Studies of the chromosome constitution of breast cancer cells in vitro have generally revealed abnormalities similar to those observed on noncultured materials.[20]

Banding studies on 58 cells from seven breast cancers indicated the existence of characteristic numerical and structural modifications, particularly changes involving the long arm of #1, #4, #5, #7, #14, #15, and the X and the short arm of #16.

Pathak, Siciliano, et al. (1979) found a number of markers identical in their origin to those present in HeLa cells in the metaphases of breast cancer cells in effusion. Inasmuch as the latter cells were examined directly and without in vitro culture, it can be speculated that the genesis of similar markers in cells of divergent origin may represent a general tendency for the human genome. Incidentally, the breast cancer cells had a modal number of 35 chromosomes vs. the very high number in the HeLa cells.

A case of EL with dyserythropoiesis developing in a patient 5 yr after irradiation for breast cancer has been reported (Chroback, Radochova, et al., 1977). Polyploidy, hyperploidy, and absence of group #17–#18 chromosomes were encountered.

Aneuploidy and structurally rearranged chromosomes were identified in most cultures of ductal epithelial cells from proliferative lesions of fibrocystic disease (E. R. Fisher and Paulson, 1978). The authors indicated that this represents some evidence in support of views that proliferative fibrocystic disease may represent a precursor lesion of cancer of the breast, if not its earliest morphologic expression.

Even though cancer of the breast in males is much rarer than in females, one center was able to report on 42 cases seen between 1938–1972 (Walach and Hochman, 1974). Apparently the prognosis is more favorable than in females. It is unfortunate that no chromosomal studies have appeared on breast cancer in males.

Tumors of the Ovary

A study of chromosome numbers in cancer of the ovary[21] shows a tendency, resembling that of carcinoma of the breast, for the tumors to fall into two categories: one with a bimodal distribution in the diploid and triploid range and another being hypodiploid. Markers, frequently as large as group A chromosomes or larger, were present in most but not all ovarian carcinomas. These markers were seen in ovarian cystadenomas and adenocarcinomas and also occurred in the peritoneal effusions of such cancers. Despite some indication of consistency in the findings of markers, the hypothesis of specificity of a long marker could not be substantiated. The finding of 57% breaks in a cystadenoma of the ovary was not observed by other authors.

In one study of five cases with *bilateral* ovarian tumors (one Q-banded), a similar aneuploid chromosome pattern was seen *in each ovary.* In one case, however, there was also a number of diploid metaphases in one of the ovarian tumors: 50% of the 60 metaphases were diploid. Because few mitoses were seen among the stromal cells in histological sections and squashes of this tumor, some of the diploid metaphases were possibly tumor cells. It is conceivable that the primary tumor was in this ovary and that only the aneuploid cells (with a mode of 38 chromosomes) had metastasized, because apart from occasional diploid cells that were probably not tumor cells, these constituted the cell population in metastases to the other ovary, in an omental metastasis, and in a malignant ascites, all of which had a similar hypodiploid pattern. It is noteworthy that the diploid cells generally had clear-cut chromosomes of good morphology, whereas those of the aneuploid cells tended to be more contracted with poor morphology and more characteristic of cancer cells. There was a tendency for a reduction of chromosomes in groups B, D, and G, with most of the cells being hypodiploid (Atkin, Baker, et al., 1974).

In a study on the cancerous effusions in 21 patients with cancer of the ovary it was shown that each case had an abnormal chromosomal picture; the chromosome counts ranged from 30–159 with the modal numbers ranging from 43 to 74–76. Marker chromosomes of various morphology were found in 17 of the cases, with most of the cells containing more than one marker (Olinici, Galatir, et al., 1973). Comparative morphologic and cytogenetic studies of eight ovarian papillary adenocarcinomas revealed that early in the disease the pattern is nearly diploid with scarce karyotypic changes. As the disease progresses from a single layer of atypical epithelium to multilayered proliferation and invasiveness, the cytogenetic pattern shifts from near-diploidy to heteroploidy with different markers. In a review of the literature, exclusive of teratomas, consisting of 79 cases of effusion material and 54 of solid ovarian malignant tumors, it was shown that the modal number in effusions was 55.2 and in solid tumors 56.4; the difference was not significant. Also, the mean number of cells with a modal chromosome number was higher in effusions (38.5%) than in solid tumors (26.2%); the difference, again, was not significant statistically.

A possible example of a metastasizing tumor with a diploid karyotype is provided by a *folliculoma*; the primary tumor, an omental metastasis, and malignant ascites were studied and all had apparently normal karyotypes.[22]

Only a few studies have appeared on benign ovarian tumors. Dermoid cysts of the ovary exhibit a normal karyotypic picture in the few studies published.[23] In a "vegetant cyst" of the ovary, abnormal metaphases were found, the cells being of two distinct modes, i.e., 45 and 88–90, the counts ranging from 36–94 in 57 cells.[24] Material from a papillary cystadenoma cultured for 12 days yielded cells with 47 and 48 chromosomes which had one and two extra C-chromosomes, respectively.[25] However, it has been suggested[26] from the published photomicrographs that this tumor was in fact a well-differentiated mucinous cystadenocarcinoma. In another report,[27] three aneuploid metaphases with markers and chromosome numbers of 60, 150, and 300 were found in uncultured material from a mucinous cystadenoma, and it was suggested that the aneuploidy implied that areas of early malignant transformation were present.

Only diploid karyotypes were found in six benign ovarian teratomas.[28] Another benign teratoma in culture yielded diploid cells and cells with 47 chromosomes including an extra member in group C.[29] A low-grade malignant teratoma in a 14-yr-old patient examined after culture for 29 and 169 days, showed on both occasions a mode of 47 chromosomes with an extra chromosome resembling #3.[30] A malignant teratoma in a 9-yr-old child had a mode of 46 chromosomes; nine cells were analyzed, all of which showed a slight departure from a diploid karyotype with a #3 being replaced by a metacentric chromosome of the size of group C.[31]

Early studies of cancer of the ovary. In 1956 Hansen-Melander, Kullander, et al. described the chromosome findings in ascites due to metastases from an ovarian cystadenocarcinoma. Two samples were taken at intervals of 10 days and in both there were two distinct populations with modes of 58 and 63 chromosomes, respectively. Subsequent to this publication a large number of papers appeared on the chromosome findings on ascitic and pleural effusions and later, as techniques improved, on primary ovarian tumors.[32] Usually, the presence of markers and other consistent karyotypic features, including changes in the number of normal or apparently normal chromosomes in the various groups, have indicated a clonal origin (Figures 154–156). Olinici, Galatir, et al. (1973) found a preponderance of diploid cells in unbanded chromosome preparations from malignant effusions. As mucin-secreting tumor cells outnumbered normal ones in the effusions, it was considered that this was one of the very small groups of malignant tumors with an apparently normal diploid stem line.

Serial studies of malignant effusions from carcinoma of the ovary generally showed little change in the chromosomal pattern, apart from a diminution in the frequency of aneuploid cells in cases that respond satisfactorily to treatment (Siracký, 1969). Occasionally, however, the pattern changes during the course of treatment (Sandberg, Yamada, et al., 1967), usually by the selective growth of a minor line which was already present in the effusion before therapy. It has been suggested that such an alteration of the karyotypic profile during chemotherapy may be an indication for a change to another chemotherapeutic agent (Visfeldt and Lundwall, 1970).

Chromosome analysis on malignant effusions secondary to a vegetating cyst of the ovary and an ovarian carcinoma revealed the presence of an acrocentric marker (the size of B-group chromsomes) in both tumors with 100% of the cells containing this marker in the former and 30% of the cells of the latter. The two tumors had modes of 45 and 48 chromosomes, respectively (Grouchy, Vallée, et al., 1963).

Because of the high frequency of large markers in cancer of the ovary, protruding arms possibly related to those large markers may often be seen in metaphases, anaphases and telophases in histologic sections and also in interphase nuclei in optimal cytological preparations (Atkin, 1969b; Atkin and Baker, 1964;

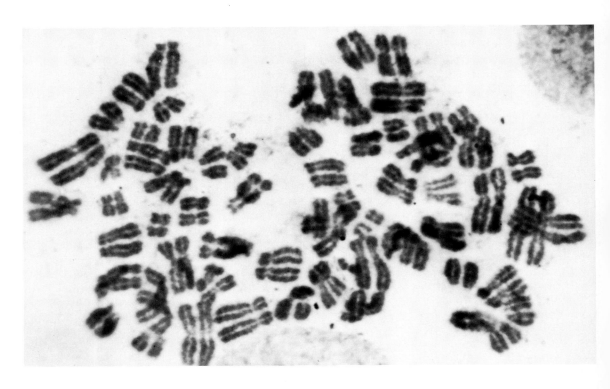

LoCurcto and Fraccaro, 1974). As in other female tumors, most near-diploid ovarian cancers have a single Barr body; those with modes in the triploid region or above are usually either chromatin negative or have two and, occasionally, three bodies (Atkin, 1976b).

Studies on modal DNA values and chromosome numbers in ovarian tumors reveal two groups: a near-diploid group and a group of high-value tumors with modes mainly in the triploid-tetraploid range (Atkin, 1971d). In five cases (Atkin, Baker, et al., 1974), both ovaries were involved and in each of these the chromosomal pattern indicated a common origin of the malignant cells, showing that the two ovaries contained primary and metastatic tumors, respectively, and not two different primaries. In one case a tumor from one ovary contained a substantial proportion of diploid cells which, in comparison with the aneuploid cells, had a better chromosome morphology, i.e., a clearer outline and less contraction (Atkin, 1976b). It was speculated that the ovary containing the diploid cells represented the primary site and that these cells were possible precursors of the aneuploid malignant cells, with a mode of 38 chromosomes and generally similar karyo-

154 A metaphase from the ascites of a patient with cancer of the ovary showing endoreduplication with each chromosome having four chromatids instead of the usual two. This endoreduplication involved the total set including a marker chromosome.

types, which were found in both ovaries, peritoneal metastases, and ascitic fluid.

On the basis of DNA measurements, the near-diploid cases showed a better average survival rate than those with high values (Atkin, 1970b, 1971d). However, among the near-diploid cases whose chromosomes have been studied, the survival rate has not been high. The author speculated that the latter may represent a selective group, in that the tumors with a favorable prognosis may be less suitable for chromosome studies. The DNA data suggested that the prognosis of tumors with modal numbers at or slightly above the diploid level was favorable compared with those that are hypodiploid.

Chromosome studies in cancer of the ovary based on banding analyses. One

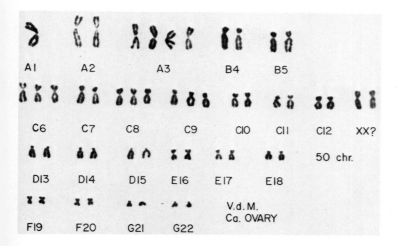

155 Karyotype of a cell from the ascites of a patient with cancer of the overy containing 50 chromosomes, all of apparently normal morphology. Of course, without banding it is difficult to be certain about this statement; however, no distinctly abnormal chromosomes are visible in this karyotype.

156 A hypodiploid karyotype (G-banded) consisting of 40 chromosomes of a cell from a cancerous ascites due to an ovarian carcinoma. For some of the markers (M_1, M_3 and M_4) Q-banding was also performed. The origin of 6 of the markers could be established, whereas 3 markers (UM_{1-3}) remained unidentified.

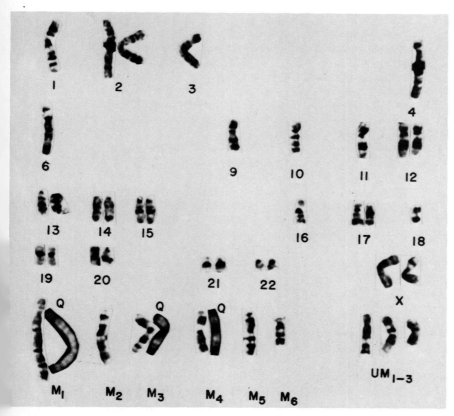

tumor in a series of 14 near-diploid carcinomas yielded adequate G- and Q-banded preparations (Atkin, Baker, et al., 1974). The modal number was 40 and in unbanded preparations 12 markers could be identified; with banding, however, 23 of the chromosomes were seen to be abnormal. The metaphases obtained in the solid tumor before the commencement of treatment showed a fairly high incidence of chromosome breakage and unstable rearrangements; a high incidence of breakage was similarly reported in solid tumor tissue and ascitic fluid, the latter in direct preparations and following culture for 9 and 25 days, from an untreated patient with an ovarian cystadenocarcinoma (Fraccaro, Gerli, et al., 1965). Similar breakage of chromosomes in ovarian cancers has been reported by others (Zara, Fraccaro, et al., 1966; Mannini, Fraccaro, et al., 1966).

Tiepolo and Zuffardi (1973) studied the Q- and G-banded chromosome patterns of malignant cells in ascitic fluid from two ovarian adenocarcinomas, which had modes of about 40 and 75 chromosomes, respectively. Only about 35–50% of the chromosomes could be recognized as normal. In one tumor, #8 was virtually absent and the abnormal chromosomes were mostly the products of translocations or deletions. Chromosomes #1 and #3 were frequently involved in structural changes that included a complex inversion of #1 involving three break points and inversion of both inner segments. Several abnormal chromosomes could not be identified; they had no major segments whose banding patterns were similar to any segment of the normal chromosomes and were presumed to be a product of multiple translocations or rearrangements involving short segments. In the second tumor, which had a mode of about 40 chromosomes, #8 was again virtually absent, and #3, #7, and the X were underrepresented. Fourteen new chromosomes were seen; 6 of these included at least part of the long arm and two part of the short arm of #1. Frequently, #3 was also included in rearrangements: three translocations and two inversions were identified. One chromosome was composed of material from #12 and an unidentified chromosome or chromosomes; a chromosome of similar appearance was seen in the first tumor.

In an omental metastasis secondary to a papillary cystadenocarcinoma in a patient who had previously received chemotherapy, a modal chromosome number of 61 was found (Berger and Lacour, 1974). Banding revealed eight markers, including 1p−, 5p−, 7q−, Xq−, and 18q+. The origin of the additional material in the latter was uncertain; the long arm contained an additional heterochromatic segment. A translocation [t(14;18)(q11;q11)], an isochromosome [i(22q)], and a large chromosome which may have been composed of 1p, 2q, and 8q were also observed.

Two effusions were studied in my laboratory (Kakati, Hayata, et al., 1975) (Figure 157). In one the modal number was 47–48 with a range of 45–49 chromosomes. Polyploidy was, however, not uncommon and comprised about 25% of the total cell population. Unbanded preparations revealed trisomy of #2 in all cells, besides an extra C- and/or G-chromosome in the modal cells. G-banding analysis, on the other hand, revealed that the extra #2 was not a homologue of the normal chromosome, but was, in fact, an abnormal marker. Five more markers were detected on the basis of banding, which involved mainly #3, #5, #9, and the X. In fact, in all the cells analyzed both #3 and one homologue of #4, #9, and X were absent. The six markers could be identified as to their origin with banding: t(3;9)(q12;p11), t(3;3)(q29;p25), del(3)(p25), 4q+, Xq+, and a small marker whose identity could not be established. In the second case, > 50% of the cells were polyploid. The modal number was 40 with a range of 38–43. By a conventional method an unusually long submetacentric marker was seen in almost every cell. However, upon G-banding many chromosomal rearrangements were observed. Generally, #2, #12–17, #19, #20, and #21 were not involved in structural rearrangements. Of all the markers, only a few were fully or partially identifiable, e.g., t(3;?)(q21;?), t(1;11)(q21;p11), t(9;11)(q11;p11), del(1)(p36), and translocations probably involving #8, 4q, and 5q. In another case (Kakati, Oshimura, et al., 1976) of an endometroid serous adenocarcinoma of both ovaries with metastases to the serosa of the uterus and omentum, cells in the ascitic fluid were shown to have a chromosome number ranging from 60 to 65 with 10 to 16 markers, only some of which could be identified. The most common karyotype was 64,XX,−1,−1,+6, +6,+6,+6,+7,+8,+8,−12,−13,−13,+14,−21,and

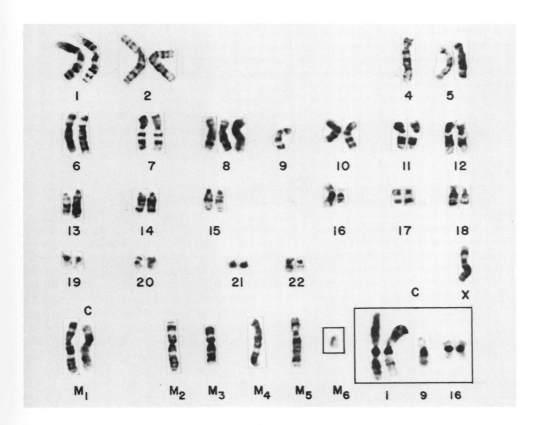

157 Karyotype (G-banded) of a cell from an ovarian cancer. Six markers (M_1–M_6), whose origins were identified, were present in the cells. The chromosomes in the bracketed area were obtained from another metaphase of the same case (C = C-banding).

(Kakati, Hayata, et al., 1975.)

16 markers. One marker was identified as an isochromosome for the long arm of #1 [i(1q)], another as an i(17q), and another marker was consistent with a modification of the i(1q).

In a study utilizing C–, G–, and Q– banding, five ovarian carcinomas were examined in detail (Van der Riet-Fox, Retief, et al., 1979). In one carcinoma biopsies taken from right and left ovaries showed major discrepancies in markers, although the modal values showed only a slight shift. Only two identified markers were common to both right and left ovaries, the right having 16 other markers common to that side and the left having another 7. In another case the chromosome counts from the sample of solid malignant ovarian tissue were slightly higher than those of the cells in ascites. Few markers were present, and there were relatively few deviations from the hypertriploid mode; three markers were common to both samples. In another case samples from the solid tumor and ascites showed no karyotypic differences. Chromosome #1 was involved most commonly in this series with most of the rearrangements involving different types of break points; second to #1 in these cancers of the ovary were #3, #11, and #16. Two of the tumors were hypodiploid (39 and 44 chromosomes) and three were near-triploid (64, 65, and 74 chromosomes). Each one of the tumors contained marker chromosomes. In some the markers could be readily identified even though they were present in large numbers (in one tumor 18 such markers could be identified), whereas in other tumors only some of the markers could be identified.

A comparative cytogenetic and histologic study on 18 mesothelial ovarian tumors revealed a normal chromosome complement in the benign lesions and definite abnormalities in the cystadenocarcinomas. However, in all borderline tumors evidence of an abnormal stem line was present, with a more or less marked tendency to polyploidization. Serous papillary cystadenomas of this group showed (in five out of the six cases) a stem line with a karyotype 47,XX,+10 present in both sides of bilateral lesions. The authors (Knoerr-Gaertner, Schuhmann, et al., 1977) concluded that a malignant change at the chromosomal level precedes histologically detectable features of malignancy. Histologic equivalents appeared when the abnormal cell line was established. The initiation of malignant transformation, therefore, may be signaled by karyotypic abnormalities before structural changes can be detected in the corresponding histologic specimens.

Atkin and Pickthall (1977) found a structurally rearranged #1 in nine out of 14 ovarian carcinomas and believe they may also have been present in 3 others. In the remaining two, pericentric inversions involving the heterochromatic regions of #1 were seen and identified in one of the chromosomes #1 in the patients normal lymphocytes. Heterochromatin variants were stressed by the authors as being present in more than half of the tumors, with some of them also being present in the normal cells of the individuals. The possibility of an association between the presence of #1 heterochromatin in variants as a constitutional anomaly, a liability to ovarian (and perhaps other forms of) cancer and structural changes involving #1 in tumor cells was proposed for consideration by these authors.

As an extension of the above, Atkin (1976b) indicated that A-group chromosomes are commonly involved in the formation of large markers which are frequently seen in ovarian cancer. This not only applied to #1 but also to #3, a conspicuous marker observed in cancer of the ovary. However, in my opinion, the involvement of these chromosomes in the formation of markers is also frequent in other cancers and may not have the significance attributed to it by Atkin and his co-workers.

Evidence is accumulating[33] that patients with ovarian cancer may develop AML after or associated with chemotherapy and/or radiation for the disease.

A benign serous cystadenoma studied with DNA analysis and karyotyping was found to have a diploid mode, whereas the cells from the cystic fluid showed both a diploid and triploid population, the latter with a marker, possibly indicative of malignant transformation (Wager, Granberg, et al., 1978).

In 16 granulosa cell tumors, no cytogenetic changes suggestive of malignancy were found (Arzapalo, Meneses, et al., 1977).

Ovarian teratomas. The common form of teratoma (dermoid cyst) is benign and cystic, and is composed of mature histologic structures of ectodermal, mesodermal and endodermal origin. Sebaceous material and hair fill the cyst. These and all other structures within the cysts arise from a nodular growth nidus. Samples from this nidus thus contain a wide variety of cell types, including connective tissue, bone, and teeth. Because of their location in the ovary and the presence of cell types that are inappropriate for the ovary, it has been speculated for a number of years that these tumors were of germ-cell derivation. Uncertainty existed whether these teratomas arose through conjugation of gametes or by parthenogenesis of a single haploid cell with chromosome reduplication. Studies on the sex-chromatin make-up of such tumors indicated the presence of such a body in the tumors of females and in about 50% of males (Hunter and Lennox, 1954), thus strengthening the concept that these tumors arise through parthenogenesis. This concept was further strengthened when it was shown that benign ovarian cystic teratomas contained two X-chromosomes (Corfman and Richart, 1964; Galton and Benirschke, 1959; Rashad, Fathalla, et al., 1966). The germ-cell hypothesis of ovarian teratomas has been tested by biochemical and cytogenetic experiments and the observations are consistent with the following conclusions:[34] (1) an ovarian teratoma originates from a germ-cell; (2) the likely mechanism for parthenogenic teratoma formation is failure of meiosis II (or equivalently, reentry of the second polar body); (3) teratomas contain two haploid chromosome complements derived only from their host; and (4) the two haploid chromosome complements are identical unless recombination recurred during meiosis.[35]

Abnormal teratomas have been examined cytogenetically. In one tumor in culture (Serr, Padeh, et al., 1969) a mixture of diploid and cells with 47 chromosomes was encountered, the extra chromosome belonging to group C. A low-grade malignant teratoma in a 14-yr-old patient yielded cells with 47 chromosomes in culture, with the extra chromosome being a #2 (Arias-Bernal and Jones, 1968). Another malignant teratoma in a 9-yr-old child had cells with 46 chromosomes in which one #3 was replaced by a medium-sized metacentric (Toews, Katayama, et al., 1968b). One teratoma from a female 35 yr of age studied by Atkin (1976b) had a modal number of 49 chromosomes with an extra C- and two extra D-group chromosomes. Thus, the tendency of teratomas may be toward pseudodiploidy or hyperdiploidy rather than hypodiploidy.

Further comments on the origin of benign or variant teratomas. To determine the origin of benign cystic teratomas of the ovary, banding studies (Linder, Kaiser-McCaw, et al., 1975) were done on normal tissues and teratomas from five patients. The normal tissues were heterozygous for #17 polymorphism at or near the centromere, whereas the teratomas were uniformly homozygous for such polymorphism and contained 46,XX karyotypes. These findings and those based on electrophoretic variants indicated that ovarian teratomas are parthenogenic tumors arising from a single germ cell after the first meiotic division. These teratomas result then from meiotic error in a germ cell homozygous for Q-fluorescence markers, most of which are located at or near the centromere. However, whereas the enzyme data showed that in some instances the teratoma cells were homozygous at some loci for which the host was heterozygous, in other situations both teratoma and host tissues were heterozygous. The heterozygosity of the cells from the teratoma is the result of crossing-over in meiosis, and the frequency of crossing-over makes it possible to estimate the genetic distance of various loci (Ott, Linder, et al., 1976).

Teratomas are usually differentiated tumors containing tissues derived from ectoderm, endoderm, and mesoderm and may arise in the gonads or elsewhere. Benign or variant teratomas are parthenogenetic tumors derived from a single germ cell after the first division of meiosis, as studies of chromosome markers and enzyme variants have shown. Extragonadal teratomas develop in a different manner. Examination of single gene products and chromosomes in extragonadal teratomas (including endodermal sinus tumors, because of their similarity to teratomas) has revealed that they originate from several cells which are of diverse origin and not related to meiotic events. Thus, an endodermal sinus tumor was shown to have a 46,XY karyotype and, interestingly, a retroperitoneal teratoma in a person with DS was shown to have a 47,XX,+21 karyotype (Linder, Kaiser-McCaw, et al., 1975).

Cancer of the Corpus Uteri and Endometrium

Compared with cancer of the cervix, the uterine body is much less frequent as a primary site of cancer. This tumor is seldom detected in its initial stages and, hence, cytogenetic analysis does not yield much information about karyotypic changes during the course of the disease. The cytogenetic data are generally inconclusive but do show a disorganized and variable karyotypic picture with modes as low as 28[36] and as high as the tetraploid range.[37] Various markers have been observed but there is no uniformity or consistency in the findings of various authors. The mean modal chromosome number of 50–51 for 39 cancers of the corpus uteri (Figure 128) was found to be significantly lower and nearer to diploidy than that in cancer of the cervix (mean: 61 chromosomes)[38]. Furthermore, the difference between modes in different specimens of cancer of corpus uteri was smaller than in cancer of the cervix, which show a rather marked scatter. One report of a benign fibroma of the uterus has appeared in which the chromosomal picture was found to be diploid.[39]

The chromosomal changes in cancer of the corpus uteri showed a bimodal pattern somewhat related to the histological picture.[40] Thus, most undifferentiated tumors had high modal numbers, with the well-differentiated tumors usually being in the near-diploid range. Among the latter, modes of 46–49 chromosomes were common and some of these tumors showed small deviations, such as a single group C-trisomy or at most small numbers of additional and perhaps missing chromosomes. These minimal deviations may be associated with the presence of slight, if any, myometrial invasion.[41] In contrast, tumors of high malignancy generally showed evidence of structural as well as more extensive numerical chromosomal changes. Such a correlation between the degree of malignancy and the extent of the chromosomal changes has been generally confirmed by other workers.[42] However, in a series of seven tumors, which included not only differentiated carcinomas but also more malignant tumors, over half of the cells karyotyped were found to be diploid.[43] Some indication was given that the origin of these diploid cells, at least in one of the tumors, were probably of cancer cell origin. One highly malignant tumor yielded predominantly diploid metaphases.[44] Two squamous cell carcinomas were cultured; one was found to be diploid and the other to have a mode of 60 chromosomes (Kang, Kim, et al., 1972).

Cancer of the endometrium develops in a vast majority of cases after menopause. It is five to seven times less frequent than cancer of the cervix, and the prognosis is usually relatively favorable, because bleeding frequently attracts the attention of the patient before extensive invasion by the tumor. Of the 67 recorded cases of carcinoma of the endometrium, 28 exhibited a clear mode with a mean modal number of 46–47, thus not deviating markedly from the normal diploid set.[45] However, about 60% of the reported cases did not show any definite modes, and chromosome counts were usually scattered around the diploid and triploid values. Analysis of solid tumors and effusions did not reveal significant differences. Normal metaphases with 46 chromosomes are usual in endometrial cancer; their proportion decreases, however, with increasing invasiveness of the tumor. Two cases of atypical endometrial hyperplasia have been examined,[46] with one exhibiting a normal karyotype and the other a pseudodiploid picture with six of the cells having 38–46 chromosomes.

A uterine cancer was found to have a pseudodiploid (46,+3,–D) stem line, whereas one of the metastases in the mesentery was hypotetraploid with a smaller population of near-triploid cells (Granberg, Gupta, et al., 1974). This observation is in keeping with the findings in my laboratory in which we have shown that primary tumors tend to have lower chromosome numbers (diploid range) than the metastases (triploid range). Whether the change in chromosome number is a consequence of the metastasis or is generated in the primary tumor, thus possibly enhancing metastatic spread, is a problem still to be solved.

Relatively few chromosome studies have appeared on cancer of the uterus. In general, chromosome studies have shown that hyperdiploidy with minimal changes, perhaps amounting only to trisomy of a single chromosome, is quite common; in this respect, cancer of the corpus uteri differs from that of other common sites in adults (Atkin, 1976b). In a series of seven near-diploid cancers studied by Baker (1968), three had modal numbers of 47, two with an extra C-group and one with an extra D-group chromosome as the only apparent change. The two with an extra C showed no residual tumor at hysterectomy a few weeks after biopsy, and therapy with radioactive cobalt. The tumor with the extra D was the only one of the three in which metaphases with more than the modal number of 47 were found: there were several with 48–51 chromosomes, the additional chromosomes being mostly C- and D-group autosomes. Residual tumor which had not infiltrated the myometrium was present at hysterectomy. The other four tumors showed more or less extensive invasion of the myometrium. One had a mode of 49 chromosomes with three extra C-group autosomes, another had 47 chromosomes with a marker and changes in various groups, and two had 46 chromosomes with an additional G and a missing #16 in one and apparently diploid karyotypes in the other. The latter was a highly malignant tumor that was found at hysterectomy to have metastasized to the peritoneum. This small series suggested that single trisomies may characterize early carcinomas that have not yet invaded the myometrium; such tumors would be classified by some as carcinoma-*in situ*. Even though one tumor appeared to be predominantly diploid, this could not be verified by banding studies. Tseng and Jones (1969) described a well-differentiated and a poorly differentiated tumor in each of which over half the metaphases appeared to be diploid.

In this regard, it may be worthwhile to mention the chromosome findings in normal endometrium. In the past, it was repeatedly found that normal endometrium contained a considerable proportion of polyploid cells.[47] However, an investigation[48] of a large number of endometrial cells in mitosis studied by a direct method has indicated that polyploidy rarely occurs and that only 5.5% of the cells were aneuploid in a sample consisting of nearly 3,000 cells, suggesting that the previously reported high percentage of polyploid cells was most likely due to technical procedures. The basic diploidy of endometrial cells has been confirmed by others, utilizing both cultured and uncultured cells.[49] A study of hyperplasia of the endometrium has shown only normal, diploid cells (Baker, 1968).

Studies of cancer of the uterus before banding. In 1962 Wakonig-Vaartaja described a well-differentiated carcinoma with a modal number of 46 chromosomes, including a giant marker that could be identified in 97% of the metaphases and 100% of anaphases. In this tumor, therefore, the marker characterized a clone that comprised virtually the whole of the dividing cell population. In a tumor of moderate to poor differentiation (Granberg, Gupta, et al., 1974) evidence was found from chromosome and DNA studies for the concurrent or sequential development of three clones: the primary tumor was pseudodiploid, but a mesenteric metastases had a hypotetraploid mode and also a small near-triploid mode. Karyotypes of tumors showing more or less extensive changes, often with markers, have been published by various authors without banding analysis.

Chromosome studies on normal hyperplastic and "atypical" endometrium have generally revealed no evidence of aneuploidy (Bonilla-Musoles, 1971). Counts higher than 46 are generally lacking, apart from occasional polyploid (probably tetraploid) metaphases. Hypodiploid metaphases can usually be regarded as artifacts resulting from the loss of chromosomes from diploid metaphases during preparation of the material (Bowey and Spriggs, 1968). However, pseudodiploid cells in curettings that histologically showed atypical cystic hyperplasia have been described (Stanley and Kirkland, 1968), and diploid and hypodiploid cells in endometrium showing atypical hyperplasia from a patient who also had cervical carcinoma *in situ* has been reported (Granberg, Traneus, et al., 1972). Among the hypodiploid cells four were found with 45 chromosomes, lacking a D-group autosome, the consistency suggesting that these were not artifacts.

Fibromyomas are common benign tumors of the uterus that are frequently multiple. Occasionally, sarcomas arise in these fibromyomas. Because of their tough texture and low mitotic index, fibromyomas are not amenable to direct chromosome studies. In culture, they have yielded diploid metaphases (Fiocchi, 1967).

The significance of endometrial chromosome aberrations (gaps, breaks, etc.) attributed to clomiphene citrate therapy (or other forms of therapy) is to be yet established (Charles, Turner, et al., 1973). A report (Bamford, Mitchell, et al., 1969) on size variation of the late-replicating X-chromosome in leukocytes of patients with hyperplastic and malignant lesions of the endometrium is difficult to evaluate.

Banding studies in cancer of the uterus. Only a few attempts have been made to study uterine cancers with banding techniques and, hence, no generalization can be made regarding the cytogenetic picture in this condition on the basis of such studies. In one study (Kuramoto and Hamano, 1977),[50] samples from tumorous tissue were cultured in vitro and the chromosome constitution determined. Out of 42 trials for culture, a little over 40% of the samples could be analyzed. The mode of chromosome number was found at the diploid range in all cases, with a majority of the endometrial carcinomas having a pseudodiploid picture with some minor structural changes. A marker was identified in only one case. No common specific chromosome change was noted, and the histologic findings did not correlate with the chromosome aberrations. The authors concluded that carcinoma of the corpus uteri consists mainly of cells with a pseudodiploid chromosome constitution. However, examination of the three published karyotypes in the article reveals definite markers in each case, the markers being of rather large size and possibly related to either #1 or #2. As discussed previously, the findings obtained after culture of various tumors have to be interpreted with some caution, because it is possible that the major portion of the cancerous cells may fail to divide in vitro and, thus, the cytogenetic findings obtained may not necessarily represent the in vivo status.

Chorionic Villi, Hydatidiform Moles, and Chorionepithelioma

Trophoblastic tumors are known to occur much less frequently in Caucasian than in Japanese and Chinese women; the frequency of hydatidiform moles has been estimated to be about 1–2 in 2,000–2,500 pregnancies in white women, whereas in a Chinese population it was about 1 in 125. In Southeast Asia and India the frequency is about 7–10 times higher than in Europe or the United States and tends to occur in mothers younger than 20 yr and older than 39 yr. The sequence of events favoring the development of chorionepithelioma through several steps of transformation is supported by chromosomal data,[51] at least the transition from villi to mole.

In an examination of 2,000 cells of *normal chorionic villi* by a direct method, 0.1–0.45% of the cells were found in mitosis.[52] Of 2,000 cells in hydatidiform moles in the 4th mo of pregnancy the proportion of mitotic figures was higher (0.9–1.7%). Furthermore, a total of 552 mitoses from three cases of apparently normal villi were analyzed and in all these the mode was clearly 46; polyploidy was found in 0.9–1% of the cells. The first apparent chromosome changes were detected in abnormal *transitional villi*. Specimens obtained from three women in the first trimester of pregnancy characterized by proliferation of trophoblasts and edema of stromal cells with absence or extreme scantiness of blood vessels were examined and in all the chromosomal count ranged from 53–71 with a distinct mode of 69. All triploid cells showed three times the haploid autosome number plus an *XXY* set of chromosomes. A total of 75 mitoses was analyzed (Makino, Sasaki, et al., 1964a).

Utilizing Q- and R-banding polymorphism as markers, Kajii and Ohama (1977) established the karyotypes in uncultured and/or cultured villi of *hydatidiform moles* and in seven of them demonstrated a 46,XX karyotype of androgenetic origin, i.e., that the 46 chromosomes were all of paternal origin. Stromal cells in villi were studied for the X-chromatin in 47 moles; all but three were X-chromatin positive. Of the three X-chromatin-negative moles, two were Y-chromatin-negative and one was Y-chromatin-positive in 56% of the stromal cells. In a woman with a balanced reciprocal translocation [t(7;8)] whose mole could be karyotyped, it was revealed to have a 46,XX constitution. The high malignancy rate in hydatidiform moles could be explained by a recessive mutation of a gene that controls cell growth. Because the homologous chromosomes in moles are genetically homozygous, they are also homozygous for the mutation and, thus, escape from normal growth control. Similar findings and views have been reported by P. A. Jacobs, Hassold, et al. (1978).

Hydatidiform moles of all stages may show abnormal karyotypes. However, the modal number does not deviate much from near-diploidy, and in 17 reported cases the mean mode was close to diploidy:[53] 9 had a modal number of 46, 1 being pseudodiploid. In all nine cases, however, the proportion of aneuploid cells was higher than 16% except for one case with 1.4% aneuploid metaphases. In four cases with chorionepithelioma, all had near-diploid but abnormal stem lines, except for one having 18 cells with normal 46 chromosomes and 4 aneuploid ones. However, there are a few reports that indicate that hydatidiform moles are usually euploid and may show both a triploid or diploid–triploid mosaicism, which may also be the chromosome constitution of the associated fetus. In fact, the placenta of triploid conceptions frequently undergoes a partial or complete hydatidiform change.

In one study of 13 abortuses showing clearly visible cystic degeneration of the majority of their chorionic villi it was found that 10 were chromosomally abnormal:[54] 9 were triploid (6-XXY;3-XXX) and 1 tetraploid (XXXX). The other three were diploid (46,XX).

In a case of chorioadenoma destruens, however, an increase in number of aneuploid cells was found[55] whereas in choriocarcinoma changes similar to those in other cancers and malignancies have been observed.[56]

Miscellaneous Gynecologic Tumors

There have been a few reports on carcinoma of the vulva.[57] A significant degree of aneuploidy was present in carcinoma *in situ* but not in condyloma acuminatum or Paget's disease.[58] Chromosome counts from an area of Bowen's disease of the vagina showed a mode of 46–49[59] and a mode of 75 was found in a carcinoma *in situ* adjacent to an early vaginal carcinoma.[60] An adenocarcinoma of the vagina yielded 54 metaphases with abnormal chromosomes and a mode of 68.[61] In three cases of squamous cell carcinoma of the vagina, one case was characterized by a modal number of 46 and the remaining two showed abnormal chromosomes without a definite mode.[62] In three carcinomas of the fallopian tube, the modal chromosome number was 74, 42, and 82–84, respectively, with all tumors containing marker chromosomes.[63]

Hydatidiform moles (Szulman and Surti, 1978*a,b*). On a morphologic and cytogenetic basis hydatidiform moles can be divided into at least two different groups, if not syndromes:

1. Complete or classic mole. This is seen without any evidence of an ascertainable embryo or fetus, contains a 46,XX karyotype, and manifests a progressive fluid engorgment of the villi as well as a gross and haphazardly distributed trophoblastic hyperplasia. No trace of embryo or of fetal erythrocytes in villous vessels is found even in the early stages of the pregnancy. Microscopically, the villi exhibit immature mesenchyme with empty *in situ* capillaries and edema leading to cyst formation. The size of the vesicles increases with gestational age, but early stages of hydatidiform changes are observed in all cases.

2. Partial or imcomplete moles. These contain either a live or macerated fetus, characterized by a triploid karyotype,[64] and exhibit a slowly progressing hydatidiform swelling in the presence of functioning villous capillaries. Trophoblastic immaturity and focal hyperplasia are rather inconspicuous. Well-preserved or macerated fetal parts are usually found or the existence of the fetus can usually be verified by erythrocytes remaining in the villous vessels. Such fetuses usually have multiple congenital anomalies including spina bifida, cleft lip or palate, syndectomy, and growth retardation. In the partial mole consisting of normal chorionic villi intermingled with hydatidiform villi there was always association with a fetus, cord and/or amniotic membranes. Most of these moles have chromosomal abnormalities. Thus, in the presence of a partial mole one must suspect a pregnancy with chromosomal abnormalities. None of these cases showed advanced trophoblastic hyperplasia or anaplasia and none had a malignant evolution. These moles are not to be regarded as potentially malignant.

On the other hand, complete moles are never associated with a fetus, cord, or amniotic membranes. As opposed to partial moles, complete moles constitute a high risk group for malignant trophoblastic neoplasia, 2 of 16 cases showing distant metastases and all showing pronounced trophoblastic hyperplasia and anaplasia. Their karyotype is usually 46,XX (Vassilakos and Kajii, 1976).

However, Stone and Bagshawe (1976) are somewhat in disagreement, indicating that they have observed patients with transitional or partial moles who developed invasive moles or choriocarcinomas as evidenced by pulmonary metastases. These authors suggested that the ability to predict which moles are likely to give rise to a trophoblastic tumor would be useful, but it is unlikely to be clinically helpful unless comparable in reliability to follow-up by human chorionic gonadotropin (HCG) radioimmunoassay levels.

Trophoblastic tumors. Usually hydatidiform moles are chromatin positive. Out of a total of 23 moles described in three papers published in 1958,[65] 48 out of 53 were chromatin positive. Chromosome studies on cultured materials from seven of the moles showed female diploid karyotypes in four positive moles and male diploid karyotypes in two negative moles; one positive mole, however, had a triploid mode with three X-chromosomes, most cytotrophoblast and stromal cells having two Barr bodies (Atkin and Klinger, 1961). A further mole that was positive in the stroma but not in the trophoblast had a modal DNA content consistent with a triploid consitution. In histological sections the preponderant percentage of moles were found to be chromatin positive.[66] These findings are not surprising because in both types of moles discussed above at least one X-chromosome is present and in incomplete moles two to four X-chromosomes have been described (Carr, 1969).

Apparently, therefore, moles arise much more often in female than in male diploid conceptuses. However, they may also arise in triploid conceptuses as shown by chromosome studies in cultured molar and fetal material from a chromatin-positive case, both of which showed a 69,XXX constitution (Beischer, Fortune, et al., 1967); and from two cases in another study (W. G. Paterson, Hobson, et al., 1971). Three placentas showing a vesicular appearance in some areas but normal villi in others (Makind, Sasaki, et al., 1964a) were all triploid with an XXY sex-chromosome constitution. It appears that such partial hydatidiform moles are particularly common in triploid conceptuses.

About 85% of triploid abortions show cystic (hydatidiform) changes; inversely, 70% of abortions showing partial hydatidiform changes, but only 12–15% complete hydatidiform moles are triploid (Carr, 1971). Histologically, a series of eight triploid moles were found to lack hyperplastic changes, and it was therefore suggested that triploidy might be a favorable sign (Levy, Chadeyron, et al., 1972). Carr (1969) found one tetraploid mole, as well as nine that were triploid, among 13 abortions showing hydatidiform changes in the majority of the placental villi. It was postulated that triploidy may result from hormonal disturbances, e.g., there is an increased incidence up to 6 mo after the discontinuation of oral contraceptives (Carr, 1969). Triploidy in moles may be revealed by DNA measurements on histological sections (Atkin, 1974b). The presence of double Barr bodies (found in some triploid moles) may also suggest triploidy. One mole has been reported that had a 47,XX,+2 constitution in culture, the extra chromosome being confirmed as #2 by Q-banding (Honoré, Dill, et al., 1974). Whereas hydatidiform moles are benign tumors, placentas may also be the seat of locally invasive or destructive moles (chorioadenoma destruens) or frankly malignant tumors, i.e., choriocarcinomas. In a series of nine destructive moles, aneuploid counts were obtained in addition to others which may have been diploid or tetraploid (Makino, Sasaki, et al., 1965). In choriocarcinoma the neoplastic cell population shows consistent aneuploidy as in other malignant tumors, modes in the hyperdiploid and hypotetraploid regions having been reported.[67]

In four reported cases of chorioepithelioma,[68] all had near-diploid abnormal stem lines except for one having 18 cells with normal 46 chromosomes and 4 aneuploid cells.

The risk of malignant development in various hydatidiform moles and the unusual behavior of some choriocarcinomas have been the subjects of several papers.[69] A mediastinal choriocarcinoma in a chromatin-positive boy, a patient with Klinefelter's syndrome, has been reported (Storm, Fallon, et al., 1976).

On the basis of an anatomical, pathologic, and cytogenetic study in 75 specimens with gross swelling of chorionic villi, Vassilakos, Riotton et al. (1977) classified them into partial and complete moles. Most cases of partial moles had chromosome abnormalities including triploidy and trisomies, whereas the complete moles showed exclusively a normal female karyotype (46,XX). Aside from the macroscopic morphology of villi, these two entities were distinct from each other; complete moles with villi consisting of pronounced hyperplastic and anaplastic trophoblasts were never accompanied by a fetus, cord, or amniotic membrane, whereas partial moles with villi consisting of slightly hyperplastic to hypoplastic trophoblasts were always associated with cord and/or amniotic membranes. Those moles that had undergone neoplastic evolution were complete moles.

In animal studies (Wake, Tagaki, et al., 1978), it was clearly ruled out that host cells, as well as normally fertilized ova, are the origin of complete moles and pointed to androgenesis as a cause of this type of chorionic lesion. Most probably the moles originate from fertilized ova, with the maternal nuclear complement either being eliminated or inactivated. The presence of paternally derived markers in duplicate suggested that the doubling of paternal haploid sets occurred either after fertilization or at the second meiotic division.

Androgenesis probably is largely, if not solely, responsible for complete moles. Kajii and Ohama (1977) made observations in seven human cases similar to the animal study just mentioned. Such observations would account for the preponderance of XX moles, because YY counterparts are probably lost during early cleavage stages. A unique case with 48,XXYY was reported (Shinohara, Sasaki, et al., 1971) and seems compatible with the androgenetic origin. Exclusive of apparently partial moles, only 3 XY cases[70] out of 95 cases have been reported in the literature. Analysis of familial polymorphic variants in such cases, if present, in the future might unveil another mechanism for the genesis of complete moles.

The incidence of hydatidiform moles rises with advanced maternal age. It may be envisioned that the incidence of meiotic disturbances leading to the total elimination of the maternal chromosome set increases in oocytes from aged females, as in the case of trisomy syndromes. Alternatively, the mishap occurs with equal frequency regardless of maternal age but androgenetic conceptuses survive better in older females. Currently, no direct evidence favors either of the possibilities.

Tumors of the Urinary Tract

Chromosomes in Cancer of the Bladder

Even though the number of comprehensive studies published on the chromosomal findings in cancer of the bladder (Figures 158 and 159) is sparse, more meaningful cytogenetic data are available for this cancer than for some other human neoplasms.[1] Furthermore, considering the discrepancies existing among pathologists in classifying bladder tumors, it is remarkable that the chromosomal findings in various laboratories have shown as much consistency as they have. Thus, it is imperative that these studies be extended to a much larger series of patients at a variety of medical centers around the world to obtain a body of data that would make it possible to establish the usefulness of the cytogenetic picture in the diagnosis, therapy, and prognosis of cancer of the bladder.

A particular need exists in the case of papillary tumors, which often may have a benign histologic appearance but behave in an invasive or recurrent manner. The application of banding techniques to the examination of the chromosomal substructure of such tumors is

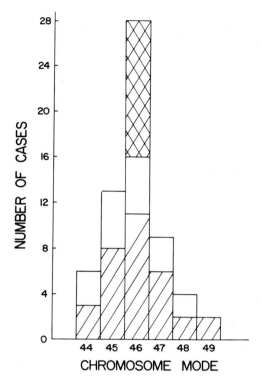

158 Distribution of chromosome numbers in cells from papillary cancers of the bladder, tumors that tend to fall in the diploid range. Of the 62 papillary tumors of the bladder, 12 were benign (cross-hatched area) and 50 cancerous. The 12 benign tumors had a diploid number of chromosomes. The papillary cancers had a chromosome number ranging from 44–49, with markers being present in 32 tumors (hatched areas) and absent in 18 (clear areas).

not only essential for the rigorous identification of the normal chromosomes, but, more importantly, in the elucidation of the origins of markers, which may play an important role in the biology, diagnosis, and therapy of bladder tumors.[2]

Sporadic karyotypic findings in bladder cancers were reported prior to 1967,[3] but the first study presenting detailed analysis on a relatively large series of patients and correlation with the pathologic picture and invasiveness of the tumors was not presented until 1967.[4]

Early studies on the chromosomes in bladder carcinoma, usually performed on perito-

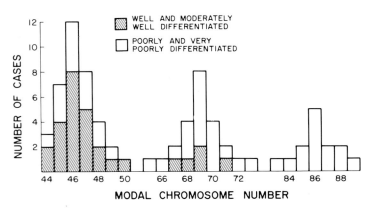

159 Modal chromosome numbers in transitional cell cancers of the bladder and their relation to histologic types of the tumors.

neal effusions, indicated that the mode was near diploid, i.e., 47–49 chromosomes.[5] We examined two cases of primary bladder cancers in 1965; 95% of the cells from one case had chromosome numbers ranging between 48 and 87 with a clear mode of 82.[6] All cells examined contained a ring marker. Extra chromosomes were present in all groups except B. The second tumor had a mode of 45 (44% of the cells), the chromosome number ranging from 40 to 49. In another study, two bladder "papillomas" yielded only diploid karyotypes, 2 carcinomas *in situ* had aneuploid cells with markers and 19 cancers were generally aneuploid (Duruman, 1969). In another case, a mode of 45 and missing chromosomes in groups B, D, and G in a carcinoma of the bladder were found (Atkin and Baker, 1969).

In a study of 15 bladder cancers, cystoscopic resection or biopsy coupled with cytogenetic analysis at intervals of 3–69 mo revealed a persistence of the chromosomal patterns with markers in four of seven noninvasive and six of eight submucosal invasive carcinomas.[7] Recurrence was limited to those carcinomas containing markers. Thus, markers have potential value as a prognostic aid in bladder cancer. Furthermore, the triad of tetraploidy, markers, and submucosal invasive, moderately well-differentiated cancer appears to carry such a poor prognosis as to militate for early radical resection.

Spectrophotometric DNA studies. Before discussing the chromosomal findings in bladder cancer, it may be worthwhile to take up those data based on the spectrophotometric determination of the DNA in cells as an index of ploidy. Thus, when the amount of DNA was found to be higher than that in the nucleus of a lymphocyte, the cell was recorded as being hyperdiploid, or tetraploid when it was found to contain twice the amount of the DNA of the lymphocyte. In other words, the allegedly nondividing DNA of the nuclei of mature lymphocytes or of neutrophils (cells that do not undergo mitosis) were assumed to have a diploid complement of chromosomes (and DNA) constituting the $2n$ (or $2c$) number of chromosomes. This approach has some serious shortcomings, because it is possible for a cell to have a diploid set of chromosomes and, yet, contain extra DNA when examined spectrophotometrically. It has been well established that the chromosomes replicate their DNA during interphase in an asynchronous fashion, i.e., some chromosomes or some of their parts replicate their DNA late, whereas others accomplish it early in the interphase period. Hence, it is possible to measure the DNA of an interphase nucleus spectrophotometrically and obtain spuriously high DNA levels of varying values depending on the point in interphase such a nucleus is being observed. Ideally, only a telophase nucleus contains, with certainty, the diploid amount of DNA, because it cannot be ruled out that the DNA has not started replicating after the cell enters interphase.

Another serious shortcoming of the spectrophotometric analysis of DNA in tumor cells is the not infrequent occurrence of pseudodiploid cells or cells with a chromosome number near the diploid region but with a definitely abnormal karyotype. These cells, however, when measured spectrophotometrically have been revealed to have a normal complement of DNA. Thus, to accurately establish the exact chromosomal picture in malignancy, it is essential that a detailed analysis of the karyotype be performed.

Tavares, Costa, et al. (1966) reported the spectrophotometric DNA measurements in 62 cases of transitional cell cancer of the bladder. These authors found 35 to be diploid or tetraploid, with 13 of the patients with diploid tumors dead 1–9 yr (average 33.6 mo) after surgery. In the group of 27 cases with tumors with a triploid-hexaploid amount of DNA, the mortality rate was higher, with 21 of the patients dead 1 mo–6 yr (average 18.1 mo) after surgery.

In a spectrophotometric study[8], allegedly based on a technique more accurate than the one employed by Tavares, Costa, et al. (1966), the DNA distribution in 31 bladder tumors was examined. In this study normal bladder was found to be polyploid, i.e., having a substantial population of tetraploid and some octoploid cells. This, it should be mentioned, has not been confirmed with cytogenetic studies, and it is possible that these authors were, in fact, measuring diploid cells with a fully replicated DNA complement. Nevertheless, from this study the authors postulated that bladder tumors arise as diploid and progress to hypotetraploidy and then to gross aneuploidy. Solid tumors were frequently very aneuploid, but a persistence of near-diploidy was observed.

In a study utilizing DNA measurement with scanning cytophotometry of the nuclei of 24 bladder tumors, which were compared to the measurements obtained with nuclei from urinary transitional cell epithelium of 13 normal tissues, it was shown that well-differentiated transitional cell carcinomas of the bladder had mainly diploid DNA stem lines comprising 85% or more of the cells. One case with a tetraploid DNA-stem line was found.[9] The moderately differentiated bladder cancers showed a DNA content ranging from diploid to hexaploid. Interestingly, four out of five poorly differentiated tumors had a diploid DNA-content. The author felt that the predominance of cells in the diploid, tetraploid, and/or octoploid range without marked frequency of DNA values in the intermediary classes seems to be a sign of clinically more "benign" bladder tumors, whereas a clinically more "malignant" tumor appears to be characterized by an increase of the DNA values in the triploid and hexaploid classes.

Further comparison (Fosså and Kaalhus, 1976) with normal transitional cell epithelium revealed nuclear enlargement and a decreased chromatin concentration in differentiated transitional cell carcinomas. Undifferentiated tumors could have the same or higher mean chromatin concentration as the control cells as a result of the many high-ploidy nuclei. It was also found (Fosså and Kaalhus, 1976) that changes in "noncondensed" and "condensed" chromatin were primarily dependent on nuclear size and total chromatin content and were not found to be a characteristic of cancer nuclei when compared with control nuclei of the same size and ploidy. Histologically atypical epithelium taken from bladder mucosa near transitional cell carcinomas showed changes similar to, though less marked than, DNA changes observed in the corresponding tumor (Fosså, 1977).

DNA measurements in 60 bladder tumors (E. H. Cooper, Levi, et al., 1969) indicated that as transitional cell carcinomas become more invasive, there is usually a progressive shift of the modal DNA content toward higher values. Chromosome analysis showed that even well-differentiated tumors had karyotypic changes. Incidentally, the authors injected colcemide (10 mg i.v.) 3 hr before resection of the bladder tissue.

A study[10] in which a noninvasive cancer of the bladder in a female was found to have more than one sex-chromatin body indicated to the authors that this may characterize an intermediate stage of clonal evolution from noninvasiveness, particularly because such extra sex-chromatin bodies were generally present in a variety of invasive tumors of females associated with a high chromosome number.

Cytogenetic studies: chromosome numbers and morphology. In contrast to statements made in the literature, we found only diploidy in the cells of normal bladder tissues.[11] The cells with < 46 chromosomes, constituting no more than 5–8% of the metaphases of any single bladder specimen, were due to random loss of chromosomes. Metaphases with a polyploid number of chromosomes, e.g., tetraploid cells with 92 chromosomes, were not observed in any of the specimens examined. Because it is possible that polyploid cells may not undergo mitosis[12] and, hence, that metaphases with 92 or more chromosomes may not be available for observation, we examined interphase cells in the bladder specimens of male patients using a quinacrine fluorescent technique. This method leads to high fluorescence of the Y, which can be readily identified as a single Y-body in diploid interphase cells. No cells with two such fluorescent bodies were observed, which would be expected if polyploid cells were present in the male bladder specimens. Thus, these observations lend further support to the lack of polyploid cells in human bladder mucosa. In contrast to the observations on elderly males, of whom a high percentage have a missing Y in their marrow cells, we could not find any significant number of cells with a missing Y in the bladder, including over 20 specimens from males over the age of 65 yr.

The first comprehensive study of chromosomes in bladder cancer appeared in 1967, though these data were briefly presented in a publication in 1965.[13] Lamb (1967) studied 29 cases with transitional cell cancer of the bladder (two tumors were examined in one case). A cell suspension was made of a piece of the tumor and incubated for 1–2 hr at 37°C with colcemide (0.5–1.0 μg/ml). The results indicated a relationship between the chromosome number and the histologic appearance and invasiveness of the tumor, i.e., that well-differentiated transitional cell carcinoma is usually diploid or near-diploid, though no karyotyping was performed and, hence, it is unknown whether the diploid cases were either truly normal chromosomally (diploid) or pseudodiploid. All noninvasive tumors had a near-diploid mode and only 2 out of 22 invasive tumors had a near-diploid chromosome count. With loss of differentiation, the chromosome number increased to near-tetraploid, with the poorly differentiated cancers being in the triploid to near-tetraploid range. Markers were not present in the well-differentiated ones. In the cases with a greatly increased number of chromosomes structural abnormalities, a widening of the range of chromosome counts and a decrease of the percentage of counts at the modal value also occurred, i.e., 54% in well-differentiated tumors vs. 17% in poorly differentiated ones. The presence of near-diploidy seems to relate more to an absence of invasiveness than to the picture of differentiation.

Examination of the sediments from irrigations of the bladder obtained by catheterization for their chromosome analysis has been used as an approach in the diagnosis of the disease of the organ.[14] The finding of normal and abnormal chromosome constitutions in the irrigated material from one patient with a papillary (pedunculated) tumor and from another with an advanced transitional cancer, respectively, correlated well with the histologic appearance of the tumors.

Shigematsu (1965) examined the chromosomes in a series of bladder tumors of all types. Even though it is difficult to compare series due to differences in histologic classification, the results agree, in general, with the trend for noninvasive tumors to be near-diploid and with less differentiated ones to have counts that fall away from the tetraploid level.

Falor (1971a) examined the chromosomes with a direct technique in seven noninvasive papillary carcinomas of the bladder obtained as biopsies during cytoscopy. In this series 23% of the cells were at the modal number (44–48 chromosomes) and all tumors had markers.

Karyotyping was performed and several abnormalities listed by the author: extra autosomes in groups A and E, missing chromosomes in groups C and D, and the presence of a long marker. These findings were compared to those in invasive cancer of the bladder, in which a wide range of modal numbers ranging from hypodiploidy to hypertetraploidy with markers and abnormal karyotypes were found and only 20% cells at the modal number.

The findings of Falor (1971a) are somewhat at variance with those previously reported, in that Lamb (1967) and Shigematsu (1965) found no markers in the six cases of noninvasive cancer of the bladder studied by them, though it must be pointed out that no detailed karyotypic analysis was established by these two authors.

In a more extensive study published by Spooner and Cooper (1972) on 64 tumors of the bladder from 58 patients, some interesting data were presented. To obtain sufficient metaphases for analysis these authors injected 10 mg I.V. of colcemide 2–4 hr before resection of the tumor. Pieces of the removed tumor were incubated for 30 min at room temperature. Squash preparations were examined.

Spooner and Cooper (1972) found that all well-differentiated (22 tumors) and moderately well-differentiated (13 tumors) cancers fell within the diploid range (42–49 chromosomes), whereas the poorly differentiated cases had a discrete distribution of modal numbers in the diploid range (20%) and widespread distribution throughout the triploid-tetraploid range (80%), with a distinct interval between the low and the high ploidy ranges of these tumors.

The frequency of polyploid metaphases (those with twice or higher multiple of the modal number) was estimated in 39 tumors. Of these, 37 were near-diploid and 2 were tetraploid with a secondary near-diploid mode present in 30–40% of the cells. In most tumors, the polyploid cells constituted < 5% of the metaphases, but in nine they made up a substantial percentage of the cell population.

A total of 32 papillary tumors with near-diploid chromosome numbers from 25 patients were analyzed (17 well differentiated, 9 moderately well differentiated, 6 poorly differentiated) in relation to invasiveness. It was shown that a third of the well-differentiated and moderately well-differentiated tumors and all but one of the poorly differentiated ones showed signs of invasion. Thus, these findings differ from those of Lamb (1967), who indicated that near-diploid tumors were usually not invasive.

Five tumors with the modal chromosome number of 46 were shown to have normal karyotypes (in one 19% of the metaphases were tetraploid) and in another five tumors 50–90% of the metaphases had normal karyotypes, and the tumors were of a well-differentiated variety. The remaining well- and moderately well-differentiated tumors and the poorly differentiated ones had preponderantly abnormal or pseudodiploid karyotypes.

A loss of predominantly A- and C-group chromosomes and a gain of D-chromosomes and little change in other groups were present in the well-differentiated tumors. The moderately well-differentiated tumors and the poorly differentiated ones showed more complicated karyotypic rearrangements, with a tendency of loss of chromosomes in group B. Thus, these findings differ from those reported by Falor (1971a). In only nine tumors were markers found by Spooner and Cooper (1972), and these consisted of a variety of morphologic types, i.e., large subtelocentric, acrocentric, and submetacentric markers. These observations also differ from those of Falor (1971a), who consistently found a large A-type marker. As observed in other human tumors, the markers in bladder cancer may be present in a small or in a preponderant percentage of cells of the tumor and in some more than one marker per cell may be present.

Spooner and Cooper (1972) essentially disagreed with Lamb's conclusion regarding the relation of ploidy to invasiveness and stressed the correlation of ploidy with differentiation of the tumor. Thus, the histologic grades of bladder tumors, which usually have a better prognosis (well- and moderately well-differentiated tumors), were all near-diploid, even though 20% of poorly differentiated tumors were only near-diploid. These authors hastened to add, however, that all of the latter tumors showed mainly a papillary growth pattern and may represent an intermediate histologic type.

In a study[15] utilizing banding techniques (G-banding) it was shown that in six moderately well-differentiated transitional cell carcinomas of the bladder, 162 metaphase spreads were suitable for counting. The counts ranged from 40 to 183 chromosomes with 18% of the cells having a hypotetraploid mode of 87 and the majority of counts in the 63-96 chromosome range. Bimodal metaphases were not seen. A total of 54 cells were karyotyped and in 36 of these a large submetacentric marker, larger than any of the chromosomes in group A, was observed. In 15 karyotypes a somewhat smaller marker of the B-group size with a submetacentric location of the centromere was also found. Even though banding methods were used, the origins of these markers were not clearly ascertained.

In a study of 53 cases of noninvasive and submucosal invasive, well-differentiated cancers of the bladder followed over a period of 4 to over 100 mo, repeated cytogenetic analysis by a direct technique revealed markers in 33 of the cases, with recurrences developing in 32.[16] In contrast, in the 20 patients without markers followed over a period up to 8 yr all but one patient have remained recurrence free. In the single case, 8 mo postdiagnosis, the mode changed from 69 to 92 with evidence of dedifferentiation and development of a new tumor in a bladder prone to neoplasia. Except for one tumor, which was in the near-diploid range, the noninvasive tumors had chromosome modes ranging from 41 to 50 with most of them in the near-diploid area. In contrast, the submucosal invasive carcinomas of the bladder were all in the tetraploid range with the modal numbers ranging from 81 to over 100.

Based on these authors' experience[17] with 165 patients with carcinoma of the bladder, they indicated that markers have potential value as a prognostic aid in bladder cancer and, furthermore, that the triad of tetraploidy, markers, and submucosal invasive moderately well-differentiated carcinoma appears to carry such a lethal prognosis as to militate for early radical resection. The markers observed most commonly consisted of a large metacentric and, less frequently, a large submetacentric. A ring chromosome was observed in two cases with submucosal invasive carcinoma. These same authors published a report on a recurrent noninvasive papillary carcinoma of the bladder, in which karyotypic abnormalities including aneuploidy and a ring and other markers, were established. Fifty-three mo later, when the tumor recurred, the karyotypic picture seemed to be very similar to the original one, e.g., the deficiency in group C was still present and the ring chromosome was present in 89% of the metaphases.

A number of reports have appeared on the chromosome constitution and DNA content of transitional cell cancers grown in culture for long periods of time.[18] In general, all have had an abnormal chromosome constitution with the chromosome numbers generally being in the triploid to the near-diploid range, with a substantial number of the cultures having marker chromosomes. Most of the tumors which were derived from male patients had identifiable Y-chromosomes in their metaphases. Also, most of the tumors from female patients were shown to contain a sex-chromatin body with occasional tumors having two such bodies. The pattern of two or more bodies in near-diploid cells seen in a noninvasive tumor may characterize an intermediate stage of clonal evolution, eventually resulting in malignancy, when the cell line has not yet achieved the ability to invade or metastasize.[19]

Utilizing conventional staining, Sekine (1976) examined with a direct technique 12 transitional cell bladder cancers and found a chromosome range of 32 to over 100, with 2 having hypodiploid modes, 3 near-diploid, 1 hyperdiploid, 3 near- or hypertriploid, and 1 hypotetraploid. In two cancers the modal counts could not be determined. Tumors of low-grade malignancy had a tendency toward near-diploidy, and those with high-grade malignancy a tendency to near-triploidy with a wider distribution of chromosome counts. In two exceptional cases, in which the histological pattern was of lower malignancy (grade 2), frequent recurrences were observed at the site of resection of these triploid tumors. The author suggested that a tendency to polyploidy and/or wide distribution of chromosome numbers in transitional cell cancer of the bladder may indicate more careful follow-up and/or more radical treatment. Markers were seen in some of the tumors.

The chromosome constitution of 62 papillary tumors and 75 invasive transitional cell carcinomas of the bladder has been examined in my laboratory (Sandberg, 1977). Of the 62 papillary tumors, 12 were benign and found to be preponderantly diploid; 2 of these tumors recurred, however, and in each a few karyotypically abnormal cells were present on the original examination (Figure 160). The modal chromosome number in the 50 papillary cancers ranged in the diploid area (44-49 chromosomes) with the karyotypic picture being different from tumor to tumor. Of the 50 papillary cancers, 32 had one to two markers (Figures 161 and 162). Twelve of the papillary tumors recurred; only 1 did not have a marker. Thus, the recurrence of the 11 papillary tumors with markers indicates that their presence may be of value in predicting the behavior of papillary cancers of the bladder. The chromosomal picture in the recurrent papillary tumors did not differ materially from the original one, although a slight modification in the chromosome number may occur. On the other hand, invasive transitional cell cancer of the bladder was accompanied by a large number of chromosomes and a relatively large number of marker chromosomes with many complicated karyotypic pictures. The presence of markers in papillary cancers may be indicative of the likelihood of recurrence and/or progression of such tumors and should attract the attention of

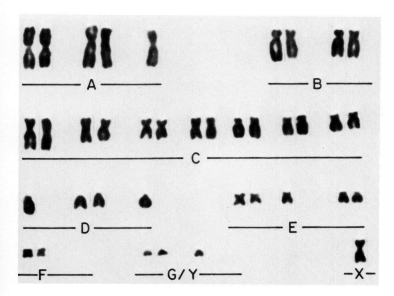

160 Unbanded karyotype from a cell of a papillary cancer of the bladder with a hypodiploid chromosome number: chromosomes are missing in groups A, D, E, F, and G.

those involved in the care of patients with these cancers.

As in other cancers in males, a Y-chromatin body may or may not be present in bladder cancers,[20] though a missing Y as the sole anomaly has not been described. The incidence of Y-bodies may correlate with the ploidy level of the tumor (Atkin, 1973a). Duplication of the body was seen in most bladder cancers, and in tumors at other sites, which had high chromosome numbers.

Studies with banding. Atkin and Baker (1977c) studied the chromosomes of 13 cancers of the bladder with C- and G-banding techniques. Heteromorphism for the centromeric heterochromatin of #1 was apparent in eight tumors, and in three of these the heteromorphism was also found in the patients' normal lymphocytes. In four tumors there were pericentric inversions of the heterochromatic regions of one or more #1 chromosomes, and major structural changes appeared to involve

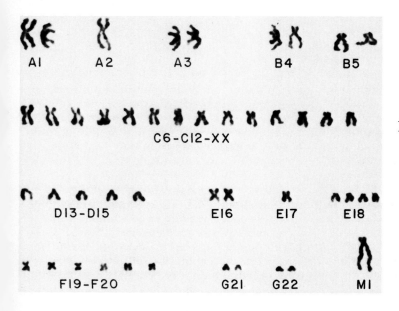

161 Unbanded pseudodiploid karyotype from a papillary cancer of the bladder containing one marker. Note abnormal distribution of chromosomes.

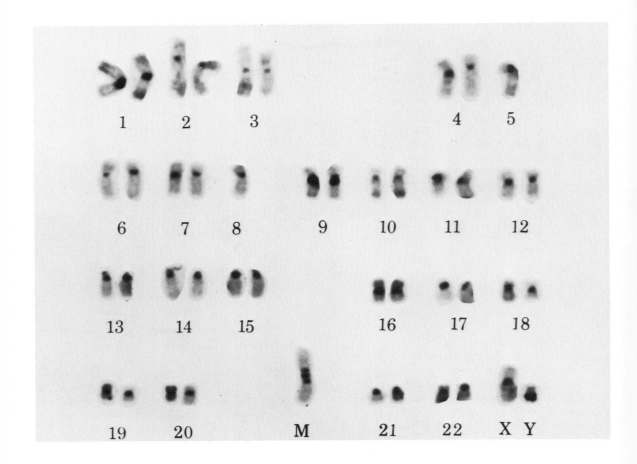

162 Hypodiploid karyotype from a papillary cancer of the bladder (C-banded) with an abnormal chromosome (M) and autosomes missing in groups #5 and #8.

that chromosome in at least seven of the tumors. In addition to the high incidence of heterochromatin variants, a presumably constitutional phenomenon, and major structural changes involving #1, one further feature common to four tumors was the presence of a heterochromatic minute chromosome. Atkin and Baker (1977c) indicated that these findings are similar to those obtained (Atkin and Pickthall, 1977) in tumors of the ovary and, although this heteromorphism is a common phenomenon, a significantly higher incidence has been found in normal cells in a series of cancer patients as compared with controls, suggesting that it is associated with an increased risk of cancer (Atkin, 1977). The results of all these studies suggested to the authors an association between heteromorphism for the size of the C-bands of #1, pericentric inversion of the heterochromatic region, the susceptibility of cancer at these and perhaps other sites, and the presence of gross structural changes involving #1 in tumor cells. Such changes are rarely found in chronic leukemias or benign tumors and may represent changes commonly found in the later stages of malignant transformation (AL and invasive tumors) irrespective of site or histological type.

A further feature of note in the series of bladder tumors was a heterochromatic minute found in four of the cancers. A similar minute was seen in bronchial cancer (Pickthall, 1976), though its mode of origin may vary in different tumors. One possibility is that it represents a product of a Robertsonian change involving two acrocentrics. The minute was often seen to associate with acrocentrics or with the centromeric regions of other chromosomes. In one of the cancers of the bladder a ring chromosome was seen. Rings have been noted by Falor and Ward (1976b) in one out of seven noninvasive and in two out of eight invasive bladder carcinomas.

An established cell line derived from a transitional cell cancer was shown to have a mode of 58 chromosomes with retention of the X and Y (G. E. Moore, Morgan, et al., 1978). In addition, the origin of 10 markers was established with banding, including abnormalities of #1, del(3), a dicentric, and several isochromosomes. A number of markers remained unidentified.

Cancer of the Kidney

Examination of seven cases of untreated primary kidney tumors (nephroblastomas) obtained from children revealed hypodiploid, pseudodiploid, and near-triploid counts with surprising consistency.[21] Neither highly polyploid cells nor abnormal markers were found. Of eight karyotypes in a tumor with 48 chromosomes, two were normal diploid and the rest contained extra chromosomes in groups A and C. Another tumor (no. 3) had a mode of 46, and all the 22 karyotypes examined were diploid except for three cells with 45 chromosomes. Each of these hypodiploid cells was obtained from a different site within the tumor. All cells counted, as well as all 13 karyotypes of another tumor (no. 4), had a normal set of 46 chromosomes; the cells were shown histologically to be tumor cells and not host cells. In the fifth tumor 46 chromosomes were found in all eight cells counted, and three cells had a normal karyotype. In the sixth tumor all 15 cells counted had a chromosome number ranging from 55 to 58, and five karyotypes showed a missing G- and extra A- and C-chromosomes. Case no. 7 was clearly pseudodiploid, all 14 karyotypes having 46 chromosomes with a missing #1 and an extra chromosome in the C group. All 44 cells contained 46 chromosomes.

Thus, this study can be summarized as follows. Except for one case in which the modal chromosome number was 55-58, in all others it was at or near 46. In fact, three patients, including the two youngest (aged 11 and 18 mo), yielded diploid karyotypes only. Two additional tumors had some hyperdiploid cells with extra C- or #1 chromosomes, whereas the tumor in the oldest patient (13 yr) had a consistent karyotype with 46 chromosomes in which a #1 was missing and an extra C-like chromosome which might have been derived from the #1.

Three renal carcinomas and two of the renal pelvis were studied; four were hypodiploid and one hyperdiploid. The range of chromosome numbers was 23-85 (Sekine, 1976). A carcinoma of the kidney was found to have 50-51 chromosomes with similar karyotypes on two occasions (Atkin, 1976b).

Riccardi, Sujansky, et al. (1978) reported an interstitial 11p− deletion in patients with the triad of aniridia, ambiguous genitalia, and mental retardation (AGR triad). In two cases a Wilms' tumor was present, and the authors indicate that the genesis of some of Wilms' tumors may be related to the chromosomal anomaly.

Chromosome studies in cancer of the urethra have been scarce. One case with 84 chromosomes has been reported.[22] In a reported case of a mesonephroma a mode of 92 with a larger marker was apparent.[23]

A biopsy specimen of a liver metastasis from a rapidly progressing renal adenocarcinoma was investigated with banding. Of the metaphases studied, 40% had a chromosome number in the triploid region. A characteristic set of markers was consistently present in these metaphases. The main part of the metaphases studied, however, had a normal diploid karyotype, and the authors postulated that these cells apparently represented normal host cells stimulated to mitotic activity by the presence of the neoplasm (Gripenberg, Ahlqvist, et al., 1977).

Tumors of the Male

Tumors of the Testis

This group of tumors constitutes only 1% of all the human cancers, affects males chiefly between the ages of 20-40 yr, and occurs rarely in blacks. Testicular tumors (Figure 163) arise in undescended testes with relatively high frequency, abdominal testes being more often involved than inguinally retained testes. Malignant testicular tumors are often found in pa-

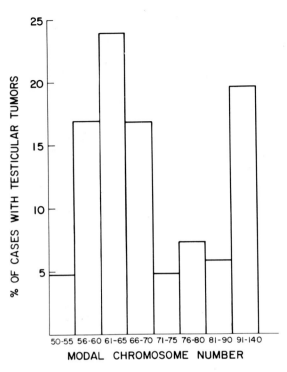

163 Distribution of modal chromosome numbers in testicular tumors. Most of the tumors have numbers in the near-triploid range.

tients with male pseudohermaphroditism, especially in cases with XO/XY mosaicism and a well-developed uterus; they have been noted in about 50% of the patients with this syndrome.

In contrast to the ovary, a salient feature of the > 100 seminomas and malignant teratomas of the testes that have been studied is the absence of tumors with chromosome numbers at or below the diploid range.[24] The seminomas (Figure 164) tend to have higher chromosome numbers than teratomas, though there was a predilection in both types of tumors for modal chromosome numbers in the hypotriploid region or above. Few of the seminomas had modes of < 60 chromosomes, and they were concentrated in the range of 60–69. There was an appreciable number with higher modes. Malignant teratomas, however, had modes of 50–59 and less frequently 60–69, and only a few tumors had higher modes. A nonrandom distribution of chromosomes in both malignant teratomas and seminomas and particularly a deficiency of group B and excess of group C chromosomes have been noted.[25] An excess of group F and a deficiency of #17–18 chromosomes in seminoma have also been reported.[26] Comparing the ratios of various pairs of groups, these authors found that in tumors combining the features of both seminomas and teratomas, the ratio of #3 to group B chromosomes was increased, whereas the ratio of groups B/F and #17–#18/F group was decreased. The ratio of #3/B group was also increased in pure teratomas and the ratios B/F groups and #17–#18/F group were decreased in pure seminomas. The pure tumors could also be distinguished by other features of the various ratios, and the tumors combining the features of seminomas and teratomas appear to occupy an intermediate position in this respect.

Five testicular tumors studied with conventional staining by Sekine (1976) yielded the following results: one seminoma and one benign teratoma appeared to be diploid, one seminoma triploid, one malignant teratoma hyperdiploid, and one embryonal cancer undetermined.

Nearly all malignant testicular tumors have at least one large or medium-sized marker; however, a seminoma without any obvious marker has been observed.[27] Testicular teratomas are frequently X-chromatin positive; seminomas uniformly lack this feature. The reason for this is unknown, but it is clear that studies of the sex-chromosomes are of special interest in teratomas. How this is related to the androgenetic origin of ovarian teratomas and hydatidiform moles in females is an intriguing question and, hopefully, data regarding the exact genetic origin of these teratomas will be forthcoming in the near future. Quinacrine fluorescence observations of metaphases and interphases have confirmed or at least suggested the presence of a Y (or Y-bodies) in most seminomas and malignant teratomas, including among the latter some that were X-chromatin positive and some in the former with double Y-bodies.[28]

Much attention has been given to a large marker consistently found in a high proportion of cells of four testicular tumors.[29] Thus, one seminoma with counts ranging from 63 to 74 (modes of 72 and 73) was found to have a long

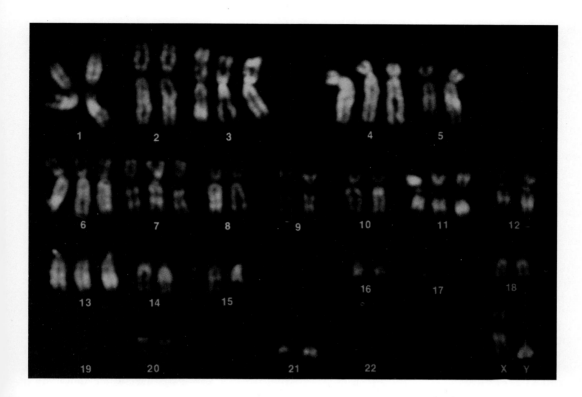

dicentric marker in 38 cells, a long acrocentric marker in 64, and a long B-type marker in 29 of 69 cells karyotyped. In a second seminoma and malignant teratoma of the right testis, the counts ranged from 45 to 124 and a long marker was noted in 58 out of 68 cells karyotyped. Most of the cells examined probably came from the seminoma part of the tumor. In a third malignant teratoma, the chromosome counts ranged from 44 to 160, and a long dicentric marker was present in 28 out of 34 cells. The fourth tumor, a malignant teratoma, had a range of chromosomes of 41 to 125, and a long marker, similar to that in the three other cases, was present in 35 out of 43 examined metaphases. The similarity in morphology and size of the marker was remarkable as was the fact that it never occurred in duplicate in any cell in all of the four tumors. However, these observations must be interpreted with caution, because without banding it is impossible to decide upon the origin and identity of the markers and, furthermore, we have shown that markers of similar morphology in different tumors, when examined with banding, may have different origins.

164 Q-banded karyotype of a seminoma of testis showing 53,XY,+3,+4,+6,+7,+11,+13, and +19.

The data in another study do not support the hypothesis of a specific testicular tumor marker.[30] A long chromosome having similar morphology was seen in only one of five tumors, a metastatic pleomorphic seminoma with chromosome numbers ranging from 59 to 75. All aneuploid cells possessed a very long submetacentric chromosome nearly twice the length of #1. In another well-differentiated seminoma with inflammatory infiltrate only metaphases with normal 46 chromosomes were found. In a third case, a metastatic teratocarcinoma with mostly near-diploid cells and two karyotypes with 50-63 chromosomes, no marker was found. In a fourth case, a metastatic embryonal carcinoma, only diploid cells were found. The examined cells were most likely of nonmalignant origin because the specimen histologically showed considerable hyperplastic lymphoid tissue.

In another study,[31] 10 testicular tumors were analyzed, and the karyotypes of all cells were diverse. Extra chromosomes were distributed irregularly among all groups, with the most prominent increase being in group C. An outstanding feature of the tumors was the presence of markers. A chromosome with a subterminal centromere was present in every tumor and in 35%–95% of the cells. A long marker with a secondary centromere or constriction was found in at least 50% of the cells of four cases.

Referring back to the long marker in testicular tumors, if there is specificity to this marker it would be site specific, because this long chromosome has been observed in testicular tumors of different histology and pathology, e.g., seminoma and malignant teratoma. The morphology of the marker is similar but not identical in different reported specimens. The proportion of the long and short arms, as well as the relative length of the chromosome varied. On the other hand, the marker has not been seen in duplicate in any cell and appears to be quite characteristic for stem line cells in the tumors described. In one case, it occurred in seminomas of both testes.[32]

The occurrence of tumors with modal numbers predominantly in the near-triploid range raises the question of how such a chromosome number is generated. The chromosome complements of tumors are probably the outcome of a series of events. Triploidy may be achieved by repeated nondisjunctions. Alternatively, this might also be reached by a combination of a complete doubling of the complement by endoreduplication followed by chromosomal loss, but not necessarily in that order. Another possibility is that testicular tumors, which are often near-triploid, commonly arise from triploid rather than diploid (or haploid?) cells. Such might be the case were the tumors to arise from the chromosomally abnormal triploid twins. Near-triploid complements may also result from a process of "triploidization" involving duplication of a haploid or near-haploid set in a diploid or near-diploid cell; that such a process can occur is suggested by the occurrence of sporadic triploid cells in cultures of normal lymphocytes and fibroblasts and the finding of a near-triploid cell, which could have arisen from a pseudodiploid cell by duplication of a haploid set, in a patient with CML and lymphadenopathy (Nava, Grouchy, et al., 1969b). It is interesting to note that Atkin (1973b), in discussing two seminomas that had modal DNA values equivalent to 52 and 80 chromosomes, respectively, and were histologically of the spermatocytic variety, raised the possibility that this type of seminoma may arise from spermatocytes.

The chromosomes have been examined in two cases of squamous cell carcinoma of the penis;[33] both had a stem line in the range of 71–80 chromosomes with no markers.

Further comments on tumors of the testes. There is uncertainty about the histogenesis of testicular tumors and, hence, there is no universally accepted classification. Most tumors, however, can be broadly classified as either teratomas or seminomas, even though mixed forms occur. Two main cytogenetic features of testicular tumors are the frequent occurrence of Barr bodies in teratomas[34] and the apparent universally high chromosome numbers (i.e., over 50) in both teratomas and seminomas.[35]

Unlike the common teratomas (dermoid cysts) of the ovary, which are benign tumors, teratomas of the testis are rather uncommon and usually malignant tumors. In the more differentiated examples, elements of all three embryonic germ layers are usually present; in the undifferentiated tumors there may only be cells of carcinomatous appearance. Even though there have been several theories as to the origin of testicular teratomas, perhaps the most generally accepted one is that they arise from pluripotential cells that have escaped the influence of organizers.

Seminomas have a uniform histological appearance and may arise from the germinal (seminiferous) epithelium of the mature or maturing testis. They are malignant tumors, however, that are radiosensitive and carry a better prognosis than teratomas. Barr bodies have not been described in seminomas.

On the basis of enzyme and chromosome studies, Linder, Hecht, et al. (1975) examined a number of extragonadal teratomas and endodermal sinus tumors and came to the conclusion that these tumors develop from mitotic cells; and that the cell origin could be either a somatic cell or a misplaced germ cell that failed to undergo meiosis and had proceeded directly into mitosis. Reports have appeared on the role of chromosome breakage in teratology, on the distribution of chromosome spiralization in testicular tumors, and on extragonadal teratomas.[36]

Barr bodies in teratomas were first described in 1954[37] and subsequently by numerous other authors. In a review[38] Barr bodies had been found in 110 out of 240 male teratomas (including some arising at extragenital sites). Sometimes they were present in some regions of the tumor but not in others; or they were confined to certain cellular components. Even though Barr bodies can occasionally be explained on other bases, e.g., in a case of Klinefelter's syndrome with a pineal teratoma,[39] this is exceptional in the great majority of cases, and almost all patients with testicular teratomas appear to be chromosomally normal. Various theories have been put forward to account for the appearance of Barr bodies in teratomas of males; some light has recently been shed by the demonstration that most chromatin-positive teratomas, as well as chromatin-negative teratomas and seminomas, contain Y-chromosomes, either as seen in metaphases, where they are particularly easy to identify with Q- and C-banding, or revealed in interphase nuclei where in Q-stained preparations they are represented by Y-chromatin bodies.[40] The presence of both X- and Y-chromosomes in chromatin-positive teratomas is not in accord with theories that postulate their development from haploid cells containing an X, with subsequent conjugation of two cells or chromosome duplication in a single cell to give XX diploid cells. Aneuploidy per se would not appear to explain the occurrence of Barr bodies in teratomas because, with extremely rare exceptions, other types of malignant tumors in males are chromatin negative. If, however, the aneuploidy first occurred in developing teratomas in cells where, like those in the early stages of fetal development, differentiation of X-chromosomes to allocyclic behavior (formation of Barr body) had not yet occurred, the subsequent development of the latter properties might result in the appearance of Barr bodies, as it does in patients with Klinefelter's syndrome. One possibility is that the precursor cells are triploid; it is known that triploid embryos with both XXX and XXY sex-chromosomes may have Barr bodies. Whether teratomas originate from diploid or triploid cells, secondary chromosome changes may, of course, involve the sex-chromosomes and could account for the duplication of the Y seen in some teratomas.[41] An origin from triploid cells might also explain the consistently high chromosome numbers of both teratomas and seminomas. Kaiser-McCaw and Latt (1977) showed that in ovarian tumors the pattern of duplication of the late-replicating X is identical to that of normal fibroblasts, but different from that usually observed in peripheral lymphocytes. Thus, if late replication is an accurate gauge of X inactivation, the data confirmed that such an activation can occur without fertilization.

Choriocarcinomas of the testes are highly malignant tumors generally considered to be a variety of teratoma. Barr bodies have been described in one tumor.[42]

A distinction between benign and well-differentiated malignant teratomas is not always easy to make on histologic criteria, especially in young adults. In infants and young children, however, a polycystic tumor with completely differentiated elements occurs and carries a good prognosis. According to a report on eight of these tumors,[43] there was a lack or a very low incidence of Barr bodies; it will be interesting if further studies confirm that these tumors are generally chromatin negative. Chromosome studies on benign teratomas have not been reported, but the modal DNA value of a presumably benign well-differentiated teratoma was compatible with a diploid complement.[44]

Cancer of the Prostate

Detailed chromosome analyses of human prostatic cancer, either primary or metastatic, are egregoriously lacking. Studies on cellular DNA content of cancer of the prostate have been published and correlated with prognosis;[45] however, these methodologies lack the specificity and morphologic detail of chromosome analysis, particularly regarding the possibility of a specific karyotypic change characterizing all or some cancers of the prostate.

We reported (Oshimura and Sandberg, 1975) the presence of an isochromosome 17 [i(17q)], established with Q- and G-banding, in the metastatic cells in the BM of a patient with prostatic cancer. Direct marrow chromosome preparations showed a mode of 70 with considerable scatter in counts and the presence (ca 15%) of normal diploid metaphases with 46 chromosomes. The latter were undoubtedly of normal BM origin.

A prostatic cancer with hypodiploidy and no markers was described by Sekine (1976). Aneuploidy has been found in long-term established cell lines from cancers of the prostate.[46]

Because studies[47] based on DNA content in prostatic carcinoma have indicated that ploidy may have some relationship in response to estrogens, i.e., all but one of the nine triploid and hexaploid tumors did not respond to estrogen therapy, sharply contrasting with 22 or 24 diploid and tetraploid tumors that did respond, it becomes important to ascertain with chromosomal studies whether this particular relationship holds up.

Tumors of the Lung

Cancer of the Lung (Bronchus) and Larynx

Most of the reports on the chromosomes of lung cancer (Figure 165) have dealt with cells in pleural effusion (Figure 166), though some information is available regarding the karyotypic picture in primary tumors (Figure 167).[1] The first report on cancer of the lung appeared over 20 years ago,[2] in which the behavior of a ring chromosome in the pleural effusion of an anaplastic adenocarcinoma of the lung was followed. The marker persisted in 100% of cells in two samples drawn 2 wk apart. This tumor was analyzed about a year later,[3] at which time the modal number was 75; five karyotypes had ring, long acrocentric, and metacentric markers. A chromosome number exceeding 140 was found in 10% of the cells. Nearly all published modal chromosome numbers in cancer of the lung, whether from primary tumors or malignant effusions, have been in the region of 60 or above; the distribution for different histological sites is uncertain. Four cases with hypo- and/or pseudodiploidy in the cells of pleural effusions resulting from lung cancer have been reported (Carlevaro, Rossi, et al., 1978). In 1963 we published[4] the results on several cases with cancer of the lung; in one of them three different samples of a pleural effusion were obtained. The first sample contained cells with a chromosome number ranging from 94 to 106; the second sample, obtained 6 hr after the injection of vincristine, had cells with 49–194 chromosomes. A third sample, drawn 24 hr after the vincristine administration, showed cells with 75–109 chromosomes. In the last specimen, two or three acrocentric markers appeared in all cells. An interesting case of a pleural effusion from a pleomorphic anaplastic carcinoma of the lung in a 58-yr-old man with a chromosome number of 75 was published.[5] In addition to the set of apparently

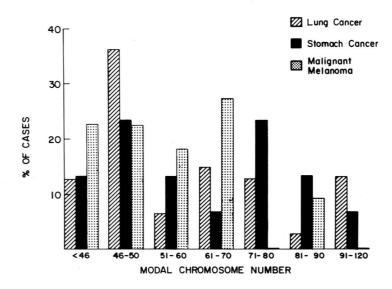

165 Distribution of the modal chromosome numbers in cancers of the lung and stomach and in malignant melanoma.

morphologically undamaged chromosomes, almost all dividing cells contained double-minute chromosomes (DMS), the size of which was near the limit of visibility; the authors suggested the association of DMS with micronuclei formation (see discussion of DMS in section on brain tumors). In another study[6] four bronchogenic carcinomas were examined: one had chromosome numbers ranging from 62 to 131, another had a mode of 60, and the third a mode of 65. Markers were observed in all three tumors with consistency. The fourth carcinoma had abnormal chromosomes but of poor quality to construct karyotypes. A carcinoid adenoma of the lung has been examined and yielded three karyotypes with 45 chromosomes without any consistent pattern (no markers were seen), and a cylindromatous adenoma of the lung showing local invasion had a mode of 46 chromosomes, including markers.

A solid primary carcinoma of the larynx has been analyzed in my laboratory and found to be bimodal, i.e., 40 (present in 22% of the cells) and 78–80 chromosomes. Karyotypes of the modal cells contained haploid numbers in

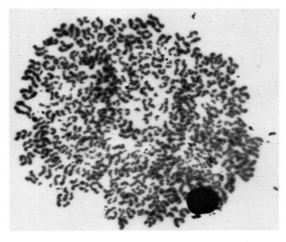

166 Cell with a large number of degenerating chromosomes from a cancer of the lung (pleural effusion).

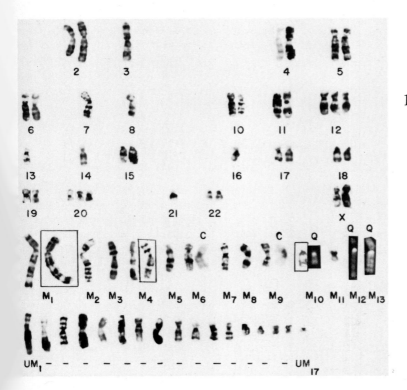

167 Karyotype (G-banded) of a cell from a lung cancer. Bracketed chromosomes and those on which C- and Q-banding had been performed are from another metaphase of the same case. This extremely complicated case karyotypically had a large number of marker chromosomes ($M_1 - M_{13}$), the origin of which could be ascertained, and a large number of unidentified markers (UM_{1-17}).
(*Kakati, Hayata, et al., 1975.*)

groups D, F, and G; one minute and three long markers were present in almost all the cells examined.

The chromosome numbers in lung and larynx cancer vary greatly, as is true of other malignant neoplasms. Modal numbers were usually near-diploid and triploid values and ranged from 40 to 114. Stem lines detected in two solid tumors did not differ from most stem-cell numbers of effusion material. In the karyotypes of malignant cells the distribution of chromosomes did not reveal any consistent patterns. However, groups E, F, and G often contained supernumerary chromosomes and group D frequently lacked several chromosomes. All these findings await confirmation with banding, preferably on primary tumors.

G- and C-banding analysis of a metastatic bronchial cancer (Pickthall, 1976) revealed a mode of 60 chromosomes and 1–3 DMS in the majority of the cells. Seven different types of markers were observed: t(5;?) (5qter→p11::?); t(6;10)(6qter→cen→10qter); t(7;11) (7qter→p22::11q13→qter); t(14;18) (14qter→cen→18qter); del(2) (pter→q21); t(6;?) (6qter→cen→?); and the DMS. Extra chromosomes were present in groups #1–#5, #10, #12, #17, #20, and #21; autosomes were missing in groups #14 and #18. Apparently, two Y-chromosomes were present.

Several authors[7] have examined the sex-chromatin incidence in cells of lung cancer in an attempt to correlate this occurrence with the phenotypic sex of the patient or with the sex-chromatin incidence of tumors of other sites. An examination of five bronchogenic carcinomas in males revealed an average of 17.4% of cells to be chromatid positive. When these data are compared with samples from 50 carcinomas of the uterine cervix, in which an average of 67.2% of cells were chromatin positive, it appears that cancer in general preserves the cytomorphologic sex characteristic of the host tissue. In one study,[8] 20 bronchogenic carcinomas in women were examined and confirmed the findings noted above. Positive sex-chromatin was found in 64.5% of the cells. In 37 carcinomas of the lung in female patients, it was found that in 20 cases 9–100% of the tumor cells contained Barr bodies. In no case did < 30% of the cells have a clearly visible sex-chromatin body.

In the oat cell variety of bronchial carcinoma, hypodiploidy may be commoner than in adenocarcinoma or squamous cell carcinoma of the lung. Among oat cell carcinomas studied (Atkin, 1976b), two had modal chromosome numbers of 43 and 45, respectively, another 75–80 chromosomes, and the modal DNA value of another was near-diploid. In contrast, only one out of nine tumors of other histological types was near-diploid. Measurement of DNA values (Greisen, 1971) yielded values for oat cell carcinomas that tended to be lower than those of other histological types.

In a bronchial adenoma in the few suitable metaphases the modal number was apparently 45, and in a second adenoma showing local invasion, pseudodiploid karyotypes with markers were found (Falor, 1971b).

In an established cell line, originating from a squamous cell cancer of the bronchus growing continuously for 6 yr, the chromosome number ranged between 58–61. Banding analysis indicated that except for minor changes the basic karyotype remained fairly stable (P. Fischer and Vetterlein, 1977). A dozen markers were identified as to their origin and included: del(1)(p13), del(1)(q11), del(3)(q11), two i(5p), del(6)(q21), t(8;15)(p23;q15), t(1;3)(q11;p13), t(3;13)(q11;p13), t(2;21)(p11;p13), 8p+? and 17p+?

Participation of #1 in marker formation in metastatic cells in the BM from small cell anaplastic cancers of the lung in 14 of 18 patients with karyotypic abnormalities has been reported (Wurster-Hill and Maurer, 1978).

Tumors of the Thyroid and Adrenal Glands

The number of cases analyzed is not too large and in some instances the cultured metastatic thyroid cancer cells were accompanied by karyotypes that were probably of host origin.[1] In one study[2] one malignant and five benign tumors of the thyroid gland were examined. The three follicular adenomas were shown to have modes of 46 in 199 cells counted. When 100 cells were karyotyped, abnormalities were detected in 20% of the cells with no consistent pattern; however, gains and losses of chromosomes were apparent. Two cases of microfollicular adenoma showed patterns similar to those of the follicular adenomas. In the one case of follicular adenocarcinoma the chromosome picture differed sharply from that in the benign tumors. Of the 100 cells counted, the majority had 52–59 chromosomes with gains in groups B, C, E, F, and G. Of 29 karyotypes, only 1 was normal. The rest displayed a variable picture with no typical markers; in 40% of the cells one or two acentric fragments were present. It should be pointed out that in this study the cells of both the benign and malignant tumors were examined after 3–8 days in culture.

In a DNA photometric analysis[3] of 371 nodular lesions of the thyroid, including 17 cases of carcinoma, irregularly increased DNA values were found in 75% of nodules with microfollicular structure, in 81% of lesions of possible malignant nature, and in 82% of frank carcinomas. The largest deviations from the diploid were found in three microfollicular lesions: 38% of the cells in one tumor and 34% in another were aneuploid, with one of these having a stem line of 60 chromosomes (53% of the cells). The third, a solid carcinoma, had a stem line of 46 chromosomes (81% of the cells).

In another study of the DNA content[4] of carcinomas of the thyroid, including the "occult sclerosing papillary carcinoma" which has a relatively good prognosis, hypodiploidy was common.

An adrenocortical carcinoma in a female infant (1 yr old) showed 47 chromosomes, the karyotype being diploid except for an additional marker chromosome (+Cp−) (Berger, 1969c). Another case in which chromosome changes were found in such a tumor has been described (Pascasio, Jesalva, et al., 1967).

Further Comments on Tumors of the Thyroid

The first reported study of human thyroid chromosomes was that of Miles and Gallagher (1961), who by direct preparations found no significant abnormality in eight cells of a papillary and follicular thyroid carcinoma metastatic to a local lymph node. In view of the paucity of data and the lack of definite abnormalities, one may wonder whether this material was representative of the malignant stem line or of normal lymphoid tissue. Ishihara, Moore, et al. (1962) analyzed the karyotypes of 81 cells in a pleural effusion of a patient with metastatic thyroid cancer. Two-thirds of the cells were found to have normal karyotypes, whereas one-third were highly abnormal, possessing 64–78 chromosomes. It was concluded that the diploid cells probably represented normal host elements, whereas the cells in the triploid range were derived from the thyroid carcinoma.

Hashimoto's thyroiditis is an autoimmune disease which may be associated with thyroid carcinoma; the thyroiditis may be secondary to the cancer rather than the reverse. It is commoner in females than males, with a relatively high incidence in patients with gonadal dysgenesis. The presence of cells with an XX,47,+C karyotype in short-term cultures of thyroid tissue removed surgically has been described.[5] Short-term cultures of thyroid tissue from 23 patients with Hashimoto's disease have been studied[6] and in 13 of these (all female) satisfactory preparations were obtained. Five had only diploid metaphases, whereas eight included some hyperdiploid cells with most of these having 47 chromosomes due to an extra C-autosome or a marker. Two cases had clear evidence of single clones but the others showed some variation, e.g., in one case some cells had an additional D-chromosome or a minute. Blood cultures from five affected glands and one from an adjacent lymph node, among those showing abnormalities in thyroid cultures, were normal. It is uncertain whether the aneuploid cells were epithelial or stromal, but it seems more likely that they were the former. The epithelium in Hashimoto's disease frequently shows atypia with polyploid cells, appearances consistent with a neoplastic condition.

An interesting case has been reported[7] of a 58-yr-old woman who in 1945 at age 30 was found to have Hashimoto's thyroiditis and possibly lymphoma in the gland, but who in 1961 was shown to be a t(14;21) carrier and the grandmother of a child with DS. In 1970 HD in the left upper neck was discovered, followed by systemic spread of the disease and death in 1973.

Normal chromosomes were found in a mucosal neuroma variant of a medullary thyroid carcinoma syndrome (Nankin, Hydovitz, et al., 1970).

DNA estimations in 12 benign pheochromocytomas revealed near-diploid DNA modes, whereas 3 other tumors of similar histologic appearance, but which had metastasized, were hyperdiploid or near-diploid.[8] The chromosomes (of blood cells?) in four patients with Sipple's syndrome (medullary thyroid carcinoma plus pheochromocytoma) were shown with G-banding to be normal.[9]

Malignant Melanoma

Chromosome studies on some 50 cases of malignant melanoma (Figure 165) have been reported.[1] The majority of specimens examined were metastatic cells either in effusion, from BM, or lymph nodes, though a number of primary tumors has also been included. A summary of representative karyotypes in 22 tumors reported in detail showed an average chromosome number of 65.[2] A proportional increase of chromosomes in groups E, #3, and a reduction in groups #1, B, D, and G were noted. In addition, the karyotypes included, on the average, two markers. Large markers with secondary constrictions have been described or illustrated in several tumors.[3] In one study cells obtained from metastases of melanoma to lymph nodes were examined with a direct method;[4] no karyotypic consistency or predominant cell lines were observed. A giant submetacentric marker, ring chromosomes, and medium-sized telocentric markers were found in several cells. There were 10 chromosomes resembling those of group F in some cells and only 2 chromosomes from the G-Y group in others. In another series of 10 melanomas (9 metastatic and 1 primary), 2 specimens were characterized by predominant stem lines of 42 and 78 chromosomes, respectively.[5] Twelve of the modal cells with 42 chromosomes were karyotyped and all contained a small condensed F-sized marker. This marker was not seen in seven karyotypes constructed from stem-line cells with 78 chromosomes, all of which had a ring chromosome. The remaining eight specimens, including a primary melanoma, contained abnormal numbers of chromosomes, variable markers, and structural rearrangements with no consistent pattern. In a more recent series of 19 cases (Berger and Lacour, 1973), in which the cells were examined either by a direct method or after short-term culture, numerical and structural chromosome abnormalities were observed with a relative deficit in groups B, D, and #17-18, a relative excess in group F, and the presence of a variety of markers. No correlation between the cytogenetic and clinical data could be established.

Inasmuch as some of the studies on malignant melanoma have been done on established culture lines, it may be worthwhile to mention some of the results; particularly, as these have been examined with various banding techniques. In one cell line a modal chromosome number of 45 with the presence of a ring, a large marker with an interstitial C-band, and at least five other structural rearrangements resulting in markers were observed.[6] Most of these changes probably occurred in vivo, because they were identifiable in the biopsy material 4 hr after the culture was initiated. They persisted in early and later subcultures. In another study,[7] analysis in six cell lines and two primary outgrowths derived from malignant melanoma were performed. Gross aneuploidy was seen in all specimens, but each culture contained at least one distinctive marker specific for the cell line in 87–100% of the metaphases. One of the primary explants contained a marker (5p+) that was demonstrable in fresh tissue and persisted for 2 wk of culture. The same marker was found in all metaphases from two different metastases; skin fibroblasts from the same patient had a normal chromosome complement. In six of the eight cultures the most frequently found marker was formed by a brightly banded chromatin addition. Relative polysomy of #7 was found in seven of the eight cultures and for #22 in five of the eight. The frequency of polysomy of #7 and #22 was significant at the 5% level.

In the above two reports, examination of the published materials reveals that #1 was more frequently involved in the formation of markers than any other chromosome, even though this was not stressed by the authors. Thus, in the report by Chen and Shaw (1973), out of the seven markers described, two were derived from #1. In the paper by McCulloch, Dent, et al. (1976), markers involving #1 were present in four lines; in two it was 1p+ and in two others 1p−.

Some Chromosome Findings in Malignant Melanoma

A report[8] on 19 tumors, some previously reported on, showed that there were numerical and structural chromosome abnormalities in all, including relative deficiencies in groups B, D, and #17–18. Modal numbers ranged from an exceptionally low 29 to about 108, but, as in other studies, near-triploid modes were favored. The tumor with a modal number of 29, a local recurrence after surgery in a 64-yr-old female, was shown to have reduced numbers of chromosomes in all groups apart from the presence of two #3 chromosomes. Frequent endomitoses were observed, and there was a secondary mode of 58–59 chromosomes and some more highly polyploid cells with 100-130 chromosomes. The majority of cells with 28 or 29 chromosomes contained one marker (a dicentric). However, this marker was only seen occasionally in the cells with higher chromosome numbers, which frequently contained other markers, including a ring. Only small changes in the karyotypes of a metastatic melanoma and lymph nodes removed on three occasions over a 5-mo period were observed.[9] The mode changed from 47 to 46 chromosomes, but the same markers persisted in each sample.

Analysis of X-chromatin frequency in primary cell cultures from malignant melanomas from the uvea showed that the frequency was in close accord with the phenotypic sex of the patients.[10]

Karyotype determination in seven malignant melanoma cell cultures showed wide numerical variability of chromosomes, ranging from 24–50 chromosomes. Most of the karyotypes had a modal number of 40.[11]

In studies of a human melanoma cell line kept in vitro, karyotypic analysis showed two distinct markers, one in the A and the other in the B group.[12]

The presence of sex-chromatin in a metastatic malignant melanoma from a 26-yr-old male patient, who showed no evidence of any constitutional chromosome anomaly, has been described.[13] A possible association between the apparently "female" origin of the tumor and its favorable response to therapy was advanced.

We published (Kakati, Song, et al., 1977) a detailed karyotypic analysis with G- and Q-banding on cells of four malignant melanomas (Figures 168–169). The modal number in two cases was in the hypodiploid range, varying from 39 to 43. These two tumors had 5–13 markers. The other two melanomas were in the polyploid range, with modal numbers of 63–157 chromosomes. The cells had a minimum of 11 and a maximum of 40 markers. Chromosome #1 was more frequently involved in aberrations than any other chromosome in this material which was directly examined without resort to culture. Frequent break points were also noticed in the centromeric region of various chromosomes, but in #1 this area did not seem to be involved. Frequent involvement of centromeric breakage during formation of complicated markers and other chromosomal abnormalities has been noted in other tumors and emphasized in one of our previous studies on CML. The most common break points on the various chromosomes were 1q21, 1q25, 1q32, 5p13, 9q13, 11q23, and 12q13. No common markers were noted among these four cases of melanoma, but were seen in unrelated tumors.

Chromosome #1, Malignant Melanoma, and Other Cancers

Several of the observations obtained by us are of interest as they relate to data previously published in the literature. Mitelman and Levan (1976b) reviewed 270 cases of human neoplasia from the literature and indicated clustering of aberrations to specific chromosomes, i.e., mainly #7, #8, #9, #14, #20, #21, and #22. We would like to add #1 to this list. This is not only based on the observations with malignant melanoma, but also on the frequent aberrations of #1 seen in cancer of the ovary, lung, and breast.[14] It should be mentioned that the break points were not the same in every case. However, from the observations with melanomas and from those of our previous studies it appeared that the breakage rate is high at region 1q21–25. The high frequency of aberrations cannot be related necessarily to the size of #1, because #2 and #3, though of relatively large size, have very few aberrations. Incidentally, in two cell lines and one outgrowth obtained from malignant melanoma, a 5p+ and involvement of #1 are apparent from an examination of the figures in the publications.[15] From our studies it appears that not only is #1 more frequently involved in structural aberrations than any other chromosome in melanoma, but we have observed a total of 11 break points in #1 in one such a case, which is the highest number of break points observed in any chromosome. Granberg, Gupta, et al. (1973) reported a karyotype of a gastric tumor that had a marker identical to one of the markers in one

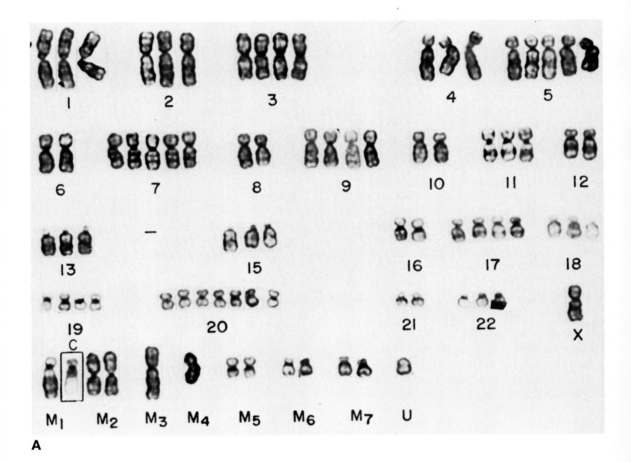

of the melanoma cases studied by us, i.e., a 1q+, having the following constituents: ins i(q31::q21→q32::q32→qtr). A similar marker was also noticed in the leukemic cells of a patient with CML and in those of the marrow of a subject with multiple myeloma. We have described the presence of an i(1)(q) in an ovarian cancer and in another malignancy of unknown origin.[16] An identical i(1)(q) was present in another case of melanoma studied.[17] An isochromosome of the long arm of #1 was also reported in a histiocytic lymphoma,[18] however, the i(1q) of this chromosome was altered by deletion of regions q22–25 and, hence, this chromosome was actually i(1)(q−). Involvement of #1 is not infrequently observed in AL and other myeloproliferative disorders which may terminate in AL.[19] A marker chromosome similar, again, to one seen in melanoma, i.e., t(21;?)(q11;?), was also present in two cases of

168 A. G-banded karyotype of a malignant melanoma cell with a large number of chromosomes including eight markers (M_1–M_7). One marker remained unidentified (U). A C-banded M_1 is shown for comparison. Note the missing Y-chromosome. B. G-banded karyotype of a malignant melanoma cell with missing chromosomes in a dozen groups and an extra chromosome in group #18. The cell contained six markers, four of which (M_1–M_4) could be identified with banding, and two of which remained unidentified (U_1 and U_2).

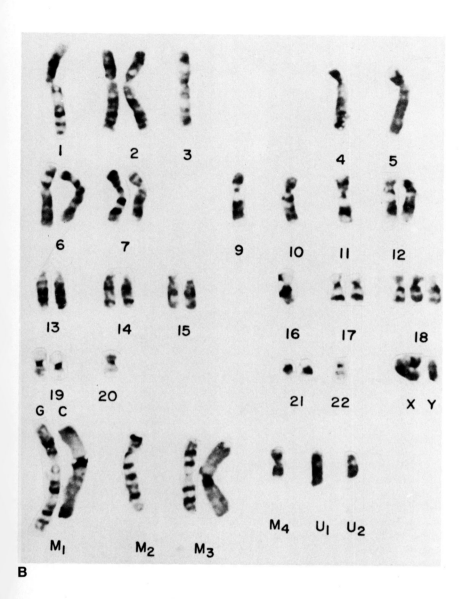

lung cancer.[20] A submetacentric marker seen in one of the melanomas, i.e., del(1p)(13), was also noted in a breast cancer and in an ovarian tumor. Thus, there seems to be little doubt that #1 is involved in a number of chromosome aberrations in an array of human tumors and should be added to the list of chromosomes frequently involved in human neoplasia. A more detailed analysis of the involvement of #1 has been published by Rowley (1977d) and is presented in another section of this book.

Miscellaneous Tumors

This group includes tumors of diverse origin and for which only sporadic cases examined karyotypically have been published. A summary is presented with some of the findings, and it is hoped that future studies will reveal a more clear-cut picture based on more extensive cytogenetic analyses.

Chromosome preparations of a solid retroperitoneal leiomyosarcoma have been ana-

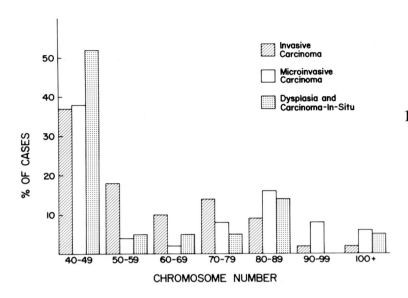

169 Chromosome number (modal) in invasive, microinvasive, and dysplasia and carcinoma *in situ* of the cervix uteri.
(*According to Spriggs, 1974.*)

lyzed;[1] three karyotypes exhibited only normal diploid cells despite the histologic examination which showed a very cellular tumor with a high mitotic rate and no apparent mixture of lymphocytes or other inflammatory components.

The chromosomal findings in rhabdomyosarcoma have been described by several authors. In one study a metastatic embryonic rhabdomyosarcoma in a male patient, age 16 yr, had a mode of about 60 chromosomes.[2] All metaphases had a number of DMS, but no other markers were seen. Two successive recurrences after radiotherapy of a rhabdomyosarcoma in a girl, age 3 yr, showed modes of 70 and 67–68, respectively;[3] among the markers was one in the second specimen which had not been present in the first. A similar tumor in the bladder (female, age 3 mo) showed only diploid karyotypes, whereas an undifferentiated sarcoma (female, age 10 yr) was aneuploid with 47 chromosomes including a large marker.[4] The cells of a rhabdomyosarcoma from a 5-yr-old girl cultured for several days were found to have a pseudodiploid mode with a karyotype 46,XX,t(Bq−;Cp+),t(Al?;Dq+) (Orye and Delbeke, 1974).

In a series of tumors of the bone,[5] six sarcomas had modes of 58 or above in direct preparations, whereas eight benign giant cell tumors appeared to be diploid, both in direct preparations and in a long-term culture. A periosteal sarcoma had a mode of 42 chromosomes, including markers.[6] A thymoma (female, age 7 yr), probably benign, yielded only diploid karyotypes,[7] whereas in another case with thymoma it was shown that after tissue culture the cells contained a stem line that was pseudodiploid with hypodiploid stem lines. A great variation of the chromosome numbers in the E (losses) and C groups (gains) was observed.[8]

A series of seven embryonic sarcomas, studied in direct preparations as well as in cultured material from patients up to 7 yr of age, yielded 54 abnormal karyotypes out of the 97 cells examined, six of the tumors having large markers.[9] In another case of embryonic sarcoma in a child a ring D-chromosome was reported.[10] An embryonic sarcoma of the urogenital sinus (male, 2.5 yr) and a highly malignant embryonic tumor of the mediastinum (female, age 8 yr) were aneuploid with the modal numbers ranging from 58 and 48, respectively.[11] In another study of eight embryonal sarcomas five were found to be aneuploid after short-term culture (chromosome numbers: 48–69), with five of the tumors containing markers, one with fragmented chromosomes and one with minutes (Rousseau, 1973).

Chromosomes were examined in cultured fibroblasts established from tumor tissue of six patients with multiple basal cell carcinoma and from one with a solitary tumor.[12] The chromosome findings in the latter were normal

Three cases showed increased rates of chromosome breakage and rearrangements (46,XY, 1q+,2q+,?2p−; 46,XY,Bq+; 46,XY,C+,F−, G−). The karyotypic aberrations noted in one patient probably reflect the long-term effect of exposure to arsenic, and in another the effects of X-ray therapy of multiple nevoid basal cell carcinomas.

A benign sacrococcygeal teratoma, when cultured for a week, yielded a few diploid karyotypes and several with 47 or 48 chromosomes, the latter having two additional group E autosomes.[13] Studies of 5 mesotheliomas revealed all of them to be aneuploid.[14]

A cell line from a salivary gland mixed tumor was shown to have a near-triploid chromosome number with a number of markers (1−5) (Kondo, Muragishi, et al., 1971).

The chromosomes of an intestinal leiomyosarcoma from a 77-yr-old female were studied with G-banding and a direct technique (Mark, 1976). The tumor had a hypodiploid stem line ($s = 42$ chromosomes) with both numerical and structural changes. The modal karyotype consisted of 42,XX,1p−,+7,11q−,−13,−14,−15, −18 and −22. The author indicated that the evolutionary pattern in the sarcoma was reminiscent of that seen in meningiomas.

Dyskeratosis congenita (Zinsser-Cole-Engmann syndrome) is characterized by poikiloderma, leukoplakia of mucosal surfaces, and nail dystrophy. The onset is in childhood, males being more frequently affected. Later there may be aplastic anemia and malignant changes frequently occurring in areas of leukoplakia. An 11-yr-old male was found to have more chromosome breakage and numerical changes than controls.[15]

A rhabdomyosarcoma had a pseudodiploid stem line with two probable translocations, with hypodiploid and polyploid variants.[16]

Two papillomas of the larynx of children were analyzed in tissue culture,[17] and only diploidy was found, although, in one, an enlarged secondary constriction was found on one #16, probably an unrelated constitutional variant. Two cell lines derived from a treated embryonal rhabdomyosarcoma (7-yr-old girl) at an interval of 5 mo had modal chromosome numbers of 51 and 49 and 45−170, respectively.[18] In the second line a consistent marker was present in most cells and in form resembled a large metacentric in group C. The same workers obtained a cell line from an osteogenic sarcoma of a 13-yr-old girl with a modal range of 58−65 chromosomes.[19] Some cells had markers of varied morphology and some minute chromosomes. Evidence was presented that these cells were of tumor rather than stromal origin. A cell line derived from an osteogenic sarcoma from a girl aged 15 was found to have a hypodiploid mode of 34−38 chromosomes.[20] Culture of fibroblasts from the patient's skin remained predominantly diploid during 65 passages.

Of eight reported mesotheliomas,[21] one was shown to have a mode of 76 chromosomes (Spriggs and Boddington, 1968), two that yielded only a few counts were mainly in the hyperdiploid region, two were pseudodiploid and in two tumors markers were seen but no chromosome numbers were reported.

Two additional mesotheliomas were studied by banding (Mark, 1978); they were hypodiploid with 5 and 8 markers, respectively; each was characterized by −14, −22, and an interstitial 13q−. In addition, #1 and #3 were involved in both tumors in the formation of other markers of dissimilar morphology.

Six malignant tumors of bone (three osteogenic sarcomas, a chondrosarcoma, a fibrosarcoma, and a reticulosarcoma) had high modes that ranged from 58 to 92 chromosomes in direct preparations.[22] No chromosome abnormality was found in direct preparations or early subculture passages from eight giant cell tumors (osteoclastomas). A periosteal sarcoma was hypodiploid (42 chromosomes) with markers.[23] A Ewing sarcoma had a modal number of 49, the same two markers being seen in 271 out of 272 metaphases.[24] It was considered that a missing #1 and B-group chromosomes had contributed to the formation of the two markers, with the labeling pattern of the markers in culture containing tritiated thymidine supporting this view.

Mixed parotid tumors are generally benign, although recurrence after surgical removal is not uncommon. In cultures from six tumors, variant cells with 45−48 chromosomes were found but a similar variation, comprising structural as well as numerical changes, was found in cultures derived from normal salivary glands from subjects with tumors of the upper respiratory or alimentary tract and from the glands of patients with chronic sialadenitis.[25] The significance of these aneuploid cells is unknown. Two cell lines derived from mixed parotid tumors[26] had modal chromosome numbers in the region of 62–66; it is uncertain whether these represented the chromosome constitution of the original tumor.

Basal cell carcinomas are common, slow growing tumors of the skin that invade underlying tissue but very rarely metastasize. Cell suspensions are difficult to prepare, and direct observations on chromosomes have not been reported. DNA estimations, however, suggest that most are near-diploid, although some are near-tetraploid, perhaps with a minor near-diploid component.[27]

Maxillary tumors have been described with modes in the region of 65 or more.[28] There have been only a few studies of squamous cell carcinoma of the skin.[29] Hyperdiploid metaphases in three condylomata acuminata and one Paget's disease of the vulva have been described.[30] A cavernous hemangioma with secondary thrombocytopenia with 47 chromosomes has been described.[31]

Epidermal *in situ* carcinomas were shown to have DNA values that ranged from diploid to hypertetraploid and higher. A high incidence of mitotic irregularities (chromosome bridges, fragments, multipolar mitoses) were encountered.[32]

A Ewing sarcoma of the ileum was shown to have a mode of 49 chromosomes. Banding revealed a 49,XX,+3,+13,+14 karyotype.[33]

In a case with multiple sebaceous adenomas with "internal malignant disease," the blood cells were found to be diploid. Unfortunately, the tumorous cells were not studied cytogenetically (Tschang, Poulos, et al., 1976).

In 1 out of 13 cell lines obtained from nasopharyngeal cancers, a hypodiploid mode due to a missing Y was observed. All other lines were diploid (Utsumi and Yoshida, 1971). It was uncertain whether the lines were derived from tumor or stromal cells.

Large marker chromosomes have been found in tumors of the skin (Paulette-Vanrell, 1966).

Cancer of the Uterine Cervix

An evaluation of the extensive literature on chromosomal findings in cervical lesions prior to the banding era is made difficult by certain facets to be discussed later. Hence, in presenting this area I have relied heavily on the publications and views of Spriggs and co-workers,[1] who admittedly have discussed the cytogenetics of cervical lesions in terms of their own experience and expertise and, thus, can be challenged by those who view the field differently. Nevertheless, in my opinion, the views expressed by Spriggs (1974) appear reasonable and cogent, considering the vicissitudes in studying chromosomes of cervical lesions (see Figures 169 and 170). Unfortunately, sparse banding studies have appeared on cervical cancer; such studies may contribute considerably to the elucidation of some of the problems existing in the cytogenetics of cervical lesions.

Lesions of the cervix afford one of the best and most direct means of studying neoplastic lesions during their various stages in human subjects, e.g., simple dysplasia, carcinoma in situ, and invasive cancer. In addition, it is the only place in the body where supposedly precancerous epithelial changes are very frequently discovered, as a result of screening female populations by cytological techniques. Also, lesions of the cervix afford the cytogeneticist an unusual opportunity to add to the diagnostic armamentarium of the pathologist and clinician, in that the chromosomal changes may be of considerable help in establishing the type of lesion present in the cervix and, hence, reflect on the diagnostic implications in such lesions. The cervix is an organ that can be readily inspected and subjected to biopsy and, indeed, the whole of the area where cancer most commonly develops can be removed without affecting the patient's child-bearing capability. The nature of the findings in cervical lesions varies with the methods used for analysis, i.e., examination of the tumor cells with or without previous in vitro culture, site from which specimens were obtained, diagnostic criteria in labeling the lesion, etc. Furthermore, in early lesions and in normal cervical tissue the number of metaphases may be very small, due to the low incidence of dividing cells and, hence, the results may have to be based on rather sparce cytogenetic data. In addition, even though there is every likelihood that in dysplasia and carcinoma *in situ* and rarely in invasive carcinoma diploid cells may truly be neoplastic and the shift to aneuploidy is only an indication of the biological change in these cells, the possibility cannot be excluded that some diploid cells are of normal origin. Admittedly, the chromosomal material very often obtained from cervical lesions is technically too poor for

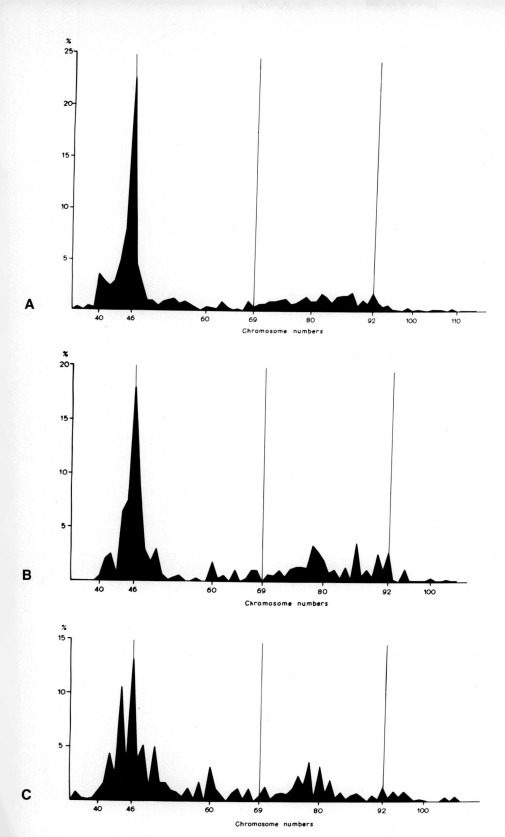

170 The percentage distribution of chromosome numbers in cervical lesions: A = carcinoma *in situ* (32 cases); B = carcinoma *in situ* with microinvasion (5 cases); and C = invasive carcinoma (10 cases). *(Granberg, 1971.)*

reliable examination, particularly that originating from normal squamous epithelium, nevertheless, favorable material for direct cytogenetic study is often obtained from abnormal epithelia of carcinoma *in situ* and invasive cancer. These often yield preparations of quality good enough for direct karyotypic analysis.

There are cogent reasons, then, why the cervix offers a unique opportunity to investigate cytogenetic changes in the developmental stages of a common spontaneous cancer in the human. It can be argued that if chromosome changes are seldom or inconsistently present in precancerous stages, then they can be dismissed as late or unessential features of the established cancer, perhaps due to adverse conditions in tumor tissue. If, on the other hand, chromosome abnormalities are regularly found in lesions that precede cancer by many years, it can be reasonably held that they play an important part, perhaps even a necessary one, in carcinogenesis of cancer of the cervix. The type and sequence of such abnormalities then become of the greatest interest for our understanding of the development of neoplasia in human subjects.

Pathology of Cervical Lesions

One of the major problems in comparing cytogenetic data on cervical lesions is the considerable variability in the interpretation of the pathology of such lesions. Despite attempts of international and national committees to establish a uniform nomenclature for cervical lesions, considerable variation continues to exist in terminology and interpretation, making comparisons difficult, if not impossible. For example, the term carcinoma *in situ* is not considered by some to be a cancerous lesion but rather a precancerous condition. However, in some gynecologic literature it is classified as squamous carcinoma, stage 0. The most common histologic type of cervical tumor is squamous cell carcinoma, which in its fullest form differs insignificantly from squamous carcinoma of other sites. The lesions believed to be precancerous are characterized by a failure of normal differentiation, so that mitotic activity occurs at levels of the epithelium well above the basal layer and sometimes near the surface. The nuclei tend to be enlarged, pleomorphic, and hyperchromatic. It is these nuclear changes that permit detection by cervical smears. Where there is little differentiation of the surface layers, the lesion is called carcinoma *in situ*; where there is surface maturation, the lesion is called dysplasia. However, this classification is not accepted by some of the major authorities on cytodiagnosis. Even though most pathologists do use the terms "dysplasia" and "carcinoma *in situ*," the exact lesions described by them have considerable variability. Thus, the cytogenetic literature on cancer of the cervix has to be evaluated with these ambiguities and conflicting aspects in mind.

There seems little reason to doubt that most cancers of the cervix are preceded by carcinoma *in situ* or dysplasia. But, these conditions need not necessarily progress to cancer. Certainly, they can persist unchanged for many years, and it is thought that they may often regress spontaneously, particularly the mild dysplasias.

It is interesting to note that well before the technical breakthroughs in cytogenetics, reports had appeared on the chromosomes of the cervix uteri. Not only was the chromosome number in endometrial tissue and uterine cervix described several years before 1956, but it was also known that abnormal divisions were frequent in cervical cancer. Thus, lagging chromosomes, stickiness, hollow spindles, colchicine effect, multipolar divisions, and structural changes were described by several groups.[2] The striking "three-group metaphases" of carcinoma *in situ* were described as early as 1951.[3] Just before the introduction of the technique of hypotonic treatment for chromosome analysis, 29 cancers of the human cervix were examined cytogenetically[4] and, considering the technical limitations, the findings were remarkably similar to those found later by other workers with more refined techniques. The cells of the normal cervix and endometrium were found to have a peak of chromosome number at the normal level (thought at that time, of course erroneously, to be 48). Considerable scatter of chromosome numbers to the left of the modal peak, but not to the right, was also encountered and was probably the result of artifactual parameters caused by cell breakage.[5] On the other hand, cervical cancers were shown to have a scatter of numbers to the right as well as to the left of the modal value, and there was variation from case to case. The distribution of the chromosomes agreed well with the stemline concept of tumor growth, which had gained acceptance in the mid-1950s. The introduction of hypotonic pretreatment made it possible to determine the normal human chromosome number and, therefore, to recognize deviations from it. Nevertheless, it is still technically difficult to make reliable direct preparation from cervical tissues. In the first significant series of solid tumors reported upon in 1959,[6] five uterine cancers were included, but whether they were of endometrial or cervical origin was not stated. In the same year,[7] studies on six cervical cancers were published and clearly established that like many other types of cancers they consisted of aneuploid populations of cells, each with unique abnormalities including markers and each with variation of counts about the stem-line mode. Soon thereafter considerable efforts to establish the karyotypic picture of various lesions of the uterine cervix were undertaken, and the first such study appeared in 1962,[8] in which it was shown that aneuploid cells were present, sometimes along with cells containing 46 or 92 chromosomes. Furthermore, in two cases in which microcarcinoma was subsequently found in another part of the cervix, there was either a distinct abnormal stem-line number or a marker common to a number of different cells. Obviously, these indicated the presence of a neoplastic clone and that clonal proliferation had occurred in lesions presumed on other grounds to be the early stages of cervical cancer.

Problems Related to Cytogenetic Studies in Cancer of the Cervix

The bulk of cytogenetic work on cancer of the cervix and its precursors during the last decades or so has been done with direct preparations, intended to display the chromosomes of cells already in division. Incubation for one to several hours in the presence of colchicine or colcemide has sometimes been used, a step intended simply to improve the presentation of cells already dividing by damaging the spindle fibers and contracting the chromosome arms and, possibly, to increase the number of cells in mitosis by using a mitotic arresting agent. In most cases the final results leave much to be desired, and new technical methods will have to be developed to increase our knowledge in the future, rather than the accumulation of more and more data using "old" methods. The new banding techniques cannot, at present, be usefully applied to these direct biopsy preparations, because they also require that the chromosomes be clearly presented without overlapping or distortion.

Some of these technical difficulties have been overcome by the use of tissue culture. However, there is certainly a differential in vitro in the growth of some cells in preference to others, and the cell line that is finally selected may not necessarily be derived from the neoplastic population. It is no wonder then that after culture overwhelmingly diploid cells have been found,[9] whereas it is well known that most neoplastic lesions of the cervix have an abnormal karyotype. In one report (Lin, Dent, et al., 1973) the cells of three invasive cervical cancers were cultured; two of the cultures had only normal karyotypes, as determined with banding, whereas the third culture consisted predominantly of cells with 47 chromosomes due to +7. The results based on microdensitometry of Feulgen-stained nuclei have been utilized by some workers in an interpretation of cervical lesions. This method of DNA assay is not accurate enough to show small chromosomal deviations, but it is of value in detecting gross abnormalities, particularly abnormal modal peaks and may confirm the abnormally wide scatter of values with abnormal modes found in carcinoma *in situ* and dysplasia of the cervix.[10]

Chromosomal Findings in Cervical Lesions

Invasive carcinoma. As in all other cancers, carcinoma of the cervix consist of populations of cells with abnormal karyotypes, each tumor usually showing a single cytogenetically unique stem line. Chromosome counts above and below the modal number are found regularly; these certainly represent genuine karyotypic variation rather than artifact. The scatter of values is least, however, when preparations are of high quality.[11] The distribution of the modal chromosome numbers in invasive cancer of the cervix published in the literature are shown in Figure 169. Chromosomal numbers below 50 are the most common, and hypodiploidy is particularly frequent. Even though modal numbers from 50–79 are also common, higher ones are relatively rare. The distribution of the modal chromosome numbers shown in Figure 169 differs somewhat from those usually shown, in that most authors tend to graph all available cell chromosome counts from many cases with the result that the curves embody two different distributions.[12] First, there is the distribution of chromosome counts within one lesion, usually showing a modal peak but sometimes with secondary peaks. Second, there is a wide variation in the position of the modal peak in different cases. Moreover, no correction is made for the fact that each case contributes a different number of counts, some with many and others with few cells examined, so that the final distribution curves have little meaning.

Even though sufficient data based on chromosomal banding have not appeared for cancer of the cervix, a number of authors[13] have examined the karyotypic data, and apparently there is some consistency in reduction of the number of chromosomes in groups B, D, and G, relative to the other groups, whereas group C usually has an increased number of chromosomes. In this respect, cancer of the cervix probably does not differ from other carcinomas.[14] It should be emphasized that the loss or gain of chromosomes does not necessarily involve every cell in the tumor. As judged by published karyotypes, of which about three dozen are available in the literature,[15] markers are frequently present in invasive cancer of the cervix and differ from case to case. Apparently, the most common markers are long metacentrics and acrocentrics, but rings, dicentrics, small metacentrics, and minute chromosomes have also been described. Again, because banding has not been applied to such lesions it is difficult to establish the origin of these markers and whether, in fact, normally appearing chromosomes may not be of abnormal derivation.

The cytogenetic findings in invasive cancer of the cervix indicate that the whole cell population of the malignant epithelium in a particular cervical tumor is of clonal derivation and is usually descended from a single abnormal cell. During the evolution of the abnormal cells from the clonal cell, all of the cells of the clone do not necessarily have identical karyotypes. In some cases of cervical cancer there is evidence of two separate stem lines, but this can usually be explained by polyploidization.[16]

Microcarcinoma (microinvasive carcinoma). The microscopic appearance of microinvasive carcinoma indicates incipient or perhaps aborted invasion, or else shows clinical evidence of invasive growth. After removal of the lesion, the prognosis is uniformly good, and some authorities, therefore, prefer to classify it with carcinoma *in situ* rather than as stage I carcinoma. The microinvasive points are isolated or occur at intervals in an epithelium that otherwise would be classified as showing carcinoma *in situ* or dysplasia, so that a piece taken for cytogenetic study does not necessarily represent the same lesion as is studied histologically from an adjacent piece. The cytogenetic evidence, therefore, concerns abnormal epithelium taken from cervices that harbor microcarcinoma in some part.

The distribution of the modal numbers of microinvasive carcinomas of the cervix published in the literature is shown in Figure 169. The number of cases is small, but there does appear to be a different chromosome distribution in microcarcinoma from that found in invasive cancer of the cervix. The most frequent higher modes are in the range of 80–90, whereas those in the diploid range almost all had 46 chromosomes or above. In this respect, the picture is very much like that for carcinoma *in situ* to be discussed later.

In two of the earliest cases of microcarcinoma described,[17] the chromosome number was mostly diploid or tetraploid but four of the cells with 46 chromosomes showed a long marker. In one case, there was a distinct clone having a modal number of 54 with a substantial proportion of the cells showing a long acrocentric marker. Nine additional cases from the same center were described,[18] seven of which showed markers of various types. In one case without a marker the chromosome numbers clustered between 75 and 88 in samples taken both from the anterior and posterior lips of the cervix. In another case, karyotypes representing anterior and posterior aspects of the lesion showed a long acrocentric chromosome similar in both cells; in the third case anterior and posterior lip samples showed a similar ring chromosome. Another case with a ring marker in two separate biopsies has also been reported.[19] In the cases with rings the microinvasive focus was, in each case, at a distance from the area sampled, and it seems certain that there were extensive areas of clonal origin that in most parts of the cervix only showed the histologic picture of carcinoma *in situ* or dysplasia. Further evidence of clonal proliferation is given by the finding of a similar pseudodiploid karyotype in most of the cells from a single sample involved in microcarcinoma.[20] In another case an area of marked dysplasia in the posterior lip and an area in the anterior lip showing microcarcinoma both had the same abnormal peak at 63–64 chromosomes; however, another case showed discordant modes in two samples of 46 and 87 chromosomes, respectively. Discordant chromosome counts were also found in one of five cases in which the samples were examined.[21]

It appears that noninvasive and microinvasive or invasive areas are derived from the same altered cell, but because the noninvasive portions all have the cytogenetic hallmarks of neoplasia and in some cases cells of undoubtedly common ancestry are distributed widely over the cervical epithelium, there is a strong likelihood that microcarcinoma does develop as a subline from carcinoma *in situ* or dysplasia and that it is a developmental stage of cervical cancer.

Carcinoma in situ and dysplasia. Because of the confusion of nomenclature, it seems best to combine the conditions of carcinoma *in situ* and dysplasia, with only those lesions that are clearly of the mild type being discussed under "mild dysplasia." Almost all observers have found chromosome abnormalities in most cases of carcinoma *in situ* and dysplasia. There is a scattering of the chromosome numbers in each case, with or without a recognizable modal peak either at 46 or in some abnormal area. Even when the modal count is 46, aneuploid cells have nearly always been encountered. The distribution of the main modal regions or clusters derived from the data presented in the literature is shown in Figure 169. It is clear that the patterns for invasive carcinoma and of carcinoma *in situ* and dysplasia are distinctly different. Among invasive cancers there are many with hypodiploid modes, the upper peak being in the 70–79 chromosomal region; modes in the 50s are common, whereas modal numbers above 80 are relatively rare. In contrast, microcarcinoma, carcinoma *in situ,* and dysplasia often have diploid and rarely hypodiploid modes; those in the 50–69 region are rare and there are many above 80. The bimodal distribution is also evident in pooled results of chromosome counts from multiple cases, which, as has been pointed out before, provide a blend rather than a distillate of information.[22]

The difference between the distribution of modal numbers of chromosomes in precancerous and invasive disease is most easily explained by postulating that invasion tends to be associated with the loss of chromosomes, the numbers in general being shifted downward. That one series[23] of cases published differs from others in that there are modal numbers of exactly 46 in the majority of cases of carcinoma *in situ* (21 out of 32) may perhaps be explained by the inclusion of many cases which, in other hands, would not have furnished any countable metaphases. Other workers[24] have found essentially diploid pictures very rarely (2 out of 23 cases of carcinoma *in situ* and dysplasia reported by one group), whereas in the study referred to some normal diploid cells were found in 16 cases, and in 2 all the cells analyzed were consistent with a diploid picture.

Because of the technical difficulties in obtaining good preparations in cervical materials, it is even more difficult to make reliable generalizations about the detailed karyotype of carcinoma *in situ* and dysplasia than about the frequency of chromosome numbers. There are about three dozen photographs of karyotypes available in the literature,[25] and it can be assumed that these have been selected either because they were the best qualitatively or because they make a particular point (e.g., contain markers). In the precancerous lesions, as in the case of invasive cancer of the cervix, chromosomes are sometimes found to be missing from groups B and D,[26] though an analysis of the data in the literature revealed only a moderate deficit in group B.[27] Most of the published karyotypes are abnormal and contain no distinct markers. This is in contrast to the findings in invasive cancer, in which markers are very common. Nevertheless, various markers have been reported in carcinoma *in situ* and dysplasia, either sporadic or occurring in a number of related cells. Markers resembling #1 or #2 with a pericentric inversion, giving a chromosome with a high arm ratio, has been given a special name (Api);[28] however, this marker has not been found consistently in the lesions. Long acrocentrics, very small metacentrics, minutes, and rings have also been seen.[29] The frequency of a stretched area in the long arm of #2 has been emphasized in one study,[30] but in the absence of control material from tissues that are equally difficult to process, it is impossible to assess the significance of this finding. There is certainly no justification for proposing any specificity to a marker in this disease, and the frequency with which aneuploid cells occur without obvious markers does not support the idea that chromosome breakage and rearrangements are of primary importance in producing the abnormal karyotypes. In any case, the derivation of the abnormal markers and the identification of all other chromosomes in lesions of the cervix with banding techniques is the minimum requisite for establishing the possibility of common marker chromosomes existing in cervical lesions.

Mild dysplasia. Because of the lack of uniformity of nomenclature, there is conflicting information about the findings in cervical dysplasia. However, in some series there is separate mention of lesions that may be regarded as having "minimum-deviation" changes. These have been referred to as dysplasia,[31] minimal dysplasia,[32] and mild dysplasia.[33]

The numerical results have been somewhat conflicting, one group reporting seven cases with counts at or near the diploid range, with some abnormal karyotypes.[34] In another study,[35] pooled chromosome distributions for nine cases was presented with a high peak at 46 and a scatter above and below that number but with no tetraploid cells. However, in the former study, in which cases of mild dysplasia were studied, chromosome numbers in three were mostly or all in the tetraploid region.

In most mild dysplasias the mitotic rate is too low to furnish sufficient cytogenetic data and, therefore, those reported cannot be considered definitive. Mild dysplasia is detected by the finding of "superficial dyskaryosis" in cervical smears, i.e., large and hyperchromatic nuclei and otherwise well-differentiated squamous cells. The enlargement and hyperchromasia are not due to DNA synthesis preparatory to cell division, because the superficial cells in mild dysplasia do not synthesize DNA or divide, and it seems most likely therefore that these nuclei are polyploid. In two published histograms[36] of two cases a wide distribution of values, mainly between diploid and tetraploid but extending above, have been reported. In another case of very mild dysplasia there were many nuclei in the tetraploid class.[37] In another study[38] based on DNA estimations of 14 cases of dysplasia, 8 of which were classified as mild, as well as three cases of carcinoma *in situ*, the former were found to have modes located in the diploid level and its multiples, and, although the full data were not given, it was concluded that "the atypical cells which characterize mild dysplasia generally reflect polyploidization of the epithelium."

The finding of normal chromosome counts, on the one hand, and polyploid DNA levels, on the other, would be explained by supposing that normal mitosis occurs in the base layer while endomitosis produces polyploid cells at a more superficial level. The information on mild dysplasia is very incomplete, and it seems almost certain that many or most of these lesions regress spontaneously so that the relationship of chromosomal changes to this reversibility is a particularly intriguing problem.

Evidence for clonal evolution in cervical lesions. Whenever there is a marked cluster of chromosome numbers at an abnormal level, a likelihood exists that the cells concerned are related by descent from a common ancestral cell with that chromosome number. In an attempt to discover whether extensive areas are occupied by the same clone, multiple samples have been studied from the same cervix, including information based on markers. In one case[39] samples from the anterior and posterior cervical lips both showed chromosome numbers clustered in the region 78–83; both lesions contained cells with a similar long acrocentric marker. In another case[40] two of three different samples showed clearly different modes, one being 47 and the other about 96; a long marker was present in the first of these only. More frequently the evidence for clonal evolution comes from a comparison of counts in two or more areas, without the help of markers. In one series multiple biopsies were taken from 10 cervices carrying moderate or severe dysplasia or carcinoma *in situ*.[41] Some of the discordant counts were explained by inclusion of areas of normal epithelium;

even allowing for this, in two cases the counts agreed and in four they did not. In another series[42] two biopsies were taken from each of three cases and in two of these modes from the separate areas were concordant (diploid and near-diploid); in the third case, they were discordant. In another study[43] concordant counts in four out of six cases were found, including the one mentioned above. Using DNA estimations on Feulgen-stained nuclei, abnormal peaks in two samples from one case were found,[44] and in another study,[45] using scrapings from a relatively wide area rather than multiple small pieces, what appeared to be a stem line was found, usually with an abnormal DNA value, in all 18 cases.

Evidently, the extensive proliferation of a clone is only demonstrable with confidence in a small proportion of cases, but it can be suspected in many more, probably the majority. At the same time, there is sometimes more than one mode as well as many counts outside the modal region, and in some cases there are no distinct modes at all. Where photographic karyotyping has been done, it is frequently impossible to recognize any real relationship between different cells from the same lesion. Even allowing for technical imperfections, it seems almost certain that some lesions contain many competing cell lines that do not necessarily have a common origin from a single chromosomally altered cell.

Certain Aspects of Chromosomal Changes in the Uterine Cervix

Polyploidy. Cervical epithelium in its normal state and in conditions of inflammation and regeneration appears to consist of diploid cells, though direct evidence of this from chromosomal studies is minimal.[46] However, DNA estimates are consistent with this supposition, and it can be safely assumed that under normal conditions the cells are diploid. The mildest and presumably earliest cytological changes believed to bear any relationship to cancer of the cervix are characterized by nuclear enlargement and hyperchromasia in otherwise well-differentiated superficial cells. Similar appearances can be produced experimentally with podophyllin which induces polyploidy;[47] such lesions are invariably reversible. Cases of more severe dysplasia and carcinoma *in situ* with modal counts near 92 are common, and those with hypotetraploid counts are most probably derived from tetraploids by chromosome loss.

Tetraploidy may be produced by endomitosis, nondisjunction, endoreduplication, or cell fusion.[48] Apparent endomitoses are common in carcinoma,[49] carcinoma *in situ*, and dysplasia, and such chromatin condensation may be seen simultaneously in many adjacent cells. DNA studies[50] have shown the content to be typical of resting nuclei rather than of nuclei that have completed DNA synthesis and, therefore, it has been proposed that the condensed appearance results from a prolonged telophase pattern (rather than from endomitosis). Perhaps both conditions occur. Endoreduplication has sometimes been demonstrated by the finding of diplochromosomes;[51] about 25% of such cells have been demonstrated in one case.

Polyploidy is then a highly probable explanation for the morphological findings of an increased nuclear size and density in some of the mildest dysplasias, and it would serve well to explain the later development of hypotetraploid clones by chromosome loss. However, direct evidence from chromosome counts is minimal, and the questions awaits better evidence.

Aneuploidy. When cells are found to have one or few chromosomes more than normal, this is a significant finding. With past techniques loss of chromosomes during preparation[52] was common and without banding the exact chromosome loss could not be identified. Under these conditions, the best evidence for a genuine aneuploidy is clustering of chromosome numbers, with a mode at a level other than 46 (or its multiples) in most cases of carcinoma *in situ*. Most authors have failed to find more than a few cells with normal karyotypes, but in some instances a significant number of such cells was encountered in carcinoma *in situ*,[53] though not in microinvasive or invasive carcinoma. The mechanisms of chromosome gain and loss are not known specifically for cervical lesions, but it is common to see abnormal mitoses in sections, and these can sometimes be observed in smears. The so-called three-group metaphases are particularly common in carcinoma *in situ*, but also occur in other tumors and in cells that have been experimentally poisoned.[54] It has been shown[55] that the misplaced chromosomes are pairs of sister chromatids, still together, but it is unknown whether particular chromosomes are preferentially involved. This and other types of abnormal division could easily result in aneuploidy, but it must be admitted that a division seen in the milder dysplasias usually appears regular and the "three-group metaphases" are more often a feature of classic carcinoma *in situ* and microcarcinoma rather than of lesser abnormalities.

Chromosome breakage. Chromosome breakage has not been a common feature of cervical lesions. Exaggerated secondary constrictions producing an unusually long #2 have been noted[56] but chromosome fragments in figures resulting from chromatid exchanges are seldom mentioned by authors. Markers are, or course, a sure sign that chromosome breakage and recombination have taken place and there are many reports of markers in cervical lesions, but they are distinctly more common in microcarcinoma and invasive carcinoma.[57] Visible chromosome breaks and recombination figures are, therefore, not a regular feature of putative precancerous states of the cervix, even if they are actually present in an occult form.

Summary of Chromosome Findings in Lesions of the Uterine Cervix

Normal and inflammatory conditions of the cervix are known to be associated with diploidy.[58] The mildest changes regarded as possibly on the road to neoplasia,[59] i.e., mild dysplasia, have in one series shown chromosome numbers near tetraploid, some of which could represent normal tetraploid cells, but most other series have had numbers near or exactly at the diploid number of 46. In severe dysplasia and carcinoma *in situ* some or all of the cells are aneuploid.[60] There is a wide scattering of the chromosome numbers within one sample, but in many cases clustering of numbers in an abnormal region suggests the formation of an aneuploid clone. Karyotypic evidence of clonal growth has also been obtained in a few cases. According to some authors, cells with normal karyotypes are commonly present along with the abnormal ones. In microcarcinoma (microinvasive carcinoma) there is good evidence from karyotype analysis and from the presence of markers that all or a substantial part of the lesions consist of a neoplastic clone. This is also the typical feature for established invasive carcinoma of the cervix.[61]

Some facets of chromosomes in cervical cancer. In more recent studies, attempts have been made to correlate the histological type of cervical lesions, survival and ploidy as revealed by DNA values.[62] Utilizing similar measurements, a differentiation between benign and precancerous lesions of the cervix was thought to exist.[63] In a study of seven cases of preinvasive carcinomas of the cervix in which biopsies were examined at two different time intervals (ranging from 1 to 8 mo), it was shown that a general karyotypic stability exists, though it is possible that these intervals were entirely too short for changes to become evident.[64]

A number of studies have been published in which correlations between the karyotypic findings (including DNA measurements) in cervical lesions and histology, response to radiotherapy, and other parameters were attempted (L. W. Cox, Stanley, et al., 1969). No concensus can be gleaned from these articles. In a series of papers[65] the general opinion was expressed that numerical and structural chromosome changes cannot differentiate between dysplasia, carcinoma *in situ*, with and without microinvasion, and invasive carcinoma of the cervix. However, triploidy occurred somewhat more often in the invasive group. The dysplasia group had, in general, fewer numerical deviations than the other groups, with chromosome loss being more frequent than chromosome gains. Chromosome gains usually involved group C and the losses, the B, and D, and G groups.

In 14 consecutive near-diploid carcinomas of the cervix, 5 had trisomy for #1, but banded preparations were too poor to reveal structural details. In three other tumors, isochromosomes of #1 were found, two for the long and one for the short arm of #1. In tumors with high chromosome numbers, structurally changed and/or a relative excess of #1 was frequently present.[66] In an anaplastic small, round-cell tumor of the vault of the vagina and base of the bladder with a modal number of 47 chromosomes, the only abnormality was an extra chromosome derived from the #1. The frequent encounter of abnormalities of #1 in these cervical lesions were related by the authors to be common abnormalities of #1 seen in cancers of the bladder, ovary, and breast (Atkin and Baker, 1978, 1979b).

Six cervical carcinomas were examined with banding and in each morphologically abnormal chromosomes varying in number from 1 to 10 were found (Van der Riet-Fox, Retief, et al., 1979). In two carcinomas abnormalities of #1 were established, and isochromosomes of #8, #13, and #14 seen in three separate cancers.

In a continuation of their studies, Atkin and Baker (1979b) performed cytogenetic studies on 26 carcinomas of the cervix and showed that #1 was consistently involved in one or more structural changes or as a relative excess of #1. Several types of structural changes were repeatedly seen, i.e., 1p− in seven tumors, i(1q) in six tumors, translocations of unidentified chromosomal material onto one of the arms of #1 in 11 of the tumors and an i(1p) in one tumor. The most consistent feature of the aneuploid complements of these tumors appeared to be an excess of the centromeric region and at least part of the adjacent heterochromatin of #1.

A few words regarding the HeLa cell line are in order. This line originally came from a cervical cancer, but whether it still remains so over the decades is open to question. Chromosomally, it is in the triploid range with a number of markers. How many of the chromosomal changes are due to contamination with mycoplasma, is difficult to say. Other cell lines originating from cancer of the cervix have differed considerably in chromosome number and constitution from HeLa, e.g., the line reported by Herz, Miller, et al. (1977).

Brain Tumors

Meningiomas

Next to the Ph[1], the description more than 10 years ago by Zang and co-workers[1] of monosomy #22 in meningiomas probably constitutes chronologically the next cytogenetic milestone in medical oncology. This finding was confirmed by Mark[2] and the bulk of the cases to be discussed has emanated from these two laboratories in West Germany and Sweden, respectively. However, much of the data on meningiomas are based on cultured materials, always raising the possibility of preferential *in vitro* growth of one cell type over another. Hence, the interpretation of the data must take this fact into consideration.

The closest human tumors to "minimal deviation" hepatomas are the meningiomas. These tumors have usually been considered to be benign in character, because they do not metastasize or invade adjacent tissues and, except for occasional sarcomatous transformation, they look "benign" histologically. When a large number of these tumors was examined, it was found that a substantial percentage of meningiomas was aneuploid, with a fairly characteristic cytogenetic anomaly being present in a significant number of these tumors. This karyotypic abnormality consists of a missing G-group autosome,[3] thus leading to the presence of only 45 chromosomes in the cells of the meningiomas. Loss of this chromosome is unrelated to the sex of the patient. The other aneuploid meningiomas have a variable chromosomal picture without consistencies in the cytogenetic profiles.[4]

The findings in meningioma confront the cytogeneticist, pathologist, clinician, and cancer biologist with a serious problem. What is the role of the chromosomal findings in the definition of cancer? Does the presence of an abnormal chromosome constitution indicate irrevocable evidence for neoplasia?

In my opinion, the types of changes found in meningioma indicate the presence of malignancy and point to a malignant transformation of the meningioma cells, however early the change may be on the basis of other cytological and clinical data. The results with meningiomas also indicate that there is no correlation between the nature of the karyotypic change and the degree of malignancy of a tumor. The meningiomas with chromosomal changes probably represent the least malignant human tumors, yet the cytogenetic aberrations may be extreme and similar to those observed in some frankly anaplastic cancers. The presence of chromosomal changes similar to those found in meningiomas indicate the presence of malignancy.

Over 200 meningiomas have been studied to date[5] and, thus, we know more about the chromosomes of this tumor than any other. Early in the studies on meningiomas attention was drawn to the fact that a strikingly uniform anomaly of the chromosome constitution found in these tumors is characterized by loss of a G-group chromosome, which has been shown with banding techniques to be consistently a #22.

A few parameters related to meningiomas are worthy of note. Generally, female patients with this tumor outnumber male patients, and the condition occurs mostly in patients over the age of 50. In some patients the tumors may be multiple. Usually, the cells have to be examined after culture; pieces of tumor tissue are usually explanted in vitro and chromosome preparations made after 5–13 days of growth. Some of the tumors were also studied in preparations from in vitro passages usually performed once a week, the number of subcultures varying between 3 and 15.

The chromosome results on the 208 meningiomas studied to date are shown in the accompanying figure (Figure 171); 43 of these tumors have been studied with banding techniques. Approximately 33% of the meningiomas had diploid stem lines (S), 60% were hypodiploid, and the remaining 7% were in the hyperdiploid-hypotriploid (4.5% hyperdiploid; 2.5% hypotriploid) region. None was found to be tetraploid. In the preferred region of variation, the hypodiploid region, the frequency of monosomic numbers exceeded all the others, and there was a gradual fall in frequency of stem lines as the S numbers decreased. Only scattered numbers were found above the diploid one and rare tumors had more than 46 chromosomes. Thus, karyotypic evolution in human meningioma is characterized by a decrease in the chromosome number and the absence of stem lines in the $4n$ region and indicates that loss of chromosomes below the diploid number is well tolerated by such tumors. This is certainly unique among human tumors.

Studies prior to banding. Of the 158 tumors studied prior to banding,[6] 62 had different stem-line karyotypes. Two karyotypes were outstanding by their frequency, i.e., monosomy of group G (41 tumors) and the presence of a normal karyotype in 46 tumors. However, the majority of the tumors with a diploid mode contained either a side line or variant cells with monosomy G (Figure 172). Conversely, the monosomy G group had, in addition, a side line or cells with a normal karyotype. An evaluation based on 104 meningiomas revealed 5 to consist exclusively of G-group monosomy, 3

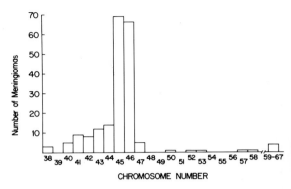

171 Modal chromosome number in meningiomas.

to have monosomy for groups D and G, 2 to have monosomy for group G in addition to other changes, and 38 variable karyotypes in only 1 tumor each (Figures 173–176).

In spite of this variability, however, the meningiomas do have important karyotypic features that they share in common and that distinguish them from other types of human tumors; the deviations affecting mainly chromosomes of groups G, C, D, and to a lesser degree those of groups #17–18 and #1.

The distribution of aberrations among the chromosome types in the 67 meningiomas with additional karyotypic deviations or changes other than monosomy G revealed a predominant involvement of groups C, D, and G (Table 86).[7] In the heteroploid tumors, groups C and D approached group G in importance, with > 60% of the stem lines deviating in groups C and D. When an analysis was made in which the three chromosomes in group A were distinguished and #16 separated from #17–18, it

172 Partial karyotypes from three meningiomas showing 22q–. No evidence of a translocation was found in these cases.
(Courtesy of Dr. J. Mark.)

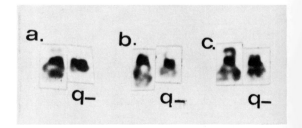

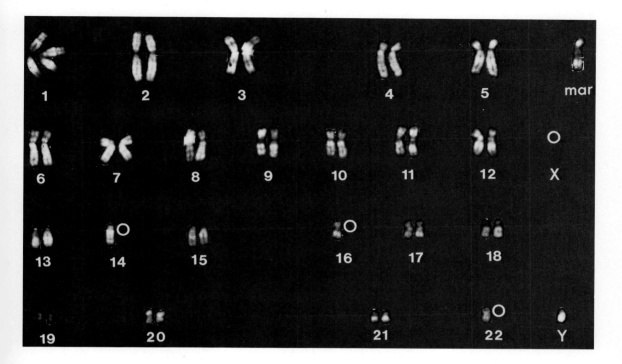

173 Karyotype (Q-banded) of a meningioma from a male patient showing the presence of a marker chromosome, loss of the X, a #14, #16, and a #22.
(Courtesy of Dr. K. Zang.)

was found that the variations in these groups affected mainly #1 and #17–18. All these data indicated a nonrandom pattern for the karyotypic evolution subsequent to the usual first step, i.e., the development of monosomy G. The high figures in groups C and D are of special interest, because these were usually involved in the few stem lines with no change in group G. It is possible that in rare cases karyotypic evolution in meningiomas may originate without those steps that follow the primary G-group change seen in the majority of the tumors. In recent studies of meningiomas using fluorescence analysis, in which all the individual pairs were identified, similar results were reported. In that analysis in addition to the selective involvement of pair #22, involvement of #1, #8, #12, #17, #7, #6, and #11 (shown in order of frequency) were also often found. The only discrepancy was in group D, which showed an infrequent involvement in the fluorescent studies; further work is needed to explain this discrepancy.

In chromosome groups with frequent involvement both losses and gains of chromosomes took place, with losses predominating, in keeping with the observation that the majority of heterodiploid stem lines were hypodiploid. The high incidence of losses and gains in specific chromosomes indicates that these chromosomes are more liable to nondisjunction than others. For unknown reasons, in meningiomas cells with gains seem usually to be inferior in viability to those with losses. The preferential pattern was also apparent from the karyotypic deviations recorded in the few meningiomas having an abnormal stem line without involvement of group G. Except for unusual

Table 86 Involvement (%) of the various chromosome groups in 67 stem lines with other or additional changes than monosomy G

Chromosome groups							
A	B	C	D	E	F	G	Markers
45	16	62	63	40	18	87	36

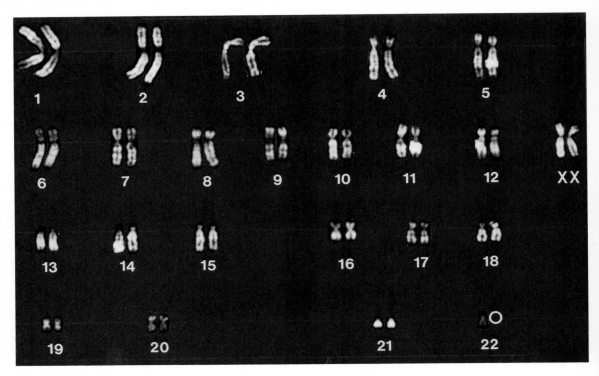

174 Karyotype (Q-banded) of a meningioma from a female patient showing loss of one #22.
(Courtesy of Dr. K. Zang.)

175 Karyotype (Q-banded) of a meningioma from a female patient showing loss of one #22 and trisomy of #7 and #9.
(Courtesy of Dr. K. Zang.)

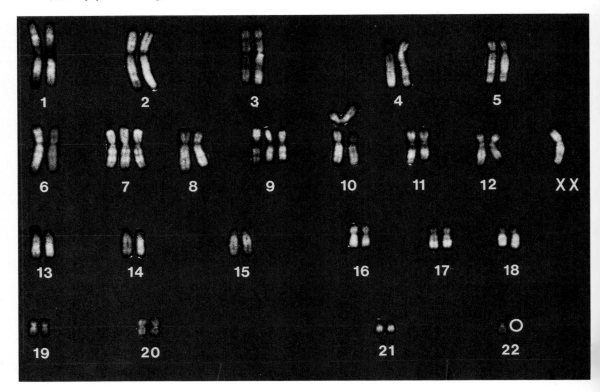

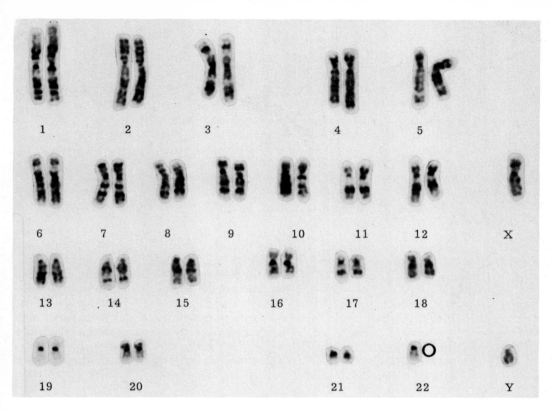

176 Karyotypes (G-banded) from two meningiomas showing the characteristic absence of one #22.
(Courtesy of Dr. K. Zang.)

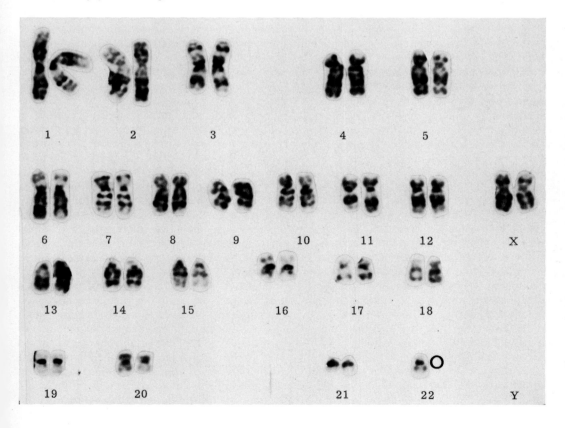

cases, most of these karyotypes involve deviations in groups C, D, or both. Furthermore, groups C and D were involved in all cases in which the only change was a gain or loss of one normal chromosome. Thus, it seems that in rare cases the karyotypic evolution of meningiomas may commence directly with those steps that usually follow the initial loss or change of a group G chromosome observed in the majority of the tumors. This conclusion is supported by the common occurrence of variant cells (making up < 10% of the tumor cell population) with monosomy of group C, as well as by the occurrence of side lines ($s=10\%$ or more of the cell population) with the loss of one C-chromosome as the only change.

Variant cells were observed in nearly all meningiomas, often in a frequency close to that seen in most other malignant tumors; usually these variant cells followed a pattern of deviations similar to that observed in stem-line or side-line cells, though the degree of involvement differed. In 13 of the meningiomas studied by the fluorescence technique, further evidence was found for the genuineness of the variant cells, which, especially in the cases with markers, demonstrated convincingly their close relation to the S cells. These findings are at variance with the conclusions of another group studying meningiomas, who found variant cells to be comparatively rare in meningiomas and viewed hypomodal cells in meningiomas as artifacts.

About one-quarter of the 158 meningiomas had marker chromosomes with 9 containing > 1. This frequency of about 25% is certainly impressive and is in opposition to the statements made that markers do not occur in meningiomas. Except for one case (Benedict, Brown, et al., 1971) all the meningiomas with markers were histologically benign. The one exception was a meningioma with structural rearrangements in which recurrence of the tumor with histological signs of change to malignancy occurred.[8] The markers varied among the tumors in which they were found. Abnormally short or long acrocentrics were the most frequent markers; such rare markers as rings and dicentrics have been observed in some meningiomas (uncommon even in malignant tumors). The ring chromosomes are an obstacle to smooth mitotic function and, hence, are of some importance. There were three meningiomas with ring markers, and the data indicate a different origin for the ring in each of these three tumors. Among the markers whose origin could be traced, the majority were wholly or partially derived from groups G and #1.

Information gathered from dicentric markers bears on the question of the origin of the ring chromosomes. Thus, in one study[9] dicentrics were found in the side lines of two tumors and were occasionally present in variant cells from the first preparation of 11 additional meningiomas, sometimes together with centric or acentric rings. In addition, dicentric markers were often seen in variant cells of all meningiomas placed in long-term culture. In another study,[10] single cells with dicentrics in several of the tumors were also reported. Even though the morphological characteristics of the dicentrics were highly variable, they may have been formed by translocations between group F or G and A or C chromosomes. This impression, gained from conventional squash preparations, was confirmed in one meningioma studied by fluorescence technique.[11] The tendency of meningiomas to form variant cells with dicentrics gives a plausible explanation for the development of stem lines with ring chromosomes. The fairly common occurrence of these variant cells might also be an important reason for the high frequency of stem lines with markers other than rings. The most frequent structural change of #1 was a short arm deletion. Deletion of the long arm, in particular, and those of both arms were prevalent in the G group. Such anomalies of group G were the only deviations in several tumors, and in a few meningiomas with a normal, diploid stem line they represented a side line or variant cells. These observations have raised the question[12] as to whether structural G-group changes might not constitute the pristine karyotypic change in all meningiomas, the structurally changed G-chromosome then rapidly being lost through mitotic disturbances in most tumors.

These evolutionary features of the meningiomas, i.e., the formation of dicentrics, and the loss of various normal chromosomes (especially acrocentrics), constitute an interesting parallel to the karyotypic change observed during the degenerated phase of cultures of normal human cells. It may be that mechanisms for chromosome behavior in some tumors are analogous to those occurring in senescence; in the case of tumors, such changes are perhaps triggered earlier than in normal cells by the influence of oncogenic factors that do not necessarily lead to cell death.

Another group of markers of special interest to meningiomas were traceable to the group G chromosomes. In one evolution, 16 of the tumors, or more than half of the stem lines with markers, displayed one or two small chromosomes recognizable as structurally changed derivatives of group G chromosomes.[13] The predominating abnormality was deletion of parts of the long arm. Because of the small size of the group G chromosomes, it is difficult to determine in each case the exact extent of the deletion. It is certain, however, that the size varied among the tumors and that in some half or more of the long arm had been lost and in others only a third or less had disappeared. A similar situation also prevails with the aberration Gp−q−, as regards the change in both the long and the short arms. The situation becomes even more complex by the finding of single tumors with an abnormally elongated short arm of a group G chromosome and others in which one group G chromosome has been replaced by an isochromosome for the short arm. To these results from S cells should be added two tumors, each of which had one side line with Gq− and another in which variant cells occurred with Gq−. Thus, structural aberrations in group G have been observed in 19 of the 105 meningiomas studied in detail, and in almost all of these cases at least a third (probably the distal third) of the long arm of one G-chromosome had been lost.

Banding studies. The results discussed above should be considered in relation to the observations in 18 meningiomas studied by means of the fluorescence and/or G-banding technique.[14] These analyses demonstrated unequivocally that the group G chromosome subjected to loss, deletion, or structural changes was consistently a #22. No evidence was obtained of a regular translocation of the missing #22, or part of it, onto any other chromosome. Thus, there is good reason to believe that the numerical and structural G-group aberrations observed in the remaining 86 stem lines (studied by conventional methods only) also affected a #22. The implications of this specificity of the "G pattern" for meningiomas has been discussed[15] and, as was pointed out, a deletion of a #22 chromosome might well be the initial change in all meningiomas. Thereafter, the aberrant #22 could either be maintained as a marker or become lost through some mitotic disturbance. If the deletion is terminal, further disturbances may be expected to lead to further rearrangements and additional markers.

Banding data are available on 43 meningiomas.[16] Chromosome #22 was consistently the one most often involved (Figures 177 and 178). Conversely, #21 and the Y were affected in only a small number of tumors and usually

177 G-banded karyotype from a meningioma. The karyotype contains the following anomalies: −1, −5, 9q+, −14, −18, −22, and a ring chromosome.
(Courtesy of Dr J. Mark.)

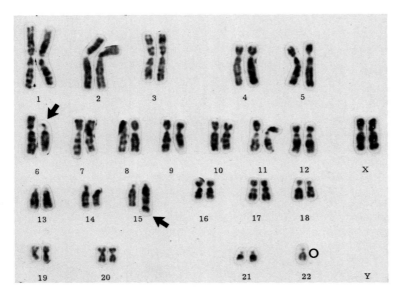

178 G-banded karyotype of a meningioma showing loss of a #22 and a t(6p−;15q+). *(Courtesy of Dr. K. Zang.)*

in a low frequency of their cells. Other chromosomes often involved were #1, #8 (most frequent), #9, #14, and #15.

The banding data suggest that #1 and #22 are most frequently involved in structural rearrangements, with the resulting markers varying in morphology from case to case. No indications were obtained that certain specific chromosome regions ("hot spots") are particularly liable to breakage.

In rare cases of meningioma near-triploid stem lines occur.[17] The mechanism for such an evolution is unknown. Q-banding of two such tumors indicates a progression toward a balanced triploid complement.

In a significant number of meningiomas, invariably accompanied by other chromosomal changes, the loss of either the X- or Y has been established.[18] However, it should be pointed out that the loss of a sex-chromosome as the sole karyotypic anomaly has not been described in meningiomas. Furthermore, the chromosomal pattern of the cultured lymphocytes and BM, in particular, from patients with meningioma has not been studied in detail to ascertain the possibility that such patients may have a hereditary predisposition to chromosomal loss.

Zankl and Huwer (1978) reported on the karyotypes of two human meningiomas in which, besides other aberrations, a deleted #1 and #6 could be observed. In these chromosomes most of the short arm was missing and after silver staining for the detection of NORs, not only the satellite regions of most of the acrocentric chromosomes were stained but also the deleted ends of #1 and #6. Though there may be other explanations for this unusual NOR-staining, the distinct possibility exists that the NORs of acrocentric chromosomes were easily translocated to the deleted chromosomes.

Summary of cytogenetic findings in meningioma. It appears that meningioma approaches CML in cytogenetic specificity. The Ph[1] that characterizes CML is an aberrant #22 also with loss of approximately half of the long arm. This marker, however, is not particularly liable to either loss or further structural rearrangement. The stability and consistency of the Ph[1] are important features of CML, in contrast to the G-group pattern of the meningiomas. One significant factor of these differences might be the oncogenic agents involved, which have recently been shown in certain groups of animal tumors to be of crucial importance to their particular cytogenetic pattern.

To summarize the data on the chromosomal findings in meningioma (Table 87), it would appear that recent findings in over 200 meningiomas indicate that these tumors have a re-

Table 87 Possible evolution of karyotypic changes in meningioma

1. Normal diploid stem line
   ↓
2. Normal diploid stem line
   plus
   variant cells with monosomy G
   ↓
3. Normal diploid stem line
   plus
   sideline with monosomy G
   ↓
4. Stem line with monosomy G
   plus
   normal diploid sideline
   ↓
5. Stem line with monosomy G
   plus
   variant cells with diploidy
   ↓
6. Stem line with monosomy G
   ↓
7. Stem line with monosomy G
   plus
   additional karyotypic deviations

markable tendency to develop hypodiploid stem lines. The initial step is usually a loss or deletion of a #22 and the subsequent steps affect chromosomes in groups C (#8 and #9) and D (#14 and #15), in particular, and also #1. Structural aberrations are frequent, often affecting #1 and #22. The formation of dicentrics is a common occurrence especially as a result of translocations of group F or G chromosomes onto group A or C. The importance of dicentrics in the origin of ring chromosomes found in several stem lines and the significance of various new markers frequently found in meningiomas are worthy of further evaluation and investigation. Whether or not there is a correlation between the histology and the cytogenetic findings of meningiomas, the chromosomal data indicate that a karyotypic evolution occurs in these tumors. The fact that meningiomas without chromosomal anomalies exist suggests that these tumors originally are diploid. Then, one or several cells probably lose a G-chromosome, as indicated by observations in which only a minor proportion of the cells were monosomic for a G-group chromosome. Finally, the cell line with 45 chromosomes and a missing G overgrows all other cells completely. With time, additional chromosome losses or changes may occur. Thus, the later the stage of the tumor, the more chromosomal anomalies may be found.

The majority of the meningiomas examined are of the arachnothelial type. This group shows the basic type of chromosome aberrations, i.e., #22 monosomy. In tumors that show a transition to the fibromatous type, accessory stem lines appear in addition to the missing #22. One group of meningiomas, characterized histologically by an endothelial picture with regressive alterations and enzymatically by absence of an alkaline phosphatase reaction, was associated with more than the usual hypodiploidy, i.e., with a chromosome number between 40–42 and similarity between the different karyotypes, consisting of losses in groups G, D, and C.[19] The relation of the karyotypic changes in meningioma to those associated with SV-40 virus, an agent capable of producing brain tumors in animals, remains to be elucidated.[20]

Malignant Gliomas (Astrocytic)

The astrocytic malignant gliomas (Figures 179 and 180) are the predominant type of gliomas in adults and are usually located in the cerebral hemisphere. The soft texture of these neoplasms and especially their high mitotic index make them favorable material for chromosome analysis. Fewer than 70 cases have been published in the literature;[21] 50 of these from one laboratory. Two of the tumors occurred in children and both were hypertriploid-hypotetraploid as far as their modal chromosome number, but no karyotypic data are given for these two cases. The distribution of the chromosome numbers in 50 tumors emanating from one laboratory (Mark, 1974 *a, b*) are shown in the accompanying Figure 179. About ¾ of the gliomas had a diploid or near-diploid modal number, but all were karyotypically abnormal; those with near tetraploid numbers were more than twice as common as those with a near triploid number. The distribution of the stem line-side line numbers show that the karyotypic evolution in these gliomas usually does not stop at a diploid number, in contrast to several types of neurogenic tumors in children, such as medulloblastomas and neu-

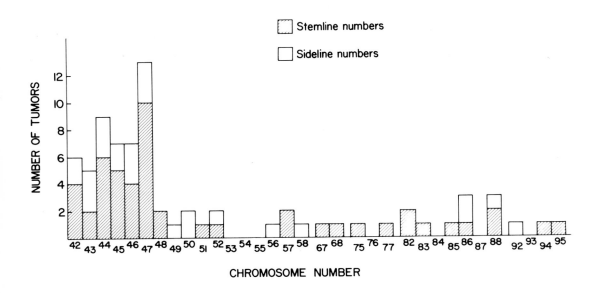

179 Stem-line and side-line numbers in astrocytomas (Mark, 1974a).

roblastomas, but proceeds with about equal frequency in both hyperdiploid and hypodiploid directions. The sharp peak at a hyperdiploid mode of 47 indicates that loss of chromosomes is a greater threat to viability than gains of chromosomes; the loss more often upsets the genetic balance necessitating compensatory changes and, hence, a greater diversity of the hypodiploid modal numbers. The chromosome pattern in the polyploid region shows that the doubling products of the modal number and related variant cells are very liable to chromosome losses. The "flattened" modes in the triploid to tetraploid range also demonstrate the increased tolerance to the numerical deviations, once a cell becomes polyploid. The small population of cells with high ploidy chromosomes ($5n-16n$) is probably composed of the giant cells frequently found during histologic examination of gliomas.

Karyotypic analysis of the 50 gliomas revealed no case with a normal diploid picture.[22] On the contrary, most tumors showed extensive deviations; markers of varying morphology were found in almost ¾ of the cases. In the modal karyotypes, the chromosome groups affected most often were C and D. Groups G and A (#1) were also involved relatively often but less than groups C and D. The nature of the chromosome groups involved was elucidated further by statistical analysis of the relative chromosome representation in the 50 stemline karyotypes as well as in the different S groups. According to the results of this analysis, the karyotypic evolution in astrocytic gliomas was characterized by a proportional loss of chromosomes particularly in groups D and C and less frequently of #1, as well as by a tendency toward gaining of #3. It should be indicated, however, that the bulk of these analyses were performed prior to the introduction of banding techniques and, hence, the identification of the various chromosomes cannot be accepted with absolute certainty.

When tumors with and without markers were compared (excluding those with DMS), loss of chromosomes in groups C, D, and #1 could be partly ascribed to their common participation in the formation of markers. Among the many different types of markers, the DMS were seen often. The nature of these markers will be discussed separately. In an established cell line originating from a parietal malignant astrocytic glioma, a 45-chromosomes stem line was found with eight different markers, whose origin could be traced almost completely with G-banding. Missing #3, #4, #7, and #15 were actually preserved, since their arms were rearranged to form four of the markers.

Analysis of the relation between the cytoge-

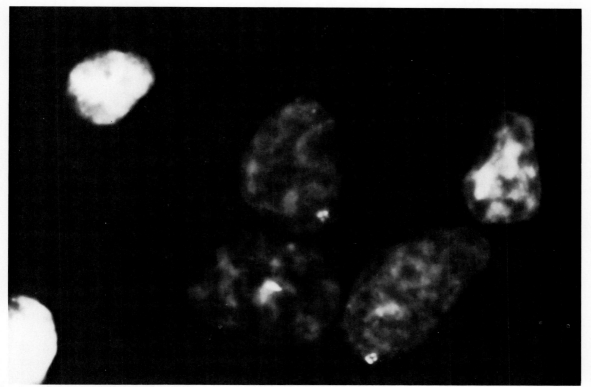

180 Interphase nuclei and a metaphase of brain tumor cells stained with quinacrine. Some of the fluorescent bodies in the upper figure probably represent Y-chromosomes. A possible Y-body is present in the nucleus shown on the left of the lower figure. A fluorescent Y-chromosome in a normal metaphase is also shown (arrow).

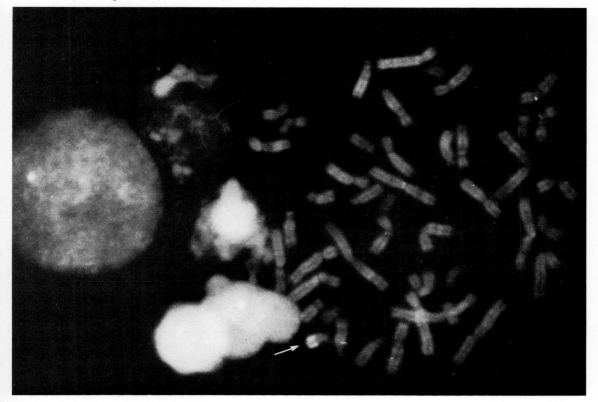

netic findings and the clinical and pathological pictures of these tumors revealed several interesting trends. The age of the patient appeared to be a factor; the results suggested that gliomas with a pseudodiploid karyotype were most likely to appear in young patients, whereas those with hyperdiploid and tetraploid stem lines tended to develop in elderly patients. Thus, the early appearance of gliomas of a certain karyotypic constitution suggests a correlation between the degree of malignancy and specific karyotypic features. The sex of the patient appeared to have an effect both on the frequency of different karyotypic S groups (hypodiploid, pseudodiploid, etc.) and on the pattern in certain groups, particularly triploid and tetraploid. When the site of the glioma and also the duration of symptoms before operation were considered, there was an influence of sex, too. Thus, there was a correlation between male sex in three tumor qualities: development of hyperdiploid stem lines, a parietal location, and a long period of symptoms before operation. This is of interest in view of the recent observations[23] that cell lines could be established from malignant gliomas much more frequently from parietally or temporally located tumors in males than those of females. No definite conclusion could be reached regarding a correlation between the histopathological characteristic of the gliomas and the chromosomal picture. The one cerebellar glioblastoma studies had features similar to those of cerebral origin.

Little is known about the cytogenetic picture of cerebral or cerebellar astrocytomas. Direct preparations from these more mature tumors have been repeatedly unsuccessful because of the low mitotic index. Many astrocytomas, especially cerebellar ones, have been studied from primary cultures. The chromosomal picture has been normal[24] in these cases but the results must be viewed with some skepticism because it was impossible to exclude stromal cells from overgrowing the tumor cells.

Oligodendrogliomas

Oligodendrogliomas are generally found in adults, but in a much lower frequency than the astrocytic malignant gliomas. This is reflected in the small number of cases studied cytogenetically. Thus, only four tumors have been reported.[25] Out of these four tumors, two were well differentiated with special chromosomal characteristics: one with 47 chromosomes differed from the normal only by a minute marker and the other had a 92-chromosome stem line which apparently was tetraploid in its distribution. This comparatively slight departure from the normal in the mature forms of oligodendrogliomas (which are usually more benign than the astrocytic types) is a noticeable feature. Recent findings in a fifth oligodendroglioma (Mark, 1974a) of the mature type studied with Giemsa banding added further support to the above results. This hypodiploid tumor had a male karyotype that differed from the normal only by the loss of the Y-chromosome in some of the cells; many cells with 46 chromosomes and a completely normal male karyotype were observed. It is difficult to ascertain the nature of the cells belonging to the tumor tissue or to the supporting tissue and other types of cells present in that location.

Two other tumors of the immature variety (oligodendroblastomas) showed chromosomal deviations as extensive of those observed in astrocytic malignant gliomas. One of the tumors had many DMS. Gain of chromosomes in group C and loss in group F in oligodendroblastomas appear to contrast with findings obtained in the astrocytic gliomas (Mark, 1974a).

Ependymomas

Ependymomas occur predominantly in children and young adults. The cytogenetic findings in five cases in children and three in adults have been reported.[26] The ependymomas in adults and two of those in the children had a normal diploid karyotype. Thus, only 3 out of the 8 tumors studied showed a stemline with chromosome abnormalities, a much lower frequency than in any other type of glioma so far analyzed.[27] The proportional losses in groups C, D and G and the low incidence of markers, were results supported by a separate analysis of the chromosomal representation in the sidelines and the variant cells found in several of the five ependymomas stemlines. The significance of the proportional gains of #1, #3 and #16 could not be clarified. Obviously, more data on ependymomas are necessary in order to clarify the karyotypic picture, though a pesimistic outlook is based on the fact that very few direct preparations have given results.

Medulloblastomas and Neuroblastomas

These tumors are highly malignant and found predominantly in children. Chromosomes have been studied in 9 primary and 2 metastatic medulloblastomas[28] and in 10 primary and 9 metastatic neuroblastomas.[29] As in the astrocytic malignant gliomas, about ¾ of the medulloblastomas are in the diploid range. In the neuroblastomas, the frequency is reduced to about ⅔ due to an increase in both triploid and hypotetraploid stem lines; metastatic neuroblastomas tended toward an increase in polyploid stem lines. Diploid stemlines were comparatively common among both medullo- and neuroblastomas. A few of these had normal karyotypes, but the others also had aneuploid ones often displaying chromosomal changes as extensive as those seen in the astrocytic gliomas, i.e., loss or gain of normal chromosomes in many and sometimes in all chromosome groups. Structural deviations were common in both tumor types. About one third of the primary and all of the metastatic medulloblastomas had one or several markers in their karyotypes and more than ¾ of the primary and about two-thirds of the metastatic neuroblastomas had markers. A statistical evaluation revealed that groups B and #17 – #18, in particular, have an evolutionary pattern that is different in these two histopathologically related tumor types. The medulloblastomas tended to gain chromosomes in group B, whereas the neuroblastomas tended to lose such chromosomes, with the former losing chromosomes in group #17 – #18 and no changes observed in the neuroblastomas. However, some tumors of both types had DMS. These markers were morphologically the same as those seen in the gliomas of adults, though they were usually somewhat larger. The explanation for this minor difference is unknown.

Six human neuroblastomas were analyzed by Giemsa and fluorescence banding techniques.[30] Two neuroblastomas were primary tumors from untreated children, and four were cell lines established from human neuroblastomas. Five of the six tumors studied were diploid or near-diploid; one was near-tetraploid. A 1p− deletion was found in three of the neuroblastomas examined. The 1p− was present in both primary tumors, and in one it was the only abnormality detected. This deletion was also found in the cells of an established line, in addition to other abnormalities. Giant marker chromosomes were present in three of the four cell lines at the time of initial study, and the fourth developed a giant marker in culture. However, three of these cell lines were from treated children and had been in culture for several years. Even though DMS have been described in human neuroblastomas, none was found in the primaries or cell lines examined. Thus, a 1p− deletion was the most consistent abnormality found in the six human neuroblastomas examined in this study.

Cytogenetic studies were performed in 21 children with neuroblastoma, and increased numbers of abnormal metaphases were found in the blood and marrow.[31] In three marrows with metastases, heteroploid (unusually hyperploid) cells with consistent markers were present.

In five established neuroblastoma cell lines the modes ranged from hypodiploid to near-tetraploid with a variety of markers, some of which were identified with banding.[32] Two markers contained HSR (see section on DMS). Three of the markers involved #1. The possible relation of DMS to spontaneous regression of neuroblastoma[33] has been discussed in one of our papers.[34]

A neuroblastoma was found in the infant daughter of parents each of whom had a child with neuroblastoma by previous marriage.[35] Banding analysis revealed a 21p−q− in the *proposita*, her mother, and maternal grandfather. An atypical #11 was found in the *proposita* and her father (11q21 was wide and 11q23 narrow). The authors suggested that family studies be performed in instances of neuroblastoma.

The exact relationship of chromosomal changes in patients with neuroblastoma and the developmental genetics of this condition[36] remain to be elucidated.

In an early study[37] of eight cases of neuroblastoma in children with BM involvement, four were found to have abnormal karyotypes. In one the chromosome constitution was pseudodiploid and in three hyperdiploid. The karyotypes were revealed to contain markers in three cases, with one containing as many as five markers.

Retinoblastomas and Optical Gliomas

These tumors are malignant neoplasms usually found in children. In one tumor a hyperdiploid stem line with 47 chromosomes (+1,+B, −C,−G, +metacentric marker) was observed.[38] Other marker chromosomes were also found in variant cells of this retinoblastoma; DMS were found in only a single cell. In a metastatic retinoblastoma (BM and spinal fluid) in a 2-yr-old boy Q-banding revealed the majority of the cells to be hyperdiploid (47 or 48 chromosomes) with at least six distinct markers (Inoue, Ravindranath, et al., 1974). One of the markers had a striking resemblence to the M_2 marker of a retinoblastoma described by Mark (1970c). No D-group abnormalities were observed.

Lele, Penrose, et al. (1963) found a chromosome abnormality, a deletion of a D-group chromosome, in cultures of conjunctiva, skin, blood, and tumor tissue from a patient with bilateral tumors. It is uncertain whether the dividing cells from the tumor tissue were neoplastic or stromal. In all probability the authors observed the congenital 13q− anomaly present in the somatic cells of patients with familial bilateral retinoblastomas (see earlier section on Retinoblastoma). In another retinoblastoma (Orye, Delbeke, et al., 1974) 44 chromosomes were found.

In a patient with bilateral retinal retinoblastomas, in one of the tumors examined karyotypically (Hossfeld and Schmidt, 1978), a 14q+ anomaly was found, in addition to other chromosomal changes.

Reports on the chromosome numbers in retinoblastomas, usually based on cultured material, ranged from hypodiploidy to tetraploidy, but no specific karyotypic changes were described. Hossfeld, Höpping, et al. (1976) successfully studied eight tumors from untreated patients (without a family history of retinoblastoma) by a direct technique. All of the tumor metaphases were abnormal, the mode being 46 in four tumors, 47 in two, and 49 and 51 in one tumor each. Banding was achieved in five cases and revealed considerable variability from tumor to tumor. One to five markers were found in each case. A specific marker characteristic of retinoblastoma could not be established. A comparison of histological and chromosomal findings suggested that the differentiated tumors tended to have a diploid and undifferentiated tumors hyperdiploid chromosome numbers.

The chromosomes have been studied in two optical gliomas,[39] one was found to have completely normal karyotype and the other modal number of 47 due to +C,−D,+E. No marker chromosomes were observed.

The relation of retinoblastoma to the congenital 13q− disorder has been discussed previously. Obviously the cytogenetic data on the tumors themselves are too scanty to draw any conclusions. It should be pointed out, however, that when the tumors are examined in the congenital disorder, in which they tend to be multiple, the 13q− anomaly does not appear to be present in the tumor. However, in the tumor of a 21-mo-old boy a deleted D(?13q−) chromosome was found, though the authors indicate that the karyotype (based on banding) most likely was 44,XY,−D,−F,t(Dq−;Bp+) (Orye and Delbeke, 1974).

Neurinomas (Schwannomas)

These tumors are found at any age, but most commonly in adults. To date, seven tumors have been examined cytogenetically,[40] with some of these tumors being located in the central nervous system: three intracranially and three in the spinal canal. The seventh neurinoma was located in the peripheral nervous system in the branches of the left brachial plexus. Preparations from primary cell cultures were used for chromosome analysis; in three a normal diploid karyotype was found with the other four having pseudodiploid or hypodiploid stem lines. The chromosomal deviations in the abnormal stem lines were restricted to groups B, C, #17−18, and G, with the three groups first mentioned also participating in the formation of markers, which were found in three of the four abnormal stem lines. No consistent karyotypic involvement of any group was found among the tumors.

Most neurosarcomas probably develop from preexisting neurinomas. One such sarcoma in a patient with von Recklinghausen's disease has been studied chromosomally;[41] because of extensive structural rearrangements in this hyperdiploid tumor, only a few karyologic similarities between the neurosarcoma and neurinoma were encountered.

In a 5-yr-old girl with neurofibromatosis, an apparently balanced translocation, t(2p−22q+) was found, and family studies showed it to be a *de novo* translocation (Alvarez and Villalpando, 1971). In another case (Ganner, 1969a) the patient had a 46,XX,t(Dq−;Bq+) karyotype. A neurosarcoma from a 38-yr-female with von Recklinghausen disease was found to have 79 chromosomes (Mark, 1972a) with karyotypic variation among the cells. Diploid and pseudodiploid cells were found in a neurofibroma (Milcu, Stanereu, et al., 1961).

A ganglioneuroma had a modal number of 53, but there were variant cells whose karyotypes suggested a pattern of clonal evolution with successive acquisition of C-, D-, and E-group chromosomes (Orye and Delbeke, 1974).

Subependymomas (subependymal gliomas) are tumors somewhat similar to ependymomas. Of three such tumors (D. Cox, 1968), two in 2-yr-old infants had diploid metaphases (one also had some "hypotetraploid" metaphases which may have been broken tetraploids) and the third tumor in a 13-yr-old boy had 47 chromosomes including trisomy of #1, a marker and loss of G-group chromosomes. In another tumor (Kucheria, 1968) five diploid and six pseudodiploid metaphases, including some with DMS, were encountered.

Pituitary Adenomas

These tumors are benign and found predominantly in adults. Some of the tumors are locally invasive but malignant transformation rarely (if ever) occurs. The chromosomes were studied in direct preparations of one eosinophilic and six chromophobic adenomas.[42] Only two of the tumors had a normal diploid karyotype. Of the remaining five tumors, three had a near-diploid stem line and two showed a near-triploid chromosome number; only one (near-diploid) tumor had a marker. Groups C and F, in particular, were involved in the heteroploid transformation, usually as a result of excessive chromosomes in these groups. Thus, the chromosomal evolutionary pattern is different from that found in meningiomas and neurinomas. It should be pointed out, however, that whereas the chromosomal findings in pituitary adenomas are usually based on direct examination of the tissues, the karyotypic examinations in neurinomas and meningiomas have usually been performed after one or another type of in vitro culture.

The cytogenetic findings in pituitary adenomas and in other "benign" tumors of the nervous system indicate that karyotypic abnormalities are by no means restricted to so-called malignant tumors. Because generally the malignancy of a tumor is based on invasiveness or metastatic spread of a tumor, on the surface the pituitary adenoma would appear to be "benign." However, the chromosomal changes do point to malignant transformation in some of the tumors, though this is not expressed by invasiveness or metastatic spread of the tumors. In this respect, the pituitary adenomas, like most meningiomas, occupy the extreme of the spectrum, as far as spread and invasiveness are concerned, though on the basis of the cytogenetic findings they have undergone a malignant transformation.

Metastatic Tumors to the Nervous System

The most common lesions that metastasize to the brain are those from cancer of the lung, kidney, breast, and, occasionally, from pancreas, colon, and malignant melanoma. About ¾ of the tumors had stem lines in the triploid region, i.e., hypotriploid zone.[43] Other cases with pentaploid or extreme hypodiploid modes (S = 35) illustrate further the advanced character of the karyotypic evolution of most of the metastatic cancers studied. Pronounced variability among the cells studied in a particular neoplasm was common; in several cases it was difficult to find even two cells of the same tumor with the same karyotype. A substantial part of the variation could be attributed to markers, which were found in all but one tumor examined. Many tumors contained one of several marker types with a notable tendency toward variation in size and morphology, indicating that these particular markers were unstable and often involved in further structural rearrangements. Also contributing to the variation were losses or gains of one or several marker types from the stem-line karyotype. This numerical deviation in the markers was not unexpected in view of the high numbers of markers found in many of the metastases (about ½ of the tumors had seven or more markers) and the common occurrence of mitotic irregularities.

The changes in chromosome numbers were accompanied by changes in the chromosomal representation of the metastases. The most striking features were the high proportional losses in groups D and G and the high positive values for markers. The brain metastases displayed more extreme numerical and structural changes in the chromosomes than did cancerous effusions. It is possible that there are neoplastic properties related to the capacity to metastasize to the central nervous system. In three patients, two different brain metastases from the same primary tumor were examined; cytogenetic analyses indicated that different subclones from the same primary tumor had a similar invasive capacity (Mark, 1974a).

A lymphoma of the brain was found to have a 14q+ and a late-replicating #22 (Yamada, Yoshioka, et al., 1977), findings characteristic of lymphomas rather than brain tumors.

Double Minute Chromosomes (DMS)

A rare finding in human cancer cells is the presence of double minute chromosomes (DMS) (Figures 181–184) described in more than a dozen tumors, primarily of neurogenic origin. However, a number of other tumors has been reported with a similar anomaly, as well as in certain tumors of animals.[44] Occasional DMS may be seen in a small number of cultured cells, in leukemic cells in the marrow, and in tumor cells, but in the tumors referred to above, a large number of DMS are present and, more importantly, in a substantial number of cells. An established cell line from a metastatic human neuroblastoma was shown to be near-diploid with several markers and containing DMS. Such an observation was also made some time ago by A. Levan, Manolov, et al., (1968) in another case of human neuroblastoma. The DMS may vary somewhat in morphology, even in the same cell, but have a general resemblance from tissue to tissue, even in different species. The exact significance of DMS in tumor cells or in normal cells is unknown. There is little doubt that they have some relationship to the neoplastic state. We have advanced a hypothesis that they may represent the chromomeres of fragmented chromosomes, but we do not know the cause for this fragmentation.

A recent report by A. Levan, Levan, et al. (1977) on DMS being replaced by chromosomes originating *de novo* in mouse tumor cells is of considerable interest. The DMS were present in about 90% of the cells of a mouse ascites tumor and varied in number from 1 to 100; occasionally > 1,000 were present with most cells having 1–50. The DMS replicated their DNA synchronously early during the S phase and were eliminated when the cells were grown in vitro; when the cells were transferred back in vivo the DMS appeared again. After 4 yr one specific subline was discovered that was almost completely devoid of DMS. This subline instead contained 1–12 new, medium-sized metacentrics in 97% of the cells. The new chromosomes differed sharply from ordinary ones, but had several properties in common with the DMS. They were completely without C-heterochromatin (similar to DMS), even though they clearly had a centromeric structure. They replicated their DNA synchronously early during the S phase and were eliminated when the cells were grown in vitro. When transferred back in vivo, cells with the new chromosomes appeared again. The authors came to the tentative conclusion that the new chromosome type had originated from the DMS by a mechanism so far unknown (A. Levan, Levan, et al., 1978) and that both the DMS and the new chromosomes may represent amplification of genetic material stimulating malignant growth.

The status of DMS has become further complicated by a recent report (Barker and Hsu, 1978) in which no centromeres were found in DMS of a human breast cancer cell line studied with Cd-banding. The authors indicated that the DMS cluster at the periphery of metaphase plates, usually encased in a matrix material; and that they move in anaphase passively with the chromosomes by attaching to the sides or ends of the chromosomes. The two "sister minutes" moved to the same pole without separation. Such anomalous mitotic behavior suggested to the authors that DMS are not chromosomes. The findings of Barker and Hsu (1978) must be reconciled with the demonstration by other authors of the incorporation of DNA precursors into DMS.

The report by A. Levan and Levan (1978), in which the authors have indicated that normally DMS are extremely difficult to observe at other mitotic stages than prophase and metaphase, is cogent to the subject discussed in the previous

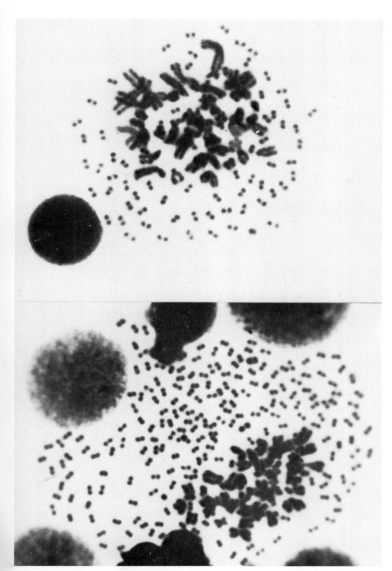

181 Metaphases of neuroblastoma cells from the marrow containing a very large number of DMS.

paragraph. In a certain clonal subline of the mouse ascites tumor SEWA, the DMS were unusually distinct and could be followed through all mitotic stages. Their behavior was extraordinary in that during metaphase-anaphase they exhibited no response whatever toward the spindle forces, and the authors concluded that the DMS were without functioning centromeres. Instead, at metaphase the DMS were attached to or enclosed by the nucleolar matter persisting around the chromosome ends and at anaphase were transported to the poles together with the nucleolar matter, sticking to the daughter chromosomes. The daughter halves of each DMS were usually carried to the same pole. Thus, most of the DMS, in spite of their lack of detectable centromeric activity, were included into the telophase nuclei. Some micronuclei were found, however, probably leading to the loss of a proportion of the DMS. The content of DMS in a certain tumor cell population therefore must depend on the bal-

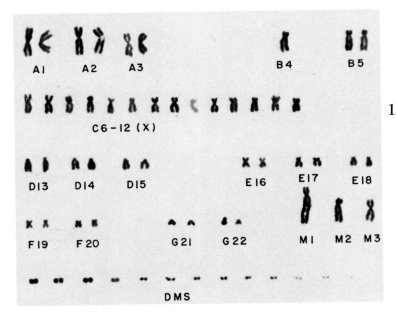

182 Karyotype of a *diploid* cell in marrow infiltrated by neuroblastoma and containing over two dozen DMS. The morphology of the DMS is somewhat different from set to set.
(Sandberg, Sakurai, et al., 1972.)

ance between this loss and the gain due to a positive selection value of cells containing DMS.

Nevertheless, the enigma of DMS remains unsolved. Their high incidence in tumors of the brain (11 of 15 reported tumors) is intriguing, indeed, and warrants further research as to their origin and significance in relation to brain tumors, particularly in view of the concepts mentioned in the preceding paragraph.

Comments on DMS Observed in Human Tumors

The karyotypic findings in human tumors with DMS described to date have shown a variable picture (Table 88). Thus, the total chromosome number in such tumors has ranged from pseudodiploid to as high as 85 chromosomes, with no consistent or characteristic picture having been obtained for such tumors. Furthermore,

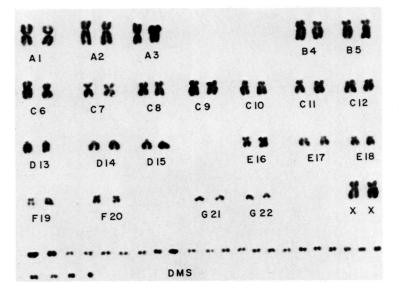

183 Karyotype (pseudodiploid) of a neuroblastoma cell in the marrow, including several markers. Note the presence of more than two dozen DM of varying sizes.

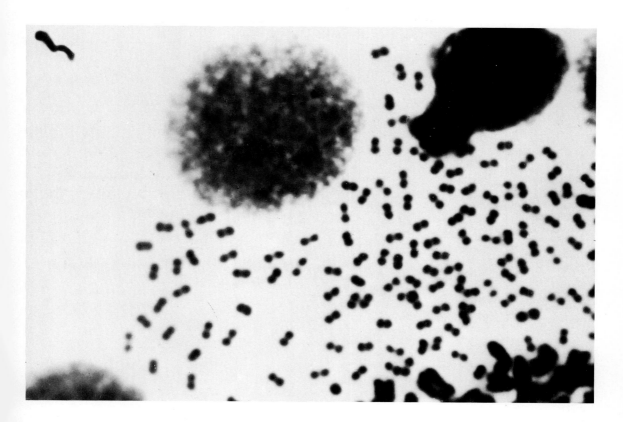

in several of the tumors there was considerable variability in the chromosome number of cells examined and no characteristic mode was obtained. In addition, most of the tumors contained abnormal markers which showed considerable variability in morphology and incidence in the various cells examined. Thus, no characteristic karyotypic picture or markers appear to characterize tumors containing DMS.

To date, the preponderant incidence of DMS has been in neurogenic tumors and rhabdomyosarcomas, and this fact remains an intriguing and interesting phenomenon. At present, it is difficult to ascertain whether it is the tissue of origin which is susceptible to DMS formation or whether a unique agent residing within such tumors leads to DMS genesis. It is possible that both mechanisms may be operative.

Even though DMS are encountered in largest number and more frquently in neurogenic tumors, particularly neuroblastomas, occasional cells with such abnormalities may be observed in other conditions, e.g., AL. The pro-

184 Close-up view of DMS in the cell of a patient with neuroblastoma. Note that the DMS differ somewhat in size and are much smaller than the chromosomes (shown in corner).

cess by which these originate may be similar to the ones to be described, though the cause for the DMS remains just as obscure.

The morphological characteristics of DMS show remarkably low variability from tumor to tumor, from case to case and, even, from species to species.[45] Nevertheless, in the same tumor the DMS may vary somewhat in size in the same cell and from one cell to another, though generally they tend to be a similar size. Thin fibrils may connect one member of DMS with the other, but the origin of these fibrils and their function remain obscure.

Even though small DMS often appear as paired and pale dots, more often they stain darker than the chromosomes. In all probability, some of the DMS represent ring chromo-

Table 88 Published cases of tumors with DMS

Diagnosis	Age, yr	Sex	Therapy	Source of cells	Chromosome findings		Reference
					Modal number	Markers	
Cancer of lung	58	M		Effusion	75		Spriggs, Boddington, et al., 1962b
Medulloblastoma	8	F	None	Tumor, BM	49,47	+	Lubs, Salmon, et al., 1966
Neuroblastoma	1 8/12	M	None	Tumor	46		Lubs and Salmon, 1965
Neuroblastoma	4	F	Radiation therapy	Tumor	54	+	D. Cox, Yuncken, et al., 1965
Neuroblastoma	1 8/12	F	None	Tumor	75		D. Cox, Yuncken, et al., 1965
Medulloblastoma	10	M	None	Tumor	75–85		D. Cox, Yuncken, et al., 1965
Rhabdomyosarcoma	3	F	Radiation therapy	Tumor	70		D. Cox, Yuncken, et al., 1965
Neuroblastoma	9/12	M	None	Tumor	46	+	A. Levan, Manolov, et al., 1968
Malignant glioma	49	M	None	Tumor	47(44)	+	Mark and Granberg, 1970
Malignant glioma	69	F	None	Tumor	47(43)	+	Mark and Granberg, 1970
Glioblastoma	40	M	None	Tumor	75–75	+	Mark and Granberg, 1970
Embryonic rhabdomyosarcoma	16	M	Radiation therapy	Axillary node	60	−	Granberg and Mark, 1971
Glioma	58	M	None	Tumor	47	+	Mark, 1971c
Ovarian cancer	65	F	None	Effusion	43–44? 65–66?		Olinici, 1971
Neuroblastoma	3	F	Radiation therapy, chemotherapy	BM	46	+	Sandberg, Sakurai, et al., 1972

somes, most commonly of the acentric form. The presence of large ring chromosomes in some of the cells of the material studied by us supports this concept (Sandberg, Sakurai, et al., 1972).

The occurrence of DMS in tumor cells or in cells grown in vitro, both normal and abnormal, has been reported.[46] We have had the opportunity to observe DMS in leukemic cells in marrow and lymphocytes grown in culture, and in some tumor cells. However, the number of cells containing such DMS is small, and the number of DMS per cell is usually also small. It is of interest that even among gliomas, Mark (1974a) found 13 tumors out of 50 which contained only small numbers of cells with DMS. Whether with time these cells with DMS would have found an appropriate milieu in the gliomas for multiplying and whether these DMS bear the same significance in relation to the behavior of the tumors remains unknown. The exact incidence of DMS in all tumors is unknown, because there is a natural tendency to report only those that contain large numbers of DMS.

Even though the morphology of the DMS is unique, it has been difficult to obtain rigorous evidence of their origin. The presence of a large number of DMS in cells with a nearly normal number of chromosomes raises many questions regarding their origin. However, in most cells, the number and morphology of DMS is of such a nature that their origin could be readily explained on the basis of disruption of some of the chromosomes.

We postulated (Sandberg, Sakurai, et al., 1972) that a number of agents may lead to DMS formation, but felt that they originate by the same mechanism, i.e., breakdown of existing chromosomes. It is possible that the general uniformity and duality of DMS within a cell are related to the architecture of the chromosomes from which they originate. The DMS may be compared to the brick or stone of a building, which, when the latter collapses, are but units of a preexisting structure, these units being of essentially uniform size and shape. Thus, depending on the morphology and size of the chromosomes, the DMS will vary in relation to the characteristics of the chromosomal origin.

Recent studies with a variety of techniques have shown that the human chromosome appears to consist of a number of chromomeres of different sizes, with the chromomeres probably being connected by thin strands of DNA. The variation in size and staining between chromomeres seems to disappear to a large extent upon contraction of the chromosomes. Thus, it is possible to visualize that the DMS are but chromomeres or groups of chromomeres which have been set free by one process or another and the variability and size among the DMS being due to the above mentioned observations.

An impressive finding in the cases with DMS is the fact that the number of DMS per cell did not bear any relationship to the total number of chromosomes. In fact, in the case of the diploid cells with DMS, the set was shown not to lack any chromosomes even when the number of DMS was very high. A plausible explanation is that the DMS replicate but remain within the parent cell as a result of nondisjunction of the pairs. A repeated process of such nondisjunction, posssibly associated with concomitant endoreduplication, would then lead to an accumulation of DMS within one cell, the number of DMS depending on the number of nondisjunction divisions undergone by them. Thus, it is possible to envision a single DMS pair being present in a diploid cell, but, because of nondisjunction and/or endoreduplication through a number of cell divisions, the DMS accumulate within the parent cell, ultimately being present in rather large numbers. This explanation would readily account for either the diploid or aneuploid cells without DMS and the lack of correlation between the chromosome number and the number of DMS within a cell.

The presence of DMS in cells that contain perfectly normal karyotypes or characteristic aneuploid karyotypes without loss of chromosomes may possibly be explained on the basis of inclusion of DMS by such cells. The DMS could originate from breakage of chromosomes in other cells, such cells exuding their DMS before dying. The incorporation of DMS by normal or cancerous cells followed by nondisjunction and/or endoreduplication could account for the findings observed. Even though attempts to incorporate DMS by cells grown in vitro have failed to demonstrate such a phenomenon (Lubs, Salmon, et al., 1966), the occurrence of such a sequence of events in vivo cannot be ruled out and should definitely be considered. Such an explanation would account for the presence of apparently totally diploid metaphases with a large number of DMS and cells with aneuploid or pseudodiploid metaphases without any evidence of fragmentation of chromosomes and containing a large number of DMS.

It is our contention that DMS originate from existing chromosomes within the cell (Figures 185 and 186). The process of DMS formation most likley occurs over a period of time, as indicated not only by the absence of dicentric chromosomes and only the occasional presence of ring chromosomes in tumor material, but also by the demonstration of early and rapid DNA synthesis in DMS and the rarity of finding a cell in which a DMS can be seen to originate from chromosomes. The absence of chromatid types of aberrations possibly indicates that the DMS originate during the G1 period. That the DMS are not of external origin has been amply demonstrated.[47] Thus, it has been shown that they are not of bacterial origin, that they do not originate outside the cell and then become incorporated by the cell, that they do consist of DNA, and that they cannot be incorporated in vitro when isolated DMS are added to a culture of cells.

The spontaneous regression of cancer in human subject, although extremely rare, has been documented (W. Boyd, 1966). The highest incidence of spontaneous regression has been described for neuroblastoma and other neurogenic tumors (W. Boyd, 1966; D'Angio, Evans, et al., 1971; Helson, 1971). Even though mechanisms for such spontaneous regression of cancer in humans have not been established, it

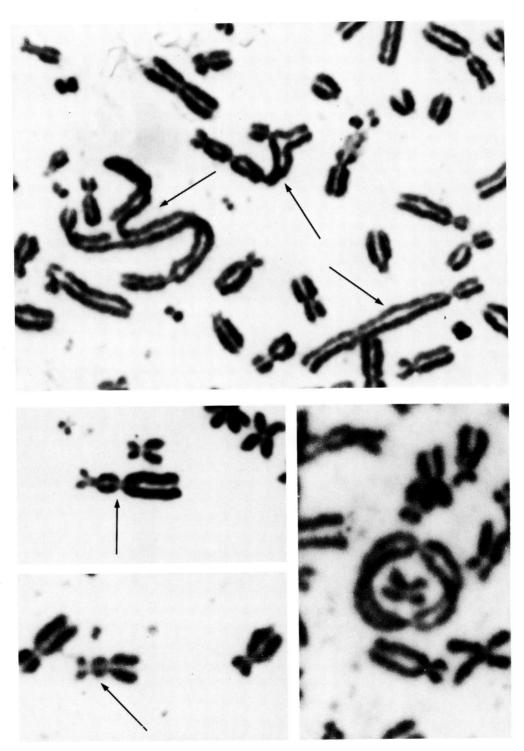

185 Various chromosome abnormalities observed in cultured cells: large markers and complicated exchange figures shown in upper figure, with an occasional DMS dispersed in the figure, a large dicentric ring in lower right figure, and dicentrics in the two figures in lower left.

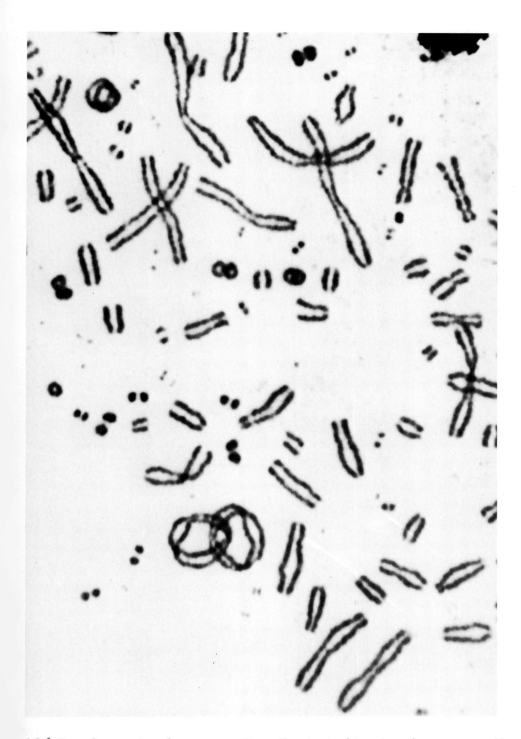

186 Part of a metaphase from a mammalian cell maintained in culture for over a year. Note the formation of double fragments of various sizes, which may be related to DMS formation. In some cases, minute units can be seen originating from association of the long arms of an acrocentric chromosome with a small metacentric chromosome. Other minute units can be seen coming off other chromosomes shown in the photograph.

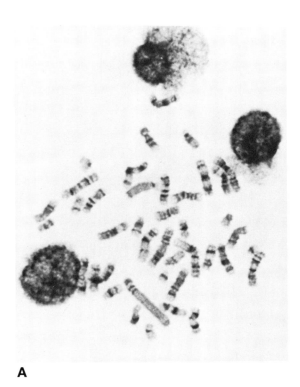

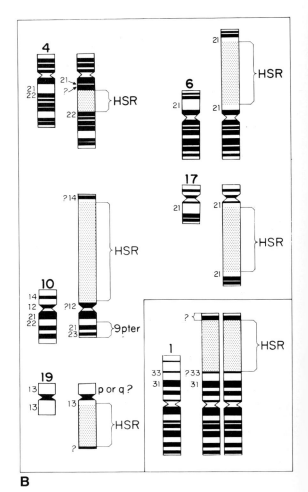

is extremely tempting to speculate that the presence of DMS may be responsible for such an unusual occurrence. The presence of DMS in tumor cells may not only interfere with normal metabolism in and survival of such cells but may, in fact, inhibit the neoplastic process. It should be stressed, though, that most neuroblastomas do not contain DMS and that other karyotypic aspects may play a role in spontaneous regression of these tumors. Nevertheless, more has to be learned about the agents that lead to DMS formation, because their application in tumor therapy would deserve a serious try, in view of the discussion just concluded.

Homogeneously Staining Regions and Double-Minute Chromosomes

The presence of large, homogeneously staining regions (HSR) (Figure 187) in chromosomes that lack the longitudinal differentiation ordinarily revealed by banding has been described in established human neuroblastoma cell lines.[48] Studies with tritiated thymidine indicated that the long, abnormally staining region

187 A. Trypsin-Giemsa banded metaphase of a human neuroblastoma cell in culture containing two chromosomes (#10 and #19) with HSR. B. A diagrammatic representation of metaphase chromosomes with HSR. One or two of these chromosomes were present per cell, never more. In those cells with an HSR on #10, a t(9;10)(p13; q22) was present. Two identical #1 chromosomes with interstitial HSR are shown in the *inset*.
(Biedler and Spengler, 1976.)

replicated fairly rapidly and synchronously and terminated replication before the midpoint of the S phase. The HSR, thus, appear to represent a novel metaphase chromosome abnormality that may be present in cells with specialized functions. The above authors could not determine whether the HSR is transcriptionally active. One of the neuroblastoma cell lines in which an HSR was ultimately observed, when first established and reported upon (Tumilowicz, Nichols, et al., 1970), was shown to be comprised of two morphologically distinct cell types but each had the same chromosome constitution characterized by hyperdiploidy and the presence of two large markers.

A most interesting study involved four neuroblastoma cell lines[49] in which, with banding techniques, all the lines were shown to contain a marker with a long nonbanding HSR. This HSR-containing chromosome differed in each line. One line contained two classes of cells: one with an HSR marker and the other with DMS. Each cell had one of these abnormalities, but no cell had both. The presence of two additional markers in all cells of this line indicated a common origin. The authors suggested that the DMS are derived from the HSR, with the location and size of the HSR differing from cell line to cell line. This was also true of the other cell lines studied (Biedler and Spengler, 1976b). For example, in one an HSR was located within the long arm of #4, another of unusually large size was located in the short arms of #6 and #10, and another in the long arms of #17 and #19. In another neuroblastoma cell line relatively large HSR were located within the short arm of #1, replicate copies of it being present in some cells.

An extensive HSR region which showed marked fluorescence with Q-banding and dark staining with C-banding has been reported in the cells of an established cell line originating from a cancer of the breast (Barker and Pathak, 1978). This suggests that HSRs may involve euchromatin or heterochromatin. The authors suggested that the large heterochromatic region resulted from tandem amplification of a small paracentric chromosomal region containing primarily heterochromatin.

References

Pages 461–468

1. Baker and Atkin, 1970; Baker in Atkin 1976a; Enterline and Arvan, 1967; Kotler and Lubs, 1967; Lubs, 1970; Lubs and Clark, 1963; Lubs and Kotler, 1967; Mark, Mitelman, et al. 1973; Messinetti, Zelli et al., 1968; Mitelman, Levan, et al., 1976a.
2. Danes, 1975, 1976a, b; Law, Hollinshead, et al., 1977; Mark, Mitelman, et al., 1973; McConnell and Parsons, 1968.
3. Danes, 1975, 1976a, b, 1978.
4. Atkin, Baker, et al., 1967a; Benedict, Brown, et al., 1967; Dallapiccola and Tataranni, 1969; Ishihara, Moore, et al., 1961; Ishihara, Kikuchi, et al., 1963; Ishihara and Sandberg, 1963; Ising and Levan, 1957; Jacob, 1961; Krompotic and Zellner, 1968; A. Levan, 1956c; A. Levan and Hauschka, 1953; Makino, Ishihara, et al., 1959a, b; Makino, Tonomura, et al., 1959; Makino, Sasaki et al., 1964b.
5. Hansson and Korsgaard, 1974; Hansteen, Hillestad, et al., 1977; Miles and Wolinska, 1973.
6. Kakati, Hayata, et al., 1975, 1976; Kakati, Oshimura, et al., 1976; Sandberg 1968a, 1973: Sandberg and Hossfeld, 1973; Sandberg, Yamada, et al., 1967; Sandberg and Yamada, 1965, 1966; Sonta, Oshimura, et al., 1977; Sonta and Sandberg, 1977c; Yamada, Takagi et al., 1977c; Yamada and Sandberg, 1966b.
7. Atkin, 1976b; Atkin and Ross, 1960; Ballesta and Baldellou, 1970; Bartram and Rüdiger, 1978; Bayreuther, 1960; Berger, Lejeune, et al., 1971; Bottomley, 1968; Brandão and Atkin, 1968; Brieux de Salum, 1972; Cadotte, 1974; Clausen and Syverton, 1962; Conen and Falk, 1967: D. Cox, 1966, 1968; Falor, Gordon, et al., 1969; E.H.R. Ford, 1973; Forteza-Bover, Báguena-Candela et al., 1964b; Ganina, Polishchuk, et al., 1976; German, 1974c; Goodlin, 1962, 1963; Gottlieb, 1969; Gripenberg, 1976; Grouchy, 1964, 1967a; Grouchy, Vallée, et al., 1963; Grouchy and Wolff, 1969; Hansteen, 1967; Hossfeld, 1971, 1972; T.C. Hsu, 1954, 1961; D.T. Hughes, 1965; Inui, 1966; Ishihara, 1959; Ishihara, Moore, et al., 1962; J.F. Jackson, 1967; Koller, 1972; Kotler and Lubs, 1967; Lancet, 1977; Legrand, 1968; Levi, Cooper, et al., 1969; Lubs, 1970; Maeda, Tabata, et al., 1965; Makino, 1975; Mannini, Fraccaro, et al., 1966; Mark, 1970b; Miles, 1966, 1967a, b, 1974; Miles and Gallagher, 1961; G.E. Moore and Sandberg, 1963; Nelsen-Rees, Flandermeyer, et al., 1975; Pogosianz, 1963, 1973; Pogosianz and Prigogina, 1972, 1976; Richards and Atkin, 1960; Rigby and Franks, 1970; Rousseau, Laurent, 1970; Ruffié, Ducos, et al., 1964; Sasaki, 1961; Schuler and Dobos, 1970; Serr, Padeh, et al., 1969; Spooner and Cooper, 1972; Spriggs, 1964, 1973, 1976; Spriggs and Boddington, 1968; Tabata, 1959a, b; Tavares, 1968; A.I. Taylor, 1970; Tiepolo and Zuffardi, 1973; Tonomura, 1959a; Wakabayshi and Ishihara, 1958; Wakonig-Vaartaja, Helson, et al., 1971; Whang-Peng, Canellos, et al., 1968; L. White and Cox,

1967; Wisniewski and Lech, 1974; Wu, 1961; Wurster-Hill, 1975; Zech, 1974.
8. Sandberg, Ishihara, et al., 1963b; Sonta and Sandberg, 1977c.
9. Sandberg and Hossfeld, 1970.
10. Mutton and Daker, 1973.
11. Atkin and Baker, 1966; P. Fischer and Golob, 1967; Galton, Benirschke, et al., 1966; Martineau, 1966.
12. Sandberg, 1966b; Sandberg, Cortner, et al. 1966.
13. Lelikova, Laskina, et al., 1971.
14. Auersperg, Corey, et al., 1967.
15. Lubs and Kotler, 1967.
16. Sandberg, 1966b.
17. Sandberg, Bross, et al., 1968.
18. Sandberg, Yamada, et al., 1967.
19. Yamada, Shinohara, et al., 1966.
20. Sandberg, Yamada, et al., 1967.
21. Sandberg and Hossfeld, 1970.
22. Sandberg, 1966b.

Pages 468–485

1. T.C. Hsu, 1961.
2. Atkin and Richards, 1956.
3. Makino, Ishihara, et al., 1959a, b.
4. Ertl, Schlegel, et al., 1970.
5. Lubs and Clark, 1963.
6. T.C. Hsu, 1954; Ising and Levan, 1957; A. Levan, 1956b.
7. Makino, Ishihara, et al., 1959a, b.
8. Ishihara, 1959; Tonomura, 1960.
9. Ishihara, 1959.
10. Makino, Ishihara, et al., 1959a, b.
11. Tonomura, 1960.
12. Makino, Sasaki, et al., 1964b.
13. Ising and Levan, 1957; A. Levan, 1956b.
14. Awano, Tsuda, et al., 1960; Awano and Tsuda, 1969; Ganina, Polishchuk, et al., 1976; Ishihara, Moore, et al., 1961; Ishihara, Kikuchi, et al., 1963; Messinetti, Zelli, et al., 1966a, b, 1968; Miles, 1967a, b; Sandberg, Yamada, et al., 1967; Sasaki, 1961; Spriggs, Boddington, et al., 1962b; Spriggs, 1964.
15. Ishihara, 1959; Makino, Ishihara, et al., 1959a, b.
16. Makino, Sasaki et al., 1964b; Tonomura, 1960.
17. Messinetti, Zelli, et al., 1968.
18. Steenis, 1966.
19. Sandberg, Bross, et al., 1968.
20. H.F. Stich, Florian, et al., 1960.
21. Atkin, 1970b; Baker and Atkin, 1970; Enterline and Arvan, 1967; Lubs and Clark, 1963; Lubs and Kotler, 1967; Mark, Mitelman, et al., 1973; Messinetti, Zelli, et al., 1968; Miles, 1967a, b; Mitelman, Mark, et al., 1974.
22. Atkin, 1973b.
23. Atkin, 1970b; Baker and Atkin, 1970.
24. Lubs and Clark, 1963.
25. Lubs and Kotler, 1967.
26. Mark, Mitelman, et al., 1973. The significance of increased tetraploidy in Gardner's syndrome has been reported by Danes, 1975, 1976a,b.
27. Mitelman, Mark, et al., 1974.
28. Enterline and Arvan, 1967.
29. Miles, 1967a, b.
30. Xavier, Prolla, et al., 1974.
31. Findings similar to these have been reported by Emerit, Emerit, et al., 1972.
32. Kotler and Lubs, 1967; Lubs and Kotler, 1967.
33. Messinetti, Zelli, et al., 1968.
34. Atkin and Baker, 1969.
35. Atkin and Baker, 1966, 1969; Enterline and Arvan, 1967; Ishihara, Moore, et al., 1961, 1962; Ishihara and Sandberg, 1963; J.F. Jackson, 1967; Legrand, 1968; Leibowitz, Stinson, et al., 1976; Lubs and Clark, 1963; Lubs and Kotler, 1967; Makino, Sasaki, et al., 1964b; Messinetti, Zelli, et al., 1968; Miles, 1967a, b; Richards and Atkin, 1960; Sandberg, Ishihara, et al., 1963b; Sandberg, Yamada, et al., 1967; Spriggs, Boddington, et al., 1962b; Yamada and Sandberg, 1966a; Yamada, Takagi, et al., 1966.
36. Koller, 1947.
37. Baker in Atkin, 1974a.
38. Atkin, 1970b.
39. Sandberg and Yamada, 1966.
40. Sandberg, Ishihara, et al., 1963b.
41. Sandberg, Ishihara, et al., 1963b.
42. Atkin and Baker, 1966.
43. Atkin, 1969a; Atkin and Baker, 1969.
44. Ishihara and Sandberg, 1963.
45. Makino, Sasaki, et al., 1964b.
46. J.F. Jackson, 1967.
47. Sandberg, Yamada, et al., 1967.
48. Ishihara and Sandberg, 1963.
49. Sandberg, Yamada, et al., 1967.
50. Sandberg and Yamada, 1966.

Pages 485–503

1. Castoldi, Scapoli, et al., 1968; Ruffié, Ducos, et al., 1964.
2. Katayama and Masukawa, 1968.
3. Toews, Katayama, et al., 1968a.
4. Emson and Kirk, 1967.
5. Cervenka and Koulischer, 1973.
6. Baker in Atkin, 1974a.
7. Toews, Katayama, et al., 1968b.
8. Miles, 1967a, b.
9. Olinici and Simu, 1970.
10. Toews, Katayama, et al., 1968b.
11. Goncalves, 1976; Ghosh and Shah, 1975; Gropp, Pera, et al., 1967; Gropp, Wolf, et al., 1965; Gros, Colin, et al., 1973; Kallenberger, 1964; Kallenberger, Hagmann, et al., 1967, 1968; Kimel, 1957; Rajeswari, Ghosh, et al., 1977; Shirley, 1967; Siracka, Simko, et al., 1970; Stanley, Bigham, et al., 1966.

12. Fritz-Niggli, 1954, 1956; Makino, Ishihara, et al., 1959*a, b*; Makino, Tonomura, et al., 1959; Wakabayashi and Ishihara, 1958.
13. Atkin, 1972; Ayraud, Lambert, et al., 1977; Brandão and Atkin, 1968; Bodor, Håkansson, et al., 1974; Castoldi, Scapoli, et al., 1968; Genes, 1970; Grouchy, Vallée, et al., 1963; Ishihara and Sandberg, 1963; Ishihara, Kikuchi, et al., 1963; Ishihara, Moore, et al., 1961; Kallenberger, 1974; Khishin, El-Zawahri, et al., 1976; Krompotic and Zellner, 1968; Legrand, 1968; Mark 1975*a*; Messinetti, Zelli, et al., 1971; Miles, 1967*a, b*; Neishtadt, 1974; Porter, Benedict, et al., 1969; Spriggs and Boddington, 1968; Spriggs, Boddington, et al., 1962*a, b*.
14. Toews, Katayama, et al. 1968*a*.
15. Bodor, Håkansson, et al., 1974.
16. Messinetti, Marcellino, et al., 1971.
17. R.K. Young, Cailleau, et al., 1974.
18. Kallenberger, 1974.
19. Khishin, El-Zawahri, et al., 1976.
20. Ayraud, Lambert, et al., 1977; Cailleau, Cruciger, et al., 1976; Seman, Hunter, et al., 1976; Trempe, 1976.
21. Atkin, 1971*a, b*; Atkin, Baker, et al., 1974; Atkin and Baker, 1964, 1966; Atkin and Pickthall, 1977; Bader, Taylor, et al., 1960; Benedict, Brown, et al., 1971; Berger, 1968*a, b*; Borch, Stanley, et al., 1969; Brigato, Franco, et al., 1965; Chudina and Pichugina, 1976; Curcio, 1966*c*; Curcio and Sartori, 1966; Forteza-Bover, Báguena-Candela, et al., 1964*a*; Fraccaro, Gerli, et al., 1965; Goodlin, 1962; Grouchy, Vallée, et al., 1963; Ishihara and Sandberg, 1963; Ishihara, Kikuchi, et al., 1963; Ishihara, Moore, et al., 1961; J.F. Jackson, 1967; Katayama and Toews, 1969; Knoerr-Gaertner, Schuhmann, et al., 1973, 1977; Legrand, 1968; Lejeune and Berger, 1966; Makino, Sasaki, et al., 1964*a, b*; Mannini, Fraccaro, et al., 1966; Messinetti, Marcellino, et al., 1971; Meuge, 1967; Miles, 1967*a*; Moraru and Fadei, 1973, 1974; Olinici, 1971; Olinici, Galatir, et al., 1973; Pogosianz, Chudina, et al., 1972; Rodkina, Pogosyants, et al., 1972; Rugiati, Ragni, et al., 1966, 1967; Ruffié, Marquès, et al., 1964; Siracký, 1969; Šlot, 1967*b*, 1970; Šlot and Frauenklin, 1968; Spriggs, Boddington, et al., 1962*a, b*; Tiepolo, Zara, et al., 1967; Tortora, 1967; Wakonig-Vaartaja and Auersperg, 1970; Witkowski and Zelli, et al., 1970; Miles, 1967*a*; Ojima, Inui, noio, 1976; Zara, Fraccaro, et al., 1966.
22. Curcio, 1966*b*.
23. Rashad, Fathalla, et al., 1966; Toews, Katayama, et al., 1968*a*.
24. Grouchy, Vallée, et al., 1963.
25. Fraccaro, Mannini, et al., 1968.
26. Richart and Ludwig, 1969.
27. Benedict, Rosen, et al., 1969.
28. Rashad, Fathalla, et al., 1966.
29. Serr, Padeh, et al., 1969.
30. Arias-Bernal and Jones, 1968.
31. Toews, Katayama, et al., 1968*a*.
32. Curcio and Sartori, 1966; Eicke, Emminger, et al., 1965; Katayama, 1969; Katayama and Jones, 1968; Katayama, Woodruff, et al., 1972; Messinetti and Moscarini, 1970; Messinetti, Zelli, et al., 1970; Moraru and Fadei, 1974; Olinici, Galatir, et al., 1973; Tagliani, Mastrangelo, et al., 1963; Tortora, 1963, 1969.
33. Einhorn, 1978; Greenspan and Tung, 1974; Reimer, Hoover, et al., 1977.
34. Kaiser-McCaw and Latt, 1977.
35. Kaiser-McCaw and Latt, 1977; Linder, Hecht et al., 1975; Linder, Kaiser-McCaw, et al., 1975; Ott, Linder, et al., 1976.
36. Richards and Atkin, 1960.
37. Makino, Yamada, et al., 1965.
38. Cervenka and Koulischer, 1973.
39. Fiocchi, 1967.
40. Atkin, 1971*a, b*.
41. Baker, 1968.
42. Atkin, 1960*a*; Atkin, Richards, et al., 1959; Cecco, Rugiati, et al., 1966; Dehnhard, Knörr-Gartner, et al., 1973; P. Fischer, Golob, et al., 1966*a, b*; Gaffuri and Bertoli, 1964; Ganina, Polishchuk, et al., 1976; Ichinoe, 1970; Kang, Kim, et al., 1968; Katayama and Jones, 1967; S.W. Kim, Kang, et al., 1967; Kirkland, 1966*a, b*; Messinetti and Moscarini, 1970; Messinetti, Zelli, et al., 1970; Miles, 1967*a*; Ojima, Inui, et al., 1960; Olinici, Galatir, et al., 1973; Paulete-Vanrell and Camacho, 1966; Spriggs, 1964; Stanley and Kirkland, 1968; Szulman, 1966; Tseng and Jones, 1969; Wakonig-Vaartaja, 1962, 1963; Wakonig-Vaartaja and Hughes, 1967.
43. Tseng and Jones, 1969.
44. Baker, 1968.
45. Cervenka and Koulischer, 1973.
46. Stanley and Kirkland, 1968.
47. E.C. Hughes and Csermely, 1966; Manna, 1954, 1957*a, b*; L. Sachs, 1953, 1954; Sherman, 1969; Takemura, 1960; Tjio and Puck, 1958; B.E. Walker and Boothroyd, 1954.
48. Rask-Madsen and Philip, 1970.
49. Bowey and Spriggs, 1967; Stanley, 1969.
50. Kuramoto, 1972; Kuramoto and Hamano, 1977.
51. Makino, Sasaki, et al., 1963, 1965; Sasaki, Fukuschima, et al., 1962.
52. Sasaki, Fukuschima, et al., 1962.
53. Kawasaki, 1968*a*; Serr, Padeh, et al., 1969; Stolte, Kessel, et al., 1960.
54. Carr, 1969.
55. Makino, Sasaki, et al., 1965.
56. Katayama, 1969; Makino, Sasaki, et al., 1965.

57. Katayama, Woodruff, et al., 1972; Messinetti, Zelli, et al., 1970; Paulete-Vanrell, Laguardia, et al., 1964; Wakonig-Vaartaja and Hughes, 1967.
58. Woodruff, Davis, et al., 1969; Von Mayenburg, Ehlers, et al., 1977.
59. Makino, Tonomura, et al., 1959.
60. Spriggs, 1964.
61. Wakonig-Vaartaja and Hughes, 1967.
62. Goodlin, 1962; Wakonig-Vaartaja and Hughes, 1967.
63. Curcio, 1966a; Goodlin, 1962; Weise and Biittner, 1972.
64. Atkin, 1974b; Beischer, Fortune, et al., 1967; W.G. Paterson, Hobson, et al., 1971.
65. Klinger, Ludwig, et al., 1958; Serr, Padeh, et al., 1969; D. Wagner, 1973.
66. Baggish, Woodruff, et al., 1968; Loke, 1969.
67. Katayama, 1969; Makino, Sasaki, et al., 1965.
68. Kawasaki, 1968a, b; Serr, Padeh, et al., 1969.
69. Hertz, 1976; Stone and Bagshawe, 1976; Vassilakos and Kajii, 1976.
70. Bourgoin, Baylet, et al., 1965; Sasaki, Fukuschima, et al., 1962; Shinohara, Sasaki, et al., 1971.

Pages 503-516

1. Sandberg and Hossfeld, 1970.
2. Falor and Ward, 1976a, 1977, 1978; Sandberg, 1977.
3. Spriggs, Boddington, et al., 1962a, b; Yamada, Takagi, et al., 1966.
4. Lamb, 1967.
5. Spriggs, Boddington, et al., 1962a, b.
6. Yamada, Takagi, et al., 1966.
7. Falor and Ward, 1976a.
8. Levi, Cooper, et al., 1969.
9. Fosså, 1975.
10. Atkin and Petković, 1973.
11. Sandberg, 1977.
12. A low mitotic index in the bladder epithelium has been reported by Pringle and Williams, 1967.
13. D.T. Hughes, 1965.
14. Nagayama and Kataumi, 1972.
15. Falor and Ward, 1973.
16. Falor and Ward, 1978.
17. Falor, 1971a; Falor and Ward 1973, 1976a, b, 1977, 1978.
18. Elliott, Cleveland, et al., 1974; Elliot, Bronson, et al., 1977; Fosså, Kaalhus, et al., 1977; Malkovský and Bubeník, 1977; Melamed, Darzynkiewicz, et al., 1977; Nayak, O'Toole, et al., 1977; Rasheed, Gardner, et al., 1977; Rigby and Franks, 1970; Traganos, Darzynkiewicz, et al., 1977.
19. Atkin and Petković, 1973.
20. Atkin, 1973a; Litton, Hollander, et al., 1972; Sellyei and Vass, 1972.
21. D. Cox, 1966. See also Rousseau, Laurent, et al., 1970.
22. Makino, Tonomura, et al., 1959.
23. Straub, Lucas, et al., 1969.
24. Atkin, 1973a, b; Atkin and Baker, 1966; Berger and Martineau, 1970; P. Fischer and Golob, 1967; Galton, Benirschke, et al., 1966; Khudr, Walsh, et al., 1973; Lelikova, Laskina, et al., 1970, 1971; Martineau, 1966, 1967, 1969; Miles, 1967a; Quiroz-Gutiérrez, Alfaro-Kofman, et al., 1967; Rigby, 1968.
25. Lelikova, Laskina, et al., 1970, 1971.
26. Berger and Martineau, 1970.
27. Atkin, 1973a.
28. Atkin, 1973b; Khudr, Walsh, et al., 1973.
29. Martineau, 1966.
30. Miles, 1967a.
31. Rigby, 1968
32. Rigby, 1968.
33. Tabata, 1959a, b.
34. Pierce and Nakane, 1967.
35. Atkin, 1973a, b.
36. Laurent, Rousseau, et al., 1968; Linder, Hecht, et al., 1975; Roux, Emerit, et al., 1971; Zakharov and Lelikova, 1972.
37. Hunter and Lennox, 1954.
38. Tavares, Costa, et al., 1966.
39. Hunter and Lennox, 1954.
40. Atkin, 1973b.
41. Atkin, 1973b.
42. Cavallero, 1958.
43. Pierce and Nakane, 1967.
44. Atkin, 1973b; Lederer, Autengruber, et al., 1976.
45. Tavares, Costa, et al., 1973.
46. Jellinghaus, Okada, et al., 1976; Okada and Schroeder, 1974; Stone, Mickey, et al., 1978.
47. Tavares, Costa, et al., 1973.

Pages 516-518

1. Atkin, 1966; Benedict, Brown, et al., 1967; E. Davidson and Bulkin, 1966; Genes, 1970; Greisen, 1971; Ishihara, Moore, et al., 1961; J.F. Jackson, 1967; Kotler and Lubs, 1967; Makino, Ishihara, et al., 1959a, b; Miles, 1967b; Sandberg, Yamada, et al., 1967, Spriggs, Boddington, et al., 1962b; Yamada, Takagi, et al., 1966.
2. A. Levan, 1956c.
3. Ising and Levan, 1957.
4. Ishihara, Kikuchi, et al., 1963.
5. D. Cox, Yuncken, et al., 1965.
6. Falor, 1971b; Falor, Gordon, et al., 1969.
7. Baradnay, Monus, et al., 1968; Cervenka and Koulischer, 1973; Hanschke and Hoffmeister, 1960.
8. Baradnay, Monus, et al., 1968.

Pages 518–520
1. Miles and Gallagher, 1961.
2. Socolow, Engel, et al., 1964.
3. Haemmerli, 1970.
4. Izuo, Okagaki, et al., 1971.
5. Atkin, 1970b; Atkin and Baker, 1965a.
6. Baker *in* Atkin, 1976a.
7. Langlands and Maclean, 1976.
8. P.D. Lewis, 1971.
9. G. Levan, Mitelman, et al., 1973.

Pages 520–524
1. Atkin, 1971c; Atkin and Baker, 1969; Benedict, Brown, et al., 1971; Berger, 1968b; Berger and Aubert, 1975; Berger and Lacour, 1973; Berger, Lejeune, et al., 1971; Forteza-Bover, Báguena-Candela, et al., 1964b; Ganina, Polishchuk, et al., 1976; Genes, 1970; Goh, 1968c; T.C. Hsu, 1954; Huang, Imamura, et al., 1969; Ishihara, Moore, et al., 1961; Kakati, Song, et al., 1977; Kanzaki, Hashimoto, et al. 1976; Katayama, Woodruff, et al., 1972; Miles, 1967b; Quiroz-Gutiérrez, Islas, et al., 1968; Sandberg, Yamada, et al., 1967; Spriggs, Boddington, et al., 1962b; Svejda, Vrba et al., 1974; Whang-Peng, Chretien, et al., 1970; Witkowski, 1970; Witkowski and Zabel, 1972.
2. Atkin, 1974a.
3. Atkin, 1971c; Ishihara, Moore, et al., 1961; Quiroz-Gutiérrez, Islas, et al., 1968; Whang-Peng, Chretien, et al., 1970.
4. Forteza-Bover, Báguena-Candela, et al., 1964b.
5. Miles, 1967b.
6. Chen and Shaw, 1973.
7. McCulloch, Dent, et al., 1976.
8. Berger and Lacour, 1973.
9. Berger, Lejeune, et al., 1971.
10. Vrba, 1974.
11. Berger and Aubert, 1975.
12. Kanzaki, Hashimoto, et al., 1976, 1977.
13. Atkin, 1971c.
14. Kakati, Hayata, et al., 1975, 1976; Kakati, Oshimura, et al., 1976.
15. Chen and Shaw, 1973; McCulloch, Dent, et al., 1976.
16. Kakati, Oshimura, et al., 1976.
17. Kakati, Song, et al., 1977.
18. Mark, 1973a.
19. Oshimura, Sonta, et al., 1976; Rowley, 1975a; Yamada and Furasawa, 1976.
20. Kakati, Oshimura, et al., 1976.

Pages 524–526
1. Miles, 1967a, b.
2. Granberg and Mark, 1971.
3. L. White and Cox, 1967.
4. D. Cox, 1968.
5. Ishii, 1965; Maunoury, Arnoult, et al., 1972.
6. Berger, 1968a.
7. D. Cox, 1968.
8. Kristoffersson, 1973.
9. Nezelof, Laurent, et al., 1967.
10. Ayraud, Szepetowski, et al., 1970; Ayraud, Rey, et al., 1971.
11. Berger, 1968a.
12. Happle and Hoehn, 1973.
13. Laurent, Rosseau, et al., 1968.
14. Ayraud and Kermarec, 1968; Legrand, 1968; Spriggs and Boddington, 1968.
15. Robledo Aguilar, Gomez, et al., 1974.
16. Orye, Delbeke, et al., 1974.
17. Prunieras, Gazzolo, et al., 1968.
18. McAllister, Melnyk, et al., 1969.
19. McAllister, Gardner, et al., 1971.
20. Pontén and Saksela, 1967.
21. Ayraud and Kermarec, 1968; Carlevaro, Rossi, et al., 1978; Legrand, 1968; Mark, 1978; Spriggs and Boddington, 1968.
22. Ishii, 1965.
23. Berger, 1968a; 1969c.
24. Porter, Benedict, et al., 1969.
25. Scappaticci, Lo Curto, et al., 1973.
26. Duran-Troise and Lustig, 1972; Kondo, Muragishi, et al., 1971.
27. Manocha, 1969.
28. Maeda, Tabata, et al., 1964, 1965; Makino, Tonomura, et al., 1959.
29. Benedict, Brown, et al., 1971; Makino, Tonomura, et al., 1959.
30. Katayama, Woodruff, et al., 1972.
31. Becak, Becak, et al., 1963.
32. H.D. Steele, Monocha, et al., 1963.
33. Sonta and Sandberg, 1978a.

Pages 526–534
1. Spriggs, 1964, 1972, 1974; Spriggs, Boddington, et al., 1962a, Spriggs, Bowey, et al., 1971.
2. Koller, 1947; Timonen, 1950.
3. Parmentier and Dustin, 1951.
4. Manna, 1955, 1957a, b, 1962.
5. Bowey and Spriggs, 1968.
6. Makino, Ishihara, et al., 1959a, b.
7. Tonomura, 1959a, b.
8. Spriggs, Boddington, et al., 1962a.
9. Biswas and Chowdhury, 1968; Richart and Wilbanks, 1966.
10. Atkin, 1969a, b; Brandão, 1969; Steele, Monocha, et al., 1963; Wilbanks, Richart, et al., 1967.
11. Atkin, 1967c.
12. Granberg, 1971; H.W. Jones, Woodruff, et al., 1970; Kirkland and Stanley, 1967, 1971; Kirkland, Stanley, et al., 1967; Wakonig-Vaartaja and Hughes, 1965.
13. Atkin and Baker, 1966, 1969; Auersperg and

Wakonig-Vaartaja, 1970; Granberg, 1971; Wakonig-Vaartaja and Hughes, 1965.
14. Atkin and Baker, 1969.
15. Atkin, 1967b, c, 1970b, 1971a, b; Atkin and Baker, 1966; Auersperg and Wakonig-Vaartaja, 1970; P. Fischer, Golob, et al., 1967; Granberg, 1971; Katayama, 1969; Salimi and Jones, 1970; Tortora, 1963, 1969; Wakonig-Vaartaja and Auersperg, 1970.
16. Atkin, 1967b.
17. Spriggs, Boddington, et al., 1962a.
18. Spriggs, Bowey, et al., 1971.
19. Spriggs and Cowdell, 1972.
20. Stanley and Kirkland, 1968, 1969.
21. Granberg, 1971.
22. Granberg, 1971; H.W. Jones, Woodruff, et al., 1970; Kirkland and Stanley, 1967; Wakonig-Vaartaja and Hughes, 1965.
23. Granberg, 1971.
24. Spriggs, Bowey, et al., 1971.
25. Auersperg, Corey, et al., 1967; Boddington, Spriggs, et al., 1965; Cecco, Rugiati, et al., 1966; Dehnhard, Breinl, et al., 1971; Granberg, Traneus, et al., 1972; H.W. Jones, Woodruff, et al., 1970; H.W. Jones, Davis, et al., 1968; Katayama, 1969; Kirlkand, Stanley, et al., 1967; Moricard and Cartier, 1959; Pescetto, 1967; Richart and Corfman, 1964; Spriggs, 1974; Spriggs, Bowey, et al., 1971; Wakonig-Vaartaja and Kirkland, 1965; Wakonig-Vaartaja, 1969; Woodruff, Davis, et al., 1969.
26. Granberg, 1971; Wakonig-Vaartaja, 1969.
27. Atkin and Baker, 1969.
28. Auersperg, Corey, et al., 1967.
29. Spriggs, Bowey, et al., 1971.
30. Granberg, 1971.
31. Kirkland, 1966a.
32. Auersperg, Corey, et al., 1966, 1967.
33. H.W. Jones, Davis, et al., 1968; Spriggs, Bowey, et al., 1971.
34. Spriggs, Bowey, et al., 1971.
35. Kirkland and Stanley, 1967.
36. Wilbanks, Richart, et al., 1967.
37. Brandão, 1969.
38. D. Wagner, Sprenger, et al., 1972.
39. Spriggs, Bowey, et al., 1971.
40. Wakonig-Vaartaja and Kirkland, 1965.
41. Auersperg, Corey, et al., 1967.
42. H.W. Jones, Davis, et al., 1968.
43. Spriggs, Bowey, et al., 1971.
44. Brandão, 1969.
45. Atkin, 1969a, b.
46. Atkin and Richards, 1962; H.W. Jones, Woodruff, et al., 1970.
47. Kaminetzky and Jagiello, 1967.
48. Oksala and Therman, 1974.
49. Atkin and Ross, 1960.
50. Atkin, 1967b.
51. Katayama, 1969; Spriggs, Bowey, et al., 1971.
52. Bowey and Spriggs, 1968.
53. Granberg, 1971.
54. Kirkland, Stanley, et al., 1967; Parmentier and Dustin, 1951.
55. Heneen, Nichols, et al., 1970.
56. Granberg, 1971.
57. Granberg, 1971; Spriggs, Bowey, et al., 1971.
58. Atkin and Baker, 1965b, 1969; Atkin and Richards, 1956; Atkin, Baker, et al., 1967b; E.C. Hughes and Csermely, 1966, Rask-Madsen and Philip, 1970; Stanley, 1969.
59. Cellier, Kirkland, et al., 1970; P. Fischer, Golob, et al., 1966a, b, 1967; H.W. Jones, Katayama, et al., 1967; Richart and Corfman, 1964; Wakonig-Vaartaja, 1963, 1970; Wakonig-Vaartaja and Hughes, 1967.
60. Auersperg and Wakonig-Vaartaja, 1970; Auersperg and Worth, 1966; Borch, Stanley, et al., 1969.
61. Atkin, 1970b; Auersperg and Hawryluk, 1962; Boddington, Spriggs, et al., 1965; Kang, Kim, et al., 1968; S.W. Kim, Kang, et al., 1967; Kirkland, 1966a, b, 1969; Tseng and Jones, 1969.
62. Atkin, 1964a, c; Dixon and Stead, 1977; Jaffurs, Marlow, et al., 1970; Ng and Atkin, 1973; Nishiya, Kikuchi, et al., 1977; Sprenger, Moore, et al., 1973; D. Wagner, 1973; Wilbanks, 1976.
63. H. Sachs and Schittko, 1975.
64. Stanley and Kirkland, 1975.
65. Dehnhard, 1973; Dehnhard, Breinl, et al., 1970, 1971, 1975a–c; Schüssler, Dehnhard, et al., 1977.
66. Atkin and Baker, 1977a.

Pages 535–559

1. Zang and Singer, 1967, 1968a, 1971.
2. Mark, 1970a, 1974c.
3. Mark 1973c; Zankl, Seidel, et al., 1975a, b; Zankl, Weiss, et al., 1975; Zankl and Zang, 1972b; Zang and Zankl, 1975.
4. Mark, 1973a.
5. Mark, 1970a, 1971a–d, 1973a–c, 1974b, c; Mark, Levan, et al., 1972; Mark, Mitelman, et al., 1972; Paul, Porter, et al., 1973; Singer and Zang, 1970; Zang and Singer, 1967, 1968a, b; Zankl and Zang, 1971b, 1972b, 1978b; Zang and Zankl, 1975; Zankl and Love, 1976; Zankl, Singer, et al., 1971; Zankl, Stengel-Rutkowski, et al., 1973; Zankl, Seidel, et al., 1975a, b; Zankl, Weiss, et al., 1975.
6. Mark, 1977a.
7. Mark, 1970b, 1974b, 1977a.
8. The tumor had modes of 38 and 39 in 1970 when examined by a direct preparation and after 6 wk in culture. After 28 mo the frozen cells were thawed and cultured for 3 mo (Paul, Porter, et al., 1973); they were shown to have two

populations: hypodiploid and near-triploid with loss of #22.
9. Mark, 1974c.
10. Singer and Zang, 1970.
11. Mark, Levan, et al., 1972; Mark, Mitelman et al., 1972.
12. Mark, 1974c; Mark, Mitelman et al., 1972.
13. Mark, 1974c.
14. Mark, Levan, et al., 1972; Mark, Mitelman et al., 1972; Zankl and Zang, 1972b.
15. Mark, Levan, et al., 1972; Mark, Mitelman et al., 1975.
16. Mark, 1972a, b, 1973a-c, 1977a; A.F. Weiss, Portmann, et al., 1975; Zankl and Zang, 1972b; Zankl, Seidel, et al., 1975a, b.
17. Mark, 1973c, 1977a.
18. Zankl, Seidel, et al., 1975a, b.
19. Singer and Zang, 1970.
20. Hollinshead, Suskind et al., 1976; A.F. Weiss, Portmann, et al., 1975; A.F. Weiss, Zang, et al., 1976; Zang, Weiss, et al., 1976.
21. Bicknell, 1967; D. Cox, 1968; Dharker, Chaurasia, et al., 1973; Erkman and Conen, 1964a; Hansteen, 1967; Lubs and Salmon, 1965; Mark 1970b, 1971a-d, 1974b; Spriggs, Boddington, et al., 1962b; C.B. Wilson, Kaufman, et al., 1970.
22. Mark, 1974a-c.
23. Pontén and McIntyre, 1968; Westermark, Ponten, et al., 1973; C.B. Wilson and Barker, 1969.
24. Mark, 1971d, 1974b.
25. Lubs and Salmon, 1965; Mark 1971a, b.
26. Conen and Falk, 1967; Erkman and Conen, 1964a; Mark, 1970b, 1971d.
27. Mark, 1970b, 1971d.
28. D. Cox, 1968; Erkman and Conen, 1964a; Lubs, Salmon, et al., 1966; Mark, 1970b, 1971d; McAllister, Isaacs, et al., 1977.
29. Berger, 1968a, b; Brewster and Garrett, 1965; Conen and Falk, 1967; D. Cox, 1968; Gagnon, Dupal, et al., 1962; A. Levan, Manolov, et al., 1968; Makino, Sofuni, et al., 1965; Mark, 1970b; Miles, 1967a; Orye, Delbeke, et al., 1974; Sandberg, Sakurai, et al., 1972; Whang-Peng and Bennett, 1968.
30. Brodeur, Sekhon, et al., 1977.
31. Wakonig-Vaartaja, Helson, et al., 1971.
32. Seeger, Rayner, et al., 1977.
33. D'Angio, Evans, et al., 1971.
34. Sandberg, Sakurai, et al., 1972.
35. Pegelow, Ebbin, et al., 1975.
36. Knudson and Meadows, 1976; Knudson, Strong, et al., 1973.
37. Nakagome, 1965.
38. Mark, 1970c.
39. Mark, 1970b.
40. Mark, 1972a.
41. Mark, 1972a.
42. Mark, 1969b, 1971c.
43. Mark, 1972b, 1974b.
44. Donner and Bubenik, 1968; G. Levan, Mandahl, et al., 1976; Mark, 1967a; Mitelman, Levan, et al., 1972.
45. Larripa, Brieux de Salum, et al., 1975; G. Levan, Mandahl, et al., 1976; Mark, 1967a; Mitelman, Levan, et al., 1972.
46. Balaban-Malenbaum and Gilbert, 1977; Barker and Hsu, 1979; Biedler, Helson, et al., 1973; L. White and Cox, 1967.
47. D. Cox, Yuncken et al., 1965; Lubs, Salmon, et al., 1966.
48. Biedler, 1975a, b; Biedler and Spengler, 1976a, b.
49. Balaban-Malenbaum and Gilbert, 1977.

18

Synoptic View of Specific Chromosome Changes in Human Cancer, Leukemia, and Gene Loci

Involvements of Certain Chromosomes in Human Neoplasia, Particularly Lympho- and Myeloproliferative Disorders

Gene Mapping in Cancer Causation

Involvements of Certain Chromosomes in Human Neoplasia, Particularly Lympho- and Myeloproliferative Disorders

Every chromosome in the human set has been described as being involved in one or another karyotypic anomaly, e.g., trisomy, monosomy, deletion, translocation, etc., in human cancer and leukemia. However, some of the involvements are more frequent (i.e., nonrandom) than others, and in describing the karyotypic aberrations for each chromosome only those significantly associated with a clinical condition or in which the abnormality was of special interest have been included. However, the presentation is by no means exhaustive, and no attempt has been made to cover every clinical case in which a particular chromosome was affected. The abnormalities listed and discussed are thought to be the most frequent and outstanding.

This chapter is intended to serve, at the most, as a short survey and a quick guide to the involvement of various chromosomes in human cancer and leukemia. The reader will have to consult the appropriate chapters and tables for more details regarding the clinical and chromosomal involvement. More space is devoted to those chromosomes (e.g., #1, #14, #17) for which a body of karyotypic information has been accumulated and very little space to those (e.g., #4) infrequently associated with a particular disease.

For some of the chromosomes I have compiled tables illustrating their involvement in various disease states (Table 89), including chromosomes (e.g., #3 and #6) usually not considered to be involved in nonrandom karyotypic changes. In my opinion, the tables indicate otherwise. In all probability, similar tables could be obtained for all other chromosomes.

Chromosome #1

The involvement of #1 (Figures 188–190) in human neoplasia, primarily hematologic disorders, has been summarized by Rowley (1977a, b, 1978 a–d), who has pointed out the susceptibility of particular regions of this autosome to be affected by duplications, deletions, and/or translocations, in addition to numerical changes, in a number of different neoplasias (Tables 90 and 91). We have also been impressed by the frequency with which this chromosome is involved in a number of different solid tumors (Table 92), such as cancer of the breast, malignant melanoma, cancer of the bladder, etc.

Rowley has compiled all patients known to her (from her laboratory and the literature) with hematological diseases and abnormalities of #1. Of the 36 patients, 34 had duplication of part or all of #1. Every patient was trisomic for the region from 1q25 to 1q32. In addition, 5 of these cases were trisomic for all of #1, and 16 had part of #1 translocated to another chromosome, usually with two other normal #1 chromosomes as well. Five patients had an additional chromosome composed only of part of #1, as well as two normal #1 chromosomes. Finally, eight patients had complex structural rearrangements involving only #1 and leading to a duplication of part of the chromosome, usually as a very large marker.

Except for the 5 patients who had an extra intact #1, 29 were trisomic for no more than the proximal one-third of 1p and for 1q. In addition to the trisomy of region 1q25–1q32 present in every patient, trisomy 1q32–qter or from 1q25 to 1q21 was present in 28 and 30 patients, respectively. Except for cases with an extra whole #1, trisomy for 1p and band p22–pter was not seen. The 16 translocation trisomies involved #1 with #4, #6, #9, #11, #13, #15, #17, #22, and the Y. All breaks in #1 were at the centromeric region or in bands q21 or q25. Translocations involving the centromere of #1 usually occurred with acrocentrics #13, #15, or #22, which were also broken at the centromeric areas. The exception involved the translocation with 9p (two cases) and the end of 1p. Six of the eight tandem duplications of 1q had the distal break point in band 1q32; half of the proximal breaks took place at q21 and the others in band q11 or q12.

Two patients had deletions of #1: In one there was a 1;11 reciprocal translocation and a deletion of the same #1 distal to 1q32 and in the other a complex three-way translocation that resulted in the loss of part of #1 from 1p32–pter.

In addition to the 36 cases compiled by Rowley, she refers to 9 with hematologic disorders with reciprocal translocations involving #1, which did not lead to either gain or loss of

Table 89 Chromosome abnormalities associated with hematologic disorders*

Disorder	% of patients abnormal	Trisomy for all or long arm of #1	5q-	-7	+8	8q+ transl.‡	+9	9q+	14q+	15/17 transl. location	17q+	20q-	Ph¹ (22q-)	Double Ph¹
Chronic myelocytic leukemia	85												+	
Acute phase	70§				+								+	+
Acute nonlymphocytic leukemia	50	+			+	+		+		+	+			
APL	100						+			+	+			
PV	25	q+			+			+						
Refractory anemia	15?			+	+							+		
Myelosclerosis with or without myeloid metaplasia	15?	q+			+							+		
Lymphoma									+ (usually 8/14)					

*Adapted from table published by Rowley (1977a–c)
‡Usually to #21; frequently associated with loss of an X- or Y-chromosome.
§Other chromosome abnormalities superimposed on Ph¹-positive cell line.

#1. Four of the 36 patients considered earlier also had balanced translocations of #1 as well as trisomy or deletion. In two cases, the reciprocal rearrangement involved the affected #1. Except for one case, the break points were located in the short arm of #1, specifically in bands p22, p34, and p36.

Of the 36 cases in which the abnormalities of #1 just described occurred, 11 were diagnosed as PV, 6 as MM, 8 as AML, and the remaining cases as lymphoma, refractory anemia, and multiple myeloma.

Based on the involvement of #1 (and #17) in hematological disorders and possibly in some other neoplastic human conditions, Rowley felt that the hematologic disorders can now be correlated with gene loci on these chromosomes or chromosomal segments, in an attempt to identify specific genes that might be related to malignancy. Utilizing the mapping of various human genes she came to the conclusion that it may be significant that chromosomes carrying gene loci related to nucleic acid metabolism are more frequently involved in hematological disorders and other malignancies than are gene loci related to intermediary or carbohydrate metabolism. Furthermore, the known virus-human chromosome associations are closely correlated with the chromosomes affected in hematological disorders. She postulated that if one of the effects of carcinogens, including viruses, is to activate genes that regulate host cell DNA synthesis and if translocations or duplications of specific chromosomal segments produce the same effect, then either of these mechanisms might provide the affected cell with proliferative advantages.

Atkin and his co-workers have stressed the importance of changes involving #1 in various cancers (e.g., cancer of the ovary, cervix, vagina, and bladder).[1] They stressed not only the participation of #1 in numerical changes, including total and partial trisomy, and isochromosome formation, but also the importance of heterochromatin variants of #1, such as variation in size of the C bands and/or the presence of a pericentric inversion in a number of different tumors. Heteromorphism for the C bands of #1, shown mostly in cultured lymphocytes, existed in 41 out of 76 (54%) cases with cancer, as compared with 22 out of 68 (32%) of controls. Pericentric inversion of #1 was present in a minimum of 15% of the patients with cancer and in only 4% of the controls. The above au-

Region Trisomic for Chromosome No. 1 in 34 Patients

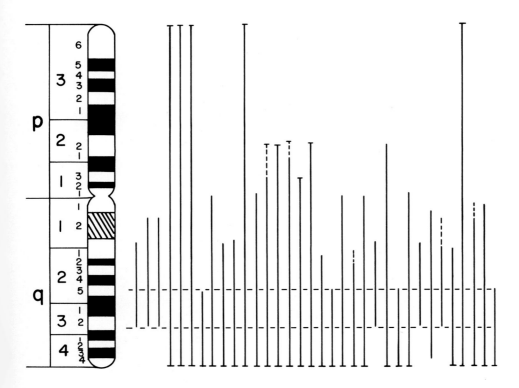

thors speculated that the heterochromatin polymorphism, pericentric inversion, and the risk of cancer are interrelated. How these observations will relate to recent demonstrations of lateral polymorphism of C bands will have to be established with data on large populations and cogent analysis.

Atkin and Baker (1977 $a-c$) stated that involvements of #1 have been found in tumors but not in the CP of CML and myeloproliferative disorders. The frequent involvement of #1 in human cancer was pointed to originally by findings in my laboratory, including cases of cancer of the breast, malignant melanoma, and some myeloproliferative disorders.[2]

In particular, trisomy, either partial or complete, of the long arm of #1 [i(1q)] is a common and nonrandom karyotypic change in solid tumors (Kovacs, 1978) (e.g., neuroblastoma, rhabdomyosarcoma, Wilm's tumor) and may occur in lymphomas.

Areas 1p22 and 1q2(1−3) are thought to be vulnerable points in AML, and a high incidence of 1qht in CML has been reported (Berger and Bernheim, 1977). 1p− and −1 may occur in lymphomas and meningiomas, respectively.

188 The part of #1 that is trisomic in each patient with a different condition, including a number of myeloproliferative and lymphoproliferative disorders, is represented by a vertical line. The dashed vertical lines indicate uncertain break points. A short horizontal bar indicates the end(s) of the chromosome. The dashed horizontal lines enclose the segment that is trisomic in every one of the 34 patients studied.

The sites of breakage show a very curious distribution. Breakage that leads directly to the trisomic state, with tandem duplication, translocation to another chromosome, or the production of an extra 1q chromosome, is confined to bands p22, cen, q12, q21, q25, and q32. All breaks in the short arm of #1 (p13 and p22) lead to an extra, separate chromosome that results in trisomy for the entire long arm. Breaks at the centromere usually involve a translocation with the long arm of an acrocentric chromosome. Of the other translocations that lead directly to trisomy for part of 1q, all have breaks at q21 or q25. All of the six breaks seen in band 1q32 were identified in patients with tandem duplications of #1.

(Rowley, 1977d.)

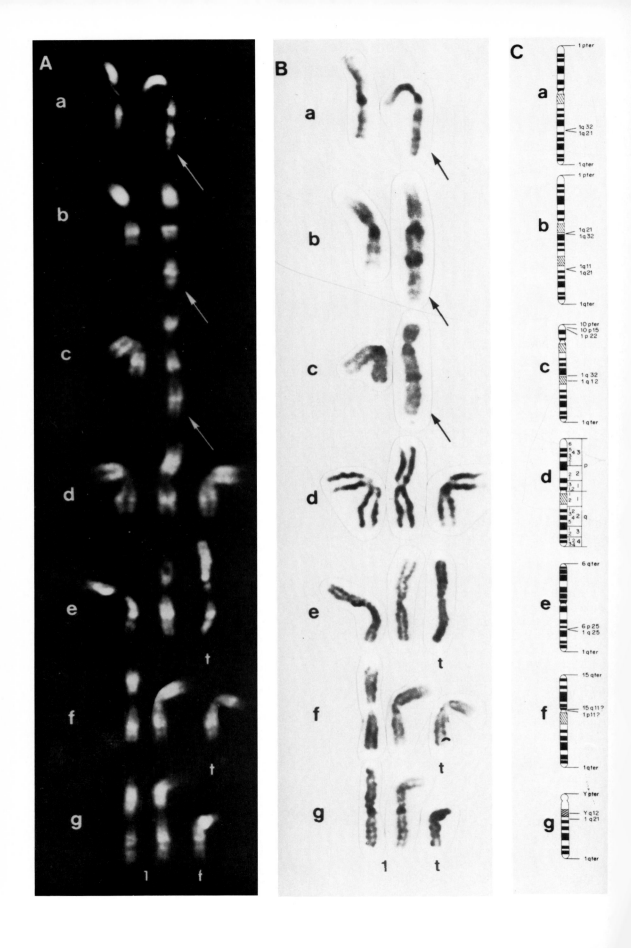

189 Chromosomes #1 from seven different patients: A. Stained with quinacrine mustard. B. Same chromosomes stained with conventional Giemsa or after sodium hydroxide (NaOH) treatment to reveal constitutive heterochromatin. C. Diagrammatic presentation of the chromosomes according to the Paris nomenclature (1972) of the arrangements involving #1. (a) Myelofibrosis; the abnormality results in the duplication of 1q21 to 1q32. There is only one block of constitutive heterochromatin at the centromere. (b) Histiocytic lymphoma; due to an inversion of the duplicated segment 1q11 to 1q32 between two blocks of constitutive heterochromatin. (c) Plasma cell leukemia; the duplicated segment of 1q11 to 1q32 is between two blocks of constitutive heterochromatin. (d) AML; an extra intact #1 is present. Diagram shows complete numbering system for a normal #1. (e) Myelofibrosis; a translocation between the end of the short arm of chromosome #6 (p25) and chromosome #1 (q25). Patients (e), (f), and (g) also had two normal chromosomes #1. (f) AML after PV; the entire long arm of #1 is translocated to the long arm of #15. (g) PV; part of the long arm of #1 (q21) has been translocated to the end of the Y-chromosome.
(Rowley, 1977d.)

190 A diagramatic representation of #1 showing the number of breaks in a particular band related to sites of breaks in adenovirus 12-infected cells (first vertical column of numbers). Chromosome #1 is the site of the highest and #4 the lowest frequency of breaks. Also shown are break points leading to partial trisomy (second vertical column) and those involved in balanced reciprocal translocations (third column). The break points for structural rearrangements noted in inborn chromosomal abnormalities are shown in the next column and that portion of #1 that is trisomic in hematological disorders is shown by a solid line and the dashed line represents trisomy in most patients. The location of selected gene loci on #1 are also shown.
(Rowley, 1977c.)

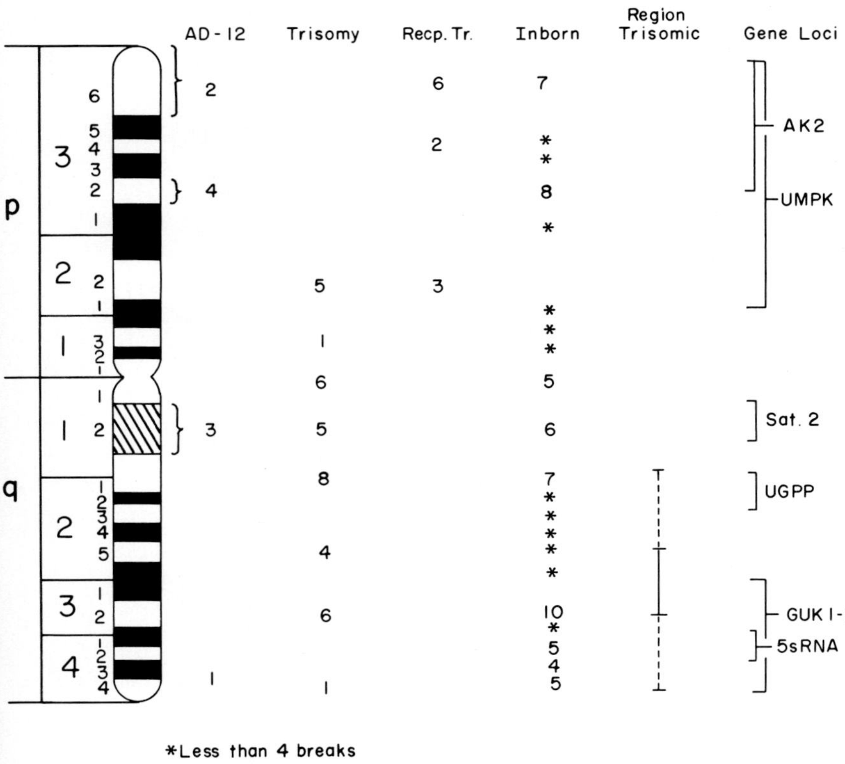

*Less than 4 breaks

Table 90 Involvement of chromosome #1 in translocation in lympho- and myeloproliferative disorders

Disorder	Karyotype	Reference
AML	t(1;4)(q22→25;p14→16)	Oshimura, Hayata, et al., 1976;
	t(1;8)(p22;q24)	Oshimura, Sonta, et al., 1976
LS → AML	t(1;9)(q22or23;p24)	
	t(1;13)(p36;q14)	
ANLL	rcp(1;3)	P. Philip, 1977a
	t(1p—;?)(1qter→1p22::?)	
	rcp(1;17)	
	t(1;1)(1pter→1q42::1q21→1qter)	
	t(1;14)(14qter→14pter::1pter→1p22)	
	t(1;15)(15qter→15pter::1pter→1p22)	
	t(1;?)(1qter→1p3::?)	
BP of CML	t(1;9)(q11;p11—13)	Oshimura, Sonta, et al., 1976
	+2 normal #1	
	t(1;9)	Sonta and Sandberg, 1978b
	t(1q;17q)	Prigogina, Fleischman, et al., 1978
ALL	t(1q+;10q—)	Castoldi, Grusovin, et al., 1976
PV	t(1;3)	Westin and Weinfeld, 1977;
	t(1;5)(p36;q31)	Westin, Wahlström, et al., 1976
	t(1;13)	L.Y.F. Hsu, Pinchiaroli, et al., 1977
CML	t(1q—;1q+)	Meisner, Inhorn, et al., 1973
Lymphoma	t(1;10)(q22;qter)	Fleischman and Prigogina, 1977
	t(1q;11p)(also 1q— and 1p—)	
	1q/i(17q)(also 1p+)	Fleischman and Prigogina, 1975
HD	t(X:1)(q28;q22)	Hossfeld and Schmidt, 1978
	t(1;5)	
AILD	t(1p—;2p+)	Goh and Bakemeier, 1978
Myelosclerosis	t(1;6)	L.Y.F. Hsu, Pinchiaroli, et al., 1977
Untreated PV	t(1;9)(q22;q13)	Westin, Wahlström, et al., 1976
Treated PV	t(1q;15q)	Wurster-Hill, Whang-Peng, et al., 1976; Rowley, 1975a
PV → CML	t(1;8)	Stavem, vander Hagen, et al., 1975
Myelofibrosis with MM	t(1;6)(q25;q25)	Rowley, 1975a
PV → AML	t(1;15)(cen;cen)	Rowley, 1975a
PV	t(1;4)(q21;q12)	Rowley, 1975a
AL in FA	t(1q+;10p—)	Dosik and Verma, 1977
AML	t(1;6;11)(q12;q23;p15)	Yamada and Furasawa, 1976
	t(1;22)(q11;p11or12)	Oshimura, Hayata, et al., 1976
	t(1;9)(q22or23;p24)	Oshimura, Hayata, et al., 1976
	t(1;13)(p36;q14)	Oshimura, Hayata, et al., 1976
	t(1;X)(p1→q2;Xq)	J.H. Ford, Pittman, et al., 1975
	t(1;13)	Kaneko and Sakurai, 1977
	t(1;3)	Barlogie, Hittelman, et al., 1977
Blast cell leukemia	t(1q+;5q—)	Soderström, 1974
Myelofibrosis	t(1p+;4q—;7q—)	Nowell, Jensen, et al., 1976a
	t(1q;9p)	Nowell, Jensen, et al., 1976a
	t(1p;9q)	Nowell, Jensen, et al., 1976a
Acute myelofibrosis	t(1q+;3p—)	Slyck, Weiss, et al., 1970
PV (treated)	t(1;Y)(q21;q12)	Rowley, 1973c
RCS of brain	t(1;14)(q23;q32)	Yamada, Yoshioka, et al., 1977
Ph¹ + CML	t(1;20)	Norman and Boucher, 1978

Table 90—*continued*

Disorder	Karyotype	Reference
Ph¹ + CML with PV	t(1;3)	Pris, Launais, et al., 1978
AML	tan(1;13)	T. Abe and Misawa, 1978
Myelofibrosis → AML	t(1;6)(q23;p21)	Gahrton, Friberg, et al., 1978a,b
Preleukemia → AMMoL	+t(1;11)(q11or12;q25)	Najfeld, Singer, et al., 1978
Myelofibrosis	+t(1q2q)	Whang-Peng, Knutsen, et al., 1978
Histiocytic lymphoma	t(1;7)(7pter→7q32::1p22→1pter)	Mark, Ekedahl, et al., 1978
	and t(1;21)(21pter→21q22::1q22→1qter)	Mark, Ekedahl, et al., 1978
	t(1;18)(1qter→1p36::18q21→18qter)	Mark, Ekedahl, et al., 1978
	t(1q17q)	Mark, Ekedahl, et al., 1978
LS	t(1;9?)(p32;q11?)	Fukuhara, Rowley, et al., 1979
	t(1;14)(p21;q13)	Fukuhara, Rowley, et al., 1979
MM	t(1;10)	Liang, Hopper, et al., 1979
ALL	t(1;13)(q12.2;p13)	Roth, Cimino, et al., 1979
BP of CML	t(1;8)(q25;qter)	Alimena, Brandt, et al., 1979
ANLL	t(1;2)(2qter-2cen::1q2-1qter)	Philip, Krogh Jensen, et al., 1978
AL	t(1;22)(q21;p11)	Humbert, Morse, et al., 1978
PV → AML	t(1;22)(p12;q13)	Berger and Bernheim, 1979
HL	t(1;17),t(1;19)	Mark, Dahlenfors, et al., 1979

Table 91 Trisomy of chromosome #1 or its total [iso(1q)] or partial long arm and other structural changes of #1 in lympho- and myeloproliferative disorders

Disorder	Reference
Myeloma	Spriggs, Holt, et al., 1976; P. Philip, 1975c; P. Philip, Müller-Berat, et al., 1976; P. Philip, 1977b(6cases); P. Philip and Drivsholm, 1974
Two cases of dysproteinemia	Elves and Israëls, 1963
Lymphoma	Miles, 1967b, Coutinho, Bottura, et al., 1971; Castoldi, 1973b; Reeves, 1973 (2 cases); Mark, Ekedahl, et al., 1976; Zech, Haglund, et al., 1976 (cell line); Fleischman and Prigogina, 1975
Mycosis fungoides	Erkman-Balis and Rappaport, 1974
BL cell line	Miles and O'Neill, 1967; Gripenberg, Levan, et al., 1969; Toshima, Takagi, et al., 1967
Histiocytic lymphoma	Mark, 1975b, 1977b; Brynes, Golomb, et al., 1978
Myelofibrosis	L.Y.F. Hsu, Pinchiaroli, et al., 1977; Rowley, 1975a; Nowell, Jensen, et al., 1976a
PV	Rowley, 1975a,b; Wurster-Hill, Whang-Peng, et al., 1976; Visfeldt, Franzén, et al., 1973; L.Y.F. Hsu, Alter, et al., 1974, Westin, Wahlström, et al., 1976; Lawler, Millard, et al., 1970
AML	Rowley, 1975a; Oshimura, Hayata, et al., 1976 (3 cases); Hungerford, 1961; Sakurai and Sandberg, 1976a (2 cases); Prieto, Badia, et al., 1978a
BP of CML	Oshimura, Sonta, et al., 1976; Hayata, Sakurai, et al., 1975; Sonta and Sandberg, 1978a (2 cases)

continued

Table 91 Trisomy of chromosome #1 or its total [iso(1q)] or partial long arm and other structural changes of #1 in lympho- and myeloproliferative disorders—*continued*

Disorder	Reference
RDA	Warburton and Bluming, 1973; Yamada and Furasawa, 1976
AL	Zuelzer, Inoue, et al., 1976
EL	P. Philip, 1975a; Sakurai and Sandberg, 1976b; Kroll and Schlesinger, 1970
ALL	Sandberg, Ishihara, et al., 1962a; Sandberg, Ishihara, et al., 1964b (2 cases); Yamada and Furasawa, 1976
HD	Hossfeld, 1978
Sézary syndrome	Lutzner, Emerit, et al., 1973
HD	Hossfeld, 1978
1p—	
1q—	
Sézary syndrome	Dewald, Spurbeck, et al., 1974
−1,−1	
ALL	Sandberg, Ishihara, et al., 1962a, 1964b
1p—	
Histiocytic lymphosarcoma	Mark, 1977b
1p—(p31p35)	
ANLL	Hagemeijer, Smit, et al., 1977
1p—	
Untreated PV	Westin, Wahlström, et al., 1976
+del(1)(p21?)	
PV → AML	Rowley, 1975a
Myelofibrosis	Nowell, Jensen, et al., 1976a
1q+,1q—	
BP of agnogenic myeloid metaplasia (1q+)	Cehreli, Ezdinli, et al., 1976
Blast cell leukemia	Soderström, 1974
1q+	
AL	J.H. Ford, Pittman, et al., 1975
1p—	
BL	Zech, Haglund, et al., 1976
Myelofibrosis → AML	Gahrton, Friberg, et al., 1978a,b
Preleukemia → AMMoL	Najfeld, Singer, et al., 1978
CML	Lilleyman, Potter, et al., 1978
1q— [del(1)(q22)] and HL	Mark, Ekedahl, et al., 1977
1p—	
LS	Fukuhara, Rowley, et al., 1979
+1p—	
LS	Fukuhara, Rowley, et al., 1979
1p+	
HL (3 cases)	Fukuhara and Rowley, 1978
1q+	
LS	Fukuhara and Rowley, 1978
1q—,+1p—	
LS (2 cases)	Fukuhara and Rowley, 1978
1p+	
MF	Fukuhara and Rowley, 1978
1p—	
MM → AML	Liang, Hopper, et al., 1979
+1	
MM	Liang, Hopper, et al., 1979
1q+	

Table 91 — *continued*

Disorder	Reference
MM (PCL) inv ins(1)	Liang, Hopper, et al., 1979
MM (2 cases) +1	Liang, Hopper, et al., 1979
MM del(1)(p36)	Liang, Hopper, et al., 1979
MM 1q−	Liang, Hopper, et al., 1979
AUL 45,XY,−1	Shiloh, Naparstek, et al., 1979

BP of CML
  del(1)(q21),
  dup(1)(q21)
	Alimena, Brandt, et al., 1979

BP of CML 1p+ (2 cases)	Alimena, Brandt, et al., 1979
BP of CML del(1)(p11)	Alimena, Brandt, et al., 1979
Acute myelofibrosis	Nowell and Finan, 1978a,b
BP of CML +1	Dallapiccola and Alimena, 1979
AL (2 cases) 46,XY,dup(1)(q31q41)	Humbert, Morse, et al., 1978
AML 1q+	Van Den Berghe, David, et al., 1978a
BP of CML del(1)(q12)	Lyall and Garson, 1978

AUL = acute undifferentiated leukemia; RDA = refractory dysplastic anemia; HL = histiocytic lymphoma; PCL = plasma cell leukemia.

Chromosome #2

This autosome has not been found to be involved with any frequency in any particular human neoplasia, though some think that the centromeric area of #2 is a vulnerable point in AML.

Chromosome #3

The vulnerability of several bands of #3 in lympho- and myeloproliferative disorders has been stressed by P. Philip (1975d.) Table 93 gives a compilation of conditions in which #3 has been involved either in translocation or other changes.

Possible involvement of #3 in conditions characterized by DIC (diffuse intravascular coagulation) has been stressed by Sweet, Golomb, et al. (1979). Trisomy of #3 may be seen in cancers and multiple myeloma, and deletions in lymphoma.

Chromosome #4

This autosome has not been described to be characteristically involved in any neoplastic condition. Translocation between #4 and other chromosomes occur in leukemia.

Chromosome #5

An interstitial deletion of #5 has been described in refractory anemias and has been shown to involve a significant number of patients with a somewhat characteristic clinical and hematological picture. More recently, several cases of AML, EL, and AMMoL with the

Table 92 Trisomy of long arm of chromosome #1 [i(1q)], translocations involving #1, q+, and other structural anomalies of #1 in various cancers

Cancer	Reference
Cancer of breast	Kakati, Hayata, et al., 1976; Kakati, Oshimura, et al., 1976; Nelson-Rees, Flandermeyer, et al., 1975 (culture 5/6)
	Jones Cruciger, Pathak, et al., 1976 (culture) (7 cell lines); Seman, Hunter, et al., 1976 (culture, 1 cell line); Cailleau, Olivé, et al., 1978 (12 cell lines with 1q+ and 1 with +1; 7 of the lines previously described)
Leiomyosarcoma	Mark, 1976
Lung cancer	Kakati, Hayata, et al., 1975; Kakati, Oshimura, et al., 1976; Pickthall, 1976
Cancer of stomach	Granberg, Gupta, et al., 1973
Cancer of ovary	Tiepolo and Zuffardi, 1973
Neuroblastoma (1p−)	Brodeur, Sekhon, et al., 1977
Cancer of rectum (1p+)	Sonta and Sandberg, 1978b
(1q+)	Sonta, Oshimura, et al., 1977
Cancer of breast del(1)(pter→q32)	Kovacs, 1978
Melanoma del(1)(q21::q25→qter) Ins 1(q32::q21→q32::q32→qter)	Kakati, Song, et al., 1977
Cancer of colon	Miles, 1967a
Cancer of breast t(1;20)(1p12→pter::20qter→pter) t(1;?)(qter→pter::?) t(1;14)(1qter→q21::14p12→qter) t(dic(1;3)(1qter→1p13::3qter→pter)	Kovacs, 1978
Undifferentiated tumor of limb t(1;?)(qter→q21::?) t(1;13)(1qter→q21::13p12→qter)	Kovacs, 1978
Cancer of colon (large bowel) t(1;14)(q11?;q24?) t(1;17)(p36;q21)	Sonta and Sandberg, 1978b
Melanoma t(1;9)(p22;q34) t(1q;7q)	Kakati, Song, et al., 1977
Cancer of colon i(1q)(qter→p11::p11→qter)	Kovacs, 1978
Cancer of breast i(1q)(qter→p11::pll→qter) i(1q)+2 normal #1 i(1q)(qter→pll::pll→qter)	Kovacs, 1978
Melanoma i(1q)	Kakati, Song, et al., 1977
Cancer of ovary i(1q)	Kakati, Oshimura, et al., 1976
Undifferentiated cancer i(1q)	Kakati, Oshimura, et al., 1976
Cancer of ovary 1q− 1q+ inv(1) 1p− 1p+	Atkin and Pickthall, 1977
Cancer of bladder 1q−,1p−,1q+,1p+,i(1q)	Atkin and Baker, 1977c
Cancer of cervix 1q+,1p+,q−,1p−,i(1q)	Atkin and Baker, 1977a

Table does not include numerical changes (losses or gains) of the whole chromosome #1, a rather frequent finding in human cancers.

Table 93 Changes involving chromosome #3, including trisomy, monosomy, deletions, additions, and translocations

Karyotypic change	Diagnosis	Reference
	TRISOMY AND MONOSOMY	
+3	AML	Sandberg, Cortner, et al., 1966 (2 cases); Mitelman, personal communication; Sakurai and Sandberg 1976a,c (2 cases); J.H. Ford, Pittman, et al., 1975
+3	CML	Mitelman, Levan, et al., 1976b; Rowley, 1977a; Prigogina, Fleischman, et al., 1978; Stoll, Oberling, et al., 1978
+3	ALL	Sandberg, Takagi, et al., 1968c (2 cases); Oshimura, Freeman, et al., 1977b
+3	Lymphoma	Prigogina and Fleischman, 1975b (2 cases); Mark, Ekedahl, et al., 1977; Van Den Berghe, Michaux, et al., 1977a
+3	EL	Sakurai and Sandberg, 1976b (2 cases)
+3	AMMoL	Zuelzer, Inoue, et al., 1976
+3	Preleukemia	Panani, Papayannis, et al., 1977
+3	AILD	Hossfeld, Höffken, et al., 1976
+3	Cancer of colon	Sonta and Sandberg, 1978b
+3	Ewing's sarcoma	Sonta and Sandberg, 1978b
+3	Seminoma of testis	Sonta and Sandberg, 1978b
+3	Liposarcoma	Sonta and Sandberg, 1978b
+3	AL	J.H. Ford, Pittman, et al., 1975
+3	ANLL	P. Philip, 1977a and personal communication
+3	LS	Fleischman and Prigogina, 1977 (3 cases)
+3	BL	Zech, Haglund, et al., 1976
+3	RCS of brain	Yamada, Yoshioka, et al., 1977
+3	Multiple myeloma	Liang, Hopper, et al., 1979
+3	Hairy cell leukemia	Golomb, 1978a
+3	PV	Nowell and Finan, 1978a,b
+3 (8 cases)	Lymphoma	Mark, Dahlenfors, et al., 1979
−3	AML	Mitelman, personal communication; Yamada and Furusawa, 1976; Rowley, 1977c; J.H. Ford, Pittman, et al., 1975
−3	EL	Sakurai and Sandberg, 1976b (3 cases)
−3	ANLL	P. Philip, 1977a and personal communication
−3	PV→AL	Lawler, Millard, et al., 1970
−3	Lymphosarcoma	Fukuhara, Rowley, et al., 1979
−3	Lymphoma→AL	Whang-Peng, Knutsen, et al., 1979
−3	PV	Nowell and Finan, 1978a,b
−3	Pancytopenia	Nowell and Finan, 1978a,b
−3	BP of CML	Lyall and Garson, 1978
iso 3q	Lymphoblastoid cell line from tumors	Zech, Haglund, et al., 1976
3q+	Lymphoma	Zech, Haglund, et al., 1976
Inv(3)	AML	Rowley, 1976d
Inv(3)(qter→q25::p13→25::p13→pter)	Cancer of colon	Sonta and Sandberg, 1978b
inv(3)(pter→p24::q13→p24::q13→qter)	Mesothelioma	Mark, 1978
Ins(3;3)	AML	Golomb, Vardiman, et al., 1978

continued

Table 93 Changes involving chromosome #3, including trisomy, monosomy, deletions, additions, and translocations—*continued*

Karyotypic change	Diagnosis	Reference
\multicolumn{3}{c}{DELETIONS AND ADDITIONS}		

Karyotypic change	Diagnosis	Reference
del(3)(p15p23)	Histiocytic LS	Mark, 1977b
del(3)	Nonendemic BL	P. Philip, Krogh-Jensen, et al., 1977
del(3)	Lymphoma	Reeves, 1973 (5 cases)
del(3)(pter→q21)	AML	Sasaki, Okada, et al., 1976
del(3)(3cen–3qter)	ANLL	P. Philip, 1977a and personal communication
del(3)(q11), del(3)(q21) del(3)(qter→p25)	Lymphoma (T cell)	Reeves and Stathopoulos, 1976
del(3)(q33)	Histiocytic lymphoma	Mark, Ekedahl, et al., 1978
del(3)(q11)	Histiocytic lymphoma	Mark, Ekedahl, et al., 1978
del(3)(p14)	LS	Mark, Dahlenfors, et al., 1979
del(3)(q21)	LS	Mark, Dahlenfors, et al., 1979
del(3)(q21–q22)	HL	Mark, Dahlenfors, et al., 1979
i(3p)	Histiocytic lymphoma	Mark, Ekedahl, et al., 1978
3q+,3q−	Histiocytic lymphoma	Mark, Ekedahl, et al., 1978
3q− and 3q+	AL	Lawler, Secker Walker, et al., 1975 (1 case each)
+3p+q−	Lymphoma→AL	Whang-Peng, Knutsen, et al., 1979
3q−	HD	Hossfeld, 1978
3q−	BP of CML	Knuutila, Helminen, et al., 1978
3p−	ANLL	Mitelman, personal communication
3p−	APL	Teerenhovi, Borgstrom, et al., 1978
3p−	Epidermoid cancer of uterus	Ayraud, 1975
3p+	Histiocytic lymphoma	Fukuhara and Rowley, 1978
3q+	Lymphosarcoma	Fukuhara, Rowley, et al., 1979

TRANSLOCATIONS

Karyotypic change	Diagnosis	Reference
ins(3;3)(q21;q21q25)	AMMoL	Rowley, 1977a–c
ins(3:3?)(3qter→3p14::3p21→3p25::3p14→3pter)	Histiocytic lymphoma	Mark, Ekedahl, et al., 1978
t(3;5)(q25;q33?)	AMMoL	Rowley, 1977a–c
t(2;3)(q31;q27)	AML	Oshimura, Hayata, et al., 1976
t(3;5)(q21;q31)	AML	Oshimura, Hayata, et al., 1976
rcp(1q−;3q+)	AML	P. Philip, 1976
t(1;3)	AML	Barlogie, Hittelman, et al., 1977
t(1;3;11)	AL	Ruutu, Ruutu, et al., 1977a,b
t(3;17)(q21;p1?)	BP of CML	Hayata, Sakurai, et al., 1975
t(3;11)	CML	Rowley, 1977a–c
t(3;6)(6qter→6p12::3p21→3pter)	Preleukemia	Panani, Papayannis, et al., 1977
t(3p−;Dq+)	Myelofibrosis	Merker, Schneider, et al., 1968
rcp(3;5)	Myeloma	P. Philip and Drivsholm, 1974; P. Philip, 1975d
t(1;3)	Myeloma	P. Philip, 1975d
t(3;5)(q29;q31)	HD	Hossfeld, 1978
t(3q+;5q−)	Myelofibrosis	Hagemeijer, Smit, et al., 1977
t(3;12)	ANLL	Mitelman, personal communication
t(2q;3p)	AL	J.H. Ford, Pittman, et al., 1975
t(3;7)	ANLL	P. Philip, 1977a, and personal communication

Table 93—continued

Karyotypic change	Diagnosis	Reference
TRANSLOCATIONS		
t(3p+;?)(p21or23?)	HD	Reeves, 1973
t(3p;13)	Lymphoblastoid cell line from tumors	Zech, Haglund, et al., 1976
t(3;3)	ALL	Barlogie, Hittelman, et al., 1977
t(4q;3p)	Sézary syndrome	Bosman, van Vloten, et al., 1976
t(6q;3q)	Sézary syndrome	Bosman, van Vloten, et al., 1976
t(1;3)	Ph¹ + CML with PV	Pris, Launais, et al., 1978
t(3;9)(q29;p21)	MM→AML	Liang, Hopper, et al., 1979
t(3;8)(3pter→3q22::8q11→8qter)	Histiocytic lymphoma	Mark, Ekedahl, et al., 1978
t(3;12)(3pter→3q13::12q15→12qter)	Histiocytic lymphoma	Mark, Ekedahl, et al., 1978
t(3;12),t(3;18)	APL	Teerenhovi, Borgstrom, et al., 1978
t(3q+;3q−)	Acute myelofibrosis	Nowell and Finan, 1978a,b
t(3q9q),t(3q9p)	HL	Mark, Dahlenfors, et al., 1979

same anomaly, either seen pristinely in the disease or as a progression from the anemia, have been described. Table 27a shows a number of other conditions with 5q− (not all interstitial). Apparently, frequent involvement of #5 (including its loss from the karyotype) has been described in AML in association with lymphoma or multiple myeloma. We have seen (S. Abe, Kohno, et al., 1979) a case of ALL with 5q− (interstitial) (Figure 190). A 5q− may be seen in PV.

A −5 is not infrequently seen in AML developing in association with other diseases, e.g., multiple myeloma, lymphosarcoma, and HD.

Whatever the cellular mechanics are that lead to long arm deletion, e.g., 5q−, 22q− and 6q−, they ultimately result in and/or are associated with the loss of the chromosome involved in not a small number of tumors or conditions. For example, the loss of #22 in meningioma is often preceded by 22q−, as is the loss of #5 by 5q− in some cases of AML and −6 by 6q− in lymphoma or ALL.

Chromosome #6

A compilation of involvement of #6 in a number of disorders is shown in Table 94. It seems that 6q− is not an infrequent anomaly in the cells of ALL and lymphoma.

Chromosome #7

Some of the conditions in which #7 is involved and a partial list of references dealing with −7 are shown in Table 95. The most common involvement is that of −7 in AML, which may be preceded by 7q−. The latter may be seen in AMMoL and PV. An i(7q) has been described in ALL and in some cancers. Occasionally, −7 may be seen in CML (BP), myelofibrosis, EL, myeloproliferative disorders in children and in patients with megaloblastic erythroid features and disturbance of myeloid maturation. 7q22 may be a vulnerable point in AML. A +7 has been observed in AML and CML (with +12).[3]

Chromosome #7 (and/or #17) has been implicated, on the basis of laboratory data, as playing a role in the development of malignancy (Croce, 1976, 1977). Faulty chemotaxis associated with monosomy of #7 has been reported by Ruutu, Ruutu, et al. (1977a).

Chromosome #8

Probably the most frequently involved chromosome in human neoplasia is #8; it is certainly true for leukemias. Thus, +8 is common in AML and CML (both Ph¹-negative and Ph¹-positive) and is seen not infrequently in myelofibro-

Table 94 Involvement of chromosome #6 in lympho- and myeloproliferative disorders

Karyotypic change	Diagnosis	Reference
TRANSLOCATIONS		
t(6q;17)	BL cell line	Zech, Haglund, et al., 1976
t(1;6)(q25;q22)	Myelosclerosis	L.Y.F. Hsu, Pinchiaroli, et al., 1977
t(3;6)	Preleukemia	Panani, Papayannis, et al., 1977
t(1;6;11)(q12;q23;p15)	AML	Yamada and Furusawa, 1976
t(6;9)(p23;q34)	AML	Rowley, 1977a (2 cases)
t(6;11)		Fleischman and Prigogina, 1975
t(6;11)	ANLL	P. Philip, 1977a
t(6;14)	ANLL	P. Philip, 1977a
t(6;9;22)(q21;q34;q11)	CML—Ph^1+	Potter, Sharp, et al., 1975
t(6;10)	CML—Ph^1+	Van Den Berghe, personal communication
t(6;22)	CML—Ph^1+	Grinblat, Mammon, et al., 1977
t(6;22)	CML—Ph^1+	Berger, Gyger, et al., 1976
t(6;21)(p25;q11)	RCS of brain	Yamada, Yoshioka, et al., 1977
t(6;7)	Histiocytic lymphoma	Mark, Ekedahl, et al., 1978
t(1;6)(q23;p21)	Myelofibrosis→AML	Gahrton, Friberg, et al., 1978a,b
t(6;18)(p25;q21)	ALL (B cell)	Roth, Cimino, et al., 1979
t(6;8)(q34;q11)	BP of CML	Alimena, Brandt, et al., 1979
t(6q;12q)	Pancytopenia	Nowell and Finan, 1978a,b
DELETIONS AND ADDITIONS		
6q—	HD	Hossfeld, 1978
6q—	Lymphoma	Reeves, 1973 (4 cases); Mark, Ekedahl, et al., 1978 (2 cases)
6q—	Undifferentiated leukemia	Lawler, Secker Walker, et al., 1975
del(6)(q13)	ALL	Lawler, Secker Walker, et al., 1975
6q—	ALL	Zuelzer, Inoue, et al., 1976
6q—	Ph^1+/CML	Prigogina and Fleischman, 1975a
6q—	Ph^1+/CML	E. Engel, McGee, et al., 1975a
6q—	ALL	Oshimura, Freeman, et al., 1977b
del(6)(q15or11)	HD→APL	Rowley, Golomb, et al., 1977b
del(6)(q13)	Lymphoma→preleukemia	Rowley, Golomb, et al., 1977b
6q—	ALL	Cimino, Kinnealey, et al., 1977
6q—	Lymphoma	Boecker, Hossfeld, et al., 1975
6q—	Histiocytic lymphoma	Fukuhara and Rowley, 1978
6q—	Lymphosarcoma	Fukuhara and Rowley, 1978
6q— (q21)	Retinoblastoma	Hossfeld, 1978
6q—	Histiocytic lymphoma	Kaiser-McCaw, Epstein, et al., 1977a
6q+	Lymphosarcoma	Fukuhara and Rowley, 1978
6q— (q25)	Multiple myeloma	Liang, Hopper, et al., 1979
inv(6)(p21q13)	Multiple myeloma	Liang, Hopper, et al., 1979
del(6)(q13)	Lymphosarcoma	Fukuhara, Rowley, et al., 1979
6p—	Lymphoma	Mark, Ekedahl, et al., 1976, 1978 (2 cases)
6p—	AML	Golomb, Vardiman, et al., 1978
6p+	Ph^1+/CML	Hagemeijer, Smit, et al., 1977
6p+	ALL	Oshimura, Freeman, et al., 1977b
6p+	Histiocytic lymphoma	Mark, Ekedahl, et al., 1978
del(6)(q13)	BP of CML	Alimena, Brandt, et al., 1979
6p+	Pancytopenia	Nowell and Finan, 1978a,b
6q—	Pancytopenia	Nowell and Finan, 1978a,b
6q— (3 cases)	Lymphoma	Mark, Dahlenfors, et al., 1979

Table 94 — continued

Karyotypic change	Diagnosis	Reference
\multicolumn{3}{c}{MONOSOMY AND TRISOMY}		
+6	Lymphoma	Reeves, 1973 (2 cases); Mark, Ekedahl, et al., 1978
+6	AA	Geraedts and Haak, 1976
+6	AML	Zuelzer, Inoue, et al., 1976
+6	Ph¹+/CML	Prigogina and Fleischman, 1975a (2 cases)
+6	Ph¹+/CML	E. Engel, McGee, et al., 1975b
−6	AML	Kohn, Manny, et al., 1975
+6	ALL	Oshimura, Freeman, et al., 1977b
−6	ALL	Oshimura, Freeman, et al., 1977b
+6	Ph¹+/CML	Sonta and Sandberg, 1978a (2 cases)
+6	Lymphoma	Van Den Berghe, David, et al., 1976a
−6	Lymphoma	Dewald, Spurbèck, et al., 1974; Mark, Ekedahl, et al., 1978 (2 cases)
−6	Lymphoma	Fleischmann, Håkansson, et al., 1976
+6	Lymphoma	Kaiser-McCaw, Epstein, et al., 1977a
6q+	Cell line (AML?)	Venuat, Dutrillaux, et al., 1977
+6	AL (undifferentiated) AML	J.H. Ford, Pittman, et al., 1975
+6	ANLL	P. Philip, 1977a (2 cases)
−6	ANLL	P. Philip, 1977a (2 cases)
+6	AML	Barlogie, Hittelman, et al., 1977
−6	Lymphosarcoma	Fukuhara, Rowley, et al., 1979
−6	Glioma cell lines	Mark, Ekedahl, et al., 1977
+6	BP of CML	Alimena, Brandt, et al., 1979
−6	Histiocytic lymphoma	Fukuhara, Rowley, et al., 1978b
+6	AUL	Prosser, Bradley, et al., 1978
−6	Pancytopenia (2 cases)	Nowell and Finan, 1978a,b
+6	AL	Humbert, Morse, et al., 1978
+6 (3 cases)	Lymphoma	Mark, Dahlenfors, et al., 1979

sis, PV, pancytopenia, AMMoL, sideroblastic anemia (SBA), essential thrombocythemia (ET) (with +9), lymphoma, and multiple myeloma; it has been described in gastric polyps and other tumors. Chromosome #8 participates in the prototypic translocation of AML and AMMoL [t(8;21)]. A missing #8, a rather rare phenomenon, has been seen in meningioma and also in rare cases of AL.[4] The 8q22 area is thought to be vulnerable in AML. Table 96 shows the various conditions in which +8 is present and references.

A t(8;14) is rather characteristic for BL, though it may be seen in other forms of lymphoma.

The induction by EB-virus of a t(8q−;14q+) in the blood lymphocytes from a 17-yr old male patient with AT has been reported (Jean, Richer, et al., 1979). Since the translocation seemed to be identical to that found in BL, the authors suggested that in vitro transformation of lymphocytes from patients with genetic predisposition to chromosome damage favors the development of a chromosome rearrangement observed in lymphoid malignancies.

Chromosome #9

Trisomy of #9 occurs in the CP or BP of CML, PV, myelofibrosis, or myelosclerosis, AMMoL, and ET (with either +8 or +10) (Table 97). In > 90% of Ph¹-positive CML cases, #9 is involved

Table 95a Some hematological conditions associated with monosomy #7 reported in the literature (CML not included)

Diagnosis	Reference
EL	Pegoraro and Rovera, 1964 (50 ♀); McClure, Thaler, et al., 1965 (5 ♂, chronic EL); Dyment, Melnyk, et al., 1968 (11 ♂); Polák and Žižka, 1970 (4 ♀); Crossen, Mellor, et al., 1972 (72 ♀); Petit, Alexander, et al., 1973 (4 1/2 ♂)*; MacDougall, Brown, et al., 1974 (4 ♀, banding)*
AML	Freireich, Whang, et al., 1964 (child); Rowley, Blaisdell, et al., 1966 (adult); J. F. Jackson and Higgins, 1967 (adult); Krogh Jensen, 1967b(49♀), 1968 (51 ♂); Teasdale, Worth, et al., 1970 (1 ♂); Holden, Garcia, et al., 1971 (child); Humbert, Hathaway, et al., 1971 (child); Rowley, 1973d (adult;banding), 1975b; U. Kaufmann, Löffler, et al., 1974 (5 ♀)*; Chapelle, Vuopio, et al., 1975; J. H. Ford, Pittman, et al., 1975 (adults); Lawler, Secker Walker, et al., 1975 (adults); Zech, Lindsten, et al. 1975 (2 adults*; Larsen and Schimke, 1976 (familial); Mitelman, Nilsson, et al., 1976 (51 ♀)*; Rowley and Potter, 1976 (44♂,45♂)*; Yamada and Furasawa, 1976 (49 ♂)*; Boetius, Hustinx, et al., 1977 (55 ♀,21 ♂)*; Oshimura and Sandberg, 1977; P. Philip, 1977b(3 adult cases); Secker Walker and Sandler, 1978 (7 ♀)*; Ruutu, Ruutu, et al., 1977a (49♂,29♀,56♀)*
PV	Rowley, 1973d (7q−); Tsuchimoto, Bühler, et al., 1974
Multiple myeloma	P. Philip, 1977b (adult)
Myeloproliferative syndrome	Krompotic, Silberman, et al. 1968 (adult); Kamiyama, Shibata, et al., 1973*; Lisker, Cobo de Gutiérrez, et al., 1973; Rowley, 1973d (adult); Nowell, Jensen, et al., 1976a (2 adults)
AMMoL	Freireich, Whang, et al., 1964 (3 children ; Ruutu, Ruutu, et al., 1977a (60 ♀)*
Refractory anemia	Van Den Berghe, Michaux, et al., 1977a (7q−)
ALL	Oshimura, Freeman, et al., 1977b
SBA	Chapelle, Vuopio, et al., 1975
Preleukemia	Chapelle, Vuopio, et al., 1975; Ruutu, Ruutu, et al., 1977a (49 ♂); Van Den Berghe, David, et al., 1976a; Pierre, 1978a
MM → AML	Liang, Hopper, et al., 1979
Lymphoma → ANLL (4 cases)	Whang-Peng, Knutsen, et al., 1979
+7 (ALL)	Roth, Cimino, et al., 1979

*Cases in which monosomy #7 was the sole karyotypic abnormality.

Table 95b Some changes related to trisomy−8, or 7−monosomy patients

Bone marrow Karyotype	Cell	Metabolic finding	Gene location	Clinical significance
8-Trisomy	Erythrocyte	Increased glutathione reductase	Glutathione reductase in chromosome 8	None
7-Monosomy	Erythrocyte	Lack of Colton blood group	Colton blood group in chromosome 7	Immunization against Colton blood group?
7-Monosomy	Granulocyte	Defective chemotaxis and chemokinesis	Chemotaxis and/or chemokinesis in chromosome 7	Infections
del(7)(q22)	Granulocyte	Defective chemotaxis and chemokinesis	Chemotaxis and/or chemokinesis in region 7q22→7qter	Infections

Table obtained through the courtesy of Dr. P. Vuopio.

Table 96 Some hematological conditions associated with trisomy #8 reported in the literature

Diagnosis	Reference
AML	Hellström, Hagenfeldt, et al., 1971; Rowley, 1973c, 1974a, 1975b; Jonasson, Gahrton, et al., 1974; Sakurai, Oshimura, et al., 1974; J.H. Ford, Pittman, et al., 1975; Lawler, Secker Walker, et al., 1975; P. Philip, 1975a, 1977a; Mitelman, Nilsson, et al., 1976; Hagemeijer, Smit, et al., 1977; Panani, Papayannis, et al., 1977
AMMoL	Chapelle, Schröder, et al., 1973a; Najfeld, Singer, et al., 1978
PV	Rowley, 1973c; T.C. Hsu, Pathak, et al., 1974; Wurster-Hill, Whang-Peng, et al., 1976; Zech, Gahrton, et al., 1976; Shabtai, Weiss, et al., 1978
CML	Rowley, 1973c; T.C. Hsu, Pathak, et al., 1974 (includes a case of Ph¹-negative CML); Hagemeijer, Smit, et al., 1977; Bondare, Barawika, et al., 1978; Lilleyman, Potter, et al., 1978; Alimena, Brandt, et al., 1979
Pancytopenias, etc.	Chapelle, Schröder, et al., 1972
SBA	Hellström, Hagenfeldt, et al., 1971; Jonasson, Gahrton, et al., 1974; Moretti, Broustet, et al., 1975; Van Den Berghe, Michaux, et al., 1977a
Myeloproliferative syndrome	Hagemeijer, Smit, et al., 1977; Jacobson, Salo, et al., 1978
Eosinophilic leukemia (Ph¹-negative)	Weinfeld, Westin, et al., 1977b
AML with familial t(7;20)(p13;p12) and mosaicism for +8	Riccardi, Humbert, et al., 1978
AMØL	Shiloh, Naparstek, et al., 1979
Lymphoma (LS, HL)	Mark, Dahlenfors, et al., 1979
Preleukemia	Pierre, 1978a

Table 97 Some hematological conditions associated with trisomy #9 reported in the literature

Diagnosis	Reference
Myelosclerosis	W. M. Davidson and Knight, 1973
ET	Rowley, 1973c
PV	Rowley, 1973c, 1975a; Gahrton, Zech, et al., 1975; Wurster-Hill, Whang-Peng, et al., 1976; Zech, Gahrton, et al., 1976; Westin, 1976; Shabtai, Weiss, et al., 1978
Myeloproliferative syndrome	Knight, Davidson, et al., 1974
AML	Rutten, Hustinx, et al., 1974; J.H. Ford, Pittman, et al., 1975; Mitelman, Levan, et al., 1976b; Mitelman, Nilsson, et al., 1976
AMMoL	Rutten, Hustinx, et al., 1974
AL	J.H. Ford, Pittman, et al., 1975
Lymphoma→ANLL	J.H. Ford, Pittman, et al., 1975; Sandberg, Ishihara, et al., 1964a
CML	Whang-Peng, Knutsen, et al., 1979
?	Hagemeijer, Smit, et al., 1979

Table 98 Anomalies of chromosome #11 in lympho- and myeloproliferative disorders

Translocation	Diagnosis	Reference
t(11;11)(pter23;q13qter)	RCS	Fleischman and Prigogina, 1977
t(6p;11q)	RCS	Fleischman and Prigogina, 1977
t(1q;11p)	RCS	Fleischman and Prigogina, 1977
t(11;17)	RCS	Fleischman and Prigogina, 1977
t(11;14)(q13;qter)	LS	Flesichman and Prigogina, 1977
t(11;14)(q13;qter)	LS	Fleischman and Prigogina, 1977
t(11;19)(q13;qter)	LS	Fleischman and Prigogina, 1977
t?D(14?)q+,del(11)(q12)	LS	Fleischman and Prigogina, 1977
t(11;17)(q21;qter)	Lymphocytic lymphoma	Fleischman and Prigogina, 1977
t(?;11)(:,q+)	LS	Reeves, 1973
t(11;14)(q14;q32?)	LS	Van Den Berghe, David, et al., 1976a
t(11;14)(q14;q32?)	LS	Van Den Berghe, David, et al., 1976a
t(11;14)(q14;q32)	LS	Van Den Berghe, David, et al., 1976a
t(4;11)(q13;q22)	ALL	Oshimura and Sandberg, 1977
t(4;11)(q13;q22)	ALL	Van Den Berghe, David, et al., 1978b
t(4;11)(q13;q22)	ALL	Van Den Berghe, David, et al., 1978b
t(4;11)(q13;q22)	ALL	Van Den Berghe, David, et al., 1978b
t(3p−;11p+)	Juvenile CML	Ionue, Ravindranath, et al., 1977
t(9q+;11q−)	Juvenile CML	Warburton and Shah, 1976
t(1p−,11p+)	ALL	Zuelzer, Inoue, et al., 1976
t(11;20)(p15;q11)	PV	Berger, 1972b, 1975
t(3;11)	HD	Hossfeld, 1978
t(3;11),t(4;11)	HD	Hossfeld, 1978
t(11;22)(q13;p11or12)	AML	Oshimura, Hayata, et al., 1976; Oshimura, Sonta, et al., 1976
t(11;13)(q23→25;q12−14)	AML	Oshimura, Hayata, et al., 1976; Oshimura, Sonta, et al., 1976
t(5;11)	AML	Mitelman, personal communication
t dic(8;11)(8qter→8p23::11p13 or 14 or 15→11qter)	ANLL	P. Philip, 1977a
t(11;13)(11qter→cen→13qter); t(11;15)(11qter→cent→15qter)	ANLL	P. Philip, 1977a
t(6;11)(6qter→6p25 or 24 or 25:: 11q13→11qter);t(8;11)(11pter→ 11q13::8q22→8qter)	ANLL	P. Philip, 1977a
t(1;11)	CML?	Rowley, 1977a
t(9;11)	Refractory anemia with 5q−	Swolin, Waldenström, et al., 1977
t(10p;11)	BL cell line	Zech, Haglund, et al., 1976
t(2?; 11)	LS	Mark, Dahlenfors, et al., 1979
t(11;12)	Ph¹ + CML → Ph¹ − BP	Hagemeijer, Smit, et al., 1979
DELETIONS AND OTHER CHANGES		
del(11)(q22)	CLL	Fleischman and Prigogina, 1977
del(11)q23	Lymphoma	Reeves, 1973
+11p−	HD	Hossfeld, 1978
11q−(q13→qter)	LS	Mark, 1977b
del(11)(q14)	AML	Mitelman, personal communication
11p−	CML(Ph¹+)	Prigogina, Fleischman, et al., 1978
11q−	PV	Shiraishi, Hayata, et al., 1975
11q+	EL (after x-ray R_x for cancer of breast)	Whang-Peng, Gralnick, et al., 1977

Table 98—*continued*

Translocation	Diagnosis	Reference
del(11)q23)	AML	Yamada and Furasawa, 1976
t(1;6;11)(q12;q23;p13)	AML	Yamada and Furasawa, 1976
t(8;11)	EL	P. Philip, 1976
t(8;11;21)	AML	Lindgren and Rowley, 1977
del(11)(q12)	LS	Fleischman and Prigogina, 1977
11p+	RCS	Fleischman and Prigogina, 1977
t(7;11)(q11;q25)	RCS of brain	Yamada, Yoshioka, et al., 1977
t(8;11)	AML	Barlogie, Hittelman, et al., 1977
t(11;9)	ALL	Barlogie, Hittelman, et al., 1977
t(11;2q)	Sézary syndrome	Bosman, van Vloten, et al., 1976
t(11;14)(q23;q32)	ALL (B cell)	Roth, Cimino, et al., 1979
t(11;14)	ALL	Cimino, Rowley, et al., 1979
t(1;11)(q11 or 12;q25)	Preleukemia→AMMol	Najfeld, Singer, et al., 1978
inv(11)	Ph[1] + CML	Lilleyman, Potter, et al., 1978
del(q13)	Histiocytic lymphoma	Mark, Ekedahl, et al., 1978
inv(11)(p15q21)	Histiocytic lymphoma	Mark, Ekedahl, et al., 1978
t(11;14)	Histiocytic lymphoma	Mark, Ekedahl, et al., 1978
+11,+11p+q−	Histiocytic lymphoma	Mark, Ekedahl, et al., 1978
11p+,+11p+	Histiocytic lymphoma	Fukuhara and Rowley, 1978
11q+,11q−	Histiocytic lymphoma	Fukuhara and Rowley, 1978
−11 (2 cases)	LS	Fukuhara and Rowley, 1978
−11	LS	Fukuhara, Rowley, et al., 1979
t(11;14)(q13;q32)	LS	Fukuhara, Rowley, et al., 1979
t(11;21)(q23;q22)	LS	Fukuhara, Rowley, et al., 1979
+11p−	Lymphoma→ANLL	Whang-Peng, Knutsen, et al., 1979
del(11)(q21)	LS	Mark, Dahlenfors, et al., 1979
11q−	AML	Prosser, Bradley, et al., 1978
del(11)(q21)	BP of CML	Lyall and Garson, 1978
11q−	Pancytopenia	Nowell and Finan, 1978a

in the Ph[1]-translocation with #22, [t(9;22)(q34;q11)]. A −9 has been observed in subacute AMMoL[5] and 9q− in lymphoma.

Chromosome #10

This autosome is rarely, if ever, involved in structural changes in AML. Trisomy of #10 has been described in AML, ET (with +9), and CML.[6] A 10q− has been seen in AML.

Chromosome #11

Abnormalities involving #11 are shown in Table 98. This chromosome is not infrequently involved in translocations, particularly with 14q in lymphoma or with #4 in ALL. Trisomy of #11 has been seen in the BP of CML[7] and 11q− in lymphoma.

Chromosome #12

This autosome is rather infrequently involved in human neoplasia. A 12q+ in lymphoma, +12 in PV,[8] and an i(12p) in leukemia have been seen in my laboratory.

Chromosome #13

13q− as a constitutional anomaly is seen in families with a high incidence of retinoblastoma; the tumor *per se* does not have this karyotypic change. However, 13q− has been seen in lymphoma tissue and the cells of CML, PV,

Table 99a Cases of myelo- and lymphoproliferative disorders with 13q−, 15q−, or Dq− reported in the literature

Group	Diagnosis	Reference
15q−	Lymphosarcoma	Fukuhara and Rowley, 1978; Mark, Dahlenfors, et al., 1979
	PV	Nowell and Hungerford, 1962
	CML	Potter, Sharp, et al., 1975; Lawler, O'Malley, et al., 1976; Geraedts, personal communication; Kohno, Van Den Berghe, et al., 1979; Kwan, Singh, et al., 1977
	BL cell line	Zech, Haglund, et al., 1976
	ALL	Zuelzer, Inoue, et al., 1976
	ANLL	P. Philip, 1977a
	APL	Van Den Berghe, Louwagie, et al., 1979b
	Myelofibrosis and MM	Whang-Peng, Lee, et al., 1978
13q−	PV → subacute myeloid leukemia	Nowell and Hungerford, 1962
	Histiocytic lymphoma (5 cases)	Mark, 1975a,b; Mark, Ekedahl, et al., 1978
	PV (treated with busulfan and ^{32}P)	Shiraishi, Hayata, et al., 1975
	PV (untreated)	Wurster-Hill, Whang-Peng, et al., 1976
	AML	Van Den Berghe (personal communication to Rowley, 1977); J.H. Ford, Pittman, et al., 1975
	Follicular lymphoma in transformation	Catovsky, Pittman, et al., 1977
	Multiple myeloma	Van Den Berghe, Louwagie, et al., 1979c
	CML	Kohno, Van Den Berghe, et al., 1979
Dq−	CML (2 Ph¹)	Hammouda, Quaglino, et al., 1964
	CML (spleen irradiated)	E. Engel, 1965a,b
	PV (2 cases; one with Dq−/Dq+)	Kay, Lawler, et al., 1966
	PV →AL	Erkman, Hazlett, et al., 1967; Lawler, Millard, et al., 1970
	CML (treated with x-ray) t(Dq−;2p+)	Nava, Grouchy, et al., 1969a
	"Preleukemia" PV (^{32}P)	Nowell, 1971b,c
	PV (4 cases)	Visfeldt, Franzén, et al., 1973
	HD (x-ray) → AML	Weiden, Lerner, et al., 1973
	Myelofibrosis (?15q−)	Ganner-Millonig, 1974; W. Engel, Merker, et al., 1968
	ALL (2 cases)	Lawler, Secker Walker, et al., 1975
	EL	Lawler, Secker Walker, et al., 1975; Sakurai and Sandberg, 1976b
	AML (+Dq−)	Sakurai and Sandberg, 1976a
	Undifferentiated leukemia (?Dq−)	Zuelzer, Inoue, et al., 1976

In collecting the data from the literature on Dq− anomalies for inclusion in this table, the verbal descriptions in the publications and examination of the published karyotypes by the present author served as the main basis for selection. No 14q− anomalies have been included. The present author has in several cases also used his judgment in deciding as to the D-group chromosomes involved, even though no banding analyses were available.

Table 99b Cases of myeloproliferative disorders with congenital D/D translocations

Chromosome abnormality	Diagnosis	Reference
45,D/D	CML	E. Engel, McGee, et al., 1965
45,D/D	AL	Prigogina, Stavrovskaja, et al., 1968
45,D/D	EL	Dallapiccola and Malacarne, 1971
45,XX,t(13q14q)	CML (Ph¹+)	Wennström and Schröder, 1973
46,Dq+/Dq−	PV	Berger, Parmentier, et al, 1974
t(13q14q)	Multiple myeloma→AML	Rowley, 1976c
t(13q14q)	Myeloproliferative disorder	Goh, Bauman, et al., 1979
t(13q14q)	Ph¹-positive CML	Kohno and Sandberg, 1979

etc. (see Table 99). A −13 has been described in meningioma.

Chromosome #14

The frequent occurrence of 14q+ in BL, lymphosarcoma, HD, and MM has been discussed previously. This is usually due to a t(8;14); however, #1, #4, #5, #10, #11, #14, (Table 100) and #18 have been described to be involved with a #14 in translocations in lymphomas. Some facets of changes affecting #14 in lymphoma have been discussed previously. Chromosome #14 appears to have donor and receptor sites commonly involved in neoplastic disorders (Figure 191). A 14q− (band q12) is seen in the somatic cells of patients with AT, where most of the long arm has been translocated to the other #14 in tandem, [t(14q;14q)], though other chromosomes may be involved in this translocation. Trisomy of #14 has been seen in CML and −14 in meningioma. #14 has been described to be involved in a Ph¹-translocation with #22 both in AML and CML. An 8;14 (8q−;14q+) translocation was seen by us in a case of CML.

Chromosome #15

15q− (see Table 99) is not an infrequent finding in a number of myeloproliferative disorders, though it has also been seen in lymphoma. Monosomy of #15 has been described in meningioma. The t(15;17) is thought to be characteristic for acute promyelocytic leukemia.

Chromosome #16

This autosome appears to be rarely involved in lymphoma. An iso (16q) has been seen in some tumors (lung cancer).

Chromosome #17

The appearance of an isochromosome for the long arm of #17 is an interesting and not uncommon karyotypic finding in a number of disorders (BP of CML, AML, lymphoma, myelofibrosis),[9] including solid tumors, i.e., cancer of the prostate (see Table 46).

One may wonder why an extra copy of 17q in tumor cells usually appears as an isochromosome rather than a trisomy 17. Generally, trisomy is more common in the evolution of neoplastic cell population than are isochromosomes (Nowell, 1976). As E. Engel, McGee, et al. (1975a, b) pointed out, no evidence exists that the centromere of #17 is unusually susceptible to abnormal division; thus, the significant cytogenetic changes may not simply be the extra 17q but rather an imbalance between the long and short arms.

The concept of chromosome "imbalance," as opposed to a single gene locus, being critical in carcinogenesis is being widely studied in vitro through cell fusion techniques. These methods have also provided evidence that a gene for thymidine kinase is located on the long arm of #17 (O.J. Miller, Miller, et al., 1971). As these approaches to the genetic basis of neoplasia are pursued with cell lines and tissue culture, the data on i(17q), and reports of monoso-

Table 100 Chromosome 14q+ in various disorders

Disorder	Reference
BL	Gripenberg, Levan, et al., 1969; Manolov and Manolova, 1972; Zech, Haglund, et al., 1976
BL cell lines	Petit, Verhest, et al., 1972; Jarvis, Ball, et al., 1974; Kaiser-McCaw, Epstein, et al., 1977a,b; Hellriegel, Diehl, et al., 1977; Ben-Bassat, Goldblum, et al., 1977
Lymphoma	
Histiocytic	Mark, 1975b; Fukuhara, Shirakawa, et al., 1976; Fleischman and Prigogina, 1977; Mark, 1977b (2 cases); Fukuhara and Rowley, 1977, 1978; Van Den Berghe, David, et al., 1976a (3 cases); Mark, Ekedahl, et al., 1978 (3 cases); Mark, Dahlenfors, et al., 1979 (3 cases).
Lymphocytic	Reeves, 1973; Prigogina and Fleischman, 1975b; Fukuhara, Shirakawa, et al., 1976; Kaiser-McCaw, Epstein, et al., 1977a,b; Mark, Dahlenfors, et al., 1979 (3 cases)
HD	Reeves, 1973; Lawler, Reeves, et al., 1975; Fukuhara, Shirakawa, et al., 1976; Hossfeld, 1978
American and Non-endemic BL	Zech, Haglund, et al., 1976; Temple, Baumiller, et al., 1976; P. Philip, Krogh Jensen, et al., 1977a; Kaiser-McCaw, Epstein, et al., 1977a,b; Kakati, Barcos, et al., 1979
Multiple myeloma	Wurster-Hill, McIntyre, et al., 1973; P. Philip, 1975c; Liang and Rowley, 1978 (3 cases); Liang, Hopper, et al., 1979 (5 cases)
AT	Hecht, McCaw, et al., 1973
AL (AML and EL)	Sakurai and Sandberg, 1976a
Undifferentiated AL	Lawler, Secker Walker, et al., 1975
RCS of brain	Yamada, Yoshioka, et al., 1977
ALL	Whang-Peng, Knutsen, et al., 1976b; Oshimura, Freeman, et al., 1977b; Van Den Berghe, David, et al., 1978b, Roth, Cimino, et al., 1979
Ph¹-negative CML (6q−;14q+)	Mintz, Pinkhas, et al., 1973
Retinoblastoma	Hossfeld, 1978
BP of CML (8;14)	Hayata, Sakurai, et al., 1975
Mycosis fungoides	Fukuhara, Rowley, et al., 1978a
Follicular lymphoma in transformation	Catovsky, Pittman, et al., 1977
Gastric cancer	Ayraud, 1975
EBV-genome-negative LN without malignancy in a patient with nasopharyngeal cancer and probable lymphoma	Mitelman, Klein, et al., 1979
AML	Van Den Berghe, David, et al., 1978a

my 17 and deletion of 17q in several leukemias and lymphomas (Mitelman, Levan, et al., 1976a; Golomb, Rowley, et al., 1976) suggest that concurrent studies in vivo may also provide clues to specific chromosome segments and chromosome sites important in human neoplasia (Nowell and Finan, 1977).

A translocation between #15 and #17, [t(15;17) (q22;q21)], has been described in acute promyelocytic leukemia. Trisomy or monosomy of #17 has been seen in CML.

Chromosome #18

Deletion of the long arm of #18 (18q−) has been seen in lymphoma and ALL and −18

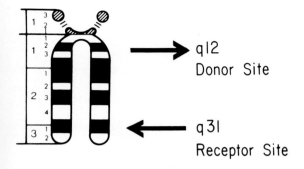

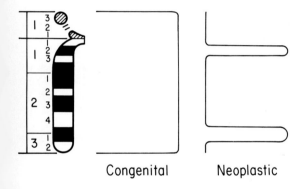

191 Schematic presentation of the segments of #14 involved in various disorders. (above) Areas most commonly affected by deletions (donor site) or translocations (receptor site). (below) Areas involved in neoplastic disorders; congenital changes affect any area of #14.

(Courtesy of Dr. B. Kaiser-McCaw.)

meningioma. A translocation between #18 and #14 leading to a 14q+ appears to be rather common in some lymphomas.

Chromosome #19

Trisomy of #19 appears to be the only consistent abnormality described to date and has been seen most frequently in the BP of CML, though it has also been seen in meningioma and gliomas. A 19;22 Ph¹-translocation has been described in AML. Even though +19 is a relatively common karyotypic change in the BP of CML and may take place concurrently with other chromosomal changes, e.g., +8, an extra Ph¹, it has not been observed to occur in conjunction with an i(17q). Exceptions have been observed by me and others (Alimena, Brandt, et al., 1979).

Chromosome #20

The 20q— anomaly is rather characteristic for PV, though it may be seen in other conditions (see Table 70). These conditions include SBA, Ph¹-negative and Ph¹-positive CML, atypical and typical myeloproliferative syndromes, smouldering leukemia, and AMMoL, but all in elderly patients.

Chromosome #21

Trisomy of this chromosome[10] is "classically" seen in all of the somatic cells of subjects with DS. Monosomy of #21 has been described in meningioma and AML; in the latter 21q22 appears to be vulnerable. Trisomy of #21 has been seen in AML and ALL and in many tumors. Translocation between #21 and #8 constitutes the prototypic change in AML.

A 21q— anomaly seen frequently in ET has been recently described, though it may also be encountered in other conditions (e.g., PV, CML, AL) when associated with thrombocythemia (Fuscaldo, Erlick, et al., 1979).

Chromosome #22

This chromosome is the source of the Ph¹ resulting from a deletion of its long arm. 22q— and —22 occur not infrequently in meningioma.

A case of a 19-yr old male with Ph¹-positive CML (the Ph¹-positivity was observed only at diagnosis and no banding analysis was performed) was found to have a normal 46,XY karyotype during the chronic phase of the disease when studied cytogenetically next more than 4 yr following the diagnosis and to have a succession of different karyotypic abnormalities (49,XY,+9,+10,+12; 46,XY, 14p+; 45, XY,t(5p;17q); 44,XY,+8,t(11;12), 17p+,—18,—22; 49,XY,t(5;17),+6,+8,+9, all without a Ph¹, during and after several episodes of the BP (Hagemeijer, Smit, et al., 1979). The suggested interpretation was that the Ph¹-positive cells had been eradicated by therapy of the disease following diagnosis and that the new abnormal clones arose from normal 46,XY cells.

A case of CML with a complex Ph¹-translocation, i.e., t(1;9;22), similar to that described by Verma and Dosik (1977) (Table 35), has been reported by Berger and Bernheim (1978). This appears to be the thirty-third case of CML with a complex translocation described to-date.

In a study in my laboratory (Kohno, Abe, et al., 1979) examination of the NOR on the Ph¹ revealed a significantly lower value than the expected one.

A list of references related to the Ph¹ (22q−), CML, and other aspects of this disease is included.[14]

X-Chromosome

Extra X-chromosomes may be seen in a number of conditions (see Table 101), including about every variety of cancer. Missing X-chromosomes have been most frequently encountered in AML (usually part of the prototypic karyotype with an 8;21 translocation), meningioma, and lymphoma. A Ph¹-translocation involving the short arm of the X and #22 has been described.

Autoradiography of the leukemic cells in a female patient with AML and a 45,X,−X, t(8;21) karyotype demonstrated that the X-chromosome was active (G. Levan and Mitelman, 1979).

Y-Chromosome

Extra Y-chromosomes may be present in leukemia (particularly CML) or in cancers (see Table 101). A missing Y has been described in various tumors, as part of the prototypic karyotype in AML (with an 8;21 translocation), meningioma, and CML (all invariably associated with other chromosomal changes).

Though a missing Y is often seen in the marrows of elderly males (>65 yr old), its occurrence as the *sole* karyotypic anomaly in leukemic and cancer cells is very rare.

Gene Mapping in Cancer Causation

The use of somatic cell hybrids, complemented by linkage data from pedigree studies, has led to the chromosomal localization of *nearly* 100 genes, coding for enzymes or other substances involved in the synthesis and/or metabolism of carbohydrates, protein, lipids, nucleic acids, etc. A correlation of specific chromosomal abnormalities with various known genetic loci holds much promise for future research in human neoplasia.

In my opinion, however, it is premature to impute certain chromosomal bands or segments in the causation of cancer of leukemia because the loci involved and the enzymes affected are entirely too numerous and heterogeneous to be implicated in the direct causation of human neoplasia. If neoplasia can be induced by a mutation in one gene, its chromosomal location, regardless of some evidence brought forth by in vitro observations, is still unknown. There is little doubt that nonrandom changes occur in certain human chromosome, e.g., in #1 and #17, in leukemia and cancer, but, in my opinion, they are a reflection of events tangential to the basic causation of these diseases, with a small number of nonrandom chromosomal lesions reflecting a susceptibility of some chromosomes or their regions to the influences of the already existing neoplastic condition. When one considers that the average chromosome band contains 5×10^6 base pairs and that a deletion or duplication of ⅓ of a band (or 2×10^6 nucleotide pairs) would be difficult if not impossible to detect with current methods, it is premature, in my opinion, to correlate such gross chromosomal changes, observed with the best of banding techniques, with causation of neoplasia in human subjects.

Most importantly, only a small fraction of the gene loci on the human chromosomes has been established and, hence, it is also premature to implicate chromosomes containing genes involved in the metabolism of nucleic acids in neoplastic proliferations, when there is the possibility that just as many loci related to nucleic acids may be present on other chromosomes, not often involved in neoplasia, and in those representing the nonrandom karyotypic changes containing gene loci for protein, lipids, etc.

A variety of acquired chromosomal abnormalities occur uniquely in the marrow and could be useful in gene mapping (Chapelle, 1976; McKusick and Ruddle, 1977). One such abnormality has been used to assign a human gene, i.e., that for glutathione reductase, to #8. High levels of glutathione reductase activity

Table 101 Extra sex chromosomes in neoplasia

Sex-chromosome complement in neoplastic cells	Diagnosis	Representative karyotypes	Reference
XYY	AML	53,XYY,−2,+22,+6M	Zuelzer, Inoue, et al., 1976
XYY	PV	45,XY,+21,−16,−17/ 48,XYY,+11,+16, −17,−17,+22	Wurster-Hill, Whang-Peng, et al., 1976
XXX	LS	48,XXX,+t(i17q/ 1q),+2r	Fleischman and Prigogina, 1975, 1977
XYY	CML	51,XYY,+3C,−22, +2Ph¹	Goh, 1974
	CML	47,XYY,t(9q+; 13q−q+;15q−;22q−) (q34;q12;q22;q11)	Potter, Sharp, et al., 1975
	CML	54,XYY,t(9;22),+6, +8,+11,+19,+21, +22q−	Prigogina, Fleischman, et al., 1978
	CML	53,XYY,2q+,+8,+10, +14,+18,+21,+2Ph¹	Van Den Berghe, personal communication
XYY	Eosinophilic leukemia	49,XYY,t(3;5),+8 +M/46,XY	Bitran, Rowley, et al., 1977
XXY	CML (BP)	49,XXY,t(9;22),+8, +22q−	Rowley, 1977a−c
XYY	BP of CML	50,XYY,+8,+19, +22q−,t(9q+;22q−)	Matsunaga Sadamori, et al., 1976a
XXX	AL	54,XXX,+3,+4,+9, +10,+19,+21,+22	J.H. Ford, Pittman, et al., 1975
XYY	Follicular lymphoma	52 chromosomes	Baker in Atkin, 1976c
XYY and XXY	Lymphoma	48,XXY,+G/49, XYY,+2G	Nassar and Khouri, 1974
XYY	RCS	47,XYY	Coutinho, Bottura, et al., 1971
XYY	RCS	47,XXY,−13,−15, +t(11;6),+t(?;13q, 14q), etc.	Fleischman and Prigogina, 1977
XYY	Follicular lymphoma		Spiers and Baikie, 1968a
XXY	AML	65,XXY,complex	Golomb, Vardiman, et al., 1978
XXX	Histiocytic lymphoma	82 chromosomes	Mark, Ekedahl, et al., 1978
XYY	BP of CML	48,XYY,+21	Alimena, Brandt, et al., 1979
XXX	BP of CML	50,XXX,+8,+9,+19	Alimena, Brandt, et al., 1979
XYY	BP of CML	51,XYY,+8,+19, i(17q),+Ph¹,t(6;8)	Alimena, Brandt, et al., 1979

were observed in the erythrocytes of two patients whose marrow cells were trisomic for an unidentified chromosome belonging to group C, and later studies showed that the mean value for glutathione reductase activity was significantly higher (approximately 150%) in patients with trisomy #8 than in control subjects. Thus, acquired abnormalities in the marrow can be used and even have some advantages over constitutional abnormalities. However, a number of difficulties and limitations exist in which blood elements and cells in other tissues offer advantages over those in the marrow.

Defective chemotaxis and abnormalities of

Table 102 The gene map of each human chromosome

Chromosome	Most consistent smallest region	Gene marker	Gene symbol
#1	p32–pter	Enolase-1 (E.C. no. 4.2.1.11)	ENO_1
	p34–pter	Phosphogluconate dehydrogenase (E.C. no. 1.1.1.44)	PGD
	p36–pter	Adenovirus-12 chromosome modification site-1p	AdV-12-CMS-$1p$
	p32–p34	α-L-Fucosidase (E.C. no. 3.2.1.51)	αFUC
	p32–pter	Rhesus blood group	Rh
	p32–pter	Adenylate kinase-2 (E.C. no. 2.7.4.3)	AK_2
	p32–pter	Uridine monophosphate kinase (E.C. no. 2.7.4.4)	$UMPK$
	p21–p33	Phosphoglucomutase-1 (E.C. no. 2.7.5.1)	PGM_1
	1p	Amylase$_1$ (salivary) (E.C. no. 3.2.1.1)	AMY_1
	1p	Amylase$_2$ (pancreatic) (E.C. no. 3.2.1.1)	AMY_2
	q21–q23	UDP glucose pyrophosphorylase (E.C. no. 2.7.7.9)	$UGPP_1$
	q32–qter	Guanylate kinase (soluble) (E.C. no. 2.7.4.8)	GUK_S
	q41–qter	Peptidase-C (E.C. no. 3.4.11.-)	$PEPC$
	q42	Adenovirus-12 chromosome modification site-1q	AdV-12-CMS-$1q$
	q32–qter	Fumarate hydratase (soluble) (E.C. no. 4.2.1.2)	FH_S
	q42	5S RNA locus	$RN5S$
	1p	Elliptocytosis	El_1
	1q	Duffy blood group	Fy
	1q	Zonular pulverulent cataract	Cae
		Scianna blood group	Sc
		Retinitis pigmentosa-1	Rp_1
		Glucose dehydrogenase (E.C. no. 1.1.1.47)	GDH
		Dombrock blood group	Do
		Antithrombin III	$AT3$
		Aniridia type II	AN-2
#2	p23	Acid phosphatase-1 (erythrocyte) (E.C. no. 3.1.3.2)	ACP_1
	p23-pter	Malate dehydrogenase (soluble) (E.C. no. 1.1.1.37)	MDH_S
	p11–p22	Galactose enzyme activator	$Ga1+act$
	1q	Isocitrate dehydrogenase (soluble) (E.C. no. 1.1.1.42)	IDH_S
		Arylhydrocarbon hydroxylase (E.C. no. 1.14.14.1)	AHH
		UDP glucose pyrophosphorylase (E.C. no. 2.7.7.9)	$UGPP_2$
		Interferon	If–1
		MNS_s blood group	MNS_s
#3		β-Galactosidase (E.C. no. 3.2.1.23)	βGAL
		Glutathione peroxidase$_1$ (E.C. no. 1.11.1.9)	GPX_1
		Herpes virus sensitivity	HVS
		Temperature-sensitive complement	$tsAF8$
#4	p14–p21	Phosphoglucomutase-2 (E.C. no. 2.7.5.1)	PGM_2
	pter–q21	Peptidase-S (E.C. no. 3.4.11.-)	$PEPS$
	q11–q13	Group-specific protein	Gc
		Phosphoribosylpyrophosphate amidotransferase	$PRPP$–ATF
		Albumin	Alb
#5	cen–q13	Hexosaminidase-B (E.C. no. 3.2.1.30)	HEX_B
		Arylsulfatase-B (E.C. no. 3.1.6.1)	ARS_B
		Diphtheria toxin sensitivity	DTS
		Interferon-2	If–2
		Leucyl-TRNA synthetase	$LEURS$
#6	pter–p12	Glyoxylase-1 (E.C. no. 4.4.1.5)	GLO_1
	p12–p22	Major histocompatibility complex HLA-A,B,C,D C2, C4, C8	MHC
		Rogers blood group	Rg

Table 102 — *continued*

Chromosome	Most consistent smallest region	Gene marker	Gene symbol
#6		Chido blood group	Ch
		Properdin factor B	Bf
	p12–qter	Phosphoglucomutase-3 (E.C. no. 2.7.5.1)	PGM_3
	q12–q15	Malic enzyme (soluble) (E.C. no. 1.1.1.40)	ME_S
	q21	Superoxide dismutase (mitochondrial) (E.C. no. 1.15.1.1)	SOD_M
		Glutamic oxaloacetic transaminase (mitochondrial) (E.C. no. 2.6.1.1)	GOT_M
	q21	Pepsinogen (E.C. no. 3.4.23)	Pg
#7	p13–cen	β-Glucuronidase (E.C. no. 3.2.1.31)	βGUS
	p22–q22	Malate dehydrogenase (mitochondrial) (E.C. no. 1.1.1.37)	MDH_M
	q35	Hageman factor	HaF
	7q	SV40 T antigen	$SV40\text{-}T$
	7q	Kidd locus	Jk
		Colton blood group	CO
		Hydroxyacyl-CoA dehydrogenase (E.C. no. 1.1.1.35)	$HADH$
		Surface antigens	$SA\text{-}7x$
			$SA\text{-}7y$
		Uridine phosphorylase (E.C. no. 2.4.2.3)	UP
#8	p11	Glutathione reductase (E.C. no. 1.6.4.2)	GSR
#9	7p	Adenylate kinase-3 (E.C. no. 2.7.4.10)	AK_3
	7p	Aconitase (soluble) (E.C. no. 4.2.1.3)	$ACON_S$
	q34	ABO blood group	ABO
	q34	Adenylate kinase-1 (E.C. no. 2.7.4.3)	AK_1
	q34	Nail-patella syndrome	NPa
		Argininosuccinate synthetase (E.C. no. 6.3.4.5)	ASS
		Galactose-1 phosphate uridyl transferase (E.C. no. 2.7.7.12)	$GALT$
#10	pter–q24	Hexokinase-1 (E.C. no. 2.7.1.1)	HK_1
	pter–q24	Pyrophosphatase (inorganic) (E.C. no. 3.6.1.1)	PP
	q24–q26	Glutamic oxaloacetic transaminase (soluble) (E.C. no. 2.6.1.1)	GOT_S
		Adenosine kinase (E.C. no. 2.7.1.20)	ADK
		Glutamate-α-semialdehyde synthetase (E.C. no. 6.3.-.-)	GSS
		Polykaryocytosis promoter	$FUSE$
		External membrane protein-130	$EMP\text{–}130$
#11	pter–p13	Lactate dehydrogenase-A (E.C. no. 1.1.1.27)	LDH_A
	pter–p13	Lethal antigen	AL
	p12–cen	Acid phosphatase-2 (tissue) (E.C. no. 3.1.3.2)	ACP_2
	cen–q22	Esterase-A_4 (E.C. no. 3.1.1.1)	ESA_4
	11p	Species antigen	$SA\text{–}1$
#12	pter–p12	Triosephosphate isomerase (E.C. no. 5.3.1.1)	TPI
	pter–p12	Glyceraldehyde-3-phosphate dehydrogenase (E.C. no. 1.2.1.21)	$GAPDH$
	p12	Lactate dehydrogenase-B (E.C. no. 1.1.1.27)	LDH_B
	p21	Peptidase-B (E.C. no. 3.4.11.-)	$PEPB$
	pter–q14	Serine hydroxymethyltransferase (E.C. no. 2.1.2.1)	$SHMT$
		Complement to glycine auxotroph	$Gly+A$
		Enolase-2 (E.C. no. 4.2.1.11)	ENO_2
		Citrate synthase (E.C. no. 4.1.3.7)	CS
#13	q3	Esterase-D (E.C. no. 3.1.1.1)	ESD
		Retinoblastoma	RB
	13p	Ribosomal RNA	$rRNA$

continued

Table 102 The gene map of each human chromosome — *continued*

Chromosome	Most consistent smallest region	Gene marker	Gene symbol
#14	q12–q20	Nucleoside phosphorylase (E.C. no. 2.4.2.1)	NP
	q21–qter	Tryptophenyl-tRNA synthetase (E.C. no. 6.1.1.2)	$TRPRS$
		External membrane protein-195	$EMP-195$
	14p	Ribosomal RNA	$rRNA$
#15	q14–q21	β_2-microglobulin	β_2-MG
	q22–qter	Hexosaminidase-A (E.C. no. 3.2.1.30)	HEX_A
	q22–qter	Mannosephosphate isomerase (E.C. no. 5.3.1.8)	MPI
	q22–qter	Pyruvate kinase-M2 (E.C. no. 2.7.1.40)	PK_{M_2}
	q21–qter	Isocitrate dehydrogenase (mitochondrial) (E.C. no. 1.1.1.42)	IDH_M
	q11–qter	α-Mannosidase-A (E.C. no. 3.2.1.24)	αMAN_A
	15p	Ribosomal RNA	$rRNA$
#16	16q	Adenine phosphoribosyl transferase (E.C. no. 2.4.2.7)	$APRT$
	16q	Thymidine kinase (mitochondrial) (E.C. no. 2.7.1.75)	TK_M
	cen–q22	Cholinesterase (serum) (E.C. no. 3.1.1.8)	$E2$
	cen–q22	α-Haptoglobin	αHp
		α-Hemoglobin	$Hb\alpha$
		Lecithin-cholesterol acyltransferase (E.C. no. 2.3.1.43)	$LCAT$
#17	q21	Thymidine kinase (soluble) (E.C. no. 2.7.1.75)	TK_S
		Acid α-glucosidase	α-GLU
	q21–q22	Galactokinase (E.C. no. 2.7.1.6)	$GALK$
	q21–q22	Adenovirus-12 chromosome modification site-17	$Adv.12-CMS-17$
		Collagen-1	COL_1
		SV40 integration site	$SV40-I$
		SV40 transformation site	$SV40-T$
#18	q23–qter	Peptidase-A (E.C. no. 3.4.11.-)	$PEPA$
#19	pter–q13	Glucosephosphate isomerase (E.C. no. 5.3.1.9)	GPI
		Poliovirus sensitivity	PVS
		Peptidase-D (E.C. no. 3.4.13.9)	$PEPD$
	pter–q13	α-Mannosidase-B (E.C. no. 3.2.1.24)	αMAN_B
#20	p11–qter	Adenosine deaminase (E.C. no. 3.5.4.4)	ADA
		Desmosterol-to-cholesterol enzyme	DCE
		Inosine triphosphate (E.C. no. 3.6.1.3)	ITP
#21	q22	Superoxide dismutase (soluble) (E.C. no. 1.15.1.1)	SOD_S
		Antiviral protein (antiviral response)	AVP
		Phosphoribosoyl glycinamide synthetase (E.C. no. 6.3.4.13)	$PRGAS$
		Glutathione peroxidase (E.C. no. 1.11.1.9)	GPX
	21p	Ribosomal RNA	$rRNA$
		Ag Lipoprotein	Ag
#22	22p	Ribosomal RNA	$rRNA$
		Arylsulfatase-A (E.C. no. 3.1.6.1)	ARS_A
		Aconitase (mitochondrial) (E.C. no. 4.2.1.3)	$ACON_M$
		NADH diaphorase (E.C. no. 1.6.2.2)	DIA_1
		α-Galactosidase B	α-$GALA$
X	q13–q22	Phosphoglycerate kinase (E.C. no. 2.7.2.3)	PGK
	q22–q24	α-Galactosidase (E.C. no. 3.2.1.22)	αGAL
	q26–qter	Hypoxanthine phosphoribosyl transferase (E.C. no. 2.4.2.8)	$HPRT$
	q26–qter	Glucose-6-phosphate dehydrogenase (E.C. no. 1.1.1.49)	$G6PD$
		Xg antigen	Xg
		Tyrosine aminotransferase regulator	$TATr$
		Phosphorylase kinase (E.C. no. 2.7.1.38)	PHK
		Ornithine transcarbamylase (E.C. no. 2.1.3.3)	OTC
		Species antigen	SAX

Table 102 — *continued*

Chromosome	Most consistent smallest region	Gene marker	Gene symbol
Y		Color blindness	cb
		Hemophilia	heA
		Y histocompatibility antigen	$H-Y$
		ADDITIONAL	
#1		Cholinesterase (serum)-1 (E.C. no. 3.1.1.8)	E_1
		Nail-patella syndrome	NPa
		Transferrin (provisional)	Tf
#2		Hemoglobin-β (also on #4 and ?#12)	Hb_β
	q11–q14	Galactose-1-phosphate uridyltransferase	$GALT$
		Phosphopyruvate hydratase	PPH
#6		Plasminogen activator (provisional)	PA
		P	P
	p12–p22	Complement component 2	C_2
	p12–p22	Complement component 4	C_4
		Surface antigens	$SA-6$
		Juvenile diabetes mellitus	JDM
		Indophenol oxidase B	$IPO-B$
		Adenosine deaminase complexing protein	$ADCP$
#7		Neutrophil differentiating factor (provisional)	NDf
		Type I Collagen	COL_1
#8		Clotting Factor VII	$F7r$
#9		Galactose-1-phosphate uridyltransferase (E.C. no. 2.7.7.12)	$GALT$
#13		Lipoprotein (provisional)	LP
#16		Hemoglobin-α (also on ?#2 and ?#4)	Hb_α
#17		Surface antigen	$SA-17$
#18		Chorionic gonadotropin (provisonal)	HCG
#19		Echoll sensitivity	$E11_s$
#21		Diaphorase (NADM) (E.C. no. 1.6.2.2)	DIA_1
		Indophenol oxidase A	$IDO-A$
X		Kell precursor controller	X_k
		α-2-macroglobulin	X_m
		Antihemophilic globulin A	$Hem-A$
		Antihemophilic globulin B	$Hem-B$
	q	Becker's muscular dystrophy	MDB
		Mental retardation	MR
	q	Ocular albinism	OA
	q	Retinoschisis	RS
		Xeroderma pigmentosum	XP

the Colton blood-groups have been shown to be associated with monosomy of #7,[11] a deficit in Factor VII in coagulation in subjects with trisomy #8,[12] and evidence for localization of genes for human α-globin on the long arm of #4.[13]

The various loci identified in individual chromosomes are shown in Table 102. The table is based on those published by Shows (1978) and Shows and McAlpine (1979).

References

1. Atkin, 1977; Atkin and Baker, 1977a–c. 1978; Atkin and Pickthall, 1977.
2. Kakati, Oshimura, et al., 1976; Kakati, Song, et al., 1977.
3. J.H. Ford, Pittman, et al., 1975; Kaffe, Hsu, et al., 1974.
4. J.H. Ford, Pittman, et al., 1974.
5. Lampert, Phebus, et al., 1972.

6. Beck and Chesney, 1973; Wurster-Hill, Whang-Peng, et al., 1976.
7. P. Philip, 1975b; Rowley, 1977a.
8. Kaffe, Hsu, et al., 1974; Wurster-Hill, Whang-Peng, et al., 1976.
9. E. Engel, McGee et al., 1975a, b; E. Engel, McKee, et al., 1975; Fitzgerald and Adams, 1965; Nowell and Hungerford, 1962; Ricci, Punturieri, et al., 1962a; Sasaki, Sofuni, et al., 1965; Spiers and Baikie, 1968a, 1970.
10. J.H. Ford, Pittman, et al., 1975; Mitelman, Levan, et al., 1976a; Rowley, 1973b; Sakurai, Oshimura, et al., 1974.
11. Chapelle, Vuopio, et al., 1975; Ruutu, Ruutu, et al., 1977a.
12. Chapelle, Schröder, et al., 1973a.
13. Gandini, Dallapiccola, et al., 1977.
14. Anthonioz, Diakhate, et al., 1974; Baillet and Jullien, 1965; Barton and Conrad, 1978; Bauke, 1970c; Bottura and Coutinho, 1968b, 1975; Brătianu, Răutu, et al., 1975; Cáp, Izakovič, et al., 1977; Cassuto, Ayraud, et al., 1972; Cawein, Lappat, et al., 1965; Chervenick, Lawson, et al., 1971; Drew, Terasaki, et al., 1977; Dubois-Ferrière, 1968; Egozcue, 1971; E. Engel, McGee, et al., 1976, 1977; Gasparini, Bersi, et al., 1972; Gebhardt, Schwanitz, et al., 1973; Gropp, 1973; Grozdéa, Colombiès, et al., 1970, 1975b; Hampel, 1967; Hellriegel, Koebke, et al., 1971; Jedrzejczak, 1972; M.H. Khan, 1974; Khouri, Shahid, et al., 1969b; Kitchin and Shaw, 1971; Korenevskaia, Bartashchuk, et al., 1970; Korenevskaia, Nevskaia, et al., 1971; Lawler, 1967; B.B. Lozzio, Lozzio, et al., 1976; Lucas, Whang, et al., 1966; Mansour, 1970; Merker, 1972; Negrini, Azzolini, et al., 1968; Pedersen, 1965, 1967d,e; Perillie, 1967; Peterlik, Pietschmann, et al., 1970; Pierre, O'Sullivan, et al., 1974; Poroshenko, 1966; Prieto Garcia, Forteza Bover, et al., 1968a,b; Re, Bagnara, et al., 1976; Rovera, Pegoraro, et al., 1965; Saffhill, Dexter, et al., 1976; Y. Sakurai and Nakao, 1973; T. Sakurai, Kudo, et al., 1975; Sei, Hanzawa, et al., 1972; Tezok, Sayli, et al., 1966; Turri, Lorand, et al., 1976; M.Y.F.W. Wang and Desforges, 1967; Winkelman, Fisher, et al., 1971.

References

Abbott, C.R.: The Christchurch chromosome. Lancet i, 1155–1156 (1966).

Abbrederis, K.: Klinische Relevanz zytochemischer Befunde bei differenzierten myelogenen Leukämien des Erwachsenen. Wien. Klin. Wochenschr. 89, 1–26 (1977).

Abbrederis, K., Schmalzl, F., Braunsteiner, H.: Zur Differentialdiagnose akuter Leukämien mittels zytochemischer Methoden. Schweiz. Med. Wochenschr. 99, 1425–1430 (1969).

Abdulla, U., Campbell, S., Dewhurst, C.J., Talbert, D., Lucas, M., Mullarkey, M.: Effect of diagnostic ultrasound on maternal and fetal chromosomes. Lancet ii, 829–831 (1971).

Abe, S., Sandberg, A.A.: Chromosomes and causation of human cancer and leukemia. XXXII. Unusual features of Ph^1-positive acute myeloblastic leukemia (AML), including a review of the literature. Cancer (in press) (1979).

Abe, S., Sasaki, M.: Studies on chromosomal aberrations and sister chromatid exchanges induced by chemicals. Proc. Jap. Acad. 53, 46–49 (1977a).

Abe, S., Sasaki, M.: Chromosome aberrations and sister chromatid exchanges in Chinese hamster cells exposed to various chemicals. J. Natl. Cancer Inst. 58, 1635–1641 (1977b).

Abe, S., Golomb, H.M., Rowley, J.D., Mitelman, F., Sandberg, A.A.: Chromosomes and causation of human cancer and leukemia. XXXV. The missing Y in acute non-lymphocytic leukemia (ANLL). Cancer (in press) (1979).

Abe, S., Kohno, S.I., Kubonishi, I., Minowada, J., Sandberg, A.A.: Chromosomes and causation of human cancer and leukemia. XXXIII. 5q- in a case of acute lymphoblastic leukemia (ALL). Am. J. Hematol. (in press) (1979).

Abe, T., Misawa, S.: Cytogenetic studies on leukemia with special reference to the relevance of chromosome abnormalities to the development of the disease (In Japanese). Clin. Hematol. 18, 771–776 (1978).

Abe, T., Morita, M., Kawai, K.: Cd banding of human chromosomes observed in the C-banding procedure. Lancet ii, 981 (1975).

Abe, T., Morita, M., Kawai, K., Misawa, S., Masuda, M.: Banding analysis of human leukemic and malignant lymphoma cell chromosomes. Jap. J. Hum. Genet. 20, 250–251 (1976).

Abe, T., Morita, M., Misawa, S., Hattori, A., Nakano, S., Hojo, H.: Breakage and reunion at the chromosome band 22q11 in somatic cells—its relevance to the genetic role of the Philadelphia chromosome. Acta Haematol. Jap. 41, 140–143 (1978).

Abrahamson, J.R., Edgington, T.S.: Sideroblastic anemia associated with cytogenetic aberrations of bone marrow cells. Am. J. Clin. Pathol. 57, 348–351 (1972).

Abramson, S., Miller, R.G., Phillips, R.A.: The identification in adult bone marrow of pluripotent and restricted stem cells of the myeloid and lymphoid systems. J. Exp. Med. 145, 1567–1569 (1977).

Abramovsky, I., Vorsanger, G., Hirschhorn, K.: Sister-chromatid exchange induced by X-ray of human lymphocytes and the effect of L-cysteine. Mutat. Res. 50, 93–100 (1978).

Absatz, M., Borgaonkar, D.S.: Chromosome arm involvement in interchanges. Lancet i, 96 (1977).

Adams, A., Fitzgerald, P.H., Gunz, F.W.: A new chromosome abnormality in chronic granulocytic leukaemia. Br. Med. J. ii, 1474–1476 (1961).

Adams, F.H., Norman, A., Bass, D., Oku, G.: Chromosome damage in infants and children after cardiac catheterization and angiocardiography. Pediatrics 62, 312–316 (1978).

Adamson, J.W., Fialkow, P.J.: Annotation: The pathogenesis of myeloproliferative syndromes. Br. J. Haematol. 38, 299–303 (1978).

Adamson, J.W., Fialkow, P.J., Murphy, S., Prchal, J.F., Steinmann, L.: Polycythemia vera: Stem-cell and probable clonal origin of the disease. N. Engl. J. Med. 295, 913–916 (1976).

Adler, K.R., Lempert, N., Scharfman, W.B.: Chronic granulocytic leukemia following successful renal transplantation. Cancer 41, 2206–2208 (1978).

Ahearn, M.J., Trujillo, J.M.: Cytogenetic and ultrastructural evidence of altered DNA metabolism in leukemic cells. Proc. Am. Assoc. Cancer Res. 13, 108 (1972).

Ahearn, M.J., Trujillo, M., Cork, A., Fowler, A., Hart, J.S.: The association of nuclear blebs with aneuploidy in human acute leukemia. Cancer Res. 34, 2887–2896 (1974).

Ahlström, U.: Chromosomes of primary carcinomas induced by 7,12-dimethylbenz (α) anthracene in the rat. Hereditas 78, 235–244, (1974).

Ahnström, G., Natarajan, A.T.: Mechanism of chromosome breakage. A new theory. Hereditas 54, 379–388 (1965).

Aksoy, M.: Leukemia in workers due to occupational exposure to benzene. N. Istanbul Contr. Clin. Sci. 12, 3–14 (1977).

Aksoy, M., Erdem, S., Erdogan, G., Dinçol, G.: Acute leukemia in two generations following chronic exposure to benzene. Hum. Hered. 24, 70–74 (1974).

Albahary, C., Grouchy, J. De, Turleau, C., Auffret, M., Martin, S., Leger, B.: A fatal case of pancytopenia during viral hepatitis. Presse Med. 79, 1055–1058 (1971).

Alfi, O.S., Menon, R.: A rapid C-band staining technique for chromosomes. J. Lab. Clin. Med. 82, 692–694 (1973).

Alimena, G., Annino, L., Balestrazzi, P., Montuoro, A., Dallapiccola, B.: Cytogenetic studies in acute leukaemias. Prognostic implications of chromosome imbalances. Acta Haematol. 58, 234–239 (1977).

Alimena, G., Annino, L., Dallapiccola, B., Ricci, N.: Ring chromosomes and leukaemia. Experientia 31, 1359–1360 (1975).

Alimena, G., Brandt, L., Dallapiccola, B., Mitelman, F., Nilsson, P.G.: Secondary chromosome changes in chronic myeloid leukemia: Relation to treatment. Cancer Genet. Cytogenet. (in press) (1979).

Allard, S., Cadotte, M.: Neoplasia and G monosomy. Union Med. Can. 101, 2116–2120 (1972).

Allderdice, P.W., Davis, J.G., Miller, O.J., Klinger, H.P., Warburton, D., Miller, D.A., Allen, F.J., Jr., Abrams, C.A.L., McGilvray, E.: The 13q-syndrome. Am. J. Hum. Genet. 21, 499–512 (1969).

Allen, J.W., Latt, S.A.: Analysis of sister chromatid exchange formation *in vivo* in mouse spermatogonia as a new test system for environmental mutagens. Nature 260, 449–451 (1976).

Allen, J.W., Shuler, C.F., Mendes, R.W., Latt, S.A.: A simplified technique for in vivo analysis of sister-chromatid exchanges using 5-bromodeoxyuridine tablets. Cytogenet. Cell Genet. 18, 231–237 (1977).

Allison, A.C., Paton, G.R.: Chromosome damage in human diploid cells following activation of lysosomal enzymes. Nature 207, 1170–1173 (1965).

Al-Saadi, A., Mizejewski, G.J.: Immunological and cytogenetic properties of developing thyroid tumors in the rat. Cancer Res. 32, 501–505 (1972).

Alter, A.A., Lee, S.L., Pourfar, M., Dobkin, G.: Studies of leukocyte alkaline phosphatase in mongolism: A possible chromosome marker. Blood 22, 165–177 (1963).

Altman, A.J., Baehner, R.L.: *In vitro* colony-forming characteristics of chronic granulocytic leukemia in childhood. J. Pediatr. 86, 221–224 (1975).

Altman, A.J., Palmer, C.G., Baehner, R.L.: Juvenile "chronic granulocytic" leukemia. A panmyelopathy with prominent monocytic involvement and circulating monocyte colony-forming cells. Blood 43, 341–350 (1974).

Alvarez, I.M., Villalpando, C.J.: Neurofibromatosis y translocacion cromosomica. Informe de un case 46,XX,t(2p–;22p+). Rev. Invest. Clin. Hosp. Enferm. Nutr. (Mex.) 23, 311–316 (1971).

Amano, K., Shigeta, S., Awano, I., Hinuma, Y.: Chromosomal aberrations of a lymphoblastoid cell line by experimental infection with Epstein-Barr virus. Gann 64, 309–311 (1973).

Amarose, A.P.: Chromosomal patterns in cancer patients during treatment. N.Y. State J. Med. 64, 2407–2413 (1964).

Amarose, A.P., Baxter, D.H.: Chromosomal changes following surgery and radiotherapy in patients with pelvic cancer. Obstet. Gynecol. 25, 828–843 (1965).

Amarose, A.P., Kyriazis, A.A., Dorus, E., Azizi, F.: Clinical, pathologic, and genetic findings in a case of 46,XY pure gonadal dysgenesis (Swyer's syndrome). Am. J. Obstet. Gynecol. 127, 824–828 (1977).

Amarose, A.P., Plotz, E.J., Stein, A.A.: Residual chromosomal aberrations in female cancer patients after irradiation therapy. Exp. Mol. Pathol. 7, 58–91 (1967).

Amarose, A.P., Tartaglia, A.P., Propp, S.: Cytogenetic findings in Blackfan-Diamond syndrome. Lancet ii, 1020 (1965).

Anday, G.J., Fishkin, B., Gabor, E.P.: Cytogenetic studies in multiple myeloma. J. Natl. Cancer Inst. 52, 1069–1079 (1974).

Andersen, E., Videbaek, A.: Stem cell leukaemia in myelomatosis. Scand. J. Haematol. 7, 201–207 (1970).

Anderson, C., Mohandas, T., Okun, D.: Chronic myelogenous leukemia with a complex translocation. American Society of Human Genetics, 29th Annual Meeting, Vancouver, B.C., October 4–7, Abstract, p. 73A (1978).

Anderson, J.W., Pearson, G., Valdmanis, A., Mann, J.D.: Duplication of the Y chromosome during myeloblastic crisis of chronic myelocytic leukemia. Ann. Genet. (Paris) 11, 135–137 (1968).

Anderson, T.F.: Techniques for the presentation of three-dimensional structure in preparing specimens for the electron microscope. Trans. N.Y. Acad. Sci. 13, 130–134 (1951).

Andersson, J., Edelman, G.M., Möller, G., Sjöberg, O.: Activation of B lymphocytes by locally concentrated concanavalin A. Eur. J. Immunol. 2, 233–235 (1972).

Andres, A.H.: Zellstudien an Menschenkrebs. Der chromosomale Bestand im Primärtumor und in der Metastase. Z. Zellforsch. 16, 88–122 (1932).

Andres, A.H.: *Introduction to Human Karyology*, C.G. Levitt, P.I. Zhivago (eds.) (in Russian). Medgis (1934).

Andres, A.H., Jiv, B.V.: Somatic chromosome complex of the human embryo. Cytologia (Tokyo) 7, 371–388 (1936).

Andres, A.H., Shiwago, P.I.: Karyologische Studien an myeloischer Leukämie des Menschen. Folia Haematol. (Leipz.) 49, 1–20 (1933).

Angell, R.R., Jacobs, P.A.: Lateral asymmetry in human constitutive heterochromatin: Frequency and inheritance. Am. J. Hum. Genet. 30, 144–152 (1978).

Anner, R.M., Drewinko, B.: Frequency and significance of bone marrow involvement by metastatic solid tumors. Cancer 39, 1337–1344 (1977).

Anthonioz, P., Diakhate, L., Arnold, J.: A propos d'un chromosome de Philadelphie. Bull. Soc. Med. Afr. Noire Lang. Fr. 19, 309–314 (1974).

Archidiacono, A., Capoa, A. de, Ferraro, M., Pelliccia, F., Rocchi, A., Rocchi, M.: Nucleolus organizer and N-band distribution in morphologic and fluorescence variants of human chromosomes. Hum. Genet. 37, 285–289 (1977).

Ardito, G., Lamberti, L., Brøgger, A.: Satellite associations of human acrocentric chromosomes identified by trypsin treatment at metaphase. Ann. Hum. Genet. 41, 455–462 (1978).

Arduino, L.J.: Carcinoma of the prostate in sex chromatin positive (XXY/XY) Klinefelter's syndrome. J. Urol. 98, 234–240 (1967).

Arias-Bernal, L., Jones, H.W., Jr.: Chromosomes of a malignant ovarian teratoma. Am. J. Obstet. Gynecol. 100, 785–789 (1968).

Arlin, Z., Chaganti, R.S., Gee, T., Clarkson, B.: Complete remission of the blastic phase of chronic myelogenous leukemia (BLCML). Proc. Am. Assoc. Cancer Res. 18, 196 (1977).

Armenta, D., Cadotte, M., Beaulieu, R., Neemeh, J., Long, L., Pretty, H., Gosselin, G.: Cytogenetic evidence of the splenic origin of chronic myelogenous leukemia. Union Med. Can. 105, 922–927 (1976).

Arnold, J.: Über feinere Struktur der Zellen unter normalen und pathologischen Bedingungen. Virchows Arch. Pathol. Anat. 77, 181–206 (1879a).

Arnold, J.: Beobachtungen über Kernteilungen in den Zellen der Geschwülste. Virchows Arch. Pathol. Anat. 78, 279–301 (1879b).

Arrighi, F.E., Hsu, T.C.: Localization of heterochromatin in human chromosomes. Cytogenetics 10, 81–86 (1971).

Arrighi, F.E., Hsu, T.C., Bergsagel, D.E.: Chromosome damage in murine and human cells following cytoxan therapy. Tex. Rep. Biol. Med. 20, 545–549 (1962).

Arzapalo, E.G., Meneses, A.C., Garay, M.E., Gurrola, J.A., Diddi, M.G., Salamanca, F.: Granulosa cell tumors. Cytogenetic and histogenetic investigations of 16 cases. Ginecol. Obstet. Mex. 41, 59–67 (1977).

Asamer, H., Schmalzl, F., Braunsteiner, H.: Immunocytological demonstration of lysozyme (muramidase) in human leukaemic cells. Br. J. Haematol. 20, 571–574 (1971).

Astaldi, G.: Chairmen's opening remarks at the section on chromosomes. *Proceedings of the 9th Congress of the European Society of Haematology*, p. 14–19, Lisbon (1963).

Astaldi, G., Airò, R., Sauli, S.: *In vitro* studies on leukaemic cells. *In* F.G.J. Hayhoe (ed.): *Current*

Research in Leukaemia, pp. 139–163. London: Cambridge University Press (1965).
Astaldi, G. Strosseli, E., Airò, R., Pollini, G.: Recherches cytogénétiques dans l'anémie pernicieuse aprés traitement. Schweiz. Med. Wochenschr. 92, 1332–1333 (1962).
Athens, J.W.: Granulocyte kinetics in health and disease. Natl. Cancer Inst. Monogr. 30, 135–155 (1969).
Atkin, N.B.: Sex chromatin and chromosomal variation in human tumours. Acta Un. Int. Cancer 16, 41–46 (1960a).
Atkin, N.B.: A single heteropycnotic chromosome in a human tumour. Exp. Cell Res. 20, 214–215 (1960b).
Atkin, N.B.: The deoxyribonucleic acid content of malignant cells in cervical smears. Acta Cytol. 8, 68–72 (1964a).
Atkin, N.B.: The chromosomal changes in malignancy; an assessment of their possible prognostic significance. Br. J. Radiol. 37, 213–218 (1964b).
Atkin, N.B.: Nuclear size in carcinoma of the cervix: Its relation to DNA content and to prognosis. Cancer 17, 1391–1399 (1964c).
Atkin, N.B.: The influence of nuclear size and chromosome complement on prognosis of carcinoma of the cervix. Proc. R. Soc. Med. 59, 979–982 (1966).
Atkin, N.B.: Triple sex chromatin, and other sex chromatin anomalies, in tumours of females. Br. J. Cancer 21, 40–47 (1967a).
Atkin, N.B.: A high incidence of cells with a condensed (telophase) chromatin pattern in human tumors and carcinoma-in-situ. Acta Cytol. 11, 81–85 (1967b).
Atkin, N.B.: A carcinoma of the cervix uteri with hypodiploid and hypotetraploid stem-lines. Europ. J. Cancer 3, 289–291 (1967c).
Atkin, N.B.: Perimodal variation of DNA values of normal and malignant cells. Acta Cytol. 13, 270–273 (1969a).
Atkin, N.B.: Variant nuclear types in gynecologic tumors: Observations on squashes and smears. Acta Cytol. 13, 569–575 (1969b).
Atkin, N.B.: The use of microspectrophotometry. Obstet. Gynecol. Surv. 24, 794–804 (1969c).
Atkin, N.B.: The chromosome number of "hypochronic" cancer cells. Acta Cytol. 14, 463–464 (1970a).
Atkin, N.B.: Cytogenetic studies on human tumors and premalignant lesions: The emergence of aneuploid cell lines and their relationship to the process of malignant transformation in man. *In: Genetic Concepts and Neoplasia,* ed. by Staff of the Univ. of Texas and M.D. Anderson Hospital, 23rd Annual Symposium on Fundamental Cancer Research, University of Texas M.D. Anderson Hospital Tumor Institute, Houston, Tex., 1969, pp. 36–56. Baltimore, Md.: Williams & Wilkins (1970b).
Atkin, N.B.: Cytogenetic factors influencing the prognosis of uterine carcinoma. *In* T.J. Deeley (ed.): *Modern Radiotherapy,* 2nd edition, pp. 138–154. London: Butterworth (1971a).
Atkin, N.B.: Cytogenetic aspects of malignant transformation in the uterus. *In* J. M. Riggott (ed.): *The Early Diagnosis of Cancer of the Cervix,* pp. 33–44. University of Hull, Hull, England (1971b).
Atkin, N.B.: Sex chromatin positive metastatic melanoma in a male with a favourable prognosis. Br. J. Cancer 25, 487–492 (1971c).
Atkin, N.B.: Modal DNA value and chromosome number in ovarian neoplasia. A clinical and histopathologic assessment. Cancer 27, 1064–1073 (1971d).
Atkin, N.B.: Modal deoxyribonucleic acid value and survival in carcinoma of the breast. Br. Med. J. i, 271–272 (1972).
Atkin, N.B.: Y bodies and similar fluorescent chromocentres in human tumours including teratomata. Br. J. Cancer 27, 183–189 (1973a).
Atkin, N.B.: High chromosome numbers of seminomata and malignant teratomata of the testis: A review of data on 103 tumours. Br. J. Cancer 28, 275–279 (1973b).
Atkin, N.B.: Chromosomes in human malignant tumors. A review and assessment. *In* J. German (ed.): *Chromosomes and Cancer,* pp. 375–422. New York, London, Sydney, Toronto: Wiley (1974a).
Atkin, N.B.: Histophotometric determination of near-triploidy in products of conception including hydatiform moles. Aust. N.Z. J. Obstet. Gynaecol. 14, 39–41 (1974b).
Atkin, N.B.: Cytogenetic aspects of malignant transformation. *In* A. Wolsky (ed.): *Experimental Biology and Medicine,* Vol. 6, pp. 1–171. Basel, New York: Karger (1976a).
Atkin, N.B.: Prognostic significance of ploidy level in human tumours. II. Extra-uterine cancers and summary of data on 1171 tumours. Cytobios 15, 233–237 (1976b).
Atkin, N.B.: *Cytogenetic Aspects of Malignant Transformation.* Basel, München, Paris, London, New York, Sydney: Karger, (1976c).
Atkin, N.B.: Chromosome 1 heteromorphism in patients with malignant disease: a constitutional marker for a high-risk group? Br. Med. J. i, 358 (1977).
Atkin, N.B., Baker, M.C.: A nuclear protrusion in a human tumor associated with an abnormal chromosome. Acta Cytol. 8, 431–433 (1964).
Atkin, N.B., Baker, M.C.: Chromosome abnormalities, neoplasia, and autoimmune disease. Lancet i, 820–821 (1965a).
Atkin, N.B., Baker, M.C.: Chromosomes in carci-

noma of the cervix. Br. Med. J. i, 522–523 (1965b).

Atkin, N.B., Baker, M.C.: Chromosome abnormalities as primary events in human malignant disease: Evidence from marker chromosomes. J. Natl. Cancer Inst. 36, 539–559 (1966).

Atkin, N.B., Baker, M.C.: Possible differences between the karyotypes of preinvasive lesions and malignant tumours. Br. J. Cancer 23, 329–336 (1969).

Atkin, N.B., Baker, M.C.: Chromosome and DNA abnormalities in ovarian cystadenomas. Lancet i, 470 (1970).

Atkin, N.B., Baker, M.C.: Chromocentres in polymorphs as interphase markers for chromosomes having increased constitutive heterochromatin. J. Med. Genet. 11, 371–373 (1974).

Atkin, N.B., Baker, M.C.: Chromosome 1 in cervical carcinoma. Lancet ii, 984 (1977a).

Atkin, N.B., Baker, M.C.: Pericentric inversion of chromosome 1: frequency and possible association with cancer. Cytogenet. Cell Genet. 19, 180–184 (1977b).

Atkin, N.B., Baker, M.C.: Abnormal chromosomes and number 1 heterochromatin variants revealed in C-banded preparations from 13 bladder carcinomas. Cytobios 18, 101–109, 1977c.

Atkin, N.B., Baker, M.C.: Duplication of the long arm of chromosome 1 in a malignant vaginal tumour. Br. J. Cancer 38, 468–471 (1978).

Atkin, N.B., Baker, M.C.: Nuclear protrusions in malignant tumours with large abnormal chromosomes: Observations on C-banded preparations. Experientia (in press) (1979a).

Atkin, N.B., Baker, M.C.: Chromosome 1 in 26 carcinomas of the cervix uteri: Structural and numerical changes. (submitted) (1979b).

Atkin, N.B., Brandão, H.J.S.: Evidence for the presence of a large marker chromosome in histological sections of a carcinoma in situ of the cervix uteri. J. Obstet. Gynaecol. Br. Commonw. 75, 211–214 (1968).

Atkin, N.B., Klinger, H.P.: The superfemale mole. Lancet ii, 727–728 (1961).

Atkin, N.B., Petković, I.: Variable sex chromatin pattern in an early carcinoma of the bladder. J. Clin. Pathol. 26, 126–129 (1973).

Atkin, N.B., Pickthall, V.J.: No. 1 chromosomes in 14 ovarian cancers: Heterochromatin variants and structural changes. Hum. Genet. 38, 25–33 (1977).

Atkin, N.B., Richards, B.M.: Deoxyribonucleic acid in human tumours as measured by microspectrophotometry of Feulgen stain. A comparison of tumours arising at different sites. Br. J. Cancer 10, 769–786 (1956).

Atkin, N.B., Richards, B.M.: Clinical significance of ploidy in carcinoma of cervix. Its relation to prognosis. Br. Med. J. ii, 1445–1446 (1962).

Atkin, N.B., Ross, A.J.: Polyploidy in human tumours. Nature 187, 579–581 (1960).

Atkin, N.B., Taylor, M.C.: A case of chronic myeloid leukaemia with a 45-chromosome cell-line in the blood. Cytogenetics 1, 97–103 (1962).

Atkin, N.B., Baker, M.C., Brandão, H.J.S.: Chromosomes in effusions. Lancet ii, 1367 (1967a).

Atkin, N.B., Baker, M.C., Robinson, R., Gaze, S.E.: Chromosome studies on 14 near-diploid carcinomas of the ovary. Eur. J. Cancer 10, 143–146 (1974).

Atkin, N.B., Baker, M.C., Wilson, S.: Stem-line karyotypes of 4 carcinomas of the cervix uteri. Am. J. Obstet. Gynecol. 99, 506–514 (1967b).

Atkin, N.B., Mattinson, G., Baker, M.C.: A comparison of the DNA content and chromosome number of fifty human tumours. Br. J. Cancer 20, 87–101 (1966).

Atkin, N.B., Richards, B.M., Ross, A.J.: The deoxyribonucleic acid content of carcinoma of the uterus: An assessment of its possible significance in relation to histopathology and clinical course, based on data from 165 cases. Br. J. Cancer 13, 773–787 (1959).

Atkins, L., Goulian, M.: Multiple clones with increase in number of chromosomes in the G group in a case of myelomonocytic leukemia. Cytogenetics 4, 321–328 (1965).

Audebert, A.A., Krulik, M., Smadja, N., Debray, J.: Monoclonal paraprotein disorders other than multiple myeloma and Waldenstrom's disease. A report of 67 observations. Ann. Med. Interne (Paris) 10, 727–732 (1977).

Auerbach, A.D., Wolman, S.R.: Susceptibility of Fanconi's anemia fibroblasts to chromosome damage by carcinogens. Nature 261, 494–496 (1976).

Auersperg, N., Hawryluk, A.P.: Chromosome observations on three epithelial-cell cultures derived from carcinomas of the human cervix, J. Natl. Cancer Inst. 28, 605–627 (1962).

Auersperg, N., Wakonig-Vaartaja, T.: Chromosome changes in invasive carcinomas of the uterine cervix. Acta Cytol. 14, 495–501 (1970).

Auersperg, N., Worth, A.: Growth patterns in vitro of invasive squamous carcinomas of the cervix—a correlation of cultural, histologic, cytogenetic and clinical features, Int. J. Cancer 1, 219–238 (1966).

Auersperg, N., Corey, M.J., Austin, G.: Chromosomes in cervical lesions, Lancet i, 604–605 (1966).

Auersperg, N., Corey, M.J., Worth, A.: Chromosomes in preinvasive lesions of the human uterine cervix. Cancer Res. 27, 1394–1401 (1967).

Aula, P., Koskull, H. von: Distribution of spontaneous chromosome breaks in human chromosomes. Hum. Genet. 32, 143–148 (1976).

Aula, P., Saksela, E.: Banding characteristics of paracentric marker constrictions in human chromosomes. Hereditas 70, 309–310 (1972).

Awa, A., Bloom, A.D.: Cytogenetics at the Atomic Bomb Casualty Commission: Report of a symposium. Jap. J. Hum. Genet. 12, 69–75 (1967).

Awa, A.A., Mitani, M., Ōkōchi, K.: Chromosomes in two cases with acute childhood leukemia. Z. Krebsforsch. 67, 23–30 (1965).

Awano, I.: Chromosomes of man, normal and cancerous. Nucleus 4, 127–144 (1961).

Awano, I., Tsuda, F.: The chromosomes of stomach cancers and myelogenous leukemias in comparison with normal human complex. Jap. J. Hum. Genet. 34, 220–225 (1969).

Awano, I., Toshima, S., Tsuda, F.: A study of the chromosomes in human myelogenous leukemic cells with a note on the occurrence of virus-like particles. Cytologia (Tokyo) 26, 20–31 (1961).

Awano, I., Tsuda, F., Toshima, S.: The chromosomes of human gastric cancer, myelogenous leukemia and somatic cells, with special reference to the cancer virus (in Japanese). Symp. Cell Chem. 10, 159–176 (1960).

Awano, I., Tsuda, F., Ito, K., Abe K., Kokubun, K., Komada, R., Watari, H.: Chromosome studies in human acute leukemia (in Japanese). Jap. J. Intern. Med. 53, 1519–1532 (1965).

Aya, T., Makino, S.: Notes on chromosome abnormalities in cultured leucocytes from serum hepatitis patients. Proc. Jap. Acad. 42, 648–653 (1966).

Aya, T., Makino, S., Hirayama, A.: Some chromosomal studies in patients with infectious hepatitis. Proc. Jap. Acad. 42, 1088–1093 (1966).

Aymé, S., Mattei, J.F., Mattei, M.G., Aurran, Y., Giraud, F.: Nonrandom distribution of chromosome breaks in cultured lymphocytes of normal subjects. Hum. Genet. 31, 161–175 (1976).

Ayraud, N.: Identification par dénaturation thermique ménagée des anomalies chromosomiques observées dans six tumeurs métastatiques humaines. Biomedicine (Paris) 23, 423–430 (1975a).

Ayraud, N.: Identification of chromosomal aberrations observed in human cancer cells by R banding. C.R. Soc. Biol. (Paris) 169, 365–373 (1975b).

Ayraud, N., Kermarec, J.: Etude cytogénique de huit tumeurs d'origine mésothéliale. Bull. Cancer (Paris) 55, 91–110 (1968).

Ayraud, N., Dujardin, P., Audoly, P.: Leucémie aiguë lymphoblastique avec chromosome Philadelphie. Nouv. Presse Med. 4, 3013 (1975).

Ayraud, N., Duplay, J., Grellier, P., Bezon, A., Martinon, J.: Caryotype 45,X/46,XX/47,XXX et tumeur du systéme nerveux. Deux observations. Nouv. Presse Med. 1, 2902 (1972).

Ayraud, N., Lambert, J.-C., Hufferman-Tribollet, K., Basteris, B.: Etude cytogénétique comparative de sept carcinomes d'origine mammaire. Ann. Genet. (Paris) 20, 171–177 (1977).

Ayraud, N., Rey, C., Roustan, J.: Chromosome D en anneau dans un cas de sarcome embryonnaire. Arch. Anat. Pathol. (Paris) 19, 91–93 (1971).

Ayraud, N., Szepetowski, G., Gares, R., Vaillaud, J.C.: Sarcome embryonnaire et chromosome d'en anneau. Ann. Genet. (Paris) 13, 199–200 (1970).

Baccarani, M., Zaccaria, A., Tura, S.: Philadelphia-chromosome-positive preleukaemic state. Lancet ii, 1094 (1973).

Baccaredda, A.: Reticulohistiocytosis cutanea hyperplastica benigna cum melanoderma. Arch. Dermatol. Syph. 179, 209–256 (1939).

Bader, S., Taylor, H.C., Jr., Engle, E.T.: Deoxyribonucleic acid (DNA) content of human ovarian tumors in relation to histological grading. Lab. Invest. 9, 443–459 (1960).

Bagby, G.C., Jr., Kaiser-McCaw, B., Hecht, F., Koler, R.D., Linman, J.W.: Clonal evolution in atypical chronic granulocytic leukemia: A non-Philadelphia translocation. Blood 51, 997–1004 (1978).

Bagby, G.C., Jr., Richert-Boe, K., Koler, R.D.: [32]P and acute leukemia: Development of leukemia in a patient with hemoglobin Yakima. Blood 52, 350–354 (1978).

Baggish, M.S., Woodruff, J.D., Tow, S.H., Jones, H.W., Jr.: Sex chromatin pattern in hydatidiform mole. Am. J. Obstet. Gynecol. 102, 362–370 (1968).

Baguena-Candela, R., Forteza-Bover, G.: Estudio citogenético de una leucosis aguda del tipo mieloblástico-promielocítico con cromosoma Filadelfia (Ph[1]). Med. Esp. 298, 1–8 (1964).

Bahr, G.F.: Human chromosome fibres. Considerations of DNA-protein packing and of looping patterns. Exp. Cell Res. 62, 39–49 (1970).

Bahr, G.F.: The fibrous structure of human chromosomes in relation to rearrangements and aberrations; a theoretical consideration. Fed. Proc. 34, 2209–2217 (1975).

Bahr, G.F.: Considerations of the structure of chromosomes and chromatin in relation to chromosomal aberrations. Virchows Arch. B Cell Pathol. 29, 3–14 (1978).

Bahr, G.F., Engler, W.F.: Association of centrioles and chromosomes observed in preparations of whole-mounted human chromosomes. Chromosoma 63, 295–303 (1977).

Bahr, G.F., Golomb, H.M.: Karyotyping of single human chromosomes from dry mass determined by electron microscopy. Proc. Natl. Acad. Sci. USA 68, 726–730 (1971).

Bahr, G.F., Golomb, H.M.: Constancy of a 200 Å fiber in human chromatin and chromosomes Chromosoma 46, 247–254 (1974).

Bahr, G.F., Larsen, P.M.: Structural "bands" in human chromosomes. Adv. Cell Mol. Biol. 3, 191–212 (1974).
Baikie, A.G.: Polymorph alkaline phosphatase and genes on the mongol chromosome. Lancet ii, 937 (1962).
Baikie, A.G.: The Philadelphia chromosome. Lancet i, 556–557 (1964).
Baikie, A.G.: Chromosomal aspects of leukaemia. XIth Congress of the International Society of Hematology, Sydney, Australia, pp. 198–210 (1966a).
Baikie, A.G.: Chromosomes and leukaemia. Acta Haematol. (Basel) 36, 157–173 (1966b).
Baikie, A.G.: Chronic granulocytic leukaemia. Med. J. Aust. 2, 12–17 (1974).
Baikie, A.G., Buckton, K.E., Court Brown, W.M., Harnden, D.G.: Two cases of leukaemia and a case of sex chromosome abnormality in the same sibship. Lancet ii, 1003–1004 (1961).
Baikie, A.G., Court Brown, W.M., Buckton, K.E., Harnden, D.G., Jacobs, P.A., Tough, I.M.: A possible specific chromosome abnormality in human chronic myeloid leukaemia. Nature 188, 1165–1166 (1960a).
Baikie, A.G., Court Brown, W.M., Jacobs, P.A.: Chromosome studies in leukaemia. Lancet i, 168 (1960b).
Baikie, A.G., Court Brown, W.M., Jacobs, P.A.: Chromosome studies in leukaemia. Lancet i, 280 (1960c).
Baikie, A.G., Court Brown, W.M., Jacobs, P.A.: Chromosome constitution of mongols with leukaemia. Lancet i, 1251 (1960d).
Baikie, A.G., Court Brown, W.M., Jacobs, P.A.: Chromosome studies in leukaemia. Lancet i, 615 (1961).
Baikie, A.G., Court Brown, W.M., Jacobs, P.A., Milne, J.S.: Chromosome studies in human leukaemia. Lancet ii, 425–428 (1959).
Baikie, A.G., Garson, O.M., Spiers, A.S.D., Ferguson, J.: Cytogenetic studies in familial leukaemias. Australas. Ann. Med. 18, 7–11 (1969).
Baikie, A.G., Jacobs, P.A., McBride, J.A., Tough, I.M.: Cytogenetic studies in acute leukaemia. Br. Med. J. i, 1564–1571 (1961).
Baillet, J., Jullien, P.: Si on me trouvait un chromosome Philadelphie ... Concours Med. 87, 6847–6853 (1965).
Bain, B., Catovsky, D., O'Brien, M., Spiers, A.S., Richards, H.G.: Megakaryoblastic transformation of chronic granulocytic leukaemia. An electron microscopy and cytochemical study. J. Clin. Pathol. 30, 235–242 (1977).
Bajer, A.: Subchromatid structure of chromosomes in the living state. Chromosoma 17, 291–302 (1965).
Bak, A.L., Zeuthen, K., Crick, F.H.C.: Higher-order structure of human mitotic chromosomes. Proc. Natl. Acad. Sci. USA 74, 1595–1599 (1977).
Baker, M.C.: A chromosome study of seven near-diploid carcinomas of the corpus uteri. Br. J. Cancer 22, 683–695 (1968).
Baker, M.C., Atkin, N.B.: Short-term culture of lymphoid tissue for chromosome studies. Lancet i, 1164 (1963).
Baker, M.C., Atkin, N.B.: Chromosomes in short-term cultures of lymphoid tissue from patients with reticulosis. Br. Med. J. i, 770–771 (1965).
Baker, M.C., Atkin, N.B.: Chromosome abnormalities in polyps and carcinomas of the large bowel. Proc. R. Soc. Med. 63 (Suppl.), 9–10 (1970).
Balaban-Malenbaum, G., Gilbert, F.: Double minute chromosomes and the homogeneously stained regions in chromosomes of a human neuroblastoma cell line. Science 198, 739–741 (1977).
Baldwin, J.P., Boseley, P.G., Bradbury, E.M., Ibel, K.: The subunit structure of the eukaryotic chromosome. Nature 253, 245–249 (1975).
Balducci, L., Weitzner, S., Beghe, C., Morrison, F.S.: Acute megakaryocytic leukemia. Description of a case initially seen as preleukemia syndrome. Arch. Intern. Med. 138, 794–795 (1978).
Balíček, P., Zizka, J., Skalská, H.: Length of human constitutive heterochromatin in relation to chromosomal contraction. Hum. Genet. 38, 189–193 (1977).
Ballesta, I., Baldellou, A.: Cytogenetics and neoplasia. Bol. Soc. Catal. Pediatr. 31, 240–249 (1970).
Bamford, S.B., Mitchell, G.W., Jr., David, J., Sperber, A., Cassin, C.: Size variation of the late replicating X-chromosome in the leukocytes of individuals with hyperplastic and malignant lesions of uterine epithelium. Acta Cytol. 13, 238–245 (1969).
Banerjee, A.R.: Chromosome abnormalities in Waldenström's macroglobulinaemia. Hum. Chrom. Newsl. 14, 2–3 (1964).
Banerjee, A., Jung, O., Huang, C.C.: Response of hematopoietic cell lines derived from patients with Down's syndrome and from normal individuals to Mitomycin C and caffeine. J. Natl. Cancer Inst. 59, 37–39 (1977).
Baradnay, G., Monus, Z., Kukla, F.: Sexchromatin. Untersuchungen in weiblichen Lungenkrebs-Fällen. Zentralbl. Allg. Pathol. 111, 275–276 (1968).
Barcinski, M.A., Abreu, M. do C.A., Almeida, J.C.C. de, Naya, J.M., Fonseca, L.G., Castro, L.E.: Cytogenetic investigation in a Brazilian population living in an area of high natural radioactivity. Am. J. Hum. Genet. 27, 802–806 (1975).
Barker, P.E., Hsu, T.C.: Are double minutes chromosomes? Exp. Cell Res. 113, 457–458 (1978).
Barker, P.E., Hsu, T.C.: Double minutes in human

carcinoma cell lines, with special reference to breast tumors. J. Natl. Cancer Inst. 62, 257–262 (1979).
Barker, P.E., Pathak, S.: A new class of chromosome change in human tumor cells. American Society on Human Genetics, 29th Annual Meeting, Vancouver, B.C., October 4–7, Abstract, p. 73A (1978).
Barlogie, B., Hittelman, W., Spitzer, G., Trujillo, J.M., Hart, J.S., Smallwood, L., Drewinko, B.: Correlation of DNA distribution abnormalities with cytogenetic findings in human adult leukemia and lymphoma. Cancer Res. 37, 4400–4407 (1977).
Barnes, C.A., Holmes, H.L., Ilbery, P.L.T.: Chromosome aberration following radiophosphorus treatment of polycythaemia. Australas. Radiol. 13, 396–417 (1969).
Barnett, R.I., MacKinnon, E.A., Romero-Sierra, C.: Delineation of human chromosome contour by heat treatment and hematoxylin staining. Chromosoma 40, 299–306 (1973).
Barnett, R.I., MacKinnon, E.A., Romero-Sierra, C.: The lability of acid-alcohol fixed human chromosomes. Cytobios 11, 115–122 (1974).
Barr, M.L., Bertram, E.G.: A morphological distinction between neurons of the male and female, and the behavior of the nucleolar satellite during accelerated nucleoprotein synthesis. Nature 163, 676–677 (1949).
Barr, M.L., Carr, D.H.: Sex chromatin, sex chromosomes and sex anomalies. Can. Med. Assoc. J. 83, 979–986 (1960).
Barr, M.L., Carr, D.H.: Correlations between sex chromatin and sex chromosomes. Acta Cytol. 6, 34–45 (1962a).
Barr, M.L., Carr, D.H.: Nuclear sex. In J.L. Hamerton (ed.): *Chromosomes in Medicine*, pp. 61–72. London: Heineman (1962b).
Barr, R.D., Fialkow, P.J.: Clonal origin of chronic myelocytic leukemia. N. Engl. J. Med. 289, 307–309 (1973).
Barr, R.D., Watt, J.: Preliminary evidence for the common origin of a lympho-myeloid complex in man. Acta Haematol. (Basel) 60, 29–35 (1978).
Bartalos, M.: Time factor in cytogenetics and neoplasia. Acta Genet. Med. Gemellol. (Roma) 20, 350–358 (1971).
Barton, J.C., Conrad, M.E.: Current status of blastic transformation in chronic myelogenous leukemia. Am. J. Hematol. 4, 281–291 (1978).
Bartram, C.R., Rüdiger, H.W.: Chromosomenanomalien bei malignen Tumoren. Klin. Wochenschr. 56, 733–741 (1978).
Bartram, C.R., Koske-Westphal, T., Passarge, E.: Chromatid exchanges in ataxia telangiectasia, Bloom's syndrome, Werner syndrome and xeroderma pigmentosum. Ann. Hum. Genet. 40, 79–86 (1976).
Baserga, A.: Il cariotipe nelle situazione preleucemiche e preneoplastiche. Haematologica (Pavia) 55, 165–172 (1970).
Baserga, A.: Studio cromosomico delle malattie immunoproliferative. Recent Prog. Med. (Roma) 53, 183–197 (1972).
Baserga, A., Castoldi, G.L.: Chromosomes in chronic lymphatic leukaemia. Lancet ii, 1299–1300 (1965).
Baserga, A., Castoldi, G.L.: Chromosome alterations in chronic myeloid leukaemia. Present aspects of the problem. In S. Tura, M. Baccarani (eds.): *Proceedings of an International Symposium on Chronic Myeloid Leukemia*, pp. 25–44. Pavia: Edizioni di Haematologica (1972a).
Baserga, A., Castoldi, G.L.: Chromosome alterations in chronic myeloid leukaemia. Present aspects of the problem. Haematologica (Pavia) 57, 621–640 (1972b).
Baserga, A., Castoldi, G.L., The Philadelphia chromosome. Biomedicine 18, 89–94 (1973).
Baserga, A., Ricci, N.: Cytogenetic studies in erythroleukemia. 10th Int. Congr. Haematol., Stockholm (1964).
Baserga, A., Bosi, L., Castoldi, G.L., Punturieri, E., Ricci, N.: Lo studio dei cromosomi nelle emoblastosi. Haematologica (Pavia) 47 (Suppl), 83–120 (1962).
Baserga, A., Castoldi, G.L., Franceschini, F.: Étude chromosomique des leucémies lymphocytaires chroniques. Schweiz. Med. Wochenschr. 96, 1220–1222 (1966).
Bath, D.W.: Surface ultrastructure of trypsin-banded chromosomes. Exp. Cell Res. 98, 262–268 (1976).
Bauchinger, M., Dresp, J., Schmid, E., Englert, N., Krause, Chr.: Chromosome analyses of children after ecological lead exposure. Mutat. Res. 56, 75–80 (1977).
Bauer, K.H.: *Mutationstheorie der Geschwulstentstehung, Übergang von Körperzellen in Geschwulstzellen durch Genänderung.* S. 1–72. Berlin: Springer (1928).
Bauke, J.: Chronic myelocytic leukemia. Chromosome studies of a patient and his nonleukemic identical twin. Cancer 24, 643–648 (1969).
Bauke, J.: Diskussionsbemerkung zur Arbeit: Hossfeld, D.K., Sandberg, A.A.: Das Philadelphia Chromosom." Klin. Wochenschr. 49, 1144–1145 (1970a).
Bauke, J.: Cytogenetic studies of a patient with chronic myelocytic leukemia and his nonleukemic identical twin. Acta Genet. Med. Gemellol. (Roma) 19, 180–182 (1970b).
Bauke, J.: Extramedullary manifestations of chronic myelogenous leukemia. Presence of hyperdip-

loid stemlines with Ph¹ disomy in lymph nodes and pleural effusions. Proceedings of the 2nd Meeting of the Asian-Pacific Division of the International Society of Haematology, Melbourne (1971).
Bauke, J.: Klonale Evolution in Spätstadium der chronischen myeloischen Leukämie. I. Beobachtungen von Stammlinien mit einen metazentrischen Marker-Chromosom bei acht Patienten. Dtsch. Med. Wochenschr. 98, 1956–1959 (1973).
Bauke, J., Bach, G.: Klonale Evolution mit Acquisation und Duplikation von Extrachromosomen bei Ph¹ negativer chronischer myeloischer Leukämie. *In* R. Gross, J. Van de Loo (eds.): *Leukämie,* S. 93–97. Berlin, Heidelberg, New York: Springer (1972).
Bauke, J., Schöffling, K.: Polyploidy in human malignancy. Hypopentaploid chromosome pattern in malignant reticulosis with secondary sideroachrestic anemia. Cancer 22, 686–694 (1968).
Bauke, J., Cremer, H.J., Heimpel, H.: Kongenitale myelomonocytäre Leukämie mit aneuploider Stammlinie. Z. Kinderheilkd. 168, 288–296 (1970).
Bauke, J., Hoerler, S., Gienger, S., Heimpel, H.: Clinical and cytogenetic investigations of preleukemic states. Verh. Dtsch. Ges. Inn. Med. 79, 419–422 (1973).
Bauke, J., Kaiser, G., Schöffling, K.: Chromosome studies in plasmacytoma and plasma cell leukemia. Verh. Dtsch. Ges. Inn. Med. 78, 122–125 (1972).
Bauke, J., Neubauer, A., Schöffling, K.: Über das Auftreten einer Stammlinie mit 55 Chromosomen und Diplo-Philadelphia-Chromosom vor der terminalen Blastenkrise einer chronischen myeloischen Leukämie. Dtsch. Med. Wochenschr. 92, 301–305 (1967).
Bauters, F., Goudemand, M.: Les débuts thrombocytémiques de leucémie myéloide chronique. Nouv. Rev. Fr. Hematol. 15, 241–255 (1975).
Bauters, F., Croquette, M.F., Delmas-Marsalet, Y., Deminatti, M., Goudemand, M.: Une forme particuliére de leucémie myélöide chez `l'-homme: évolution prolongée et présence du chromosome Philadelphie avec perte du chromosome Y dans les cellules myélöides. A propos de trois obsérvations. Nouv. Rev. Fr. Hematol. 10, 697–707 (1970).
Bayle, R., Linhard, J.: Leucose myéloblastique á deux populations de cellules sanguines (cellules normales et cellules haplo 21–22). Med. Afrique Noire 13, 311 (1966).
Bayreuther, K.: Chromosomes in primary neoplastic growth. Nature 186, 6–9 (1960).

Beard, M.E.: Fanconi's anaemia. Ciba Found. Symp. 37, 103–114 (1976).
Beard, M.E.J., Durrant, J., Catovsky, D., Wiltshaw, E., Amess, J.L., Brearley, R.L., Kirk, B., Wrigley, P.F.M., Janossy, G., Greaves, M.F., Galton, D.A.G.: Blast crisis of chronic myeloid leukaemia (CML). 1. Presentation simulating acute lymphoblastic leukaemia (ALL). Br. J. Haematol. 34, 169–181 (1976).
Beatty-de Sana, J.W., Hoggard, M.J., Cooledge, J.W.: Non-random occurrence of 7-14 translocation in human lymphocyte cultures. Nature 255, 242–243 (1975).
Becak, W., Becak, M.L., Andrade, J.D., Manissadjian, A.: Extra acrocentric chromosome in a case of giant cavernous haemangioma with secondary thrombocytopenia. Lancet ii, 468–469 (1963).
Becak, W., Becak, M.L., Saraiva, L.G.: Cromosoma acrocéntrico gigante en un caso de sindrome de Di Guglielmo. Sangre (Barc.) 12, 65–70 (1967).
Beck, W.S., Chesney, T. McC: Myelocytic leukemia with an unusual chromosomal pattern. N. Engl. J. Med. 288, 957–963 (1973).
Beischer, N.A., Fortune, D.W., Fitzgerald, M.G.: Hydatidiform mole and coexistent foetus, both with triploid chromosome constitution. Br. Med. J. iii, 476–478 (1967).
Bekkum, D.W. van, Noord, M.Y. van, Maat, E., Dicke, K.A.: Attempts at identification of hematopoietic stem cell in mouse. Blood 38, 547–558 (1971).
Bell, W.R., Whang, J.J., Carbone, P.P., Brecher, G., Block, J.B.: Cytogenetic and morphologic abnormalities in human bone marrow cells during cytosine arabinoside therapy. Blood 27, 771–781 (1966).
Belling, J.: The number of chromosomes in the cells of cancerous and other human tumors. J.A.M.A. 88, 396 (1927).
Ben-Bassat, H., Goldblum, N., Mitrani, S., Goldblum, T., Yoffey, J.M., Cohen, M.M., Bentwich, Z., Ramot, B., Klein, E., Klein, G.: Establishment in continuous culture of a new type of lymphocyte from a "Burkitt-like" malignant lymphoma (line D.G.-75). Int. J. Cancer 19, 27–33 (1977).
Bender, M.A.: Chromosome aberrations in irradiated human subjects. Ann. N.Y. Acad. Sci. 114, 249–251 (1964).
Bender, M.A., Gooch, P.C.: Persistent chromosome aberrations in irradiated human subjects. II. Three and one-half year investigation. Radiat. Res. 18, 389–396 (1963).
Bender, M.A., Kastenbaum, M.A., Lever, C.A.: Chromosome 16: A specific chromosomal pathway for the origin of human malignancy? Br. J. Cancer 26, 34–42 (1972).

Benedict, W.F., Porter, I.H.: The cytogenetic diagnosis of malignancy in effusions. Acta Cytol. 16, 304–306 (1972).
Benedict, W.F., Brown, C.D., Porter, I.H.: Chromosomes in effusions. Lancet ii, 1146 (1967).
Benedict, W.F., Brown, C.D., Porter, I.H.: Long acrocentric marker chromosomes in malignant effusions and solid tumors. N.Y. State J. Med. 71, 952–955 (1971).
Benedict, W.F., Porter, I.H., Brown, C.D., Florentin, R.A.: Cytogenetic diagnosis of malignancy in recurrent meningioma. Lancet i, 971–973 (1970).
Benedict, W.F., Rosen, W.C., Brown, C.D., Porter, I.H.: Chromosomal aberrations in an ovarian cystadenoma. Lancet ii, 640 (1969).
Benedict, W.F., Rucker, N., Mark, C., Kouri, R.E.: Correlation between balance of specific chromosomes and expression of malignancy in hamster cells. J. Natl. Cancer Inst. 54, 157–161 (1975).
Benirschke, K., Brownhill, L., Ebaugh, F.G.: Chromosomal abnormalities in Waldenström's macroglobulinaemia. Lancet i, 594–595 (1962).
Benn, P.A., Harnden, D.G., Fairburn, E.A.: Subpopulations of cytogenetically abnormal cells in fibroblast cultures derived from workers exposed to mineral oil. J. Natl. Cancer Inst. 60, 45–50 (1978).
Bennett, J.M.: Plasma cell myeloma terminating in erythremic myelosis. Proc. Am. Assoc. Cancer Res. 12, 16 (1971).
Bennett, J.M., Reed, C.E.: Acute leukemia cytochemical profile. Diagnostic and clinical implications. Blood Cells 1, 101 (1975).
Bennett, J.M., Catovsky, D., Daniel, M.-T., Flandrin, G., Galton, D.A.G., Gralnick, H.R., Sultan, C.: Proposals for the classification of the acute leukaemias. Br. J. Haematol. 33, 451–458 (1976).
Bentley, D.: A case of Down's syndrome complicated by retinoblastoma and celiac disease. Pediatrics 56, 131–133 (1975).
Benvenisti, D.S., Ultman, J.E.: Eosinophilic leukemia: Report of five cases and review of literature. Ann. Intern. Med. 71, 731–745 (1969).
Benyush, V.A., Luckash, V.G., Shtannikov, A.V.: Quantitative analysis of C-bands based on optical density profiles in human chromosomes. Hum. Genet. 39, 169–175 (1977).
Berard, C., O'Conor, G.T., Thomas, L.B., Torloni, H.: Histopathological definition of Burkitt's tumour. Bull. WHO 40, 601–607 (1969).
Berard, C.W., Jaffe, E.S., Braylan, R.C., Mann, R.B., Nanba, K.: Immunologic aspects and pathology of the malignant lymphomas. Cancer 42, 911–921 (1978).
Berger, R.: Chromosomes et leucémies humaines. La notion d'évolution clonale. Ann. Genet. (Paris) 8, 70–82 (1965).
Berger, R.: Sur la méthodologie de l'analyse des chromosomes des tumeurs. Paris, Faculty of Science, Thèses (1968a).
Berger, R.: Leucémies, cancers et anomalies chromosomiques. Gaz. Med. France 75, 961–976 (1968b).
Berger, R.: Chromosomes et tumeurs. I. Techniques de préparation. Pathol. Biol. (Paris) 17, 1051–1058 (1969a).
Berger, R.: Chromosomes et tumeurs humaines. II. Méthodes d'analyse des anomalies des chromosomes dans les tumeurs. Pathol. Biol. (Paris) 17, 1059–1068 (1969b).
Berger, R.: Chromosomes et tumeurs humaines. Pathol. Biol. (Paris) 17, 1133–1151 (1969c).
Berger, R.: Anomalies chromosomiques constitutionnelles et leucémies. Nouv. Rev. Fr. Hematol. 10, 99–107 (1970a).
Berger, R.: Leucémie myélöide chronique et chromosomes. Rev. Europ. Etud. Clin. Biol. 15, 1000–1007 (1970b).
Berger, R.: Anomalies chromosomiques constitutionnelles et néoplasies. Presse Med. 79, 1107–1109 (1971).
Berger, R.: Leucémie myélomonocytaire de l'enfant. Etude chromosomique. Pathol. Biol. (Paris) 20, 479–484 (1972a).
Berger, R.: Cytogénétique et hémopathies. Nouv. Rev. Fr. Hematol. 12, 510–515 (1972b).
Berger, R.: Chromosome Philadelphie. Nouv. Presse Med. 2, 3121 (1973).
Berger, R.: Translocation t(11;20) et polyglobulie primitive. Nouv. Presse Med. 4, 1972 (1975).
Berger, R.: Unusual chromosomal segregations in hematologic disorder. Helsinki Chromosome Conference, p. 192, August 29–31 (abstract) (1977).
Berger, R., Aubert, C.: Transformation of a malignant melanoma in vitro. Chromosomal study. C.R. Acad. Sci. [D.] (Paris) 280, 2409–2412 (1975).
Berger, R., Bernheim, A.: Anomalie de fréquence du chromosome 1 qh+ dans la leucémie myélöide chronique. C. R. Acad. Sci. [D] (Paris) 285, 1183–1185 (1977).
Berger, R., Bernheim, A.: Reciprocal translocation and the Philadelphia chromosome. Hum. Genet. 44, 357–358 (1978a).
Berger, R., Bernheim, A.: Non-randomness in complex translocations of chronic myeloid leukaemia. Scand. J. Haematol. 21, 418–420 (1978b).
Berger, R., Bernheim, A.: Y chromosome loss in leukemias. Cancer Genet. Cytogenet. (in press) (1979).
Berger, R., Lacour, J.: Etude chromosomique de mélanomes malins. Biomedicine 19, 22–27 (1973).
Berger, R., Lacour, J.: Anomalies chromosomiques

dans un cancer de l'ovarire. Pathol. Biol. (Paris) 22, 603–606 (1974).
Berger, R., Martineau, M.: Analyse caryotypique de tumeurs testiculaires. Eur. J. Cancer 6, 61–66 (1970).
Berger, R., Parmentier, C.: Leucémie lymphoide chronique et chromosomes. Nouv. Rev. Fr. Hematol. 11, 261–277 (1971).
Berger, R., Bussel, A., Schenmetzler, C.: Anomalies chromosomiques et anémie de Fanconi. Etude de 4 cas. Nouv. Rev. Fr. Hematol. 15, 539–550 (1975).
Berger, R., Bussel, A., Schenmetzler, C.: Somatic segregation and Fanconi anemia. Clin. Genet. 11, 409–415 (1977).
Berger, R., Gyger, M., Bussel, A.: Anomalie chromosomique nouvelle dans une leucémie myélöide chronique. Nouv. Rev. Fr. Hematol. 16, 309–320 (1976).
Berger, R., Lejeune, J., Lacour, J.: Evolution chromosomique d'un mélanome malin. Rev. Eur. Etud. Clin. Biol. 16, 476–481 (1971).
Berger, R., May-Lewin, F., Contesso, G.: Cancer du sein et syndrome de Klinefelter. Nouv. Presse Med. 2, 111 (1973).
Berger, R., Parmentier, C., Droz, J.P.: Karyotype studies in polycythaemia vera. Nouv. Rev. Fr. Hematol. 14, 705–712 (1974).
Berger, R., Weisgerber, C., Bernard, J.: Evolution clonale au cours d'une leucémie aiguë chez un enfant mongolien. Nouv. Rev. Fr. Hematol. 13, 229–236 (1973).
Berghoff, D., Passarge, E.: ^{3}H-thymidine incorporation at the end of the S phase in cultured human lymphocytes. Humangenetik 24, 141–144 (1974).
Bergsman, K.L., Slyck, E.J. van: Acute myelofibrosis. An accelerated variant of agnogenic myeloid metaplasia. Ann. Intern. Med. 74, 232–235 (1971).
Bernard, C., Tanzer, J.: Chromosomes et leucémies. Bull. Assoc. Fr. Cancer 51, 7–24 (1964).
Bernard, C., Geraldes, A., Boiron, M.: Effects of phytohaemagglutinin on blood-cultures of chronic lymphocytic leukaemias. Lancet i, 667–668 (1964a).
Bernard, C., Geraldes, A., Boiron, M.: Action de la phytohémagglutinine "in vitro" sur les lymphocytes de leucémie lymphoides chronique. Nouv. Rev. Fr. Hematol. 4, 69–76 (1964b).
Berns, M.W., Cheng, W.K.: Are chromosome secondary constrictions nucleolar organizers? Exp. Cell Res. 69, 185–192 (1971).
Berry, E. W., Desforges, J.F.: Changing cytogenetic picture in an acute myeloproliferative disorder. Am. J. Med. 47, 299–305 (1969).
Bersi, M., Gasparini, C.: Anomalie cromosomiche in un caso di anemia di Fanconi. Minerva Med. 64, 1633–1637 (1973).
Bersi, M., Gasparini, C., Cardini, G.: Studio cromosomico su di un caso di leucemia emocitoblastica. Minerva Med. 62, 1981–1984 (1971a).
Bersi, M., Gasparini, C., Cardini, G.: Su di un corredo cromosomico alterato in un caso di istioleucemia acuta. Minerva Med. 62, 1120–1124 (1971b).
Bessis, M.: Ultrastructure of normal and leukemic granulocytes. In C.J.D. Zarafonetis (ed.) Proceedings of the International Conference on Leukemia–Lymphoma, pp. 281–303. Philadelphia: Lea & Febiger (1968).
Better, O., Brandstaetter, S., Padeh, R., Bianu, G.: Myeloid metaplasia. Clinical, laboratory and cytogenetic observations. Israel Med. 23, 162–175 (1964).
Beutler, E., Goldenburg, E.W., Ohno, S., Yettra, M.: Chromosome-21 and paroxysmal nocturnal hemoglobinuria. Blood 24, 160–163 (1964).
Bey, E., Alexander, J., Whitcutt, J.M., Hunt, J.A., Gear, J.H.S.: Carcinoma of the esophagus in Africans: Establishment of a continuously growing cell line from a tumor specimen. In Vitro 12, 102–114 (1976).
Bicknell, J.M.: Chromosome studies of human brain tumors. Neurology 17, 485–490 (1967).
Biedler, J.L.: A novel chromosomal abnormality in human neuroblastoma cells. Presented at the Symposium of the American Cancer Society on Advances in Neuroblastoma Research, Philadelphia (1975a).
Biedler, J.L.: Chromosome abnormalities in human tumor cells in culture. In J. Fogh (ed.): Human Tumor Cells in Vitro, pp. 359–394. New York: Plenum Press (1975b).
Biedler, J.L., Spengler, B.A.: Metaphase chromosome anomaly: Association with drug resistance and cell-specific products. Science 191, 185–187 (1976a).
Biedler, J.L., Spengler, B.A.: A novel chromosome abnormality in human neuroblastoma and antifolate-resistant Chinese hamster cell lines in culture. J. Natl. Cancer Inst. 57, 683–695 (1976b).
Biedler, J.L., Helson, L., Spengler, B.A.: Morphology and growth, tumorigenicity, and cytogenetics of human neuroblastoma cells in continuous culture. Cancer Res. 33, 2643–2652 (1973).
Biervliet, J.P. van, Hemel, J. van, Guerts, K., Punt, K., de Boer-van Wering, E.: Philadelphia chromosome in acute lymphocytic leukaemia. Lancet ii, 617 (1975).
Bigger, T.L.R., Savage, J.R.K., Watson, G.E.: A scheme for characterizing ASG banding and an illustration of its use in identifying complex chromosomal rearrangements in irradiated human skin. Chromosoma 39, 297–310 (1972).
Binder, R.A., Jencks, J.A., Chun, B., Rath, C.E.: "B" cell origin of malignant cells in a case of Ameri-

can Burkitt's lymphoma. Characterization of cells from a pleural effusion. Cancer 36, 161–168 (1975).

Biscatti, G., Vaccaro, R.: Chromosoma Ph¹ in un caso di leucemia mioloide cronica in ita infantile. Pediatria (Napoli) 73, 938–945 (1965).

Bishop, J.O., Madson, E.C.: Retinoblastoma: Review of the current status. Surv. Ophthalmol. 19, 342–366 (1975).

Bishop, R.P., Young, I.T.: The automated classification of mitotic phase for human chromosome spreads. J. Histochem. Cytochem. 25, 730–740 (1977).

Bishun, N.P.: The cytogenetic effect of cyclophosphamide on a Burkitt tumor cell line (EB_4) in vitro. Mutat. Res. 11, 258–260 (1971).

Bishun, N.P., Morton, W.R.M.: Endoreduplicated chromosomes. Lancet i, 1169–1170 (1965).

Bishun, N.P., Sutton, R.N.P.: Cytogenetic and other studies on the EB4 line of Burkitt tumour cells. Br. J. Cancer 21, 675–678 (1967).

Bishun, N.P., Eddie, H., Williams, D.C.: Cytogenetic studies with diethyl stilboestrol. Clin. Oncol. 4, 159–165 (1978).

Bishun, N.P., Morton, W.R.M., Rashad, M.N.: Chromosome endoreduplication. Lancet i, 1452 (1964).

Bishun, N.P., Raven, R.W., Williams, D.C.: Chromosomes and breast cancer. Cytobios 10, 7–10 (1974).

Bishun, N.P., Williams, D.C., Mills, J., Lloyd, N., Raven, R.W., Parke, D.V.: Chromosome damage induced by chemicals. Chem. Biol. Interact. 6, 375–392 (1973).

Biswas, S., Chowdhury, J.R.: Evaluation of chromosome morphology in human normal and cancer cervix cells *in vitro*. Indian J. Med. Res. 56, 1595–1599 (1968).

Bitran, J., Golomb, H.M., Rowley, J.D.: Idiopathic acquired sideroblastic anemia: Banded chromosomal analysis in six patients. Acta Haematol. (Basel) 57, 15–23 (1977).

Bitran, J.D., Rowley, J.D., Plapp, F., Golomb, H.M., Ultmann, J.E.: Chromosomal aneuploidy in a patient with hypereosinophilic syndrome: Evidence for a malignant disease. Am. J. Med. 63, 1010–1014 (1977).

Björnness, H., Bühler, E.M., Fricker, H., Gugler, E.: Kongenitale Leukämie mit Chromosomenveränderungen (Trisomie G) bei einem nichtmongoloiden Kinde. Helv. Paediatr. Acta 29, 457–470 (1974).

Blackburn, E.K., Callender, S.T., Dacie, J.V., Doll, R., Girdwood, R.H., Mollin, D.L., Saracci, R., Stafford, J.L., Thompson, R.B., Varadi, S., Wetherley-Mein, G.: Possible association between pernicious anaemia and leukaemia: A prospective study of 1,625 patients with a note on the very high incidence of stomach cancer. Int. J. Cancer 2, 163–170 (1968).

Blackstock, A.M., Garson, O.M.: Direct evidence for involvement of erythroid cells in acute myeloblastic leukaemia. Lancet ii, 1178–1179 (1974).

Blake, M.N.: A chromosome abnormality associated with macroglobulinaemia. Aust. N.Z. J. Med. 15, 162–168 (1966).

Blatnik, D., Jung, F., Jung, M.: Klinični in laboratorijsky študij kemoterapije hemoblastoz. I. Analize kromosomov pri nekaterih bolnikih z malignimi obolenji. Zdravstveni Vestnik 31, 1–11 (1962).

Blattner, R.J.: Chromosomes in chronic myeloid leukemia and in acute leukemia associated with mongolism. J. Pediatr. 59, 145–158 (1961).

Blein, J.P., Garnier, D., Maraud, R.: Influence sur le caryotype lymphocytaire, du traitement par le radioiode ^{131}I au cours de diverses affections thyroidiennes. Ann. Endocrinol. (Paris) 33, 231–242 (1972).

Blij-Philipsen, M. van der, Breed, W.P.M., Hustinx, T.W.J.: A case of chronic myeloid leukemia with a translocation (12;22) (p13;q11). Hum. Genet. 39, 229–231 (1977).

Bloch-Shtacher, N., Sachs, L.: Chromosome balance and the control of malignancy. J. Cell Physiol. 87, 89–100 (1976).

Bloch-Shtacher, N., Goodman, R.M., Lev, T., Slor, H.: Biochemical and chromosomal changes in xeroderma pigmentosum. Harefuah 83, 536–537 (1972).

Block, M.: Chromosome studies in chronic myeloid leukaemia. Br. Med. J. i, 994 (1965).

Bloom, A.D., Tjio, J.H.: *In vivo* effects of diagnostic X-irradiation on human chromosomes. N. Engl. J. Med. 270, 1341–1344 (1964).

Bloom, A.D., Nakagome, Y., Awa, A.A., Neriishi, S.: Chromosome aberrations and malignant disease among A-bomb survivors. Am. J. Public Health 60, 641–644 (1970).

Bloom, A.D., Neriishi, S., Awa, A.A., Honda, T., Archer, P.G.: Chromosome aberrations in leucocytes of older survivors of the atomic bombings of Hiroshima and Nagasaki. Lancet ii, 802–805 (1967).

Bloom, A.D., Neriishi, S., Kamada, N., Iseki, T., Keehn, R.J.: Cytogenetic investigation of survivors of the atomic bombings of Hiroshima and Nagasaki. Lancet ii, 672–674 (1966).

Bloom, G.E.: Leukocyte adenosine deaminase phenotypes in acute leukemia. Cancer 29, 1357–1360 (1972).

Bloom, G.E., Diamond, L.K.: Chromosome abnormalities in anemias. Ann. N.Y. Acad. Sci. 155, 770–776 (1968).

Bloom, G.E., Gerald, P.S., Diamond, L.K.: Chronic myelogenous leukemia in an infant: Serial cytogenetic and fetal hemoglobin studies. Pediatrics 38, 295 – 299 (1966).

Bloom, G.E., Warner, S., Gerald, P.S., Diamond, L.K.: Chromosome abnormalities in constitutional aplastic anemia. N. Engl. J. Med. 274, 8 – 14 (1966).

Bloom, S.E.: Marek's disease: chromosome studies of resistant and susceptible strains. Avian Dis. 14, 478 – 490 (1970).

Bloom, S.E., Goodpasture, C.: An improved technique for selective staining of nucleolar organizer regions in human chromosomes. Hum. Genet. 34, 199 – 206 (1976).

Bloomfield, C.D., Brunning, R.D.: Acute leukemia as a terminal event in nonleukemic hematopoietic disorders. Semin. Oncol. 3, 297 – 317 (1976).

Bloomfield, C.D., Theologides, A.: Acute granulocytic leukemia in elderly patients. J.A.M.A. 226, 1190 – 1193 (1973).

Bloomfield, C.D., Lindquist, L.L., Brunning, R.D., Yunis, J.J., Coccia, P.F.: The Philadelphia chromosome in acute leukemia. Virchows Arch. B Cell Pathol. 29, 81 – 91 (1978).

Bloomfield, C.D., Peterson, L.C., Yunis, J.J., Brunning, R.D.: The Philadelphia chromosome (Ph¹) in adults presenting with acute leukaemia: a comparison of Ph¹+ and Ph¹− patients. Br. J. Haematol. 36, 347 – 358 (1977).

Bobrow, M., Collacott, H.E.A.C., Madan, K.: Chromosome banding with acridine orange. Lancet ii, 1311 (1972).

Bobrow, M., Emerson, P.M., Spriggs, A.I., Ellis, H.L.: Ring-1 chromosome, microcephalic dwarfism and acute myeloid leukemia. Am. J. Dis. Child. 126, 257 – 260 (1973).

Bobrow, M., Madan, K., Pearson, P.L.: Staining of some specific regions of human chromosomes, particularly the secondary constriction of No. 9. Nature [New Biol.] 238, 122 – 124 (1972).

Bobrow, M., Pearson, P.L., Pike, M.C., El Alfi, O.S.: Length variation in the quinacrine-binding segment of human Y chromosomes of different sizes. Cytogenetics 10, 190 – 198 (1971).

Bobzien, W.F.: Chromosomes in multiple myeloma. Georgetown Med. Bull. 21, 48 – 49 (1967).

Bochkov, N.P., Lopukhin, Y.M., Kuleshov, N.P., Kovalchuk, L.V.: Cytogenetic study of patients with ataxia-telangiectasia. Humangenetik 24, 115 – 128 (1974).

Bock, H.E., Haberlandt, W., Ritter, H.: Chromosomal and cytochemical findings in osteomyelosclerosis. Blut 20, 205 – 213 (1970).

Boczkowski, K.: Familial occurrence of gonadal tumors in XY females with breast development. Hum. Genet. 33, 289 – 294 (1976).

Boczkowski, K., Teter, J., Tomaszenska, H., Philip, J.: Gonadoblastoma (gonocytoma III) in a boy with XO/XY mosaicism. Case report with a survey of literature. Acta Pathol. Microbiol. Scand. 71, 46 – 58 (1967).

Boddington, M.M., Spriggs, A.I., Wolfendale, M.R.: Cytogenetic abnormalities in carcinoma-in-situ and dysplasias of the uterine cervix. Br. Med. J. i, 154 – 158 (1965).

Bodor, F., Fleischmann, T.: Anomalies in the banding pattern of chromosomes with acute myelocytic leukemia. Mag. Onkol. 19, 199 – 202 (1975).

Bodor, F., Håkansson, C. H., Norgren, A.: Psuedodiploid karyotype in a breast carcinoma. Acta Pathol. Microbiol. Scand. 82, 386 – 388 (1974).

Boecker, W.R., Hossfeld, D.K., Gallmeier, W.M., Schmidt, C.G.: Clonal growth of Hodgkin cells. Nature 258, 235 – 236 (1975).

Boetius, G., Hustinx, T.W.J., Smits, A.P.T., Scheres, J.M.J.C., Rutten, F.J., Haanen, C.: Monosomy 7 in two patients with a myeloproliferative disorder. Br. J. Haematol. 37, 101 – 109 (1977).

Boggs, D.R.: Hematopoietic stem cell theory in relation to possible lymphoblastic conversion of chronic myeloid leukemia. Blood 44, 449 – 453 (1974).

Boggs, D.R.: The pathogenesis and clinical patterns of blastic crisis of chronic myeloid leukemia. Semin. Oncol. 3, 289 – 296 (1976).

Boisseau, M., Le Menn, R.: Ultrastructure des plaquettes sanguines chez deux nouveau-nés atteints de trisomie-21 avec hyper-plaquettose et syndrome leucémique. Nouv. Rev. Fr. Hematol. 14, 371 – 382 (1974).

Boll, I., Mersch, G.: Morphologische Untersuchungen zur Proliferationskinetik der normalen und pathologischen Granulozytopoese in vitro. Blut 19, 257 – 274 (1969).

Bond, H.E., Flamm, W.G., Burr, H.E., Bond, S.B.: Mouse satellite DNA. Further studies on its biological and physical characteristics and its intracellular localization. J. Mol. Biol. 27, 289 – 302 (1967).

Bondare, D., Barawika, I., Kolodjasnaja, I.: Untersuchungen über den Karyotyp der Knochenmarkzellen in der präleukämischen Phase der chronischen myeloischen Leukämie. Folia Haematol. (Leipz) 105, 58 – 65 (1978).

Bonilla-Musoles, F.: Chromosomen bei adenomatüser Hyperplasie und Adenocarcinom des Endometrium. Z. Geburtshilfe Gynakol. 175, 218 – 225 (1971).

Boque, A., Wilson, R.G.: Chronic myeloid leukaemia of adult type in a 5-month-old infant. Br. Med. J. 2, 1397 (1977).

Borch, S. von der, Stanley, M.A., Kirkland, J.A.: Chromosome studies on direct and cultured preparations from malignant tissues. Pathology i, 243 – 250 (1969).

Bordelon, M.R., Stubblefield, E.: Human tumours in mice confirmed by chromosomal analysis. Nature 252, 324–326 (1974).
Borgaonkar, D.S.: 47, XYY Bibliography. Ann. Genet. (Paris) 12, 67–70 (1968).
Borgaonkar, D.S.: Cytogenetic studies in myeloproliferative disorders. Cytologia (Tokyo) 37, 271–280 (1972).
Borgaonkar, D.S.: Philadelphia-chromosome translocation and chronic myeloid leukaemia. Lancet i, 1250 (1973).
Borgaonkar, D.S., Hollander, D.H.: Quinacrine fluorescence of the human Y chromosome. Nature 230, 52 (1971).
Borgaonkar, D.S., Bias, W.B., Sertt, C.I., Wadea, R.S., Bozkawf, S.P.: IgA and abnormal chromosome 18. Lancet i, 206–207 (1969).
Borgaonkar, D.S., Trips, L., Krush, A.J., Murphy, E.A.: The identification of individuals at high risk for large bowel cancer and the application of preventive measures in their management. Cancer 40, 2531–2533 (1977).
Borges, W.H., Nicklas, J.W., Hamm, C.W.: Prezygotic determinants in childhood leukemia. J. Pediatr. 68, 837–838 (1966).
Borges, W.H., Nicklas, J.W., Hamm, C.W.: Prezygotic determinants in acute leukemia. J. Pediatr. 70, 180–184 (1967).
Borges, W.H., Wald, N., Kim, J.: Non-specificity of chromosomal abnormalities in human leukemia. Clin. Res. 10, 211 (1962).
Borghi, A., Moutali, E., Bigozzi, U., Giusti, V.: XO/XY mosaicism in a phenotypic female with gonadoblastoma. Helv. Paediatr. Acta 20, 185–192 (1965).
Borgström, G.H. Vuopio, P., Chapelle, A. de la: Polyploidy of the bone marrow. Scand. J. Haematol. 17, 123–131 (1976).
Bornstein, R.S., Nesbit, M., Kennedy, B.J.: Chronic myelogenous leukemia presenting in blastic crisis. Cancer 30, 939–941 (1972).
Bosi, L., Castoldi, G.L., Punturieri, E., Ricci, N.: Alterazioni cromosomiche nelle leucemie mieloidi croniche. Prog. Med. (Napoli) 18, 225–228 (1962).
Bosman, F.T., van der Ploeg, M., Geraedts, J.P.M.: Influence of Q- and G-banding on the Feulgen-stainability of human metaphase chromosomes. Histochem. J. 9, 31–42 (1977a).
Bosman, F.T., van der Ploeg, M., van Duijn, P., Schaberg, A.: Photometric determination of the DNA distribution in the 24 human chromosomes. Exp. Cell Res. 105, 301–311 (1977b).
Bosman, F.T., van Vloten, W.A., Pearson, P.L.: Cytogenetic and cytochemical observations in patients with the Sézary syndrome. Chromosomes Today 5, 235–239 (1976).
Bostock, C.J., Sumner, A.T.: *The Eukaryotic Chromosome.* Amsterdam, New York, Oxford: North-Holland Publishing Co. (1978).
Bottomley, R.H.: Cytogenetic studies in human malignant disease. J. Arkansas Med. Soc. 64, 417–419 (1968).
Bottomley, R.H.: Cytogenetic heterogeneity of the acute leukemias. Semin. Oncol. 3, 253–257 (1976).
Bottomley, R.H., Trainer, A.L., Condit, P.T.: Chromosome studies in a "cancer family." Cancer 28, 519–528 (1971).
Bottura, C.: Chromosome analysis. Lancet ii, 1092–1093 (1960).
Bottura, C.: Chromosome abnormalities in multiple myeloma. Acta Haematol. (Basel) 30, 274–279 (1963).
Bottura, C., Coutinho, V.: The chromosome anomalies of the megaloblastic anemia. Blut 16, 193–199 (1968a).
Bottura, C., Coutinho, V.: Cromossomos e leucemias. I. Leucemia mielóide crónica. Rev. Paul. Med. 72, 55–60 (1968b).
Bottura, C., Coutinho, V.: A possible explanation for the origin of the Philadelphia chromosome. Blut 22, 273–276 (1971a).
Bottura, C., Coutinho, V.: Chromosome and leukemia. II. Acute leukemia. Rev. Paul. Med. 78, 5–10 (1971b).
Bottura, C., Coutinho, V.: G/G translocation and chronic myelocytic leukaemia. Blut 29, 216–218 (1974).
Bottura, C., Coutinho, V.: Morfogênese do cromossomo Philadelphia (Ph). Rev. Assoc. Med. Bras. 21, 251–252 (1975).
Bottura, C., Ferrari, I.: Simplified technique for examination of chromosomes in the bone marrow of man. Can. Med. Assoc. J. 85, 381 (1961).
Bottura, C., Ferrari, I.: Chromosome abnormalities in multiple myeloma. Hum. Chrom. Newsl. 7, 12 (1962a).
Bottura, C., Ferrari, I.: Études chromosomiques dans quelques anémies hémolytiques. Acta Haematol. (Basel) 28, 20–24 (1962b).
Bottura, C., Ferrari, I.: Endoreduplication in acute leukaemia. Blood 21, 207–212 (1963).
Bottura, C., Ferrari, I., Veiga, A.A.: Caryotype anormal dans la leucémie aiguë. Acta Haematol. (Basel) 26, 44–49 (1961a).
Bottura, C., Ferrari, I., Veiga, A.A.: Chromosome abnormalities in Waldenström's macroglobulinaemia. Lancet i, 1170 (1961b).
Bourgeois, C.A., Hill, F.G.H.: Fanconi anemia leading to acute myelomonocytic leukemia: Cytogenetic studies. Cancer 39, 1163–1167 (1977).
Bourgeois, C.A., Calverley, M.H., Forman, L., Polani, P.E.: Bloom's syndrome: A probable new case with cytogenetic findings. J. Med. Genet. 12, 423 (1975).

Bourgoin, P., Baylet, R., Ballon, C.: Exploration d'une hypothèse sur l'étiopathogénie des môles hydatidiformes. Étude chromosomique. Dakar. Rev. Fr. Gynecol. 60, 673–684 (1965).

Bouroncle, B.A., Clausen, K.P., Aschenbrand, J.F.: Studies of the delayed response of phytohemagglutinin (PHA) stimulated lymphocytes in 25 chronic lymphatic leukemia patients before and during therapy. Blood 34, 166–178 (1969).

Bousser, J., Tanzer, J.: Syndrome de Klinefelter et leucémie aiguë. A propos d'un cas. Nouv. Rev. Fr. Hematol. 70, 194–197 (1963).

Bousser, J., Zittoun, R., Casin, I., Turleau, C., Grouchy, J. de: Appearance of Ph¹ chromosome during the course of polycythemia vera transformed to subacute myeloid leukemia. Nouv. Presse Med. 6, 753–754 (1977).

Bouton, M.J., Phillips, H.J., Smithells, R.W., Walker, S.: Congenital leukaemia with parental consanguinity. Case report with chromosome studies. Br. Med. J. ii, 866–869 (1961).

Boveri, T.: Über Differenzierung der Zellkerne während der Furchung des Eies von *ascaris megalocephala*. Anat. Anz. 2, 688–693 (1887).

Boveri, T.: Über mehrpolige Mitosen als Mittel zur Analyse des Zellkerns. Verh. Phys. Med. Ges. Wurzb. (1902).

Boveri, T.: Veitrag zum Studium des Chromatins in den Epithelzellen der Carcinome. Beitr. Pathol. 14, 249 (1912).

Boveri, T.: *Zur Frage der Entstehung maligner Tumoren*, S. 1–64. Jena: Gustav Fischer, 1914.

Boveri, T.: *The Origin of Malignant Tumors*. London: Baillière, Tindall & Cox, 1929.

Bowen, P., Lee, C.S.N.: Ph¹ chromosome in the diagnosis of chronic myeloid leukemia: Report of a case with features simulating myelofibrosis. Bull. Johns Hopkins Hosp. 113, 1–12 (1963).

Bowey, C.E., Spriggs, A.I.: Chromosomes of human endometrium. J. Med. Genet. 4, 91–95 (1967).

Bowev, C.E., Spriggs, A.I.: Lost chromosomes in endometrial cells. J. Med. Genet. 5, 58–59 (1968).

Boyd, E., Abdulla, U., Donald, I., Fleming, J.E.E., Hall, A.J., Ferguson-Smith, M.A.: Chromosome breakage and ultrasound. Br. Med. J. ii, 501–502 (1971).

Boyd, E., Buchanan, W.W., Lennox, B.: Damage to chromosomes by therapeutic doses of radioiodine. Lancet i, 977–978 (1961).

Boyd, E., Pinkerton, P.H., Hutchison, H.E.: Chromosomes in aleukaemic leukaemia. Lancet ii, 444–445 (1965).

Boyd, J.T., Court Brown, W.M., Vennart, J., Woodcock, G.E.: Chromosome studies on women formerly employed as luminous-dial painters. Br. Med. J. i, 377–382 (1966).

Boyd, W.: *The Spontaneous Regression of Cancer*, Springfield, Ill.: Charles C. Thomas (1966).

Bram, S.: The function of the structure of DNA in chromosomes. Biochimie 54, 1005–1011 (1972).

Brandão, H.J.S.: DNA content in epithelial cells of dysplasias of the uterine cervix. Histologic and microspectrophotometric observations. Acta Cytol. 13, 232–237 (1969).

Brandão, H.J.S., Atkin, N.B.: Protruding chromosome arms in histological sections of tumours with larger marker chromosomes. Br. J. Cancer 22, 184–191 (1968).

Brandom, W.F., Bloom, A.D., Bristline, R.W., Archer, P.G.: Radiation cytogenetics of plutonium workers. Radiat. Res. 67, 588 (1976).

Brandt, L.: Differences in the proliferative activity of myelocytes from bone marrow, spleen and peripheral blood in chronic myeloid leukemia. Scand. J. Haematol. 6, 105–112 (1969).

Brandt, L., Mitelman, F.: Splenic biopsy and early detection of blastic crisis in chronic granulocytic leukaemia. Excerpta Medica (in press) (1979).

Brandt, L., Levan, G., Mitelman, F., Olsson, I., Sjögren, U.: Trisomy G-21 in adult myelomonocytic leukaemia. An abnormality common to granulocytic and monocytic cells. Scand. J. Haematol. 12, 117–122 (1974*a*).

Brandt, L., Levan, G., Mitelman, F., Sjögren, U.: Defective differentiation of megakaryocytes in acute myeloid leukemia. Acta Med. Scand. 196, 227–230 (1974*b*).

Brandt, L., Mitelman, F., Beckman, G., Laurell, H., Nordenson, I.: Different composition of the eosinophilic bone marrow pool in reactive eosinophilia and eosinophilic leukaemia. Acta Med. Scand. 201, 177–180, (1977).

Brandt, L., Mitelman, F., Nilsson, P.G., Sjögren, U.: Low mitotic activity of granulopoietic precursor cells in Ph¹-chromosome-negative chronic myeloid leukaemia. Lancet ii, 719–720 (1974).

Brandt, L., Mitelman, F., Panani, A.: Cytogenetic differences between bone marrow and spleen in a case of agnogenic myeloid metaplasia developing blast crisis. Scand. J. Haematol. 15, 187–191 (1975*a*).

Brandt, L., Mitelman, F., Panani, A., Lenner, H.C.: Extremely long duration of chronic myeloid leukaemia with Ph¹ negative and Ph¹ positive bone marrow cells. Scand. J. Haematol. 16, 321–325 (1976).

Brandt, L., Mitelman, F., Sjögren, U.: Relation between chromosomes and mitotic activity in acute myeloid leukemia. Hereditas 79, 305–306 (1975*b*).

Brandt, L., Mitelman, F., Sjögren, U.: Megaloblastic changes and chromosome abnormalities of eryth-

ropoietic cells in acute myeloid leukaemia. Acta Haematol. (Basel) 54, 280–283 (1975c).
Brandt, N.J., Froland, A., Mikkelsen, M., Mielsen, A., Tolstrup, N.: Galactosaemia locus and the Down's syndrome chromosome. Lancet ii, 700–703 (1963).
Brătianu, A., Răutu, I., Mărgineau, I., Burdea, M.: A case of chronic myeloid leukemia, associated with karyotype changes, in a child (In Romanian). Rev. Med. Chir. Soc. Med. Nat. Isai 79, 597–600 (1975).
Braun-Falco, O., Marhgescu, S.: Bloom-Syndrome. A disease with relatively high leukemia morbidity. Münch. Med. Wochenschr. 111, 65–69 (1969).
Braunsteiner, H., Rothenbuchner, G., Schober, B.: Chromosomenuntersuchungen bei Morbus Waldenström. Wien. Klin. Wochenschr. 76, 502–503 (1964).
Breckenridge, R.L., Nash, E., Litz, L.: Gonadoblastoma in male pseudohermaphroditism. Exp. Med. Surg. 27, 330–335 (1970).
Breg, W.R.: Quinacrine fluorescence for identifying metaphase chromosomes, with special reference to photomicrography. Stain Technol. 47, 87–93 (1972).
Breg, W.R., Miller, D.A., Allderdice, P.W., Miller, O.J.: Identification of translocation chromosomes by quinacrine fluorescence. Am. J. Dis. Child. 123, 561–564 (1972).
Brewster, D.J., Garrett, J.V.: Chromosome abnormalities in neuroblastoma. J. Clin. Pathol. 18, 167–169 (1965).
Bridge, M.F., Melamed, M.R.: Leukocyte chromosome abnormalities in advanced nonhematopoietic cancer. Cancer Res. 32, 2212–2220 (1972).
Brière, J., Castro-Malaspina, H., Tanzer, J., Bernard, J.: Subacute myeloid leukemias with Philadelphia chromosome. Nouv. Rev. Fr. Hematol. 15, 407–424 (1975).
Brieux de Salum, S.: Citogenética en hematologia. Medicina (B. Aires) 28 (Suppl. 1), 96–102 (1968).
Brieux de Salum, S.: Anormalidades cromosomicas de neoplasias humanas. Medicina (B. Aires) 32, 530–534 (1972).
Brieux de Salum, S., Larripa, I.: Minute chromatin bodies in a murine *in vitro* cell line. J. Natl. Cancer Inst. 55, 717–720 (1975).
Brieux de Salum, S., Pavlovsky, A.: Estudios de citogenética en hiperplasias ganglionares. Medicina (B. Aires) 28, 192–202 (1968).
Brieux de Salum, S., Suárez, H.G., Pavlovsky, A.: Chromosome serial studies of cultured cell line (GH_7) from a human lymphosarcoma. Rev. Eur. Etudes Clin. Biol. 16, 711–714 (1971).
Brigato, G., Franco, G., Cipani, F., Zanoio, L.: Studio cromosomico sulle cellule del liquido ascitico nei tumori ovarici maligni. Attual. Obstet. Ginecol. 11, 666–675 (1965).
Brile, A.B., Tomanaga, M., Heyssel, R.M.: Leukemia in man following exposure to ionizing radiation. A summary of the findings in Hiroshima and Nagasaki and a comparison with other human experience. Ann. Intern. Med. 56, 590–609 (1962).
British Medical Journal: Chromosome abnormality in chronic myeloid leukaemia (Annotation). Vol. 1, 347 (1961).
British Medical Journal: Chromosomes and leukaemia (Leading article). Vol. 2, 719–720 (1966).
Broder, S., Edelson, R.L., Lutzner, M.A., Nelson, D.L., MacDermott, R.P., Durm, M.E., Goldman, C.K., Meade, B.D., Waldmann, T.A.: The Sézary syndrome. A malignant proliferation of helper T cells. J. Clin. Invest. 58, 1297–1306 (1976).
Brodeur, G.M., Sekhon, G.S., Goldstein, M.N.: Chromosomal aberrations in human neuroblastomas. Cancer 40, 2256–2263 (1977).
Brody, J.I., Burningham, R.A., Nowell, P.C., Rowlands, D.T., Jr., Freiburg, P., Daniele, R.P.: Persistent lymphocytosis with chromosomal evidence of malignancy. Am. J. Med. 58, 547–552 (1975).
Brøgger, A.: Apparently spontaneous chromosome damage in human leukocytes and the nature of chromatid gaps. Humangenetik 13, 1–14 (1971).
Brøgger, A.: Iso-labelling in BrdU-substituted chromosomes. Hereditas 79, 311–313 (1975).
Brouet, J.-C., Flandrin, G., Seligmann, M.: Indications of the thymus-derived nature of the proliferating cells in six patients with Sézary's syndrome. N. Engl. J. Med. 289, 341–344 (1973).
Brouet, J.C., Labaume, S., Seligmann, M.: Evaluation of T and B lymphocyte membrane markers in human non-Hodgkin malignant lymphomata. Br. J. Cancer 31, 121 (1975).
Brouet, J.C., Preud'Homme, J.L., Flandrin, G., Chelloul, N., Seligmann, M.: Membrane markers in "histiocytic" lymphomas (reticulum cell sarcomas). J. Natl. Cancer Inst. 56, 631–633 (1976).
Brouet, J.C., Preud'Homme, J.-L., Seligmann, M.: The use of B and T membrane markers in the classification of human leukemias with special reference to acute lymphoblastic leukemia. Blood Cells 1, 81 (1975).
Broustet, A., Hartmann, L., Moulinier, J., Staeffen, J., Moretti, G: Etude cytogénétique de dix-huit cas de maladie de Waldenström. Nouv. Rev. Fr. Hematol. 7, 809–826 (1967).
Broustet, A., Meuge, C., Legrand, E.: Anomalies chromosomiques au cours de la leucémie lymphoide chronique, des processus lymphoréticulaires malins et de la maladie de Kahler. Nouv. Rev. Fr. Hematol. 10, 91–99 (1970).
Brown, A.K., Elves, M.W., Gunson, H.H., Pell-Il-

derton, R.: Waldenström's macroglobulinaemia. Acta Haematol. (Basel) 38, 184–192 (1967).
Brown, C.D., Propp, S.: Trisomy of chromosome 21 and tetraploid metaphases in congenital acute granulocytic leukemia in mongolism. Hum. Chrom. Newsl. 1, 12 (1962).
Brown, J.K., McNeill, J.R.: Aberrations in leukocyte chromosomes of personnel occupationally exposed to low levels of radiation. Radiat. Res. 40, 534–543 (1969).
Brown, W.M.C., Tough, I.M.: Cytogenetic studies in chronic myeloid leukemia. Adv. Cancer Res. 7, 351–381 (1963).
Brun, A., Sköld, G.: CNS malformations in Turner's syndrome. An integral part of the syndrome? Acta Neuropathol. (Berl.) 10, 159–161 (1968).
Brunel, D., Donnadio, D., Emberger, J.M., Astruc, J., Manassero, J.: Bloom's syndrome – Discussion of the diagnosis concerning two cases of terminal leukemia in a sibship. J. Genet. Hum. 25, 177–188 (1977).
Brynes, R.K., Golomb, H.M., Desser, R.K., Recant, W., Reese, C., Rowley, J.: Acute monocytic leukemia. Am. J. Clin. Pathol. 65, 471–482 (1976).
Brynes, R.K., Golomb, H.M., Gelder, F., Desser, R.K., Rowley, J.D.: The leukemic phase of histiocytic lymphoma. Histologic, cytologic, cytochemical, ultrastructural, immunologic and cytogenetic observations in a case. Am. J. Clin. Pathol. 69, 550–558 (1978).
Buchanan, J.G., Becroft, D.M.O.: Down's syndrome and acute leukaemia: A cytogenetic study. J. Med. Genet. 7, 57–69 (1970).
Buchanan, J.G., Scott, P.J., McLachlan, E.M., Smith, F., Richmond, D.E., North, J.D.K.: A chromosome translocation in association with periarteritis nodosa and macroglobulinaemia. Am. J. Med. 42, 1003–1010 (1967).
Büchner, T., Gatermann, B.: Zur Chromosomen-Replikation in Zellen der chronischen myeloischen Leukämie. Klin. Wochenschr. 49, 644–648 (1971).
Buckton, K.E., O'Riordan, M.L.: Easy identification of chromosome translocation involved in chronic myeloid leukaemia. Lancet ii, 1064 (1976).
Buckton, K.E., Court Brown, W.M., Smith, P.G.: Lymphocyte survival in men treated with X-rays for ankylosing spondylitis. Nature 214, 470–473 (1967).
Buckton, K.E., Harnden, D.G., Baikie, A.G., Woody, G.E.: Mongolism and leukaemia in the same sibship. Lancet i, 171–172 (1961).
Buckton, K.E., Jacobs, P.A., Court Brown, W.M., Doll, R.: A study of the chromosome damage persisting after X-ray therapy for ankylosing spondylitis. Lancet ii, 676–682 (1962a).
Buckton, K.E., Jacobs, P.A., Court Brown, W.M., Doll, R.: Cancer subjects and abnormal cell division. Nature 193, 591 (1962b).
Buckton, K.E., O'Riordan, M.L., Jacobs, P.A., Robinson, J.A., Hill, R., Evans, H.J.: C- and Q-band polymorphisms in the chromosomes of three human populations. Ann. Hum. Genet. 40, 99–112 (1976).
Buhler, E.M., Jurik, L.P., Voyame, M., Bühler, U.K.: Presumptive evidence of two active X chromosomes in somatic cells of a human female. Nature 265, 142–144 (1977).
Bühler, E.M., Tsuchimoto, T., Korztolànyi, G., Jurik, L.P.: Q-banding of human chromosomes after BUdR and BCdR treatment. Hum. Genet. 31, 309–316 (1976).
Buick, R.N., Till, J.E., McCulloch, E.A.: Colony assay for proliferative blast cells circulating in myeloblastic leukaemia. Lancet i, 862–863 (1977).
Bukowski, R.M., Weick, J.K., Reimer, R.R., Groppe, G.W., Hewlett, J.S.: Characteristics of acute leukemia in patients with non-malignant diseases receiving alkylating agent therapy. Blood 50 (Suppl. 1), 185 (abstract) (1977).
Bull, J.: Cytogenetic studies of marrow and peripheral blood granulocyte colonies in treated chronic myelogenous leukemia. Blood Cells 1, 161–162 (1975).
Burgdorf, W., Kurvink, K., Cervenka, J.: Elevated sister chromatid exchange rate in lymphocytes of subjects treated with arsenic. Hum. Genet. 37, 69–72 (1977).
Burgess, M.A., Garson, O.M.: Homologous leucocyte transfusion in acute leukaemia with cytogenetic evidence of myeloid graft. Med. J. Aust. 1, 1243–1246 (1969).
Burk, P.G., Lutzner, M.A., Robbins, J.H.: Decreased incorporation of thymidine into the DNA of lymphocytes from patients with xeroderma pigmentosum after UV-irradiation *in vitro*. Clin. Res. 17, 614 (1969).
Burkhardt, R., Pabst, W., Kleber, A.: Knochenmark-Histologie und Klinik der Polycythaemia vera. Arch. Klin. Med. 216, 64–104 (1969).
Burkholder, G.D.: Reciprocal Giemsa staining of late DNA replicating regions produced by low and high pH sodium phosphate. Exp. Cell Res. 111, 489–492 (1978).
Burkholder, G.D., Weaver, M.G.: DNA-protein interactions and chromosome banding. Exp. Cell Res. 110, 251–262 (1977).
Burns, G.F., Cawley, J.C., Barker, C.R., Goldstone, A.H., Hayhoe, F.G.J.: New evidence relating to the nature and origin of the hairy cell of leukaemic reticuloendotheliosis. Br. J. Haematol. 36, 71–84 (1977).
Burns, G.F., Cawley, J.C., Higgy, K.E., Barker, C.R., Edwards, M., Rees, J.K.H., Hayhoe, F.G.J.:

Hairy-cell leukaemia: A B-cell neoplasm with a severe deficiency of circulating normal B lymphocytes. Leukemia Res. 2, 33–40 (1978).

Bushkell, L.L., Kersey, J.H., Cervenka, J.: Chromosomal breaks in T and B lymphocytes in Fanconi's anemia. Clin. Genet. 9, 583–587 (1976).

Buyssens, N., Bourgeois, N.H.: Chronic myelocytic leukemia versus idiopathic myelofibrosis. A diagnostic problem on bone marrow biopsies. Cancer 40, 1548–1561 (1977).

Cabrol, C., Abele, R.: Chromosome 5q− dans les cellules medullaires d'un sujet atteint d'anémie ayant évolué en leucémie aiguë indifferenciée. J. Genet. Hum. 26, 195–202 (1978).

Cabrol, C., Peytremann, R., Maurice, P.-A., Klein, D.: Triple chromosome Philadelphie dans la phase accelerée d'une leucémie myéloïde chronique. Schweiz. Med. Wochenschr. 106, 1381 (1976).

Cadiou, M., Ruff, F., Meunier, F., Attalah, N., Bernadou, A., Zittoun, R., Parrot, J.L., Bousser, J.: L'histaminémie dans les syndromes myéloproliferatifs. Nouv. Rev. Fr. Hematol. 15, 261–269 (1975).

Cadotte, M.: Chromosome anomalies in cancer cells. Can. J. Med. Technol. 36, 221–225 (1974).

Cailleau, R., Cruciger, Q., Hokanson, K.M., Olive, M., Blumenschein, G.: Morphological, biochemical and chromosomal characterization of breast tumor lines from pleural effusions. In Vitro 12, 331 (1976).

Cailleau, R., Olivé, M., Cruciger, Q.V.: Long-term breast carcinoma cell lines of metastatic origin: Preliminary characterization. In Vitro 14, 911–915 (1978).

Calculli, G., Donisi, C., Sacchi, G.: A case of eosinophilic leukemia: Clinical and morphological follow-up. 2nd International Symposium on Therapy of Acute Leukemias, Rome, Italy, Dec. 7–10, p. 103 (abstract) (1977).

Callender, S.T., Kay, H.E.M., Lawler, S.D., Millard, R.E., Sanger, R., Tippett, P.A.: Two populations of Rh groups together with chromosomally abnormal cell lines in the bone marrow. Br. Med. J. i, 131–133 (1971).

Canellos, G.P.: Second malignancies complicating Hodgkin's disease in remission. Lancet i, 1294 (1975).

Canellos, G.P., Whang-Peng, J.: Philadelphia-chromosome-positive preleukaemic state. Lancet ii, 1227–1229 (1972).

Canellos, G.P., DeVita, V.T., Arseneau, J.C., Whang-Peng, J., Johnson, R.E.C.: Second malignancies complicating Hodgkin's disease in remission. Lancet i, 947 (1975).

Canellos, G.P., DeVita, V.T., Whang-Peng, J., Carbone, P.: Chronic granulocytic leukemia: clinical and cytogenetic correlations. Clin. Res. 18, 531 (1970).

Canellos, G.P., DeVita, V.T., Whang-Peng, J., Carbone, P.P.: Hematologic and cytogenetic remission of blastic transformation in chronic granulocytic leukemia. Blood 38, 671–679 (1971).

Canellos, G.P., DeVita, V.T., Whang-Peng, J., Chabner, B.A., Schein, P.S., Young, R.C.: Chemotherapy of the blastic phase of chronic granulocytic leukemia: Hypodiploidy and response to therapy. Blood 47, 1003–1009 (1976).

Canellos, G.P., Whang-Peng, J., DeVita, V.T.: Chronic granulocytic leukemia without the Philadelphia chromosome. Am. J. Clin. Pathol. 65, 467–470 (1976).

Canellos, G.P., Whang-Peng, J., Schnipper, L., Brown, C.H. III: Prolonged cytogenetic and hematologic remission of blastic transformation in chronic granulocytic leukemia. Cancer 30, 288–293 (1972).

Cantolino, S.J., Schmickel, R.D., Ball, M., Cisar, C.F.: Persistent chromosomal aberrations following radioiodine therapy for thyrotoxicosis. N. Engl. J. Med. 275, 739–745 (1966).

Cantu, J.M., Castillo, V. del, Jimenez, M., Ruiz-Barquin, E.: Chromosomal instability in incontinentia pigmenti. Ann. Genet. (Paris) 16, 117–120 (1973).

Cao, A., Leone, P., Trabalza, N., Virgiliis, S. de: Considerazioni clinico-ematologiche su di un caso di leucemia mieloide cronica "tipo giovanile." Minerva Pediatr. 21, 1646–1656 (1969).

Cáp, J., Izakovič, V., Vojtassák, J.: Juvenile type of chronic myelogenous leukemia in a 3 6/12 year old boy with trisomy 21 mosaicism but no symptoms of Down's syndrome (in German). Monatsschr. Kinderheilkd. 125, 740–743 (1977).

Capoa, A. de, Ferraro, N., Archidiacono, N., Pelliccia, F., Rocchi, M., Rocchi, A.: Nucleolus organizer and satellite association in a variant D-group chromosome. Hum. Genet. 34, 13–16 (1976).

Carbone, P.P., Tjio, J.H., Whang, J., Block, J.B., Kremer, W.B., Frei, E., III: The effect of treatment in patients with chronic myelogenous leukemia. Hematologic and cytogenetic studies. Ann. Intern. Med. 59, 622–628 (1963).

Cardillo, M., Krause, E., Sciorra, L.J.: Human chromosome preparations obtained from immunoadsorbent separated T- and B-cells. Clin. Genet. 13, 271–277 (1978).

Cardini, G., Bersi, M., Gasparini, C.: Anomalies chromosomiques dans une leucémie á plasmocytes. Nouv. Rev. Fr. Hematol. 10, 787–791 (1970).

Cardini, G., Clemente, R., Bersi, M.: Su di un corredo cromosomico particolare in un caso di leucemia acuta mieloblastica comparsa in un sog-

geto portatore di un mosaico XY/XO. Arch. Sci. Med. (Torino) 125, 433–438 (1968).
Carlevaro, C., Rossi, G.A., Cerri, E., Pelucco, D.: Cytogenetic study of pleural effusions. Tumori 64, 335–344 (1978).
Carr, D.H.: Cytogenetics and the pathology of hydatidiform degeneration. Obstet. Gynecol. 33, 333–341 (1969).
Carr, D.H.: Chromosomes and abortion. In H. Harris, K. Hirschhorn (eds.) Advances in Human Genetics, Vol. 2, pp. 201–257, New York, London: Plenum Press (1971).
Carrano, A.V., Heddle, J.A.: The fate of chromosome aberrations. J. Theor. Biol. 38, 289–304 (1973).
Carrano, A.V., Minkler, J., Piluso, D.: On the fate of stable chromosomal aberrations. Mutat. Res. 30, 153–156 (1975).
Carrano, A.V., Thompson, L.H., Lindl, P.A., Minkler, J.L.: Sister chromatid exchange as an indicator of mutagenesis. Nature 271, 551–553 (1978).
Carpentier, S., Lejeune, J.: Analyse chromosomique des tumeurs malignes. Technique d'examen en culture organotypique. Pathol. Biol. (Paris) 21, 665–669 (1973).
Cascos, A.S., Barreiro, E.: Cromosoma Ph¹ y translocacion 5–12 en un caso de leucemia mieloide cronica. Revta. Clin. Esp. 94, 10–13 (1964).
Caspersson, T., Farber, S., Folwy, G.E., Kudynowski, J., Modest, E.J., Simonsson, E., Wagh, U., Zech, L.: Chemical differentiation along metaphase chromosomes. Exp. Cell Res. 49, 219–222 (1968).
Caspersson, T., Gahrton, G., Lindsten, J., Zech, L.: Identification of the Philadelphia chromosome as a number 22 by quinacrine mustard fluorescence analysis. Exp. Cell Res. 63, 238–240 (1970).
Caspersson, T., Lindston, J., Zech, L.: The nature of the structural X chromosome aberrations in Turner's syndrome as revealed by quinacrine mustard fluorescence analysis. Hereditas 66, 287–292 (1970).
Caspersson, T., Lomakka, G., Lindston, J.: Die Identifikation menschlicher Chromosomen mit Helfe der Fluoreszenmethode. Triangle 11, 73–80 (1973).
Caspersson, T., Lomakka, G., Møller, A.: Computerized chromosome identification by the aid of the quinacrine mustard fluorescence technique. Hereditas 67, 103–110 (1971a).
Caspersson, T., Lomakka, G., Zech, L.: The 24 fluorescence patterns of the human metaphase chromosomes. Distinguishing characters and variability. Hereditas 67, 89–102 (1971b).
Caspersson, T., Zech, L., Johansson, C.: Differential banding of alkylating fluorochromes in human chromosomes. Exp. Cell Res. 60, 315–319 (1970a).

Caspersson, T., Zech, L., Johansson, C., Modest, E.J.: Identification of human chromosomes by DNA-binding fluorescent agents. Chromosoma 30, 215–227 (1970b).
Caspersson, T., Zech, L., Modest, E.J.: Fluorescent labeling of chromosomal DNA: superiority of guinacrine mustard to quinacrine. Science 170, 762 (1970c).
Caspersson, T., Zech, L., Modest, E.J., Foley, G.E., Wagh, U., Simonsson, E.: Chemical differentiation with fluorescent alkylating agents in vicia fabia metaphase chromosomes. Exp. Cell Res. 58, 128–140 (1969a).
Caspersson, T., Zech, L., Modest, E.J., Foley, G.E., Wagh, U., Simonsson, W.: DNA-binding fluorochromes for the study of the organization of the metaphase nucleus. Exp. Cell Res. 58, 141–152 (1969b).
Cassidy, S.B., McGee, B.J., Van Eys, J., Nance, W.E., Engel, E.: Trisomy 8 syndrome, Pediatrics 56, 826–831 (1975).
Cassuto, P., Ayraud, N., Dujardin, P., Audoly, P.: Ph¹ chromosome and B-12 hypervitaminemia without obvious symptoms of chronic myeloid leukemia (in French). Nouv. Presse Med. 1, 2769 (1972).
Castel, Y., Riviere, D., Roche, J., Daridon, I., Toudie, L., Clavier, M., Benardeau, M.: Neonatal acute leukemia in a 21 trisomy infant: Case report, Cah. Méd. 12, 1305–1312 (1971).
Castelman, B., Scully, R.E., McNeely, B.U.: Case records of the Massachusetts General Hospital. Case 18-1973. N. Engl. J. Med. 288, 957–963 (1973).
Castelman, K.R., Melnyk, J., Frieden, H.J., Persinger, G.W., Wall, R.J.: Karyotype analysis by computer and its application to mutagenicity testing of environmental chemicals. Mutat. Res. 41, 153–162 (1976).
Castoldi, G.L.: Anomalie cromosomiche nel mieloma multiplo. Haematologica (Pavia) 49, 751–766 (1964).
Castoldi, G.L.: I cromosomi nelle reticulo-linfopatie. Haematologica (Pavia) 55, 185–224 (1970).
Castoldi, G.L.: Cytogenetics of Hodgkin's disease. Recent Prog. Med. (Roma) 55, 4 (1973a).
Castoldi, G.L.: Malattie Linfoproliferative. In A. Baserga, G.L. Castoldi, B. Dallapiccola (eds.): La Patologia Cromosomica. Rome: Pozzi (1973b).
Castoldi, G.L., Mitus, W.J.: Chromosome vacuolization and breakage. Arch. Intern. Med. 121, 177–179 (1968).
Castoldi, G.L., Grusovin, G.D., Gualandi, M., Scapoli, G.L.: Nuclear projections in tumour cells and large chromosome markers. Experientia 32, 856–857 (1976).
Castoldi, G.L., Grusovin, G.D., Scapoli, G.L.: Consecutive cytochemical staining for the analysis

of the blastic population in the acute phase of chronic myeloid leukemia. Biomedicine 23, 12-16 (1975).

Castoldi, G.L., Grusovin, G.D., Scapoli, G., Gualandi, M., Spanedda, R., Anzanel, D.: Acute myelomonocytic leukemia terminating in histiocytic medullary reticulosis. Cytochemical, cytogenetic and electron microscopic studies. Cancer 40, 1735-1747 (1977).

Castoldi, G.L., Grusovin, G.D., Scapoli, G.L., Spanedda, R.: Association of multiple haematological disorders (acute myeloblastic leukaemia, paraproteinaenemia and thalassaemia) in a 46,XX/46,XXqi female. Acta Haematol. (Basel) 46, 294-306 (1971).

Castoldi, G.L., Grusovin, G.D., Scapoli, G.L., Spanedda, R.: Differential acridine orange staining of human chromosomes. Acta Genet. Med. Gemmellol. 21, 319-326 (1972).

Castoldi, G.L., Ricci, N., Punturieri, E., Bosi, L.: Chromosomal imbalance in plasmacytoma. Lancet i, 829 (1963).

Castoldi, G.L., Scapoli, G.L., Dallapiccola, B., Spanedda, R.: Analyse des anomalies chromosomiques dans l'anémie pernicieuse. Nouv. Rev. Fr. Hematol. 9, 769-782 (1969).

Castoldi, G., Scapoli, G., Grusovin, G.D., Gualandi, M., Spanedda, R., Cavazzini, L., Anzanel, D.: Chromosomal abnormalities in angio-immunoblastic lymphadenopathy. La Ricerca Clin. Lab. 6, 145-165 (1976).

Castoldi, G.L., Scapoli, G.L., Spanedda, R.: Sull' evoluzione clonale del cariotipodelle cellule neoplastiche in versamento pleurico da adenocarcinoma mammario. Tumori 11, 3-22 (1968).

Castoldi, G., Spanedda, R., Scapoli, G., Grusovin, G.D., Baserga, M., Gualandi, M., Bariani, L., Bertocco, S.: Multiple chromosomal associations and paracentromeric region instability in a case of acute leukemia. La Ricerca Clin. Lab. 5, 234-247 (1975).

Castoldi, G., Yam, L.T., Mitus, W.J., Crosby, W.H.: Chromosomal studies in erythroleukemia and chronic erythremic myelosis. Blood 31, 202-215 (1968).

Castro-Sierra, E., Gorman, L.Z., Merker, H., Obrecht, P., Wolf, U.: Clinical and cytogenetic findings in the terminal phase of chronic myelogenous leukemia. Humangenetik 4, 62-73 (1967).

Catovsky, D.: T-cell origin of acid-phophatase-positive lymphoblasts. Lancet ii, 327 (1975).

Catovsky, D., Holt, P.J.L., Galton, D.A.G.: Lymphocyte transformation in immunoproliferative disorders. Br. J. Cancer 26, 154-163 (1972).

Catovsky, D., Pittman, S., Lewis, D., Pearse, E.: Marker chromosome 14q+ in follicular lymphoma in transformation. Lancet ii, 934 (1977).

Catovsky, D., Shaw, M.T., Hoffbrand, A.V., Dacie, J.V.: Sideroblastic anaemia and its association with leukaemia and myelomatosis: A report of five cases. Br. J. Haematol. 20, 385-393 (1971).

Catti, A.: Aberrations chromosomiques dans les maladies du sang. Schweiz. Med. Wochenschr. 101, 1646-1649 (1971).

Catti, A.A., Cork, A., Trujillo, J.M.: Application of newer cytogenetic techniques to the study of human leukemia bone marrow. Mammal. Chrom. Newsl. 14, 37 (1973).

Cavallin-Ståhl, E., Landberg, T., Ottow, Z., Mitelman, F.: Hodgkin's disease and acute leukaemia: A clinical and cytogenetic study. Scand. J. Haematol. 19, 273-280 (1977).

Cave, M.D.: Reverse patterns of thymidine-H^3 incorporation in human chromosomes. Hereditas 54, 338-355 (1966).

Cavellero, G.: Sulla frequenza della cromatina sessuale in un caso di corionepithelioma del testicolo. Pathologica 50, 215-218 (1958).

Cawein, M.J., Lappat, E.J.: A case of Down's syndrome with the Philadephia chromosome, chronic myelogenous leukemia, and low leukocyte alkaline phosphatase. Clin. Res. 12, 32 (1964).

Cawein, M., Lappat, E.J., Rackley, J.W.: Down's syndrome and chronic myelogenous leukemia. Arch. Intern. Med. 116, 505-508 (1965).

Cecco, L. de, Rugiati, S., Sbernini, R.: Aspetti citogenetici dei carcinomi uterini. Quad. Clin. Obstet. Ginecol. (Parma) 21, 562-573 (1966).

Cehreli, C., Ezdinli, E.Z., Lee, C.Y., Krmpotic, E.: Blastic phase of agnogenic myeloid metaplasia simulating malignant lymphoma. Cancer 38, 1297-1305 (1976).

Cellier, K.M., Kirkland, J.A., Stanley, M.A.: Statistical analysis of cytogenetic data in cervical neoplasia. J. Natl. Cancer Inst. 44, 1221-1230 (1970).

Ceppelini, R., Celada, F., Franceschini, P., Gavosto, F., Pileri, A., Pegoraro, L., Zanalda, A.: Anomalia del cariotipo midollare e del genotipo eritrocitario in un caso de ertremia cronica (malattia di Di Guglielmo). Atti. Assoc. Genet. Ital. (Pavia) 9, 258 (1964).

Cernea, P., Teodorescu, F., Angheloni, T.: Rétinoblastome á mosaicisme chromosomien 46XXX/47XX,G+. Ann. Oculist. 206, 607-611 (1973).

Černy, M., Šimánková, N., Misarová, Z., Černá, M.: Chromosomal study in childhood leukemia. Cesk. Pediatr. 17, 976-979 (1962).

Cervenka, J., Koulischer, L.: *Chromosomes in Human Cancer.* Springfield, Ill.: Charles C. Thomas (1973).

Cervenka, J., Anderson, R.S., Nesbit, M.E., Krivit, W.: Familial leukemia and inherited chromosomal aberration. Int. J. Cancer 19, 783-788 (1977).

Cervenka, J., Thorn, H.L., Gorlin, R.J.: Structural basis of banding pattern of human chromosomes. Cytogenet. Cell Genet. 12, 81–86 (1973).

Chaganti, R.S.K., Schonberg, S., German, J.: A manyfold increase in sister chromatid exchanges in Bloom's syndrome lymphocytes. Proc. Natl. Acad. Sci. USA 71, 4508–4512 (1974).

Chamla, Y., Ruffié, M.: Production of C and T bands in human mitotic chromosomes after heat treatment. Hum. Genet. 34, 213–216 (1976).

Chandra, H.S.: A genetic test for multiplicity of origin of the Ph¹ chromosome in human chronic granulocytic leukaemia. Ann. Genet. (Paris) 11, 3–5 (1968).

Chang, K.S.S.: Susceptibility of xeroderma pigmentosum cells to transformation by murine and feline sarcoma viruses. Cancer Res. 36, 3294–3299 (1976).

Chang, T.D., Biedler, J.L., Stockert, E., Old, L.J.: Trisomy of chromosome 15 in X-ray induced mouse leukemia. Proc. Am. Assoc. Cancer Res. 18, 225 (1977).

Chapelle, A. de la: The use of cytogenetic abnormalities in bone marrow cells for mapping by gene dosage. Scand. J. Haematol. 17, 81–88 (1976).

Chapelle, A. de la, Gräsbeck, R.: Normal mitotic activity and karyotype of leucocytes from pernicious anaemia patients cultured in vitamin B₁₂-deficient medium. Nature 197, 607–608 (1963).

Chapelle, A. de la, Anla, P., Kivalo, E.: Enlarged short arm or satellite region. A heritable trait probably unassociated with developmental disorder. Cytogenetics 2, 129–139 (1963).

Chapelle, A. de la, Erickson, A.W., Kirgarinta, M., Knutar, F.: Glutathione reductase activity in haematological disorders associated with C trisomy. Eur. J. Clin. Invest. 1, 366 (1971).

Chapelle, A. de la, Schröder, J., Ikkala, E.: Significance of 8-trisomy in the bone marrow. Bull. Eur. Soc. Hum. Genet. 6, 63–64 (1973a).

Chapelle, A. de la, Schröder, J., Selander, R.: *In situ* localization and characterization of different classes of chromosomal DNA. Acridine orange and quinacrine mustard fluorescence. Chromosoma 40, 347–360 (1973b).

Chapelle, A. de la, Schröder, J., Selander, R., Stenstrand, K.: Differences in DNA composition along mammalian metaphase chromosomes. Chromosoma 42, 365–382 (1973c).

Chapelle, A. de la, Schröder, J., Vuopio, P.: 8-trisomy in the bone marrow. Report of two cases. Clin. Genet. 3, 470–476 (1972).

Chapelle, A. de la, Vuopio, P., Borgstrom, G.H.: The origin of bone marrow fibroblasts. Blood 41, 783–787 (1973).

Chapelle, A. de la Vuopio, P., Icén, A.: Trisomy 8 in the bone marrow associated with high red cell glutathione reductase activity. Blood 47, 815–826 (1976).

Chapelle, A. de la, Vuopio, P., Sanger, R., Teesdale, P.: Monosomy-7 and the Colton blood-groups. Lancet ii, 817 (1975).

Chapelle, A. de la, Wennström, J., Wasastjerna, C., Knutar, F., Stenman, U.H., Weber, T.H.: Apparent C trisomy in bone marrow cells. Report of two cases. Scand. J. Haematol. 7, 112–122 (1970).

Charles, D., Turner, J.H., Redmond, C.: The endometrial karyotypic profiles of women after clomiphene citrate therapy. J. Obstet. Gynecol. Br. Commonw. 80, 264–270 (1973).

Chatterjea, J.B.: Chromosomal abnormalities in leukaemias. J. Indian Med. Assoc. 45, 558–559 (1965).

Chaudhuri, A., Roy, S.: The Philadelphia (Ph¹) chromosome in chronic myeloid leukaemia. Indian J. Cancer 4, 165–168 (1967).

Chaudhuri, J.P., Ludwig, E., Labouvie, C.: Human lymphocyte culture: Intrinsic factors influencing the quality of chromosome preparations. Blut 35, 223–228 (1977).

Chaudhuri, J.P., Vogel, W., Voiculescu, I., Wolf, U.: A simplified method of demonstrating Giemsa-band pattern in human chromosomes. Humangenetik 14, 83–84 (1971).

Chechik, B.E., Pyke, K.W., Gelfand, E.W.: Human thymus/leukemia-associated antigen in normal and leukemic cells. Int. J. Cancer 18, 551–556 (1976).

Chen, T.R.: A simple method to sequentially reveal Q- and C- bands on the same metaphase chromosomes. Chromosoma 47, 147–156 (1974).

Chen, T.R., Ruddle, F.H.: Karyotype analysis utilizing differentially stained constitutive heterochromatin of human and murine chromosomes. Chromosoma 34, 51–72 (1971).

Chen, T.R., Shaw, M.W.: Stable chromosome changes in a human malignant melanoma. Cancer Res. 33, 2042–2047 (1973).

Cheng, W.-S., Tarone, R.E., Andrews, A.D., Whang-Peng, J.S., Robbins, J.H.: Ultraviolet light-induced sister chromatid exchanges in xeroderma pigmentosum and in Cockayne's syndrome lymphocyte cell lines. Cancer Res. 38, 1601–1609 (1978).

Chernay, P.R., Kardon, N.B., Hsu, L.Y., Shapiro, L.R., Beratis, N.G., Kerr, J., Hirschhorn, K.: A differential staining technique for chromosome identification and its comparison with fluorescence technique. Clin. Genet. 3, 347–356 (1972).

Chervenick, P.A., Ellis, L.D., Pan, S.F., Lawson, A.L.: Human leukemic cells: *In vitro* growth of colonies containing the Philadelphia (Ph¹) chromosome. Science 174, 1134–1136 (1971).

Chervenick, P.A., Lawson, A.L., Ellis, L.D., Pan, S.F.: In vitro growth of leukemic cells containing the Philadelphia (Ph¹) chromosome. J. Lab. Clin. Med. 78, 838–839 (1971).

Chessells, J.M., Janossy, G., Lawler, S.D., Secker

Walker, L.M.: The Ph¹ chromosome in childhood leukaemia. Br. J. Haematol. 41, 25–41 (1979).

Chicago Conference: Standardization in human cytogenetics, D. Bergsma (ed.). New York: The National Foundation—March of Dimes (1966). (Birth Defects: Orig. Art. Ser. Vol. 2, No. 2).

Chitham, R.G., MacIver, E.J.: Chromosome abnormality with lymphoid leukaemia. Lancet i, 1044–1045 (1964).

Christenhuss, R., Büchner, TH., Pfeiffer, R.A.: Visualization of human somatic chromosomes by scanning electron microscopy. Nature 216, 379–380 (1967).

Christensen, L.P.: Applied photography in chromosome studies. *In* J.J. Yunis (ed.): *Human Chromosome Methodology* pp. 129–153. New York: Academic Press (1965).

Chroback, L., Radochova, D., Zizka, J., Smetana, K., Mirova, S., Kerekes, Z.: Di Guglielmo syndrome (erythroleukemia) with dyserythropoiesis in a patient with mammary carcinoma treated by irradiation. Vnitr. Lek. 23, 688–694 (1977).

Chrustchov, G.K., Berlin, E.A.: Cytological investigations on cultures of normal human blood. J. Genet. 31, 243–261 (1935).

Chrustchov, G.K., Andres, A.G., Iljina-Kakujeva, W.: Die Kulturen der Leukozyten des Blutes als eine Methode für das Studium von Menschenkaryotypen. J. Biol. Exp. (Moscow) 7, 455–561 (1931).

Chu, E.H.Y.: The chromosome complements of human somatic cells. Am. J. Hum. Genet. 12, 97–103 (1960).

Chu, E.W., Whang, J.J.K., Rabson, A.S.: Cytogenetic studies of lymphoma cells from an American patient with a tumor similar to Burkitt's tumors in African children. J. Natl. Cancer Inst. 37, 885–891 (1966).

Chu, J.Y., O'Connor, D.M., Blair, J., Roodman, S., McElfresh, A.E.: Congenital leukemia. Proc. Am. Soc. Clin. Oncol. 18, 299 (1977).

Chudina, A.P., Pichugina, M.N.: Chromosome abnormalities in ovarian cancer. Vopr. Onkol. 22, 8–12 (1976).

Chusid, M.J., Dale, D.C.: Eosinophilic leukemia. Remission with vincristine and hydroxyurea. Am. J. Med. 59, 297–300 (1975).

Chusid, M.J., Dale, D.C., West, B.C., Wolff, S.M.: The hypereosinophilic syndrome: Analysis of fourteen cases with review of literature. Medicine (Baltimore) 54, 1–27 (1975).

Cimino, M.C., Kinnealey, A., Variakojis, D., Golomb, H.M., Rowley, J.D.: Chromosomal abnormalities in patients with acute lymphocytic leukemia (ALL). *Program of the American Society of Human Genetics*, 28th Annual Meeting, October 19–22, San Diego, Calif., p. 32 Abstract (1977).

Cimino, M.C., Roth, D.G., Golomb, H.M., Rowley, J.D.: A chromosome marker for B-cell cancers. N. Engl. J. Med. 298, 1422 (1978).

Cimino, M.C., Rowley, J.D., Kinnealey, A., Variakojis, D., Golomb, H.M.: Banding studies of chromosomal abnormalities in patients with acute lymphocytic leukemia. Cancer Res. 39, 227–238 (1979).

Clarkson, B.D., Dowling, M.D., Gee, T.S., Cunningham, I., Hopfan, S., Knapper, W.H., Vaartaja, T., Haghbin, M.: Radical therapy for chronic granulocytic leukemia. *15th Congress of the International Society of Hematology* (Jerusalem), Sept. 1–6, p. 136 (1974).

Clausen, J.J., Syverton, J.T.: Comparative chromosomal study of 31 cultured mammalian cell lines. J. Natl. Cancer Inst. 28, 117–145 (1962).

Cleary, B., Binder, R.A., Kales, A.N., Veltri, B.J.: Simultaneous presentation of acute myelomonocytic leukemia and multiple myeloma. Cancer 41, 1381–1386 (1978).

Cleaver, J.E.: Xeroderma pigmentosum. A human disease in which an initial stage of DNA repair is defective. Proc. Natl. Acad. Sci. USA 63, 428–435 (1969).

Cleaver, J.E.: DNA repair and radiation sensitivity in human (xeroderma pigmentosum) cells. Int. J. Radiat. Biol. 18, 557–565 (1970).

Cleaver, J.E.: Xeroderma pigmentosum. Variants with normal DNA repair and normal sensitivity to ultraviolet light. J. Invest. Dermatol. 58, 124–128 (1972).

Cleaver, J.E.: DNA repair with purine and pyrimidines in radiation and carcinogen-damaged normal and xeroderma pigmentosum human cells. Cancer Res. 33, 362–369 (1973).

Cleaver, J.E.: DNA repair processes and their impairment in some human diseases. *In* D. Scott, B.A. Bridges, F.H. Sobels (eds.): *Progress in Genetic Toxicology*. New York: Elsevier/North-Holland Biomedical Press, pp. 29–42 (1977a).

Cleaver, J.E.: Nucleosome structure controls rates of excision repair in DNA of human cells. Nature 270, 451–453 (1977b).

Clein, G.P., Flemans, R.J.: Involvement of the erythroid series in blastic crisis of chronic myeloid leukaemia. Further evidence for the presence of Philadelphia chromosome in erythroblasts. Br. J. Haematol. 12, 754–758 (1966).

Clifford, P., Gripenberg, U., Klein, E., Fenyö, E.M. Manolov, G.: Treatment of Burkitt's lymphoma. Lancet ii, 517–518 (1968).

Cobo, A., Lisker, R., Cordova, S., Pizzuto, J.: Cytogenetic findings in acquired aplastic anaemia. Acta Haematol. (Basel) 44, 32–36 (1970).

Codish, S.D., Paul, B.: Reversible appearance of specific chromosome which suppresses malignancy. Nature 252, 610–612 (1974).

Cohen, H.J., Huang, A.: A marker chromosom

abnormality occurrence in chloramphenicol-associated acute leukemia. Arch. Intern. Med. 132, 440–443 (1973).
Cohen, M.M., Shaw, M.W.: Two XY siblings with gonadal dysgenesis and a female karyotype. N. Engl. J. Med. 272, 1083–1088 (1965).
Cohen, M.M., Ariel, I., Dagan, J.: Chromosome deletion (46,XX,20q-) in sideroblastic anemia. Israel J. Med. 10, 1393–1396 (1974).
Cohen, M.M., Kohn, G., Dagan, J.: Chromosomes in ataxia-telangiectasia. Lancet ii, 1500 (1973).
Cohen, M.M., Shaham, M., Dagan, J., Shmueli, E., Kohn, G.: Cytogenetic investigations in families with ataxia-telangiectasia. Cytogenet. Cell Genet. 15, 338–356 (1975).
Cohen, S.M.: Chronic myelogenous leukemia with myelofibrosis. Four years after auto-immune hemolytic anemia. Arch. Intern. Med. 119, 620–625 (1967).
Coleman, C.N., Williams, C.J., Flint, A., Glatstein, E.J., Rosenberg, S.A., Kaplan, H.S.: Hematologic neoplasia in patients treated for Hodgkin's disease. N. Engl. J. Med. 297, 1249–1252 (1977).
Coley, G.M., Otis, R.D., Clark, W.E. II: Multiple primary tumors including bilateral breast cancers in a man with Klinefelter's syndrome. Cancer 27, 1476–1481 (1971).
Collins, A.J., Bloomfield, C.D., Peterson, B.A., McKenna, R.W.: Acute nonlymphocytic leukemia in patients with nodular lymphoma. Cancer 40, 1748–1754 (1977).
Colombiès, P., Ducos, J., Ruffié, J., Salles-Mourlan, A.M.: Existe-t-il des anomalies chromosomiques au cours des leucémies lymphocytaires chroniques? Étude de 16 cas. Rev. Fr. Etud. Clin. Biol. 10, 525–529 (1965).
Coltman, C.A.: Multiple myeloma without a paraprotein. Report of a case with observations on chromosomal composition. Arch. Intern. Med. 120, 687–696 (1967).
Comings, D.E.: The rationale for an ordered arrangement of chromatin in the interphase nucleus. Am. J. Hum. Genet. 20, 440–460 (1968).
Comings, D.E.: The structure and function of chromatin. Adv. Hum. Genet. 3, 237–446 (1972).
Comings, D.E.: A general theory of carcinogenesis. Proc. Natl. Acad. Sci. USA 70, 3324–3328 (1973a).
Comings, D.E.: Model for evolutionary origin of chromosome bands. Nature 244, 576–577 (1973b).
Comings, D.E.: What is a chromosome break? In J. German (ed.) Chromosomes and Cancer, pp. 95–133. New York: Wiley (1974).
Comings, D.E.: Mechanisms of chromosome banding. IV. Optical properties of the Giemsa dyes. Chromosoma 50, 89–110 (1975a).
Comings, D.E.: Implications of somatic recombination and sister chromatid exchange in Bloom's syndrome and cells treated with mitomycin C. Humangenetik 28, 191–196 (1975b).
Comings, D.E.: Mammalian chromosome structure. Chromosomes Today 6, 19–26 (1977).
Comings, D.E.: Lateral asymmetry of human chromosomes. Am. J. Hum. Genet. 30, 223–226 (1978).
Comings, D.E., Avelino, E.: Mechanisms of chromosome banding. VII. Interaction of methylene blue with DNA and chromatin. Chromosoma 51, 365–379 (1975).
Comings, D.E., Drets, M.E.: Mechanisms of chromosome banding. IX. Are variations in DNA base composition adequate to account for quinacrine, Hoechst 33258 and Daunomycin banding? Chromosoma 56, 199–211 (1976).
Comings, D.E., Okada, T.A.: Whole-mount electron microscopy of the centromere region of metacentric and telocentric mammalian chromosomes. Cytogenetics 9, 436–449 (1970).
Comings, D.E., Okada, T.A.: Fine structure of kinetochore in Indian Muntjac. Exp. Cell Res. 67, 97–110 (1971).
Comings, D.E., Okada, T.A.: Some aspects of chromosome structure in eukaryotes. Cold Spring Harbor Symp. Quant. Biol. 38, 145–153 (1973).
Comings, D.E., Okada, T.A.: The fibrillar nature of the nuclear matrix. In S. Armendares, R. Lisker (eds.) V International Congress of Human Genetics, p. 117, Abstract 299, October 10–15, Mexico (1976).
Comings, D.E., Avelino, E., Okada, T.A., Wyandt, H.E.: The mechanism of C- and G-banding of chromosomes. Exp. Cell Res. 77, 469–493 (1973).
Comings, D.E., Kovacs, B.W., Avelino, E., Harris, D.C.: Mechanisms of chromosome banding. V. Quinacrine banding. Chromosoma 50, 111–145 (1975).
Conen, P.E.: Chromosome studies in leukemia. Can. Med. Assoc. J. 96, 1599–1605 (1967).
Conen, P.E., Erkman, B.: Combined mongolism and leukemia. Report of eight cases with chromosome studies. Am. J. Dis. Child. 112, 429–443 (1966).
Conen, P.E., Erkman, B.: Chromosome studies in tumours and leukaemia. Can. Med. Assoc. J. 99, 348–353 (1968).
Conen, P.E., Falk, R.E.: Chromosome studies on cultured tumors of nervous tissue origin. Acta Cytol. 11, 86–91 (1967).
Conen, P.E., Lansky, G.S.: Chromosome damage during nitrogen mustard therapy. A case report. Br. Med. J. ii, 1055–1057 (1961).
Conen, P.E., Erkman, B., Laski, B.: Chromosome studies on a radiographer and her family. Report of one case of leukemia and two cases of

Down's syndrome. Arch. Intern. Med. 117, 125–132 (1966).
Conrad, R.A.: Acute myelogenous leukemia following fallout radiation exposure. J.A.M.A. 232, 1356–1357 (1975).
Conroy, J.F., Wieczoreck, R.C., Fuscaldo, K.E.: Trisomy D in a patient with Philadelphia negative chronic leukemia. Proc. Am. Assoc. Cancer Res. 16, 193 (1975).
Contrafatto, G.: Marker chromosome of macroglobulinemia identified by G-banding. Cytogenet. Cell Genet. 18, 370–373 (1977).
Cooke, P.: Non-random participation of chromosomes 13, 14 and 15 in acrocentric associations. Humangenetik 13, 309–314 (1974).
Cooper, E.H., Hughes, D.T., Topping, N.E.: Kinetics and chromosome analyses of tissue culture lines derived from Burkitt lymphomata. Br. J. Cancer 20, 102–113 (1966).
Cooper, E.H., Levi, P.E., Anderson, C.K., Williams, R.E.: The evolution of tumour cell populations in human bladder cancer. Br. J. Urol. 41, 714–717 (1969).
Cooper, H.L., Hirschhorn, K.: Enlarged satellites as a familial chromosome marker. Am. J. Hum. Genet. 14, 107–124 (1962a).
Cooper, H.L., Hirschhorn, K.: Improvements in white cell culture including differential leukocyte separation. Blood 20, 101 (1962b).
Corfman, P.A., Richart, R.M.: Chromosome number and morphology of benign ovarian cystic teratomas. N. Engl. J. Med. 271, 1241–1244 (1964).
Cosson, A., Depres, P., Gazengel, C., Breton-Gorius, J., Prieur, M., Josso, F.: Syndrome leucémique singulier chez un nouveau-né trisomique 21. Proliferation megacaryocyto-plaquettaire; syndrome de coagulation intravasculaire diffuse. Nouv. Rev. Fr. Hematol 14, 181–198 (1974).
Countryman, P.I., Heddle, J.A., Crawford, E.: The repair of X-ray induced chromosomal damage in trisomy 21 and normal diploid lymphocytes. Cancer Res. 37, 52–58 (1977).
Court Brown, W.M.: Role of genetic change in neoplasia. Br. Med. J. i, 961–964 (1962).
Court Brown, W.M.: Chromosomal abnormality and chronic lymphatic leukaemia. Lancet i, 986 (1964).
Court Brown, W.M.: Human population cytogenetics. *Frontiers of Biology,* Vol. 5. Amsterdam: North Holland (1967).
Court Brown, W.M., Tough, I.M.: Cytogenetic studies in chronic myeloid leukemia. Adv. Cancer Res. 7, 351–381 (1963).
Court Brown, W.M., Buckton, K.E., Jacobs, P.A., Tough, I.M., Kuenssberg, E.V., Knox, J.D.E.: Chromosome studies on adults. Eugenics laboratory memoirs, 42, p. 91. London: Cambridge University Press (1966).
Court Brown, W.M., Buckton, K.E., Langlands, A.O., Woodcock, G.E.: The identification of lymphocyte clones, with chromosome structural aberrations in irradiated men and women. Int. J. Radiat. Biol. 13, 155–168 (1967).
Court Brown, W.M., Buckton, K.E., McLean, A.S.: Quantitative studies of chromosome aberrations in man following acute and chronic exposure to X rays and gamma rays. Lancet i, 1239–1241 (1965).
Court Brown, W.M., Jacobs, P.A., Doll, R.: Interpretation of chromosome counts made on bone marrow cells. Lancet i, 160–163 (1960).
Court Brown, W.M., Jacobs, P.A., Tough, I.M., Baikie, A.G.: Manifold chromosome abnormalities in leukaemia. Lancet i, 1242 (1962).
Coutinho, V., Bottura, C., Falcao, R.P.: Cytogenetic studies in malignant lymphomas: a study of 28 cases. Br. J. Cancer 25, 789–801 (1971).
Coutinho, V., Falcao, R.P., Bottura, C.: Cytogenetic observations in congenital familial panmyelopathy (Fanconi's syndrome). Nouv. Rev. Fr. Hematol. 11, 781–790 (1971).
Cox, D.: Chromosome constitution of nephroblastomas. Cancer 19, 1217–1224 (1966).
Cox, D.: Chromosome studies in 12 solid tumours from children. Br. J. Cancer 22, 402–414 (1968).
Cox, D., Yuncken, C., Spriggs, A.I.: Minute chromatin bodies in malignant tumours of childhood. Lancet ii, 55–58 (1965).
Cox, L.W., Stanley, M.A., Harvey, N.D.M.: Cytogenetic assessment of radiosensitivity of carcinoma of the uterine cervix. Obstet. Gynecol. 33, 82–91 (1969).
Cox, R.P.: Activity and regulation of alkaline phosphatase in cells from patients with abnormal chromosome complements. Ann. N.Y. Acad. Sci. 166, 406–416 (1969).
Cox, R.P.: Genetics and cancer. *In* R. Andrade, S.L. Gumport, G.L. Popkin, T.D. Rees (eds.) *Cancer of the Skin: Biology-Diagnosis-Management,* Vol. 1, pp. 137–162. Philadelphia: W.B. Saunders, (1976).
Craig, A.P., Shaw, M.W.: Autoradiographic studies of the human Y chromosome. Chromosoma 32, 364–377 (1971a).
Craig-Holmes, A.P., Shaw, M.W.: Polymorphism of human constitutive heterochromatin. Science 174, 702–704 (1971b).
Craig-Holmes, A.P., Shaw, M.W.: Effects of six carcinogens on SCE frequency and cell kinetics in cultured human lymphocytes. Mutat. Res. 46, 375–384 (1977).
Craig-Holmes, A.P., Moore, F.B., Shaw, M.W.: Polymorphism of human C-band heterochromatin. I. Frequency of variants. Am. J. Hum. Genet. 25, 181–192 (1973).
Crawfurd, M., Pegrum, G.: Chronic granulocytic leukaemia. Lancet i, 1044 (1964).
Crick, F.: General model for the chromosomes of higher organisms. Nature 234, 25–27 (1971).

Crick, F., Klug, A.: Kinky helix. Nature 255, 530–533 (1975).
Crisalli, M., Monteverde, R., Dagna-Briccarelli, F.: Chromosomal aneuploidy during the course of acute lymphatic leukemia in a subject with trisomy 21. Minerva Pediatr. 23, 1791–1797 (1971).
Crist, W.M., Ragab, A.H., Ducos, R.: Lymphoblastic conversion in chronic myelogenous leukemia. Pediatrics 61, 560–563 (1978a).
Crist, W.M., Ragab, A., Moreno, H., Pereira, F., Foster, J.C.: Granulopoiesis in chronic myeloproliferative disorders in children. Pediatrics 61, 889–893 (1978b).
Croce, C.M.: Assignment of gene(s) for cell transformation and malignancy to human chromosome 7 carrying the SV40 genome. *In* R.L. Crowell, H. Friedman, J.E. Prier (eds.): *Tumor Virus Infections and Immunity*, pp. 223–230. Baltimore: University Park Press (1976).
Croce, C.M.: Assignment of the integration site for Simian virus 40 to chromosome 17 in GM54 VA, a human cell line transformed by Simian virus 40. Proc. Natl. Acad. Sci. USA 74, 315–318 (1977).
Crosby, W.H.: Splenectomy in hematologic disorders. N. Engl. J. Med. 286, 1252–1254 (1972).
Cross, H.E., Hansen, R.C., Morrow, G., Davis, J.R.: Retinoblastoma in a patient with a 13qxp translocation. Am. J. Ophthalmol. 84, 548–554 (1977).
Crossen, P.E.: Giemsa banding patterns of human chromosomes. Clin. Genet. 3, 169–179 (1972).
Crossen, P.E.: Unusual chromosome bands revealed by aging. Humangenetik 21, 197–202 (1974).
Crossen, P.E.: Giemsa banding patterns in chronic lymphocytic leukaemia. Humangenetik 27, 151–156 (1975).
Crossen, P.E., Morgan, W.F.: Analysis of human lymphocyte cell cycle time in culture measured by sister chromatid differential staining. Exp. Cell Res. 104, 453–457 (1977a).
Crossen, P.E., Morgan, W.F.: Proliferation of PHA- and PWM-stimulated lymphocytes measured by sister chromatid differential staining. Cell Immunol. 32, 432–438 (1977b).
Crossen, P.E., Drets, M.E., Arrighi, F.E., Johnston, D.A.: Analysis of the frequency and distribution of sister chromatid exchanges in cultured human lymphocytes. Hum. Genet. 35, 345–352 (1977).
Crossen, P.E., Fitzgerald, P.H., Menzies, R.C., Brehaut, L.A.: Chromosomal abnormality, megaloblastosis, and arrested DNA synthesis in erythroleukaemia. J. Med. Genet. 6, 95–104 (1969).
Crossen, P.E., Mellor, J.E.L., Adams, A.C., Gunz, F.W.: Chromosome studies in Fanconi's anaemia before and after treatment with oxymetholone. Pathology 4, 27–33 (1972).
Crossen, P.E., Mellor, J.E.L., Finley, A.G., Ravich, R.B.M., Vincent, P.C., Gunz, F.W.: The Sézary syndrome. Cytogenetic studies and identification of the Sézary cell as an abnormal lymphocyte. Am. J. Med. 50, 24–34 (1971a).
Crossen, P.E., Mellor, J.E.L., Vincent, P.E., Gunz, F.W.: Clonal evolution in human leukaemia. Cytobios 4, 29–48 (1971b).
Cuenca, C.R., Becker, K.L.: Klinefelter's syndrome and cancer of the breast. Arch. Intern. Med. 121, 159–162 (1968).
Curcio, S.: Studio citogenetico di un carcinoma primitivo della salpinge. Arch. Ostet. Ginecol. 71, 450–456 (1966a).
Curcio, S.: Analisi cromosomica di un folliculoma. Estratto dagli atti del III Congresso della Società Italiana di Citologia Clinica et Sociale. Arch. Ostet. Ginecol. 71, 139–143 (1966b).
Curcio, S.: Analisi cromosomica di un caso di ca. ovarico aspetti della replicazione del DNA. Arch. Ostet. Ginecol. 71, 436–449 (1966c).
Curcio, S., Sartori, R.: Citogenetica delle neoplasie ginecologiche. Arch. Ostet. Ginecol. 71, 423–435 (1966).
Curnow, R.N., Franklin, M.F.: Some further problems in the classification of human chromosomes. Biometrics 29, 429–440 (1973).
Czeizel, A., Csösz, L., Gárdonyi, J., Remenár, L., Ruzicska, P.: Chromosome studies in twelve patients with retinoblastoma. Humangenetik 22, 159–166 (1974).

Daguillard, F., Fontaine, L., Tardieu, M.: B and T lymphocytes in chronic lymphatic leukaemia. Lancet i, 308–309 (1974).
Dahlke, M.B., Nowell, P.C.: Chromosomal abnormalities and dyserythropoiesis in the preleukaemic phase of multiple myeloma. Br. J. Haematol. 31, 111–116 (1975).
Dallapiccola, B.: Il cariotipo nelle leucemie eosinofile. Haematologica (Pavia) 55, 358–370 (1970).
Dallapiccola, B., Alimena, G.: Inactivated normal X in a female leukaemic patient with an acquired X/autosome translocation. Hum. Genet. 48, 169–177 (1979).
Dallapiccola, B., Malacarne, P.: Di Guglielmo syndrome in a t(Dq Dq) heterozygote. J. Med. Genet. 8, 209–214 (1971).
Dallapiccola, B., Ricci, N.: Observations on specific Giemsa staining of the Y and on selective destaining of the chromosomes. Humangenetik 26, 251–255 (1975).
Dallapiccola, B., Tataranni, G.: Modello di evoluzione clonale del cariotipo in versamento pleurico neoplastico. Arch. Ital. Pathol. Clin. Tumori 12, 3–21 (1969).
Daly, M.B.: Epidemiologic factors as related to specific neoplasms in ataxia-telangiectasia families. Am. J. Epidemiol. 104, 331–332 (1976).

D'Ambrosio, S.M., Setlow, R.B.: Defective and enhanced postreplication repair in classical and variant xeroderma pigmentosum cells treated with N-acetoxy-2-acetylaminofluorene. Cancer Res. 38, 1147–1153 (1978).

Dameshek, W.: Some speculations on the myeloproliferative syndromes. Blood 6, 372–375 (1951).

Dameshek, W.: Foreward and a proposal for considering paroxysmal nocturnal hemoglobinuria (PNH) as a "candidate" myeloproliferative disorder. Blood 33, 263–264 (1969).

Dameshek, W., Gunz, F.W.: *Leukemia*, 2nd edition. New York: Grune and Stratton (1964).

Dammacco, F., Trizio, D., Bonomo, L.: A case of IgAK-myelomatosis with two urinary Bence-Jones proteins (BJK and BJL) and multiple chromosomal abnormalities. Acta Haematol. (Basel) 41, 309–320 (1969).

Danes, B.S.: The Gardner syndrome. A study in cell culture. Cancer 36, 2327–2333 (1975).

Danes, B.S.: Increased tetraploidy: Cell-specific for the Gardner gene in the cultured cell. Cancer 38, 1983–1988 (1976a).

Danes, B.S.: The Gardner syndrome. Increased tetraploidy in cultured skin fibroblast. J. Med. Genet. 13, 52–56 (1976b).

Danes, B.S.: Increased in vitro tetraploidy: Tissue specific within the heritable colorectal cancer syndromes with polyposis coli. Cancer 41, 2330–2334 (1978).

Daneshvar-Alavi, B., Lutcher, C.L., Welter, D.: Multiple myeloma in acute myelogenous leukemia with the presence of a Philadelphia (Ph[1]) chromosome. South. Med. J. 70, 1477–1479 (1977).

D'Angio, G.J., Evans, A.E., Koop, C.E.: Special pattern of widespread neuroblastoma with a favourable prognosis. Lancet i, 1046–1049 (1971).

Daniel, A., Lam-poo-tang, P.R.: Mechanism for the chromosome banding phenomenon. Nature 244, 358–359 (1973).

Daniel, A., Francis, S.E., Stewart, L.A., Barber, S.: A near haploid clone: 24,XY,t(9;22)(q34;q11) from a patient in blast crisis of chronic myeloid leukaemia. Scand. J. Haematol. 21, 99–103 (1978).

Darai, G., Braun, R., Flugel, R.M., Munk, K.: Malignant transformation of rat embryo fibroblasts by herpes simplex virus types 1 and 2 at suboptimal temperature. Nature 265, 744–746 (1977).

Darlington, C.D.: *Recent Advances in Cytology*, 2nd edition. Philadelphia: P. Blakiston's Son (1937).

Dartnall, J.A., Mundy, G.R., Baikie, A.G.: Cytogenetic studies in myeloma. Blood 42, 229–239 (1973).

Das, K.C., Aikat, B.K.: Chromosomal abnormalities in multiple myeloma. Blood 30, 738–748 (1967).

Das, K.C., Nayak, J., Mohanty, D., Garewal, G.: Cytology and cytogenetics in chronic myelogenous leukemias. Indian J. Med. Res. 68, 148–163 (1978).

Davidenkova, E.F., Kolosova, N.N.: Chromosomal anomalies in acute and chronic leukemias (in Russian). Vopr. Onkol. 10 (6), 3–7 (1964).

Davidson, E., Bulkin, W.: Long marker chromosome in bronchogenic carcinoma. Lancet ii, 227 (1966).

Davidson, N.R.: Photographic techniques for recording chromosome banding patterns, J. Med. Genet. 10, 122–126 (1973).

Davidson, R.J.L., Walker, W., Watt, J.L., Page, B.M.: Familial erythroleukaemia: A cytogenetic and haematological study. Scand. J. Haematol. 20, 351–359 (1978).

Davidson, W.M., Knight, L.A.: Acquired trisomy 9. Lancet i, 1510 (1973).

Davis-Lawas, D., Lawas, I.: The Ph[1] chromosome in three cases of chronic myelogenous leukemia in Filipino children. Philippine J. Pediatr. 14, 289 (1965).

Day, R.S.: Xeroderma pigmentosum variants have decreased repair of ultraviolet-damaged DNA. Nature 253, 748–749 (1975).

Day, R.W., Wright, S.W., Koons, A., Quigley, M.: XXX 21-trisomy and retinoblastoma. Lancet ii, 154–155 (1963).

Debray, J., Cheymol, G., Krulik, M., Schmitt, S., Audebert, A.: Étude préliminaire de la valeur diagnostique et pronostique de l'histaminémie au cours de la leucémie myéloïde chronique. Nouv. Rev. Fr. Hematol. 15, 250–260 (1975).

Dehnhard, F.: Zytogenetische Untershungen an Praekanzerosen der Cervix Uteri. Verh. Dtsch. Ges. Pathol. 57, 223–228 (1973).

Dehnhard, F., Breinl, H., Knörr-Gartner, H.: Chromosomenveränderungen bei Krebsvorstadien der Cervix Uteri. Geburtshilfe Frauenheilkd. 30, 602–612 (1970).

Dehnhard, F., Breinl, H., Knörr-Gartner, H.: Chromosomenbefunde bei Krebsvorstadien der Cervix Uteri (Zusammenfassende Darstellung erster Ergebnisse). Minerva Ginecol. 23, 78–81 (1971).

Dehnhard, F., Breinl, H., Schüssler, J., Wehler, V.: Cytogenetische Untersuchungen an Praecancerosen und Carcinomen der Cervix Uteri. I. Numerische Chromosomenanomalien. Arch. Gynaekol. 220, 123–138 (1975a).

Dehnhard, F., Breinl, H., Schüssler, J., Wehler, V.: Cytogenetische Untersuchungen an Praecancerosen und Carcinomen der Cervix Uteri. II Strukterelle Chromosomenanomalien Ver änderungen im Karyotypus. Arch. Gynaekol. 220, 139–159 (1975b).

Dehnhard, F., Breinl, H., Schüssler, J., Wehler, V. Cytogenetische Untersuchungen an Praecancer osen und Carcinomen der Cervix Uteri. III. Literaturübersicht. Diskussion der bisher mitgeteilten Befunde. Arch. Gynaekol. 220, 161–177 (1975c).

Dehnhard, F., Knörr-Gartner, H., Breinl, H.: Zur Chromosomenpathologie des Korpus-Karzinoms. Geburtshilfe Frauenheilkd. 33, 98–106 (1973).

Deisseroth, A., Nienhuis, A., Ruddle, F., Lawrence, J., Turner, P.: Chromosomal localization of the human β globin gene to human chromosome 11. Blood 50 (Suppl. 1), 105 (abstract) (1977).

DeMayo, A.P., Kiossoglou, K.A., Erlandson, M.E., Notterman, R.F., German, J.: A marrow chromosomal abnormality preceding clinical leukemia in Down's syndrome. Blood 29, 233–241 (1967).

Demin, A.A., Radjabli, S.I., Loseva, M.I., Metelkina, N.V.: Cytogenetic studies in cases of Hodgkin's disease and some reticuloses. (in Russian). Genetika 8 (1), 125–141 (1972).

Denegri, J.F., Naiman, S.C., Gillen, J., Thomas, J.W.: In vitro growth of basophils containing the Philadelphia chromosome in the acute phase of chronic myelogenous leukaemia. Brit. J. Haematol. 40, 351–356 (1978).

Denton, T.E., Howell, W.M., Barrett, J.V.: Human nucleolar organizer chromosomes: Satellite associations. Chromosoma 55, 81–84 (1976).

Denver Conference: A proposed standard system of nomenclature of human mitotic chromosomes. Lancet i, 1063–1065 (1960); J.A.M.A. 174, 159–162 (1960).

Desbois, J.C., Coicadan, L., Hagege, N., Allaneau, C.: Gondoblastomas (review of the literature and the report of one observation). Med. Infant. 84, 525–547 (1977).

Dev, V.G., Warburton, D., Miller, O.J., Erlanger, B.F.: Consistent pattern of binding of anti adenosine antibodies to human metaphase chromosomes. Exp. Cell Res. 74, 288–293 (1972).

Dewald, G., Dines, D.E., Weiland, L.H., Gordon, H.: Usefulness of chromosome examination in the diagnosis of malignant pleural effusions. N. Engl. J. Med. 295, 1494–1500 (1976).

Dewald, G., Spurbeck, J.L., Vitek, H.A.: Chromosomes in a patient with the Sezary syndrome. Mayo Clin. Proc. 49, 553–557 (1974).

Dewurst, C.J., Ferreira, H.P., Gillett, P.G.: Gonadal malignancy in XY females. J. Obstet. Gynaecol. Br. Commonw. 78, 1077–1083 (1971).

Dharker, R.S., Chaurasia, B.D., Goswami, H.K.: Hypoploidy in brain tumours, Acta Biol. Acad. Sci. Hung. 24, 233–235 (1973).

Diekmann, L. Von, Rickers, H., Pfeiffer, R.A., Bäumer, A.: Morphologische, immunozytologische und zytogenetische Untersuchungen bei einem Kind mit Erythroleukämie. Blut 18, 321–333 (1969).

Dieśka, D., Izaković, V., Gajdoś, M., Gōcǎrova, K.: Karyotyp leukemickych buniek u chronickej myeloickej leukémie v štàdiu akútneho zvratu. Vnitr. Lek. 13, 209–216 (1967a).

Dieśka, D., Izaković, V., Gajdoś, M.: Filadelfsky chromozóm (Ph^1) — cytogeneticka charakteristika chronickej myeloickej leukémie. Cas. Lek. Cesk. 106, 92–97 (1967b).

DiGrado, F., Mendes, F.T., Schroeder, E.: Ring chromosome in a case of Di Guglielmo syndrome. Lancet ii, 1243–1244 (1964).

DiLeo, P.E., Müller, Hj., Obrecht, J.-P., Speck, B., Bühler, E.M., Stalder, G.R.: Loss of the Y chromosome from bone marrow cells of males with myeloproliferative disorders. Acta Haematol. (Basel) 57, 310–320 (1977).

Dinauer, M.C., Pierre, R.V.: Philadelphia chromosome by translocation. Lancet ii, 971 (1973).

Ding, J.C., Adams, P.B., Patison, M., Cooper, I.A.: Thymic origin of abnormal lymphoid cells in Sézary syndrome. Cancer 35, 1325–1332 (1975).

Dingman, C.W., Kakunaga, T.: DNA strand breaking and rejoining in response to ultraviolet light in normal human and xeroderma pigmentosum cells. Int. J. Radiat. Biol. 30, 55–66 (1976).

DiPaolo, J.A.: Karyological instability of neoplastic somatic cells. In Vitro 11, 89–96 (1975).

DiPaolo, J.A.: Chromosomal alterations in carcinogen transformed mammalian cells. *In* P.O. Ts'o (ed.): *Molecular Biology of the Mammalian Genetic Apparatus*, Vol. 2, Amsterdam: Elsevier-North Holland Biomedical Press, pp. 205–227 (1977).

DiPaolo, J.A., Popescu, N.C.: Distribution of chromosome constitutive heterochromatin of Syrian hamster cells transformed by chemical carcinogens. Cancer Res. 33, 3259–3265 (1973).

DiPaolo, J.A., Popescu, N.C.: Relationship of chromosome changes to neoplastic cell transformation. Am. J. Pathol. 85, 709–738 (1976).

DiPaolo, J.A., Nelson, R.L., Donovan, P.J.: Morphological, oncogenic, and karyological characteristics of Syrian hamster embryo cells transformed *in vitro* by carcinogenic polycyclic hydrocarbons. Cancer Res. 31, 1118–1127 (1971).

Dische, M.R., Gardner, H.A.: Mixed teratoid tumors of the liver and neck in trisomy 13. Am. J. Clin. Pathol. 69, 631–637 (1978).

Distéche, C., Bontemps, J.: Chromosome regions containing DNA of known base composition, specifically evidenced by 2,7-di-t-butyl proflavine. Comparison with the Q-banding and relation to dye-DNA interactions. Chromosoma 47, 263–281 (1974).

Disteche, C., Bontemps, J.: Method for the determination of mean densitometric profiles of chromosomes. Chromosoma (Berl.) 54, 39–59 (1976).

Dixon, B., Stead, R.H.: Feulgen microdensitometry and analysis of S-phase cells in cervical tumour biopsies. J. Clin. Pathol. 30, 907–913 (1977).

Djaldetti, M., Padeh, B., Pinkhas, J., De Vries, A.: Prolonged remission in chronic myeloid leukemia after one course of Busulfan. Blood 27, 103–109 (1966).

Dodge, O.G., Jackson, A.W., Muldal, S.: Breast cancer and interstitial-cell tumor in a patient with Klinefelter's syndrome. Cancer 24, 1027–1032 (1969).

Dofuku, R., Biedler, J.L., Spengler, B.A., Old, L.J.: Trisomy of chromosome 15 in spontaneous leukemia of AKR mice. Proc. Natl. Acad. Sci. USA 72, 1515–1517 (1975).

Doida, Y., Hoke, C., Hempelmann, L.H.: Chromosome damage in thyroid cells of adults irradiated with X-rays in infancy. Radiat. Res. 45, 645–656 (1971).

Doll, D.C., Kandzari, S., Jenkins, J.J., III, Amato, S., Jones, B.: Seminoma in a 12 year old male with 46,XY/45,XO karyotype. Cancer 42, 1823–1825 (1978).

Doll, D.C., Weiss, R.B., Evans, H.: Klinefelter's syndrome and extragenital seminoma. J. Urol. 116, 675–676 (1976).

Dominguez, C.J., Greenblatt, R.B.: Dysgerminoma of the ovary in a patient with Turner's syndrome. Am. J. Obstet. Gynecol. 83, 674–677 (1962).

Donlon, J.A., Jaffe, E.S., Braylan, R.C.: Terminal deoxynucleotidyl transferase activity in malignant lymphomas. N. Engl. J. Med. 297, 461–464 (1977).

Donner, L., Bubenik, J.: Minute chromatin bodies in two mouse tumours induced *in vivo* by Rous sarcoma virus. Folia Biol. (Praha) 14, 86–88 (1968).

Dor, J.F., Mattéi, J.F., Mattéi, M.G., Giraud, F., Mongin, M.: Acquired idiopathic sideroblastic anemia: 3 cases with the same extra chromosome (47 mar+). Blood Cells 18, 235–236 (1977a).

Dor, J.F., Mattéi, J.F., Mattéi, M.G., Giraud, F., Mongin, M.: Subacute myeloid leukemia with the Philadelphia chromosome and an additional translocation: 9–12. Nouv. Rev. Fr. Hematol. Blood Cells 18, 245–246 (1977b).

Dorus, E., Amarose, A.P., Koo, G.C., Wachtel, S.S.: Clinical, pathologic, and genetic findings in a case of 46,XY pure gonadal dysgenesis (Swyer's syndrome). Am. J. Obstet. Gynecol. 127, 829–831 (1977).

Dosik, H., Verma, R.S.: New cytogenetic findings in patients with hematologic disorders. *American Society of Human Genetics, 28th Annual Meeting*, San Diego, Calif., Oct. 19–22, p. 38a (abstract) (1977).

Dosik, H., Hsu, L.Y., Todaro, G.J., Lee, S.L., Hirschhorn, K., Selirio, E.S., Alter, A.A.: Leukemia in Fanconi's anemia: Cytogenetic and tumor virus susceptibility studies. Blood 36, 341–352 (1970).

Dosik, H., Verma, R.S., Wilson, C., Miotti, A.B.: Fanconi's anemia and a familial stable chromosome abnormality in a family with multiple malignancies. Blood 50 (Suppl. 1), 190 (abstract) (1977).

Dougan, L., Woodliff, H.J.: Presence of two Ph[1] chromosomes in cells from a patient with chronic granulocytic leukaemia. Nature 205, 405–406 (1965).

Dougan, L., Onesti, P., Woodliff, H.J.: Cytogenetic studies in chronic granulocytic leukaemia. Australas. Ann. Med. 16, 52–61 (1967).

Dougan, L., Scott, I.D., Woodliff, H.J.: A pair of twins, one of whom has chronic granulocytic leukaemia. J. Med. Genet. 3, 217–219 (1966).

Dougan, L., Woodliff, H.J., Onesti, P.: Cytogenetic studies in megakaryocytic myelosis. Med. J. Aust. 1, 62–65 (1967).

Dowsett, J.W.: Corpus carcinoma developing in a patient with Turner's syndrome treated with estrogen. Am. J. Obstet. Gynecol. 86, 622–625 (1963).

Dresp, J., Schmid, E., Bauchinger, M.: The cytogenetic effect of bleomycin on human peripheral lymphocytes in vitro and in vivo. Mutat. Res. 56, 341–353 (1978).

Drets, M.E., Shaw, M.W.: Specific banding pattern of human chromosomes. Proc. Natl. Acad. Sci. USA 68, 2073–2077 (1971).

Drew, S.I., Terasaki, P.I., Billing, R.J., Bergh, O.J., Minowada, J., Klein, E.: Group-specific human granulocytic antigens on a chronic myelogenous leukemia cell line with a Philadelphia chromosome marker. Blood 49, 715–718 (1977).

Dreyfus, B., Rochant, H., Sultan, C., Varet, B., Breton-Gorius, J., Yvart, J.: Anémies réfractanes ou dysérythropoïeses acquises. Etude clinique et biologique. Rev. Prat. Paris 9, 1341–1356 (1971).

Dreyfus, B., Sultan, C., Rochant, H., Salmon, C., Mannoni, P., Cartron, J.P., Boivin, P., Galand, C.: Anomalies of blood group antigens and erythrocyte enzymes in two types of chronic refractory anemia. Br. J. Haematol. 16, 303–312 (1969).

Dubois-Ferrière, H.: Etude cytologique de la transformation aiguë d'une leucémie myélocytaire chronique. Acta Haematol. (Basel) 39, 249–256 (1968).

Dubrova, S.E., Dygin, V.P., Ushakova, E.A.: Cytogenetical and clinico-hematological changes in myeloma disease (in Russian). Tsitologiia 8, 241–249 (1966).

Ducatman, A., Hirschhorn, K., Selikoff, I.J.: Vinyl chloride exposure and human chromosome aberrations. Mutat. Res. 31, 163–168 (1975).

Ducos, J., Colombiés, P.: Chromosomes in chronic lymphocytic leukaemia. Lancet i, 1038 (1968).
Ducos, J., Ruffié, J., Marty, Y., Salles-Mourlan, A.M., Colombiés, P.: Does a connection exist between blood group modifications observed in leukaemia and certain chromosomal alterations? Nature 203, 432–433 (1964).
Dumars, K.W., Kitzmiller, N., Gaskill, C.: Cancer chromosomes and congenital abnormalities. Cancer 20, 1006–1014 (1967).
DuPraw, E.J.: Evidence for a "folder fibre" organization in human chromosomes. Nature 209, 577–581 (1966).
DuPraw, E.J.: *DNA and Chromosomes.* New York: Holt, Rinehart and Winston (1970).
DuPraw, E.J., Bahr, G.F.: The arrangement of DNA in human chromosomes, as investigated by quantitative electron microscopy. Acta Cytol. 13, 188–205 (1969).
Durant, J.R., Tassoni, E.M.: Coexistent Di Guglielmo's leukemia and Hodgkin's disease. A case report with cytogenetic studies. Am. J. Med. Sci. 254, 824–830 (1967).
Duran-Troise, G., Lustig, E.S. de: Cytogenetic studies on a cell line from a mixed parotid tumor. Rev. Eur. Clin. Biol. 17, 605–610 (1972).
Duruman, N.A.: Chromosome abnormalities in bladder tumors. Hacettepe Bull. Med. Surg. 2, 15–23 (1969).
Dutrillaux, B.: Application to the normal karyotype of R-band and G-band techniques involving protolytic digestion. Nobel Symp. 23, 38–42 (1973a).
Dutrillaux, B.: New system of chromosome banding. The T bands. Chromosoma 41, 395–402 (1973b).
Dutrillaux, B.: Obtention simultanée de plusieurs marguages chromosomiques dur les mêmes préparations, aprés traitement par le BrdU. Humangenetik 30, 297–306 (1975).
Dutrillaux, B.: The relationship between DNA replication and chromosome structure. Hum. Genet. 35, 247–253 (1977).
Dutrillaux, B., Lejeune, J.: Sur une nouvelle technique d'analyse du caryotype humain. C.R. Acad. Sci. [D] (Paris) 272, 2638–2640 (1971).
Dutrillaux, B., Aurias, A., Fosse, A.M.: Différenciation des mécanismes induisant la segmentation et l'asymétrie des chromatides, aprés traitement par le 5-bromodéoxyuridine. Exp. Cell Res. 97, 313–321 (1976).
Dutrillaux, B., Croquette, M.F., Viegas-Pequignot, E., Aurias, A., Coget, J., Couturier, J., Lejeune, J.: Human somatic chromosome chains and rings. A preliminary note on end-to-end fusions. Cytogenet. Cell Genet. 20, 70–77 (1978).
Dutrillaux, B., Couturier, J., Viegas-Péquingnot, E., Schaison, G.: Localization of chromatid breaks in Fanconi's anemia, using three consecutive stains. Hum. Genet. 37, 65–71 (1977).
Dutrillaux, B., Grouchy, J. de, Defianz, C., Lejeune, J.: Cytogénétique humaine. Mise en évidence de la structure fine des chromosomes humains par digistion enzymatique (pronase en partrentaier). C.R. Acad. Sci. [D] (Paris) 273, 587–588 (1971).
Duttera, M., Bull, J.M.C., Whang-Peng, J., Carbone, P.P.: Cytogenetically abnormal cells *in vitro* in acute leukaemia. Lancet i, 715–718 (1972).
Dutta-Choudhuri, R., Choudhuri, A.: Chromosome findings in leucocytes of patients with cancer of the throat region. Indian J. Cancer 4, 180–184 (1967).
Duvall, C.P., Carbone, P.P., Bell, W.R., Whang, J., Tjio, J.H., Perry, S.: Chronic myelocytic leukemia with two Philadelphia chromosomes and prominent peripheral lymphadenopathy. Blood 29, 652–666 (1967).
Dygin, V.P.: *Cytogenetical Investigations in Systemic Diseases of Blood* (in Russian). Leningrad: Meditsina (1976).
Dyment, P.G., Melnyk, J., Brubaker, C.A.: A cytogenetic study of acute erythroleukemia in children. Blood 32, 997–1002 (1968).

Ebbin, A.J., Heather, C.W., Jr., Moldow, R.E., Lee, J.: Down's syndrome and leukemia in a family. J. Pediatr. 73, 917–920 (1968).
Egozcue, J.: Philadelphia (Ph-1) chromosome. Nature 231, 405 (1971).
Ehrlich, P., Apolaut, M.: Beobachtungen über maligne Mausetumoren. Z. Naturforsch. [C] 42, 871–874 (1905).
Eiberg, H.: G, R, and C banding patterns of human chromosomes produced by heat treatment in organic and inorganic salt solutions. Clin. Genet. 4, 556–562 (1973).
Eicke, J., Emminger, A., Strauss, Ch., Mohr, U., Wrba, H.: Cytogenetisch-karyologische Studien an klinisch behandelten gynäkologischen Tumoren. Z. Krebsforsch. 67, 205–212 (1965).
Einhorn, N.: Acute leukemia after chemotherapy (melphalan). Cancer 41, 444–447 (1978).
Elejalde, R., Restrepo, A.: Cytogenetic studies in patients with leukemia. Antioquia Med. 22, 5–22 (1972).
Elfenbein, G.J., Borgaonkar, D.S., Bias, W.B., Burns, W.H., Saral, R., Sensenbrenner, L.L., Tutschka, P.J., Zaczek, B.S., Zander, A.R., Epstein, R.B., Rowley, J.D., Santos, G.W.: Cytogenetic evidence for recurrence of acute myelogenous leukemia after allogenic bone marrow transplantation in donor hematopoietic cells. Blood 52, 627–636 (1978).
Elliott, A.Y., Bronson, D.L., Cervenka, J., Stein, N., Fraley, E.E.: Properties of cell lines established

from transitional cell cancers of the human urinary tract. Cancer Res. 37, 1279–1289 (1977).
Elliott, A.Y., Cleveland, P., Cervenka, J., Castro, A.E., Stein, N., Hakala, T.R., Fraley, E.E.: Characterization of a cell line from human transitional cell cancer of the urinary tract. J. Natl. Cancer Inst. 53, 1341–1349 (1974).
Ellis, J.R., Penrose, L.S.: Enlarged satellites and multiple malformations in the same pedigree. Ann. Hum. Genet. 25, 159–162 (1960).
Ellison, J.R., Barr, H.J.: Quinacrine fluorescence of specific chromosome regions. Late replication and high A:T content in *samoaia leonensis*. Chromosoma 36, 375–390 (1972).
Ellman, L.: Bone marrow biopsy in the evaluation of lymphoma, carcinoma and granulomatous disorders. Am. J. Med. 60, 1–7 (1976).
Ellman, L., McChesney, T.: Dyspnea and lymphadenopathy in a patient with two Ph¹ chromosomes. Case reports of the Massachusetts General Hospital (Case 36, 1973). N. Engl. J. Med. 289, 524–530 (1973).
Elves, M.W., Brown, A.K.: Cytogenetic studies in a family with Waldenström's macroglobulinemia. J. Med. Genet. 5, 118–122 (1968).
Elves, M.W., Israëls, M.C.G.: Chromosomes and serum proteins; a linked abnormality. Br. Med. J. ii, 1024–1026 (1963).
Elves, M.W., Israëls, M.C.G.: Cytogenetic studies in unusual forms of chronic myeloid leukaemia. Acta Haematol. (Basel) 38, 129–141 (1967).
Elves, M.W., Buttoo, A.S., Israëls, M.C.G., Wilkinson, J.F.: Chromosome changes caused by 6-azauridine during treatment of acute myeloblastic leukaemia. Br. Med. J. i, 156–159 (1963).
Elves, M.W., Buttoo, A.S., Israëls, M.C.G., Wilkinson, J.F.: Chromosome changes caused by 6-azauridine during treatment of acute myeloblastic leukaemia. Br. Med. J. i, 156–159 (1963).
Emanuel, B.S.: Compound lateral asymmetry in human chromosome 6: BrdU-dye studies of 6q12→6q14. Am. J. Hum. Genet. 30, 153–159 (1978).
Emberger, J.M., Taib, J., Izarn, P.: Étude d'une évolution clonale au cours d'une leucémie myeloblastique. Lyon Med. 233, 219–222 (1975).
Emerit, I.: Chromosomal breakage in systemic sclerosis and related disorders. Dermatologica 153, 145–156 (1976).
Emerit, I., Housset, E.: Chromosome studies on bone marrow from patients with systemic sclerosis. Evidence for chromosomal breakage *in vivo*. Biomedicine 19, 550–554 (1973).
Emerit, I., Marteau, R.: Chromosome studies in 14 patients with disseminated sclerosis. Hum. Genet. 13, 25–33 (1971).
Emerit, I., Emerit, J., Tosoni-Pittoni, A., Bousquet, O., Sarrazin, A.: Chromosome studies in patients with ulcerative colitis. Humangenetik 16, 313–322 (1972).
Emerit, I., Housset, E., Feingold, J.: Chromosomal breakage and scleroderma: Studies in family members. J. Lab. Clin. Med. 88, 81–86 (1976).
Emerit, I., Housset, E., Grouchy, J. de, Camus, J.-P: Chromosome breakage in diffuse scleroderma. A study of 27 patients. Rev. Eur. Clin. Biol. 16, 684–694 (1971).
Emerit, I., Levy, A., Housset, E.: Sclérodermie généralisée et cassures chromosomiques. Mise en évidence d'une "facteur cassant" dans le serum des malades. Ann. Genet. (Paris) 16, 135–138 (1973).
Emson, H.E., Kirk, H.: Value of desoxyribonucleic acid (DNA) in evolution of carcinomas of the human breast. Cancer 20, 1248–1252 (1967).
Endo, A., Yamamoto, M., Watanabe, G.I., Suzuki, Y., Sakai, K.: "Antimongolism" syndrome. Br. Med. J. iv, 148–149 (1969).
Engel, E.: Chromosomes in aleukaemic leukaemia. Lancet ii, 1242 (1965*a*).
Engel, E.: X-rays and Philadelphia chromosome. Lancet ii, 291–292 (1965*b*).
Engel, E.: Les facteurs chromosomiques de la leucémogenése. Schweiz. Med. Wochenschr. 107, 1426–1436 (1977).
Engel, E., Forbes, A.P.: Cytogenetic and clinical findings in 48 patients with congenitally defective or absent ovaries. Medicine (Baltimore) 44, 135–164 (1965).
Engel, E., McKee, L.C.: Double Ph¹ chromosomes in leukaemia. Lancet ii, 337 (1966).
Engel, E., Flexner, J.M., Engel-de Montmollin, M., Frank, H.E.: Blood and skin chromosomal alterations of a clonal type in a leukemic man previously irradiated for a lung carcinoma. Cytogenetics 3, 228–251 (1964).
Engel, E., Jenkins, D.E., Tipton, RE., McGee, B.J., Engel-de Montmollin, M.: Ph¹-positive chronic myelogenous leukemia, with absence of another G chromosome, in a male. N. Engl. J. Med. 273, 738–742 (1965).
Engel, E., McGee, B.J., Flexner, J.M., Krantz, S.B.: Translocation of the Philadelphia chromosome onto the 17 short arm in chronic myeloid leukemia. A second example. N. Engl. J. Med. 293, 666–667 (1975*a*).
Engel, E., McGee, B.J., Flexner, J.M., Krantz, S.B.: Chromosome band analysis in 19 cases of chronic myeloid leukemia: 9 chronic, 10 blastic, two with Ph¹ (22q–) translocation on 17 short arm. Ann. Genet. (Paris) 18, 239–240 (1975*b*).
Engel, E., McGee, B.J., Flexner, J.M., Russell, M.T., Evans, B.J.: Philadelphia chromosome (Ph¹) translocation in an apparently Ph¹ negative, minus G22, case of chronic myeloid leukemia. N. Engl. J. Med. 291, 154 (1974).
Engel, E., McGee, B.J., Hartmann, R.C., Engel-de

Montmollin, M.: Two leukemic peripheral blood stemlines during acute transformation of chronic myelogenous leukemia in a D/D translocation carrier. Cytogenetics 4, 157–170 (1965).

Engel, E., McGee, B.J., Myers, B.J., Flexner, J.M., Krantz, S.B.: Chromosome banding patterns of 49 cases of chronic myelocytic leukemia. N. Engl. J. Med. 296, 1295 (1977).

Engel, E., McGee, B.J., Russell, M.H., Cassidy, P.S., Flexner, J.M., Engel-de Montmollin, M.: Transformation of mouse cells by fusion with chronic granulocytic leukemia cells: Possible role of human chromosome. Ann. Genet. (Paris) 19, 249–252 (1976).

Engel, E., McKee, L.C., Bunting, K.W.: Chromosomes 17–18 in leukaemias. Lancet ii, 42–43 (1967).

Engel, E., McKee, L.C., Engel-de Montmollin, M.: Aberrations chromosomiques dans les maladies malignes du sang. Union Med. Can. 97, 901–906 (1968).

Engel, E., McKee, L.C., Flexner, J.M., McGee, B.J.: 17 long arm isochromosome. A common anomaly in malignant blood disorders. Ann. Genet. (Paris) 18, 56–60 (1975).

Engel, R.R., Hammond, D., Eitzman, D.V., Pearson, H., Krivit, W.: Transient congenital leukemia in 7 infants with mongolism. J. Pediatr. 65, 303–305 (1964).

Engel, W., Merker, H., Schneider, G., Wolf, U.: Clonal occurrence of a chromosome Dq− in myelosclerosis with myeloid metaplasia. Humangenetik 6, 335–337 (1968).

Enterline, H.T., Arvan, D.A.: Chromosome constitution of adenoma and adenocarcinoma of the colon. Cancer 20, 1746–1759 (1967).

Epstein, A.L., Henle, W., Henle, G., Hewetson, J.F., Kaplan, H.S.: Surface marker characteristics and Epstein-Barr virus studies of two established North American Burkitt's lymphoma cell lines. Proc. Natl. Acad. Sci. USA 73, 228–232 (1976).

Epstein, M.A., Barr, Y.M.: Cultivation in vitro of human lymphoblasts from Burkitt's malignant lymphoma. Lancet i, 252–253 (1964).

Epstein, M.A., Achong, B.G., Barr, Y.M.: Virus particles in cultured lymphoblasts from Burkitt's lymphoma. Lancet i, 702–703 (1964).

Epstein, M.A., Barr, Y.M., Achong, B.G.: Preliminary observations on new lymphoblast strains (EB4, EB5) from Burkitt tumours in a British and Ugandan patient. Br. J. Cancer 20, 475 (1966).

Erdogan, G., Aksoy, M.: Cytogenetic studies in thirteen patients with pancytopenia and leukemia associated with long term exposure to benzene. N. Istanbul Contrib. Clin. Sci. 10, 230 (1973).

Erkman, B., Conen, P.E.: Consistent pseudodiploid and near diploid karyotypes in three intracranial tumors. Am. J. Pathol. 44, 18a (abstract) (1964a).

Erkman, B., Conen, P.E.: Chromosome constitution of 14 malignant tumors. Proc. Am. Assoc. Cancer Res. 5, 17 (1964b).

Erkman, B., Crookston, J.H., Conen, P.E.: Ph¹ chromosomes in leukaemia. Lancet i, 368–369 (1966).

Erkman, B., Crookston, J.H., Conen, P.E.: Double Ph¹ chromosomes in chronic granulocytic leukemia. Cancer 20, 1963–1975 (1967).

Erkman, B., Hazlett, B., Crookston, J.H., Conen, P.E.: Hypodiploid chromosome pattern in acute leukemia following polycythemia vera. Cancer 20, 1318–1325 (1967).

Erkman-Balis, B., Conen, P.E.: Consistent chromosome abnormalities in each of three cases of childhood lymphosarcoma. Eur. J. Cancer 8, 683–688 (1972).

Erkman-Balis, B., Rappaport, H.: Cytogenetic studies in mycosis fungoides. Cancer 34, 626–633 (1974).

Ertl, M., Schlegel, D., Wiesner, O.: Zytogenetische Untersuchungen an autochtonen, menschlichen Tumoren der Mundhöhle. Dtsch. Zahnaerztl. Z. 25, 407–409 (1970).

Evans, H.J.: Molecular architecture of human chromosomes. Br. Med. Bull. 29, 196–202 (1973).

Evans, H.J.: Effects of ionizing radiation on mammalian chromosomes. In J. German (ed.): Chromosomes and Cancer, pp. 191–237. New York: Wiley (1974).

Evans, H.J.: Molecular mechanisms in the induction of chromosome aberrations. In D. Scott, B.A. Bridges, F.H. Sobels: Progress in Genetic Toxicology. Elsevier/North-Holland Biomedical Press, New York, pp. 57–74 (1977).

Evans, H.J., O'Riordan, M.L.: Human peripheral blood lymphocytes for the analysis of chromosome aberrations in mutagen tests. Mutat. Res. 31, 135–148 (1975).

Evans, H.J., Adams, A.C., Clarkson, J.M., German, J.: Chromosome aberrations and unscheduled DNA synthesis in X- and UV-irradiated lymphocytes from a boy with Bloom's syndrome and a man with xeroderma pigmentosum. Cytogenet. Cell Genet. 20, 124–140 (1978).

Evans, H.J., Buckland, R.A., Pardue, M.L.: Location of the genes coding for 18S and 28S ribosomal RNA in the human genome. Chromosoma 48, 405–426 (1974).

Evans, H.J., Buckton, K.E., Sumner, A.T.: Cytological mapping of human chromosomes: results obtained with quinacrine fluorescence and the acetic-saline-Giemsa techniques. Chromosoma 35, 310–325 (1971).

Evans, H.J., Gosden, J.R., Mitchell, A.R., Buckland, R.A.: Location of human satellite DNAs on the

Y chromosome. Nature 251, 346–347 (1974).
Evans, H.M., Swezy, O.: The chromosomes in man. Sex and somatic. Mem. Univ. Calif. 9, 1–64 (1929).
Ezdinli, E.Z., Sandberg, A.A., Sokal, J.E.: Definition of chronic myelocytic leukemia (CML) according to the Ph¹ chromosome. Ann. Intern. Med. 70, 1083 (1969).
Ezdinli, E.Z., Sokal, J.E., Aungst, C.W., Kim, U., Sandberg, A.A.: Myeloid leukemia in Hodgkin's disease: chromosomal abnormalities. Ann. Intern. Med. 71, 1097–1104 (1969).
Ezdinli, E.Z., Sokal, J.E., Crosswhite, L., Sandberg, A.A.: Philadelphia-chromosome-positive and -negative chronic myelocytic leukemia. Ann. Intern. Med. 72, 175–182 (1970).

Faed, M.J.W., Mourelatos, D.: Enhancement by caffeine of sister-chromatid exchange frequency in lymphocytes from normal subjects after treatment by mutagens. Mutat. Res. 49, 437–440 (1978).
Falls, H.F., Neel, J.V.: Genetics of retinoblastoma. Arch. Ophthalmol. 46, 367–389 (1951).
Falor, W.H.: Chromosomes in noninvasive papillary carcinoma of the bladder. J.A.M.A. 216, 791–794 (1971a).
Falor, W.H.: Chromosomes in bronchial adenomas and in bronchogenic carcinomas. Am. Rev. Respir. Dis. 104, 198–205 (1971b).
Falor, W.H., Ward, R.M.: DNA banding patterns in carcinoma of the bladder. J.A.M.A. 226, 1322–1327 (1973).
Falor, W.H., Ward, R.M.: Cytogenetic analysis. A potential index for recurrence of early carcinoma of the bladder. J. Urol. 115, 49–52 (1976a).
Falor, W.H., Ward, R.M.: Fifty-three month persistence of ring chromosome in noninvasive bladder carcinoma. Acta Cytol. 20, 272–275 (1976b).
Falor, W.H., Ward, R.M.: Prognosis in well differentiated noninvasive carcinoma of the bladder based on chromosomal analysis. Surg. Gynecol. Obstet. 144, 515–518 (1977).
Falor, W.H., Ward, R.M.: Prognosis in early carcinoma of the bladder based on chromosomal analysis. J. Urol. 119, 44–48 (1978).
Falor, W.H., Gordon, M., Kaczala, O.A.: Chromosomes in bronchoscopic biopsies from patients with bronchial adenoma, bronchogenic carcinoma, and from heavy smokers. Cancer 24, 198–209 (1969).
Farber, R.A., Davidson, R.L.: Differences in the order of termination of DNA replication in human chromosomes in peripheral blood lymphocytes and skin fibroblasts from the same individual. Cytogenet. Cell Genet. 18, 349–363 (1977).

Farmer, J.B., More, J.E.S., Walker, G.E.: On the cytology of malignant growth. Proc. R. Soc. B 77, 336–353 (1906).
Faust, J., Vogel, W.: Are "N bands" selective staining of specific heterochromatin? Nature 249, 352–353 (1974).
Fefer, A., Cheever, M., Thomas, E.D., Boyd, C., Ramberg, R., Glucksberg, H., Buckner, C.D., Sanders, J., Storb, R.: Disappearance of Ph¹-positive cells from marrows of 4 CML patients after chemotherapy, radiation, and marrow transplantation from an identical twin. Blood 50 (Suppl. 1), 232 (abstract) (1977).
Fefer, A., Thomas, E.D., Buckner, C.D., Storb, R., Neiman, P., Glucksberg, H., Clift, R.A., Lerner, K.G.: Marrow transplants in aplastic anemia and leukemia. Semin. Hematol. 11, 353–367 (1974).
Feig, S.A., Falk, P.M., Neerhout, R.C., Sparkes, R., Gale, R.P., Opelz, G., Cline, M.J., Fahey, J., Smith, G., Sarna, G., Territo, M., Young, L., Langdon, E.A., Fawzi, F.: Experience with incompatible maternal donors for bone marrow transplantation. Blut 34, 1–10 (1977).
Feldges, A.J., Aur, R.J.A., Verzosa, M.S., Daniels, S.: Periodic acid-Schiff reaction, a useful index of duration of complete remission in acute childhood lymphocytic leukemia. Acta Haematol. (Basel) 52, 8–13 (1974).
Ferenczy, A., Richart, R.M., Miller, O.J.: Gonadoblastoma occurring in a female with XO/XY fragment gonadal dysgenesis. Am. J. Obstet. Gynecol. 109, 564–569 (1971).
Ferguson, J., Mackay, I.R.: Macroglobulinaemia with chromosomal anomaly. Aust. Ann. Med. 12, 197–201 (1963).
Ferguson-Smith, M.A.: The identification of human chromosomes. Proc. R. Soc. Med. 55, 471–475 (1962).
Ferguson-Smith, M.A.: The sites of nucleolus formation in human pachytene chromosomes. Cytogenetics 3, 124–134 (1964).
Ferguson-Smith, M.A., Handmaker, S.D.: Observations on the satellited human chromosomes. Lancet i, 638–640 (1961a).
Ferguson-Smith, M.A., Handmaker, S.D.: Observations on the satellited human chromosomes. Lancet ii, 1362 (1961b).
Ferguson-Smith, M.A., Handmaker, S.D.: The association of satellited chromosomes with specific chromosomal regions in cultured human somatic cells. Ann. Hum. Genet. 27, 143–156 (1963).
Ferguson-Smith, M.A., Ferguson-Smith, M.E., Ellis, P.M., Dickson, M.: The sites and relative frequencies of secondary constrictions in human somatic chromosomes. Cytogenetics 1, 325–343 (1962).
Ferreyra, M.E., Dondo Lascano, F.: Alteraciones

cromosómicas en algunas hemopatías malignes. Prensa Med. Argent. 56, 148 – 163 (1969).
Ferrier, P.E., Ferrier, S.A., Schärer, K.O., Genton, N., Hedinger, C., Klein, D.J.: Disturbed gonadal differentiation in a child with XO,XY,XYY mosaicism: relationship with gonadoblastoma. Helv. Paediatr. Acta 22, 479 – 490 (1967).
Festa, R.S., Meadows, A.T., Boshes, R.A.: Leukemia in a black child with Bloom's syndrome. Somatic recombination as a possible mechanism for neoplasia. Cancer (submitted).
Fettig, O., Schröter, R., Wolf, U., Schneider, J.: Dysgerminom bei einem Fall von Gonadendysgenesie mit 47 Chromosomen (XY + Fragment). Arch. Gynaekol. 205, 309 – 324 (1968).
Fialkow, P.J.: Autoimmunity: A predisposing factor to chromosomal aberrations. Lancet i, 474 – 475 (1964).
Fialkow, P.J.: Clonal origin of human tumors. Biochim. Biophys. Acta 458, 283 – 321 (1976a).
Fialkow, P.J.: Human tumors studied with genetic markers. Birth Defects 12, 123 – 132 (1976b).
Fialkow, P.J.: Clonal origin and stem cell evolution of human tumors. Prog. Cancer Res. Ther. 3, 439 – 453 (1977).
Fialkow, P.J., Gartler, S.M., Yoshida, A.: Clonal origin of chronic myelocytic leukemia in man. Proc. Natl. Acad. Sci. USA 58, 1468 – 1471 (1967).
Fialkow, P.J., Klein, G., Gartler, S.M., Clifford, P.: Clonal origin for individual Burkitt tumours. Lancet i, 384 – 386 (1970).
Fialkow, P.J., Thomas, E.D., Bryant, J.I., Neiman, P.E.: Leukaemic transformation of engrafted human marrow cells *in vivo*. Lancet I, 251 – 255 (1971).
Field, E.O.: Chromosomal loss and cancer. Lancet ii, 87 – 88 (1972).
Filip, D.A., Gilly, C., Mouriquand, C.: The metaphase chromosome ultrastructure. 1. Acute angle metal deposition technique as an appropriate use of shadow casting in chromosome structure investigation. Exp. Cell Res. 92, 245 – 252 (1975a).
Filip, D.A., Gilly, C., Mouriquand, C.: The metaphase chromosome ultrastructure. II. Helical organization of the basic chromosome fiber as revealed by acute angle metal deposition. Humangenetik 30, 155 – 165 (1975b).
Finan, J.B., Daniele, R.P., Rowlands, S.T., Jr., Nowell, P.C.: Cytogenetics of chronic T cell leukemia; including two patients with a 14q+ translocation. Virchows Arch. B Cell Pathol. 29, 121 – 127 (1978).
Finaz, C., Grouchy, J.: Le caryotype humain après traitment par l'α-chymotrypsine. Ann. Genet. (Paris) 14, 309 – 311 (1971).
Finaz, C., Grouchy, J.: Identification of individual chromosomes in the human karyotype by their banding pattern after proteolytic digestion. Humangenetik 15, 249 – 252 (1972).
Finch, J.T., Lutter, L.C., Rhodes, D., Brown, R.S., Rushton, B., Levitt, M., Klug, A.: Structure of nucleosome core particles of chromatin. Nature 269, 29 – 36 (1977).
Finney, R., McDonald, G.A., Baikie, A.G., Douglas, A.S.: Chronic granulocytic leukemia with Ph[1] negative cells in bone marrow and a ten year remission after busulphan hypoplasia. Br. J. Haematol. 23, 283 – 288 (1972).
Fiocchi, E.: Studies of the chromosome patterns of some types of tumors of the female genital system. Folia. Hered. Pathol. (Milano) 16, 157 – 164 (1967).
First International Workshop on Chromosomes in Leukemia. Helsinki, Finland, Aug. 25 – 28, 1977. Cancer Res. 38, 867 – 868 (1978a).
First International Workshop on Chromosomes in Leukaemia (1977): Chromosomes in Ph[1]-positive chronic granulocytic leukaemia. Br. J. Haematol. 39, 305 – 309 (1978b).
First International Workshop on Chromosomes in Leukaemia (1977): Chromosomes in acute non-lymphocytic leukaemia. Br. J. Haematol. 39, 311 – 316 (1978c).
Fischer, P., Golob, E.: Similar marker chromosomes in testicular tumours. Lancet i, 216 (1967).
Fischer, P., Hebrard, E.: High risk groups, definition and recognition by chromosome aberrations. Cancer Cytol. 14, 16 – 20 (1974).
Fischer, P., Vetterlein, M.: Establishment and cytogenetic analysis of a cell line derived from a human epithelioma of the lung. Oncology 34, 205 – 208 (1977).
Fischer, P., Golob, E., Holzner, J.H.: Chromosomenzahl und DNS-wert bei malignen Tumoren des weiblichen Genitaltraktes. Z. Krebsforsch 68, 200 – 208 (1966a).
Fischer, P., Golob, E., Holzner, J.H.: Die praktische Bedeutung zytogenetischer und zytochemischer Untersuchungen an malignen Tumoren des weiblichen Genitaltraktes. Wien. Klin. Wochenschr. 78, 284 – 286 (1966b).
Fischer, P., Golob, E., Holzner, J.H.: Zytogenetische Untersuchungen am Portioepithel bei positivem und bei zweifelhaftem Abstrichbefund. Krebsarzt 22, 289 – 294 (1967).
Fischer, P., Golob, E., Holzner, H.: XY gonadal dysgenesis and malignancy. Lancet ii, 110 (1969).
Fischer, P., Golob, E., Kunze-Mühl, E., Haim, A.B., Dudley, R.A., Müllmer, T., Parr, R.M., Vetter, H.: Chromosome aberrations in peripheral blood cells in man following chronic irradiation from internal deposits of thorotrast. Radiat. Res. 29, 505 – 515 (1966c).
Fischer, P., Vetterlein, M., Pohl-Rüling, J., Krepler, P.: Cytogenetic effects of chemotherapy and cranial irradiation on the peripheral lympho-

cytes of children with malignant disease. Oncology 34, 224–228 (1977).

Fischer, R., Hennekeuser, H.H., Schaefer, H.E.: Extramedulläre Blutbildung in der Milz, insbesondere bei Knochenmarkmetastasierung. In K. Lennert, D. Harms (eds.): Die Milz, pp. 81–92. Berlin, Heidelberg, New York: Springer (1970).

Fischer, W., Fölsch, E.: Chronisch-myeloische Leukämie und Osteomyelofibrose – zwei verschiedene Erkrankungen? Dtsch. Med. Wochenschr. 100, 1025–1028 (1975).

Fisher, B.L., Lyons, R.M., Sears, D.A.: Development of chronic myelocytic leukemia during the course of acute lymphocytic leukemia in an adult. Am. J. Hematol. 2, 291–297 (1977).

Fisher, E.R., Paulson, J.D.: Karyotypic abnormalities in precursor lesions of human cancer of the breast. Am. J. Clin. Pathol. 69, 284–288 (1978).

Fitzgerald, P.H.: The Ph^1 chromosome in uncultured leukocyte and marrow cells from human chronic granulocytic leukaemia. Exp. Cell Res. 26, 220–222 (1961a).

Fitzgerald, P.H.: Cytogenetic studies in chronic granulocytic leukaemia. Proc. Univ. Otago Med. Sch. 39, 38–40 (1961b).

Fitzgerald, P.H.: Chromosomes of two cases of human chronic myeloid leukaemia. Nature 194, 393 (1962).

Fitzgerald, P.H.: Abnormal length of the small acrocentric chromosomes in chronic lymphocytic leukemia. Cancer Res. 25, 1904–1909 (1965).

Fitzgerald, P.H.: A complex pattern of chromosome abnormalities in the acute phase of chronic granulocytic leukaemia. J. Med. Genet. 3, 258–264 (1966).

Fitzgerald, P.H.: The life-span and role of the small lymphocyte. In H.J. Evans, W.M. Court Brown, A.S. McLean (eds.): Human Radiation Cytogenetics, pp. 94–98. Amsterdam: North-Holland (1967).

Fitzgerald, P.H.: Autoradiography of terminal DNA replication of the G group autosomes of man. Homologue asynchrony of the later-replicating G pair. Am. J. Hum. Genet. 23, 390–402 (1971).

Fitzgerald, P.H.: Evidence that chromosome band 22q12 is concerned with cell proliferation in chronic myeloid leukaemia. Hum. Genet. 33, 269–274 (1976).

Fitzgerald, P.H.: Scientific method and the Philadelphia chromosome. Hum. Genet. 39, 257–259 (1977).

Fitzgerald, P.H., Adams, A.: Chromosome studies in chronic lymphocytic leukemia and lymphosarcoma. J. Natl. Cancer Inst. 34, 827–839 (1965).

Fitzgerald, P.H., Gunz, F.W.: Chromosomal abnormality and chronic lymphocytic leukaemia. Lancet ii, 150 (1964).

Fitzgerald, P.H., Hamer, J.W.: Third case of chronic lymphocytic leukaemia in a carrier of the inherited Ch^1 chromosome. Br. Med. J. iii, 752–754 (1969).

Fitzgerald, P.H., Hamer, J.W.: Primary acquired red cell hypoplasia associated with a clonal chromosomal abnormality and disturbed erythroid proliferation. Blood 38, 325–335 (1971).

Fitzgerald, P.H., Hamer, J.W.: Karyotype and survival in human acute leukemia. J. Natl. Cancer Inst. 56, 459–462 (1976).

Fitzgerald, P.H., Adams, A., Gunz, F.W.: Chronic granulocytic leukemia and the Philadelphia chromosome. Blood 21, 183–196 (1963).

Fitzgerald, P.H., Adams, A., Gunz, F.W.: Chromosome studies in adult acute leukemia. J. Natl. Cancer Inst. 32, 395–417 (1964).

Fitzgerald, P.H., Crossen, P.E., Adams, A., Sharman, C.V., Gunz, F.W.: Chromosome studies in familial leukaemia. J. Med. Genet. 3, 96–100 (1966).

Fitzgerald, P.H., Crossen, P.E., Hamer, J.W.: Abnormal karyotypic clones in human acute leukemia. Their nature and clinical significance. Cancer 31, 1069–1077 (1973).

Fitzgerald, P.H., Pickering, A.F., Eiby, J.R.: Clonal origin of the Philadelphia chromosome and chronic myeloid leukaemia: Evidence from a sex chromosome mosaic. Br. J. Haematol. 21, 473–480 (1971).

Fitzgerald, P.H., Pickering, A.F., Mercer, J.M., Miethke, P.M.: Premature centromere division. A mechanism of non-disjunction causing X chromosome aneuploidy in somatic cells of man. Ann. Hum. Genet. 38, 417 (1975).

Fitzgerald, P.H., Rastrick, J.M., Hamer, J.W.: Acute plasma cell leukaemia following chronic lymphatic leukaemia. Transformation of two separate diseases? Br. J. Haematol. 25, 171–177 (1973).

Flannery, E.P., Corder, M.P.: The Philadelphia or short Y chromosome. J.A.M.A. 228, 286 (1974).

Flannery, E.P., Dillon, D.E., Freeman, M.V.R., Levy, J.D., D'Ambrosio, U., Bedynek, J.L.: Eosinophilic leukemia with fibrosing endocarditis and short Y chromosome. Ann. Intern. Med. 77, 223–228 (1972).

Fleig, I., Thiess, A.M.: Mutagenicity of vinyl chloride. External chromosome studies on persons with and without VC illness, and on VC exposed animals. J. Occup. Med. 20, 557–561 (1978)

Fleischman, E.V., Prigogina, E.L., Platonova, G.M., Chudina, A.P., Kruglova, G.V., Kaverzneva,

M.M.: Comparative chromosome characteristics of reticulosarcomas and Hodgkin's disease. Neoplasma 21, 51 – 61 (1974).
Fleischman, E.W., Prigogina, E.L.: G banding in cytogenetic study of hemoblastoses. Humangenetik 26, 335 – 342 (1975).
Fleischman, E.W., Prigogina, E.L.: Karyotype peculiarities of malignant lymphomas. Hum. Genet. 35, 269 – 279 (1977).
Fleischman, E.W., Prigogina, E.L., Volkova, M.A., Petkovitch, I.: Unusual translocation (10;22) in chronic myelogenous leukemia. Hum. Genet. 39, 127 – 129 (1977).
Fleischmann, T., Bodor, F.: Aneuploidy in paroxysmal nocturnal haemoglobinuria. Acta Haematol. (Basel) 44, 251 – 253 (1970).
Fleischmann, T., Bodor, F.: Chromosome banding anomalies in acute myeloid leukaemia. Int. J. Cancer 15, 980 – 984 (1975).
Fleischmann, T., Krizsa, F.: Marker chromosome in myeloproliferative syndrome. Acta Haematol. (Basel) 54, 59 – 63 (1975).
Fleischmann, T., Krizsa, F.: Chromosomes in malignant lymphomas. Orv. Hetil. 117, 1263 – 1266 (1976).
Fleischmann, T., Gustafsson, T., Håkansson, C.H.: Computer-display of the chromosomal fluorescence pattern. Hereditas 68, 325 – 328 (1971).
Fleischmann, T., Gustafsson, T., Håkansson, C.H., Levan, A.: The fluorescent pattern of normal chromosomes in biopsies of malignant lymphomas, and its computer display. Hereditas 70, 75 – 88 (1972).
Fleischmann, T., Håkansson, C.H., Levan, A.: Fluorescent marker chromosomes in malignant lymphomas. Hereditas 69, 311 – 314 (1971).
Fleischmann, T., Håkansson, C.H., Levan, A.: Chromosomes of malignant lymphomas. Studies in short-term cultures from lymph nodes of twenty cases. Hereditas 83, 47 – 56 (1976).
Fleischmann, T., Håkansson, C.H., Levan, A., Möller, T.: Multiple chromosome aberrations in a lymphosarcomatous tumor. Hereditas 70, 243 – 258 (1972).
Fleishman, E.V., Prigozhina, E.L., Kruglova, G.V., Volkova, M.A.: Features specific to karyotype of the lymphatic system neoplasm. Probl. Gematol. Pereliv. Krovi 22, 16 – 22 (1977).
Fleishman, E.W., Wolkowa, M.A.: Cytogenetic studies in the dynamics of chronic myeloid leukaemia. In K. Stampfli (ed.): Proceedings of the 12th Congress, International Society of Blood Transfusion, Moscow, 1969, pp. 536 – 540. Basel: Karger (Bibliotheque Haematologica No. 31, Pt. 1) (1971).
Fleishman, Y.V., Volkova, M.A.: Aneuploid lines of cells in chronic myeloleukemia not during blastic crisis. Probl. Gematol. Pereliv. Krovi 16, 10 – 16 (1971).
Flemming, W.: Beitrage zur Kenntniss der Zelle und ihre Lebenserscheinungen. Arch. Mikr. Anat. 20, 1 – 86 (1882).
Flemming, W.: Über die Chromosomenzahl beim Menschen. Anat. Anz. 14, 171 – 174 (1889).
Fliedner, T.M., Thomas, E.D., Meyer, L.M., Cronkite, E.P.: The fate of transfused H^3 thymidine-labeled bone marrow cells in irradiated recipients. Ann. N.Y. Acad. Sci. 114, 510 – 527 (1964).
Foadi, M.D.: Tetraploid cell line in a girl with acute leukaemia. Acta Haematol. (Basel) 57, 55 – 64 (1977).
Foadi, M.D., Slater, A.M., Pegrum, G.D.: Kinetic features of a lymphoid population in the blastic crisis of chronic granulocytic leukaemia. Europ. J. Cancer 14, 271 – 277 (1977).
Focan, Ch., Brictieux, N., Lamaire, M., Hughes, J.: Neoplasies secondaires compliquant une maladie de Hodgkin. Nouv. Presse Med. 25, 1385 (1974).
Foerster, W., Medau, H.J., Löffler, H.: Chronische myeloische Leukamie mit Philadelphia-Chromosom und Tandem-Translokation am 2. Chromosom Nr. 22; 46,xx,tan (22q+;22q−). Klin. Wochenschr. 52, 123 – 126 (1974).
Ford, C.E.: The chromosomes of normal human somatic and leukaemic cells. Proc. R. Soc. Med. 53, 491 – 493 (1960).
Ford, C.E.: Chromosomes et leucémies. Nouv. Rev. Fr. Hematol. 1, 165 – 171 (1961).
Ford, C.E., Clarke, C.M.: Cytogenetic evidence of clonal proliferation in primary reticular neoplasms. Can. Cancer Conf. 5, 129 – 146 (1963).
Ford, C.E., Clegg, H.M.: Reciprocal translocations. Br. Med. Bull. 25, 110 – 114 (1969).
Ford, C.E., Hamerton, J.L.: The chromosomes of man. Nature 178, 1020 – 1023 (1956).
Ford, C.E., Mole, R.H.: Chromosome studies in human leukaemia. Lancet ii, 732 (1959).
Ford, C.E., Jacobs, P.A., Lajtha, L.G.: Human somatic chromosomes. Nature 181, 1565 – 1568 (1958).
Ford, C.E., Jones, K.W., Polani, P.E., De Almeida, J.C., Briggs, J.H.: A sex-chromosome anomaly in a case of gonadal dysgenesis (Turner's syndrome). Lancet i, 711 – 713 (1959).
Ford, C.E.: Chromosomal abnormalities in leukaemic cells. Proceedings, Eighth Congress of the European Society of Haematology, Wien, 1961. Abstract #104 (1962).
Ford, E.H.R.: Human Chromosomes London-New York: Academic Press (1973).
Ford, E.H.R., Woollam, D.H.M.: Significance of variation in satellite incidence in normal human mitotic chromosomes. Lancet ii, 26 – 27 (1967).
Ford, J.H., Pittman, S.M.: Duplication of 21 or 8/21

translocation in acute leukaemia. Lancet ii, 1458 (1974).

Ford, J.H., Pittman, S.M., Gunz, F.W.: Consistent chromosome abnormalities in acute leukaemia. Br. Med. J. iv, 227–228 (1974).

Ford, J.H., Pittman, S.M., Singh, S., Wass, E.J., Vincent, P.C., Gunz, F.W.: Cytogenetic basis of acute myeloid leukemia. J. Natl. Cancer Inst. 55, 761–765 (1975).

Ford Bainton, D., Farquhar, M.G.: Differences in enzyme content of azurophil and specific granules of polymorphonuclear leukocytes. II. Cytochemistry and electron microscopy of bone marrow cells. J. Cell Biol. 39, 299–317 (1968).

Forman, E.N., Padre-Mendoza, T., Smith, P.S., Barker, B.E., Farnes, P.: Ph[1]-positive childhood leukemias: Spectrum of lymphoid-myeloid expressions. Blood 49, 548–558 (1977).

Fornace, A.J., Jr., Kohn, K.W., Kann, H.E., Jr.: DNA single-strand breaks during repair of UV damage in human fibroblasts and abnormalities of repair in xeroderma pigmentosum. Proc. Natl. Acad. Sci. USA 73, 39–43 (1976).

Forni, A.: Chromosome changes due to chronic exposure to benzene. *Proceedings of the 15th International Congress on Occupational Health* (Vienna), Vol. 2, pp. 437–439 (1966).

Forni, A., Moreo, L.: Cytogenetic studies in a case of benzene leukaemia. Eur. J. Cancer 3, 251–255 (1967).

Forni, A., Moreo, L.: Chromosome studies in a case of benzene-induced erythroleukemia. Eur. J. Cancer 5, 459–463 (1969).

Forni, A., Baroni, M., Pacifico, E.: Chromosomal findings in a case of leukemic reticuloendotheliosis. Acta Cytol. 15, 173–178 (1971).

Forni, A., Cappellini, A., Pacifico, E., Vigliani, E.C.: Chromosome changes and their evolution in subjects with past exposure to benzene. Arch. Environ. Health 23, 385–391 (1971).

Forni, A., Pacifico, E., Limonta, A.: Chromosome studies in workers exposed to benzene or toluene or both. Arch. Environ. Health 22, 373–378 (1971).

Forrester, R.H., Louro, J.M.: Philadelphia chromosome abnormality in agnogenic myeloid metaplasia. Ann. Intern. Med. 64, 622–627 (1966).

Forteza-Bover, G., Báguena-Candela, R.: Anomalias cromosomicas en la leucemia mieloside cronica. Rev. Inf. Med. Terap. 7, 426 (1962).

Forteza-Bover, G., Báguena-Candela, R.: Analisis cytogenetica de un caso de anemia perniciosa autes Y· despues del tratamiento. Rev. Clin. Esp. 88, 251–254 (1963).

Forteza-Bover, G., Báguena-Candela, R.: Anàlisis cromosómico de las celulás metastáticas de un melanoblastoma maligno obtenidas mediante punción ganglionar. Sangre (Barc.) 11, 161 (1968).

Forteza-Bover, G., Báguena-Candela, R., Bordon, A.J.: Un caso de leucosis aguda de paramieloblastos promielocitoides con células sanguineas trisomicas D, haplosomicas C y haplosomicas G. Med. Esp. 50, 357–361 (1963).

Forteza-Bover, G., Báguena-Candela, R., Tortajada Martínez, M.: Citogenética de les ascitis carcinomatosas en ginecologia. Rev. Esp. Obstet. Ginecol. 23, 301–316 (1964a).

Forteza-Bover, G., Báguena-Candela, R., Tortajada Martínez, M. Oncocitogenetica. *In: Symposium Sobre Mitoses e Inhibidores en la Quimioterapia Oncologica,* p. 3. Barcelona: Sandoz S.A.E. (1964b).

Fortune, D.W., Lewis, F.J.W., Poulding, R.H.: Chromosome pattern· in myeloid leukaemia in a child. Lancet i, 537 (1962).

Fosså, S.D.: Feulgen-DNA-values in transitional cell carcinoma of the human urinary bladder. Beitr. Pathol. 155, 44 (1975).

Fosså, S.D.: DNA-variations in neighbouring epithelium in patients with bladder carcinoma. Acta Pathol. Microbiol. Scand. [A] 85, 603–610 (1977).

Fosså, S.D., Kaalhus, O.: "Non-condensed" and "condensed" chromatin in transitional cell carcinoma of the human urinary bladder. Beitr. Pathol. 158, 241–254 (1976).

Fosså, S.D., Kaalhus, O., Scott-Knudsen, O.: The clinical and histopathological significance of Feulgen DNA-values in transitional cell carcinoma of the human urinary bladder. Eur. J. Cancer 13, 1155–1162 (1977).

Fournier, J.L., Saint-Aubert, P., Ponte, C., Gaudier, B., Walbaum, R., Farriaux, J.P., Fontaine, G.: Nosological study of true hermaphroditism and mixed asymmetrical gonadal dysgenesis. Ann. Pediatr. (Paris) 23, 763–775 (1976).

Fraccaro, M., Gerli, M., Tiepolo, L., Zara, C.: Analisi della variabilita cariotipica in un caso de·neoplasia ovarica. Minerva Ginecol. 17, 485–492 (1965).

Fraccaro, M., Mannini, A., Tiepolo, L., Gerli, M., Zara, C.: Karyotypic clonal evolution in a cystic adenoma of the ovary. Lancet i, 613–614 (1968).

Fraccaro, M., Mannini, A., Tiepolo, L., Zara, C.: High frequency of spontaneous recurrent chromosome breakage in an untreated human tumour. Mutat. Res. 2, 559–561 (1965).

Fraccaro, M., Tiepolo, L., Gerli, M., Zara, C.: Analysis of karyotype changes in ovarian malignancies. Panminerva Med. 8, 1–19 (1966).

France, H.F. de, Bijlsma, J.B., Bond, C.P.: Direct Giemsa-banding pattern analysis of human chromosomes by means of a television microdensitometer. The Quantimet 720D. Humangenetik 22, 167–170 (1974).

Francesconi, D., Pasquali, F.: Three chromosomes'

(7;9;22) rearrangement and the origin of the Philadelphia chromosome. Hum. Genet. 43, 133–137 (1978a).
Francesconi, D., Pasquali, F.: 8/21 translocation, loss of the Y chromosome and Philadelphia chromosome. Br. J. Haematol. 38, 149–150 (1978b).
Francke, C., Robert, W.N.: Chromosomal aberrations in monoclonal gammapathies without clinical features of plasma cell myeloma or macroglobulinaemia. Folia Med. Neerl. 13, 2–9 (1970).
Francke, U.: Retinoblastoma and chromosome 13. In Baltimore Conference (1975): Third International Workshop on Human Gene Mapping, Birth Defects, Orig. Art. Ser. 12 (7), 131–137 (1976).
Francke, U., Kung, F.: Sporadic bilateral retinoblastoma and 13q− chromosomal deletion. Med. Pediatr. Oncol. 2, 379–385 (1976).
Francois, J., Matton, M.T., de Bie, S., Tanaka, Y., Vandenbulcke, D.: Genesis and genetics of retinoblastoma. Ophthalmologica 170, 405–425 (1975).
Frank, D.W., Trzos, R.J., Good, P.I.: A comparison of two methods for evaluating drug-induced chromosome alterations. Mutat. Res. 56, 311–317 (1978).
Frasier, S.D., Bashore, R., Mosier, D.: Chromatin-negative twins with female phenotype, gonadal dysgenesis, gonadoblastoma: chromosome evaluation. Am. J. Dis. Child. 102, 582 (1961).
Frasier, S.D., Bashore, R., Mosier, H.D.: Gonadoblastoma associated with pure gonadal dysgenesis in monozygous twins. J. Pediatr. 64, 740–745 (1964).
Fraumeni, J.F., Jr.: Constitutional disorders of man predisposing to leukemia and lymphoma. Natl. Cancer Inst. Monogr. 32, 221–232 (1969).
Fraumeni, J.F., Jr., Vogel, C.L., DeVita, V.T.: Familial chronic lymphocytic leukemia. Ann. Intern. Med. 71, 279–284 (1969).
Freeman, A.I., Edwards, J.A., Cohen, M., Sinks, L.F.: Morphologic studies in a recently described dyserythropoietic state. Blood 40, 473–486 (1972).
Freeman, A.I., Sinks, L.F., Cohen, M.M.: Lymphosarcoma in siblings, associated with cytogenetic abnormalities, immune deficiency, and abnormal erythropoiesis. J. Pediatr. 77, 996–1003 (1970).
Freeman, M.V.R., Miller, O.J.: XY gonadal dysgenesis and gonadoblastoma. Obstet. Gynecol. 34, 478–483 (1969).
Freese, E.: The arrangement of DNA in the chromosomes. Cold Spring Harbor Symp. Quant. Biol. 23, 13–18 (1958).
Frei, E., III, Tjio, J.H., Whang, J., Carbone, P.P.: Studies of the Philadelphia chromosome in patients with chronic myelogenous leukemia. Ann. N.Y. Acad. Sci. 113, 1073–1080 (1964).
Freireich, E.J., Whang, J., Tjio, J.H., Levin, R.H., Brittin, G.M., Frei, E., III: Refractory anemia, granulocytic hyperplasia of bone marrow, and a missing chromosome in marrow cells. A new clinical syndrome? Clin. Res. 12, 284 (1964).
Frey, I., Siebner, H.: Osteomyelofibrose mit Philadelphia- (Ph[1])-Chromosom. Med. Welt 19, 2274–2279 (1968).
Frick, P.G.: Primary thrombocythemia. Clinical, haematological and chromosomal studies of 13 patients. Helv. Med. Acta 35, 20–29 (1969).
Friedman, B.I., Saenger, E.L., Kreindler, M.S.: Endoreduplication in leucocyte chromosomes. Lancet ii, 494–495 (1964).
Fritz-Niggli, H.: Analyse menschlicher Chromosomen. I. Karyotyp eines Mammakarzinoms. Pathol. Microbiol. (Basel) 17, 340–351 (1954).
Fritz-Niggli, H.: Chromosomenanalysen bei Karzinomen des Menschen. Oncologia 8, 121–135 (1955).
Fritz-Niggli, H.: Die Chromosomen in menschlichen Mammakarzinom. Acta Int. Cancer 12, 623–636 (1956).
Frizzera, G., Moran, E.M., Rappaport, H.: Angioimmunoblastic lymphadenopathy. Am. J. Med. 59, 803–818 (1975).
Frolov, A.K., Slysuarev, A.A., Dement'ev, I.V., Sokhnin, A.A., Frolov, V.K., Lebedinsky, A.P., Lysakova, V.I., Bol'shinskaya, Zh. I.: Cytogenetic investigation of peripheral blood lymphocytes in children re-immunized with the smallpox vaccine. Genetika (Moskva) 11, 142–146 (1975)
Fujita, K., Fujita, H.M.: Klinefelter's syndrome and bladder cancer. J. Urol. 116, 836–837 (1976).
Fukuhara, S.: Significance of 14q translocations in non-Hodgkin lymphomas. Virchows Arch. B Cell Pathol. 29, 99–106 (1978).
Fukuhara, S., Rowley, J.D.: Origin of donor chromosome in the 14q+ marker chromosome in histiocytic lymphoma. Blood 50 (Suppl. 1), 218 (abstract) (1977).
Fukuhara, S., Rowley, J.D.: Chromosome 14 translocations in non-Burkitt lymphomas. Int. J. Cancer 22, 14–21 (1978).
Fukuhara, S., Rowley, J.D., Variakojis, D.: Banding studies of chromosomes in a patient with mycosis fungoides. Cancer 42, 2262–2268 (1978a).
Fukuhara, S., Shirakawa, S., Uchino, H.: Specific marker chromosomes 14 in malignant lymphomas. Nature 259, 210–211 (1976).
Fukuhara, S., Rowley, J.D., Variakojis, D., Golomb, H.M.: Correlation of 14q+ marker chromosome with histopathology in malignant lymphoma, poorly differentiated lymphocytic type. Cancer Res. (submitted).

Fukuhara, S., Rowley, J.D., Variakojis, D., Sweet, D.L., Jr.: Banding studies in diffuse "histiocytic" lymphomas: Correlation of 14q+ marker chromosome with cytology. Blood 52, 989–1001 (1978b).

Funaki, K., Matsui, S.-I., Sasaki, M.: Location of nucleolar organizers in animal and plant chromosomes by means of an improved N-banding technique. Chromosoma 49, 357–370 (1975).

Funes-Cravioto, F., Kolmondin-Hedman, B., Lindsten, J., Nordenskjöld, M., Zapata-Gayon, C., Lambert, B., Norberg, E., Olin, R., Swensson, Å.: Chromosome aberrations and sister chromatid exchange in workers in chemical laboratories and a rotoprinting factory and in children of women laboratory workers. Lancet ii, 322–325 (1977).

Funes-Cravioto, F., Lambert, B., Lindsten, J., Ehrenberg, L., Natarajan, A.T., Osterman-Golkar, S.: Chromosome aberrations in workers exposed to vinyl chloride. Lancet i, 459 (1975).

Furukawa, M., Huang, C.C.: Enhanced effect of cyclophosphamide on Burkitt lymphoma cell lines in vivo. Proc. Soc. Exp. Biol. Med. 153, 536–538 (1976).

Furusawa, S., Adachi, Y., Komatsu, F.: Loss of the Y chromosome in various hematologic disorders and its significance. Acta Haematol. Jap. 36, 120–128 (1973).

Furusawa, S., Kawada, K., Adachi, Y., Komiya, M., Yamada, K.: Cytological and cytogenetical studies in preleukemia. Rinsho Ketsueki (Jap. J. Clin. Hematol.) 12, 559–566 (1971).

Fuscaldo, K.E., Brodsky, I., Conroy, J.F.: CGL: Effect of sequential chromosomal analysis, splenectomy and intensive chemotherapy on survival. Proc. Am. Soc. Clin. Oncol. 18, 294 (1977).

Fuscaldo, K.E., Erlick, B.J., Fuscaldo, A.A., Brodsky, I.: Chromosomal markers, retroviral indicators and thrombocythemia. Blood 52 (suppl. 1), 250 (abstr. #527) (1978).

Fuscaldo, K.E., Erlick, B.J., Fuscaldo, A.A., Brodsky, I.: Correlation of a specific chromosomal marker, 21q−, and retroviral indicators in patients with thrombocythemia. Cancer Letters 6, 51–56 (1979).

Gaffuri, S., Bertoli, S.: Analisi cromosomica in alcuni tumori maligni dell'apparato genitale feminile. Minerva Ginecol. 16, 607–612 (1964).

Gafter, U., Shabtai, F., Kahn, Y., Halbrecht, I., Djaldetti, M.: Aplastic anemia followed by leukemia in congenital trisomy 8 mosaicism: Ultrastructural studies of polymorphonuclear cells in peripheral blood. Clin. Genet. 9, 134–142 (1976).

Gagnon, J., Dupal, M.-F., Katyk-Longtin, N.: Anomalies chromosomiques dans une observation de sympathome congenital. Rev. Can. Biol. 21, 145–155 (1962).

Gahrton, G., Brandt, L., Franzén, S., Nordén, A.: Cytochemical variants of neutrophil leukocyte population in chronic myelocytic leukaemia. A microspectrophotometric study of the change in the periodic acid-Schiff reaction in blood and bone marrow neutrophils during busulfan treatment. Scand. J. Haematol. 6, 365–372 (1969).

Gahrton, G., Friberg, K., Lindsten, J., Zech, L.: Duplication of part of the long arm of chromosome 1 in myelofibrosis terminating in acute myeloblastic leukemia. Hereditas 88, 1–5 (1978b).

Gahrton, G., Friberg, K., Zech, L.: A new translocation involving three chromosomes in chronic myelocytic leukemia, 46,XY,t(9;11;22). Cytogenet. Cell Genet. 18, 75–81 (1977).

Gahrton, G., Friberg, K., Zech, L., Lindsten, J.: Duplication of part of chromosome No. 1 in myeloproliferative diseases. Lancet i, 96–97 (1978a).

Gahrton, G., Lindsten, J., Zech, L.: Origin of the Philadelphia chromosome. Exp. Cell Res. 79, 246–247 (1973).

Gahrton, G., Lindsten, J., Zech, L.: The Philadelphia chromosome and chronic myelocytic leukemia (CML) — Still a complex relationship? Acta Med. Scand. 196, 353–354 (1974a).

Gahrton, G., Lindsten, J., Zech, L.: Clonal origin of the Philadelphia chromosome from either the paternal or the maternal chromosome number 22. Blood 43, 837–840 (1974b).

Gahrton, G., Lindsten, J., Zech, L.: Involvement of chromosomes 8, 9, 19 and 22 in Ph¹ positive and Ph¹ negative chronic myelocytic leukemia in the chronic or blastic stage. Acta Med. Scand. 196, 355–360 (1974c).

Gahrton, G., Zech, L., Friberg, K., Lundgren, G., Möller, E., Groth, C.-G.: Chromosomal satellites as markers in human bone-marrow transplantation. Hereditas 84, 15–18 (1976).

Gahrton, G., Zech, L., Lindsten, J.: A new variant translocation (19q+,22q−) in chronic myelocytic leukemia. Exp. Cell Res. 86, 214–216 (1974).

Gahrton, G., Zech, L., Lindsten, J.: Significance of chromosomes 7, 8 and 9 abnormalities in myeloproliferative disease. *International Society of Haematology*, Third Meeting, London, p. 5:03 (abstract) (1975).

Galan, H.M., Lida, E.J., Kleisner, E.H.: Chromosomes of Sternberg–Reed cells. Lancet i, 335 (1963).

Galbraith, P.R.: Granulocyte kinetic studies in chronic myelogenous leukemia. Natl. Cancer Inst. Monogr. 30, 121–134 (1969).

Galeotti, G.: Beitrag zum Studium der Chromatins

in der Epithelzellen der Carcinome. Beitr. Pathol. Anat. 14, 249–271 (1893).
Gall, J.: Chromosome fibers from an interphase nucleus. Science 139, 120–121 (1963).
Gall, J.A., Boggs, D.R., Chervenick, P.A., Pan, S., Fleming, R.B.: Discordant patterns of chromosome changes and myeloblast proliferation during the terminal phase of chronic myeloid leukemia. Blood 47, 347–353 (1976).
Gallo, R.C.: Viruses and the pathogenesis of human leukemia. Schweiz. Med. Wochenschr. 107, 1436–1440 (1977).
Gallo, R., Bhattacharyya, J., Anderson, P.: Specific markers of chronic myelogenous leukemia (CML) cells (Ph¹ chromosome), thymic-derived cells (terminal transferase), and type-C RNA tumor virus (reverse transcriptase) in blastic leukemia. J. Clin. Invest. 53, 26a (1974).
Galloway, S.M.: Ataxia telangiectasia: The effects of chemical mutagens and X-rays on sister chromatid exchanges in blood lymphocytes. Mutat. Res. 45, 343–349 (1977).
Galloway, S.M., Buckton, K.E.: Aneuploidy and ageing: Chromosome studies on a random sample of the population using G-banding. Cytogenet. Cell Genet. 20, 78–95 (1978).
Galloway, S.M., Evans, H.J.: Asymmetrical C-bands and satellite DNA in man. Exp. Cell Res. 94, 454–459 (1975).
Galton, D.A.G.: Problems in the management of the myeloproliferative states. Scand. J. Haematol. 1, 37–46 (1965).
Galton, M., Benirschke, K.: Forty-six chromosomes in an ovarian teratoma. Lancet ii, 761–762 (1959).
Galton, M., Benirschke, K., Baker, M.C., Atkin, N.B.: Chromosomes of testicular teratomas. Cytogenetics 5, 261–275 (1966).
Gandini, E., Dallapiccola, B., Laurent, C., Suerinck, E.F., Forabosco, A., Conconi, F., Del Senno, L.: Evidence for localisation of genes for human α-globin on the long arm of chromosome 4. Nature 265, 65–66 (1977).
Gange, R., Tanguay, R., Laberge, C.: Differential staining patterns of heterochromatin in man. Nature [New Biol.] 232, 29–30 (1971).
Ganina, K.P.: The diagnostic significance of sex chromatin in hormone-dependent preneoplastic and neoplastic processes. Vopr. Onkol. 18, 46–50 (1972).
Ganina, K.P., Polishchuk, L.Z., Kireeva, S.S., Gritsenko, A.P.: The state of karyotype of tumor cells in patients with stomach and uterus cancer and malignant melanoma. *In* S. Armendares, R. Lisker (eds.): *V International Congress of Human Genetics*, pp. 125–126, Abstract 324, Oct. 10–15, Mexico (1976).
Ganner, E.: Chromosomenuntersuchungen bei Morbus Waldenström. Wien. Klin. Wochenschr. 79, 20–21 (1967).
Ganner, E.: Eine Patientin mit Translokation 46, XX, t(Dg–;Bq+) und Neurofibromatose, Schwachsinn sowie Aortenisthmusstenose. Schweiz. Med. Wochenschr. 99, 182–186 (1969a).
Ganner, E.: Monoklonales abnormes Karyogramm: 45,XX,2D–, (17–18)–, 2C+, Gp+ in den Tumorzellen eines Rethothelsarkoms. Blut 19, 416–419 (1969b).
Ganner-Millonig, E.: Ungewöhnlicher Chromosomensatz (46,XX,Dq–) bei Osteomyelofibrose. Blut 28, 411–414 (1974).
Garfinkel, L.S., Bennett, D.E.: Extramedullary myeloblastic transformation in chronic myelocytic leukemia simulating a coexistent malignant lymphoma. Am. J. Clin. Pathol. 51, 638–645 (1969).
Gariepy, G., Cadotte, M.: Lymphome malin caractérisé par une trisomie C. Ann. Genet. (Paris) 13, 112–114 (1970).
Garrison, F.H.: *An Introduction to the History of Medicine*, 4th edition, p. 519. Philadelphia, London: W.B. Saunders (1929).
Garson, O.M., Burgess, M.A., Stanley, L.G.: Cytogenetic remission in acute transformation of chronic granulocytic leukemia. Br. Med. J. ii, 556 (1969).
Garson, O.M., Gruchy G.C. de: Chromosome changes in the blastic transformation stage of chronic granulocytic leukaemia. Haematologia (Budap.) 8, 21–27 (1975).
Garson, O.M., Milligan, W.J.: The 45,XO,Ph¹ subgroup of chronic myelocytic leukemia. Scand. J. Haematol. 9, 186–192 (1972).
Garson, O.M., Milligan, W.J.: Acute leukaemia associated with an abnormal genotype. Scand. J. Haematol. 12, 256–262 (1974).
Garson, O.M., Milligan, W.J., Hurdle, A.D.F.: Chromosome abnormalities in chronic myelomonocytic leukaemia. *4th International Congress on Human Genetics*, Paris, 1971. Abstracts of papers presented, p. 71. Amsterdam: Excerpta Medica (1971) (Excerpta Med. Int. Congr. Ser. No. 233).
Gartler, S.M.: Patterns of cellular proliferation in normal and tumor cell populations. Am. J. Pathol. 86, 685–691 (1977).
Garvin, A.J., Pratt-Thomas, H.R., Spector, M., Spicer, S.S., Williamson, H.: Gonadoblastoma: Histologic, ultrastructural, and histochemical observations in five cases. Am. J. Obstet. Gynecol. 125, 459–471 (1976).
Gasparini, C., Bersi, M., Cardini, G.: Unusual chromosome changes in a case of chronic myeloid leukemia in the blastic phase (in Italian). Arch. Sci. Med. (Torino) 129, 56–63 (1972).
Gastearena Erice, J., Lasa Doria, E., Martinez-Peñue-

la Garcia, J.M.: Estudio cromosomico de un caso de anemia de Fanconi. Rev. Clin. Esp. 125, 169–172 (1972).

Gavosto, F.: Citogenetica e problemi attuali delle leucemie umane. Rass. Med. Sarda 68, 373–385 (1965a).

Gavosto, F.: Recenti progressi nell 'impiego dei radioisotopi per lo studio delle cellule leucemiche. Minerva Nucl. 9, 227–231 (1965b).

Gavosto, F., Ghemi, F., Pegoraro, L., Pileri, A.: Sintesi dell'ADN in cromosomi di leucemia acuta umana. Atti Assoc. Genet. It. 9, 280–289 (1964).

Gavosto, F., Pegoraro, L., Pileri, A.: Thymidine incorporation in the chromosomes of human acute leukaemia. In F.G.J. Hayhoe (ed.): Current Research in Leukaemia, pp. 177–191. London: Cambridge University Press (1964).

Gavosto, F., Pegoraro, L., Pileri, A.: Possibilité de marquer, a l'aide de précurseurs tritiés, les chromosomes de cellules leucémiques chez l'homme. Rev. Fr. Clin. Biol. 8, 920–922 (1963a).

Gavosto, F., Pileri, A., Pegoraro, L.: X-rays and Philadelphia chromosome. Lancet i, 1336–1337 (1965).

Gavosto, F., Pileri, A., Pegoraro, L., Momigliano, A.: In vivo incorporation of tritiated thymidine in acute leukaemia chromosomes. Nature 200, 807–809 (1963b).

Gebhart, E.: Comparative studies on the distribution of aberrations of human chromosomes treated with busulfan in vivo and in vitro. Humangenetik 21, 263–272 (1974).

Gebhart, E., Schwanitz, G., Hartwich, G.: Zytogenetische Wirkung von Vincristin auf menschliche Leukozyten in vivo und in vitro. Med. Klin. 64, 2366–2371 (1969).

Gebhart, E., Schwanitz, G., Hartwich, G.: Chromosomenveränderungen bei Myelerantherapie. Verh. Dtsch. Ges. Inn. Med. 79, 1389–1391 (1973).

Gebhart, E., Schwanitz, G., Hartwich, G.: Chromosomenaberrationen bei Busulfan-Behandlung. Dtsch. Med. Wochenschr. 99, 52–56 (1974).

Geiser, C.F., Schindler, A.M.: Long survival in a male with 18-trisomy syndrome and Wilms' tumor. Pediatrics 44, 111–116 (1969).

Geneix, A., Jaffray, J.-Y., Malet, P., Turchini, J.-P.: Ultrastructure du chromosome humain: fibre chromosomique et chromomère. C.R. Acad. Sci. [D] (Paris) 279, 327–329 (1974).

Genes, I.S.: Chromosome constitution of cells in ascitic and pleural fluids in cancer. Vopr. Onkol. 16, 39–45 (1970).

Geraedts, J.P.M., Haak, H.L.: Trisomy 6 associated with aplastic anemia. Hum. Genet. 35, 113–115 (1976).

Geraedts, J.P.M., Mol, A., Den Ottolander, G.J., van der Ploeg, M., Pearson, P.L.: Variation in the chromosomes of CML patients. Helsinki Chromosome Conference, Aug. 29-31, p. 194 (abstract) (1977).

Geraedts, J.P.M., Pearson, P.L., van der Ploeg, M., Vossepoel, A.M.: Polymorphisms for human chromosomes 1 and Y. Exp. Cell Res. 95, 9–14 (1975).

Gerber, P., Whang-Peng, J., Monroe, J.H.: Transformation and chromosome changes induced by Epstein-Barr virus in normal human leukocyte cultures. Proc. Natl. Acad. Sci. USA 63, 740–747 (1969).

Germain, D., Requin, Ch., Robert, J.M., Viala, J.-L.: Les anomalies chromosomiques dans l'anemie de Fanconi. Pediatrie 23, 153–167 (1968).

German, J.L.: DNA synthesis in human chromosomes. Trans. N.Y. Acad. Sci. 24, 395 (1962).

German, J.: The pattern of DNA synthesis in the chromosomes of human blood cells. J. Cell Biol. 20, 37–55 (1964).

German, J.: Bloom's syndrome. I. Genetical and clinical observations in the first twenty-seven patients. Am. J. Hum. Genet. 21, 196–277 (1969a).

German, J.L.: Chromosomal breakage syndromes. Birth Defects. Orig. Art. Ser. 5, 117–131 (1969b).

German, J.L.: Oncogenic implications of chromosomal instability. Hosp. Prac. 8, 93–104 (1972a).

German, J.: Genes which increase chromosomal instability in somatic cells and predispose to cancer. Prog. Med. Genet. 8, 61–101 (1972b).

German, J.: Genetic disorders associated with chromosomal instability and cancer. J. Invest. Dermatol. 60, 427–434 (1973).

German, J.L.: An advance in cytogenetics. Science 183, 647–648 (1974a).

German, J.: Bloom's syndrome. II. The prototype of human genetic disorders predisposing to chromosome instability and cancer. In J. German (ed.): Chromosomes and Cancer, pp. 616–636. New York: Wiley (1974b).

German, J. (ed.): Chromosomes and Cancer, New York, London, Sydney, Toronto: Wiley (1974c).

German, J.L., III: To cure leukemia and related conditions. In Vitro 14, 402–404 (1978).

German, J.L., Bearn, A.G.: Asynchronous thymidine uptake by human chromosomes. J. Clin. Invest. 40, 1041–1042 (1961).

German, J.L., Crippa, L.P.: Chromosomal breakage in a diploid cell line from Bloom's syndrome and Fanconi's anemia. Ann. Genet. (Paris) 9, 143–154 (1966).

German, J., Archibald, R., Bloom, D.: Chromosomal breakage in a rare and probably genetically de-

termined syndrome in man. Science 148, 506–507 (1965).
German, J.L., Biro, C.E., Bearn, A.G.: Chromosomal abnormalities in Waldenström's macroglobulinaemia. Lancet ii, 48 (1961).
German, J., Crippa, L.P., Bloom, D.: Bloom's syndrome. III. Analysis of the chromosome aberration characteristic of this disorder. Chromosoma 48, 361–366 (1974).
German, J.L., Demayo, A.P., Bearn, A.G.: Inheritance of an abnormal chromosome in Down's syndrome (mongolism) with leukemia. Am. J. Hum. Genet. 14, 31–43 (1962).
German, J., Gilleran, T.G., Setlow, R.B., Regan, J.D.: Mutant karyotypes in a culture of cells from a man with xeroderma pigmentosum. Ann. Genet. (Paris) 16, 23–28 (1973).
German, J.L., Passarge, E., Bloom, D.: Surveillance for cancer in congenital telangiectatic erythrema and stunted growth. *4th International Congress on Human Genetics,* Paris, 1971, Amsterdam: Excerpta Medica (1971) (Excerpta Medica Int. Congr. Ser. No. 233).
German, J., Schonberg, S., Louie, E., Chaganti, R.S.K.: Bloom's syndrome. IV. Sister-chromatid exchanges in lymphocytes. Am. J. Hum. Genet. 29, 248–255 (1977).
Gey, W.: Dq−, multiple Missbildungen und Retinoblastom. Humangenetik 10, 362–365 (1970).
Ghosh, M.L.: Primary haemorrhagic thrombocythaemia with Philadelphia chromosome. Postgrad. Med. J. 48, 686–688 (1972).
Ghosh, P.K., Singh, I.P.: Morphologic variability of human chromosomes: Polymorphism of constitutive heterochromatin. Hum. Genet. 32, 149–154 (1976).
Ghosh, S.N., Shah, P.N.: Prognosis and incidence of sex chromatin in breast cancer. A preliminary report. Acta Cytol. 19, 58–61 (1975).
Giangiacomo, J., Penchansky, L., Monteleone, P.L., Thompson, J.: Bilateral neonatal Wilms' tumor with B-C chromosomal translocation. J. Pediatr. 86, 98–102 (1975).
Giannelli, F., Howlett, R.M.: The identification of the chromosomes of the D-group (13-15) Denver: An autoradiographic and measurement study. Cytogenetics 5, 186–205 (1966).
Gibbs, T.J., Wheeler, M.V., Bellingham, A.J., Walker, S.: The significance of the Philadelphia chromosome in acute lymphoblastic leukaemia: A report of two cases. Br. J. Haematol. 37, 447–453 (1977).
Gilbert, C.W., Lajtha, L.G., Muldal, S., Ockey, C.H.: Synchrony of chromosome duplication. Nature 209, 537–538 (1966).
Gilgenkrantz, S., Alexandre, P., Baue, G., Streiff, F.: Chromosome X et replication tardive. Essai d'interpretation à l'aide de cas pathologiques et des techniques cytogénétiques avec autoradiographie et B.U.D.R. Lyon Med. 233, 231–240 (1975).
Gilman, P.A., Jackson, D.P., Guild, H.G.: Congenital agranulocytosis. Prolonged survival and terminal acute leukemia. Blood 36, 575–585 (1970).
Gingold, N., Oproiu, C.D., Comanescu, N.: Cytochemical and cytogenetic findings in chronic neutrophilic leukaemia of mature cell type. Lancet ii, 1123 (1964).
Gmyrek, U., Witkowski, R., Sylim-Rapoport, I., Jacobasch, G.: Chromosomenaberrationen und Stoffwechselstörungen der Blutzellen bei Fanconi-Anämie vor und nach Übergang in Leukose am Beispiel einer Patientin. Dtsch. Med. Wochenschr. 92, 1701–1707 (1967) [in English in Germ. Med. Mth. 13, 105–111 (1968)].
Goerttler, E.A., Jung, E.G.: Parakeratosis Mibelli and skin carcinoma. A critical review. Humangenetik 26, 291–296 (1975).
Gofman, J.W., Minkler, J.L., Tandy, R.K.: A specific common chromosomal pathway for the origin of human malignancy. (University of California at Livermore, Lawrence Radiation Lab) Springfield, Va.: Clearinghouse for Federal Scientific and Technical Information, Report UCRL-50356 (1967).
Goh, K.O.: Studies of lymphocytes frm patients with chronic myelocytic leukemia. Clin. Res. 12, 449 (1964).
Goh, K.O.: Cytogenetic interrelation of the myeloproliferative disorders. Clin. Res. 13, 273 (1965).
Goh, K.O.: Smaller G chromosome in irradiated man. Lancet i, 659–660 (1966).
Goh, K.O.: The *in vitro* autoradiographic studies of the Ph¹ chromosomes cells. Clin. Res. 15, 278 (1967a).
Goh, K.O.: Cytogenetic studies in blastic crisis of chronic myelocytic leukemia. Arch. Intern. Med. 120, 315–320 (1967b).
Goh, K.O.: Pseudodiploid chromosomal pattern in chronic lymphocytic leukemia. J. Lab. Clin. Med. 69, 938–949 (1967c).
Goh, K.O.: Autoradiographic studies in chronic myelocytic leukaemia. *Proceedings, Symposium held at Oak Ridge Association Universities, Nov. 13–16,* pp. 695–715 (1967d).
Goh, K.O.: Chromosomes in chronic lymphocytic leukaemia. Lancet ii, 104 (1968a).
Goh, K.O.: Total-body irradiation and human chromosomes. Cytogenetic studies of the peripheral blood and bone marrow leukocytes seven years after total-body irradiation. Radiat. Res. 35, 155–170 (1968b).

Goh, K.O.: Large abnormal acrocentric chromosome associated with human malignancies. Arch. Intern. Med. 122, 241–248 (1968c).

Goh, K.O.: Smaller G (Gp–) and t(Gp–;Dp+) chromosomes. A familial study with one member having acute leukemia. Am. J. Dis. Child. 115, 732–738 (1968d).

Goh, K.O.: Chloramphenicol, acute leukemia and chromosomal vacuolizations. South. Med. J. 64, 815–819 (1971a).

Goh, K.O.: Total body irradiation and human chromosomes. II. Cytogenetic studies of the cultured bone marrow cells seven years after total body irradiation. Am. J. Med. Sci. 262, 43–49 (1971b).

Goh, K.O.: Total body irradiation and human chromosomes. III. Cytogenetic studies in patients with malignant hematologic diseases treated with total-body irradiation. Am. J. Med. Sci. 266, 179–186 (1973).

Goh, K.O.: Additional Philadelphia chromosomes in acute blastic crisis of chronic myelocytic leukemia: Possible mechanims of producing additional chromosomal abnormalities. Am. J. Med. Sci. 267, 229–240 (1974).

Goh, K.O.: Cytogenetic evidence of *in-vivo* leukaemic transformation of engrafted marrow. Lancet i, 1338–1339 (1975).

Goh, K.-O., Anderson, F.W.: Cytogenetic studies in basophilic chronic myelocytic leukemia. Arch. Pathol. Lab. Med. (in press) (1979).

Goh, K.O., Bakemeier, R.F.: Is angioimmunoblastic lymphadenopathy with dysproteinemia a malignant disease? J. Am. Med. Women's Assoc. 33, 38–40 (1978).

Goh, K.O., Joiner, E.: Autoradiographic studies of the abnormal chromosome in human malignancies. Res. Rep. AEC-ORAU 107, 99–102 (1968).

Goh, K.O., Klemperer, M.R.: In vivo leukemic transformation: Cytogenetic evidence of in vivo leukemic transformation of engrafted marrow cells. Am. J. Hematol. 2, 283–290 (1977).

Goh, K.O., Swisher, S.N.: Chromosomal studies in patients with chronic myelocytic leukemia and myeloid metaplasia. Clin. Res. 11, 194 (1963).

Goh, K.O., Swisher, S.N.: Studies of the Philadelphia chromosome. Ann. Intern. Med. 60, 729–730 (1964a).

Goh, K.O., Swisher, S.N.: Specificity of the Philadelphia chromosome. Cytogenetic studies in cases of chronic myelocytic leukemia and myeloid metaplasia. Ann. Intern. Med. 61, 609–624 (1964b).

Goh, K.O., Swisher, S.N.: Identical twins and chronic myelocytic leukemia. Chromosomal studies of a patient with chronic myelocytic leukemia and his normal identical twin. Arch. Intern. Med. 115, 475–478 (1965).

Goh, K.O., Swisher, S.N.: Macroglobulinemia of Waldenström and the chromosomal morphology. Am. J. Med. Sci. 260, 237–244 (1970).

Goh, K.O., Bauman, A.W., Townes, P.L.: Myeloproliferative disorder in a t (13q14q) carrier. Cancer (in press)(1979).

Goh, K., Lee, H., Miller, G.: Down's syndrome and leukemia: Mechanism of additional chromosomal abnormalities. Am. J. Ment. Defic. 82, 542–548 (1978).

Goh, K.O., Reddy, M.M., Joishy, S.K.: Chromosomes and B and T cells in mycosis fungoides. Am. J. Med. Sci. 276, 197–204 (1978).

Goh, K.O., Reddy, M.M., Webb, D.R.: Cancer in a familial IgA deficiency patient: Abnormal chromosomes and B lymphocytes. Oncology 33, 237–240 (1976).

Goh, K.O., Swisher, S.N., Herman, E.C.: Chronic myelocytic leukemia and identical twins. Additional evidence of the Philadelphia chromosome as postzygotic abnormality. Arch. Intern. Med. 120, 214–219 (1967).

Goh, K.O., Swisher, S.N., Rosenberg, C.A.: Cytogenetic studies in eosinophilic leukemia. The relationship of eosinophilic leukemia and chronic myelocytic leukemia. Ann. Intern. Med. 62, 80–86 (1965).

Goh, K.O., Swisher, S.N., Troup, S.B.: Submetacentric chromosome in chronic myelocytic leukemia. Arch. Intern. Med. 114, 439–443 (1964).

Goh, K.O., Bauman, A., Bakemeier, R., Lee, H., Woll, J.E.: Leukemia in radiation-treated patients: cytogenetic studies in eight cases. Am. J. Med. Sci. 276, 189–195 (1978).

Goldberg, L.S., Winkelstein, A., Sparkes, R.S.: Acquired G-group trisomy in acute monomyeloblastic leukemia. Cancer 21, 613–618 (1968).

Goldberg, M.B., Scully, A.L.: Gonadal malignancy in gonadal dysgenesis: Papillary pseudomucinous cystadenocarcinoma in a patient with Turner's syndrome. J. Clin. Endocrinol. Metab. 27, 341–347 (1967).

Goldberg, M.B., Scully, A.L., Solomon, I.L., Steinbach, H.L.: Gonadal dysgenesis in phenotypic female subjects. A review of eighty-seven cases, with cytogenetic studies in fifty-three. Am. J. Med. 45, 529–543 (1968).

Goldblum, N.: Establishment in continuous culture and characterization of cell surface markers, and other immunologic and virologic properties of lymphoblastoid cells derived from patients with different types of leukemia. Isr. J. Med. Sci. 13, 725–730 (1977).

Golde, D.W., Bersch, N.L., Sparkes, R.S.: Chromosomal mosaicism associated with prolonged remission in chronic myelogenous leukemia. Cancer 37, 1849–1852 (1976).

Golde, D.W., Burgaleta, C., Sparkes, R.S., Cline, M.J.: The Philadelphia (Ph[1]) chromosome in

human macrophages. Blood 49, 367–370 (1977).
Golde, D.W., Cline, M.J.: Human preleukemia. Identification of a maturation defect in vitro. N. Engl. J. Med. 288, 1083–1086 (1973).
Golde, D.W., Byers, L.A., Cline, M.J.: Chronic myelogenous leukemia cell growth and maturation in liquid culture. Cancer Res. 34, 419–423 (1974).
Golde, D.W., Quan, S.G., Cline, M.J.: Hairy cell leukemia: In-vitro culture studies. Ann. Intern. Med. 85, 78–79 (1976).
Goldman, J.M.: Acute promyelocytic leukaemia. Br. Med. J. i, 380–382 (1974).
Goldman, J.M., Najfeld, V., Th'ng, K.H.: Agar culture and chromosome analysis for diagnosis of eosinophilic leukemia. International Society of Hematology meeting, London, Aug. 24–28, Abstract 1, 7:01 (1975).
Goldner, F., Hale, M., Engel, E.: Virilization superimposed on Turner's syndrome. Cytogenetic studies of a gonadal tumor in a patient with XO/XY mixed gonadal dysgenesis. South. Med. J. 66, 129–134 (1973).
Goldschmidt, R., Fischer, A.: Chromosomenstudien an Carcinomzellen in vitro. Z. Krebsforsch. 30, 281–285 (1930).
Goldsmith, C.I., Hart, W.R.: Ataxia-telangiectasia with ovarian gonadoblastoma and contralateral dysgerminoma. Cancer 36, 1838–1842 (1975).
Golob, N.E., Klearchou, N., Baumung, H.: Fluoreszenmikroskopischer Nachweis des menschlichen Y-chromosoms in Interphasezellkernen. Wien. Klin. Wochenschr. 85, 11–12 (1973).
Golomb, H.M.: Human chromatin and chromosomes studied by scanning electron microscopy. Progress and perspectives. J. Reprod. Med. 17, 29–35 (1976).
Golomb, H.M.: Hairy cell leukemia: An unusual lymphoproliferative disease. A study of 24 patients. Cancer 42, 946–956 (1978a).
Golomb, H.M.: Clinical implications of chromosome abnormalities in acute non-lymphocytic leukemia: Current status. Virchows Arch. B Cell Pathol. 29, 73–79 (1978b).
Golomb, H.M., Bahr, G.F.: Analysis of an isolated metaphase plate by quantitative electron microscopy. Exp. Cell Res. 68, 65–74 (1971a).
Golomb, H.M., Bahr, G.F.: Scanning electron microscopic observations of surface structure of isolated human chromosomes. Science 171, 1024–1026 (1971b).
Golomb, H.M., Bahr, G.F.: Correlation of the fluorescent banding pattern and ultrastructure of a human chromosome. Exp. Cell Res. 84, 121–126 (1974).
Golomb, H.M., Lindgren, V., Rowley, J.D.: Chromosome abnormalities in patients with hairy cell leukemia. Cancer 41, 1374–1380 (1978a).
Golomb, H.M., Lindgren, V., Rowley, J.D.: Hairy cell leukemia: An analysis of the chromosomes of 26 patients. Virchows Arch. B Cell Pathol. 29, 113–120 (1978b).
Golomb, H.M., Rowley, J., Vardiman, J., Baron, J., Locker, G., Krasnow, S.: Partial deletion of long arm of chromosome 17. A specific abnormality in acute promyelocytic leukemia? Arch. Intern. Med. 136, 825–828 (1976).
Golomb, H.M., Vardiman, J., Rowley, J.D.: Acute nonlymphocytic leukemia in adults: Correlations with Q-banded chromosomes. Blood 48, 9–21 (1976).
Golomb, H.M., Vardiman, J.W., Rowley, J.D., Testa, J.R., Mintz, U.: Correlation of clinical findings with quinacrine-banded chromosomes in 90 adults with acute nonlymphocytic leukemia. An eight-year study (1970–1977). N. Engl. J. Med. 299, 613–619 (1978).
Gomez, G., Hossfeld, D.K., Sokal, J.E.: Removal of abnormal clone of leukaemic cells by splenectomy. Br. Med. J. ii, 421–423 (1975).
Gomez, G.A., Sokal, J.E., Aungst, C.W.: Chemotherapy of the terminal phase of chronic myelocytic leukemia (CML) with combinations of colchicine derivatives and purine analogs. Proc. Am. Assoc. Cancer Res. 18, 194 (1977).
Goncalves, A.: Cromatina sexual e carcinoma mamário. Rev. Bras. Med. 33, 295–296 (1976).
Gonzalez, F., Trujillo, J.M., Alexanian, R.: Acute leukemia in multiple myeloma. Ann. Intern. Med. 86, 440–443 (1977).
Goodall, H.B., Robertson, J.: The chromosomes in sideroblastic anemia. In W.J. Clarke, E.B. Howard, P.L. Hackett (eds.): *Myeloproliferative Disorders of Animals and Man. Proceedings, 8th Annual Hanford Biology Symposium, Richland, Wash., May, 1968.* Springfield, Va.: Clearinghouse for Federal Scientific and Technical Information (AEC symposium series No. 19), U.S.A.E.C. Conf. 680529, pp. 253–271 (1970).
Goodlin, R.C.: Karyotype analysis of gynecologic malignant tumors. Am. J. Obstet. Gynecol. 84, 493–500 (1962).
Goodlin, R.C.: Utilization of cell chromosome number for diagnosing cancer cells in effusion. Nature 197, 507 (1963).
Goodman, R.M., Bouroncle, B.A., Miller, F., North, C.: Unusual chromosomal findings in a case of myelofibrosis. J. Hered. 59, 348–350 (1968).
Goodman, W.N., Cooper, W.C., Kessler, G.B., Fischer, M.S., Gardner, M.B.: Ataxia-telangiectasia: A report of two cases in siblings presenting a picture of progressive spinal muscular atrophy. Bull. Los Angeles Neurol. Soc. 34, 1–22 (1969).
Goodpasture, C., Bloom, S.E.: Visualization of nucleolar organizer regions in mammalian chro-

mosomes using silver staining. Chromosoma 53, 37–50 (1975).

Goodpasture, C., Bloom, S.E., Hsu, T.C., Arrighi, F.E.: Human nucleolus organizers: The satellites or the stalks? Am. J. Hum. Genet. 28, 559–566 (1976).

Goradia, R.Y., Davis, B.K.: Banding and spiralization of human metaphase chromosomes. Hum. Genet. 36, 155–160 (1977).

Gormley, I.P., Ross, A.: Studies on the relationship of a collapsed chromosomal morphology to the production of Q- and G-bands. Exp. Cell Res. 98, 152–158 (1976).

Gosden, J.R., Mitchell, A.R., Buckland, R.A., Clayton, R.P., Evans, H.J.: The location of four human satellite DNAs on human chromosomes. Exp. Cell Res. 92, 148–158 (1975).

Gött, E., Vitéz, L.: Zur Zytogenetik des Morbus Waldenström. Wien. Klin. Wochenschr. 82, 143–145 (1970).

Gottlieb, S.K.: Chromosomal abnormalities in certain human malignancies. J.A.M.A. 209, 1063–1066 (1969).

Govan, A.D.T., Woodcock, A.S., Gowing, N.F.C., Langley, F.A., Neville, A.M., Anderson, M.C.: A clinico-pathological study of gonadoblastoma. Br. J. Obstet. Gynaecol. 84, 222–228 (1977).

Goyanes, V.J.: Differential silver carbonate staining of sister chromatids in BrdU-substituted chromosomes. Hum. Genet. 40, 205–208 (1978).

Grace, E., Bain, A.D.: A simple method for chromosome banding. J. Clin. Pathol. 25, 910 (1972).

Grace, E., Drennan, J., Culver, D., Gordon, R.R.: The 13q deletion syndrome. J. Med. Genet. 8, 351–357 (1971).

Gralnick, H.R., Harbor, J., Vogel, C.: Myelofibrosis in chronic granulocytic leukemia. Blood 37, 152–162 (1971).

Granberg, I.: Chromosomes in preinvasive, microinvasive and invasive cervical carcinoma. Hereditas 68, 165–218 (1971).

Granberg, I., Mark, J.: The chromosomal aberration of double-minutes in a human embryonic rhabdomyosarcoma. Acta Cytol. 15, 42–45 (1971).

Granberg, I., Gupta, S., Joelsson, I., Sprenger, E.: Chromosome and nuclear DNA study of a uterine adenocarcinoma and its metastases. Acta Pathol. Microbiol. Scand. [A] 82, 1–6 (1974).

Granberg, I., Gupta, S., Zech, L.: Chromosome analyses of a metastatic gastric carcinoma including quinacrine fluorescence. Hereditas 75, 189–194 (1973).

Granberg, I., Traneus, A., Silfversward, C.: Chromosome pattern in a patient with cervical carcinoma in situ and atypical hyperplasia of the endometrium. Acta Obstet. Gynecol. Scand. 51, 47–53 (1972).

Granlund, G.H., Zack, G.W., Young, I.T., Eden, M.: A technique for multiple-cell chromosome karyotyping. J. Histochem. Cytochem. 24, 160–167 (1976).

Grant, M.D., Horowitz, H.L., Lezarevic, B.M., Spielvogel, A.R.: Eosinophilic leukemia and myelofibrosis with acute blastic termination. Report of a case with review of literature. Conn. Med. 38, 227–240 (1974).

Gras, L., Cowling, D.C., Sullivan, J., Hurley, T.H.: Unusual chromosomal changes in polycythaemia vera. Med. J. Aust. 1, 36–37 (1975).

Gray, J.W., Carrano, A.V., Steinmetz, L.L., van Dilla, M.A., Moore, D.H. II, Mayall, B.H., Mendelsohn, M.L.: Chromosome measurement and sorting by flow systems. Proc. Natl. Acad. Sci. USA 72, 1231–1234 (1975).

Graze, P., Sparkes, R., Como, R., Gale, R.P.: Hematopoietic engraftment following transplantation of bone marrow cells carrying a Philadelphia (Ph[1])-like chromosome. Am. J. Hematol. 3, 137–142 (1977).

Greaves, M.F.: Clinical applications of cell surface markers. *In* Elmer B. Brown (ed.): *Progress in Hematology*, Vol. 9. pp. 255–303. New York: Grune & Stratton (1975).

Greaves, M.F., Bauminger, S.: Activation of T and B lymphocytes by insoluble phytomitogens. Nature [New Biol.] 235, 67–70 (1972).

Greaves, M.F., Roitt, T.M.: The effect of phytohemagglutinin and other lymphocyte mitogens on immunoglobulin synthesis by human peripheral blood lymphocytes in vitro. Clin. Exp. Immunol. 3, 393–412 (1968).

Greenberg, B.R., Ikeda, R.M., Lewis, J.P.: Eradication of the intramedullary Ph[1] positive cell line without accompanying improved survival in chronic myelogenous leukemia. Cancer 42, 2115–2122 (1978b).

Greenberg, B.R., Wilson, F.D., Woo, L., Jenks, H.M.: Cytogenetics of fibroblastic colonies in Ph[1]-positive chronic myelogenous leukemia. Blood 51, 1039–1044 (1978a).

Greenspan, E.M., Tung, B.G.: Acute myeloblastic leukemia after cure of ovarian cancer. J.A.M.A. 230, 418–420 (1974).

Greisen, O.: The bronchial epithelium. Nucleic acid content in morphologically normal, metaplastic and neoplastic bronchial mucosae. A microspectrophotometric study of biopsy material. Acta Otolaryngol., [Suppl.] (Stockh) 276, 1–110 (1971).

Griffith, J.D.: Chromatin structure: Deduced from a minichromosome. Science 187, 1202–1203 (1975).

Grinblat, J., Mammon, Z., Lewitus, Z., Joshua, H.: Chronic myelogenous leukemia with elevated leukocyte alkaline phosphatase, positive indirect Coombs' test, neutrophilic leukocytosis and unusual cytogenetical findings. Acta Haematol. (Basel) 57, 298–304 (1977).

Gripenberg, U.: Chromosome studies in some virus infections. Hereditas 54, 1–18 (1965).

Gripenberg, U.: Size variation of exceptionally large marker chromosomes in human tumors. Annual Meeting of the European Society of Human Genetics on "Genetic Polymorphism," Athens, Greece. May 8–9 (1976).

Gripenberg, U., Ahlqvist, J., Stenström, R., Gripenberg, L.: Two chromosomally different cell populations in a human neoplasm. Hereditas 87, 51–56 (1977).

Gripenberg, U., Levan, A., Clifford, P.: Chromosomes in Burkitt lymphomas. I. Serial studies in a case with bilateral tumors showing different chromosomal stemlines. Int. J. Cancer 4, 334–349 (1969).

Gromults, J.M., Jr., Hirschhorn, K.: Satellites of acrocentric chromosomes. Lancet ii, 54 (1962).

Gropp, A.: Klonale Evolution im Spätstadium der chronischen myeloischen Leukämie. Dtsch. Med. Wochenschr. 98, 2219 (1973).

Gropp, A., Flatz, G.: Chromosome breakage and blastic transformation of lymphocytes in ataxia-telangiectasia. Humangenetik 5, 77–79 (1967).

Gropp, A., Fischer, R., Niederalt, G., Klesse, M.P., Hensen, S.: Philadelphia-Chromosom und alkalische Leukozyten-Phosphatase bei chronischer Myelose. Klin. Wochenschr. 46, 177–187 (1968).

Gropp, H., Pera, F., Wolf, U.: Untersuchungen über die Anzahl der X-Chromosomen beim Mammacarcinom. Z. Krebsforsch. 69, 326–334 (1967).

Gropp, H., Wolf, U., Pera, F.: Chromatin und Chromosomenstatus beim Mammakarzinom. Dtsch. Med. Wochenschr. 90, 637–642 (1965).

Gros, Ch., Colin, C., Haehnel, P.: Sex chromatin (Barr body) and inflammatory carcinoma of the breast. Biomedicine 19, 65–67 (1973).

Grossbard, L., Rosen, D., McGilvray, E., de Capoa, A., Miller, O., Bank, A.: Acute leukemia with Ph[1]-like chromosome in an LSD user. J.A.M.A. 205, 791–793 (1968).

Grouchy, J. de: Chromosome studies in leukaemia. Lancet i, 615–616 (1961).

Grouchy, J. de: Verleichende Chromosomenuntersuchungen von Krebszellen und *in vitro* bestrahlten Blut und Knochenmarkszellen. Z. Menschl. Vereb. Konstitutionsl. 37, 410–425 (1964).

Grouchy, J. de: Genetic diseases, chromosome rearrangements, and malignancy. Ann. Intern. Med. 65, 603–607 (1966).

Grouchy, J. de: Le fond permanent de remaniements chromosomiques. Frequence de survenue de chromosomes marqueurs dans les cellules normales. Ann. Genet. (Paris) 10, 46–48 (1967a).

Grouchy, J. de: Chromosomes in neoplastic tissues. In J.F. Crow, J.V. Neel (eds.) *Proceedings, 3rd International Congress on Human Genetics, University of Chicago, 1966*, pp. 137–149. Baltimore, Md.: Johns Hopkins Press, (1967b).

Grouchy, J. de: Cancer and the evolution of species. A ransom. Biomedicine 18, 6–8 (1973).

Grouchy, J. de, Lamy, M.: Délétion partielle d'un chromosome moyen dans une leucémie aiguë lymphoblastique. Rev. Fr. Etud. Clin. Biol. 7, 639–643 (1962).

Grouchy, J. de, Nava, C. de: A chromosomal theory of carcinogenesis. Ann. Intern. Med. 69, 381–391 (1968).

Grouchy, J. de, Turleau, C.: Clonal evolution in the myeloid leukemias. In J. German (ed.): *Chromosomes and Cancer*, pp. 287–311. New York: Wiley (1974).

Grouchy, J. de, Wolff, E.: Analyse chromosomique d'une tumeur canceréuse humaine en culture organotypique. Eur. J. Cancer 5, 159–163 (1969).

Grouchy, J. de, Bonnette, J., Brussieux, J., Roidot, M., Begin, P.: Cassures chromosomique dans *l'incontinentia pigmenti*. Etude d'une familie. Ann. Genet. (Paris) 15, 61–65 (1972).

Grouchy, J. de, Josso, F., Beguin, S., Turleau, C., Jalbert, P., Laurent, C.: Déficit en facteur VII de la coagulation chez trois sujets trisomiques 8. Ann. Genet. (Paris) 17, 105–108 (1974).

Grouchy, J. de, Nava, C. de, Bilski-Pasquier, G.: Duplication d'une Ph[1] et suggestion d'une évolution clonale dans une leucémie myéloïde chronique en transformation aiguë. Nouv. Rev. Fr. Hematol. 5, 69–78 (1965a).

Grouchy, J. de, Nava, C. de, Bilski-Pasquier, G.: Analyse chromosomique d'une évolution clonale dans une leucémie myéloïde. Nouv. Rev. Fr. Hematol. 5, 565–590 (1965b).

Grouchy, J. de, Nava, C. de, Bilski-Pasquier, G., Bousser, J.: Chromosome Ph[1] et perte de petits acrocentriques dans une leucémie myéloïde chronïque a évolution prolongée chez un homme. Ann. Genet. (Paris) 9, 73–77 (1966a).

Grouchy, J. de, Nava, C. de, Bilski-Pasquier, G., Zittoun, R., Bernadou, A.: Endoréduplication sélective d'un chromosome surnuméraire dans un cas de myélome multiple (maladie de Kahler). Ann. Genet. (Paris) 10, 43–45 (1967a).

Grouchy, J. de, Nava, C. de, Cantu, J.M., Bilski-Pasquier, G., Bousser, J.: Models for clonal evolutions: a study of chronic myelogenous leukemia. Am. J. Hum. Genet. 18, 485–503 (1966b).

Grouchy, J. de, Nava, C. de, Feingold, J., Bilski-Pasquier, G., Bousser, J.: Onze observations d'une modèle précis d'évolution caryotypique au cours de la leucémie myéloïde chronique. Eur. J. Cancer 4, 481–492 (1968).

Grouchy, J. de, Nava, C. de, Feingold, J., Frézal, J., Lamy, M.: Asynchronie chromosomique dans un cas de xeroderma pigmentosum. Ann. Genet. (Paris) 10, 224–225 (1967b).

Grouchy, J. de, Nava, C. de, Marchand, J.-C., Feingold, J., Turleau, C.: Études cytogénétique et biochemique de huit cas d'anémie de Fanconi. Ann. Genet. (Paris) 15, 29–40 (1972).

Grouchy, J. de, Nava, C. de, Zittoun, R., Bousser, J.: Analyses chromosomiques dans l'anémie sidéroblastique idiopathique acquise. Une étude de six cas. Nouv. Rev. Fr. Hematol. 6, 367–387 (1966c).

Grouchy, J. de, Tudela, V., Feingold, J.: Études cytogénétiques *in vivo* et *in vitro* après infections virales et après vaccination anti-amarile. Pathol. Biol. (Paris) 15, 870–885 (1967).

Grouchy, J. de, Vallée, G., Lamy, M.: Analyse chromosomique directe de deux tumeurs malignes. C.R. Acad. Sci. [D] (Paris) 256, 2046–2048 (1963).

Grouchy, J. de, Valée, C., Nava, V.C., Lamy, M.: Analyse chromosomique de cellules cancéruses et de cellules médullaires et sanguines irradiées "in vitro." Ann. Genet. (Paris) 6, 9–20 (1963).

Grozdea, J., Colombiès, P., Kessous, A.: Correlations between clone Ph^1 and the leukocyte alkaline phosphatase in the association of chronic myeloid leukemia and pregnancy. C.R. Soc. Biol. (Paris) 169, 1376–1379 (1975a).

Grozdéa, J., Colombiès, P., Kessous, A.: Chronic myeloid leukemia and pregnancy. Clone Ph^1 and levels of leucocyte alkaline phosphatases (in French). Nouv. Presse Med. 4, 1596 (1975b).

Grozdea, J., Kessous, A., Colombiès, P.: Leukemia and loss of Y chromosome. Lancet ii, 506 (1973).

Grozdéa, J., Colombiès, P., Bierme, R., Ducos, J., Kessous, A.: Cytochemical and chromosomal studies in hemopathies. I. Chronic myelocytic leukemia (in French). Nouv. Rev. Fr. Hematol. 10, 535–540 (1979).

Gruenwald, H., Kiossoglou, K.A., Mitus, W.J., Dameshek, W.: Philadelphia chromosome in eosinophilic leukemia. Am. J. Med. 39, 1003–1010 (1965).

Grusovin, G.D., Castoldi, G.L.: Characterization of blast cells in acute nonlymphoid leukemias by consecutive cytochemical reactions. Acta Haematol. (Basel) 55, 338–345 (1976).

Guanti, C., Petrinelli, P., Schettini, F.: Cytogenetical and clinical investigations in splenic anemia (Fanconi's type). Hum. Genet. 13, 222–233 (1971).

Guillan, R.A., Ranjini, R., Zelman, S., Hocker, E.V., Smalley, R.L.: Multiple myeloma with hypogammaglobulinemia. Electron microscopic and chromosome studies. Cancer 25, 1187–1192 (1970).

Guinet, P., Eyraud, M.-T.: Les dysgénésies gonadiques avec gonocytome. Rev. Lyon Med. Decembre, 975–994 (1968).

Gunz, F.W., Fitzgerald, P.H.: Chromosomes and leukemia. Blood 23, 394–400 (1964).

Gunz, F.W., Bach, B.I., Crossen, P.E., Mellor, J.E.L., Singh, S., Vincent, P.C.: Relevance of the cytogenetic status in acute leukemia in adults. J. Natl. Cancer Inst. 50, 55–61 (1973).

Gunz, F.W., Fitzgerald, P.H., Adams, A.: An abnormal chromosome in chronic lymphocytic leukaemia. Br. Med. J. ii, 1097–1099 (1962).

Gunz, F.W., Ravich, R.B.M., Vincent, P.C., Stewart, J.H., Crossen, P.E., Mellor, J.: A case of acute leukemia with a rapidly changing chromosome constitution. Ann. Genet. (Paris) 13, 79–86 (1970).

Gustavsson, A., Mitelman, F., Olsson, I.: Acute myeloid leukaemia with the Philadelphia chromosome. Scand. J. Haematol. 19, 449–452 (1977).

Gutierrez, A. Cobo de, Lisker, R., Uribe, M.: Estudios citogéneticos en médula ósea en 185 pacientes con padecimientos hematológicos diversos. Rev. Invest. Clin. Hosp. Enferm. Nutr., (Mex.) 26, 153–168 (1974).

Haberlandt, W.: Aberrations of chromosome number and structure in industrial workers exposed to benzene. Zentralbl. Arbeitsmed. 21, 338–341 (1971).

Haemmerli, G.: Zytophotometrische und zytogenetische Untersuchungen an knotigen Veränderungen des menschlichen Schilddrüse. Schweiz. Med. Wochenschr. 100, 633–641 (1970).

Haemmerli, G., Zweidler, A., Sträuli, P.: Transplantation behavior and cytogenetic characteristics of a spontaneous reticulum cell sarcoma in the Golden hamster. Int. J. Cancer 1, 599–612 (1966).

Haerer, A.F., Jackson, J.F., Evers, C.G.: Ataxia telangiectasia with gastric adenocarcinoma. J.A.M.A. 210, 1884–1887 (1969).

Hagemeijer, A., Smit, E.M.E., Bootsma, D.: Chromosome analysis in myeloproliferative disorders. Has trisomy 8 such a poor prognosis? *Program of Helsinki Chromosome Conference,* August 29–31, p. 196 (abstract) (1977).

Hagemeijer, A., Smit, E.M.E., Löwenberg, B., Abels, J.: Chronic myeloid leukemia with permanent disappearance of the Ph^1 chromosome and development of new clonal subpopulations. Blood 53, 1–14 (1979).

Haglund, U., Lundell, G., Zech, L., Ohlin, J.: Radioiodine administration in hyperthyroidism – a cytogenetic study. Hereditas 87, 85–98 (1977).

Hahner, U., Burkhardt, R.: Knochenmarksdiagnos-

tik bei Haarzell-Leukämie. Klin. Wochenschr. 55, 933–944 (1977).
Haim, A.B., Dudley, R.A., Müllner, T., Parr, R.M., Vetter, H.: Chromosome aberrations in peripheral blood cells in man following chronic irradiation from internal deposits of thorotrast. Radiat. Res. 29, 505–515 (1966).
Haines, M.: Autoradiographic studies of the chromosomes in chronic granulocytic leukaemia. Nature 207, 552–553 (1965).
Halkka, O., Meynadier, G., Vago, C., Brummer-Korvenkontio, M.: Rickettsial induction of chromosome aberrations. Hereditas 64, 126–128 (1970).
Hall, B.: Down's syndrome (mongolism) with a morphological Philadelphia chromosome. Lancet i, 558 (1963).
Hall, W.T., Schidlovsky, G.: Typical type-C virus in human leukemia. J. Natl. Cancer Inst. 56, 639–642 (1976).
Hamann, W., Oehlert, W., Musshoff, K., Nuss, A., Schnellbacher, B.: Histologische Einteilung des Lymphoma malignum Hodgkin und seine Bedeutung für die Prognose. Dtsch. Med. Wochenschr. 95, 112–116 (1970).
Hamerton, J.L.: Chromosomes and neoplastic disease. Clin. Genet. 2, 407–441 (1971a).
Hamerton, J.L.: *Human Cytogenetics*, Vol. 2. New York: Academic Press (1971b).
Hammouda, F.: Chromosome abnormality in acute leukaemia. Lancet ii, 410–411 (1963).
Hammouda, F., Quaglino, D., Hayhoe, F.G.J.: Blastic crisis in chronic granulocytic leukaemia. Cytochemical, cytogenetic, and autoradiographic studies in four cases. Br. Med. J. i, 1275–1281 (1964).
Hampel, K.E.: Endoreduplikation bei der chronischen myeloischen Leukämie. Naturwissenschaften 50, 619–620 (1963).
Hampel, K.E.: Diplo-Ph¹-Chromosom bei der myeloischen Leukämie. Klin. Wochenschr. 42, 522–524 (1964).
Hampel, K.E.: Differentialdiagnostische Bedeutung des Ph-Chromosoms. Verh. Dtsch. Ges. Inn. Med. 73, 457–460 (1967).
Hampel, K.E., Levan, A.: Breakage in human chromosomes induced by low temperature. Hereditas 51, 315–343 (1964).
Hampel, K.E., Palme, G.: Chromosomenstudien bei akuter und chronischer myeloischer Leukämie. Neoplasma 11, 113–122 (1964).
Hampel, K.E., Lohr, G.W., Blume, K.G., Rudiger, H.W.: Spontane und chloramphenicolinduzierte Chromosomemutationen und biochemische Befunde bei zwei Fallen mit Glutathionreduktasemangel (NAD(P)H glutathione oxidoreductase, E.C.I. 6.4.2.). Humangenetik 7, 305–313 (1969).

Han, T., Moayeri, H., Minowada, J.: T- and B-lymphocytes in chronic lymphocytic leukemia: Correlation with clinical and immunologic status of the disease. J. Natl. Cancer Inst. 57, 477–481 (1976).
Hand, R.: Human DNA replication: Fiber autoradiographic analysis of diploid cells from normal adults and from Fanconi's anemia and ataxia telangiectasia. Hum. Genet. 37, 55–64 (1977).
Hand, R., German, J.: A retarded rate of DNA chain growth in Bloom's syndrome. Proc. Natl. Acad. Sci. USA 72, 758–762 (1975).
Hand, R., German, J.: Bloom's syndrome: DNA replication in cultured fibroblasts and lymphocytes. Hum. Genet. 38, 297–306 (1977).
Handmaker, S.D.: The satellited chromosomes of man with reference to the Marfan syndrome. Am. J. Hum. Genet. 15, 11–18 (1963).
Hanschke, H.J., Hoffmeister, H.: Die zellernmorphologische Geschlechtsbestimmung beim Bronchialcarzinom der Frau. Zentralbl. Allg. Pathol. 101, 99 (1960).
Hansemann, D. von: Über asymmetrische Zellteilung in Epithelkrebsen und deren biologische Bedeutung. Arch. Pathol. Anat. Physiol. 119, 299–326 (1890).
Hansen-Melander, E., Kullander, S., Melander, Y.: Chromosome analysis of a human ovarian cysto-carcinoma in the ascites form. J. Natl. Cancer Inst. 16, 1067–1081 (1956).
Hansson, A., Korsgaard, R.: Cytogenetical diagnosis of malignant pleural effusions. Scand. J. Respir. Dis. 55, 301–308 (1974).
Hansteen, I.-L.: Chromosome studies in glial tumours. Eur. J. Cancer 3, 183–191 (1967).
Hansteen, I.-L., Hillestad, L., Thiis-Evensen, E., Heldaas, S.S.: Effects of vinyl chloride in man. A cytogenetic follow-up study. Mutat. Res. 51, 271–278 (1978).
Hansteen, I.-L., Hillestad, L., Thomassen, O.K.: Chromosome analysis and cell cytology in effusions. A comparative study. Scand. J. Respir. Dis. 58, 51–56 (1977).
Happle, R., Hoehn, H.: Cytogenetic studies on cultured fibroblast-like cells derived from basal cell carcinoma tissue. Clin. Genet. 4, 17–24 (1973).
Happle, R., Kupferschmid, A.: A further case of basal cell nevus syndrome and structural chromosome abnormalities. Humangenetik 15, 287–288 (1972).
Happle, R., Mehrle, G., Sander, L.Z., Hoehn, H.: Basalzellnävus-Syndrom mit Retinopathia pigmentosa, rezidivierender Glaskörperblutung und Chromosomenveränderungen. Arch. Dermatol. Forsch. 241, 96–114 (1971).
Hardisty, R.M., Speed, D.E., Till, M.: Granulocytic

leukaemia in childhood. Br. J. Haematol. 10, 551–566 (1964).
Hare, W.C.D., Yang, T.-J., McFeely, R.A.: A survey of chromosome findings in 47 cases of bovine lymphosarcoma (leukemia). J. Natl. Cancer Inst. 38, 383–392 (1967).
Harnden, D.G.: The role of genetic change in neoplasia. Constitutional abnormalities. In: *Genetic Concepts and Neoplasia, 23rd Annual Symposium on Fundamental Cancer Research, M.D. Anderson Hospital*, pp. 31–35. Houston, Baltimore: Williams and Wilkins (1970).
Harnden, D.G.: Viruses, chromosomes and tumors. The interaction between viruses and chromosomes. *In* J. German (ed.): *Chromosomes and Cancer*, pp. 151–190. New York: Wiley (1974a).
Harnden, D.G.: Ataxia telangiectasia. Cytogenetic and cancer aspects. *In* J. German (ed.): *Chromosomes and Cancer*, pp. 616–636. New York: Wiley (1974b).
Harnden, D.G.: Chromosome abnormalities and predisposition towards cancer. Proc. R. Soc. Med. 69, 41–43 (1976a).
Harnden, D.G.: The genetic component in oncological diseases. *In* T. Symington, R.L. Carter (eds.): *Scientific Foundations of Oncology*, pp. 181–190. London: William Heinemann Medical Books (1976b).
Harnden, D.G.: Cytogenetics of human neoplasia. *In* J.J. Mulvhill, R.W. Miller, J.F. Fraumeni, Jr. (eds.): *Progress in Cancer Research and Therapy*, Vol. 3, *Genetics of Human Cancer*, pp. 87–104. New York: Raven Press (1977).
Harnden, D.G., Benn, P.A., Oxford, J.M., Taylor, A.M.R., Webb, T.P.: Cytogenetically marked clones in human fibroblasts cultured from normal subjects. Somat. Cell Genet. 2, 55–62 (1976).
Harnden, D.G., Langlands, A.O., McBeath, S., O'Riordan, M., Faed, M.J.W.: The frequency of constitutional chromosome abnormalities in patients with malignant disease. Eur. J. Cancer 5, 605–614 (1969).
Harnden, D.G., Maclean, N., Langlands, A.O.: Carcinoma of the breast and Klinefelter's syndrome. J. Med. Genet. 8, 460–461 (1971).
Harousseau, J.-L., Smadja, N., Krulik, M., Audebert, A.-A., Debray, J.: Transformation lymphoblastique inaugurale d'une leucémie myéloïde chronique. Étude clinique et cytogénétique d'une cas. Sem. Hop. Paris 53, 2437–2444 (1977).
Harris, V.J., Seeler, R.A.: Ataxia-telangiectasia and Hodgkin's disease. Cancer 32, 1415–1420 (1973).
Harrisson, C.M.H.: The arrangement of chromatin in the interphase nucleus with reference to cell differentiation and repression in higher organisms. Tissue Cell 3, 523–550 (1971).

Hart, J.S., Trujillo, J.M., Freireich, E.J., George, S.L., Frei, E., III: Cytogenetic studies and their clinical correlates in adults with acute leukemia. Ann. Intern. Med. 75, 353–360 (1971).
Hartmann, R.C., Jenkins, D.E., Jr.: Paroxysmal nocturnal hemoglobinuria: Current concepts of certain pathophysiologic features. Blood 25, 850–865 (1965).
Hartwich, G., Schricker, K.T.: Pilzsepsis im Verlauf einer Benzol-Leukämie mit Chromosomenveränderungen. Fortschr. Med. 87, 563–565 (1969b).
Hartwich, G., Schwanitz, G.: Chromosomenuntersuchungen nach chronischer Benzol-Exposition. Dtsch. Med. Wochenschr. 97, 45–49 (1972).
Hartwich, G., Schwanitz, G., Becker, J.: Chromosomenaberrationen bei einer Benzol-Leukämie. Dtsch. Med. Wochenschr. 94, 1228–1229 (1969).
Hashem, N., Khalifa, S.: Retinoblastoma: A model of hereditary fragile chromosome regions. Hum. Hered. 25, 35–49 (1975).
Hashimoto, Y., Takaku, F., Kozaka, K.: Damaged DNA in lymphocytes of aplastic anemia. Blood 46, 735–742 (1975).
Hastings, J., Freedman, S., Rendon, O., Cooper, H.L., Hirschhorn, K.: Culture of human white cells using differential leucocyte separation. Nature 192, 1214–1215 (1961).
Hatcher, N.H., Pollara, B., Hook, E.B.: Chromosome breakage in two siblings with ataxia-telangiectasia: A search for intrafamilial similarities. Am. J. Hum. Genet. 26, 39A (1974).
Hauschka, T.S.: Relationship between chromosome ploidy and histocompatibility in mouse ascites tumors. Cancer Res. 12, 269 (1952).
Hauschka, T.S.: Cell population studies on mouse ascites tumors. Trans. N.Y. Acad. Sci. 16, 64–73 (1953).
Hauschka, T.S.: The chromosomes in ontogeny and oncogeny. Cancer Res. 21, 957–974 (1961).
Hauschka, T.S.: Chromosome patterns in primary neoplasia. Exp. Cell Res. (Suppl.) 9, 86–98 (1963).
Hauschka, T.S., Levan, A.: Characterization of five ascites tumors with respect to chromosome ploidy. Anat. Rec. 111, 467 (1951).
Hauschka, T.S., Hasson, J.E., Goldstein, M.N., Koepf, G.F., Sandberg, A.A.: An XYY man with progeny indicating familial tendency to nondisjunction. Am. J. Hum. Genet. 14, 22–30 (1962).
Havemann, K.: Delayed reaction of lymphocytes in Hodgkin's disease to phytohaemagglutinin (PHA). Ger. Med. Mon. 14, 243–246 (1969).
Hayashi, K., Schmid, W.: Tandem duplication q14 and dicentric formation by end-to-end chromosome fusions in ataxia telangiectasia (AT). Clin-

ical and cytogenetic findings in 5 patients. Humangenetik 30, 135–141 (1975).
Hayata, I., Sasaki, M.: A case of Ph^1-positive chronic myelocytic leukemia associated with complex translocations. Proc. Jap. Acad. 52, 29–32 (1976).
Hayata, I., Kakati, S., Sandberg, A.A.: A new translocation related to the Philadelphia chromosome. Lancet ii, 1385 (1973).
Hayata, I., Kakati, S., Sandberg, A.A.: The monoclonal origin of chronic myelocytic leukemia. Proc. Jap. Acad. 50, 381–385 (1974).
Hayata, I., Kakati, S., Sandberg, A.A.: Another translocation related to the Ph^1 chromosome. Lancet i, 1300 (1975).
Hayata, I., Sakurai, M., Kakati, S., Sandberg, A.A.: Chromosomes and causation of human cancer and leukemia. XVI. Banding studies in CML, including five unusual Ph^1-translocations. Cancer 36, 1177–1191 (1975).
Hayata, I., Oshimura, M., Sandberg, A.A.: N-band polymorphism of human acrocentric chromosomes and its relevance to satellite association. Hum. Genet. 36, 55–61 (1977).
Hayhoe, F.G.J.: *Leukaemia – Research and Clinical Practice.* London: Churchill (1960).
Hayhoe, F.G.J., Hammouda, F.: Cytogenetic and metabolic observations in leukaemias and allied states. *In* F.G.J. Hayhoe (ed.): *Current Research in Leukaemia,* pp. 55–76. London: Cambridge University Press (1965).
Hayhoe, F.G.J., Sinks, L.F., Flemans, R.J.: Studies on the transformation in vitro of lymphocytes from chronic lymphocytic leukaemia. *In* J.M. Yoffey (ed.): *The Lymphocyte in Immunology and Haemopoiesis.* (Sympos. Bristol, 1966), p. 66. London: Edward Arnold, 1966.
Hays, T., Humbert, J.R., Peakman, D.C., Hutter, J.J., Morse, H.G., Robinson, A., August, C.S.: Missing Y chromosome in juvenile chronic myelogenous leukemia. Humangenetik 29, 259–264 (1975).
Hays, T., Morse, H., Peakman, D., Rose, B., Robinson, A.: Cytogenetic studies in childhood chronic myelocytic leukemia. Cancer (submitted).
Heath, C.W.: Cytogenetic observations in vitamin B$_{12}$ and folate deficiency. Blood 27, 800–815 (1966).
Heath, C.W., Moloney, W.C.: Cytogenetic observations in a case of erythremic myelosis. Cancer 18, 1495–1504 (1965a).
Heath, C.W., Moloney, W.C.: The Philadelphia chromosome in an unusual case of myeloproliferative disease. Blood 26, 471–478 (1965b).
Heath, C.W., Bennett, J.M., Whang-Peng, J., Berry, E.W., Wiernick, P.H.: Cytogenetic findings in erythroleukemia. Blood 33, 453–467 (1969).
Hecht, F., Case, M.: Emergence of a clone of lymphocytes in ataxia-telangiectasia. Annual Meeting, American Society on Human Genetics, San Francisco, Calif. (1969).
Hecht, F., Kaiser-McCaw, B.: The consequences of the Philadelphia chromosome rearrangement in chronic myeloid leukemia. Hum. Genet. 36, 127–128 (1977a).
Hecht, F., Kaiser-McCaw, B.: Chromosomes and genes in human cancer cells: Multidisciplinary approaches to a unitary genodemographic hypothesis. Chromosomes Today 6, 357–361 (1977b).
Hecht, F., McCaw, B.K.: Leukemia and chromosomes. Science 185, 735 (1974).
Hecht, F., McCaw, B.K.: Social contacts and leukaemia-lymphoma: The Philadelphia chromosome rearrangement. Lancet i, 1031 (1975).
Hecht, F., McCaw, B.K.: Chromosome instability syndromes. *In* J.J. Mulvhill, R.W. Miller, J.F. Fraumeni, Jr. (eds.): *Progress in Cancer Research and Therapy,* Vol. 3, *Genetics of Human Cancer,* pp. 105–123. New York: Raven Press (1977).
Hecht, F., Koler, R.D., Rigas, D.A., Dahula, G.S., Case, M.P., Tudale, V., Miller, R.W.: Leukemia and lymphocytes in ataxia-telangiectasia. Lancet ii, 1193 (1966).
Hecht, F., McCaw, B.K., Koler, R.D.: Ataxia-telangiectasia. Clonal growth of translocation lymphocytes. N. Engl. J. Med. 289, 286–291 (1973).
Hecht, F., McCaw, B.K., Peakman, D., Robinson, A.: Non-random occurrence of 7–14 translocations in human lymphocyte cultures. Nature 255, 243–244 (1975).
Heiberg, K.H., Kemp, T.: Über die Zahl der Chromosomen in Carcinomzellen beim Menschen. Virchows Arch. Pathol. Anat. 273, 693–700 (1929).
Heimpel, H.: Präleukämie: klinisches und pathogenetisches Konzept. Dtsch. Arztebl. 69, 2705–2707 (1972).
Heller, A., Gross, R.: Cytochemical observations in preleukemic states and remission of akute leukemia. Blut 34, 465–469 (1977).
Hellriegel, K.P.: Chromosomenbefunde bei myeloproliferativen Erkrankungen. Internist 9, 465–470 (1968).
Hellriegel, K.P.: Chromosomenbefunde bei monoklonalen Gammapathien. Internist 12, 13–23 (1971).
Hellriegel, K.P.: Cytogenetic studies in potentially preleukemic stages. Proc. Int. Soc. Hematol. (London) Abstract 5, p. 11 (1975).
Hellriegel, K.P.: Chromosome findings in monoclonal gammapathies. Haematol. Bluttransfus. 18, 369–375 (1976).
Hellriegel, K.P., Gross, R.: Die Frühdiagnose der chronischen myeloischen Leukämie mit Hilfe

der Chromosomenanalyse. Verh. Dtsch. Ges. Inn. Med. 75, 505–507 (1969).
Hellriegel, K.P., Koebke, J.: Chromosomenbefunde in Blastenschub chronischer Myelosen. Med. Welt 22, 200–202 (1971).
Hellriegel, K.P., Koebke, J., Gross, R.: Differenzierung von Myeloblastenleukämien und Blastenschüben chronischer Myelosen mit cytogenetischen Untersuchungen. Verh. Dtsch. Ges. Inn. Med. 77, 83–87 (1971).
Hellriegel, K.P., Borberg, H., Reitz, H., Gross, R.: Evaluation of the efficacy of granulocyte transfusions by Y-chromatin studies. In J.M. Goldman, R.M. Goldman (eds.): Leukocytes: Separation, Collection and Transfusion, pp. 415–423. New York: Academic Press (1975).
Hellriegel, K.P., Diehl, V., Krause, P.H., Meier, S., Blankenstein, M., Busche, W.: The significance of chromosomal findings for the differentiation between lymphoma and lymphoblastoid cell lines. Haematol. Bluttransfus. 20, 307–313 (1977).
Hellriegel, K.-P., Heit, W., Byrne, P.: Effekt von Kurzzeit-Suspensionskulturen hämatopoetischer Zellen auf die Mitoserate bei Frühstadien unreifzelliger Leukämien: Eine Methode zur Verbesserung der Präparations-bedingungen für zytogenetische Untersuchungen. Blut 34, 398–402 (1977).
Hellriegel, K.P., Pfeiffer, R.A., Seiler, R., Schütz, C., Rickers, H.J.: Mongolismus und Leukämie. Klinische und zytogenetische Befunde bei vier mongoloiden Kindern mit Leukämie. Münch. Med. Wochenschr. 111, 1522–1528 (1969).
Hellström, K., Hagenfeldt, L., Larsson, A., Lindsten, J., Sundelin, P., Tiepolo, L.: An extra C chromosome and various metabolic abnormalities in the bone marrow from a patient with refractory sideroblastic anemia. Scand. J. Haematol. 8, 293–306 (1971).
Hellweg-Fründ, S., Koske-Westphal, T., Fuchs-Mecke, S., Passarge, E.: Fluoreszenmikroskopische Identifizierung von Anomalien des Y-Chromosoms. Dtsch. Med. Wochenschr. 97, 1650–1661 (1972).
Helson, L.: Regression of neuroblastomas. Lancet i, 1075 (1971).
Henderson, A.S., Warburton, D., Atwood, K.C.: Location of ribosomal DNA in human chromosome complement. Proc. Natl. Acad. Sci. USA 69, 3394–3398 (1972).
Henderson, A.S., Warburton, D., Atwood, K.C.: Ribosomal DNA connectives between human acrocentric chromosomes. Nature 245, 95–97 (1973).
Heneen, W.K.: Silver staining and nucleolar patterns in human heteroploid and measles-carrier cells. Hereditas 88, 213–227 (1978).
Heneen, W.K., Nichols, W.W., Norby, E.: Polykaryocytosis and mitosis in a human cell line after treatment with measles virus. Hereditas 64, 53–84 (1970).
Heni, F., Siebner, H.: Chromosomenveränderungen bei der Makroglobulinämie Waldenström. Dtsch. Med. Wochenschr. 88, 1781–1782 (1963) [in English in Ger. Med. Mon. 9, 72–73 (1964)].
Heni, F., Siebner, H.: Chromosomal abnormality and chronic lymphocytic leukaemia. Lancet i, 1109–1110 (1964).
Henle, W., Diehl, V., Kohn, G., Hausen, H. zur, Henle, G.: Herpes-type virus and chromosome marker in normal leukocytes after growth with irradiated Burkitt cells. Science 157, 1064–1065 (1967).
Hennekeuser, H.H., Citoler, P., Niemczyk, H., Gropp, A.: Klinische, histologische und cytogenetische Befunde bei Patienten mit Thorotrastschaden. Klin. Wochenschr. 48, 895–906 (1970).
Hennekeuser, H.H., Mobius, W.: Untersuchungen zur Bedeutung des Peroxidase-Nachweises bei akuter myeloischer Leukämie. Blut 29, 317–322 (1974).
Hentel, J., Hirschhorn, K.: The origin of some bone marrow fibroblasts. Blood 38, 81–86 (1971).
Hersh, E.M., Whitecar, J.P., McCredie, K.B., Bodey, G.P., Freireich, E.J.: Chemotherapy, immunocompetence, immunosuppression and prognosis in acute leukemia. N. Engl. J. Med. 285, 1211–1216 (1971).
Hertz, R.: Spontaneous regression in choriocarcinoma and related gestational trophoblastic neoplasms. Natl. Cancer Inst. Monogr. 44, 59–60 (1976).
Herz, F., Miller, O.J., Miller, D.A., Auersperg, N., Koss, L.G.: Chromosome analysis and alkaline phosphatase of C41, a cell line of human cervical origin distinct from HeLa. Cancer Res. 37, 3209–3213 (1977).
Hester, J.P., Kwaan, H.C., Moake, J., Trujillo, J., Hart, J., McBride, C., McCredit, K.B., Freireich, E.J.: Cytokinetic, cytogenetic, coagulation profiles of chronic myelogenous leukemia (CML) patients undergoing splenectomy. Proc. Am. Soc. Clin. Oncol. 18, 280 (1977).
Heston, W.E.: The genetic aspects of human cancer. Adv. Cancer Res. 23, 1–21 (1976).
Higurashi, M., Conen, P.E.: In vitro chromosomal radiosensitivity in Fanconi's anemia. Blood 38, 336–342 (1971).
Higurashi, M., Conen, P.E.: In vitro chromosomal radiosensitivity in patients and in carriers with abnormal non-Down's syndrome karyotypes. Pediatr. Res. 6, 514–520 (1972).
Higurashi, M., Conen, P.E.: In vitro chromosomal radiosensitivity in "chromosomal breakage syndromes." Cancer 32, 380–383 (1973).

Higurashi, M., Nakagome, Y., Matsui, I., Naganuma, M.: Cytogenetic observations in children with leukemia. Paediatr. Univ. Tokyo 18, 36–40 (1970).

Higurashi, M., Nakagome, Y., Matsui, I., Nagao, T.: On the DNA replication pattern of the Ph¹ chromosome. Paediatr. Univ. Tokyo 15, 37–40 (1968).

Higurashi, M., Tamura, T., Nakatake, T.: Cytogenetic observations in cultured lymphocytes from patients with Down's syndrome and measles. Pediatr. Res. 7, 582–587 (1973).

Hill, J.A., McKenna, H.: Pure gonadal dysgenesis. An XY female with gonadoblastoma and adenofibroma. Aust. N.Z. J. Obstet. Gynecol. 14, 50–52 (1974).

Hillman, E.A., Charamella, L.J., Temple, M.J., Elser, J.F.: Biological characterization of an Epstein-Barr nuclear antigen-positive American Burkitt's tumor-derived cell line. Cancer Res. 37, 4546–4558 (1977).

Hilton, H.B., Lewis, I.C., Trowell, H.R.: C group trisomy in identical twins with acute leukemia. Blood 35, 222–226 (1970).

Hinkes, E., Crandall, B.F., Weber, F., Craddock, C.G.: Acute leukemia with C-G chromosome translocation. Blood 41, 259–263 (1973).

Hiraki, S., Miyoshi, I., Kubonishi, I., Matsuda, Y., Nakayama, T., Kishimoto, H., Masuji, H.: Human leukemic "null" cell line (NALL-1). Cancer 40, 2131–2135 (1977).

Hirschhorn, K.: Cytogenetic alterations in leukemia. In W. Dameshek, R.M. Dutcher (eds.): Perspectives in Leukemia, pp. 113–120. New York, London: Grune and Stratton (1968).

Hirschhorn, K.: Discussion paper: The role of cytogenetics in mutagenesis testing. Ann. N.Y. Acad. Sci. 269, 12–15 (1975).

Hirschhorn, K.: Chromosomes and cancer. In Daniel Bergsma (ed.): Cancer and Genetics, pp. 113–121. The National Foundation—March of Dimes, Birth Defects: Orig. Art. Ser., Vol. XII, New York: Alan R. Liss, Inc. (1976).

Hirschman, R.J., Shulman, N.R., Abuelo, J.G., Whang-Peng, J.: Chromosomal aberrations in two cases of inherited aplastic anemia with unusual clinical features. Ann. Intern. Med. 71, 107–117 (1969).

Hittelman, W.N., Rao, P.N.: Predicting response and progression of human leukemia by premature chromosome condensation of bone marrow cells. Cancer Res. 38, 416–423 (1978).

Hitzeroth, H.W., Bender, K., Ropers, H.-H., Geerthsen, J.M.P.: Tentative evidence for 3–4 haematopoietic stem cells in man. Hum. Genet. 35, 175–183 (1977).

Hoch, S.O., Longmire, R.L., Hoch, J.A.: Unique DNA-banding-protein in the serum of patients with various neoplasms. Nature 255, 560–562 (1975).

Hoefnagel, D., Sullivan, M., McIntyre, O.R., Gray, J.A., Storrs, R.C.: Panmyelopathy with congenital anomalies (Fanconi) in two cousins. Helv. Paediatr. Acta 21, 230–238 (1966).

Hoehn, H., Au, K., Karp, L.E., Martin, G.M.: Somatic stability of variant C-band heterochromatin. Hum. Genet. 35, 163–168 (1977).

Hoffbrand, A.V., Ganeshaguru, K., Janossy, G., Greaves, M.F., Catovsky, D., Woodruff, R.K.: Terminal deoxynucleotidyltransferase levels and membrane phenotypes in diagnosis of acute leukaemia. Lancet ii, 520–523 (1977).

Holden, D., Lichtman, H.: Paroxysmal nocturnal hemoglobinuria with acute leukemia. Blood 33, 283–286 (1969).

Holden, J.D., Garcia, U.G., Samuels, M., Dupin, C., Stallworth, B., Anderson, E.: Myelofibrosis with C monosomy of marrow elements in a child. Am. J. Clin. Pathol. 55, 573–579 (1971).

Hollander, D.H., Borgaonkar, D.S.: The quinacrine fluorescence method of Y-chromosome identification. Acta Cytol. 15, 452–454 (1971).

Hollander, D.H., Tockman, M.S., Liang, Y.W., Borgaonkar, D.S., Frost, J.K.: Sister chromatid exchanges in the peripheral blood of cigarette smokers and in lung cancer patients; and the effect of chemotherapy. Hum. Genet. 44, 165–171 (1978).

Hollinshead, A.C.: Cell membrane antigens associated with human adult acute leukemia. Blood Cells 3, 257–265 (1976).

Hollinshead, A., Suskind, G., Jacobson, G., Randall, J.: R-type virus particles in primary human brain tumor cell cultures associated with defined chromosome loss and membrane antigens. Proc. Am. Assoc. Cancer Res. 17, 211 (abstract) (1976).

Holton, C.P., Johnson, W.W.: Chronic myelocytic leukemia in infant siblings. J. Pediatr. 72, 377–383 (1968).

Honda, F., Punnett, H.H., Charney, E., Miller, G., Thiede, H.A.: Serial cytogenetic and hematologic studies on a mongol with trisomy-21 and acute congenital leukemia. J. Pediatr. 65, 880–887 (1964).

Honoré, L.H., Dill, F.J., Poland, B.J.: The association of hydatidiform mole and trisomy 2. Obstet. Gynecol. 43, 232–237 (1974).

Hook, E.B., Engel, R.R.: Leukocyte life-span, leukocyte alkaline phosphatase, and the 21st chromosome. Lancet i, 112 (1964).

Hook, E.B., Hatcher, N.H., Calka, O.J.: Apparent "in situ" clone of cytogenetically marked ataxia-telangiectasia lymphocytes. Humangenetik 30, 251–257 (1975).

Hoppin, E.C., Lewis, J.P.: Polycythemia rubra vera progressing to Ph¹-positive chronic myelogenous leukemia. Ann. Intern. Med. 83, 820–823 (1975).

Horland, A.A., Wolman, S.R., Distenfeld, A., Cohen, T.: Another variant translocation in chronic myelogenous leukemia. N. Engl. J. Med. 294, 164–165 (1976).

Horvat, D.: Chromosomal aberrations in persons occupationally exposed to ionizing radiation. Arh. Hig. Rada Toksikol. 26, 139–146 (1975).

Hossfeld, D.K.: Chromosomal anomalies in neoplasia. 11th Meeting of the German Cancer Association, Hannover (Sept.), pp. 71–72 (1971).

Hossfeld, D.K.: Chromosomal features occurring in human neoplastic diseases. Z. Krebsforsch. 78, 123–128 (1972).

Hossfeld, D.K.: Identification of chromosome anomalies in the blastic phase of chronic myelocytic leukemia (CML) by Giemsa- and quinacrine-banding techniques. Humangenetik 23, 111–118 (1974a).

Hossfeld, D.K.: No chromosome 9p+ in Ph¹-negative CML. Nature 249, 864 (1974b).

Hossfeld, D.K.: Additional chromosomal indication for the unicellular origin of chronic myelocytic leukemia. Z. Krebsforsch. 83, 269–273 (1975a).

Hossfeld, D.K.: Chronic myelocytic leukemia: Cytogenetic findings and their relations to pathogenesis and clinic. Ser. Haematol. 8, 53–72 (1975b).

Hossfeld, D.K.: Chromosome 14q+ in a retinoblastoma. Int. J. Cancer 21, 720–723 (1978).

Hossfeld, D.K., Cohnen, G.: Die chronische myeloische Leukämie. In H. Begemann (ed.): Handbuch der Inneren Medizin, Vol. II, 3rd ed. Berlin: Springer Verlag, pp. 443–518 (1978).

Hossfeld, D.K., Hirche, H.: Relations between hematological-clinical findings and prognosis in chronic myelocytic leukemia (CML). In A. Stacher, P. Höcker (eds.): Erkrankungen der Myelopoese, pp. 324–325. Munchen-Berlin-Wien: Urban and Schwarzenberg (1976).

Hossfeld, D.K., Köhler, S.: New translocations in chronic granulocytic leukaemia: t(X;22) (p22;q11) and t(15;22) (q26;q11). Br. J. Haemat. 41, 185–191 (1979).

Hossfeld, D.K., Sandberg, A.A.: Das Philadelphia Chromosom. Klin. Wochenschr. 48, 1431–1441 (1970a).

Hossfeld, D.K., Sandberg, A.A.: Chromosomes of marrow cells in acute leukemia: direct method vs. culture. Proceedings 13th Annual Meeting of the American Society of Hematology, San Juan (1970b).

Hossfeld, D.K., Schmidt, C.G.: Chromosomal data suggesting a primary role of the spleen in the pathogenesis of chronic myelocytic leukemia (CML) and blastic phase of CML. In S. Garratini, G. Franchi (eds.): Chemotherapy and Cancer Dissemination and Metastasis, pp. 223–235. New York: Raven Press (1973).

Hossfeld, D.K., Schmidt, C.G.: Chromosome findings in effusions from patients with Hodgkin's disease. Int. J. Cancer 21, 147–156 (1978).

Hossfeld, D.K., Wendehorst, E.: Ph¹-negative chronic myelocytic leukemia with a missing Y chromosome. Acta Haematol. (Basel) 52, 232–237 (1974).

Hossfeld, D.K., Bremer, K., Meusers, P., Wendehorst, E., Reis, H.E.: Extramedullary manifestation of the blastic phase of chronic myelocytic leukemia. A chromosome study. Z. Krebsforsch. 84, 49–57 (1975).

Hossfeld, D.K., Ezdinli, E.Z., Han, T., Aungst, C.W.: Prognostic parameters in chronic myelocytic leukemia (CML). Proc. Am. Assoc. Cancer Res. 12, 86, Abstract 342 (1971).

Hossfeld, D.K., Han, T., Holdsworth, R.N., Sandberg, A.A.: Chromosomes and causation of human cancer and leukemia: VII. The significance of the Ph¹ in conditions other than CML. Cancer 27, 186–192 (1971).

Hossfeld, D.K., Höffken, K., Schmidt, C.G., Diedrichs, H.: Chromosome abnormalities in angioimmunoblastic lymphadenopathy. Lancet i, 198 (1976).

Hossfeld, D.K., Höpping, W., Vogel, M.: Chromosomal patterns in retinoblastomas. In S. Armendares, R. Lisker (eds.): V International Congress of Human Genetics, pp. 130–131, Abstract 338. Held in Mexico, Oct. 10–15. Amsterdam: Excerpta Medica (1976).

Hossfeld, D.K., Holland, J.F., Cooper, R.G., Ellison, R.R.: Chromosome studies in acute leukemias developing in patients with multiple myeloma. Cancer Res. 35, 2808–2813 (1975).

Hossfeld, D.K., Schmidt, C.G., Sandberg, A.A.: Die 'F'-Chromosomenanomalie in Erkrankungen des erythropoetischen Systems. Verh. Dtsch. Ges. Inn. Med. 78, 126–129 (1972a).

Hossfeld, D.K., Schmidt, C.G., Sandberg, A.A.: Chromosomenanalysen an Knochenmark- und Milzgewebe von Patienten mit chronisch myeloischer Leukämie in Blastenphase. In R. Gross, J. van de Loo (eds.): Leukämie (Verh. 15. Kongr. Dtsch. Ges. Hämat., Köln, 1971), pp. 89–92. Berlin, Heidelberg, New York: Springer (1972b).

Hossfeld, D.K., Tormey, D., Ellison, R.R.: Ph¹-positive megakaryoblastic leukemia. Cancer 36, 576–581 (1975).

Housset, E., Emerit, I., Baulon, A., Grouchy, J.de: Anomalies chromosomiques dans le sclérodermie généralisée. Une étude de dix malades. C.R. Acad. Sci. [D] (Paris) 269, 413–416 (1969).

Houston, E.W., Hoshino, T., Kawasaki, S., Nahayama, S.: Chromosome abnormality and its sig-

nificance in human multiple myeloma. Acta Haematol. Jap. 33, 54 – 66 (1970).
Houston, E.W., Levin, W.C., Ritzmann, S.E.: Endoreduplication in untreated early leukaemia. Lancet ii, 496 – 497 (1964a).
Houston, E.W., Levin, W.C., Ritzmann, S.E.: Untreated chronic myelocytic leukemia associated with an unusual chromosome pattern. Ann. Intern. Med. 61, 696 – 702 (1964b).
Houston, E.W., Ritzmann, S.E., Levin, W.C.: Chromosomal aberrations common to three types of monoclonal gammopathies. Blood 29, 214 – 232 (1967).
Howard, R.O., Breg, W.R., Albert, D.M., Lesser, R.L.: Retinoblastoma and chromosome abnormality. Partial deletion of the long arm of chromosome 13. Arch. Ophthalmol. 92, 490 – 493 (1974).
Howell, W.M., Denton, T.E.: An ammoniacal-silver stain technique specific for satellite III DNA regions on human chromosomes. Experientia 30, 1364 – 1366 (1974).
Howell, W.M., Denton, T.E.: Negative silver staining in A-T and satellite DNA-rich regions of human chromosomes. Chromosoma 57, 165 – 169 (1976).
Howell, W.M., Denton, T.E., Diamond, J.R.: Differential staining of the satellite regions of human acrocentric chromosomes. Experientia 31, 260 – 262 (1975).
Hsu, L.Y.F., Alter, A., Hirschhorn, K.: Trisomy 8 in bone marrow cells of patients with polycythemia vera and myelogenous leukemia. Clin. Genet. 6, 258 – 264 (1974).
Hsu, L.Y.F., Papenhausen, P., Greenberg, M.L., Hirschhorn, K.: Trisomy D in bone marrow cells in a patient with chronic myelogenous leukemia. Acta Haematol. (Basel) 52, 61 – 64 (1974).
Hsu, L.Y.F., Pinchiaroli, D., Gilbert, H.S., Wittman, R., Hirschhorn, K.: Partial trisomy of the long arm of chromosome 1 in myelofibrosis and polycythemia vera. Am. J. Hematol. 2, 375 – 383 (1977).
Hsu, T.C.: Mammalian chromosomes *in vitro*. I. The karyotype of man. J. Hered. 43, 167 – 172 (1952).
Hsu, T.C.: Mammalian chromosomes *in vitro*. IV. Some human neoplasms. J. Natl. Cancer Inst. 14, 905 – 917 (1954).
Hsu, T.C.: Chromosomal evolution in cell population. Int. Rev. Cytol. 12, 69 – 161 (1961).
Hsu, T.C.: Longitudinal differentiation of chromosomes. Annu. Rev. Genet. 7, 153 – 176 (1974).
Hsu, T.C., Pathak, S.: Differential rates of sister chromatid exchanges between euchromatin and heterochromatin. Chromosoma 258, 269 – 273 (1976).
Hsu, T.C., Pomerat, C.M.: Mammalian chromosomes *in vitro*. III. Somatic aneuploidy. J. Morphol. 93, 301 – 329 (1953).
Hsu, T.C., Somers, C.E.: Effect of 5-bromodeoxyuridine on mammalian chromosomes. Proc. Natl. Acad. Sci. USA 47, 396 – 403 (1961).
Hsu, T.C., Collie, C.J., Lusby, A.F., Johnston, D.A.: Cytogenetic assays of chemical clastogens using mammalian cells in culture. Mutat. Res. 45, 233 – 247 (1977).
Hsu, T.C., Pathak, S., Cailleau, R., Cowles, S.R.: Nature of nuclear projections in an adenocarcinoma of the breast. Lancet ii, 413 – 414 (1974).
Hsu, T.C., Pathak, S., Chen, T.R.: The possibility of latent centromeres and a proposed nomenclature system for total chromosome and whole arm translocations. Cytogenet. Cell Genet. 15, 41 – 49 (1975).
Hsu, T.C., Pathak, S., Shafer, D.A.: Induction of chromosome crossbanding by treating cells with chemical agents before fixation. Exp. Cell Res. 79, 484 – 487 (1973).
Huang, C.C., Banerjee, A., Hou, Y.: Chromosomal instability in cell lines derived from patients with xeroderma pigmentosum. Proc. Soc. Exp. Biol. Med. 148, 1244 – 1248 (1975).
Huang, C.C., Banerjee, A., Tan, J.C., Hou, Y.: Comparison of radiosensitivity between human hematopoietic cell lines derived from patients with Down's syndrome and from normal persons. J. Natl. Cancer Inst. 59, 33 – 36 (1977).
Huang, C.C., Hou, Y., Woods, L.K., Moore, G.E., Minowada, J.: Cytogenetic study of human lymphoid T-cell lines derived from lymphocytic leukemia. J. Natl. Cancer Inst. 53, 655 – 660 (1974).
Huang, C.C., Imamura, T., Moore, G.E.: Chromosomes and cloning efficiencies of hematopoietic cell lines derived from patients with leukemia, melanoma, myeloma and Burkitt lymphoma. J. Natl. Cancer Inst. 43, 1129 – 1146 (1969).
Huang, C.C., Minowada, J., Smith, R.T., Osunkoya, B.O.: Reevaluation of relationship between C chromosome marker and Epstein-Barr virus: chromosome and immunofluorescence analyses of 16 human hematopoietic cell lines. J. Natl. Cancer Inst. 45, 815 – 829 (1970).
Huang, C.S., Gomez, G.A., Kohno, S., Sokal, J.E., Sandberg, A.A.: Chromosomes and causation of human cancer and leukemia. XXXIV. "Hypereosinophilic syndrome" with unusual cytogenetic findings terminating in blastic transformation and CNS leukemia. Cancer (in press) (1979).
Hubbell, H.R., Hsu, T.C.: Identification of nucleolus organizer regions (NORs) in normal and neoplastic human cells by the silver-staining technique. Cytogenet. Cell Genet. 19, 185 – 196 (1977).
Huber, C., Huber, H., Schmalzl, F., Braunsteiner,

H.: Decreased proliferative activity of erythroblasts in granulocytic stem cell leukaemia. Nature 29, 113–114 (1971).

Huberman, J.A.: Structure of chromosome fibers and chromosomes. Annu. Rev. Biochem. 42, 355–378 (1973).

Hübner, K.F., Littlefield, L.G.: Burkitt lymphoma in three American children: Clinical and cytogenetic observations. Am. J. Dis. Child. 129, 1219–1223 (1975).

Hughes, A.F.W.: Some effects of abnormal tonicity on dividing cells in chick tissue cultures. Q. J. Microsc. Sci. 93, 207–219 (1952).

Hughes, D.T.: The role of chromosomes in the characterisation of human neoplasms. Eur. J. Cancer 1, 233–243 (1965).

Hughes, D.T.: Cytogenetical polymorphism and evolution in mammalian somatic cell populations *in vivo* and *in vitro*. Nature 217, 518–523 (1968).

Hughes, E.C., Csermely, T.V.: Chromosome constitution of human endometrium. Am. J. Obstet. Gynecol. 93, 777–792 (1965); Nature 209, 326 (1966).

Huhn, D., Rodt, H., Thiel, E., Grosse-Wilde, H., Fink, U., Theml, H., Jäger, G., Steidle, C., Thierfelder, S.: T-Zell-Leukämien des Erwachsenen. Blut 33, 141–160 (1976).

Hultén, M., Weerd-Kastelein, E.A. de, Bootsma, D., Solari, A.J., Skakkebaek, N.E., Swanbeck, G.: Normal chiasma formation in a male with xeroderma pigmentosum. Hereditas 78, 117–124 (1974).

Humbert, J.R., Hathaway, W.E., Robinson, A., Peakman, D.C., Githens, J.H.: Pre-leukemia in children with a missing bone marrow C chromosome and a myeloproliferative disorder. Br. J. Haematol. 21, 705–716 (1971).

Humbert, J.R., Morse, H.G., Hutter, J.J., Jr., Rose, B., Robinson, A.: Non-leukemic dividing cells in the blood of leukemic patients. Am. J. Hematol. 4, 217–224 (1978).

Hungerford, D.A.: Chromosome studies in human leukemia. I. Acute leukemia in children. J. Natl. Cancer Inst. 27, 983–1011 (1961).

Hungerford, D.A.: The Philadelphia chromosome and some others. Ann. Intern. Med. 61, 789–793 (1964).

Hungerford, D.A.: Some early studies of human chromosomes, 1879–1955. Cytogenet. Cell Genet. 20, 1–11 (1978).

Hungerford, D.A., Nowell, P.C.: Chromosome studies in human leukemia. III. Acute granulocytic leukemia. J. Natl. Cancer Inst. 29, 545–565 (1962).

Hunter, W.F., Lennox, B.: The sex of teratomata. Lancet ii, 633–634 (1954).

Hurdle, A.D.F., Garson, O.M., Buist, D.G.P.: Clinical and cytogenetic studies in chronic myelomonocytic leukaemia. Br. J. Haematol. 22, 773–782 (1972).

Huskins, C.L.: The internal structure of chromosomes – A statement of opinion. Cytologia, Fujii Jub. 2, 1015–1022 (1937).

Hustinx, T.W.J., Burghouts, J.T.M., Scheres, J.M.J.C., Smits, A.P.T.: A case of AMMoL with "prototypic" 8/21 translocation and loss of the Y as probably secondary events. Cancer (in press) (1979).

Hustinx, T.W.J., Scheres, J.M.J.C., Rutten, F.J., Burghouts, J.T.M., Smits, A.P.T.: Serial bone marrow karyotyping in an AMMoL patient, before and during a drug-induced remission. *Helsinki Chromosome Conference, August 29–31,* p. 199 (abstract) (1977).

Hütteroth, T.H., Litwin, S.D., German, J.: Abnormal immune responses of Bloom's syndrome lymphocytes *in vitro*. J. Clin. Invest. 56, 1–7 (1975).

Ichinoe, K.: Chromosomal abnormalities in gynecologic neoplasms. The invited lecture on the theme commissioned by Japanese Obstetrical and Gynecological Society (in Japanese). J. Jap. Obstet. Gynecol. Soc. 22, 843–852 (1970).

Iinuma, K., Nakagome, Y.: Y-chromatin in aged males. Jap. J. Hum. Genet. 17, 57–61 (1972).

Iinuma, K., Nakagome, Y.: Fluorescence of Barr body in human amniotic-fluid cells. Lancet i, 436–437 (1973).

Ikeuchi, T., Honda, T.: Cytologic studies of tumors. 48. Chromosomes of nine primary rat hepatomas induced by administration of 3'-methyl-4-dimethylaminoazobenzene. Cytologia (Tokyo) 36, 173–182 (1971).

Ikeuchi, T., Kawasaki, T.: Spontaneous chromosome breakages in leucocytes from patients with hereditary spinocerebellar ataxia. Chrom. Inform. Serv. 14, 9–11 (1973).

Ikeuchi, T., Minowada, J., Sandberg, A.A.: Chromosomal variability in ten cloned sublines of newly established Burkitt's lymphoma cell line. Cancer 28, 499–512 (1971).

Ikeuchi, T., Sonta, S., Sasaki, M., Hujita, M., Tsunematsu, K.: Chromosomal banding patterns in an infant with 13q– syndrome. Humangenetik 21, 309–314 (1974).

Ikushima, T.: Role of sister chromatid exchanges in chromatid aberration formation. Nature 268, 235–236 (1977).

Ilbery, P.L.T., Ahmad, A.: An extra small acrocentric chromosome in a case of acute monocytic leukaemia. Med. J. Aust. 2, 330–332 (1965).

Ilbery, P.L.T., Louer, C.S.: Ph^1 chromosome in the differential diagnosis of a case of ascites. Australas. Radiol. 10, 135–138 (1966).

Inoue, S., Ravindranath, Y., Ottenbreit, M.J.,

Thompson, R.I., Zuelzer, W.W.: Chromosomal analysis of metastatic retinoblastoma cells. Humangenetik 25, 111–118 (1974).

Inoue, S., Ravindranath, Y., Thompson, R.I., Zuelzer, W.W., Ottenbreit, M.J.: Cytogenetics of juvenile type chronic granulocytic leukemia. Cancer 39, 2017–2024 (1977).

Inoue, S., Ravindranath, Y., Zuelzer, W.W.: Cytogenetic analysis of erythroleukemia in two children – Evidence of nonmalignant nature of erythron. Scand. J. Haematol. 14, 129–139 (1975).

Introzzi, P., Buscarini, L.: Contributo allo studio malattie eritremica cronica. Haematologica (Pavia) 5, 1 (1966).

Inui, N.: Histological and chromosomal studies in two human gastric carcinomas and in their metastatic lesions. Jap. J. Genet. 41, 115–120 (1966).

Iriki, S.: A preliminary report on the human chromosomes. Zool. Mag. (Tokyo) 48, 184 (1936).

ISCN (1978): An International System for Human Cytogenetic Nomenclature (1978). Birth Defects: Original Article Series, Vol. XIV, No. 8 (The National Foundation, New York, 1978).

Ishida, T., Tagatz, G.E., Okagaki, T.: Gonadoblastoma: Ultrastructural evidence for testicular origin. Cancer 37, 1770–1781 (1976).

Ishihara, T. (1958) cf. Wakabayashi, M.

Ishihara, T.: Cytological studies of tumors. XXXI. A chromosome study in a human gastric carcinoma. Gann 50, 403–408 (1959).

Ishihara, T., Kumatori, T.: Chromosome aberrations in human leukocytes irradiated *in vivo* and *in vitro*. Acta Haematol. Jap. 28, 291–307 (1965).

Ishihara, T., Kumatori, T.: Chromosome studies on Japanese exposed to radiation resulting from nuclear bomb explosions. In H.J. Evans, W.M. Court Brown, A.S. McLean (eds.): *Human Radiation Cytogenetics*. (Proc. Int. Sympos., Edinburgh, 1966), pp. 144–166. Amsterdam: North-Holland (1967).

Ishihara, T., Kumatori, T.: Cytogenetic studies on fisherman exposed to fallout radiation in 1954. Jap. J. Genet. 44, 241–251 (1969).

Ishihara, T., Makino, S.: Chromosomal conditions in some human subjects with nonmalignant diseases. Tex. Rep. Biol. Med. 18, 427–437 (1960).

Ishihara, T., Sandberg, A.A.: Chromosome constitution of diploid and pseudodiploid cells in effusion of cancer patients. Cancer 16, 885–895 (1963).

Ishihara, T., Kikuchi, Y. Sandberg, A.A.: Chromosomes of twenty cancer effusions: correlation of karyotypic, clinical and pathologic aspects. J. Natl. Cancer Inst. 30, 1303–1361 (1963).

Ishihara, T., Kohno, S.-I., Hayata, I., Kumatori, T.: A nine-year cytogenetic follow-up of a patient injected with thorotrast. Hum. Genet. 42, 99–108 (1978).

Ishihara, T., Kohno, S., Hirashima, K., Kumatori, T., Sugiyama, H., Kurisu, A.: Chromosome aberrations in persons accidentally exposed to ^{192}IR gamma-rays. J. Radiat. Res. (Tokyo) 14, 328–335 (1973).

Ishihara, T., Kohno, S.-I., Kumatori, T.: Chromosome analysis in chronic myelocytic leukemia by banding techniques. Jap. J. Hum. Genet. 19, 75–76 (1974a).

Ishihara, T., Kohno, S.-I., Kumatori, T.: Ph^1 translocation involving chromosomes 21 and 22. Br. J. Cancer 29, 340–342 (1974b).

Ishihara, T., Kohno, S.-I., Minamihisamatsu, M., Kumatori, T.: Banding analysis on Ph^1 chromosome translocation: A hypothetic Ph^1 region relating to the development of CML. Natl. Inst. Radiol. Sci. Ann. Rept., p. 50, NIRS-15 (1975).

Ishihara, T., Moore, G.E., Sandberg, A.A.: Chromosome constitutions of cells in effusions of cancer patients. J. Natl. Cancer Inst. 27, 893–933 (1961).

Ishihara, T., Moore, G.E., Sandberg, A.A.: The *in vitro* chromosome constitution of cells from human tumors. Cancer Res. 22, 375–379 (1962).

Ishii, S.: Chromosome studies on human bone tumors in *in vivo* and *in vitro*. Gann 56, 251–260 (1965).

Ising, U., Levan, A.: The chromosomes of two highly malignant human tumours. Acta Pathol. Microbiol. Scand. [A] 40, 13–24 (1957).

Isurugi, K., Imao, S., Hirose, K., Aoki, H.: Seminoma in Klinefelter's syndrome with 47,XXY,15s+ karyotype. Cancer 39, 2041–2047 (1977).

Itani, S., Hoshino, T., Kawasaki, S., Nakayama, S.: Chromosome abnormality and its significance in human multiple myeloma. Acta Haematol. Jap. 33, 54–66 (1970).

Izaković, V., Šteruská, M., Vojtaššák, J.: Karyotype 45,XX,−11,2q+ of bone marrow cells in a case of di Guglielmo syndrome. Folia Haematol. (Leipz.) 104, 533 (1977).

Izuo, M., Okagaki, T., Richart, R.M., Lattes, R.: Nuclear DNA content of occult sclerosing and frank papillary carcinoma of the thyroid. Cancer 27, 902–909 (1971).

Jackson, A.W., Muldal, S., Ockey, C.H., O'Connor, P.J.: Carcinoma of male breast in association with the Klinefelter syndrome. Br. Med. J. i, 223–225 (1965).

Jackson, E.W., Norris, F.D., Klauber, M.R.: Childhood leukemia in California-born twins. Cancer 23, 913–919 (1969).

Jackson, H., Parker, F.: Hodgkin's disease. I. Gen-

eral considerations. N. Engl. J. Med. 230, 1–8 (1944a).
Jackson, H., Parker, F.: Hodgkin's disease. II. Pathology. N. Engl. J. Med. 231, 35–44 (1944b).
Jackson, J.F.: Chromosome analysis of cells in effusions from cancer patients. Cancer 20, 537–540 (1967).
Jackson, J.F., Clement, E.G.: Nuclear projections and chromosome abnormalities. Lancet ii, 1270 (1974).
Jackson, J.F., Higgins, L.C.: Group C monosomy in myelofibrosis with myeloid metaplasia. Arch. Intern. Med. 119, 403–406 (1967).
Jackson, L.G.: Chromosomes and cancer: Current aspects. Semin. Oncol. 5, 3–10 (1978).
Jackson, S.J.: Chromosome studies in hematology. Blood 18, 783 (1961).
Jacob, G.F.: Diagnosis of malignancy by chromosome counts. Lancet ii, 724 (1961).
Jacobs, E.M., Luce, J.K., Cailleau, R.: Chromosome abnormalities in human cancer. Report of a patient with chronic myelocytic leukemia and his nonleukemic monozygotic twin. Cancer 19, 869–876 (1966).
Jacobs, P.A., Court Brown, W.M.: Age and chromosomes. Nature 212, 823–824 (1966).
Jacobs, P.A., Strong, J.A.: A case of human intersexuality having a possible XXY sex-determining mechanism. Nature 183, 302–303 (1959).
Jacobs, P.A., Baikie, A.G., Court Brown, W.M., Forrest, H., Roy, J.R., Stewart, J.S.S., Lennox, B.: Chromosomal sex in the syndrome of testicular feminisation. Lancet ii, 591–592 (1959a).
Jacobs, P.A., Baikie, A.G., Court Brown, W.M., MacGregor, T.N., Maclean, N., Harnden, D.G.: Evidence for the existence of the human "super female." Lancet ii, 423–425 (1959b).
Jacobs, P.A., Brunton, M., Court Brown, W.M.: Cytogenetic studies in leucocytes on the general population: Subjects of ages 65 years and more. Ann. Hum. Genet. 27, 353–365 (1964).
Jacobs, P.A., Brunton, M., Court Brown, W.M., Doll, R., Goldstein, H.: Change of human chromosome count distributions with age. Evidence for a sex difference. Nature 197, 1080–1081 (1963).
Jacobs, P.A., Buckton, K.E., Cunningham, C., Newton, M.: An analysis of the break points of structural rearrangements in man. J. Med. Genet. 11, 50–64 (1974).
Jacobs, P.A., Court Brown, W.M., Doll, R.: Distribution of human chromosome counts in relation to age. Nature 191, 1178–1180 (1961).
Jacobs, P.A., Hassold, T.J., Matsuyama, A.M., Newlands, I.M.: Chromosome constitution of gestational trophoblastic disease. Lancet ii, 49 (1978).
Jacobs, P.A., Mayer, M., Morton, N.E.: Acrocentric chromosome associations in man. Am. J. Hum. Genet. 28, 567–576 (1976).
Jacobs, P.A., Tough, I.M., Wright, D.H.: Cytogenetic studies in Burkitt's lymphoma. Lancet ii, 1144–1146 (1963).
Jacobson, R.J., Salo, A., Fialkow, P.J.: Agnogenic myeloid metaplasia: A clonal proliferation of hematopoietic stem cells with secondary myelofibrosis. Blood 51, 189–194 (1978).
Jacquillat, Cl., Belpomme, D., Weil, M., Anclerc, G., Teillet, F., Weisberger, C., Tanzer, J., Boiron, M., Bernard, J.: Les néoplasies simultanées et successives. A propos de 18 observations d'affections malignes compliquant l'évolution de la maladie de Hodgkin. Nouv. Presse Med. 2, 3089–3092 (1973).
Jacquillat, Cl., Chastang, Cl., Tanzer, J., Briére, J., Weil, M., Pereira-Neto, M., Gemon-Auclerc, M.F., Schaison, G., Domingo, A., Boiron, M., Bernard, J.: Facteurs de pronostic de la leucémie myéloïde chronique. A propos de 798 observations. Nouv. Rev. Fr. Hematol. 15, 229–240 (1975).
Jaffurs, W.J., Marlow, J.L., Turner, T.R., Khanizadeh, A.: Chromosomal aberrations correlated with the cytologic-histologic findings of carcinoma of the cervix uteri. Cancer Cytol. 10, 27–36 (1970).
Jalal, S.M., Pfeiffer, R.A., Pathak, S., Hsu, T.C.: Subdivision of the human Y-chromosome. Humangenetik 24, 59–65 (1974).
Jami, J., Aviles, D.: Malignancy of hybrids derived from in vivo fusion of tumor cells with host cells. J. Cell Biol. 70, 34a (1976).
Janossy, G., Greaves, M.F., Capellaro, D., Minowada, J., Rosenfeld, C.: Membrane antigens of leukaemic cells and lymphoid cell lines. In H. Peeters (ed.): Protides of the Biologic Fluids, pp. 591–600. Oxford, New York: Pergamon Press (1978).
Janossy, G., Greaves, M.F., Revesz, T., Lister, T.A., Roberts, M., Durrant, J., Kirk, B., Catovsky, D., Beard, M.E.J.: Blast crisis of chronic myeloid leukaemia (CML). II. Cell surface marker analysis of "lymphoid" and myeloid cases. Br. J. Haematol. 34, 179–192 (1976).
Janossy, G., Greaves, M.F., Sutherland, R., Durrant, J., Lewis, C.: Comparative analysis of membrane phenotypes in acute lymphoid leukaemia and in lymphoid blast crisis of chronic myeloid leukaemia. Leukemia Res. 1, 289–300 (1977).
Janossy, G., Roberts, M., Greaves, M.F.: Target cell in chronic myeloid leukaemia and its relationship to acute lymphoid leukaemia. Lancet ii, 1058–1061 (1976).
Janossy, G., Roberts, M., Greaves, M.F., Woodruff, R., Pippard, M., Prentice, G., Hoffbrand, A.V.: Lymphoid blast crisis in chronic myeloid leu-

kaemia and Philadelphia positive acute lymphoid leukaemia. 2nd International Symposium on Therapy of Acute Leukemias. Held in Rome, Italy, Dec. 7–10 (1977).

Janossy, G., Woodruff, R.K., Paxton, A., Greaves, M.F., Capellaro, D., Kirk, B., Innes, E.M., Eden, O.B., Lewis, C., Catovsky, D., Hoffbrand, A.V.: Membrane marker and cell separation studies in Ph¹-positive leukemia. Blood 51, 861–877 (1978).

Janson, K.L., Roberts, J.A., Varela, M.: Multiple endocrine adenomatosis: In support of the common origin theories. J. Urol. 119, 161–165 (1978).

Jarvik, L.F., Yen, F.S.: A comment to the paper. Variations in normal human chromosomes. Humangenetik 24, 337–338 (1974).

Jarvik, L.F., Yen, F.S., Moralishvili, E.M.: Chromosome examinations in aging institutionalized women. J. Gerontol. 29, 269–276 (1974).

Jarvis, J.E., Ball, G., Rickinson, A.B., Epstein, M.A.: Cytogenetic studies on human lymphoblastoid cell lines from Burkitt's lymphomas and other sources. Int. J. Cancer 14, 716–721 (1974).

Jean, P., Richer, C.-L., Murer-Orlando, M., Luu, D.H., Joncas, J.H.: Translocation 8;14 in a ataxia telangiectasia-derived cell line. Nature 277, 56–57 (1979).

Jędranzejczak, W.: Diagnostic value of Philadelphia chromosome (in Polish). Wiad. Lek. 25, 1865–1867 (1972).

Jellinghaus, W., Okada, K., Ragg, C., Gerhard, H., Schröder, F.H.: Chromosomal studies of human prostatic tumors *in vitro*. Invest. Urol. 14, 16–19 (1976).

Jenkins, D.E., Hartmann, R.C.: Paroxysmal nocturnal hemoglobinuria terminating in acute myeloblastic leukemia. Blood 33, 274–282 (1969).

Jenkins, D.E., Rivera, H.P., Coltman, C.A.: Case report: acute lymphatic leukemia followed by a clinical picture indistinguishable from chronic granulocytic leukemia in the same patient. Am. J. Med. Sci. 264, 329–333 (1972).

Jensen, K., Christensen, K.R., Jacobsen, P., Nielsen, J., Friedrich, U., Tsuboi, T.: Ring chromosome 18 and gamma-M-globulin abnormality. Lancet ii, 497–498 (1969).

Jensen, R.D., Miller, R.W.: Retinoblastoma: Epidemiologic characteristics. N. Engl. J. Med. 285, 307–311 (1971).

Jonasson, J., Gahrton, G., Lindsten, J., Simonsson-Lindemalm, C., Zech, L.: Trisomy 8 in acute myeloblastic leukemia and sideroachrestic anemia. Blood 43, 557–563 (1974).

Johnson, F.L., Hartmann, J.R., Thomas, E.D., Chard, R.L., Hersman, J.A., Buckner, C.D., Clift, R.A., Storb, R.: Marrow transplantation in treatment of children with aplastic anemia or acute leukemia. Arch. Dis. Child. 51, 403–410 (1976).

Johnston, A.W.: The chromosomes in a child with mongolism and acute leukemia. N. Engl. J. Med. 264, 591–594 (1961).

Jones, H.W., Davis, H.J., Frost, J.K., Park, I.-J., Salimi, R., Tseng, P.-Y., Woodruff, J.D.: The value of the assay of chromosomes in the diagnosis of cervical neoplasia. Am. J. Obstet. Gynecol. 102, 624–640 (1968).

Jones, H.W., Katayama, K.P., Stafl, A., Davis, H.J.: Chromosomes of cervical atypia, carcinoma *in situ*, and epidermoid carcinoma of the cervix. Obstet. Gynecol. 30, 790–805 (1967).

Jones, H.W., Woodruff, J.D., Davis, H.J., Katayama, K.P., Salimi, R., Park, I.-J., Tseng, P.-Y., Preston, E.: The evolution of chromosomal aneuploidy in cervical atypia, carcinoma *in situ* and invasive carcinoma of the uterine cervix. Johns Hopkins Med. J. 127, 125–135 (1970).

Jones, K.W.: Chromosomal and nuclear location of mouse satellites DNA in individual cells. Nature 225, 912–915 (1970).

Jones, K.W.: Satellite DNA. J. Med. Genet. 10, 273–281 (1973).

Jones, K.W.: Chromosomes and malignancy. Nature 252, 525 (1974).

Jones, K.W., Corneo, G.: Location of satellite and homogeneous DNA sequences on human chromosomes. Nature [New Biol.] 233, 268–271 (1971).

Jones, K.W., Prosser, J.: The chromosomal location of human satellite DNA III. Chromosoma 42, 445–451 (1973).

Jones Cruciger, Q.V., Pathak, S., Cailleau, R.: Human breast carcinomas: Marker chromosomes involving lq in seven cases. Cytogenet. Cell Genet. 17, 231–235 (1976).

Joseph, R.R., Belpomme, D.: T and B lymphocytes in spleen in Hodgkin's disease. Lancet i, 747–748 (1975).

Joseph, R.R., Zarafonetis, C.J.D., Durant, J.R.: "Lymphoma" in chronic granulocytic leukemia. Am. J. Med. Sci. 251, 417–427 (1966).

Josso, N., Nezelof, C., Picon, R., Grouchy, J. de, Dray, F., Rappaport, R.: Gonadoblastoma in gonadal dysgenesis. A report of two cases with 46,XY/45,X mosaicism. J. Pediatr. 74, 425–437 (1969).

Jotterand-Bellomo, M.: One case of a Ph¹ chromosome resulting from translocation of the distal end of 22q onto the short arm of chromosome 15. Cytogenet. Cell Genet. 21, 168–169 (1978).

Juberg, R.C., Jones, B.: The Christchurch chromosome (Gp−), mongolism, erythroleukemia and an inherited Gp− chromosome (Christchurch). N. Engl. J. Med. 282, 292–297 (1970).

Jung, F., Blatnik, D., Jung, M.: A cytogenetic study on the therapeutic effect of myleran in a case of chronic myeloid leukaemia. Acta Med. Iugosl. 17, 321–335 (1963).

Jung, M., Blatnik, D., Jung, F.: Determination of sensitivity against chemotherapeutics in effusion cells of cancer patients. Hum. Chrom. Newsl. 13, 27–28 (1964a).

Jung, M., Blatnik, D., Jung, F.: Sensitivity testing of malignant cells in pleural effusions of cancer patients during chemotherapeutic treatment. Lijec. Vjesn. 86, 807 (1964b).

Kaback, M.M., Saksela, E., Mellman, W.J.: The effect of 5-bromodeoxyuridine on human chromosomes. Exp. Cell Res. 34, 182–212 (1964).

Kaffe, S., Hsu, L.Y.F., Hirschhorn, K.: Acquired trisomies 12 and 7. Lancet i, 261–262 (1974).

Kaffe, S., Hsu, L.Y.F., Hoffman, R., Hirschhorn, K.: Association of 5q− and refractory anemia. Am. J. Hematol. 4, 269–272 (1978).

Kahn, H.: The evolution of the Ph^1 trisomic clone in a case of chronic myeloid leukemia. Humangenetik 24, 207–212 (1974).

Kaiser-McCaw, B., Latt, S.A.: X-chromosome replication in parthenogenic benign ovarian teratomas. Hum. Genet. 38, 163–168 (1977).

Kaiser-McCaw, B., Epstein, A.L., Kaplan, H.S., Hecht, F.: Chromosome 14 translocation in African and North American Burkitt's lymphoma. Int. J. Cancer 19, 482–486 (1977a).

Kaiser-McCaw, B., Epstein, A.L., Overton, K.M., Kaplan, H.S., Hecht, F.: The cytogenetics of human lymphomas: Chromosome 14 in Burkitt's, diffuse histiocytic and related neoplasms. Chromosomes Today 6, 383–390 (1977b).

Kajii, T., Ohama, K.: Androgenetic origin of hydatidiform mole. Nature 268, 633–634 (1977).

Kajii, T., Neu, R.L., Gardner, L.I.: Chromosome abnormalities in lymph node cells from patient with familial lymphoma. Loss of No. 3 chromosome and presence of large submetacentric chromosome in reticulum cell sarcoma tissue. Cancer 22, 218–224 (1968).

Kakati, S., Sandberg, A.A.: Chromosomes in solid tumors. Virch. Arch. B Cell Pathol. 29, 129–137 (1978).

Kakati, S., Sinha, A.K.: Induction of distinctive chromosomal bands in selected human subjects with D, G and Y chromosome anomalies. Hum. Hered. 23, 313–330 (1973).

Kakati, S., Abe, S., Sandberg, A.A.: Sister chromatid exchange (SCE) in Philadelphia-positive leukemia. Cancer Res. 38, 2918–2921 (1978).

Kakati, S., Barcos, M., Sandberg, A.A.: Chromosomes and causation of human cancer and leukemia. XXXVI. 14q+ anomaly in an American Burkitt lymphoma and its value in the definition of lymphoproliferative disorders. Med. Pediatr. Oncol. 6, 121–129 (1979).

Kakati, S., Hayata, I., Oshimura, M., Sandberg, A.A.: Chromosomes and causation of human cancer and leukemia. X. Banding patterns in cancerous effusions. Cancer 36, 1729–1738 (1975).

Kakati, S., Hayata, I., Sandberg, A.A.: Chromosomes and causation of human cancer and leukemia. XIV. Origin of a large number of markers in a cancer. Cancer 37, 776–782 (1976).

Kakati, S., Oshimura, M., Sandberg, A.A.: The chromosomes and causation of human cancer and leukemia. XIX. Common markers in various tumors. Cancer 38, 770–777 (1976).

Kakati, S., Song, S.Y., Sandberg, A.A.: Chromosomes and causation of human cancer and leukemia. XXII. Karyotypic changes in malignant melanoma. Cancer 40, 1173–1181 (1977).

Kalden, J.R., Peter, H.H., Odriozola, J., Richter, W., Richter, R.: Sézary syndrome. Lancet i, 688 (1974).

Kallenberger, A.: Geschlechts Chromatin bei Mammakarzinom. Schweiz. Med. Wochenschr. 94, 1450–1458 (1964).

Kallenberger, A.: Chromosomal anomalies in benign and premalignant breast lesions. *Proceedings, 5th International Symposium on Biological Characterization of Human Tumors, Bologna, April 4–6, 1973*, i, 171–179 (1974).

Kallenberger, A., Hagmann, A., Descoeudres, C.: The interpretation of abnormal sex chromatin incidence in human breast tumors on the basis of DNA measurements. Eur. J. Cancer 3, 439–448 (1968).

Kallenberger, A., Hagmann, A., Meier-Ruge, W., Descoeudres, C.: Beziehungen zwischen Sexchromatinvorkommen, Kerngrösse und DNS-Werten in Mammatumoren und ihre Bedeutung für die Überlebenszeit. Schweiz. Med. Wochenschr. 97, 678–682 (1967).

Kamada, N.: The effects of radiation on chromosomes of bone marrow cells. III. Cytogenetic studies on leukemia in atomic bomb survivors. Acta Haematol. Jap. 32, 249–274 (1969a).

Kamada, N.: The effects of radiation on chromosomes of bone marrow cells. II. Studies on bone marrow chromosomes of atomic bomb survivors in Hiroshima. Acta Haematol. Jap. 32, 236–248 (1969b).

Kamada, N., Uchino, H.: Double Ph^1 chromosomes in leukaemia. Lancet i, 1107 (1967).

Kamada, N., Uchino, H.: Preleukemic states in atomic bomb survivors in Japan. Blood Cells 2, 57–65 (1976).

Kamada, N., Uchino, H.: Chronologic sequence in appearance of clinical and laboratory findings characteristic of chronic myelocytic leukemia. Blood 51, 843–850 (1978).

Kamada, N., Kuramoto, A., Katsuki, T., Hinuma, T.: Chromosome aberrations of B lymphocytes of atomic bomb survivors. Blood 53, 1140–1147 (1979).

Kamada, N., Oguma, N., Mikami, M., Tanaka, R., Ishii, Y., Uchino, H.: Serial cytogenetic studies in acute leukemia. Jap. J. Med. 15, 18–25 (1976).

Kamada, N., Okada, K., Ito, T., Nakatsui, T., Tomonaga, M.: Chromosome aberrations and neutrophil alkaline phosphatase in forty-three cases of leukemia, including fourteen cases of leukemia in atomic bomb survivors. J. Kyushu Hematol. Soc. 17, 115–142 (1967).

Kamada, N., Okada, K., Ito, T., Nakatsui, T., Uchino, H.: Chromosomes 21–22 and neutrophil alkaline phosphatase in leukaemia. Lancet i, 364 (1968).

Kamada, N., Okada, K., Oguma, N., Tanaka, R., Uchino, H.: C-G translocation in acute myelocytic leukemia with low neutrophil alkaline phosphatase activity. Cancer 37, 2380–2387 (1976).

Kamarov, E.: Chromosome aberrations as a biological indicator of the effects of radiation and other environmental hazards. WHO Chron. 27, 463–465 (1973).

Kaminetzky, H.A., Jagiello, G.M.: Differential chromosomal effects of carcinogenic and noncarcinogenic substances. Am. J. Obstet. Gynecol. 98, 349–355 (1967).

Kamiyama, R., Shibata, T., Mori, W.: Two autopsy cases of atypical myeloproliferative disorder with group C monosomy occurring in siblings. Acta Pathol. Jap. 23, 815–835 (1973).

Kanda, N.: Banding pattern observed in human chromosomes by the modified BSG technique. Hum. Genet. 31, 283–292 (1976).

Kaneko, Y., Sakurai, M.: 15/17 translocation in promyelocytic leukaemia. Lancet i, 961 (1977).

Kaneko, Y., Sakurai, M., Hattori, M.: A case of acute myelogenous leukemia with an 8-21 translocation, missing Y, and additional karyotypic abnormalities. Am. J. Hematol. 4, 273–280 (1978a).

Kaneko, Y., Sakurai, M., Hattori, M.: Childhood acute myelogenous leukemia with an 8–21 chromosome translocation. J. Pediatr. 93, 1066–1067 (1978b).

Kang, Y.S., Kim, S.W., Lee, C.K.: Chromosome analysis of cultured uterine carcinoma. J. Korean Cancer Res. Assoc. 3, (1968).

Kang, Y.S., Kim, S.W., Park, E.H.: On the chromosome distribution of uterine carcinoma in culture. Korean J. Zool. 15, 15–24 (1972).

Kanzaki, T., Hashimoto, K., Bath, D.: Human melanoma: Biologic and ultrastructural studies in vivo. J. Invest. Dermatol. 66, 280 (1976).

Kanzaki, T., Hashimoto, K., Bath, D.W.: Human malignant melanoma in vivo and in vitro. J. Natl. Cancer Inst. 59, 775–785 (1977).

Kanzow, U., Lange, B., Niederlat, G., Gropp, A.: Chromosomenuntersuchungen bei Paraproteinämien. Klin. Wochenschr. 45, 1076–1084 (1967).

Kaplan, J.C., Zamansky, G.B., Black, P.H.: Parallel induction of sister chromatid exchanges and infectious virus from SV40-transformed cells by alkylating agents. Nature 271, 662–663 (1978).

Kaplow, L.S.: Leukocyte alkaline phosphatase cytochemistry: applications and methods. Ann. N.Y. Acad. Sci. 155, 911–947 (1968).

Kardinal, C.G., Bateman, J.R., Weiner, J.: Chronic granulocytic leukemia. Review of 536 cases. Arch. Intern. Med. 136, 305–313 (1976).

Kariminejad, M.H., Movlavi, M.A., Nasserghodssi, M.A., Gharoorzadeh, D., Behjatnia, Y.: Gonadoblastoma associated with mixed gonadal dysgenesis. Am. J. Obstet. Gynecol. 113, 410–414 (1972).

Karpas, A., Khalid, G., Burns, G.F., Hayhoe, F.G.J.: Continuous culture of malignant haemic cells from human acute myelomonocytic leukaemia: Cytological, cytochemical, cytogenetic and immunological studies. Br. J. Cancer 37, 308–315 (1978).

Katayama, K.P.: Chromosomes and pelvic cancer. Clin. Obstet. Gynecol. 12, 435–458 (1969).

Katayama, K.P., Jones, H.W.: Chromosomes of atypical (adenomatous) hyperplasia and carcinoma of the endometrium. Am. J. Obstet. Gynecol. 97, 978–983 (1967).

Katayama, K.P., Jones, H.W., Jr.: The chromosomes of normal and hyperplastic endometrium. Johns Hopkins Med. J. 122, 84–86 (1968).

Katayama, K.P., Masukawa, T.: Ring chromosomes in a breast cancer. Acta Cytol. 12, 159–161 (1968).

Katayama, K.P., Toews, H.A.: Chromosomes of metastatic ovarian carcinoma treated with a progestogen and alkylating agents. Am. J. Obstet. Gynecol. 104, 997–1003 (1969).

Katayama, K.P., Woodruff, J.D., Jones, H.W., Jr., Preston, E.: Chromosomes of condyloma acuminatum, Paget's disease, *in situ* carcinoma, invasive squamous cell carcinoma, and malignant melanoma of the human vulva. Obstet. Gynecol. 39, 346–356 (1972).

Kato, H.: Induction of sister chromatid exchanges by chemical mutations and its possible relevance to DNA repair. Exp. Cell Res. 85, 239–247 (1974a).

Kato, H.: Spontaneous sister chromatid exchanges detected by a BUdR-labelling method. Nature 251, 70–72 (1974b).

Kato, H.: Possible role of DNA synthesis in formation of sister chromatid exchanges. Nature 252, 739–741 (1974c).
Kato, H.: Spontaneous and induced sister chromatid exchanges as revealed by BUdR-labeling method. In G.H. Bourne, J.F. Danielli, K.W. Jeon (eds.): International Review of Cytology, Vol. 49, pp. 55–97. New York, San Francisco, London: Academic Press (1977a).
Kato, H.: Mechanisms for sister chromatid exchanges and their relation to the production of chromosomal aberrations. Chromosoma 59, 179–191 (1977b).
Kato, H., Moriwaki, K.: Factors involved in the production of banded structures in mammalian chromosomes. Chromosoma 38, 105–120 (1972).
Kato, H., Sandberg, A.A.: Chromosome pulverization in human binucleate cells following colcemid treatment. J. Cell Biol. 34, 35–45 (1967).
Kato, H., Sandberg, A.A.: Cellular phase of chromosome pulverization induced by Sendai virus. J. Natl. Cancer Inst. 41, 1125–1131 (1968).
Kato, H., Sandberg, A.A.: Effects of herpes simplex virus on sister chromatid exchange and chromosome abnormalities in human diploid fibroblasts. Exp. Cell Res. 109, 423–427 (1977a).
Kato, H., Sandberg, A.A.: The effect of sera on sister chromatid exchanges in vitro. Exp. Cell Res. 109, 445–448 (1977b).
Kato, H., Stich, H.F.: Sister chromatid exchanges in ageing and repair-deficient human fibroblasts. Nature 260, 447–448 (1976).
Kato, H., Yosida, T.H.: Banding patterns of Chinese hamster chromosomes revealed by new techniques. Chromosoma 36, 272–280 (1972).
Kato, R.: The chromosomes of forty-two primary Rous sarcomas of the Chinese hamster. Hereditas 59, 63–119 (1968).
Kato, R., Levan, A.: Chromosomal aberrations in carcinogenesis. International Symposium on Genetic Effects of Radiation and Radiometric Chemistry, Kyoto. Held in August (1968).
Kauer, G.L., Jr., Engle, R.L., Jr.: Eosinophilic leukaemia with Ph¹-positive cells. Lancet ii, 1340 (1964).
Kaufmann, B.P., Gay, H., McDonald, M.R.: Organizational patterns within chromosomes. Int. Rev. Cytol. 9, 77–127 (1960).
Kaufmann, R.W., Schechter, G.P., McFarland, W.: Paroxysmal nocturnal hemoglobinuria terminating in acute granulocytic leukemia. Blood 33, 287–291 (1969).
Kaufmann, U., Löffler, H., Foerster, W., Desaga, J.F., Koch, F.: Fehlendes Chromosom NR. 7 in der präleukämischen Phase einer Myeloblastenleukose bei einem Kind. Blut 29, 50–61 (1974).

Kaung, D.T., Swartzendruber, A.: Effect of chemotherapeutic agents on chromosomes of patients with lung cancer. Dis. Chest 55, 98–100 (1966).
Kaur, J., Catovsky, D., Valdimarsson, H., Jensson, O., Spiers, A.S.D.: Familial acute myeloid leukaemia with acquired Pelger-Huët anomaly and aneuploidy of C group. Br. Med. J. iv, 327–331 (1972).
Kaur, J., Spiers, A.S.D., Galton, D.A.G.: T and B lymphocytes in spleens of patients with malignant diseases. Lancet i, 747 (1975).
Kawamura, N.: Chromosome studies on mouse leukemias. Acta Haematol. Jap. 28, 195–205 (1965).
Kawasaki, M., Alteration of chromosomes in the peripheral leukocytes by administration of antineoplastic drugs to patients with gyneco-obstetric tumors. J. Jap. Obstet. Gynecol. Soc. 20, 413–422 (1968a).
Kawasaki, M.: Chromosomal studies of trophoblastic tumors. J. Jap. Obstet. Gynecol. Soc. 20, 699–707 (1968b).
Kawasaki, M.: Peripheral leukocyte chromosome anomaly in cases of irradiation therapy after radical surgery of uterine cancer. J. Jap. Obstet. Gynecol. Soc. 20, 491–499 (1968c).
Kay, H.E.M., Lawler, S.D., Millard, R.E.: The chromosomes in polycythaemia vera. Br. J. Haematol. 12, 507–527 (1966).
Kay, H.E.M., Millard, R.E., Lawler, S.D.: Aneuploidy of C-group chromosomes in chronic myeloproliferative disorders In: J.L. Ambrus (ed.): Hematologic Reviews, Vol. 2, pp. 19–40. New York: M. Dekker (1970).
Kegel, J., Conen, P.E.: Nuclear sex identification in human tissues. A histologic study using quinacrine fluorescence. Am. J. Clin. Pathol. 57, 425–430 (1972).
Keller, R., Nordén, A.: Chromosome observations in bone marrow cells and lymphocytes of peripheral blood of patients with megaloblastic anemia. Hereditas 58, 265–283 (1967).
Keller, R., Lindstrand, K., Nordén, A.: Disappearance of chromosomal abnormalities in megaloblastic anemia after treatment. Scand. J. Haematol. 7, 478–485 (1970).
Kemp, N.H.: Cytogenetic studies in acute leukaemia. Br. Med. J. i, 48 (1961).
Kemp, N.H., Stafford, J.L., Tanner, R.K.: Acute leukaemia and Klinefelter's syndrome. Lancet ii 434–435 (1961).
Kemp, N.H., Stafford, J.L., Tanner, R.: Chromosome studies during the development of human leukaemia. In: Proceedings of the 9th Congress of the European Society of Haematology, Lisbon, pp. 33–38. Basel, New York: Karger (1963).
Kemp, N.H., Stafford, J.L., Tanner, R.: Chromosom

studies during early and terminal chronic myeloid leukaemia. Br. Med. J. i, 1010–1014 (1964).
Kemp, T.: Ueber des verhälten der Chromosomen in den somatischen Zellen des Menschen. Z. Mikrosk. Anat. Forsch. 16, 1–20 (1929).
Kemp, T.: Ueber die somatischen Mitosen bei Menschen und warmblütigen Tieren unter normalen und pathologischen Verhältnissen. Z. Zellforsch. Mikrosk. Anat. 11, 429–444 (1930).
Kenis, Y., Koulischer, L.: Étude clinique et cytogénétique de 21 patients atteints de leucémie myéloïde chronique. Eur. J. Cancer 3, 83–93 (1967).
Keogh, E.J., Kretser, D.M. de, Fitzgerald, M.G.: Isochromosome for the short arm of X with primary amenorrhoea and a pituitary tumour. Aust. N.Z. J. Med. 3, 617–619 (1973).
Kersey, J.H., Sabad, A., Vance, J.C., White, J.G., Neely, A.N.: Calcium ionophore as a mitogen: Response of normal T and B and immunodeficient lymphocytes. Clin. Res. 24, 411 (1975).
Kessous, A., Colombiès, P.: Hypodiploidy and cellular survival. Biomedicine 23, 108–111 (1975).
Kessous, A., Corberand, J., Grozdea, J., Colombiès, P.: Clone cellulaire a 27 chromosomes dans une leucémie aiguë humaine. Nouv. Rev. Fr. Hematol. 15, 73–81 (1975a).
Kessous, A., Corberand, J., Robert, A., Colombiès, P.: Caractéres cytogénétiques des leucémies aiguës de l'enfant. Lyon Med. 233, 253–260 (1975b).
Keusch, F., Rüttner, J.R., Pedio, G., Gut, D.: Occurrence of plasmocytoid cells in long-term blood-cell cultures from a patient with "hairy cell" leukemia. Exp. Cell Biol. 44, 20–27 (1976).
Keutel, J.: Cytogenetische, immunologische und cytologische Familienuntersuchungen bei Bloom-Syndrom. Humangenetik 8, 142–157 (1969).
Keutel, J., Maigheseu, I., Teller, W.: Bloom-Syndrom. Bericht über einen Fall mit dermatohistologischen, endokrinologischen und cytogenetischen Untersuchungen. Z. Kinderheilkd. 101, 165–180 (1967).
Khaleeli, M., Keane, W.M., Lee, G.R.: Sideroblastic anemia in multiple myeloma. A preleukemic change. Blood 41, 17–25 (1973).
Khan, H., Khan, M.H.: Cytogenetische Untersuchungen bei chronischer Bensolex-position. Arch. Toxicol. 31, 39–49 (1973).
Khan, M.H.: C trisomy in bone marrow cells in a case of preleukaemic acute myelogenous leukaemia. Remarks on the karyotypic analysis and chemotherapy. Humangenetik 16, 323–327 (1972a).

Khan, M.H.: Heteromorphic pair of metacentric chromosomes with fused arms and the Philadelphia chromosome in a case of acute myeloid leukemia. Acta Haematol. (Basel) 48, 312–319 (1972b).
Khan, M.H.: Acute myeloid leukemia with two Philadelphia chromosomes in forty-six stemline. Remarks on the karyotypic analysis and chemotherapy. Humangenetik 18, 55–62 (1973a).
Khan, M.H.: Forty-three chromosome-clone in chronic myeloid leukaemia. Z. Krebsforsch 77, 6–10 (1973b).
Khan, M.H.: The evolution of the Ph^1-trisomic clone, in a case of chronic myeloid leukemia. Humangenetik 24, 207–212 (1974).
Khan, M.H., Martin, H.: Myeloblastenleukämie mit Philadelphia-Chromosom. Klin. Wochenschr. 45, 821–824 (1967a).
Khan, M.H., Martin, H.: Two Ph^1 chromosomes in blastic crisis of a granulocytic leukaemia. Acta Haematol. (Basel) 38, 391–396 (1967b).
Khan, M.H., Martin, H.: G21 trisomy in a case of acute myeloblastic leukaemia. Acta Haematol. (Basel) 38, 142–146 (1967c).
Khan, M.H., Martin, H.: Multiple chromosomal aberrations in a case of malignant myelosclerosis. Acta Haematol. (Basel) 39, 299–308 (1968a).
Khan, M.H., Martin, H.: Zytogenetische Untersuchungen bei perniziöser Anämie. Blut 18, 129–141 (1968b).
Khan, M.H., Martin, H.: Presence of two Ph^1 chromosomes in cells with 49 clone from patient in blast crisis of granulocytic leukaemia. Acta Haematol. (Basel) 42, 357–360 (1969).
Khan, M.H., Martin, H.: Chromosomal aberrations in a case of erythroleukaemia. Blut 21, 29–34 (1970a).
Khan, M.H., Martin, H.: G-Trisomie bei akuter Erythroleukämie. Klin. Wochenschr. 48, 445–447 (1970b).
Khare, A.G., Bhisey, A.N., Advani, S.H., Ranadive, K.J.: Cytogenetic studies in chronic myeloid leukemia. Indian J. Med. Res. 67, 413–424 (1978).
Khishin, A.F., El-Zawahri, M.M., Hassan, A.A., Morad, M.M.: Cytogenetic studies of benign and malignant tumours of the breast. In S. Armendares, R. Lisker (eds.): *International Congress of Human Genetics,* pp. 133–134, Abstract 347. Held in Mexico, Oct. 10–15 (1976).
Khondkarian, O.A., Burak, A.I., Skolbesky, M.D.: Chromosome complex in patients with disseminated sclerosis. Vestn. Akad. Med. Nauk. SSSR 22, 85–88 (1967).
Khouri, F.P., Nassar, V.H.: Non-Hodgkin's lymphoma: A cytogenetic study. J. Med. Liban. 27, 599–604 (1974).

Khouri, F.P., Shahid, M.J., Kronfol, N.: Chromosomal pattern in the progression of chronic granulocytic leukemia. Cancer 24, 807–809 (1969a).

Khouri, F.P., Shahid, M.J., Kronfol, N.: The Philadelphia (Ph¹) chromosome and chronic granulocytic leukemia. J. Med. Liban. 22, 603–608 (1969b).

Khouri, F.P., Shahid, M.J., Yehikomshian, S.: Cytogenetic studies in acute leukemia. Acta Haematol. (Basel) 40, 192–199 (1968).

Khudr, G., Walsh, P.C., Benirschke, K.: Quinacrine fluorescence of testicular tumors. Urology 2, 162–164 (1973).

Kihlman, B.A.: Molecular mechanisms of chromosome breakage and rejoining. Adv. Cell Biol. 1, 59–107 (1971).

Kikuchi, Y., Sandberg, A.A.: Chronology and pattern of human chromosome replication. I. Blood leukocytes of normal subjects. J. Natl. Cancer Inst. 32, 1109–1143 (1964).

Kikuchi, Y., Sandberg, A.A.: Chronology and pattern of human chromosome replication. II. Autoradiographic behavior of various Y and X chromosomes. J. Natl. Cancer Inst. 34, 795–813 (1965).

Kilian, D.J., Picciano, D.J., Jacobson, C.B.: Industrial monitoring: A cytogenetic approach. Ann. N.Y. Acad. Sci. 269, 4–11 (1975).

Killmann, S.-A.: Chronic myelogenous leukemia. Preleukemia or leukemia? *In* S. Tura, M. Baccarini (eds.): *Chronic Myeloid Leukemia*, pp. 45–53. Pavia: Edizioni di Haematologica (1972); Haematologia 57, 641–649 (1972).

Killmann, S.-A.: Preleukemia: Does it exist? Blood Cells 2, 81–105 (1976).

Killmann, S.-A., Philip, P., Baccarani, M.: Rapid blastic transformation and early ectopic proliferation of hyperdiploid myeloblasts in chronic myeloid leukemia. Eur. J. Cancer 12, 763–766 (1976).

Kim, My. A.: Identification and characterization of heterochromatic regions in the human metaphase and interphase nucleus. Humangenetik 21, 331–340 (1974).

Kim, My. A.: Fluorometrical detection of thymine base differences in complementary strands of satellite DNA in human metaphase chromosomes. Humangenetik 28, 57–63 (1975).

Kim, My. A., Bier, L.: Identification and location of the Y-body in interphase by quinacrine and Giemsa. Humangenetik 16, 261–265 (1972).

Kim, My. A., Bier, L., Majewski, F., Pfeiffer, R.A.: Fluorochromierung menschlicher Chromosomen mit Atebrin-Essigsäure. Humangenetik 12, 257–260 (1971).

Kim, My. A., Johannsmann, R., Grzeschik, K.-H.: Giemsa staining of the sites replicating DNA early in human lymphocyte chromosomes. Cytogenet. Cell Genet. 15, 363–371 (1975).

Kim, S.W., Kang, Y.S., Kim, S.R.: A chromosome study on uterine carcinoma. J. Korean Cancer Res. Assoc. 2, (1967).

Kimel, V.M.: Clinical-cytological correlations of mammary carcinoma based upon sex chromatin counts. Cancer 10, 922–927 (1957).

King, M.J., Gillis, E.M., Baikie, A.G.: The polymorph alkaline phosphatase in mongolism. Lancet ii, 1302–1305 (1962).

Kinlough, M.A., Robson, H.N.: Study of chromosomes in human leukaemia by a direct method. Br. Med. J. ii, 1052–1055 (1961).

Kint, A., de Weert, J., de Smet, M.: Nosology of the Sézary syndrome. Dermatologica (Basel) 153, 277–283 (1976).

Kiossoglou, K.A., Mitus, W.J.: Chromosomal aberration in acute leukemia. Blood 22, 839–840 (1963).

Kiossoglou, K.A., Mitus, W.J.: Cytogenetic studies in pernicious anemia, megaloblastic and Di Guglielmo syndrome. Clin. Res. 12, 217 (1964).

Kiossoglou, K.A., Mitus, W.J.: Chromosomal studies in chronic myeloproliferative syndromes. Clin. Res. 13, 276 (1965).

Kiossoglou, K.A., Mitus, W.J., Dameshek, W.: A direct method for chromosome studies of human bone marrow. Am. J. Clin. Pathol. 41, 183–187 (1964).

Kiossoglou, K.A., Mitus, W.J., Dameshek, W.: Chromosomal aberrations in acute leukemia. Blood 26, 610–641 (1965a).

Kiossoglou, K.A., Mitus, W.J., Dameshek, W.: Two Ph¹ chromosomes in acute granulocytic leukaemia. A study of two cases. Lancet ii, 665–668 (1965b).

Kiossoglou, K.A., Mitus, W.J., Dameshek, W.: Chromosomal aberrations in pernicious anemia: study of three cases before and after therapy. Blood 25, 662–682 (1965c).

Kiossoglou, K.A., Mitus, W.J., Dameshek, W.: Cytogenetic studies in the chronic myeloproliferative syndrome. Blood 28, 241–252 (1966a).

Kiossoglou, K.A., Mitus, W.J., Dameshek, W.: Double Ph¹ chromosomes in leukaemia. Lancet ii, 590–591 (1966b).

Kiossoglou, K.A., Rosenbaum, E., Mitus, W.J., Dameshek, W.: Multiple chromosome aberrations in Down's syndrome associated with twinning and acute granulocytic leukaemia. Lancet ii, 944–945 (1963).

Kiossoglou, K.A., Rosenbaum, E.H., Mitus, W.J., Dameshek, W.: Multiple chromosomal aberrations in a patient with acute granulocytic leukemia associated with Down's syndrome and twinning. Study of a family with possible tendency to nondisjunction. Blood 24, 134–159 (1964a).

Kiossoglou, K.A., Rosenbaum, E., Mitus, W.J., Dameshek, W.: Multiple chromosome aberrations. Lancet i, 1066–1067 (1964b).

Kirchner, M., Franck, A., Appenroth, D.: Zytogenetische Untersuchungen nach Kontakt zu kindlichen Parablastenleukosen. Folia Haematol. (Leipz.) 103, 37–42 (1976).

Kirchner, M., Hofmann, A.: Zytogenetische Untersuchungen als Bietrag zur Diagnostik von Präleukamie und Leukämierezidiv im Kindesalter. Folia Haematol. (Leipz.) 103, 831–843 (1976).

Kirkland, D.J., Lawler, S.D., Venitt, S.: Chromosomal damage and hair dyes. Lancet ii, 124–127 (1978).

Kirkland, J.A.: Mitotic and chromosomal abnormalities in carcinoma *in situ* of the uterine cervix. Acta Cytol. 10, 80–86 (1966a).

Kirkland, J.A.: Chromosomes in uterine cancer. Lancet i, 152 (1966b).

Kirkland, J.A.: Study of chromosomes in cervical neoplasia. Obstet. Gynecol. Surv. 24, 784–794 (1969).

Kirkland, J.A., Stanley, M.A.: The cytogenetics of carcinoma of the cervix. Aust. N.Z. J. Obstet. Gynaecol. 7, 189–193 (1967).

Kirkland, J.A., Stanley, M.A.: Chromosomes of cancer cells. Nature 232, 632–633 (1971).

Kirkland, J.A., Stanley, M.A., Cellier, K.M.: Comparative study of histologic and chromosomal abnormalities in cervical neoplasia. Cancer 20, 1934–1952 (1967).

Kishimoto, H.: Studies of Japanese Burkitt's lymphoma. Part I. Establishment of a Japanese Burkitt's lymphoma cell line. Jap. J. Clin. Hematol. 19, 93–99 (1978).

Kisliak, N.S.: Acute lymphoid leukemia in children. Pediatriia 8, 17–26 (1976).

Kissling, M., Speck, B.: Chromosome aberrations in experimental benzene intoxication. Helv. Med. Acta (Basel) 36, 59–66 (1972).

Kitchin, F.D.: Genetics of retinoblastoma. In A.B. Reese (ed.): *Tumors of the Eye,* third edition, pp. 125–132. Hagarstown, Md.: Harper and Row (1976).

Kitchin, R.M., Loudenslager, E.J.: An *in vivo* Giemsa banding technique. Stain Techol. 50, 371–374 (1976).

Kitchin, R.M., Shaw, M.W.: The association pattern of the Ph[1] chromosome. Cytogenetics 10, 235–243 (1971).

Klein, E., Ben-Bassat, H., Naumann, G., Ralph, P., Zeuthen, J., Polliack, A., Vanky, F.: Properties of the K562 cell line, derived from a patient with chronic myeloid leukemia. Int. J. Cancer 18, 421–431 (1976).

Klein, E., Klein, G., Nadkarni, J.S., Nadkarni, J.H., Wigzell, H., Clifford, P.: Surface IgM kappa specificity on a Burkitt lymphoma cell in vivo and in derived culture lines. Cancer Res. 28, 1300–1310 (1968).

Klein, G.: Comparative studies of mouse tumors with respect to their capacity for growth as "ascites tumors" and their average nucleic acid content per cell. Exp. Cell Res. 2, 518–573 (1951).

Klein, G., Bregula, U., Wiener, R., Harris, H.: The analysis of malignancy by cell fusion. I. Hybrids between tumour cells and L cells derivatives. J. Cell Sci. 8, 659–672 (1971).

Klein, V.A.E.: Erste Beobachtung einer Ph[1]-positive chronischen Myelose mit cytochemisch gesicherten terminalen proerythroblasten Schulb. Haematol. Blattransfus. 8, 116–118 (1969).

Kleinschmidt, A.K., Lang, D.K.: Intrazellulare Desoxyribonucleinsaure von Bakterien. 5th Int. Congr. Electron Microsc. 2 (1962).

Klinger, H.P., Glasser, M., Kava, H.W.: Contraceptives and the conceptus. I. Chromosome abnormalities of the fetus and neonate related to maternal contraceptive history. Obstet. Gynecol. 48, 40–48 (1976).

Klinger, H.P., Ludwig, K.S., Schwarzacher, H.G., Hauser, G.A.: Neue Ergrbnisse der Sex-Chromatin-Forschung an Placenta und Eihaüten. Gynaecologia 146, 328–335 (1958).

Knight, L.A., Davidson, W.M., Cuddigan, B.J.: Acquired trisomy 9. Lancet i, 688 (1974).

Knight, L.A., Gardner, H.A., Gallie, B.L.: Segregation of chromosome 13 in retinoblastoma. Lancet i, 989 (1978).

Knoerr-Gaertner, H., Schuhmann, R., Kraus, H., Uebele-Kallhardt, B.: Comparative cytogenetic and histologic studies on early malignant transformation in mesothelial tumors of the ovary. Hum. Genet. 35, 281–297 (1977).

Knörr-Gärtner, H., Schuhmann, R., Uebele-Kallhardt, B.: Zytogenetische befunde bei Ovarialtumoren. Verh. Dtsch. Ges. Pathol. 57, 203–207 (1973).

Knospe, W.H., Gregory, S.A.: Smoldering acute leukemia. Clinical and cytogenetic studies in six patients. Arch. Intern. Med. 127, 910–918 (1971).

Knospe, W.H., Klatt, R.W., Bergin, J.W., Jacobson, C.B., Conrad, M.E.: Cytogenetic changes in chronic granulocytic leukemia during blast crisis: Two Ph[1] chromosomes and hyperdiploidy. Am. J. Med. Sci. 254, 816–823 (1967).

Knudson, A.G., Jr.: Mutation and cancer. Statistical study of retinoblastoma. Proc. Natl. Acad. Sci. USA 68, 820–823 (1971).

Knudson, A.G., Jr.: Genetic predisposition to cancer. In P. Bucalossi, U. Veronesi, N. Cascinelli (eds.): *Proceedings of the International Cancer Congress, 11th,* Vol. 4. *Cancer Campaigns, Detection, Rehabilitation, Clinical Classification,* pp. 183–187. Held in Florence, Italy, Oct. 20–26, 1974. New York: American Elsevier, (1975).

Knudson, A.G.: Genetics and the etiology of childhood cancer. Pediatr. Res. 10, 513–517 (1976).

Knudson, A.G., Jr.: Mutation and cancer in man. Cancer 39, 1882–1886 (1977).

Knudson, A.G., Jr.: Retinoblastoma: A prototypic hereditary neoplasm. Semin. Oncol. 5, 57–60 (1978).

Knudson, A.G., Meadows, A.T.: Developmental genetics of neuroblastoma. J. Natl. Cancer Inst. 57, 675–682 (1976).

Knudson, A.G., Jr., Strong, L.C.: Mutation and cancer: Neuroblastoma and pheochromocytoma. Am. J. Hum. Genet. 24, 514–532 (1972).

Knudson, A.G., Meadows, A.T., Nichols, W.W., Hill, R.: Chromosomal deletion and retinoblastoma. N. Engl. J. Med. 295, 1120–1122 (1976).

Knudson, A.G., Jr., Strong, L.C., Anderson, D.E.: Heredity and cancer in man. In A.G. Steinberg, A.G. Bearn (eds.): *Progress in Medicine,* Vol. 9, pp. 113–158. New York: Grune and Stratton (1973).

Knudtzon, S.: In vitro culture of leukaemic cells from 81 patients with acute leukaemia. Scand. J. Haematol. 18, 377–394 (1977).

Knuutila, S.: Polyploid mitoses in human bone-marrow cells. Hereditas 82, 263–265 (1976).

Knuutila, S., Helminen, E., Vuopio, P., de la Chapelle, A.: Sister chromatid exchanges in human bone marrow cells. I. Control subjects and patients with leukaemia. Hereditas 88, 189–196 (1978).

Knuutila, S., Laasonen, M., Siimes, M., Simell, O.: Chromosomal abnormalities in bone-marrow cells and in cultured lymphocytes in a variety of human viral infections. Hereditas 85, 250–254 (1977).

Knuutila, S., Mäki-Paakkanen, J., Kähkönen, M., Hokkanen, E.: An increased frequency of chromosomal changes and SCE's in cultured blood lymphocytes of 12 subjects vaccinated against smallpox. Hum. Genet. 41, 89–96 (1978).

Knuutila, S., Simell, O., Lipponen, P., Saarinen, I.: Bone-marrow chromosomes in healthy subjects. Hereditas 82, 29–36 (1976).

Kodani, M.: Three chromosome numbers in whites and Japanese. Science 127, 1339–1340 (1958).

Koeffler, H.P., Golde, D.W.: Cellular maturation in human preleukemia. Blood 52, 355–361 (1978).

Kohler, P.O., Bridson, W.E., Hammond, J.M., Weintraub, B., Kirschner, M.A., van Thiel, D.H.: Clonal lines of human choriocarcinoma cells in culture. *Karolinska Symposia on Research Methods in Reproductive Endocrinology, 3rd Symposium. In Vitro Methods in Reproductive Cell Biology,* Jan. 25–27, pp. 137–153 (1971).

Kohn, G., Manny, N., Eldor, A., Cohen, M.M.: De novo appearance of the Ph[1] chromosome in a previously monosomic bone marrow (45,XX,–6). Conversion of a myeloproliferative disorder to acute myelogenous leukemia. Blood 45, 653–657 (1975).

Kohn, G., Mellmann, W.J., Moorhead, P.S., Loftus, J., Henle, G.: Involvement of C group chromosomes in five Burkitt lymphoma cell lines. J. Natl. Cancer Inst. 38, 209–222 (1967).

Kohno, S., Sandberg, A.A.: Ph[1]-positive CML in a 13;14 translocation carrier. Med. Pediatr. Oncol. 5, 61–64 (1978).

Kohno, S., Abe, S., Matsui, S., Sandberg, A.A.: Chromosomes and causation of human cancer and leukemia. XXXVII. Nucleolus organizers on the Ph[1] chromosome in CML. Cancer Genet. Cytogenet. (in press) (1979).

Kohno, S., Minowada, J., Sandberg, A.A.: Chromosomal evolution of near-haploid cell clones in an established ALL cell line (NALM 16). Program of the American Society on Human Genetics, 29th Annual Meeting, Vancouver, B.C., Canada, Oct. 4–7, p. 86A (1978).

Kohno, S., Van Den Berghe, H., Sandberg, A.A.: Chromosomes and causation of human cancer and leukemia. XXXI. Dq– deletions and their significance in proliferative disorders. Cancer 43, 1350–1357.

Kolb, H.-J., Wündisch, G.F., Bender-Götze, Ch., Spitzer, I., Brehm, G., Rodt, H., Lieven, H., Grosse-Wilde, H., Albert, E.D., Thiel, E., Ruppelt, W., Balk, O., Thierfelder, S.: Bone marrow transplantation in children with aplastic anemia and acute lymphatic leukemia. Blut 31, 343–346 (1975).

Koller, P.C.: The genetical and mechanical properties of sex chromosomes. III. Man. Proc. R. Soc. Edinburgh [Biol] 57, 194–214 (1937).

Koller, P.C.: Abnormal mitosis in tumours. Br. J. Cancer 1, 38–47 (1947).

Koller, P.C.: Chromosome behavior in tumors: Readjustments to Boveri's theory. In R.M. Cumley, M. Abbott, J. McCay (eds.): *Cell Physiology of Neoplasia.* pps. 9–48. Austin Texas: University Texas Press (1960).

Koller, P.C.: Chromosomes in neoplasia. In P. Emmelot, O. Mühlbock (eds.): *Cellular Control Mechanisms and Cancer,* pp. 174–189. Amsterdam: Elsevier (1964).

Koller, P.C.: *The Role of Chromosomes in Cancer Biology.* New York, Heidelberg, Berlin: Springer-Verlag (1972).

Koller, P.C.: Incidence of cancer in subjects with congenital and acquired chromosome aberrations. Recent Progr. Med. (Roma) 54, 383–395 (1973).

Kondo, T., Muragishi, H., Imaizumi, M.: A cell line from a human salivary gland mixed tumor. Cancer 27, 403–410 (1971).

Kondo, K., Sasaki, M., Mikuni, C.: A complex translocation involving chromosomes 1, 8, and 21 i

acute myeloblastic leukemia. Proc. Jap. Acad. 54, 21 – 24 (1978).
Konietzko, H., Haberlandt, W., Heilbronner, H., Reill, G., Weichardt, H.: Chromosome studies of trichloroethylene workers. Arch. Toxicol. (Berl.) 40, 201 – 206 (1978).
König, E., Cohnen, G., Brittinger, G., Douglas, S.D.: Response to phytohaemagglutinin and pokeweed mitogen in chronic lymphocytic leukaemia. Lancet i, 795 (1972).
Konstantinowa, B., Bratanowa, N.N.: Chromosomal aberrations in patients with utero-hemorrhagic colitis. Digestion 2, 329 – 337 (1969).
Kontras, S.B., Robbins, M., Ambuel, J.P.: Morphological Philadelphia chromosome: Occurrence in a child with congenital anomalies. Am. J. Dis. Child. 111, 324 – 326 (1966).
Korenvskaia, M.I., Bartashchuk, E.I., Dul'Tsina, S.M., Terenteva, E.I.: Case of blast crisis in chronic myeloleukemia with di- and trisomy Ph-chromosome (in Russian). Probl. Gematol. Pereliv. Krovi 15, 51 – 53 (1970).
Korenvskaia, M.I., Nevskaia, T.P., Cherntsova, T.A., Terenteva, E.I.: Cytogenetic studies at different stages of development of chronic myeloleukemia (in Russian). Probl. Gematol. Pereliv. Krovi 16, 23 – 29 (1971).
Korf, B.R., Schuh, B.E., Salwen, M.J., Warburton, D., Miller, O.J.: The role of trypsin in the pretreatment of chromosomes for Giemsa banding. Hum. Genet. 31, 27 – 33 (1976).
Kornberg, R.D.: Chromatin structure: A repeating unit of histones and DNA. Science 184, 868 – 871 (1974).
Kornberg, R.D.: Structure of chromatin. Annu. Rev. Biochem. 46, 931 – 954 (1977).
Korsgaard, R.: Chromosome analysis as diagnostic tool in malignant pleural effusions. I.R.C.S. Suppl. 2, 1349 (1974).
Kosenow, W., Pfeiffer, R.A.: Chronisch-myeloische Leukämie bei eineiigen Zwillingen. Dtsch. Med. Wochenschr. 94, 1170 – 1176 (1969).
Koskull, H. von, Aula, P.: Nonrandom distribution of chromosome breaks in Fanconi's anemia. Cytogenet. Cell Genet. 12, 423 – 434 (1973).
Koskull, H. von, Aula, P.: Distribution of chromosome breaks in measles, Fanconi's anemia and controls. Hereditas 87, 1 – 10 (1977).
Kosmidis, P.A., Axelrod, A.R., Palacas, C., Stahl, M.: Angioimmunoblastic lymphadenopathy. A T-cell deficiency. Cancer 42, 447 – 452 (1978).
Kotler, S., Lubs, H.A.: Comparison of direct and short-term tissue culture techniques in determining solid tumor karyotypes. Cancer Res. 27, 1861 – 1866 (1967).
Koulischer, L.: Mitotic chromatid separation in a case of acute leukemia. Eur. J. Cancer 2, 347 – 352 (1966).
Koulischer, L., Frühling, J., Henry, J.: Obsérvations cytogénétiques dans la maladie de Vaquez. Eur. J. Cancer 3, 193 – 201 (1967).
Kovacs, G.: Abnormalities of chromosome No. 1 in haematological malignancies. Lancet i, 555 (1978).
Kovacs, Gy.: Abnormalities of chromosome No. 1 in human solid malignant tumours. Int. J. Cancer 21, 688 – 694 (1978).
Kovacs, M., Vass, L., Sellyei, M.: Detection of Y chromatin and Barr bodies in histological sections by quinacrine fluorescence. Stain Techol. 48, 94 – 96 (1973).
Kovary, P.M., Lonauer, G., Niedorf, H., Nautsch, C., Pawlowitzki, I.H.: Pyoderma gangrenosum with Philadelphia chromosome-negative chronic myeloid leukemia. Dermatologica (Basel) 154, 360 – 366 (1977).
Kowalski, Ch. J., Nasjleti, C.E., Harris, J.E.: Human chromosomes. Evidence for autosomal sexual dimorphism. Exp. Cell Res. 100, 56 – 62 (1976).
Koziner, B., Filippa, D.A., Mertelsmann, R., Gupta, S., Clarkson, B., Good, R.A., Siegel, F.P.: Characterization of malignant lymphomas in leukemic phase by multiple differentiation markers of mononuclear cells. Am. J. Med. 63, 556 – 567 (1977).
Koziner, B., McKenzie, S., Straus, D., Clarkson, B., Good, R.A., Siegel, F.P.: Cell marker analysis in acute monocytic leukemias. Blood 49, 895 – 901 (1977).
Kraemer, P.M., Peterson, D.F., van Dilla, M.A.: DNA constancy in heteroploidy and the stem line theory of tumors. Science 174, 714 – 717 (1971).
Krajinčanić, B., Lazarov, A., Žunić, Z., Radojičić, B.: Chromosome changes in patients treated with "myleran." Strahlentherapie 151, 459 – 462 (1976).
Krauss, S.: Chronic myelocytic leukemia with features simulating myelofibrosis with myeloid metaplasia. Cancer 19, 1321 – 1332 (1966).
Krauss, S.: The Philadelphia chromosome and leukocyte alkaline phosphatase in chronic myelocytic leukemia and related disorders. Ann. N.Y. Acad. Sci. 155, 983 – 991 (1968).
Krauss, S., Sokal, J.E., Sandberg, A.A.: Comparison of Philadelphia chromosome-positive and -negative patients with chronic myelocytic leukemia. Ann. Intern. Med. 61, 625 – 635 (1964).
Kristoffersson, U.: Chromosome studies in a thymoma *in vitro*. Humangenetik 20, 191 – 192 (1973).
Krmpotic, E., Vykoupil, K.F., Stangel, W.: The role of chromosome studies in polycythaemia vera. International Society on Haematology, European African Division, 4th Meeting. Held in Istanbul, Sep . 5 – 9, Abstract #164 (1977).
Krogh Jensen, M.: Chromosomal findings in two

cases of acute erythroleukaemia. Acta Med. Scand. 180, 245–252 (1966).

Krogh Jensen, M.: Chromosome studies in patients treated with azathioprine and amethopterin. Acta Med. Scand. 182, 445–455 (1967a).

Krogh Jensen, M.: Chromosome studies in acute leukaemia. III. Chromosome constitution of bone marrow cells in 30 cases. Acta Med. Scand. 182, 629–644 (1967b).

Krogh Jensen, M.: Chromosome studies in acute leukaemia. II. A comparison between the chromosome patterns of bone marrow cells and cells from the peripheral blood. Acta Med. Scand. 182, 157–165 (1967c).

Krogh Jensen, M.: Chromosome studies in potentially leukaemic myeloid disorders. Acta Med. Scand. 183, 535–542 (1968).

Krogh Jensen, M.: Chromosome studies in acute leukemia. Copenhagen: Munksgaard. Doctoral thesis (1969).

Krogh Jensen, M.: Cytogenetic studies in acute myeloid leukaemia. Acta Med. Scand. 190, 429–434 (1971).

Krogh Jensen, M.: Cytogenetic findings in pernicious anaemia. Comparison between results obtained with chromosome studies and the micronucleus test. Mutat. Res. 45, 249–252 (1977).

Krogh Jensen, M., Friis-Møller, A.: Chromosome studies in pernicious anemia. Acta Med. Scand. 181, 571–576 (1967).

Krogh Jensen, M., Hüttel, M.S.: Assessment of the effect of azathioprine on human bone marrow cells in vivo, combining chromosome studies and the micronucleus test. Danish Med. Bull. 23, 152–155 (1976).

Krogh Jensen, M., Killmann, S.-A.: Chromosome studies in acute leukaemia. Evidence for chromosomal abnormalities common to erythroblasts and leukaemic white cells. Acta Med. Scand. 181, 47–53 (1967).

Krogh Jensen, M., Killmann, S.-A.: Additional evidence for chromosome abnormalities in the erythroid precursors in acute leukaemia. Acta Med. Scand. 189, 97–100 (1971).

Krogh Jensen, M., Mikkelsen, M.: Cytogenetic studies in sideroblastic anemia. Cancer 37, 271–274 (1976).

Krogh Jensen, M., Philip, P.: Cytogenetic studies in haematological disorders which may terminate in acute leukaemia. Acta Med. Scand. 193, 353–357 (1973).

Krogh Jensen, M., Philip, P.: Cytogenetic studies in myeloproliferative disorders during transformation into leukaemia. Scand. J. Haematol. 7, 330–335 (1970).

Krogh Jensen, M., Søborg, M.: Chromosome aberrations in human cells following treatment with Imuran. Preliminary report. Acta Med. Scand. 179, 249–250 (1966).

Krogh Jensen, M., Eriksen, J., Djernes, B.W.: Cytogenetic studies in myelomatosis. Scand. J. Haematol. 14, 201–209 (1975).

Kroll, W., Schlesinger, K.: Chromosome studies in an infant with acute erythremic myelosis. Blood 35, 282–285 (1970).

Krompotic, E., Zellner, J.M.: Medical cytogenetics. Part XV. Chromosomes in malignancies. Chic. Med. Sch. Quart. 27, 171–185 (1968).

Krompotic, E., Lewis, J.P., Donnelly, W.J.: Chromosome aberrations in two patients with chronic granulocytic leukemia undergoing acute transformation. Am. J. Clin. Pathol. 49, 161–170 (1968).

Krompotic, E., Silberman, S., Einhorn, M., Uy, E.S., Chernay, P.R.: Clonal evolution in Di Guglielmo syndrome. Ann. Genet. (Paris) 11, 225–229 (1968).

Kucheria, K.: Double minute chromatin bodies in a sub-ependymal glioma. Br. J. Cancer 22, 696–697 (1968).

Kuhn, E.M.: Localization by Q-banding of mitotic chiasmata in cases of Bloom's syndrome. Chromosoma 57, 1–11 (1976).

Kuhn, E.M.: Mitotic chiasmata and other quadriradials in mitomycin C-treated Bloom's syndrome lymphocytes. Chromosoma (Berl.) 66, 287–297 (1978).

Kumatori, T., Ishihara, T., Ueda, T., Miyoshi, K.: Medical survey of Japanese exposed to fall-out radiation in 1954. A report after 10 years. Natl. Inst. Radiol. Sci. 4, 1–18 (1965).

Kundel, D.W., Tanaka, Y., Tjio, J.H., Whang, J., Fishbein, W.N.: Chromosome Philadelphie, inclusions filamenteuses et corps d'Auer dans un cas de transformation aiguë de leucémie myéloïde chronique. Nouv. Rev. Fr. Hematol. 3, 844–849 (1963).

Kunze, J., Frenzel, U.H., Hüttig, E., Grosse, F.-R., Wiedemann, H.-R.: Klinefelter's syndrome and incontinentia pigmenti Bloch-Sulzberger. Hum. Genet. 35, 237–240 (1977).

Kuramoto, H.: Studies of the growth and cytogenetic properties of human endometrial adenocarcinoma in culture and its development into an established line. Acta Obstet. Gynaecol. Jap. 19, 47–58 (1972).

Kuramoto, H., Hamano, M.: Cytogenetic studies of human endometrial carcinomas by means of tissue culture. Acta Cytol. 21, 559–565 (1977).

Kurita, S., Kamei, Y., Ota, K.: Genetic studies on familial leukemia. Cancer 34, 1098–1101 (1974).

Kurnit, D.M.: DNA helical content during the C-banding procedure. Cytogenet. Cell Genet. 13, 313–329 (1974).

Kurvink, K., Bloomfield, C.D., Cervenka, J.: Sister chromatid exchange in patients with viral disease. Exp. Cell Res. 113, 450–453 (1978a).
Kurvink, K., Bloomfield, C.D., Keenan, K.M., Levitt, S., Cervenka, J.: Sister chromatid exchange in lymphocytes from patients with malignant lymphoma. Hum. Genet. 44, 137–144 (1978b).
Kurz, R., Haas, H.: Value of the combined cytological and cytochemical classification in the management of acute childhood leukemia. Acta Haematol. (Basel) 52, 1–7 (1974).
Kushner, J.P., Lee, G.R., Wintrobe, M.M.: Idiopathic refractory sideroblastic anemia. Clinical and laboratory investigation of 17 patients and a review of the literature. Medicine (Baltimore) 50, 139–159 (1971).
Kwan, Y.-L., Singh, S., Vincent, P.C., Gunz, F.W.: Metamorphosis of chronic granulocytic leukaemia arising in an extramedullary site. Leukemia Res. 1, 301–307 (1977).
Kyle, R.A., Pease, G.L.: Basophilic leukemia. Arch. Intern Med. 118, 205–210 (1966).

Ladda, R., Atkins, L., Littlefield, J., Neurath, P., Marimuthu, K.M.: Computer-assisted analysis of chromosomal abnormalities: Detection of a deletion in aniridia/Wilms' tumor syndrome. Science 185, 784–787 (1974).
Ladda, R., Atkins, L., Littlefield, J., Pruett, R.: Retinoblastoma: Chromosome banding in patients with heritable tumour. Lancet ii, 506 (1973).
Lahey, M.E., Beier, F.R., Wilson, J.F.: Leukemia in Down's syndrome. J. Pediatr. 63, 189–190 (1963).
Lamb, D.: Correlation of chromosome counts with histological appearances and prognosis of transitional-cell carcinoma of bladder. Br. Med. J. i, 273–277 (1967).
Lambert, B., Hansson, K., Bui, T.H., Funes-Cravioto, F., Lindsten, J., Holmberg, M., Strausmanis, R.: DNA repair and frequency of X-ray and U.V.-light induced chromosome aberrations in leukocytes from patients with Down's syndrome. Ann. Hum. Genet. 39, 293–303 (1976a).
Lambert, B., Hansson, K., Lindsten, J., Sten, M., Werelius, B.: Bromodeoxyuridine-induced sister chromatid exchanges in human lymphocytes. Hereditas 83, 163–174 (1976b).
Lambert, B., Lindblad, A., Nordenskjöld, M., Werelius, B.: Increased frequency of sister chromatid exchanges in cigarette smokers. Hereditas 88, 147–149 (1978).
Lambert, B., Ringborg, U., Harper, E., Lindblad, A.: Sister chromatid exchanges in patients receiving chemotherapy against malignant disorders. Cancer Treat. Rep. 62, 1413–1419 (1978).
Lampert, F.: Cellulärer DNS-Gehalt und Chromosomenzahl bei der akuten Leukämie im Kindesalter un ihre Bedeutung für Chemotherapie und Prognose. Klin. Wochenschr. 45, 763–768 (1967a).
Lampert, F.: DNS (Feulgen)–und Chromosomenuntersuchungen bei Mongolismus mit akuter Leukämie. Klin. Wochenschr. 45, 512–516 (1967b).
Lampert, F.: Akute lymphoblastische Leukämie bei Geschwistern mit progressiver Kleinhirnataxie (Louis-Bar-Syndrom). Dtsch. Med. Wochenschr. 94, 217–220 (1969).
Lampert, F.: Coiled supercoiled DNA in critical point dried and thin sectioned human chromosome fibres. Nature [New Biol.] 234, 187–188 (1971).
Lampert, F., Gauger, J.U.: Chromosomen der Zellen der akuten Leukämie im Kindesalter. Klin. Wochenschr. 46, 882–888 (1968).
Lampert, F., Lampert, P.: Ultrastructure of the human chromosome fiber. Humangenetik 11, 9–17 (1970).
Lampert, F., Bahr, G.F., Dupraw, E.J.: Ultrastructure of a Burkitt's lymphoma marker chromosome, as investigated by quantitative electron microscopy. Cancer 24, 367–376 (1969).
Lampert, F., Phebus, C.K., Huhn, D., Myer, G., Greifenegger, M.: Leukemic xanthomatosis with a missing No. 9 chromosome. Z. Kinderheilkd. 112, 251–260 (1972).
Lam-Poo-Tang, P.R.L.C.: An improved method of processing bone marrow for chromosomes. Scand. J. Haematol. 5, 158–160 (1968).
Lancet: New thoughts on leukaemia. ii, 447 (1959).
Lancet: A proposed standard system of nomenclature of human mitotic chromosomes (special article). i, 1063–1065 (1960).
Lancet: Leading article: The Philadelphia chromosome. i, 433–435 (1961a).
Lancet: The numbering of chromosomes (annotation). 928–929 (1961b).
Lancet: Endoreduplication, polyploidy and leukaemia (leading article). ii 511–513 (1964).
Lancet: The changing pattern of retinoblastoma (editorial). ii, 1016–1017 (1971a).
Lancet: Dyeing the Y chromosome (editorial). i, 275 (1971b).
Lancet: Chromosomes and cancer (editorial). ii, 227–228 (1977).
Landau, J.W., Sasaki, M.S., Newcomer, V.D., Norman, A.: Bloom's syndrome. The syndrome of telangiectatic erythrema and growth retardation. Arch. Dermatol. Syph. (Chic.) 94, 687–694 (1966).
Landaw, S.A.: Acute leukemia in polycythemia vera. Semin. Hematol. 13, 33–48 (1976).
Lange, M., Alfi, O.S., Benedict, W., Derencsenyi, A.: Prognostic value of bone marrow cytogenetics

in children with acute myelogenous leukemia. *American Society on Human Genetics, 28th Annual Meeting,* p. 66a. Held in San Diego, Calif., Oct. 19–22 (abstract) (1977).

Langlands, A.O., Maclean, N.: Lymphoma of the thyroid. An unusual clinical course in a patient possessing a 14/21 translocation. Cancer 38, 259–267 (1976).

Larripa, I., Brieux de Salum, S., Pavlovsky, S.: Hematological and genetic study of a neonatal acute myeloblastic leukemia. Sangre (Barc.) 20, 69–73 (1975).

Larsen, W.E., Schimke, R.N.: Familial acute myelogenous leukemia with associated C-monosomy in two affected members. Cancer 38, 841–845 (1976).

Latt, S.A.: Microfluorometric detection of deoxyribonucleic acid replication in human metaphase chromosomes. Proc. Natl. Acad. Sci. USA 70, 3395–3399 (1973).

Latt, S.A.: Sister chromatid exchanges, indices of human chromosome damage and repair: Detection by fluorescence and induction by mitomycin C. Proc. Natl. Acad. Sci. USA 71, 3162–3166 (1974a).

Latt, S.A.: Localization of sister chromatid exchanges in human chromosomes. Science 185, 74–76 (1974b).

Latt, S.A.: Fluorescence analysis of late DNA replication in human metaphase chromosomes. Somat. Cell Genet. 1, 293–321 (1975).

Latt, S.A.: Analysis of human chromosome structure, replication and repair, using BrdU-33258 Hoechst techniques. J. Reprod. Med. 17, 41–52 (1976).

Latt, S.A.: Analysis of sister chromatid exchange and chromosome replication kinetics using BrdU-dye techniques. Virchows Arch. B Cell Pathol. 29, 19–27 (1978).

Latt, S.A., Loveday, K.S.: Characterization of sister chromatid exchange induction by 8-methoxypsoralen plus near UV light. Cytogenet. Cell Genet. 21, 184–200 (1978).

Latt, S.A., Davidson, R.L., Lin, M.S., Gerald, P.S.: Lateral asymmetry in the fluorescence of human Y chromosomes stained with 33258 Hoechst. Exp. Cell Res. 87, 425–429 (1974).

Latt, S.A., Stetten, G., Juergens, L.A., Buchanan, G.R., Gerald, P.S.: Induction by alkylating agents of sister chromatid exchanges and chromatid breaks in Fanconi's anemia. Proc. Natl. Acad. Sci. USA 72, 4066–4070 (1975).

Lau, Y.-F., Arrighi, F.E.: Comparative studies of N-banding and silver staining of NORs in human chromosomes. Seminar–Workshop, Montevideo, Uruguay, February (1977).

Lau, Y.-F., Pfeiffer, R.A., Arrighi, F.E., Hsu, T.C.: Combination of silver and fluorescent staining for metaphase chromosomes. Am. J. Hum. Genet. 30, 76–79 (1978).

Laurent, M., Rosseau, M.-F., Nezelof, C.: Étude caryotypique d'une tératome sacro-coccygien. Ann. Anat. Pathol. (Paris) 13, 413–422 (1968).

Law, I.P., Hollinshead, A.C., Whang-Peng, J., Dean, J.H., Oldham, R.K., Heberman, R.B., Rhode, M.C.: Familial occurrence of colon and uterine carcinoma and of lymphoproliferative malignancies. II. Chromosomal and immunologic abnormalities. Cancer 39, 1229–1236 (1977).

Lawler, S.D.: Chromosomes and acute transformation of chronic myeloid leukemia (in French). Nouv. Rev. Fr. Hematol. 7, 529–535 (1967).

Lawler, S.D.: Chromosomes in haematology. Br. J. Haematol. 17, 139–143 (1969).

Lawler, S.D.: Cytogenetic studies. *In: Smithers Hodgkin's Disease,* pp. 55–63. London: Churchill, Livingstone (1973).

Lawler, S.D.: Can cytogenetic investigation of malignant lymphomas have practical applications? Acta Cytol. 19, 489 (1975).

Lawler, S.D.: New thoughts about the Philadelphia chromosome. Br. J. Cancer 34, 318 (1976).

Lawler, S.D.: The cytogenetics of chronic granulocytic leukaemia. Clin. Hematol. 6, 55–75 (1977a).

Lawler, S.D.: Chromosomes in haematology. Br. J. Haematol. 36, 455–460 (1977b).

Lawler, S.D., Galton, D.A.G.: Chromosome changes in the terminal stages of chronic granulocytic leukaemia. Acta Med. Scand. [Suppl.] 445, 312–318 (1966).

Lawler, S.D., Lele, K.P.: Chromosomal damage induced by chlorambucil in chronic lymphocytic leukaemia. Scand. J. Haematol. 9, 603–612 (1972).

Lawler, S.D., Millard, R.E.: Chromosome breakage and leukaemia. Lancet i, 160 (1967).

Lawler, S.D., Reeves, B.R.: Chromosome studies in man: past achievements and recent advances. J. Clin. Pathol. 29, 569–582 (1976).

Lawler, S.D., Sanger, R.: Xg blood-groups and clonal-origin theory of chronic myeloid leukaemia. Lancet i, 584–585 (1970).

Lawler, S.D., Kay, H.E.M., Birbeck, M.S.C.: Marrow dysplasia with C trisomy and anomalies of the granulocyte nuclei. J. Clin. Pathol. 19, 214–219 (1966).

Lawler, S.D., Lobb, D.S., Wiltshaw, E.: Philadelphia-chromosome positive bone-marrow cells showing loss of the Y in males with chronic myeloid leukaemia. Br. J. Haematol. 27, 247–252 (1974).

Lawler, S.D., Millard, R.E., Kay, H.E.M.: Further cytogenetical investigations in polycythaemia vera. Eur. J. Cancer 6, 223–233 (1970).

Lawler, S.D., O'Malley, F., Lobb, D.S.: Chromosome

banding studies in Philadelphia chromosome positive myeloid leukaemia. Scand. J. Haematol. 17, 17–28 (1976).

Lawler, S.D., Pentycross, C.R., Reeves, B.R.: Lymphocyte transformation and chromosome studies in Hodgkin's disease. Br. Med. J. iii, 704–708 (1967).

Lawler, S.D., Pentycross, C.R., Reeves, B.R.: Chromosomes and transformation of lymphocytes in lymphoproliferative disorders. Br. Med. J. iv, 213–219 (1968).

Lawler, S.D., Reeves, B.R., Hamlin, I.M.E.: A comparison of cytogenetics and histopathology in the malignant lymphomata. Br. J. Cancer 31, 162–167 (1975).

Lawler, S.D., Roberts, P.D., Hoffbrand, A.V.: Chromosome studies in megaloblastic anemia before and after treatment. Scand. J. Haematol. 8, 309–320 (1971).

Lawler, S.D., Secker Walker, L.M., Summersgill, B.M., Reeves, B.R., Lewis, J., Kay, H.E.M., Hardisty, R.M.: Chromosome banding studies in acute leukaemia at diagnosis. Scand. J. Haematol. 15, 312–320 (1975).

Lawrence, J.H., Winchell, H.S., Donald, W.G.: Leukemia in polycythemia vera. Relationship to splenic myeloid metaplasia and therapeutic radiation dose. Ann. Intern. Med. 70, 763–771 (1969).

Lech, H., Polaniecka, B., Wisniewski, L.: X-trisomy associated with Hodgkin's disease. Br. Med. J. i, 517 (1974).

Lederer, B., Autengruber, M., Mikuz, G.: Statistical analysis of cytophotometric DNA measurements demonstrated on malignant testicular teratoma. Acta Cytol. 20, 5–6 (1976).

Lederlin, P., Puchelle, J.C., Aymard, J.P., Thibaut, G., Guerci, O., Herbeuval, R.: Evolution of chronic myeloid leukemias turned "acute." Ann. Med. Nancy 14, 191–193 (1975).

Leeksma, C.H.W., Friden-Kill, L., Brommer, E.J.P., Neuberg, C.W., Kerkhofs, H.: Chromosomes in premyeloid leukaemia. Lancet ii, 1299 (1965).

Legrand, E.: Étude cytogénétique de neuf épanchements néoplastiques (Considérations sur l'aneuploïde et les chromosomes anormaux des cellules cancéreuses). Travail de la Fondation Bergonié, Centre Anti-cancéreux de Bordeaux, p. 176. Bordeaux: Baillet (1968).

Lehmann, A.R., Arlett, C.F.: Human genetic disorders with defects in the repair of deoxyribonucleic acid. Biochem. Soc. Trans. 5, 1199–1203 (1977).

Lehmann, A.R., Stevens, S.: The production and repair of double strand breaks in cells from normal humans and from patients with ataxia telangiectasia. Biochim. Biophys. Acta 474, 49–60 (1977).

Leibovitz, A., Stinson, J.C., McCombs, W.B. III, McCoy, C.E., Mazur, K.C., Mabry, N.D.: Classification of human colorectal adenocarcinoma cell lines. Cancer Res. 36, 4562–4569 (1976).

Leibowitz, M.R., Derman, D.P., Jacobson, R., Stevens, K., Katz, J.: Chronic myeloid leukaemia in South African blacks. S. Afr. Med. J. 50, 2035–2037 (1976).

Lejeune, J.: Le Mongolisme, maladie chromosomique. La Nature 3296, 521–523 (1959).

Lejeune, J.: Aberrations chromosomiques et cancer. In R.J.C. Harris (ed.): Proceedings of the 9th International Cancer Congress, pp. 71–85. Berlin: Springer Verlag (UICC Monograph Ser. Vol. 9) (1967).

Lejeune, J.: Cytogenetics of clonal growths. N. Engl. J. Med. 289, 320–321 (1973).

Lejeune, J., Berger, R.: Sur une méthode de recherche d'un variant common des tumeurs de l'ovarie. C.R. Acad. Sci. [D] (Paris) 262, 1885–1887 (1966).

Lejeune, J., Berger, R., Cailie, B., Turpin, R.: Évolution chromosomique d'une leucémie myéloïde chronique. Ann. Genet. (Paris) 8, 44–49 (1965).

Lejeune, J., Berger, R., Haines, M., Lafourcade, J., Vialatte, J., Satge, P., Turpin, R.: Constitution d'un clone à 54 chromosomes au cours d'une leucoblastose congénitale chez une enfant mongolienne. C.R. Acad. Sci. [D] (Paris) 256, 1195–1197 (1963).

Lejeune, J., Berger, R., Rethore, M.-O.: Sur l'endoréduplication selective de certains segments du génome. C.R. Acad. Sci. [D] (Paris) 263, 1880–1882 (1966).

Lejeune, J., Gautier, M., Turpin, R.: Étude des chromosomes somatiques de neuf enfants mongoliens. C.R. Acad. Sci. [D] (Paris) 248, 1721–1722 (1959).

Lejeune, J., Turpin, R., Gautier, M.: Le Mongolisme, premier exemple d'aberration autosomique humaine. Ann. Genet. (Paris) 1, 41–49 (1959).

Lele, K.P., Penrose, L.S., Stallard, H.B.: Chromosome deletion in a case of retinoblastoma. Ann. Hum. Genet. 27, 171–174 (1963).

Lelikova, G.P., Laskina, A.V., Zacharov, A.F., Pogosianz, H.E.: Cytogenetic study of teratoid testicular tumors in man. Vopr. Onkol. 16, 32–38 (1970).

Lelikova, G.P., Laskina, A.V., Zacharov, A.F., Pogosianz, H.E.: Cytogenetic studies of human seminomas (in Russian) Vopr. Onkol. 17, 20–28 (1971).

Leon, N., Reichhardt Epps, D., Becak, M.L., Becak, W.: Discrepancies between bone marrow and peripheral-blood chromosomal constitution. Lancet ii, 880–881 (1961).

Levan, A.: Chromosomes number and structure of

malignant cells. The Swedish Cancer Society Yearbook 3, 94-98.
Levan, A.: The effect of colchicine on root mitoses in allium. Hereditas 24, 471-486 (1938).
Levan, A.: Chromosome studies on some human tumors and tissues of normal origin grown *in vivo* and *in vitro* at the Sloan-Kettering Institute. Cancer 9, 648-663 (1956*a*).
Levan, A.: Chromosomes in cancer tissue. Ann. N.Y. Acad. Sci. 63, 774-789 (1956*b*).
Levan, A.: Self-perpetuating ring chromosomes in two human tumors. Hereditas 42, 366-372 (1956*c*).
Levan, A.: Cancerogenesis: A genetic adaptation on the cellular level. *Eighth Yearbook of Cancer Examination and Cancer Treatment,* pp. 110-126. Netherlands (1958).
Levan, A.: Non-random representation of chromosome types in human tumor stemlines. Hereditas 55, 28-38 (1966).
Levan, A.: Some current problems of cancer cytogenetics. Hereditas 57, 343-355 (1967).
Levan, A.: Chromosome abnormalities and carcinogenesis. In A. Lima-de-Faria (ed.): *Handbook of Molecular Cytology,* pp. 717-731. Amsterdam, London: North-Holland (1969).
Levan, A.: Chromosome patterns in tumours. In T. Caspersson, L. Zech, V. Runnstrom-Reco (eds.): *Nobel Symposium. 23. Chromosome Identification,* pp. 55-61. New York: Academic Press (1973).
Levan, A., Hauschka, T.S.: Endomitotic reduplication mechanisms in ascites tumors of the mouse. J. Natl. Cancer Inst. 14, 1-43 (1953).
Levan, A., Hsu, T.C.: The human ideogram. Hereditas 45, 665-674 (1959).
Levan, A., Levan, G.: Have double minutes functioning centromeres? Hereditas 88, 81-92 (1978).
Levan, A., Müntzing, A.: Terminology of chromosome number. Port. Acta Biol. 7, 1-16 (1963).
Levan, A., Fredga, K., Sandberg, A.A.: Nomenclature for centromeric position on chromosomes. Hereditas 52, 201-220 (1964).
Levan, A., Levan, G., Mandahl, N.: A new chromosome type replacing the double minutes in a mouse tumor. Cytogenet. Cell Genet. 20, 12-23 (1978).
Levan, A., Levan, G., Mitelman, F.: Chromosomes and cancer. Hereditas 86, 15-30 (1977).
Levan, A., Manolov, G., Clifford, P.: Chromosomes of a human neuroblastoma: a new case with accessory minute chromosomes. J. Natl. Cancer Inst. 41, 1377-1387 (1968).
Levan, A., Nichols, W.W., Hall, B., Löw, B., Nilsson, S.B., Nordén, A.: Mixture of Rh positive and Rh negative erythrocytes and chromosomal abnormalities in a case of polycythemia. Hereditas 52, 89-105 (1964).

Levan, A., Nichols, W.W., Nordén, Å: A case of chronic myeloid leukemia with two leukemic stemlines in the blood. Hereditas 49, 433-441 (1963).
Levan, G.: The detailed chromosome constitution of a benzpyrene-induced rat sarcoma. A tentative model for G-band analysis in solid tumors. Hereditas 78, 273-290 (1974).
Levan, G.: Cytogenetic studies in experimental rat sarcomas. Hereditas 79, 1-20 (1975).
Levan, G., Levan, A.: Specific chromosome changes in malignancy: Studies in rat sarcomas induced by two polycyclic hydrocarbons. Hereditas 79, 161-198 (1975).
Levan, G., Mandahl, N., Bengtsson, B.O., Levan, A.: Experimental elimination and recovery of double minute chromosomes in malignant cell populations. Hereditas 86, 75-90 (1977).
Levan, G., Mitelman, F.: Clustering of aberrations to specific chromosomes in human neoplasms. Hereditas 79, 156-160 (1975).
Levan, G., Mitelman, F.: G-banding in Rous rat sarcomas during serial transfer: Significant chromosome aberrations and incidence of stromal mitoses. Hereditas 84, 1-14 (1976).
Levan, G., Mitelman, F.: Chromosomes and the etiology of cancer. Chromosomes Today 6, 363-371 (1977).
Levan, G., Mitelman, F.: Absence of late-replicating X-chromosome in a female patient with acute myeloid leukemia and the 8;21 translocation. J. Natl. Cancer Inst. 62, 273-275 (1979).
Levan, G., Ahlström, U., Mitelman, F.: The specificity of chromosome A2 involvement in DMBA-induced rat sarcomas. Hereditas 77, 263-280 (1974).
Levan, G., Mandahl, N., Bregula, U., Klein, G., Levan, A.: Double minute chromosomes are not centromeric regions of the host chromosome. Hereditas 83, 83-90 (1976).
Levan, G., Mitelman, F., Telenius, M.: Chromosomes in Sipple's syndrome. Lancet i, 1510 (1973).
Levi, P.E., Cooper, E.H., Anderson, C.K., Path, M.C., Williams, R.E.: Analyses of DNA content, nuclear size, and cell proliferation of transitional cell carcinoma in man. Cancer 23, 1074-1085 (1969).
Levin, W.C., Houston, E.W., Ritzmann, S.E.: Polycythemia vera with Ph[1] chromosomes in two brothers. Blood 30, 503-512 (1967).
Levin, W.C., Ritzmann, S.E., Houston, E.W.: Reply to Summitt. Blood 32, 181-183 (1968).
Levine, M.: The chromosome number in cancer tissue of man, of rodent, of bird and in crown gall tissue of plants. J. Cancer Res. 14, 400-425 (1930).

Levine, M.: Studies in the cytology of cancer. Am. J. Cancer 15, 144–211 (1931).

Levitt, R., Pierre, R.V., White, W.L., Siekert, R.G.: Atypical lymphoid leukemia in ataxia-telangiectasia. Blood 52, 1003–1011 (1978).

Levy, J., Chadeyron, D.-A., Fonty, B.: Avortements spontanés et triploidie. Rev. Fr. Gynecol. Obstet. 67, 327–342 (1972).

Lewis, E.B.: Leukemia, multiple myeloma, and aplastic anemia in American radiologists. Science 142, 1492–1494 (1963).

Lewis, F.J.W., MacTaggart, M.: Chromosome counts in myeloma. Hum. Chrom. Newsl. 8, 20–21 (1962).

Lewis, F.J.W., Fraser, I.L., MacTaggart, M.: An abnormal chromosomal pattern in myelomatosis. Lancet ii, 1013–1015 (1963).

Lewis, F.J.W., MacTaggart, M., Andrews, M.I.J.: Chromosome studies in acute leukaemia. J. Clin. Pathol. 17, 475–476 (1964).

Lewis, F.J.W., MacTaggart, M., Crow, R.S., Wills, M.R.: Chromosomal abnormalities in multiple myeloma. Lancet i, 1183–1184 (1963).

Lewis, F.J.W., MacTaggart, M., Poulding, R.H., Stevenson, P.: A malignant hypodiploid cell line in a presumptive case of myelomatosis presenting as an acute leukaemia. Hum. Chrom. Newsl. 19, 26–28 (1966).

Lewis, F.J.W., Poulding, R.H., Eastham, R.D.: Acute leukaemia in an XO/XXX mosaic. Lancet ii, 306 (1963).

Lewis, P.D.: A cytophotometric study of benign and malignant phaeochromocytomas. Virchows Arch. [Zellpathol.] 9, 371–376 (1971).

Lewis, R.M., Lockwood, J.: The tetraploid number of chromosomes in the malignant cells of the Walker rat sarcoma No. 1. Bull. Johns Hopkins Hosp. 44, 187–199 (1929).

Li, F.P., Fraumeni, J.F., Jr.: Soft-tissue sarcomas, breast cancer, and other neoplasms. A familial syndrome? Ann. Intern. Med. 71, 747–752 (1969).

Liang, W., Hopper, J.E., Rowley, J.D.: Karyotypic abnormalities and clinical aspects of patients with multiple myeloma and related paraproteinemic disorders. Cancer (submitted) (1979).

Liang, W., McLean, F.: Karyotypic abnormalities in multiple myeloma and plasma cell leukemia. Proc. Am. Assoc. Cancer Res., p. 212, Abstract #848 (1978).

Liang, W., Rowley, J.D.: 14q+ marker chromosomes in multiple myeloma and plasma-cell leukaemia. Lancet i, 96 (1978).

Liaudet, J., Combaz, M.: Chronic myeloid leukemia in a 35-year-old petroleum chemist who had been exposed to benzene since the age of 18. Eur. J. Toxicol. 6, 309–313 (1973).

Libre, E.P., McFarland, W.: Chronic myelogenous leukemia. Possible association with reticulum cell sarcoma. Arch. Intern. Med. 119, 626–630 (1967).

Lieber, E., Hsu, L., Spitler, L., Fudenberg, H.H.: Cytogenetic findings in a parent of a patient with Fanconi's anemia. Clin. Genet. 3, 357–363 (1972).

Lilleyman, J.S., Potter, A.M., Watmore, A.E., Cooke, P., Sokol, R.J., Wood, J.K. Myeloid karyotype and the malignant phase of chronic granulocytic leukaemia. Br. J. Haematol. 39, 317–323 (1978).

Lima-de-Faria, A.: Incorporation of tritiated thymidine into meiotic chromosomes. Science 130, 503–504 (1959).

Lima-de-Faria, A., Bianchi, N.O., Nowell, P.C.: Patterns of chromosome replication in a patient with chronic granulocytic leukemia. Hereditas 58, 31–62 (1967).

Lin, C.C., Dent, P.B., Ward, E.J., McCulloch, P.B.: Extra chromosome 7 in cells cultured from cervical carcinoma. J. Natl. Cancer Inst. 50, 1399–1401 (1973).

Lin, C.C., van de Sande, H., Smink, W.K., Newton, D.R.: Quinacrine fluorescence and Q-banding patterns of human chromosomes. I. Effects of various factors. Can. J. Genet. Cytol. 17, 81–92 (1975).

Lin, M.S., Alfi, O.S.: Detection of sister chromatid exchanges by 4'-6-diamidino-2-phenylindole fluorescence. Chromosoma 57, 219–225 (1976).

Lin, M.S., Alfi, O.S.: Variation in lateral asymmetry of human chromosome 1. Cytogenet. Cell Genet. 21, 243–250 (1978).

Linder, D., Hecht, F., Kaiser-McCaw, B., Campbell, J.R.: Origin of extragonadal teratomas and endodermal sinus tumours. Nature 254, 597–598 (1975).

Linder, D., Kaiser-McCaw, B., Hecht, F.: Parthenogenic origin of benign ovarian teratomas. N. Engl. J. Med. 292, 63–66 (1975).

Lindgren, V., Rowley, J.D.: Comparable complex rearrangements involving the 8;21 and 9;22 translocations in leukaemia. Nature 266, 744–745 (1977).

Lindquist, R., Gahrton, G., Friberg, K., Zech, L.: Trisomy 8 in the chronic phase of Philadelphia negative chronic myelocytic leukaemia. Scand. J. Haematol. 21, 109–114 (1978).

Linman, J.W., Bagby, Y.C., Jr.: The preleukemic syndrome: Clinical and laboratory features, natural course, and management. Blood Cells 2, 11–31 (1976).

Lisker, R., Cobo, A.: Chromosome breakage in ataxia-telangiectasia. Lancet i, 618 (1970).

Lisker, R., Cobo de Gutierrez, A.: Cytogenetic studies in Fanconi's anemia. Description of a case with bone marrow clonal evolution. Clin. Genet. 5, 72–76 (1974).

Lisker, R., Cobo de Gutiérrez, A., Valázquez-Ferrari, M.: Longitudinal bone marrow chromosome studies in potential leukemic myeloid disorders. Cancer 31, 509–515 (1973).

Littlefield, L.G.: An abnormal cell-line in a patient with acute granulocytic leukemia. Cytogenetic studies before and after an allogenic marrow transplant. Cancer 29, 1281–1286 (1972).

Littlefield, L.G., Goh, K.O.: Cytogenetic studies in control men and women. I. Variations in aberration frequencies in 29,709 metaphases from 305 cultures obtained over a three-year period. Cytogenet. Cell Genet. 12, 17–34 (1973).

Littlefield, L.G., Mailhes, J.B.: Observations of *de novo* clones of cytogenetically aberrant cells in primary fibroblast cell strains from phenotypically normal women. Am. J. Hum. Genet. 27, 190–197 (1975).

Littlefield, L.G., Vodopick, H.A.: Cytogenetic studies in a patient with acute granulocytic leukemia of seven and one-half years duration. Blood 46, 783–789 (1975).

Littlefield, L.G., Lever, W.E., Miller, F.L., Goh, K.O.: Chromosome breakage studies in lymphocytes from normal women, pregnant women and women taking oral contraceptives. Am. J. Obstet. Gynecol. 121, 976–980 (1975).

Litton, L.E., Hollander, D.H., Borgaonkar, D.S., Frost, J.K.: Y-chromatin of interphase cancer cells, a preliminary study. Acta Cytol. 16, 404–407 (1972).

Lloyd, D.C., Puttott, R.J., Dolphin, G.W., Horton, P.W., Halman, K.E., Scott, J.S., Mair, G.: A comparison of physical and cytogenetic estimates of radiation dose in patients treated with iodine-131 for thyroid carcinoma. Int. J. Radiat. Biol. 30, 473–485 (1976).

Lobb, D.S., Reeves, B.R., Lawler, S.D.: Identification of isochromosome 17 in myeloid leukemia. Lancet i, 849–850 (1972).

Lo Curto, F., Fraccaro, M.: Nuclear projections in tumour cells. Lancet ii, 847 (1974).

Löffler, H., Meyhöfer, W., Lange, R.H., Ehlers, G., Remmele, W.: Sézary-Syndrom: Eine leukämische Variante der Mykosis Fungoides. Dtsch. Med. Wochenschr. 99, 429–434 (1974).

Loke, Y.W.: Sex chromatin of hydatidiform moles. J. Med. Genet. 6, 22–25 (1969).

London Conference: The normal human karyotype. Ann. Hum. Genet. 27, 294–296 (1963); Cytogenetics 2, 264–268 (1963).

Long, J.C., Zamecnik, P.C., Aisenberg, A.C., Atkins, L.: Tissue culture studies in Hodgkin's disease. Morphologic, cytogenetic, cell surface, and enzymatic properties of cultures derived from splenic tumors. J. Exp. Med. 145, 1484–1500 (1977).

Longo, D.L., Whang-Peng, J., Jaffe, E., Triche, T.J., Young, R.C.: Myeloproliferative syndromes: A unique presentation of chronic myelogenous leukemia (CML) as a primary tumor of bone. Blood 52, 793–801 (1978).

Louwagie, A.C., Desmet, V.J., van den Berghe, H.: Coexistence of a myelo- and lymphoproliferative disorder. Scand. J. Haematol. 11, 350–355 (1973).

Lowry, W.S.B., Halmos, P.B.: Malignant tumour of brown fat in patient with Turner's syndrome. Br. Med. J. 4, 720–721 (1967).

Lozzio, C.B., Lozzio, B.B.: Human chronic myelogenous leukemia cell-line with positive Philadelphia chromosome. Blood 45, 321–334 (1975).

Lozzio, C.B., Lozzio, B.B., Wust, C.J., Kim, J.: Correlation of leukemia-associated antigens and Ph^1 chromosome in fibroblastlike cells derived from bone marrow. Blood 52, 673–680 (1978).

Lozzio, C.B., Lozzio, B.B., Yang, W.-K., Ichiki, A.T., Bamberger, E.G.: Absence of thymus-derived lymphocyte markers in myelogenous leukemia (Ph^1+) cell line K-562. Cancer Res. 36, 4657–4662 (1976).

Lozzio, B.B., Lozzio, C.B., Machado, E.: Human myelogenous (Ph^1+) leukemia cell line: Transplantation into athymic mice. J. Natl. Cancer Inst. 56, 627–628 (1976).

Lozzio, B.B., Machado, E.A., Lozzio, C.B., Lair, S.: Hereditary asplenic-athymic mice: Transplantation of human myelogenous leukemic cells. J. Exp. Med. 143, 225–231 (1976).

Lubiniecki, A.S., Blattner, W.A., Dosik, H., Sun, C., Fraumeni, J.F.: SV40 T-antigen expression in skin fibroblasts from clinically normal individuals and from ten cases of Fanconi anemia. Am. J. Hematol. 2, 33–40 (1977).

Lubs, H.A., Jr.: Testicular size in Klinefelter's syndrome in men over fifty. Report of a case with XXY/XY mosaicism. N. Engl. J. Med. 267, 326–331 (1962).

Lubs, H.A.: The clinical significance of chromosomal abnormalities in colonic cancer. *In* W.J. Burdette (ed.): *Carcinoma of the Colon and Antecedent Epithelium,* pp. 319–332. Springfield, Ill.: C.C. Thomas (1970).

Lubs, H.A., Clark, R.: The chromosome complement of human solid tumors. I. Gastrointestinal tumors and technique. N. Engl. J. Med. 268, 907–911 (1963).

Lubs, H.A., Kotler, S.: The prognostic significance of chromosome abnormalities in colon tumors. Ann. Intern. Med. 67, 328–336 (1967).

Lubs, H.A., Lurie-Blitman, S.: The diagnostic signifi-

cance of the Ph¹ chromosome. Conn. Med. 29, 498–500 (1965).
Lubs, H.A., Salmon, J.H.: The chromosomal complement of human solid tumors. II. Karyotypes of glial tumors. J. Neurosurg. 22, 160–168 (1965).
Lubs, H.A., McKenzie, W.H., Patil, S.R., Merrick, S.: New staining methods for chromosomes. *In* D.M. Prescott (ed.): *Methods in Cell Biology*, Vol. VI, pp. 345–380. New York: Academic Press (1973).
Lubs, H.A., Jr., Salmon, J.H., Flanigan, S.: Studies of a glial tumor with multiple minute chromosomes. Cancer 19, 591–599 (1966).
Lubs, H.A., Verma, R.S., Summitt, R.L., Hecht, F.: Re-evaluation of the effect of spray adhesives on human chromosomes. Clin. Genet. 9, 302–306 (1976).
Lucas, L.S., Whang, J.J.K., Tjio, J.H., Manaker, R.A., Zeve, V.H.: Continuous cell culture from a patient with chronic myelogenous leukemia. I. Propagation and presence of Philadelphia chromosome. J. Natl. Cancer Inst. 37, 753–756 (1966).
Luciani, J.M.: Le contenu genetique du chromosome Y de l'homme. Rev. Fr. Gynecol. 68, 661–664 (1973).
Luckasen, J.R., White, J.G., Kersey, J.H.: Mitogenic properties of a calcium ionophore, A23187. Proc. Natl. Acad. Sci. USA 71, 5088–5090 (1974).
Lüers, Th.: Die Chromosomen des Menschen. Z. Naturwiss.-Med. Grundlag. 2, 1-21 (1964).
Lüers, Th., Struck, E., Albrecht, M.: Strukturanomalien der Chromosomen bei einer akuten paramyeloblasten Leukämie. Med. Welt 17, 948–951; 955–956 (1963).
Lüers, Th., Struck, E., Boll, I.: Über eine spezifische Chromosomenanomalie bei Leukämie (das "minute" oder Ph¹-Chromosom). Münch. Med. Wochenschr. 104, 1493–1496 (1962).
Lukes, R.J.: The pathologic manifestations of Hodgkin's disease. Z. Krebsforsch. 78, 129–136 (1972).
Lukes, R.J., Butler, J.J., Hicks, E.B.: Natural history of Hodgkin's disease as related to its pathologic picture. Cancer 19, 317–344 (1966).
Lukes, R.J., Collins, R.D.: Immunologic characterization of human malignant lymphomas. Cancer 34, 1488–1503 (1974).
Lukes, R.J., Collins, R.D.: Lukes-Collins classification and its significance. Cancer Treat. Rep. 61, 971–979 (1977).
Lundh, B., Mitelman, F., Nilsson, P.G., Stenstam, M., Söderström, N.: Chromosome abnormalities identified by banding technique in a patient with acute myeloid leukaemia complicating Hodgkin's disease. Scand. J. Haematol. 14, 303–307 (1975).

Lundsteen, C., Kristoffersen, L., Ottosen, F.: Studies on the mechanism of chromosome banding with trypsin. Humangenetik 24, 67–69 (1974).
Lundsteen, C., Lind, A.-M., Granum, E.: Visual classification of banded human chromosomes. I. Karyotyping compared with classification of isolated chromosomes. Ann. Hum. Genet. 40, 87–97 (1976).
Lustman, F., Stoffes-de Saint Georges, A., Ardichvili, D., Koulischer, L., Deinol, H.: La macroglobulinemie de Waldenström. Acta Clin. Belg. 23, 67 (1968).
Lutzner, M.A., Edelson, R.L., Smith, A.W., Shevach, E.M., Green, I.: Two varieties of Sézary syndrome, both bearing T-cell markers. Lancet ii, 207 (1973).
Lutzner, M.A., Emerit, I., Durepaire, R., Flandrin, G., Grupper, Ch., Prunieras, M.: Cytogenetic, cytophotometric, and ultrastructural study of large cerebriform cells of the Sézary syndrome and description of a small-cell variant. J. Natl. Cancer Inst. 50, 1145–1162 (1973).
Lyall, J.M., Garson, O.M.: Non-random chromosome changes in the blastic transformation stage of Ph¹-positive chronic granulocytic leukaemia. Leukemia Res. 2, 213–222 (1978).
Lyall, J.M., Brodie, G.N., Garson, O.M.: A variant chromosomal translocation found in a series of 24 patients with Philadelphia positive chronic granulocytic leukaemia. Aust. N.Z. J. Med. 8, 288–289 (1978).
Lynch, H.T., Kaplan, A.R., Lynch, J.F.: Klinefelter syndrome and cancer: a family study. J.A.M.A. 229, 809–811 (1974).
Lyon, M.F.: Possible mechanisms of X chromosome inactivation. Nature [New Biol.] 232, 229–232 (1971).

Macavei, I., Halmos, S.: High T-cell counts in chronic lymphatic leukaemia. Lancet i, 220–221 (1975).
Macciotta, A., Cao, A., Corda, R., Zorcolo, G., Pala, P.: Ulteriore studio sulla sindrome anemica panemocitopenica di Fanconi, con particolare signardo a possibili aspetti di alterazione cromosomica. Ann. Ital. Pediatr. 18, 1–22 (1965).
MacDiarmid, W.D.: Chromosomal changes following treatment of polycythaemia with radioactive phosphorus. Q. J. Med. 34, 133–143 (1965).
Macdougall, L.G., Brown, J.A., Cohen, M.M., Judisch, J.M.: C-monosomy myeloproliferative syndrome: A case of 7-monosomy. J. Pediatr. 84, 256–259 (1974).
Mace, M.L., Jr., Tevethia, S.S., Brinkley, B.R.: Differential immunofluorescent labeling of chromosomes with antisera specific for single strand DNA. Exp. Cell Res. 75, 521–523 (1972).

Macek, M., Benyesh-Melnick, M.: Chromosomal analysis of lymphoblastoid cell lines from patients with leukaemia, infectious mononucleosis, or Burkitt lymphoma. Neoplasma 19, 51–56 (1972).

Macek, M., Seidel, E.H., Lewis, R.T., Brunschwig, J.P., Wimberly, I., Benyesh-Melnick, M.: Cytogenetic studies of EB virus-positive and EB virus-negative lymphoblastoid cell lines. Cancer Res. 31, 308–321 (1971).

MacSween, R.N.M.: Reticulum-cell sarcoma and rheumatoid arthritis in a patient with XX/XXY/XXXY Klinefelter's syndrome and normal intelligence. Lancet i, 460–461 (1965).

Maeda, T., Tabata, T., Kusayama, S., Kimura, T.: Chromosomes of human tumors. V. Chromosomal alternatives in a metastatic lesion of a maxillary cancer. Wakayama Med. Rep. 8, 85–91 (1964).

Maeda, T., Tabata, T., Saito, H.: Chromosomes of human tumors. VI. A comparative chromosome survey of the primary and its secondary tumors. Wakayama Med. Rep. 9, 225–234 (1965).

Maino, V.C., Green, N.M., Crumpton, M.J.: The role of calcium ions in initiating transformation of lymphocytes. Nature (Lond.) 251, 324–327 (1974).

Makino, S.: Some observations on the chromosomes of the Yoshida sarcoma cells based on the homoplastic and heteroplastic transplantations. A preliminary report. Gann 42, 87–90 (1951).

Makino, S.: The concept of stemline-cells as progenitors of a neoplastic population. Proc. Int. Genet. Symposia, Tokyo, pp. 177–181 (1956). Cytologia [Suppl.] (1957a).

Makino, S.: The chromosome cytology of the ascites tumors of rats, with special reference to the concept of the stem-line cell. Int. Rev. Cytol. 6, 26–84 (1957b).

Makino, S.: The role of tumor stem-cells in regrowth of the tumor following drastic applications. ACTA 15, 196–198 (1959).

Makino, S.: Historic sketches of human cytogenetics. Heredity (Japan) 27, 7–10 (1973).

Makino, S.: Cytogenetics of canine veneral tumors: Worldwide distribution and a common karyotype. *In* J. German (ed.): *Chromosomes and Cancer*, pp. 335–372. New York: Wiley (1974).

Makino, S.: *Human Chromosomes.* Tokyo: Igaku Shoin Ltd. (1975).

Makino, S., Awa, A.A.: Chromosome breakage in four chronic myeloid leukemia patients. Hereditas 52, 253–264 (1964).

Makino, S., Ishihara, T.: Chromosomes of some human subjects with non-malignant diseases. Proc. Jap. Acad. 36, 162–167 (1960).

Makino, S., Kano, K.: Cytologic studies on cancer. II. Daily observations on the mitotic frequency and the variation of the chromosome number in tumor cells of the Yoshida sarcoma through a transplant generation. J. Fac. Sci. Hokkaido Univ., Ser. VI. Zool. 10, 225–242 (1951).

Makino, S., Muramoto, J.-I.: Some observations on the variability of the human Y chromosome. Proc. Jap. Acad. 40, 757–761 (1964).

Makino, S., Nishimura, I.: Water pretreatment squash technique. Stain Technol. 27, 1–7 (1952).

Makino, S., Sasaki, M.: On the chromosome number of man. A preliminary note. Proc. Jap. Acad. 35, 99–104 (1959).

Makino, S., Sasaki, M.: A study of the somatic chromosomes in the Japanese. Jap. J. Genet. 35, 228–237 (1960).

Makino, S., Sasaki, M.S.: A chromosomal abnormality in a myelocytic aleukaemic leukaemia. Lancet i, 851–852 (1964).

Makino, S., Ishihara, T., Tonomura, A.: Cytological studies of tumors. XXVII. The chromosomes of thirty human tumors. Z. Krebsforsch. 63, 184–208 (1959a).

Makino, S., Ishihara, T., Tonomura, A.: Chromosome conditions in thirty human tumors. Proc. Jap. Acad. 35, 252–256 (1959b).

Makino, S., Kikuchi, Y., Sasaki, M.S., Sasaki, M., Yoshida, M.: A further survey of the chromosomes in the Japanese. Chromosoma 13, 148–162 (1962).

Makino, S., Obara, Y., Sasaki, M., Ohshima, M., Mikuni, C.: Cytologic studies on tumors XLVII. Acute myelogenous leukemia with C/G translocation and differential response to PHA of normal and leukemic cells of blood and marrow. Cancer 24, 758–763 (1969).

Makino, S., Sasaki, M., Fukuschima, T.: Preliminary notes on the chromosomes of human chorionic lesions. Proc. Jap. Acad. 39, 54–57 (1963).

Makino, S., Sasaki, M.S., Fukuschima, T.: Cytological studies of tumors. XLI. Chromosomal instability in human chorionic lesions. Okajimas Folia Anat. Jap. 40, 439–465 (1965).

Makino, S., Sasaki, M.S., Fukuschima, T.: Triploid chromosome constitution in human chorionic lesions. Lancet ii, 1273–1275 (1964a).

Makino, S., Sasaki, M.S., Tonomura, A.: Cytological studies of tumors. XL. Chromosome studies in fifty-two human tumors. J. Natl. Cancer Inst. 32, 741–763 (1964b).

Makino, S., Sofuni, T., Mitani, M.: Cytological studies on tumors. XLVIII. A chromosome condition in effusion cells from a patient with neuroblastoma. Gann 56, 127–133 (1965).

Makino, S., Tonomura, A., Ishihara, T.: Studies on the chromosomes of some types of human cancer cells. Dobkutsaku Zasshi (Tokyo) 68, 142 (1959).

Makino, S., Yamada, K., Kajii, T.: Chromosome aberrations in leukocytes of patients with aseptic meningitis. Chromosoma 16, 372–380 (1965).

Maldonado, J.E., Pierre, R.V.: The platelets in preleukemia and myelomonocytic leukemia: Ultrastructural cytochemistry and cytogenetics. Mayo Clin. Proc. 50, 573–587 (1975).

Males, J.L., Lain, K.C.: Epithelioid sarcoma in XO/XX Turner's syndrome. Arch. Pathol. 94, 214–216 (1972).

Málková, J., Michalová, K., Chrz, R., Kobilková, J., Motlík, K., Stárka, L.: Dicentric Yp chromosome in a patient with the gonadal dysgenesis and gonadoblastoma. Humangenetik 27, 251–253 (1975).

Malkovský, M., Bubeník, J.: Human urinary bladder carcinoma cell line (T24) in long-term culture: Chromosomal studies on a wild population and derived sublines. Neoplasma 24, 319–326 (1977).

Mamaev, N.N., Gerchak, D., Mamaeva, S.E., Afanasjev, B.V., Gaponov, G.G.: Particularities of DNA replication of the chromosomes in human leucaemia cells. Tsitologiia 18, 718–723 (1976).

Mammon, Z., Grinblat, J., Joshua, H.: Philadelphia chromosome with t(6;22)(q25;q12). N. Engl. J. Med. 294, 827–828 (1976).

Mamunes, P., Lapidus, P.H., Abbott, J.A., Roath, S.: Acute leukaemia and Klinefelter's syndrome. Lancet ii, 26–27 (1961).

Mancinelli, S., Durant, J.R., Hammack, W.J.: Cytogenetic abnormalities in a plasmacytoma. Blood 33, 225–233 (1969).

Mandel, E.M., Shabtai, F., Gafter, U., Klein, B., Halbrecht, I., Djaldetti, M.: Ph¹-positive acute lymphocytic leukemia with chromosome 7 abnormalities. Blood 49, 281–287 (1977).

Mandelli, F., Amadori, S., Alimena, G., Annino, L., Nardelli, S., Papa, G.: Experience on the treatment of chronic myelocytic leukaemia (CML) in blastic crisis. Scand. J. Haematol. 19, 496–502 (1977).

Maniatis, A.K., Amsel, S., Mitus, W.J., Coleman, N.: Chromosome pattern of bone marrow fibroblasts in patients with chronic granulocytic leukaemia. Nature 222, 1278–1279 (1969).

Manna, G.K.: Chromosome number of human endometrium. Nature 173, 271–272 (1954).

Manna, G.K.: Chromosome number of human cervix uteri. Nature 176, 354–355 (1955).

Manna, G.K.: A study on chromosomes of the human nonneoplastic and neoplastic uterine tissues. *Proceedings of the International Genetics Symposium,* pp. 182–187 (1957a).

Manna, G.K.: A study on the chromosome number of human neoplastic uterine cervix tissue. *Proceedings, Zoological Society of Calcutta,* Mookerjee Memorial Volume, pp. 95–112 (1957b).

Manna, G.K.: The relative frequencies of different mitotic stages with some of their abnormalities in nonneoplastic and neoplastic human cervix uteri. *Proceedings, Zoological Society of Calcutta,* 15, 1–10 (1962).

Mannini, A., Fraccaro, M., Gerli, M., Tiepolo, L., Zara, C.: Studi citogenetici in una serie di carcinomi ovarici prima del trattamento. Clin. Ostet. Ginecol. 71, 1–15 (1966).

Manocha, S.L.: The diploid deoxyribonucleic acid (DNA) content to basal cell carcinomas in man. Experientia 25, 201–203 (1969).

Manoharan, A., Garson, O.M.: Familial polycythaemia vera: A study of 3 sisters. Scand. J. Haematol. 17, 10–16 (1976).

Manolov, G., Manolova, Y.: A marker band in one chromosome 14 in Burkitt lymphomas. Hereditas 69, 300 (1971).

Manolov, G., Manolova, Y.: Marker band in one chromosome 14 from Burkitt lymphomas. Nature 237, 33–34 (1972).

Manolov, G., Levan, A., Nadkarni, J.S., Clifford, P.: Burkitt's lymphoma with female karyotype in an African male child. Hereditas 66, 79–100 (1970).

Manolov, G., Manolova, Y., Fiskejö, G., Levan, A.: The complexity of the fluorescent pattern of the human Y chromosome. Hereditas 68, 328–331 (1971a).

Manolov, G., Manolova, Y., Levan, A.: The fluorescence pattern of the human karyotype. Hereditas 69, 273–286 (1971b).

Manolov, G., Manolova, Y., Levan, A., Klein, G.: Experiments with fluorescent chromosome staining in Burkitt tumors. Hereditas 68, 235–244 (1971c).

Manolov, G., Manolova, Y., Levan, A., Klein, G.: Fluorescent pattern of apparently normal chromosomes in Burkitt lymphomas. Hereditas 68, 160–163 (1971d).

Mansour, N.J.: Philadelphia chromosome and leukemia. Ann. Intern. Med. 72, 960–961 (1970).

Marimuthu, K.M., Selles, W.D., Neurath, P.W.: Computer analysis of Giemsa banding patterns and automatic classification of human chromosomes. Am. J. Hum. Genet. 26, 369–377 (1974).

Marcovitch, H., Cain, F., Havard, C.W.H.: Eosinophilic leukaemia. Br. J. Clin. Pract. 27, 185–188 (1973).

Mark, H.F.L., Mendoza, T.: R banding. Am. J. Hum. Genet. 28, 191–192 (1976).

Mark, J.: Double minutes—a chromosomal aberration in Rous sarcomas in mice. Hereditas 57, 1–22 (1967a).

Mark, J.: Chromosomal analysis of ninety-one pri-

mary Rous sarcoma virus-induced tumours in the mouse. Hereditas 57, 1–22 (1967b).
Mark, J.: Rous sarcomas in mice: The chromosomal progression in primary tumours. Eur. J. Cancer 5, 307–315 (1969a).
Mark, J.: Two benign intracranial tumours with an abnormal chromosomal picture. Acta Neuropathol. (Berl.) 14, 174–184 (1969b).
Mark, J.: Chromosomal patterns in human meningiomas. Eur. J. Cancer 6, 489–498 (1970a).
Mark, J.: Chromosomal characteristics of neurogenic tumors in children. Acta Cytol. 14, 510–518 (1970b).
Mark, J.: Chromosomal analysis of a human retinoblastoma. Acta Ophthalmol. [Kbh.] 48, 124–135 (1970c).
Mark, J.: Chromosomal characteristics of human pituitary adenomas. Acta Neuropathol. (Berl.) 19, 99–109 (1971a).
Mark, J.: Chromosomal aberrations and their relation to malignancy in meningiomas: a meningioma with ring chromosomes. Acta Pathol. Microbiol. Scand. [A] 79, 193–200 (1971b).
Mark, J.: Chromosomal characteristics of neurogenic tumours in adults. Hereditas 68, 61–100 (1971c).
Mark, J.: The chromosomes in two oligodendrogliomas and three ependymomas in adults. Hereditas 69, 145–149 (1971d).
Mark, J.: The chromosomal findings in seven human neurinomas and one neurosarcoma. Acta Pathol. Microbiol. Scand. [A] 80, 61–70 (1972a).
Mark, J.: Chromosomal characteristics of secondary human brain tumours. Eur. J. Cancer 8, 399–407 (1972b).
Mark, J.: The fluorescence karyotypes of three human meningiomas with hyperdiploid-hypotriploid stemlines. Acta Neuropathol. (Berl.) 25, 46–53 (1973a).
Mark, J.: Origin of the ring chromosome in a human recurrent meningioma studied with G-band technique. Acta Pathol. Microbiol. Scand. [A] 81, 588–590 (1973b).
Mark, J.: Karyotype patterns in human meningiomas. A comparison between studies with G- and Q-banding techniques. Hereditas 75, 213–220 (1973c).
Mark, J.: G-band analyses of an established cell line of a human malignant glioma. Humangenetik 22, 323–326 (1974a).
Mark, J.: Chromosome patterns in benign and malignant tumors in the human nervous system. In J. German (ed.): Chromosomes and Cancer, pp. 481–495. New York: Wiley (1974b).
Mark, J.: The human meningioma: A benign tumor with specific chromosome characteristics. In J. German (ed.): Chromosomes and Cancer, pp. 497–517. New York: Wiley (1974c).

Mark, J.: Two pseudodiploid human breast carcinomas studied with G-band technique. Eur. J. Cancer 11, 815–819 (1975a).
Mark, J.: Histiocytic lymphomas with the marker chromosome 14q+. Hereditas 81, 289–292 (1975b).
Mark, J.: G-band analyses of a human intestinal leiomyosarcoma. Acta Pathol. Microbiol. Scand. [A] 84, 538–540 (1976).
Mark, J.: Chromosomal abnormalities and their specificity in human neoplasms: An assessment of recent observations by banding techniques. Adv. Cancer Res. 24, 165–222 (1977a).
Mark, J.: On the specificity of medium-sized isomarker chromosomes in non-Burkitt lymphomas. Acta Pathol. Microbiol. Scand. [A] 85, 557–558 (1977b).
Mark, J.: Monosomy 14, monosomy 22 and 13q−. Three chromosomal abnormalities observed in cells of two malignant mesotheliomas studied by banding techniques. Acta Cytol. 22, 398–401 (1978).
Mark, J., Granberg, I.: The chromosomal aberration of double-minutes in three gliomas. Acta Neuropathol. (Berl.) 16, 194–204 (1970).
Mark, J., Ekedahl, C., Arenander, E.: Cytogenetical observations in two cases of clinico-pathologically suspected malignant lymphomas. Hereditas 84, 225–230 (1976).
Mark, J., Ekedahl, C., Dahlenfors, R.: Characteristics of the banding patterns in non-Hodgkin and non-Burkitt lymphomas. Hereditas 88, 229–242 (1978).
Mark, J., Ekedahl, C., Hagman, A.: Origin of the translocated segment of the 14q+ marker in non-Burkitt lymphomas. Hum. Genet. 36, 277–282 (1977).
Mark, J., Levan, G., Mitelman, F.: Identification by fluorescence of the G chromosome lost in human meningiomas. Hereditas 71, 163–168 (1972).
Mark, J., Mitelman, F., Dencker, H., Norryd, C., Tranberg, K.-G.: The specificity of the chromosomal abnormalities in human colonic polyps. A cytogenetic study of multiple polyps in a case of Gardner's syndrome. Acta Pathol. Microbiol. Scand. 81, [A] 85–90 (1973).
Mark, J., Mitelman, F., Levan, G.: On the specificity of the G abnormality in human meningiomas studied by the fluorescence technique. Acta Pathol. Microbiol. Scand. [A] 80, 812–820 (1972).
Mark, J., Pontén, J., Westermark, B.: G-band analyses of an established cell line of a human malignant glioma. Humangenetik 22, 323–326 (1974a).
Mark, J., Pontén, J., Westermark, B.: Origin of the marker chromosomes in an established hypo-

triploid glioma cell line studied with G-band technique. Acta Neuropathol. (Berl.) 29, 223–228 (1974b).
Mark, J., Pontén, J., Westermark, B.: Cytogenetical studies with G-band technique of established cell lines of human malignant gliomas. Hereditas 78, 304–307 (1974c).
Mark, J., Westermark, B., Pontén, J., Hugosson, R.: Banding patterns in human glioma cell lines. Hereditas 87, 243–260 (1977).
Marković, V.D., Worton, R.G., Berg, J.M.: Evidence for the inheritance of silver-stained nucleolus organizer regions. Hum. Genet. 41, 181–187 (1978).
Marks, S.M., Baltimore, D., McCaffrey, R.: Terminal transferase as a predictor of initial responsiveness to vincristine and prednisone in blastic chronic myelogenous leukemia. N. Engl. J. Med. 298, 812–814 (1978).
Marks, S.M., McCaffrey, R., Rosenthal, D.S., Moloney, W.C.: Blastic transformation in chronic myelogenous leukemia: Experience with 50 patients. Med. Pediatr. Oncol. 4, 159–167 (1978).
Martin, P., Levin, B., Golomb, H.M., Riddell, R.H.: Chromosome analysis of primary large bowel tumors: A new method for improving the yield of analyzable metaphases. Cancer (submitted).
Martineau, M.: A similar marker chromosome in testicular tumours. Lancet i, 839–842 (1966).
Martineau, M.: Chromosomes in testicular tumors. Lancet i, 386 (1967).
Martineau, M.: Chromosomes in human testicular tumours. J. Pathol. 99, 271–282 (1969).
Masera, P., Pogoraro, L., Rovera, G.: Anomalie del cromosoma 21 e fosfatasi alcalina nelle leucemie. Bull. Soc. Ital. Biol. Exp. 41, 748–751 (1965).
Mastrangelo, R., Zuelzer, W.W., Ecklund, P.S., Thompson, R.I.: Chromosomes in the spinal fluid: evidence for metastatic origin of meningeal leukemia. Blood 35, 227–235 (1970).
Mastrangelo, R., Zuelzer, W.W., Thompson, R.I.: The significance of the Ph[1] chromosome in acute myeloblastic leukemia: serial cytogenetic studies in a critical case. Pediatrics 40, 834–841 (1967).
Matre, R., Talstad, I., Haugen, Å.: Surface markers in nonphagocytic hairy cell leukemia. Acta Pathol. Microbiol. Scand. [A] 85, 406–412 (1977).
Matsaniotis, N., Kiossoglou, K.A., Karpouzas, J., Anastaśea-Vlachou, K.: Chromosomes in Kostmann's disease. Lancet ii, 104 (1966a).
Matsaniotis, N., Kiossoglou, K.A., Maounis, F., Anagnostakis, D.E.: Chromosomes in infectious hepatitis. Lancet ii, 1421 (1966b).
Matsaniotis, N., Kiossoglou, K.A., Maounis, F., Basti-Maouni, B.: Chromosomal study in megaloblastic anaemia of children. Acta Haematol. (Basel) 39, 29–35 (1968).
Matsui, S., Sasaki, M.: Differential staining of nucleolus organisers in mammalian chromosomes. Nature 246, 148–150 (1973).
Matsui, S., Weinfeld, H., Sandberg, A.A.: Dependence of chromosome pulverization in virus fused cells on events of the G2-period. J. Natl. Cancer Inst. 47, 401–411 (1971).
Matsui, S., Weinfeld, H., Sandberg, A.A.: Fate of chromatin of interphase nuclei subjected to prophasing in virus-fused cells. J. Natl. Cancer Inst. 49, 1621–1630 (1972).
Matsui, S., Yoshida, H., Weinfeld, H., Sandberg, A.A.: Induction of prophase in interphase nuclei by fusion with metaphase cells. J. Cell Biol. 54, 120–132 (1972).
Matsunaga, M., Sadamori, N., Tomonaga, Y., Tagawa, M., Ichimaru, M.: Cytogenetic study on 22 patients with chronic myelogenous leukemia. Jap. J. Hum. Genet. 20, 267–268 (1976a).
Matsunaga, M., Sadamori, N., Tomonaga, Y., Tagawa, M., Ichimaru, M.: Chronic myelogenous leukemia with an unusual karyotype: 46,XY,t(17q+; 22q−). N. Engl. J. Med. 295, 1537 (1976b).
Mattei, J.F., Ayme, S., Mattei, M.G., Gouvernet, J., Giraud, F.: Quantitative and qualitative study of acrocentric associations in 109 normal subjects. Hum. Genet. 34, 185–194 (1976).
Maunoury, R., Arnoult, J., Vedrenne, Cl.: Establishing and characterization of a cell line derived from a human osteosarcoma. Pathol. Biol. (Paris) 20, 369–376 (1972).
Mauri, C., Torelli, U., di Prisco, U., Silingardi, V., Artusi, T., Emilia, G.: Lymphoid blastic crisis at the onset of chronic granulocytic leukemia. Report of 2 cases. Cancer 40, 865–870 (1977).
Maurice, P.A., Alberto, P., Ferrier, S., Freund, M.: Leucémie myélocytaire chronique: "guérison" apparente depuis plus de 9 ans, consécutive á une hyperplasie médullaire thérapeutique. Schweiz. Med. Wochenschr. 101, 1781–1782 (1971).
Mayall, B.M., Carrano, A.V., Moore, D.H., Rowley, J.D.: Quantification by DNA-based cytophotometry of the 9q+/22q− chromosomal translocation associated with chronic myelogenous leukemia. Cancer Res. 37, 3590–3593 (1977).
Mayall, B.H., Carrano, A.V., Rowley, J.D.: DNA cytophotometry of chromosomes in a case of chronic myelogenous leukemia. Clin. Chem. 20, 1080–1085 (1974).
McAllister, R.M., Gardner, M.B., Greene, A.E., Bradt, C., Nichols, W.W., Landing, B.H.: Cultivation *in vitro* of cells derived from a human osteosarcoma. Cancer 27, 397–402 (1971).
McAllister, R.M., Isaacs, H., Rongey, R., Peer, M., Au, W., Soukup, S.W., Gardner, M.B.: Establish-

ment of a human medulloblastoma cell line. Int. J. Cancer 20, 206–212 (1977).

McAllister, R.M., Melnyk, J., Finklestein, J.Z., Adams, E.C., Jr., Gardner, M.B.: Cultivation *in vitro* of cells derived from a human rhabdomyosarcoma. Cancer 24, 520–526 (1969).

McBride, J.A.: Acute leukaemia after treatment for hyperthyroidism with radioactive iodine. Br. Med. J. ii, 736 (1964).

McCaffrey, R., Harrison, T.A., Parkman, R., Baltimore, D.: Terminal deoxynucleotidyl transferase activity in human leukemic cells and in normal human thymocytes. N. Engl. J. Med. 292, 775–780 (1975).

McCaw, B.K., Hecht, F., Harnden, D.G., Teplitz, R.L.: Somatic rearrangement of chromosome 14 in human lymphocytes. Proc. Natl. Acad. Sci. USA 72, 2071–2075 (1975).

McClain, D.A., Wang, J.L., Edelman, G.M.: The effect of sodium metaperiodate on T and B lymphocytes. Cell Immunol. 15, 287–293 (1975).

McClure, P.D., Thaler, M.M., Conen, P.E.: Chronic erythroleukemia with chromosome mosaicism. Report of a case in a 5-year-old boy. Arch. Intern. Med. 115, 697–703 (1965).

McConnell, T.S., Parsons, L.: Chromosome evaluation in familial polyposis of the colon. Rocky Mt. Med. J. 65, 51–53 (1968).

McCormack, K.R., Sheline, G.E.: Leukemia after radioiodine therapy for hyperthyroidism. Calif. Med. 98, 207–209 (1963).

McCormick, D.P., Amman, A.J., Kimishige, I., Miller, D.G., Hong, R.: A study of allergy in patients with malign lymphoma and CLL. Cancer 27, 93–99 (1971).

McCulloch, P.B., Dent, P.B., Hayes, P.R., Liao, S.-K.: Common and individually specific chromosomal characteristics of cultured melanoma. Cancer Res. 36, 398–404 (1976).

McDermott, A., Romain, D., Fraser, I.D., Scott, G.L.: Isochromosome 17q in two cases of acute blastic transformation in myeloproliferative disorders. Hum. Genet. 45, 215–218 (1978).

McDonough, P.G., Byrd, J.R., Tho, P.T., Otken, L.: Gonadoblastoma in a true hermaphrodite with a 46,XX karyotype. Obstet. Gynecol. 47, 355–358 (1976).

McDougall, J.K.: Spontaneous and adenovirus type 12-induced chromosome aberrations in Fanconi's anaemia fibroblasts. Int. J. Cancer 7, 526–534 (1971*a*).

McDougall, J.K.: Adenovirus-induced chromosome aberrations in human cells. J. Genet. Virol. 12, 43–51 (1971*b*).

McIntyre, O.R.: C group chromosome anomaly in polycythemia vera. Proc. Am. Soc. Hematol., p. 282 (abstract) (1970).

McKay, R.D.G.: The mechanism of G and C banding in mammalian metaphase chromosomes. Chromosoma 44, 1–14 (1973).

McKenzie, M., Perrotta, A.: Philadelphia chromosome positive (Ph¹+) chronic basophilic myelogenous leukemia—Report of a second case. *Program of the 18th Annual Meeting of the American Society of Hematology,* P. 140. Held in Dallas, Tex., Dec. 6–9 (1975).

McKusick, V.A.: Chromosomes in Marfan's syndrome. Lancet i, 1194 (1960).

McKusick, V.A., Ruddle, F.H.: The status of the gene map of the human chromosomes. Science 196, 390–405 (1977).

McLaren, G.D., Tebbi, K., Muir, W.A.: Cytogenetic analysis of granulocyte colonies formed *in vitro* in chronic granulocytic leukemia. American Society of Hematology, Nineteenth Annual Meeting, Abstract #293. Held in Boston, Mass., Dec. 4–7 (1976).

McLaughlin, H., Wetherly-Mein, S., Pitcher, C., Hobbs, R.J.: Nonimmunoglobulin-bearing "B" lymphocytes in chronic lymphatic leukaemia? Br. J. Haematol. 25, 7–14 (1973).

Meighan, S., Stitch, H.F.: Simplified technique for examination of chromosomes in the bone marrow of man. Can. Med. Assoc. J. 84, 1004–1006 (1961).

Meisner, L.F., Chuprevich, T.W., Inhorn, S.L.: Giemsa banding specificity. Nature [New Biol.] 245, 145–147 (1973).

Meisner, L.F., Chuprevich, T.W., Inhorn, S.L.: Mechanisms of chromatid breakage in human lymphocyte cultures. Acta Cytol. 21, 555–558 (1977).

Meisner, L.F., Inhorn, S.L., Chuprevich, T.W.: Cytogenetic analysis as a diagnostic aid in leukemia. Am. J. Clin. Pathol. 60, 435–444 (1973).

Meisner, L., Inhorn, S.L., Nielsen, P.: Karyotype evolution of cells with the Philadelphia chromosome. Acta Cytol. 14, 192–199 (1970).

Melamed, M.R., Darzynkiewicz, Z., Traganos, F., Sharpless, T.K.: Nucleic acid content and nuclear chromatin structure of human bladder cell culture lines as studied by flow cytofluorometry. Cancer Res. 37, 1227–1231 (1977).

Meme, J.S., Oduori, M.L., Gripenberg, U.: Fanconi's aplastic anaemia: A case report of an affected African child and a review of the literature. East Afr. Med. J. 52, 462–466 (1975).

Mende, S.: Y-Chromosomen-Verlust bei hämatologischen Erkrankungen älterer Männer. Blut 33, 219 (1976).

Mende, S., Aveissenfels, I., Pribilla, W.: Cytogenetic follow-up studies in idiopathic refractory sideroblastic anemia. International Society of Haematology, European African Division, 4th Meeting. Held in Istanbul, Sept. 5–9, Abstract #717 (1977).

Mende, S., Fülle, H.-H., Knuth, A., Weissenfels, I.: Myelomonozytäre Leukämie: Klinische, zytologische und zytogenetische Studien bei akuten, subakuten und chronischen Verlaufsformen. Blut 35, 21 – 34 (1977).

Mendelsohn, M.L., Mayall, B.H., Bogart, E., Moore, D.H., III, Perry, B.H.: DNA content and DNA-based centromeric index of the 24 human chromosomes. Science 179, 1126 – 1129 (1973).

Mendes, N.F., Musatti, C.C., Tolani, E.A.: T and B lymphocyte membrane markers in cells from patients with leukemia and lymphoma. Int. Arch. Allergy Appl. Immunol. 46, 695 – 706 (1974).

Menzies, R.C., Crossen, P.E., Fitzgerald, P.H., Gunz, F.W.: Cytogenetic and cytochemical studies on marrow cells in B_{12} and folate deficiency. Blood 28, 581 – 594 (1966).

Mercer, R.D., Keller, M.K., Lonsdale, D.: An extra abnormal chromosome in a child with mongolism and acute myeloblastic leukemia. Cleve. Clin. Q. 30, 215 – 224 (1963).

Merker, H.: The significance of cytochemical and cytogenetic findings in chronic granulocytic leukemia and related diseases. In F.G.J. Hayhoe (ed.): *Current Research in Leukaemia*, pp. 1 – 15. London: Cambridge University Press (1965).

Merker, H.: Zum Stand der Diskussion über das Philadelphia-Chromosom. Dtsch. Med. Wochenschr. 96, 296 – 298 (1971).

Merker, H.: Stato attuale delle discussioni sul cromosoma Fidelfia. Minerva Med. 63, 4154 – 4155 (1972).

Merker, H., Schneider, G., Burmeister, P., Wolf, U.: Chromosomentranslokation bei chronisch-myeloproliferativem Syndrom. Bericht über zwei Beobachtungen unter besonderer Berücksichtigung pathogenetischer Gesichtspunkte. Klin. Wochenschr. 46, 593 – 600 (1968).

Mertelsmann, R., Mertelsmann, I., Koziner, B., Moore, M.A.S., Clarkson, B.D.: Improved biochemical assay for terminal deoxynucleotidyl transferase in human blood cells: Results in 89 adult patients with lymphoid leukemias and malignant lymphomas in leukemic phase. Leukemia Res. 2, 57 – 69 (1978).

Merz, T., El-Mahdi, A.M., Prempree, T.: Unusual chromosomes and malignant disease. Lancet i, 337 – 339 (1968).

Messinetti, S., Moscarini, M.: Chromosome analysis in the study and diagnosis of benign and malignant epithelial tumors of the gastrointestinal tract. Prog. Med. (Napoli) 26, 721 – 724 (1970).

Messinetti, S., Marcellino, L.R., Zelli, G.P., Moscarini, M.: Osservazioni citogenetiche sui carcinomi ovarici in fase ascitica sottoposti a terapia antiblastica. Prog. Med. (Napoli) 27, 130 (1971).

Messinetti, S., Zelli, G.P., Marcellino, L.R., Alcini, E.: Benign and malignant epithelial tumors of the gastro-enteric tract. Cancer 21, 1000 – 1010 (1968).

Messinetti, S., Zelli, G.P., Marcellino, L.R., Moscarini, M.: L'analisi cromosomica nello studio di ulcune lesioni proliferative e neoplastiche dell'apparatogenitale femminile. Prog. Med. (Napoli) 26, 382 – 401 (1970).

Messinetti, S., Zelli, G. P., Marcellino, L.R., Tumino, G.: L'analisi cromosomica nelli citodiagnostica dei versamenti arcitici di carcinomi gastroenterici. Ann. Ital. Chir. 42, 800 – 816 (1966a).

Messinetti, S., Zelli, G.P., Marcellino, L.R., Tumino, G.: L'analisi cromosomica nello studio e nella diagnostica dei tumori epitheliali benigni e maligni del tubo gastroenterico. Ann. Ital. Chir. 42, 817 – 839 (1966b).

Messinetti, S., Zelli, G.P., Moscarini, M.: In tema de cancro della mammella femminile: un particolare caso di pseudidiploidia. Prog. Med. (Roma) 27, 711 – 714 (1971).

Meuge, C.: Étude cytogénétique de trois epanchements neoplasiques chez des malades atteintes de tumeurs ovariennes. Travail de la Fondation Bergonié, p. 127. Bordeaux: Baillet (1967).

Meytes, D., Akstein, E., Modan, B.: Cytogenetic findings in polycythemia vera. A review and reevaluation. Israel J. Med. Sci. 133, 1226 – 1239 (1977).

Meytes, D., Seligsohn, U., Ramot, B.: Multiple myeloma with terminal erythroleukemia. Acta Haematol. (Basel) 55, 358 – 362 (1976).

Michalová, K., Málková, J., Činátl, J., Placerová, J.: Chromosomal characteristics and non-random distribution of sister chromatid exchanges in lymphoblastoid cell lines isolated from acute leukemias. Neoplasma 24, 537 – 545 (1977).

Michaux, J.-L., Van Den Berghe, H., Rodhain, J., Sokal, G., David, G., Hulhoven, R.: Étude simultanée du caryotype et de l'histologie médullaire dans la leucémie myéloïde chronique a chromosome Ph[1]. Nouv. Rev. Fr. Hematol. 15, 575 – 588 (1975).

Miescher, P.A., Farquet, J.J.: Chronic myelomonocytic leukemia in adults. Semin. Hematol. 11, 129 – 139 (1974).

Mikelsaar, A.-V., Schmid, M., Krone, W., Schwarzacher, H.G., Schnedl, W.: Frequency of Ag-stained nucleolus organizer regions in the acrocentric chromosomes of man. Hum. Genet. 37, 73 – 77 (1977).

Mikkelsen, M., Petersen, G.S., Bøgh, A.: Chromosome studies of two patients with Down's syndrome (mongolism) and leukemia (in Danish). Ugeskr. Laeger 126, 1365 – 1368 (1964).

Milcu, St.-M., Ionescu, B., Strihan, P., Iliescu, I., Augustin, M., Maximilian, C.: Sindrom Turner cu adenom hipofizar si cariotip XO. Stud. Cercet. Endocrinol. 15, 257–262 (1964).

Milcu, St.-M., Ionescu, B., Strihan, P., Parhan, C. C., Iliescu, J., Maximilian, C.: A rare form of intersexuality with karyotype XO/XY. Hum. Chrom. Newsl. 15, 17 (1965).

Milcu, St.-M., Stanereu, V., Ionereu, V., Maximilian, C.: Mozaic cromozomial: 46 cromozomi si poliploide neurofibromatizo. Stud. Cercet. Endocrinol. 12, 745 (1961).

Miles, C.P.: Sex chromatin in cultured normal and cancerous human tissues. Cancer 12, 299–305 (1959).

Miles, C.P.: Chromosomal alterations in cancer. Med. Clin. North Am. 50, 875–885 (1966).

Miles, C.P.: Chromosome analysis of solid tumors. II. Twenty-six epithelial tumors. Cancer 20, 1274–1287 (1967a).

Miles, C.P.: Chromosome analysis of solid tumors. I. Twenty-eight nonepithelial tumors. Cancer 20, 1253–1273 (1967b).

Miles, C.P.: Chromosome changes in Hodgkin's disease. Natl. Cancer Inst. Monogr. 36, 197–201 (1973).

Miles, C.P.: Non-random chromosome changes in human cancer. Br. J. Cancer 30, 73–85 (1974).

Miles, C. P., Gallagher, R.E.: Chromosomes of a metastatic human cancer. Lancet ii, 1145–1146 (1961).

Miles, C.P., O'Neill, F.: Chromosome studies of 8 *in vitro* lines of Burkitt's lymphoma. Cancer Res. 27, 392–402 (1967).

Miles, C.P., Wolinska, W.: A comparative analysis of chromosomes and diagnostic cytology in effusions from 58 cancer patients. Cancer 32, 1458–1469 (1973).

Miles, C.P., Geller, W., O'Neill, F.: Chromosomes in Hodgkin's disease and other malignant lymphomas. Cancer 19, 1103–1116 (1966).

Millard, R.E.: Abnormalities of human chromosomes following therapeutic irradiation. Cytogenetics 4, 277–294 (1965).

Millard, R.E.: Chromosome abnormalities in the malignant lymphomas. Eur. J. Cancer 4, 97–105 (1968).

Millard, R.E., Seif, G.: Chromosomes in malignant lymphomas. Lancet i, 781 (1967).

Millard, R.E., Lawler, S.D., Kay, H.E.M., Cameron, C.B.: Further observations on patients with a chromosomal abnormality associated with polycythaemia vera. Br. J. Haematol. 14, 363–374 (1968).

Miller, D.A., Allerdice, P.W., Miller, O.J., Breg, W.R.: Quinacrine fluorescence patterns of human D group chromosomes. Nature 232, 24–27 (1971).

Miller, D.A., Dev, V.G., Tantravahi, R., Croce, C.M., Miller, O.J.: Human tumor and rodent-human hybrid cells with an increased number of active human NORs. Cytogenet. Cell Genet. 21, 33–41 (1978).

Miller, O.J.: Sex chromosome anomalies. Am. J. Obstet. Gynecol. 90, 1078–1139 (1964).

Miller O. J., Allerdice, P.W., Capoa, A. de, Miller, D.A.: Asynchronous duplication of homologous autosomes in man associated with differences in length. J. Cell Biol. 39, 91a–92a, Abstract #222 (1968).

Miller, O.J., Breg, W.R., Schmickel, G., Tretter, W.: A family with an XXXY male, a leukemic male and two 21-trisomic mongoloid females. Lancet ii, 78–79 (1961).

Miller, O.J., Breg, W.R., Warburton, D., Miller, D.A., Firchein, I.L., Hirschhorn, K.: Alternative DNA replication patterns associated with long arm length of chromosomes 4 and 5 in the *Cri-du-Chat* syndrome. Cytogenetics 5, 137–151 (1966).

Miller, O.J., Miller, D.A., Allerdice, P.W., Dev, V.G., Grewal, M.S.: Quinacrine fluorescent karyotypes of human diploid and heterodiploid cells lines. Cytogenetics 10, 338–346 (1971).

Miller, O.J., Miller, D.A., Tantravahi, R., Dev, V.G.: Nucleolus organizer activity and the origin of Robertsonian translocations. Cytogenet. Cell Genet. 20, 40–50 (1978).

Miller, O.J., Mukherjee, B.B., Breg, W.R.: Normal variations in the human karyotype. Trans. N.Y. Acad. Sci. 24, 372–382 (1962).

Miller, O.J., Schreck, R.R., Beiser, S.M., Erlanger, B.R.: Immunofluorescent studies of chromosome banding with antinucleoside antibodies. Nobel Symp. 23, 43–48 (1973).

Miller, R.C., Aronson, M.M., Nichols, W.W.: Effects of treatment on differential staining of BrdU labeled metaphase chromosomes: Three-way differentiation of M3 chromosomes. Chromosoma 55, 1–11 (1976).

Miller, R.W.: Down's syndrome (mongolism), other congenital malformations and cancers among sibs of leukemic children. N. Engl. J. Med. 268, 393–401 (1963).

Miller, R.W.: Radiation, chromosomes and viruses in the etiology of leukemia. Evidence from epidemiologic research. N. Engl. J. Med. 271, 30–36 (1964).

Miller, R.W.: Relation between cancer and congenital defects in man. N. Engl. J. Med. 275, 88–93 (1966).

Miller, R.W.: Relation between cancer and congenital defects: an epidemiologic evaluation. J. Natl. Cancer Inst. 40, 1079–1085 (1968).

Miller, R.W.: Neoplasia and Down's syndrome. Ann. N.Y. Acad. Sci. 171, 637–644 (1970).

Miller, R.W.: Overview: Host factors. *In* J.F. Fraumeni, Jr. (ed.): *Persons at High Risk of Cancer.*

An Approach to Cancer Etiology and Control, pp. 121–128 New York: Academic Press (1975).
Miller, R.W., Todaro, G.J.: Viral transformation of cells from persons at high risk of cancer. Lancet i, 81–82 (1969).
Miller, R.W., Fraumeni, J.F., Jr., Hill, J.A.: Neuroblastoma: Epidemiologic approach to its origin. Am. J. Dis. Child. 115, 253–261 (1968).
Milligan, W.J., Garson, O.M.: Giesma banding of "normal" leukaemic chromosomes: a preliminary report. Pathology 6, 143 (1974).
Milner, G.R., Testa, N.Y., Geary, C.G., Dexter, T.M., Muldal, S., MacIver, J.E., Lajtha, L.G.: Bone marrow culture studies in refractory cytopenia and smouldering leukaemia. Br. J. Haematol. 35, 251–261 (1976).
Minkler, J.L., Gofman, J.W., Tandy, R.K.: A specific common chromosomal pathway for the origin of human malignancy. Br. J. Cancer 24, 726–740 (1970a).
Minkler, J.L., Gofman, J.W., Tandy, R.K.: A specific common chromosomal pathway for the origin of human malignancy. II. Evaluation of longterm human hazards of potential environmental carcinogens. Adv. Biol. Med. Phys. 13, 107–151 (1970b).
Minouchi, O., Ohta, T.: The chromosome number of man. Zool. Mag. (Tokyo) 44, 85 (1932).
Minouchi, O., Ohta, T.: On the number of chromosomes and the type of sex-chromosomes in man. Cytologica 5, 472–490 (1934).
Minowada, J., Moore, G.E., Gerner, R.E., Toshima, S., Takagi, N., Sandberg, A.A.: Comparative studies of Burkitt's lymphoma cell lines with cell lines derived from patients with leukemia, other malignancies and normal humans. East Afr. Med. J. Monogr. "Cancer in Africa," pp. 166–179 (1968).
Minowada, J., Oshimura, M., Tsubota, T., Higby, D.J., Sandberg, A.A.: Cytogenetic and immunoglobulin markers of human leukemic B-cell lines. Cancer Res. 37, 3096–3099 (1977).
Minowada, J., Tsubota, T., Nakazawa, S., Srivastava, B.I.S., Huang, C.C., Oshimura, M., Sonta, S., Han, T., Sinks, L.F., Sandberg, A.A.: Establishment and characterization of leukemic T-cell lines, B-cell lines, and null-cell lines: A progress report on surface antigen study of fresh lymphatic leukemias in man. *In* S. Thierfelder, H. Rodt, E. Thiel (eds.): *Haematology and Blood Transfusion*, Vol. 20, *Immunological Diagnosis of Leukemias and Lymphomas*, pp. 241–251. Berlin, Heidelberg: Springer-Verlag (1977).
Mintz, U., Pinkhas, J., Pick, A.I., Vries, A. de: Philadelphia chromosom-negative, lysozyme-positive chronic myeloid leukemia. Haematologica (Pavia) 7, 3–6 (1973).
Misawa, S., Takino, T., Morira, M., Abe, T., Ashihara, T.: Staining properties of a benzimidazol derivative "33258 Hoechst" and a simplified staining method for chromosome banding. Jap. J. Hum. Genet. 22, 1–9 (1977).
Misawa, S., Takino, T., Sugishima, K., Abe, T.: Quinacrine-banding analysis of the Philadelphia chromosome from five patients with chronic myelogenous leukemia. Acta Haematol. Jap. 38, 87–90 (1975).
Misset, J.-L., Venuat, A.-M., Nevarez, L., Mathé, G.: Crises blastiques révélatrices de leucémies myéloïdes chroniques et simulant des leucémies aiguës primaires, ou leucémies aiguës primaires à chromosome Philadelphie? Nouv. Presse Med. 6, 2409–2413 (1977).
Mitani, M., Okochi, K.: A H^3-thymidine autoradiographic study of cell proliferation in patients with acute leukemia in childhood. Tumor Res. 2, 67–89 (1967).
Mitelman, F.: The chromosomes of fifty primary Rous rat sarcomas. Hereditas 69, 155–186 (1971).
Mitelman, F.: Comparative chromosome analysis of primary and metastatic Rous sarcomas in rats. Hereditas 70, 1–14 (1972a).
Mitelman, F.: Predetermined sequential chromosome changes in serial transplantation of Rous rat sarcomas. Acta Pathol. Microbiol. Scand [A] 80, 313–328 (1972b).
Mitelman, F.: Kromosomförändringar och cancer. Lakartidningen (Stockholm) 70, 1651–1654 (1973).
Mitelman, F.: The Rous sarcoma virus story: Cytogenetics of tumors induced by RSV. *In* J. German (ed.): *Chromosomes and Cancer*, pp. 675–693. New York: Wiley (1974a).
Mitelman, F.: Different chromosome morphology of diploid and aneuploid malignant cells. J. Natl. Cancer Inst. 52, 561–564 (1974b).
Mitelman, F.: Heterogeneity of Ph^1 in chronic myeloid leukaemia. Hereditas 76, 315–316 (1974c).
Mitelman, F.: Cytogenetic findings in leukemia. Lakartidningen (Stockholm) 72, 3417–3421 (1975a).
Mitelman, F.: Comparative cytogenetic studies of bone marrow and extramedullary tissues in chronic myeloid leukemia. Ser. Haematol. 8, 113–117 (1975b).
Mitelman, F., Brandt, L.: Chromosome banding pattern in acute myeloid leukaemia. Scand. J. Haematol. 13, 321–330 (1974).
Mitelman, F., Levan, G.: The chromosomes of primary 7,12-dimethylbenz (α) anthracene-induced rat sarcomas. Hereditas 71, 325–334 (1972).
Mitelman, F., Levan, G.: Do only a few chromosomes carry genes of prime importance for malignant transformation? Lancet ii, 264 (1976a).
Mitelman, F., Levan, G.: Clustering of aberrations to specific chromosomes in human neoplasms. II.

A survey of 287 neoplasms. Hereditas 82, 167–174 (1976b).
Mitelman, F., Levan, G.: Clustering of aberrations to specific chromosomes in human neoplasms. III. Incidence and geographic distribution of chromosome aberrations in 856 cases. Hereditas 89, 207–232 (1978).
Mitelman, F., Levan, G.: Chromosomes in neoplasia: An appeal for unpublished data. Cancer Genet. Cytogenet. (in press) (1979).
Mitelman, F., Mark, J.: Chromosomal analysis of primary and metastatic Rous sarcomas in the rat. Hereditas 65, 227–235 (1970).
Mitelman, F., Brandt, L., Levan, G.: Identification of isochromosome 17 in acute myeloid leukaemia. Lancet ii, 972 (1973).
Mitelman, F., Brandt, L., Nilsson, P.G.: The banding pattern in Philadelphia-chromosome negative chronic myeloid leukemia. Hereditas 78, 302–304 (1974a).
Mitelman, F., Brandt, L., Nilsson, P.G.: Cytogenetic evidence for splenic origin of blastic transformation in chronic myeloid leukaemia. Scand. J. Haematol. 13, 87–92 (1974b).
Mitelman, F., Brandt, L., Nilsson, P.G.: Relation among occupational exposure to potential mutagenic/carcinogenic agents, clinical findings and bone marrow chromosomes in acute non-lymphocytic leukemia. Blood 52, 1229–1237 (1978).
Mitelman, F., Hartley-Asp, B., Ursing, B.: Chromosome aberrations and metronidazole. Lancet ii, 802 (1976).
Mitelman, F., Klein, G., Andersson-Anvret, M., Forsby, N., Johansson, B.: 14q+ marker chromosome in an EBV-genome-negative lymph node without signs of malignancy in a patient with EBV-genome-positive nasopharyngeal carcinoma. Int. J. Cancer 23, 32–36 (1979).
Mitelman, F., Levan, G., Brandt, L.: Highly malignant cells with normal karyotype in G-banding. Hereditas 80, 291–293 (1975).
Mitelman, F., Levan, G., Brandt, L.: Specific chromosome changes in neoplasms of man. In S. Armendares, R. Lisker (eds.): *V International Congress of Human Genetics,* p. 141, abstract 368. Held in Mexico, Oct. 10–15 (1976a).
Mitelman, F., Levan, G., Mark, J.: The origin of double-minutes in Rous rat sarcoma. Acta Pathol. Microbiol. Scand. [A] 80, 428–429 (1972).
Mitelman, F., Levan, G., Nilsson, P.G., Brandt, L.: Non-random karyotypic evolution in chronic myeloid leukemia. Int. J. Cancer 18, 24–30 (1976b).
Mitelman, F., Mark, J., Levan, G.: Chromosomes of six primary sarcomas induced in the Chinese hamster by 7,12-dimethylbenz (α) anthracene. Hereditas 72, 311–318 (1972a).

Mitelman, F., Mark, J., Levan, G., Levan, A.: Tumor etiology and chromosome pattern. Science 176, 1340–1341 (1972b).
Mitelman, F., Mark, J., Nilsson, P.G., Dencker, H., Norryd, C., Tranberg, K.-G.: Chromosome banding pattern in human colonic polyps. Hereditas 78, 63–68 (1974).
Mitelman, F., Nilsson, P.G., Brandt, L.: Abnormal clones resembling those seen in blast crisis arising in the spleen in chronic myelocytic leukemia. J. Natl. Cancer Inst. 54, 1319–1321 (1975).
Mitelman, F., Nilsson, P.G., Levan, G., Brandt, L.: Non-random chromosome changes in acute myeloid leukemia. Chromosome banding examination of 30 cases at diagnosis. Int. J. Cancer 18, 31–38 (1976).
Mitelman, F., Panani, A., Brandt, L.: Isochromosome 17 in a case of eosinophilic leukaemia. An abnormality common to eosinophilic and neutrophilic cells. Scand. J. Haematol. 14, 308–312 (1975).
Mittwoch, U.: Sex chromatin. J. Med. Genet. 1, 50–76 (1964).
Mitus, W.J., Kiossoglou, K.A.: Leukocytic alkaline phosphatase in myeloproliferative syndrome. Ann. N.Y. Acad. Sci. 155, 976–979 (1968).
Mitus, W.J., Coleman, N., Kiossoglou, K.A.: Abnormal (marker) chromosomes in two patients with acute myelofibrosis. Arch. Intern. Med. 123, 192–197 (1969).
Miyamoto, N.: A long-term cytogenetic survey on fifty patients with chronic myelogenous leukemia. Acta Med. (Fukuoka) 65, 601–621 (1974).
Miyamoto, N., Takita, A.: Three abnormal cell lines (with a Ph[1], two Ph[1]'s and without Ph[1]) in acute phase of chronic myelogenous leukemia. Jap. J. Hum. Genet. 18, 288–293 (1973).
Miyoshi, I., Kubonishi, I., Uchida, H., Hiraki, S., Toki, H., Tanaka, T., Masuji, H., Hiraki, K.: Direct implantation of Ph[1] chromosome-positive myeloblasts into newborn hamsters. Blood 47, 355–361 (1976).
Moake, J.L., Lebos, H., Warren, R. J.: Chromosomal abnormalities in a patient with adolescent myelofibrosis. Acta Haematol. (Basel) 52, 173–179 (1974).
Modan, B., Lilienfeld, A.M.: Polycythemia vera and leukemia—the role of radiation treatment, a study of 1222 patients. Medicine (Baltimore) 44, 305–344 (1965).
Modan, B., Padeh, B., Kallner, H., Akstein, E., Meytes, D., Czerniak, P., Ramot, B., Pinkhas, J., Modan, M.: Chromosomal aberrations in polycythemia vera. Blood 35, 28–38 (1970).
Moloney, W.C.: Natural history of chronic granulocytic leukaemia. Clin. Hematol. 6, 41–53 (1977).

Montaldo, G., Zucca, G.: Anomalies of the chromosomal system and their possible relationship with the viral etiology of malignant tumors in humans. Arch. De Vecchi Anat. Patol. 56, 411–428 (1970).

Moore, G.E., Sandberg, A.A.: Chromosomes and surgeons. Surg. Gynecol. Obstet. 113, 777–778 (1961).

Moore, G.E., Sandberg, A.A.: Studies of a human tumor cell line with a diploid karyotype. Cancer 17, 170–175 (1963).

Moore, G.E., Fjelde, A., Huang, C.C.: Established hyperdiploid hematopoietic cell line with a minute marker chromosome persisting both in culture and in the "normal" donor. Cytogenetics 8, 332–336 (1969).

Moore, G.E., Ishihara, T., Koepf, G.F., Sandberg, A.A.: The importance to radiologists of recent advances in cytology and cytogenetics. Am. J. Roentgenol. Radium Ther. Nucl. Med. 89, 584–589 (1963).

Moore, G.E., Morgan, R.T., Quinn, L.A., Woods, L.K.: A transitional cell carcinoma cell line. In Vitro 14, 301–306 (1978).

Moore, M.A.S.: Marrow culture—a new approach to classification of leukemias. Blood Cells 1, 149–158 (1975).

Moore, M.A.S.: Prediction of relapse and remission in ANL by marrow culture criteria. Blood Cells 2, 109–124 (1976).

Moore, M.A.S., Metcalf, D.: Cytogenetic analysis of human acute and chronic myeloid leukemic cells cloned in agar culture. Int. J. Cancer 11, 143–152 (1973).

Moore, M.A.S., Ekert, H., Fitzgerald, M.G., Carmichael, A.: Evidence for the clonal origin of chronic myeloid leukemia from a sex chromosome mosaic: Clinical, cytogenetic, and marrow culture studies. Blood 43, 15–22 (1974a).

Moore, M.A.S., Ekert, H., Fitzgerald, M.G., Carmichael, A.: Correspondence: To the Editor. Blood 44, 768–769 (1974b).

Moore, M.A.S., Spitzer, N., Williams, N., Metcalf, D., Buckley, J.: Agar culture studies in 127 cases of untreated acute leukemia: The prognostic value of reclassification of leukemia according to *in vitro* growth characteristics. Blood 44, 1–18 (1974).

Moore, R.: Ionizing radiations and chromosomes. J. Coll. Radiol. Aust. 9, 272–283 (1965).

Moorhead, P.S., Nowell, P.C., Mellman, W.J., Battips, D.M., Hungerford, D.A.: Chromosome preparations of leukocytes cultured from human peripheral blood. Exp. Cell Res. 20, 613–616 (1960).

Moraine, P.C., Brémond, J.L., Despert, F., Leroy, J., Leroux, M.E., Maillet, M.: Chromosome Philadelphie et marqueur du groupe C dans un cas de leucose aiguë découverte sous traitement par noramidopyrine. Nouv. Rev. Fr. Hematol. 14, 461–470 (1974).

Moraru, I., Fadei, L.: Morphological and cytogenetical characters of human ovarian tumors in short-term tissue cultures. Arch. Geschwulstforsch. 42, 163–171 (1973).

Moraru, I., Fadei, L.: Comparative morphological and cytogenetical studies on human ovarian papillary adenocarcinomas. Oncology 30, 113–124 (1974).

Moretti, G., Broustet, A., Moulinier, J., Beylot, J., Serville, F., Veyret, V., Bonnin, H., Bourdeau, M. J., Gachet, M.: Anomalies chromosomiques et dyserythropoieses acquises (à propos de 12 obsérvations). Lyon Med. 233, 281–286 (1975).

Moretti, G., Broustet, A., Moulinier, J., Veyret, V., Beylot, J., Paccalin, J., Garcia, J.: Anémies refractaires et anomalies chromosomiques une nouvelle approche de ebat préleucemiques. Bord. Med. 5, 661–668 (1972).

Moretti, G., Hartmann, L., Staeffen, J., Grouchy, J. de, Catanzano, G., Broustet, A.: L'anomalie chromosomique de la maladie de Waldenström. Ann. Genet. (Paris) 8, 55–59 (1965).

Morgan, W.F., Crossen, P.E.: The frequency and distribution of sister chromatid exchanges in human chromosomes. Hum. Genet. 38, 271–278 (1977a).

Morgan, W.F., Crossen, P.E.: The incidence of sister chromatid exchanges in cultured human lymphocytes. Mutat. Res. 42, 305–312 (1977b).

Mori, M., Sasaki, M.: Chromosome studies on rat leukemias and lymphomas, with special attention to fluorescent karyotype analysis. J. Natl. Cancer Inst. 52, 153–160 (1974).

Moricard, R., Cartier, R.: Chromosomal and cytoplasmic cytopathology of intraepithelial squamous cell epithelioma of the cervix uteri. Ciba Found. Study Group 3, 28 (1959).

Morse, H.G., Ducore, J.M., Hays, T., Peakman, D., Robinson, A.: Multiple leukemic clones in acute leukemia of childhood. Hum. Genet. 40, 269–278 (1978).

Morse, H., Hays, T., Peakman, D., Rose, B., Robinson, A.: Acute nonlymphocytic leukemia in childhood: High incidence of clonal abnormalities and non-random changes. Cancer (submitted).

Morse, H.G., Humbert, J.R., Hutter, J.J., Robinson, A.: Karyotyping of bone-marrow cells in hematologic diseases. Hum. Genet. 37, 33–39 (1977).

Motoiu-Raileanu, I.: Chromosomal aberrations in human malignant blood diseases. Stud. Cercet. Med. Interna 12, 423–433 (1971).

Motoiu-Raileanu, I., Bercaneanu, S.: Ph[1] chromosome associated with changes of clonal evolution in acute leukemia. Stud. Cercet. Med. Interna 12, 537 – 548 (1971).

Motoiu-Raileanu, I., Bercaneanu, S., Gogiu, M., Sighetea, E.: Study of chromosomal abnormalities in polycythemia vera. Rev. Roum. Med. Intern. 7, 223 (1970).

Motomura, S., Yamamoto, K.: A case of Fanconi's anemia with multiple chromosome aberrations. Rinsho Ketsueki 12, 598 – 603 (1971).

Motomura, S., Misutake, T., Wakasugi, H.: Successive karyotype evolution in a case of acute myelogenous leukemia. J. Kyushu Hematol. Soc. 22, 41 – 47 (1972).

Motomura, S., Ogi, K., Horie, M.: Monoclonal origin of acute transformation of chronic myelogenous leukemia. Acta Hematol. (Basel) 49, 300 – 305 (1973).

Motomura, S., Ogi, K., Kaneko, S.: Progression and prognosis of chronic myelogenous leukemia in the acute stage: Hematologic and cytogenetic aspects. Acta Haematol. (Basel) 56, 78 – 83 (1976).

Mourelatos, D., Faed, M.J.W., Gould, P.W., Johnson, B.E., Frainbell, W.: Sister chromatid exchanges in lymphocytes of psoriatics after treatment with 8-methoxypsoralen and long wave ultraviolet radiation. Br. J. Dermatol. 97, 649 – 654 (1977).

Mouriquand, C.: Chromosomes, leucémies et cancers. Grenoble Med. Chir. 6, 31 – 40 (1968).

Mouriquand, C., Gilly, C., Wolff, C.: Données recentes sur l'ultrastructure du chromosome entier. Lyon Med. 233, 287 – 295 (1975).

Muir, P.D., Occomore, M.A., Thornley, B., Singh. S., Gunz, F.W.: The value of chromosome banding methods in the study of adult acute leukemia. Pathology 9, 323 – 330 (1977).

Mukerjee, D., Bowen, J., Anderson, D.E.: Simian papovavirus 40 transformation of cells from cancer patient with XY/XXY mosaic Klinefelter's syndrome. Cancer Res. 30, 1769 – 1772 (1970).

Mukerjee, D., Bowen, J.M., Trujillo, J.M., Cork, A.: Increased susceptibility of cells from cancer patients with XY-gonadal dysgenesis to Simian papovavirus 40 transformation. Cancer Res. 32, 1518 – 1520 (1972).

Mukerjee, D., Trujillo, J.M., Cork, A., Bowen, J.M.: Genetic susceptibility of human cells to transformation by oncogenic viruses. *4th International Congress on Human Genetics, Paris, 1971.* Abstracts of papers presented, p. 128. Amsterdam: Excerpta Medica (1971). (Excerpta Medica Int. Congr. Ser. No. 233)

Mukherjee, A.B., Moser, G.C., Nitowsky, H.M.: Fluorescence of X and Y chromatin in human interphase cells. Cytogenetics 11, 216 – 227 (1972).

Muldal, S., Lajtha, L.G.: Chromosomes and leukemia. In J. German (ed.): *Chromosomes and Cancer,* pp. 451 – 480. New York: Wiley (1974).

Muldal, S., Elejalde, R., Harvey, P.W.: Specific chromosome anomaly associated with autonomous and cancerous development in man. Nature 229, 48 – 49 (1971).

Muldal, S., Lajtha, L.G., Gilbert, C.W.: Time sequence of human chromosome duplication. In F.G.H. Hayhoe (ed.): *Current Research in Leukemia,* p. 86. London: Cambridge University Press (1965).

Muldal, S., Mir, M.A., Freeman, C.B., Geary, C.G.: A new translocation associated with the Ph[1] chromosome and an acute course of chronic granulocytic leukaemia.. Br. J. Cancer 31, 364 – 368 (1975).

Muldal, S., Taylor, J.J., Asquith, P.: Non-random karyotype progression in chronic myeloid leukaemia. Int. J. Radiat. Biol. 12, 219 – 226 (1967).

Müller, D., Haberlandt, W.: Chromosomale und zytochemische Befunde bei Osteomyelosklerose. Blut 20, 205 – 213 (1970).

Müller, Hj., Stalder, G.R.: Chromosomes and human neoplasms. Achievements using new staining techniques. Eur. J. Pediatr. 123, 1 – 13 (1976).

Müller, Hj., Buhler, E.E., Sartorius, J.A., Stalder, G.R.: Chromosomen bei Leukämien. Krebsinformation 11, 1 – 9 (1976).

Muller, P., Clavert, J., Phillippe, E., Levy, G., Klein, M., Muller, R., Doerr, R.: La dysgénésie gonadique mixte: Sa place parmi les syndromes dysgénétiques et ses relations avec le gonocytome. Ann. Endocrinol. (Paris) 31, 1014 – 1021 (1970).

Müller, V.D., Orywall, D., Hübner, N.: Vergleichende zytogenetische und quantitativ-zytochemische DNS-untersuchungen bei akuten Leukämiem. Blut 23, 287 – 301 (1971).

Müller, W., Rosenkranz, W.: Rapid banding technique for human and mammalian chromosomes. Lancet i, 898 (1972a).

Müller, W., Rosenkranz, W.: Identifizierung numerischer und struktureller Anomalien menschlicher G-Chromosomen mit dem Pankreatin-Giesma-Banden-Muster. Klin. Paediatr. 184, 265 – 271 (1972b).

Mulvihill, J.J., Wade, W.M., Miller, R.W.: Gonadoblastoma in dysgenetic gonads with a Y chromosome. Lancet i, 863 (1975).

Mutton, D.E.: Origin of the trisomic 21 chromosome. Lancet I, 375 (1975).

Mutton, D.E., Daker, M.G.: Pericentric inversion of

chromosome 9. Nature [New Biol.] 241, 80 (1973).
Nadel, M., Koss, L.G.: Klinefelter's syndrome and male breast cancer. Lancet i, 366 (1967).
Nadkarni, J.S., Nadkarni, J.J., Clifford, P., Manolov, G., Fenyo, E.M., Klein, E.: Characteristics of new cell lines derived from Burkitt lymphomas. Cancer 23, 64–79 (1969).
Nagao, K., Yonemitsu, H., Yamaguchi, K., Okuda, K.: A case of acute myeloblastic leukemia with Ph[1] chromosome showing translocation 9q+; 22q−. Blood 50, 259–262 (1977).
Nagayama, T., Kataumi, S.: Exofoliative cytology and chromosomal analysis in tumor of the bladder by pumping method. Acta Urol. Jap. 18, 5–11 (1972).
Nagy, G., Yurgutis, R.P.: Chromosome studies on patients with polycythaemia vera. Haematologica (Pavia) 2, 179–186 (1968).
Najfeld, V.: Isochromosome 17 in a case of chronic erythroleukaemia. Scand. J. Haematol. 17, 101–104 (1976).
Najfeld, V., Price, T.H., Adamson, J.W., Fialkow, P.J.: Myelofibrosis with complex chromosome abnormality in a patient with erythrocytosis due to hemoglobin Rainier and treated with ^{32}P. Am. J. Hematol. 5, 63–69 (1978).
Najfeld, V., Singer, J.V., James, M.C., Fialkow, P.J.: Trisomy of 1q in preleukaemia with progression to acute leukaemia. Scand. J. Haematol. 21, 24–28 (1978).
Nakagome, Y.: Chromosome studies in malignant diseases in children with special reference to bone-marrow metastasis of neuroblastoma. Acta Paediatr. Jap. 69, 651–667 (1965).
Nakagome, Y.: Initiation of DNA replication in human chromosomes. Exp. Cell Res. 106, 457–461 (1977).
Nakagome, Y., Kudo, H., Watanabe, Y., Matsui, I.: On the congenital leukemic disease. A review. Acta Paediatr. Jap. 68, 153–162 (1964).
Nakagome, Y., Oka, S., Higurashi, M.: Quinacrine and acridine R banding without a fluorescence microscope. Hum. Genet. 40, 171–176 (1978).
Nakagome, Y., Oka, S., Matsunaga, E.: LBA technique in the detection of chromosome variants. II. Chromosomes except for those with Q variants. Hum. Genet. 38, 307–314 (1977).
Nakamuro, K., Yoshikawa, K., Sayato, Y., Kurata, H., Tonomura, M., Tonomura, A.: Studies on selenium-related compounds. V. Cytogenetic effect and reactivity with DNA. Mutat. Res. 40, 177–184 (1976).
Naman, R., Cadotte, M., Gosselin, G., Long, L.A.: Extramedullary blastic crisis resembling malignant lymphoma and Ph[1] duplication in chronic myeloid leukemia. Union Med. Can. 101, 884–887 (1972).
Naman, R., Cadotte, M., Long, L.A.: Les chromosomes dans les érythro-leucémies. Étude d'un cas et revue de la littérature. Nouv. Rev. Fr. Hematol. 11, 211–218 (1971).
Nanba, K., Jaffe, E.S., Soban, E.J., Braylan, R.C., Berard, C.W.: Hairy cell leukemia. Enzyme histochemical characterization, with special reference to splenic stromal changes. Cancer 39, 2323–2336 (1977).
Nankin, H., Hydovitz, J., Sapira, J.: Normal chromosomes in mucosal neuroma variant of medullary thyroid carcinoma syndrome. J. Med. Genet. 7, 374–378 (1970).
Nasjleti, C.E., Spencer, H.H.: Chromosome damage and polyploidization induced in human peripheral leukocytes *in vivo* and *in vitro* with nitrogen mustard, 6-mercaptopurine, and A-649. Cancer Res. 26, 2437–2443 (1966).
Nassar, V.H., Khouri, F.P.: Malignant lymphomas. Tissue karyotyping in doubtful cases. Arch. Pathol. 98, 367–369 (1974).
Nathwani, B.N., Rappaport, H., Moran, E.M., Pangalis, G.A., Kim, H.: Malignant lymphoma arising in angio-immunoblastic lymphadenopathy. Cancer 41, 578–606 (1978).
Nau, R.C., Hoagland, H.C.: A myeloproliferative disorder manifested by persistent basophilia, granulocytic leukemia and erythroleukemic phases. Cancer 28, 662–665 (1971).
Nava, C. de: Les anomalies chromosomiques au cours des hémopathies malignes et nonmalignes. Une étude de 171 cas. Monogr. Ann. Genet. (Paris) 1, 89 (1969).
Nava, C. de, Grouchy, J. de, Thoyer, C., Bousser, J., Bilski-Pasquier, J., Freteaux, J.: Évolutions clonales et duplication du Ph[1]; présentation de deux cas. Ann. Genet. (Paris) 12, 83–93 (1969a).
Nava, C. de, Grouchy, J. de, Thoyer, C., Turelau, C., Siguier, F.: Polyploidisation et évolutions clonales. Ann. Genet. (Paris) 12, 237–241 (1969b).
Nava, C. de, Zittoun, R., Grouchy, J. de: Étude chromosomiques dans 5 cas d'ànémie sideroblastique idiopathique acquise. *Proceedings of the 10th Congress of the European Society of Haematology, Strasburg*, p. 546–550 (1965).
Nayak, S.K., O'Toole, C., Price, Z.H.: A cell line from an anaplastic transitional cell carcinoma of human urinary bladder. Br. J. Cancer 35, 142–151 (1977).
Nayar, S.: Chromosomal and DNA analysis of human lymphoma cases. Indian J. Med. Res. 64, 1543–1551 (1976).
Nayar, S., Sharma, G.P.: Nuclear variations in cells

of twenty human solid neoplastic tumors. Indian J. Cancer 12, 400–404 (1975).
Nebel, B.R.: Chromosome structure. Bot. Rev. 5, 563–626 (1939).
Neerhout, R.C.: Chronic granulocytic leukaemia. Early blast crisis simulating acute leukaemia. Am. J. Dis. Child. 115, 66–70 (1968).
Negrini, A.C., Azzolini, A., Bertocchi, I., Dolcino, G.: Hyperlymphocytosis in a case of busulfan-induced medullar aplasia in chronic myeloid leukemia. Chromosome and functional study (in Italian). Pathologica 59, 407–410 (1968).
Neishtadt, E.L.: Chromosome anomalies and pathological mitoses in fibroadenomatosis and cancer of the breast. Vopr. Onkol. 20, 3–8 (1974).
Nelson, M.M., Blom, A., Arens, L.: Chromosomes in ataxia-telangiectasia. Lancet i, 518–519 (1975).
Nelson-Rees, W.A., Flandermeyer, R.R., Hawthorne, P.K.: Distinctive banded marker chromosomes of human tumor cell lines. Int. J. Cancer 16, 74–82 (1975).
Neurath, P., Remer, K. DE, Bell, B., Jarvik, L., Kato, T.: Chromosome loss compared with chromosome size, age and sex of subjects. Nature 225, 281–282 (1970).
Nevin, N.C., Dodge, J.A., Allen, I.V.: Two cases of trisomy D associated with adrenal tumours. J. Med. Genet. 9, 119–122 (1972).
Nevstad, N.P.: Sister chromatid exchanges and chromosomal aberrations induced in human lymphocytes by the cytostatic drug adriamycin in vivo and in vitro. Mutat. Res. 57, 253–258 (1978).
Newsome, Y.L., Singh, D.N.: Cytologic effects of adriamycin on human peripheral lymphocytes. Acta Cytol. 21, 137–140 (1977).
Nezelof, C., Laurent, M., Rosseau, M.-F., Ayraud, N., Urano, Y.: Le sarcome embryonnaire. Étude caryotypique de 7 observations. Bull. Cancer (Paris) 54, 423–446 (1967).
Ng, A.B.P., Atkin, N.B.: Histological cell type and DNA value in the prognosis of squamous cell cancer of uterine cervix. Br. J. Cancer 28, 322–331 (1973).
Nichols, W.W.: Relationships of viruses, chromosomes and carcinogenesis. Hereditas 50, 53–80 (1963).
Nichols, W.W.: Interaction between viruses and chromosomes. In A. Lima-de-Faria (ed.): Handbook of Molecular Cytology, pp. 733–750. Amsterdam: North-Holland (1969).
Nichols, W.W.: Viruses and chromosomal abnormalities. Ann. N.Y. Acad. Sci. 171, 478–485 (1970).
Nichols, W.W.: Chromosomal changes due to viruses. Triangle 11, 103–106 (1972a).
Nichols, W.W.: Genetic hazards of drugs of abuse. In C.J.D. Zarafonetis (ed.): Proceedings of the International Conference, Drug Abuse, pp. 93–100. Philadelphia: Lea and Febiger (1972b).
Nichols, W.W.: Cytogenetic techniques in mutagenicity testing. Agents Actions 3, 86–92 (1973).
Nichols, W.W.: Somatic mutation in biologic research. Hereditas 81, 225–236 (1975).
Nichols, W.W., Bradt, C.I., Toji, L.H., Godley, M., Segawa, M.: Induction of sister chromatid exchanges by transformation with simian virus 40. Cancer Res. 38, 960–964 (1978).
Nichols, W.W., Levan, A., Hall, B., Östergren, G.: Measles-associated chromosome breakage. A preliminary communication. Hereditas 48, 367–370 (1962).
Nichols, W.W., Nordén, Å., Bradt, C., Berg, B., Peluse, M.: Cytogenetic studies in a case of erythroleukemia. Scand. J. Haematol. 7, 32–36 (1970).
Nicoara, S., Butoianu, E., Brosteanu, R.: Specificity of the Ph[1] chromosome. Lancet ii, 1312–1313 (1967).
Nicolau, C.T., Nicoară, S., Wechsler, B., Munteanu, N., Enache, F.: Studiu asupra cromozomilor in leucemie acute. Duc. Haematol. 2, 125–142 (1966).
Nicolau, C.T., Popescu, E.R., Nicoară, S.T., Wechsler, B., Butoianu, E., Taigăr, S.: A study of chromosomes in chronic granulocytic leukemia. Med. Interna (Bucur.) 16, 775–782 (1964).
Nieburgs, H.E., Herman, B.E., Reisman, H.: Buccal cell changes in patients with malignant tumors. Lab. Invest. 11, 80–88 (1962).
Nielsen, V.G., Krogh Jensen, M.: Pernicious anaemia and acute leukaemia. Scand. J. Haematol. 7, 26–31 (1970).
Nigam, R., Dosik, H.: Chronic myelogenous leukemia presenting in the blastic phase and its association with a 45 XO Ph[1] karyotype. Blood 47, 223–227 (1976).
Nilsson, B.: A bibliography of literature concerning chromosome identification — with special reference to fluorescence and Giemsa staining techniques. Hereditas 73, 259–270 (1973).
Nilsson, C., Hansson, A., Milsson, G.: Influence of thyroid hormones on satellite association in man and the origin of chromosome abnormalities. Hereditas 80, 157–166 (1975).
Nilsson, P.G., Brandt, L., Mitelman, F.: Prognostic implications of chromosome analysis in acute non-lymphocytic leukemia. Leukemia Res. 1, 31–34 (1977a).
Nilsson, P.G., Brandt, L., Mitelman, F.: Relation between age and chromosomal aberrations at diagnosis of adult non-lymphocytic leukemia. Leukemia Res. 1, 385–386 (1977b).
Nishiya, I., Kikuchi, T., Moriya, S., Shimotamai, K., Sakamura, I.: Cytophotometric study of prema-

lignant and malignant cells of the cervix in an approach towards automated cytology. Acta Cytol. 21, 271–275 (1977).

Noël, B., Quack, B., Rethoré, M.-O.: Partial deletions and trisomies of chromosome 13: Mapping of bands associated with particular malformations. Clin. Genet. 9, 593–602 (1976).

Nordenson, I.: Effect of superoxide dismutase and catalase on spontaneously occurring chromosome breaks in patients with Fanconi's anemia. Hereditas 86, 147–150 (1977).

Nordenson, I., Beckman, G., Beckman, L., Nordström, S.: Occupational and environmental risks in and around a smelter in northern Sweden. II. Chromosomal aberrations in workers exposed to arsenic. Hereditas 88, 47–50 (1978a).

Nordenson, I., Beckman, G., Beckman, L., Nordström, S.: Occupational and environmental risks in and around a smelter in Northern Sweden. IV. Chromosomal aberrations in workers exposed to lead. Hereditas 88, 263–267 (1978b).

Norman, C.S., Boucher, B.J.: Atypical chronic myelogenous leukemia with Philadelphia (Ph¹) chromosome and an additional translocation. Cancer 41, 1123–1127 (1978).

Novogrodsky, A.: Selective activation of mouse T and B lymphocytes by periodate, galactose oxidase and soybean agglutinin. Eur. J. Immunol. 4, 646–648 (1974).

Nowell, P.C.: Phytohemagglutinin: An initiator of mitosis in cultures of normal human leukocytes. Cancer Res. 20, 462–466 (1960).

Nowell, P.C.: The minute chromosome (Ph¹) in chronic granulocytic leukaemia. Blut 8, 65–66 (1962).

Nowell, P.C.: Chromosome changes in primary tumors. Prog. Exp. Tumor Res. 7, 83–103 (1965a).

Nowell, P.C.: Prognostic value of marrow chromosome studies in human "preleukemia." Arch. Pathol. 80, 205–208 (1965b).

Nowell, P.C.: Chromosome abnormalities in human leukemia and lymphoma. *In* C.J.D. Zarafonetis (ed.): *Proceedings of the International Conference on Leukemia-Lymphoma*, pp. 47–53. Philadelphia: Lea and Febiger (1968).

Nowell, P.C.: Biological significance of induced human chromosome aberrations, Fed. Proc. 28, 1797–1803 (1969).

Nowell, P.C.: Commentary on chromosomal abnormalities in neoplasia. *In* W.J. Burdette (ed.): *Carcinoma of the Colon and Antecedent Epithelium*, pp. 333–337. Springfield, Ill.: C.C. Thomas (1970).

Nowell, P.C.: Genetic changes in cancer: Cause or effect? Hum. Pathol. 2, 347–348 (1971a).

Nowell, P.C.: Marrow chromosome studies in "preleukemia." Further correlation with clinical course. Cancer 28, 513–518 (1971b).

Nowell, P.C.: Marrow chromosome studies in "preleukemia." Proc. Am. Assoc. Cancer Res. 12, 37, Abstract 148 (1971c).

Nowell, P.C.: Diagnostic and prognostic value of chromosome studies in cancer. Ann. Clin. Lab. Sci. 4, 234–240 (1974a).

Nowell, P.C.: Chromosome changes and the clonal evolution of cancer. *In* J. German (ed.): *Chromosomes and Cancer*, pp. 267–285. New York, Wiley (1974b).

Nowell, P.C.: The clonal evolution of tumor cell populations. Science 194, 23–28 (1976).

Nowell, P.C.: Preleukemia. Cytogenetic clues in some confusing disorders. Am. J. Pathol. 89, 459–476 (1977).

Nowell, P.C.: Tumors as clonal proliferation. Virchows Arch. B Cell Pathol. 29, 145–150 (1978).

Nowell, P.C., Finan, J.B.: Isochromosome 17 in atypical myeloproliferative and lymphoproliferative disorders. J. Natl. Cancer Inst. 59, 329–333 (1977).

Nowell, P.C., Finan, J.: Chromosome studies in preleukemic states. IV. Myeloproliferative versus cytopenic disorders. Cancer 42, 2254–2261 (1978a).

Nowell, P.C., Finan, J.B.: Cytogenetics of acute and chronic myelofibrosis. Virchows Arch. B Cell Pathol. 29, 45–50 (1978b).

Nowell, P.C., Hungerford, D.A.: Chromosome studies on normal and leukemic human leukocytes. J. Natl. Cancer Inst. 25, 85–109 (1960a).

Nowell, P.C., Hungerford, D.A.: A minute chromosome in human chronic granulocytic leukemia. Science 132, 1497 (1960b).

Nowell, P.C., Hungerford, D.A.: Chromosome studies in human leukemia. II. Chronic granulocytic leukemia. J. Natl. Cancer Inst. 27, 1013–1035 (1961).

Nowell, P.C., Hungerford, D.A.: Chromosome studies in human leukemia. IV. Myeloproliferative syndrome and other atypical myeloid disorders. J. Natl. Cancer Inst. 29, 911–931 (1962).

Nowell, P.C., Hungerford, D.A.: Chromosome changes in human leukemia and a tentative assessment of their significance. Ann. N.Y. Acad. Sci. 113, 654–662 (1964).

Nowell, P.C., Hungerford, D.A.: Chromosome in hematologic disorders, *In* R. Philip Custer (ed.): *An Atlas of the Blood and Bone Marrow*, Chapter 21, pp. 427–439. Philadelphia: W.B. Saunders (1974).

Nowell, P.C., Daniele, R., Winger, L., Rowlands, D.T., Jr.: T cells in chronic lymphocytic leukaemia. Lancet i, 915 (1975).

Nowell, P.C., Hungerford, D.A., Brooke, C.D.: Chromosomal characteristics of normal and leukemic human leukocytes after short-term tissue culture. Proc. Am. Assoc. Cancer Res. 2, 331–332 (1958).

Nowell, P.C., Jensen, J., Gardner, F.: Two complex translocations in chronic granulocytic leukemia involving chromosomes 22, 9 and a third chromosome. Humangenetik 30, 13–21 (1975).

Nowell, P.C., Jensen, J., Gardner, F., Murphy, S., Chaganti, R.S.K., German, J.: Chromosome studies in "preleukemia." III. Myelofibrosis. Cancer 38, 1873–1881 (1976a).

Nowell, P., Jensen, J., Winger, L., Daniele, R., Growney, P.: T cell variant of chronic lymphocytic leukaemia with chromosome abnormality and defective response to mitogens. Br. J. Haematol. 33, 459–468 (1976b).

Nowell, P.C., Morris, H.P., Potter, V.R.: Chromosomes of "minimal deviation" hepatomas and some other transplantable rat tumors. Cancer Res. 27, 1565–1579 (1967).

Nusbacher, J., Hirschhorn, K., Cooper, L.Z.: Chromosomal abnormalities in congenital rubella. N. Engl. J. Med. 276, 1409–1413 (1967).

Obara, Y.: A short arm deletion of a G group chromosome found in a patient with chronic myelogenous leukemia with notes on the familial transmission and the disease state. Zool. Mag. (Tokyo) (Dobutsugaku Zasshi) 77, 317–321 (1968).

Obara, Y., Makino, S., Mikuni, C.: Cytological studies of tumors. XLIX. Chronic lymphocytic leukemia with A_1/G chromosome translocation and high serum γ-globulin production. Gann 61, 1–6 (1970a).

Obara, Y., Makino, S., Mikuni, C.: Cytologic studies of tumors. LI. Notes on some abnormal chromosome features in a case of mycosis fungoides. Proc. Jap. Acad. 46, 561–566 (1970b).

Obara, Y., Makino, S., Oshima, M., Mikuni, C.: Cytological studies of tumors. XLV. The short arm deletion of a G-group chromosome in a patient with chronic myelogenous leukemia and in his family. J. Facult. Sci., Hokkaido Univ. 16, 632–639 (1968).

Obara, Y., Makino, S., Sasaki, M.: A chromosome survey in 50 cases of human hematopoietic disorders. Proc. Jap. Acad. 45, 495–500 (1969).

Obara, Y., Sasaki, M., Makino, S.: Cytologic studies of tumors. LIII. Manifold chromosome observations in acute myelogenous leukemia. Gann 62, 301–308 (1971a).

Obara, Y., Sasaki, M., Makino, S., Mikuni, C.: Cytologic studies of tumors. L. Clonal proliferation of four stemlines in three hematopoietic tissues of a patient with reticulosarcoma. Blood 37, 87–95 (1971b).

Obe, G.: Chromosome endoreduplication in normal controls. Lancet i, 59 (1965).

Obe, G., Herha, J.: Chromosomal aberrations in heavy smokers. Hum. Genet. 41, 259–263 (1978).

Obe, G., Lüdeke, J.B.P., Waldenmaier, K., Sperling, K.: Premature chromosome condensation in a case of Fanconi's anemia. Humangenetik 28, 159–162 (1975).

Oberling, F., Stoll, C., Lang, J.M., Giron, C., Batzenschlager, A., Mayer, G., Waitz, R.: Leucémie myeloïde chronique, adénopathies et duplication du chromosome Ph[1]. Acutisation extramédullaire. Nouv. Rev. Fr. Hematol. 15, 279–283 (1975a).

Oberling, F., Stoll, C., Lang, J.M., Mayer, G.: Duplication of Philadelphia chromosome in acute transition of chronic granulocytic leukemia. Ann. Intern. Med. 83, 231 (1975b).

O'Brien, R.L., Poon, P., Kline, E., Parker, J.W.: Susceptibility of chromosomes from patients with Down's syndrome to 7,12-dimethylbenz (α) anthracene-induced aberrations in vitro. Int. J. Cancer 8, 202–210 (1971).

O'Carroll, D.I., McKenna, R.W., Brunning, R.D.: Bone marrow manifestations of Hodgkin's disease. Cancer 38, 1717–1728 (1976).

Ogawa, M., Fried, J., Sakai, Y., Strife, A., Clarkson, B.D.: Studies of cellular proliferation in human leukemia. VI. The proliferative activity, generation time, and emergence time of neutrophilic granulocytes in chronic granulocytic leukemia. Cancer 25, 1031–1049 (1970).

Ogawa, M., Wurster, D.H., McIntyre, O.R.: Multiple myeloma in one of a pair of monozygotic twins. Acta Haematol. 44, 295–304 (1970).

O'Grady, R.B., Ruthstein, T.B., Romano, P.E.: D-group deletion syndromes and retinoblastoma. Am. J. Ophthalmol. 77, 40–45 (1974).

Oguma, K.: A further study on the human chromosomes. Arch. Biol. (Liege) 40, 205–226 (1930).

Oguma, K.: The segmentary structure of the human X-chromosome compared with that of rodents. J. Morphol. 61, 59–93 (1937).

Oguma, K., Kihara, H.: A preliminary report on the human chromosomes. Zool. Mag. (Tokyo) (Dobutsugaku Zasshi) 34, 424–435 (1922).

Oguma, K., Kihara, H.: Étude ses chromosomes chez l'homme. Arch. Biol. (Liege) 33, 493–514 (1923).

Ohno, S.: Genetic implication of karyological instability of malignant somatic cells. Physiol. Rev. 51, 496–526 (1971).

Ohno, S., Trujillo, J.M., Kaplan, W.D., Riojun, K.: Nucleolus-organisers in the causation of chromosomal anomalies in man. Lancet ii, 123–125 (1961).

Ohnuki, Y.: Structure of chromosomes. I. Morphological studies of the spiral structure of human somatic chromosomes. Chromosoma 25, 402–428 (1968).

Ojima, Y., Inui, N., Makino, S.: Cytochemical studies on tumor cells. V. Measurement of desoxyribonucleic acid (DNA) by Feulgen microspectrophotometry in some human uterine tumors. Gann 51, 371–376 (1960).

Oka, S., Nakagome, Y., Matsunaga, E., Arima, M.: LBA technique in the detection of chromosome variants. I. Chromosomes with known sites of Q variants. Hum. Genet. 39, 31–37 (1977).

Okada, H., Hattori, Y., Kanoh, T., Okada, T., Takahashi, T., Furukawa, H., Ohga, T.: An autopsied case of plasma cell leukemia with long term survival. Jap. J. Clin. Hematol. 18, 145–152 (1977).

Okada, H., Lin, P.I., Hoshino, T., Yamamoto, T., Yamaoka, H., Murakami, M.: Down's syndrome associated with myelocytic and megakaryocytic leukemia. Hiroshima ABCC Tech. Rep. 1–10, 33–70 (1970).

Okada, K., Schroeder, F.H.: Human prostatic carcinoma in cell culture: Preliminary report on the development and characterization of an epithelial cell line (EB33). Urol. Res. 2, 111–121 (1974).

Okada, M., Miyazaki, T., Kumota, K.: 15/17 translocation in acute promyelocytic leukaemia. Lancet i, 961 (1977).

Okada, T.A., Comings, D.E.: Mechanisms of chromosome banding. III. Similarity between G-bands of mitotic chromosomes and chromomeres of meiotic chromosomes. Chromosoma 48, 65–71 (1974).

Oksala, T., Therman, E.: Mitotic abnormalities and cancer. In J. German (ed.): Chromosomes and Cancer, pp. 239–263. New York: Wiley (1974).

Olinici, C.D.: Double minute chromatin bodies in a case of ovarian ascitic carcinoma. Br. J. Cancer 25, 350–353 (1971).

Olinici, C.D.: Cytogenetic observations of different cell lines in Hodgkin's disease. Acta Haematol. (Basel) 48, 283–287 (1972).

Olinici, C.D.: Further data on the common origin of various stem-lines in human tumors. Cytologia (Tokyo) 38, 271–276 (1973).

Olinici, C.D., Simu, G.: Histologic and cytogenetic observations on a case of giant fibroadenoma. Neoplasma 17, 663–665 (1970).

Olinici, C.D., Dobáy, O., Almasan, M.: Constitutive heterochromatin in Ehrlich ascites carcinoma cells. Neoplasma 23, 71–76 (1976).

Olinici, C.D., Galatir, N., Lazarov, P., Giurgiuman, M.: Chromosomes in malignant gynaecological effusions. Neoplasma 20, 311–324 (1973).

Olinici, C.D., Marinca, E., Macavei, I., Dobay, O.: Missing X chromosome and ring chromosome in a case of acute myelomonocytic leukemia. Arch. Geschwulstforsch. 48, 202–204 (1978).

Olinici, C.D., Petrov, L., Macavi, I., Dobay, O.: Different cell clones in bone marrow and spleen of a patient with chronic myelocytic leukemia (CML) in blastic phase. Cancer 42, 2707–2709 (1978).

Olins, A.L., Olins, D.E.: Sperhoid chromatin units (ν bodies). Science 183, 330–332 (1974).

Oliver, R.T.D., Pillai, A., Klouda, P.T., Lawler, S.D.: HLA linked resistance factors and survival in acute myelogenous leukemia. Cancer 39, 2337–2341 (1977).

Onesti, P., Woodliff, H.J.: Cytogenetic studies in leukemia and allied disorders in Western Australia during the period 1963/65. Med. J. Aust. 2, 1176–1182 (1968).

Onesti, P., Woodliff, H.J.: The Philadelphia chromosome in a patient with acute leukaemia. Med. J. Aust. 1, 544–546 (1970).

Oni, S.B., Osunkoya, B.O., Luzzatto, L.: Paroxysmal nocturnal hemaglobinuria: Evidence for monoclonal origin of abnormal red cells. Blood 36, 145–152 (1970).

Oppenheim, J.J., Whang, J., Frei E.M., III,: Immunologic and cytogenetic studies of chronic lymphocytic leukemic cells. Blood 26, 121–132 (1965).

O'Riordan, M.L., Berry, E.W., Tough, I.M.: Chromosome studies on bone marrow from a male control population. Br. J. Haematol. 19, 83–90 (1970).

O'Riordan, M.L., Langlands, A.O., Harnden, D.G.: Further studies on the frequency of constitutional chromosome abnormalities in patients with malignant disease. Eur. J. Cancer 8, 373–379 (1972).

O'Riordan, M.L., Robinson, J.A., Buckton, K.E., Evans, H.J.: Distinguishing between the chromosomes involved in Down's syndrome (trisomy-21) and chronic myeloid leukaemia (Ph[1]) by fluorescence. Nature 230, 167–168 (1971).

Ortega Arambura, J.J., Garcia, J.: Chromosomal study of Fanconi's aplastic anemia. Apropos of 2 cases. Sangre (Barc.) 17, 87–100 (1972).

Orye, E.: Satellite association and variations in length of the nucleolar constriction of normal and variant human G chromosomes. Humangenetik 22, 299–309 (1974).

Orye, E., Delbeke, M.J.: Clonal karyotype evolution in solid tumours in children. Oncology 29, 520–533 (1974).

Orye, E., Delbeke, M.J., Vandenabeele, B.: Retinoblastoma and D-chromosome deletions. Lancet ii, 1376 (1971).

Orye, E., Delbeke, M.J., Vandenabeele, B.: Retinoblastoma and long arm deletion of chromosome 13. Attempts to define the deleted segment. Clin. Genet. 5, 457–464 (1974).

Orywall, D., Muller, D.: Comparative chromosome findings and DNA determination in acute leukemia. Artzl. Forsch. 25, 268–271 (1971).

Osgood, E.E., Brooke, J.H.: Continuous tissue culture of leukocytes from human bloods by application of "gradient" principle. Blood 10, 1010–1022 (1955).

Osgood, E.E., Krippaehne, M.L.: The gradient tissue culture method. Exp. Cell Res. 9, 116–127 (1955).

Oshimura, M., Sandberg, A.A.: Isochromosome #17 in a prostatic cancer. J. Urol. 114, 249–250 (1975).

Oshimura, M., Sandberg, A.A.: Chromosomal 6q– anomaly in acute lymphoblastic leukaemia. Lancet ii, 1045–1046 (1976).

Oshimura, M., Sandberg, A.A.: Chromosomes and causation of human cancer and leukemia. XXV. Significance of the Ph¹ (including unusual translocations) in various acute leukemias. Cancer 40, 1149–1160 (1977).

Oshimura, M., Freeman, A. I., Sandberg, A.A.: Chromosomes and causation of human cancer and leukemia. XXIII. Near-haploidy in acute leukemia. Cancer 40, 1143–1148 (1977a).

Oshimura, M., Freeman, A. I., Sandberg, A.A.: Chromosomes and causation of human cancer and leukemia. XXVI. Banding studies in acute lymphoblastic leukemia (ALL). Cancer 40, 1161–1171 (1977b).

Oshimura, M., Hayata, I., Kakati, S., Sandberg, A.A.: Chromosomes and causation of human cancer and leukemia. XVII. Banding studies in acute myeloblastic leukemia (AML). Cancer 38, 748–761 (1976).

Oshimura, M., Kakati, S., Sandberg, A.A.: Chromosomes and causation of human cancer and leukemia. XXVII. Possible mechanisms for the genesis of common chromosome abnormalities, including isochromosomes and the Ph¹. Cancer Res. 37, 3501–3507 (1977).

Oshimura, M., Sasaki, M., Makino, S.: Chromosomal banding patterns in primary and transplanted venereal tumors of the dog. J. Natl. Cancer Inst. 51, 1197–1203 (1973).

Oshimura, M., Sonta, S.-I., Sandberg, A.A.: Trisomy of the long arm of chromosome #1 in human leukemia. J. Natl. Cancer Inst. 56, 183–184 (1976).

Osserman, E.F.: Plasma cell dyscrasia. General considerations. In P.B. Beeson, W. McDermott (eds.) Cecil-Loeb Textbook of Medicine. 12th edition, pp. 1106–1116, Philadelphia: W.B. Saunders (1968).

Østergaard, P.A.: A girl with recurrent infections, low IgM and an abnormal chromosome number 1. Acta Paediatr. Scand. 62, 211–215 (1973).

Osztovics, M., Bühler, E.M., Müller, H., Stalder, G.R.: Banding techniques in the evaluation of human chromosomal variants. Humangenetik 18, 123–128 (1973).

Osztovics, M., Ivady, Gy., Ruzicska, I.P., Bühler, E.M., Király, L.: 45,X/46,XY/47,XYY mosaicism in a phenotypic female with gonadoblastoma. Acta Paediatr. Acad. Sci. Hung. 15, 295–299 (1974).

O'Toole, C., Price, Z.H., Ohnuki, Y., Unsgaard, B.: Ultrastructure, karyology and immunology of a cell line originated from a human transitional-cell carcinoma. Br. J. Cancer 38, 64–76 (1978).

Ott, J., Linder, D., Kaiser-McCaw, B., Lovrien, E.W., Hecht, F.: Estimating distances from the centromere by means of benign ovarian teratomas in man. Ann. Hum. Genet. 40, 191–196 (1976).

Overzier, C.: Ein XX/XY-Hermaphrodit mit einem "intratubulären Ei" und einem Gonadoblastom Gonocytom 3. Klin. Wochenschr. 42, 1052–1060 (1964).

Owen, J.J.T., Moore, M.A.S., Biggs, P.M.: Chromosome studies in Marek's disease. J. Natl. Cancer Inst. 37, 199–209 (1966).

Oxford, J.M., Harnden, D.G., Parrington, J.M., Delhanty, J.D.A.: Specific chromosome aberrations in ataxia telangiectasia. J. Med. Genet. 12, 251–262 (1975).

Pacheco, J., Gabuzda, T.G., Jackson, L.: Erythroleukemia with Philadelphia chromosome J.A.M.A. 226, 787 (1973).

Pachmann, U., Rigler, R.: Quantum yield of acridines interacting with DNA of defined base sequence. A basis for the explanation of acridine bands in chromosomes. Exp. Cell Res. 72, 602–608 (1972).

Padeh, B., Bianu, G., Akstein, E., Schaki, R.: Cytogenetic studies in chronic myeloid leukemia. Israel J. Med. Sci. 1, 791–793 (1965a).

Padeh, B., Bianu, G., Schaki, R., Akstein, E.: Cytogenetic studies in proliferative disorders. Israel J. Med. Sci. 1, 795–797 (1965b).

Padre-Mendoza, T., Farnes, P., Barker, B.E., Smith, P.S., Forman, E.N.: Y chromosome loss in childhood leukaemias. Br. J. Haematol. 41, 43–48 (1979).

Padre-Mendoza, T., Forman, E.N., Farnes, P., Barker, B.E., Smith, P.S.: Short Y chromosome and Ph¹ chromosome in acute monomyelocytic leukaemia. Lancet i, 667 (1978).

Painter, T.S.: The Y-chromosome in mammals. Science 53, 503–504 (1921).

Painter, T.S.: Studies in mammalian spermatogenesis. II. The spermatogenesis of man. J. Exp. Zool. 37, 291–335 (1923).

Painter, T.S.: The sex chromosomes of man. Am. Naturalist 58, 506–524 (1924).

Painter, T.S.: A comparative study of the chromosomes of mammals. Am. Naturalist 59, 385–409 (1925).

Painter, T.S.: Recent work on human chromosomes. A review. J. Hered. 21, 61–64 (1930).

Pajares, J.M., Espinos, D.: Alteraciones cromosomicas en la Eritromeilosis. Analisis citogenetico en cinco casos. Sangre (Barc.) 15, 98–110 (1970).

Palade, C., Postelmieu, R., Nocoara, S.: Anomalies chromosomiques dans un cas de syndrome de Fanconi. Nouv. Rev. Fr. Hematol. 10, 266–269 (1970).

Palmer, C.G., Funderburk, K.S.: Secondary constrictions in human chromosomes. Cytogenetics 4, 261–276 (1965).

Pan, S.F., Rodnan, G.P., Deutsch, M., Wald, N.: Chromosomal abnormalities in progressive systemic sclerosis (sceleroderma) with consideration of radiation effects. J. Lab. Clin. Med. 86, 300–308 (1975).

Panani, A., Papayannis, A.G., Kyrkou, K., Gardikas, C.: Cytogenetic studies in preleukaemia using the G-banding staining technique. Scand. J. Haematol. 18, 301–308 (1977).

Pant, G.S., Kamada, N., Tanaka, R.: Sister chromatid exchanges in peripheral lymphocytes of atomic bomb survivors and of normal individuals exposed to radiation and chemical agents. Hiroshima J. Med. Sci. 25, 99–105 (1976).

Papamichail, M., Brown, J.C., Holborow, E.J.: Immunoglobulins on the surface of human lymphocytes. Lancet ii, 850–852 (1971).

Paquin, L.A., Purtilo, D.T.: Genetic studies of the X-linked recessive lymphoproliferative syndrome (XLRLS). Genetics 88, (Suppl. 4, Pt. 2), s76 (1978).

Pardon, J.F., Wilkins, M.H.F.: A super coil model for nucleohistone. J. Mol. Biol. 68, 115–124 (1972).

Pardon, J.F., Wilkins, M.H.F., Richards, B.M.: Molecular structure: Super helical model for nucleohistone. Nature 215, 508–509 (1967).

Pardue, M.L., Gall, J.G.: Chromosomal localization of mouse satellite DNA. Science 168, 1356–1358 (1970).

Paris Conference (1971): Standardization in Human Cytogenetics. Birth Defects: Orig. Art. Ser., Vol. VIII, No. 7. New York: The National Foundation (1972).

Paris Conference (1971), Supplement (1975): Standardization in Human Cytogenetics. Birth Defects: Orig. Art. Ser., Vol. XI, No. 9. New York: The National Foundation (1975).

Park, I.J., Heller, R.H., Jones, H.W., Jr., Woodruff, J.D.: Apparent pseudopuberty in a phenotypic female with a gonadal tumor and an autosome/Y chromosome translocation. Am. J. Obstet. Gynecol. 119, 661–668 (1974).

Parker, J.W., O'Brien, R.L., Luke, R.J., Steiner, J.: Transformation of human lymphocytes by sodium periodate. Lancet i, 103–104 (1972).

Parker, J.W., O'Brien, R.L., Steiner, J., Paolilli, P.: Periodate-induced lymphocyte transformation. II. Character of response and comparison with phytohemagglutinin and pokeweed mitogen stimulation. Exp. Cell Res. 78, 279–286 (1973).

Parmentier, R., Dustin, P.: Reproduction experimentale d'une anomalie particulie de la metaphase de celules maliques. ("metaphase a trois groupes"). Caryologica 4, 98–109 (1951).

Parrington, J.M.: Chromosome aberrations induced by UV in cultured fibroblasts from patients with xeroderma pigmentosum. Excerpta Medica International Congress Series 233, 138, Abstract No. 504 (1971).

Parrington, J.M., Casey, G., West, L., de Vasconcelos Maia, V.: Frequency of chromosome aberrations and chromatid exchange in cultured fibroblasts from patients with xeroderma pigmentosum, Huntington's chorea and normal controls. Mutat. Res. 46, 146 (1977).

Parrington, J.M., Delhanty, J.D.A., Baden, H.P.: Unscheduled DNA synthesis, UV induced chromosome aberrations and SV_{40} transformation in cultured cells from xeroderma pigmentosum. Ann. Hum. Genet. 35, 149–160 (1971).

Pascasio, F.M., Jesalva, P.S., Cruz, E.P., Alikpala, S.: Adrenocortical carcinoma with cytogenetic findings. Acta Med. Phillip. 3, 266–268 (1967).

Pascoe, H.R.: Tumors composed of immature granulocytes occurring in the breast in chronic granulocytic leukemia. Cancer 25, 697–704 (1970).

Passarge, E.: Spontaneous chromosome instability. Humangenetik 16, 151–157 (1972).

Patau, K.: The identification of individual chromosomes, especially in man. Am. J. Hum. Genet. 12, 250–276 (1960).

Patau, K.: Chromosomal abnormalities in Waldenström's macroglobulinaemia. Lancet ii, 600–601 (1961a).

Patau, K.: Chromosome identification and the Denver report. Lancet i, 933–934 (1961b).

Patau, K.: Identification of chromosomes. In J.J. Yunis (ed.): *Human Chromosome Methodology,* p. 155–186. New York, London: Academic Press (1965).

Paterson, M.C., Smith, B.P., Lohman, P.H.M., Anderson, A.K., Fishman, L.: Defective excision repair of γ-ray damaged DNA in human (ataxia telangiectasia) fibroblasts. Nature 260, 444–446 (1976).

Paterson, W.G., Hobson, B.M., Smart, G.E., Bain, A.D.: Two cases of hydatidiform degeneration of the placenta with fetal abnormality and trip-

loid chromosome constitution. J. Obstet. Gynecol. Br. Commonw. 78, 136–142 (1971).
Pathak, S., Arrighi, F.E.: Loss of DNA following C-banding procedures. Cytogenet. Cell Genet. 12, 414–422 (1973).
Pathak, S., Siciliano, M.J., Cailleau, R., Wiseman, C.L., Hsu, T.C.: A human breast adenocarcinoma with chromosome and isozyme markers similar to those of the HeLa line. J. Natl. Cancer Inst. 62, 263–271 (1979).
Patil, S.R., Lubs, H.A.: Non-random association of human acrocentric chromosomes. Humangenetik 13, 157–159 (1971).
Patil, S.R., Lubs, H.A.: Classification of qh regions in human chromosomes 1, 9, and 16 by C-banding. Hum. Genet. 38, 35–38 (1977).
Patil, S.R., Merrick, S., Lubs, H.A.: Identification of each human chromosome with a modified Giemsa stain. Science 173, 821–822 (1971).
Paul, B., Porter, I.H., Benedict, W.F.: Giemsa banding in an established line of a human malignant meningioma. Humangenetik 18, 185–187 (1973).
Paulete-Vanrell, J.: DNA content and chromosome number in twenty-five human carcinomas. Oncology 24, 48–57 (1970).
Paulete-Vanrell, J.: Análisis del cariotipo de dos tumores dermatológicos. Rev. Assoc. Méd. Argent. 80, 372–377 (1966).
Paulete-Vanrell, J., Camacho, D.O.O.: Estudio cromosómico de un carcinoma de endometrio. Actas Ginecot. (Montevideo) 20, 298–324 (1966).
Paulete-Vanrell, J., Camacho, D.O.O.: Cariotipo de un adenocarcinoma cervical primitivo. Actas Ginecot. (Montevideo) 20, 24–34 (1964).
Paulete-Vanrell, J., Laguardia, A., Camacho, D.O.O.: Determinacion del numero de cromosomas de un carcinoma vulvar humano tratado con "Trenimón." Actas Ginecot. (Montevideo) 18, 98–112 (1964).
Pawelski, S., Maj, St., Topolska, P.: Chromosomal abnormalities of spleen cells in osteomyelosclerosis. Acta Haematol. 38, 397–402 (1967).
Pawelski, S., Topolska, P., Maj, S.: Nieprawidłowości chromosomalne w zespole di Guglielmo. Nowotwory 15, 293–300 (1965).
Pawliger, D.F., Barrow, M., Noyes, W.D.: Acute leukaemia and Turner's syndrome. Lancet i, 1345 (1970).
Pearson, M.A., Grello, F.W., Cone, T.E.: Leukemia in identical twins. N. Engl. J. Med. 268, 1153–1156 (1963).
Pearson, P.L., Bobrow, M., Vosa, C.G.: Technique for identifying Y chromosomes in human interphase nuclei. Nature 226, 78–80 (1970).
Pearson, P.L., Bobrow, M., Vosa, C.G., Barlow, P.W.: Quinacrine fluorescence in mammalian chromosomes. Nature 231, 326–329 (1971).
Peckham, M.J., Cooper, E.H.: Proliferation characteristics of the various classes of cells in Hodgkin's disease. Cancer 24, 135–146 (1969).
Pedersen, B.: Two cases of chronic myeloid leukaemia with presumably identical 47-chromosome cell-lines in the blood. Acta Pathol. Microbiol. Scand [A] 61, 497–502 (1964a).
Pedersen, B.: Chromosome aberrations in blood, bone marrow, and skin from a patient with acute leukaemia treated with 6-mercaptopurine. Acta Pathol. Microbiol. Scand. [A] 61, 261–267 (1964b).
Pedersen, B.: Development and possible significance of abnormal cell-lines in the acute stage of chronic myelogenous leukaemia. Scand. J. Haematol. 2, 167–173 (1965).
Pedersen, B.: The aneuploid Ph^1 positive cell population during progression and treatment of chronic myelogenous leukaemia. Acta Pathol. Microbiol. Scand. [A] 67, 451–462 (1966a).
Pedersen, B.: Karyotype profiles in chronic myelogenous leukaemia. Influence of therapy and progression of disease. Acta Pathol. Microbiol. Scand. [A] 67, 463–478 (1966b).
Pedersen, B.: Studies concerning the cytogenetic relationship between *in vivo* and corresponding *in vitro* cell populations from patients with chronic myelogenous leukaemia. Acta Pathol. Microbiol. Scand. [A] 68, 408–420 (1966c).
Pedersen, B.: Ph^1 prevalence, peripheral blood picture and cytostatic therapy. Acta Pathol. Microbiol. Scand. [A] 69, 35–49 (1967a).
Pedersen, B.: Cytogenetic structure of aneuploid blood culture cell populations during progression and treatment of chronic myelogenous leukaemia. Acta Pathol. Microbiol. Scand. [A] 69, 192–204 (1967b).
Pedersen, B.: Evolutionary trends of aneuploid blood culture cell populations during progression and treatment of chronic myelogenous leukaemia. Acta Pathol. Microbiol. Scand. [A] 69, 184–191 (1967c).
Pedersen, B.: The Philadelphia chromosome and chronic myelogenous leukemia (in Danish). Ugeskr. Laeger 129, 1289–1292 (1967d).
Pedersen, B.: The role of chromosomal conditions in progression of chronic myelogenous leukemia (in Danish). Ugeskr. Laeger 129, 1293–1297 (1967e).
Pedersen, B.: Males with XO Ph^1-positive cells: a cytogenetic and clinical subgroup of chronic myelogenous leukaemia? Report of a case. Acta Pathol. Microbiol. Scand. [A] 72, 360–366 (1968a).
Pedersen, B.: Influence of hyperdiploidy on Ph^1 prevalence response to therapy in chronic myelogenous leukaemia. Br. J. Haematol. 14, 507–512 (1968b).
Pedersen, B.: Ph^1-disomy and prognosis in chronic

myelogenous leukaemia. Acta Haematol. 39, 102–111 (1968c).
Pedersen, B.: Cytogenetic evolution in chronic myelogenous leukaemia. Relation of chromosomes to progression and treatment of the disease. Copenhagen, Faculty of Medicine Dissertation. Copenhagen: Munksgaard (1969).
Pedersen, B.: Relation between karyotype and cytology in chronic myelogenous leukaemia. Scand. J. Haematol. 8, 494–504 (1971).
Pedersen, B.: The blastic crisis of chronic myeloid leukaemia: acute transformation of a preleukaemic condition? Br. J. Haematol. 25, 141–145 (1973a).
Pedersen, B.: The karyotype evolution in chronic granulocytic leukaemia. I. The chromosomes gained and lost during initiation of the evolution. Eur. J. Cancer 9, 503–507 (1973b).
Pedersen, B.: The karyotype evolution in chronic granulocytic leukaemia. II. The chromosome and karyotype pattern of advanced evolution. Eur. J. Cancer 9, 509–513 (1973c).
Pedersen, B.: Periodic acid-Schiff positive myeloblasts in chronic myelogenous leukaemia: Relation to karyotype evolution. Scand. J. Haematol. 11, 112–121 (1973d).
Pedersen, B.: Clonal evolution and progression in chronic myeloid leukaemia. Blood Cells 1, 227–234 (1975a).
Pedersen, B.: The pathogenesis of granulopoietic hyperplasia in chronic myeloid leukaemia and the humoral regulating factors: A hypothesis. Scand. J. Haematol. 14, 108–113 (1975b).
Pedersen, B.: Pathogenesis and blastic transformation of chronic myeloid leukemia as consequences of Ph^1-positive stem cell hyperplasia: A unifying concept. Blood Cells 3, 535–551 (1977).
Pedersen, B., Hayhoe, F.G.J.: Relation between phagocytic activity and alkaline phosphatase content of neutrophils in chronic myeloid leukaemia. Br. J. Haematol. 21, 257–260 (1971a).
Pedersen, B., Hayhoe, F.G.J.: Cellular changes in chronic myeloid leukaemia. Br. J. Haematol. 21, 251–256 (1971b).
Pedersen, B., Killmann, S.-A.: Leukaemic subclones in chronic myelogenous leukaemia. Acta Med. Scand. 190, 61–69 (1971).
Pedersen, B., Videbaek, A.: Several cell-lines with abnormal karyotypes in a patient with chronic myelogenous leukaemia. Scand. J. Haematol. 1, 129–137 (1964).
Pedersen-Bjergaard, J., Worm, A.-M., Hainau, B.: Blastic transformation of chronic myelocytic leukaemia. Scand. J. Haematol. 18, 292–300 (1977).
Pegelow, C.H., Ebbin, A.J., Powars, D., Towner, J.W.: Familial neuroblastoma. J. Pediatr. 87, 763–765 (1975).

Pegoraro, L., Rovera, G.: Anomalie cromosomiche nelle leucemie acuta umane. Prog. Med. (Napoli) 20, 587–593 (1964).
Pegoraro, L., Jakšić, B., Gavosto, F.: T-cells in chronic lymphatic leukaemia. Lancet ii, 909 (1973).
Pegoraro, L., Pileri, A., Gavosto, F.: Anomalia cromosomica in un caso di leucemia acuta con aplasia midollare. Haematologica (Pavia) 48, 713–719 (1963).
Pegoraro, L., Pileri, A., Rovera, G., Gavosto, F.: Trisomia del cromosoma Filadelfia (Ph^1) nella crisi blastica di un caso di leucemia mieloide cronica. Tumori 53, 315–321 (1967).
Pegoraro, L., Rovera, G., Masera, P., Gavosto, F.: Chromosome size and aneuploidy. Lancet ii, 618 (1967).
Pelc, S.R., Appelton, T.C., Welton, M.E.: State of light autoradiography. In C.P. Leblond, K.B. Warren (eds.): *The Use of Radioautography in Investigating Protein Synthesis*, pp. 9–22. New York: Academic Press (1965).
Perebra, D.J.B., Pegrum, G.D.: The lymphocyte in chronic lymphatic leukaemia. Lancet i, 1207–1209 (1974).
Perillie, P.E.: Studies of the changes in leucocyte alkaline phosphatase following pyrogen stimulation in chronic granulocytic leukemia. Blood 29, 401–406 (1967).
Perillie, P.E., Finch, S.C.: Muramidase studies in Philadelphia-chromosome-positive and chromosome-negative chronic granulocytic leukemia. N. Engl. J. Med. 283, 456–459 (1970).
Perkins, J., Timson, J., Emery, A.E.H.: Clinical and chromosome studies in Fanconi's aplastic anaemia. J. Med. Genet. 6, 28–33 (1969).
Perlin, E., Granatir, R.G., Roche, J., Moquin, R.B.: Remission of leukemia associated with polycythemia vera. J.A.M.A. 223, 192 (1973).
Pero, R.W., Bryngelsson, C., Mitelman, F., Thulin, T., Nordén, Å.: High blood pressure related to carcinogen-induced unscheduled DNA synthesis, DNA carcinogen binding, and chromosomal aberrations in human lymphocytes. Proc. Natl. Acad. Sci. USA 73, 2496–2500 (1976).
Perona, G., Testolin, R.: Fanconi-Anämie bei einem Erwachsenen. Blut 13, 90–99 (1966).
Perrin, E.V., Landing, B.H.: Evidence for a relation between gonadoblastoma and gonadal dysgenesis. Am. J. Dis. Child. 102, 575 (1961).
Perry, P., Evans, H.J.: Cytological detection of mutagen-carcinogen exposure by sister chromatid exchange. Nature 258, 121–125 (1975).
Perry, P., Wolff, S.: New Giemsa method for the differential staining of sister chromatids. Nature 251, 156–158 (1974).
Perry, S.: Leukocyte kinetics in leukemia. In C.J.D. Zarafonetis (ed.): Proceedings of the International Conference on Leukemia-Lymphoma,

pp. 229–244. Philadelphia: Lea and Febiger (1968).
Perry, S., Moxley, J.H., III, Weiss, G.H., Zelen, M.: Studies of leukocyte kinetics by scintillation counting in normal individuals and in patients with chronic myelocytic leukemia. J. Clin. Invest. 45, 1388–1399 (1966).
Pescetto, G.: Considerazioni e rilievi sugli aspetti citogenetici delle neoplasie ginecologiche. Rass. Clin. Sci. 43, 329–337 (1967).
Peterlik, M., Pietschmann, H., Vormittag, W.: Isoenzyme der alkalischen Leukozytenphosphatase bei einer chronischen Myelose ohne Philadelphia-Chromosom und mit erhöhtem Phosphataseindex. Folia Haematol. (Leipz.) 93, 24–34 (1970).
Peterson, L.C., Bloomfield, C.D., Brunning, R.D.: Philadelphia chromosome positive (Ph$^+$) acute leukemia: Morphology, clinical features, and survival of nine patients. Lab. Invest. 34, 346–347 (1976a).
Peterson, L.C., Bloomfield, C.D., Brunning, R.D.: Blast crisis as an initial or terminal manifestation of chronic myeloid leukemia. Am. J. Med. 60, 209–220 (1976b).
Peterson, R.D.A., Cooper, M.D., Good, R.A.: Lymphoid tissue abnormalities associated with ataxia-telangiectasia. Am. J. Med. 41, 342–359 (1966).
Peterson, R.D.A., Kelly, W.D., Good, R.A.: Ataxia-telangiectasia. Its association with defective thymus, immunological-deficiency disease and malignancy. Lancet i, 1189–1193 (1964).
Petit, P., Cauchie, C.H.: Philadelphia chromosome by translocation. Lancet ii, 94 (1973).
Petit, P., Alexander, M., Fondu, P.: Monosomy 7 in erythroleukaemia. Lancet ii, 1326–1327 (1973).
Petit, P., Maurus, R., Richard, J., Koulischer, L.: Chromosome du cri du chat chez un trisomique 21 leucémique. Ann. Genet. (Paris) 11, 125–128 (1968).
Petit, P., Verhest, A., van der Bilt, F.L., Jongsma, A.: The chromosomes of the EB virus-positive Burkitt cell line P3J.HR1K studied by the fluorescent staining technique. Pathol. Eur. 7, 17–21 (1972).
Petit, P., Vryens, R., Canchie, C., Koulischer, L.: Anomalie chromosomique du tissue gangionnaire dans la macroglobulinine de Waldenström. Acta Clin. Belg. 23, 182–190 (1968).
Petite, J.P.: La signification clinique d'une macroglobulinémie. Les rapports entre maladie de Waldenström, myélome et leucose. Presse Med. 72, 2400 (1964).
Petres, J., Schmid-Ullrich, K., Wolf, U.: Chromosomenaberrationen an menschlichen Lymphozyten bei chronischen Arsenschäden. Dtsch. Med. Wochenschr. 95, 79–80 (1970).

Pfeiffer, R.A.: Chromosomal abnormalities in ataxia-telangiectasia (Louis-Barr's syndrome). Humangenetik 8, 302–306 (1970).
Pfeiffer, R.A., Kim, A.: Leukemia in patients with a "chromosomal breakage syndrome." Excerpta Medica International Congress Series 233, 142 (1971).
Pfeiffer, R.A., Kosenow, W.: Further cases with apparently normal pattern. Hum. Chrom. Newsl. 6, 3 (1962).
Pfeiffer, R.A., Kosenow, W., Bäumer, A.: Chromosomenuntersuchungen an Blutzellen eines Patienten mit Makroglobulinämie Waldenström. Klin. Wochenschr. 40, 342–344 (1962).
Pfitzer, P., Pape, H.D.: Characterization of tumor cell populations by DNA-measurements. Acta Cytol. 17, 19–26 (1973).
Philip, J.: Gonadoblastoma in dysgenetic gonads with a Y chromosome. Lancet i, 1244 (1975).
Philip, J., Teter, J.: Significance of chromosomal investigation of somatic cells to determine the genetic origin of gonadoblastoma. Acta Pathol. Microbiol. Scand. [A] 61, 543–550 (1964).
Philip, J., Hansen, M.K., Reintoft, I.: A cytogenetic study of gonadoblastoma tissue in two cases. Acta Pathol. Microbiol. Scand. [A] 83, 559–567 (1975).
Philip, P.: Trisomy 8 in acute myeloid leukemia. Scand. J. Haematol. 14, 140–147 (1975a).
Philip, P.: Trisomy 11 in acute phase of chronic myeloid leukemia. Acta Haematol. 54, 188–191 (1975b).
Philip, P.: Marker chromosome 14q+ in multiple myeloma. Hereditas 80, 155–156 (1975c).
Philip, P.: A vulnerable point on human chromosome 3 in myeloproliferative disorders? Hereditas 81, 124–125 (1975d).
Philip, P.: G banding analysis of complex aneuploidy in a case of erythroleukaemia. Scand. J. Haematol. 16, 365–368 (1976).
Philip, P.: Chromosomal aberrations in acute non-lymphocytic leukemia. Marker chromosomes and vulnerable points. *Helsinki Chromosome Conference*, Aug. 29–31, p. 206 (abstract) (1977a).
Philip, P.: Chromosomal aberrations in multiple myeloma. International Society of Haematology, European African Division, 4th Meeting, Held in Istanbul, Sept. 5–9, Abstract #82 (1977b).
Philip, P., Drivsholm, A.A.: G-banding of 3 multiple myeloma marker chromosomes. Biomedicine 21, 429–430 (1974).
Philip, P., Drivsholm, A.A.: G-banding analysis of complex aneuploidy in multiple myeloma bone marrow cells. Blood 47, 69–77 (1976).
Philip, P., Ernst, P., Lange Wantzin, G.: Karyotypes in infectious mononucleosis. Scand. J. Haematol. 15, 201–206 (1975).

Philip, P., Jensen, M.K., Killmann, S.-Aa., Drivsholm, A., Hansen, N.E.: Chromosomal banding patterns in 88 cases of acute nonlymphocytic leukemia. Leukemia Res. 2, 201–212 (1978).

Philip, P., Krogh-Jensen, M., Pallesen, G.: Marker chromosome 14q+ in non-endemic Burkitt's lymphoma. Cancer 39, 1495–1499 (1977).

Philip, P., Müller-Berat, N., Killmann, S.-A.: Philadelphia chromosome in acute lymphocytic leukaemia. Hereditas 84, 231–232 (1976).

Philip, P., Wantzin, G.L., Jensen, K.G., Drivsholm, A.: Trisomy 8 in acute myeloid leukaemia: A non-random event. Scand. J. Haematol. 18, 163–169 (1977).

Pickthall, V.J.: Detailed cytogenetic study of a metastatic bronchial carcinoma. Br. J. Cancer 34, 272–278 (1976).

Pierce, G.B., Nakane, P.K.: Nuclear sex of testicular teratomas of infants. Nature 214, 820–821 (1967).

Pierre, R.V.: Preleukemic states. Semin. Hematol. 11, 73–92 (1974).

Pierre, R.V.: Cytogenetic studies in preleukemia: Studies before and after transition to acute leukemia in 17 subjects. Blood Cells 1, 163–170 (1975).

Pierre, R.V.: Preleukemic syndromes. Virchows Arch. B Cell Pathol. 29, 29–37 (1978a).

Pierre, R.V.: Cytogenetics in malignant lymphoma. Virchows Arch. B Cell Pathol. 29, 107–112 (1978b).

Pierre, R.V., Hoagland, H.C.: 45,X cell lines in adult men: loss of Y chromosome, a normal aging phenomenon? Mayo Clin. Proc. 46, 52–55 (1971).

Pierre, R.V., Hoagland, H.C.: Age-associated aneuploidy: loss of Y chromosome from human bone marrow cells with aging. Cancer 30, 889–894 (1972).

Pierre, R.V., Hoagland, H.C.: The missing Y chromosome and human leukaemia. Lancet i, 1008–1009 (1973).

Pierre, R.V., Hoagland, H.C., Linman, J.W.: Microchromosomes in human preleukemia and leukemia. Cancer 27, 160–175 (1971).

Pierre, R.V., O'Sullivan, M.B., Hilton, P.K.: Philadelphia chromosome studies of mailed-in blood samples. Am. J. Clin. Pathol. 62, 713–714 (1974).

Pierseus, W.F., Shur, P.H., Molony, W.H., Churchill, W.M.: Lymphocyte surface immunoglobulins: Distribution and frequency in lymphoproliferative disorders. N. Engl. J. Med. 288, 176–180 (1973).

Pileri, A., Masera, P., Pegoraro, L., Gavosto, F.: Indagini fisiopatologiche e citogenetiche su un caso di mielofibrosi trattato con successo con androgeni. Haematologica (Pavia) 51, 57–72 (1966).

Pinedo, H.M., van Hemel, J.O., Vrede, M.A., van der Sluys Veer, J.: Acute myelofibrosis and chromosome damage after procarbazine treatment. Br. Med. J. 3, 525 (1974).

Pisano, D.: Chronic myelocytic leukemia: Report of case and review of current literature. J. Am. Osteopath. Assoc. 75, 971–978 (1976).

Pogosiants, E.E.: Chromosome variability and carcinogenesis. Vestn. Akad. Med. Nauk. SSSR 1, 49–54 (1973).

Pogosiants, E.E., Prigozhina, E.L.: Progress in the cytogenetic study of malignant neoplasias. Vopr. Onkol. 22, 85–93 (1976).

Pogosianz, H.E., (Pogosyantz, E.E.): Cytogenetics of tumors (in Russian) Zh. Vses. Khim. Obshchestva Imeni D.I. Mendeleeva 8(4), 449–457 (1963).

Pogosianz, H.E., Prigogina, E.L.: Chromosome abnormalities and carcinogenesis. Neoplasma 19, 319–325 (1972).

Pogosianz, H.E., Chudina, A.P., Rodkina, R.A.: Karyotypic abnormalities in human ovarian cancers (in Russian). Vopr. Onkol. 18, 3–9 (1972).

Pogosianz, H.E., Prigogina, E.L., Fleishman, E.V.: Chromosome changes in human haemoblastoses. In S. Armendares, R. Lisker (eds.): V International Congress of Human Genetics, p. 147, Abstract 385. Held in Mexico, Oct. 10–15 (1976).

Polák, J., Žižka, J.: Myeloproliferative disease in a child with monosomia of a C group chromosome. Acta Paediatr. Scand. 59, 591–595 (1970).

Polani, P.E.: Cytogenetics of Fanconi anaemia and related chromosome disorders. Ciba Found. Symp. 37, 261–306 (1976).

Pollini, G., Colombi, R.: Il danno cromosomico midollare nell'anemia aplastica benzolica. Med. Lav. 55, 241–255 (1964a).

Pollini, G., Colombi, R.: Il danno cromosomico dei linfociti nell'emopatia benzenica. Med. Lav. 55, 641–654 (1964b).

Polliack, A., Douglas, S.D.: Surface features of human eosinophils. A scanning and transmission electron microscopic study of a case of eosinophilia. Br. J. Haematol. 30, 303–306 (1975).

Pontén, J., McIntyre, E.H.: Long term culture of normal and neoplastic human glia. Acta Pathol. Microbiol. Scand. [A] 74, 465–486 (1968).

Pontén, J., Saksela, E.: Two established in vitro cell lines from human mesenchymal tumours. Int. J. Cancer 2, 434–447 (1967).

Ponti, G.B., Valentini, R., Carrara, P.M., Eridani, S.: Investigations on the chromosome complement in some myeloproliferative disorders. Acta Haematol. (Basel) 34, 36–43 (1965).

Ponzone, A., de Sanctis, C., Fabris, G., Ciriotti, G., Franceschini, P.: Blast cell proliferation in perinatal leukaemia with chromosomal transloca-

tion (Bq+; Dq−). Helv. Paediatr. Acta 27, 3−13 (1972).
Ponzone, A., Rovera, G., Tarocco, R.P.: Leucemia mieloide cronica Ph¹ positive in una bambina di 5 anni. Minerva Pediatr. 19, 1458−1490 (1967).
Poon, P.K., O'Brien, R.L., Parker, J.W.: Defective DNA repair in Fanconi's anaemia. Nature 250, 223−225 (1974).
Popescu, H.I., Stephanescu, D.T.: Cytogenetic investigation of industrial workers occupationally exposed to gamma rays. Radiat. Res. 47, 562−570 (1971).
Popescu, N.C., Olinici, C.D., Castro, B.C., DiPaolo, J.A.: Random chromosome changes following SA7 transformation of Syrian hamster cells. Int. J. Cancer 14, 461−472 (1974).
Popescu, N.C., Turnbull, D., DiPaolo, J.A.: Sister chromatid exchange and chromosome aberration analysis with the use of several carcinogens and noncarcinogens: Brief communication. J. Natl. Cancer Inst. 59, 289−293 (1977).
Poroshenko, G.G.: Chromosome endoreduplication in a patient with chronic myeloid leukemia (in Russian). Tsitologiia 8, 316−319 (1966).
Porter, I.H., Benedict, W.F., Brown, C.D., Paul, B.: Recent advances in molecular pathology: a review. Some aspects of chromosome changes in cancer. Exp. Mol. Pathol. 11, 340−367 (1969).
Potter, A.M., Sharp, J.C., Brown, M.J., Sokol, R.J.: Structural rearrangements associated with the Ph¹ chromosome in chronic granulocytic leukemia. Humangenetik 29, 223−228 (1975).
Powsner, E.R.: Frequency of endoreduplication in short-term cultures of human blood cells. J. Lab. Clin. Med. 67, 610−614 (1966).
Powsner, E.R., Berman, L.: The chromosomes of human bone marrow cells. Morphologic study of material prepared promptly after aspiration. Blood 17, 360 (1961).
Powsner, W.R., Berman, L.: Human bone marrow chromosomes in megaloblastic anemia. Blood 26, 784−789 (1965).
Pravtcheva, D., Andreeva, P., Tsaneva, R.: A new translocation in chronic myelogenous leukemia. Hum. Genet. 32, 229−232 (1976).
Pravtcheva, D., Manolov, G.: Genesis of the Philadelphia chromosome: Possible points of breakage in chromosome No. 22. Hereditas 79, 301−303 (1975).
Prescott, D.M.: Structure and replication of eukaryotic chromosomes. Adv. Cell Biol. 1, 57−118 (1970).
Preud'Homme, J.L., Hurez, D., Seligmann, M.: Immunofluorescence studies in Waldenström's macroglobulinemia. Rev. Eur. Etud. Clin. Biol. 15, 1127−1131 (1970).
Pridie, G., Dimitrescu-Purvu, D.: Leucemie acutasi sindrom Bonnene-Ulrich la un nou-naseut. Pediatria (Bucur.) 10, 345−349 (1961).
Prieto, F., Badia, L., Calabuig, J.R., Mayans, J., Perez-Sirvent, M.L., Martinez, J., Luno, E., Marty, M.L.: Chromosome abnormalities in refractory anemia with partial myeloblastosis. Sangre (Barc.) 21, 701−712 (1976).
Prieto, F., Badia, L., Mayans, J., Besalduch, J., Marty, M.L.: Trisomía de los brazos largos del cromosoma 1 e inversíon del cromosoma 9 en células leucémicas. Sangre (Barc.) 23, 64−68 (1978a).
Prieto, F., Badia, L., Mayans, J., Gomis, F., Marty, M.L.: Twenty-six chromosomes hypodiploidy in acute lymphoblastic leukemia (in Spanish). Sangre (Barc.) 23, 484−488 (1978b).
Prieto, F., Egozcue, J., Forteza, G., Marco, F.: Identification of the Philadelphia (Ph¹) chromosome. Blood 35, 23−38 (1970).
Prieto Garcia, F., Forteza Bover, G., Baguena Candela, R., Marco Orts, F.: Double Philadelphia chromosome (Ph¹) and hyperdiploidy in the acute stage of a chronic myeloid leukemia (in Spanish). Sangre (Barc.) 13, 193−202 (1968a).
Prieto Garcia, F., Forteza Bover, G., Baguena Candela, R., Marco Orts, F.: Hyperdiploidy and Philadelphia chromosome (Ph¹) in the blastic crisis of a chronic myelogenous leukemia (in Spanish). Sangre (Barc.) 13, 443−448 (1968b).
Prigogina, E.L., Fleischman, E.W.: Certain patterns of karyotype evolution in chronic myelogenous leukaemia. Humangenetik 30, 113−119 (1975a).
Prigogina, E.L., Fleischman, E.W.: Marker chromosome 14q+ in two non-Burkitt lymphomas. Humangenetik 30, 109−112 (1975b).
Prigogina, E.L., Fleischman, E.W., Volvoka, M.A., Frenkel, M.A.: Chromosome abnormalities and clinical and morphologic manifestations of chronic myeloid leukemia. Hum. Genet. 41, 143−156 (1978).
Prigogina, E.L., Stavrovskaja, A.A., Kakpakova, E.S., Streljuchina, N.V., Zakharov, A.F., Lelikova, G.P., Chudina, A., Pogosianz, E.E.: Congenital chromosome abnormalities and leukaemia. Lancet ii, 524 (1970).
Prigogina, E.L., Stavrovskaja, A.A., Zakharov, A.F., Lelikova, G.P., Streljukhina, N.V.: Congenital anomalies of the karyotype in human acute leukaemia. Vopr. Onkol. 14, 58−64 (1968).
Pringle, J.A.S., Williams, R.F.: Mitosis in human bladder biopsies. Annu. Rep. Br. Empire Cancer Campaign 11, 237 (1967).
Pris, J., Launais, B., Capdeville, J., Monnier, J., Bourrouilloux, G., Duchayne, E., Colombies, P.: Leucémie myéloïde chronique avec chromosome Philadelphie compliquant l'évolution d'une maladie de Vaquez. Nouv. Presse Med. 7, 1114−1115 (1978).
Prokofjeva-Belgovskaya, A.A., Kosmachevskaya,

G.A., Terentyeva, E.I., Veshneva, I.V.: Study of human chromosomes in the cases of chronic myeloleucosis and acute hemocytoblastosis. Hum. Chrom. Newsl. 46, 16–18 (1964).

Propp, S., Lizzi, F.A.: Philadelphia chromosome in acute lymphocytic leukemia. Blood 36, 353–360 (1970).

Propp, S., Brown, Ch.D., Tartaglia, A.P.: Down syndrome and congenital leukemia. N.Y. State J. Med. 66, 3067–3071 (1966).

Prosser, J., Bradley, M.L., Muir, P.D., Vincent, P.C., Gunz, F.W.: Satellite III DNA hybridised to chromosomes from patients with acute leukemia. Leukemia Res. 2, 151–161 (1978).

Pruett, R.C., Atkins, L.: Chromosome studies in patients with retinoblastoma. Arch. Ophthalmol. 82, 177–181 (1969).

Prunieras, M.: Le syndrome de Sézary existe-t-il? Nouv. Rev. Fr. Hematol. 13, 237–242 (1973).

Prunieras, M.: DNA content and cytogenetics of the Sezary cell. Mayo Clin. Proc. 49, 548–552 (1974).

Prunieras, M., Gazzolo, L., Delescluse, C., Charachon, J., Bouchayer, M.: Étude caryologique et ultrastructurale de deux cas de papillome du larynx de l'enfant. Pathol. Biol. (Paris) 16, 277–285 (1968).

Psaroudakis, A., Oettinger, M., Byrd, J.R., Greenblatt, R.B.: Cytogenetic studies in gonadal dysgenesis with dysgerminoma. Am. J. Obstet. Gynecol. 126, 508–510 (1976).

Punnett, T., Punnett, H.H., Kaufmann, B.N.: Preparation of a crude human leukocyte growth factor from Phaseolis vulgaris. Lancet i, 1359–1360 (1962).

Purchase, I.F.H., Richardson, C., Anderson, D.: Chromosomal effects in peripheral lymphocytes. Proc. R. Soc. Med. 69, 290–292 (1976).

Purchase, I.F.H., Richardson, C.R., Anderson, D., Paddle, G.M., Adams, W.G.F.: Chromosomal analyses in vinyl chloride-exposed workers. Mutat. Res. 57, 325–334 (1978).

Purtilo, D.T., Bhawan, J., Hutt, L.M., Denicola, L., Szymanski, I., Yang, J.P.S., Boto, W., Maier, R., Thorley-Lawson, D.: Epstein-Barr virus infections in the X-linked recessive lymphoproliferative syndrome. Lancet i, 798–801 (1978).

Purtilo, D.T., DeFlorio, D., Jr., Hutt, L.M., Bhawan, J., Yang, J.P.S., Otto, R., Edwards, W.: Variable phenotypic expression of an X-linked recessive lymphoproliferative syndrome. N. Eng. J. Med. 297, 1077–1081 (1977).

Quaglino, D., Cowling, D.C.: Cytochemical studies on cells from chronic lymphocytic leukaemia and lymphosarcoma cultured with phytohaemagglutinin. Br. J. Haematol. 10, 358–364 (1964).

Queisser, W., Queisser, U., Ansmann, M., Brunner, G., Hoelzer, D., Heimpel, H.: Megakaryocyte polyploidization in acute leukaemia and pre-leukaemia. Br. J. Haematol. 28, 261–270 (1974).

Quinlan, M.F., Scopa, J.: Thorotrast-induced haemangioendothelial sarcoma: A lesson from the past. Aust. N.Z. J. Med. 6, 329–335 (1976).

Quiroz-Gutiérrez, A., Alfaro-Kofman, S., Marquez-Monter, H.: Chromosome markers in a seminoma. Lancet ii, 306 (1967).

Quiroz-Gutiérrez, A., Islas, G.M., Robles, I.N.H.: Observaciones sobre la accion de la methilhidracina en algunos melanomas. II. Estudios cromosomicos. Rev. Med. Hosp. Gen. 31, 645–651 (1968).

Rabson, A.S., O'Conor, G.T., Baron, S., Whang, J.J., Legallais, F.Y.: Morphologic, cytogenetic and virologic studies in vitro of a malignant lymphoma from an African child. Int. J. Cancer 1, 89–106 (1966).

Ragen, P.A., McGuire, P., Antonius, J.I.: Decreased formation of erythrocyte antigen A and a consistent chromosome abnormality in a patient with myelomonocytic leukemia. Acta Haematol. (Basel) 39, 309–319 (1968).

Raich, P.C., Carr, R.M., Meisner, L.F., Korst, D.R.: Acute granulocytic leukemia in Hodgkin's disease. Am. J. Med. Sci. 269, 237–241 (1975).

Rajeswari, S., Ghosh, S.N., Shah, P.N., Borah, V.J.: Barr body frequency in the human breast cancer tissue. A prospective study on its prognostic value and its correlation with specific oestradiol receptors. Eur. J. Cancer 13, 99–102 (1977).

Ranjini, R.: Chromosomal abnormalities in multiple myeloma. J. Kans. Med. Soc. 72, 435–437 (1971).

Raposa, T.: Sister chromatid exchange studies for monitoring DNA damage and repair capacity after cytostatics in vitro and in lymphocytes of leukaemic patients under cytostatic therapy. Mutat. Res. 57, 241–251 (1978).

Rarposa, T., Natarajan, A.T.: Fluorescence banding pattern of human and mouse chromosomes with a benzimidazol derivative (Hoechst 33258). Humangenetik 21, 221–226 (1974).

Raposa, T., Natarajan, A.T., Granberg, I.: Identification of Ph¹ chromosome and associated translocation in chronic myelogenous leukemia by Hoechst 33258. J. Natl. Cancer Inst. 52, 1935–1938 (1974).

Rappaport, H.: Tumors of the hematopoietic systems. In: Atlas of Tumor Pathology, Washington, D.C., Armed Forces Institute of Pathology, Sect. III, Fasc. 8, p. 442 (1966).

Rapparpot, H., Braylan, R.C.: Changing concepts in the classification of malignant neoplasms of the hematopoietic system. In D.W. Rebuck, C.W.

Beard, M.R. Abell (eds.): *International Academy of Pathology, Monogr. 16: The Reticuloendothelial System*, pp. 1–19. Baltimore: Williams and Wilkins (1975).

Rary, J.M., Bender, M.A., Kelly, T.E.: Cytogenetic studies of ataxia telangiectasia. Am. J. Hum. Genet. 26, 70A (abstract) (1974).

Rary, J.M., Bender, M.A., Kelly, T.E.: A 14/14 marker chromosome lymphocyte clone in ataxia telangiectasia. J. Hered. 66, 33–35 (1975).

Rashad, M.N., Fathalla, M.F., Kerr, M.G.: Sex chromatin and chromosome analysis in ovarian teratomas. Am. J. Obstet. Gynecol. 96, 461–465 (1966).

Rashad, W.N., Morton, W.R.M.: Chromosomes and serum proteins. Br. Med. J. i, 181 (1964).

Rasheed, S., Gardner, M.B., Rongey, R.W., Nelson-Rees, W.A., Arnstein, P.: Human bladder carcinoma: Characterization of two new tumor cell lines and search for tumor viruses. J. Natl. Cancer Inst. 58, 881–890 (1977).

Rask-Madsen, J., Philip, J.: The chromosome complement of human endometrium. Cytogenetics 9, 24–41 (1970).

Rastrick, J.M.: A method for the positive identification of erythropoietic cells in chromosome preparations of bone marrow. Br. J. Haematol. 16, 185–191 (1969).

Rastrick, J.M., Fitzgerald, P.K., Gunz, F.W.: Direct evidence for presence of Ph^1 chromosome in erythroid cells. Br. Med. J. i, 96–98 (1968).

Rattner, J.B., Branch, A., Hamkalo, B.A.: Electron microscopy of whole mount metaphase chromosomes. Chromosoma 52, 329–338 (1975).

Rauh, J.L., Soukup, S.W.: Bloom's syndrome. Am. J. Dis. Child. 116, 409–413 (1968).

Rausen, A.R., Kim, H.J., Burstein, Y., Rand, S., McCaffrey, R.M., Kung, P.C.: Philadelphia chromosome in acute lymphatic leukaemia of childhood. Lancet i, 432 (1977).

Ravindranath, Y., Inoue, S., Considine, B., Lusher, J.: New leukemia in the course of therapy of acute lymphoblastic leukemia (ALL). Proc. Am. Assoc. Cancer Res. 18, 203 (1977).

Ray-Chandhuri, S.P., Kakati, S., Sharma, T.: Split and enlarged satellites in man. Heredity 23, 146–149 (1968).

Re, G., Bagnara, G.P., Brunelli, M.A.: Cytogenetic studies of the peripheral blood in chronic myeloid leukemia cultured in diffusion chambers (in Italian). Boll. Soc. Ital. Biol. Sper. 52, 503–506 (1976).

Reed, W.B., Landing, B., Sugarman, G., Cleaver, J.E., Melnyk, J.: Xeroderma pigmentosum. J.A.M.A. 207, 2073–2079 (1969).

Reeves, B.R.: Cytogenetics of malignant lymphomas. Studies utilizing a Giemsa-banding technique. Humangenetik 20, 231–250 (1973).

Reeves, B.R., Houghton, J.A.: Serial cytogenetic studies of human colonic tumour xenografts. Br. J. Cancer 37, 612–619 (1978).

Reeves, B.R., Lawler, S.D.: Preferential breakage of sensitive regions of human chromosomes. Humangenetik 8, 295–301 (1970).

Reeves, B.R., Lawler, S.D.: Identification of isochromosome 17 in myeloid leukaemia. Lancet i, 849–850 (1972).

Reeves, B.R., Margoles, C.: Preferential location of chlorambucil-induced breakage in the chromosomes of normal human lymphocytes. Mutat. Res. 26, 205–208 (1974).

Reeves, B.R., Stathopoulos, G.: Cytogenetic and cell-surface marker studies in two non-Hodgkin's lymphomata of T-cell origin. Hum. Genet. 31, 203–210 (1976).

Reeves, B.R., Lobb, D.S., Lawler, S.D.: Identity of the abnormal F-group chromosome associated with polycythaemia vera. Humangenetik 14, 159–161 (1972).

Reeves, B.R., Pickup, V.L., Lawler, S.D., Dinning, W.J., Perkins, E.S.: A chromosome study of patients with uveitis treated with chlorambucil. Br. Med. J. iv, 22–24 (1974).

Refsum, S.B., Hansteen, I.-L.: The relationship between the cell population kinetics and chromosome pattern in human tumours. Acta Pathol. Microbiol. Scand. [A] (Suppl.) 248, 153–155 (1974).

Reimann, F., Endogan, G., Tangün, Y.: Ein Fall von sideroblastichen Anämie mit besonderen Chromosomenbefund. *Proceedings of the 10th Congress of the European Society of Haematology*, pp. 551–557 (1965).

Reimer, R.R., Hoover, R., Fraumeni, J.F., Jr., Young, R.C.: Acute leukemia after alkylating-agent therapy of ovarian cancer. N. Engl. J. Med. 297, 177–181 (1977).

Reisman, L.E., Trujillo, J.M.: Chronic granulocytic leukemia of childhood. Clinical and cytogenetic studies. J. Pediatr. 62, 710–723 (1963).

Reisman, L.E., Mitani, M., Zuelzer, W.W.: Chromosome studies in leukemia. I. Evidence for the origin of leukemic stem lines from aneuploid mutants. N. Engl. J. Med. 270, 591–597 (1964).

Reisman, L.E., Zuelzer, W.W., Mitani, M.: Endoreduplication in a patient with acute monocytic leukaemia. Lancet ii, 1038–1039 (1963).

Reisman, L.E., Zuelzer, W.W., Thompson, R.I.: Further observation on the role of aneuploidy in acute leukemia. Cancer Res. 24, 1448–1456 (1964).

Reitalu, J., Bergman, S., Ekwall, B., Hall, B.: Correlation between Y chromosome length and fluorescence intensity of Y chromatin on interphase nuclei. Hereditas 72, 261–268 (1972).

Remy, D.: Chromosomen-Anomalien bei Hämoblastosen. Dtsch. Med. Wochenschr. 89, 684–688 (1964).

Rethoré, M.-O., Prieur-Lecuyer, A.-M., Griscelli, C., Mozziconacci, P., Lejeune, J.: Évolution clonale au cours d'une leucémie aiguë myéloblastique chez un enfant trisomique 21. Ann. Genet. (Paris) 14, 193–198 (1971).

Rethoré, M.O., Saraux, H., Prieur, M., Dutrillaux, B., Meer, J.J., Lejeune, J.: 48,XXY,+21 syndrome associated with retinoblastoma. Arch. Fr. Pediatr. 29, 533–538 (1972).

Retief, A.E., Rüchel, R.: Histones removed by fixation. Their role in the mechanism of chromosomal banding. Exp. Cell Res. 106, 233–237 (1977).

Rhodes, C.A., Robinson, W.A., Entringer, M.A.: Granulocyte colony formation in chronic granulocytic leukemia during stable, accelerated and blastic disease. Proc. Soc. Exp. Biol. Med. 157, 337–341 (1978).

Riccardi, V.M.: Trisomy 8 mosaicism in the skin of a patient with leukemia. Birth Defects: Original Art. Ser., Vol. 12, No. 1, p. 187 (1976).

Riccardi, V.M., Humbert, J.R., Peakman, D.: Acute leukemia associated with trisomy 8 mosaicism and a familial translocation 46,XY,t(7;20) (p13; p12). Am. J. Hum. Genet. 2, 15–21 (1978).

Riccardi, V.M., Sujansky, E., Smith, A.C., Francke, U.: Chromosomal imbalance in the aniridia-Wilms' tumor association: 11p interstitial deletion. Pediatrics 61, 604–610 (1978).

Ricci, N.: Anomalia cromosomica in un caso di sindrome eritoleucemica. Prog. Med. (Napoli) 21, 309–317 (1965).

Ricci, N., Dallapiccola, B., Preto, G.: Familial transmission of a Gq– Ph¹-like chromosome. Ann. Genet. (Paris) 13, 263–264 (1970).

Ricci, N., Punturieri, E., Bosi, L., Castoldi, G.L.: Su di una particolare alterazione cromosomica in corso di emopatia acuta. Prog. Med. (Napoli) 18, 297–301 (1962a).

Ricci, N., Punturieri, E., Bosi, L., Castoldi, G.L.: Chromosomes of Sternberg-Reed cells. Lancet ii, 564 (1962b).

Richards, B.M., Atkin, N.B.: The difference between normal and cancerous tissues with respect to the ratio of DNA content by chromosome number. Acta Un. Int. Cancer 16, 124–128 (1960).

Richards, B.M., Pardon, J.F.: The molecular structure of nucleo-histone (DNH). Exp. Cell Res. 62, 184–196 (1970).

Richart, R.M., Corfman, P.A.: Chromosome number and morphology of a human preinvasive neoplasm. Science 144, 65–67 (1964).

Richart, R.M., Ludwig, A.S., Jr.: Alterations in chromosomes and DNA content in gynecologic neoplasms. Am. J. Obstet. Gynecol. 104, 463–471 (1969).

Richart, R.M., Wilbanks, G.D.: The chromosomes of human intraepithelial neoplasia: Report of 14 cases of cervical intraepithelial neoplasia and review. Cancer Res. 26, 60–74 (1966).

Richmond, H.G., Ohnuki, Y., Awa, A.A., Pomerat, C.M.: Multiple myeloma—an in vitro study. Br. J. Cancer 15, 692–700 (1961).

Rieger, R., Michaelis, A., Green, M.M.: *A Glossary of Genetics and Cytogenetics. Classical and Molecular,* 4th edition. Berlin, Heidelberg, New York: Springer-Verlag (1976).

Rigby, C.C.: Chromosome studies in ten testicular tumours. Br. J. Cancer 22, 480–485 (1968).

Rigby, C.C., Franks, L.M.: A human tissue culture cell line from a transitional cell tumour of the urinary bladder: Growth, chromosome pattern and ultrastructure. Br. J. Cancer 24, 746–754 (1970).

Rigo, S.J., Stannard, M., Cowling, D.C.: Chronic myeloid leukaemia associated with multiple chromosome abnormalities. Med. J. Aust. 2, 70–72 (1966).

Ris, H.: Chromosome structure. *In* W.D. McElroy, B. Glass (eds.): *Symposium on the Chemical Basis of Heredity,* p. 23–69. Baltimore: Johns Hopkins Press (1957).

Ris, H.: Chromosomes and genes: Fine structure of chromosomes. Proc. R. Soc. Biol. 164, 246–257 (1966).

Ris, H., Kubai, D.F.: Chromosome structure. Annu. Rev. Genet. 4, 263–294 (1970).

Ritzman, S.E., Stoufflet, E.J., Houston, E.W., Levin, W.C.: Coexistent chronic myelocytic leukemia, monoclonal gammopathy and multiple chromosomal abnormalities. Am. J. Med. 41, 981–989 (1966).

Riva, G., Spengler, G.A., Siebner, H.: Se cariotipo nelle paraproteinemia. Minerva Med. 62, 2081–2084 (1971).

Robbins, J.H.: Significance of repair of human DNA: Evidence from studies of xeroderma pigmentosum. J. Natl. Cancer Inst. 61, 645–656 (1978).

Robbins, J.H., Kraemer, K.H., Lutzner, M.A., Festoff, B.W., Coon, H.G.: Xeroderma pigmentosum: An inherited disease with sun sensitivity, multiple cutaneous neoplasms, and abnormal DNA repair. Ann. Intern. Med. 80, 221–248 (1974).

Roberts, M., Greaves, M., Janossy, G., Sutherland, R., Pain, C.: Acute lymphoblastic leukaemia (ALL) associated antigen—I. Expression in different haematopoietic malignancies. Leukemia Res. 2, 105–114 (1978).

Robinson, A., Priest, R.E., Bigler, P.C.: Male pseudohermaphrodite with XY/XO mosaicism and bilateral gonadoblastomas. Lancet i, 111–112 (1964).

Robinson, J.C., Pierce, J.E., Goldstein, D.P.: Leukocyte alkaline phosphatase: electrophoretic variants associated with chronic myelogenous leukemia. Science 150, 58–60 (1965).

Robinson, K.M.: Karyotype studies on a carcinoma of the oesophagus over 80 passages in vitro and on peripheral lymphocytes of several patients. Helsinki Chromosome Conference, Aug. 29–31, p. 208 (abstract) (1977).

Robledo Aguilar, A., Gomez, F., Tojo Sierra, R., Larripa, P.: Dyskeratosis congenita Zinsser-Cole-Engmann form with abnormal karyotype. Dermatologica 148, 98–103 (1974).

Robson, M.C., Santiago, Q., Huang. T.W.: Bilateral carcinoma of the breast in a patient with Klinefelter's syndrome. J. Clin. Endocrinol. Metab. 28, 897–902 (1968).

Rochon, M., Cadotte, M., Pretty, H.M., Long, L.A.: Les anomalies chromosomiques du myélome multiple: Expérience personnelle et revue de la littérature. Union Med. Can. 100, 1750–1754 (1971).

Rodkina, R.A., Pogosyants, E.E., Chudina, A.P.: Cytogenetic investigation of cancer of the ovary. Akush. Ginekol. (Mosk.) 6, 46–53 (1972).

Rodman, T.C.: Human chromosome banding by Fuelgen stain aids in localizing classes of chromatin. Science 184, 171–173 (1974).

Rolović, Z., Ćirić, M.: Letter to Editor. Blood 44, 623–624 (1974).

Rolović, Z., Markovic, M., Jangic, M., Kalieanin, P., Boskovic, D., Rudicic, R.: Double Ph^1 chromosomes in chronic granulocytic leukemia. Lijec. Vjesn. 92, 141.5–1423 (1970).

Romano, N., Comes, R., Valentino, L.: Chromosome changes in human diploid cells infected by mycoplasmas. J. Microbiol. 18, 33–46 (1970).

Rønne, M.: Induction of uncoiled chromosomes with RNA'se. Hereditas 86, 245–250 (1977).

Rønne, M., Bøye, H. A., Sandermann, J.: A mounting medium for banded chromosomes. Hereditas 86, 155–158 (1977).

Rønne, N., Sanderman, J.: Simple methods to induce banding in human chromosomes. Hereditas 86, 151–154 (1977).

Rook, A., Hsu, L.Y., Gertner, M., Hirschhorn, K.: Identification of Y and X chromosomes in amniotic fluid cells. Nature 230, 53 (1971).

Ros, Y.: Étude de la transformation de lymphocytes *in vitro* dans le cas de deux sujets atteints d'ataxie-telangiectasie et recherche d'anomalies chromosomiques eventuelles. Humangenetik 26, 223–230 (1975).

Rosano, M., Delellis, M., Massara, B., Ditondo, U., Casini, C.: Cariotipo XYY e medulloblastoma. Acta Genet. Med. Gemellol. (Roma) 23, 259–263 (1970).

Rosen, R.B., Nishiyama, H.: Leukocyte alkaline phosphatase in chronic granulocytic leukemia of childhood. Ann. N.Y. Acad. Sci. 155, 992–1002 (1968).

Rosen, R.B., Teplitz, R.L.: Chronic granulocytic leukemia complicated by ulcerative colitis: Elevated leukocyte alkaline phosphatase and possible modifier gene deletion. Blood 26, 148–156 (1965).

Rosenfeld, C., Venuat, A.M.: Cytogenetic studies of 7 patients with chronic myeloid leukaemia before and after splenectomy. In S. Armendares, R. Lisker (eds.): *V International Congress of Human Genetics,* p. 149, Abstract 390. Held in Mexico, Oct. 10–15, (1976).

Rosenfeld, C., Goutner, A., Venuat, A.M., Choquet, C., Pico, J.L., Dore, J.F., Liabeuf, A., Durandy, A., Desgrange, C., de The, G.: An effective human leukaemic cell line: REH. Eur. J. Cancer 13, 377–379 (1977).

Rosenkranz, W., Holzer, S.: Satellite association. A possible cause of chromosome aberrations. Humangenetik 16, 147–150 (1972).

Rosenszajn, L.A., Radnay, J.: The lysosomal nature of the anomalous granules and chromosome aberrations in cultures of peripheral blood in Chediak-Higashi syndrome. Br. J. Haematol. 18, 683–689 (1970).

Rosenthal, D.S., Moloney, W.C.: Myeloid metaplasia: a study of 98 cases. Postgrad. Med. 45, 136–142 (1969).

Rosenthal, D.S., Moloney, W.C.: Occurrence of acute leukaemia in myeloproliferative disorders. Br. J. Haematol. 36, 373–382 (1977).

Rosenthal, I.R., Makowitz, A.D., Medenis, R.: Immunologic incompetence in ataxia telangiectasia. Am. J. Dis. Child. 110, 69–75 (1965).

Rosenthal, S., Canellos, G.P., Devita, V.T., Gralnick, H.R.: Characteristics of blast crisis in chronic granulocytic leukemia. Blood 49, 705–714 (1977).

Rosenthal, S., Schwartz, J.H., Canellos, G.P.: Basophilic chronic granulocytic leukaemia with hyperhistaminaemia. Br. J. Haematol. 36, 367–372 (1977).

Rosner, F., Lee, S.L.: Down's syndrome and acute leukemia: myeloblastic or lymphoblastic? Report of forty-three cases and review of the literature. Am. J. Med. 53, 203–218 (1972).

Rosner, F., Schreiber, Z.R., Parise, F.: Leukocyte alkaline phosphatase. Fluctuations with disease status in chronic granulocytic leukemia. Arch. Intern. Med. 130, 892–894 (1972).

Ross, D.D., Wiernik, P.H., Sarin, P.S., Whang-Peng, J.: Loss of terminal deoxynucleotidyltransferase (TdT) activity as a predictor of emergence of resistance to chemotherapy in a case of chronic myelogenous leukemia in blast crisis. Cancer (submitted) (1979).

Ross, J.D., Atkins, L.: Chromosomal anomaly in a mongol with leukaemia. Lancet ii, 612–613 (1962).

Ross, J.D., Rosenbaum, E.: Paroxysmal nocturnal hemoglobinuria presenting as aplastic anemia in a child. Am. J. Med. 37, 130–139 (1964).

Roth, D.G., Cimino, M.C., Variakojis, D., Golomb, H.M., Rowley, J.D.: B-cell acute lymphoblastic leukemia (ALL) with a 14q+ chromosome abnormality. Blood 53, 235–243 (1979).

Rothfels, K.H., Siminovitch, L.: An air-drying technique for flattening chromosomes in mammalian cells grown *in vitro*. Stain Technol. 33, 73–77 (1958).

Rouesse, J., Berger, R., Amiel, J.L.: Myelomonocytic leukemia patient with abnormal clone cells bearing a chromosome marker. Possible cure. Ann. Med. Interne 129, 215–218 (1978).

Rousseau, M.F.: Etude chromosomique de 20 tumeurs embryonnaires aprés culture à court terme. Biomedicine [Express] 19, 275–280 (1973).

Rousseau, M.F., Laurent, M., Nezelof, C.: Le néphroblastome: étude chromosomique. Ann. Anat. Pathol. (Paris) 15, 399–414 (1970).

Roux, C., Emerit, I., Taillemite, J.L.: Chromosomal breakage and teratogenesis. Teratology 4, 303–315 (1971).

Rovera, G., Pegoraro, L.: Clonal evolution and characteristics of the distribution of the extra chromosomes in three cases of acute transformation in chronic myeloid leukemia. Haematologica (Pavia) 53, 465–480 (1968).

Rovera, G., Pegoraro, L., Masera, P.: Development of a cellular clone with 3 Ph^1 chromosomes in the terminal stage of a case of chronic myeloid leukemia (in Italian). Boll. Soc. Ital. Biol. Sper. 41, 741–743 (1965).

Rowley, J.D.: Multiple chromosome aberrations in Down's syndrome associated with twinning and acute granulocytic leukemia. Lancet i, 664–665 (1964).

Rowley, J.D.: Cytogenetics in clinical medicine. J.A.M.A. 207, 914–919 (1969).

Rowley, J.D.: Loss of the Y chromosome in myelodysplasia: a report of three cases studied with quinacrine fluoescence. Br. J. Haematol. 21, 717–728 (1971).

Rowley, J.D.: Chromosomal patterns in myelocytic leukemia. N. Engl. J. Med. 289, 220–221 (1973a).

Rowley, J.D.: Identification of a translocation with quinacrine fluorescence in a patient with acute leukemia. Ann. Genet. (Paris) 16, 109–112 (1973b).

Rowley, J.D.: Acquired trisomy 9. Lancet ii, 390 (1973c).

Rowley, J.D.: Deletions of chromosome 7 in haematological disorders. Lancet ii, 1385–1386 (1973d).

Rowley, J.D.: A new consistent chromosomal abnormality in chronic myelogenous leukaemia identified by quinacrine fluorescence and Giemsa staining. Nature 243, 290–293 (1973e).

Rowley, J.D.: Do human tumors show a chromosome pattern specific for each etiologic agent? J. Natl. Cancer Inst. 52, 315–320 (1974a).

Rowley, J.D.: Missing sex chromosomes and translocations in acute leukaemia. Lancet ii, 835–836 (1974b).

Rowley, J.D.: Absence of the 9q+ chromosome in Ph^1 negative chronic myelogenous leukaemia. J. Med. Genet. 11, 166–170 (1974c).

Rowley, J.D.: Abnormalities of chromosome 1 in myeloproliferative disorders. Cancer 36, 1748–1757 (1975a).

Rowley, J.D.: Nonrandom chromosomal abnormalities in hematologic disorders of man. Proc. Natl. Acad. Sci. USA 72, 152–156 (1975b).

Rowley, J.D.: Chromosomes in malignancy. 5th International Congress on Human Genetics, p. 1–25. Excerpta Medica (1976a).

Rowley, J.D.: Chromosomes in human cancer. J. Reprod. Med. 17, 36–40 (1976b).

Rowley, J.D.: 5q− acute myelogenous leukemia. Reply. Blood 48, 626 (1976c).

Rowley, J.D.: The role of cytogenetics in hematology. Blood 48, 1–7 (1976d).

Rowley, J.D.: Are chromosomal changes related to etiologic agents? *Proceedings of the 11th Canadian Cancer Conference,* pp. 124–133 (1976e).

Rowley, J.D.: The relationship of chromosomal abnormalities to neoplasia. Adv. Pathobiol. 4, 67–73 (1976f).

Rowley, J.D.: Population cytogenetics of leukemia. *In* E.B. Hook, I.H. Porter (eds.): *Population Cytogenetics,* pp. 189–216. New York, San Francisco, London: Academic Press (1977a).

Rowley, J.D.: Are nonrandom karyotypic changes related to etiologic agents? *In* J.J. Mulville, R.W. Miller, J.F. Fraumeni, Jr. (eds.): *Progress in Cancer Research and Therapy,* Vol. 3, *Genetics of Human Cancer,* pp. 125–136. New York: Raven Press (1977b).

Rowley, J.D.: Nonrandom chromosomal changes in human malignant cells. *In* R.S. Sparks, D.E. Comings, C.F. Fox (eds.): *Molecular Human Cytogenetics* (ICN-UCLA Symposia on Molecular and Cellular Biology, Vol. VII). New York: Academic Press (1977c).

Rowley, J.D.: Mapping of human chromosomal regions related to neoplasia: Evidence from chromosomes 1 and 17. Proc. Natl. Acad. Sci. USA 74, 5729–5733 (1977d).

Rowley, J.D.: Abnormalities of chromosome No. 1 in haematological malignancies. Lancet i, 554–555 (1978a).

Rowley, J.D.: Chromosomes in leukemia and lymphoma. Semin. Hematol. 15, 301–319 (1978b).

Rowley, J.D.: Chromosome abnormalities in the

acute phase of CML. Virchows Arch. B Cell Pathol. 29, 57–63 (1978c).
Rowley, J.D.: Abnormalities of chromosome No. 1: Significance in malignant transformation. Virchows Arch. B Cell Pathol. 29, 139–144 (1978d).
Rowley, J.D.: The cytogenetics of acute leukemia. Clin. Haematol. 7, 385–406 (1978e).
Rowley, J.D., Blaisdell, R.K.: Karyotype of treated thrombocythaemia. Lancet ii, 104–105 (1966).
Rowley, J.D., Bodmer, W.F.: Relationship of centromeric heterochromatin to fluorescent banding patterns of metaphase chromosomes in the mouse. Nature 231, 503–506 (1971).
Rowley, J.D., Potter, D.: Chromosomal banding patterns in acute nonlymphocytic leukemia. Blood 47, 705–721 (1976).
Rowley, J.D., Blaisdell, R.K., Jacobson, L.O.: Chromosome studies in preleukemia. I. Aneuploidy of group C chromosomes in three patients. Blood 27, 782–799 (1966).
Rowley, J.D., Golomb, H.M., Dougherty, C.: 15/17 translocation, a consistent chromosomal change in acute promyelocytic leukaemia. Lancet i, 549–550 (1977a).
Rowley, J., Golomb, H., Vardiman, J.: Nonrandom chromosomal abnormalities in acute nonlymphocytic leukemia in patients treated for Hodgkin disease and non-Hodgkin lymphomas. Blood 50, 759–770 (1977b).
Rowley, J.D., Golomb, H.M., Vardiman, J.: Acute leukemia after treatment of lymphoma. N. Engl. J. Med. 297, 1013 (1977c).
Rowley, J.D., Golomb, H.M., Vardiman, J., Fukuhara, S., Dougherty, C., Potter, D.: Further evidence for a non-random chromosomal abnormality in acute promyelocytic leukemia. Int. J. Cancer 20, 869–872 (1977d).
Rowley, J.D., Potter, D., Mikita, J.: Reuse of chromosome preparations for fluorescent staining. Stain Technol. 46, 97–99 (1971).
Rowley, J.D., Wolman, S.R., Horland, A.A.: Another variant translocation in chronic myelogenous leukemia—revisited. N. Engl. J. Med. 295, 900–901 (1976).
Rozenszajn, A.L., Radnai, J., Tatarski, A., Benderlei, A.: Blood cell culture and chromosomal findings in Chediak-Higashi syndrome. Isr. J. Med. Sci. 5, 1087 (1969).
Rozman, C., Sans-Sabrafen, J., Woessner, S.: Panmielopatia de Fanconi con cariotipo normal. Med. Clin. (Barc.) 11, 325–327 (1963).
Rożynkowa, D., Marczak, T.: Chromosomes in chronic lymphatic leukemia. A survey of seven cases. Genet. Pol. 11, 415–421 (1970).
Rożynkowa, D., Stepień, J.: New achievements in cytogenetics of chronic myeloid leukemia. Pol. Arch. Med. Wewn. 54, 39–45 (1975a).
Rożynkowa, D., Stepień, J.: Culture of leukemic cells. II. New achievements in cytogenetics of chronic myeloid leukemia. Pol. Arch. Med. Wewn. 54, 39–45 (1975b).
Rożynkowa, D., Marczak, T., Rupniewska, Z.: E-1 chromosome abnormality in lymphatic leukaemia. Humangenetik 6, 300–302 (1968).
Rożynkowa, D., Stepień, J., Kowalewski, J., Nowakowski, A.: Nonrandom chromosomal rearrangements in 27 cases of human myeloid leukemia. Hum. Genet. 39, 293–301 (1977).
Rubenstein, C.T., Verma, R.S., Dosik, H.: Centromeric banding (C) of sequentially Q- and R-banded human chromosomes. Hum. Genet. 40, 279–283 (1978).
Rubin, A.D., Havemann, K., Dameshek, W.: Studies in chronic lymphocytic leukemia. Further studies of the proliferative abnormality of the blood lymphocyte. Blood 33, 313–328 (1969).
Rudders, R.A., Kilcoyne, R.F.: Myeloproliferative disorder with lytic osseous lesions and chromosomal anomalies. Am. J. Clin. Pathol. 61, 673–679 (1974).
Rudkin, G.T., Hungerford, D.A., Nowell, P.C.: DNA contents of chromosome Ph[1] and chromosome 21 in human chronic granulocytic leukemia. Science 144, 1229–1232 (1964).
Ruffié, J.: Les chromosomes des cellules du sang au cours des hemopathies. Toulouse Med. 63, 249–256 (1962).
Ruffie, J.: Modifications des chromosomes dans les cellules des leucémies aiguës. Soc. Fr. Hematol., pp. 830–844, May (1963).
Ruffié, J., Lejeune, J.: Deux cas de leucémie aiguë myéloblastique avec cellules sanguines normales et cellules haplo (21 ou 22). Rev. Fr. Etud. Clin. Biol. 7, 644–647 (1962).
Ruffié, J., Biermé, R., Ducos, J., Colombiès, P., Salles-/Mourlan, A.M., Sendrail, A.: Haplosomie 21 associée à une trisomi 6–12 chez un tuberculeux présentant une myéloblastose sanguine et médullaire. Nouv. Rev. Fr. Hematol. 4, 719–725 (1964).
Ruffié, J., Colombiés, P., Combes, P.-F., Ducos, J.: Leucémie lymphoblastique chez un porteur d'une anomalie congénitale complexe (type XXY probable). Bull. Acad. Natl. Med. (Paris) 150, 342–346 (1966).
Ruffié, J., Ducos, J., Biermé, R., Colombiès, P., Salles-Mourlan, A.M.: Multiple chromosomal abnormalities in an acute exacerbation of myeloid leukaemia. Lancet i, 609–610 (1965).
Ruffié, J., Ducos, J., Biermé, P., Colombiès, P., Salles-Mourlan, A.M.: Chromosomes in chronic lymphocytic leukaemia. Lancet ii, 227–228 (1966).
Ruffié, J., Ducos, L., Biermé, R., Salles-Mourlan, A.M., Colombiès, P., Quilici, J.C.: Chromosomal abnormalities in leukaemia. Lancet ii, 589–590 (1964).

Ruffié, J., Ducos, J., Colombiès, P., Carles-Trochain, E.: Les anomalies chromosomiques dans les leucoses aiguës. Nouv. Rev. Fr. Hematol. 10, 84–91 (1970).

Ruffié, J., Marquès, P., Mourlan, A.M.: Étude cytogénétique de deux tumeurs cancéreuses. C.R. Acad. Sci. (Paris) 258, 1935–1937 (1964).

Rugiati, S., Ragni, N., Cecco, L. de: Analisi citogenetica di un gruppo di tumori maligni dell'ovario. Quad. Clin. Ostet. Ginecol. 21, 617–627 (1966).

Rugiati, S., Ragni, S., Sbernini, R., de Cecco, L.: Quadri citogenetici in corso di terapia antiblastica. Minerva Ginecol. 19, 840 (1967).

Rundles, R.W., Moore, J.O.: Chronic lymphocytic leukemia. Cancer 42, 941–945 (1978).

Rutovitz, D.: Machines to classify chromosomes. In H.J. Evans, W.M. Court Brown, A.S. McLean (eds.) *Human Radiation Cytogenetics*. Proceedings of an International Symposium held in Edinburgh, Oct. 12–15, 1966. Amsterdam: North-Holland (1967).

Rutten, F.J., Hustinx, T.W.J., Scheres, J.M.J.C., Wagener, D.J.T.: Trisomy-9 in the bone marrow of a patient with acute myelomonoblastic leukaemia. Br. J. Haematol. 26, 391–394 (1974).

Ruutu, P., Ruutu, T., Vuopio, P., Kosunen, T.U., de la Chapelle, A.: Defective chemotaxis in monosomy-7. Nature 265, 146–147 (1977a).

Ruutu, P., Ruutu, T., Vuopio, P., Kosunen, T.U., de la Chapelle, A.: Function of neutrophils in preleukaemia. Scand. J. Haematol. 18, 317–325 (1977b).

Ruvalcaba, R.H.A., Thuline, H.C.: IgA absence associated with short arm deletion of chromosome No. 18. J. Pediatr. 74, 964–965 (1969).

Ruzicka, F.: Effect of G-banding techniques on the ultrastructure of human chromosomes. Humangenetik 22, 119–126 (1974a).

Ruzicka, F.: Organization of human mitotic chromosomes. Humangenetik 23, 1–22 (1974b).

Ruzicka, F., Pawlowsky, J., Erber, A., Nowotny, H.: Drei Fälle von Eosinophilenleukämie mit atypischer Granulation in Eosinophilen und Neutrophilen. Blut 32, 337–346 (1976).

Ryan, T.J., Boddington, M.M., Spriggs, A.I.: Chromosomal abnormalities produced by folic acid antagonists. Br. J. Dermatol. 77, 541–555 (1965).

Saarni, M.I., Linman, J.W.: Myelomonocytic leukemia: Disorderly proliferation of all marrow cells. Cancer 27, 1221–1230 (1971).

Saarni, M.I., Linman, J.W.: Preleukemia: The hematologic syndrome preceding acute leukemia. Am. J. Med. 55, 38–48 (1973).

Sachs, H., Schittko, G.: Zytofotometrisch ermittelte DNA-Muster in präkanzerösen Dysplasien der Cervix uteri. Beitrag zur morphologischen Abgrenzung und klinischen Wertigkeit dieser Plattenepithelveränderungen. Arch. Gynaekol. 218, 95–112 (1975).

Sachs, L.: Subdiploid chromosome variation in man and other mammals. Nature 172, 205–206 (1953).

Sachs, L.: The chromosome constancy of the normal mammalian uterus. Heredity 8, 117–124 (1954).

Saffhill, R., Dexter, T.M., Muldal, S., Testa, N.G., Jones, P.M., Joseph, A.: Terminal deoxynucleotidyl transferase in a case of Ph^1 positive infant chronic myelogenous leukemia. Br. J. Cancer 33, 664–667 (1976).

Saksela, E., Moorhead, P.S.: Enhancement of secondary constrictions and the heterochromatic X in human cells. Cytogenetics 1, 225–244 (1962).

Sakurai, M.: Chromosome studies in hematological disorders. III. Chromosome findings in "preleukemia" and related diseases. Acta Haematol. Jap. 33, 127–136 (1970a).

Sakurai, M.: Chromosome studies in hematological disorders. II. Chromosome findings in acute leukemia. Acta Haematol. Jap. 33, 116–126 (1970b).

Sakurai, M.: Chromosome studies in hematological disorders. I. Chromosome findings in chronic myelogenous leukemia with special reference to those after blastic transformation. Acta Haematol. Jap. 33, 103–115 (1970c).

Sakurai, M., Sandberg, A.A.: Prognosis in acute myeloblastic leukemia: chromosomal correlation. Blood 41, 93–104 (1973).

Sakurai, M., Sandberg, A.A.: Chromosomes and causation of human cancer and leukemia. IX. Prognostic and therapeutic value of chromosomal findings in acute myeloblastic leukemia. Cancer 33, 1548–1557 (1974).

Sakurai, M., Sandberg, A.A.: Chromosomes and causation of human cancer and leukemia. XI. Correlation of karyotypes with clinical features of acute myeloblastic leukemia. Cancer 37, 285–299 (1976a).

Sakurai, M., Sandberg, A.A.: Chromosomes and causation of human cancer and leukemia. XIII. An evaluation of karyotypic findings in erythroleukemia. Cancer 37, 790–804 (1976b).

Sakurai, M., Sandberg, A.A.: The chromosomes and causation of human cancer and leukemia. XVIII. The missing Y in acute myeloblastic leukemia (AML) and Ph^1-positive chronic myelocytic leukemia (CML). Cancer 38, 762–769 (1976c).

Sakurai, M., Hayata, I., Sandberg, A.A.: Chromosomes and causation of human cancer and leukemia. XV. Prognostic value of chromosomal findings in Ph^1-positive CML. Cancer Res. 36, 313–318 (1976).

Sakurai, M., Oshimura, M., Kakati, S., Sandberg,

A.A.: 8-21 translocation and missing sex chromosomes in acute leukaemia. Lancet ii, 227–228 (1974).
Sakurai, T., Kudo, H., Nagasawa, T., Yoda, Y., Kawada, K.: Correlation of marrow morphology and cytogenetics after blastic transformation in a case with chronic myelocytic leukemia (in Japanese). Jpn. J. Clin. Hematol. 16, 636–643 (1975).
Sakurai, Y., Nakao, I.: C monosomy changes in chronic myelocytic leukemia with special reference to response to therapy (in Japanese). Jpn. J. Clin. Genet. 14, 1109–1113 (1973).
Salamanca, F., Armendares, S.: C-bands in human metaphase chromosomes treated by barium hydroxide. Ann. Genet. (Paris) 17, 135–136 (1974).
Salberg, D., Kurtides, E.S., McKeever, W.P.: Monomyelocytic leukemia in an untreated case of Waldenström macroglobulinemia. Arch. Intern. Med. 137, 514–516 (1977).
Salimi, R., Jones, H.W.: Chromosomes of adenocarcinoma of the cervix uteri with a ring and a minute marker chromosome. J. Surg. Oncol. 2, 17–22 (1970).
Salkinder, M., Gear, J.H.S.: Fluorescent staining of chromosomes. Lancet i, 107 (1962).
Salmon, S.E., Fudenberg, H.H.: Abnormal nucleic acid metabolism of lymphocytes in plasma cell myeloma and macroglobulinemia. Blood 33, 300–312 (1969).
Samad, F.U., Engel, E., Hartmann, R.C.: Hypoplastic anemia, Friedreich's ataxia and chromosomal breakage: Case report and review of similar disorders. South. Med. J. (Nashville) 66, 135–140 (1973).
Sanchez, O., Yunis, J.J.: The relationship between repetitive DNA and chromosomal bands in man. Chromosoma 48, 191–202 (1974).
Sandberg, A.A.: Chromosomes and leukemia. CA 15, 2–13; 42–44 (1965).
Sandberg, A.A.: Chromosomes and leukemia. In: Atti Convegni Farmitalia, pp. 146–162. Torino, Italy: Edizioni Minerva Medica, (1966a).
Sandberg, A.A.: The chromosomes and causation of human cancer and leukemia. Cancer Res. 26, 2064–2081 (1966b).
Sandberg, A.A.: Chromosome abnormalities in human leukemia. Geriatrics 23, 124–136 (1968a).
Sandberg, A.A.: Chromosomes in human cancer and leukemia. In: Cancer Management. A special graduate course on cancer, sponsored by American Cancer Society, pp. 60–65. Philadelphia: Lippincott (1968b).
Sandberg, A.A.: Chromosomes in clinical oncology. In G.P. Murphy (ed.): Perspectives in Cancer Research and Treatment, pp. 223–235. New York: Alan R. Liss, Inc. (1973).
Sandberg, A.A.: Chromosome changes in human malignant tumors: an evaluation. In E. Grundmann (ed.): Recent Results in Cancer Research, Vol. 44, pp. 75–85. Berlin, Heidelberg, New York: Springer-Verlag (1974).
Sandberg, A.A.: Chromosome markers and progression in bladder cancer. Cancer Res. 37, 222–229 (1977).
Sandberg, A.A.: Cytogenetic "staging" of chronic myelocytic leukemia (CML). Boll. 1st. Sieroter. Milan 57, 247–256 (1978a).
Sandberg, A.A.: Cytogenetic data and prognosis in acute leukemia. In F. Mandelli (ed.): Therapy of Acute Leukemias. 2nd International Symposium, Rome, Italy, Dec. 7–10, 1977, pp. 186–192. (1978b).
Sandberg, A.A.: Chromosomes in the chronic phase of CML. Virchows Arch. B Cell Path. 29, 51–55 (1978c).
Sandberg, A.A.: Some comments regarding chromosome pulverization (premature chromosome consensation or PCC, prophasing). Virchows Arch. B Cell Path. 29, 15–18 (1978d).
Sandberg, A.A., Hossfeld, D.K.: Chromosomal abnormalities in human neoplasia. Annu. Rev. Med. 21, 379–408 (1970).
Sandberg, A.A., Hossfeld, D.K.: Chromosomes in the pathogenesis of human cancer and leukemia. In J.F. Holland, E. Frei, III (eds.): Cancer Medicine, pp. 151–177. Philadelphia: Lea and Febiger (1973).
Sandberg, A.A., Hossfeld, D.K.: Chromosomal changes in human tumors and leukemias. In H.W. Altmann, F. Büchner, H. Cottier, E. Grundmann, G. Holle, E. Letterer, W. Masshoff, H. Meessen, F. Roulet, G. Seifert, G. Siebert (eds.): Handbuch der allgemeinen Pathologie, Vol. VI, pp. 141–287. Berlin, Heidelberg, New York: Springer-Verlag (1974).
Sandberg, A.A., Hossfeld, D.K.: Chromosomes in the pathogenesis of human cancer and leukemia. In: J.F. Holland, E. Frei (eds.): Cancer Medicine, Philadelphia: Lea & Febiger. (in press) (1979).
Sandberg, A.A., Sakurai, M.: The role of chromosomal studies in cancer epidemiology. In W. Nakahara, T. Kirayama, K. Nishioka, H. Sugano (eds.): Analytical and Experimental Epidemiology of Cancer, pp. 297–328 (paper presented at the 4th International Symposium of the Princess Takamatsu Cancer Research Fund). Tokyo, Japan: University of Tokyo Press (1973a).
Sandberg, A.A., Sakurai, M.: The missing Y chromosome and human leukaemia. Lancet i, 375 (1973b).
Sandberg, A.A., Sakurai, M.: Chromosomes in the causation and progression of cancer and leukemia. In H. Busch (ed.): The Molecular Biology of Cancer, pp. 81–106. New York: Academic Press (1974).
Sandberg, A.A., Sakurai, M.: The role of chromosomal studies in cancer epidemiology. In H.T. Lynch (ed.): Cancer Genetics, pp. 122–

145. Springfield, Ill.: C.C. Thomas (1976).
Sandberg, A.A., Sonta, S.-I.: Chromosomal findings in relation to the blastic phase of chronic myelocytic leukemia (CML). Boll. Ist. Sieroter. Milan 57, 334–343 (1978).
Sandberg, A.A., Yamada, K.: Chromosomes in human tumors. CA 15, 58–74 (1965).
Sandberg, A.A., Yamada, K.: Chromosomes and causation of human cancer and leukemia. I. Karyotypic diversity in a single cancer. Cancer 19, 1869–1878 (1966).
Sandberg, A.A., Aya, T., Ikeuchi, T., Weinfeld, H.: Definition and morphologic features of chromosome pulverization: a hypothesis to explain the phenomenon. J. Natl. Cancer Inst. 45, 615–621 (1970).
Sandberg, A.A., Bross, I.D.J., Takagi, N., Schmidt, M.L.: Chromosomes and causation of human cancer and leukemia. IV. Vectorial analysis. Cancer 21, 77–82 (1968).
Sandberg, A.A., Cohen, M.M., Rimm, A.A., Levin, M.L.: Aneuploidy and age in a population survey. Am. J. Hum. Genet. 19, 633–643 (1967).
Sandberg, A.A., Cortner, J., Takagi, M., Moghadam, M.A., Crosswhite, L.H.: Differences in chromosome constitution of twins with acute leukemia. N. Engl. J. Med. 275, 809–812 (1966).
Sandberg, A.A., Hossfeld, D.K., Ezdinli, E.Z., Crosswhite, L.H.: Chromosomes and causation of human cancer and leukemia. VI. Blastic phase, cellular origin, and the Ph¹ in CML. Cancer 27, 176–185 (1971).
Sandberg, A.A., Ishihara, T., Crosswhite, L.H.: Group-C trisomy in myeloid metaplasia with possible leukemia. Blood 24, 716–725 (1964a).
Sandberg, A.A., Ishihara, T., Crosswhite, L.H., Hauschka, T.S.: Chromosomal dichotomy in blood and marrow of acute leukemia. Cancer Res. 22, 748–756 (1962a).
Sandberg, A.A., Ishihara, T., Crosswhite, L.H., Hauschka, T.S.: Comparison of chromosome constitution in chronic myelocytic leukemia and other myeloproliferative disorders. Blood 20, 393–423 (1962b).
Sandberg, A.A., Ishihara, T., Kikuchi, Y., Crosswhite, L.H.: Chromosomal differences among the acute leukemias. Ann. N.Y. Acad. Sci. 113, 663–716 (1964b).
Sandberg, A.A., Ishihara, T., Kikuchi, Y., Crosswhite, L.H.: Chromosomes of lymphosarcoma and cancer cells in bone marrow. Cancer 17, 738–746 (1964c).
Sandberg, A.A., Ishihara, T., Crosswhite, L.H., Koepf, G.F.: XYY genotype. N. Engl. J. Med. 268, 585 (1963a).
Sandberg, A.A., Ishihara, T., Miwa, T., Hauschka, T.S.: The in vivo chromosome constitution of marrow from 34 human leukemias and 60 nonleukemic controls. Cancer Res. 21, 678–689 (1961).
Sandberg, A.A., Ishihara, T., Moore, G.E., Pickren, J.W.: Unusually high polyploidy in a human cancer. Cancer 16, 1246–1254 (1963b).
Sandberg, A.A., Kikuchi, Y., Crosswhite, L.H.: Mitotic ability of leukemic leukocytes in chronic myelocytic leukemia. Cancer Res. 24, 1468–1473 (1964).
Sandberg, A.A., Koepf, G.F., Crosswhite, L.H., Hauschka, T.S.: The chromosome constitution of human marrow in various developmental and blood disorders. Am. J. Hum. Genet. 12, 231–249 (1960).
Sandberg, A.A., Koepf, G.F., Ishihara, T., Hauschka, T.S.: An XYY human male. Lancet ii, 488–489 (1961).
Sandberg, A.A., Sakurai, M., Holdsworth, R.N.: Chromosomes and causation of human cancer and leukemia. VIII. DMS chromosomes in a neuroblastoma. Cancer 29, 1671–1679 (1972).
Sandberg, A.A., Sakurai, M., Hossfeld, D.K.: An appraisal of cytogenetics in leukemia including a correlation with the evolution and prognosis of acute leukemia. In: Proceedings of the Seventh National Cancer Conference, pp. 333–343. Philadelphia, Toronto: Lippincott (1973).
Sandberg, A.A., Sofuni, T., Takagi, N., Moore, G.E.: Chronology and pattern of human chromosome replication. IV. Autoradiographic studies of binucleate cells. Proc. Natl. Acad. Sci. USA 56, 105 (1966).
Sandberg, A.A., Takagi, N., Kato, H.: Cytogenetic studies of normal and neoplastic cells in vitro. In: The Proliferation and Spread of Neoplastic Cells (21st Annual Symposium on Fundamental Research, University of Texas M.D. Anderson Hospital and Tumor Institute, Houston, Tex. 1967), Ed. by staff of M.D. Anderson Hospital and Tumor Institute, pp. 99–136. Baltimore: Williams and Wilkins (1968a).
Sandberg, A.A., Takagi, N., Schmidt, M.L., Bross, I.D.J.: Chronology and pattern of human chromosome replication. IX. Metasynchronous DNA replication in homologs. Cytogenetics 7, 298–332 (1968b).
Sandberg, A.A., Takagi, N., Sofuni, T., Crosswhite, L.H.: Chromosomes and causation of human cancer and leukemia. V. Karyotypic aspects of acute leukemia. Cancer 22, 1268–1282 (1968c).
Sandberg, A.A., Yamada, K., Kikuchi, Y., Takagi, M.: Chromosomes and causation of human cancer and leukemia. III. Karyotypes of cancerous effusions. Cancer 20, 1099–1116 (1967).
Sande J.H. van de, Lin, C.C., Jorgenson, K.F.: Reverse banding on chromosomes produced by a guanosine-cytosine specific DNA binding antibiotic: Olivomycin. Science 195, 400–402 (1976).
Sandhofer, M., Tuschl, H., Kovac, R., Altmann, H.: Xeroderma pigmentosum mit intaktem Exzisionsrepair-Mechanismus und herabgesetzter

UV-Toleranz. Wien. Klin. Wochenschr. 88, 296–299 (1976).
Sarin, P.S., Anderson, P.N., Gallo, R.C.: Terminal deoxynucleotidyl transferase activities in human blood leukocytes and lymphoblast cell lines: High levels in lymphoblast cell lines and in blast cells of some patients with chronic myelogenous leukemia in acute phase. Blood 47, 11–20 (1976).
Sarna, G., Tomasulo, P., Lotz, M.J., Bubinak, J.F., Shulman, N.R.: Multiple neoplasms in two siblings with a variant form of Fanconi's anemia. Cancer 36, 1029–1033 (1975).
Sasaki, M.S.: Cytological effect of chemicals on tumors. XII. A chromosome study in a human gastric tumor following radioactive colloid gold (Au^{198}) treatment. J. Fac. Sci. Hokkaido Univ., Ser. VI, Zool. 14, 566–575 (1961).
Sasaki, M.S.: DNA repair capacity and susceptibility to chromosome breakage in xeroderma pigmentosum cells. Mutat. Res. 20, 291–293 (1973).
Sasaki, M.S.: Is Fanconi's anaemia defective in a process essential to the repair of DNA cross links? Nature 257, 501–503 (1975).
Sasaki, M.S., Makino, S.: The demonstration of secondary constrictions in human chromosomes by means of a new technique. Am. J. Hum. Genet. 15, 24–33 (1963).
Sasaki, M.S., Tonomura, A.: Chromosomal radiosensitivity in Down's syndrome. Jap. J. Hum. Genet. 14, 81–92 (1969).
Sasaki, M.S., Tonomura, A.: A high susceptibility of Fanconi's anemia to chromosome breakage by DNA cross-linking agents. Cancer Res. 33, 1829–1836 (1973).
Sasaki, M., Fukuschima, T., Makino, S.: Some aspects of the chromosome constitution of hydatidiform moles and normal chorionic villi. Gann 53, 101–106 (1962).
Sasaki, M., Mori, M., Kobayashi, H., Takeichi, N.: Cytologic studies of tumors. LII. Persistence of diploid stemlines in Friend virus-induced ascites tumors of rats. Proc. Jap. Acad. 46, 792–797 (1970).
Sasaki, M., Muramoto, J., Makino, S., Hara, Y., Okada, M., Tanaka, E.: Two cases of acute myeloblastic leukemia associated with a 9/22 translocation. Proc. Jap. Acad. 51, 193–197 (1975).
Sasaki, M., Okada, M., Kondo, I., Muramoto, J.-I.: Chromosome banding patterns in 27 cases of acute myeloblastic leukemia. Proc. Jap. Acad. 52, 505–508 (1976).
Sasaki, M., Oshimura, M., Makino, S., Toshio Koike, M.J.A., Itoh, M., Watanabe, F., Tanaka, N.: Further karyological evidence for contagiousness and common origin of canine venereal tumors. Proc. Jap. Acad. 50, 636–640 (1974).
Sasaki, M.S., Sofuni, T., Makino, S.: Cytological studies of tumors. XLII. Chromosome abnormalities in malignant lymphomas of man. Cancer 18, 1007–1013 (1965).
Sasaki, M.S., Tonomura, A., Matsubara, S.: Chromosome constitution and its bearing on the chromosomal radiosensitivity in man. Mutat. Res. 10, 617–633 (1970).
Saunders, G.F., Hsu, T.C., Getz, M.J., Simes, E.L., Arrighi, F.E.: Locations of human satellite DNA in human chromosomes. Nature [New Biol.] 236, 244–246 (1972).
Saunders, G.F., Shirakawa, S., Saunders, P.P., Arrighi, F.E., Hsu, T.C.: Populations of repeated DNA sequences in the human genome. J. Mol. Biol. 63, 323–334 (1972).
Savage, J.R.K.: Chromosomal aberrations as tests for mutagenicity. Nature 258, 103–104 (1975).
Savage, J.R.K., Bigger, T.R.L., Watson, G.E.: Location of quinacrine mustard-induced chromatid exchange points in relation to ASG bands in human chromosomes. In P.L. Pearson, K.R. Lewis (eds.): *Chromosomes Today,* Vol. 5, pp. 281–291. New York: Wiley (1976).
Sawitsky, A., Bloom, D., German, J.: Chromosomal breakage and acute leukemia in congenital telangiectatic erythema and stunted growth. Ann. Intern. Med. 65, 487–495 (1966).
Say, B., Balci, S., Tuncbilek, E.: 45, XO Turner's syndrome, Wilms' tumor and imperforate anus. Humangenetik 12, 348–350 (1971).
Say, B., Tuncbilek, E., Yamak, B., Balci, S.: An unusual chromosomal aberration in a case of Chediak-Higashi syndrome. J. Med. Genet. 7, 417–421 (1970).
Scappaticci, S., LoCurto, F., Mira, E.: Karyotypic variation in benign pleomorphic adenoma of the parotid and in normal salivary glands. Acta Otolaryngol. 76, 221–228 (1973).
Schade, H., Schoeller, L., Schultze, K.W.: D-Trisomie (Patau-Syndrom) mit kongenitaler myeloischer Leukamie, Med. Welt 2, 2690–2692 (1962).
Schappert-Kimmijser, J., Hammes, G.D., Nijland, R.: The heredity of retinoblastoma. Ophthalmologica 151, 197–213 (1966).
Scharfman, W.B., Amarose, A.P., Propp, P.: Primary erythrocytosis of childhood. J.A.M.A. 210, 2274–2276 (1969).
Scheike, O., Visfeldt, J., Petersen, B.: Male breast cancer. 3. Breast carcinoma in association with the Klinefelter syndrome. Acta Pathol. Microbiol. Scand. [A] 81, 352–358 (1973).
Schellhas, H.F., Trujillo, J.M., Rutledge, F.N., Cork, A.: Germ cell tumors associated with XY gonadal dysgenesis. Am. J. Obstet. Gynecol. 109, 1197–1204 (1971).
Scheres, J.M.J.C.: CT banding of human chromosomes. The role of cations in the alkaline pre-

treatment. Hum. Genet. 33, 167–174 (1976a).
Scheres, J.M.J.C.: CT banding of human chromosomes. Description of the banding technique and some of its modifications. Hum. Genet. 31, 293–307 (1976b).
Scheres, J.M.J.C.: The effect of cations on C-band formation in human chromosomes. Cytogenet. Cell Genet. 18, 2–12 (1977).
Scheres, J.M.J.C., Hustinx, T.W.J., de Vaan, G.A.M., Rutten, F.J.: 15/17 translocation in acute promyelocytic leukaemia. Hum. Genet. 43, 115–117 (1978).
Scherz, R.G., Louro, J.M.: A simple method for making chromosome slides. Preparation from short-term cultures of peripheral blood. J. Clin. Pathol. 40, 222–225 (1963).
Schettini, F., Guanti, G., Violante, N., Santoro, A.: Rilievi citogenetici nella leucemia mieloide cronica nella infanzia. Minerva Pediatr. 21, 1–11 (1969).
Schiffer, L.M.: Kinetics of chronic lymphocytic leukemia. Ser. Haematol. 1, 3–23 (1968).
Schiffer, L.M., Vaharu, T., Gardner, L.I.: Acridine orange as a chromosome stain. Lancet ii, 1362–1363 (1961).
Schinzel, A., Schmid, W.: Lymphocyte chromosome studies in humans exposed to chemical mutagens. The validity of the method in 67 patients under cytostatic therapy. Mutat. Res. 40, 139–166 (1976).
Schleicher, W.: Die Knorpelzellteilung. Ein Beitrag zur Lehre der Teilung von Gewebezellen. Arch. Mikr. Anat. 16, 248–300 (1879).
Schleiermacher, E., Kroll, W.: A constant chromosome aberration in two children with acute myeloid leukaemia. Humangenetik 5, 80–82 (1967).
Schmiady, H., Sperling, K.: Length of human C-bands in relation to the degree of chromosome condensation. Hum. Genet. 35, 107–111 (1976).
Schmiady, H., Wegner, R.D., Sperling, K.: Relative DNA content of human euchromatin and heterochromatin after G, C and Giemsa 11 banding. Humangenetik 29, 85–89 (1975).
Schmid, M., Krone, W., Vogel, W.: On the relationship between the frequency of association and the nucleolar constriction of individual acrocentric chromosomes. Humangenetik 23, 267–277 (1974).
Schmid, W.: DNA replication patterns of human chromosomes. Cytogenetics 2, 175–193 (1963).
Schmid, W.: Autoradiography of human chromosomes. In J.J. Yunis (ed.): *Human Chromosome Methodology*. pp. 91–110. New York: Academic Press (1965).
Schmid, W.: Familial constitutional panmyelocytopathy, Fanconi's anemia. II. A discussion of the cytogenetic findings in Fanconi's anemia. Semin. Hematol. 4, 241–249 (1967).
Schmid, W., Fanconi, G.: Fragility and spiralization anomalies of the chromosomes in three cases, including fraternal twins, with Fanconi's anemia, type Estren-Dameshek. Cytogenet. Cell Genet. 20, 141–149 (1978).
Schmid, W., Jerusalem, F.: Cytogenetic findings in two brothers with ataxia telangiectasia (Louis-Barr syndrome). Arch. Genet. (Zur.) 45, 49–52 (1972).
Schmid, W., Schärer, K., Baumann, T., Fanconi, G.: Chromosomenbrüchigkeit bei der familiären Panmyelopathie (Typus Fanconi). Schweiz. Med. Wochenschr. 95, 1461–1464 (1965).
Schmidt, R., Dar, H., Santorineou, M., Sekine, I.: Ph[1] chromosome and loss and reappearance of the Y chromosome in acute lymphocytic leukaemia. Lancet i, 1145 (1975).
Schnedl, W.: Banding pattern of human chromosomes. Nature [New Biol.] 233, 93–94 (1971a).
Schnedl, W.: Analysis of the human karyotype using a reassociation technique. Chromosoma 34, 448–454 (1971b).
Schnedl, W.: Observations on the mechanisms of Giemsa staining methods. Nobel Symp., Chromosome Identification, 23, 342–345 (1973).
Schnedl, W.: Banding patterns in chromosomes. Int. Rev. Cytol. (Suppl.) 4, 237–272 (1974a).
Schnedl, W.: Der Polymorphismus des menschlichen Chromosomensatzeseine Möglichkeit für den Vaterschaftsnachweis. Z. Rechtsmedizin 74, 17–23 (1974b).
Schnedl, W.: Structure and variability of human chromosomes analyzed by recent techniques. Hum. Genet. 41, 1–9 (1978).
Schnedl, W., Mikelsaar, A.-V., Breitenbach, M., Dann, O.: DIPI and DAPI: Fluorescence banding with only negligible fading. Hum. Genet. 36, 167–172 (1977).
Schnedl, W., Roscher, U., Czaker, R.: A photometric method for quantifying the polymorphisms in human acrocentric chromosomes. Hum. Genet. 35, 185–191 (1977).
Schneider, G., Stecher, G., Obrecht, P., Merker, H.: Atypical Ph[1] chromosome by pericentric inversion. Lancet ii, 1367–1368 (1967).
Schoen, E.J., Shearn, M.A.: Immunoglobin deficiency in Bloom's syndrome. Am. J. Dis. Child. 113, 594–596 (1967).
Schoyer, N.H.: The Philadelphia chromosome. Lancet i, 1045 (1964).
Schreck, R.R., Erlanger, B.F., Miller, O.J.: The use of antinucleoside antibodies to probe the organization of chromosomes denatured by ultraviolet irradiation. Exp. Cell Res. 88, 31–39 (1974).

Schreck, R.R., Warburton, D., Miller, O.J., Beiser, S.M., Erlanger, B.F.: Chromosome structure as revealed by a combined chemical and immunochemical procedure. Proc. Natl. Acad. Sci. 70, 804–807 (1973).

Schrek, R.: Effect of phytohaemagglutinin on lymphocytes from patients with chronic lymphocytic leukaemia. Arch. Pathol. 83, 58–63 (1967).

Schrek, R., Rabinowitz, Y.: Effect of phytohaemagglutinin on rat and normal and leukemic human blood cells. Proc. Soc. Exp. Biol. Med. 113, 191–194 (1963).

Schroeder, T.M.: Cytogenetischer Befund und Ätiologie bei Fanconi-Anämie. Ein Fall von Fanconi-Anämie ohne Hexokinasedefekt. Humangenetik 3, 76–81 (1966a).

Schroeder, T.M.: Cytogenetische und cytologische Befunde bei enzymopenischen Panmyelopathien. Familiäre Panmyelopathie Typ Fanconi, Glutathionreduktasemangel-Anämie und megaloblastäre Vitamin B_{12}-Mangel-Anämie. Humangenetik 2, 287–316 (1966b).

Schroeder, T.M.: Genetische Faktoren der Krebsentstehung. Fortschr. Med. 90, 603–608 (1972).

Schroeder, T.M.: Chromosomal instability and malignancy. In P. Bucalossi, U. Veronesi, N. Cascinelli (eds.): Proceedings of the International Cancer Congress, 11th, Vol. 2, Chemical and Viral Oncogenesis, pp. 62–65. Held in Florence, Italy, Oct. 20–26, 1974. New York: American Elsevier (1975a).

Schroeder, T.M.: Sister chromatid exchanges and chromatid interchanges in Bloom's syndrome. Humangenetik 30, 317–323 (1975b).

Schroeder, T.M., Bock, H.E.: Trisomie des Ph^1-Chromosoms in Myeloblasten während der terminalen Phase einer chronisch myeloischen Leukämie. Humangenetik 1, 681–685 (1965).

Schroeder, T.M., Drings, P.: Verlaufsbeobachtung einer Fanconi-Anämie bei einen Erwachsenen. Verh. Dtsch. Ges. Inn. Med. 79, 477 (1973).

Schroeder, T.M., German, J.: A comparative study of the patterns of chromosomal instability in Fanconi's anemia and Bloom's syndrome. Excerpta Medica International Congress Ser. No. 233, 4th Int. Congr. Hum. Genet. Vol. 231, p. 161 (1971).

Schroeder, T.M., German, J.: Bloom's syndrome and Fanconi's anemia: Demonstration of two distinctive patterns of chromosome disruption and rearrangement. Humangenetik 25, 299–306 (1974).

Schroeder, T.M., Kurth, R.: Spontaneous chromosomal breakage and high incidence of leukemia in inherited disease. Blood 37, 96–112 (1971).

Schroeder, T.M., Passarge, E.: Spontaneous chromosomal instability. Humangenetik 17, 276 (1973).

Schroeder, T.M., Stahl Mauge, C.: Spontaneous chromosome instability, chromosome reparation and recombination in Fanconi's anemia and Bloom's syndrome. In H. Altman (ed.): DNA-Repair and Late Effects, pp. 35–50. International Symposium of the "IGEM," held in Vienna, Dec. 1–2, 1975. Institut für Biologie am Forschungszentrum, Seibersdor/Wien, Switzerland. Edition Roetzer, Einstadt (1976).

Schroeder, T.M., Anschütz, F., Knopp, A.: Spontane Chromosomenaberrationen bei familiärer Panmyelopathie. Humangenetik 1, 194–196 (1964).

Schroeder, T.M., Drings, P., Beilner, P., Buchinger, G.: Clinical and cytogenetic observations during a six-year period in an adult with Fanconi's anaemia. Blut 34, 119–132 (1976).

Schroeder, T.M., Tilgen, D., Krüger, J., Vogel, F.: Formal genetics of Fanconi's anemia. Hum. Genet. 32, 257–288 (1976).

Schuh, B.E., Korf, B.R., Salwen, M.J.: Dynamic aspects of trypsin-Giemsa banding. Humangenetik 28, 233–237 (1975).

Schuler, D., Dobos, M.: Malignant tumors and chromosomal aberrations. Acta Paediatr. Acad. Sci. Hung. 11, 3–10 (1970).

Schuler, D., Kiss, S.: Investigations on chromosomes in leukemic children. Folia Haematol. (Leipz.) 80, 419–424 (1963).

Schuler, D., Dobos, M., Fekete, G.: Investigation of chromosomal mutability in patients with high risk of malignancy. Rev. Cubano Pediatr. 47, 343–351 (1975).

Schuler, D., Dobos, M., Fekete, G., Machay, T., Nemeskéri, A.: Down's syndrome and malignancy. Acta Paediatr. Acad. Sci. Hung. 13, 245–252 (1972).

Schuler, D., Kiss, A., Fábián, F.: Chromosome studies in Fanconi's anemia. Orv. Hetil. 110, 713–720 (1969a).

Schuler, D., Kiss, A., Fábián, F.: Chromosomal peculiarities and in vitro examination in Fanconi's anaemia. Humangenetik 7, 314–322 (1969b).

Schuler, D., Schöngut, L., Cserháti, E., Siegler, J., Gács, G.: Lymphoblastic transformation, chromosome pattern and delayed-type skin reaction in ataxia telangiectasia. Acta Paediatr. Scand. 60, 66–72 (1971).

Schuller, J.L., van de Merwe, J.P.: Sezary's syndrome. Ned. Tijdschr. Geneeskd. 120, 639 (1976).

Schüssler, J., Dehnhard, F., Breinl, H.: Zur Frage der Ploidiereduktion bei der Carcinogenese an der Cervix uteri: Numerische Chromosomenanomalien beim Carcinoma in situ. Arch. Gynaekol. 222, 285–293 (1977).

Schuster, J., Hart, Z., Stimson, C.W., Brought, A.J., Poulik, M.D.: Ataxia telangiectasia with cerebella tumour. Pediatrics 37, 776–786 (1966).

Schwanitz, G., Gebhart, E., Rott, H.-D., Schaller,

K.-H., Essing, H.-G., Lauer, O., Prestele, H.: Chromosomenuntersuchungen bei Personen mit beruflicher Bleiexposition. Dtsch. Med. Wochenschr. 100, 1007 – 1011 (1975).

Schwanitz, G., Lehnert, G., Gebhart, E.: Chromosomenschäden bei beruflicher Bleibelastung. Dtsch. Med. Wochenschr. 95, 1636 – 1641 (1970).

Schwartz, D.: Studies on crossing over in maize and drosophilia. J. Cell Comp. Physiol. (Suppl. 2) 5, 171 – 188 (1955).

Schwarzacher, H.G.: Modern ideas on chromosome structure. Pathol. Eur. 11, 5 – 13 (1976a).

Schwarzacher, H.G.: *Chromosomes in Mitosis and Interphase.* Berlin, Heidelberg, New York: Springer-Verlag (1976b).

Schwarzacher, H.G., Wolf, U. (eds.): *Methods in Human Cytogenetics* (trans. from German by E. Passarge). Berlin: Springer-Verlag (1974).

Schwarzacher, H.G., Bielek, E., Ruzicka, F.: Neue befunde zur Struktur der Chromosomen. Hum. Genet. 35, 125 – 135 (1977).

Schwarzacher, H.G., Mikelsaar, A.-V., Schnedl, W.: The nature of the Ag-staining of nucleolus organizer regions. Electron- and light-microscopic studies on human cells in interphase, mitosis, and meiosis. Cytogenet. Cell Genet. 20, 24 – 39 (1978).

Schwarzacher, H.G., Ruzicka, F., Sperling, K.: Electron microscopy of human banded and prematurely condensed chromosomes. *In* P.L. Pearson, K.R. Lewis (eds.): *Chromosomes Today,* Vol. 5, pp. 227 – 234. New York: Wiley (1976).

Schwarze, E.W., Schwalbe, P., Klein, U.E.: Pathoanatomical features of so-called Ph¹-chronic myeloid leukemia. Virchows Arch. [Pathol. Anat.] 367, 137 – 148 (1975).

Schwinger, E.: Non-fluorescent Y chromosome. Lancet i, 437 (1973).

Schwinger, E., Sperling, H., Schwieder, R.: Strongly fluorescent chromosome bands in metaphase and interphase. Humangenetik 22, 127 – 132 (1974).

Scott, T.: Turner's syndrome and vermiform phlebectasia of the bowel. Trans. Am. Clin. Climatol. Assoc. 79, 45 – 50 (1968).

Scully, R.E., Galdabini, J.J., McNeely, B.U.: Case records of the Massachusetts General Hospital: Case 14-1976. N. Engl. J. Med. 294, 772 – 777 (1976).

Seabright, M.: A rapid banding technique for human chromosomes. Lancet ii, 971 – 972 (1971).

Seabright, M.: The use of proteolytic enzymes for the mapping of structural rearrangements in the chromosomes of man. Chromosoma 36, 204 – 210 (1972).

Seabright, M., Cooke, P., Wheeler, M.: Variation of trypsin banding at different stages of contraction in human chromosomes and the definition, by measurement, of the "average" karyotype. Hum. Genet. 29, 35 – 40 (1975).

Sebahoun, G., Gratecos, N., Foa, J., Drony, S., Carcassonne, Y.: Successive acute transformations of chronic myeloid leukemia. "Lymphoblastic" and granulous transformations. Nouv. Presse Med. 5, 2616 – 2618 (1976).

Secker Walker, L.M.: The chromosomes of bonemarrow cells of haematologically normal men and women Br. J. Haematol. 21, 455 – 461 (1971).

Secker Walker, L.M., Hardy, J.D.: Philadelphia chromosome in acute leukemia: Case report. Cancer 38, 1619 – 1624 (1978).

Secker Walker, L.M., Sandler, R.M.: Acute myeloid leukaemia with monosomy-7 follows acute lymphoblastic leukaemia. Br. J. Haematol. 38, 359 – 366 (1978).

Secker Walker, L.M., Lawler, S.D., Hardisty, R.M.: Prognostic implications of chromosomal findings in acute lymphoblastic leukemia at diagnosis. Br. Med. J. 2, 1529 – 1530 (1978).

Secker Walker, L.M., Summersgill, B.M., Swansbury, G.J., Lawler, S.D., Chessells, J.M., Hardisty, R.M.: Philadelphia-positive blast crisis masquerading as acute lymphoblastic leukaemia in children. Lancet ii, 1405 (1976).

Séé, G., Dayras, J.-C., Naudet, M.-C., Orlowski, J.-C.: Trisomie 21 et leucoblastose néonatale. Un cas de rémission spontanée. Ann. Pediatr. 21, 203 – 209 (1974).

Seeger, R.C., Rayner, S.A., Banerjee, A., Chung, H., Laug, W.E., Neustein, H.B., Benedict, W.F.: Morphology, growth, chromosomal pattern, and fibrinolytic activity of two new human neuroblastoma cell lines. Cancer Res. 37, 1364 – 1371 (1977).

Segall, M., Shapiro, L.R., Freedman, W., Boone, J.A.: XO/XY gonadal dysgenesis and gonadoblastoma in childhood. Obstet. Gynecol. 41, 536 – 541 (1973).

Sehested, J.: A simple method for R banding of human chromosomes, showing a pH-dependent connection between R and G bands. Humangenetik 21, 55 – 58 (1974).

Sei, M., Hanzawa, M., Kimoto, M., Ueda, T., Mikami, M.: A case of chronic myelocytic leukemia observed in an early stage. Some consideration of low LAP and Philadelphia chromosome in CML (in Japanese). Jpn. J. Clin. Hematol. 13, 976 – 980 (1972).

Seidenfeld, A.M., Smythe, H.A., Ogryzlo, M.A., Urowitz, M.B., Dotten, D.A.: Acute leukemia in rheumatoid arthritis treated with cytotoxic agents. J. Rheumatol. 3, 295 – 304 (1976).

Seif, G.S.F., Spriggs, A.I.: Chromosome changes in Hodgkin's disease. J. Natl. Cancer Inst. 39, 557 – 570 (1967).

Sekine, S.: Cytogenetic observations in tumours of

the urinary tract and male genitals. Jap. J. Urol. 67, 452–464 (1976).

Sekine, I., Alva, J.D.: Philadelphia chromosome positive (Ph¹+) chronic myelocytic leukemia (CML) in infancy—a case report. American Society of Hematology, Nineteenth Annual Meeting. Held in Boston, Mass., Dec. 4–7. Abstract #256 (1976).

Selander, R.: Interaction of quinacrine mustard with mononucleosides and polynucleosides. Biochem. J. 131, 749–755 (1973).

Selander, R., Chapelle, A. de la: The fluorescence of quinacrine mustard with nucleic acids. Nature [New Biol.] 245, 240–244 (1973).

Sele, B., Jalbert, P., van Custem, B., Lucas, M., Mouriquand, C., Bouchez, R.: Distribution of human chromosomes on the metaphase plate using banding techniques. Hum. Genet. 39, 39–61 (1977).

Selles, W.D., Marimuthu, K.M., Neurath, P.W.: Variations in normal human chromosomes. Humangenetik 22, 1–15 (1974).

Sellyei, M., Kelemen, E.: Chromosome study in a case of granulocytic leukaemia with "Pelgerisation" 7 years after benzene pancytopenia. Eur. J. Cancer 7, 83–85 (1971).

Sellyei, M., Vass, L.: Absence of fluorescent Y bodies in some malignant epithelial tumours in human males: preliminary results. Eur. J. Cancer 8, 557–559 (1972).

Sellyei, M., Vass, L.: Sex-chromosome loss in human tumours. Lancet i, 1041 (1975).

Sellyei, M., Tury, E., Fellner, F.: Neue Angaben zur chromosomalen Struktur des Sticker-Sarkoms. Z. Krebsforsch. 74, 7–14 (1970).

Sellyei, M., Vass, L., Krausz, T.: Non-random appearance of Y-chromatin-like fluorescence in the nuclei of thyroid and brain and its chromosomal background. Humangenetik 27, 339–342 (1975).

Seman, G., Hunter, S.J., Miller, R.C., Dmochowski, L.: Characterization of an established cell line (SH-3) derived from pleural effusion of patient with breast cancer. Cancer 37, 1814–1824 (1976).

Senn, H.J., Rhomberg, W.U.: Muramidaseaktivität in Serum und Urin bei akuten und chronischen Leukämien. Schweiz. Med. Wochenschr. 100, 1993–1995 (1970).

Serr, D.M., Padeh, B., Mashiach, S., Shaki, R.: Chromosomal studies in tumors of embryonic origin. Obstet. Gynecol. 33, 324–332 (1969).

Serra, A., Moneta, E., Patrona, V., Pizzolato, G.: A phenotypically Turner-like female with karyotype 45,X/46,XY gonadoblastoma and fluorescent Y. Humangenetik 24, 309–318 (1974).

Serra, A., Sargentini, S., Patrono, C., Ferrara, A., Laghi, V.: 45,X,G−,Ph¹+ cell line in bone marrow and blood of a CML affected male. Ann. Genet. (Paris) 13, 239–243 (1970).

Seshadri, R.S., Brown, E.J., Zipursky, A.: Leukemic reticuloendotheliosis. A failure of monocyte production. N. Engl. J. Med. 295, 181–184 (1976).

Setlow, R.B.: Repair deficient human disorders and cancer. Nature 271, 713–717 (1978).

Setlow, R.B., Regan, J.D., German, J., Carrier, W.L.: Evidence that xeroderma pigmentosum cells do not perform the first step in the repair of ultraviolet damage of their DNA. Proc. Natl. Acad. Sci. USA 64, 1035–1041 (1969).

Sewell, R.L.: Neutrophil alkaline phosphatase, the Philadelphia chromosome and chronic myeloid leukemia: a critical review. Med. Lab. Technol. 29, 152–159 (1972).

Sézary, A., Bouvrain, Y.: Erythrodermie avec presence de cellules monstruses dans le derme et dans le sang circulant. Bull. Soc. Fr. Dermatol. Syphiligr. 45, 254–260 (1938).

Shabtai, F., Weiss, S., van der Lijn, E., Lewinski, U., Djaldetti, M., Halbrecht, I.: A new cytogenetic aspect of polycythemia vera. Hum. Genet. 41, 281–287 (1978).

Shafer, D.A.: Replication bypass model of sister chromatid exchanges and implications for Bloom's syndrome and Fanconi's anemia. Hum. Genet. 39, 177–190 (1977).

Shahid, M.J., Khouri, F.P., Ballas, S.K.: Fanconi's anemia: Report of a patient with significant chromosomal abnormalities in bone marrow cells. J. Med. Genet. 9, 474–478 (1972).

Shapot, V.S., Krechetova, G.D., Likhtenshtein, A.V., Shliankwvich, M.A.: Molecular genetic nature of neoplastic transformation. Patol. Fiziol. Eksp. Ter., pp. 3–11 (1976).

Sharma, A.K., Sharma, A.: *Chromosome Techniques,* 2nd edition. Baltimore: University Park Press; London: Butterworths (1972).

Sharma, G.P., Parshad, R., Agnish, N.D.: Chromosome number in some malignant tumors. Res. Bull. Panjab Univ. 14, 99–101 (1963).

Sharman, C., Crossen, P.E., Fitzgerald, P.H.: Lymphocyte number and response to phytohaemagglutinin in chronic lymphocytic leukaemia. Scand. J. Haematol. 3, 375–382 (1966).

Sharp, J.C.: Karyotypic transformation of chronic granulocytic leukaemia. J. Clin. Pathol. 29, 87 (1976).

Sharp, J.C., Potter, A.M., Blackburn, E.K.: Clinical course and karyotype: The relationship in chronic granulocytic leukaemia. Br. J. Haematol. 33, 614 (1976a).

Sharp, J.C., Potter, A.M., Guyer, R.J.: Chromosome changes in congenital lymphoblastic leukaemia. Lancet ii, 1448 (1973).

Sharp, J.C., Potter, A.M., Guyer, R.J.: Karyotypic abnormalities in transformed chronic granulocytic leukaemia. Br. J. Haematol. 29, 587–591 (1975).

Sharp, J.C., Potter, A.M., Wood, J.K.: Non-random and random chromosomal abnormalities in transformed chronic granulocytic leukaemia. Scand. J. Haematol. 16, 5–12 (1976b).

Sharp, J.C., Wayne, A.W., Joynes, M.V.: Ph¹+ve acute leukaemia. *Program Helsinki Chromosome Conference,* Aug. 29–31, p. 211 (abstract) (1977).

Shaw, M.T.: The cytochemistry of acute leukemia: A diagnostic and prognostic evaluation. Semin. Oncol. 3, 219–228 (1976).

Shaw, M.T., Bottomley, R.H., Grozea, P.N., Nordquist, R.E.: Heterogeneity of morphological, cytochemical, and cytogenetic features in the blastic phase of chronic granulocytic leukemia. Cancer 35, 199–207 (1975).

Shaw, M.W.: Familial mongolism. Cytogenetics 1, 141–179 (1962).

Shaw, M.W.: Human chromosome damage by chemical agents. Annu. Rev. Med. 21, 409–432 (1970).

Shaw, M.W., Chen, T.R.: The application of banding techniques to tumor chromosomes. *In* J. German (ed.): *Chromosomes and Cancer,* pp. 135–150. New York: Wiley (1974).

Shaw, M.W., Craig, A.P., Ricciuti, F.C.: Random association of human acrocentric chromosomes. Am. J. Hum. Genet. 21, 513–515 (1969).

Sherman, A.I.: Chromosome constitution of endometrium. Am. J. Obstet. Gynecol. 34, 753–766 (1969).

Shettles, L.B.: Use of the Y chromosome in prenatal sex determination. Nature 230, 52–53 (1971).

Shevach, E., Edelson, R., Frank, M., Lutzner, M., Green, I.: A human leukemia cell with both B and T cell surface receptors. Proc. Natl. Acad. Sci. USA 71, 863–866 (1974).

Shiffman, N.J., Stecker, E., Conen, P.E., Gardner, H.A.: Males with chronic myeloid leukemia and the 45,XO,Ph¹ chromosome pattern. Can. Med. Assoc. J. 18, 1151–1154 (1974).

Shigematsu, S.: Significance of the chromosome in vesical cancer. *International Society of Urology, 13th Congress,* Vol. 2, pp. 111–121. London: E & S Livingstone (1965).

Shiloh, Y., Cohen, M.M.: An improved technique of preparing bone-marrow specimens for cytogenetic analysis. In Vitro 14, 510–515 (1978).

Shiloh, Y., Naparstek, E., Cohen, M.M.: Cytogenetic investigation of leukemic and preleukemic disorders. Isr. J. Med. Sci. 15, 500–506 (1979).

Shimada, K., Oda, I., Kobayashi, M.: Chromosome studies in retinoblastoma. Acta Soc. Ophthalmol. Jap. 71, 2014–2019 (1960).

Shinohara, T., Sasaki, M.S., Tonomura, A., Shimamine, T., Yakoyama, T., Masegawa, T.: Cytogenetic studies in human chorionic lesions. Jap. J. Hum. Genet. 16, 111–112 (1971).

Shiraishi, Y.: Differential reactivity in mammalian chromosomes. I. Relation between special segments and late-labelling patterns of DNA synthesis in human X-chromosomes. Chromosoma 36, 211–220 (1972).

Shiraishi, Y.: Cytogenetic studies in 12 patients with itai-itai disease. Humangenetik 27, 31–44 (1975).

Shiraishi, Y., Sandberg, A.A.: Studies on sister chromatid exchange in Bloom's syndrome. Proc. Jap. Acad. 52, 375–379 (1976a).

Shiraishi, Y., Sandberg, A.A.: Caffeine and sister chromatid exchange. Proc. Jap. Acad. 52, 379–382 (1976b).

Shiraishi, Y., Sandberg, A.A.: The relationship between sister chromatid exchange and chromosome aberrations in Bloom's syndrome. Cytogenet. Cell Gener. 18, 13–23 (1977).

Shiraishi, Y., Sandberg, A.A.: Evaluation of sister chromatid exhanges (SCE) in Bloom's syndrome. Cytobios 21, 174–185 (1978a).

Shiraishi, Y., Sandberg, A.A.: Effects of mitomycin C on sister chromatid exchange in normal and Bloom's syndrome cells. Mutat. Res. 49, 233–238 (1978b).

Shiraishi, Y., Sandberg, A.A.: Effects of various chemical agents on SCE, chromosome aberrations and DNA repair in normal and abnormal human lymphoid cell lines. J. Natl. Cancer Inst. 62, 27–35 (1979).

Shiraishi, Y., Yosida, T.H.: Banding pattern analysis of human chromosomes by use of a urea treatment technique. Chromosoma 37, 75–83 (1972a).

Shiraishi, Y., Yosida, T.H.: Chromosomal abnormalities in cultured leucocyte cells from itai-itai disease patients. Proc. Jap. Acad. 48, 248–251 (1972b).

Shiraishi, Y., Freeman, A.I., Sandberg, A.A.: Increased sister chromatid exchange in bone marrow and blood cells from Bloom's syndrome. Cytogenet. Cell Genet. 17, 162–173 (1976).

Shiraishi, Y., Hayata, I., Sakurai, M., Sandberg, A.A.: Chromosomes and causation of human cancer and leukemia. XII. Banding analysis of abnormal chromosomes in polycythemia vera. Cancer 36, 199–202 (1975).

Shiraishi, Y., Holdsworth, R., Minowada, J., Sandberg, A.A.: Specificity of chromosomal changes induced with X-ray in a human T-cell line. Radiat. Res. 73, 452–463 (1978).

Shiraishi, Y., Minowada, J., Sandberg, A.A.: Differential response of SCE and chromosome aberrations to mitomycin C of normal and abnormal human lymphocytic cell lines. Oncology (in press) (1978).

Shirley, R.L.: The nuclear sex of breast cancer. Surg. Gynecol. Obstet. 125, 737–740 (1967).

Shoemaker, R.H.: X chromatin and aging. Acta Cytol. 21, 127–131 (1977).

Shohet, S.B., Blum, S.F.: Coincident basophilic chronic myelogenous leukemia and pulmonary tuberculosis associated with extreme elevations of blood histamine levels and maturity onset asthma. Cancer 22, 173–174 (1968).

Shows, T.B.: Mapping of the human genome and metabolic disease. In J.W. Littlefield, J. de Grouchy (eds.): *Proceedings of the 5th International Conference on Birth Defects (1978)*, International Congress Ser. 432, pp. 66–84. Excerpta Medica. Held in Montreal, Canada, Aug. 21–27 (1977).

Shows, T.B., McAlpine, P.J.: The catalog of human gene symbols and chromosome assignments. A report on human genetic nomenclature and genes that have been mapped in man. Fourth International Workshop on Human Gene Mapping. Birth Defects, Original Article Series, The National Foundation, New York; Cytogenet. Cell Genet. 22, 132–145 (1978).

Siebenmann, R.E.: Pseudohemaphroditismus masculinus mit Gonadoblastom – Eine besondere Intersexform. Pathol. Microbiol. (Basel) 24, 233–238 (1961).

Sieber, S.M., Adamson, R.H.: Toxicity of antineoplastic agents in man: Chromosomal aberrations, antifertility effects, congenital malformations, and carcinogenic potential. Adv. Cancer Res. 22, 57–155 (1975).

Siebner, H.: Chromosomenuntersuchungen bei malignen Erkrankungen des lymphoplasmoretikulären Systems mit besonderer Berücksichtigung der Paraproteinämien. Arztl. Forsch. 21, 359–381 (1967).

Siebner, H.: Über Chromosomenuntersuchungen bei Paraproteinämien. Fortschr. Med. 87, 30–32 (1969).

Siebner, H., Aly, F.W., Braun, H.J.: Chromosomenbefund bei γ-D-Plasmocytom. Klin. Wochenschr. 47, 884–885 (1969).

Siebner, H.V., Spengler, G.A., Bütler, R., Heni, F., Riva, G.: Chromosomenanomalien bei Paraproteinämie. Schweiz. Med. Wochenschr. 95, 1767–1777 (1965).

Siegal, F.P., Voss, R., Al-Mondhiry, H., Polliack, A., Hansen, J.A., Siegal, M., Good, R.A.: Association of a chromosomal abnormality with lymphocytes having both T and B markers in a patient with lymphoproliferative disease. Am. J. Med. 60, 157–166 (1976).

Siegler, D.: Gastric carcinoma and Turner's syndrome. Postgrad. Med. J. 51, 411–412 (1975).

Silberman, S., Krmpotic, E.: Refractory anemia with leukemic transformation and chromosomal changes. A case report. Acta Haematol. (Basel) 41, 186–192 (1969).

Silver, L.M., Elgin, S.C.R.: A method for determination of the *in situ* distribution of chromosomal proteins. Proc. Natl. Acad. Sci. USA 73, 423–427 (1976).

Silver, R.T., Schleider, M.A., Senterfit, L.B., Rothe, D.J.: Early recognition of terminal phase CML by serial chromosome analysis. Proc. Am. Assoc. Cancer Res. 18, 220 (1977).

Silvergleid, A.J., Schrier, S.L.: Acute myelogenous leukemia in two patients treated with azathioprine for nonmalignant diseases. Am. J. Med. 57, 885–888 (1974).

Similä, S., Jukarainen, E., Herva, R., Heikkinen, E.S.: Gonadoblastoma associated with pure gonadal dysgenesis. Clin. Pediatr. (Phila.) 13, 177–180 (1974).

Simons, J.W.I.M.: Genome mutation and carcinogenesis. Nature 209, 818–819 (1966).

Simpson, J.L., Martin, A.O.: Cytogenetic nomenclature. Am. J. Obstet. Gynecol. 128, 167–172 (1977).

Simpson, J.L., Falk, C.T., German, J.: Autoradiographic studies of human chromosomes. VIII. The arm ratio of the late-replicating X. Ann. Genet. (Paris) 17, 19–22 (1974).

Singer, H., Zang, K.D.: Cytologische und cytogenetische Untersuchungen an Hirntumoren. I. Die Chromosomenpathologie des menschlichen Meningeoms. Humangenetik 9, 172–184 (1970).

Singh, H., Boyd, E., Hutton, M.M., Wilkinson, P.C., Peebles Brown, D.A., Ferguson-Smith, M.A.: Chromosomal mutation in bone-marrow as cause of acquired granulomatous disease and refractory macrocytic anaemia. Lancet i, 873–879 (1972).

Singh, I.P., Ghosh, P.K.: Giemsa banding analysis in myeloproliferative and lymphoproliferative disorders. In S. Armendares, R. Lisker (eds.): *V International Congress of Human Genetics*, p. 153, Abstract 404. Held in Mexico, Oct. 10–15 (1976).

Singh, R.P.: Hygroma of the neck in XO abortuses. Am. J. Clin. Pathol. 53, 104–107 (1970).

Sinha, A.K.: Spontaneous occurrence of tetraploidy and nearhaploidy in mammalian peripheral blood. Exp. Cell Res. 47, 443–448 (1967).

Sinkovics, J.G., Drewinko, B., Thornell, E.: Immunoresistant tetraploid lymphoma cells. Lancet i, 139–140 (1970).

Sinks, L.F., Clein, G.P.: The cytogenetics and cell metabolism of circulating Reed-Sternberg cells. Br. J. Haematol. 12, 447–453 (1966).

Siracka, E., Simko, I., Siracky, J.: The importance of

sex chromatin determination in the hormonal treatment of breast cancer. Neoplasma 17, 625–629 (1970).
Siracký, J.: Ploidy studies in ovarian cancer during cytostatic treatment. Neoplasma 16, 427–433 (1969).
Sjögren, U., Brandt, L.: Different composition and mitotic activity of the haematopoietic tissue in bone marrow, spleen and liver in chronic myeloid leukaemia. Acta Haematol. (Basel) 55, 73–80 (1976).
Sjögren, U., Brandt, L., Mitelman, F.: Relation between life expectancy and composition of the bone marrow at diagnosis of chronic myeloid leukemia. Scand. J. Haematol. 12, 369–373 (1974).
Skerfving, S., Hansson, K., Lindsten, J.: Chromosome breakage in humans exposed to methyl mercury through fish comsumption. Preliminary communication. Arch. Environ. Health 21, 133–139 (1970).
Skovby, F.: Nomenclature: Additional chromosome bands. Clin. Genet. 7, 21–28 (1975).
Slater, R.M., Philip, P., Badsberg, E., Behrendt, H., Hansen, N.E., Van Heerde, P.: A 14q+ chromosome in a B-cell acute lymphocytic leukemia and a case of leukemic non-endemic Burkitt's lymphoma. Int. J. Cancer (in press) (1979).
Šlot, E.: Spontaneous changes of the stemline cells in human carcinomas. Neoplasma 14, 629–639 (1967a).
Šlot, E.: A karyologic study of the cancer of the ovary and the cancer cells in the ascitic effusions. Neoplasma 14, 3–10 (1967b).
Šlot, E.: Spontaneous structural aberrations of chromosomes in human tumour cells of the effusions. Neoplasma 17, 189–195 (1970).
Šlot, E., Frauenklin, J.E.: The chromosome count in the stem cells of ovarian carcinoma and its relation to chemotherapeutic results. Zentralbl. Gynaekol. 90, 210–213 (1968).
Slowikowska, M.G., Grzymala, W.J.: Chromosome aberrations in lymphocytes of peripheral blood of roentgenologists. Nukleonika 17, 241–248 (1972).
Slyck, E.J. van, Weiss, L., Dully, M.: Chromosomal evidence for the secondary role of fibroblastic proliferation in acute myelofibrosis. Blood 36, 729–735 (1970).
Smalley, R.V.: Double Ph[1] chromosomes in leukaemia. Lancet ii, 591 (1966).
Smalley, R.V.: Chronic myelogenous leukemia. The value of chromosome analysis and a discussion of therapy. Med. Times 96, 373–381 (1968).
Smalley, R.V., Bouroncle, B.S.: Hyperdiploidy in a patient with Di Guglielmo's syndrome. Arch. Intern. Med. 120, 599–601 (1967).
Smalley, R.V., Vogel, J., Huguley, C.M., Jr., Miller,

D.: Chronic granulocytic leukemia: Cytogenetic conversion of the bone marrow with cycle-specific chemotherapy. Blood 50, 107–113 (1977).
Smetana, K., Daskal, Y., Gyorkey, F., Gyorkey, P., Lehane, D.E., Rudolph, A.H., Busch, H.: Nuclear and nucleolar ultrastructure of Sézary cells. Cancer Res. 37, 2036–2042 (1977).
Smith, K.L., Johnson, W.: Classification of chronic myelocytic leukemia in children. Cancer 34, 670–679 (1974).
Smith, W.I., Jr., Zidar, B.L., Winkelstein, A., Whiteside, T.L., Shadduck, R.K., Ziegler, Z., Brietfeld, V., Silverberg, J.H., Rosenbach, L.M., Rabin, B.S.: Hairy cell leukemia: A case with B-lymphocyte origin. Am. J. Clin. Pathol. 68, 778–786 (1977).
Socolow, E.L., Engel, E., Mantooth, L., Stanbury, J.B.: Chromosomes of human thyroid tumors. Cytogenetics 3, 394–413 (1964).
Soderström, V.B.: Giant A1 chromosome in human blast-cell leukaemia. Lancet i, 751–752 (1974).
Sofuni, T., Okada, H.: Autoradiographic and fluorescent staining studies of bone marrow chromosomes from a patient with acute granulocytic leukemia. Cancer 35, 378–384 (1975).
Sofuni, T., Sandberg, A.A.: Chronology and pattern of human chromosome replication. VI. Further studies including autoradiographic behavior of normal and abnormal A1 autosomes. Cytogenetics 6, 357–370 (1967).
Sofuni, T., Kikuchi, Y., Sandberg, A.A.: Chronology and pattern of human chromosome replication. V. Blood leukocytes of chronic myelocytic leukemia. J. Natl. Cancer Inst. 38, 141–156 (1967).
Sofuni, T., Makino, S., Kobayashi, H.: A study of chromosomes in Friend virus-induced mouse leukemias. Proc. Jap. Acad. 43, 389–394 (1967).
Sohn, K.-Y., Boggs, D.R.: Klinefelter's syndrome, LSD usage and acute lymphoblastic leukemia. Clin. Genet. 6, 20–22 (1974).
Sokal, G., Michaux, J.L., Berghe, H. van den, Cordier, A., Rodhain, J., Ferrant, A., Moriau, M., Bruyere, M. de, Sonnet, J.: A new hematologic syndrome with a distinct karyotype: the 5q− chromosome. Blood 46, 519–533 (1975).
Sokal, J.E.: Evaluation of survival data for chronic myelocytic leukemia. Am. J. Hematol. 1, 493–500 (1976).
Solari, A.J.: Experimental changes in the width of the chromatin fibers from chicken erythrocytes. Exp. Cell Res. 67, 161–170 (1971).
Solari, A.J., Sverdlick, A.B., Viola, E.R.: Chromosome abnormality in myeloid metaplasia. Lancet ii, 613 (1962).
Sonta, S.I., Sandberg, A.A.: A new complex Ph[1]-

translocation involving 3 chromosomes. J. Natl. Cancer Inst. 58, 1583–1586 (1977a).

Sonta, S.I., Sandberg, A.A.: Chromosomes and causation of human cancer and leukemia. XXIV. Unusual and complex Ph¹-translocations and their clinical significance. Blood 50, 691–697 (1977b).

Sonta, S.I., Sandberg, A.A.: Chromosomes and causation of human cancer and leukemia. XXVIII. Value of detailed chromosome studies on large numbers of cells in CML. Am. J. Hematol. 3, 121–126 (1977c).

Sonta, S.I., Sandberg, A.A.: Chromosomes and causation of human cancer and leukemia. XXIX. Further studies on karyotypic progression in CML. Cancer 41, 153–163 (1978a).

Sonta, S.I., Sandberg, A.A.: Chromosomes and causation of human cancer and leukemia. XXX. Banding studies of primary intestinal tumors. Cancer 41, 164–173 (1978b).

Sonta, S.I., Minowada, J., Tsubota, T., Sandberg, A.A.: Cytogenetic study of a new Ph¹-positive cell line (NALM-1). J. Natl. Cancer Inst. 59, 833–837 (1977).

Sonta, S.I., Oshimura, M., Evans, J.T., Sandberg, A.A.: Chromosomes and causation of human cancer and leukemia. XX. Banding patterns of primary tumors. J. Natl. Cancer Inst. 58, 49–59 (1977).

Sonta, S.I., Oshimura, M., Sakurai, M., Freeman, A.I., Sandberg, A.A.: Chromosomes and causation of human cancer and leukemia. XXI. Cytogenetically unusual cases of leukemia. Blood 48, 697–705 (1976).

Soudek, D., Laraya, P.: C and Q bands in long arm of Y chromosomes; are they identical? Hum. Genet. 32, 339–341 (1976).

Soudek, D., Langmuir, V., Stewart, D.J.: Variation in the nonfluorescent segment of long Y chromosome. Humangenetik 18, 285–290 (1973).

Sparkes, R.S., Motulsky, A.G.: The Turner syndrome with isochromosome X and Hashimoto's thyroiditis. Ann. Intern. Med. 67, 132–144 (1967).

Sparkes, R.S., Comings, D.E., Fox, C.F. (eds.): *Molecular Human Cytogenetics*. New York, San Francisco, London: Academic Press, (1977).

Speed, D.E., Lawler, S.D.: Chronic granulocytic leukaemia. The chromosomes and the disease. Lancet i, 403–408 (1964).

Speight, J.W., Smith, E., Baba, W.I., Wilson, G.M.: Lymphocyte chromosomes in untreated and ¹³¹I-treated thyrotoxic patients. J. Endocrinol. 42, 277–282 (1968).

Spence, M.A., Francke, U., Forsythe, A.B.: Evidence against the peripheral location of the Y chromosome in human metaphase cells. Cytogenet. Cell Genet. 12, 49–52 (1973).

Spengler, G.A., Siebner, H., Riva, G.: Chromosomal abnormalities in macroglobulinemia Waldenström: Discordant findings in uniovular twins. Acta Med. Scand. [Suppl.] 445, 132–139 (1966).

Sperling, K., Wegner, R.D., Riehm, H., Obe, G.: Frequency and distribution of sister-chromatid exchanges in a case of Fanconi's anemia. Humangenetik 27, 227–230 (1975).

Sperling, K., Goll, U., Kunze, J., Lüdtke, E.-K., Tolksdorf, M., Obe, G.: Cytogenetic investigations in a new case of Bloom's syndrome. Hum. Genet. 31, 47–52 (1976).

Spiers, A.S.D.: Chromosomes in haematology. In A.V. Hoffbrand, S.M. Lewis (eds.): *Haematology*, pp. 550–565. London: Heinemann (1972).

Spiers, A.S.D., Baikie, A.G.: Chronic granulocytic leukaemia. Demonstration of the Philadelphia chromosome in cultures of spleen cells. Nature 208, 497 (1965a).

Spiers, A.S.D., Baikie, A.G.: Chromosomal abnormalities in lymphoma. Br. Med. J. i, 1613 (1965b).

Spiers, A.S.D., Baikie, A.G.: Cytogenetic studies in the malignant lymphomas. Lancet i, 506–509 (1966).

Spiers, A.S.D., Baikie, A.G.: Reticulum cell sarcoma: Demonstration of chromosomal changes analogous to those in SV40-transformed cells. Br. J. Cancer 21, 679–683 (1967).

Spiers, A.S.D., Baikie, A.G.: Cytogenetic studies in the malignant lymphomas and related neoplasms: Results in twenty-seven cases. Cancer 22, 193–217 (1968a).

Spiers, A.S.D., Baikie, A.G.: Cytogentic evolution and clonal proliferation in acute transformation of chronic granulocytic leukaemia. Br. J. Cancer 22, 192–204 (1968b).

Spiers, A.S.D., Baikie, A.G.: A special role of the group 17, 18 chromosomes in reticuloendothelial neoplasia. Br. J. Cancer 24, 77–91 (1970).

Spiers, A.S.D., Baikie, A.G.: Anomalies of the small acrocentric chromosomes in human tumour cells. Aust. N.Z. J. Med. 2, 188–202 (1972).

Spiers, A.S.D., Baikie, A.G., Dartnall, J.A., Cox, J.I.: Cytogenetic studies of the spleen in chronic granulocytic leukaemia. Aust. N.Z. J. Med. 5, 295–305 (1975a).

Spiers, A.S.D., Baikie, A.G., Galton, D.A.G., Richards, H.G.H., Wiltshaw, E., Goldman, J.M., Catovsky, D., Spencer, J., Peto, R.: Chronic granulocytic leukaemia: Effect of elective splenectomy on the course of disease. Br. Med. J. i, 175–179 (1975b).

Spiers, A.S., Bain, B.J., Turner, J.E.: The peripheral blood in chronic granulocytic leukaemia. Study of 50 untreated Philadelphia-positive cases. Scand. J. Haematol. 18, 25–38 (1977).

Spiers, A.S.D., Janis, M.G., Lord, K.E.: Chromosome

studies in malignant lymphomas. Blood 50 (Suppl. 1), 231 (abstract) (1977).
Spira, J., Povey, S., Wiener, F., Klein, G., Andersson-Anvret, M: Chromosome banding, isoenzyme studies and determination of Epstein-Barr virus DNA content of human Burkitt lymphoma/mouse hybrids. Int. J. Cancer 20, 849–853 (1977).
Spitzer, G., Schwarz, M.A., Dicke, K.A., Trujillo, J.M., McCredie, K.B.: Significance of PHA induced clonogenic cells in chronic myeloid leukemia and early acute myeloid leukemia. Blood Cells 2, 149–159 (1976).
Spooner, M.E., Cooper, E.H.: Chromosome constitution of transitional cell carcinoma of the urinary bladder. Cancer 29, 1401–1412 (1972).
Sprenger, E., Moore, G.W., Naujoks, H., Schluter, G., Sandritter, W.: DNA content and chromatin pattern analysis on cervical carcinoma *in situ*. Acta Cytol. 17, 27–31 (1973).
Spriggs, A.I.: Karyotype changes in human tumour cells. Br. J. Radiol. 37, 210–212 (1964).
Spriggs, A.I.: Clonal proliferation in Hodgkin's disease. Lancet i, 857 (1971).
Spriggs, A.I.: Population screening by the cervical smear. Nature 238, 135–137 (1972).
Spriggs, A.I.: Zytogenetik und Zytologie bei malignen Erkrankungen. Verh. Dtsch. Ges. Pathol. 57, 53–61 (1973).
Spriggs, A.I.: Cytogenetics of cancer and precancerous states of the cervix uteri. In J. German (ed.): *Chromosomes and Cancer*, pp. 423–450. New York: Wiley (1974).
Spriggs, A.I.: Chromosomes in human neoplastic diseases. In T. Symington, R.L. Carter (eds.): *Scientific Foundations of Oncology*, pp. 147–155. London: William Heinemann (1976).
Spriggs, A.I., Boddington, M.M.: Chromosomes of Sternberg-Reed cells. Lancet ii, 153 (1962).
Spriggs, A.I., Boddington, M.M.: Karyotype analysis in the diagnosis of malignancy. In A.I. Spriggs, M.M. Boddington (eds.): *The Cytology of Effusions in the Pleural, Pericardial and Peritoneal Cavities and of Cerebrospinal Fluid*, 2nd ed., pp. 40–41. London: William Heinemann (1968).
Spriggs, A.I., Cowdell, R.H.: Ring chromosomes in carcinoma *in situ* of the cervix as evidence of clonal growth. J. Obstet. Gynecol. Br. Commonw. 79, 833–840 (1972).
Spriggs, A.I., Boddington, M.M., Clarke, C.M.: Carcinoma-*in-situ* of the cervix uteri. Some cytogenetic observations. Lancet i, 1383–1384 (1962*a*).
Spriggs, A.I., Boddington, M.M., Clarke, C.M.: Chromosomes of human cancer cells. Br. Med. J. ii, 1431–1435 (1962*b*).
Spriggs, A.I., Bowey, C.E., Cowdell, R.H.: Chromosomes of precancerous lesions of the cervix uteri. New data and a review. Cancer 27, 1239–1254 (1971).
Spriggs, A.I., Holt, J.M., Bedford, J.: Duplication of part of the long arm of chromosome 1 in marrow cells of a treated case of myelomatosis. Blood 48, 595–599 (1976).
Srivastava, B.I.S., Khan, S.A., Minowada, J., Freeman, A.: High terminal deoxynucleotidyl transferase activity in pediatric patients with acute lymphocytic and acute myelocytic leukemias. Int. J. Cancer 22, 4–9 (1978).
Srivastava, B.I.S., Khan, S.A., Minowada, J., Gomez, G.A., Rakowski, I.: Terminal deoxynucleotidyl transferase activity in blastic phase of chronic myelogenous leukemia. Cancer Res. 37, 3612–3618 (1977).
Srivastava, P.K., Lucas, F.V.: Evolution of human cytogenetics: An encyclopedic essay. J. Genet. Hum. 24, 235–246 (1976).
Srivastava, P.K., Miles, J.H., Lucas, F.V.: Y-chromatin fluorescence in human buccal smear. Int. J. Clin. Genet. 6, 201–204 (1974).
Srodes, C.H., Hyde, E.F., Pan, S.F., Chervenick, P.A., Boggs, D.R.: Cytogenetic studies during remission of blastic crisis in a patient with chronic myelocytic leukaemia. Scand. J. Haematol. 10, 130–135 (1973).
Stahl, A., Luciani, J.M.: Les aberrations chromosomiques provoquées par les substances chimiques. Nouv. Rev. Fr. Hematol. 10, 128–139 (1970).
Stahl, A., Papy, M.C., Muratore, R., Mongin, M., Olmer, J.: Erythromyélose avec pseudo-syndrome de Pelger-Huët et anomalies chromosomiques complexes. Nouv. Rev. Fr. Hematol. 5, 879–882 (1965).
Stanley, M.A.: Chromosome constitution of human endometrium. Am. J. Obstet. Gynecol. 104, 99–103 (1969).
Stanley, M.A., Kirkland, J.A.: Cytogenetic studies of endometrial carcinoma. Am. J. Obstet. Gynecol. 102, 1070–1079 (1968).
Stanley, M.A., Kirkland, J.A.: Chromosome analysis in a progressive lesion of the cervix. Acta Cytol. 13, 76–80 (1969).
Stanley, M.A., Kirkland, J.A.: Chromosome and histologic patterns in pre-invasive lesions of the cervix. Acta Cytol. 19, 142–147 (1975).
Stanley, M.A., Bigham, D.A., Cox, R.I., Kirkland, J.A., Opit, L.J.: Sex-chromatin anomalies in female patients with breast carcinoma. Lancet i, 690–691 (1966).
Stavem, P., Hagen, C.B., Vogt, E., Sandnes, K.: Polycythemia vera treated with ^{32}P and mylleran: Development of chronic granulocytic leukemia with chromosomal abnormalities in one patient. Clin. Genet. 7, 227–231 (1975).
Stebbings, J.H.: Cytogenetic effects of oxidant air pollution and oxides of nitrogen: Prelim-

inary results from environmental protection agency studies. Am. J. Epidemiol. 104, 360 (1976).
Steel, C.M.: Non-identity of apparently similar chromosome aberrations in human lymphoblastoid cell lines. Nature 233, 555–556 (1971).
Steel, C.M., Woodward, M.A., Davidson, C., Philipson, J., Arthur, E.: Non-random chromosome gains in human lymphoblastoid cell lines. Nature 270, 349–351 (1977).
Steele, H.D., Monocha, S.L., Stich, H.F.: Desoxyribonucleic acid content of epidermal *in-situ* carcinomas. Br. Med. J. ii, 1314–1315 (1963).
Steele, M.W.: Autoradiography may be unreliable for identifying human chromosomes. Nature 221, 1114–1116 (1969).
Steenis, H.: Chromosomes and cancer. Nature 209, 819–821 (1966).
Steffensen, D.: A comparative view of the chromosome. Brookhaven Symp. Biol. 12, 103–124 (1959).
Steffensen, D.M., Szabo, P., McDougall, J.K.: Adenovirus 12 uncoiler regions of human chromosome 1 in relation to the 5S rRNA genes. Exp. Cell Res. 100, 436–439 (1976).
Stefos, K., Arrighi, F.E.: Heterochromatic nature of W chromosome in birds. Exp. Cell Res. 68, 228–231 (1971).
Stein, H., Lennert, K., Parwaresch, M.R.: Malignant lymphomas of B-cell type. Lancet ii, 855–857 (1972).
Steinberg, M.H., Geary, C.G., Crosby, W.H.: Acute granulocytic leukemia complicating Hodgkin's disease. Arch. Intern. Med. 125, 496–498 (1970).
Steruska, M., Hrubisko, N., Izakovic, V.: The use of cytochemical and cytogenetic examination in the diagnosis of di Guglielmo's disease and erythroleukemia. Vnitr. Lek. 23, 712–719 (1977).
Stetka, D.G., Wolff, S.: Sister chromatid exchange as an assay for genetic damage induced by mutagen-carcinogens. I. In vivo test for compounds requiring metabolic activation. Mutat. Res. 41, 333–342 (1976a).
Stetka, D.G., Wolff, S.: Sister chromatid exchange as an assay for genetic damage induced by mutagen-carcinogens. II. In vitro test for compounds requiring metabolic activation. Mutat. Res. 41, 343–350 (1976b).
Stevenson, A.C., Bedford, J., Hill, A.G.S., Hill, H.: Chromosome damage in patients who have had intra-articular injections of radioactive gold. Lancet i, 837–839 (1971a).
Stevenson, A.C., Bedford, J., Hill, A.G.S., Hill, H.F.H.: Chromosomal studies in patients taking phenylbutazone. Ann. Rheum. Dis. 30, 487–500 (1971b).
Stewart, A., Pennybacker, W., Barber, R.: Adult leukaemias and diagnostic X rays. Br. Med. J. ii, 882–890 (1962).
Stewart, J.S.S.: Chromosome analysis. Lancet ii, 651 (1960).
Stewart, S.E., Lovelace, E., Whang, J.J., Ngu, V.A.: Burkitt tumor: tissue culture, cytogenetic and virus studies. J. Natl. Cancer Inst. 34, 319–327 (1965).
Stich, H.F.: Oncogenic and nononcogenic mutants of adenovirus 12: Induction of chromosome aberrations and cell divisions. Prog. Exp. Tumor Res. 18, 260–272 (1973).
Stich, H.F., Emson, H.E.: Aneuploid deoxyribonucleic acid content of human carcinomas. Nature 184, 290–291 (1959).
Stich, H.F., Steele, H.D.: The DNA content of tumor cells. III. Mosaic composition of sarcomas and carcinomas in man. J. Natl. Cancer Inst. 28, 1207–1218 (1962).
Stich, H.F., Florian, S.F., Emson, H.E.: The DNA content of tumor cells. I. Polyps and adenocarcinomas of the large intestine of man. J. Natl. Cancer Inst. 24, 471–482 (1960).
Stich, H.F., Stich, W., Lam, P.: Susceptibility of xeroderma pigmentosum cells to chromosome breakage by adenovirus type 12. Nature 250, 599–601 (1974).
Stich, H.F., Stich, W., San, R.H.C.: Chromosome aberrations in xeroderma pigmentosum cells exposed to the carcinogens, 4-nitro-quinoline-1-oxide and N-methyl-N'-nitro-nitrosoguanidine. Proc. Soc. Exp. Biol. Med. 142, 1141–1144 (1973).
Stich, W., Back, F., Dörmer, P., Tsirimbas, A.: Doppel-Philadelphia-Chromosom und Isochromosom 17 in der terminalen Phase der chronischen myeloischen Leukämie. Klin. Wochenschr. 44, 334–337 (1966).
Stobo, J.D., Rosenthal, A.S., Paul, W.E.: Functional heterogeneity of murine lymphoid cells. 1. Responsiveness to and surface binding of concanavalin A and phytohemagglutinin. J. Immunol. 108, 1–17 (1972).
Stockert, J.C., Lisanti, J.A.: Acridine-orange differential fluorescence of fast- and slow-reassociating DNA after *in situ* DNA denaturation and reassociation. Chromosoma 37, 117–130 (1972).
Stoll, C., Borgaonkar, D.S., Levy, J.-M.: Effect of vincristine on sister chromatid exchanges of normal human lymphocytes. Cancer Res. 36, 2710–2713 (1976).
Stoll, C., Oberling, F., Flori, E.: Chromosome analysis of spleen and/or lymph nodes of patients with chronic myeloid leukemia (CML). Blood 52, 828–838 (1978).
Stolte, L.A.M., Kessel, M.I.A.M., Sellen, J.C., Tijd-

ink, G.A.J.: Chromosomes in hydatidiform moles. Lancet ii, 1144–1145 (1960).
Stone, K.R., Mickey, D.D., Wundereli, H., Mickey, G.H., Paulson, D.F.: Isolation of a human prostate carcinoma cell line (DU 145). Int. J. Cancer 21, 274–281 (1978).
Stone, M., Bagshawe, K.D.: Hydatidiform mole: Two entities. Lancet i, 535–536 (1976).
Storm, P.B., Fallon, B., Bunge, R.G.: Mediastinal choriocarcinoma in a chromatin-positive boy. J. Urol. 116, 838–840 (1976).
Stowens, D.: Cytogenetic studies in children with acute leukemia. South. Med. J. 56, 1447–1448 (1963).
Straub, D.G., Lucas, L.A., McMahon, N.J., Pellett, O.L., Teplitz, R.L.: Apparent reversal of X-condensation mechanism of tumors of the female. Cancer Res. 29, 1233–1243 (1969).
Streiff, F., Peters, A., Gilgenkrantz, S.: Anomalies chromosomiques au cours de la transformation blastique terminale d'une leucémie myéloïde chronique: prédominance d'un clone a 48 chromosomes avec deux chromosomes Ph¹. Nouv. Rev. Fr. Hematol. 6, 416–422 (1966a).
Streiff, F., Peters, A., Gilgenkrantz, S.: Double Ph¹ chromosomes in leukaemia. Lancet ii, 1193–1194 (1966b).
Stroebe, H.: Zur Kenntnis verschiedener cellulärer Vorgänge und Erscheinungen in Geschwülsten. Beitr. Pathol. Anat. 11, 1–38 (1892).
Strosseli, E., Bernadelli, E.: Il cariotipo nelle' eritremia acuta. Bull. Soc. Ital. Biol. Exp. 40, 1362–1363 (1964).
Strumpf, I.J.: Gonadoblastoma in a patient with gonadal dysgenesis. Am. J. Obstet. Gynecol. 92, 992–995 (1965).
Stryckmans, P.A.: Current concepts in chronic myelogenous leukemia. Semin. Hematol. 11, 101–127 (1974).
Stubblefield, E.: The structure of mammalian chromosomes. Int. Rev. Cytol. 35, 1–60 (1973).
Stubblefield, E., Wray, W.: Architecture of the Chinese hamster metaphase chromosome. Chromosoma 32, 262–294 (1971).
Stubblefield, E., Cram, S., Deaven, L.: Flow microfluorometric analysis of isolated Chinese hamster chromosomes. Exp. Cell Res. 94, 464–468 (1975).
Suárez, H.G., Brieux de Salum, S., Pavlovsky, S., Ruibal, B., Pavlovsky, A.: Culture *in vitro* d'une lignée cellulaire provenant d'un lymphosarcome humain. I. Cytochimie, cytogénétique et ultrastructure cellulaire. Int. J. Cancer 4, 880–890 (1969).
Sugiyama, H., Ishihara, T.: Chromosome studies on bone-marrow from ageing humans. Natl. Inst. Radiol. Sci. Ann. Rep., p. 51, NIRS-15 (1975).
Sugiyama, T., Goto, K., Kano, Y.: Mechanism of differential Giemsa method for sister chromatids. Nature 259, 59–60 (1976).
Sulica, L.O., Borgaonkar, D.S., Shah, S.A.: Accurate identification of the human Y chromosome. Clin. Genet. 5, 17–27 (1974).
Sultan, C., Marquet, M., Yoffroy, Y.: Study of chronic myeloid leukemia by marrow culture *in vitro*. Nouv. Rev. Fr. Hematol. 15, 161–164 (1975).
Sumiya, M., Mizoguchi, H., Kosaka, K., Miura, Y., Takaku, F., Yata, J.: Chronic lymphocytic leukaemia of T-cell origin? Lancet ii, 910 (1973).
Summitt, R.L.: Ph¹ chromosome: Letter to the editor. Blood 32, 180 (1968).
Sumner, A.T.: A simple technique for demonstrating centric heterochromatin. Exp. Cell Res. 75, 304–306 (1972).
Sumner, A.T.: Changes in elemental composition of human chromosomes during a G-banding (ASG) and a C-banding (BSG) procedure. Histochem. J. 10, 201–211 (1978).
Sumner, A.T., Evans, H.J., Buckland, R.A.: New technique for distinguishing between human chromosomes. Nature [New Biol.] 232, 31–32 (1971).
Sumner, A.T., Evans, H.J., Buckland, R.A.: Mechanisms involved in the banding of chromosomes with quinacrine and Giemsa. 1. The effects of fixation in methanol-acetic acid. Exp. Cell Res. 81, 214–222 (1973).
Sumner, A.T., Robinson, J.A., Evans, H.J.: Distinguishing between X, Y and YY-bearing human spermatozoa by fluorescence and DNA content. Nature [New Biol.] 229, 231–233 (1971).
Sun, N.C., Chu, E.H.Y., Chang, C.C.: Staining method for the banding patterns of human mitotic chromosomes. Caryologia 27, 315–324 (1974).
Suñé, M.V., Centeno, J.V., Salzano, F.M.: Gonadoblastoma in a phenotypic female with 45,X/47,XYY mosaicism. J. Med. Genet. 7, 410–412 (1970).
Sutherland, R., Smart, J., Niaudet, P., Greaves, M.: Acute lymphoblastic leukaemia associated with antigen—II. Isolation and partial characterisation. Leukemia Res. 2, 115–126 (1978).
Sutou, Sh., Arai, Y.: Possible mechanisms of endoreduplication induction. Exp. Cell Res. 92, 15–22 (1975).
Sutton, R.N.P., Bishun, N.P., Soothill, J.F.: Immunological and chromosomal studies in first-degree relatives of children with acute lymphoblastic leukaemia. Br. J. Haematol. 17, 113–119 (1969).
Sutton, W.S.: On the morphology of the chromosome group in *Brachystola Magna*. Biol. Bull. 4, 24–39 (1902a).
Sutton, W.S.: The chromosomes in heredity. Biol. Bull. 4, 231–251 (1902b).

Svarch, E., de la Torre, E.: Myelomonocytic leukaemia with a preleukaemic syndrome and Ph¹ chromosome in monozygotic twins. Arch. Dis. Child. 52, 72–74 (1977).

Svejda, J., Vrba, M., Hornak, O.: Karyotype and ultrastructure of a malignant human melanoma. Arch. Geschwulstforsch. 43, 186–192 (1974).

Sweet, D.L., Golomb, H.M., Rowley, J.D., Vardiman, J.M.: Acute myelogenous leukemia and thrombocytosis associated with an abnormality of chromosome No. 3. Cancer Genet. Cytogenet. (in press) (1979).

Swift, M.: Fanconi's anemia in the genetics of neoplasia. Nature 230, 370–373 (1971).

Swift, M.: Fanconi's anaemia: Cellular abnormalities and clinical predisposition to malignant disease. Ciba Found. Symp. 37, 115–134 (1976).

Swift, M., Zimmermann, D., McDonough, E.R.: Squamous cell carcinomas in Fanconi's anemia. J.A.M.A. 216, 325–326 (1971).

Swift, M.R., Hirschhorn, K.: Fanconi's anemia. Inherited susceptibility to chromosome breakage in various tissues. Ann. Intern. Med. 65, 496–503 (1966).

Swolin, B., Waldenström, J., Weinfeld, A., Westin, J.: Deletion of chromosome 5 – a possible preleukemic marker? *Helsinki Chromosome Conference*, Aug. 29–31, p. 214 (abstract) (1977).

Szokol, M., Kondrai, G., Papp, Z.: Gonadal malignancy and 46,XY karyotype in a true hermaphrodite. Obstet. Gynecol. 49, 358–360 (1977).

Szulman, A.E.: Histology of endometrial carcinoma including some cytogenetical considerations. *In* G.C. Lewis, Jr., W.B. Wentz, R.M. Jaffe (eds.): *New Concepts in Gynecological Oncology*, pp. 225–228. Philadelphia: F.A. Davis (1966).

Szulman, A.E., Surti, U.: The syndromes of hydatidiform mole: I. Cytogenetic and morphologic conditions. Am. J. Obstet. Gynecol. 131, 665–671 (1978a).

Szulman, A.E., Surti, U.: The syndromes of hydatidiform mole. II. Morphologic evolution of the complete and partial mole. Am. J. Obstet. Gynecol. 132, 20–27 (1978b).

Tabata, T.: Karyological studies of human tumours. Wakayawa Igaku 9, 963 (1959a).

Tabata, T.: A chromosome study in some malignant human tumors. Cytologia (Tokyo) 24, 367–377 (1959b).

Taft, P.D., Brooks, S.E.H.: Late labelling of iso-X chromosome. Lancet ii, 1069 (1963).

Tagliani, L., Mastrangelo, C., Curcio, S.: Studio citogenetico dei tumori utero-ovarici. Nota preventive. Arch. Ostet. Ginecol. 68, 1–13 (1963).

Takagi, N., Sandberg, A.A.: Chronology and pattern of human chromosome replication. VII. Cellular and chromosomal DNA behavior. Cytogenetics 7, 118–134 (1968a).

Takagi, N., Sandberg, A.A.: Chronology and pattern of human chromosome replication. VIII. Behavior of the X and Y in the early S-phase. Cytogenetics 7, 135–143 (1968b).

Takagi, N., Sandberg, A.A.: Cytogenetics of hematopoietic cells in long term culture. *In* G.L. Tritsch (ed.): *Axenic Mammalian Cell Reactions*, pp. 27–57. New York: Dekker (1969).

Takagi, N., Aya, T., Kato, H., Sandberg, A.A.: RElation of virus induced cell fusion and chromosome pulverization to mitotic events. J. Natl. Cancer Inst. 43, 335–347 (1969).

Takahashi, M.: A behavior of Y chromatin in cancer cells of males. Acta Cytol. 21, 132–136 (1977).

Takatsuki, N.: Plasma cell myeloma and related diseases in Japan. Clinical and immunological studies on myeloma compounds. Acta Haematol. Jap. 31, 636–664 (1968).

Takebe, H.: Genetic complementation tests of Japanese xeroderma pigmentosum patients, and their skin cancers and DNA repair characteristics. *In* P.N. Magee (ed.): *Fundamentals in Cancer Prevention*, pp. 383–395. Tokyo: University of Tokyo Press; Baltimore: University Park Press (1976).

Takebe, H., Miki, Y., Kozuka, T., Furuyama, J.-I., Tanaka, K., Sasaki, M.S., Fujiwara, Y., Akiba, H.: DNA repair characteristics and skin cancers of xeroderma pigmentosum patients in Japan. Cancer Res. 37, 490–495 (1977).

Takebe, H., Nii, S., Ishii, M.I., Utsumi, H.: Comparative studies of host-cell reactivation, colony forming ability and excision repair after UV irradiation of xeroderma pigmentosum, normal human and some other mammalian cells. Mutat. Res. 25, 383–390 (1974).

Takemura, T.: Chromosome survey of normal human endometrium and endometrial carcinoma. J. Jap. Obstet. Gynecol. Soc. 7, 300–322 (1960).

Takino, T., Nakazima, K.: Chromosome abnormalities in hematological diseases other than leukemia. Jap. J. Clin. Hematol. 7, 174–184 (1966).

Taktikos, A.: Association of retinoblastoma with mental defect and other pathological manifestations. Br. J. Ophthalmol. 48, 495–498 (1964).

Takumura, H., Sakurai, M., Sugahara, T.: A study on time- and radiation-dependent changes in the frequency of lymphocytes with chromosome aberrations in the circulation, following multiple exposures of the pelvis to γ-irradiation. Blood 36, 43–51 (1970).

Tanaka, K.: Cytogenetic studies on bone marrow in chronic granulocytic leukemia. Tokyo Ika Daigaku Zasshi 32, 417–441 (1974).

Tanzer, J.: Chromosomes et hemopathies malignes humaines. In J. Bernard (ed.): Actualités Hematologiques, p. 123. Paris: Mansson Publishers (1967).

Tanzer, J., Harel, P., Boiron, M., Bernard, J.: Cytochemical and cytogenetic findings in a case of chronic neutrophilic leukaemia of mature cell type. Lancet i, 387–388 (1964).

Tanzer, J., Jacquillat, C.L., Boiron, M., Bernard, J.: Leucémie aiguë lymphoblastique suivie trois ans plus tard d'une leucémie myéloïde chronique a chromosome Philadelphie. Lyon Med. 233, 311–315 (1975).

Tanzer, J., Jacquillat, C., Levy, D.: La leucémie myéloïde chronique. Diagnostic et formes particulières: données récentes. Actualites Hematol. 3, 69–79 (1969).

Tanzer, J., Levy, A., Perrotez, C.: Chromosomes des polyglobulies. In J. Bernard (ed.): Actualités Hematologiques, p. 123. Paris: Mansson Publishers (1967).

Tanzer, J., Najean, Y., Frocrain, C., Bernheim, A.: Chronic myelocytic leukemia with a masked Ph^1 chromosome. N. Engl. J. Med. 296, 571–572 (1977).

Tanzer, J., Stoitchkov, Y., Harel, P., Boiron, M.: Chromosomal abnormalities in measles. Lancet ii, 1070–1071 (1963).

Tartaglia, A.P., Propp, S., Amarose, A.P., Propp, R.P., Hall, C.A.: Chromosome abnormality and hypocalcemia in congenital erythroid hypoplasia (Blackfan-Diamond syndrome). Am. J. Med. 41, 990–999 (1966).

Tassoni, E.M., Durant, J.R., Becker, S., Kravitz, B.: Cytogenetic studies in multiple myeloma: A study of fourteen cases. Cancer Res. 27, 806–810 (1967).

Tates, A.D.: A search for storage effects on chromosome aberrations induced in human normal type- and xeroderma pigmentosum fibroblasts by tetra-ethylene-imino-1,4-benzochinon and N-acetoxy-N-2-acetylaminofluorene. Mutat. Res. 34, 299–312 (1976).

Tavares, A.S.: Sex chromatin in tumor cells. Acta Cytol. 6, 90–94 (1962).

Tavares, A.S.: Ploidy and histological types of mammary carcinomas. Eur. J. Cancer 3, 449–455 (1968).

Tavares, A.S., Costa, J., Carvalho, A. de, Reis, M.: Tumour ploidy and prognosis in carcinomas of the bladder and prostate. Br. J. Cancer 20, 438–441 (1966).

Tavares, A.S., Costa, J., Costa-Maia, J.C.: Correlation between ploidy and prognosis in prostatic carcinoma. J. Urol. 109, 676–679 (1973).

Taylor, A.I.: Dq–, Dr and retinoblastoma. Humangenetik 10, 209–217 (1970).

Taylor, A.M.R.: Unrepaired DNA strand breaks in irradiated ataxia telangiectasia lymphocytes suggested from cytogenetic observations. Mutat. Res. 50, 407–418 (1978).

Taylor, A.M.R., Harnden, D.G., Arlett, C.F., Harcourt, S.A., Lehmann, A.R., Stevens, S., Bridges, B.A.: Ataxia telangiectasia: A human mutation with abnormal radiation sensitivity. Nature 258, 427–429 (1975).

Taylor, A.M.R., Harnden, D.G., Fairburn, E.A.: Chromosomal instability associated with susceptibility to malignant disease in patients with porokeratosis of Mibelli. J. Natl. Cancer Inst. 51, 371–378 (1973).

Taylor, A.M.R., Metcalfe, J., Oxford, J.M., Arlett, C.: Cell survival and chromosome damage in ataxia telangiectasia following X-irradiation. Heredity 36, 283 (1976a).

Taylor, A.M.R., Metcalfe, J.A., Oxford, J.M., Harnden, D.G.: Is chromatid-type damage in ataxia telangiectasia after irradiation at G_0 a consequence of defective repair? Nature 260, 441–443 (1967b).

Taylor, J.H.: The organization and duplication of genetic material. Proceedings of the 10th International Congress on Genetics, Vol. 1. Held at McGill University, Montreal, Canada (1958).

Taylor, J.H.: DNA synthesis in relation to chromosome reproduction and the reunion of breaks. J. Cell Compar. Physiol. (Suppl. 1) 62, 73–86 (1963).

Taylor, J.H.: The duplication of chromosomes. In P. Silte (ed.): 3. Wissenschaftliche Konferenz der Geselischaft Deutscher Naturforscher und Arzte, Semmerling bei Wien. pp. 9–28. Heidelberg: Springer-Verlag (1966).

Taylor, J.H., Adams, A.G., Kurek, M.P.: Replication of DNA in mammalian chromosomes. II. Kinetics of ^{3}H-thymidine incorporation and the isolation and partial characterization of labeled subunits at the growing point. Chromosoma 41, 361–384 (1973).

Taylor, J.H., Woods, P.S., Hughes, W.L.: The organization and duplication of chromosomes as revealed by autoradiographic studies using tritium-labeled thymidine. Proc. Natl. Acad. Sci. USA 43, 122–127 (1957).

Tchernia, G., Mielot, F., Subtil, E., Parmentier, C.: Acute myeloblastic leukemia after immunodepressive therapy for primary nonmalignant disease. Blood Cells 2, 67–80 (1976).

Teasdale, J.M., Worth, A.J., Corey, M.J.: A missing group C chromosome in the bone marrow cells of three children with myeloproliferative disease. Cancer 25, 1468–1477 (1970).

Teerenhovi, L., Borgström, G.H., Mitelman, F., Brandt, L., Vuopio, P., Timonen, T., Almqvist,

A., Chapelle, A. de la: Uneven geographical distribution of 15;17-translocation in acute promyelocytic leukemia. Lancet ii, 797 (1978).

Temple, M.J., Baumiller, R.C., Feller, W.C.: Marker-14 chromosome in an American Burkitt lymphoma cell line. *In* S. Armendares, R. Lisker (eds.): *V. International Congress of Human Genetics,* p. 160, Abstract 420. Held in Mexico, Oct. 10–15 (1976).

Teplitz, R.L.: Regulation of leukocyte alkaline phosphatase and the Philadelphia chromosome. Nature 209, 821–822 (1966).

Teplitz, R.L.: Cytogenetics. *In* G.D. Amromin (ed.): *Pathology of Leukaemia,* pp. 161–176. New York: Hoeber (1968).

Teplitz, R.L., Rosen, R.B., Teplitz, M.R.: Granulocytic leukaemia, Philadelphia chromosome, and leukocyte alkaline phosphatase. Lancet ii, 418–419 (1964).

Tessarscaia, T.P., Osencencaia, G.V., Hrustaler, S.A.: Modificarea cromozomilor in leucemie umane. Raportul II, leucemia acuta. Probl. Gematol. Pereliv. Krovi. 6, 10 (1964).

Testa, J.R., Rowley, J.D.: Cytogenetic patterns in acute non-lymphocytic leukemia. Virchows Arch. B Cell Pathol. 29, 65–72 (1978).

Testa, J.R., Golomb, H.M., Rowley, J.D., Vardiman, J.W., Sweet, D.L.: Acute promyelocytic leukemia (APL): Association with a consistent chromosomal abnormality. Blood 50 (Suppl. 1), 232 (abstract) (1977).

Testa, J.R., Golomb, H.M., Rowley, J.D., Vardiman, J.W., Sweet, D.L., Jr.: Hypergranular promyelocytic leukemia (APL): Cytogenetic and ultrastructural specificity. Blood 52, 272–280 (1978).

Testa, J.R., Kinnealey, A., Rowley, J.D., Golde, D.W., Potter, D.: Deletion of the long arm of chromosome 20 [del(20)(q11)] in myeloid disorders. Blood 52, 868–877 (1978).

Testa, J.R., Kinnealey, A., Rowley, J.D., Potter, D., Golde, D.W., McFarland, J.: The 20q− chromosome in myeloproliferative disorders. *Proceedings, American Society of Human Genetics,* 28th Annual Meeting, p. 106a. Held in San Diego, Calif., Oct. 19–22 (abstract) (1977).

Teter, J., Boczkowski, K.: Occurrence of tumors in dysgenetic gonads. Cancer 20, 1301–1310 (1967).

Teter, J., Philip, J., Wecewicz, G., Potocki, J.: A masculinizing mixed germ cell tumor (Gonocytoma III). Acta Endocrinol. (Kbh) 46, 1–11 (1964).

Tezok, O.F., Sayli, B.S., Utku, S.: Ph¹ chromosome in a case of chronic myeloid leukemia and other chromosomal aberrations induced by X-ray (in Turkish). Tip. Fak. Mecm. 29, 413–423 (1966).

Therman, E., Kuhn, E.M.: Cytological demonstration of mitotic crossing-over in man. Cytogenet. Cell Genet. 17, 254–267 (1976).

Therman, E., Sarto, G.E., Distèche, C., Denniston, C.: A possible active segment on the inactive human X chromosome. Chromosoma 59, 137–145 (1976).

Thiel, E., Bauchinger, M., Rodt, H., Huhn, D., Theml, H., Thierfelder, S.: Evidence for monoclonal proliferation in prolymphocytic leukemia of T-cell origin. A cytogenetic and quantitative immunoautoradiographic analysis. Blut 35, 427–436 (1977).

Thomas, E.D., Bryant, J.I., Buckner, C.D., Clift, R.A., Fefer, A., Neiman, P., Ramberg, R.E., Storb, R.: Leukemic transformation of engrafted bone marrow. Transplant. Proc. 4, 567–570 (1972).

Thomas, E.D., Clift, R.A., Fefer, A., Storb, R., Buckner, C.D.: Bone marrow transplantation. Resident and Staff Physician, pp. 53–63, June (1976).

Thomas, E.D., Storb, R., Clift, R.A., Fefer, A., Johnson, F.L., Neiman, P.E., Lerner, K.G., Glucksberg, H., Buckner, C.D.: Medical Progress: Bone-marrow transplantation. N. Engl. J. Med. 292, 832–843, 895–902 (1975).

Thompson, H., Lyons, R.B.: Retinoblastoma and multiple congenital anomalies, associated with complex mosaicism with deletion of D-chromosome and probably D/C translocation. Hum. Chrom. Newsl. 15, 21 (1965).

Thompson, M.W., Bell, R.E., Little, A.S.: Familial 21-trisomic mongolism coexistent with leukemia. Can. Med. Assoc. J. 88, 893–894 (1963).

Thomson, A.E.R., Robinson, M.A., Wetherley-Mein, G.: Heterogeneity of lymphocytes in chronic lymphocytic leukaemia. Lancet ii, 200–202 (1966).

Tiepolo, L., Zara, C., Fraccaro, M.: Chromosome DNA replication in human tumour cells labelled *in vivo* and *in vitro.* Eur. J. Cancer 3, 355–360 (1967).

Tiepolo, T., Zuffardi, O.: Identification of normal and abnormal chromosomes in tumor cells. Cytogenet. Cell Genet. 12, 8–16 (1973).

Timonen, S.: Mitosis in normal endometrium and genital cancer. Acta Obstet. Gynaecol. Scand. [Suppl. 2] 31, 1–88 (1950).

Timonen, S., Therman, E.: The changes in the mitotic mechanism of human cancer cells. Cancer Res. 10, 431–439 (1950).

Tischendorf, F.W., Ledderose, G., Müller, D., Orywall, D., Wilmanns, W.: Chronische myelosen mit massiver Lysozymurie unter Milzbestrahlung. Klin. Wochenschr. 50, 250–257 (1972).

Tishler, P.V., Rosner, B., Lamborot-Manzur, M., Atkins, L.: Studies on the location of the Y fluorescent body in human interphase nuclei. Humangenetik 22, 275–286 (1974).

Tjio, J.H., Levan, A.: The chromosome number of man. Hereditas 42, 1–6 (1956).

Tjio, J.H., Puck, T.T.: The somatic chromosomes of man. Proc. Natl. Acad. Sci. USA 44, 1229–1237 (1958).

Tjio, J.H., Whang, J.: Chromosome preparations of bone marrow cells without prior *in vitro* culture or *in vivo* colchicine administration. Stain Technol. 37, 17–20 (1962).

Tjio, J.H., Whang, J.: Direct chromosome preparations of bone marrow cells. In J.J. Yunis (ed.): *Human Chromosome Methodology*. New York: Academic Press (1965).

Tjio, J.H., Carbone, P.P., Whang, J., Frei, E., III: The Philadelphia chromosome and chronic myelogenous leukemia. J. Natl. Cancer Inst. 36, 567–584 (1966).

Tjio, J.H., Marsh, J.C., Whang, J., Frei, E., III: Abnormal karyotype findings in bone marrow and lymph node aspirates of a patient with malignant lymphoma. Blood 22, 178–190 (1963).

Tjio, J.H., Puck, T.T., Robinson, A.: The human chromosomal satellites in normal persons and two patients with Marfan's syndrome. Proc. Natl. Acad. Sci. USA 46, 532–539 (1960).

Todaro, G.J., Martin, G.M.: Increased susceptibility of Down's syndrome fibroblasts to transformation by SV_{40}. Proc. Soc. Exp. Biol. Med. 124, 1232–1236 (1967).

Todaro, G.J., Green, H., Swift, M.R.: Susceptibility of human diploid fibroblast strains to transformation by SV40 virus. Science 153, 1252–1254 (1966).

Todd, N., Wood, S.M., Robertson, J., Brown, R.A.G.: A case of leukaemia showing mixed myeloid-lymphoid characteristics and an unusual chromosome pattern. J. Clin. Pathol. 22, 743 (1969).

Toews, H.A., Katayama, K.P., Jones, H.W., Jr.: Chromosomes of normal and neoplastic ovarian tissue. Obstet. Gynecol. 32, 465–476 (1968a).

Toews, H.A., Katayama, K.P., Masukawa, T., Lewison, E.F.: Chromosomes of benign and malignant lesions of the breast. Cancer 22, 1296–1307 (1968b).

Tokuhata, G.K., Neely, C.L., Williams, D.L.: Chronic myelocytic leukemia in identical twins and a sibling. Blood 31, 216–225 (1968).

Tomkins, G.A.: Chromosome studies on cultured lymphoblast cell lines from cases of New Guinea Burkitt lymphoma, myeloblastic and lymphoblastic leukaemia and infectious mononucleosis. Int. J. Cancer 3, 644–653 (1968).

Tomonaga, M., Ichimaru, M., Danno, H., Inove, A., Okabe, N., Tomiyasu, T., Toyomasu, S., Tamari, K., Kawamoto, N.: Leukemia in atomic bomb survivors from 1946–1965 and some aspects of epidemiology of leukemia in Japan. J. Kyushu Hematol. Soc. 17, 1097 (1967).

Tonomura, A.: Cytological studies of tumors. XXXII. Chromosome analyses in stomach and uterine carcinomas. J. Fac. Sci. Hokkaido Univ. 14, 149–156 (1959a).

Tonomura, A.: A chromosome survey in six cases of human uterine cervix carcinomas. Jap. J. Genet. 34, 401–406 (1959b).

Tonomura, A.: The cytological effect of chemicals on tumors. VIII. Observations on chromosomes in a gastric carcinoma treated with carzinophilin. Gann 51, 47–53 (1960).

Tormey, D.C., Ellison, R.R., Hossfeld, D.K.: Concurrent monoclonal IgM and IgA proteins in lymphocytic leukemia. Cancer 36, 1321–1326 (1975).

Tortora, M.: Chromosome studies in gynecologic cancer. Arch. Obstet. Ginecol. 68:437–444 (1963).

Tortora, M.: Chromosome analysis of four ovarian tumors. Acta Cytol. 11, 225–228 (1967).

Tortora, M.: Cytology and chromosomes in female genital tumors. Proc. Int. Symp. Obstet. Gynecol. (Milan) pp. 99–109 (1969).

Tosato, G., Whang-Peng, J., Levine, A.S., Poplack, D.G.: Acute lymphoblastic leukemia followed by chronic myelocytic leukemia. Blood 52, 1033–1036 (1978).

Toshima, S., Takagi, N., Minowada, J., Moore, G.E., Sandberg, A.A.: Electron microscopic and cytogenetic studies of cells derived from Burkitt's lymphoma. Cancer Res. 27, 753–771 (1967).

Tough, I.M.: Cytogenetic studies in cases of chronic myeloid leukaemia with a previous history of radiation. In F.G.J. Hayhoe (ed.): *Current Research in Leukaemia*, pp. 47–54. London: Cambridge University Press (1965).

Tough, I.M., Court Brown, W.M.: Chromosome aberrations and exposure to ambient benzene. Lancet i, 684 (1965).

Tough, I.M., Buckton, K.E., Baikie, A.G., Court Brown, W.M.: X-ray-induced chromosome damage in man. Lancet ii, 849–851 (1960).

Tough, I.M., Court Brown, W.M., Baikie, A.G., Buckton, K.E., Harnden, D.G., Jacobs, P.A., King, M.J., McBride, J.A.: Cytogenetic studies in chronic myeloid leukaemia and acute leukaemia associated with mongolism. Lancet i, 411–417 (1961).

Tough, I.M., Court Brown, W.M., Baikie, A.G., Buckton, K.E., Harnden, D.G., Jacobs, P.A., Williams, J.A.: Chronic myeloid leukaemia: cytogenetic studies before and after splenic irradiation. Lancet ii, 115–120 (1962).

Tough, I.M., Harnden, D.G., Epstein, M.A.: Chromosome markers in cultured cells from Bur-

kitt's lymphomas. Eur. J. Cancer 4, 637–646 (1968).
Tough, I.M., Jacobs, P.A., Court Brown, W.M., Baikie, A.G., Williamson, E.R.D.: Cytogenetic studies on bone-marrow in chronic myeloid leukaemia. Lancet i, 844–846 (1963).
Tough, I.M., Smith, P.G., Court Brown, W.M., Harnden, D.G.: Chromosome studies on workers exposed to atmospheric benzene. The possible influence of age. Eur. J. Cancer 6, 49–55 (1970).
Traczyk, Z.: Chromosome studies of bone marrow in polycythaemia vera. Pol. Med. J. 121–127 (1963).
Traganos, F., Darzynkiewicz, Z., Sharpless, T., Melamed, M.R.: Nucleic acid content and cell cycle distribution of five human bladder cell lines analysed by flow cytofluorometry. Int. J. Cancer 20, 30–36 (1977).
Trempe, G.L.: Human breast cancer in culture. Recent results. Cancer Res. 57, 33–41 (1976).
Trujillo, J.M.: Cytogenetics in human leukemia. Pathobiol. Ann. 6, 203–220 (1976).
Trujillo, J.M., Ohno, S.: Chromosomal alteration of erythropoietic cells in chronic myeloid leukemia. Acta Haematol. (Basel) 29, 311–316 (1963).
Trujillo, J.M., Ahearn, M.J., Cork, A.: General implications of chromosomal alterations in human leukemia. Hum. Pathol. 5, 675–686 (1974).
Trujillo, J.M., Ahearn, M.J., Cork, A.: Correlated cytogenetic and ultrastructural studies in acute leukemia. Blood Cells 1, 173–175 (1975).
Trujillo, J.M., Butler, J.J., Ahearn, M.J., Shullenberger, C.C., List-Young, B., Gott, C., Anstall, H.B., Shively, J.A.: Long-term culture of lymph node tissue from a patient with lymphocytic lymphoma. II. Preliminary ultrastructural, immunofluorescence and cytogenetic studies. Cancer 20, 215–224 (1967).
Trujillo, J.M., Ahearn, M.J., Cork, A., Youness, E.L., McCredie, K.B.: Hematological and cytological characterization of the 8/21 translocation in acute leukemia. Blood 53, 695–706 (1979).
Trujillo, J.M., Cork, A., Drewinko, B., Hart, J.S., Freireich, E.J.: Case report: Tetraploid leukemia. Blood 38, 632–637 (1971).
Trujillo, J.M., Cork, A., Hart, J.S., George, S.L., Freireich, E.J.: Clinical implications of aneuploid cytogenetic profiles in adult acute leukemia. Cancer 33, 824–834 (1974).
Trujillo, J.M., Fernandez, M.M., Shullenberger, C.C., Rodriguez, L.H., Cork, A.: Cytogenetic contributions to the study of human leukemias. In: Leukemia-Lymphoma, pp. 105–123. 14th Annual Clinical Conference on Cancer, held at M.D. Anderson Hospital and Tumor Institute, Houston, Texas, 1969. Chicago, Ill.: Yearbook Medical (1970).

Tschang, T.-P., Poulos, E., Ho, C.-K., Kuo, T.-T.: Multiple sebaceous adenomas and internal malignant disease: A case report with chromosomal analysis. Hum. Pathol. 7, 589–594 (1976).
Tseng, P.Y., Jones, H.W.: Chromosome constitution of carcinoma of the endometrium. Obstet. Gynecol. 33, 741–752 (1969).
Tsessarskaya, T.P., Osechenskaya, G.V., Khrustalev, S.A.: Chromosome changes in human leukoses. I. Chronic myeloid leukosis (in Russian). Probl. Gematol. Pereliv. Krovi 9, 3–10 (1964a).
Tsessarskaya, T.P., Osechenskaya, G.V., Khrustalev, S.A.: Chromosome changes in human leukemia. Second report (in Russian). Probl. Gematol. Pereliv. Krovi 9, 10–15 (1964b).
Tsubota, T., Minowada, J., Nakazawa, S., Sinks, L.F., Han, T., Higby, R.J., Pressman, D.: Correlation of surface markers of cells of human lymphatic leukemias with disease type. J. Natl. Cancer Inst. 59, 845–850 (1977).
Tsuchimoto, T., Bühler, E.M., Stalder, G.R., Mayr, A.C., Obrecht, J.P.: Deletion of chromosome 7 in polycythaemia vera. Lancet i, 566 (1974).
Tsuchimoto, T., Ishii, Y., Uchino, H., Inoue, S.: Paroxysmal nocturnal haemoglobinuria with chromosome abnormalities: possible preleukaemia. Lancet i, 617–618 (1970).
Tsuchimoto, T., Kamada, N., Kawakami, H., Shimuzu, N., Ozono, N., Yunki, I., Ishii, Y., Uchino, H.: Blood protein anomaly and chromosome abnormality with special reference to a case of IgA type of myeloma with chromosome abnormalities (in Japanese). Saishin Igaku 23, 2678–2684 (1968).
Tsung, S.H., Heckman, M.G.: Klinefelter syndrome, immunological disorders, and malignant neoplasm. Arch. Pathol. 98, 351–354 (1974).
Tumilowicz, J.J., Nichols, W.W., Cholon, J.J., Greene, A.E.: Definition of a continuous human cell line derived from neuroblastoma. Cancer Res. 30, 2110–2118 (1970).
Turleau, C., Trebuchet, C., Finaz, C., de Nava, C., Grouchy, J. de: Étude cytogénetique do 200 cas de leucémie myéloïde chronique avec Ph[1]. Proceedings of the IV International Congress on Human Genetics, Paris, p. 179 (1971).
Turpin, R., Lejeune, J.: Leucemies et cancers. In R. Turpin, J. Lejeune (eds.): Les Chromosomes Humains, pp. 181–215. Paris: Gunthier Villars (1965).
Turri, C.M., Lorand, I.G., Ribeiro, A.F., Brandalise, S.R.: Leucémia mieloide crónica en un lactante. Sangre (Barc.) 21, 854–859 (1976).
Tursz, T., Flandrin, G., Brouet, J.C., Seligmann, M.: Coexistence d'un myélome et d'une leucemie granuleuse en l'absence de tout traitment. Nouv. Rev. Fr. Hematol. 14, 693–704 (1974).
Twomey, J.J., Levin, W.C., Melnick, M.B., Trobaugh,

F.E., Alligood, J.W.: Laboratory studies on a family with a father and son affected by acute leukaemia. Blood 29, 920–930 (1967).

Uchino, H.: Cytogenetic study on leukemia in atomic bomb survivors. Acta Haematol. Jap. 31, 818–824 (1968).

Ueda, N., Uenaka, H., Akematsu, T., Sugiyama, T.: Parallel distribution of sister chromatid exchanges and chromosome aberrations. Nature 262, 581–583 (1976).

Ueyama, Y., Morita, K., Kondo, Y., Sato, N., Asano, S., Ohsawa, N., Sakurai, M., Nagumo, F., Iijima, K., Tamaoki, N.: Direct and serial transplantation of a Ph^1+ve human myeloblastoid tumour into nude mice. Br. J. Cancer 36, 523–527 (1977).

Ulrich, H.: The incidence of leukemia in radiologists. N. Engl. J. Med. 234, 45–46 (1946).

Urasinski, I.: Proliferation of bone marrow cells in Di Guglielmo's disease. Folia Haematol. (Leipz.) 103, 211–215 (1976).

Utakoji, T.: Differential staining patterns of human chromosomes treated with potassium permanganate. Nature 239, 168–169 (1972).

Utian, H.L., Plit, M.: Ataxia telangiectasia. J. Neurol. Neurosurg. Psychiatry 27, 38–40 (1964).

Utsinger, P.D., Yount, W.J., Fuller, C.R., Logue, M.J., Orringer, E.P.: Hairy cell leukemia: B-lymphocyte and phagocytic properties. Blood 49, 19–27 (1977).

Utsumi, K.R., Yoshida, T.O.: A chromosome survey on tissue culture cells derived from nasopharyngeal cancer. Gann 10, 291–295 (1971).

Uyeda, C.K., Davis, H.J., Jones, H.W., Jr.: Nuclear protrusions and giant chromosome anomalies in cervical neoplasia. Acta Cytol. 10, 331–334 (1966).

Vagner-Capodano, A.M.: Aspects chromosomiques des leucémies myéloïdes chroniques en acutisation. Nouv. Rev. Fr. Hematol. 12, 87–100 (1972).

Vagner-Capodano, A.M., Detolle, P., Daumas, B., Arroyo, H., Aubert, L.: Syndrome lympho-proliferatif avec dysglobulinémie atypique et anomalies chromosomiques. Nouv. Rev. Fr. Hematol. 10, 541–551 (1970).

Vallejos, C.S., Trujillo, J.M., Cork, A., Bodey, G.P., McCredie, K.B., Freireich, E.J.: Blastic crisis in chronic granulocytic leukemia: Experience in 39 patients. Cancer 34, 1806–1812 (1974).

Van Bao, T., Szabo, I., Ruzicska, P., Czeizel, A.: Chromosome aberrations in patients suffering acute organic phosphate insecticide intoxication. Humangenetik 24, 33–57 (1974).

Van Beneden, E.: Contribution à l'historie de la vésicule germinative et du premier noyau embryonnaire. Bull. Acad. R. Belg. 41, (1883).

Vance, J.C., Cervenka, J., Ullman, S., Kersey, J.H., Sabad, A., Green, N.: Immunological and chromosomal studies in a patient with Sézary syndrome. Arch. Dermatol. 113, 1417–1423 (1977).

Van Den Berghe, H.: The Ph^1-chromosome: Translocation to chromosome 9. Lancet ii, 1030–1031 (1973).

Van Den Berghe, H., Fryns, J.P., Verresen, H.: Congenital leukaemia with 46,XX,t (Bq+,Cq−) cells. J. Med. Genet. 9, 468–470 (1972).

Van Den Berghe, H., Cassiman, J.-J., David, G., Fryns, J.P., Michaux, J.-L., Sokal, G.: Distinct haematological disorder with deletion of long arm of No. 5 chromosome. Nature 251, 437–438 (1974).

Van Den Berghe, H., David, G., Broeckaert-van Orshoven, A., Louwagie, A., Verwilghen, R.: Unusual Ph^1 translocation in acute myelocytic leukemia. N. Engl. J. Med. 299, 360 (1978a).

Van Den Berghe, H., David, G., Broeckaert-van Orshoven, A., Louwagie, A., Verwilghen, R., Casteels-van Daele, M., Eggermont, E., Eeckels, R.: A new chromosome anomaly in acute lymphoblastic leukemia (ALL). Hum. Genet. 46, 173–180 (1979).

Van Den Berghe, H., David, G., Michaux, J.L., Sokal, G.: Non random chromosome anomalies in myeloproliferative disorders in man. In S. Armendares, R. Lisker (eds.): V International Congress of Human Genetics, pp. 163–164, Abstract 431. Held in Mexico, Oct. 10–15 (1976a).

Van Den Berghe, H., David, G., Michaux, J.-L., Sokal, G., Verwilghen, R.: 5q− acute myelogenous leukemia. Blood 48, 624–625 (1976b).

Van Den Berghe, H., Louwagie, A., Broeckaert-van Orshoven, A., David, G., Verwilghen, R.: Chromosome analysis in two unusual malignant blood disorders presumably induced by benzene. Blood 53, 558–566 (1979a).

Van Den Berghe, H., Louwagie, A., Broeckaert-van Orshoven, A., David, G., Verwilghen, R., Michaux, J.L., Ferrant, A., Sokal, G.: Chromosome abnormalities in acute promyelocytic leukemia (APL). Cancer 43, 558–562 (1979b).

Van Den Berghe, H., Louwagie, A., Broeckaert-van Orshoven, A., David, G., Verwilghen, R., Michaux, J.L., Ferrant, A., Sokal, G.: Philadelphia chromosome in multiple myeloma. J. Natl. Cancer Inst. (in press) (1979c).

Van Den Berghe, H., Michaux, J.L., Sokal, G.: Chromosome anomalies in malignant lymphoproliferative disorders. Helsinki Chromosome Conference, Aug. 29–31, p. 215 (abstract) (1977a).

Van Den Berghe, H., Michaux, J.-L., Sokal, G., Verwilghen, R.: The 5q−anomaly: A short synopsis.

Helsinki Chromosome Conference, Aug. 29–31, p. 216 (abstract) (1977b).
Van Den Berghe, H., Parloir, C., David, G., Michaux, J.L., Ferrant, A., Sokal, G.: A new characteristic karyotypic anomaly in lymphoproliferative disorders. Cancer (in press) (1979a).
Van Den Berghe, H., Parloir, C., Gosseye, S., Englebienne, V., Cornu, G., Sokal, G.: Variant translocation in Burkitt lymphoma. Cancer Genet. Cytogenet. (in press) (1979b).
Van der Riet-Fox, M.F., Retief, A.E., van Niekerk, W.A.: Chromosome changes in twenty human neoplasms studied with banding. Cancer (submitted) (1979).
Van Hemel, J.O.: Chromosomes in patients with ataxia telangiectasia (AT). In S. Armendares, R. Lisker (eds.): V International Congress of Human Genetics, p. 164, Abstract 433. Held in Mexico, Oct. 10–15 (1976).
Van Kempen, C.: A case of retinoblastoma, combined with severe mental retardation and a few other congenital anomalies, associated with complex aberrations of the karyotype. Maandschr. Kindergeneeskd. 34, 92–95 (1966).
Varela, M.A., Sternberg, W.H.: Preanaemic state in Fanconi's anaemia. Lancet ii, 566–567 (1967).
Vass, L., Sellyei, M.: The missing Y chromosome and human leukaemia. Lancet i, 550–551 (1973a).
Vass, L., Sellyei, M.: Heated Giemsa solution for producing more consistent bands on mammalian chromosomes. Humangenetik 18, 81–83 (1973b).
Vass, L., Sellyei, M., Katalin, G.M., Bordás, E.: Tumorok és Y chromosoma Y-chromatin és Y chromosoma hiánya férfiak szolíd daganataiban. Orv. Hetil. 114, 2770–2774 (1973).
Vassilakos, P., Kajii, T.: Hydatidiform mole: Two entities. Lancet i, 259 (1976).
Vassilakos, P., Riotton, G., Kajii, T.: Hydatidiform mole: Two entities. Am. J. Obstet. Gynecol. 127, 167–170 (1977).
Valazquez, S., Arechavala, E., Márquez-Monter, H.: Translocation 21-14 in a case of chronic myeloid leukemia with chromosome Ph^1. Vth International Congress of Human Genetics, Mexico City, October 10–15, p. 164 (abstract) (1976).
Venuat, A.M., Dutrillaux, B., Rosenfeld, C.: A late clonal evolution of a human leukemia line: Sequential cytogenetic studies. Eur. J. Cancer 13, 123–130 (1977).
Venuat, A.M., Rosenfeld, C., Maman, M.T., Mathe, G.: Ten cases of acute leukaemia revealed by cytogenetic studies to be the blastic crisis of CML with Ph^1+. In S. Armendares, R. Lisker (eds.): V International Congress of Human Genetics, p. 165, Abstract 435. Held in Mexico, Oct. 10–15 (1976).

Verresen, H., Van Den Berghe, H., Creemers, J.: Mosaic trisomy in phenotypically normal mother of mongol. Lancet i, 526 (1964).
Verhest, A., Lustman, F., Wittek, M., van Schoubroeck, F., Naets, P.: Cytogenetic evidence of clonal evolution in 5q- anemia. Biomedicine 27, 211–212 (1977).
Verhest, A., van Schoubroeck, F.: Philadelphia-chromosome-positive preleukaemic state. Lancet ii, 1386 (1973).
Verhest, A., van Schoubroeck, F., Gangji, D.: Primary or secondary blastic involvement of extramedullary sites in acute CGL. In S. Armendares, R. Lisker (eds.): V International Congress of Human Genetics, p. 165, Abstract 436. Held in Mexico, Oct. 10–15 (1976a).
Verhest, A., van Schoubroeck, F., Wittek, M., Naets, J.P., Denolin-Reubens, R.: Specificity of the 5q– chromosome in a distinct type of refractory anemia. J. Natl. Cancer Inst. 56, 1053–1054 (1976b).
Verma, R.S., Dosik, H.: An improved method for photographing fluorescent human chromosomes. J. Microsc. 108, (1976).
Verma, R.S., Dosik, H.: The value of reverse banding in detecting bone marrow chromosomal abnormalities: Translocation between chromosomes 1, 9 and 22 in a case of chronic myelogenous leukemia (CML). Am. J. Hematol. 3, 171–175 (1977).
Verma, R.S., Lubs, H.A.: A simple R banding technique. Am. J. Hum. Genet. 27, 110–117 (1975a).
Verma, R.S., Lubs, H.A.: Variation in human acrocentric chromosomes with acridine orange reverse banding. Humangenetik 30, 225–235 (1975b).
Verma, R.S., Lubs, H.A.: Description of the banding patterns of human chromosomes by acridine orange reverse banding (RFA) and comparison with the Paris banding diagram. Clin. Genet. 9, 553–557 (1976a).
Verma, R.S., Lubs, H.A.: Additional observations on the preparation of R banded human chromosomes with acridine orange. Can. J. Genet. Cytol. 18, 45–50 (1976b).
Verma, R.S., Dosik, H., Lubs, H.A.: Size variation and polymorphisms of the short arm of human acrocentric chromosomes determined by R-banding by fluorescence using acridine orange (RFA). Hum. Genet. 38, 231–244 (1977a).
Verma, R.S., Dosik, H., Lubs, H.A., Jr.: Demonstration of color and size polymorphisms in human acrocentric chromosomes by acridine orange reverse banding. J. Hered. 68, 262–263 (1977b).
Verma, R.S., Dosik, H., Lubs, H.A.: Frequency of RFA colour polymorphisms of human acrocentric chromosomes in Caucasians: Interrela-

tionship with QGQ polymorphisms. Ann. Hum. Genet. 41, 257–267 (1977c).
Verma, R.S., Peakman, D.C., Robinson, A., Lubs, H.A.: Comparison of G-, Q-, and R-banding in 28 cases of chromosomal abnormalities. Cytogenet. Cell Genet. 16, 479–486 (1976).
Verschaeve, L., Kirsch-Volders, M., Hens, L., Susanne, C.: Chromosome distribution in phenyl mercury acetate exposed subjects and in age-related controls. Mutat. Res. 57, 335–347 (1978).
Verwilghen, R., Van Den Berghe, H.: Spontaneous rejection of an abnormal cell clone monitored by chromosome analysis in two patients with exposure to organic solvents. International Society of Hematology, European African Division, 4th Meeting, Istanbul, Abstract #241 (1977).
Vigliani, E.C., Forni, A.: Benzene chromosome changes and leukemia. J. Occup. Med. 11, 148–149 (1969).
Vincent, P.C., Sinha, S., Neate, R., den Dulk, G., Turner, B.: Chromosome abnormalities in a mongol with acute myeloid leukaemia. Lancet i, 1328–1329 (1963).
Vincent, P.C., Sutherland, R., Bradley, M., Lind, D., Gunz, F.W.: Marrow culture studies in adult acute leukemia at presentation and during remission. Blood 49, 903–912 (1977).
Vincent, P.C., Vandenburg, R.A., Neate, R., Nicholis, A.: Chromosome analysis in the diagnosis of malignant effusions: Report of a case. Med. J. Aust. 1, 155–157 (1964).
Visfeldt, J.: Primary polycythaemia. 2. Types of chromosome aberrations in 21 clones found in bone marrow samples from 50 patients. Acta Pathol. Microbiol. Scand. [A] 79, 513–523 (1971).
Visfeldt, J., Lundwall, F.: Alteration of karyotypic profiles in human cancerous effusion following treatment with antineoplastic drug. Acta Pathol. Microbiol. Scand. [A] 78, 551–555 (1970).
Visfeldt, J., Mortensen, E.: Chromosome aberrations in Fanconi's anaemia. Acta Pathol. Microbiol. Scand. [A] 78, 545–550 (1970).
Visfeldt, J., Franzén, S., Nielsen, A., Tribukait, B.: Primary polycythaemia. 3. Studies on the significance of the history of the disease and of the treatment for the development of clones in bone marrow cells. Acta Pathol. Microbiol. Scand. [A] 81, 195–203 (1973).
Visfeldt, J., Franzén, S., Tribukait, B.: Cytogenetic studies in myeloproliferative syndrome and some atypical myeloid disorders. Acta Pathol. Microbiol. Scand. [A] 78, 80–84 (1970).
Visfeldt, J., Franzén, S., Tribukait, B.: Primary polycythaemia: Correlations between the histologic appearances and the chromosome pattern of the bone marrow cells during the disease. Acta Radiol. [Ther.] (Stockh) 10, 86–114 (1971).
Visfeldt, J., Jensen, G., Hippe, E.: On thorotrast leukemia: Evolution of clone of bone marrow cells with radiation-induced chromosome aberrations. Acta Pathol. Microbiol. Scand. [A] 83, 373–378 (1975).
Visfeldt, J., Povlsen, C.O., Rygaard, J.: Chromosome analyses of human tumours following heterotransplantation to the mouse mutant *nude*. Acta Pathol. Microbiol. Scand. [A] 80, 169–176 (1972).
Vogel, W., Bauknecht, T.: Differential chromatid staining by *in vivo* treatment as a mutagenicity test system. Nature 260, 448–449 (1976).
Vogel, W., Faust, J., Schmid, M., Siebers, J.-W.: On the relevance of non-histone proteins to the production of Giemsa banding patterns on chromosomes. Humangenetik 21, 227–236 (1973).
Vogel, W., Schempp, W., Puel, V.: Silver-staining specificity in metaphases after incorporation of 5-bromodeoxyuridine (BUDR). Hum. Genet. 40, 199–203 (1978).
Volk, S.L.R., Monteleone, P.L., Knight, W.A.K.: Chromosomes in AILD. N. Engl. J. Med. 292, 975 (1975).
Volkova, M.A., Fleischman, E.W.: Blastic crisis of chronic myeloid leukemia. Probl. Gematol. Pereliv. Krovi 4, 12–18 (1973).
Volpe, R., Knowlton, T.G., Foster, A.D., Conen, P.E.: Testicular feminization: A study of two cases, one with a seminoma. Can. Med. Assoc. J. 98, 438–445 (1968).
Von Mayenburg, J., Ehlers, G., Muhlbauer, W., Steuer, G.: Condylomata acuminata gigantea (Buschke-Loewenstein-tumors) with carcinoma of the vulva. Z. Hautkr. 52, 869–884 (1977).
Vormittag, W.: Chromosomenuntersuchungen bei akuter Virushepatitis. Humangenetik 15, 44–65 (1972).
Vormittag, V.M., Kuhbock, J., Peskar, B.M.: Myeloproliferative disorder with hyperdiploid stemline (47,XY,mar+). Wien. Klin. Wochenschr. 85, 747–749 (1973).
Vrba, M.: The X-chromatin and the chromosome number of the cells from human malignant melanomas of the eye. Neoplasma 21, 577–581 (1974).

Waardenburg, P.J.: Mongolismus (Mongoloid idiotie). Das menschliche Auge und seine Erbanlagen. Bibliogr. Genet. 7, 44–48 (1932).
Waddell, C.C., Brown, J.A., Zelnick, P.W.: Chronic myelogenous leukemia with 45 XO Philadelphia chromosome karyotype and prolonged survival. Ann. Intern. Med. (in press) (1979).
Wager, J., Granberg, I., Vass, L., Auer, G.: A benign serous ovarian cystadenoma studied by chromo-

some and quantitative DNA analysis. Acta Cytol. 21, 774–776 (1978).
Wagner, D.: DNA-messungen an mitotischen Zellen bei Dysplasie und Carcinoma *in situ* der Cervix uteri. Verh. Dtsch. Ges. Pathol. 57, 211–213 (1973).
Wagner, D., Sprenger, E., Blank, M.H.: DNA-content of dysplastic cells of the uterine cervix. Acta Cytol. 16, 517–522 (1972).
Wagner, H.P., Tönz, O., Greyerz-Gloor, R.D.: Congenital lymphoid leukaemia. Case report with chromosomal studies. Helv. Paediatr. Acta 23, 591–610 (1968).
Wahrman, J., Schaap, T., Robinson, E.: Manifold chromosome abnormalities in leukaemia. Lancet i, 1098–1100 (1962).
Wahrman, J., Schaap, T., Robinson, E.: Chromosome studies in leukemia. *In* E. Goldschmidt, (ed.): *The Genetics of Migrant and Isolate Populations*, pp. 304–306, Baltimore: Williams and Wilkins (1963).
Wahrman, J., Voss, R., Shapiro, T., Ashkenazi, A.: The Philadelphia chromosome in two children with chronic myeloid leukemia. Israel J. Med. Sci. 3, 380–391 (1967).
Wakabayashi, M., Ishihara, T.: Cytological studies of tumors. XXVI. Chromosome analysis of a human mammary carcinoma. Cytologia (Tokyo) 23, 341–348 (1958).
Wake, N., Takagi, N., Sasaki, M.: Androgenesis as a cause of hydatidiform mole. J. Natl. Cancer Inst. 60, 55–57 (1978).
Wakonig-Vaartaja, R.: A human tumour with identifiable cells as evidence for the mutation theory. Br. J. Cancer 16, 616–618 (1962).
Wakonig-Vaartaja, R.: Chromosomes in gynaecological malignant tumours. Aust. N.Z. J. Obstet. Gynaecol. 3, 170–177 (1963).
Wakonig-Vaartaja, R.: Chromosomes in precancerous lesions and in carcinoma of the uterine cervix. Bull. N.Y. Acad. Med. 45, 22–38 (1969).
Wakonig-Vaartaja, T.: Chromosomes in invasive carcinoma and related lesions of the uterine cervix. *In* H.R.K. Barber, E.A. Graber (eds.): *Gynecological Oncology*, Proceedings, Symposium, Lenox Hill Hospital, New York, N.Y., 19–23 May, 1969, pp. 80–88. Baltimore: Williams and Wilkins (1970).
Wakonig-Vaartaja, R., Auersperg, N.: Cytogenetics of gynecologic neoplasms. Clin. Obstet. Gynecol. 13, 813–830 (1970).
Wakonig-Vaartaja, R., Hughes, D.T.: Chromosomal anomalies in dysplasia, carcinoma-in-situ, and carcinoma of cervix uteri. Lancet ii, 756–759 (1965).
Wakonig-Vaartaja, R., Hughes, D.T.: Chromosome studies in 36 gynaecological tumours: of the cervix, corpus uteri, ovary, vagina and vulva. Eur. J. Cancer 3, 263–277 (1967).
Wakonig-Vaartaja, R., Kirkland, J.A.: A correlated chromosomal and histopathologic study of preinvasive lesions of the cervix. Cancer 18, 1101–1112 (1965).
Wakonig-Vaartaja, T., Helson, L., Baren, A., Koss, L.G., Murphy, M.L.: Cytogenetic observations in children with neuroblastoma. Pediatrics 47, 839–843 (1971).
Walach, N., Hochman, A.: Male breast cancer. Oncology 29, 181–189 (1974).
Walbaum, R., François, P., Farriaux, J.-P., Woillez, M.: Un cas de rétinoblastome bilatéral avec monosomie 13 partielle (q12→q14). Hum. Genet. 44, 219–226 (1978).
Wald, N., Fatora, S.R., Herron, J.M., Preston, K., Jr., Li, C.C., Davis, L.: Status report on automated chromosome aberration detection. J. Histochem. Cytochem. 24, 156–159 (1976).
Waldeyer, W.: Karyokinesis and its relation to the process of fertilization. Q. J. Microsc. Sci. 30, 159–281 (1890).
Walker, B.E., Boothroyd, E.R.: Chromosome numbers in somatic tissues of mouse and man. Genetics 39, 210–219 (1954).
Walker, C.E., Whittingham, H.E.: Further observations upon the resemblance between the cells of malignant growths and those of normal gametogenic tissue. J. Pathol. Bacteriol. 16, 185–198 (1911).
Walker, S., Wheeler, M.V., Hatton, S., Gibbs, T.J.: Chromosome studies in acute leukaemia. *Helsinki Chromosome Conference*, Aug. 29–31, p. 218 (abstract) (1977).
Wallace, C.: Chromosomal analysis of two cases of macroglobulinaemia. South Afr. Med. J. 37, 1096 (1963).
Walther, J.-U., Stengel-Rutkowski, S., Murken, J.-D.: Observations with G-banding of human chromosomes. Humangenetik 25, 49–51 (1974).
Wang, H.C., Fedoroff, S.: Banding in human chromosomes treated with trypsin. Nature [New Biol.] 235, 52–53 (1972).
Wang, M.Y.F., Desforges, J.F.: The Philadelphia chromosome and galactose-1-phosphate uridyl transferase. Blood 29, 790–799 (1967).
Wantzin, G.L., Jensen, M.K.: The induction of chromosome abnormalities by melphalen in rat bone marrow cells. Scand. J. Haematol. 11, 135–139 (1973).
Warburton, D., Bluming, A.: A "Philadelphia-like" chromosome derived from the Y in a patient with refractory dysplastic anemia. Blood 42, 799–804 (1973).
Warburton, D., Shah, N.: A 9/11 translocation in a child with Ph^1-negative chronic myelogenous leukemia. J. Pediatr. 88, 599–601 (1976).
Warburton, D., Miller, D.A., Miller, O.J., Breg, W.R., de Capoa, A., Shaw, M.W.: Distinction between chromosome 4 and chromosome 5 by

replication pattern and length of long and short arms. Am. J. Hum. Genet. 19, 399–415 (1967).

Warkany, J., Schubert, W.K., Thompson, J.N.: Chromosome analysis in mongolism (Langdon-Down syndrome) associated with leukemia. N. Engl. J. Med. 268, 1–4 (1963).

Warkany, J., Weinstein, E.D., Soukup, S.W., Rubenstein, J.H., Curless, M.D.: Chromosomal analysis in a children's hospital. Pediatrics 33, 290–305 (1964).

Warren, J.C., Erkman, B., Cheatum, S., Holman, G.: Hilus-cell adenoma in a dysgenetic gonad with XX/XO mosaicism. Lancet i, 141–143 (1964).

Warren, S., Meisner, L.: Chromosomal changes in leukocytes of patients receiving irradiation therapy. J.A.M.A. 193, 351–358 (1965).

Wasastjerna, C., Vuorinen, E., Lehtinen, M., Ikkala, E., Huhtula, E.: The cytochemical β-glucuronidase reaction in the differential diagnosis of acute leukaemias. Acta Haematol (Basel) 53, 277–284 (1975).

Wasser, J.S., Yolken, R., Miller, D.R., Diamond, L.: Congenital hypoplastic anemia (Diamond-Blackfan syndrome) terminating in acute myelogenous leukemia. Blood 51, 991–995 (1978).

Watanabe, S.: Present status of somatic effects in atomic bomb survivors living in Hiroshima. Acta Haematol. Jap. 27, 121–130 (1964).

Watson, W.A.F., Petrie, J.C., Galloway, D.B., Bullock, I., Gilbert, J.C.: In vivo cytogenetic activity of sulphonylurea drugs in man. Mutat. Res. 38, 71–80 (1976).

Watson-Williams, E.J., Lewis, J.P., Taplett, J.K., Walling, P.A.: Chromosomal variation in chronic granulocytic leukemia (CGL): Effect on prognosis. Proc. Am. Soc. Clin. Oncol. 18, 293 (1977).

Watt, J.L., Page, B.M.: Reciprocal translocation and the Philadelphia chromosome. Hum. Genet. 42, 163–170 (1978).

Watt, J.L., Hamilton, P.J., Page, B.M.: Variation in the Philadelphia chromosome. Hum. Genet. 37, 141–148 (1977).

Watt, J.L., Page, B.M., Davidson, R.J.L.: Cytogenetic study of 10 cases of infectious mononucleosis. Clin. Genet. 12, 267–274 (1977).

Waxdale, M.J., Basham, T.Y.: B and T-cell stimulatory activities of multiple mitogens from pokeweed. Nature 251, 163–164 (1974).

Weatherall, D.J., Walker, S.: Changes in the chromosome and haemoglobin patterns in a patient with erythro-leukaemia. J. Med. Genet. 2, 212–219 (1965).

Webb, T., Harding, M.: Chromosome complement and SV40 transformation of cells from patients susceptible to malignant disease. Br. J. Cancer 36, 583–591 (1977).

Webb, T., Harnden, D.G., Harding, M.: The chromosome analysis and susceptibility to transformation by Simian virus 40 of fibroblasts from ataxia-telangiectasia. Cancer Res. 37, 997–1002 (1977).

Weber, G.: Enzymology of cancer cells. N. Engl. J. Med. 296, 486–493; 541–551 (1977).

Weerd-Kastelein, E.A. de, Keijzer, W., Rainaldi, G., Bootsma, D.: Induction of sister chromatid exchanges in xeroderma pigmentosum cells after exposure to ultraviolet light. Mutat. Res. 45, 253–261 (1977).

Wehinger, H., Niederhoff, H., Bethge, D., Kunzer, W., Hoehn, H., Cohen, G.: Myelomonocytic leukemia with Philadelphia-positive and Philadelphia-negative cell lines in early childhood. Klin. Wochenschr. 53, 431–436 (1975).

Weichselbaum, R.R., Nove, J., Little, J.B.: Skin fibroblasts from a D-deletion type retinoblastoma patient are abnormally X-ray sensitive. Nature 266, 726–727 (1977).

Weichselbaum, R.R., Nove, J., Little, J.B.: Deficient recovery from potentially lethal radiation damage in ataxia telangiectasia and xeroderma pigmentosum. Nature 271, 261–262 (1978a).

Weichselbaum, R.R., Nove, J., Little, J.B.: X-ray sensitivity of diploid fibroblasts from patients with hereditary or sporadic retinoblastoma. Proc. Natl. Acad. Sci. USA 75, 3962–3964 (1978b).

Weiden, P.L., Lerner, K.G., Gerdes, A., Heywood, J.D., Fefer, A., Thomas, E.D.: Pancytopenia and leukemia in Hodgkin's disease: Report of three cases. Blood 42, 571–577 (1973).

Weiner, L.: A family with high incidence leukemia and unique Ph¹ chromosome findings. Blood 26, 871 (1965).

Weinfeld, A., Westin, J., Ridell, B., Swolin, B.: Polycythaemia vera terminating in acute leukaemia. A clinical, cytogenetic and morphologic study in 8 patients treated with alkylating agents. Scand. J. Haematol. 19, 255–272 (1977a).

Weinfeld, A., Westin, J., Swolin, B.: Ph¹-negative eosinophilic leukaemia with trisomy 8. Case report and review of cytogenetic studies. Scand. J. Haematol. 18, 413–420 (1977b).

Weinstein, A.W., Weinstein, E.D.: A chromosomal abnormality in acute myeloblastic leukemia. N. Engl. J. Med. 268, 253–255 (1963a).

Weinstein, E.D., Weinstein, A.W.: Chromosomal abnormalities in children with acute leukemia. J. Pediatr. 63, 473–474 (1963b).

Weintraub, H., Groudine, M.: Chromosomal subunits in active genes have an altered conformation. Science 193, 848–856 (1976).

Weisblum, B.: Why centric regions of quinacrine-treated mouse chromosomes show diminished fluorescence. Nature 246, 150–151 (1973).

Weisblum, B., Haseth, P.L. de: Quinacrine, a chromosome stain specific for deoxyadenylate-

deoxythymidylate-rich regions in DNA. Proc. Natl. Acad. Sci. USA 69, 629–632 (1972).
Weisblum, B., Haseth, P.L. de: Nucleotide specificity of the quinacrine staining reaction for chromosomes. Chromosomes Today 4, 35–51 (1973).
Weise, W., Biittner, H.H.: Chromosome analysis of a primary carcinoma of the fallopian tube. Humangenetik 15, 196–197 (1972).
Weiss, A.F., Portmann, R., Fischer, H., Simon, J., Zang, K.D.: Simian virus 40-related antigens in three human meningiomas with defined chromosome loss. Proc. Natl. Acad. Sci. USA 72, 609–613 (1975).
Weiss, A.F., Zang, K.D., Birkmayer, G., Miller, F.: SV40 related Papova viruses in human meningiomas. Acta Neuropathol. (Berl.) 34, 171–174 (1976).
Weiss, R.B., Brunning, R.D., Kennedy, B.J.: Lymphosarcoma terminating in acute myelogenous leukemia. Cancer 30, 1275–1278 (1972).
Weisswichert, P., Schroeder, T.M., Stahl-Mauge, Ch.: Temperature dependent chromosome instability in human lymphocytes from healthy individuals. In H. Altmann (ed.): *International Symposium of the "IGEM,"* pp. 21–34. Held in Vienna, Dec. 1–2, 1975, "DNA-Repair and Late Effects." Instut für Biologie, Forschungszentrum Seibersdorf./Wien, Switzerland (1976).
Welch, J.C., Fiander, D.C., Main, M.: Evidence for chromosomal instability in retinoblastoma patients and band assignment of the occasionally associated chromosome 13 deletion. *Program of the American Society on Human Genetics,* 28th Annual Meeting, p. 113. Held in San Diego, Calif., Oct. 19–22 (abstract) (1977).
Welch, J.P., Lee, C.L.Y.: Non-random occurrence of 7–14 translocations in human lymphocyte cultures. Nature 255, 241–242 (1975).
Wennström, J., Schröder, J.: A t(13q14q) family with the translocation and a Philadelphia chromosome in one member. Humangenetik 20, 71–73 (1973).
Wertelecki, W., Shapiro, J.R.: 45,XO Turner's syndrome and leukaemia. Lancet i, 789–790 (1970).
Wertelecki, W., Fraumeni, J.F., Jr., Mulvihill, J.J.: Nongonadal neoplasia in Turner's syndrome. Cancer 26, 485–488 (1970).
Westermark, B., Ponten, J., Hugosson, R.: Determinants for the establishment of permanent tissue culture lines from human gliomas. Acta Path. Microbiol. Scand. [A] 81, 791–805 (1973).
Westin, J.: Chromosome abnormalities after chlorambucil therapy of polycythaemia vera. Scand. J. Haematol. 17, 197–204 (1976).
Westin, J., Weinfeld, A.: The development of chromosome abnormalities in polycythemia vera (PV). Results from a follow-up study. *Helsinki Chromosome Conference,* Aug. 29–31, p. 219 (abstract) (1977).
Westin, J., Wahlström, J., Swolin, B.: Chromosome studies in untreated polycythaemia vera. Scand. J. Haematol. 17, 183–196 (1976).
Wetter, O., Reis, H.E., Weert, M.v.d.: T-cell properties of leukaemic plasma cells. Lancet ii, 909–910 (1973).
Whang, J., Frei, E., III, Tjio, J.H., Carbone, P.P., Brecher, G.: The distribution of the Philadelphia chromosome in patients with chronic myelogenous leukemia. Blood 22, 664–673 (1963).
Whang-Peng, J.: Chromosome studies in various neoplasms of domestic animals. Natl. Cancer Inst. Monogr. 32, 117–120 (1969).
Whang-Peng, J.: Banding in leukemia: Techniques and implications. J. Natl. Cancer Inst. 58, 3–8 (1977).
Whang-Peng, J., Bennett, J.M.: Cytogenetic studies in metastatic neuroblastoma. Am. J. Dis. Child. 115, 703–708 (1968).
Whang-Peng, J., Broder, S., Lee, E., Young, R.C.: Unusual clonal evolution in a case of chronic myelogenous leukemia. Acta Haematol. (Basel) 56, 345–354 (1976).
Whang-Peng, J., Canellos, G.P., Carbone, P.P., Tjio, J.H.: Clinical implications of cytogenetic variants in chronic myelocytic leukemia (CML). Blood 32, 755–766 (1968).
Whang-Peng, J., Chretien, P., Knutsen, T.: Polyploidy in malignant melanoma. Cancer 25, 1216–1223 (1970).
Whang-Peng, J., Freireich, E.J., Oppenheim, J.J., Frei, E., III, Tjio, J.H.: Cytogenetic studies in 45 patients with acute lymphocytic leukemia. J. Natl. Cancer Inst. 42, 881–897 (1969).
Whang-Peng, J., Gerber, P., Knutsen, T.: So-called C marker chromosome and Epstein-Barr virus. J. Natl. Cancer Inst. 45, 831–839 (1970).
Whang-Peng, J., Gralnick, H.R., Johnson, R.E., Lee, E.C., Lear, A.: Chronic granulocytic leukemia (CGL) during the course of chronic lymphocytic leukemia (CLL): Correlation of blood, marrow, and spleen morphology and cytogenetics. Blood 43, 333–339 (1974).
Whang-Peng, J., Gralnick, H.R., Knutsen, T., Brereton, H., Chang, P., Schechter, G.P., Lessin, L.: Small F chromosome in myelo- and lymphoproliferative diseases. Leukemia Res. 1, 19–30 (1977).
Whang-Peng, J., Henderson, E.S., Knutsen, T., Freireich, E.J., Gart, J.J.: Cytogenetic studies in acute myelocytic leukemia with special emphasis on the occurrence of the Ph^1 chromosome. Blood 36, 448–457 (1970).
Whang-Peng, J., Knutsen, T.A., Lee, E.C.: Dicentric Ph^1 chromosome. J. Natl. Cancer Inst. 51, 2009–2012 (1973).

Whang-Peng, J., Knutsen, T., Lee, E.C., Leventhal, B.: Acquired XO/XY clones in bone marrow of a patient with paroxysmal nocturnal hemoglobinuria (PNH). Blood 47, 611–619 (1976a).

Whang-Peng, J., Knutsen, T., O'Donnell, J.F., Brereton, H.D.: Acute non-lymphocytic leukemia and acute myeloproliferative syndrome following radiation therapy for non-Hodgkin's lymphoma and chronic lymphocytic leukemia: Cytogenetic studies. Cancer (submitted) (1979).

Whang-Peng, J., Knutsen, T., Ziegler, J., Leventhal, B.: Cytogenetic studies in acute lymphocytic leukemia: Special emphasis in long-term survival. Med. Pediatr. Oncol. 2, 333–351 (1976b).

Whang-Peng, J., Lee, E.C., Knutsen, T.A.: Genesis of the Ph^1 chromosome. J. Natl. Cancer Inst. 52, 1035–1036 (1974).

Whang-Peng, J., Lee, E., Knutsen, T., Canellos, G.: The significance of chromosomal abnormalities in myelofibrosis with or without myeloid metaplasia. American Society of Hematology, Nineteenth Annual Meeting, Boston, Mass., Dec. 4–7, Abstract #270 (1976).

Whang-Peng, J., Lee, E., Knutsen, T., Chang, P., Nienhuis, A.: Cytogenetic studies in patients with myelofibrosis and myeloid metaplasia. Leukemia Res. 2, 41–56 (1978).

Whang-Peng, J., Lutzner, M., Edelson, R., Knutsen, T.A.: Cytogenetic studies and clinical implications in patients with Sézary syndrome. Cancer 38, 861–867 (1976).

White, L., Cox, D.: Chromosome changes in a rhabdomyosarcoma during recurrence and in cell culture. Br. J. Cancer 21, 684–693 (1967).

White, M.J.D.: *The Chromosomes,* 5th edition. New York: Methuen; Wiley, (1961).

Whitelaw, D.M.: Chromosome complement of lymph node cells in Hodgkin's disease. Can. Med. Assoc. J. 101, 74–81 (1969).

Whittaker, J.A., Davies, P., Khurshid, M.: Absence of the Y chromosome in patients with chronic granulocytic leukaemia. Acta Haematol. (Basel) 54, 350–357 (1975).

Wieman, H.L.: Chromosomes in man. Am. J. Anat. 14, 461–471 (1912).

Wiener, F., Fenyo, E.M., Klein, G.: Fusion of tumour cells with host cells. Nature [New Biol.] 238, 155–159 (1972).

Wiener, F., Klein, G., Harris, H.: The analysis of malignancy by cell fusion. IV. Hybrids between tumour cells and a malignant L cell derivative. J. Cell Sci. 12, 253–261 (1973).

Wiener, F., Klein, G., Harris, H.: The analysis of malignancy by cell fusion. V. Further evidence of the ability of normal diploid cells to suppress malignancy. J. Cell Sci. 15, 177–183 (1974).

Wiener, F., Ohno, S., Spira, J., Haran-Ghera, N., Klein, G.: Chromosome changes (trisomy 15) associated with tumor progression in T-cell leukemia induced by 17,12-dimethylbenz(a)anthracene (DMBA). Int. J. Cancer 22, 447–453 (1978a).

Wiener, F., Ohno, S., Spira, J., Haran-Ghera, N., Klein, G.: Chromosome changes (trisomies #15 and #17) associated with tumor progression in leukemias induced by radiation leukemia virus. J. Natl. Cancer Inst. 61, 227–237 (1978b).

Wiener, S., Reese, A.B., Hyman, G.A.: Chromosome studies in retinoblastoma. Arch. Ophthalmol. 69, 311–313 (1963).

Wiggans, R.G., Jacobson, R.J., Fialkow, P.J., Woolley, P.V., III, MacDonald, J.S., Schein, P.S.: Probable clonal origin of acute myeloblastic leukemia following radiation and chemotherapy of colon cancer. Blood 52, 659–663 (1978).

Wiik, A.S., Paulson, O.B., Sørensen, C.M., Visfeldt, J.: A study of ^{51}Cr-labelled platelets and the chromosomal pattern in a case of primary haemorrhagic thrombocythaemia. Acta Haematol. (Basel) 46, 177–187 (1971).

Wilbanks, G.D.: In vivo and in vitro "markers" of human cervical intraepithelial neoplasia. Cancer Res. 36, 2485–2494 (1976).

Wilbanks, G.D., Richart, R.M., Terner, J.Y.: DNA content of cervical intraepithelial neoplasia studied by two-wavelength Feulgen cytophotometry. Am. J. Obstet. Gynecol. 98, 792–799 (1967).

Willemse, C.H.: A patient suffering from Turner's syndrome and acromegaly. Acta Endocrinol. (Kbh) 39, 204–212 (1962).

Williams, D.M., Scott, C.D., Beck, T.M.: Premature chromosome condensation in human leukemia. Blood 47, 687–693 (1976).

Williamson, H.O., Underwood, P.B., Kreutner, A., Jr., Rogers, J.F., Mathur, R.S., Pratt-Thomas, H.R.: Gonadoblastoma: Clinicopathologic correlation in six patients. Am. J. Obstet. Gynecol. 126, 579–585 (1976).

Wilson, C.B., Barker, M.: Studies of malignant brain tumors in cell culture. Ann. N.Y. Acad. Sci. 159, 480–489 (1969).

Wilson, C.B., Kaufman, L., Barker, M.: Chromosome analysis of glioblastoma multiforme. Neurology 20, 821–828 (1970).

Wilson, J.D., Nossal, G.J.V.: Identification of human T and B lymphocytes in normal peripheral blood and in chronic lymphocytic leukaemia. Lancet ii, 788–791 (1971).

Wilson, M.G., Ebbin, A.J., Towner, J.W., Spencer, W.H.: Chromosome anomalies in patients with retinoblastoma. Clin. Genet. 12, 1–8 (1977).

Wilson, M.G., Melnyk, J., Towner, J.W.: Retinoblastoma and deletion D(14) syndrome. J. Med. Genet. 6, 322–327 (1969).

Wilson, M.G., Towner, J.W., Fujimoto, A.: Retinoblastoma and D-chromosome deletions. Am. J. Hum. Genet. 25, 57–61 (1973).
Winge, Ö.: Zytologische Untersuchungen über die Natur maligner Tumoren. I. "Crown Gall" der Zuckerrübe. Z. Zellforsch. Mikrosk. Anat. 6, 397–423 (1927).
Winge, Ö.: Zytologische Untersuchungen über die Natur maligner Tumoren. II. Teerkarzinome bei Mäusen. Z. Zellforsch. Mikrosk. Anat. 10, 683–735 (1930).
Winiwarter, H. von: Études sur la spermatogénèse humaine. I. Cellule de Sertoli. Arch. Biol. (Liege) 27, 91–127 (1912a).
Winiwarter, H. von: Étude sur la spermatogénèse humaine. II. L'Hétérochromosome et les mitoses de l'épithélium séminal. Arch. Biol. (Liege) 27, 128–188 (1912b).
Winiwarter, H. de, Oguma, K.: Recherches sur quelques points controversés de la spermatogénèse humaine. C.R. Assoc. Anat. Turin, pp. 1–8 (1925).
Winiwarter, H. de, Oguma, K.: Nouvelles recherches sur la spermatogénèse humaine. Arch. Biol. (Liege) 36, 99–166 (1926).
Winiwarter, H. de, Oguma, K.: La formule chromosomiale humaine (à propos de deux travaux récents). Arch. Biol. (Liege) 40, 541–553 (1930).
Winkelman, J., Fisher, H., Melnyk, J.: Testing for Ph[1] chromosome on mailed specimens of peripheral blood. Vox Sang. 20, 335–339 (1971).
Winkelstein, A., Goldberg, L.S., Tishkoff, G.H., Sparkes, R.S.: Leukocyte alkaline phosphatase and the Philadelphia chromosome. Arch. Intern. Med. 119, 291–296 (1967).
Winkelstein, A., Sparkes, R.S., Craddock, C.G.: Trisomy of group C in a myeloproliferative disorder. Report of a case. Blood 27, 722–733 (1966).
Winter, G.C.B., Osmond, C.B., Yoffey, J.M., Mahy, D.J.: Leucocyte cultures with phytohaemagglutinin in chronic lymphocytic leukaemia. Lancet ii, 563–565 (1964).
Winters, A.J., Benirschke, K., Whalley, P., MacDonald, P.C.: Mosaicism and lack of fluorescence of Y chromosome. Obstet. Gynecol. 46, 367–370 (1975).
Wintrobe, M.M.: *Clinical Hematology*. Philadelphia: Lea and Febiger (1974).
Wisniewski, L., Korsak, E.: Cytogenetic analysis in two cases of lymphoma. Comparison between lymphosarcoma and reticulosarcoma. Cancer 25, 1081–1086 (1970).
Wisniewski, L., Lech, H.: Cytogenetics of the neoplastic cell. Folia Med. Bialostoc. 3, 17–21 (1974).
Witkowski, R.: Chromosomenbefunde bei Melanomzellen. Derm. Wochenschr. 156, 345–347 (1970).
Witkowski, R., Anger, H.: Premature chromosome condensation in irradiated man. Hum. Genet. 34, 65–68 (1976).
Witkowski, R., Zabel, H.: Chromosomenaberrationen der Tumorzellen aus dem Aszites bei Ovarial-Karzinom. Acta Biol. Med. Ger. 16, 95–105 (1966).
Witkowski, R., Zabel, H.: Chromosomenuntersuchungen an Metastasen eines malignen Melanoms beim Menschen. Derm. Wochenschr. 158, 255–260 (1972).
Wolf, U., Merker, H., Böckelmann, W.: Chromosomenuntersuchungen bei chronisch-myeloischer Leukämie. Klin. Wochenschr. 44, 12–19 (1966).
Wolfe, S.L.: The fine structure of isolated chromosomes. J. Ultrastruct. Res. 12, 104–112 (1965a).
Wolfe, S.L.: Fine structure of isolated metaphase chromosomes. Exp. Cell Res. 37, 45–53 (1965b).
Wolff, C., Gilly, C., Mouriquand, C.: Collecting human chromosomes for whole-mount electron microscopy. Stain Technol. 49, 133–136 (1974).
Wolff, L.J., Richardson, S.T., Neiburger, J.B., Neiburger, R.G., Irwin, D.S., Baehner, R.L.: Poor prognosis of children with acute lymphocytic leukemia and increased B cell markers. J. Pediatr. 89, 956–958 (1976).
Wolff, S.: Sister chromatid exchange. Annu. Rev. Genet. 11, 183–201 (1977).
Wolff, S., Rodin, B., Cleaver, J.E.: Sister chromatid exchanges induced by mutagenic carcinogens in normal and xeroderma pigmentosum cells. Nature 265, 347–349 (1977).
Wolman, S.R., Horland, A.A.: Genetics of tumor cells. In F.F. Becker (ed.): *Cancer—A Comprehensive Treatise,* Vol. III, *The Biology of Cancer,* Chapter 7, pp. 155–198. New York: Plenum Press (1975).
Wolman, S.R., Swift, M.: Bone marrow chromosomes in Fanconi's anaemia. Case report. J. Med. Genet. 9, 473–474 (1972).
Woodliff, H.J.: *Leukaemia Cytogenetics.* London: Lloyd-Luke; (1971).
Woodliff, H.J.: Leukaemia cytogenetics. Med. J. Aust. 1, 495–499 (1975).
Woodliff, H.J., Cohen, G.: Cytogenetic studies in chronic lymphocytic leukemia. I. A study of 40 patients. Med. J. Aust. 1, 970–974 (1972).
Woodliff, H.J., Dougan, L.: Double Ph[1] chromosomes in leukaemia. Lancet i, 771 (1966).
Woodliff, H.J., Onesti, P.: Chronic granulocytic leukaemia: Studies of a patient and his twin. Med. J. Aust. 3, 397–403 (1967).
Woodliff, H.J., Chipper, L., Gallon, W.: Cytogenetic

studies in myelofibrosis. Experience in Western Australia, 1963–1970. Med. J. Aust. 1, 1075–1078 (1972).
Woodliff, H.J., Dougan, L., Goodall, D.W.: A statistical approach to the Philadelphia chromosome. Nature 207, 504–505 (1965).
Woodliff, H.J., Dougan, L., Onesti, P.: Cytogenetic studies in twins, one with chronic granulocytic leukaemia. Nature 211, 533 (1966).
Woodliff, H.J., Leong, A., Stenhouse, N.S.: Cytogenetic studies in chronic lymphocytic leukaemia. 2. Statistical studies of G-group chromosomes. Med. J. Aust. 1, 1027–1030 (1972).
Woodliff, H.J., Onesti, P., Dougan, L.: Karyotypes in thrombocythaemia. Lancet i, 114–115 (1967a).
Woodliff, H.J., Onesti, P., Goodall, D.W.: Further statistical studies on the human G-group chromosome with particular reference to chronic granulocytic leukaemia. Med. J. Aust. 2, 159–162 (1967b).
Woodruff, J.D., Davis, H.J., Jones, H.J., Recio, R.G., Salini, R., Park, I.J.: Correlative investigative technics of multiple anaplasias in the lower genital canal. Obstet. Gynecol. 33, 609–616 (1969).
Worm, A.-M., Pedersen-Bjergaard, J.: Chronic myelocytic leukaemia presenting in blastic transformation. Scand. J. Haematol. 18, 288–291 (1977).
Wu, M.: Chromosomes of cancer cells. Vopr. Onkol. 7, 9–16 (1961).
Wurster-Hill, D.H.: Chromosome banding and its application to cancer research. *In* H. Busch (ed.): *Methods in Cancer Research*, pp. 1–41. New York: Academic Press (1975).
Wurster-Hill, D.H., Maurer, L.H.: Cytogenetic diagnosis of cancer: Abnormalities of chromosomes and polyploid levels in the bone marrow of patients with small cell anaplastic carcinoma of the lung. J. Natl. Cancer Inst. 61, 1065–1075 (1978).
Wurster-Hill, McIntyre, O.R.: Chromosome studies in polycythemia vera. Virchows Arch. B Cell Pathol. 29, 39–44 (1978).
Wurster-Hill, D.H., Cornwell, G.G., McIntyre, O.R.: Chromosomal aberrations and neoplasm—a family study. Cancer 33, 72–81 (1974).
Wurster-Hill, D.H., McIntyre, O.R., Cornwell, G.G., III., Maurer, L.H.: Marker-chromosome 14 in multiple myeloma and plasma-cell leukaemia. Lancet ii, 1031 (1973).
Wurster-Hill, D.H., McIntyre, O.R., Cornwell, G.G., III: Chromosome studies in myelomatosis. Virchows Arch. B Cell Pathol. 29, 93–97 (1978).
Wurster-Hill, D., Whang-Peng, J., McIntyre, O.R., Hsu, L.Y.F., Hirschhorn, K., Modan, B., Pisciotta, A.V., Pierce, R., Balcerzak, S.P., Weinfeld, A., Murphy, S.: Cytogenetic studies in polycythemia vera. Semin. Hematol. 13, 13–32 (1976).
Wyandt, H.E., Hecht, F.: Human Y-chromatin. I. Dispersion and condensation. Exp. Cell Res. 81, 453–461 (1973a).
Wyandt, H.E., Hecht, F.: Human Y-chromatin. II. DNA replication. Exp. Cell Res. 81, 462–467 (1973b).
Wyandt, H.E., Iorio, R.J.: Human Y-chromatin. III. The nucleolus. Exp. Cell Res. 81, 468–473 (1973).
Wyandt, H.E., Wysham, D.G., Minden, S.K., Anderson, R.S., Hecht, F.: Mechanisms of Giemsa banding of chromosomes. I. Giemsa-11 banding with azure and eosin. Exp. Cell Res. 102, 85–94 (1976).

Xavier, R.G., Prolla, J.C., Bemvenuti, G.A., Kirsner, J.B.: Tissue cytogenetic studies in chronic ulcerative colitis and carcinoma of the colon. Cancer 34, 684–695 (1974).

Yam, L.T., Castoldi, G.L., Garvey, M.B., Mitus, W.J.: Functional cytogenetic and cytochemical study of the leukemic reticulum cells. Blood 32, 90–101 (1968).
Yam, L.T., Ki, C.Y., Necheles, T.F., Katayama, I.: Pseudoeosinophilia, eosinophilic endocarditis and eosinophilic leukemia. Am. J. Med. 53, 193–202 (1972).
Yamada, K., Furasawa, S.: Preferential involvement of chromosomes No. 8 and No. 21 in acute leukemia and preleukemia. Blood 47, 679–686 (1976).
Yamada, K., Sandberg, A.A.: Preliminary notes on the chromosomes of eleven primary tumors of colon. Proc. Jap. Acad. 42, 168 (1966a).
Yamada, K., Sandberg, A.A.: Chronology and pattern of human chromosome replication. III. Autoradiographic studies on cells from cancer effusions. J. Natl. Cancer Inst. 36, 1057–1073 (1966b).
Yamada, K., Shinohara, T., Furasawa, S.: Absence of the Y chromosome in a case of acute leukemia. Chrom. Inf. Serv. 7, 20–22 (1966).
Yamada, K., Takagi, N., Sandberg, A.A.: Chromosomes and causation of human cancer and leukemia. II. Karyotypes of human solid tumors. Cancer 19, 1879–1890 (1966).
Yamada, K., Yoshioka, M., Oami, H.: A 14q+ marker and a late replicating chromosome #22 in a brain tumor. J. Natl. Cancer Inst. 59, 1193–1195 (1977).
Yamamoto, K., Mizuno, F., Matsuo, T., Tanaka, A., Nonoyama, M., Osato, T.: Epstein-Barr virus and human chromosomes: Close association of the resident viral genome and the expression of the virus-determined nuclear antigen (EBNA) with the presence of chromosome 14 in human/

mouse hybrid cells. Proc. Natl. Acad. Sci. USA 75, 5155–5159 (1978).
Yamamoto, T., Rabinowitz, Z., Sachs, L.: Identification of chromosomes that control malignancy. Nature [New Biol.] 243, 247–250 (1973).
Yeung, K.-Y., Trowbridge, A.A.: Idiopathic acquired sideroblastic anemia terminating in acute myelofibrosis. Cancer 39, 359–365 (1977).
Yoneda, C., Herick, W.V.: Tissue culture cell strain derived from retinoblastoma. Am. J. Ophthalmol. 55, 987–992 (1963).
Yoshida, M.C., Ikeuchi, T., Sasaki, M.: Differential staining of parental chromosomes in interspecific cell hybrids with a combined quinacrine and 33258 Hoechst technique. Proc. Jap. Acad. 51, 184–187 (1975).
Yoshida, Tomizo: The Yoshida sarcoma, an ascites tumor. Gann 40, 1–21 (1949).
Yoshida, Toshihide: Cytologic studies on cancer. V. Heteroplastic transplantation of the Yoshida sarcoma with special regard to the behavior of tumor cells. Gann 43, 35–43 (1952).
Yosida, T.H.: Relation between chromosomal alteration and development of tumors. Jap. J. Genet. 41, 439–451 (1966).
Yosida, T.H.: Is aging of tumor cells related to the alteration of stemline karyotypes? Proc. Jap. Acad. 48, 268–273 (1972).
Yosida, T.H.: Chromosomal alterations and development of experimental tumors. In E. Grundmann (ed.): Handbuch der Allgemeinen Pathologie, pp. 677–753, Berlin, Heidelberg, New York: Springer-Verlag (1975).
Yosida, T.H., Imai, H.T., Moriwaki, K.: Chromosomal alteration and development of tumors. XXI. Cytogenetic studies of primary plasma-cell neoplasms induced in BALB/c mice. J. Natl. Cancer Inst. 45, 411–418 (1970).
Yosida, T.H., Tabata, T.: Karyological study on the normal somatic and malignant tumor cells in the human. Natl. Inst. Genet. Jap. Ann. Rep. 8, 25–26 (1957).
Young, D.: SV40 transformation of cells from patients with Fanconi's anaemia. Lancet i, 294–295 (1971a).
Young, D.: The susceptibility to SV40 virus transformation of fibroblasts obtained from patients with Down's syndrome. Eur. J. Cancer 7, 337–339 (1971b).
Young, R.K., Cailleau, R.M., Mackay, B., Reeves, W.J., Jr.: Establishment of epithelial cell line MDA-MB-157 from metastatic pleural effusion of human breast carcinoma. In Vitro 9, 239–245 (1974).
Young, R.R., Austen, K.F., Moser, H.W.: Abnormalities of serum γ-Ia globulin and ataxia telangiectasia. Medicine (Baltimore) 43, 423–433 (1964).

Yu, C.W., Borgaonkar, D.S., Bolling, D.R.: Break points in human chromosomes. Hum. Hered. 28, 210–225 (1978).
Yung, M., Blatnik, D., Yung, F.: Chromosomes of malignant tissue. Hum. Chrom. Newsl. 13, 27–28 (1964).
Yunis, A.A., Arimura, G.K., Russin, D.J.: Human pancreatic carcinoma (MIA PACA-2) in continuous culture: Sensitivity to asparaginase. Int. J. Cancer 19, 128–135 (1977).
Yunis, J.J. (ed.): Human Chromosome Methodology, 2nd edition. New York: Academic Press, (1974).
Yunis, J.J.: High resolution of human chromosomes. Science 191, 1268–1270 (1976).
Yunis, J.J. (ed.): Molecular Structure of Human Chromosomes. New York, San Francisco, London: Academic Press (1977).
Yunis, J.J., Chandler, M.E.: The chromosomes of man—clinical and biologic significance. A review. Am. J. Pathol. 88, 466–495 (1977a).
Yunis, J.J., Chandler, M.E.: High-resolution chromosome analysis in clinical medicine. In M. Stefanini, A. Hossaini (eds.): Progress in Clinical Pathology, Vol. VII, pp. 267–288. New York: Grune and Stratton (1977b).
Yunis, J.J. Ramsay, N.: Retinoblastoma and subband deletion of chromosome 13. Am. J. Dis. Child. 132, 161–163 (1978).
Yunis, J.J., Sanchez, O.: G-banding and chromosome structure. Chromosoma 44, 15–23 (1973).
Yunis, J.J., Sanchez, O.: The G-banded prophase chromosomes of man. Humangenetik 27, 167–172 (1975).
Yunis, J.J., Yasmineh, W.G.: Heterochromatin, satellite DNA, and cell function. Science 174, 1200–1209 (1971).
Yunis, J.J., Hook, E.B., Mayer, M.: Deoxyribonucleic-acid replication pattern in trisomy D_1. Lancet ii, 935–937 (1964).
Yunis, J.J., Hook, E.B., Mayer, M.: Identification of the mongolism chromosome by DNA replication analysis. Am. J. Hum. Genet. 17, 191–201 (1965).
Yunis, J.J., Kuo, M.T., Saunders, G.F.: Localization of sequences specifying messenger RNA to light-staining G-bands of human chromosomes. Chromosoma 61, 335–344 (1977).
Yunis, J.J., Sawyer, J.R., Ball, D.W.: The characterization of high-resolution G-banded chromosomes in man. Chromosoma 67, 293–307 (1978a).
Yunis, J.J., Sawyer, J.R., Ball, D.W.: Characterization of banding patterns of metaphase-prophase G-banded chromosomes and their use in gene mapping. IV International Conference on Gene Mapping, Winnipeg, 1977; Cytogenet. Cell Genet. 22, 679–683 (1978b).
Yunis, J.J., Tsai, M.Y., Willey, A.M.: Molecular orga-

nization and function of the human genome. *In* J.J. Yunis: *Molecular Structure of Human Chromosomes*, pp. 1–33. New York, San Francisco, London: Academic Press (1977).

Zaccaria, A., Tura, S.: A chromosomal abnormality in primary thrombocythemia. N. Engl. J. Med. 298, 1422–1423 (1978).

Zaccaria, A., Baccarani, M., Barbieri, E., Tura, S.: Differences in marrow and spleen cell karyotype in early chronic myeloid leukaemia. Eur. J. Cancer 11, 123–126 (1974).

Zaccaria, A., Barbieri, E., Mantovani, W., Tura, S.: Chromosome radiation-induced aberrations in patients with Hodgkin's disease. Possible correlation with second malignancy? Boll. 1st Sieroter. Milan. 57, 76–83 (1978).

Zaccaria, A., Ricci, P., Baccarani, M., Tura, S.: Chromosome studies in paroxysmal nocturnal haemoglobinuria. Acta Haematol. (Basel) 50, 350–356 (1973).

Zack, G.W., Rogers, W.E., Latt, S.A.: Automatic measurement of sister chromatid exchange frequency. J. Histochem. Cytochem. 25, 741–753 (1977).

Zack, G.W., Spriet, J.A., Latt, S.A., Granlung, G.H., Young, I.T.: Automatic detection and localization of sister chromatid exchanges. J. Histochem. Cytochem. 24, 168–177 (1976).

Zajac, B.A., Kohn, G.: Epstein-Barr virus antigens, marker chromosome, and interferon production in clones derived from cultured Burkitt tumor cells. J. Natl. Cancer Inst. 45, 399–406 (1970).

Zakharov, A.F., Lelikova, G.P.: Local disturbance of chromosome spiralization in human testicular tumors. Tsitologiia 14, 1092–1097 (1972).

Zakharova, A.V., Korenevskaya, M.I., Mokeyeva, R.A., Klimova, N.F., Kovaleva, L.G., Bartaschuk, Y.I., Terent'Yeva, E.I.: Chromosomal patterns of patients with leukemia, erythremia or myeloic fibrosis (in Russian). Probl. Gematatol. Pereliv. Krovi 16 (5), 3–8 (1971).

Zang, K.D., Back, E.: Quantitative studies on the arrangement of human metaphase chromosomes. I. Individual features in the association pattern of the acrocentric chromosomes of normal males and females. Cytogenetics 7, 455–470 (1968).

Zang, K.D., Singer, H.: Chromosomal constitution of meningiomas. Nature 216, 84–85 (1967).

Zang, K.D., Singer, H.: The cytogenetics of human tumors. Angew. Chem. [Engl.] 7, 709–718 (1968a).

Zang, K.D., Singer, H.: Verleichende morphologische und zytogenetische Untersuchungen an Meningeomen. Zbl. Neurol. Psychiatr. 192, 131 (1968b).

Zang, K.D., Singer, H.: Uniform chromosome loss in human primary meningiomas. *In: 4th International Congress on Human Genetics, Paris,* 1971; abstracts of papers presented. Amsterdam: Excerpta Medica (1971) (Excerpta Med. Int. Congr. Ser. No. 233).

Zang, K.D., Weiss, A., Zankl, H.: Cytogenetic and immunological evidence for SV-40 related Papova viruses in human meningiomas. *In* S. Armendares, R. Lisker (eds.): *V International Congress on Human Genetics*, p. 170, Abstract 449. Held in Mexico, Oct. 10–15 (1976).

Zang, K.D., Zankl, H.: True monosomy 22 in human meningiomas. Am. J. Hum. Genet. 27, 96A (1975).

Zang, K.D., Zankl, H., Weiss, A.F.: Correlation of clinical, cytological and cytogenetical data on a consecutive series of 150 human meningiomas. Helsinki Chromosome Conference, Helsinki, Finland, August 29–31, p. 220 (abstract) (1977).

Zankl, H., Bernhardt, S.: Combined silver staining of the nucleolus organizing regions and Giemsa banding in human chromosomes. Hum. Genet. 37, 79–80 (1977).

Zankl, H., Huwer, H.: Are NORs easily translocated to deleted chromosomes? Hum. Genet. 42, 137–142 (1978).

Zankl, H., Love, R.: Isonucleolinosis in cell cultures of human meningiomas. Cancer Res. 36, 1074–1076 (1976).

Zankl, H., Zang, K.D.: Structural variability of the normal human karyotype. Humangenetik 13, 160–162 (1971a).

Zankl, H., Zang, K.D.: Cytological and cytogenetical studies on brain tumors. III. Ph¹-like chromosomes in human meningiomas. Humangenetik 12, 42–49 (1971b).

Zankl, H., Zang, K.D.: Chromosomenanomalien und Tumorentstehung. Klin. Wochenschr. 56, 7–16 (1978a).

Zankl, H., Zang, K.D.: Quantitative studies on the arrangement of human metaphase chromosomes. V. The association pattern of acrocentric chromosomes in human meningiomas after the loss of G and D chromosomes. Hum. Genet. 40, 149–155 (1978b).

Zankl, H., Zang, K.D.: The role of acrocentric chromosomes in nucleolar organization. I. Correlation between the loss of acrocentric chromosomes and a decrease in the number of nucleoli in meningioma cell cultures. Virchows Arch. [Zellpathol.] 11, 251–256 (1972a).

Zankl, H., Zang, K.D.: Cytological and cytogenetical studies on brain tumors. IV. Identification of the missing G chromosome in human meningiomas as No. 22 by fluorescence technique. Humangenetik 14, 167–169 (1972b).

Zankl, H., Zang, K.D.: Quantitative studies on the arrangement of human metaphase chromo-

somes. IV. The association frequency of human acrocentric marker chromosomes. Humangenetik 23, 259–265 (1974).
Zankl, H., Seidel, H., Zang, K.D.: Cytological and cytogenetical studies on brain tumors. V. Preferential loss of sex chromosomes in human meningiomas. Humangenetik 27, 119–128 (1975a).
Zankl, H., Seidel, H., Zang, K.D.: Sex-chromosome loss in human tumours. Lancet i, 221 (1975b).
Zankl, H., Singer, H., Zang, K.D., Büscher, H., Kofler, W.: Cytological and cytogenetical studies on brain tumors. II. Hyperdiploidy, a rare event in human primary meningiomas. Humangenetik 11, 253–257 (1971).
Zankl, H., Stengel-Rutkowski, S., Zang, K.D.: The role of acrocentric chromosomes in nucleolar organization. III. Constancy of the nucleolar area independent of the number of acrocentric chromosomes in meningioma cell cultures. Virchows Arch. [Zellpathol.] 13, 113–118 (1973).
Zankl, H., Weiss, A.F., Zang, K.D.: Cytological and cytogenetical studies on brain tumors. VI. No evidence for a translocation in 22-monosomic meningiomas. Humangenetik 30, 343–348 (1975).
Zanoio, L.: Karyological study of ascitic liquid of ovarian tumor after antineoplastic drug treatment. Attual. Obstet. Ginecol. 22, 45–48 (1976).
Zara, C., Fraccaro, M., Gerli, A., Mannini, A., Tiepolo, L.: L'effetto del trattamento con ciclofosfamide sulla dinamica cromosomica di popolazioni di cellule tumorali in vivo. Clin. Obstet. Ginecol. 71 (Suppl.), 103 (1966).
Zárate, A., Karchmer, S., Esteves, R., Teter, J.: Gonadal dysgenesis with isolated clitoromegaly, XY karyotype, and bilateral gonocytoma with massive Leydig cell hyperplasia. Am. J. Obstet. Gynecol. 110, 875–878 (1971).
Zarrabi, M.H., Rosner, F., Bennet, J.M.: Acute leukemia and non-Hodgkin lymphoma. N. Engl. J. Med. 298, 280 (1978).
Zech, L.: Investigation of metaphase chromosomes with DNA-binding fluorochromes. Exp. Cell Res. 58, 463 (1969).
Zech, L.: Non-random distribution of chromosome abnormalities in tissues of neoplastic origin. *Abstract, 11th International Cancer Congress,* Florence, p. 644 (1974).
Zech, L., Haglund, U.: A recurrent structural aberration, t(7;14), in phytohemagglutinin-stimulated lymphocytes. Hereditas 89, 69–73 (1978).
Zech, L., Gahrton, G., Killander, D., Franzén, S., Haglund, U.: Specific chromosomal aberrations in polycythemia vera. Blood 48, 687–696 (1976).

Zech, L., Haglund, U., Nilsson, K., Klein, G.: Characteristic chromosomal abnormalities in biopsies and lymphoid-cell lines from patients with Burkitt and non-Burkitt lymphomas. Int. J. Cancer 17, 47–56 (1976).
Zech, L., Lindsten, J., Udén, A.-M., Gahrton, G.: Monosomy 7 in two adult patients with acute myeloblastic leukaemia. Scand. J. Haematol. 15, 251–255 (1975).
Zellweger, H., Khalifeh, R.R.: Ataxia telangiectasia. Report of two cases. Helv. Pediatr. Acta 18, 267–279 (1963).
Zidar, B.L., Winkelstein, A., Whiteside, T.L., Shadduck, R.K., Zeigler, Z., Smith, W.I., Rabin, B.S., Krause, J.R., Lee, R.E.: Hairy cell leukaemia: Seven cases with probable B-lymphocyte origin. Br. J. Haematol. 37, 455–465 (1977).
Zittoun, R.: Subacute and chronic myelomonocytic leukaemia: A distinct haematological entity. Br. J. Haematol. 31, 1–7 (1976).
Zittoun, R., Bernadou, A., Bilski-Pasquier, G., Bousser, J.: Les leucémies myélo-monocytaires subaigues. Étude de 27 cas et revue de la littérature. Semin. Hop. Paris 48, 1943–1956 (1972).
Zuelzer, W.W.: The limits of the cytogenetic method in acute leukemia. Blood Cells 3, 581–588 (1977).
Zuelzer, W.W.: Childhood leukemia—a perspective. Johns Hopkins Med. J. 142, 115–127 (1978).
Zuelzer, W.W., Inoue, S., Thompson, R.I., Ottenbreit, M.J.: Long-term cytogenetic studies in acute leukemia of children: The nature of relapse. Am. J. Hematol. 1, 143–190 (1976).
Zuelzer, W.W., Thompson, R.I., Mastrangelo, R.: Evidence for a genetic factor related to leukemogenesis and congenital anomalies: Chromosomal aberrations in pedigree of an infant with partial D trisomy and leukemia. J. Pediatr. 72, 367–376 (1968).
Zur Hausen, H.: Chromosomal changes of similar nature in seven established cell lines derived from the peripheral blood of patients with leukemia. J. Natl. Cancer Inst. 38, 683–696 (1967).
Zur Hausen, H., Schulte-Holthausen, H.: Presence of EB virus nucleic acid homology in a "virus-free" line of Burkitt tumour cells. Nature 227, 245–248 (1970).
Zur Hausen, H., Diehl, V., Wolf, H., Schulte-Holthausen, H., Schneider, U.: Occurrence of Epstein-Barr virus in genomes in human lymphoblastoid cell lines. Nature [New Biol.] 237, 189–190 (1972).
Zussman, W.V., Khan, A., Shayesteh, P.: Congenital leukemia: Report of a case with chromosome abnormalities. Cancer 20, 1227–1233 (1967).

Subject Index

A

A. *See* page 182
AA. *See* page 192
Abbreviations
  clinical, 182
  methods, 99
Aberrations, chromatid type, 57; *see also* Chromosome aberrations
  chromosome type, 52–53, 58–60
Abnormal mitosis in cervix uteri, 528
Acentric chromosome, 11, 54; *see also* Fragment
Achromatic lesion. *See* Gap
Acrocentric, 11, 42, 76
  autosomes, 28
Actinomycin D, 115–116
Acute granulocytic leukemia. *See* Acute myeloblastic leukemia (AML)
Acute leukemia (AL), 119, 263–276; *see also* Leukemia, acute
  aneuploidy in, 276
  cellular markers, 276
  chromosomes, morphology, 263
    modal number, 264
  complicating other conditions, 319–320, 381–383
  in congenital disorders, 266–268
  in Fanconi's anemia, 157–158
  in Hodgkin's disease, 381–383 *(Table 72)*
  marrow transplantation in, 275–276
  near-haploidy in, 341–344
  nuclear blebs in, 276
  Ph^1 in, 267–275 *(Table 49)*
Acute lymphoblastic leukemia (ALL), 268–270, 276–294
  chromosome abnormalities, 284–289 *(Table 54)*
  congenital abnormalities, 292–294
  karyotypic evolution, 290
  modal number, 278 *(Table 53* and *Figure 81)*
  Ph^1-positive, 268–270
  6q− in, 228–290
  surface markers, 277 *(Table 52)*
Acute lymphocytic leukemia. *See* Acute lymphoblastic leukemia (ALL)
Acute megakaryocytic leukemia, 340
Acute monocytic leukemia (AMoL), 340
Acute myeloblastic leukemia (AML), 270, 294–322
  chromosomal changes in, 294–308 *(Tables 55 and 56)*, 310–314
  chromosome patterns in, 310–319
  complicating other conditions, 319–320
  genesis of, 309–310
  karyotypic classification (MAKA, MIKA, AA, AN), 314–319, 327–329
  missing Y in, 320–322
  monosomy-7 in, 322
  Ph^1 in, 270, 308–309
  translocation 8;21, 295, 302 *(Table 57)*

Subject Index

[cont.]
  trisomy-8 in, 322
Acute myelomonocytic leukemia (AMMoL), 337–340 (Table 65)
Acute promyelocytic leukemia (APL), 335 (Table 64), 337
Adenocarcinoma. See Cancer
Adenoma
  breast, 462
  bronchus, 462, 518
  colon, 473
  multiple sebaceous, 526
  ovary, 462
  pituitary, 549
  stomach, 470
  thyroid, 461
Adenovirus, 145
  SV40, 146–147
Adjacent segregation, 195, 431 (Figures 58 and 119)
Adrenal glands, tumors of, 518–520
  pheochromocytoma, DNA value, 519–520
  Sipple's syndrome, 520
African lymphoma. See Burkitt lymphoma
AIL. See Angioimmunoblastic lymphadenopathy
AL. See Acute leukemia
Aleukemic leukemia. See Acute leukemia
Alimentary tract, tumors of, 468–485
ALL. See Acute lymphoblastic leukemia
AML. See Acute myeloblastic leukemia
AMMoL. See Acute myelomonocytic leukemia
AMoL. See Acute monocytic leukemia
AN. See page 182
Anaphase, 11, 75
  lagging, 7, 17
Anemia, 155–158, 174–180
  Fanconi's, 155–158; see also Fanconi's anemia (FA)
  idiopathic aplastic (IAA), 174
  megaloblastic, 363–365
  pure red cell aplasia, 175–176
  refractory, 176–179
  sideroblastic, 179–180
Aneuploidy, 11, 76, 276
  in cervix uteri, 533–534
Angioimmunoblastic lymphadenopathy (AILD), 409 (Table 77)
Animal tumors, 3–4, 440–447; see also Tumors, animal
  chemical carcinogens, 441–442
  contagious tumors, 444
  DMBA-induced sarcomas, 445–447
  minimal deviation hepatoma, 444–445
  oncogenic viruses, 442–444
  spontaneous tumors and leukemia, 440–441
Aniridia. See Wilms' tumor
ANLL. See Acute nonlymphocytic leukemia
Anomalies, autosomal, 119–124
  in cancer patients, 149–151
  congenital and leukemia, 266–268 (Table 47)
  relation to cancer and leukemia, 132–134
  sex chromosome, 124–132; see also Sex chromosome anomalies
Antigen, EB viral, 146
APL. See Acute promyelocytic leukemia
Appendix, cancer of, 473–484
Ascites, 3
  sarcoma, 3–4
AT. See Ataxia telangiectasia
Ataxia telangiectasia (AT), 161–165
  chromosome abnormalities, 161–165
  chromosome #14 in, 164
  clinical features, 161
  clone, 163–164
  cytogenetic studies on malignant cells, 161–163
  fibroblasts, 162
  immunological deficiencies, 161
  lymphocytes, 161–162
  marrow cells, 162
  occurrence of cancer, 161
Atom bomb survivors, 139 (Table 17a)
Autoradiography, 4, 33, 80–83, 109
Autosome, 11
Avian sarcoma virus, 147–148

B

Band, definition of, 35, 43
Banding
  definition of, 11, 43
  high-resolution, 109–116
  pattern, 34
  rearrangements, detailed system, 47–51
    short system, 47–51
  techniques, 4, 29–30, 33–41, 51, 103
    code, 51
    fluorescent, 35
  variations, 39 (Table 2a)
Basic cell carcinoma, 524–525, 526
Basal cell nevus syndrome, 168
Basophilic leukemia, 230–231
Benign tumors, 459–462
Benzene, leukemia, 141–142
Binucleate cells, 11
Bivalents, 11
BL. See Burkitt lymphoma
Bladder, cancer of, 503–511
  chromosomal findings in, 503–504
  chromosome number and morphology, 505–509
  heteromorphism, 509–511
  spectrophotometric DNA studies, 504–505, 508
  studies with banding, 509–511
  Y-chromosome, 509
Blastic phase (or crisis) (CML), 235–249 (Tables 45 and 46)

double Ph¹, 248–249
Blastic transformation. *See* Chronic myelocytic leukemia (CML)
Bloom's syndrome (BS), 158–161
  cancer in, 158
  chromosome breakage, 158–160
  clinical features, 158
  cytogenetics, 158
  DNA replication in, 161
  leukemia, 158
  SCE frequency, 160–161
  skin, 160
BM. *See* Bone marrow
Bone, tumors of, 524, 525
  giant cell, 461
Bone marrow (BM), 102–103, 184–185
  chromosome preparation, 102–103
  harvesting of cells, 102–103
  transplantation and chromosomes, 275–276
Boveri theory, 5–6
Bowen's disease of vagina, 501
BP. *See* Blastic phase of CML
Brain tumors, 535–550
  double minute chromosomes (DMS), 550–558 *(Table 88)*
  ependymomas, 546–547
  homogeneously staining regions (HSR), 558–559
  malignant gliomas (astrocytic), 543–546
  medulloblastomas, 547
  meningiomas, 535–543
  metastatic tumors, nervous system, 549–550
  neurinoma, 548–549
  neuroblastomas, 547
  oligodendrogliomas, 546
  optical gliomas, 548
  pituitary adenomas, 549
  retinoblastoma, 548
Break, chromatid, 11, 12 *(Figure 6)*
  chromosome, 13, 20, 45, 46–47, 58
  isochromatid, 54
  isolocus, 58
Breakage syndromes, 153–158
  ataxia telangiectasia (AT), 161–165
  basal cell nevus syndrome, 168
  Bloom's syndrome (BS), 158–161
  Fanconi's anemia (FA), 155–158
  glutathione reductase deficiency anemia, 168
  incontinentia pigmenti, 167
  Kostmann's agranulocytosis, 168
  porokeratosis of mibelli, 168
  scleroderma, 167–168
  xeroderma pigmentosum (XP), 165–167
Breast, tumors of, 485–490
  adenocarcinoma, 486
  adenoma, 462
  chromosome #1, 486
  cystic, 461
  DNA measurement in, 485
  EL in, 490
  fibroadenoma, 462, 485–486, 489
  lobular carcinoma-in-situ, 485
  in males, 490
  sex chromatin, 486
Bronchus
  adenoma, 462, 518
  cancer.
  See Lung, cancer of
Bromodeoxyuridine (BUdR), 4
  dye methods, 106–108
BS. *See* Bloom's syndrome
BUdR. *See* Bromodeoxyuridine
Burkitt cells, 399–400
Burkitt lines, 399–400
Burkitt lymphoma (BL), 395–405
  14q+ anomaly in, 400–404
  cell lines, 399–400
  chromosome changes in, 404–405
  nonendemic, 398–399, 404–405
Burkitt tumor. *See* Burkitt lymphoma (BL)

C

Cancer; *see also* Tumors
  age specific incidence of, 28
  causation and chromosomes, 439–453
  chromosomes, hyperdiploid, 436
    in diagnosis, therapy, and prognosis, 453–455
    hypodiploid, 436
    nonrandom changes, 450–453 *(Table 81b)*
    "practical aspects," 427–432
    in preneoplasia, 455–456
    primary and metastatic, 462–468 *(Tables 84 and 85)*
    in progression of, 447–453
    psuedodiploid, 436
    significance, 432–437
    tetraploid, 436
    triploid, 436
  DNA measurement, 439
  nuclear blebs, 437
  sex chromatin, 436–437
Carcinogenesis, chromosomal theory of, 5–7
Carcinogenic factors, 137–151
  radiation-related tumors, 137–140
  virus-associated tumors, 143–149
Carcinogens, chemical, 140–143, 441–442
Carcinoma
  breast, 485–490
  bronchus, 516–518
  of cervix, 526–529
    invasive, 529–530
  corpus uteri, 497–500
  epidermoid, 526
  -in-situ. *See* Carcinoma-in-situ
  invasive, 529–530

[cont.]
    laryngeal, 517–518
    microinvasive, 530–531
    nasopharyngeal, 526
Carcinoma-in-situ, 531–532
    of cervix uteri, 526–528
    and dysplasia, 531–532
C-banding, 34, 39–41 *(Figure 16)*, 88–90, 104
Cecum, cancer of, 473–484
Cell, cycle, 75 *(Figure 33)*
    division, 75
    fusion, 19, 331, 332
    transformation, 437–439
Centric fusion, 11
Centromere (kinetochore), 24, 76
    definition, 13
Cervix. *See* Uterine cervix, cancer of
Ch¹. *See* Christchurch chromosome
Chediak-Higashi syndrome, 150–151
Chemicals, 441–442
    benzene, 141–142
    effects on chromosomes, 140–143
Chemical oncogenesis, 140–143, 441–442
Chiasma, abbreviation, 29 *(Table 1)*
Childhood tumors; *see* Wilms' tumor, Monosomy #7, Teratomas
    CML in children, 225–227
Chimera, 55
Chinese hamster, 67
Chorioadenoma destruens, 500
Chorionepithelioma, 500
Chorionic villi, 500
Christchurch chromosome, 351
Chromatid
    aberrations, 57–58
    break, 13, 57
    exchange, 57
    gap, 57
Chromatin, 13
Chromosome, 13, 24
    abnormal, 56
    acentric, 54
    bands, 35, 43–45
    break, 13, 20, 45, 46–47, 58
    breakage syndromes, 153–170
    Christchurch (Ch¹), 351
    deletion, 46
    descriptions, 24–26 *(Figures 11–13)*
    designations, 48–51, 51–59
    dicentric, 54
    endoreduplication, 14
    etiological role in cancer, 6
    exchanges, 58
    fragments, 15
    gap, 13, 15, 58
    group A, 30–31, 81
    group B, 31, 81–82
    group C, 30–31, 82
    group D, 31, 82
    group E, 31, 82
    group F, 31, 82
    group G, 31, 82–83
    lag, 7, 17
    landmark, 43–45
    length, 32 *(Table 2)*
    marker, 30, 55, 59; *see also* Marker chromosome
    minute, 58
    number of, 2–3, 7, 59
    No. 1, 35, 567–569 *(Tables 90–92)*
    No. 2, 35, 575
    No. 3, 37, 575 *(Table 93)*
    No. 4, 37, 575
    No. 5, 37, 575–579
    No. 6, 37, 579 *(Table 94)*
    No. 7, 37–38, 579 *(Tables 95a and b)*
    No. 8, 38, 579–581 *(Tables 95b and 96)*
    No. 9, 38, 581–585 *(Table 97)*
    No. 10, 38, 585
    No. 11, 38, 585 *(Table 98)*
    No. 12, 38, 586
    No. 13, 38, 586–587 *(Table 99a)*
    No. 14, 38, 587 *(Table 100)*
    No. 15, 38, 587 *(Table 99a)*
    No. 16, 38, 587
    No. 17, 38, 587–588
    No. 18, 38, 588–589
    No. 19, 38, 589
    No. 20, 38, 589
    No. 21, 38, 589
    No. 22, 589–590
    Philadelphia. *See* Philadelphia chromosome (Ph¹)
    pulverization, 58
    rearrangements, 46–50
    region, 44–45
    ring, 48, 54, 326
    short and long arms, 24, 44–45, 53
    structure, 54, 64–73
    ultrastructure of, 64–73
Chromosome aberrations, 52, 53, 58–60
    mosaicism, 53, 55
    numerical, description of, 52–53
    structural, 53–54
Chromosomes, in cancer, 427–432
    aneuploidy, 436
    in diagnosis, therapy, and prognosis, 453–455
    nonrandom changes, 450–453 *(Table 81b)*
    in preneoplasia, 455–456
    primary and metastatic, 462–468 *(Tables 84 and 85)*
    significance, 432–437
Chronic granulocytic leukemia. *See* Chronic myelocytic leukemia (CML)
Chronic lymphocytic leukemia (CLL), 349–352
Chronic myelocytic leukemia (CML), 185–257
    basophilic, 230–231

blastic phase, 235–249 *(Tables 45 and 46)*
cell markers, 232–234
in children, 225–227 *(Table 40)*
chromosome changes, 222–225
chronic phase, 222–225
clinical implications of chromosome changes, 205–206
eosinophilic, 227–230 *(Table 41)*
extramedullary origin, 200, 254–256
features, 186 *(Table 32)*
genesis and Ph¹, 195–196
karyotypic evolution, 252–254
karyotypic staging, 250–252
LAP, 231
missing Y in, 216–222 *(Table 36)*
muramidase, 232
Ph¹-negative, 201–204
Ph¹-negative cells in, 249–250
in twins, 190 *(Table 33)*
Chronic myelocytic leukemia, Ph¹-negative, 201–204
missing Y in, 220–221
Chronic myelomonocytic leukemia (CMMoL), 257
Chronic phase of CML, 222–225
CLL. *See* Chronic lymphocytic leukemia
Clonal evolution, 13
in AML, 309–310
in carcinoma, of cervix, 532–533
in CML, 204–205, 252–254
Clonal growth. *See* Clonal evolution
Clonal origin, 60, 198–200, 253
Clone, 60
CML. *See* Chronic myelocytic leukemia
CMMoL. *See* Chronic myelomonocytic leukemia
Colchicine, 3, 7
Colon, cancer of, 473–484
adenoma, 473
banding studies, 481–484
DNA measurement in, 472
Gardner's syndrome, 462
polyps, 462, 473–473
Colitis, ulcerative, 143 *(Table 18)*, 473
Concanavalin-A, 116 *(Table 5)*
Congenital leukemia, 322–323 *(Table 59)*
Constitutive heterochromatin. *See* Heterochromatin, constitutive
Constrictions, secondary. *See* Secondary constrictions
Corpus uteri, cancer of, 497–500
CP. *See* Chronic phase of CML
Crohn's disease, chromosomes in, 473
Cystadenocarcinoma. *See* Ovary, tumors of
Cytogenetic analysis, normal lymphocytic cell lines, 101–105

D

Deletion, 12 *(Figure 7)*, 14

interestitial, 12 *(Figure 7)*, 48
terminal, 48
Derivative chromosome. *See* Marker chromosome
Designations, chromosomal, 48–51, 51–59
chromatid aberrations, 57–58
chromosome aberrations, 58–59
dicentric chromosome, 48
direct insertion, within or between chromosomes, 49
duplication of segment, 50
four-break rearrangements, 50
interstitial deletion, 48
inverted insertion, within or between chromosomes, 49
isochromosome, 48
paracentric deletion, 48
pericentric inversion, 48
reciprocal translocation, 48
ring chromosome, 48
Robertsonian translocation, 49
terminal deletion, 48
terminal rearrangements, 50
Dicentric chromosome, 12 *(Figure 6)*, 48
Dicentrics, 54
DiGuglielmo's syndrome. *See* Erythroleukemia (EL)
7, 12-Dimethylbenz (α)-anthracene (DMBA), 442 *(Table 80)*
leukemias, 442 *(Table 80)*
sarcomas, 442 *(Table 80)*, 445
thymic lymphoma, 442 *(Table 80)*
Diploid, 14, 76
number, 27
Disorders, myeloproliferative, 355–375; *see also* Myeloproliferative disorders (MD)
DMBA. *See* 7, 12-Dimethylbenz (α)-anthracene
DMS. *See* Double minute chromosomes
DNA, 33, 67–71
in bladder tumors, 504–505, 508
in breast tumors, 485
in colon, 472
epidermal carcinomas, 526
measurement of, 10
in ovarian tumors, 492
packing ratio, 66–67
pheochromocytoma, 519–520
in prostate, 516
replication, 30, 68, 80–81
satellite, 92–93 *(Figure 41)*
in testicular tumors, 514–516
in thyroid tumors, 518–519
in tumors, 439
Double minute chromosomes (DMS), 550–558 *(Table 88)*
Down's syndrome (DS) (mongolism), 4, 119–120, 332–337
leukemia in, 332–337 *(Table 63)*
leukemia incidence, 119 *(Table 6)*

[cont.]
   trisomy #21, 4, 119–120
DS. See Down's syndrome
Duplication, 14
   chromosome segment, 51
   chromosome structure, 54
   reversed, 14
   tandem, 14
Dysgerminoma, 127–129
Dyskeratosis congenita (Zinsser-Cole-Engmann), 525
Dysplasia, cervical, 531–532
Dysproteinemias. See Plasma cell dyscrasias

E

EB virus. See Epstein-Barr (EB) virus
Effusions, chromosome preparation, 103
EL. See Erythroleukemia
Electron microscopy (EM), of chromosomes, 64–66
EM. See Electron microscopy
Embryonic sarcoma, 524
Endometrium, cancer of, 497–500
Endoreduplication, 14
Eosinophilic leukemia, 227–230 (Table 41)
Ependymoma, 546–547
Epidermal carcinomas, DNA values in, 526
Epstein-Barr (EB) virus, 146
Erythroleukemia (EL), 270–274, 323–332
   chromosome changes in, 323–327 (Tables 60 and 61)
   classification (MIKA, MAKA), 327–331
   polyploidy, 329
   prophasing, 331–332
Esophagus, tumors of, 468–469
ET. See Thrombocythemia
Euchromatin, 14
Ewing sarcoma, 525, 526
Exchange, 12, 15
   chromatid, 106–109
   complete, 15
   incomplete, 15

F

FA. See Fanconi's anemia
Fallopian tube, 501
Fanconi's anemia (FA), 155–158
   acute leukemia, 157–158
Feulgen, 30
Fibrils, 65–66
Fibrosarcoma. See Sarcoma
FL. See Follicular lymphoma
Follicular lymphoma (FL), 383–393
   14q+ anomaly in, 394–395
   chromosome changes, 384–393 (Table 73a)
Fragment, 15
   acentric, 11

G

G_1-period, 15, 75
G_2-period, 15, 75
Gammopathies. See Plasma cell dyscrasias
Gap
   chromatid, 12 (Figure 6)
   chromosome, 12 (Figure 6)
Gardner's syndrome, 462
Gastric adenocarcinoma. See Stomach, cancer of
G-banding, 4, 34, 40–41 (Figures 17 and 18), 90–91, 104, 115
Gene mapping, 590–595 (Table 102)
Genome, 15
Giemsa, 30, 90–91
Glioma
   malignant (astrocytic), 543–546
   optical, 548
Glossary, 11–21
Glutathione reductase deficiency anemia, 168
Gonadal dysgenesis, 125–127
Gonadoblastoma, 125–127
Gonosomes, 15

H

"Hairy cell leukemia" (HCL), 340–341
Haploid, number, 27
Haploidy, 15, 341–344
HCL. See "Hairy cell leukemia"
HD. See Hodgkin's disease
HeLa cell line, 535
Helical coiling, 24
Hepatitis, infectious, 143 (Table 18)
Herpes viruses, 145–146
Heterochromatin, 15, 71–73
   constitutive, 15, 34, 73
   facultative, 15, 71
Heteromorphic chromosomes, 51–55
Heteromorphism, in bladder cancer, 509–511
High-resolution banding, 109–116
Histiocytic medullary reticulosis (HMR), 409–410
Historical, 1–10
HMR. See Histiocytic medullary reticulosis
Hodgkin's disease (HD), 378–383, 399
   acute leukemia in, 381–383
   chromosome changes, 380–381 (Table 72)
   chromosome number, 379 (Table 71)
Homogeneously staining regions (HSR), 558–559
Homologs, 27–30
HSR. See Homogeneously staining regions
Hydatidiform moles, 462, 500–503
   complete, 501
   partial, 501
Hyperdiploid, 16
   tetraploid, 21

triploid, 21
tumors, 436
Hypodiploid, 16
tumors, 436

I

IAA. See Idiopathic aplastic anemia
Idiogram, 16, 76
Idiopathic aplastic anemia (IAA), 174 (Table 29)
Idiopathic sideroblastic anemia, 179–181
Immunodeficiency syndromes, 416
Incontinentia pigmenti (Bloch-Sulzberger syndrome), 167
Insertion, 16, 49
  direct, 49
  inverted, 49
Intercalary, 16
Interchanges, 16
Interphase, 16–17, 75
Interstitial deletion, 12 (Figure 7), 17, 48
  5q−, 175–177 (Table 30)
  segment, 17
Invasive carcinoma, 529–530
Inversion, 12 (Figure 8), 17
  interstitial, 12 (Figure 8)
  paracentric, 17, 48
  pericentric, 17 (Figure 8), 48
Isochromosome, 12 (Figure 7), 48, 54
Isolocus break, 58
Itai-Itai disease, 167

K

Karyotype, 17, 27–30, 76
Karyotypic evolution, ALL, 290
  CML, 252–254
  meningioma, 543
Kidney, cancer of, 511
  Wilms' tumor, 511
Kinetochores, 13
Klinefelter's syndrome, 4, 129–130
  neoplasms in, 130–131 (Table 13)
Kostmann's agranulocytosis, 168

L

LA. See Lateral asymmetry
Landmark of a chromosome, 43–45
LAP. See Leukocyte alkaline phosphatase
Large bowel, 481–484
Larynx, cancer of, 518–519
  papilloma, 525
Lateral asymmetry (LA), 94–95 (Figure 42)
Leiomyosarcoma, 523–524, 525
  intestinal, 479–480
  retroperitoneal, 523–524
Leukemia
  acute, 119, 263–276
  congenital, 322–323
  "hairy cell", 340–341
  lymphoblastic, acute, 268–270, 276–294
  lymphocytic, chronic, 349–352
  megakaryocytic, acute, 340
  monocytic, 340
  murine, 148
  myeloblastic, acute, 270, 294–322
  myelomonocytic, acute, 337–340
    chronic, 257
  promyelocytic, 337
Leukocyte alkaline phosphatase (LAP), 231
  in CML, 231
  in myeloproliferative disorders, 355
Leukocyte culture, 101–102
  amethopterin synchronization, 114–115
  macrotechnique, 101–102
  microtechnique, 101
Liver, tumors of, 484–485
Long arm, 24
Loss of chromosomes. See Monosomy
Louis-Bar Syndrome, 161–165; see also Ataxia telangiectasia (AT)
LS. See Lymphosarcoma
Lung, cancer of, 516–518
  chromosome #1, 518
  oat cell, 518
  sex chromatin, 518
Lymph node, 103
Lymphoma, 387–411
  acute leukemia in, 393–394 (Table 72)
  14q+ anomaly in, 394–395 (Tables 74a and b)
  African; see Burkitt lymphoma
  angioimmunoblastic lymphadenopathy, 409 (Table 77)
  banding studies in, 387–392
  Burkitt, 395–405
  follicular, 392–393
  histiocytic medullary reticulosis, 409–410
  Hodgkin's disease, 378–383, 399
  mycosis fungoides, 405–409 (Table 76)
  reticulum cell and other lymphomas, 383–395
    chromosome changes, 384–393 (Table 73a)
  Sézary, 405–409 (Table 76)
Lymphocyte, culture, 101–102
Lymphosarcoma (LS), 383–387
  14q+ anomaly in, 394–395
  chromosome changes, 384–393 (Table 73a)
Lymphoproliferative disease. See Lymphoma
Lysosomes, 232

M

MA. See Megaloblastic anemia
MAKA, 182, 314–319, 327–331
Malignancy. See Cancer
Malignant melanoma, 520–523

738 Subject Index

[cont.]
   chromosomal findings, 520–521
   chromosome #1 in, 521–523
Malignant transformation, 437–439
   chromosomes in, 437–439
Mammary adenocarcinoma. *See* Breast, cancer of
Marek's disease, 440
Marker chromosome, 17, 30, 59
   Api in cervix, 531–532
   cell surface, 232–234, 290
   frequency in tumors, 463–465
   "W" chromosome in WM, 414
Markers
   in AL, 276
   cell surface in CML, 232–234
Maternal origin. *See* Polymorphism
Maxillary tumors, 526
MD. *See* Myeloproliferative disorders
Measles, 147
Medulloblastoma, 547
Megaloblastic anemia (MA), 363–365
Meiosis, 17
Mendelian diseases, 134
Meningiomas, 535–543
   cytogenetic findings, 542–543
   karyotypic evolution, 543 *(Table 87)*
   NOR, 542
   prior to banding, 536–542
Meningitis, aseptic, 143 *(Table 18)*
Mesotheliomas, 525
Metacentric (median), 17, 76
Metaphase, 17–18, 75
Metastases, brain, 549–550; *see also* Cancer
Metastatic tumors, brain, 549–550
Methods, 99–116
   abbreviations used, 99
   autoradiography, 109
   bone marrow, 102–103
   C-banding, 104
   effusions, 103
   G-banding, 104
   high-resolution banding, 109–116
   leukocyte culture, 101–102
   lymph nodes and spleen, 103
   N-banding, 105
   Q-banding, 103
   R-banding, 104
   SCE, 106–108
   silver NOR staining, 105
   T-banding, 104–105
   tumors, 103
MF. *See* Myelofibrosis
Microinvasive carcinoma, cervix, 530–531
Micronuclei. *See* Mitosis
MIKA, 182, 314–319, 327–328
Minute chromosomes, 58
   double, 550–558 *(Table 88)*
Mitogens, 116 *(Table 5)*

Mitomycin C (MMC), 194–195
Mitosis, 18, 75
Mitotic cycle, 75
MM. *See* Multiple myeloma
MMC. *See* Mitomycin C
MMM. *See* Myeloid metaplasia with myelofibrosis
Modal number, 18, 60
   leukemia vs. cancer, 465
Mongolism. *See* Down's syndrome (DS)
Monoclonal origin. *See* Clonal origin
Mononucleosis, infectious, 143 *(Table 18)*
Monosomy, 18, 322
   in AML, 322
   in children, #7, 362–363
Mosaicism, 53, 55
Mouse ascites, 3–4
MS. *See* page 182
Multiple myeloma (MM), 416–423
   banding studies in, 419–420
   chromosome studies in, 417–419
   leukemia complicating, 420–423 *(Table 78)*
Multipolar mitotic spindle, 18
Mumps, 147
Muramidase (lysozyme), 232
   Ph^1-positive CML, 232
Murine leukemia, 148
   Friend virus, 148
Mycosis fungoides, 405–409
Myelofibrosis (MF), 358–363
   acute, 362
Myeloid metaplasia, 358–362
   with myelofibrosis (MMM), 358–362
Myeloproliferative disorders (MD), 355–375
   LAP, 355
   myelofibrosis, 358–362
   myeloid metaplasia, 358–362
   monosomy #7 in children, 362–363
Myelosclerosis. *See* Myeloproliferative disorders (MD)

N

N. *See* page 182
Nasopharyngeal carcinoma, 526
N-banding, 4, 42–43 *(Figures 19 and 20),* 105
Near-haploidy, in AL, 341–344
Neoplasm. *See* Cancer, Tumors
Nephroblastomas, 511
Nervous system. *See* Brain, tumors of
Neuroblastomas, 547
Neurinoma, 548–549
Nomenclature, symbols, 28–29 *(Table 1),* 47
Nondisjunction, 18
NOR. *See* Nucleolar organizer regions
Nuclear blebs, acute leukemia, 276
   in tumors, 437
Nuclear membrane, in prophasing, 19
Nuclear protrusions. *See* Nuclear blebs

Nucleolar organizers (NOR), 18, 42–43, 91–94, 105–106
  in meningiomas, 542
  technique, 42–43

O

Oligodendroglioma, 546
Oncogenesis, 140–143; *see also* Carcinogenesis
  chemical, 140–143, 441–442
  viral, 442
Ovary, tumors of, 490–497
  benign tumors, 491
  bilateral tumors, 490
  chromosome #1 in, 496
  chromosome studies with banding, 494–496
  cystadenoma, 462, 491
  dermoid cyst, 491, 496
  DNA values in, 492
  folliculoma, 491
  granulosa cell, 496
  mesothelial, 496
  teratomas (benign of variant), 491, 496–497

P

Pairing, 18
Pancreas, tumors of, 484–485
Papilloma, larynx, 525
  verruca, 461
Papovaviruses, 146
Paracentric inversion. *See* Inversion, paracentric
Paramyxovirus, 147
Paraproteinemia. *See* Plasma cell dyscrasias
Parotid tumors, 525–526
Paroxysmal nocturnal hemoglobinuria (PNG), 362–363
Paternal origin; *see also* Polymorphism
  androgenic origin of hydatidiform moles, 500
PB. *See* page 182
PCC. *See* Premature chromosome condensation, Prophasing
Pericentric inversion. *See* Inversion, pericentric
Peritoneum, 484–485
PHA. *See* Phytohemagglutinin
Ph[1]. *See* Philadelphia chromosome
Philadelphia chromosome (Ph[1]), 186–204, 234–235
  in acute leukemia, 274–275
  in AML, 270, 308–309
  complex translocations, 207–214 (Table 35)
  definition of, 187–190
  in disorders other than CML, 234–235
  double Ph[1], 248–249
  genesis of CML, 195–196, 198–200
  nature of, 190–194
  novel views of, 196–197
  pathogenetic aspects of, 198
  in polycythemia vera, 371–372
  therapy and, 200–201
  without translocation, 214–216
  translocations, 207–214
  unusual translocations, 207–214 *(Table 35)*
Phytohemagglutinin (PHA), 4
Pituitary adenomas, 549
Plasma cell dyscrasias, 414–423
  multiple myeloma, 416–423
  Waldenström's macroglobinemia, 414–416
Ploidy
  aneuploid, 76
  definition, 18, 76
  diploid, 76
  low in tumors, 463 *(Table 83)*
  polyploid, 76
  pseudodiploid, 76
  tetraploid, 76
  triploid, 76
PNH. *See* Paroxysmal nocturnal hemoglobinemia
Polycythemia vera (PV), 365–375
  banding studies, 372–375
  chromosome changes in, 367–369 *(Tables 68 and 69)*
  Ph[1] in, 371–372
  progression of, 369–371
Polymorphism, 18, 94–95
Polyoma virus, 146
Polyploid, 19, 76, 198
Polyploidy
  in cancer of cervix, 533
  in EL, 329
  and Ph[1], 198
Polyploidization, 18–19
Polyps
  colon, 462, 472–473
  stomach, 470
Porokeratosis of mibelli, 168
Poxviruses, 147
Precancerous lesions
  cervix, 526–529
  colon, 473
Predisposition. *See* Breakage syndromes, Down's syndrome, Preleukemia
Preleukemia, 172–180 *(Table 27)*
  idiopathic aplastic anemia (IAA), 174–175 *(Table 29)*
  pure red cell aplasia, 175–176
Premature chromosome condensation (PCC), 19; *see also* Prophasing
Primary tumors. *See* Cancer
Prophase, 75
Prophasing, 19, 58, 331–332
  in EL, 331–332
Prostate, cancer of, 515–516
Pseudodiploid, 60
Pseudodiploidy, 19, 76

Pulverization, 58
Pure red cell aplasia, 175–176
PV. See Polycythemia vera

Q

Q-banding, 4, 35–39, 83–88, 103
    normal pattern, 35–39 (Figure 15)
Quadriradial (Qr), 158, 159 (Figure 52)
Quinacrine fluorescence, 83–88
Qr. See Quadriradial
Quinacrine mustard, 34

R

RA. See Refractory anemia
Radiation effect. See X-rays
Radiation exposure; see also X-rays
    atom bomb, 139 (Table 17a)
R-banding, 4, 35, 41, 91, 104
RC or RCS. See Reticulum cell sarcoma
Rearrangements, 46–50
    three-break, 46, 47, 50
    two-break, 47
Reciprocal translocation, 21, 46, 48, 50
Rectum, cancer of, 473–484
Refractory anemia, 176–179
Region, 37 (Figure 15B), 44–45
Reticulosarcoma. See Reticulum cell sarcoma (RCS)
Reticulum cell sarcoma (RCS), 383–392
    14q+ anomaly in, 394–395
    chromosome changes, 384–393 (Table 73a)
Retinoblastoma, 121–124 (Tables 7 and 8), 548
    congenital conditions associated with, 122 (Table 7)
Rhabdomyosarcoma, 524–525
Ring chromosome, 12 (Figure 9), 19, 48
    in leukemia and other disorders, 326 (Table 62)
Rous sarcoma virus (RSV), 442–443
RSV. See Rous sarcoma virus

S

Salivary gland, 525–526
    mixed tumor, 525
Sarcomas
    bone, 524–525
    embryonic, 524
    Ewing, 525, 526
    periosteal, 524
Satellite association, 19, 80
Satellite DNA, 19, 92–93 (Figure 41)
Satellites, 19, 42, 77–80 (Figures 34 and 35)
SCE. See Sister chromatid exchange
Scleroderma, 167–168
Sea urchin egg, 5–6

Secondary constriction, 12 (Figure 6), 19, 24, 42, 56, 77
    symbols, 56
Segregation, 195, 431
Seminomas, testicular, 512, 514
Sendai virus, 147
Sex chromatin, 7–10, 82
    in breast, 486
    in lung, 518
    in testicular tumors, 515
    in tumors, 436–437
Sex chromosome anomalies, 124–132
    in neoplasia, 131 (Table 14), 591 (Table 101)
Sézary syndrome, 405–409
Short arm, 24
Sideline, 60–61
Sideroblastic anemia, 179–180
Silver NOR staining, 105
Sister chromatid exchange (SCE), 20, 58, 95–96, 106–108
Solid tumor, 103
S-period, 20, 75
Stalk, 20, 42
Stem line, 60
    cells, 20
Sternberg-Reed cells. See Hodgkin's disease (HD)
Stomach, tumors of, 469–471
    adenoma, 470
Structure, chromosome, 64–73
Sub-band of a chromosome, 45
Sub-line, 60
Submetacentric, 20, 76
Subtelocentric, 20
SV-40 virus, 146–147
    in meningiomas, 543
Syrian (golden) hamster, RCS in, 440, 444

T

Tandem fusion or translocation, 21
T-banding, 4, 41 (Table 3), 91, 104–105
TdR (thymidine). See Autoradiography
TdT. See page 182
Telocentric (terminal), 76
Telomeres, 21
Telophase, 21, 75
Teratomas
    ovarian, 491, 496–497
    sacrococcygeal, 525
    testicular, 512, 514–515
Terminal, deletion, 21, 48
    rearrangements, 50
Testicular tumors, 511–515
    DNA measurements, 514–515
    seminomas, 512, 514
    sex chromatin, 515
    teratomas, 512, 514–515
Tetraploid, 21, 76

Thorotrast, 139 *(Table 17a)*
Thrombocythemia (essential, primary or idiopathic), 363
Thymoma, 524
Thyroid, tumors of, 518–520
    adenoma, 461
    DNA analysis, 518–519
    Hashimoto's thyroiditis, 519
    Sipple's syndrome, 520
Transformation, malignant, of cells, 148–149; *see also* Malignant transformation
Translocation, 12, 21, 46, 48, 50
    abbreviation, 46
    asymmetrical, 21
    complex, 49–50
    reciprocal, 12 *(Figure 10)*, 21, 46, 48, 50
    Robertsonian, 21, 46, 49
    symmetrical, 21
    tandem, 46
    unusual and complex Ph^1, 207–214 *(Table 35)*
    whole-arm, 50
Triploid, 76
Trisomy, 21
    #8, 322
    #21, 119–120
Trophoblastic tumors, 501–503; *see also* Hydatidiform moles, Chorionic villi, Chorioepithelioma
Trypsin. *See* G-banding
Tumors, 3–4, 143–149
    adenovirus, 145
    adrenal, 518–520
    alimentary tract, 468–485
    animal, 3–4, 440–447
    benign, 459–462
    brain, 535–559
    breast, 485–490
    Burkitt; *see* Burkitt lymphoma (BL)
    chorionic, 500–503
    colon, cecum, appendix, rectum, and anus, 471–481
    esophagus, 468–469
    liver, pancreas, and peritoneum, 484–485
    lung and larynx, 516–518
    metastatic, 462–468 *(Table 82)*
    nervous system; *see* Brain, tumors of
    oral cavity, 468–469
    ovary, 490–497
    primary, 462–468 *(Table 82)*
    spontaneous, 440
    stomach, 469–471
    testicular, 511–515
    thyroid, 518–520
    trophoblastic, 501–503
    uterus and endometrium, 497–500
    urethra, 511
    urinary tract, 503–511
    virus-induced, 143–149, 442–444

Wilms', 511
    with low ploidy, 463 *(Table 83)*
Turner's syndrome, 4, 125–127; *see also* Gonadal dysgenesis
    extragonadal neoplasms, 125 *(Table 9)*

U

Ulcerative colitis, 143 *(Table 18)*, 473
Unicellular origin of cancer, 250
Urethra, 511
Urinary tract, tumors of, 503–511
    bladder, 503–511
    kidney, 511
Uterine cervix, cancer of, 526–535
    abnormal mitosis, 528
    aneuploidy in, 533–534
    Api-marker, 531–532
    carcinoma-in-situ, 531–532
    chromosome #1, 534–535
    chromosomal findings, 529–530, 534–535
    chromosome breakage in, 534
    clonal evolution, 532–533
    cytogenetic studies, 529
    dysplasia, 531–532
    invasive, 529–530
    microcarcinoma, 530–531
    mild dysplasia, 532
    pathology of, 528–529
    polyploidy in, 533
Uterus, cancer of, 497–500
    fibromyoma, 461
UV-irradiation, 157, 165–167

V

Vagina, 501
Venereal tumor, of dog, 444
Viral oncogenesis, 442
Virus, 143–149
    adenovirus, 145
    avian sarcoma, 147–148
    chromosome breakage, 153–170
    DNA, 443
    EB; *see* Epstein–Barr (EB) virus
    hepatitis, 143
    herpes simplex, 145–146
    measles, 147
    mumps, 147
    murine leukemia, 148
    oncornaviruses, 147–148
    papilloma, 461, 525
    papovavirus, 146
    paramyxovirus, 147
    polyoma, 146
    poxviruses, 147
    RNA, 442
    Rous sarcoma; *see* Rous sarcoma virus (RSV)

[cont.]
    Sendai, 147
    SV-40, 146–147
    transformation of cells, 148–149
    UV-irradiated, 157
Vulva, 501
    Bowen's disease, 501
    condyloma acuminatum, 501

W

Waldenström's macroglobulinemia (WM), 414–416
    "W" chromosome, 414
Wilms' tumor, 511
WM. *See* Waldenström's macroglobinemia
Wright stain, 115

X

X-chromatin, 21
X-chromosome (X-body), 7–10, 28, 30, 38, 57, 83, 590
Xeroderma pigmentosum (XP), 165–167
    DNA defects in, 166–167
XLRLS = X-linked recessive lymphoproliferative syndrome, 409
XP. *See* Xeroderma pigmentosum
X-rays, 137–140
XYY male, 4

Y

Y-chromatin (Y-body), 9, 21, 57
Y-chromosome, 9, 39, 83, 590
    in bladder cancer, 509
    missing in AML, 320–322
    missing in CML, 216–222 *(Table 36)*
    missing in males, 27, 30, 216–222

Table Index

Table 1 Nomenclature symbols, 28–29

Table 2 Measurements of relative length and centromere index, 32

Table 2a Variation in staining of specific bands with various techniques, 39

Table 3 Representation of the intensity of the fluorescence of the telomeric areas in the various chromosomes as demonstrated by T-banding, 41

Table 4 Bands serving as landmarks that divide the chromosomes into cytologically defined regions, 44

Table 4a Sample descriptions of sequential observations in a single case of chronic myeloid leukemia, 57

Table 5 Lymphocyte mitogens, 116

Table 6 Summary of morphologic types of acute leukemia in Down's syndrome, 119

Table 7 Retinoblastoma and various congenital chromosomal anomalies, 122

Table 8 Retinoblastomas and 13q−, 124

Table 9 Extragonadal neoplasia in Turner syndrome (gonadal dysgenesis) with 45,XO or 45,XO/46,XX, 125

Table 10 Gonadoblastoma cases reported in the literature, 126–127

Table 11 Mixed tumors (mainly gonadoblastoma and dysgerminoma), 128

Table 12 Dysgerminomas in patients with sex chromosome abnormalities, 128

Table 13 Cases of Klinefelter's syndrome, their chromosome anomalies and neoplasia reported in the literature, 130–131

Table 14 Females with sex chromosome anomalies and neoplasia, 131

Table 15 Transformation frequency of fibroblasts by SV40 virus in controls and patients with Klinefelter's syndrome, 131

Table 16 Some Mendelian inherited diseases frequently associated with cancer or leukemia, 134

Table 17a Effects of biological, physical, and chemical agents on human chromosomes in vivo, 138–139

Table 17b Increased SCE frequency resulting from in vivo effects, 140

Table 18 Reported chromosome changes in infections and other conditions, 143

Table 19 Frequency of cell transformation in vitro by SV40, 148

Table 20 Chromosomal changes (dicentrics, rings, acentrics) induced by X-ray in cultured cells, 149

[cont.]
Table 21 Incidence of autosomal abnormalities in control and cancer subjects, 150
Table 22 Type of constitutional chromosomal anomalies in 4,543 cancer patients, 150
Table 23 Incidence of constitutional chromosomal anomalies in cancer and leukemia patients, 150
Table 24 Chromosomal breakage syndromes, 154
Table 25 Major cytogenetic features of the four main chromosome instability syndromes, 154
Table 26 Lymphocyte clones containing abnormal D-group chromosomes in patients affected with AT, 164
Table 27 Karyotypic findings in various preleukemic conditions, 173
Table 28 Some criteria for the possible diagnosis of so-called preleukemia, 174
Table 29 Chromosome findings in aplastic anemia and other refractory anemias and pancytopenias, 175
Table 30 5q— anomaly in various conditions, 177
Table 31 Karyotypic changes associated with refractory idiopathic sideroblastic anemia (SBA), 179
Table 32 Some salient cytogenetic, clinical, and survival features of CML, 186
Table 33 Reports on chromosome studies in twins with leukemia, 190
Table 34a Ph[1]-negative CML cases with karyotypic abnormalities (exclusive of juvenile type), 203
Table 34b Some further cytogenetic changes observed in the chronic phase of CML (Ph[1]-positive), 204
Table 35 Simple and complex Ph[1]-translocations in CML involving chromosome #22 and those listed in the table, 208–209
Table 36 List of papers that include cases of Ph[1]-positive CML with a missing Y, 216
Table 37a Hematological conditions associated with a missing Y (conditions other than CML), 218
Table 37b Age, karyotypic, and prognostic features of the ANLL patients with a missing Y in the literature, 219
Table 37c Survival in months (mo) of ANLL patients with missing Y, 219
Table 38 Missing Y-chromosome in Ph[1]-negative CML, 220
Table 39 Cytogenetic features of the CP of CML, 222
Table 40 Chromosome findings in childhood leukemia of the CML type, 226–227
Table 41 Summary of chromosome findings in peripheral blood or bone marrow cells from cases of eosinophilic leukemia from the literature, 228–229
Table 42 Some possible differentiating points between Ph[1]-positive AML and the BP of CML, 237
Table 43 Chromosome changes in CML in BP, 239
Table 44 Summarized information on the patients with Ph[1]-positive CML and with unusual karyotypes, 240–241
Table 45 Karyotypic patterns in Ph[1]-positive CML in the BP—determined with banding techniques, 242–245
Table 46 Iso-17[i(17)] chromosome in conditions other than CML, 246
Table 47 Leukemia and other proliferative disorders in patients with congenital chromosome abnormalities (exclusive of DS), 266–267
Table 48 Published case reports of Ph[1]-positive ALL, 268–269
Table 49 Ph[1]-positive cases of ANLL (AML, EL, and AMMoL) described in the literature, 271–273
Table 50 Ph[1]-positive cases with EL, 274
Table 51 Unusual Ph[1]-translocations in AL, 274
Table 52 Markers in leukemia, 277
Table 53 Distribution of modal chromosome number in 716 cases of AL, 278
Table 54 Published chromosome anomalies in ALL subsequent to the introduction of banding techniques, 285–286
Table 55 Chromosome numbers in acute myeloblastic leukemia and EL from major papers published since 1968, 294–295
Table 56 Chromosome abnormalities observed in AML without banding, 296–298
Table 57 Published cases of AML with probable and proved (with banding)8;21 translocations, 302–303
Table 58a Chromosome abnormalities in AML as shown by banding, 304–306
Table 58b Abnormal karyotypes in bone marrow cells from 37 ANLL cases with clonal chromosomal aberrations, 306–307
Table 58c Median survival time and cyto-

genetic pattern in 241 patients with ANLL, 308
Table 59 Congenital leukemia, 323
Table 60 Chromosome number in 125 cases of EL reported in the literature, 323
Table 61 Karyotypes in EL, 324–325
Table 62 Ring chromosome in lympho- and myeloproliferative disorders, 326
Table 63 DS and AL, 334
Table 64 Chromosome findings in APL, 335
Table 65 Chromosome findings in AMMoL, 338–339
Table 66 Chromosome constitutions of karyotypically abnormal clones in CLL, 350
Table 67 Cytogenetic changes observed in various myeloproliferative syndromes, 357–358
Table 68 Cytogenetically abnormal clones in the BM of untreated patients with PV, 366
Table 69 Chromosome findings in treated patients with PV, 368–369
Table 70 20q− anomaly in conditions other than PV, 370
Table 71 Summary of chromosomal findings in Hodgkin's disease, 379
Table 72 Leukemias complicating lymphomas, 382–383
Table 73a Karyotypes of published cases with various types of lymphoma, based primarily on banding studies, 388–391
Table 73b Karyotypic findings obtained with banding in nonendemic Burkitt lymphoma, 393
Table 74a 14q+ due to various translocations in lymphoma, 396–397
Table 74b Donor chromosomes involved in the 14q+ translocation in various lymphomas and/or their cell lines, 397
Table 74 Rearrangements of chromosome #14, 403
Table 76 Chromosome findings in Sézary syndrome and mycosis fungoides, 406
Table 77 Chromosomal changes in angioimmunoblastic lymphadenopathy, 409
Table 78 Leukemias complicating multiple myeloma, 421
Table 79 Chromosomal findings in thyroid tumors of rats, 440
Table 80 Chromosomal changes in tumors induced by chemical carcinogens, 442
Table 81a Chromosomal abnormalities associated with human neoplasia, primarily hemopoietic, 451
Table 81b Chromosomes preferentially involved in various human neoplasias, 452
Table 82 Distribution of modal number of chromosomes in 209 cases of human primary cancer and in 305 cancerous effusions reported in the literature, 463
Table 83 Unusually low ploidy in neoplasia (40 chromosomes or less), 463
Table 84 Summarized information on patients with malignant solid tumors, 464
Table 85 Distribution of chromosome numbers and banding karyotypes in patients with primary tumors, 466–467
Table 86 Involvement (%) of the various chromosome groups in 67 stem lines with other or additional changes than monosomy G, 537
Table 87 Possible evolution of karyotypic changes in meningioma, 543
Table 88 Published cases of tumors with DMS, 554
Table 89 Chromosome abnormalities associated with hematologic disorders, 568
Table 90 Involvement of chromosome #1 in translocation in lympho- and myeloproliferative disorders, 572–573
Table 91 Trisomy of chromosome #1 or its total [iso(1q)] or partial long arm and other structural changes of #1 in lympho- and myeloproliferative disorders, 573–575
Table 92 Trisomy of long arm of chromosome #1 [i(1q)], translocations involving #1, q+, and other structural anomalies of #1 in various cancers, 576
Table 93 Changes involving chromosome #3, including trisomy, monosomy, deletions, additions, and translocation, 577–579
Table 94 Involvement of chromosome #6 in lympho- and myeloproliferative disorders, 580–581
Table 95a Some hematological conditions associated with monosomy #7 reported in the literature (CML not included), 582
Table 95b Some changes related to trisomy-8, or 7-monosomy patients, 582
Table 96 Some hematological conditions associated with trisomy #8 reported in the literature, 583
Table 97 Some hematological conditions associated with trisomy #9 reported in the literature, 583

[cont.]
Table 98 Anomalies of chromosome #11 in lympho- and myeloproliferative disorders, 584–585
Table 99a Cases of myelo- and lymphoproliferative disorders with 13q—, 15q—, or Dq— reported in the literature, 586
Table 99b Cases with myeloproliferative disorders with congenital D/D translocations, 587
Table 100 Chromosome 14q+ in various disorders, 588
Table 101 Extra sex chromosomes in neoplasia, 591
Table 102 The gene map of each human chromosome, 592–595

Illustration and Table Credits

I wish to thank all the authors who have kindly supplied me with published and/or unpublished materials, photographs, and information, and have given me permission to reproduce, incorporate, or quote these in the text. Without them, this book would be woefully out of date at the time of its appearance. In addition, the following have given us permission to reproduce published textual, pictorial, or tabular materials in the book:

Figure 1A from Vol. 42, p. 2, 1956; Figure 13 from Vol. 52, p. 203, 1964; and Figure 170 from Vol. 68, p. 211, 1971, of Hereditas; The Distribution of Hereditas.

Figure 2 from Journal of the American Medical Association, Vol. 174, p. 224, 1960. "Copyright © 1960, American Medical Association."

Figure 11 from Vol. 26, p. 2074, 1966; Figure 34 from Vol. 26, p. 2075, 1966; Figures 16, 49, and 50 from Vol. 26, p. 2081, 1966; Figure 62 from Vol. 36, p. 315, 1976; Figure 86 from Vol. 22, p. 761, 1962; Figures 85 and 98 from Vol. 26, p. 2077, 1966; Figures 58 and 119 from Vol. 37, pp. 3502 and 3504, 1977; Figure 127 from Vol. 26, p. 2065, 1966; Figures 130 and 131 from Vol. 37, pp. 3506 and 3507, 1977; Figure 133 from Vol. 26, p. 2080, 1966; and Figures 117A and 153 from Vol. 26, p. 2079, 1966, of Cancer Research; Cancer Research, Inc.

Figures 12 and 55 from Vol. 20, pp. 399 and 402, 1962; and Figure 61 from Vol. 50, p. 695, 1977, of Blood; Grune & Stratton, Inc.

Figures 15A and B from Paris Conference (1971): *Standardization in Human Cytogenetics,* in Bergsma, D. (ed.). White Plains: The National Foundation–March of Dimes, BD:OAS, VIII(7), 1972.

Figure 19 from Vol. 36, p. 59, 1977, of *Human Genetics;* Springer-Verlag New York, Inc.

Figure 22 from Vol. 234, p. 187, 1971, of Nature New Biology; Macmillan Journals, Ltd.

Figures 25, 28, and 29 from Vol. 3, pp. 253, 376, and 377 of *Advances in Human Genetics,* David E. Comings/H. Harris and K. Hirschhorn (eds.), 1972; Plenum Publishing Corporation.

Figure 26 from *International Review of Cytology,* p. 30, 1973; Academic Press, Inc.

Figure 27 from Vol. 253, p. 248, 1975, of Nature; Macmillan Journals, Ltd.

Figure 31 from *Human Chromosome Methodology* (Chapter 1), p. 5, Second Edition, 1974; Academic Press, Inc.

Figure 33 from Cancer, Vol. 25, No. 1/Jan./Feb. Copyright © 1965 American Cancer Society, Inc., New York, N.Y.

Figure 41 from Vol. 92, p. 152, 1975, of Experimental Cell Research; Academic Press, Inc.

Figure 42 from Vol. 30, p. 147, 1978; and Figure 44 from Vol. 12, p. 234, 1962, of American Journal of Human Genetics; The University of Chicago Press.

Figure 47 from American Journal of Diseases of Children, Vol. 132, p. 162, 1978. "Copyright © 1978, American Medical Association."

Figure 48A from *The Molecular Biology of Cancer,* Chapter 3, p. 102, 1974; Academic Press, Inc.

Figure 48B, Reprinted by permission. From The New England Journal of Medicine, Vol. 275, p. 811, 1966.

Figure 56 from Vol. ii, p. 1385, 1973, of Lancet; The Lancet, Ltd.

Figure 70 from Vol. 30, p. 16, 1975 of Humangenetik; Springer-Verlag New York, Inc.

Figures 80, 84, 132, 154, and 166 from Vol. VI, pp. 191, 192, 234, 237, 1974, of *Handbuch der Allgemeinen Pathologie;* Springer-Verlag New York, Inc.

Figures 89 and 93 from Vol. 2, pp. 337 and 341, 1976, of Medical and Pediatric Oncology; Alan R. Liss, Inc.

Figure 98 from Vol. 113, p. 688, 1964, of Annals of the New York Academy of Sciences; The New York Academy of Sciences.

Figures 106A and B from Vol. 34, pp. 42 and 43, 1967, of Journal of Cell Biology; The Rockefeller University Press.

Figures 114A, B, C and D from Vol. 19, pp. 483–485, 1977, and Figure 108 from Vol. 20, pp. 870 and 871, 1977, of International Journal of Cancer; International Union Against Cancer.

Figures 187A and B from Biedler, J. L., and Spengler, B.A., Science, Vol. 191, pp. 185–187, 1976; Copyright © 1976 by the American Association for the Advancement of Science.

Tables 1, 2, and 2A from Paris Conference (1971): *Standardization of Human Cytogenetics,* in Bergsma, D. (ed.). White Plains: The National Foundation—March of Dimes, BD:OAS, VIII(7), 1972.

Text material in Chapter 2; from Paris Conference (1971): *Standardization of Human Cytogenetics,* in Bergsma, D. (ed.). White Plains: The National Foundation—March of Dimes, BD:OAS, VIII(7), 1972; and from Paris Conference (1971), Supplement (1975): *Standardization in Human Cytogenetics,* in Bergsma, D. (ed.). White Plains: The National Foundation—March of Dimes, BD:OAS, XI(9), 1975.

Text material in Chapters 7, 16, and 17; from *Chromosomes and Cancer,* J. German (ed.), chapters by Atkin, N.B., Harnden, D.G., Mark, J., and Spriggs, A.I.: John Wiley and Sons, 1974.

I would also like to thank Cancer, The Proceedings of the National Academy of Sciences USA, and The Journal of the National Cancer Institute for pictorial materials reproduced from these journals.

Series 4128

**THE LIBRARY
UNIVERSITY OF CALIFORNIA
San Francisco
476-2334**

THIS BOOK IS DUE ON THE LAST DATE STAMPED BELOW

Books not returned on time are subject to fines according to the Library Lending Code. A renewal may be made on certain materials. For details consult Lending Code.

| 14 DAY
FEB 16 1986
RETURNED
MAR 15 1986
14 DAY
JUL 15 1987
RETURNED
JUL 17 1987
NOV 16 1987 | RETURNED
DEC 15 1987
14 DAY
MAY 4 1988
RETURNED
APR 21 1988
14 DAY
MAY 28 1988
RETURNED
MAY 19 1988 | 14 DAY
NOV 26 1991
RETURNED
NOV 19 1991 |

INTERLIBRARY LOAN
14 DAYS AFTER RECEIPT
VA LA

Series 4128